Grant's METHOD OF ANATOMY

A CLINICAL PROBLEM-SOLVING APPROACH

Eleventh Edition

Grant's METHOD OF ANATOMY

A CLINICAL PROBLEM-SOLVING APPROACH

Eleventh Edition

JOHN V. BASMAJIAN
M.D., F.A.C.A., F.R.C.P.(C), F.S.B.M., F.A.C.R.M., F.A.B.M.R.
Emeritus Professor of Medicine and Anatomy, McMaster University; Former Director, Rehabilitation Centre, Chedoke-McMaster Hospital, Hamilton, Ontario, Canada

CHARLES E. SLONECKER
D.D.S., Ph.D.
Professor and Chairman, Department of Anatomy, University of British Columbia, Vancouver, British Columbia, Canada

WILLIAMS & WILKINS
Baltimore • Hong Kong • London • Sydney

Editor: Kimberly Kist
Associate Editor: Victoria M. Vaughn
Design: Bob Och
Illustration Planning: Lorraine Wrzosek
Copy Editor: Teresa A. Tamargo
Production: Theda Harris

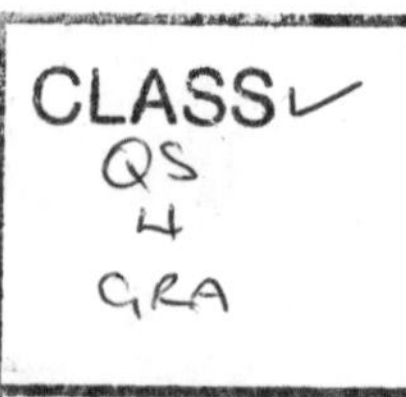

Printed in the United States of America

First Edition, 1937
Reprinted, 1938
Second Edition, 1940
Reprinted, 1941, 1943, 1944
Third Edition, 1944
Reprinted, 1945, 1946, 1947
Fourth Edition, 1948
Reprinted, 1949, 1951
Fifth Edition, 1952
Reprinted, 1953
Sixth Edition, 1958
Reprinted, 1962, 1963, 1964
Seventh Edition, 1965
Reprinted, 1966
Eighth Edition, 1971
Reprinted, 1972, 1974
Ninth Edition, 1975
Reprinted, Japanese Edition, 1979
Tenth Edition, 1980
Reprinted, 1981, 1983, 1985, 1986

Library of Congress Cataloging in Publication Data

Grant, J.C. Boileau (John Charles Boileau), 1886–1973.
[Method of anatomy]
Grant's method of anatomy.—11th ed. / John V. Basmajian.
Charles E. Slonecker.
p. cm.
Bibliography:
Includes index.
ISBN 0 683-09888-8
1. Anatomy, Human. 2. Anatomy, Surgical and topographical.
I. Basmajian, John V., 1921– . II. Slonecker, Charles E.
III. Title. IV. Title: Method of anatomy.
[DNLM: 1. Anatomy. QS 4 G762m]
QM23.2.G7 1989
611—dc19
DNLM/DLC
for Library of Congress

Printed in Egypt by Elias Modern Press
under supervision of MASS Publishing Co.

To Dora and Jan

Preface to the Eleventh Edition

This student's textbook emphasizes how to learn human anatomy—a subject fundamental to clinical practice. Although this edition includes clinical case presentations and problem solving, your limited time will not be wasted by distracting descriptions of medical conditions. Our intent is not to teach you clinical practice but to guide you in learning and appreciating the anatomical language and principles that you will need. Thus, in the Regional Sections of the book you are given clear and concise samples of *Clinical Cases* that may soon face you in the hospitals and clinics. These case descriptions will help you to discover and understand by deduction and reasoning and to learn the pertinent anatomical knowledge in the pages that follow.

This book represents a *new method* of learning, preferably to be used in conjunction with *Grant's Atlas*, with which it has been closely integrated. It is a frank response to the *GPEP Report of the American Association of Medical Colleges*, which reflected many of the *Problem-Solving Methods* that are being called for in the new basic science curricula. There is no question that this is the "wave of the future" in professional education. It also can be used in conjunction with dissection manuals, e.g., *Grant's Dissector*, in more traditional courses that require dissection of a cadaver and/or lecture demonstrations. The approach is descriptive and deductive and relies on the student's learning by problem solving wherever possible, supplemented by access to dissected material. Heavily emphasized are the relevant facts and ideas that underlie the understanding and retention of the significant anatomical knowledge for general medicine and surgery. Much irrelevant material has been deleted, but the illustrations that helped make previous editions famous have been improved and supplemented. The text has been thoroughly rewritten for today's students who have fewer hours than their counterparts in the past to devote to the study of human structure.

We hope that you will use the Mini-problems at the end of chapters and Summaries at the end of sections to reinforce your new knowledge. While the book is primarily directed to the beginning medical student, we feel it will also be interesting and useful to students in Graduate Studies, Dentistry, Rehabilitation Medicine, and most paramedical courses that require anatomical training.

Grant's Method of Anatomy is intended more than ever to make anatomy rational, interesting, and directly applicable to the clinical problems encountered in the health professions. The text and illustrations are designed to aid the student in understanding and correlating this knowledge rather than memorizing the factual information that is presented. If anatomy is simply memorized and not understood, it will soon be forgotten. The practice of good medicine requires a significant knowledge of anatomy. It is best said in the old adage: "It is not what you learn, but what you remember, that makes you wise." The challenge of this book is that it will assist you in learning your anatomy in a logical and stimulating fashion so that it can be readily recalled when reinforced in subsequent training and practice.

Medical education is always undergoing change in the face of the growing knowledge in biomedical sciences and its applicability to clinical practice. The art of progress in an institution was defined by Alfred North Whitehead as the preservation of order in an environment of change. Anatomy must reflect both order and change in medical education, training, and practice. We sincerely hope that this textbook of anatomy will serve you well.

John V. Basmajian

McMaster University
Hamilton, Ontario, Canada
1989

Charles E. Slonecker

University of British Columbia
Vancouver, British Columbia, Canada

Preface to the Tenth Edition

Courses in human anatomy continue to change not only from decade to decade but also from year to year and even semester to semester. In medical schools, both students and professors have reawakened to the need for recognizing priorities. Clinical emphasis has been rediscovered; hence, anatomy textbooks that present the subject as "basic facts" are out of favor. Fortunately, this book always relied heavily on clinical significance as the major justification for the inclusion or exclusion of details. Hence, a new edition did not dictate a complete revamping. However, this is not a book on clinical examination; it is a book with a clinical foundation, *i.e.*, an anatomical basis for clinical practice. Fortunately, I have been in the very middle of some of the most innovative changes in the teaching and learning of gross anatomy in North America. First at Emory University and since 1977 at McMaster University, I have dealt first with revolutionary modes of studying the subject. At the latter university, students acquire their anatomical knowledge while solving problems of patients in a clinical setting. More than ever, the basic approach of *Grant's Method* seems to be the touchstone of learning all that is important and essential in human gross anatomy, regardless of the educational mode, philosophy, or time allotment. This book is a friend and counselor; it relies on logic, explanation, and rational thought to lead the reader to the learning and retention of principles.

As before, this particular sequence of chapters is not of critical importance in day-to-day work because each section has been fashioned to stand alone. Hence, any sequence of dissections or lectures in specific programs can be adjusted to easily. As before, students should first read (and then reread, whenever possible) the section "General Considerations," which emphasizes systematic anatomy.

While future clinicians studying anatomy for the first time are the chief users of this book, an effort has been made to make it highly usable for biologists, graduate students, and practicing surgeons and physicians. Every attempt has been made to be up-to-date with new developments without increasing the details. Readers will soon note that clear devices are used to increase and decrease the emphasis on specific materials. Small type sections are not for memorization!

Readers will soon find that this book no longer is designed to fit any particular course of teaching. Whether the student dissects or does not is not important here; the book is planned to stand alone *if necessary*. It should make good reading on a desert island. It is a helper so that students can understand "why" and not just "what." Thus the many line drawings emphasize *concepts* and the text relies heavily on *rational explanation* and *deduction*. The human body is a logical system and not just a jumble of parts. Intelligent students want to know "why" and they soon find that understanding the embryologic and functional logic used in this book permits them to understand the body, not just memorize facts about it.

Professor Grant conceived this book in the 1930's and was its sole spirit and author until the end of the 1950's. Since then I have labored with devotion and pleasure in placing the sixth to this tenth edition at the disposal of all students. Always I have strived to maintain the essence and most of the "method" of the original—logical deduction—while shaping the size and content of the volume to the current needs. The fact that teachers and students have found the approach useful and perhaps exciting has provided the fuel that has kept this lamp burning bright. Once more, it is an honor to dedicate this volume to the memory of its founder, Professor J. C. Boileau Grant, one of the greatest teachers of human anatomy in the English language since John Hunter.

John V. Basmajian

1980

Preface to the First Edition

The study of human anatomy may be attempted in either of two ways. One consists in collecting facts and memorizing them. This demands a memory which is wax to receive impressions and marble to retain them. Even so endowed a student will not master the infinite complexities of the subject. The other way consists in correlating facts, that is studying them in their mutual relationships. This leads inevitably to the apprehending of the underlying principles involved, and the *raison d'être* of such relationships. The student will thus learn to reason anatomically and will find the acquisition of new and related facts an easier task. It is the purpose of this book to lead the student to approach the subject from this viewpoint, and it involves certain departures from tradition.

The human body is here considered by regions. In most regions some feature predominates. It may be a muscle, a vessel, a nerve, a bony landmark, or other palpable structure, or it may be a viscus. The regions are for the most part built up around the dominant or central feature.

The markings, lines and ridges, depressions and excrescences on a bone tell a story, as do the scars and irregularities of the earth's surface. Because they are in the main to be interpreted by reference to the soft parts that surround and find attachment to them, the bones are not described together under the heading "osteology," as though they were things apart. The shafts of the bones are considered with the surrounding soft parts, the ends with the joints into which they enter. The bones of the foot are primarily considered as a single mechanism—so are those of the hand and of the skull. The correct orientation of certain bones is given in cases where, without this information, the actions of certain muscles (e.g., *gluteus medius, teres major*) could not be understood.

It is not the mere presence of a ligament or its name that is of interest, but the functions it serves. These depend commonly on the direction of the fibers of the ligament, occasionally on their precise attachments. Many fibrous bands bearing individual names are really members of a community. A challenge thrown at one must be taken up by all. They act in unison, and therefore they are considered together as a unit.

In the consideration of viscera the subject is elucidated by reference to comparative anatomy and to embryology. These are cognate sciences which throw light about the existing structure of man. The positions of the viscera are referred to selected vertebral levels, the vertebral column being an ever present and ever ready measuring rod.

Illustrations, to be of value, must be simple and accurate, and must convey a definite idea. It is for these reasons that they consist entirely of line drawings. Their simplicity encourages the student to reproduce them and, though diagrammatic in nature, they are based on measurements and observations of a great deal of carefully dissected material. Their accuracy, therefore, in those details they are intended to illuminate, has been the object of very considerable work.

The book is meant to be a working instrument designed to make anatomy rational, interesting, and of direct application to the problems of medicine and surgery. The bare, dry, and unrelated facts of anatomy tend rapidly to disappear into forgetfulness. That is largely because its guiding principles are not grasped so as to capture the imagination. Once they are grasped it will be found that details and relationships will remain within certain and easy recall.

J. C. Boileau Grant

September 1937

Acknowledgments

We owe a debt of thanks to many colleagues, students and staff who assisted us in preparing this book. We also thank the editorial staff at Williams and Wilkins, especially their professional consultants who so ably reviewed and criticized the various stages of the manuscript.

Contents

Contents

Introduction and Descriptive Terms

There are few words with a longer history than the word *anatomy*. If we write anatomy, we use the name that Aristotle gave to the science of anatomy 2300 years ago. He made the first approach to accurate knowledge of the subject, although it was derived from dissections of the lower animals only. The word means cutting up—the method by which the study of the structure of living things is made possible.

The boundaries of the subject have widened. Through the use of the microscope and with the aid of stains, the field of Anatomy has come to include microscopical anatomy, or *histology*, and the study of development before birth, or *embryology*. The study of the anatomy of other animals, *comparative anatomy*, has been pursued exhaustively, partly in an endeavor to explain the changes in form, *morphology*, of different animals, including man. *Physical Anthropology*, or the branch of the study of mankind that deals chiefly with the external features and the measurements of different races and groups of people and with the study of prehistoric remains, commands interest of the anatomist. The hereditary, nutritional, chemical, and other factors controlling and modifying the growth of the embryo, of the child, and of animals are within his legitimate field; so also is the growth of tissues in test tubes, *tissue culture*. Feeding and other experiments on animals play leading parts in many investigations.

Individuals differ in outward form and features; for example, how varied are fingerprints and the arrangement of the veins visible through the skin; individuals differ also in their internal makeup. Textbooks, for the most part, describe average conditions where weights and measures are concerned and the commonest conditions where arrangements and patterns are concerned. Owing to the variety of these, the commonest may have less than a 50% incidence; therefore, it may not be truly representative. As data on variations accumulated, the subject of *statistical anatomy* emerged. Some variations are so rare as to be abnormalities or *anomalies*. Among the different races of mankind there are percentage differences in the form and arrangement of structures, just as there are among the different races of other animals—*racial anatomy*, a branch of physical anthropology.

The human body is generally dissected by regions, *regional anatomy*, and described by systems, *systematic anatomy*. The regions of the body comprise (1) the head and neck, (2) the trunk, and (3) the limbs. These can be divided and subdivided indefinitely. The trunk is divisible into thorax, abdomen, and pelvis. The systems of the body comprise the skeleton (study of which is osteology), the joints (arthrology), the muscles (myology), the nervous system (neurology, which includes the brain, spinal cord, organs of special sense, the nerves, and the autonomic nervous system), and the cardiovascular system, which includes the heart, blood vessels, and lymph vessels. The organs or viscera of the body (exclusive of the heart and parts of the nervous system) comprise four tubular systems—the digestive, respiratory, urinary, and genital—and the ductless or endocrine glands. All these are wrapped up in the skin and connective tissues.

Anatomy considered with special reference to its medical and surgical bearing is called *applied anatomy*. Anatomy can be studied profitably by means of cross-sections, *cross-section anatomy*. In the living subject, a great deal can be learned by inspection and palpation of surface parts. This and the relating of deeper parts to the skin surface, *surface anatomy*, are a necessary part of a medical education. *Radiographic anatomy* relies on the X-ray to reveal much that cannot be investigated by other means.

Regarding nomenclature, it may be said briefly that over 30,000 anatomical terms were in use in the various textbooks of Anatomy and in the journals when, in 1895, the German Anatomical Society, meeting in Basle, approved a list of about 5,000 terms (in Latin) known as the Basle Nomina Anatomica (B.N.A.). In 1933, The Anatomical Society of Great Britain and Ireland, meeting in Birmingham, adopted a revision of the B.N.A., known as the B.R.; in 1935, the German Anatomical Society, meeting in Jena, likewise adopted a revision, the J.N.A. or I.N.A. Despite their many excellent points, these found only local and restricted acceptance. In 1955, the Sixth International Congress of Anatomists, meeting in Paris, gave approval to a somewhat conservative revision of the B.N.A. which was submitted to it, and which contained many B.R. and I.N.A. terms. Subsequently minor revisions and corrections were made at a series of International Congresses and the Nomina Anatomica (the Fifth Edition of which was published in 1983 by Williams & Wilkins) is now widely accepted as "official." Translit-

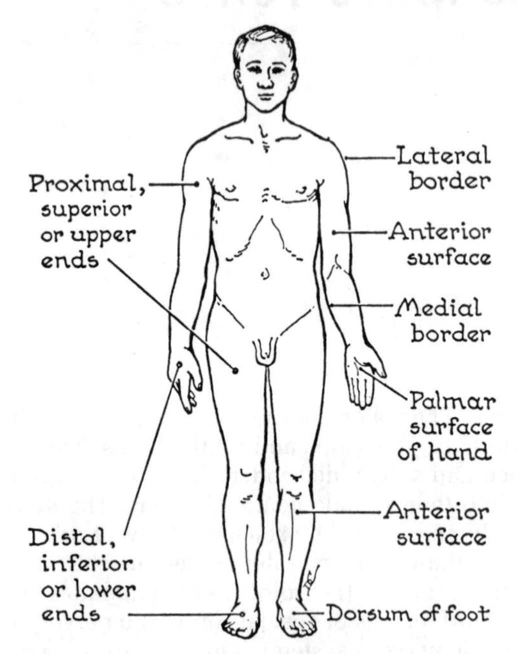

Figure I.1. The subject in the Anatomical Position—except for the right forearm, which is pronated.

eration of the Latin terms into local languages is permitted; indeed it is unavoidable in practice.

DESCRIPTIVE TERMS

For descriptive purposes the human body is regarded as standing erect, the eyes looking forward to the horizon, the arms by the sides, and the palms of the hands and the toes directed forward; this is the **anatomical position**. A person may be lying supine, prone, or whatever, but for descriptive purposes it is assumed that the body is standing erect in the anatomical position. The palm of the hand is understood to be the anterior surface of the hand (*fig. I.1*).

The body is divided into two halves, a right and a left, by the *median or midsagittal plane.* The anterior and posterior borders of this plane reach the skin surface at the front and back of the body at the *median line or midline.*

Terms of Relationship. Three pairs of relative terms suffice to express the relationship of any given structure to another (*fig. I.2*). They are:

1. Anterior or in front = nearer the front surface of the body.
   Posterior or behind = nearer the back surface of the body.
2. Superior or above = nearer the crown of the head.
   Inferior or below = nearer the soles of the feet.
3. Medial = nearer the median plane of the body.
   Lateral = farther from the median plane of the body.

The foregoing terms are applicable to all regions and all parts of the body—always provided that the body is, or is assumed to be, in the anatomical position.

Terms of Comparison. When it is desired to compare the relationship of some human structure with the same structure in, for example, a dog, it is necessary to use a different set of terms, terms related not to space but to parts of the body, such as the head, tail, belly, and back. For example, in man standing erect, the heart lies above the diaphragm; in the dog standing on all fours, it lies in front of the diaphragm; however, in both instances its position relative to other parts of the body is the same; so, *speaking comparatively*, both in man and in the dog the heart is on the head, cranial, or cephalic side of the diaphragm (*fig. I.3*).

Hence, the terms *ventral* and *dorsal, cranial* and *caudal*, as well as *medial* and *lateral* are applicable to the trunk or torso (thorax, abdomen, and pelvis) irrespective of the position assumed by the body. Moreover, it is desirable to employ these terms in embryology and comparative embryology, and it is quite correct to employ them in human anatomy—for no misunderstanding can arise from their use as synonyms for anterior, posterior, superior, and inferior.

In the limbs, terms are coupled with reference to (1) the proximity to the trunk, *proximal* = near the trunk and is synonymous with superior; *distal* = farther from the trunk and is synonymous with inferior; (2) the mor-

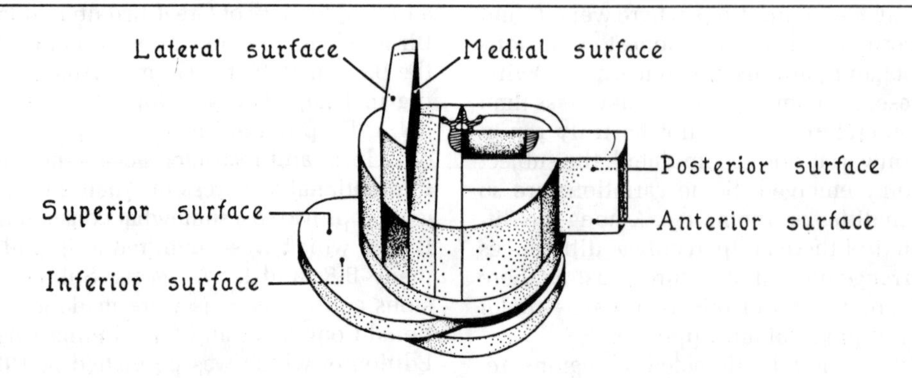

Figure I.2. Three pairs of surfaces involving six essential descriptive terms. They are related to the three fundamental planes in the body.

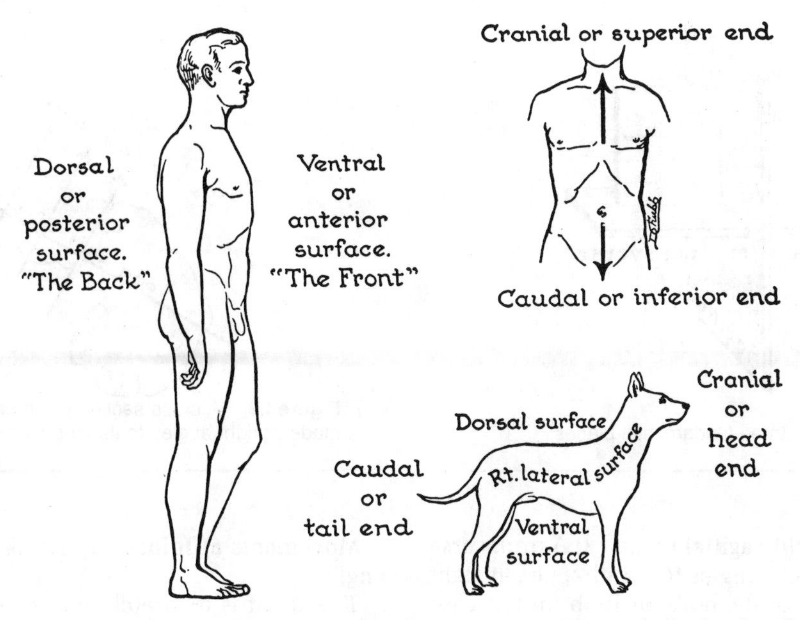

Figure I.3. Three pairs of terms necessary to comparative anatomy and of more general application than those given in *Figure I.2*.

phological borders, *preaxial* = the lateral or radial border (i.e., thumb side) of the upper limb and the medial or tibial border (i.e., big toe side) of the lower limb; *postaxial* = the medial or ulnar border of the upper limb and the lateral or fibular border of the lower limb; and (3) the functional surface, *flexor* and *extensor*, the flexor surface being anterior in the upper limb and posterior in the lower limb.

The anterior surface of the hand is generally called the *palmar* (or volar) surface, and the inferior surface of the foot is called the *plantar* surface. The opposite surfaces are called the *dorsum* of the hand and foot.

Other Terms. *Inside, interior,* or *internal* and *outside, exterior,* or *external,* are reserved (1) for bony cavities, such as the pelvic, thoracic, cranial, and orbital, and (2) for hollow organs, such as the heart, mouth, bladder, and intestine (*fig. I.4*). (Clinicians who use these terms loosely for "medial" and "lateral" cause confusion.)

An *invagination* and an *evagination* (L. *vagina* = sheath or scabbard) are inward and outward bulgings of the wall of a cavity (*fig. I.5*).

Superficial and *deep* denote nearness to and remoteness from the skin surface irrespective of whether at the front, side, or back. These two may be applied to organs, such as the liver and lung (*fig. I.6*).

On, over, and *under* are terms to beware of. They should be used in a general sense and without specific regard to the anatomical position. Carefully avoid using them loosely in place of "superior to" and "inferior to," for such misuse is the cause of much misunderstanding.

Ipsilateral refers to the same side of the body, e.g., the right arm and the right leg. *Contralateral* refers to opposite sides of the body.

Planes. (1) A *sagittal plane* is any vertical anteroposterior plane parallel to and including the median plane. (2) A *coronal* or *frontal plane* is any vertical side-to-side

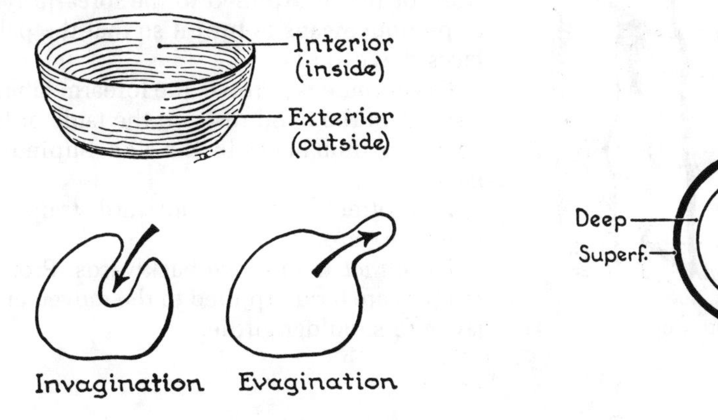

Figures I.4, I.5, and I.6. Three pairs of contrasting terms.

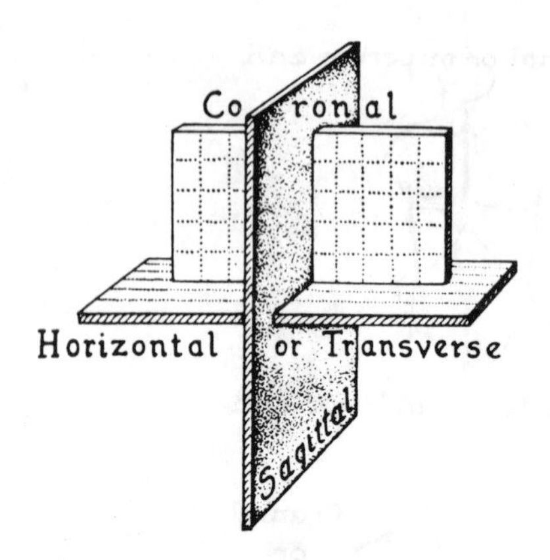

Figure I.7. The three fundamental planes.

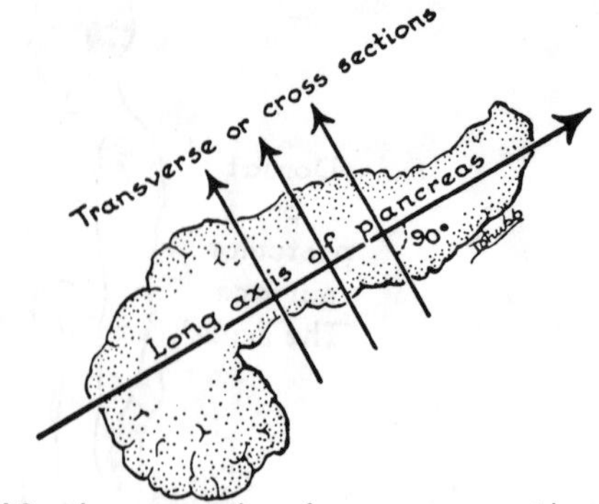

Figure I.9. A cross-section of an organ or part is a section made at right angles to its long axis.

plane at right angles to the sagittal plane. (3) A *transverse plane* is any plane at right angles to 1 and 2, i.e., at right angles to the long axis of the body or limb. In the case of an organ or other structure, a *transverse* or *cross-section* is a section at right angles to the long axis of that organ or structure. (4) An *oblique plane* may lie at any other angle (*figs. I.7–I.9*).

Attachments of Muscles. Muscles are attached at both ends. The proximal attachment of a limb muscle is its *origin*; the distal end is its *insertion*. No function is implied by these terms. When applied to muscles not associated with the limbs, the terms are arbitrarily assigned, historical precedence being the chief determinant.

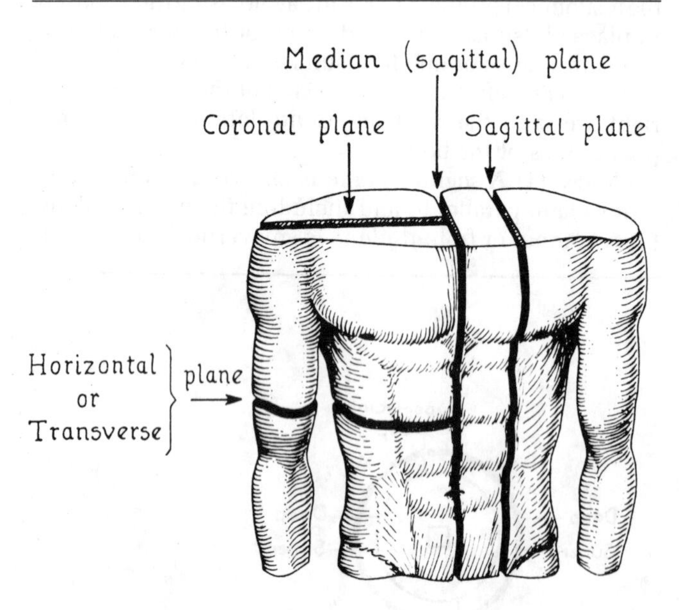

Figure I.8. Fundamental planes in the body.

Movements at Joints. *To flex* is to bend or to make an angle.

To extend is to stretch out or to straighten.

To abduct is to draw away laterally from the median plane of the body.

To adduct is the opposite movement in the same plane (L. *ab* = from; *ad* = to; *duco* = I lead). Movements of abduction and adduction, as well as of flexion and extension, take place at the wrist joint.

The middle finger is regarded as lying in the *axial line of the hand*, and the 2nd toe as lying in the *axial line of the foot*. Abduction and adduction of the fingers and toes are movements from and toward these axial lines, although, as discussed elsewhere, movements of the thumb are named differently.

To circumduct (L. *circum* = around) is to perform the movements of flexion, abduction, extension, and adduction in sequence, thereby describing a cone, as can be done at the shoulder, hip, wrist, and metacarpophalangeal (knuckle) joints.

To rotate is to turn or revolve on a long axis, as the arm at the shoulder joint, the femur at the hip joint, and certain vertebrae on each other.

To pronate was originally to bend or flex the body forward, as in obeisance in prayer, that is, to face downward or prone. Applied to the forearm lying on a table, *to pronate* means to turn it so that the palm of the hand faces downwards.

To supinate is to rotate the forearm laterally so that the dorsum of the hand rests on the table or faces backward when the limb hangs by the side. Supine = lying on the back.

To protract (L. *pro* = forward; *traho* = I pull) is to move forward.

To retract is to move backwards. Protraction and retraction are terms applied to the movements of the lower jaw and shoulder girdle.

General Considerations 1

1

Locomotor Systems

Bone

This chapter mainly concerns the bone as a tissue and its general characteristics, but the vertebrae are described in detail because an understanding of them is fundamental to several regions. The student should not proceed with a study of regional anatomy until this chapter and the succeeding ones are read and understood.

A bone of a living creature is itself a living thing. It has blood vessels, lymph vessels, and nerves. It grows and is subject to disease. When fractured it heals itself, and even if the fracture is not set perfectly, its internal structure undergoes compensatory remodeling in order to withstand strains and stresses as it did before. Unnecessary bone is resorbed. For example, the bones of a paralyzed limb atrophy (become thinner and weaker) from disuse. Conversely, when bones have increased weight to support, they hypertrophy (become thicker and stronger).

Bones have an *organic framework* of fibrous tissue and cells, among which *inorganic salts*—notably, calcium phosphate—are deposited in a characteristic fashion. The fibrous tissue gives the bones resilience and toughness; the salts give them hardness and rigidity and make them opaque to X-rays. One-third is organic; two-thirds are inorganic.

PROPERTIES OF BONE

Physical Properties

By submerging a bone in a mineral acid, the salts are removed, but the organic material remains and still displays in detail the shape of the untreated bone. Such a specimen is flexible (*fig. 1.1*).

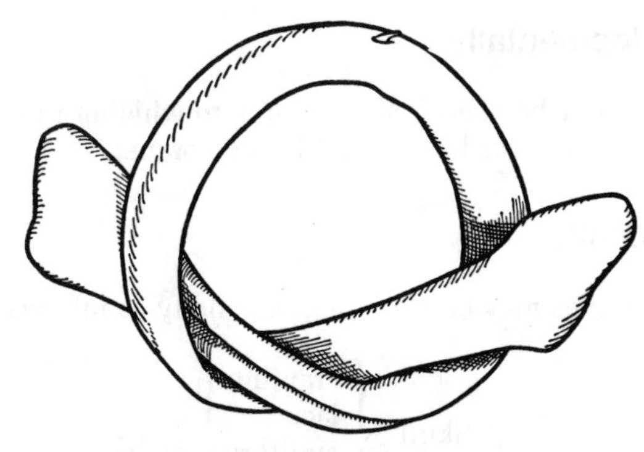

Figure 1.1. A decalcified fibula can be tied in a knot. When untied it springs back to its original shape.

The organic material of a buried bone is removed by bacterial action (i.e., decomposition), and only the salts remain. Being more brittle than porcelain, the bone will crumble and fracture easily. But bones that have lain buried in a limestone cave become petrified (i.e., calcium carbonate replaces the organic material) and, so, they endure; so do those that are mineralized through lying in soils containing iron, lead, or zinc, etc.

Function of Bones

In addition to being (1) the rigid supporting framework of the body, bones serve as (2) levers for muscles; (3) they afford protection to certain viscera (e.g., brain and spinal cord, heart and lungs, liver and bladder); (4) they contain marrow, which is the factory for blood cells; and (5) they are the storehouses of calcium and phosphorus, essential for many functions, e.g., muscle contraction.

Structure

The structure of a dried bone seen on section is shown in *Figure 1.2*. Macroscopically, there are 2 forms of bony tissue: (1) *spongy* (or "cancellous") and (2) *compact* (or dense).

All bones have a complete outer casing of compact bone; the interior is filled with spongy or cancellous bone except when replaced by a medullary cavity or an air sinus (see below). In a long bone, such as the humerus, the compact bone is thickest near the middle of the shaft, and it becomes progressively thinner as the bone expands towards its articular ends, these being covered with a thin shell of compact bone. Conversely, spongy bone fills the expanded ends and extends for a variable distance along the shaft but leaves a tubular space, the *medullary cavity*. The *lamellae* or plates of the spongework are arranged in lines of pressure and of tension, and in an X-ray photograph, the pressure lines are seen to pass across joints from bone to bone (*Plate 1.1*).

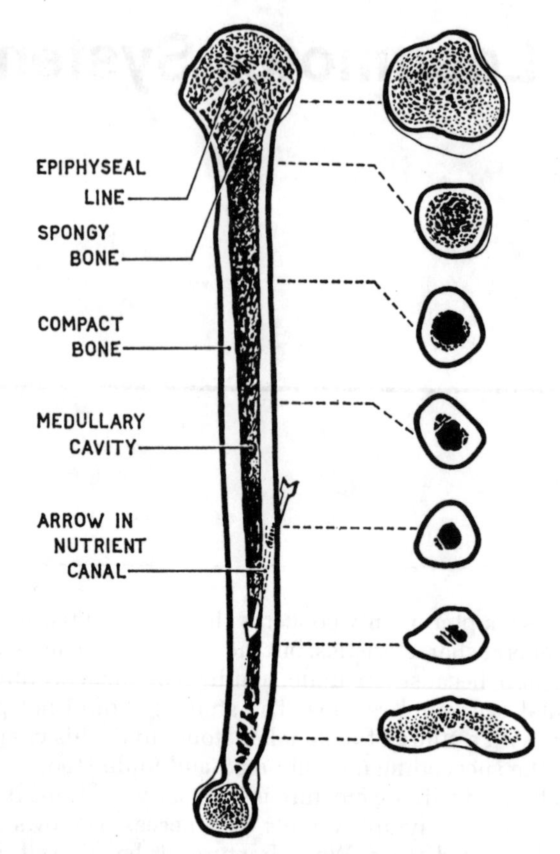

Figure 1.2. The structure of a dried bone as shown by longitudinal and transverse sections of a humerus.

CLASSIFICATION OF BONES

The bones of the body may be classified developmentally, regionally, or according to shape.

Developmentally

They may be classified according to whether they developed (1) in cartilage, or (2) in membrane.

Regionally

The bones may be classified regionally as follows.

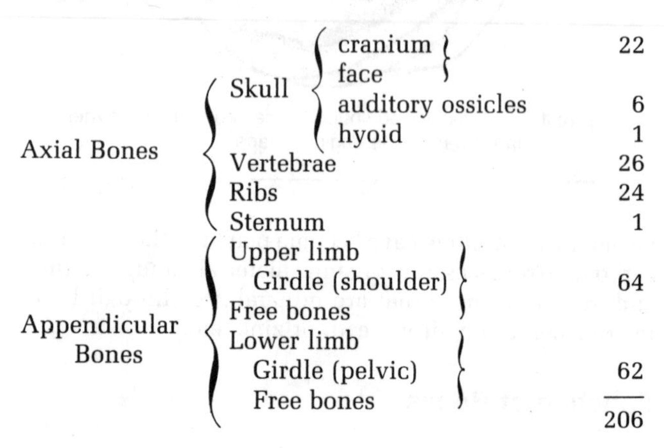

Axial Bones	Skull	cranium, face	22
		auditory ossicles	6
		hyoid	1
	Vertebrae		26
	Ribs		24
	Sternum		1
Appendicular Bones	Upper limb: Girdle (shoulder), Free bones		64
	Lower limb: Girdle (pelvic), Free bones		62
			206

This number is neither important nor exact. It varies with age and with the individual, being larger in youth while the various parts of compound bones are not yet fused together.

According to Shape

1. Long, 2. Short — found in the limbs.
3. Flat, 4. Irregular — found in the axial skeleton and the girdles.
5. Pneumatic—found in the skull.
6. Sesamoid—found in certain tendons.

1. *Long bones* are tubular, confined to the limbs, and serve as levers for muscles. A long bone has a body or shaft and two ends. The *ends*, usually being specialized for joints, are smooth, covered with articular (joint) cartilage, either convex or concave, and enlarged. The *body* is hollow (medullary cavity), thus providing maximum strength with minimum material and weight. Typically, a long bone has 3 borders that separate 3 surfaces, so on cross-section it is triangular rather than circular (*fig. 1.3*).

Long bones develop (are preformed) in cartilage. The

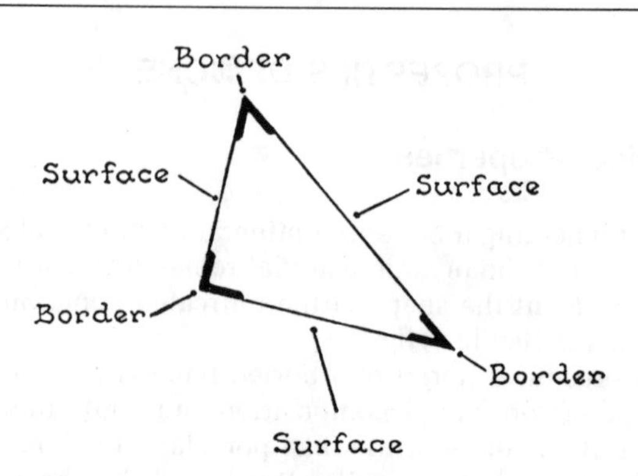

Figure 1.3. The 3 borders of the body (shaft) of many long bones are unbendable, like angle-iron.

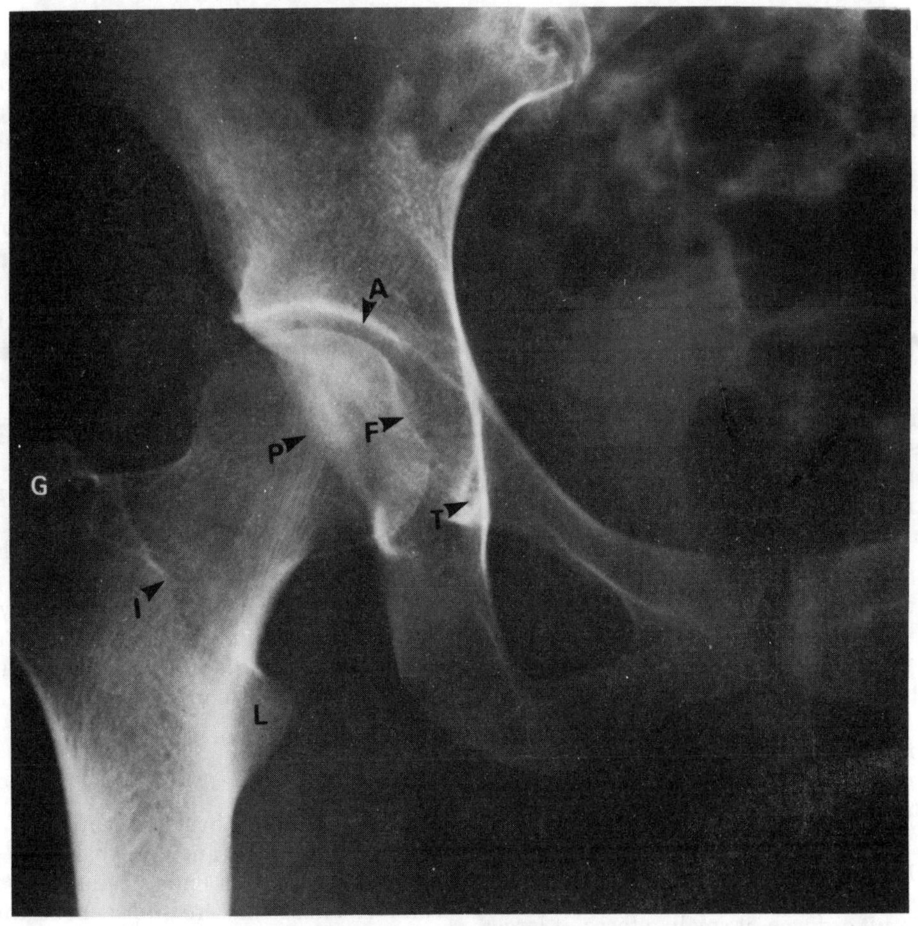

Plate 1.1. Radiograph of hip.

body of every long bone begins to ossify near its middle (primary center) about the 2nd to 3rd month of intrauterine life. One or both ends begin to ossify (secondary centers) soon after birth.

2. *Short bones* are cubical or modified cubes and are confined to the carpus and tarsus. They develop in cartilage, and they begin to ossify soon after birth.

Of the short bones, 3 (calcaneus, talus, and cuboid) start ossifying before birth; so do the epiphyses of 3 long bones (knee end of femur and of tibia and commonly the shoulder end of humerus).

3. *Flat bones* resemble sandwiches. They consist of 2 layers or plates of compact bone with spongy bone and marrow between them. Many of the skull bones, the sternum, scapulae, and parts of other bones, are of the flat type. Most flat bones help to form the walls of rounded cavities and therefore are curved. At birth a flat bone consists of a single plate. In the flat bones of the skull, the spongy bone, here called *diploe* (pronounced "dip-lo-ee"), and its contained marrow appear some years later and split the plate into inner and outer tables of compact bone.

4. *Irregular bones* have any irregular or mixed shape. All skull bones not of the flat type are irregular (e.g., sphenoid, maxilla); so too are the vertebrae and the hip bones. They are composed of spongy bone and marrow within a compact covering.

5. *Pneumatic bones* are formed by the outward expansions of the mucous lining of the nasal cavities and of the middle ear and mastoid antrum that invade the diploe of certain flat and irregular bones of the skull, thereby producing *paranasal air cells* or *air sinuses*. This pneumatic method of construction may be economical in bony material, but it invites infections of the nose that often extend to these sinuses.

6. *Sesamoid bones* are nodules of bone that develop in certain tendons where they rub on convex bony surfaces. ("Sesamoid" is of Arabic origin, meaning like a seed.) The rubbing surface of the nodule is covered with articular cartilage to form true joints; the rest is buried in the tendon. The largest, the *patella* or *kneecap*, occurs in the quadriceps femoris tendon.

Accessory bones may sometimes result from the ossification of connective tissue. Most bones normally ossify from several centers, and it sometimes happens that one or more of these centers fails to unite with the main mass of the bone; again, an abnormal or extra center of ossification may make its appearance and the resulting bone may remain discrete. In either case, the result is an accessory bone, which in an X-ray photograph could be misdiagnosed as a fracture.

MARKINGS ON A DRIED BONE

The surface of a dried bone is smooth, in fact almost polished, over areas covered with cartilage and where tendons play in grooves. Near the ends of a long bone, there are large foramina for veins, arteries and nerves. Piercing the cortical bone of the body (shaft) obliquely, the nutrient canal carries the nutrient vessels and may be 5 cm long.

Markings occur wherever fibrous tissue is attached—no matter whether it is a ligament, tendon, aponeurosis, fascia, or intermuscular septum. Fibrous tissue markings are, however, not present at birth nor in the young. They appear at about puberty and become progressively better marked. The fleshy fibers of a muscle make no mark on a bone.

Terms

Markings take the form of (1) elevations, (2) facets, and (3) depressions.

Elevations, in order of prominence, are as follows: a linear elevation is a *line*, *ridge*, or *crest*; a rounded elevation is a *tubercle*, *tuberosity*, *malleolus*, or *trochanter*; a sharp elevation is a *spine* or a *styloid process*.

Small, smooth, flat areas are called *facets* (*cf.* the facet of a diamond).

A *depression* is a *pit* or *fovea*, if small; a *fossa*, if large; a *groove* or *sulcus*, if it has length. A *notch* or *incisura*, when bridged by a ligament or by bone, is a *foramen* (i.e., a perforation or hole), and a foramen that has length is a *canal* or *meatus*. A canal has an *orifice* (os or *ostium*) at each end.

> *Os* is Latin for both "mouth" and "bone," but it is declined quite differently in each case; thus "of mouth" = *oris* and "of bone" = *ossis*. The related Greek word *osteon* for "bone" gives us words like periosteum and osteoblast (to be discussed soon).

The portion of a notch, foramen, or orifice of a canal over which an emerging vessel or nerve rolls is rounded, but elsewhere it is sharp. Therefore, even on a dried bone the direction taken by the emerging occupant is evident.

Areas covered with articular cartilage are called *articular facets*, if approximately flat. Certain rounded articular areas are called *heads*; others are called *condyles* (= knuckles). A *trochlea* is a pulley.

A LIVING BONE OR A FRESH DISSECTING-ROOM SPECIMEN

The articular parts are coated with a thin veneer of *hyaline* (articular) *cartilage*, which is smooth and lubricated.

Periosteum coats all of the external surface of a bone except in those regions that are covered with cartilage or anchoring ligaments and tendons. It consists of 2 layers: (1) an outer, fibrous membrane and (2) an inner, vascular one lined with bone-forming cells, the *osteoblasts*. The periosteum is easily scraped off with the handle of the scalpel, leaving, however, many osteoblasts adhering to the bone. *Fibrocartilage* lines the grooves where tendons exert pressure. Some elevations seen in the cleaned and dried bone are but shadows of what they were before maceration, because in life they had fibrocartilaginous extensions.

THE PARTS OF A YOUNG BONE

At birth both ends of a long bone are cartilaginous, known as *cartilaginous epiphyses* (*fig. 1.4*). The body of the bone between the cartilaginous ends is the *diaphysis* (Gk. dia = in between, across). It comprises a casing of compact bone that encloses a medullary cavity at its middle and spongy bone at each end, and all are filled with red marrow. The diaphysis is clothed in *periosteum* (Gk. peri = around; osteon = bone).

Epiphyses

(Gk. epi = upon, physis = growth).

(1) During the 1st and 2nd years, one (or both) portions of the cartilaginous ends begins to ossify as secondary centers. Those that are subjacent to the site of articula-

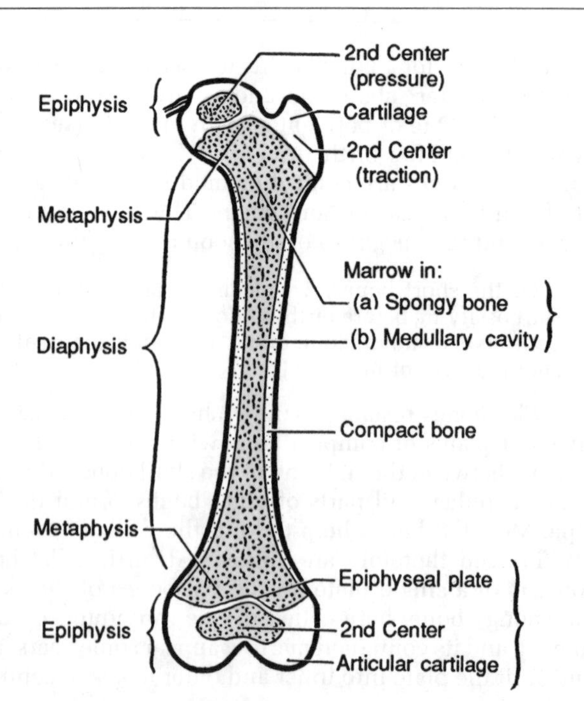

Figure 1.4. The parts of a young bone as shown by a longitudinal section of a femur.

tion, constitute *a pressure epiphysis (fig. 1.4)*. (2) Later, generally about puberty, independent ossific centers appear in the cartilage at the sites of attachment of certain tendons, constituting *traction epiphyses*. (3) A third type of epiphysis is the *atavistic epiphysis*. Atavistic epiphyses phylogenetically were independent bones that later grafted onto other bones.

The epiphyseal cartilage between an epiphysis and a diaphysis is also known as an *epiphyseal "plate."* The region of the diaphysis adjacent to the plate, the *metaphysis* (Gk. meta = beyond), is the site where growth in length takes place.

Where there are 2 epiphyseal ends, the end that has more work to do is the first to start work (ossifying) and the last to stop (to fuse with the diaphysis). When fusion (synostosis) takes place, growth in length ceases.

Nutrient Artery and Canal

The blood supply of living bones comes from many small vessels in the periosteum and from a large *nutrient artery* that enters the body through a *nutrient foramen* and is a constant feature. The nutrient canal (which earlier ran transversely) increasingly occupies an oblique position directed away from the epiphyseal end (*fig. 1.2*). Where there is an epiphysis at both ends, the canal is directed away from the more actively growing end.

Ossification

Except for certain bones of the skull and the clavicle, all the bones of the body pass through a cartilaginous stage. At about the 8th intrauterine week, ossification of the long bones begins. There are 2 types of ossification: (1) intracartilaginous or (endochondral) and (2) periosteal or intramembranous.

After birth, at the center of one or both cartilaginous ends the process of ossification begins, as shown in *Figure 1.4*, and a bony epiphysis takes form. Ossification progresses in the epiphysis until only 2 sheets of cartilage remain: (1) the *articular cartilage*, which covers the end of the bone and persists throughout life, and (2) a residual plate, the *epiphyseal cartilage*, which is placed between the diaphysis and the bony epiphysis (forming a synchondrosis). Ultimately, when the bone has attained its adult length, the plate also ossifies—the site commonly being marked by an *epiphyseal line* (*fig. 1.2*). In technical terms, a synchondrosis has been converted into a synostosis.

Short bones (i.e., carpal and tarsal) ossify endochondrally like epiphyses.

The bones of the skull, except those of the base, do not pass through a cartilaginous stage but ossify directly from membrane; hence, the term *membrane bone*.

Sexual Difference. Ossification starts earlier in females than in males, and it is completed earlier—even by as much as 2 to 3 years.

Bone Marrow

Blood cells are mass-produced by red bone marrow. At birth, spongy bone, which at this age is limited in quantity, and the medullary cavities of the long bones are filled with red (blood-forming) marrow. By the 7th year, the amount of spongy bone has increased simultaneously and the red marrow has extended into it but has receded from the medullary cavities only to be replaced there by yellow (fatty) marrow. About the 18th year, red marrow is almost entirely replaced by yellow in the limb bones; thereafter, it is confined to the axial skeleton—skull, vertebrae, ribs, sternum, hip bones, and upper ends of femur and humerus.

In certain conditions (e.g., some types of anemia) excessive death of the red cells stimulates the yellow marrow to revert to red in order to increase red cell production.

Vessels and Nerves

Arteries supply long bones thus: (1) *periosteal twigs* enter the shaft at many points, run in the small, longitudinal (haversian) canals, and supply the outer part of the compact bone of the shaft (*fig. 1.5*); (2) twigs from *articular arteries*, which anastomose around the joint

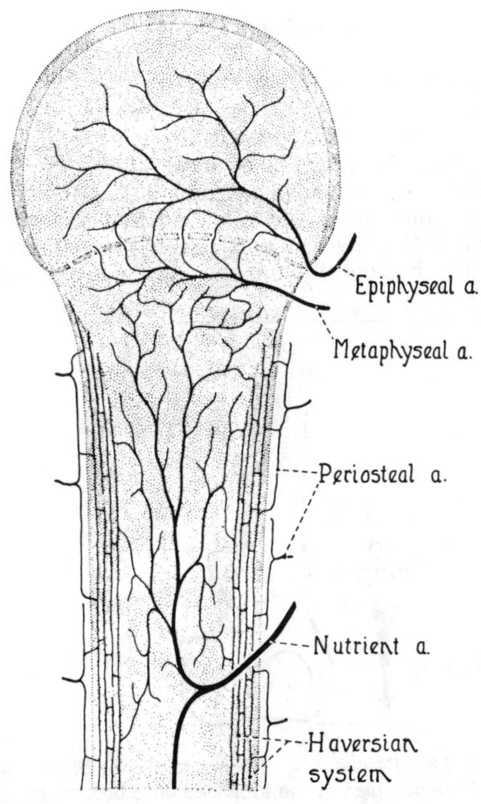

Figure 1.5. Blood supply to a long bone (schematic).

usually between the bone and the reflexion of the synovial membrane, supply the epiphyses, the metaphyseal region, and the capsule; (3) the *nutrient artery* (medullary artery), on entering the medullary cavity, divides into a proximal and a distal branch, each of which supplies the inner part of the compact bone, the marrow, and the metaphyseal region. It is the main artery of the shaft (Trueta and Cavadias). The blood flow through the cortex runs in a centrifugal and not in a centripetal direction (Brookes *et al.*).

The anastomoses between the branches of the nutrient and periosteal arteries seem to be feeble. Though many of the metaphyseal branches of the nutrient artery are end arteries, some of them anastomose with the metaphyseal branches of the articular arteries. Indeed, when the shaft of a long bone is fractured, one or other branch of the nutrient artery is necessarily torn across. Then the anastomoses with the articular arteries must provide collateral flow to the area supplied by the torn nutrient artery with blood.

There are periosteal *veins* and nutrient veins, but the chief veins, enriched with young blood cells, are believed to escape by the large foramina near the ends of the bone.

Lymph vessels exist in the periosteum and in the perivascular lymph spaces in haversian canals.

Sensory *nerves* are plentiful in the periosteum, and nerves (probably to supply blood vessels) accompany the nutrient artery.

BONE GROWTH

About 1764, John Hunter—employing experiments on pigs and using controls—proved conclusively that the growth of long bones depends on both deposition and absorption (*fig. 1.6*). The bones of the base of the skull are cartilaginous bones, but the bones of the vault are membranous bones. There are 2 conflicting views regarding their growth. In one view, the essential mode of growth is by deposition of bone on the exterior and resorption from the interior, modeling taking place as for long bones (J. C. Brash). The other view is that growth is mainly sutural in all parts of the skull, with depositions and resorptions taking place in various areas both inside and out (J. P. Weinmann and H. Sicher; L. W. Mednick and S. L. Washburn; R. K. Rau; B. H. Dawson and D. A. N. Hoyte).

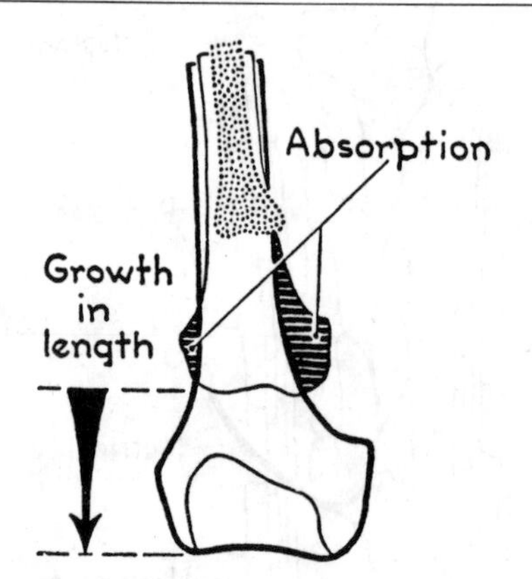

Figure 1.6. Remodelling of bone. As a long bone grows, sites once occupied by the expanded ends become parts of the more slender shaft.

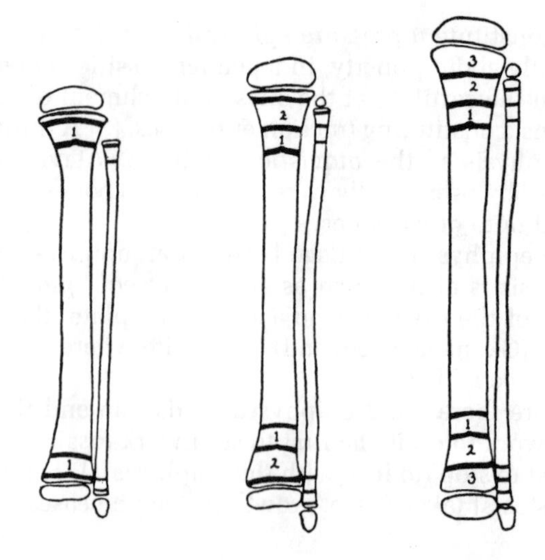

Figure 1.7. Outlines of 3 radiograms of the leg bones of a young girl taken over a period of 2 years. Observe that the 3 lines of arrested growth, denoting 3 successive illnesses, remain equidistant. (After H.A. Harris.)

Shadows of Past Events

The growing skeleton is sensitive to relatively slight and transient illnesses and to periods of malnutrition. When a child is ill or starved, the epiphyseal plates, ceasing to proliferate, become heavily calcified. When growth is resumed, this line of arrested growth appears as a veritable scar (*fig. 1.7*).

Cartilage

HYALINE CARTILAGE, FIBROCARTILAGE, AND ELASTIC CARTILAGE

Cartilage or gristle is a connective tissue in which a solid ground substance (more resilient than bone) forms the matrix. It has no blood vessels, lymph vessels, or nerves, so it is insensitive and dependent on surrounding tissues for its maintenance. There are 3 types of cartilage: (1) hyaline, (2) fibro-, and (3) elastic.

Hyaline Cartilage

Hyaline cartilage (Gk. (h)ualos = a transparent stone) is white and resilient. It is potential bone; in fact, all the bones, except certain skull bones and the clavicle, were preformed in hyaline cartilage.

Hyaline cartilage persists in the adult only at the articular ends of bones as articular cartilage, at the sternal ends of the ribs as costal cartilage, and as the cartilages of the nose, larynx, trachea, and bronchi.

Fibrocartilage

Fibrocartilage has the same structure as fibrous tissue with the addition of cartilage ground substance (*fig. 4.10*). Fibrocartilage bears the same resemblance to fibrous tissue as a starched collar bears to a soft collar. Wherever fibrous tissue is subjected to great pressure, it is replaced by fibrocartilage, which is tough, strong, and resilient. For example, it occurs in intervertebral discs and articular discs (e.g., the 2 menisci of the knee).

Elastic Cartilage

Here cartilage cells are numerous, and the solid ground substance is pervaded by yellow elastic fibers, making it more pliable. It is found only in the external ear, auditory tube, and small cartilages guarding the entrance to the larynx.

Vertebral Column

PARTS OF A TYPICAL VERTEBRA AND THEIR FUNCTIONS

The vertebral column is made up of 33 vertebrae: 7 cervical, 12 thoracic, 5 lumbar, 5 sacral, and 4 coccygeal. The sacral and the coccygeal vertebrae unite to form composite bones, called the os sacrum and os coccyx. The 5 sacral vertebrae completely fuse to form a single mass by the 23rd year; a gap, however, often persists between the 1st and 2nd sacral bodies until the 32nd year (McKern and Stewart). The last 3 pieces of the coccyx fuse together in middle life, and these in turn fuse with the first piece still later.

The bones of each region not only have features characteristic of their particular region, but also every bone in each region has one or more distinguishing features of its own.

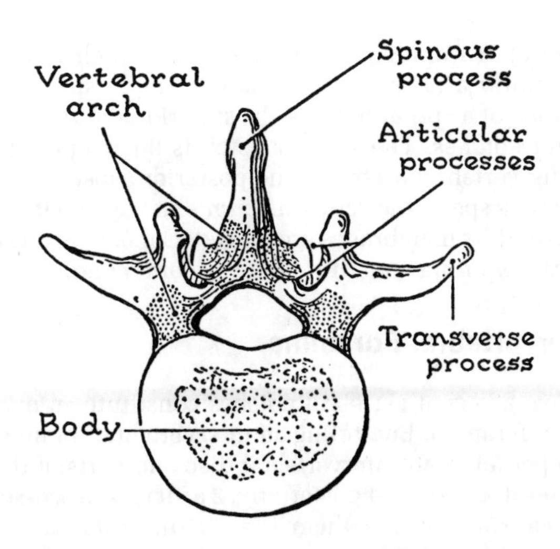

Figure 1.8. The parts of a vertebra, from above.

A vertebra is composed of the following parts (*fig. 1.8*):

1. A weight-bearing part—the *body*.
2. A part that protects the spinal cord—*the vertebral arch*, the series of which form the *vertebral canal*.
3. Three levers on which muscles pull— the *spinous process* and the right and left *transverse processes*.
4. Four projections which restrict movements—2 superior and 2 inferior *articular processes*.

The Body

The body of a vertebra (*fig. 1.9*) is constricted about its "waist," and enlarged at its 2 ends, which are articular (although rather flat). Large vascular foramina are found on its dorsal and lateral aspects. It has a primary center of ossification for the "diaphysis," which appears early, and secondary centers for the epiphyses at its ends.

The Vertebral Arch

The vertebral arch protects the spinal cord from injury. Two broad plates, the *laminae*, meet in the midline. Each is attached to the posterolateral aspect of the vertebral

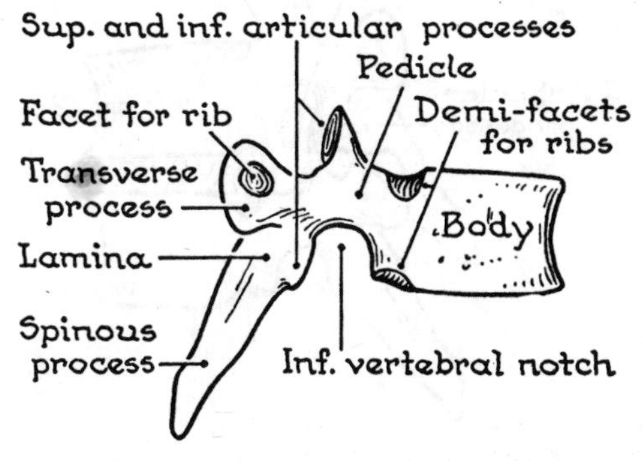

Figure 1.9. A typical vertebra (side view).

body by a rounded bar, the *pedicle*, which is notched superiorly and inferiorly to allow ample space for the passage of a spinal nerve—the *superior* and *inferior vertebral notches*. The inferior notch is the deeper one.

The vertebral arch and the posterior aspect of a body enclose a space, the *vertebral foramen*, in which the spinal cord and its membranes are lodged. Collectively, the vertebral foramina constitute the *vertebral canal*.

Intervertebral Foramina

Two adjacent vertebral notches constitute an intervertebral foramen. Encircling an intervertebral foramen are: two pedicles; an intervertebral disc and parts of the two bodies it unites; and posteriorly, 2 articular processes and the capsule uniting them (*fig. 1.10*). (Foramina is the plural of foramen.)

Transverse and Spinous Processes

The movement of one body on another is effected in part through the actions of muscles on the lever-like transverse and spinous processes. The transverse processes project laterally on each side from the junction of a pedicle and a lamina; the spinous process or "spine" projects posteriorly in the median plane from the site of union of a right and a left lamina.

Articular Processes

Articular processes form synovial joints. They arise near the junction of pedicle and lamina. The superior processes project rather more from pedicles, and their articular surfaces face in a backward direction (backward and upward in the cervical region; backward and laterally in the thoracic; backward and medially in the lumbar), whereas inferior articular processes spring from laminae and face in opposite directions. It is evident that in all regions the contact between superior and inferior articular processes prevents anterior displacement of a vertebra on its companion.

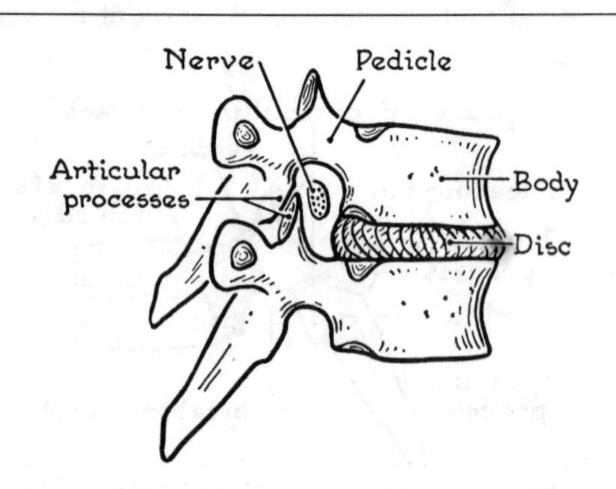

Figure 1.10. Composition of an intervertebral foramen.

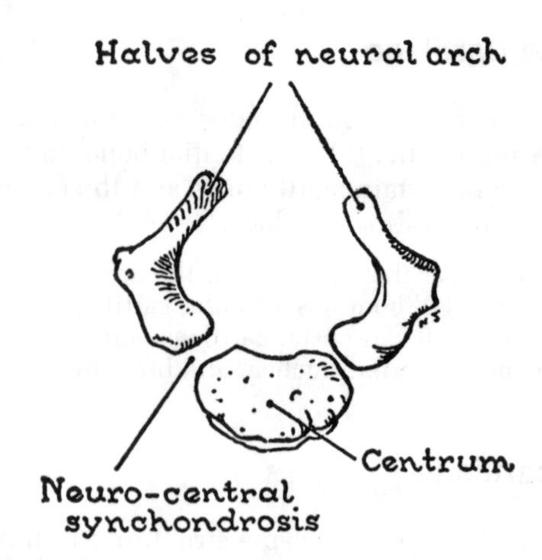

Figure 1.11. Bony parts of a vertebra at birth.

The superior and inferior surfaces of the bodies are the main articular surfaces of the vertebrae, joined together by intervertebral discs (to be described below). The articular processes (except those of the atlas and axis) transmit little or no weight. They limit the unrestricted mobility the bodies might otherwise have and decree in what direction movements between 2 adjacent vertebrae are permitted. There are, however, circumstances in which they do bear weight, e.g., on rising from the stooping position.

Ossification

At birth a vertebra is in 3 parts—a *centrum* and the right and left sides of a *neural arch*, united to each other by hyaline cartilage (*fig. 1.11*). The site of union of a centrum and a neural arch is a *neurocentral synchondrosis*.

Synostosis of the 2 halves of the arch takes place posteriorly during the 1st year, and of the arch and centrum between the 3rd and 6th years.

Epiphyses

Pressure and traction epiphyses appear at about puberty and fuse no later than the 24th year. In most mammals, the pressure epiphyses take the form of plates, but in man they are rings (*fig. 1.19*). Scale-like traction epiphyses appear on the tips of the spinous and transverse processes.

The advanced student requiring details should consult the following: Cervical vertebrae (Chap. 46), thoracic vertebrae (Chap. 5), lumbar vertebrae (Chap. 16), sacrum and coccyx (Chap. 18).

ARTICULATED VERTEBRAL COLUMN

The bodies of the vertabrae contribute three-fourths to the total length of the presacral portion of the articulated column; the intervertebral discs contribute one-fourth (*fig. 1.12*).

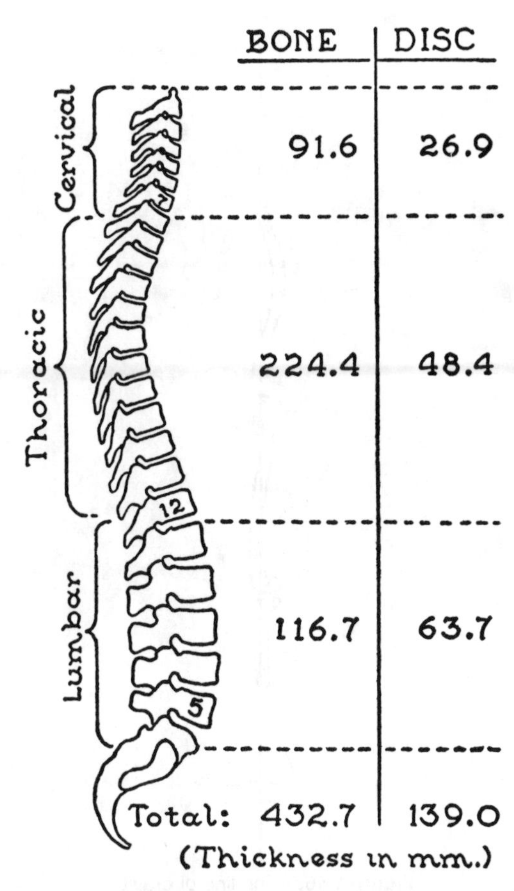

Figure 1.12. Proportions of bone and disc in the presacral parts of the vertebral column (with the use of Todd's data).

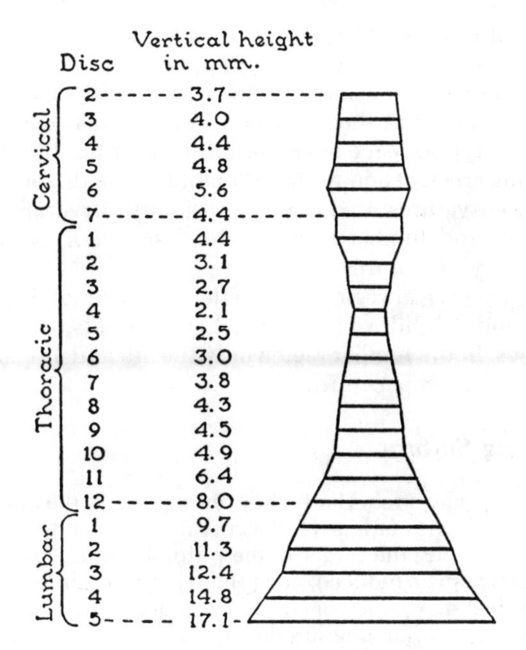

Figure 1.13. Graph of the vertical heights of the intervertebral discs (with the use of Todd's data).

Intervertebral Discs

Details of the structure and function of these joints between adjacent bodies are given in the section of this chapter devoted to joints on p. 14. Movement between 2 vertebrae is most free where the disc is thickest (vertical height greatest), namely, in the cervical and lumbar regions, where the vertebral column is convex forward. Conversely, movement is least where the disc is thinnest— in the midthoracic region (T2–6) (*fig. 1.13*).

Furthermore, in the cervical and lumbar regions each disc is thicker ventrally than dorsally, whereas in the thoracic region, the converse is the case; hence, in each region the disc contributes to the curvature of the column.

Bodies of the Vertebrae

As might be expected, the weight-bearing surfaces increase progressively to the first piece of the sacrum. From there to the tip of the coccyx, they diminish rapidly because the weight is transferred from the first 3 pieces of the sacrum laterally to the pelvis.

The 1st and 2nd cervical vertebrae—the atlas and axis—are highly specialized and their bodies are modified (see p. 000). The support of the skull is shifted to a pair of concave facets on the atlas.

The superior and inferior surfaces of the body of a *cervical* vertebra are oblong; those of the *thoracic* are heart-shaped with long anteroposterior diameters and those of the *lumbar* are kidney-shaped with long transverse diameters.

Curvatures

In prenatal life, the vertebral column is uniformly curved so as to be concave ventrally (*fig. 1.14*). In the thoracic

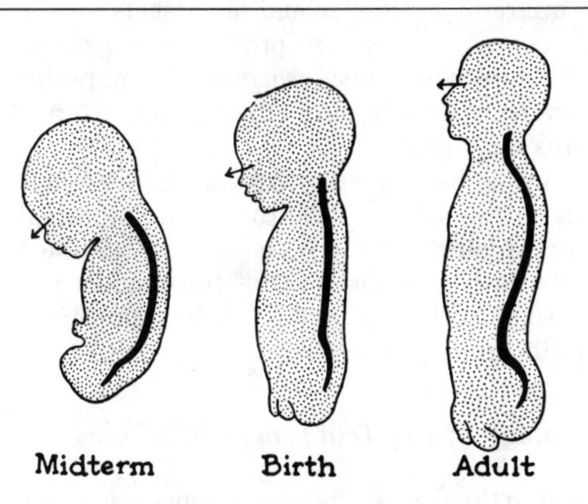

Figure 1.14. Development of the curvatures of the spine. The thoracic and sacral curvatures are primary, and the cervical and lumbar are secondary.

and sacrococcygeal regions, these concavities persist. The cervical curvature (convexity) appears when the infant learns to hold its head erect and to direct its visual axes forward, at about the 3rd month. The lumbar curvature (convexity) appears after the child acquires the art of walking erect, at about the 18th month. The thoracic and sacral curvatures, therefore, are *primary curvatures*; the cervical and lumbar curvatures are *secondary* or *compensatory* curvatures.

In the cervical region, the bodies are of equal depth in front and behind, so the cervical curvature is due solely to discs. In the lumbar region only the 5th and 4th bodies are always deeper in front.

Varying Stature

One may be shorter in the evening than in the morning because with fatigue (1) the curvatures of the spine may increase, (2) the turgor of the pulp of the intervertebral discs may be reduced, and (3) the height of the arches of the feet may be lessened. On the other hand, the stature increases when one lies down.

The Line of Gravity

The line of gravity passes through the body of the axis, just anterior to the sacrum, posterior to the centers of the hip joints, and anterior to the knee and ankle joints (*fig. 1.15*).

Figure 1.15. The line of gravity.

Transverse Processes

Transverse processes arise between superior and inferior articular processes at the junctions of pedicles with laminae, and project laterally. In the *thoracic* region, they act not only as levers for muscles but also as struts for the ribs, so they are strong and stout and, in conformity with the posterior curving of the ribs. They are inclined posterosuperiorly. Their tips bear articular facets for the ribs, except for the 11th and 12th ribs, which are floating; thus the transverse processes of the 11th and 12th vertebrae are reduced in size and lack facets.

Each *cervical* transverse process has a circular foramen, the *foramen transversarium*. The superior 6 foramina transmit the important *vertebral artery*; the 7th transmits only veins.

The *lumbar* transverse processes are thin and flat. Conforming to the shape of the rounded abdominal cavity, they are directed slightly posteriorly. Since the 4th lumbar vertebra lies at the level of the highest part of the iliac crest, it follows that the 5th lumbar vertebra must lie inferior to the highest part of the crest.

Morphology of the Transverse Processes

Except in the thoracic region, a transverse process comprises 2 elements—a *costal* or rib element and a true or *morphological* transverse process, as *Figure 1.16* makes clear.

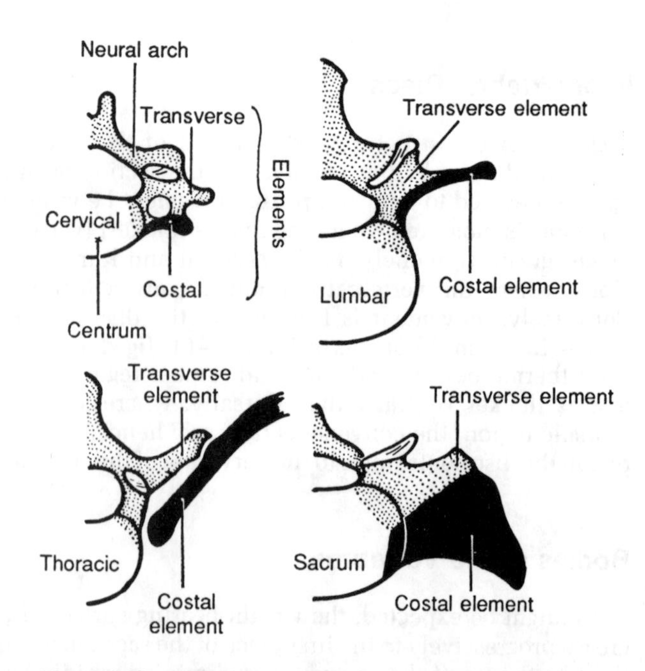

Figure 1.16. Homologous parts of cervical, thoracic, lumbar, and sacral vertebrae. (*Clear* = *centrum; Stippled* = neural arch, transverse and spinous processes; *black* = costal element.)

Pedicles

Pedicles spring from the superior half of the sides of the bodies; vertebral notches mainly lie posterior to the inferior half. *Intervertebral foramina* generally increase in size from superior to inferior (*figs. 1.10 and 1.12*).

Laminae

Laminae overlap like shingles in the thoracic region and slightly in the cervical region. In the lumbar region, there are **interlaminar gaps**, and also in the cervical region when the neck is bent. The largest gaps are between skull and atlas, atlas and axis, 4th and 5th lumbar vertebrae, and 5th lumbar vertebra and sacrum. Through these gaps, a physician can insert a spinal-puncture needle into the vertebral canal and the cerebrospinal fluid within the subarachnoid space, which surrounds the spinal cord and spinal nerve roots.

Vertebral Foramina and Vertebral Canal

In the thoracic region, the vertebral canal is circular and has the diameter of a finger ring—circular because the spinal cord is cylindrical in the thoracic region. But in the regions from which the largest nerve roots for the limbs arise, the canal is larger and triangular, or rather it is expanded transversely in adaptation to the more laterally expanded cord.

In cervical vertebrae 1–3, the vertebral canal is very roomy—so roomy that free movement between the head and the neck does not compromise the spinal cord.

Articular Processes

In all 3 regions—cervical, thoracic, lumbar—the articular processes prevent the vertebrae from slipping forwards, and they allow flexion and extension. In addition, the *cervical articular processes* allow one to look sideways and upward, because their superior facets mostly face obliquely, superiorly, laterally, and posteriorly. The *thoracic* processes allow rotation, but *lumbar* processes prevent it, allowing side bending.

Spinous Processes

The spinous processes become more massive as they are followed inferiorly. The pull on each, as in rising from the stooping posture, is mainly a caudalward one; hence, each is directed caudalward. That of the 1st *cervical* is reduced to a tubercle. In modern man, cervical spines 2–6 are bifid. The 7th ends in a tubercle and is prominent, but not so prominent as the 1st thoracic. Both are easily felt. *Thoracic* spines are long and sloping (*fig. 1.12*), while the *lumbar* spines are thick oblong plates with thickened ends (*fig. 1.8*). The 5th (and 4th) *sacral* spines and laminae are absent, and the sacral canal is exposed. The sacral articular processes on each side fuse to form an irregular crest which ends below in a cornu (horn; plural, cornua). The paired cornua articulate with the cornua of the coccyx.

Vertebral Ligaments

Between adjacent vertebrae and longitudinally along the whole length of the column, various important ligaments join the bones. They are described in the following section on *Joints* on p. 15.

Articulations or Joints

DEFINITION AND CLASSIFICATIONS

A joint is a junction between 2 or more bones. The formal classification is given below for reference:

Formal Classification

Fibrous Joints
- Syndesmoses
- Sutures
- Gomphoses

Cartilage Joints
- Synchondroses
- Symphyses

Synovial Joints

Joints also may be classified as immovable, slightly movable, and freely movable, but perhaps it is more helpful to consider them as: (1) the *skull type* (immovable or temporary joint); (2) the *vertebral type* (slightly movable or secure joint); (3) the *limb type* (freely movable, insecure, or synovial joint).

The skull type is either a suture or a synchondrosis, depending on whether the bones concerned ossify in

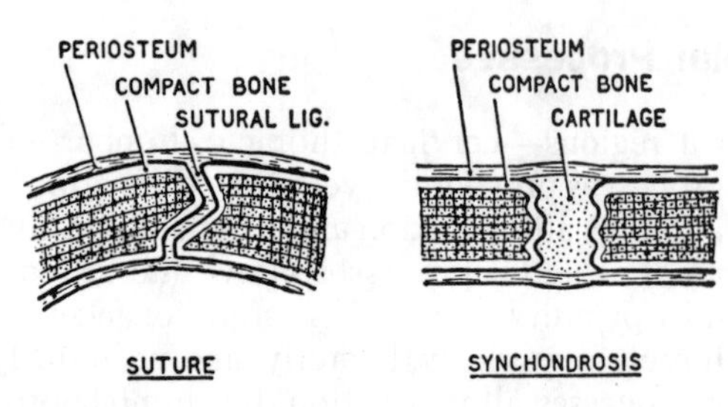

Figure 1.17. A suture and a synchondrosis.

membrane or in cartilage (*fig. 1.17*). **Gomphoses** are the joints of the roots of the teeth with their sockets (periodontal ligaments).

Suture

When the growing edges of 2 bones (or ossific centers) developing in membrane come together, a thin layer of fibrous tissue ("ligament") may persist unossified between them until middle age, or indefinitely. Such a union is called a suture, and sutures are confined to the skull (*fig. 1.17*). The edges may interlock in jigsaw fashion, or like the teeth of a saw (*fig. 26.1*). They may be beveled and overlapping or relatively flat and abutting. A ridge may fit into a groove, as it does between the sphenoid and vomer bones.

Synchondrosis

Similarly, when the growing edges of 2 bones (or ossific centers) developing in a single mass of cartilage come together, a residual plate of cartilage may persist unossified between them for a number of years. Such a union is a synchondrosis and is most typically seen as epiphyseal plates in long bones. Synchondroses also occur at the base of the skull between the various bones that develop from cartilage.

Synostosis

Synostosis is the obliteration of a suture or a synchondrosis by bone. It is associated with cessation of growth locally.

Symphysis

A symphysis is a joint where 2 opposed bony surfaces are coated with hyaline cartilage, are united by fibrocartilage, and are further united in front and behind by ligamentous bands. There is no joint cavity, but a small cleft may be present. They occur (1) between the bodies of vertebrae, (2) between the pubic bones, and (3) between the manubrium and body of the sternum. They are all situated in the median plane.

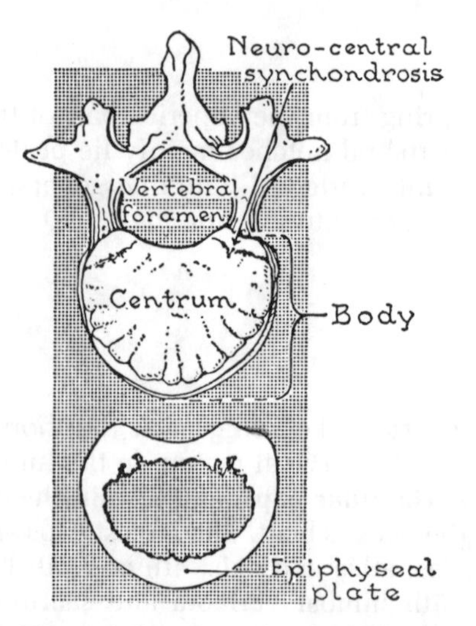

Figure 1.18. A vertebra in childhood.

INTERVERTEBRAL DISCS

Symphysis between 2 Vertebral Bodies

The upper and lower surfaces of the body of a vertebra each consist of hyaline cartilage. In the periphery of the cartilage a ring-shaped bony epiphysis appears (*fig. 1.18*) and finally fuses with the rest of the body in early adult life. The cartilaginous plate persists, helping to enclose an intervertebral disc, of which it might be considered a part.

Intervertebral Discs (fig. 1.19)

Adjacent bodies are united by a fibrocartilaginous disc whose peripheral part is composed of about a dozen con-

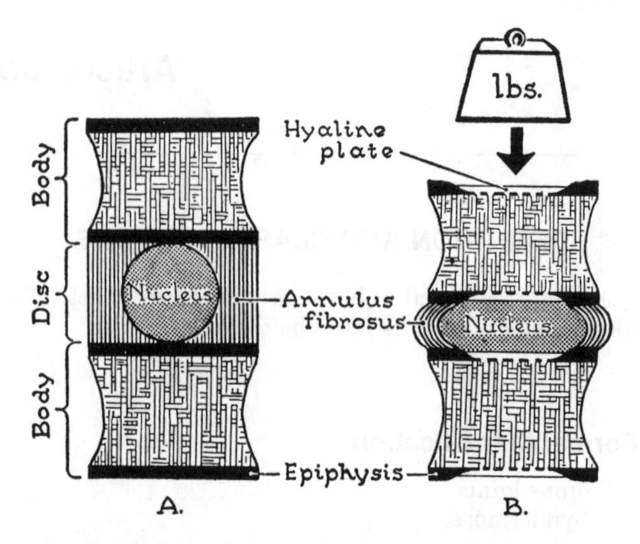

Figure 1.19. Scheme of an intervertebral disc.

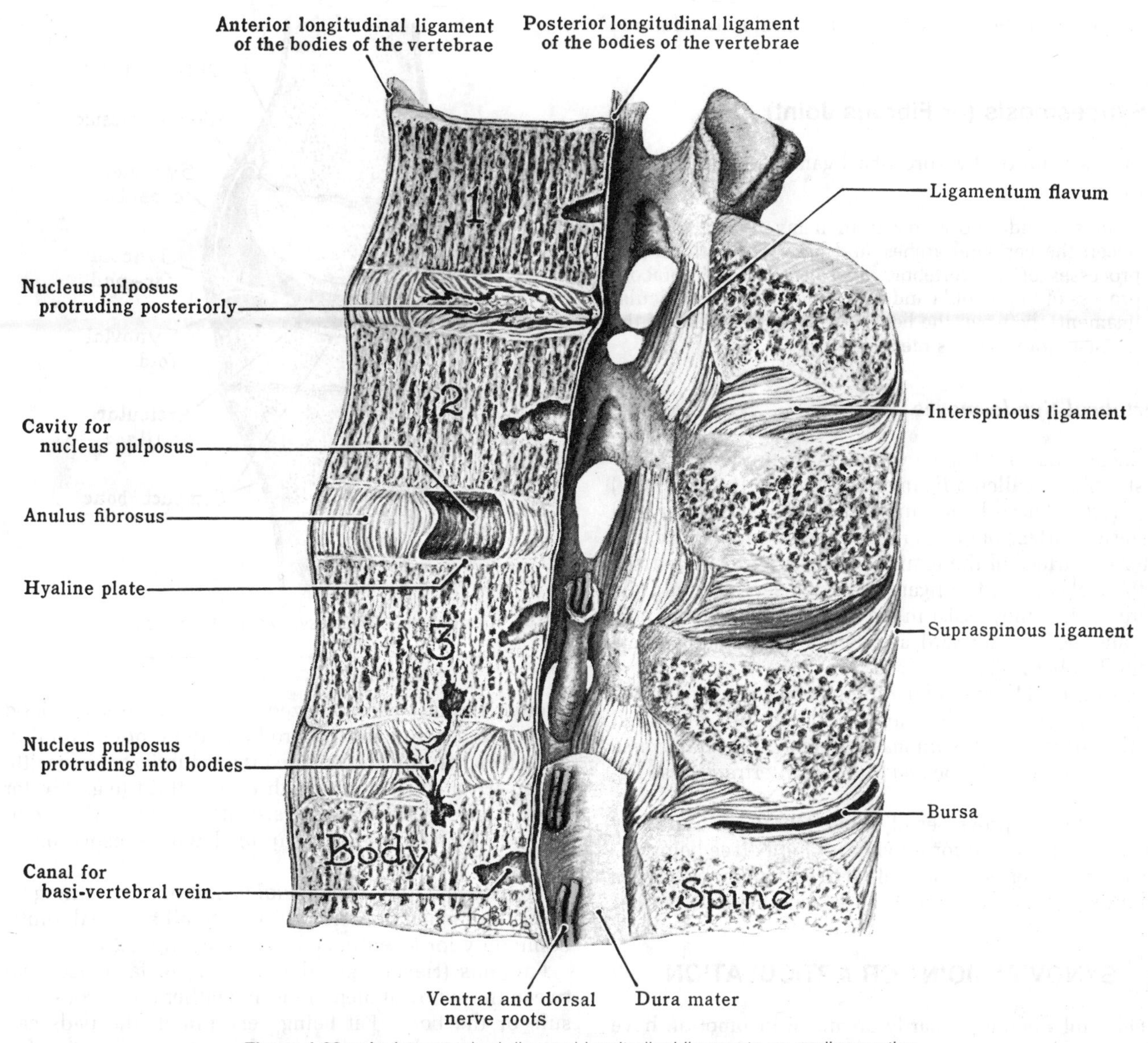

Figure 1.20. An intervertebral disc and longitudinal ligaments on median section.

centric layers of fibers, the *anulus fibrosus*. The fibers of each layer cross those of the next one obliquely, like the limbs of the letter X. The center of the disc is filled with a fibrogelatinous pulp, the *nucleus pulposus*, which acts as a cushion or shock absorber.

Protrusion of the nucleus material through a tear in the anulus is erroneously called a "slipped" disc. The pressure on spinal nerves and their sensitive coverings that may result will cause severe back pain and even partial paralysis. Sometimes when the bodies of the vertebrae are fractured, the hyaline plate also is apt to crack; then the nucleus herniates through it into the spongy bone in the body of the vertebra, forming a nodule that may be detected in a radiograph ("Schmorl's node") (*fig. 1.20*).

Longitudinal Ligaments of the Bodies (fig. 1.20)

An *anterior* and a *posterior* extend from the sacrum to the base of skull: the one is attached to the intervertebral discs and adjacent margins of the vertebral bodies anteriorly; the other is attached to them posteriorly, within the vertebral canal. The anterior ligament is a broad, strong band. The posterior ligament is weak and narrow, but it widens where it reinforces the backs of the discs.

Vessels and Nerves. Small blood vessels from the marrow spaces pass through the hyaline plate to supply the disc until the 8th year, and some of these may persist until the 20th or 30th years (Coventry *et al.*; and others). Branches of the spinal nerves have been traced to the longitudinal ligaments and to the anulus (Roofe; and others).

A Syndesmosis (or Fibrous Joint)

This is a union by cord-like ligamentous fibers (*fig. 1.20*).

Sites. Syndesmoses occur in many places, e.g., between the vertebral arches, and between the lever-like processes of the vertebrae, also between the coracoid process of the scapula and the clavicle (coracoclavicular ligament), between the bones of the forearm and of the leg (e.g., interosseous membranes).

Vertebral Syndesmoses

The laminae of adjacent vertebrae are united by yellow elastic fibers, called a *ligamentum flavum* (L. = yellow) (*fig. 1.20*). These bands unite the superior border and posterior surface of one lamina to the inferior border and anterior surface of the lamina that is superior. By virtue of their elasticity, the ligamenta flava serve as "muscle sparers," i.e., they assist in the recovery to the erect posture after bending forward, and they are particularly strong in the lumbar region.

The adjacent borders of the spinous processes are united by weak *interspinous ligaments*, and their tips are united by the strong *supraspinous ligament*. The transverse processes may be connected by weak *intertransverse ligaments*.

The articular processes of the vertebrae are united by articular capsules to form *Synovial Joints* (see below).

So, typical vertebrae are united to each other by 3 types of joints: symphyses, syndesmoses, and synovial joints.

SYNOVIAL JOINT OR ARTICULATION

The limbs being primarily organs of locomotion have joints that permit free movement. The ends of 2 bones rubbing against each other are capped with hyaline cartilage and are enclosed and united by an *articular capsule* (*fig. 1.21*). This consists of a short sleeve of fibrous tissue, called the *fibrous capsule*, which extends well beyond the articular (= joint) cartilage, and it is lined with an inner thin layer of synovial membrane (the synovial capsule). The synovial membrane is reflected from the fibrous capsule onto the bone, which it covers right up to the articular cartilage.

Synovia or synovial fluid, slippery like egg-white (hence its name—ova, L. = eggs), is a dialysate of blood plasma plus a mucin, called *hyaluronic acid*. The viscosity and lubricating properties of the synovia far exceed those of commercial lubricants.

A synovial or joint cavity develops as a cleft between the ends of the primitive bones. It is lined everywhere either with articular cartilage or with synovial membrane.

A *synovial membrane* is a thin adherent sheet of connective tissues, characterized by its richness in blood vessels and lymphatics; it produces the synovial fluid in sufficient quantity to keep the surfaces lubricated in health. When irritated, it pours forth excess fluid (e.g., "water on the knee"). Rheumatoid arthritis inflames the membrane, which is liberally supplied with sensory nerve endings.

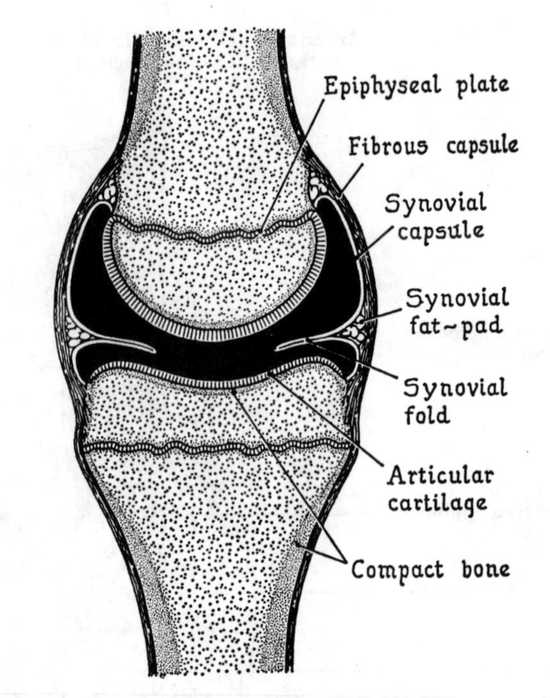

Figure 1.21. A synovial joint, on section.

Transparent *folds of synovial membrane* containing fat at their attached borders project into all synovial joints, commonly for a centimeter or farther (*fig. 1.21*).

Fat-pads (Haver's glands) are pads of fat placed between the synovial membrane and either the fibrous capsule or the bone. Fat being very pliant, the pads can accommodate themselves to changing conditions (*fig. 1.21*).

Articular discs, which are pads of fibrocartilage (or of condensed fibrous tissue) interposed between the articular surfaces of 2 bones, are found in joints where uniquely different movements take place (e.g., in the jaw joint—*fig. 1.22*). On the proximal surface of the disc, one type of movement takes place (e.g., flexion and extension); on the distal surface another type of movement (e.g., rotation or gliding). Discs are nonvascular and nonnervous, except at their peripheral attachments. They are nourished by synovial fluid. Discs are found in the temporomandibular, sternoclavicular, acromioclavicular, radioulnar, and knee joints.

The fibrous capsule is thickened in specialized parts to form cords and bands, called *ligaments*, which withstand temporary strains. Other ligaments are independent of the capsule.

Articular cartilage, having neither blood vessels nor

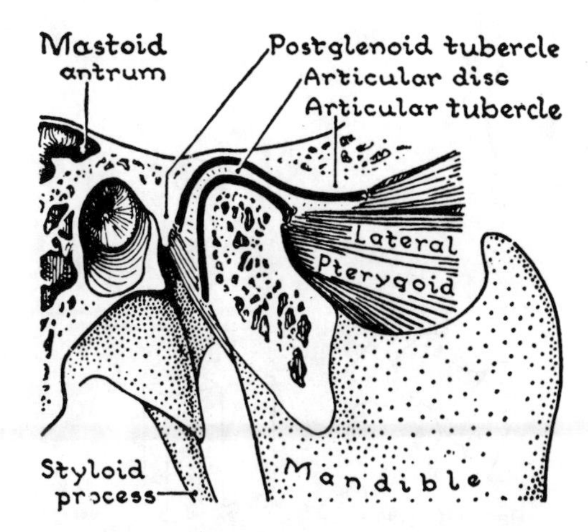

Figure 1.22. Mandibular joint (on sagittal section).

lymphatics (nor has it any nerves), must receive its nourishment by diffusion.The synovial fluid nourishes the cartilage from its free surface and the epiphyseal vessels from its attached surface (Ekholm). Detached pieces of articular cartilage can live, and even grow, in synovial fluid ("joint mice").

Lubrication

The articular components of a synovial joint are incongruous; that is, they do not fit each other reciprocally, as do structures made by machine. If apposing surfaces are parallel or congruous, there can be no self-lubrication. Self-lubrication seems to be dependent upon the presence of (1) a thin, convergent, or wedge-shaped space, (2) a viscous fluid, and (3) a certain speed of movement. Given these 3 factors, 2 bearing surfaces will be completely separated by a film of lubricant (MacConaill). This theory is challenged by others, most notably by "boundary lubrication" (Charnley; Little *et al.*), which proposes a 1-molecule-thick layer separating the surfaces; "weeping lubrication" (McCutcheon), where pressure is believed to squeeze lubricant out of the cartilage onto the rubbing surfaces; "electrohydrodynamic lubrication," in which deformation of cartilage surface is believed to generate wide, thin, fluid films capable of supporting large loads (Dintenfass); and "boosted lubrication," which depends on the trapping of thin pools of mucin between undulating surfaces (Marondas). None of these theories is fully satisfactory.

Articular discs, menisci, and synovial fat-pads and folds are important aids to the formation of thin wedge-shaped films. When at rest, there is no film between weight-bearing surfaces. Indeed, Charnley holds that such a film plays no important role. He emphasizes "boundary lubrication," which depends upon entrapped hyaluronic acid in the spongy articular cartilage. (For a lucid discussion, see Radin and Paul.)

The shoulder and hip joint each has its socket deepened by a pliable ring of fibrocartilage called its *labrum* (L., lip).

Nerve endings are found in the fibrous capsule and synovial membrane. Those in the fibrous capsule are of a type associated with position-sense or proprioception (Ruffini corpuscles) and with pain (free endings). Those in the synovial membrane and its prolongations are believed to be pain receptors and, being associated with blood vessels, they probably supply them (E. Gardner). Articular cartilage has no nerves. Articular discs are also insensitive, except at attached margins.

Blood and lymph vessels are a feature of synovial membrane, but they are absent from articular cartilage. For a discussion of *bursae*, see p. 20.

Classification of Synovial Joints (*fig. 1.23*)

Synovial joints are subdivided into *simple* and *composite* or *complex* (to be discussed); but they may be subdivided also as—

Plane

In *arthrodial or gliding joints*, the apposed bony surfaces are approximately flat, e.g., carpal joints and joints of the small tarsals.

Uniaxial

In (1) *hinge or ginglymus joints*, one surface is concave, the other convex, and movement takes place on a horizontal axis, e.g., the elbow and ankle. (2) In *pivot or trochoid joints*, a ring encircles a pivot set on a vertical axis, and rotation takes place as with a neck in a collar, *viz.*, atlantoaxial and proximal radioulnar joints.

Biaxial

Circumduction is permitted, i.e., on performing the movements of flexion, abduction, extension, and adduction in sequence, a cone is described. In (1) *condyloid joints*, one bony surface is rather round and the other like a socket, but rotation is not a conspicuous feature, e.g., the metacarpophalangeal (knuckle) joints. In (2) *ellipsoid joints*, one surface is an oval and the other a socket, e.g., the radiocarpal (wrist) joint.

Multiaxial

Movements of circumduction and of axial rotation are permitted. In *ball and socket joints*, a ball fits into a socket and provides a universal joint, e.g., the shoulder and hip joints. In *saddle joints*, the surfaces are reciprocally saddle-shaped, e.g., the carpometacarpal joint of the thumb.

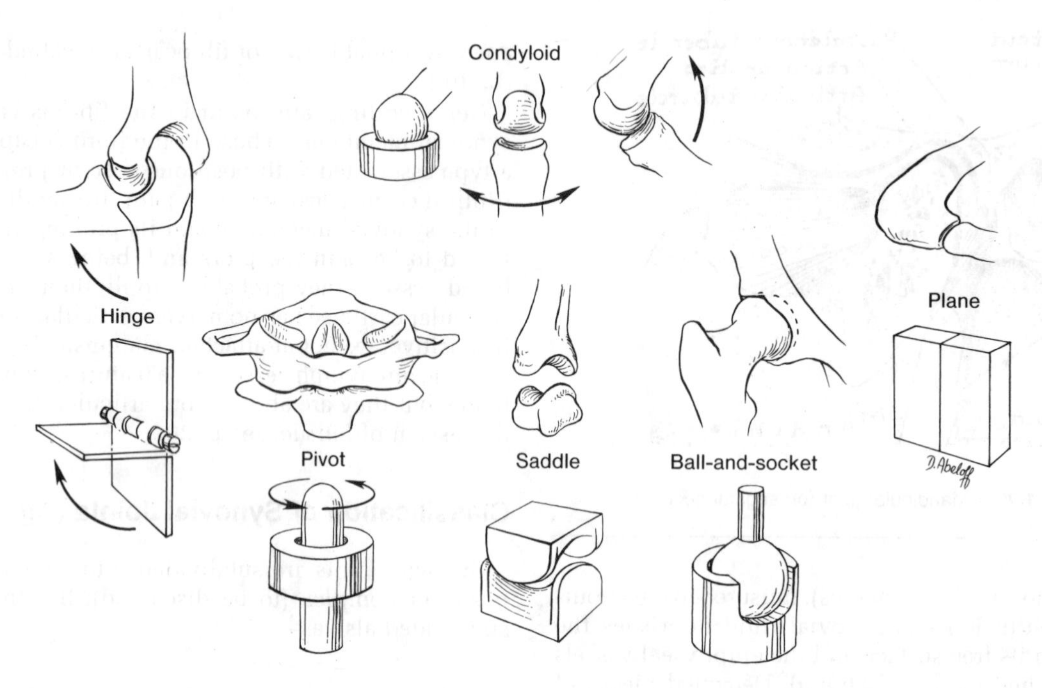

Figure 1.23. Types of joints: hinge (uniaxial, e.g., elbow; condyloid (biaxial), e.g., knuckle; plane (multiaxial), e.g., acromioclavicular; ball-and-socket (multiaxial), e.g., hip joint; saddle (multiaxial), e.g., base of thumb, base of 1st metacarpal and trapezium; pivot (uniaxial), e.g., atlas-axis. (From Basmajian, J.V.: *Primary Anatomy,* 8th edition, Williams & Wilkins, Baltimore, 1982.)

Shape of Articular Surface

No articular surface is flat; neither is any surface like a flat surface that has been rolled up to form part of the surface of a cylinder or cone. What are called "plane" surfaces in the standard textbooks are always slightly curved in 1 of 2 ways: *ovoid surfaces* and *sellar surfaces* (from the Latin word for "saddle" (*sella*).

Congruence of Articular Surfaces

Every (synovial) joint contains at least 1 *mating pair* of articular surfaces, one of these being male, the other female. In all positions of a mating pair except one, the mating pairs fit each other badly—they are loose-packed. The position in which they are fully congruent is called the *close-packed position.* The chief ligaments of the joint are so arranged that they spin and force the surfaces together in the close-packed position in such a way that they cannot be pulled apart. In the close-packed position, we no longer have 2 bones functionally, but only one, because they have been temporarily screwed together (MacConaill and Basmajian).

Muscles

It is from the fancied resemblance certain muscles bear to mice—the tendons presumably being their tails—that the diminutive term "muscle" is derived (L. *mus* = a mouse).

There are 3 types of muscular tissue: (1) *striated skeletal* (or voluntary), such as occurs in the muscles of the limbs, body wall, and face; (2) *heart* or cardiac, which is confined to the heart; and (3) *nonstriated* or *smooth,* visceral (or involuntary), such as is found in the stomach, bladder, and blood vessels.

Skeletal muscles are under the control of the will; hence, they are alternatively called voluntary muscles. Histologically, their fibers possess light and dark cross-striations. Heart muscle also is striated, but neither heart muscle nor smooth muscle is under voluntary control; they are both involuntary.

To contract and to relax is the function of all 3 types of muscle. Skeletal muscles mostly pass from one bone across a joint (or joints) to another bone, and by contracting they attempt to approximate their sites of attachment;

hence, they act upon joints. Heart muscle and smooth muscle mostly form the walls of cavities and tubes, and by contracting they expel the contents.

The distinction between voluntary and smooth muscle on the basis of their ability to be controlled by the will is not always clear. Thus, the diaphragm is structurally a voluntary muscle like the biceps, and though it can be controlled voluntarily, as on taking a deep breath or on holding the breath, ordinarily it works automatically. Again, the upper part of the esophagus is supplied with voluntary muscle and the lower part with smooth; yet, voluntary control cannot be exercised over either part under ordinary circumstances.

It would appear that the distribution of voluntary and smooth muscle is determined not so much by the type of control required as by the character of the contraction required, voluntary (skeletal) muscle having the property of rapid contraction and smooth muscle, of slow sustained contraction without fatigue.

SKELETAL MUSCLE

Skeletal muscle, the subject of the remainder of this section, forms about 43% of the total body weight—it is the "red meat."

Fibers

Muscle "fibers" are in reality thread-like cells consisting of protoplasma or sarcoplasma, multiple nuclei, and a cell membrane, the *sarcolemma* (Gk. sarx = flesh; lemma = a husk or skin.) The fibers range from about 1 mm to 41 mm or more in length. Around each individual fiber, there is some loose areolar tissue (endomysium); around a collection of fibers there is more (perimysium), and around the entire muscle still more (the muscle sheath or epimysium). This areolar tissue permits swelling, gliding and, indeed, independent action of the enclosed fiber or collection of fibers. In a long muscle, e.g., the sartorius (which runs the length of the thigh) several fibers may be arranged more or less end to end.

In some animals muscle fibers are *red* or *dark*, as in the leg of a chicken; others are *white* or *pale*, as in the breast. In man, red and white fibers are thoroughly mixed in different proportions in different muscles (E. W. Walls).

Exceptions

Though skeletal muscles typically cross at least 1 joint and are attached at both ends to bone, certain *cutaneous muscles*, particularly those of the face, are by one end attached to *skin*—through them we express our emotions; others, articular muscles, are attached to the synovial capsules of joints—an *articular muscle*, by with-

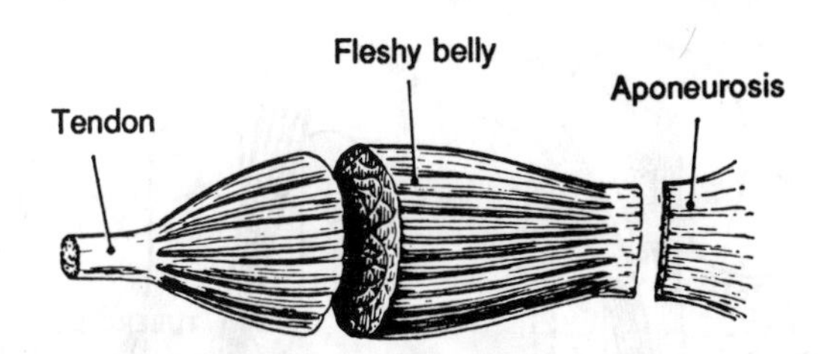

Figure 1.24. Cross-sections of tendons and fleshy belly compared. An aponeurosis is a flat tendon.

drawing the capsule, saves it from being nipped; still others form rings or *sphincter muscles* around the entrance to the orbital cavity, mouth, and anal canal—these close the eyelids, lips, and anus, respectively. The *constrictors* of the pharynx approximate the functions of a tubular muscle; the striated muscle of the esophagus actually is tubular.

The Parts of a Muscle

The proximal attachment of a limb muscle is called the *origin*, and the distal attachment, the *insertion*. The fleshy part is sometimes referred to as the *fleshy belly*. Some muscles are fleshy from end to end, but most are fibrous at one end or at both. The fibrous end has the same histological structure as ligament. When rounded it is called a *tendon*, when flattened and membranous, an *aponeurosis* (*fig. 1.24*). That name suggests it is nervous—but the ancients confused nerves with ligaments and tendinous structures.

The two chief component parts of a voluntary muscle, then, are (1) fleshy and (2) fibrous (tendon or aponeurosis). These have contrasting properties (*Table 1.1*). Tendons are designed to withstand pressure, but owing to their meager blood supply, they readily die (slough) when exposed to infection. Where a muscle presses on bone, ligament, tendon, or other unyielding structure, the fleshy fibers always give place to tendon (*fig. 1.25*). Furthermore, if the tendon is subjected to friction, a lubricating device—a *synovial bursa* or a *synovial sheath*—is always interposed (p. 20).

A muscle that arises by fleshy fibers and is inserted by tendon has a much more extensive origin than insertion (*fig. 1.24*). At fleshy attachments forces are dissipated and make no mark on the bone, but at tendinous attach-

Table 1.1. Contrasting properties of fleshy and tendinous parts of muscles

Property	Fleshy Part	Tendon
Specialization	high	low
Contractility	high	low
Elasticity	high	low
Vascularity	high	low
Upkeep Cost	high	low
Resistance to Infection	high	low

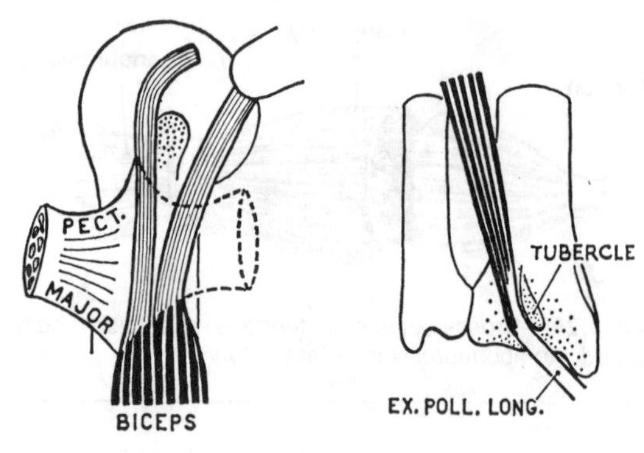

Figure 1.25. Two examples demonstrating a principle: where a muscle is subjected to pressure or friction its fleshy fibers are replaced by tendon or aponeurosis.

ments forces are concentrated and do. Thus, tendinous attachments create ridges, tubercles, and facets, and if large they may produce traction epiphyses.

Tendons are immensely strong; a tendon whose cross-sectional area is 6 cm^2 is capable of supporting a weight of from 4,000 to 8,000 kg (Cronkite).

The fibers of a tendon are not strictly parallel, but plaited; they twine about each other in such a manner that fibers from any given point at the fleshy end of the tendon are represented at all points at the insertional end (*fig. 1.26, A*); hence, the pull of the whole muscle can be transmitted to any part of the insertion (*fig. 1.26, B and C*).

The fibers of tendons, ligaments, and other fibrous structures commonly pass through a pad of fibrocartilage before plunging into their bony attachment. Like the rubber pad employed by the electrician at the junction of the free and the fixed point of a wire, the pads help to prevent fraying from frequent flexing.

Insertions

Muscles are usually inserted near the proximal end of a bone (or lever), close to an axis of movement. So, they help to retain in apposition the ends of the bones taking part in the joint, and thereby give it strength. By contracting unopposed, they produce rapid movement of the distal end (MacConaill and Basmajian).

Some muscles are inserted near the middle of the shaft of a bone (e.g., deltoid), and a few are inserted near the distal end (e.g., brachioradialis).

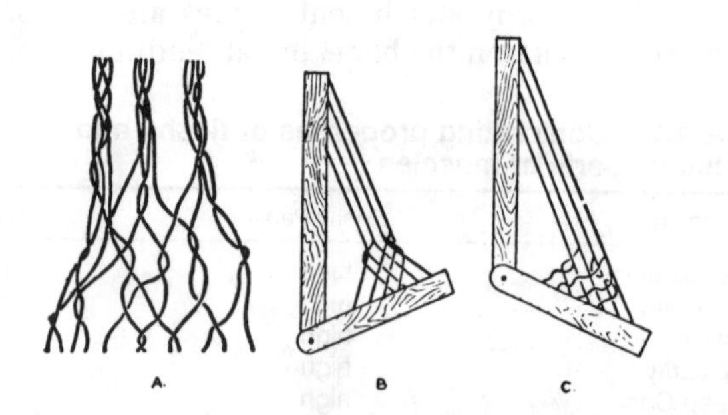

Figure 1.26. *A*, the fibers of a tendon are plaited. *B*, and *C*, in different positions of a joint different fibers take the strain.

A Synovial Bursa

A synovial bursa (L. *bursa* = a purse) is a closed sac differentiated out of areolar tissue. It is roughly the size and shape of a coin. Its delicate walls are separated from each other merely by a film of slippery fluid, like the white of an egg. As a lubricating device, diminishing friction and allowing free movement, a bursa is even more effective that areolar tissue.

Bursae may be classified thus—subtendinous, articular, and subcutaneous.

Subtendinous bursae are found wherever tendons rub against resistant structures, such as bone, cartilage, ligament, or other tendons; hence they are commonest in the limbs. Certain subtendinous bursae constantly communicate with synovial cavities, e.g., the biceps and subcapularis bursae at the shoulder.

"Articular" bursae play the part of accessory joint cavities where they separate hard structures from neighboring capsules, e.g., the subacromial bursa.

Subcutaneous bursae are present (1) at the convex surface joints, which undergo acute flexion, because here the skin requires free movement, e.g., behind the elbow (olecranon bursa), in front of the knee (prepatellar bursae); (2) over bony and ligamentous points subjected to considerable pressure and friction (e.g., under the heel). Most of these, being acquired or occupational, are inconstant.

A Synovial Sheath

This is a tubular bursa that envelops a tendon. In fact, it is 2 tubes, one within the other that are continuous with each other at the ends. The inner or *visceral tube* adheres so closely to the tendon that it is really part of the tendon; it is separated from the outer or *parietal tube* by the potential synovial cavity. The visceral and parietal tubes are united longitudinally, along the surface least subjected to pressure, by a synovial fold, the *mesotendon*, which transmits vessels to the tendon (*fig. 1.24*). If the

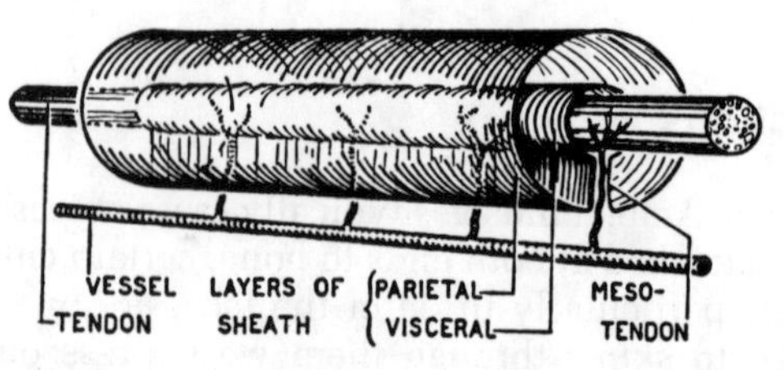

Figure 1.27. Diagram of a short section of a tendon and its synovial sheath.

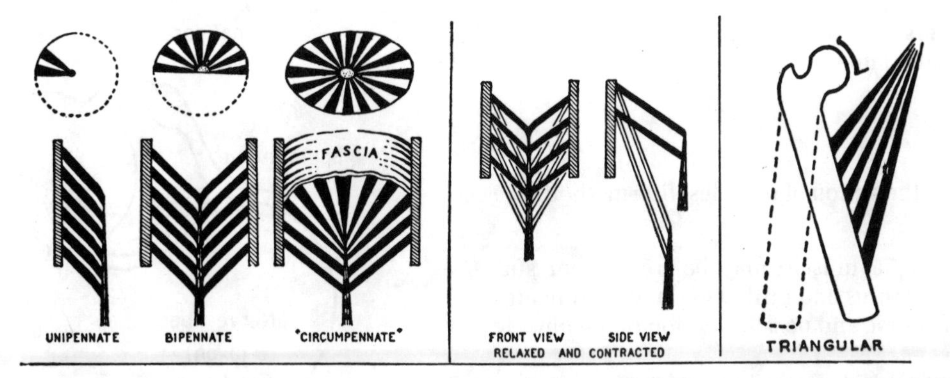

Figures 1.28. Diagrams, showing the architecture of internal structure of pennate and triangular muscles. Pennate muscles, obviously, are powerful muscles.

range of movement of the tendon is considerable, the mesotendon may disappear or be represented by a thread or *vincula* (as in the long digital flexors).

A synovial sheath is required only where a tendon is subjected to friction or pressure on 2 or more surfaces (front and back). This condition occurs only at the hand, foot, and shoulder. In all instances, it so happens that the friction results from the presence of bone on one surface and of an enclosing ligament (retinaculum) on the other. In order to allow ample play, such sheaths extend about 1 cm above and below the sites of friction.

Architecture and Functions of Internal Structure (*fig. 1.28*)

The fleshy fibers of a muscle may be disposed either (1) parallel to the long axis of the muscle, (2) obliquely, like the barbs of a feather, or (3) radially like a fan.

1. *Fibers Parallel to the Long Axis of the Muscle or Approximately so* (*fig. 1.29*). The fleshy fibers may be parallel from end to end, perhaps having a short tendon or aponeurosis at one end or at both ends. This includes the strap-like and flat muscles (e.g., sartorius). The *fusiform type* narrows to a tendon at both ends (*fig. 1.24*).
2. *Fibers Oblique to the Long Axis of the Muscle* (*fig. 1.28*). From their resemblance to feathers, these muscles are called *pennate*, the fleshy fibers corresponding to the bars of the feather and the tendon to the shaft, for they are all inserted by tendon. They are (1) *unipennate* when the fleshy fibers have a linear or narrow origin and have the appearance of one-half of a feather; (2) *bipennate* when the arrangement of the fibers is that of a whole feather; (3) *multipennate* when septa (partitions) extend into the origin and the insertion (e.g., deltoid) the appearance being that of many feathers; and finally, (4) muscles (e.g., tibialis anterior) whose fibers converge from the walls of a cylindrical space to a buried central tendon may be spoken of as *"circumpennate"*.
3. *Radial, Triangular or Fan-Shaped Muscle* (*fig. 1.28*). In these muscles, the fleshy fibers converge from a wide origin or base to an apex, which is necessarily fibrous. And, it creates a rough mark, line, ridge, or process on the bone.

Contraction

When muscle fibers contract or shorten, they necessarily increase in circumference (*fig. 1.29*)—they swell, as exemplified in gross form by the biceps brachii in changing from its relaxed to its contracted state.

On contracting, the fleshy fibers of a muscle shorten between a third and a half (57%, Haines) of their resting length. Being, so to speak, expensive in upkeep, they are never longer than necessary—those of a muscle composed of parallel fibers being 2 or 3 times the length of the distance through which the site of insertion can move. If the distance between origin and insertion is greater than the length of fiber required for full action, the surplus length is fibrous, i.e, tendinous or aponeurotic. (For example, you can easily feel the tendinous portion of the quadriceps muscle as it encompasses the patella ("knee cap") in the lower limb.) A tendon can be any length.

As a corollary, knowing the range of movement of the bony point into which a muscle is inserted, the length

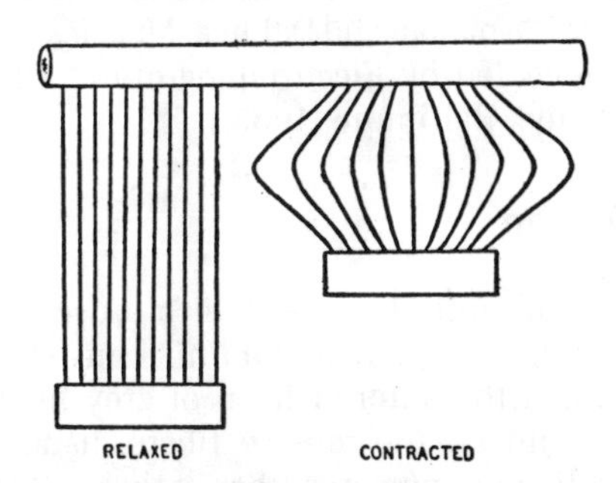

Figure 1.29. On contracting, fleshy fibers shorten by one-third to one-half of their resting length and swell correspondingly.

of the fleshy fibers (or chains of fibers) of that muscle can usually be calculated.

Investigation

In investigating the action of muscles, five methods are available.

1. In the cadaver a muscle may be freed from surrounding structures and pulled upon; this indicates the general course and possible action of the muscle in life.
2. In the living subject a muscle may be stimulated to contract by the suitable application of electricity. This was Duchenne's method of approach, and his work in the 19th century forms the basis of our knowledge of the actions of muscles (Kaplan).
3. Inspection and palpation of the muscles of the region may determine which are in action. Precautions must be taken to avoid confusing synergists with prime movers (see movements below).
4. Clinical information may be gained from the study of the effects (1) of nerve injuries—and these are plentiful during each war—and (2) of transplanting tendons surgically in cases of paralysis. These fall into the category of "natural" experiments.
5. *Electromyography* would seem to be the ultimate method. It consists of placing electrodes in or over the muscle of a living human subject and having him perform a motion. The differences in the electrical action potentials of the muscle are amplified and recorded. There is a direct relationship between the tension developed in a muscle and the electrical activity; so, by this procedure one can analyze the functions of an individual muscle during motion, noting, for example, during what stage of a movement it comes into action and during what stage it exhibits its greatest activity. Furthermore, by using a number of amplifiers, the simultaneous actions of a group of muscles can be studied.

Methods 1 and 2 give information as to what an individual muscle acting alone might do and have less practical importance than methods 3 and 4. Method 5 is throwing much new light on the subject. (For more information, see Basmajian and DeLuca, *Muscles Alive: Their Functions Revealed by Electromyography*, 5th Ed. Williams & Wilkins, Baltimore, 1985.)

Muscle Action

The structural unit of a muscle is a muscle fiber. The functional unit, known as a *motor unit*, consists of a nerve cell, situated in the anterior horn of grey matter of the spinal cord, and all the muscle fibers, usually 100 or more, controlled by the nerve fiber of that cell (*fig. 1.30*).

When an impulse is carried by the nerve fiber to its muscle fibers, they all contract almost simultaneously and for a total time that is quite brief (measured in milliseconds). The result is a twitch of a tiny volume of the whole muscle.

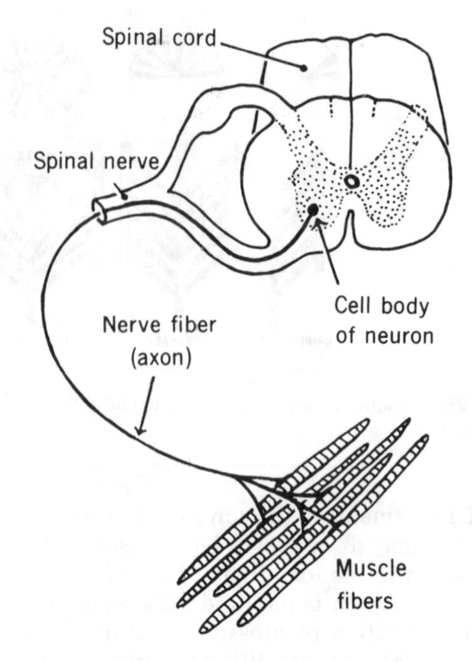

Figure 1.30. Diagram of a motor unit.

Where great precision is required, as with the extrinsic muscles of the eyeball, 1 nerve fiber controls only about a dozen muscle fibers. Only a small proportion of motor units are in action at a given moment—at least they are in different phases of activity, because impulses are discharged by the central nervous system asynchronously.

Movements

By throwing an increasing number of motor units into action and at the same time relaxing the antagonistic muscles (reflex inhibition), movements are produced.

Most bodily actions, even ordinary ones, call into play few principles and more assistants. The principal muscles, *prime movers* or agonists, by actively contracting (shortening) produce the desired movement. Muscles that are so situated that they would usually produce movements in the opposite direction are called *antagonists*.

Again, when a prime mover passes over more than one joint, certain muscles are called upon to steady the unstable joints; such muscles are *synergists* (Gk. syn = together; ergon = work). When antagonists do contract during a movement, their role is either to be a synergist or to act as a brake for deceleration. Still other muscles, *fixators*, are called upon to steady the more proximal parts of the limb or trunk. Obviously, the same muscle may act as prime mover, antagonist, synergist, or fixator under different circumstances as the nervous system dictates.

Examples. The long flexors of the fingers can also flex the wrist joint. So, on clenching the fist or on closing the hand, say on a broom handle, the tendons of the wrist extensors can be felt to contract. They act as synergists.

Gravity is a valuable aid to some movements, depending upon the position of the limb or of the body. For example, on raising the arm from the side, the deltoid is the prime mover at the shoulder; gravity (certainly if there is a weight in the hand) is sufficient to lower the arm. If resistance is encountered, the pectoral muscles and the latissimus dorsi come into play as prime movers. Similarly, in walking, the raised limb tends of its own weight to fall to the ground (MacConaill and Basmajian).

Spurt and Shunt Muscles. While acceleration of a movement is an obvious function of most muscles for moving joints (spurt muscles), in other situations muscles act by pulling along the long axis of the bones and so impart no torque and so no movement (shunt muscles). Special architectural arrangements and accessory mechanisms (e.g., fibrous tunnels enclosing the flexor tendons in the fingers) permit the shunting function of muscles across the early joints, while simultaneously providing spurt impetus across the final joint crossed (see MacConaill and Basmajian for details).

Blood Supply

The chief vessels enter a muscle with the nerve constantly in some muscles and with varying frequencies in others; others again have multiple entries. Accessory vessels, unaccompanied by nerves, are present in many muscles, forming good anastomoses in some and poor ones in others (J. C. Brash). The vessels in a muscle anastomose to form a rectangular network of large and small interconnecting branches. This assures every muscle cell of adequate nutrition. (*Anastomosis* (Gk. stoma = a mouth) means the junction or intercommunication of hollow tubes—usually blood vessels—by means of open mouths.) (Surgeons have expropriated the term for the connection of nerves end to end.)

The veins in the muscles, like those in the limbs in general, have valves; so, muscular exercise, by massaging these veins, aids in circulating the blood.

Lymph Vessels run with the blood vessels.

Nerve Supply

The nerve to a muscle, called a motor nerve, contains fibers from 2 or more consecutive spinal cord segments (e.g., the nerve to deltoid arises from cervical segments 5 and 6). Every motor nerve in a mixed nerve, being about three-fifths efferent (motor), two-fifths afferent (sensory) and containing sympathetic fibers.

1. The efferent fibers pass to *motor end-plates* (i.e., areas of sarcoplasm under the sarcolemma in which a nerve fibril branches like an open hand).
2. The afferent fibers begin as: *free endings* on the muscle fibers; *encapsulated endings* in the connective tissue; *muscle-spindles* (a spindle being a fusiform swelling, 1 to 4 mm long, containing poorly developed muscle fibers over which an afferent nerve fiber spreads or around which it forms a spiral); and *tendon-spindles* at musculotendinous junctions. (They are similar to muscle-spindles, but collagen fibers replace muscle fibers (Bridgman).) It is through the sensory nerves in muscles, tendons, and joints that one is kept informed of the position of the parts of one's body in space; so, even with the eyes shut, one can walk without stumbling and know whether a joint is extended or, if flexed, the degree.
3. The sympathetic fibers supply the vessels.

Motor Point

Muscles placed near the surface may be made to contract on applying an electrode to the skin near, but not at, the point of entry of the nerve. The most effective point of electrical stimulation for each muscle (which overlies the greatest concentration of nerve endings on muscle fibers) is called the *motor point* of that muscle.

Nomenclature

The names given to muscles may be descriptive of their *shape* (e.g., triangularis, quadratus); of their *general form* (e.g., longus, serratus [like a saw], vasti); of the number of heads or *bellies* (e.g., biceps, triceps, quadriceps, digastric), of their *structure* (e.g., semitendinosus. semimembranosus); of their *location* (e.g., supraspinatus, tibialis anterior); of their *attachments* (e.g., brachioradialis); of their *action* (e.g., flexor and extensor carpi ulnaris, abductor and adductor hallucis); of their *direction* (e.g., transversus abdominis); of *contrasting features* (e.g., pectoralis major and minor, gluteus maximus, medius, and minimus).

Variations and Accessory Muscles

Muscles are subject to variation, and books have been written about their variations (Le Double). The muscles that vary most are muscles that are either "coming" or "going," that is, muscles that are appearing in the species and muscles that are disappearing in evolution. Few have any importance. The following paragraphs are offered for interest, not for memorizing.

The *palmaris longus* in the forearm and *plantaris* in the sole of the foot are "disappearing" (the one is absent in 13.7% of 771 limbs, the other in 6.6% of 740 limbs) and when present the sites of origin and insertion and the size of the fleshy bellies vary greatly. The *peroneus tertius* in the leg is "appearing" (it is absent in 6.25% of 400 limbs), and when present it is inserted on metatarsal bones of the foot in a variable manner. The *peroneus longus* has migrated across the sole of the foot from the lateral edge to be attached to bones on the medial side.

The short muscles to the little finger and little toe are commonly in part suppressed (J.C.B.G.).

The widespread fleshy origin of a muscle may spread either a little more or less without affecting the action of the muscle. Thus, the pectoralis minor, typically arising from ribs 3, 4, and 5, may extend upward to the 2nd rib or downward to the 6th more strikingly, the lower portions of the trapezius and pectoralis major may be absent.

Accessory muscles when discovered by students mostly lend themselves to explanation on morphological grounds, e.g., *sternalis* in the front of the chest and the *axillary arch* in the armpit or axilla; but they are not clinically important. Sometimes the biceps brachii has 3 origins (instead of the 2 that its name implies) and the coracobrachialis, 3 insertions (as in some primates), and occasionally they have 4.

2

Cardiovascular and Nervous Systems and Endocrine Glands

Cardiovascular System

HEART AND BLOOD VESSELS

The blood vascular system comprises (1) the heart, and (2) the blood vessels (arteries, capillaries, veins). These form a tubular system which, with few exceptions, is closed, lined throughout with *endothelium* (a single layer of flat cells), and filled with blood.

Blood

The volume of blood in the body is about 5 liters, and it equals about 7% of the total body weight. It is composed of plasma and cells. The *plasma* or fluid portion of the blood is 91% water and 9% solids (e.g., proteins, salts, products of digestion, and waste products), and it contains also respiratory gases (O_2 and CO_2), internal secretions, enzymes, etc. The *cells* are red (erythrocytes) or white (leukocytes), and blood platelet.

The fluid products of digestion are absorbed from the digestive tract into the bloodstream for distribution to the various tissues and organs. Oxygen from the air in the lungs passes into the bloodstream, the plasma dissolving a small amount, the red cells combining with 60 times as much. This then is distributed to the tissues. The waste products of metabolism, formed in the tissues, enter the bloodstream and are brought by it to the excretory organs (kidneys, bowel, lungs, and skin). It is through the thin walls of the capillaries, which consist only of endothelium, that these various interchanges take place. Their walls are semipermeable membranes—permeable to water and crystalloids, but impermeable to the proteins of the blood plasma and to other substances of large molecular composition. Nutritive materials and O_2 pass from the blood to the tissues; waste products and CO_2 return from the tissues to the blood.

Arteries are vessels that conduct blood from the heart to the capillaries. *Veins* are vessels that conduct it back from the capillaries to the heart.

Circulation of Blood

The blood traverses 2 separate circuits (*fig. 2.1*), the *pulmonary* and the *systemic*, each with its own pump. The right side of the heart pumps the blood through the vessels of the lungs—these constitute the pulmonary or

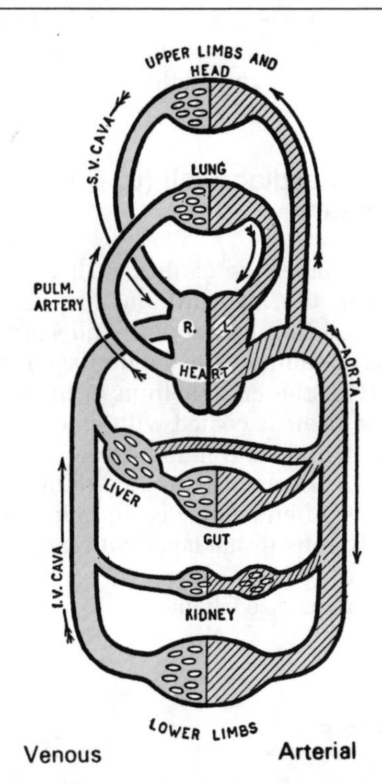

Figures 2.1. Diagram of the circulatory system. Red vessels conduct oxygenated blood; blue vessels conduct blood laden with carbon dioxide. Note the pulmonary and systemic circuits and the double set of capillaries on the subsidiary digestive and renal circuits.

lesser circuit. The left side of the heart pumps the blood through the vessels of the body generally—these constitute the systemic or greater circuit.

To serve these 2 circuits, the heart has 4 chambers, 2 on the right side and 2 on the left. One of these on each side is a thin-walled receiving chamber, the *atrium*; the other is a thick-walled distributing or pumping chamber, the *ventricle*.

The Pulmonary Circuit. The blood entering the right atrium passes to the right ventricle, which pumps it through the pulmonary arteries to the capillaries of the lungs, thence through the pulmonary veins to the left atrium.

The Systemic Circuit. From the left atrium the blood passes to the left ventricle, which pumps it through the aorta and its various arterial branches to the capillaries of the rest of the body. The capillaries join to form veins, which become increasingly larger and fewer till finally the superior and inferior venae cavae and the cardiac veins return to the right atrium the same volume of blood as the left ventricle ejects into the aorta, and at the same rate (*fig. 2.2*).

Normally blood is red when the oxygen content is high and bluish when it is low. Therefore, the blood is *red* in the pulmonary veins, left side of the heart, and systemic arteries and "*blue*" in the systemic veins, right side of the heart, and pulmonary arteries. It is customary in illustrations to color the parts red and blue accordingly.

Though the capacity of each of the 4 heart chambers is practically the same (60 to 70 cc), the resistance offered by the pulmonary vessels is obviously much less than that offered by the systemic ones; accordingly the wall of the right ventricle is much thinner than that of the left.

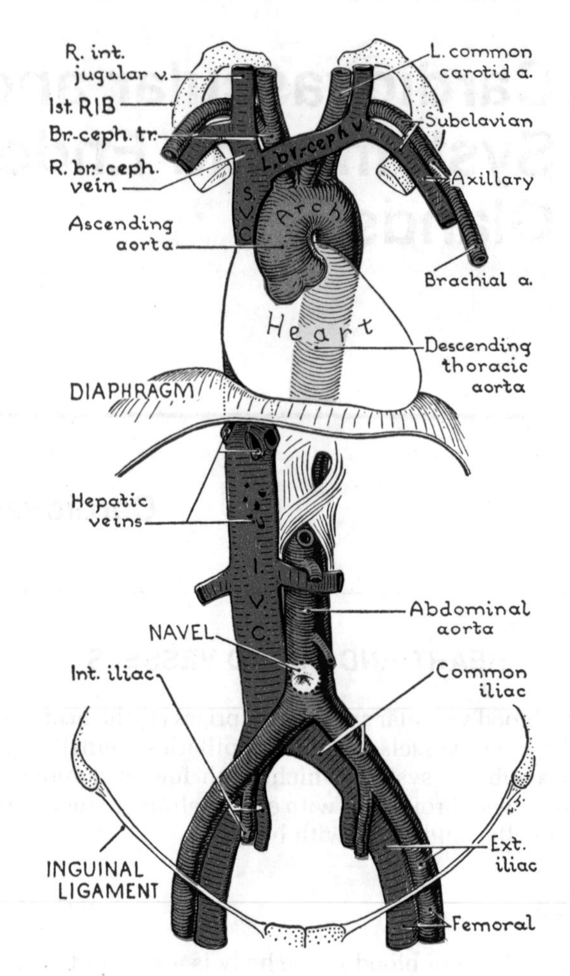

Figure 2.2. The great arteries and veins of the systemic circuit.

Structure of Cardiac Wall (L. *Cor* and Gk. Kardia = Heart)

The main layer of the cardiac wall is the middle or muscular layer, the *myocardium*. Internally, the myocardium is lined throughout with *endocardium*. This is a thin fibrous membrane, lined with flat endothelial cells continuous with the endothelium of the blood vessels. Externally the heart is coated with *epicardium* (visceral pericardium), which, like endocardium, has an extremely smooth surface. Deep to the epicardium run the main cardiac vessels, often embedded in much fat, which develops in its fibrous tissue layer with increasing age and corpulence.

For **heart valves**, see Chapter 8.

Blood Vessels

The main systemic artery (the *aorta*) and the *pulmonary trunk*, each about 30 mm in diameter, branch and rebranch like a tree, the branches becoming smaller as they become progressively more numerous. When reduced to a diameter of about 0.3 mm and just visible to the naked eye, they are called *arterioles*. These break up into a number of *capillaries* (L. *capillus* = a hair), each about ½ to 1 mm long and large enough (7 μm or 0.007 mm) to allow the passage of the red blood cells in single file.

The capillaries form an anastomosing network or *rete* (pronounced *ree*-tee). The cross-sectional area of the entire systemic rete or capillary bed is about 800 times greater than that of the aorta. From this rete, the smallest veins, called *venules*, have their source.

Veins accompany arteries and have the same tree-like pattern, the branches being called *tributaries*.

Distal to the elbow and knee, and elsewhere, arteries are closely accompanied by paired veins, *venae comitantes*, one on each side. These veins are united to each other by short branches which form a network around their artery (*fig. 2.3*). Veins also run independently of arteries (e.g., subcutaneous veins); veins are more numerous than arteries, their caliber is greater, and the pressure within them is much lower.

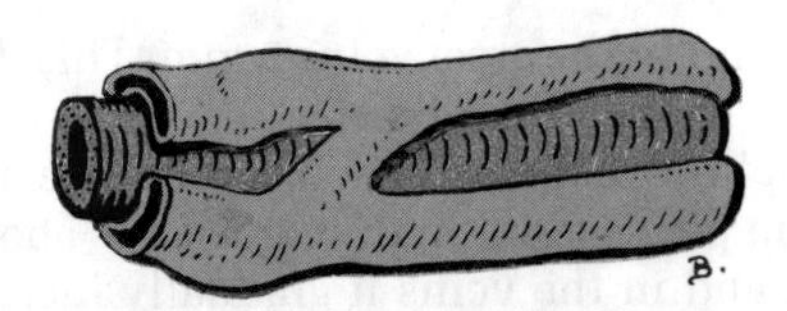

Figure 2.3. Venae Comitantes cling to their artery. (From Basmajian, J.V.: *Primary Anatomy*, 8th edition, Williams & Wilkins, Baltimore, 1982.)

Venous Valves

The inner lining of most medium and small veins is thrown at intervals into delicate semilunar folds, called *valvules*. With the wall of the vein, each valvule forms a semilunar bulging pocket or *sinus*. These pockets are mostly arranged in pairs, facing each other, to form valves. Like the gates of a canal lock, they open in one direction, namely, toward the heart (*fig. 2.4*).

The valves nearest the heart lie at the terminations of the internal jugular, subclavian, and femoral veins (*fig. 2.5*). They imprison the venous blood within the trunk, and during periods of increased intra-abdominal pressure (e.g., during defecation) and during increased intrathoracic pressure (e.g., during deep inspiration), they prevent it from being forced back into the limbs, head, and neck.

There are no functioning valves in the *portal system*, that is, in the veins that bring blood from the stomach and intestines (guts) to the liver; only rarely do those (2 or 3) in the cardiac veins function. Valves are most numerous in the veins of the limbs, and they are commonly placed just before the mouth of a tributary.

To demonstrate valves in the superficial veins of the forearm: circumduct your limb vigorously at the shoulder joint in order to cause the veins to fill; then, keep the forearm below the level of the heart so that the veins do not empty; with the tip of your index finger, obstruct a prominent vein about the middle of the forearm and, by stroking proximally with your thumbnail, empty a long segment of the vein. Note that on removing the thumb the vein refills from above as far as the nearest valve, and that on removing the index finger the empty segment fills from below (*fig. 2.6*).

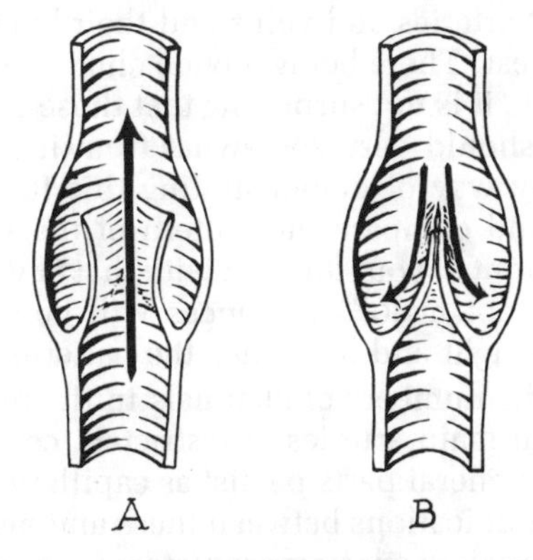

Figure 2.4. A venous bicuspid valve open (*A*) and closed by backpressure (*B*). (From Basmajian, J.V.: *Primary Anatomy*, 8th edition, Williams & Wilkins, Baltimore, 1982.)

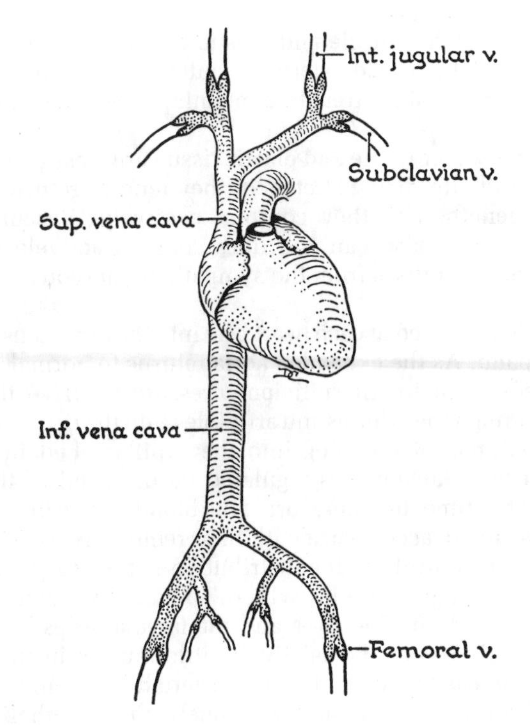

Figure 2.5. The venous blood is imprisoned in the trunk by valves.

Structure

Arteries typically have 3 coats or tunics: (1) the *tunica intima* or inner coat has a lining of endothelial cells, with a little subendothelial fibrous tissue and, outside this, a tube of elastic tissue (the internal elastic membrane); (2) the *tunica media* or middle coat is composed of alternate

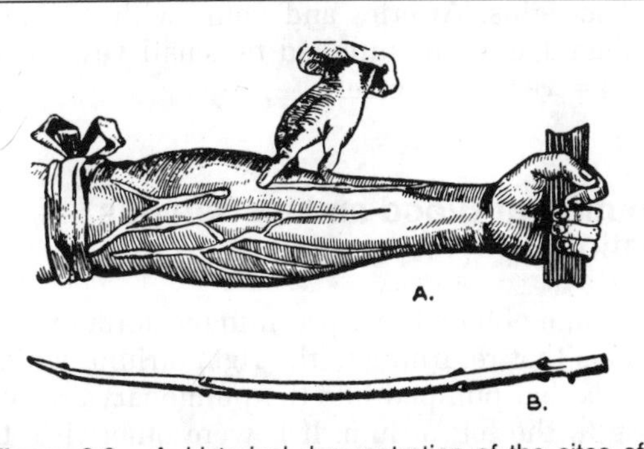

Figure 2.6. *A*, historical demonstration of the sites of the valves in the veins of the forearm during life (after William Harvey—*The motion of the heart and blood in animals*, 1628). *B*, a vein turned inside out, imperfectly showing the valves (after Fabricius, 1603).

layers of smooth muscle and elastic tissue in a fibrous bedding; (3) the *tunica externa* or outer coat is a fibrous tube of considerable strength containing some elastic fibers.

Both muscular tissue and elastic tissue are elastic, like rubber, and after being stretched, they tend to return to normal length, i.e., they contract passively. Vascular smooth muscle also can and does contract actively in response to impulses from the sympathetic nervous system.

The blood ejected at each heart beat into the aorta causes it to expand. As the expanded aorta returns to normal, it continues to maintain sufficient pressure to drive the blood through the arteries and arterioles into the capillary bed. The amount escaping into the capillary bed in a given region, however, is regulated by the need of the tissue at the time. In other words, the blood is distributed to the organs in accordance with their requirements. The mechanism controlling the distribution is the muscle in the walls of the arterioles which, so to speak, can be turned on and off. The aorta and the large arteries have much elastic tissue and relatively little muscle in their walls, whereas the arterioles are essentially muscular.

Capillaries have but one coat, namely, the endothelial lining common to all vessels and the heart.

Veins have the same general structure as their companion arteries. Their caliber, however, is greater; the blood pressure within them is lower; they have less muscle and much less elastic tissue. A vein may be described as an endothelially lined fibrous-tissue tube in which a thin muscle coat is embedded between the endothelium and the adventitia.

Some hours after death, all the muscular tissue in the body contracts; hence, the corpse becomes rigid, in *rigor mortis*, and the blood is driven from the arteries and the more muscular veins. Even after the rigor passes off and the vessels relax, the arteries are empty and the veins are filled with blood. Arteries (Gk. and L. arteria = an air tube) derive their name from the notion that they were conductors of air like the trachea, which was once known as the *arteria aspera* or rough air tube.

The heart wall is supplied by the coronary arteries and the cardiac veins. Arteries and veins with a diameter greater than 1 mm are supplied by small vessels, *vasa vasorum* (= vessel of vessels).

Distribution of Blood and Its Nervous Regulation

The volume of blood pumped into the aorta in a given time equals that returning to the right atrium, which in turn equals that pumped into the pulmonary artery and returning to the left atrium. If it were otherwise, there would be temporary or permanent local congestion—which, in fact, occurs with heart failure. The blood pressure within the aorta is about 120 mm of mercury, at the arterial end of the capillaries about 30 mm, at the venous end 12 mm, and in the great veins 5 mm. This determines the direction of the blood flow.

The rate of flow in the aorta, when the subject is resting, is about 0.5m per second, in the capillaries about 0.5 mm per second, and in the veins it gradually increases until in the venae cavae it nearly equals the rate in the aorta.

The Heart

Efferent vagal and sympathetic fibers supply the heart. Stimulation of the vagus results in slowing of the heart; stimulation of the sympathetic in acceleration. When the blood pressure in the arteries rises, afferent (sensory) fibers are stimulated. These are the vagus (10th cranial nerve) distributed to the aortic arch (aortic nerve) and the glossopharyngeal (9th cranial) nerve distributed to a swelling on the internal carotid artery, the *carotid sinus* (sinus nerve). This reflexly brings about slowing of the heart and vasodilatation of the vessels with consequent fall in blood pressure.

The Blood Vessels

All vessels, particularly the arterioles, are supplied by efferent (vasomotor) nerve fibers of the sympathetic nervous system—which almost always cause vasoconstriction—and perhaps in some locations by parasympathetic fibers having an opposite effect (vasodilatation). Various degrees of vasoconstriction allow the blood to be partly shut off from one part of the body and diverted to another. For example, during digestion, it is diverted to the abdominal viscera; during hot weather, it is diverted to the skin so that heat may be lost.

Anastomoses and Variations

During development, networks of vessels sprout into actively growing parts (e.g., organs and limbs) and, as the parts enlarge, the networks advance farther into them. Certain channels through these networks are chosen to be permanent arteries and veins and their branches; the others disappear. There being a wide choice of channels in the network, it is not surprising that those selected for permanency should vary somewhat from individual to individual. By way of demonstrating this fact, pull up your sleeve and compare the pattern of the superficial veins on the front of your forearm with that of your neighbor; minor if not major differences will be seen; even between your right and left limbs the patterns differ.

Although the numbers of channels in the network retained to form main arteries is restricted, commonly to 1 or 2, the peripheral parts persist as capillary channels, and the communications between these and neighboring capillaries are called anastomoses (*stoma*, Gk. = mouth). Anastomoses also occur between certain large vessels (e.g., coronary arteries of the heart, arterial circle at the base of the brain) and between many small vessels and precapillary vessels. In cases of obstruction of the larger arteries, it is by the enlargement of these anastomoses that a collateral circulation may be established and the vitality of distant (or neighboring) parts preserved. Failure to achieve an adequate supply has serious results.

An arterial channel, which under ordinary circumstances disappears, may persist as an *accessory* or *supernumerary* artery; if the normal artery disappears the persisting artery will act as a substitute for it. This applies to veins, too.

Arteries are *sinuous* or *tortuous* when supplying parts that are highly mobile like the cheeks and lips, protrudable like the tongue, or expansile like the uterus and colon.

Arteriovenous Anastomoses

In some regions, arterioles communicate directly with venules, e.g., in the skin of the palm of the hand and finger-tips, in the nail bed, in the skin of the lips, nose, and eyelids, and at the tip of the tongue. This is in addition to the regular communications by capillaries.

End arteries are arteries that do not anastomose with neighboring arteries except through terminal capillaries. Obstruction of such an artery is much more likely to lead to local death, resulting in the case (1) of a cerebral artery, in paralysis, (2) of the central artery to the retina, in blindness, (3) of a branch of the renal or splenic artery, in death of a segment of the kidney or spleen, and (4) of several adjacent end arteries of the gut, in gangrene of the gut.

Sinusoids

In parts of certain organs (e.g., liver, spleen, bone marrow, suprarenal glands), the place of capillaries is taken by irregularly wide, tubular spaces, called *sinusoids*. The lining cells differ somewhat from endothelial cells, many of them being phagocytic, i.e., capable of engulfing circulating particulate matter.

Cavernous Tissue

In the erectile tissue of the external genitals (penis in the male, clitoris in the female), there are innumerable venous spaces lined with endothelium and separated by partitions containing smooth muscle. In the mucous membrane of the nasal cavities, arterioles open into wide and abundant venous spaces, too. Therefore, the mucous membrane of the nasal passages also is erectile. When one has a "cold in the head," the venous spaces become engorged and may obstruct the airway.

Vascularity

Cellular tissues are highly vascular, e.g., muscles being the cellular engines of the body require much fuel; glands (kidney, thyroid, suprarenal, and liver) are also very vascular; so, obviously, are the lungs. *Connective tissues* are only slightly vascular; thus, loose and dense fibrous tissues (e.g., deep fascia, tendons, ligaments) have a very meager blood supply; fat and bone have a fair supply, but hyaline cartilage, the cornea, and the epidermis are nonvascular, receiving their nutrients by seepage from their neighbors. *Nervous tissues* vary. The gray matter of the brain and spinal medulla, being cellular, is more vascular than the white matter and the peripheral nerves.

LYMPHATIC SYSTEM

Tissue fluid is the fluid that bathes the cells (and fibers) of the tissues of the body, and it resembles blood plasma in chemical composition. From this fluid the cells get their nutritive material; to it they give their waste products, and through it they respire. Between the tissue fluid and the plasma of the circulating blood, a constant interchange of fluid and dissolved substances takes place. However, some fluid transudes into the *lymph capillaries*, whence it is drained by *lymph vessels* through *lymph nodes* to the great veins at the root of the neck where it rejoins the bloodstream.

The fluid in the lymphatics is clear and colorless and is called *lymph* (L. *lympha* = pure, clear water); lymphocytes are added to it as it passes through the lymph nodes, but otherwise it is free from blood cells.

Lymph Capillaries

These occur only where there are blood capillaries, and like blood capillaries, they form a closed network, but with a larger mesh. They are especially numerous in the skin and in mucous membranes.

The Lymph Vessels

Draining the networks are thin-walled lymph vessels, ½ to 1 mm in diameter, They are beaded due to numerous valves and run in layers of areolar tissue. Though more plentiful than veins, they tend to accompany veins and to drain corresponding territories.The propensity for spread of cancer cells along lymphatics is well known.

Lymph Nodes

Lymph nodes interrupt the flow of lymph. They vary *in size* from a pinhead to an olive and are somewhat flattened. *In color*, they are pink in life and brownish in the embalmed cadaver, but those draining the lungs are black from inhaled carbon, and after a meal, those draining the intestines are white from emulsified fat. The nodes act as filters for lymph and factories for lymphocytes. Both bacteria and cancer cells may be stopped and destroyed by phagocytic cells. Otherwise continued spread leads to more serious results.

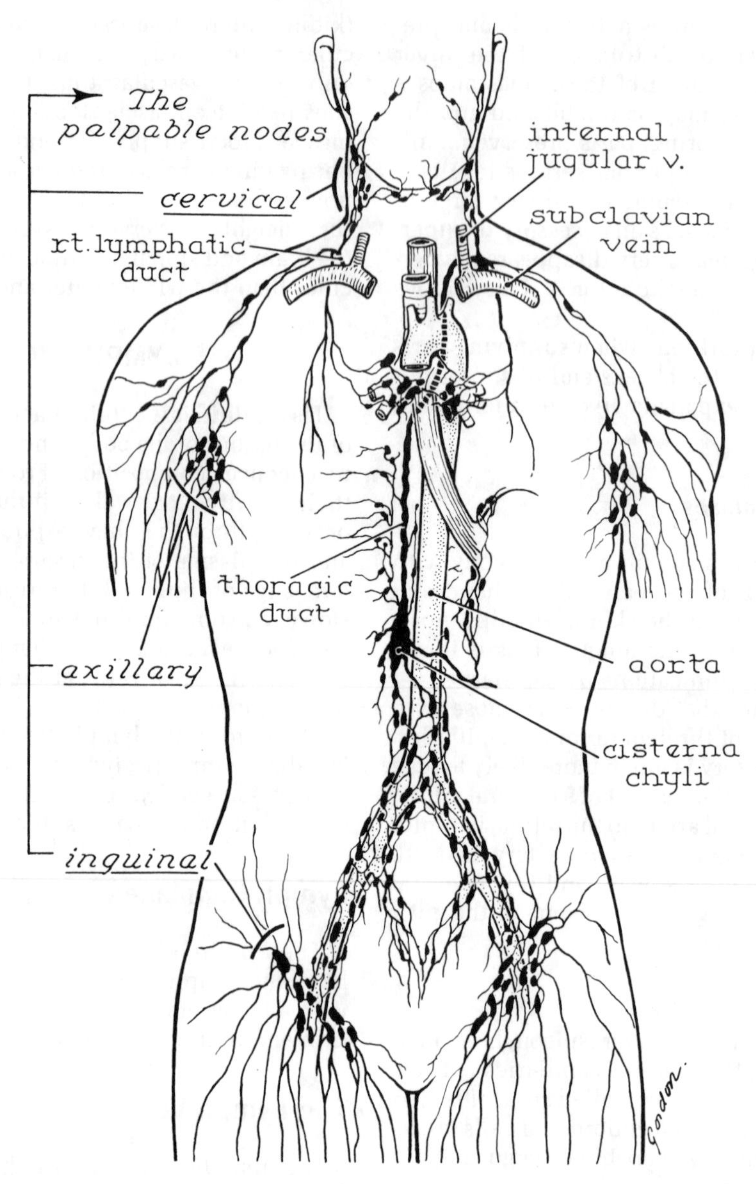

Figure 2.7. Scheme of the thoracic duct and lymphatics of the body. (From Basmajian, J.V.: *Primary Anatomy*, 8th edition, Williams & Wilkins, Baltimore, 1982.)

The flow of lymph in an immobile limb is almost negligible, but during muscular activity it is very active due probably to (1) rhythmical contraction of the vessels, (2) intermittent pressure (e.g., muscular action) on the valved vessels, (3) negative pressure or suction within the thorax, and (4) positive pressure within the abdomen during inspiration. Swelling of the feet in anyone who has sat immobile for hours (as in a long airplane trip) attests to the importance of activity to overcome the effects of gravity.

Main Channels

There are three paired lymph trunks: (1) the *jugular* accompanying the internal jugular vein in the neck, (2) the *subclavian* accompanying the subclavian vein in the upper limb, and (3) the *bronchomediastinal* from thoracic viscera. They end either separately or together near the angle of confluence of the internal jugular and subclavian veins. A larger, fourth channel, the *thoracic duct* (*fig. 2.7*), drains the chest wall and the territory below the diaphragm (the liver in part excepted); it is described in Chapter 9.

Lymphaticovenous communications exist between lymphatics and veins (Rusznyák *et al.*; Pressman and Simon).

Exceptions

Lymph capillaries are not present in epithelium (e.g., epidermis), cartilage, or other tissues devoid of blood vessels, neither are they present in the bone marrow and the main tissues of the spleen and liver. In the spleen and liver, the blood passing between the cells in irregular spaces (called sinusoids) appears to provide the sole drainage of the active tissues. However, the fibrous tissues of both organs have lymphatics. There are no lymphatics in the brain.

Lymph capillaries do not accompany the blood capillaries between the alveoli of the lungs, or in the glomeruli of the kidneys, since these capillaries are not nutritive in function, but in the lungs, lymph vessels do accompany the bronchial vessels.

Lymphoid Tissues

Lymphoid tissues are collections of enormous numbers of migratory and semimigratory scavenger cells called lymphocytes. Lymphocytes are usually classed with the white blood cells, for they are discharged with lymph into the blood stream. Lymphoid tissue occurs in lymph nodes, the palatine, lingual, and pharyngeal tonsils, the solitary and aggregated follicles of the intestine, the appendix, the thymus, and the corpuscles of the spleen. It is best developed in youth, and diseases of this tissue are commonest in youth. Lymphocytes play a vital part in the immune reactions of the body.

FETAL CIRCULATION

Red blood cells appear very early in the embryo, but until the 2nd month they are immature in type (containing nuclei). Nucleated red cells are found until a few days after birth, but they cannot be recovered from the maternal blood because the 2 circulations–fetal and maternal—are separate and closed. In the placenta, the 2 circulations come into close apposition with each other, being separated by semipermeable walls. Nutritive material and oxygen permeate from mother to fetus; waste products and carbon dioxide permeate from fetus to mother. The functions of the placenta, therefore, are concerned with nutrition, excretion, and respiration. In the fetus, the pulmonary, portal, and renal circulations are of little account; the placental circulation is paramount.

Fetal blood, charged with nutritive material and oxygen, leaves the placenta in the *umbilical vein*. This vein traverses the umbilical cord outside the fetus and the free edge of a membrane in the abdominal cavity, the falciform ligament of the liver, to end in the left portal vein. Some of the blood, thus brought to the left portal vein, then flows through the liver, leaving it via the hepatic veins, and enters the inferior vena cava; however, most of the blood bypasses the liver, being short-circuited by a special, temporary vein, the ductus venosus (*fig. 2.8*).

The *ductus venosus*, which occupies a sulcus behind the liver, connects the left portal vein to the inferior vena cava just below the diaphragm (*fig. 2.8*). Hence, the purest blood to reach the fetal heart travels through the terminal part of the inferior vena cava to the right atrium.

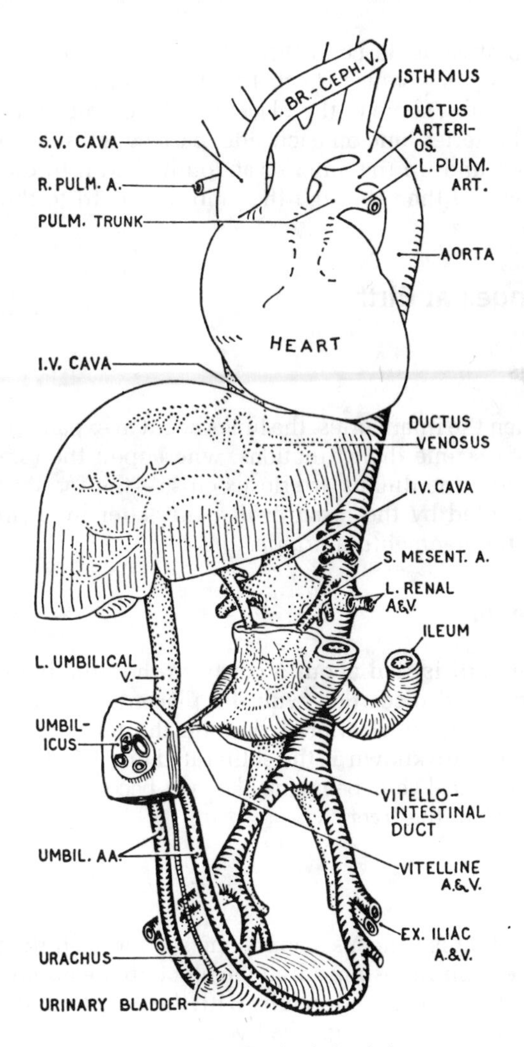

Figure 2.8. The fetal circulation.

In the fetus the blood from the inferior vena cava is directed across the right atrium and shunted through an opening in the interatrial septum, the *foramen ovale*, into the left atrium, thereby short-circuiting the pulmonary circuit. Thence it passes through the left ventricle into the ascending aorta and aortic arch, and by their branches it is distributed to the walls of the heart, the head and neck (including the brain), and the upper limbs, which therefore receive "pure" blood. It is blood that has been purified and oxygenated in the placenta and returned by the umbilical vein and ductus venosus to the inferior vena cava and so to the right atrium (*fig. 2.8*).

The blood from the head and neck and upper limbs returns via the superior vena cava to the right atrium, whence it passes through the right ventricle into the pulmonary trunk. Some of this blood then follows the pulmonary circuit through the lungs to the left atrium, but most of it is short-circuited to the aorta by a temporary artery, the *ductus arteriosus*, that connects the left pulmonary artery to the aortic arch just beyond the origin of the left subclavian artery. Beyond this connection, the united streams descend through the aorta and common iliac arteries—some to be distributed to the abdomen and lower limbs, some to the placenta. The umbilical or placental arteries, one on each side, pass by the sides of the bladder and up the anterior abdominal wall to the umbilicus and then through the umbilical cord to the placenta.

Changes at Birth

Lungs

When the infant cries, the lungs begin to expand (Chap. 7) and assume their functions, whereupon the foramen ovale and the ductus arteriosus close—the former to be represented by the *fossa ovalis*, the latter by a fibrous cord, the *ligamentum arteriosum*.

Placenta

The cord is tied about 5 cm from the umbilicus, cut beyond, and discarded with the placenta; so, the right and left umbilical arteries become thrombosed and fibrous and are known as the obliterated umbilical arteries. The umbilical vein, behaving likewise, becomes the *round ligament* of the liver.

Liver

The ductus venosus also becomes a fibrous thread, the *ligamentum venosum*; so, all the blood in the portal vein, which begins to function fully with the infant's first meal, must now pass through the liver.

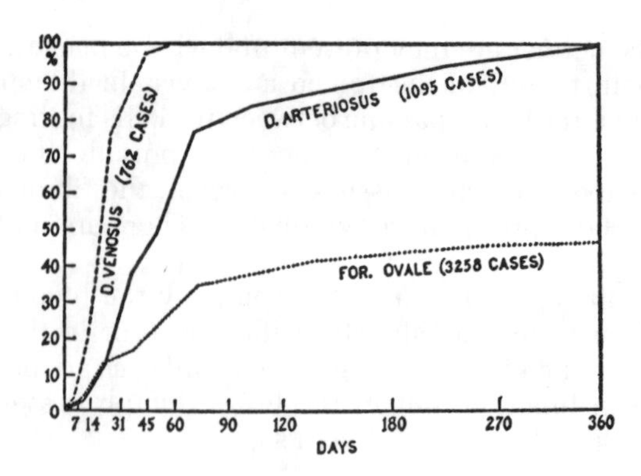

Figure 2.9. Dates of permanent closure of the ductus venosus, ductus arteriosus, and foramen ovale during the 1st year of life. (After Scammon and Norris.)

X-ray Findings

Radiopaque substances injected into veins of living fetal sheep reveal that (1) blood in the superior vena cava follows the adult pattern through the heart to the ductus arteriosus into the descending aorta, (2) most of the inferior caval blood passes through the foramen ovale, (3) the ductus arteriosus closes functionally a few minutes after delivery, and (4) the functional closure of the foramen ovale is apparently dependent on the onset of respiration (Barcroft *et al.*). However, permanent structural changes are more gradual (Patten), and the times of complete obliteration of the orifices are highly variable (*fig. 2.9*). The foramen ovale, though closed functionally, remains unobliterated (i.e., admits a probe) in about 25% of adults.

Nervous System

The species *Homo sapiens* is not the largest of the animals, nor the strongest and fastest; nor are its vision, hearing, or sense of smell the most acute. Even our marvelous hand is not quite unique—but what is unique is the superlative quality of the central nervous system that controls it.

The brain requires preferential treatment found in textbooks of neuroanatomy, but the rest of the nervous system will be described.

Subdivisions of the Nervous System

- Central Nervous System
  - Brain
  - Spinal medulla or cord
- Peripheral Nervous System
  - Peripheral nerves
    - Cranial nerves—12 pairs
    - Spinal nerves—31 pairs
  - Autonomic nervous system
    - Sympathetic
    - Parasympathetic

The central nervous system controls the voluntary muscles of the body and is concerned with functions both above and below the level of consciousness. The autonomic nervous system, also called the involuntary or visceral nervous system, controls the parts over which we do not exercise voluntary control and of which, for the most part, we are unconscious (e.g., action of the heart, movements of the viscera, the state of the blood vessels, and the secretion of glands).

NEURON

The structural unit of the nervous system is a nerve cell or neuron (shown diagrammatically in *figs. 2.10, 2.11*). A neuron has the following parts: (1) the cell body; (2) a process or *dendrite* (or processes) which transmits impulses to the cell body; and (3) a process or *axon* which transmits impulses from the cell body. An entire neuron may be microscopic in size; on the other hand, a process may approach a meter in length (e.g., a fiber of the sciatic nerve).

SPINAL MEDULLA OR CORD

The spinal cord lies within the vertebral canal, which has the diameter of a finger; thus the cord must be of smaller diameter than a finger, but in length (45 cm), it equals the femur.

The cord is surrounded by 3 **membranes:** (1) the *pia mater* (L. = tender mother) which, like a delicate skin, clings tenderly to the cord; (2) the *arachnoid mater* (Gk. = like a cobweb), which is joined to the pia by threads and is flimsy; and (3) the *dura mater*, which is tough (L., *dura*) and fibrous, but is easily split longitudinally, owing to the direction of its fibers. The arachnoid mater is separated from the pia by a space, the *subarachnoid space*

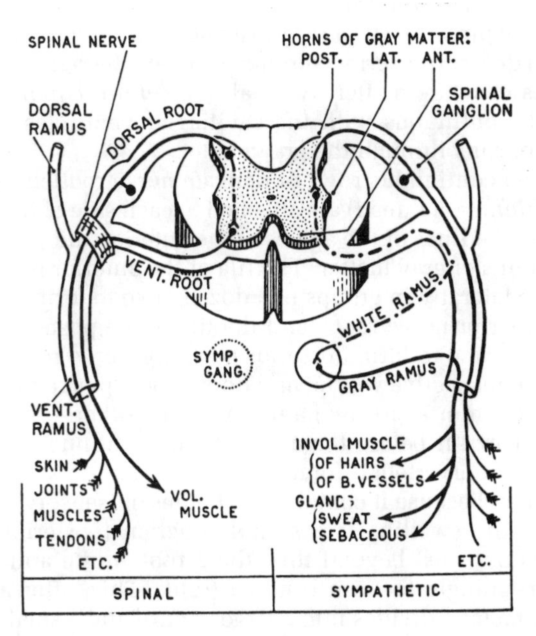

Figure 2.10. The composition of a peripheral nerve and its neurons.

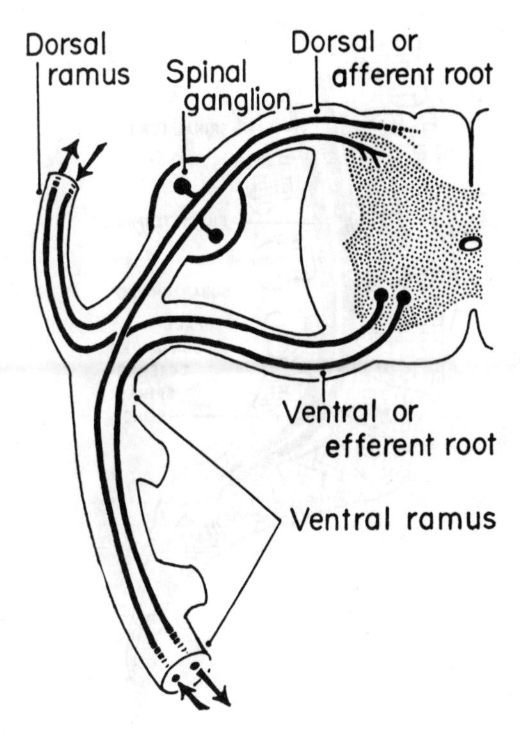

Figure 2.11. The fibers in the 2 roots and primary rami of a spinal nerve and contained axons and dendrites. The spinal ganglion contains the cell bodies of sensory neurons.

(*fig. 2.12*), filled with cerebrospinal fluid. This fluid presses the arachnoid against the dura, thereby obliterating a potential space, the *subdural space*.

In the fetus the spinal cord extends caudally to the coccyx, but as development proceeds, owing to the greater growth of the vertebral column, it fails to keep up, so that at birth it extends only to vertebra L3 and in the adult to the upper part of L2 (*fig. 2.12*). Flexing the column draws the cord temporarily higher. These facts permit a physician to insert a needle between the laminae of L3 and L4 into the subarachnoid fluid without striking the cord. A strong, glistening thread, the *filum terminale*, largely composed of pia mater, attaches the tapered end of the cord to the posterior surface of the coccyx even in the adult.

A broad band of pia, the *ligamentum denticulatum*, projects like a long lateral fin from each side of the cord and lies between the ventral and dorsal nerve roots. From the free margin of this ligament strong "tooth-like" processes, one for each segment, pass through the arachnoid to become firmly attached to the dura from the level of the foramen magnum to T12, L1, or L2 (I. B. Macdonald).

The *spinal dura* is free within the vertebral canal. It is, however, adherent to the margin of the foramen magnum and is there continuous with the cranial dura. It is fixed caudally to the coccyx by the filum terminale which, as it passes through the closed end of the dural sac, acquires an adherent dural covering. It is fixed on each side by the ventral and dorsal spinal nerve roots (described

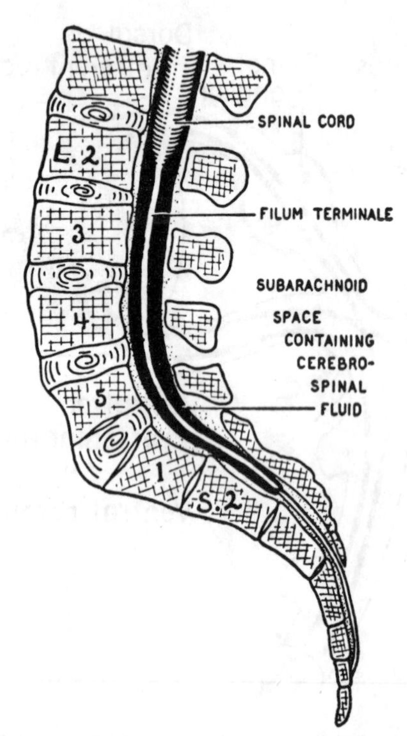

Figure 2.12. The spinal cord ends at lumbar vertebra 1 or 2, the subarachnoid space at sacral vertebra 2 (spinal nerve roots are not shown).

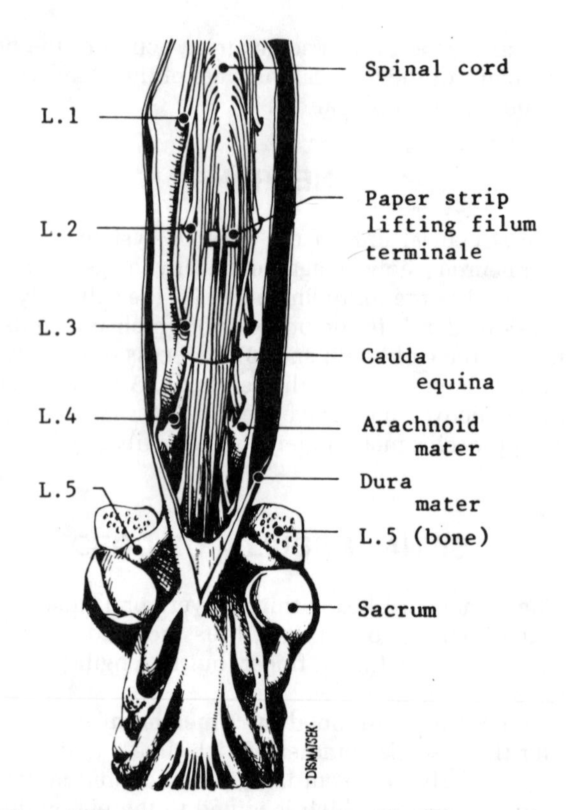

Figure 2.13. The cauda equina exposed by slitting open the dura mater and arachnoid mater. L1–5 are the (left) lumbar spinal nerves as they exit.

below). These likewise acquire adherent dural coverings that add to their thickness and give them strength.

At each segmental level, sensory and sympathetic *meningeal nerves* pass through the intervertebral foramina to the dura (Kimmel).

In the adult, the subarachnoid space and the contained cerebrospinal fluid extend to the body of sacral vertebra 2. So, between L2 and S2 in place of cord, there is filum terminale surrounded by the roots of the lower spinal nerves, which have been drawn cranially with the cord. These loosely occupy the sac of cerebrospinal fluid, here somewhat dilated, in which they lie, and from their resemblance to a horse's tail they are called the *cauda equina* (*fig. 2.13*). Fortunately, a spinal-puncture needle inserted into the CSF rarely if ever strikes these nerves, which apparently are too slippery.

On *flexing the spinal column*, the dural sac stretches, and nerves L1 and 2 are drawn from 2 to 5 mm within the intervertebral foramina. L3 is withdrawn less, and L4 negligibly. On the other hand, "*straight-leg raising*," by pulling on the sciatic nerve, can pull L5, S1, and S2 caudally by 2 to 5 mm, but again the effect on L4 is negligible (Inman and Saunders). Similarly, traction on the outstretched hand pulls the roots of C6 and C7 (Smith).

Structure

The spinal cord (in Latin, *spinal medulla*), as seen on cross-section, has an H-shaped field of gray matter enveloped in a zone of white matter (*figs. 2.10 and 2.11*). The anterior and posterior limbs of the H, called the *anterior* and *posterior horns* or *gray columns*, divide the white matter of each side into *anterior, lateral, and posterior white columns.* An anterior groove, the *median ventral (anterior) fissure*, and a posterior septum, the *median dorsal (posterior) septum*, separate the right and left sides of white matter. A canal, the *central canal of the cord*, continuous with the cavities or *ventricles* of the brain, runs through the gray matter.

Two continuous rows of delicate *nerve rootlets* or *fila* (L. *filum* = a thread) are attached to each side of the cord along the lines of the apices of the anterior and posterior columns of gray matter. The fila of the anterior row converge laterally in groups of a dozen or so to form *ventral* or *anterior nerve roots*, and the fila of the posterior row do likewise to form *dorsal* or *posterior nerve roots.*

The respective ventral and dorsal roots pierce the dura about 2 mm apart, one anterior to the other—or rather they carry it before them. They then continue laterally to the intervertebral foramina where each dorsal root is swollen because it contains the bodies of (sensory) nerve cells; the swelling is a *spinal ganglion* (posterior root ganglion). Just beyond this, the 2 roots unite and their fibers mingle to form a *nerve trunk.* This, after a few millimeters, divides into a large *ventral* and a small *dorsal ramus* (L. *ramus* = branch) (*figs. 2.10, 2.11*).

The length of cord to which the fila of one pair of spinal nerves is attached is called a *spinal segment.* Since there are 31 pairs of spinal nerves, the cord is said to have 31 segments.

Blood Supply

The spinal cord receives a series of small arteries that enter the vertebral canal as local branches of the arteries that supply the tissues near the vertebrae. Thus, there are a varying number of branches of the vertebral, deep cervical, intercostal, and lumbar arteries that travel to the cord along the spinal nerves and their roots. They form a plexus on the surface of the cord along with a trio of *spinal arteries*, which descend on the cord—the unpaired *anterior* and the paired *posterior*. At most levels, branches of the anterior spinal artery supply about ⅔ of the cross-sectional area of the cord. The companion veins form a very complex plexus that communicates through the foramen magnum with the dural venous sinuses and profusely with veins of the vertebral column.

PERIPHERAL NERVES

Macroscopic or Gross Structure

The peripheral nerves are made up of fibrous tissue enclosing bundles of nerve fibers, both afferent and efferent (*fig. 2.10*). A loose sheath (epineurium), continuous with the surrounding areolar tissue, encases the entire nerve and unites the individual bundles. Each bundle has a thicker sheath, the *perineurium*, which branches when the bundle branches and which accompanies every fiber to its destination. Between the fibers are some delicate partitions (endoneurium).

Microscopic Structure

A nerve fiber consists of a central thread, the *axon*, encased in a tube of white fatty material of varying thickness, the *myelin sheath*, around which there is a thin, but tough, nucleated membrane, the *neurilemma*. If a fiber is pinched, the axon and the myelin sheath will rupture, but the neurilemma may remain intact.

White versus Gray

It is the myelin sheath that gives whiteness to nerves and to nerve tracts in the spinal cord and brain; in its absence, nerve tissue looks gray.

Nerves transmit impulses quite rapidly, but the speed is variable, being about 60 m per second in motor nerves (Smorto and Basmajian).

Functions

A dorsal nerve root transmits impulses to the cord; it is *afferent* or sensory. A ventral nerve root transmits impulses from the cord; it is *efferent* or motor. (L. *ad-ferens* = bringing to; *ex-ferens* = bringing from, the CNS here being understood.)

The cell bodies of the afferent nerve fibers lie within the spinal ganglia. Those of the efferent nerve fibers lie in the ventral (anterior) horn of the gray matter of the cord (*fig. 2.11*).

Afferent or sensory spinal nerve fibers bring impulses from the endings in the skin, muscles, joints, etc. to the spinal ganglia, and thence, via the fila of the dorsal roots to the cord. Within the cord, the fibers branch: some branches ascend in the cord; others descend; still others at once turn into the gray matter of the dorsal horn, where they synapse (i.e., make contact) with the dendrites of short *connector* (*intercalated*) cells whose axons in turn synapse with the dendrites of large multipolar cells in the ventral horn. The axons of these motor cells emerge in the ventral root fila and end in motor endplates of voluntary muscles.

These neurons—afferent, connector, and efferent—constitute a simple spinal *reflex arc*. If one's finger is touched, even during sleep, one moves it; if it is pricked or burnt, one withdraws the limb. In the one case, a simple reflex arc is invoked; in the other, afferent impulses spread more widely, even to the opposite side of the body, perhaps arousing consciousness, due to the ascending and descending and crossing fibers by which impulses spread to other parts of the cord and to the brain.

Typical Spinal Nerve

Details of the organization and ramifications are given on p. 000 and in Figure 27.21.

Nerve Plexuses

Ventral nerve rami and their branches communicate with adjacent rami and branches to form plexuses, thereby widening the influence of the individual segments of the spinal cord. Close to the vertebral column, the ventral rami of nerves C1–4 form the cervical plexus, those of C5–T1 the brachial plexus, L1–4 the lumbar plexus, L4–S4 the sacral plexus, and S4–Co1 the coccygeal plexus.

Nerves communicate more peripherally also. Cutaneous nerves communicate freely with adjacent nerves and overlap them with the result that severing a cutaneous nerve diminishes sensation in its territory without as a rule abolishing it (*figs. 2.14 and 2.15*).

Peripheral nerves are themselves plexuses: their fibers do not run parallel courses but are plaited, not unlike the tendon fibers in Figure 1.26, *A*. The branch of a nerve is usually bound to the parent stem for 1 to 6 cm by areolar tissue, and the level at which it leaves the parent stem varies within this range. Proximal to this, the strands of the branch take part in the plait, and attempts to free the branch more proximally lead to its destruction (*fig. 2.16*).

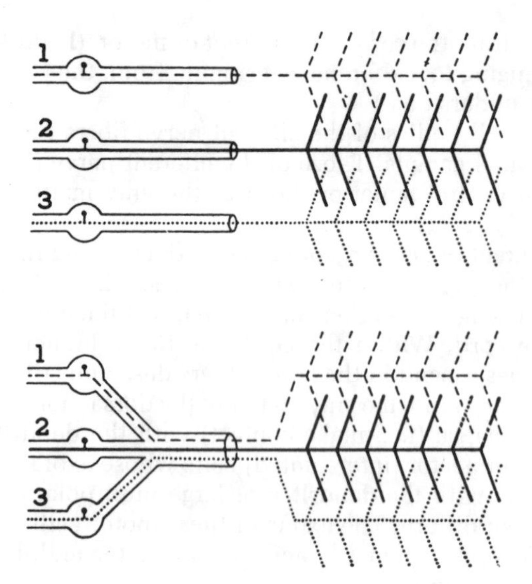

Figure 2.14. The distribution of a series of spinal nerves overlap.

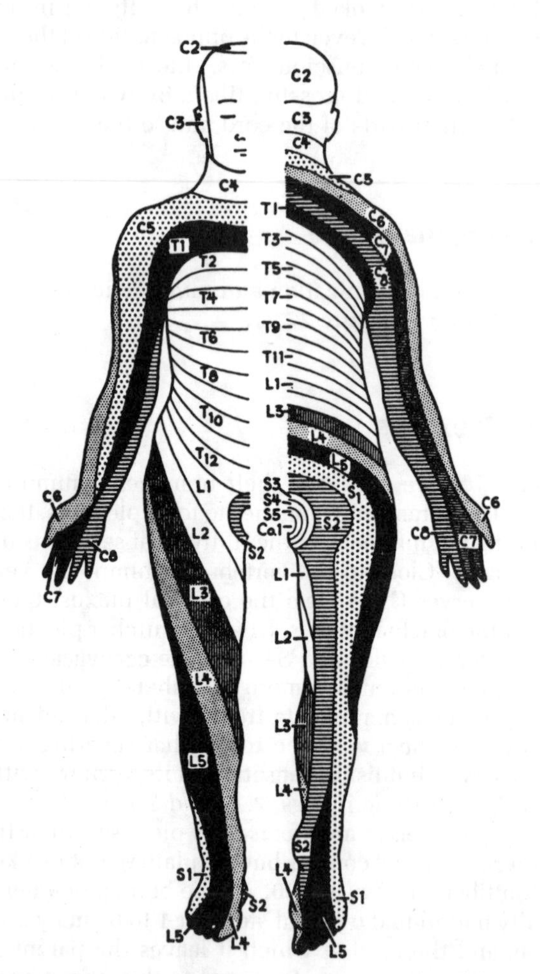

Figure 2.15. Dermatomes: the strips of skin supplied by the various levels or segments of the spinal cord.

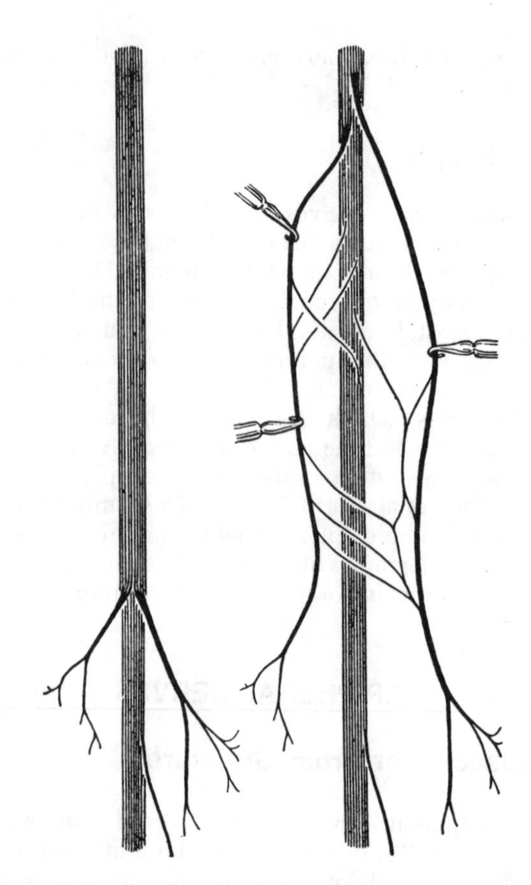

Figure 2.16. Two branches of a nerve dissected proximally.

Blood Supply

The peripheral nerves are supplied by a succession of anastomosing nutrient vessels, *arteriae nervorum* derived from the nearest arteries, but their number, size, and origin are inconstant. On reaching a nerve, the artery usually divides into ascending and descending branches, which anastomose with longitudinal chains. The veins, *venae nervorum*, have a similar pattern (Sunderland).

AUTONOMIC NERVOUS SYSTEM

This system of nerves and ganglia distributes efferent impulses to (1) the heart, (2) smooth muscle fibers everywhere, and (3) glands. A visceral afferent system collects sensory impulses from the organs innervated by the autonomic fibers. These afferent nerves accompany the autonomic fibers back to the spinal cord. The cell bodies for the visceral afferent neurons are in the dorsal root ganglion. The autonomic nervous system is a VISCERAL MOTOR system and it consists of 2 parts:

(1) the sympathetic system and
(2) the parasympathetic system.

The *sympathetic system* (*fig. 2.17*) has central connections with the thoracolumbar part of the spinal cord

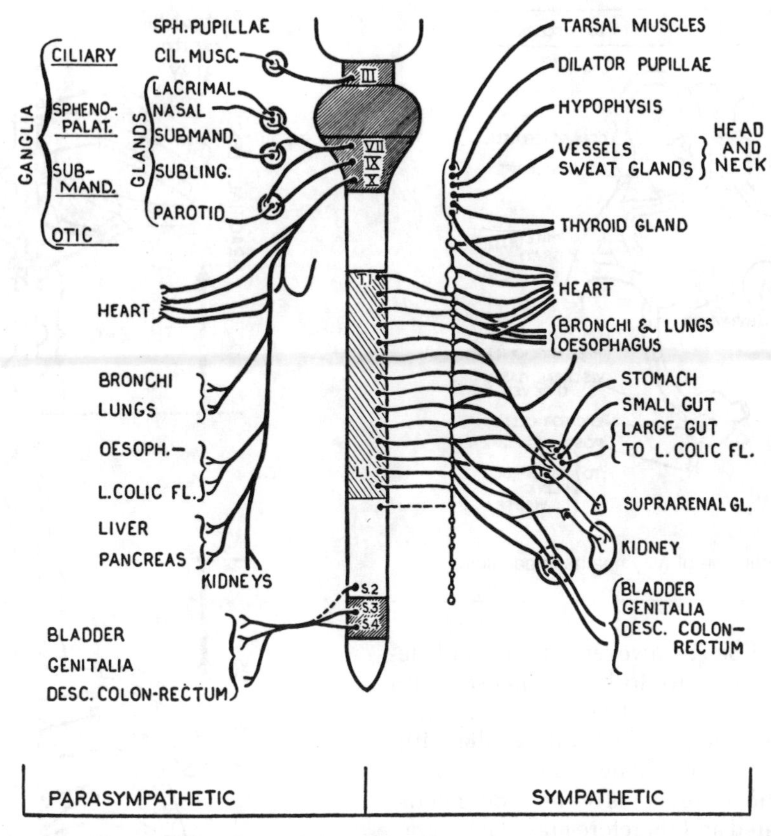

Figure 2.17. General plan of the autonomic nervous system.

from the 1st thoracic to the 2nd (or 3rd) lumbar segment. The *parasympathetic system*, which has (1) a cranial part and (2) a sacral part, has central connections with the brain through cranial nerves III, VII, IX, and X, and with the spinal cord at sacral segments 2, 3, and 4.

Sympathetic System

In the *sympathetic system*, there are neurons corresponding to the efferent, connector, and afferent neurons of the voluntary system (*figs. 2.18 and 2.19*). The efferent cells occur (1) in the *ganglia of the sympathetic trunk* (or *paravertebral ganglia*), (2) in the *prevertebral ganglia* (or *visceral ganglia*—cardiac, celiac, intermesenteric, hypogastric, and subsidiary ganglia), which are detached parts of paravertebral ganglia, and (3) in the medulla of the suprarenal gland.

The gray matter of the spinal cord from segments T1–L2 possesses an intermediolateral column (horn), and it is in this intermediolateral column that the sympathetic *connector cells* (preganglionic neurons) are lodged (*fig. 2.19*). Their axons pass as fine, medullated *preganglionic fibers* (2.6 μm) via the ventral nerve roots and white rami communicantes (T1–L2) to the paravertebral ganglia, where they form synapses with the excitor cells (postganglionic neurons).

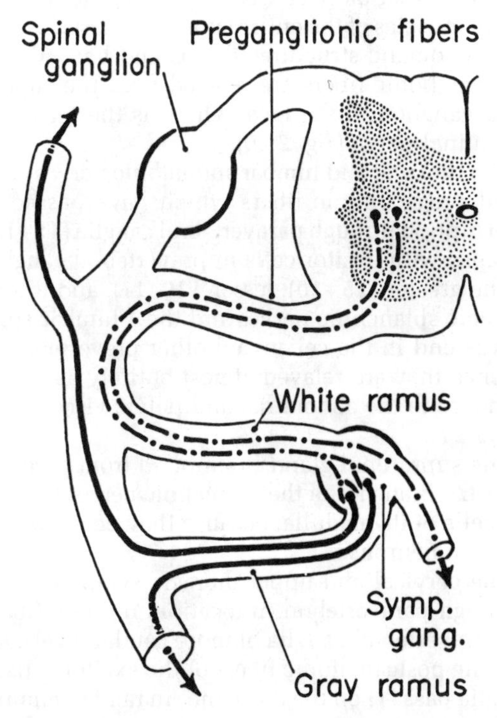

Figure 2.18. The typical sympathetic contribution to a spinal nerve.

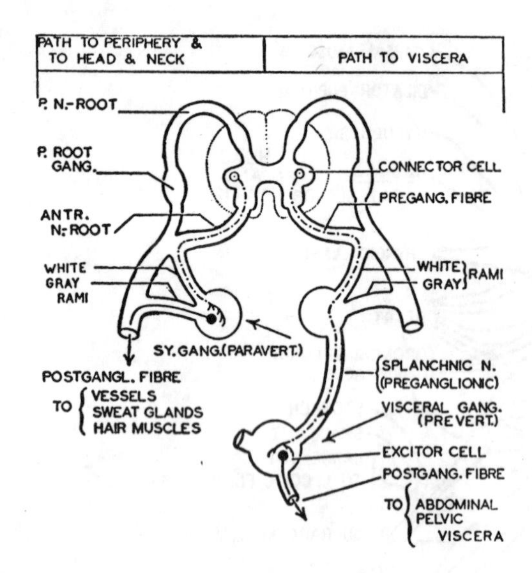

Figure 2.19. General plan of a sympathetic ganglion.

Some preganglionic fibers, however, ascend and descend in the sympathetic trunk to form synapses with the cells in the ganglia at various levels, whereas other preganglionic fibers pass through the trunk ganglia without synapsing, to form *splanchnic nerves* (*fig. 2.19*).

The *postganglionic fibers*, or axons of the excitor cells, are mostly nonmyelinated and therefore gray. These gray fibers, carried in a short nerve, a *ramus communicans*, pass laterally from the ganglia of the sympathetic trunk to each and every spinal nerve. They reach the blood vessels, sweat and sebaceous glands, and arrectores pilorum muscles of the entire cutaneous surface of the body and of somatic structures (limbs and body wall). Other postganglionic fibers are relayed from the superior cervical ganglion to the face, which is the territory of the trigeminal nerve (*fig. 2.20*).

The thoracic and lumbar *splanchnic nerves* carry elongated preganglionic fibers which have passed without interruption through paravertebral ganglia (T5–L2). They synapse with excitor cells in prevertebral ganglia.

The greater (T5–10), lesser T10, 11), and lowest (T12) thoracic splanchnic nerves and the 4 lumbar splanchnic nerves end in the celiac and other prevertebral ganglia, whence they are relayed almost entirely as perivascular branches to the abdominal and pelvic viscera (*figs. 2.19 and 2.21*).

The *suprarenal gland* is supplied from cord segments T10–L2. Branches of the splanchnic nerves ramify among the cells of its medulla, because they develop from sympathetic neurons.

The cervical and upper thoracic sympathetic ganglia, although paravertebral in location, represent both para- and prevertebral ganglia of more caudal levels; that is to say, the postganglionic fibers of the excitor cells in these ganglia pass (1) on the one hand, in rami communicantes to the spinal nerves and, so, to somatic structures, and

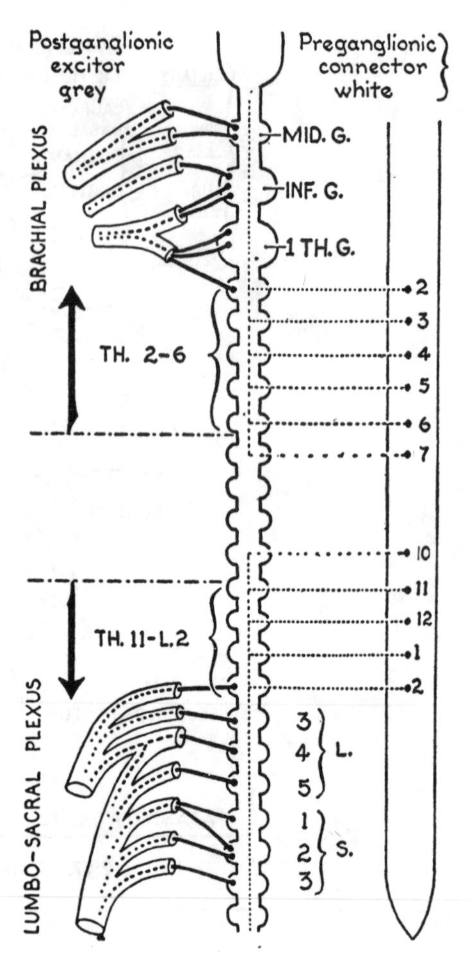

Figure 2.20. The sympathetic supply to the upper and lower limbs.

(2) on the other hand, as visceral fibers to such structures as the eye, salivary glands, heart, and lungs.

The *visceral afferent* (sensory) fibers of the sympathetic system travel with the visceral efferent fibers. They pass via the white rami communicantes to the spinal ganglia (dorsal root ganglia) where, like the afferent fibers of the voluntary system, they have their cell stations. They enter the spinal cord through the dorsal roots, mainly of T1–L2, and synapse with the connector cells in the intermediolateral column of gray matter. Many, however, first ascend or descend in the sympathetic trunk.

The Sympathetic Trunk

The trunk itself is composed of ascending and descending fibers, some of which are preganglionic efferent, postganglionic efferent, and also afferent fibers. The paravertebral ganglia (ganglia of the trunk) are formed by synapses between preganglionic efferent neurons and the cell bodies of postganglionic neurons. The influence of a single preganglionic neuron is diffused over a wide area

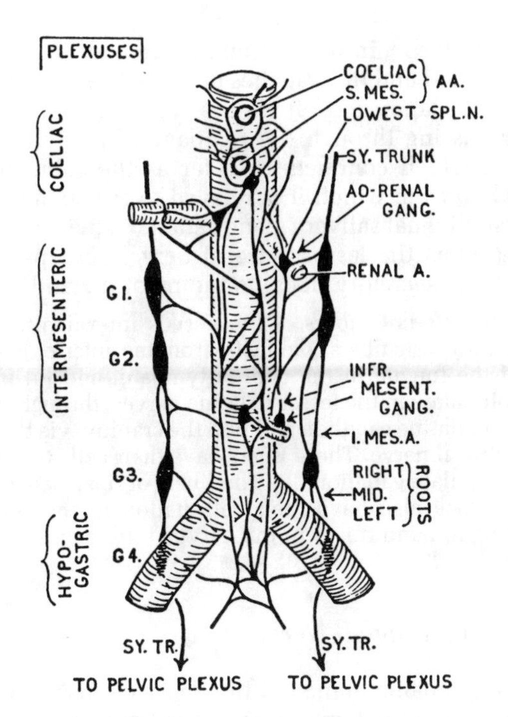

Figure 2.21. The intermesenteric and superior hypogastric plexuses. (From a dissection by K. Baldwin.)

due to the fact that each preganglionic fiber synapses with a number of postganglionic neurons.

Cerebral Control

The autonomic system is controlled by centers in the hypothalamic region, which lies toward the lower and front part of the 3rd ventricle. From here descending tracts influence the connector cells.

The Sympathetic Supply of Individual Regions and Organs

This subject will be considered by diagram and comment. *Figures 2.17–2.21* indicate for the different regions and organs (1) the probable segmental locations of the connector cells in the intermediolateral column of gray matter; (2) the ganglia in which they synapse with effector cells; and (3) the ultimate distribution of the postganglionic fibers.

Upper Limb (*fig. 2.20*). Preganglionic fibers from cord segments T2–7 ascend in the sympathetic trunk to the upper thoracic, inferior cervical, and middle cervical ganglia. A dozen or so postganglionic gray rami then pass to the roots of the brachial plexus to be distributed to the limb. Most of these fibers travel in the lower trunk of the plexus (where they may be subjected to pressure by the 1st rib) and in the median and ulnar nerves.

Lower Limb (*fig. 2.20*). Preganglionic fibers from cord segments T10, 11, 12, and L1 and 2 descend in the sympathetic trunk to ganglia L2–S3. Thence, the postganglionic fibers pass in gray rami to the nerves of the lumbar and sacral plexuses. The upper part of the femoral artery is supplied by an extension from the aortic plexuses along the common and external iliac arteries; but as in the upper limb, so in the lower, most of the femoral artery and the arteries distal to it are supplied locally.

Head and Neck (*fig. 2.17*). Cord segments are mainly T1 and 2, i.e., connector cells are situated in segments T1 and 2, though some may extend more inferiorly. The excitor cells lie mainly in the superior cervical ganglion.

Postganglionic fibers pass to the arrectores pilorum (smooth muscles that make the hairs stand up), sweat glands, and vessels of the skin; to the heart; and, via the internal carotid nerve, to the vessels of the nasal cavity (through the deep petrosal nerve), to the dura, to the cerebral vessels, and to smooth muscles in the orbital cavity (tarsal muscles, dilator pupillae, and vasoconstrictors).

The orbital fibers arise mainly in cord segment T1, pass to the 1st thoracic ganglion, and ascend to the excitor cells in the superior cervical ganglion.

Thorax. Cord segments are mainly T2, 3, and 4. For the *heart*, the cord segments are T1–5. From relay stations in the 3 cervical and superior 4 or 5 thoracic ganglia, cardiac nerves pass to the cardiac plexus (*fig. 2.22*). Pain impulses travel in the afferent fibers (but apparently not

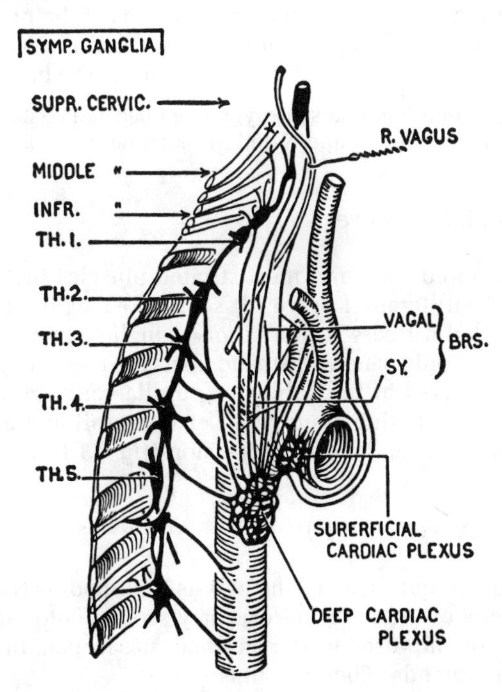

Figure 2.22. Cardiac plexuses. (From White, after Kuntz and Morehouse.)

by way of the superior cervical ganglion) and have their cell stations in the superior 4, (5) thoracic spinal ganglia.

For the lungs, the cord segments are T2–7; for the esophagus, T4, 5, and 6.

Abdomen and Pelvis. Cord segments are T5–L2. For the stomach, liver, and pancreas they are T6–10, and for the gall bladder T4–10.

For the small intestine the cord segments are T8, 9, 10, 11; for the cecum and appendix T10, 11, and 12; for the colon to the left colic flexure T11, 12, and L1; and for the descending colon to the rectum L1 and 2.

For the kidney the cord segments are T11, 12, L1, 2; for the ureter L1 and 2; and for the bladder (? T11, 12), L1 and 2. (The foregoing data are largely from Mitchell.)

From the cord segments, connector fibers pass through the thoracic and lumbar *trunk* ganglia, as the thoracic and lumbar splanchnic nerves, to be relayed in the celiac, renal, mesenteric, and superior and inferior hypogastric ganglia.

For other pelvic viscera see p. 52 and *Figures 2.21 and 4.3*.

Parasympathetic System

Parasympathetic nerve fibers (efferent) are contained in cranial nerves III, VII, IX, and X and in sacral nerves (2), 3, and 4. Details are given here for convenience and especially for review. The beginner cannot be expected to appreciate, let alone learn, them fully until the cranial nerves have been studied thoroughly.

Many branches of the trigeminal nerve (cranial nerve V) are accompanied in the terminal parts of their courses by efferent parasympathetic fibers from other nerves (e.g., the lingual nerve by the chorda tympani, see below).

Afferent (visceral sensory) nerve fibers are contained in cranial nerves IX and X and in sacral nerves 2, 3, and 4.

Oculomotor Nerve

This motor nerve to most striated muscles that move the eyeball (cranial nerve III) sends preganglionic parasympathetic fibers to the *ciliary ganglion* in the orbit. Postganglionic parasympathetic fibers are relayed by *short ciliary nerves* to the sphincter pupillae and the ciliary muscle, mediating contraction of the pupil and accommodation of the lens to near vision (*fig. 43.10*).

Facial Nerve (n. VII)

The *efferent fibers* of the nervus intermedius (pars intermedia of the facial nerve) carry secretomotor and vasodilator impulses to the lacrimal, nasal, palatine, and salivary glands. They run thus:

(1) Via the *greater petrosal nerve* to the pterygopalatine ganglion, whence as postganglionic fibers they accompany (a) the zygomatic nerve to the orbit, where they branch off to join the lacrimal nerve and run to the lacrimal gland and (b) branches of nerve V^2 to the nasal and palatine glands and (2) via the *chorda tympani*, which, after passing through the tympanum, joins the lingual nerve, which conducts it as far as the submandibular ganglion from which it is relayed to the submandibular and sublingual salivary glands and (3) a *twig of the facial nerve* joins the lesser petrosal nerve, perhaps bringing accessory secretory fibers to the parotid gland.

The *Afferent Fibers* of the nervus intermedius are: (1) mainly taste fibers coming (a) from the anterior two-thirds of the tongue in the chorda tympani, and (b) from the soft palate in the lesser palatine nerves, through the pterygopalatine ganglion and into the cranium via the greater petrosal nerve. These fibers have their cell station in the geniculate ganglion. (2) The fibers of deep sensibility in the face also have their cell station in the geniculate ganglion and travel in the nervus intermedius.

Glossopharyngeal Nerve

The tympanic branch of this mostly sensory nerve (n. IX) traverses the tympanum in the tympanic plexus, receives a twig from the facial nerve, and, as the lesser petrosal nerve, runs in the cranium to the otic ganglion, and thence by the auriculotemporal nerve to the parotid gland. If the twig from the facial nerve is secretory (secretomotor), then the parotid gland has a double nerve supply—from IX and from VII.

The *afferent fibers* are (1) fibers of taste from the posterior one-third of the tongue, (2) fibers of general sensation from the posterior one-third of the tongue, the fauces, pharynx, and tympanum, and (3) pressor fibers in the sinus nerve. Their cell station is in the ganglia on the root of nerve IX.

Vagus Nerve

This mixed nerve (n. X) supplies the digestive and respiratory passages and the heart, as described in Chapter 9 and shown in Figure 43.6. The relay stations of the efferent fibers of the vagus are in terminal ganglia, situated in (or near) the walls of the organ or part it supplies. The vagus causes hollow organs to contract and their sphincters to relax. It is secretomotor. The cell stations of its afferent fibers are in the ganglia on the root of the vagus. Its afferent fibers do not carry impulses of pain from the heart or from abdominal organs.

Pelvic Splanchnic Nerves

These nerves (S2, 3, and 4) behave like vagal fibers, and are described on p. 224. The vagus supplies the gastrointestinal tract as far as the left colic flexure, via a branch of the posterior vagal trunk (*fig. 13.3*); the pelvic splanchnic nerves supply the gut distal to that point. Their afferent fibers are conductors of pain impulses.

Endocrine Glands

An endocrine or ductless gland is one that produces an internal secretion or *hormone*, i.e., a secretion carried off in the venous blood stream to influence the activities of another part of the body. If it also produces a secretion carried off by a duct, it is both an exocrine and an endocrine gland.

HYPOPHYSIS (PITUITARY GLAND)

This is a small gland about the size of a pea attached by a stalk, the infundibulum, to the hypothalamus above, which forms the floor of the third ventricle of the brain. It rests on the deeply concave body of the sphenoid bone in the region behind the nasal and orbital cavities (*fig. 2.23*). It is made up of 2 lobes, an anterior and a posterior, entirely different from one another functionally.

The *adenohypophysis* or *anterior lobe* forms most of the gland. It produces several hormones that, because of their influence on other endocrine glands, have caused the hypophysis cerebri to be referred to as the "master gland." Disturbances of the lobe result in dwarfism, gigantism, infantilism, excessive obesity, and other manifestations associated with growth and sex.

The *neurohypophysis* or *posterior lobe* is a downgrowth from the brain and produces secretions that stimulate contractions of the uterus and the production of urine by the kidneys and that influence the production of insulin by the pancreas. This lobe also affects blood pressure. It is possibly associated with sympathetic and (or) parasympathetic functions because of its neural and vascular connections with the hypothalamus. Its portal system of veins carries the secretions first to the hypothalamus before the blood is returned to the systemic veins that drain that part of the brain.

PINEAL BODY OR GLAND

The pineal body, so called because in shape and appearance it resembles a miniature pine cone, is deeply

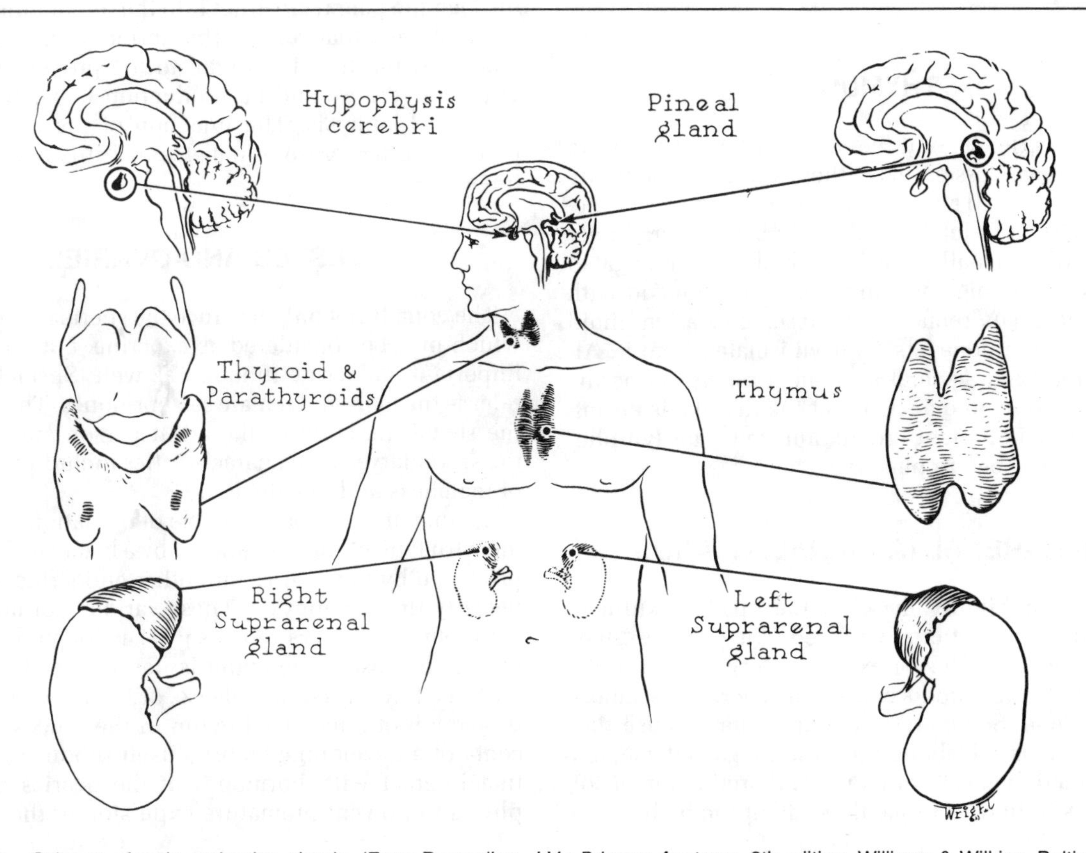

Figure 2.23. Scheme of main endocrine glands. (From Basmajian, J.V.: *Primary Anatomy*, 8th edition, Williams & Wilkins, Baltimore, 1982.)

placed within the brain. It projects backwards from the back of the roof of the 3rd ventricle. Little is known about it and its status as an endocrine gland.

THYROID GLAND

This lies in the neck (*fig. 2.23*) and is fully described in Chapter 40. The thyroid gland consists of irregular masses of tissue (follicles) filled with a colloid containing the thyroid hormone. This hormone (thyroxin) is concerned with a great many bodily functions and, when produced in excessive quantities, disturbs the normal rate at which the cells of the body function. This disturbance of the basal metabolic rate (BMR) is a notable feature of *goiter*. Congenital deficiency or absence of the thyroid gland results in a mentally defective dwarf known as a cretin. The condition is alleviated by the administration of thyroxin.

PARATHYROID GLANDS

The parathyroid glands—2 on each side—lie on the back of the thyroid gland. They are small ovoid bodies; yet, they are extremely important since they regulate the relative amounts of calcium in the blood and in the bones. Certain distressing diseases in which the bones become soft or extremely brittle are the result of malfunctioning parathyroids.

THYMUS

The thymus lies behind the manubrium sterni and in front of the great vessels above the heart. It is essentially an elongated bilobed gland whose period of greatest functional activity is in fetal life. After puberty, it gradually shrinks until it usually consists merely of 2 elongated fatty masses reaching down to the pericardium and with little thymic tissue remaining. It resembles a lymphoid organ and no hormone has yet been isolated from it. Although there is no general agreement on its endocrine function, there is no question that the thymus is an immunologically important production center for lymphocytes, especially before puberty.

SUPRARENAL (ADRENAL) GLANDS

As their names imply, these cap the top of the kidneys, but they are not functionally related. They are described fully in Chapter 15. The *cortex* of each forms a thick outer layer of the gland. It produces a group of hormones, among them cortisone. Some are essential to life because they regulate various metabolic processes, e.g., salt metabolism, production of sex hormones, and production of collagen fibers of fibrous tissue throughout the body.

The *medulla* (L. = marrow) of the adrenal is closely associated in developmental origin with the sympathetic ganglia. It is in the medullary part of the gland that adrenalin is produced.

PARAGANGLIA

Scattered along the line of the sympathetic chain and the abdominal aorta are many tiny bodies identical in structure with the suprarenal medulla. They are also known as para-aortic bodies or "chromaffin bodies," and like the suprarenal medulla, are associated with the activity of the sympathetic system.

The 2 para-aortic bodies are largest at about the age of 3, when they measure only about 1 cm in length. They atrophy and disappear by the mid-teens. The paraganglia, which are close associates of the sympathetic trunk and are quite tiny, atrophy to microscopic size in the adult. Their chromaffin cells (so named because they are easily stained by yellow-colored chromic acid) secrete adrenalin just as the suprarenal medulla does.

PANCREAS

As noted before, the pancreas has both an exocrine and an endocrine function. Scattered among the larger clumps of glandular tissue that pour their exocrine secretions through the pancreatic duct into the duodenum are small islets of cells that release the hormone *insulin* into the blood stream. Insulin is vital to the proper metabolism of carbohydrates, and its production is partly regulated by the hypophysis. The injection of the hormone into patients suffering from diabetes mellitus is a life-saving treatment.

TESTES AND OVARIES

The gonads not only produce the spermatozoa and ova, which may be considered as exocrine, but they have an important endocrine function as well. Specialized cells release the male and female sex hormones. These regulate the sexual function of the adult person and determine the secondary sexual characteristics including all aspects of maleness and femaleness.

In women, the monthly cycle of ovulation followed by menstrual bleeding is regulated by a balance of hormones produced by the ovaries and other endocrine glands, especially the hypophysis. These various hormones interact in such a way as to cause: (1) a sloughing off of the uterine mucosa if the ovum is not fertilized by a sperm within a few days of its release, or (2) an embedding and protection of a fertilized ovum in the mucosa. The placenta of a developing embryo itself produces hormones that interact with hormones of the ovaries and hypophysis to prevent premature expulsion of the child.

3

Digestive and Respiratory Systems

Digestive System

This system is essentially (1) a long hollow tube, the digestive passage, through which food is propelled, and (2) certain accessory glands (*fig. 3.1*).

The *digestive passage* (alimentary canal) begins at the lips, traverses the neck, thorax, abdomen, and pelvis, and ends at the anus.

Its successive **parts** are:

Mouth (Oral cavity)
Pharynx
Esophagus
Stomach
Small intestine
  duodenum, jejunum, and ileum.
Large intestine
  appendix, cecum, colon, rectum, anal
  canal, and anus.

The stomach and intestines comprise the *gastrointestinal tract*.

The **accessory glands** are:

Glands of the mouth
  parotid
  submandibular
  sublingual
Liver
Pancreas, with which the Spleen is associated.

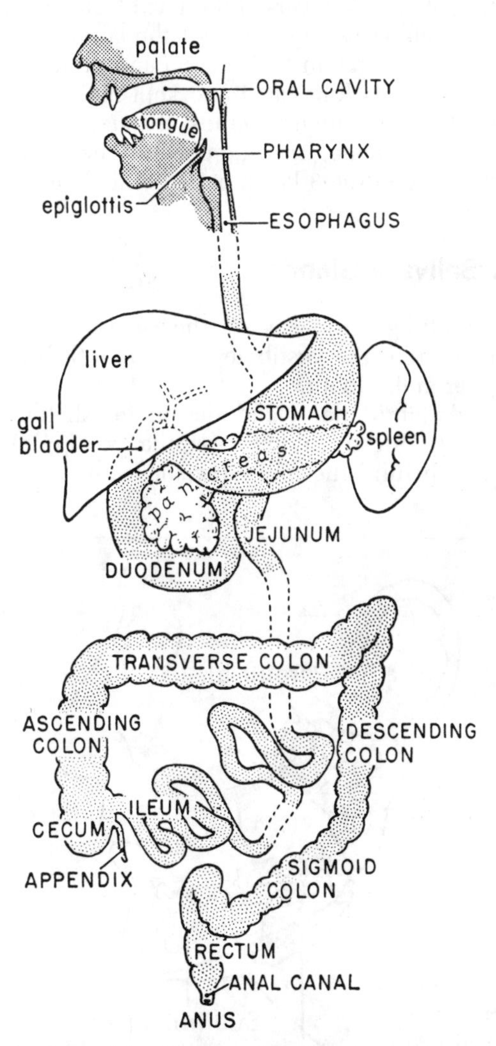

Figure 3.1. Diagram of the digestive tract.

MOUTH (ORAL CAVITY)

To survive, animals must eat, and so their mouths lead in the search for food. Around the mouth, the organs of special sense are developed: the eyes (and ears) to locate food by *sight* (and sound), the nose by *smell*, and the lips and tongue by *touch* and *taste*. Hence, the brain grows at the mouth end of the body.

The *oral cavity* is bounded externally by the cheeks and lips. The cleft between the upper and lower lips is the *aperture* of the mouth. Within, the *teeth* form 2 dental arches, one set in the upper jaw or *maxillae* and the other in the lower jaw or *mandible*. The horseshoe-shaped space external to the teeth and gums is the *vestibule* of the mouth. The space surrounded by the teeth and gums is the *mouth proper*. From the floor of the mouth rises the *tongue*; the roof comprises the *hard* and *soft palates*. The soft palate ends in the midline as a little hanging process, the *uvula* (L., "grape").

Mastication

The incisor teeth (L. *incidere* = to cut) bite off pieces of food, and the molar teeth (L. *mola* = millstone) grind them. The food is commonly coarse; so, the mouth requires a protective lining. It is lined with an epithelium consisting of special layers of flattened (squamous) cells resembling epidermis (p. 56), but the surface cells, although flattened, retain their nuclei and do not cornify. The epithelium is lubricated and kept moist by the secretions of *minor salivary glands*, the size of pinheads, which line the palate, lips, and cheeks. These secretions augment those of the 3 large, paired oral glands.

Major Salivary Glands

The *parotid gland* lies below the ear (*fig. 3.2*). Its long duct opens into the vestibule of the mouth beside the 2nd upper molar tooth.

The *submandibular gland* lies under shelter of the mandible; its duct opens onto a papilla (= nipple) beside its fellow, behind the lower incisor teeth. On opening your mouth and raising the tip of the tongue, watery secretion may be seen in a mirror to be welling up from the orifices.

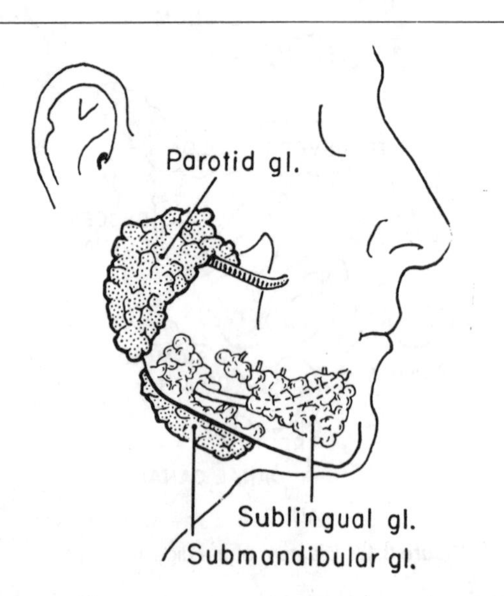

Figure 3.2. The oral (salivary) glands of the right side.

Each *sublingual gland* produces a ridge on the side of the floor of the mouth beneath the tongue. Its many fine ducts open onto the ridge (*fig. 3.2*).

The salivary glands have either a mucous or a serous secretion or both; the serous secretion (ptyalin) would digest free starch, but it cannot break down the cellulose enclosing starch in most foods.

The saliva moistens the food which has been ground into small particles by the teeth. This the tongue rolls into a bolus or lubricated mass, which can be easily swallowed; dry food is swallowed with difficulty. Saliva keeps the lips and mouth pliable when speaking.

PHARYNX AND ESOPHAGUS

The pharynx and esophagus are merely passages and, like the mouth, are protected by stratified squamous epithelium. The muscular coat of the pharynx and upper half of the esophagus, though voluntary in structure, are not under voluntary control; in the lower half of the esophagus (and therefore throughout the entire "g-i" tract to the anus), the muscles are involuntary or smooth.

STOMACH (GASTER)

Because the stomach is a receptacle for food and must expand, contract, and move about, it possesses a serous (peritoneal) coat. Its inner lining of mucosa secretes a protective mucus as well as hydrochloric acid, which is necessary for the action of the pepsin secreted by the gastric mucosa. These initiate the digestion of proteins. The outlet of the stomach, the **pylorus** (Gk. = a gatekeeper) is guarded by a strong sphincter of circular fibers, the *pyloric sphincter*, which opens in response to neural control to discharge partly digested and liquefied stomach contents into the duodenum.

INTESTINE

The intestine or gut is divisible into 2 parts: (1) the small intestine, about 7 m long, and (2) the large intestine, about 2 m long, as measured at autopsy (B.M.L. Underhill). Yet during life, a tube 3–4 m long swallowed by an adult may project from both mouth and anus. After death the gut relaxes and lengthens greatly and progressively.

The partially digested food leaving the stomach is acidic and sterile. In the first part of the small gut (10 cm beyond the pylorus) the enzymes of the pancreas and the bile are added. Here digestion continues and absorption of water and digested products begins. These processes are most active in the duodenum, and they diminish as the large gut is approached. The essential function of the large gut is to dehydrate the intestinal contents. The bacteria, present in truly enormous numbers, exemplify symbiosis, for they make an important contribution to the welfare of their host. In addition to the production of vitamin B, they convert bile salts into molecules that are resorbed and used for producing hemoglobin. They also in part break down cellulose, which is almost undigested in the small intestine.

As the intestinal contents become less fluid in the large gut, more lubricating mucus is both needed and produced by special cells.

LIVER (HEPAR)

The liver is the largest organ in the body. Attached to its inferior surface is the *gallbladder*; this extends posteriorly to a transverse fissure, the *porta hepatis* (L. = door of liver), through which the *hepatic ducts* conduct bile from the liver and through which the *portal vein* (conducting blood laden with products of digestion) and the *hepatic artery* (conducting oxygenated blood) enter the liver. After the blood has circulated through the liver, it leaves the posterior surface via the *hepatic veins* to enter the inferior vena cava.

The liver is a gland of compound tubular design. The cells of the tubules elaborate an *exocrine (external) secretion*, called bile, into the tubules. Each tubule, called a *bile canaliculus*, is really a series of narrow spaces between rows or sheets of liver cells. It drains into a system of ducts, the *bile passages*, which emerge from the liver to communicate with the gallbladder and with the duodenum (*figs. 13.1* and *13.2*).

The same liver cells also elaborate an *endocrine (internal) secretion* into the blood in the *sinusoids* (p. 29). Helping to line the sinusoids are *stellate macrophagocytes (Kupffer's cells)*, which help to dispose of effete red blood cells and are concerned with immune mechanisms.

Blood Flow through the Liver

In traditional textbook descriptions, the liver is composed of hexagonal lobules (*fig. 3.3*) surrounding a receiving (central) radicle of the hepatic vein, placed centrally. More correctly (*fig. 3.4*), a distributing (preterminal portal vein is placed centrally in a *liver acinus*, sending diverging sinusoids to 3 receiving hepatic veins, placed peripherally (Rappaport).

The *bile passages* conduct bile in the reverse direction (i.e., toward the porta). *Lymph vessels* run both with the portal vein and the hepatic veins.

The cells of the liver regulate the amount of glucose in the blood by removing the excess after a meal and temporarily storing it as glycogen. They also store vitamin A. They elaborate fibrinogen and heparin, both vital in blood-clotting mechanisms.

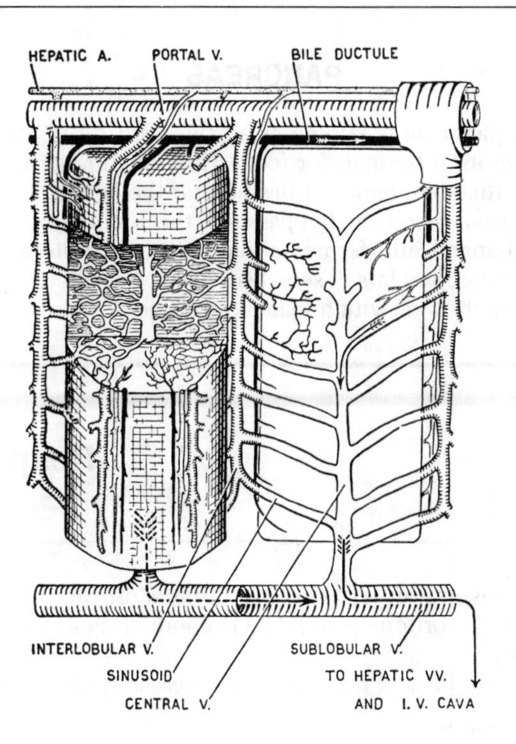

Figure 3.3. Diagram of the blood flow through 2 lobules of the liver and of the course of bile, according to the traditional description.

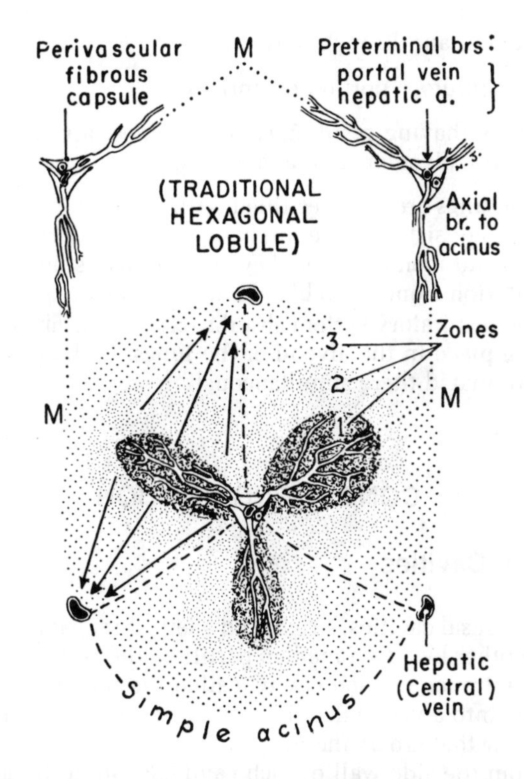

Figure 3.4. According to the modern view, the acinus is the liver unit. It incorporates sectors of several (3) adjacent lobules, a preterminal branch of the portal vein and of the hepatic artery being central and several (3) initial hepatic veins being peripheral. *Zone 1* is supplied earlier with blood than *Zone 2*, and 2 before 3. (Courtesy of Rappaport *et al.*)

PANCREAS

The pancreas is known as a "sweetbread" in animals that are used by man for food. This long gland extends across the abdomen and lies behind the stomach. It has an external secretion that passes to the duodenum, where it is changed into ferments that act upon proteins, carbohydrates, and fats. It secretes a hormone called insulin, the reduction of which leads to diabetes.

SPLEEN (LIEN)

The spleen is mentioned here for convenience, but its association with the digestive system is incidental. It is a soft sponge filled with blood, about the size of a clenched fist. For a discussion of the *structure and function* of the spleen, see p. 150.

Respiratory System

The respiratory system, which buds off the primitive gut, has 2 portions (*fig. 3.5*): (1) conducting and (2) respiratory.

The *conducting portion*, or air passages, comprises:

External nose
Nasal cavities
Pharynx
Larynx
Trachea
Bronchi and Bronchioles

The *respiratory portion* comprises:

Lungs (having respiratory bronchioles, alveolar ductules, alveolar sacs, and alveoli)

The lungs are sponges filled with air; their septa (partitions) consist almost entirely of blood capillaries. Thin, moist, and membranous, they allow transfusion of gases in solution from air to blood and *vice versa*.

The respiratory system absorbs oxygen, the absorption taking place in the alveoli of the lungs. Carbon dioxide is eliminated there simultaneously.

CONDUCTING PORTION

Nasal Cavities

The nasal cavities, a right and a left, are separated from each other by a thin median partition, the *nasal septum*. The entrance to each cavity, called the *nostril* or naris, opens into a *vestibule* which is lined with skin—hence the hair that grows there.

From the side wall of each cavity, 3 inferiorly curved shelves, the *conchae*, overhang 3 anteroposteriorly running passages, the *meatuses*. Opening into the inferior meatus inferior to the inferior concha is the *tear duct* (nasolacrimal duct); opening into the other meatuses are the orifices of large *air sinuses*. These sinuses, inflated with air, invade the surrounding bones, causing them to be large enough to carry the upper teeth and to form the framework of a large face in the adult, without adding great weight.

The mucous membrane covering the inferior and middle conchae contains dilatable *venous sinuses* that warm and humidify the inhaled air. It has a ciliated epithelium and also mucous and serous glands, diffuse lymphoid tissue, and lymphoid follicles. The mucus catches inhaled dust and bacteria and acts as a sterilizing agent. The cilia, which are microscopic "hairs," waft the mucus and the foreign matter entangled in it posteriorly to the pharynx.

The uppermost part (2 cm^2 in each nasal cavity) of the medial and lateral walls where they meet to form a narrow roof, is the *olfactory area*. Here the olfactory nerves, which stream through the thin perforated roof (*cribriform plate*) of the nasal cavity, end freely among the epithelial cells.

Pharynx

The pharynx, shared by digestive and respiratory systems, is a fibromuscular chamber only 15 cm long. It is attached superiorly to the base of the skull; it is continuous inferiorly with the esophagus. Communicating with it anteriorly are the nasal, oral, and laryngeal cavities. Accordingly, it is divisible into 3 parts (superior, middle, and inferior), called the nasal, oral, and laryngeal parts, respectively (*fig. 3.5*).

The *nasal part* (or *nasopharynx*) is the posterior extension of the nasal cavities. While it cannot be shut off from these cavities, it can be, and is, shut off from the oral part by the soft palate and uvula during the act of swallowing. Were it not so, food would be forced from the oral part into the nasal part and on to the nasal cavities; a person with a paralyzed soft palate may find that this may happen.

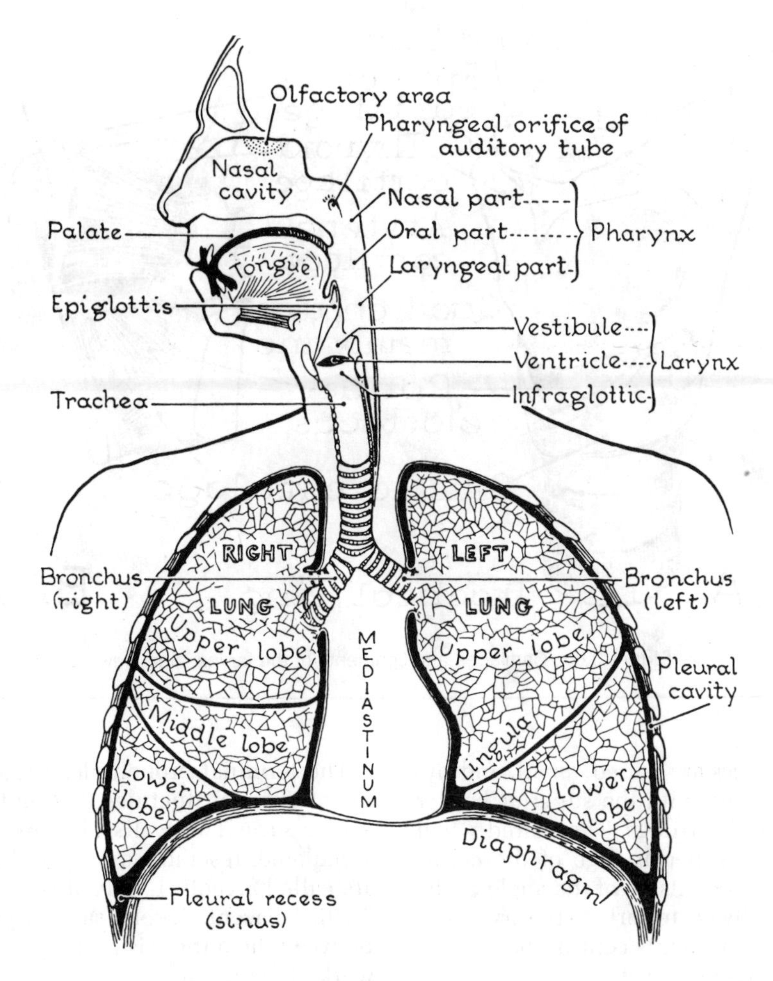

Figure 3.5. Diagram of the respiratory system.

Auditory Tube

An air duct, the auditory tube, opens into each side of the nasopharynx (*fig. 3.5*). Each tube, by bringing the pharynx into communication with the tympanic cavity (middle ear) of the corresponding side, serves to keep the air pressure on the 2 sides of the tympanic membrane (ear drum) equal under changing atmospheric conditions. Normally the tubes are closed, but the act of swallowing opens them; hence, the discomfort in the ears while ascending or descending in an airplane is relieved by swallowing. The tubes can be shut for days by inflammation of the nasal mucosa, causing temporary partial deafness; more serious, they provide a passage for infection to spread from the nasopharynx to the middle ear.

On each side of the entrance to the oral pharynx, and visible from the mouth, is a mass of lymphoid tissue, "*the tonsil*" or (more correctly) the *palatine tonsil*. Its superior pole extends from the side of the tongue superiorly into the soft palate; its inferior pole cannot be seen unless the tongue is depressed. In the roof of the nasopharynx of children, the mucosa forms another bulging mass of lymphoid tissue, the *pharyngeal tonsils* (or *adenoids*).

Larynx

The larynx or "voice box" opens off the inferior part of the pharynx and is continuous with the trachea inferiorly. This box is kept rigid by a number of hyaline and elastic cartilages that are united by membranes. It is lined with mucous membrane internally and covered with voluntary muscles externally.

The chief cartilages of the larynx are as follows.

(1) The *thyroid cartilage*, which resembles an angular shield, has 2 perpendicular *laminae* which meet anteriorly in the median plane, the prominent superior end of the *laryngeal prominence* ("Adam's apple"). Below, the thyroid cartilage grips the cricoid cartilage (*fig. 3.6*) as the knees of a horseman grip a saddle.

(2) The *cricoid cartilage* is a complete ring expanded posteriorly into a lamina or plate and so resembles a signet ring; it keeps the inferior part of the larynx perpetually open.

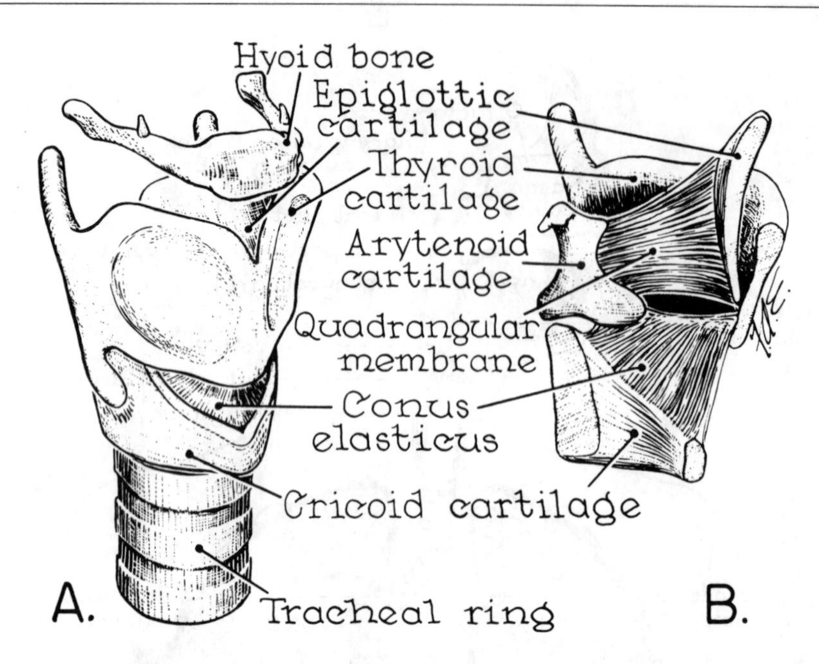

Figure 3.6. Cartilages and ligaments of the larynx, side view.

(3) The *arytenoid cartilages* are paired, small, and pyramidal; their bases articulate with the superior border of the lamina of the cricoid cartilage. The paired *vocal folds* or *cords* are the free superior edge of the *conus elasticus*; they extend from the inside of the angle of the thyroid cartilage horizontally posteriorly to the arytenoid cartilages whose various movements control the tension and distance apart of the vocal folds.

(4) The *epiglottic cartilage* is shaped like an elm leaf, its stalk being attached to the angle of the thyroid cartilage just superior to the vocal cords.

Trachea

The trachea or "windpipe" is an elastic tube about 10 cm long, with a caliber equal to the root of the index finger. It is kept patent by about 20 U-shaped rings of hyaline cartilage which are open posteriorly (*figs. 3.5* and *3.6*).

At the level of the sternal angle, 5 cm inferior to the jugular notch, the trachea bifurcates into a right and a left bronchus.

Bronchi

The bronchi have the same structure as the trachea. After an oblique course of 5 cm, each enters the respective lung at the hilus and descends toward the base, giving off branches that in turn branch and rebranch like a tree (*figs. 7.6–7.9*). Within the lung, the U-shaped rings give place to plates of hyaline cartilage, which surround the tube and hold it open. When the bronchi are reduced to the diameter of 1.0 mm, they are called *bronchioles*.

The terminal bronchioles divide into a number (2 to 11) of alveolar ductules that end in dilated air sacs, *alveolar sacs*. The walls of these sacs, being themselves sacculated, resemble a bunch of grapes, and hence they are called alveoli (L. *alveolus* = a bunch of grapes) (*fig. 7.10*). Adjacent *alveolar* sacs are practically contiguous—between them there is room only for a close-meshed network of capillaries.

RESPIRATORY PORTION

Lungs

The right and left lungs (*fig. 3.5*) are sponge-works of elastic tissue which feel like rubber sponges. In this highly elastic framework, a bronchus and a pulmonary artery and pulmonary vein branch out.

The right lung is divided by 2 complete fissures into 3 separate lobes; the left is divided by 1 fissure into 2 lobes.

Pleura

Each lobe of each lung has a delicate and inseparable "skin," the *visceral* (or *pulmonary*) *pleura* (L. *pulmo* = lung). This is a perfectly smooth, moist, serous membrane, identical in structure and in origin with peritoneum in the abdomen, i.e., a fine areolar sheet with a thin surface of pavement (mesothelial) cells. Another layer of pleura lines the ribs, diaphragm, and mediastinum (containing the heart); this is *parietal pleura.* Between the pulmonary and parietal layers of pleura there is a potential space, the *pleural cavity* (*fig.* 3.5). It allows the lung to expand and contract without friction.

The Respiratory Act

Respiration has 2 phases—inspiration and expiration. On inspiration, the diaphragm descends, thereby increasing the vertical diameter of the thorax, and the ribs rise from a sloping to a more horizontal position, thereby increasing both the anteroposterior and the side-to-side diameter, as you can determine by palpation. Air rushes down the trachea and bronchi into the lungs, which must expand to avoid formation of a vacuum in the pleural cavities. Expiration is largely a matter of elastic recoil, that is, the highly elastic tissue contracts; the stretched abdominal muscles act like an elastic belt on the abdominal contents, forcing the diaphragm upward; and the cartilages of the ribs, which underwent twisting during inspiration, now untwist.

As the table that follows shows, during quiet respiration about 500 cc of air are inspired and expired. On full inspiration, about an extra 2500 cc can be drawn in. On full expiration an extra 1000 cc can be forced out. Even then there remain in the alveolar sacs, trachea, and bronchi about 1000 cc that cannot be expelled.

	cc		
Total Capacity	2500	Complemental air	Vital Capacity
	500	Tidal air	
	1000	Supplemental air	
	1000	Residual air	

Epithelial Surfaces

The mucous membrane of the respiratory passages as far as the bronchioles of 1 mm in diameter are lined with ciliated epithelium. It has mucous and serous glands, and lymphoid tissue both diffuse and aggregated.

In the protective mucus that lines the larynx, trachea, and bronchi, inhaled dust and other foreign material are caught and entangled. The cilia cause the layer of mucus to move ever upward, like a moving carpet or escalator, to the entrance of the larynx, where it spills over into the pharynx and is swallowed. The other method of expelling foreign material is by coughing. The lymphoid tissue also is defensive against foreign invasion.

The cilia in the nasal cavities sweep backward toward the pharynx, those in the trachea and bronchi upward toward the pharynx, and those in the air sinuses spirally around the walls to the orifices.

Exceptional Areas. (1) The vestibule of the nose is lined with skin, possessing hairs, sweat glands, and sebaceous glands. (2) Areas subjected to pressure or friction, where cilia could not survive, are protected by stratified squamous epithelium. These areas are: (a) the parts of the upper surface of the soft palate and uvula which are applied to the pharyngeal wall during swallowing; (b) areas against which the food comes into contact, namely, the entire oral and laryngeal parts of the pharynx and also the dorsal aspect of the upper half of the epiglottis; and (c) the vocal cords, which vibrate and strike each other. (3) The terminal and respiratory bronchioles have epithelium, which is cubical. (4) The alveoli, whose continuous lining of epithelial cells is so extremely thin that its existence was doubted until it was shown by electron microscopy. (5) The olfactory (smell) epithelium is composed of special olfactory receptor and supportive cells.

4

Urogenital System and the Skin

Urinary Organs

The urinary organs and the genital organs develop together as the *urogenital apparatus* or *system*. The male urethra serves as a common outlet.

The urinary organs include the following (*fig. 4.1*):

Kidneys, Ureters } bilateral and paired

Bladder, Urethra } median and unpaired

The *kidneys* excrete urine. This passes down 2 muscular tubes, the *ureters*, 1 on each side, into a muscular reservoir, the *urinary bladder*, where it is stored until it can be conveniently discharged through the *urethra*.

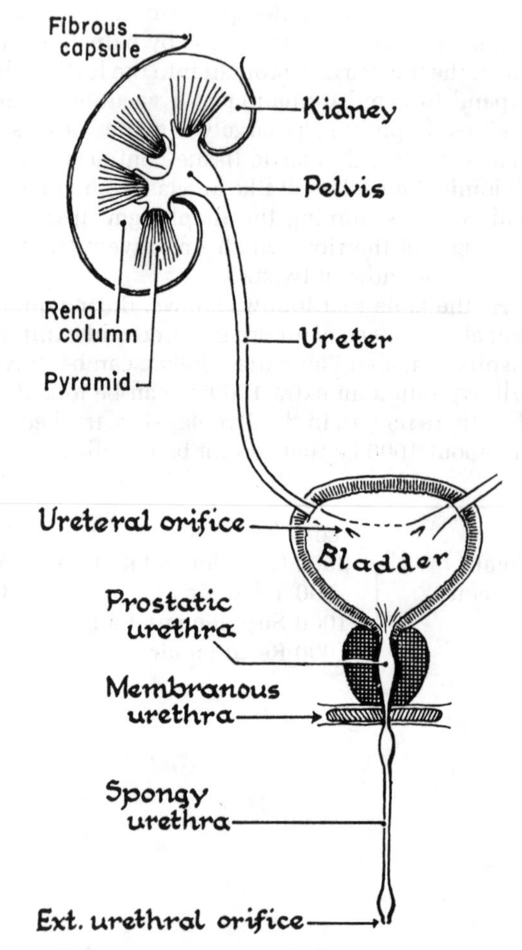

Figure 4.1. The urinary system (male).

KIDNEYS

Each of the 2 kidneys is about 11 cm long and weighs about 130–150 g. Its function is to keep the composition of the blood plasma constant by removing the excess of water and various waste products. It also maintains the acid-base balance of the blood by selective elimination of certain electrolytes.

Structure (*fig. 4.2*)

Occupying the inner two-thirds of the cut surface of a kidney are dark, striated areas, the *pyramids*. The *papillae* or tips of the pyramids project into the calices of the pelvis. The outer one-third of the kidney, i.e., the part lying external to the bases of the pyramids, is *cortex*. *Renal columns*, similar to cortical tissue, extend between the pyramids. These columns and the pyramids constitute the *medulla* (Gk. *calix*, = cup, singular).

A kidney contains about 1,000,000 microscopical units. Each unit or *nephron* has 2 parts: (1) a *glomerulus* and (2) a *uriniferous tubule*.

A glomerulus (L. = a small ball) is a spherical bunch of looped capillaries which invaginates the expanded blind end of a uriniferous tubule, called a *glomerular capsule* or "*Bowman's capsule*." The surface area of all the glomeruli of each kidney is estimated to be 0.3813 m^2 (M.H. Book). The 2 layers of glomerular capsule, outer and inner or invaginated, plus the glomerulus are known as a *renal* or *Malpighian corpuscle*.

The capsule is succeeded by the proximal convoluted tubule, the proximal and distal straight tubules of "Henle's loop," the distal convoluted tubule, and finally the junc-

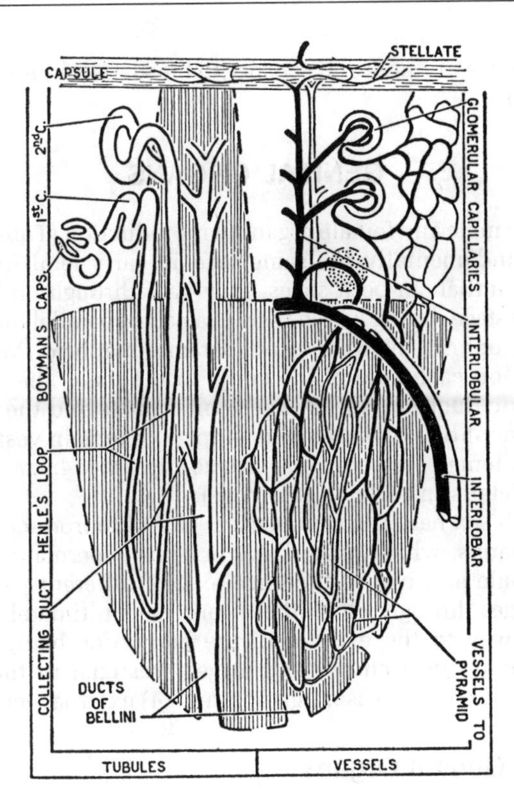

Figure 4.2. Diagram of the tubules and blood supply of the kidney (see also *fig. 14.5*).

tional tubule, which discharges into a system of collecting tubules. About a dozen collecting tubules open onto the papilla of each pyramid (*fig. 4.2*) and discharge their contents into the pelvis of the kidney.

Each named part of the tubule has a distinctive epithelium and, accordingly, a different function. The glomeruli and the convoluted tubules occupy the cortex and renal columns; the straight tubules of "Henle's loops" and the collecting tubules occupy the pyramids.

Details of Arteries and Veins. The **renal arteries**, a right and a left, carry far more blood to the kidneys than is needed for their nourishment, and all the blood first passes through the glomeruli.

Each renal artery typically divides into 5 *segmental arteries* (commonly arising irregularly) which, after entering the renal sinus, send branches up along the sides of the pyramids and, because the pyramids are spoken of as lobes of the kidney, the arteries are called *interlobar arteries*. At the junction between medulla and cortex, the interlobar arteries divide into many *arcuate arteries* that curve between these 2 parts, forming arcs but not arches.

From the arcuate arteries, *interlobular arteries* pass radially toward the capsule, each giving off many branches from which short arterioles, the *afferent* arteries, pass to the glomeruli.

Each *afferent artery* enters a glomerulus and there forms capillary loops, as noted. These unite and leave the glomerulus as an *efferent artery*, which breaks up into capillaries that ramify throughout the renal substance. Thus (1) the *efferent arteries* from the outermost glomeruli (those nearest the capsule) provide a capillary network among the convoluted tubules; (2) those from the innermost glomeruli send long meshed capillaries, arteriolae rectae, to the pyramids; and (3) the intermediate vasa do both. The 'arcuate' arteries are end arteries—they do not anastomose with their fellows to form complete arches.

The **interlobular veins** begin under the capsule as stellate venules and may be seen on stripping it off. The tributaries of the renal veins anastomose freely.

URETERS, URINARY BLADDER, AND URETHRA

From the kidney, the urine is propelled by peristaltic action along a 25-cm muscular tube, the *ureter*, into a hollow muscular reservoir, the *urinary bladder*. Through a cystoscope (a lighted tube used by urologists), jets of urine are seen to squirt into the bladder from the ureteral orifices 2 or 3 times a minute.

The bladder has a widely varying capacity (average about ½ liter). From its lowest part a fibromuscular tube, the *urethra*, leads to the exterior of the body. It is about 20 cm long in the male, 4 cm in the female.

At the junction of the bladder and urethra (i.e., at the neck of the bladder) involuntary muscle forms specialized bundles, but these do not constitute a true involuntary sphincter, as formerly taught (Woodburne). Beyond this (i.e., between the fasciae of the urogenital diaphragm), urinary control is maintained by a sphincter of voluntary muscle, the *sphincter urethrae*.

Parts of the Male Urethra (fig. 4.1) include: (1) prostatic (3 cm); (2) membranous (surrounded by a thin voluntary muscle and its membranes)—quite short but elastic; and (3) penile or spongy urethra, traversing the entire length of the spongy body of the penis—in the nonerect state, about 15–18 cm, but stretching proportionately during penile erection.

Nerve Supply

Kidney

Cutting dorsal nerve roots of T12, L1, and L2 may relieve renal pain (White and Smithwick). A denervated kidney continues to excrete normal urine.

The Ureter

Its spinal or cord segments are L1 and 2. The peristalsis of the ureter is not disturbed by lumbar sympathectomy; in fact, by the withdrawal of inhibitory sympathetic influences the functions of a dilated (hydronephrotic) ureter and of a dilated colon (megacolon) may be improved.

The Bladder and Urethra

These structures receive both parasympathetic and sympathetic nerves (*fig. 4.3*).

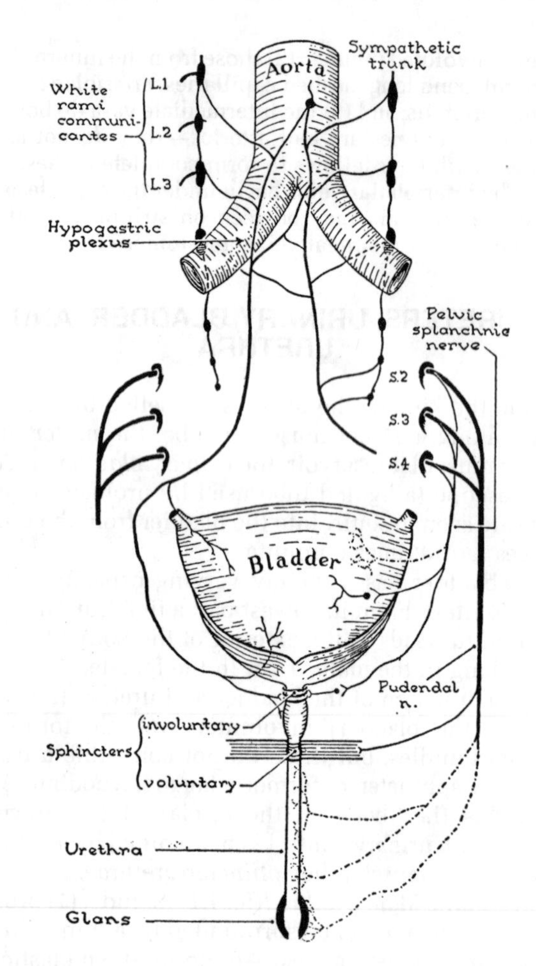

Figure 4.3. Diagram of the nerve supply of the bladder and urethra.

Parasympathetic. The pelvic splanchnic nerves (S2, 3, 4) are the motor nerves to the bladder; when they are stimulated, the bladder empties, the blood vessels dilate, and the penis becomes erect. They are also the sensory nerves to the bladder.

Sympathetic. The superior hypogastric plexus (lower thoracic and lumbar 1, 2, 3), is motor to a continuous muscle sheet comprising the ureteric musculature, the trigonal muscle, and the muscle of the urethral crest. It also supplies the muscle of the epididymis, ductus (vas) deferens, seminal vesicle, and prostate. When the plexus is stimulated, the seminal fluid is ejaculated into the urethra but is hindered from entering the bladder, perhaps by the muscle sheet that is drawn towards the internal urethral orifice. The sympathetic is also a vasoconstrictor, and to some slight extent it is sensory to the trigonal region.

It would seem that the sympathetic supply to the bladder has a vasoconstrictor and a sexual effect and that as regards micturition, it is not antagonistic to the parasympathetic supply (Learmonth; Langworthy *et al.*).

Somatic Nerves. The *pudendal nerve* is motor to the sphincter urethrae and sensory to the glans penis and the urethra.

GENITAL ORGANS

The male and female organs of reproduction (*Table 4.1*) are fundamentally the same, and in early fetal life are very similar. In both sexes, they pass through an indifferent stage during which there is a pair of parallel ducts—mesonephric (Wolffian) and paramesonephric (Mullerian)—on each side of the body. In the male, the mesonephric ducts are utilized as genital ducts and the paramesonephric ducts largely disappear or remain vestigial; in the female, the converse is true (*figs. 4.4, 4.5,* and *4.6* and details in table on this page).

Each sex has (1) a symmetrical pair of reproductory or sex glands, which produce germ cells—spermatozoa in the male and ova in the female; (2) 2 different pairs of passages through which these germ cells ultimately find their way to the exterior of the body, one being well developed in each sex and largely vestigial in the opposite sex; (3) accessory glands; and (4) external genitals.

Male Genital Organs

The male reproductive organs are:

Testis
Epididymis
Ductus Deferens
Seminal Vesicle
Ejaculatory Duct
Prostate
Bulbourethral Gland

Table 4.1. Homologous Male and Female Parts

	Male	Female
Sexual Gland	Testis	
MALE PASSAGES		Appendix (W) (?)
Mesonephric tubules & duct	Epididymis	*Epoophoron*
	paradidymis	*Paroophoron*
	Ductus Deferens	*Duct of Gartner* (*of epoophoron*)
	Urethra:	
	prostatic	Urethra
	penile	Labia Minora, enclosing vestibule
FEMALE PASSAGES	*Appendix* (M) (?)	Uterine tube
Paramesonephric duct	*Prostatic Utricle* (?)	Uterus
		Vagina
	Seminal vesicle	—
Accessory glands	Prostate	Para-urethal ducts
	Bulbo-urethral gland	Greater vestibular glands
	Urethral glands	Lesser vestibular glands
External genitals	Penis	Clitoris
	bulb of penis	Bulbs of vestibule
	Scrotum	Labia Majora

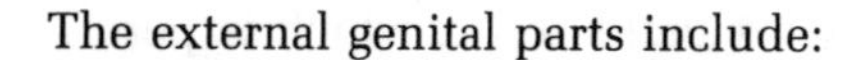

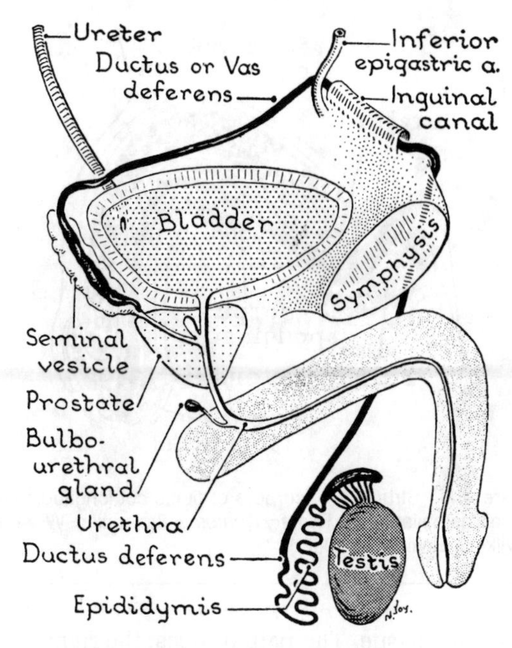

Figure 4.4. The male genital system.

The external genital parts include:

- Penis
- Urethra
- Scrotum

Rudimentary or vestigial structures:

- Prostatic utricle
- Paradidymis
- Aberrant ductules

Testes

The testes or male sex glands, one on each side, lie in the scrotum (L. = a leather bag) (*figs. 4.4, 11.16*). Each testis is ovoid and 4 cm long. Like the eyeball, it has a thick, white, inelastic, fibrous outer coat, the *tunica albuginea*. Within this covering are numerous delicate, threadlike macroscopic *seminiferous tubules*, the linings of which produce enormous numbers of microscopic *spermatozoa*. These are provided with long whiplike tails which later provide propulsive power. Specialized cells in the testis produce the male sex hormone.

The testis is covered anteriorly and at the sides with a

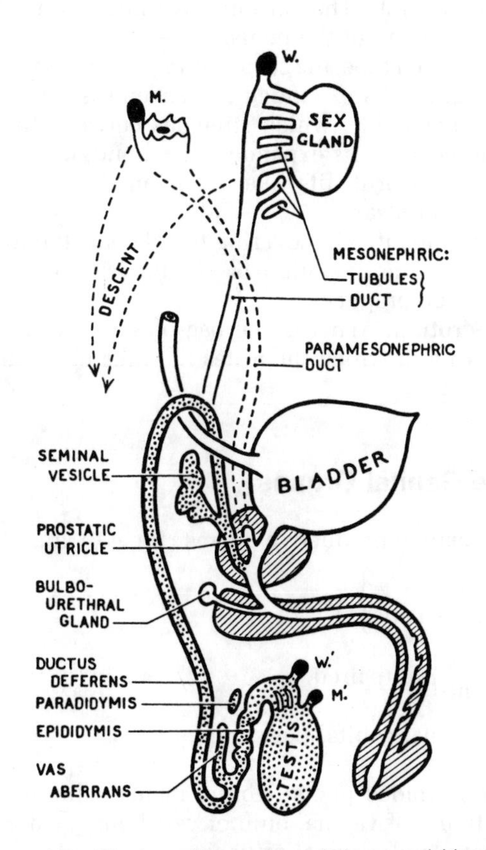

Figure 4.5. Diagram of the male genital system (side view).

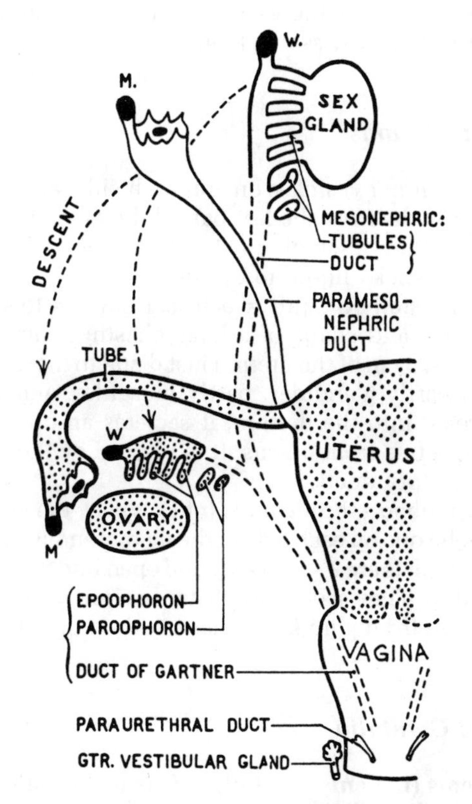

Figure 4.6. Diagram of the female genital system (front view).

In the upper halves of both figures, the parts are shown in their early or indifferent state. Most of the paramesonephric (Mullerian) structures disappear in the male, and most of the mesonephric disappear (Wolffian) in the female. (*M* = Wolffian appendage.)

bursal sac, the *tunica vaginalis testis*, which is peritoneum that was continuous with the peritoneal cavity until about the time of birth.

The Ducts of the Testis. From 6 to 12 fine *efferent ductules* lead out of the superior part of the testis into the *duct of the epididymis*. This threadlike duct, although about 6 m long, is so folded as to form a compact body, the *epididymis* (Gk. epi = upon, didumos = a twin; i.e., the testes are twins), which caps the superior pole of the testis and is applied to its posterior border. The cells lining the duct of the epididymis discharge a mucoid secretion to carry the spermatozoa.

The *ductus* (or *vas*) *deferens* connects the duct of the epididymis to the urethra. It is about 45 cm long. It ascends through the superior part of the scrotum to the abdominal wall. This it pierces obliquely in a tunnel, the *inguinal canal*. Continuing, it runs under cover of the peritoneum to the base of the bladder and then descends between the bladder and the rectum. Its terminal 2 cm, the *ejaculatory duct*, pierces the prostate and opens into the urethra close to its fellow, about 3 cm distal to the bladder. From this point onward, the male urethra is the common duct of the urinary and genital systems. It is also the royal road for gonorrheal infection throughout the mucous membranes of the male genital tract.

Semen (L. = seed). Semen is composed of spermatozoa suspended in the secretions of the ducts of the testes and of the accessory glands.

Accessory Glands

1. *The seminal vesicles*, one on each side, are tubular outgrowths from the last part of the deferent ducts, which they resemble in structure. They add a yellowish sticky liquid to the semen.
2. *The prostate* (Gk. pro = before, istanai = to stand), the size and shape of a large chestnut, surrounds the first 3 cm of the urethra just distal to the bladder. It is partly glandular, partly muscular, and partly fibrous. Into the urethra, it secretes an opalescent liquid, free from mucus. It is pierced by the paired ejaculatory ducts.
3. *The bulbourethral glands* (of Cowper), the size of a pea, lie one on each side of the membranous urethra. Their ducts are 2–3 cm long and open into the spongy urethra. They secrete a substance similar to mucus.
4. *The urethral glands* are discussed on p. 218.

External Genitalia

The penis (L. *penis* = a tail). This is the male organ of copulation (see *fig. 21.6*). It comprises 3 parallel fibrous tubes, 2 paired and 1 unpaired (*fig. 4.7*). These have innumerable cavernous spaces, filled with blood. They are enclosed in a single loosely fitting tube of skin and

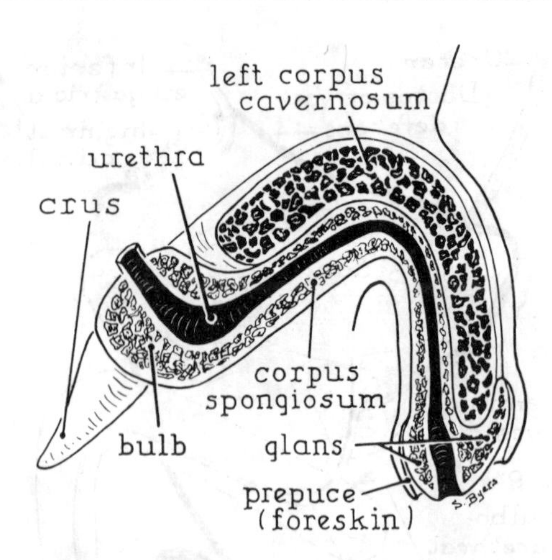

Figure 4.7. Structure of corpora of penis cut longitudinally. (From Basmajian, J.V.: *Primary Anatomy*, 8th edition, Williams & Wilkins, Baltimore, 1982.)

subcutaneous tissue. The paired tubes, the right and left *corpora cavernosa*, are fused side by side. Distally, they present a rounded end; proximally, they diverge into right and left *crura* (L. *crus* = a leg) that are firmly attached to the pubic arch. The corpora cavernosa are the "supporting skeleton" of the penis.

The unpaired tube, the *corpus spongiosum*, is traversed by the urethra. Its expanded hinder end is fixed to the perineal membrane, which stretches between the sides of the pubic arch; its expanded distal end, the *glans* (L. glans = an acorn), fits like a cap on the ends of the corpora cavernosa.

The redundant skin covering the glans is the foreskin or *prepuce*. The operation of circumcision consists of removing the prepuce.

The Scrotum. This is the bag of skin and subcutaneous tissue in which the testes lie. Like the penis it is free from fat.

Female Genital Organs

The female reproductive organs (*fig. 4.8*) are:

Ovary, Uterine Tube } paired

Uterus, Vagina } unpaired

The external genital parts include:
Clitoris
Pudenda (mons pubis, labium majus, labium minus, vestibule of vagina, bulb of vestibule, greater vestibular gland, vaginal orifice)
Rudimentary or vestigial structures:
Epoophoron and duct
Paroophoron

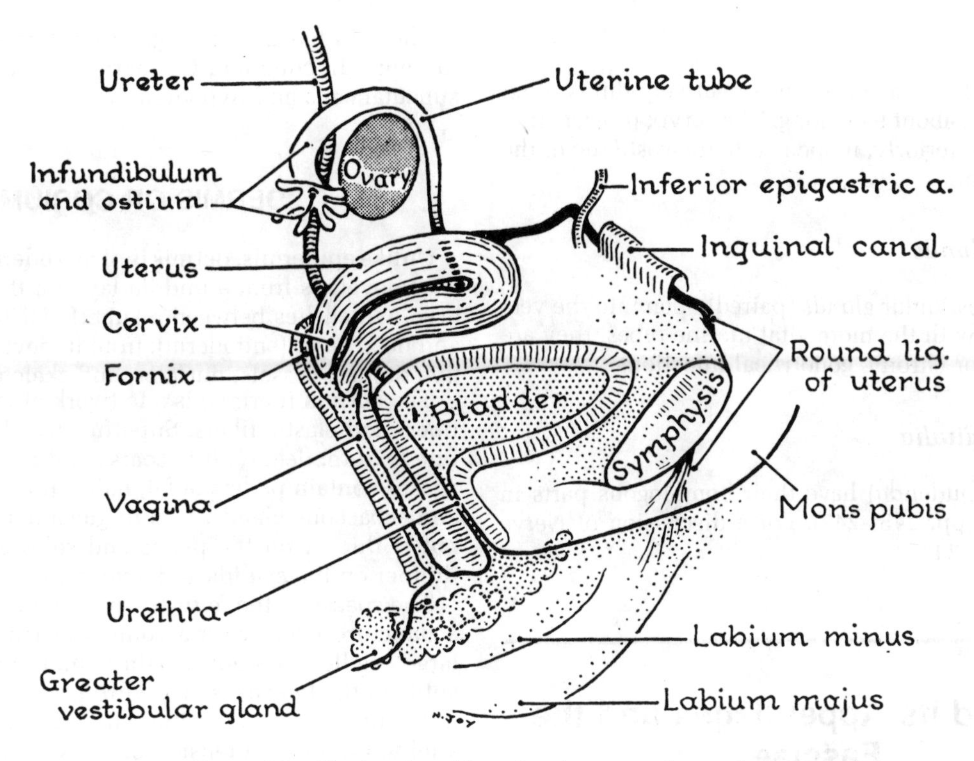

Figure 4.8. The female pelvis and perineum, median section.

Ovaries

The ovaries or female sex glands, one on each side, lie on the lateral walls of the pelvis (*fig. 4.8*). Each ovary is a solid body about half the size of a testis. It has an attached border; otherwise it lies free in the peritoneal cavity. However, it is not covered with peritoneal cells but with cubical cells, some of which become the ova (eggs). Its surface is scarred and pitted due to the shedding of ova. At birth each ovary contains about 200,000 (immature) ova. From puberty to the end of the reproductive period (15th to 50th year), an ovum is shed into the peritoneal cavity about once a lunar month. In all, about 400 ova are shed; the rest are absorbed in the ovary. Ovarian cells also release female sex hormones into the bloodstream.

Uterine Tubes

The uterine tube (*fig. 4.8*) is about 10 cm long, is as large as a pencil, lies in the free edge of a fold of peritoneum, the *broad ligament*, and takes a twisted course from ovary to uterus. Its ovarian end, the *infundibulum*, is funnel shaped, and at the bottom of the funnel is the abdominal mouth or *ostium*, which lies open to the peritoneal cavity, beside the ovary. Its other end opens into the cavity of the uterus.

When an ovum is about to be shed, the mouth of the tube apparently lies ready to receive the ovum. Peristaltic action and a carpet of cilia in the mucous membrane lining the tube propel it along. In the tube, the ovum may be met and fertilized by a spermatozoon, which is able to propel itself from the vagina and through the uterus in about 6 hours.

Uterus

The uterus is a thick-walled, hollow, muscular organ placed near the center of the pelvis and projecting superiorly into the peritoneal cavity between the bladder and the rectum (see *fig. 19.2*). It is shaped like an inverted pear, somewhat flattened anteroposteriorly so that its cavity is collapsed, and is 8 cm long. Superiorly, a uterine tube opens into it on each side; inferiorly, it opens into the vagina. The superior two-thirds (5 cm) are the *body*; the inferior one-third is the *cervix* (= neck); the part of the body superior to the entrance to the tubes is the *fundus*. The peritoneum covering the body and fundus stretches from each side of the uterus as a fold, the *broad ligament*, which rises from the pelvic floor, extends to the side wall of the pelvis, and contains the uterine tube in its upper free edge. The function of the uterus is to harbor a fertilized ovum as it becomes a child for 9 months. During the first 2 months, the ova are in the indifferent or embryonic stage; the last 7, the formed or fetal stage.

Vagina

The *vagina* (L. = a sheath) is a relatively thin-walled collapsed tube, about 9 cm long. The cervix projects into it superiorly; inferiorly, it opens into the *vestibule* of the external genitalia.

Accessory Glands

The great vestibular glands (paired) open into the vestibule. Along with the more vital uterine tubes, they are prone to harbor chronic gonorrheal infections.

External Genitalia

These (the *pudenda*) have their homologous parts in the male (see pp. 245–247). For a discussion of *Nerve Supply*, see p. 248.

Skin and its Appendages and the Fasciae

The skin is no mere envelope wrapped around our bodies like paper around a parcel. It is, indeed, a wrapping but it is more than a wrapping—it is one of our most versatile organs.

As an *envelope*, it has admirable properties: being waterproof, it prevents the evaporation and escape of tissue fluids; it becomes thick where it is subject to rough treatment; it is fastened down where it is most likely to be pulled off; and it has friction ridges where it is most liable to slip. "Even with our ingenious modern machinery we cannot create a tough but highly elastic fabric that will withstand heat and cold, wet and drought, acid and alkali, microbic invasion, and the wear and tear of three score years and ten, yet effect its own repairs throughout and even present a seasonable protection of pigment against the sun's rays. It is indeed the finest fighting tissue" (Whitnall).

As an *organ* it is the regulator of the body temperature; it is an excretory organ capable of relieving the kidneys in time of need; it is a storehouse for chlorides; it is the factory for antirachitic vitamin D (ergosterol) formed by the action of the ultraviolet rays of the sun on the sterols in the skin and necessary for the mineralization of bones and teeth; and it is the most extensive and varied of the sense organs.

In an average adult, the skin covers a body surface of 1.7 m^2. At the orifices of the body, it is continuous with the mucous membranes. The skin, somewhat modified, forms the conjunctiva and the outer layer of the ear drum.

The skin has 2 parts (*fig. 4.9*): (1) dermis, and (2) epidermis. (L. cutis and Gk. derma = skin; *cf.* the terms subcutaneous and hypodermic.)

DERMIS OR CORIUM

Unlike epidermis, dermis is of mesodermal origin. That is, it develops from a middle layer of the primitive embryo, which lies between a superficial layer (ectoderm) and deep layer (endoderm); from it, develop the connective tissues, vessels, muscles, and skeleton.

The corium (dermis) is a feltwork of bundles of white fibers and elastic fibers. Superficially, the feltwork is of fine texture; deeply, it is coarse and more open, and its spaces contain pellets of fat, hair follicles, sweat glands, and sebaceous glands. It is in general 1 to 2 mm thick, but is thicker on the palms and soles and back and is thinner on the eyelids and external genital parts.

The spaces in the feltwork are lozenge-shaped; hence, a puncture made with a conical instrument, such as a large needle, does not remain a round hole, for the skin splits in the long axes of the lozenges. The long axes are differently directed in different parts, usually being parallel to the lines of tension of the skin. A cut across the long axes of these **lines of cleavage** (Langer's Lines) will gape (Cox).

It is due to the presence of elastic fibers that the skin, after being stretched or pinched into a fold, returns to normal.

THE EPIDERMIS

The epidermis is a nonvascular stratified epithelium of ectodermal origin—the other major derivative of embryonic ectoderm being the central nervous system. The deeper layer, the "*germinative layer*," is living. It consists of several strata of polyhedral cells resting on a single stratum of columnar (basal) cells. The superficial layer, the horny layer or *stratum corneum*, is dead. It consists of several strata of dry, flattened, scaly cells without visible nuclei. The surface cells are perpetually being rubbed away and are perpetually being replaced by cells of the germinative layer.

Ridges of dermis, which on cross-section have the appearance of nipple-like processes and are therefore called *papillae*, project into the epidermis.

For *finger prints*, see *Figure 32.13*, and pp. 000–000, and consult Cummins and Midlo.

THE NAILS

The nails are thickenings of the deeper layers of the stratum corneum. A nail has a free end which projects, a root which extends proximally deep to the overhanging

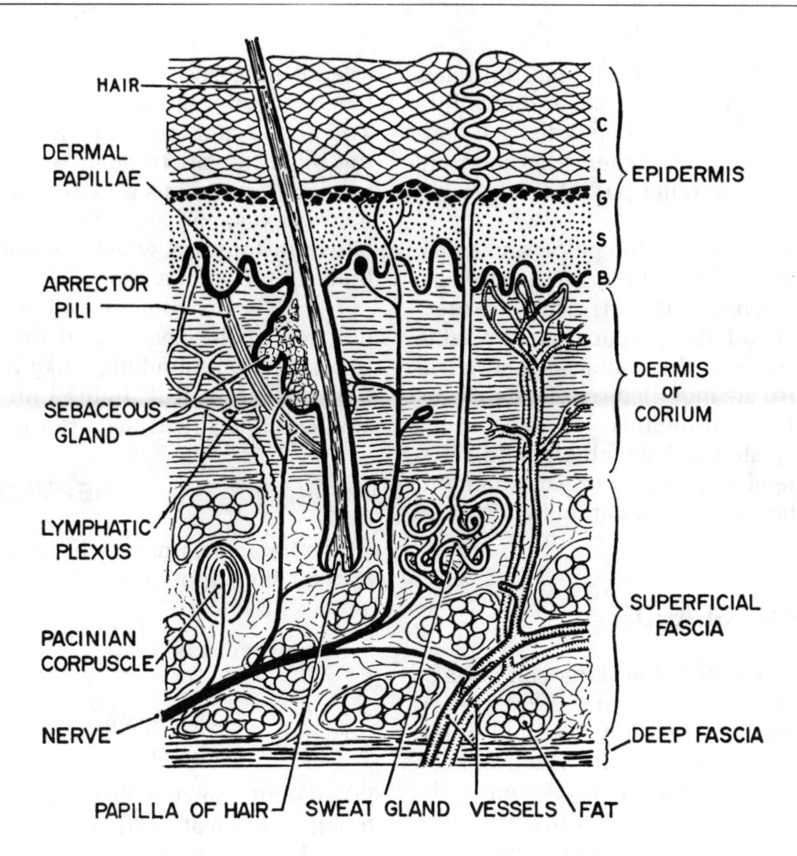

Figure 4.9. Section of thick skin, superficial fascia, and deep fascia. (*C, L, G, S, B* = Strata Corneum, Lucidum, Granulosum, Spinosum, and Basal layer, respectively.)

nail fold, 2 lateral borders, a free surface, and a deep one. The white crescent appearing distal to the nail fold is the *lunule.* The deep surface rests on and adheres to the *nail bed.* This largely consists of white fibrous tissue which attaches the nail to the periosteum.

Growth takes place at the root and from the bed as far distally as the lunule; beyond this, the nail probably slides distally on its bed, adhering to it. Poisons, formed during an acute illness, temporarily arrest the growth of the nails (as they do of the bones, see *fig. 1.7*), and a transverse ridge appearing on each nail when growth is resumed is evidence of a past illness. Since the average rate of nail growth is 0.1 mm per day or 3 mm a month (i.e., about 3–5 cm per year), the date of a past illness can be estimated. The toenails grow much more slowly than fingernails, of which the nail of digit III consistently grows fastest, that of digit V, slowest. Nails grow rapidly in "nail biters"; in immobilized limbs, they grow slowly (Le Gros Clark and Buxton).

THE HAIRS

Hair is distributed over the entire surface of the body, except for the palms and soles, dorsum of the last segment of the digits of the hand and foot, red of the lips, and parts of the external genitals and the conjunctiva. Hairs are also present in the vestibule of the nasal cavity and in the outer part of the external acoustic meatus.

Hairs may be short (a few mm) or long. Long hairs are present in the scalp, eyebrows, margins of the eyelids, vestibule of the nose, and outer part of the external acoustic meatus. At puberty, they appear on the pubes, external genitals, axillae and, in the male, on the face.

A hair has a *shaft* or part that projects beyond the skin surface, a *root* that lies in a follicle of the skin, and at the end of the root there is a swelling, the *bulb,* which is moulded over a dermal papilla. The life of a hair on the head is about 2 to 4 years; of an eyelash about 3 to 5 months. Old hairs are constantly falling out and new ones taking their place.

The *arrectores pilorum muscles* are bundles of smooth muscle that pass obliquely from the epidermis to the slanting surface of the hair follicles deep to the sebaceous glands. By contracting they cause the hairs to stand erect. In birds, by erecting the feathers, they increase the air spaces between them, thereby preserving heat; hence, sparrows look plump in cold weather. In man, spasm of the arrectores produces "goose skin."

SEBACEOUS GLANDS

Sebaceous glands are simple alveolar glands, bottle-shaped and filled with polyhedral cells, which break down into a fatty secretion called sebum in the hair follicles. The glands develop as outgrowths of hair follicles into the dermis, one or more being associated with each hair. Commonly, the glands are largest where the hairs are shortest (e.g., end of nose and outer part of the external acoustic meatus).

Fortunately there are no hairs on the palms and soles; neither are there sebaceous glands to make the surfaces greasy. Independently of hairs, sebaceous glands are present on the inner surface of the prepuce, on the labia minora, and on the areolae of the mammae. The tarsal glands of the eyelids also are modified sebaceous glands which waterproof the edges of the lids.

Boils (and carbuncles) start in hair follicles and sebaceous glands and are therefore possible, indeed common, in the vestibule of the nose and outer part of the external acoustic meatus.

SWEAT GLANDS

The sweat glands are present in the skin of all parts of the body (except the red of the lips and glans penis), being most numerous on the palms and soles and in the axillae. They are simple tubular glands. The secretory part of each is coiled to form a ball (0.3–0.4 mm in diameter) situated in the fat deep to the dermis. The duct runs tortuously through the dermis, enters the epidermis between 2 ridges, and proceeds spirally to the skin surface. In the stratum corneum it is represented merely by a cleft between the cells. The resemblance to the "intestines of a fairy" was fancifully suggested by Oliver Wendell Holmes.

The *ceruminous* (= wax) *glands* in the outer parts of the external acoustic meatus and the ciliary glands of the eyelids are modified sweat glands. In the axilla, about the external genitals, and around the anus are long, large (3–5 mm in diameter) modified sweat glands that produce an odor. They are spoken of as "sexual skin glands."

Sweating lowers the temperature. In man at rest, sweating is observed to begin abruptly when the body temperature is elevated a fraction of a degree. This is due to the action of the heated blood on the brain centers. Since the autonomic nerve fibers to sweat glands travel to the skin in the ordinary cutaneous nerves, if such a nerve is cut the area is not only deprived of sensation but it also cannot sweat.

Sweat glands are excretory organs—accessory to the kidneys. The salt taste of sweat is due to sodium chloride. The normal sweat secretion is important in keeping the thick horny layers of the palms and soles supple, and it increases the friction between the skin and an object grasped. In dogs, sweat glands are confined to the foot pads; so, being unable to sweat, dogs pant.

VESSELS

The vessels for the supply of the skin run in the subcutaneous fatty tissue. From these, the dermis receives 2 arterial plexuses; one is deeply seated near the subcutaneous tissue, and the other is in the subpapillary layer. This sends capillary loops into the papillae. The returning blood passes through several layers of thin-walled subpapillary venous plexuses, thence through a deep venous plexus, and so to the superficial veins. Arteriovenous anastomoses, which are sometimes open and sometimes closed, connect some of these arterioles and venules.

The *lymph vessels* of the skin form a plexus at the junction of the dermis and the superficial fascia. This plexus receives blind fingerlike vessels (or networks) from the papillae, and it drains into lymph vessels that accompany the superficial arteries and veins.

NERVES

The cutaneous nerves have (1) *efferent* autonomic fibers for the supply of:

Smooth muscle { of hairs (Arrectores)
of blood vessels }

Glands { sweat glands
sebaceous glands }

(2) *afferent* somatic fibers of general sensation, namely touch, pain, heat, cold, and pressure.

As *Figure 4.9* indicates, free fibers end between the cells of the germinative layer (hence, intraepidermal injections may cause pain), and around the hair follicle and beside it (probably touch fibers); tactile corpuscles occupy occasional papillae (for touch); Pacinian corpuscles lie in the superficial fascia and are plentiful along the sides of the digits (for pressure).

SUPERFICIAL FASCIA OR TELA SUBCUTANEA

Superficial fascia is a subcutaneous layer of **loose areolar tissue**, which unites the corium of the skin to the underlying deep fascia. It consists of (1) bundles of *white* or *collagenous fibers*, which by branching and uniting with other bundles, form an open webbing, filled with (2) *tissue fluid*; (3) a slender network of *yellow elastic fibers*, and scattered among all this (4) *connective tissue cells* (*fig. 4.10*).

When areolar tissue is exposed to the air, held fluids rapidly evaporate with consequent drying and shrinking. Fortunately, the addition of an appropriate laboratory solution will restore to areolar tissue its original fluffy texture.

Areolar tissue is derived from those portions of mesoderm that remain after bones, ligaments, tendons, muscles, and vessels have taken form. It is, therefore, not confined to the superficial fascia but is diffusely spread; for example, it forms the sheaths of muscles, vessels, nerves, and viscera, and it fills up the spaces between them; it forms the basis of the mucous, submucous, and subserous coats of the hollow viscera; the serous mem-

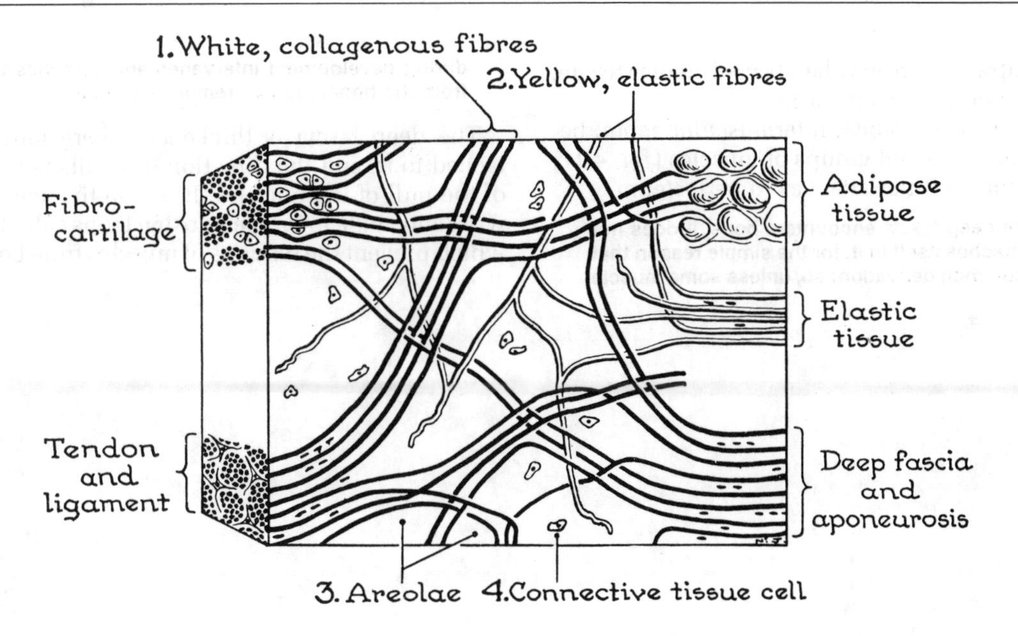

Figure 4.10. Scheme to indicate that the 4 ingredients of areolar tissue (*viz.*, collagenous fibers, elastic fibers, areolar spaces, and cells), when blended in different proportions, from other tissues (e.g., adipose, elastic, collagenous, ligamentous, and fibrocartilaginous) and that one may merge into another.

branes (i.e., peritoneum, pleura, pericardium, and tunica vaginalis testis) are but areolar membranes lined with flat mesothelial cells.

Areolar tissue is potentially **Adipose Tissue** or fat, and wherever areolar tissue occurs, there fat also may occur. The fat accumulates in the connective tissue cells. Fat is fluid at body temperature, but because each drop of fat is imprisoned in a cell, it does not "run" away, nor can it be massaged away normally.

Distribution

The superficial fascia almost everywhere contains fat—except in the eyelids, external ear, penis, scrotum, and at the flexion creases of the digits. In the palms and soles, it forms a protective cushion; here and still more so in the breast and scalp, it is loculated, i.e., imprisoned in loculi. It is most abundant in the buttocks.

In *women*, in whom special fat deposits are a normal secondary sexual characteristic, the typical areas are in the gluteal and lumbar regions, front of the thigh, anterior abdominal wall below the naval, mammae, postdeltoid region, and cervicothoracic regions.

Adipose or fatty tissue, as mentioned, is modified areolar tissue and is notably present in the superficial fascia, in the subserous layer of the abdominal wall, and in the omenta and mesenteries. It covers parts of the urinary tract (e.g., kidney, ureter, and sides of the bladder), but it leaves free the gastrointestinal tract, liver, spleen, testis, ovary, uterus, and lung. It is odd, then, that it should be present in the sulci of the heart. There is no fat within the cranial cavity to dispute possession with the brain, nor within the dura mater covering the spinal cord. The loose fatty tissue of the orbital cavity provides the eyeball with a soft and pliant bed.

Subcutaneous Bursae. See p. 20.

DEEP FASCIA

Deep fascia is the membranous investment of the structures deep to the superficial fascia (*figs. 4.10* and *4.11*). Like tendons, aponeuroses, and ligaments, it contains the same 4 ingredients as areolar tissue, but in different proportions; and, like them, being subjected to tensile strains, the white collagenous fibers form parallel bundles; the tissue fluids are at a minimum, and the cells are flattened (stellate on cross-section) from pressure.

Deep fascia is best marked in the limbs and neck, where it is wrapped around the muscles, vessels, and nerves like a bandage, the fibers being chiefly circularly arranged. Around the thorax and abdomen, which must

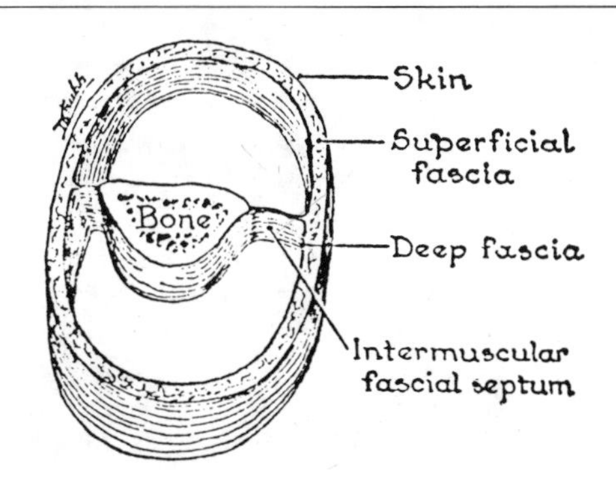

Figure 4.11. An intermuscular septum passing from deep fascia to bone.

expand and contract, there may be a film of areolar tissue, but there can be no true deep fascia.

The deep fascia sends septa, *intermuscular septa*, between various muscles and groups of muscles (*fig. 4.11*) and usually blends with or attaches to periosteum.

AXIOM. Where deep fascia encounters bone, it does not cross it but attaches itself to it, for the simple reason that both have a common derivation; so, unless some muscle during development intervenes and detaches the fascia from the bone, the two remain attached.

The deep fascia is thickened where muscles are attached to it, and the direction of its fibers takes the line of the pull of the muscles. It is also thickened about the wrist and ankle to form enclosing loops called *retinacula*, which prevent the tendons of muscles from bowstringing.

Thorax 2

5

Walls of the Thorax

Clinical Case 5.1

Patient Arthur S. This 63-year-old man injured his chest when he was forcibly thrown against the steering wheel of his car during an accident. When he was removed from the car, he showed some signs of respiratory distress, and he was brought by ambulance to the hospital emergency ward where you are an assistant. Lateral and anteroposterior radiographs of his thorax reveal fractures in the bodies of the 4th, 5th and 6th ribs over the region of the midaxillary line, and fractures in the right 4th, 5th and 6th costal cartilages near their union with the sternum. These double fractures in the bony and cartilaginous elements of the right side of his rib cage create an unstable portion in the chest wall, producing a "flail chest." Upon inspiration for chest expansion, the unstable wall appears to "sink" into the thorax. When the patient exhales to decrease the thoracic volume, the affected portion of the chest wall expands paradoxically. A surgeon has been called to perform a mechanical fixation of the mobilized portion of the rib cage—the subject of *Clinical Case 5.2* that comes later in this chapter.

Skin

The skin over the thoracic cage is innervated by nerves from the cervical and thoracic segments of the spinal cord. The dermatomes or areas of skin supplied by the sensory root of a single spinal nerve are illustrated in *Figure 5.1*. While dermatomes C3 and 4 are present over the clavicular and scapular regions of the upper thorax, the dermatomes of C5, 6, 7, 8 and T1 are found primarily on the upper extremity. The T1 dermatome is therefore adjacent to the C5 dermatome. This point must be remembered in examining the chest wall in patients suspected of having neck injuries. The T6 dermatome is found anteriorly over the xiphoid process (solar plexus region) and dermatomes from T7–T12 overlie portions of the posterior thoracic cage as well as the abdominal wall anteriorly and laterally.

The skin of the thorax is modified to produce the nipple of the breast. This is located over the 4th intercostal space in most individuals, but this is variable depending on the sex and weight of the individual. The female breast is described in the section on the axilla but is a very important structure of the thoracic wall. The relationship of the breast to the underlying tissues, blood vessels, nerves, and lymphatics is of paramount importance to all physicians who examine female patients for breast disease.

The thoracic skin overlies a number of bony elements that are easily palpable and identifiable beneath the skin (*fig. 5.2*). These surface landmarks include the subcutaneously located clavicle (collar bone) with its associated muscles, the sternum, the articulations of the sternum with the clavicle and upper 7 ribs, and the sternal angle (of Louis) where the second rib cartilage articulates with the lower manubrium and upper body of the sternum. The inferior end of the body of the sternum also articulates with the cartilaginous xiphoid process. The cartilaginous costal margin projects (as a vest border) inferiorly, laterally, and posteriorly to provide protection to the thoracic contents as well as the viscera in the superior quadrants of the abdomen.

The skin of the posterior aspect of the thorax is thicker and less sensitive to cutaneous stimuli, particularly in the region overlying the spinal column and head, neck and angle of each rib. The superficial back muscles are demonstrable under the skin of many patients, and their relationship to the subcutaneous scapula is easily determined in surface examination. The scapula is located between the 2nd and the 7th ribs and its medial border parallels the 6th rib when patients protract the scapula by placing their hand on the back of their head. This

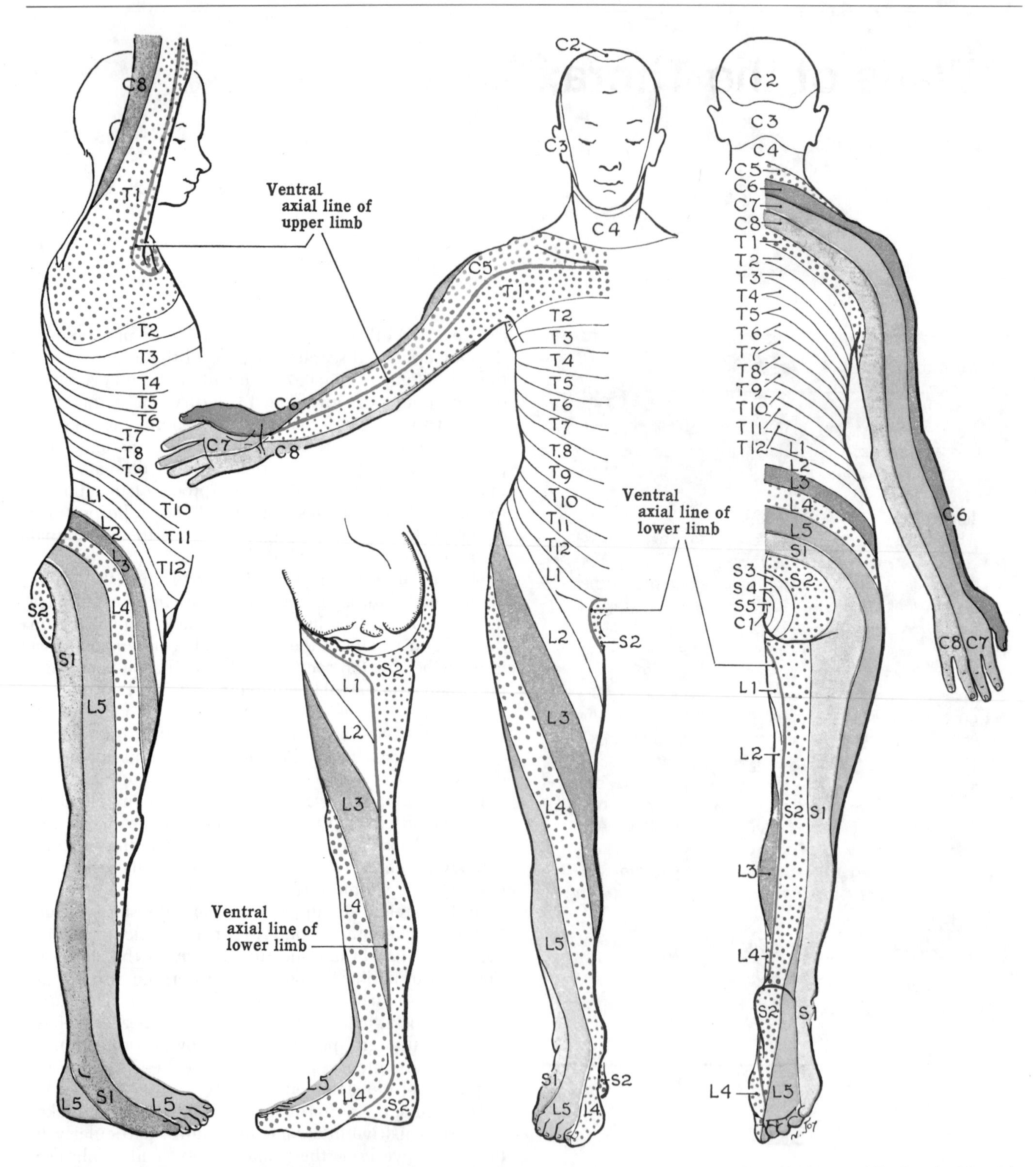

Figure 5.1. A dermatome is an area of skin supplied by the dorsal (sensory) root of a spinal nerve. (From Anderson, J.E., *Grant's Atlas*, 8th edition, Williams & Wilkins, Baltimore, 1983.)

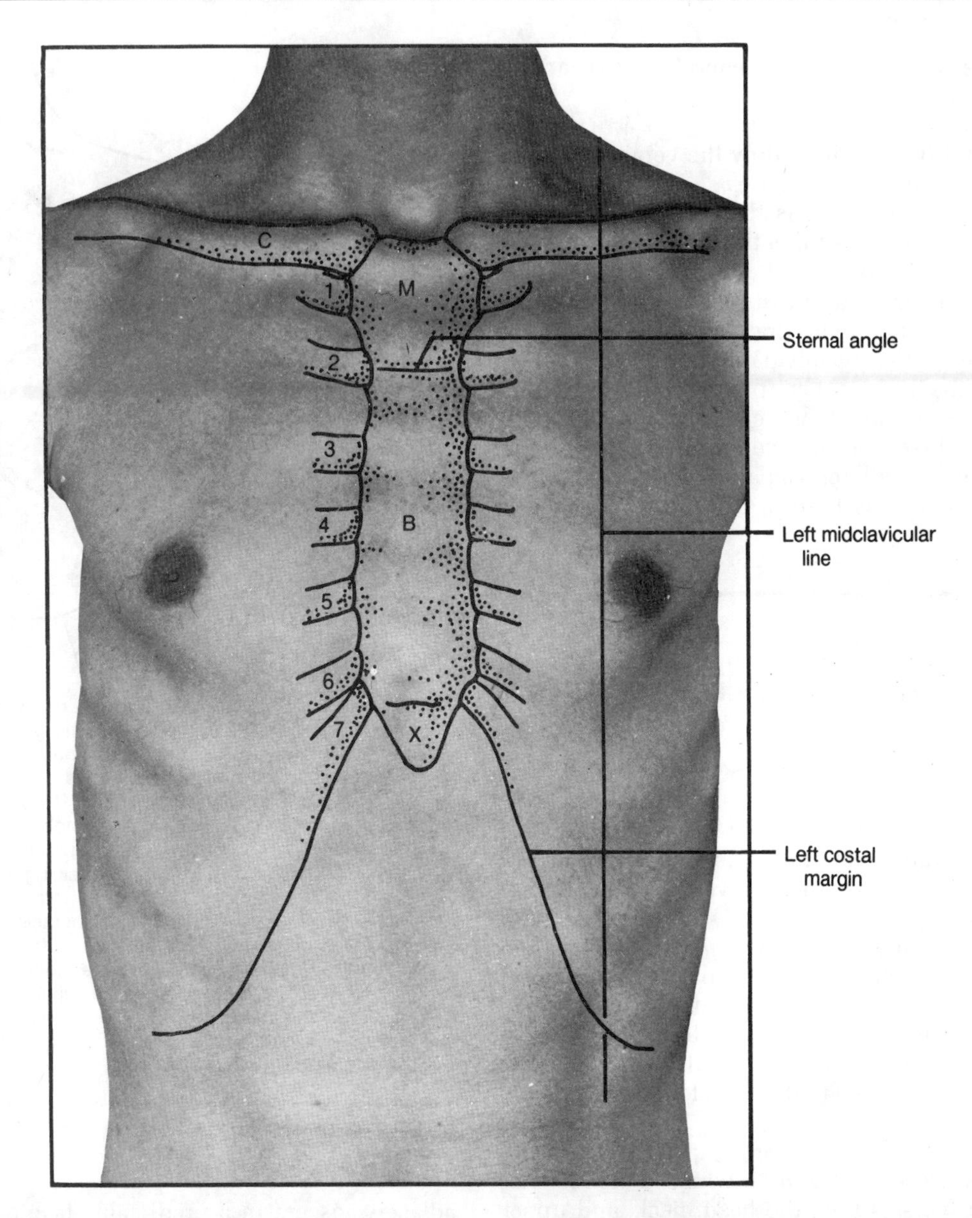

Figure 5.2. Photograph of an anterior view of the thorax of a 27-year-old man on which the outlines of the clavicle, costal cartilages, and sternum have been drawn. *C*, clavicle; *M*, manubrium of sternum; *B*, body of sternum; and *X*, xiphoid process of sternum. The sternal angle is the best guide for numbering the ribs. Note that it lies adjacent to the second costal cartilage. (From Moore, K.L., *Clinically Oriented Anatomy*, 2nd edition, Williams & Wilkins, Baltimore, 1985.)

maneuver is commonly used in the examination of the lungs. It allows one to place a stethoscope over the *triangle of auscultation* to maximize the acoustics for listening to lung sounds in the inferior lobe of the lung.

Subcutaneous Tissue

The underlying subcutaneous tissue is quite variable in its thickness over the thoracic cage. In the breast (especially in the postpubertal female) and the axilla, it is quite thick. This tissue is rich in blood vessels, nerves, and lymphatic tissues. While these elements are difficult to visualize at the gross level of examination, understanding of their normal positions and function is extremely important for physical examination. This illustrates the point that clinicians need to conceptualize the underlying basis for innervation of a dermatome and the lymphatic drainage of a superficial tissue, even though it is not apparent by direct observation in an anatomical preparation. Lymphatic distribution in the body is always a perplexing problem for beginning students. This is due to the fact that in the nondiseased tissues very few of these elements are apparent. **The general rules regarding lymphatics are:**

a) **Superficial lymphatics follow the veins, and deep lymphatics follow the arteries.**
b) **Lymphatic channels pass through a lymph node before the lymph re-enters the venous system.**

These rules illustrate why a student should pay particular attention to the venous drainage of the breast. One can then *deduce* the lymphatic drainage and subsequently locate the important lymph node groups that parallel the axillary vein and its branches, draining the breast region (*fig. 5.3*). Likewise, the arterial supply to the lungs and thoracic viscera will form a pattern for the lymphatic drainage in these deeper tissues.

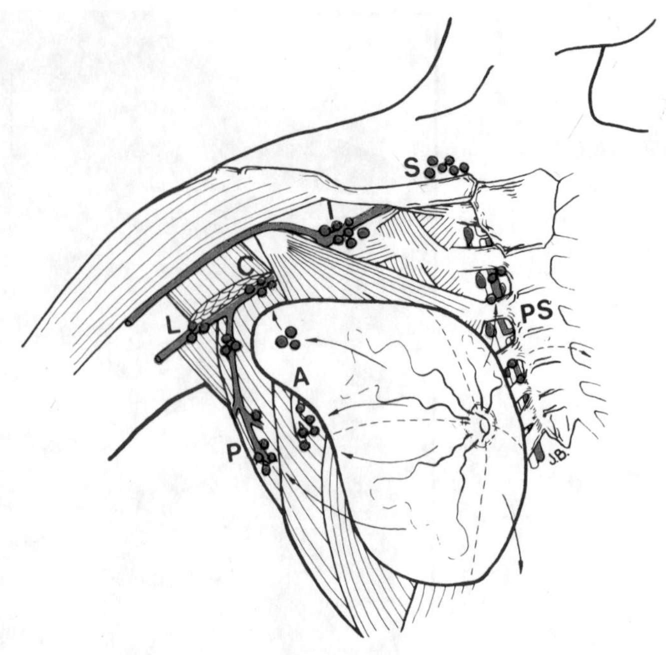

Figure 5.3. The lymphatic drainage of the breast parallels the venous drainage of the breast. The lymphatic drainage of the lateral one-half of the breast and areolar region courses along the axillary vein and its branches. *A* = Anterior or Pectoral axillary lymph nodes, *P* = Posterior axillary lymph nodes; *L* = Lateral axillary lymph nodes; *C* = Central axillary lymph nodes; *I* = Infraclavicular or apical axillary lymph nodes. The lymphatics draining the medial portion of the breast drain into the (*PS*) parasternal lymph nodes, which surround the internal thoracic vessels. The supraclavicular lymph nodes (*S*) are in the base of the neck where the lymphatic channels of the thorax, upper extremity, and neck converge to enter the venous system near the junction of the subclavian and internal jugular veins. (Illustration from Anderson, J.E.: *Grant's Atlas*, 8th edition, Williams & Wilkins, Baltimore, 1983.)

Bony Thorax and Its Diaphragm

The bony thorax is comprised of 12 thoracic vertebrae, 12 pairs of ribs and a sternum (*fig. 5.4*). The upper six ribs articulate directly with thoracic vertebrae posteriorly and the sternum anteriorly. Ribs 7–10 articulate with their respective vertebral elements posteriorly and then fuse their anterior costal cartilages into the flexible costal margin. The 11th and 12th ribs articulate with thoracic vertebrae but are unattached to the anterior elements of the rib cage. They are referred to as "floating" ribs, but in effect, are well invested in the muscles of the thoracic and abdominal walls.

The thoracic cage has a superior aperture (opening) called the thoracic inlet (*fig. 5.5*), which transmits viscera and blood vessels from the head, neck, and upper extremity to the thoracic cavity. There is also an inferior aperture (thoracic outlet), which is closed in life by a muscular and moveable diaphragm. While this diaphragm separates the thoracic and abdominal cavities, it also allows transmission of the inferior vena cava, the aorta, and the esophagus to allow for blood flow and gastrointestinal communication between these two body cavities. The diaphragm is attached to the periphery of the inferior aperture as well as to portions of the vertebral column in the thoracic and lumbar segments. The diaphragm is dome-shaped on the right and left sides with an intervening central tendon. The height of the diaphragmatic domes extends superiorly to the level of the 5th or 6th ribs (*Plate 5.1*). This permits the thoracic cage to protect the thoracic viscera (heart and lungs) above the diaphragm and the abdominal viscera (liver, stomach, spleen and kidneys) below the diaphragm. Normal movement of the diaphragm during contraction will displace these adjacent viscera in a predictable fashion. The enclosed thoracic cavity is therefore capable of increasing and decreasing its volume by the moveable diaphragm. Control of this contractile phenomenon is by the *phrenic nerves*. These are very important nerves that arise on the right and left sides of the neck and enter the thorax through the superior aperture. The phrenic nerves traverse the thoracic cavity and pierce the diaphragm before they innervate the muscle. They also provide sensory input to the membranous linings of the diaphragm (pleura and peritoneum). Neck injuries that do not affect the nerve fibers contained in the phrenic nerve (C3, 4, 5) will permit the patients to breathe even though movement of the arm, chest, abdomen, and lower extremity may be lost. **Patients may be** *paraplegic* **(2-limb paralysis) or** *quadraplegic* **(4-limb paralysis) and still have innervation of their diaphragm. When the diaphragm is also involved, the patients may be described as pentaplegic; they require a mechanical respirator for life support.**

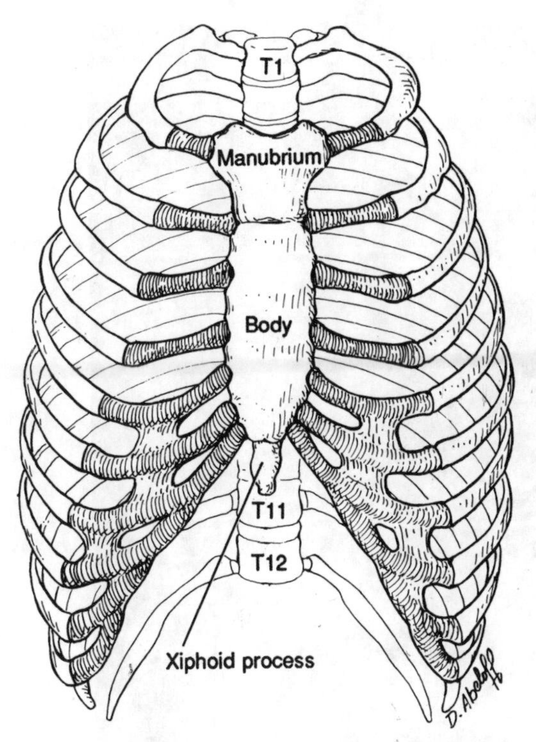

Figure 5.4. Thoracic cage. Three parts of sternum: manubrium; body; and xiphoid process. Costal cartilages are shaded dark. (From Basmajian, J.V.: *Primary Anatomy*, 8th edition, William & Wilkins, Baltimore, 1982.)

THORACIC VERTEBRAE

For typical features, see *Figures 5.6* and *5.7*. The bodies of the thoracic vertebrae have articulations for the corresponding heads of the ribs. These articulations are characteristic of the thoracic vertebrae and are a distinguishing feature not found in vertebrae of the cervical, lumbar,

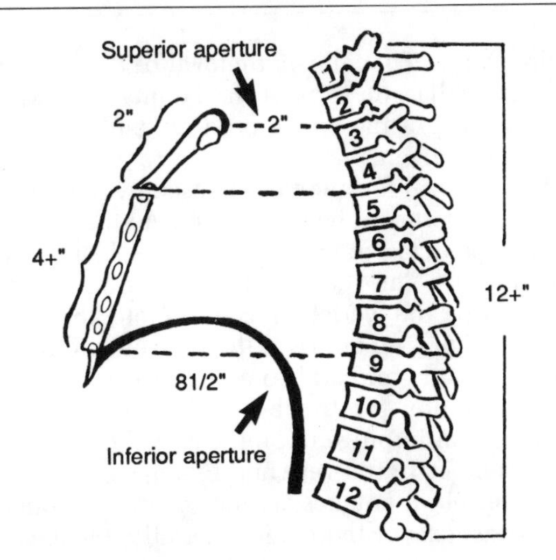

Figure 5.5. The bony thorax in median section—levels and lengths (1″ = 2.5 cm).

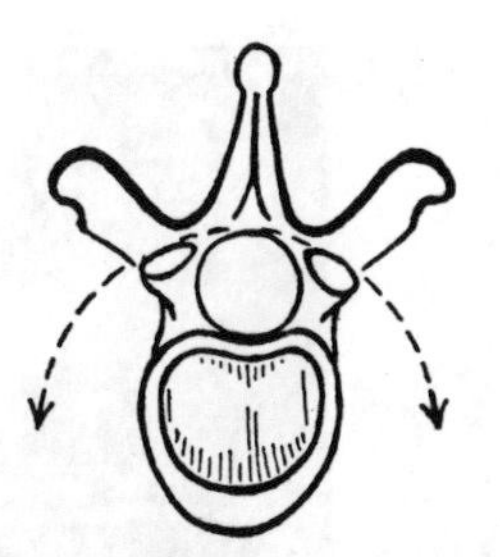

Figure 5.6. The thoracic articular processes are set on an arc, so they permit rotation.

or sacral regions. Another feature of the thoracic vertebral bodies is their heart-shaped appearance, particularly in the mid-thoracic range (*fig. 5.6*). The bodies of these central thoracic vertebrae are askewed to the right. During development, the pulsating aorta lies on the left side of the vertebral column and influences the shape of the developing and calcifying vertebral bodies. The size of the thoracic vertebral bodies increases on downward inspection of the vertebral column. The increased weight of the bipedal organism requires a greater support base. *Figure 5.8* shows the primary curvature that exists in the thoracic segment of the vertebral column. This is a very stable configuration in the vertebral column and helps in confirming a greater degree of stability for this part of the column. The bodies are separated by intervertebral discs. They constitute approximately ¼ of the length of the thoracic segment, while the vertebral bodies constitute ¾ of the linear dimension. The intervertebral discs (described on page 14) act as joints, stabilizers, and shock absorbers within the vertebral column. The **pedicles** project from the posterior aspect of the vertebral body to form the lateral side of the neural arch of bone that surrounds the spinal cord.

The **transverse processes** act like a support for the neck and tubercle of the ribs as the ribs project posterolaterally behind the vertebral body. Each transverse process has an articulating facet on the anterior surface for its corresponding rib. The transverse processes also serve as attachments for ligaments and muscles related to the rib cage and back.

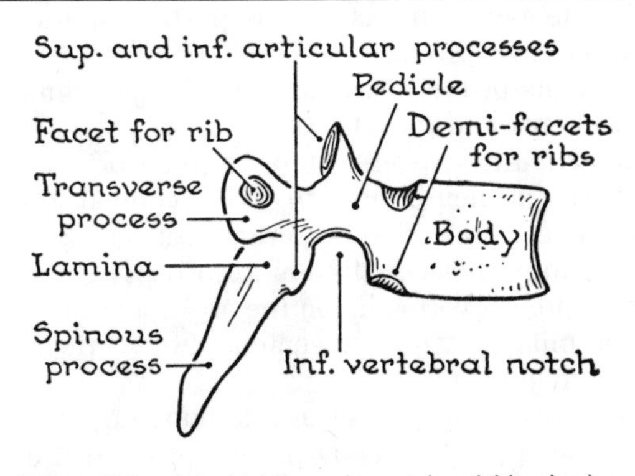

Figure 5.7. A typical thoracic vertebra (side view).

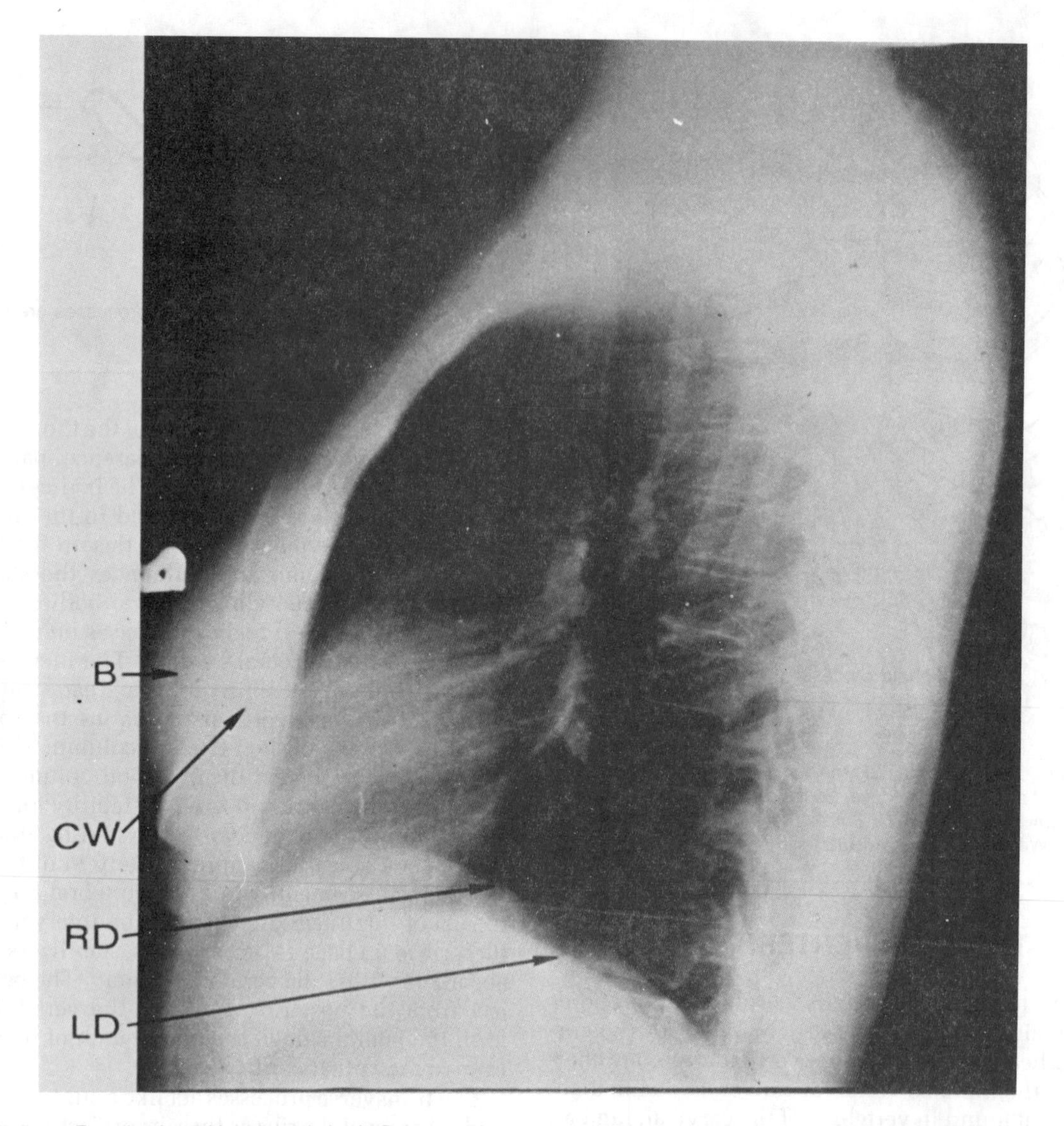

Plate 5.1. Female chest (right lateral view). *B*, breast shadow; *CW*, chest wall; *LD*, left dome of diaphragm; *RD*, right dome.

The spinous processes are subcutaneous and palpable. The T1 spinous process is the most prominent and horizontal in its projection at the base of the neck. *Figure 5.8* shows how the spinous processes of the lower thoracic vertebrae are directed more postero-inferiorly. Like the transverse processes, the spinous processes serve as components of the neural arch and as attachment sites for ligaments and muscles.

The articular processes project superiorly and inferiorly from the neural arch to form synovial joints with adjacent vertebrae. Their superior facets face posteriorly, and the inferior facets face anteriorly. The articulating surfaces are set vertically on the arc of a circle (*fig. 5.6*). This permits a rotatory movement of the vertebrae in the thoracic region.

The laminae are the sheet-like bony connections between the transverse and spinous processes, and they complete the posterior aspect of the neural arch. Access to the underlying spinal cord may be gained by doing a *laminectomy*. This allows removal of the spinous processes and adjacent parts of the lamina and exposes the dorsal aspect of the spinal cord or cauda equina in the vertebral foramen.

The vertebral foramen is enclosed by the posterior aspect of the vertebral body and the neural arch formed by the bony pedicles, transverse processes, laminae, and spinous process. The vertebral foramina from adjacent vertebrae form the vertebral canal. Note that the intervertebral discs are between the vertebral bodies and are anterior to the vertebral foramen of each vertebra (*fig. 5.9*). Herniation (protrusion) of the disc in a posterior direction can impinge upon the spinal cord and its segmental nerves in the vertebral foramen.

The vertebral notches are located on the superior and inferior aspects of the pedicles of all vertebrae. The inferior vertebral notch is most prominent and together with the superior vertebral notch of the vertebra below forms an *intervertebral foramen*. The *intervertebral fora-*

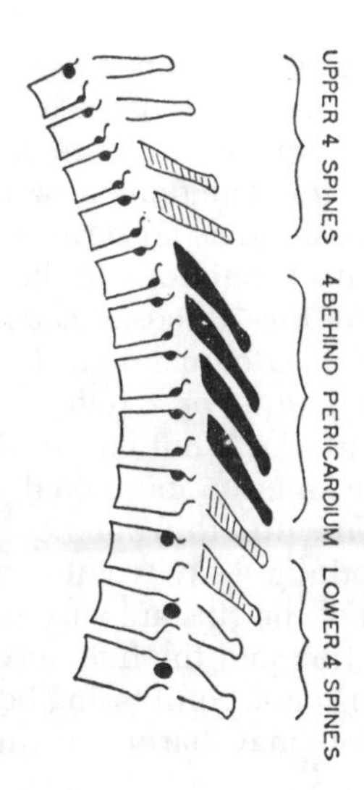

Figure 5.8. Thoracic vertebrae, showing costal facets (for heads of ribs) and the inclination of the spinous processes.

mina are the exit points for the spinal nerves that leave the vertebral canal. Each thoracic spinal nerve exits the vertebral canal through a specific intervertebral foramen on the right and left side of the vertebral column (*fig. 5.9*).

Transitions

There is a gradual transition in the form of the thoracic vertebrae as one descends in the thoracic series of the spinal column. The 1st thoracic vertebra has characteristics of the C7 vertebra, but it is uniquely thoracic by virtue of its rib articulations and its lack of *foramina* in the *transverse* processes seen in cervical vertebrae. The 12th thoracic vertebra is thoracic-like in its superior portion and lumbar-like in its inferior portion. T12 possesses a single articulating facet on each side of the body for the 12th rib. The superior articulating facet of T12 is more in the coronal plane like the other thoracic vertebral interarticulations, while the inferior articulating facet is in the sagittal plane like those of the lumbar vertebrae.

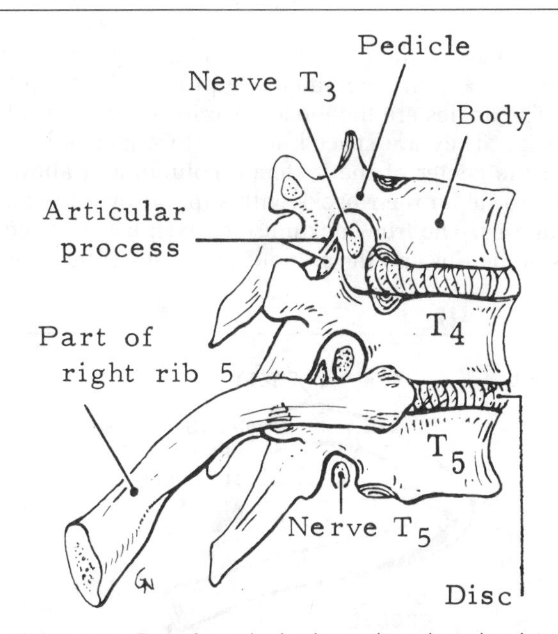

Figure 5.9. Part of vertebral column, thoracic region: intervertebral disc, intervertebral foramen with spinal nerve, rib attachment. (From Sauerland, E.K.: *Grant's Dissector,* 9th edition, Williams & Wilkins, Baltimore, 1984.)

Sternum

The sternum, or breast bone, is subcutaneous and palpable. It forms the anterior aspect of the rib cage in the midline of the chest. The sternum has articulations with the upper seven ribs and the clavicles on each side. It is composed of three parts: the manubrium, the body, and the xiphoid process (*fig. 5.10*).

The manubrium has a superior concave surface, the **jugular notch**, in the base of the neck. The head of clavicle, the costal cartilage of the 1st rib and the upper aspects of the 2nd costal cartilage on each side articulate with the manubrium. The manubrium in turn articulates with the superior aspect of the body of the sternum at the **sternal angle** (of Louis). *Figure 5.5* illustrates how the manubrium and the sternum articulate to create this sternal angle. This is a very important surface landmark for identifying the level of the 2nd rib in a patient and the 4th and 5th thoracic vertebrae in a radiograph. The horizontal plane of section from the sternal angle to the T4/5 vertebral level is of major importance in thoracic radiological interpretation. At this level, the trachea bifurcates, the arch of the aorta traverses the thorax from right to left in a posterolateral direction. The sternal angle is also used to locate the 2nd intercostal space. When a physician listens for heart sounds in this location, the aortic valves are heard over the right 2nd intercostal space, and the pulmonary valves are heard over the 2nd intercostal space on the left side of the sternum. The manu-

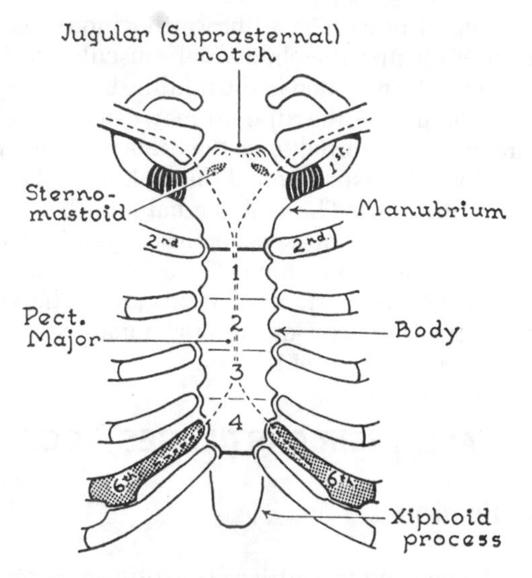

Figure 5.10. The anterior surface of the sternum. *Dashed lines* indicate origins of the *pectorales majores*.

brium sternum also serves as a site for muscle attachments. *Figure 5.10* shows the attachments for the *pectoralis major* and the *sternomastoid* muscles. The sternohyoid and sternothyroid muscles of the neck also arise from the posterior surface of the manubrium.

The *sternoclavicular joint* is at the manubrium. It is a pivotal joint for the upper extremity and represents the only joint between the axial skeleton and the upper extremity. The tremendous freedom of movement in the upper extremity is partially accounted for by this multiaxial saddle joint. To withstand the forces transmitted from the upper extremity to the axial skeleton (rib cage and spinal column), the very strong sternoclavicular ligaments and costoclavicular ligaments reinforce the capsule of the sternoclavicular joint. This joint also contains an intra-articular disc to help cushion the forces transmitted from the clavicle to the manubrium sternum.

The body of the sternum is made up of 4 fused sternebrae. Ossification of these units occurs during the 3rd trimester of intrauterine life, and the sternal body is fused into a single bone from the 6th rib to the 2nd rib in the young adult. It articulates with the costal cartilages of the 2nd, 3rd, 4th, 5th, 6th and 7th ribs. The intervening intercostal spaces can be palpated and identified by locating the sternal angle and counting the rib attachments from the 2nd rib to the 7th rib. The body of the sternum also possesses an abundant amount of red marrow in its medullary cavity. The thin bony cortex of the body and its subcutaneous position allow easy but painful access to bone marrow samples for examination and transplantation procedures. The body of the sternum also provides an attachment site for part of the pectoralis major on the external surface and transversus thoracic muscle on the internal surface.

The *manubriosternal joint* allows movement during deep respiration, as the rib cage is elevated and extended forward. This is a symphysis type joint and may become ossified (synostosis) in 10% of adults after the age of 30 (Trotter).

The xiphoid process is a fibrocartilaginous extension of the sternum into the abdominal musculature (rectus sheath). It can be palpated in the midline below the costal margin. Pressure on the xiphoid may cause discomfort. This area is also referred to as the "solar plexus" region of the abdomen. Ossification of the xiphoid makes it visible on radiographs. The xiphisternal joint between the xiphoid and the inferior aspect of the body of the sternum is a synchondrosis type union. The costal cartilage of the 7th rib attaches to the sternum at this joint. The xiphoid process marks the level of the 6th thoracic dermatome on the anterior surface of the body.

RIBS AND THEIR CARTILAGES—COSTAE

Classification

The rib bone and its cartilage constitute a *costa*. There are 12 pairs of ribs that articulate with the vertebral column. Their anterior ends are connected to a costal cartilage (*fig. 5.4*). The cartilages of the upper 7 ribs connect directly to the sternum, and they are classified as "true" or *vertebrosternal ribs*. The remaining five ribs (8, 9, 10, 11, and 12) are classified as false ribs. The 8th, 9th, and 10th ribs interconnect indirectly to the sternum through the costal margin of the 7th costal cartilage. These three false ribs are termed *vertebrochondral ribs*. The final two ribs inferiorly are *floating or vertebral* ribs. They have a cartilaginous tip arteriorly but end in the investment of the muscles of the anterior abdominal wall.

Ribs are flattened with a thin outer cortex of bone. They possess a large medullary cavity with an abundant blood supply. Fractures of the ribs are very painful due to the endosteal innervation and the free nerve endings in the surrounding periosteum. Splintering bone fragments are also dangerous and may harm the underlying viscera (pleura and lungs).

A typical rib consists of the following parts beginning at the articulation with the vertebral column (*fig. 5.11*). The *head* has two facets that articulate with the corresponding vertebra and the vertebra that lies supra-adjacent. A bony crest separates the inferior and superior articulating facets on the head of the rib. This crest is attached to the intervertebral disc that lies between the body of the corresponding vertebra and the body of the vertebra above the rib. The *neck* of the rib parallels the transverse process of the corresponding thoracic vertebra. It has an *articulating tubercle* on its posterior surface for articulation with the transverse process of the vertebra. The neck of the rib also has a *crest* that provides attachments for the superior costotransverse ligaments. The dorsal rami of the intercostal nerves pass into the muscles of the back below the neck of the ribs, and the sympathetic trunk lies on the anterior surface of the neck of the rib.

The *angle of the rib* is formed as the posteriorly projecting neck joins the anteriorly projecting body of the rib. The angles are the most posterior projections of the rib cage. They are spaced about 3 to 4 inches laterally from the center of the vertebral column and allow humans to lie on their backs with support over the angles of the ribs. The *true back muscles* exist between the angles of the ribs and the spinous processes of the thoracic

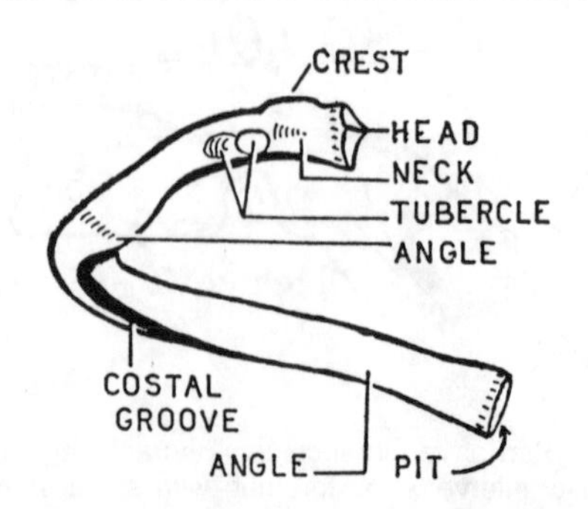

Figure 5.11. A typical rib viewed obliquely from behind.

vertebrae. These muscles are innervated by the dorsal rami of the intercostal nerves.

The *body of the rib* projects laterally, anteriorly, and finally medially, as it circumscribes the lateral and anterior thoracic wall. The external surface is subcutaneous and provides attachments for the pectoral muscles (major and minor), the *serratus anterior muscle* and the muscles of the abdominal wall. The internal surface of the body of the rib supports the costal parietal pleura of the thoracic cavity. A *costal groove* lies on the undersurface of each rib and contains an intercostal vein, artery, and nerve. The innermost intercostal muscles attach to the internal lip of this intercostal groove, and the external and internal intercostal muscles attach to the external lip. The body of the rib slopes downward, as the rib projects from the posterior to the anterior aspects of the rib cage. The ribs continue to the sternum through articulations with the costal cartilages. Elevation of the upper ribs therefore elevates the sternum and projects it anteriorly to increase the chest volume within the rib cage (*fig. 5.12*).

Examination of the Rib Cage

With the articulated skeleton before you, confirm the following facts, because they have an important bearing on the mechanism of respiration.

1. The typical rib takes a downward slope; the cartilage, an upward one (*fig. 5.12*).
2. The 1st rib slopes downward throughout its entire course.
3. The sternal end of each arch lies at a lower level than the vertebral end.
4. The middle of each arch (except the 1st) lies at a lower level than a straight line joining the 2 ends of the costa.
5. Both ribs and cartilages increase in length progressively from 1st to 7th.
6. The transverse diameter of the thorax increases progressively from the 1st to the 8th rib, the 8th rib having the greatest lateral projection.
7. The ribs increase in obliquity progressively from 1st to 9th; the 9th rib is the most obliquely placed.
8. The anterior ends of the 11th and 12th ribs, not being subjected to terminal pressure, taper.

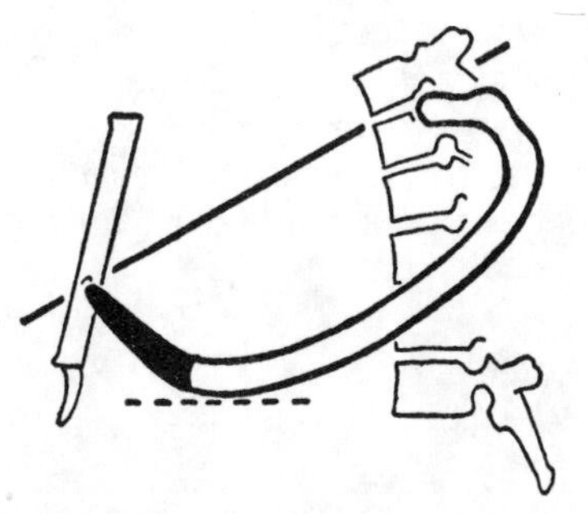

Figure 5.12. A costal arch (side view).

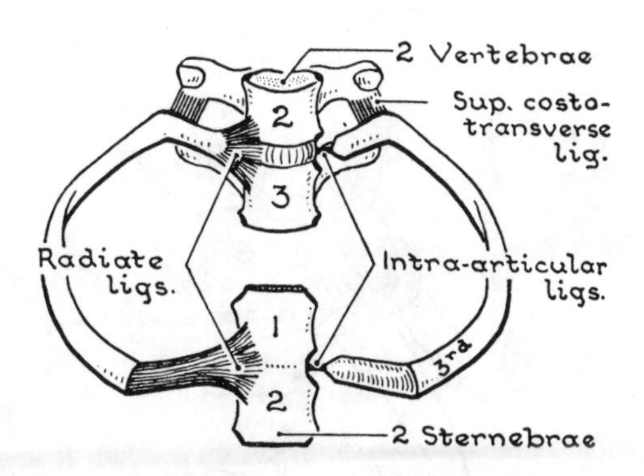

Figure 5.13. The articulations at the dorsal and ventral ends of a costal arch, compared.

COSTAL JOINTS

Typically, the head of a rib articulates with the sides of the bodies of 2 vertebrae; the tubercle of a rib articulates with the tip of a transverse process, and the costal cartilage articulates with the sides of 2 sternebrae. The joints are classified as follows:

1. Costovertebral articulation:
   a. Joint of the head of a rib.
   b. Joint of a tubercle of a rib.
2. Sternocostal articulation.
3. Interchondral joints, between costal cartilages 7, 8, and 9.

Vertebrocostal Articulations

The head of each typical rib (2nd to 10th) articulates with each demifacet of 2 adjacent vertebrae and with the intervertebral disc between these vertebrae (*fig. 5.13*). It is attached to the intervertebral disc by a transversely placed *intra-articular ligament*. The *capsule* is strongest in front, where its fibers *radiate* from the anterior margin of the head.

The heads of the ribs 1, (10), 11, and 12, being confined to single vertebrae, are rounded, and their joints have no intra-articular ligaments.

Sternocostal Articulations (*fig. 5.13*)

Like the posterior joints, each joint cavity is divided into two by an intra-articular ligament, and it is closed ventrally by a ligament that radiates from the perichondrium to the sternum. The cavity may be obliterated by fibrous union.

Joint of the Tubercle of a Rib

The *tubercle of a rib* articulates with the facet at the tip of the transverse process of its own vertebra (except

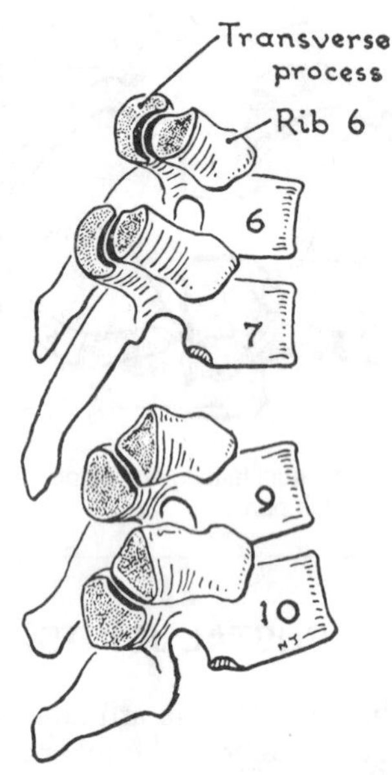

Figure 5.14. To demonstrate why upper ribs rotate on transverse processes and lower ribs glide.

11 and 12) to form a synovial joint, called a *costotransverse joint* (*fig. 5.14*).

The superior costotransverse joints permit rotational movements of the rib. The inferior costotransverse joints permit a gliding movement for the articulated rib.

Ligaments

Figure 5.15 shows that the strong ligamentous fibers that bind a rib to a transverse process are divided into a medial and a lateral group by the cavity of the joint. They are the *(medial) costotransverse ligament* (ligament of the neck) and the *lateral costotransverse ligament* (ligament of the tubercle).

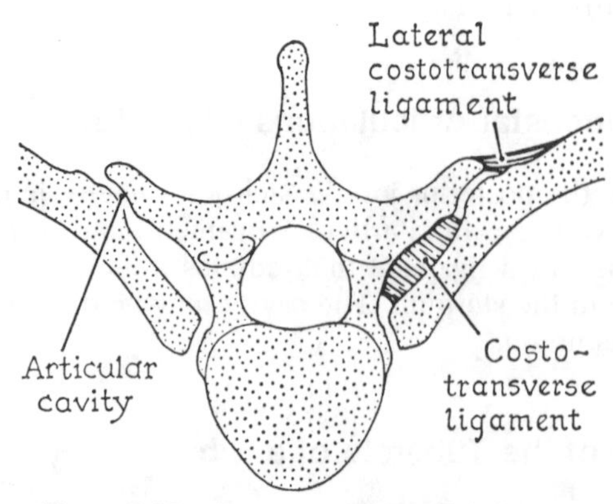

Figure 5.15. Costotransverse articulation.

A band, the *superior costotransverse ligament* (*fig. 5.13*), descends from a transverse process to the upper border of the neck of the next rib below, producing a sharp *crest of the neck* (*fig. 5.11*).

Ossification begins near the angle (about the 9th prenatal week) and spreads in both directions but fails to reach the sternal end, hence, the costal cartilages. Scale-like epiphyses, which cap the head and tubercle, are in all cases fused by the 24th year.

Variations (See *figs. 5.16* and *5.17*). Accessory ribs may appear in the lower cervical or upper lumbar regions. Ribs may also be bifid.

Clinical Case 5.2

Patient Arthur S. (cont'd from Clinical Case 5.1). Now we are in the operating room, assisting the surgeon, and have the following additional information about the chest injury. The intercostal muscles suspending the broken skeletal elements are still attached and limit the degree of movement of the unstable portion of the chest wall. The right lung underlying the traumatized chest wall is not collapsed but is only capable of minimal expansion during inspiration. The patient's airway is cleared using a endotracheal suction tube, and the patient has been intubated for a general anesthetic.

The skin and subcutaneous fascia over the mid-axillary region is incised and wire fixation done on the fracture sites of each rib. Care is taken to avoid damaging the long thoracic nerve and the intercostal nerves and vessels that traverse the costal groove on the inferior surface of each rib. The anterior fracture sites through the costal cartilages are exposed by a skin incision 1 cm lateral to the right margin of the sternum. Wire fixation is placed with care to avoid damaging or occluding the underlying internal thoracic artery and vein. The costal pleura underlying the fracture sites is found to be intact. The skin incisions are sutured and dressed and a compression bandage is placed around the chest. The patient is sent to the recovery ward where you follow his progress. Respiratory function becomes stabilized, permitting discharge from the hospital after six days. He breathes without discomfort while at rest but experiences considerable intercostal pain on the right side if he inhales deeply, coughs or forcibly moves his right arm. On follow-up over the next 5 weeks, this pain has diminished, and gradually he regains a full range of movement in both his rib cage and right arm.

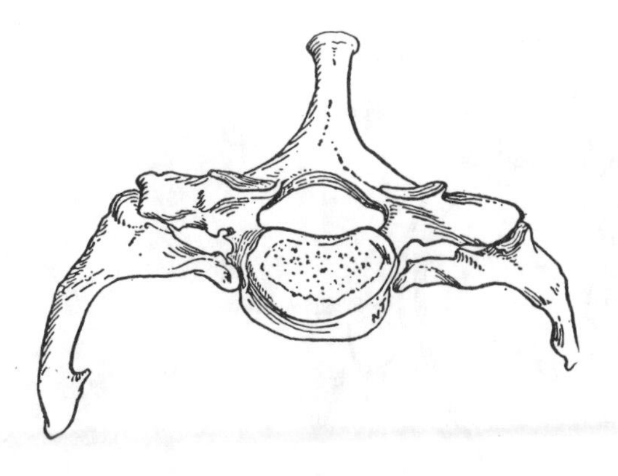

Figure 5.16. Cervical rib.

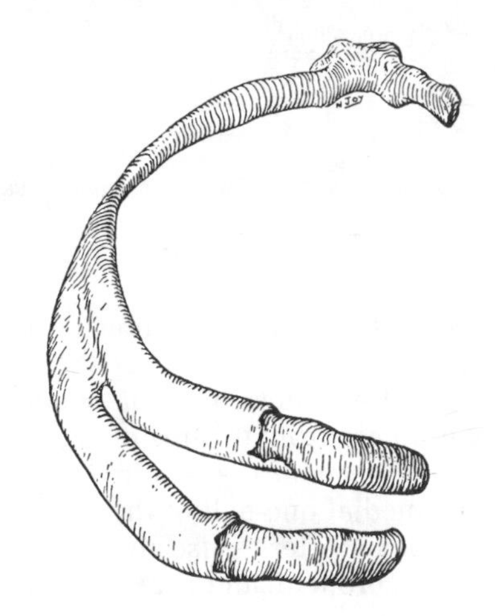

Figure 5.17. A bifid rib.

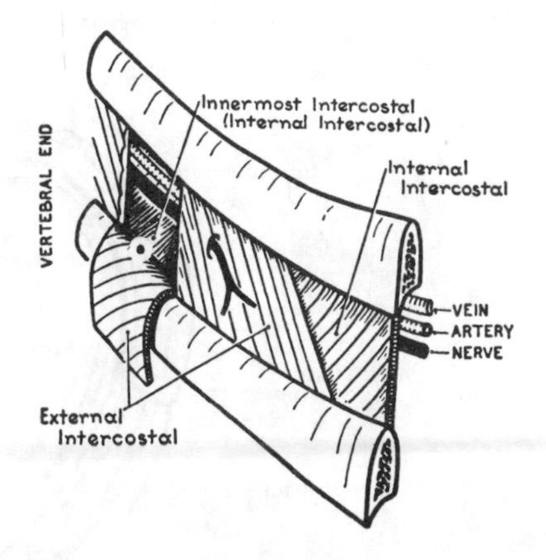

Figure 5.18. Contents of intercostal space.

Intercostal Spaces

Muscles of the Thoracic Wall

The intercostal muscles are arranged in a similar manner to the muscles of the anterior abdominal wall. There are three layers of flattened muscles attached to the ribs and cartilages that bound the intercostal spaces. *Figure 5.18* shows how the fibers of the *external intercostal muscles* project inferiorly in the posterior to anterior direction (the same way one's fingers project when they are placed in one's front pocket). The underlying *internal intercostal muscle* fibers run perpendicular to the external fibers (the same way one's fingers would project if inserted into one's back pocket). The deepest layer of muscle fibers, the *innermost intercostals* (*fig. 5.19*), are incomplete. This group includes the *transversus thoracis* muscle, which arises from the posterior aspects of the xiphoid and sternum and fans out to insert onto the 3rd to 6th costochondral junctions (*fig. 5.20*). The innermost intercostal muscles tend to bridge more than one intercostal space. The *intercostal vein, artery*, and *nerves* lie between the internal and innermost intercostal muscle groups (*figs. 5.18, 5.19, 5.20*). A transverse section of the idealized intercostal space is shown in *Figure 5.20*. This illustrates the relationships of the three muscles groups and nerves on the right side and the innermost intercostal muscles and the intercostal artery arising from the aorta on the left side. Due to the slope of the ribs and therefore the intercostal vessels and nerves, one would only see the entire nerve distribution if the "transverse" section were sloping in a posterior and anterior direction as shown in *Figure 5.12*.

The *intercostal nerves* are mixed nerves and contain both motor and sensory fibers. The posterior (dorsal) rami innervate the true back muscles between the angles of the ribs and the spinous processes of the vertebrae. Cutaneous branches also innervate the skin overlying these muscles. The anterior (ventral) rami continue in the plane between the innermost and internal intercostal muscles to innervate the intercostal musculature, the periosteum of the ribs, and the skin of the thorax. Each intercostal nerve distributes sensory fibers to the area of skin that constitutes that specific thoracic dermatome. The sensory fibers of each of these intercostal nerves have their cell bodies in a single *dorsal root ganglion*. The dorsal root ganglion is located in the dorsal root of the spinal nerve in its intervertebral foramen.

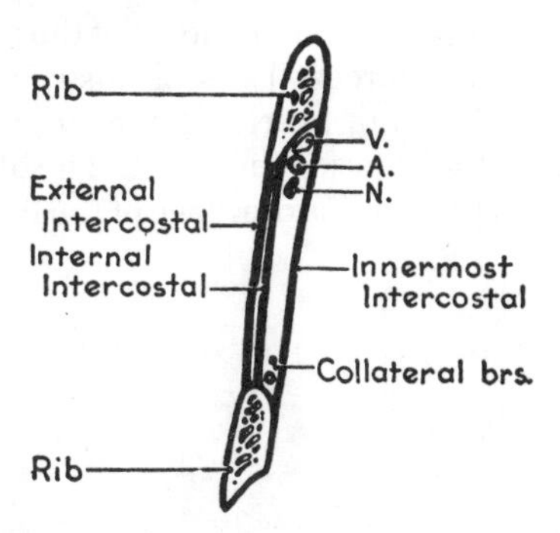

Figure 5.19. Intercostal muscles, coronal section.

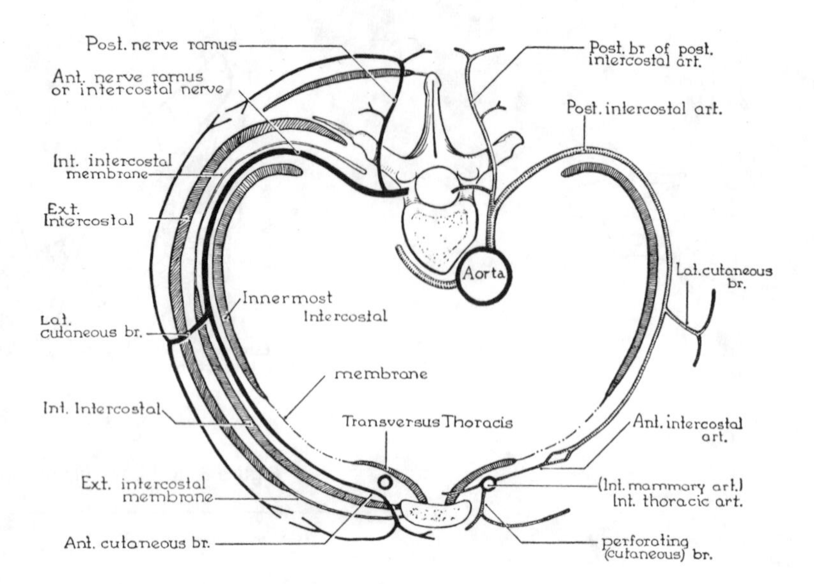

Figure 5.20. The contents of an intercostal space (horizontal section). (*Nerve rami: posterior and anterior = dorsal and ventral.*)

The *intercostal arteries* and *veins* have two origins. The posterior intercostal arteries (3 to 12) arise from the aorta. *Posterior intercostal arteries* in the first two intercostal spaces arise from the costocervical trunk, which branches from the subclavian artery (*fig. 9.14*). The *posterior intercostal veins* drain into the azygos and hemiazygos system on the right and left side of the thorax respectively. Venous drainage of the posterior intercostal system drain into the brachiocephalic veins (*fig. 9.15*). The intercostal vessels bifurcate in the intercostal spaces and have branches running in the intercostal groove under the body of the rib as well as collateral branches above the body of the subjacent rib (*fig. 5.19*).

Anterior intercostal arteries and *veins* arise from the *internal thoracic vessels* (*fig. 5.20*). The vessels anastomose with the comparable posterior intercostal vessels in the intercostal spaces near the midclavicular line. The internal thoracic artery is a branch of the subclavian artery that descends posteriorly to the costal cartilages about a finger's breadth lateral to the sternum. In the lower aspect of the thoracic cavity, the artery lies between the sternum and the transversus thoracis muscle. It terminates by dividing into the *superior epigastric artery* and the *musculophrenic artery* at the level of the xiphosternal joint. The anterior intercostal vessels also send perforating branches to the skin overlying the sternum and the medial aspect of the anterior chest wall. The blood supply to the breast and the venous drainage of the medial aspect of the breast are via these internal thoracic vessels. Lymphatic vessels draining the medial aspect of the breast drain into a chain of *parasternal lymph nodes* that filters the lymphatic drainage from this region. Examination of the parasternal lymph nodes is required in tumors that are found in the medial one-half of the breast.

The internal thoracic arteries also give off branches that supply the thymus, bronchi, and pericardium within the thoracic cavity.

Clinical Mini-Problems

1. **a. At what level of the sternum is the attachment of the 2nd costal cartilage?**
   b. Which dermatome overlies this region?
2. **Which general grouping of lymph nodes would likely be stimulated by a tumor in the upper medial quadrant of the right breast?**
3. **Which spinal nerves would be associated with the 1st thoracic vertebra?**
4. **What is the origin for the blood supply to the 5th rib?**

(Answers to these questions can be found on p. 585.)

6

Pleurae

Clinical Case 6.1

Patient Ted B. During a winter sailboat race, this crewman was thrown against the side rail. He suffered a severe and painful soft tissue injury to his chest over the right 4th rib. He tried to continue with his duties, but the pain drove him to a bunk until they returned to port that evening. As a junior member of the medical team, you note that he is somewhat cyanotic (bluish) around his lips and that his chest is extremely sensitive to pressure over the 4th rib. An obvious hyperresonant sound is produced when the physician in charge percusses the right chest wall below the 6th intercostal space. A chest radiograph reveals a broken and fragmented right 4th rib, a collapsed right lung, and a pneumothorax. The left lung is functioning normally, but the mediastinum and trachea are both deviating to the left side of the thoracic cavity on inspiration. A thoracic surgeon has been called to handle the emergency, which is continued as Clinical Case 7.1 in the next chapter.

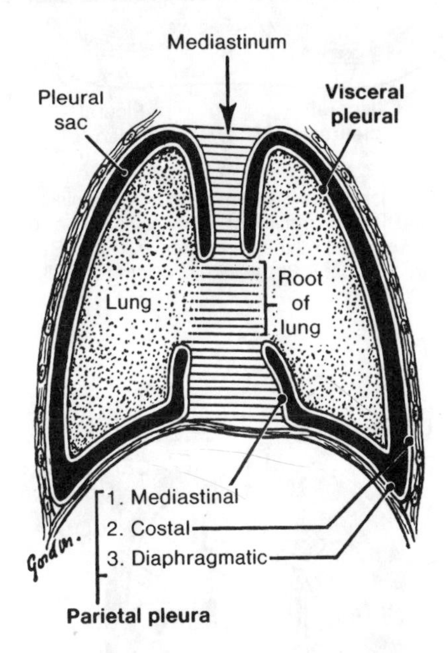

Figure 6.1. Schematic coronal section of thorax to show parts of pleura and separation of right and left pleural sacs by mediastinum. (From Basmajian, J.V.: *Primary Anatomy*, 8th edition, Williams & Wilkins, Baltimore, 1982.)

Subdivisions of the Thoracic Cavity

The thoracic cavity is within the thoracic cage. It contains the right and left pleural cavities, the right and left lungs, and a mass of tissues and organs that separates the two lungs, the mediastinum (*figs. 6.1* and *6.2*). The prominent structures within the mediastinum are (1) the heart with its pericardium; (2) the great vessels that leave and enter the heart; (3) the trachea; and (4) the structures that traverse the thorax in passing from the neck to the abdomen, e.g., esophagus, the vagus nerves, phrenic nerves, and thoracic duct.

The thorax in the embryo becomes partially cavitated by the extension of the intraembryonic celom. This celomic cavity is subsequently divided into 2 pleural and 1 pericardial cavity in the developing thorax. The developing endodermally derived lungs and mesodermally derived heart arise from tissues adjacent to these membraned-lined celomic cavities. As the lungs and heart enlarge, they invaginate into these thoracic cavities (*figs. 6.3* and *6.4*). The resultant effect is that the lungs and heart of the adult are contained within separate pleural and pericardial cavities early in fetal life. The pleural cavities are subsequently separated from the abdominal component of the celomic cavity by the development of the diaphragm.

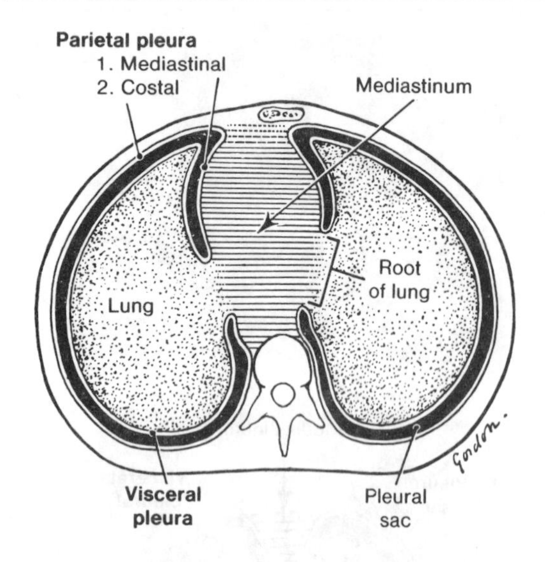

Figure 6.2. Horizontal section of thorax (semischematic). The right and left pleural sacs, each enclosing a lung, are completely separated by the mediastinum. (From Basmajian, J.V.: *Primary Anatomy*, 8th edition, Williams & Wilkins, Baltimore, 1982.)

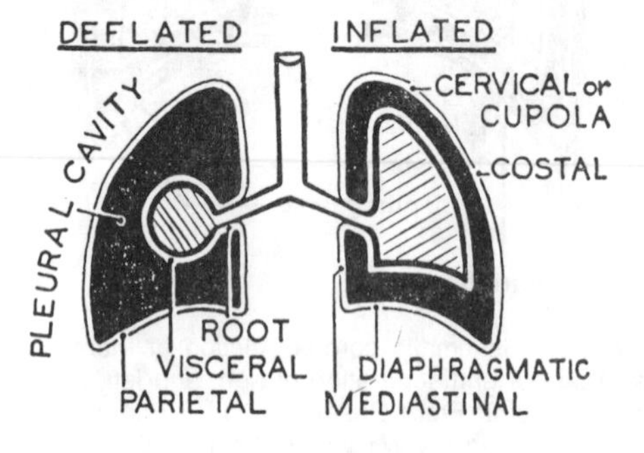

Figure 6.3. The pleura. The lung represented as a balloon with a stalk (highly schematic).

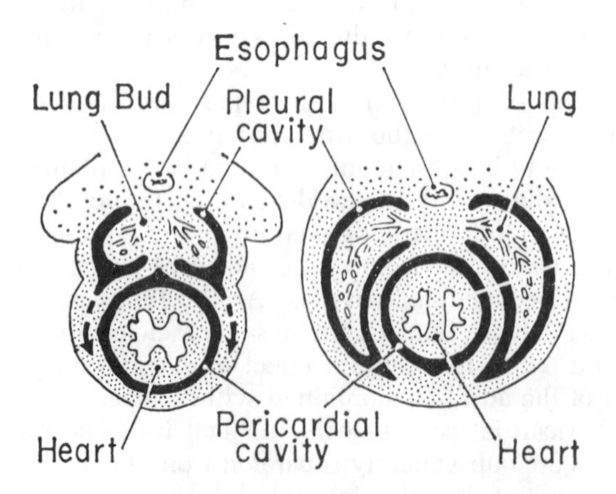

Figure 6.4. The lung buds expand in their pleural cavities to embrace the heart in its pericardial sac.

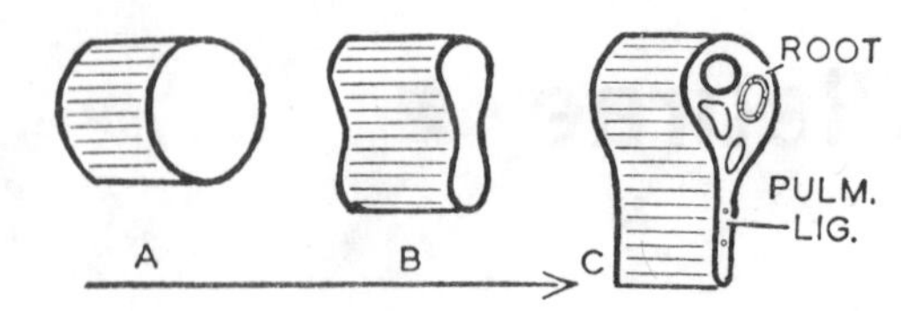

Figure 6.5. Diagram of pleura at root of lung.

THE PLEURAE

The pleurae are the membranes lining the two pleural cavities that surround the developing right and left lungs. *Figures 6.1, 6.2,* and *6.3* demonstrate how this pleural lining becomes related to the lung surface. The pleura that coats and so forms the surface of the lung is termed *visceral pleura.* The pleura that maintains its relationship to the walls of the thoracic cavity and the sides of the mediastinum is termed *parietal pleura.* The parietal and visceral pleura are continuous at the root of each lung (*fig. 6.3*). A short segment of the pleura inferior to the root of each lung is called the *pulmonary ligament* (*fig. 6.5*). The use of the term ligament in this context is confusing. Ligament is commonly used to describe a connective tissue support between bony elements. However, ligament is also used to describe certain membrane-lined conduits of vascular and neural projections that are important in surgical procedures. The pulmonary ligament is an example of the latter.

The parietal pleura lines the entire pleural cavity and is attached to the costal, diaphragmatic, and mediastinal surfaces of the pleural cavity by a loose connective tissue, the *endothoracic fascia.* The dome of the parietal pleura, which projects into the root of the neck on each side, is called the *cupola. Figure 6.6* shows how the cupola rises superior to the level of the clavicle on the anterior aspect of the neck. Caution must therefore be used in penetrating this area with anesthetic needles and exploring wounds in the base of the neck. Perforation of the parietal pleura in cases involving trauma of the thorax, lower neck, or upper abdomen may result in the equalization between the pleural cavity pressure and atmospheric pressure, and thus cause the lung in this space to collapse (*a pneumothorax*). The parietal pleura is innervated by the free nerve endings of sensory fibers in the intercostal and phrenic nerves. Irritations and inflammation of the pari-

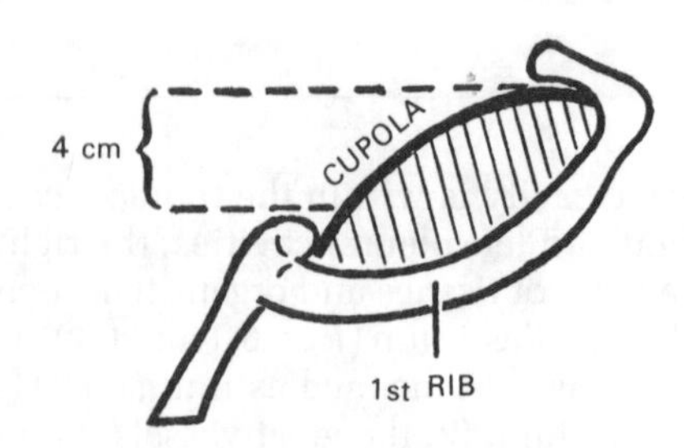

Figure 6.6. The cupola of the pleura.

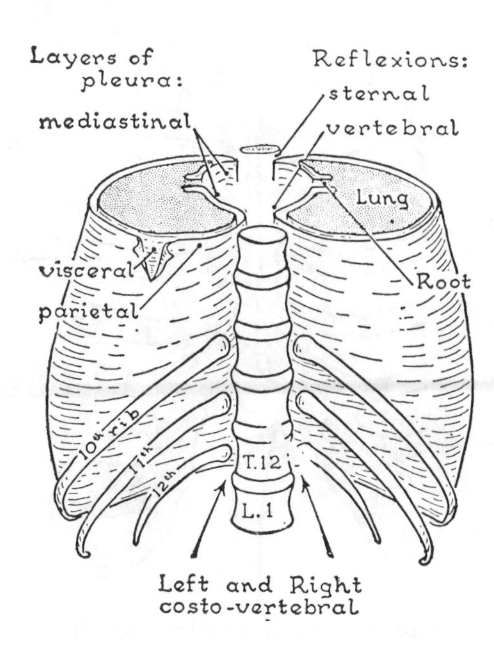

Figure 6.7. Pleural reflexions, from behind.

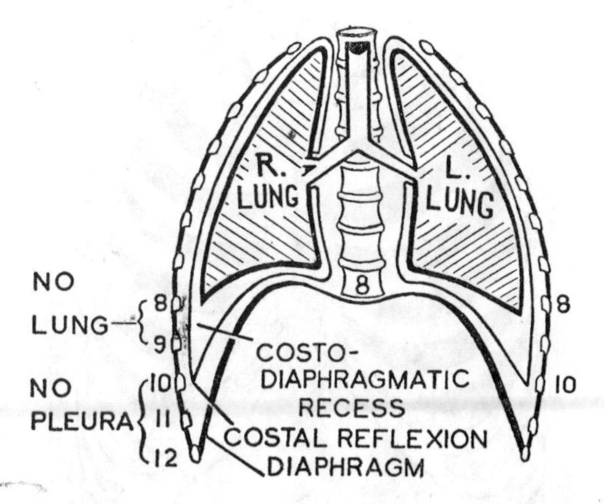

Figure 6.8. Coronal section of thorax (semischematic).

etal pleura can refer pain to the dermatomes that are served by specific thoracic intercostal or phrenic nerves (C3 and 4).

The *visceral pleura* is a simple squamous layer of epithelium over the surface of the lung. *It is insensitive to pain* and provides a moistened and lubricated surface so that the lung can move freely within the pleural cavity. Infections, inflammatory reactions, and lung immobility can promote adhesions between the parietal and visceral pleural linings. Because the parietal pleura is sensitive to pain, lung movement in these conditions may then produce severe pain (*pleurisy*), as these adhesions are stretched and disrupted.

Lines of Pleural Reflexion (*figs. 6.7 and 6.8*)

The parietal pleura on the costal surface is continuous with (1) the mediastinal pleura anterior to the vertebral column—the vertebral reflexion; (2) the mediastinal pleura posterior to the sternum—the sternal reflexion; and (3) the diaphragmatic pleura near the thoracic wall—the costal reflexion (*fig. 6.8*).

Surface Anatomy

The visualization of the sternal and costal reflexion on the external chest wall is extremely important in clinical examination. One plots these reflexions by remembering the positions on the *even* numbered ribs—2, 4, 6, 8, 10 and 12.

Figures 6.9 and 6.10 show how the right and left sternal reflexions pass behind the sternoclavicular joints and meet each other in the median plane at the sternal angle (2nd rib). The right reflexion continues inferiorly in the midline to the xiphoid process (rib 6). The left reflexion also descends in the midline but separates from it and the right reflexion at the 4th rib level. This is due to the positioning of the heart and pericardium on the left side of the midline at this level. The left reflexion then descends over the midposition of the left 6th costal cartilage.

The pleural reflexions then reach the vicinity of the 8th rib in the midclavicular line. *Figure 6.7* shows how the pleural reflexions continue onto the back and cross the 10th rib in the midaxillary line and the neck of the 12th rib at the vertebral column. The differences in the heights of the right and left domes of the diaphragm also contribute to the slight asymmetry in the pleural reflexion on the right and left anterior chest surface. The left is slightly lower between the xiphoid area and the midaxillary line (*fig. 6.10*).

Figures 6.7 and *6.10* show three areas where the pleural reflexions descend below the costal margins. Abdominal surgery could therefore cause a pneumothorax if the parietal pleura were ruptured at (1) the right xiphicostal margin, (2) the right costovertebral angle, or (3) the left costovertebral angle.

Pleural Recesses

The space between the parietal and visceral pleura is normally only of molecular thickness and filled with a serous pleural fluid. However, in the areas of pleural reflexion onto the diaphragm and mediastinum, the distance between the parietal and visceral pleura is greatly expanded in the normal patient. These expanded pleural cavity regions are called *pleural recesses*, and they are physiologically important for lung expansion in forced inspiration. Their presence and recognition is also important in clinical examination of the thorax and clinical treatment for thoracic disorders that require assessment or entry into the pleural cavity. These recesses are termed the *costodiaphragmatic recesses* and the *sternocostal re-*

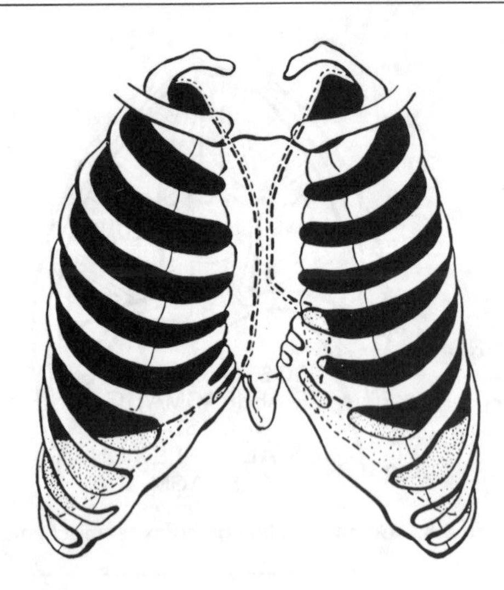

Figure 6.9. The two lungs (*solid black*) fail to fill lower parts of two pleural sacs (*stippled*). (From Basmajian, J.V.: *Primary Anatomy*, 8th edition, Williams & Wilkins, Baltimore, 1982, p. 235.)

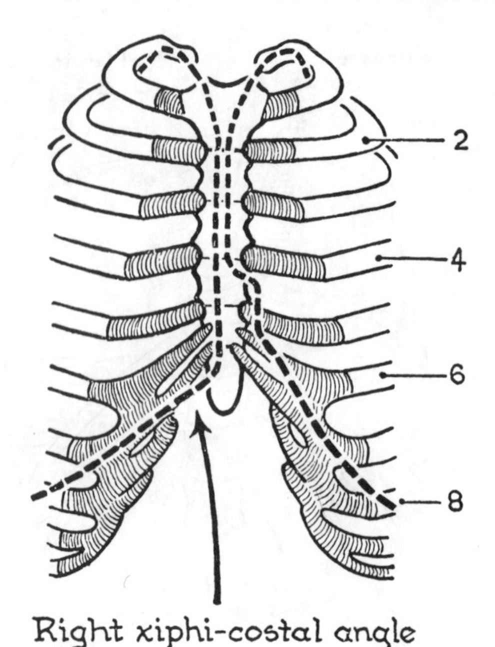

Figure 6.10. Sternocostal reflexion of pleura.

cesses. *Figures 6.8, 6.9,* and *6.10* show their location in relation to the thoracic cage and lungs. These locations can be confirmed clinically be percussing (tapping) the chest wall and listening for changes in tone as one moves from the area overlying the unoccupied recess to an area of underlying lung tissue, or sternum and mediastinum, or diaphragm and abdominal viscera. *Figure 6.8* shows the relationship of the lung to the 8th, 9th and 10th ribs at the midaxillary line. One can see how the pleural cavity can be entered through the 9th and 10th intercostal spaces without penetrating the lungs in a patient with "shallow" breathing. Since gravity would favor accumulation of fluids in this part of the pleura cavity, sampling pleural effusions following diseases or thoracic surgery is frequently done in the costodiaphragmatic recess. The costomediastinal (sternal) recesses (*fig. 6.6*) overlie the heart and pericardium and are infrequently opened in clinical procedures.

Clinical Mini-Problems

1. **Which layer of pleura is firmly attached to the surface of each lung?**
2. **Which pleura lines the costodiaphragmatic recess?**
3. **a. What is the inferior level of the resting lung in the midaxillary line?**
   b. What is the level of the parietal pleural reflexion that forms the costodiaphragmatic recess in the midaxillary line?

(Answers to these questions can be found on p. 585.)

7

Lungs

Clinical Case 7.1

Patient Ted B. (continued from *Clinical Case 6.1*, the case of the sailor with the severely contused chest and collapse of the right lung, now being treated in the operating room). Under anesthesia, the right 4th rib has been surgically exposed. A sharp bony fragment had been driven through the parietal pleura and had penetrated the visceral pleura of the right lung. A major segmental bronchus was also lacerated by the fragment and air from the bronchial tree had entered the pleural cavity and produced the "closed" pneumothorax. The bronchus and lung are surgically repaired and the fragmented rib debris removed. Then the parietal pleura, periosteal lining of the 4th rib, and skin are repaired. A large needle (trocar) is placed above the 9th rib in the 8th intercostal space at the midaxillary line, and a drainage tube is inserted into the greatly expanded costodiaphragmatic recess. This tube will drain the inflammatory exudate from the pleural space that developed in response to the intrapleural blood accumulation following both the injury and surgery.

Followup. The tube is removed 5 days later. The right lung has reinflated, and no radiological signs of mediastinal shift or tracheal deviation remain. Two weeks following the injury, the patient has normal respiratory findings and normal percussion sounds over the rib cage. He still experiences some occasional pain over his healing 4th rib but this is well controlled by analgesic drugs (pain relievers).

Gross Features

The lungs are conical and reflect the space of the pleural cavities except for the costodiaphragmic and costomediastinal recesses (*fig. 7.1*). Each lung has a base and an apex, costal and medial surfaces, anterior and inferior borders, and a hilus (point of entry for vessels, nerves, and bronchial components on the medial surface) (*fig. 7.3*).

The base of the lung is concave and conforms to the shape of the dome of the diaphragm. The right dome of the diaphragm ascends to a higher level (mid level of the T8 vertebra) than the dome of the left diaphragm (T8/9 vertebral disc level). The right lung is therefore shorter than the left. Since the heart occupies a position somewhat left of the midline, the base of the right lung is broader than the base of the left lung. The apex forms the cupola that rises anteriorly above the first rib and clavicle under the fascia of the muscles in the neck (Scaleus muscles).

The medial surface has two parts: (1) vertebral and (2) mediastinal. The vertebral part is posterior and rounded to blend imperceptibly with the costal surface. This area of the lung occupies the thoracic gutters at the side of the vertebral column. The mediastinal part of the medial surface lies anterior to the vertebral column. Its most distinctive feature is the *hilus* or *root* of the lung (*fig. 7.3*). The cardiac impression is anterior to the root of both lungs and is much more prominent on the left lung due to the heart's position in the thoracic cavity. The root of the lung and its relationship to the surrounding structures is a *key* to understanding the anatomical relationship of the thoracic viscera.

Except for the phrenic nerves, all important structures that traverse the thorax from the neck to the abdomen do so posterior to the root of the lung. The root of each lung is "bridged" by a major vascular structure (*fig. 7.3*). On the venous side of the heart (right), the azygos arch passes from the posterior to the anterior side of the root of the right lung as it enters the superior vena cava. On the arterial side of the heart (left), the arch of the aorta passes anterior to posterior above the root of the left lung. A lateral radiographic view of the thorax demonstrates the radiopaque root of the lung centrally in the radiolucent field of the lung tissue (*Plate 7.1*). The clinically important viscera are noted in the root of the lung and posteriorly in the mediastinum.

The broader but shorter right lung has three distinct lobes (*fig. 7.2B*). These lobes are demarcated by two fissures. The *oblique fissure* separates the *superior* and the *inferior lobes* of the lung. It passes from the posterior surface along the pathway of the 6th rib. When the arm is raised above the head, the medial border of the scapula

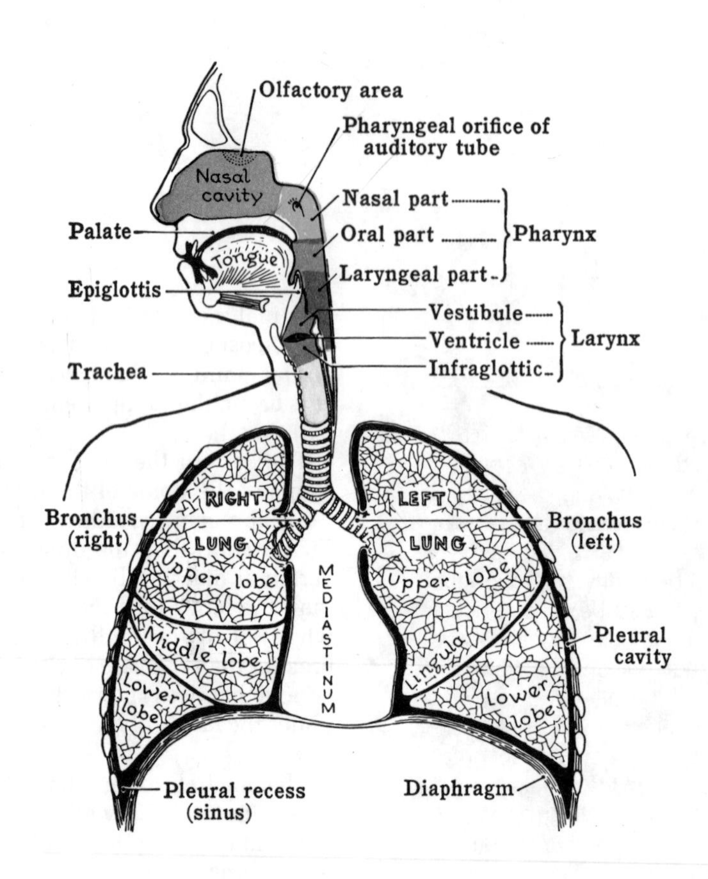

Figure 7.1. The respiratory system consists of the upper airway from the nasal apertures to the trachea, and a lower airway is within the thorax and includes the trachea, the bronchial tree, and the lungs. (From Anderson, J.E.: *Grant's Atlas*, 8th edition, Williams & Wilkins, Baltimore, 1983, Section 1, *Fig. 1-20*.)

practically overlies this oblique fissure. Examination of the superior lobe of the lung is therefore done on the anterior chest wall, as it is covered by the scapula posteriorly. The inferior lobe is likewise best examined posteriorly below the scapula. Very little of inferior lobe projects to the anterior chest wall (*fig. 7.2A*). The superior lobe of the right lung is subdivided by a second *horizontal fissure*. The surface anatomy reference for the *in situ* position of horizontal fissure is the 4th rib. It lies in a horizontal plane from the midaxillary line to the attachment of the 4th costocartilage to the sternum.

The left lung has two distinct lobes, a *superior* and an *inferior lobe*. These lobes are separated by an *oblique fissure* that is comparable in its position and extent to the oblique fissure found in the right lung. The superior lobe of the *left lung* rarely contains a horizontal fissure and this is one of its most distinguishing features when examined grossly. Another distinguishing feature in the left lung is the extensive *cardiac notch* (*fig. 7.2A*) (impression) that appears on its medial (mediastinal) surface. This pronounced impression is formed by the heart as it projects from its centrosternal position into the left side of the thoracic cavity. The heart abuts the superior lobe of the left lung anterior to the hilus. The heart, therefore, occupies the space of the left lung that would be analogous, in large part, to the space occupied by the middle lobe of the right lung. The left lung lacks a middle lobe but has a tongue-like projection of the superior lobe, *the lingula*, which overlies much of the anterior surface of the heart. The lingula of the left lung is therefore somewhat analogous to the middle lobe in the right lung when the two organs are compared or described. In a physical examination of the left lung with a stethoscope, the clinician would examine the superior lobe from the anterior side of the thorax and the inferior lobe from the posterior side, inferior to the 6th rib.

Variations occur that produce additional fissures in some cases, but the typical arrangement is that the right lung has three lobes (superior, middle, and inferior) separated by the oblique and horizontal fissures (*fig. 7.1*). The left lobe has two lobes separated by an oblique fissure. The superior lobe of the left lung contains the lingula that projects over the anterior aspect of the pericardium covering the heart.

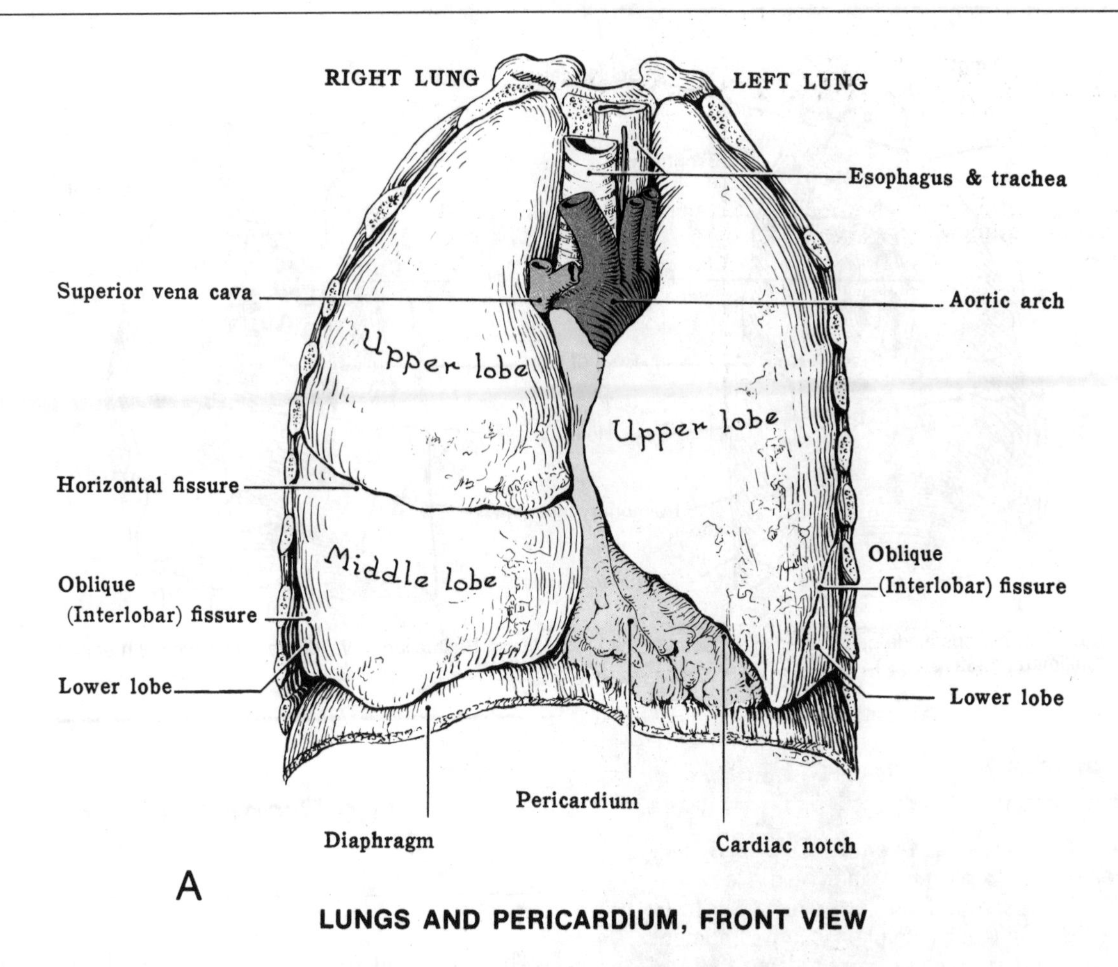

LUNGS AND PERICARDIUM, FRONT VIEW

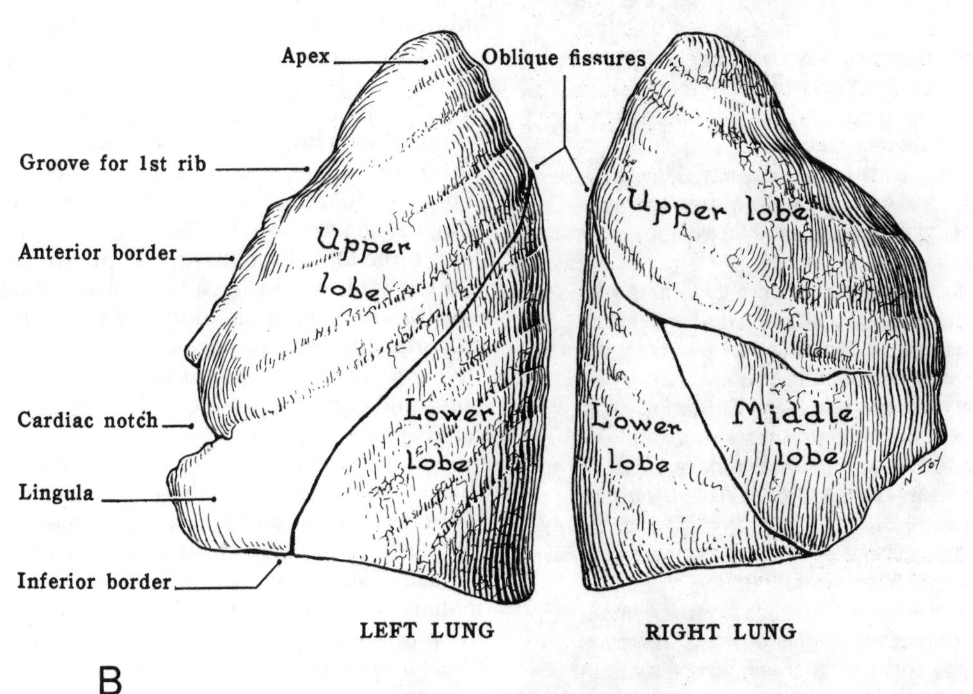

LUNGS, LATERAL VIEWS

Figure 7.2. *A*, Lungs and pericardium. An anterior view showing superior and middle lobes on the right lung and a single superior lobe on the left lung. *B*, Lungs. These lateral views show that both lungs have an equivalent oblique fissure on the posterior and lateral aspect. (From Anderson, J.E.: *Grant's Atlas*, 8th edition, Williams & Wilkins, Baltimore, 1983.)

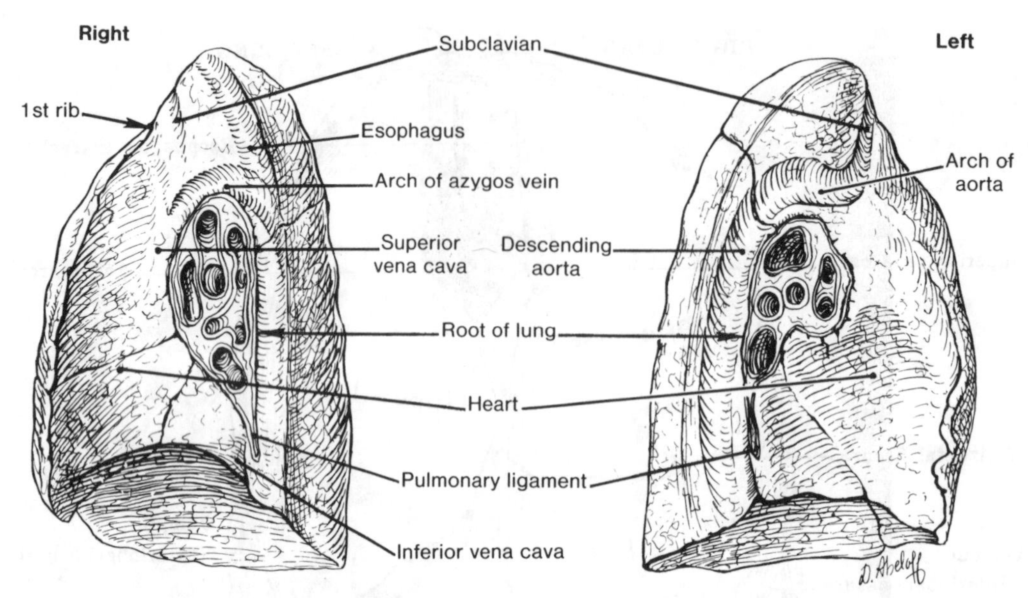

Figure 7.3. Impressions made on the medial surfaces of the lungs. (From Basmajian, J.V.: *Primary Anatomy*, 8th edition, Williams & Wilkins, Baltimore, 1982, p. 232.)

Clinical Case 7.2

Patient Donna F. This 21-year-old college student was seated in a dental chair and given a local anesthetic for the removal of a lower 3rd molar (wisdom tooth). During the extraction, a large fragment of the tooth crown fractured and entered the oral pharynx. The patient began coughing and the chair was tilted in a more upright position. She continued to cough and complained that her "throat" was quite sore. The dentist gave her a mild sedative intravenously. He finished extracting the remaining parts of the tooth and closed the oral tissue over the "open" tooth socket. He took the patient to a radiologist in the clinic and had an X-ray of the upper and lower respiratory system and an X-ray of the upper gastrointestinal tract taken. Although the radiograph of the abdomen showed no evidence of the tooth crown in the stomach, it showed up in the posterior basal region of the right lung. (Why?)

The patient is brought to the emergency clinic where you are a student and under a general anesthetic, bronchoscopy is used to remove the tooth fragment from the bronchial tree. The patient is placed on prophylactic antibiotic therapy and observed for two days before being released from the clinic. At follow-up five days later at the clinic, you find no signs of respiratory dysfunction or pneumonia in the right lung. After the dentist removes the sutures from the extraction site she returns to school the next day.

The Bronchial Tree

The bronchial tree is formed by the branching of the midline tracheal tube in the thorax. The initial branching into two primary bronchi (right and left) occurs at the level of the T4/5 intervertebral disc. The primary bronchi (*fig. 7.1*) are contained in the mediastinum and enter the hilum of each lung at a slightly lower level than the initial bifurcation. The primary bronchus then divides into secondary or lobar bronchi (*fig. 7.4*), which extend into the lung tissue of each lobe. The right lung would therefore have three lobar bronchi (upper, middle, and lower), and the left lung would have two lobar bronchi (upper and lower). The lobar bronchi then divide into the *extremely important* tertiary or segmental bronchi. *These tertiary divisions of the bronchial tree are the basis for understanding and identifying the bronchopulmonary segments of the right and left lungs.*

A *bronchopulmonary segment* is the portion of lung that is aerated by the subsequent division of a tertiary bronchus. These wedge-shaped segments of lung tissue extend from an apex at the hilum to a base on the lung surface. The tertiary bronchus within each bronchopulmonary segment is also joined by a tertiary division of the pulmonary arteries. Their terminal divisions constitute an anatomical division of lung tissue that can be surgically resected with less physiological loss than a lobectomy (lobe removal) or *pneumonectomy* (lung removal). The pattern of bronchopulmonary segments are shown in *Figures 7.4* and *7.5*. The right lung has 10 bronchopulmonary segments, and the left lung has 8 bron-

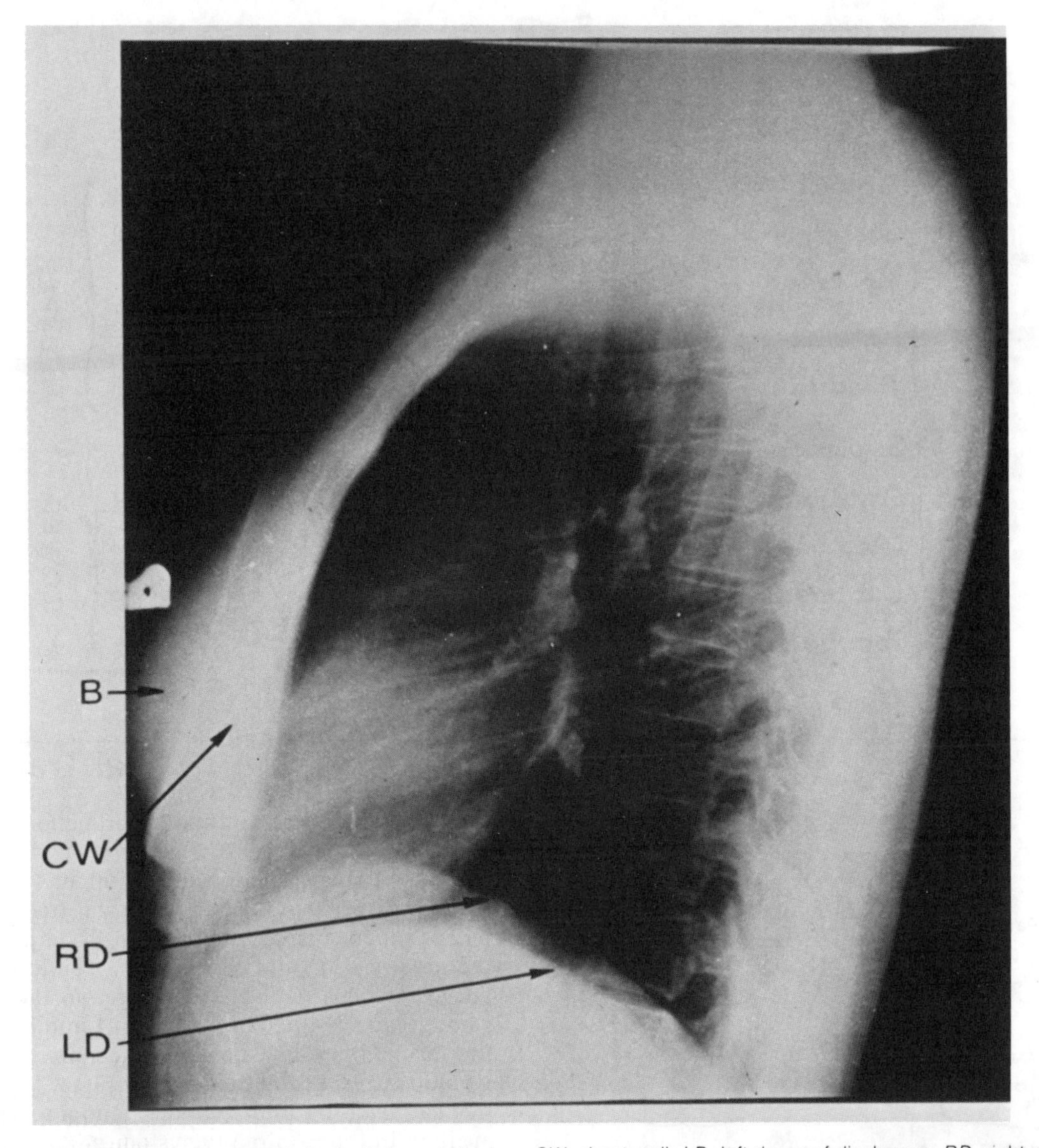

Plate 7.1. Female chest (right lateral view). *B*, breast shadow; *CW*, chest wall; *LD*, left dome of diaphragm; *RD*, right dome.

chopulmonary segments. The left differs from the right in that the apical and posterior bronchopulmonary segments of the right lung are considered to be a single segment in the upper lobe of the left lung, and the medial and anterior basal bronchopulmonary segments of the right lobe are fused into a single segment in the left lung for surgical purposes. Since the heart is occupying much of the medial basal area of the left lower lobe, it is logical that this segment would be least likely to be found in the left lung. The bronchopulmonary segments can be visualized very readily in a *bronchogram* (*Plate 7.2*) (special enhanced radiograph of the bronchial tree). Foreign bodies and pathological lesions can be localized and visualized in individual bronchopulmonary segments. The localization of aspirated material could be predicted if one knew the position of the patient at the time of aspiration. Gravity is a principal factor in localizing the aspirated material. If a patient is sitting up, the likely route of a foreign body would be to the right lung (least angled from the midline at the primary site of tracheal bifurcation) and then into the posterior basal segment. If the patient were lying on the right side, the right posterior or middle lobe bronchopulmonary segments would be affected (*figs.* 7.6 and 7.7). If the patient were lying on the left side, the superior and inferior bronchopulmonary segments of the lingula would be most favored by gravitational force. Likewise, a patient lying on the back would most likely localize aspirated material in the superior bronchopulmonary segment of the lower lobe (*figs.* 7.7 and 7.8). Direct visualization of these bronchopulmonary segments can be made through the oral pharynx and trachea (*fig.* 7.1) using a bronchoscope. Fiber optics transmit an image of the tissue surfaces, and biopsies or foreign object removals can be done to confirm radiological find-

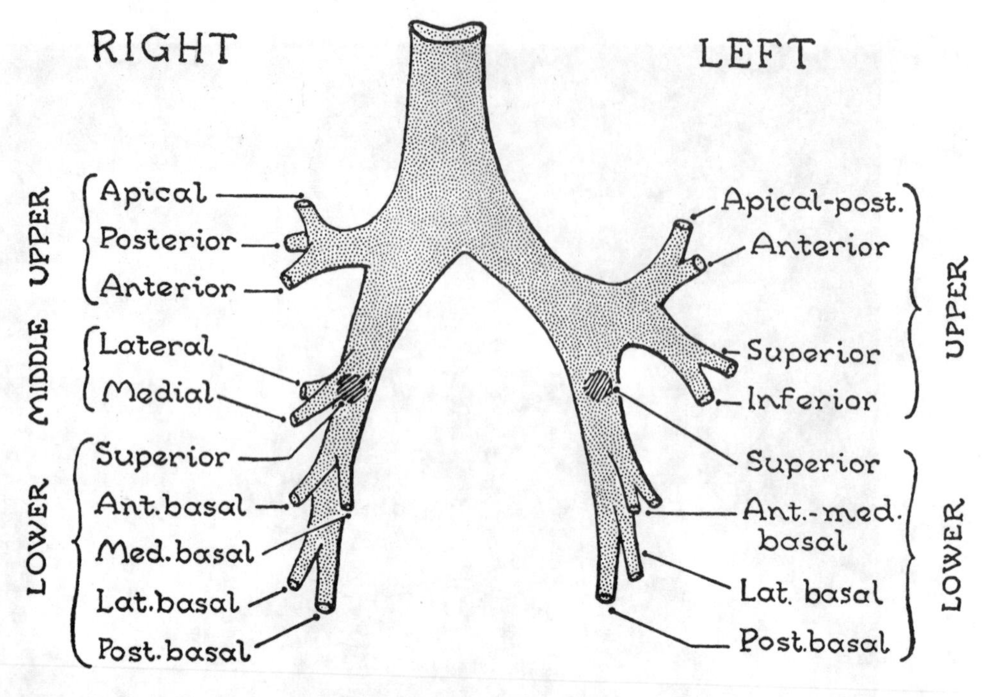

Figure 7.4. The 10 right and 8 left segmental bronchi. (After Jackson and Huber.)

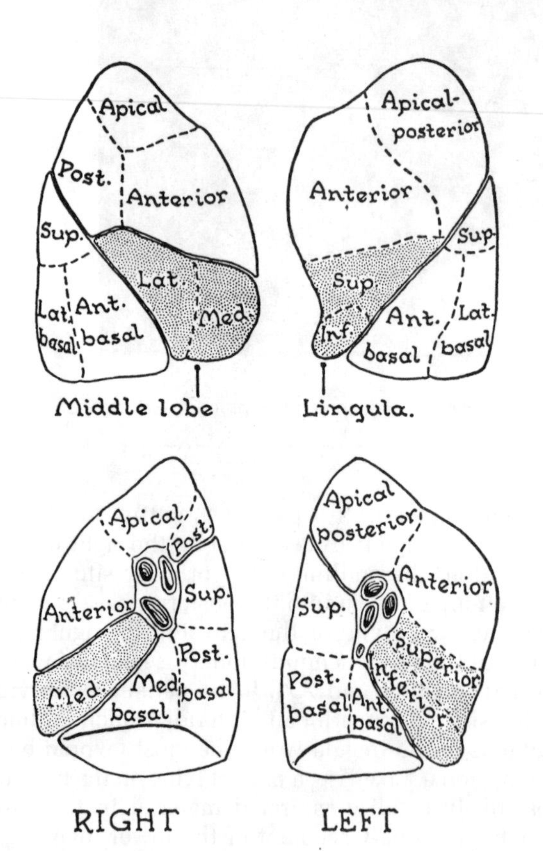

Figure 7.5. The 10 right and 8 left bronchopulmonary segments. (After Jackson and Huber.)

ings of disorders in individual bronchopulmonary segments.

The principal function of the upper respiratory tract and the larger parts of the bronchial tree is to ensure a patent (open) respiratory system. Bony passageways in the nasal cavity, and cartilaginous reinforcement of the larynx, trachea, and bronchial tree to the levels of the smooth muscle lined bronchioles (1 mm in diameter) are the means by which patency is ensured. Lack of cartilaginous rings in the trachea and plates in the bronchial tree can lead to tubular collapse and a resistance to free air flow in the respiratory system.

Vessels and Nerves of the Lungs

PULMONARY ARTERIES

The pulmonary arteries are derived from the bifurcating *pulmonary trunk* (*fig. 7.9*). The *left* and *right pulmonary* arteries lie anterior to the primary bronchi as they enter the hilus of their respective lungs. The right pulmonary artery is crossed superiorly by the arch of the azygos vein, while the left pulmonary artery is inferior to the arch of the aorta at the T5 vertebral level of the thorax. The pulmonary arteries form *lobar branches* in

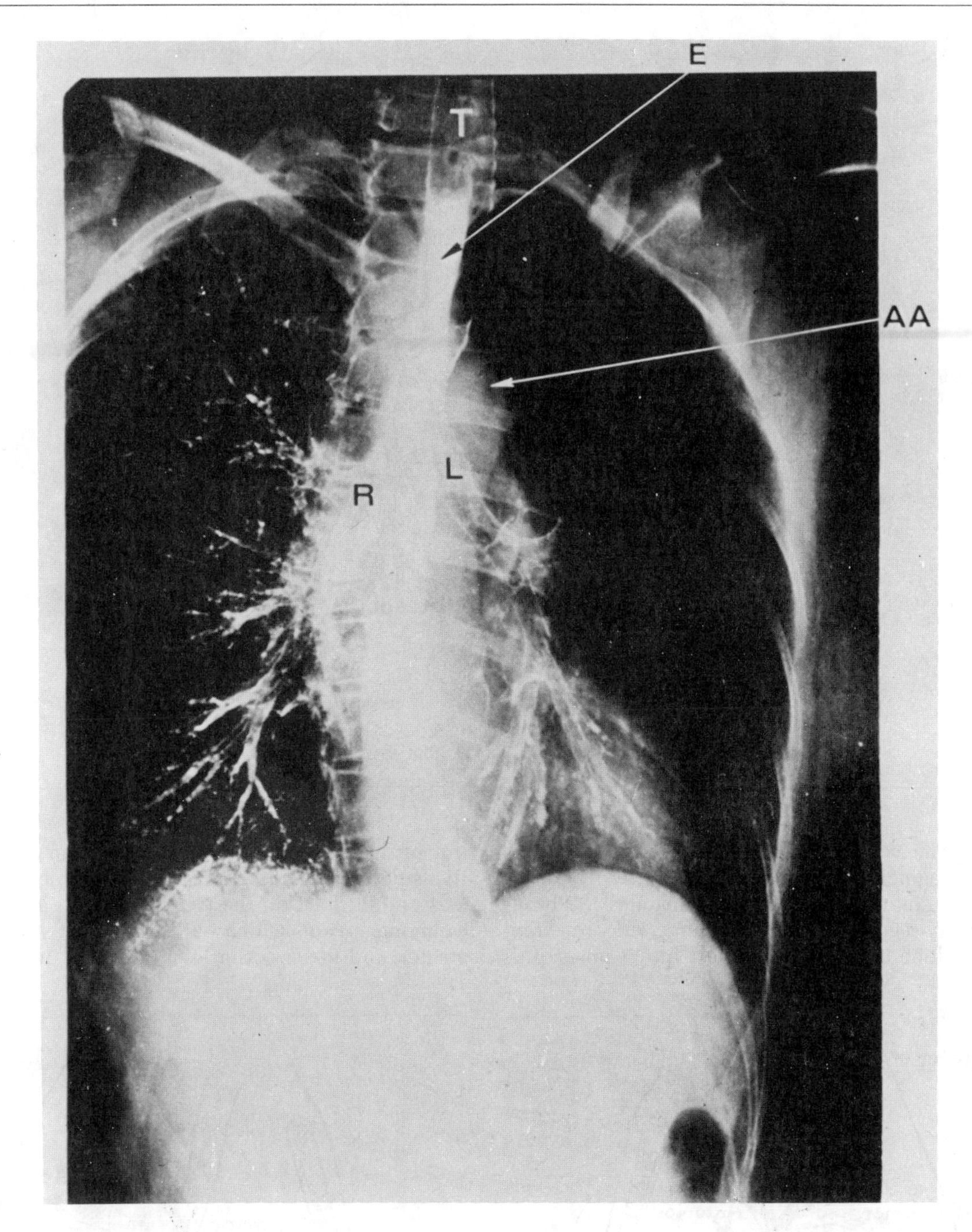

Plate 7.2. Bronchogram and esophagogram (P-A view). *AA*, aortic arch; *E*, esophagus; *L*, left main bronchus; *R*, right main bronchus; *T*, trachea.

the hilus of each lung and subsequently divide into *tertiary branches* (segmental divisions), which correspondingly branch successively with each branching division of the bronchial tree. This intimate relationship of the pulmonary artery and the bronchial tree beyond the tertiary division of each structure is the anatomical basis for the bronchopulmonary segmentation of the lung. The pulmonary arteries are carrying deoxygenated blood from the right side of the heart to the terminal alveolar ducts and sacs of the bronchial tree for exchange of CO_2 and O_2 in the hemoglobin of the red blood cells. The oxygenated blood leaves the capillary beds associated with bronchial alveolar sacs and returns to the left side of the heart through the pulmonary veins (*fig. 7.10*).

PULMONARY VEINS

The pulmonary venous system is anatomically separated from the bronchial and pulmonary arterial system

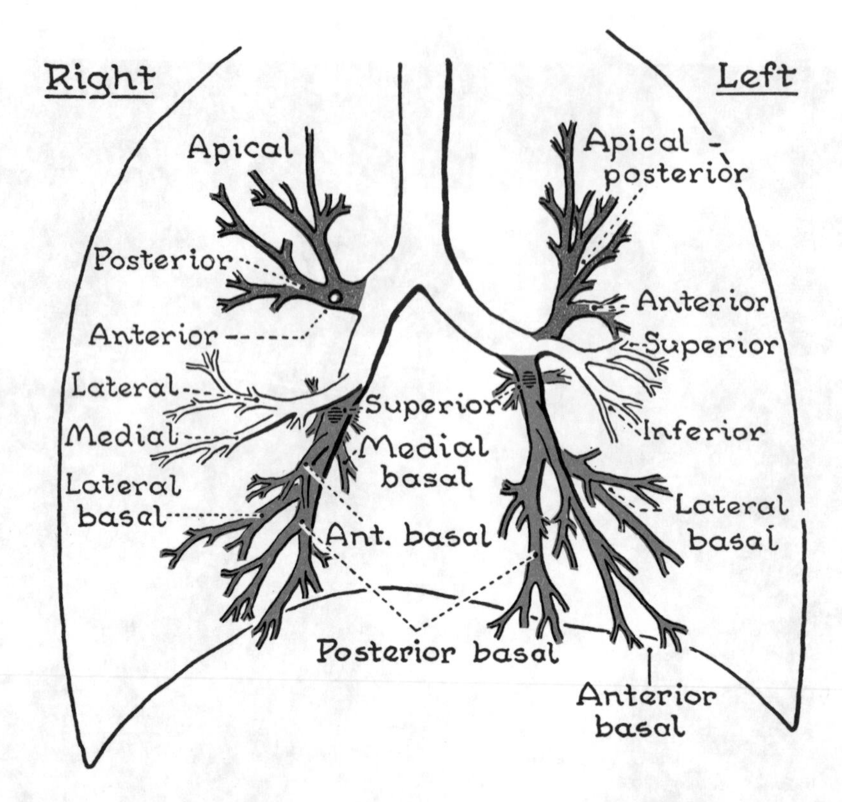

Figure 7.6. The distribution of the bronchi, front view. (After Nelson, modified.)

(*fig. 7.10*). The veins actually drain adjacent bronchopulmonary segments and readily come to lie in the intersegmental connective tissue that separates the bronchopulmonary segments within each lung. The veins from the inferior lobe of each lung collect into a *lower pulmonary* vein which enters the left atrium on the respective side of the left atrium. The veins from the superior and middle lobe on the right lung form an *upper right pulmonary vein*, which enters the right side of the left atrium, and the superior lobe of the left lung also has a

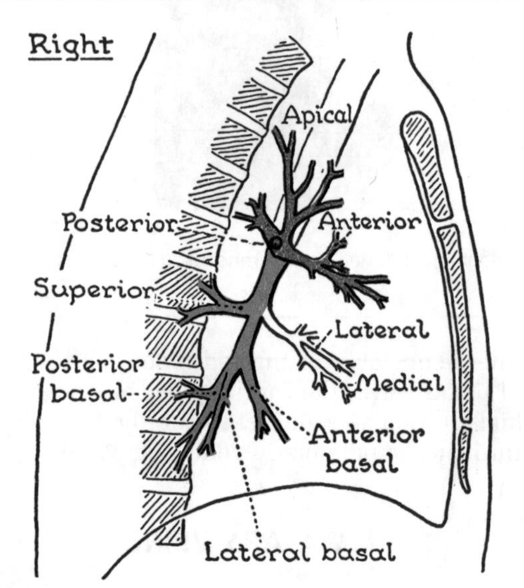

Figure 7.7. The distribution of the right bronchus, side view. (After Nelson, modified.)

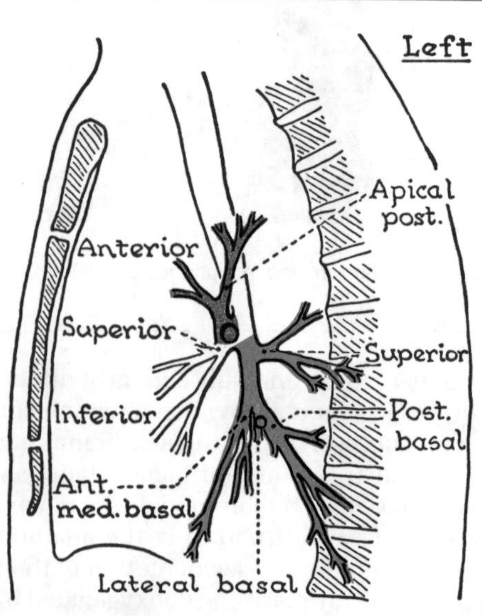

Figure 7.8. The distribution of the left bronchus, side view. (After Nelson, modified.)

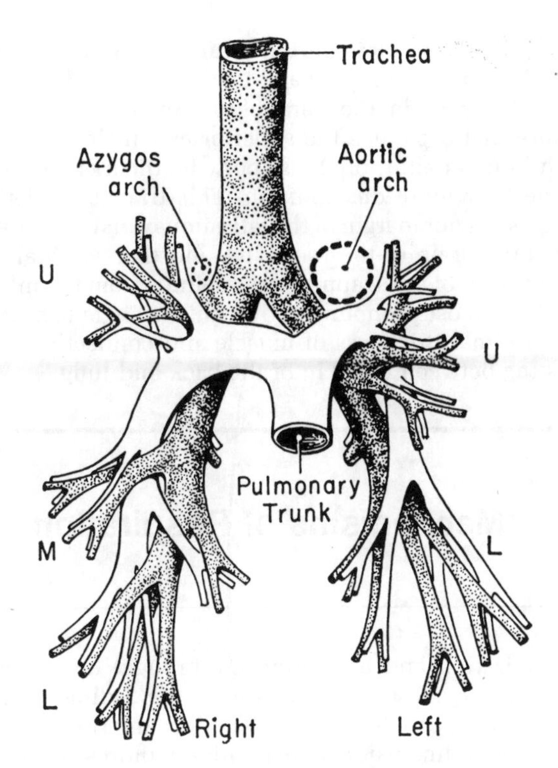

Figure 7.9. To show the relationship of pulmonary arteries to bronchi, front view. *U. M. L.*, branches to upper, middle, and lower lobes. (After Hayek, translated by V.E. Krahl.)

separate *upper left pulmonary vein* that enters the left atrium. The veins, in general, lie anterior to the pulmonary arteries and bronchi in the hilus of the lung.

THE BRONCHIAL ARTERIES

The bronchial arteries (*fig. 7.10*) arise from the descending aorta or the 3rd intercostal branch of the aorta. They carry oxygenated blood to the structures in the hilum, bronchial tree, and the supportive tissues of the lung. This blood supply is eventually drained from the lung through the pulmonary venous system. The lungs therefore have two arterial systems, one carrying deoxygenated blood and one oxygenated blood, and a single venous system that drains into the left atrium of the heart.

LYMPHATICS (*Fig. 7.10*)

The lung has a rich lymphatic drainage system that is readily apparent by the surface pattern of carbon particles under the visceral pleura in the cadaver. These lymphatics follow the vascular tree and the intersegmental tissues to reach the hilus where the draining lymph is filtered by the pulmonary lymph nodes. These pulmonary lymph nodes are usually very evident due to their carbon content in most cadaver or autopsy specimens. The lymphatic drainage from the hilum of the right lung enters the *right lymphatic duct* and eventually enters the venous system at the junction of the right subclavian and right internal jugular veins in the neck. The hilum of the left lung is drained by lymphatic vessels that connect with the *thoracic duct*, which in turn enters the venous system in the root of the neck on the left side. Lung

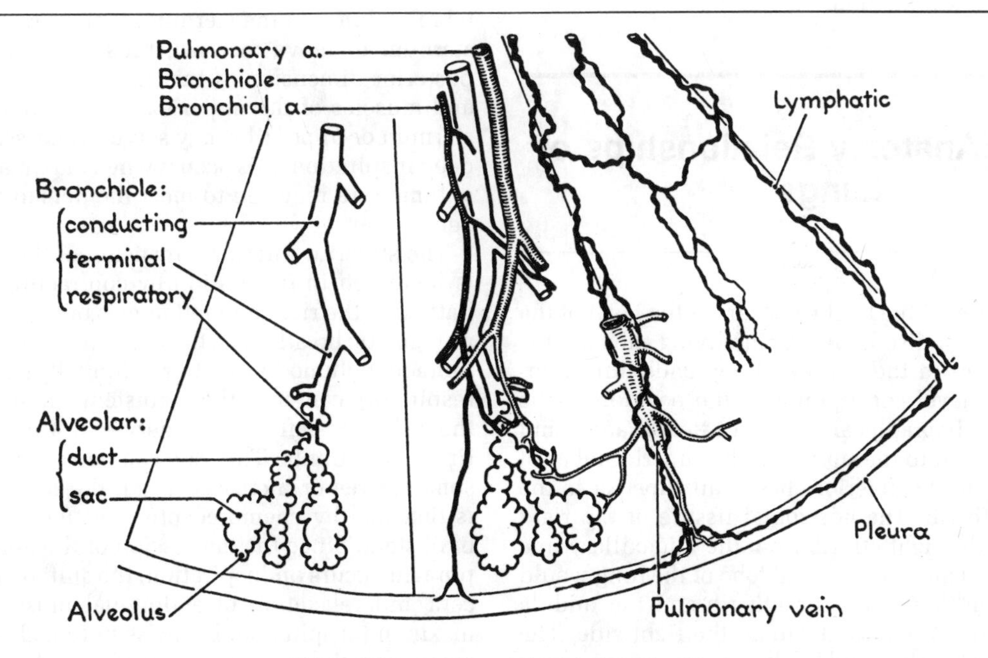

Figure 7.10. Structure of a lobule of the lung.

tumors can therefore spread via the lymphatic system to the neck and systemic circulation once they bypass the lymph nodes in the hilum of their respective lung.

NERVES OF THE LUNGS AND PLEURA

The *pulmonary plexus* contains both parasympathetic and sympathetic nerves to the bronchial and vascular "trees" of each lung. The parasympathetic vagal fibers are preganglionic and *secretomotor* to the glands in the bronchial mucosa. The sympathetic postganglionic fibers are *vasomotor* to the arterial system. Control of the smooth muscle of the pulmonary bronchioles is controversial. It is known that drugs that mimic the sympathetic nervous system, such as epinephrine, cause bronchodilatation. Patients with asthma can relieve bronchiole constriction by inhaling epinephrine-like drugs. Severe bronchiole constriction can also be relieved by injecting epinephrine intravenously.

Sensory nerves also innervate the lung. These may have a function in limiting the amount of expansion permitted in inhalation through stretch receptors in the pleura. This regulatory feedback system is transmitted by the vagal branches. Pain sensations are also felt in the lung, and they usually relate to insufficiency in the vascular supply to tissue. These visceral pain pathways usually follow the distribution of the sympathetic fibers, which innervate vascular smooth muscle. Severe pain in the lungs can also be associated with the parietal pleura. This pleural lining of the thoracic cage is innervated by the intercostal nerves in the adjacent rib cage and the phrenic nerves that innervate the diaphragm. The pain of pleurisy is most commonly associated with inflammation or adhesions to the parietal pleura. The visceral pleura is insensitive to painful stimulation.

Surface Anatomy Relationships of Lungs

The bifurcation of the trachea occurs at the level of the T4/5 intervertebral disc in the supine living subject. It is somewhat higher in the child and descends during inspiration. This level corresponds to the position of the palpable and radiographically distinct sternal angle and 2nd rib attachment to the sternum. The superior lobes of the lung relate to the anterior chest wall superior to the level of the 6th rib. The horizontal fissure of the right lung parallels the right 4th rib from the midaxillary line to the sternum. The superior right lobe of the lung would therefore be audible above the 4th rib and the middle lobe between the 4th and 6th rib on the right side. The inferior lobes of the right and left lungs are best examined on the back, inferior to the 6th rib and the vertebral border of the scapula when the hand is placed behind the head. These bony landmarks parallel the course of the oblique fissure that separates the superior and inferior lobes of each lung. In addition, these bony landmarks also relate to the *triangle of auscultation*. This triangle is defined by the superior margin of the latissimus dorsi, the inferior (lateral) margin of the trapezius, and the vertebral (medial) border of the scapula. Lung sounds from the inferior lobes are most pronounced over this triangle because of the minimal thickness of muscle and connective tissue existing between the skin of the back and lung.

Mechanisms of Respiration

Inspiration and expiration are brought about by the alternate increase and decrease in the 3 dimensions of the thoracic cavity (inferior-superior, anterior-posterior, and lateral dimensions). Inspiration requires an increase in the intrathoracic volume. The attached parietal pleura will move with the expanding thoracic wall and descending diaphragmatic domes, creating a more negative pressure in the pleural space. These pressure decreases will expand the lungs and draw air through the trachea and into the bronchial tree. At the same time, the capillaries in the lung dilate to facilitate the pulmonary circulation.

Increases in the anteroposterior dimensions of the rib cage occur when the upper six ribs are elevated. Since these ribs are usually obliquely oriented from the vertebral column to the sternum, their elevation causes the sternum to move forward, thus increasing the anteroposterior dimension of the thoracic cage. Neck muscles and muscles of the upper extremities that insert on the sternum or upper ribs may serve as accessory muscles in deep inspiration; this occurs when maximal intrathoracic volumes are required to meet the demands of increased ventilation.

The superior-inferior dimension of the thoracic cavity is increased in inspiration by contracting and therefore flattening the right and left domes of the diaphragm. Diaphragmatic breathing is the main means of expanding the thoracic volume in quiet breathing. It is controlled by a respiratory center in the brainstem and effected through the right and left phrenic nerves at a normal rate of 16 times per minute. This rate can be voluntarily altered to some degree or increased when the demand for oxygen is detected by chemoreceptors in the great vessels and brainstem. Since the depression of the domes of the diaphragm occurs on inspiration, the shift of abdominal viscera and relaxation of abdominal muscles also aid in maximum inspiration. Likewise in forced expiration (e.g., coughing), the abdominal muscles will contract to force the abdominal viscera against the undersurface of the diaphragm and drive it superiorly into the thoracic cavity. The subsequent decrease in the inferior-superior dimension of the thorax will expel the air from the bronchial tree as the thoracic volume decreases and the intrathoracic pressure rises.

The lateral dimension of the thorax can be increased when the lower ribs (7–10) are elevated. Since these ribs descend obliquely from the spinal column to the mid-axillary line and then ascend with the rising costal margin to attach to the lower part of the sternum, their elevation expands the thoracic cage in the lateral dimension. It is analogous to raising a bucket handle and observing the handle's movement away from the side of the bucket prior to swinging above the bucket. This lateral movement is evident in an individual who is breathing deeply in forced inspiration.

The intercostal muscles of the rib cage act to stabilize the soft tissue walls between the ribs and maintain the intercostal spacing during inspiration and expiration. Without these muscles non-bony walls of the chest wall would tend to depress on inspiration and expand on expiration. Patients with spinal cord injuries in the cervical region below the C4 spinal cord segment demonstrate the characteristics of diaphragmatic breathing. The paralyzed abdominal and intercostal muscles in these patients move in a passive fashion away from increased pressures in the abdominal and thoracic cavities, respectively. The loss of these muscles greatly decreases the respiratory capacity of these patients.

Clinical Mini-Problems

1. **At which vertebral levels does one normally find the bifurcation of the trachea?**
2. **What major thoracic structure passes anterior to the root of the lung?**
3. **Where would one place a stethoscope to listen to lung sounds in the middle lobe?**
4. **In which bronchopulmonary segment would one suspect an object aspirated by a patient who was sitting in a chair?**
5. **a. Which nerve is sensory to the parietal pleura overlying the right dome of the diaphragm?**
   b. Which nerve ennervates the costal parietal pleura underlining the right sixth rib?

(Answers to these questions can be found on p. 585.)

8

Heart and Pericardium

Clinical Case 8.1

Patient Bernard W. This 55-year-old pharmacist complained of pain in his left arm and axilla while climbing the back stairs of his home after a brisk walk. His physician suspected that the pain was due to cardiac ischemia (lack of blood supply to the heart muscle). This patient has been referred to the University Clinic where you are a student. After a physical examination, he is scheduled for a radiological examination of the coronary blood supply (angiography). *Follow-up:* The angiogram reveals a constricted right coronary artery 1 cm distal to the ostium in the aortic sinus of the right aortic valve. The patient is now scheduled for coronary bypass surgery to improve the blood supply to the right ventricle of his heart.

Pericardium

Pericardium (Gr. Peri = around; Kardia = heart) consists of a tough outer fibrous sac, lined within by a serous sac (*fig. 8.1*). The heart and great vessels lie inside the fibrous sac and invaginate the serous sac from behind during development (*fig. 8.1*). The resultant effect is that the external surface of the heart and the internal surface of the fibrous percardium are covered by a layer of **serous percardium**. The part covering the surface of the heart is called the **visceral pericardium** or **epicardium** and the part covering the inner aspect of the fibrous pericardium is called the **parietal pericardium**. The intervening space between these two serous pericardial linings is the **pericardial cavity**. This space contains a small amount of serous fluid that lubricates the serous linings and allows the heart to move freely within the pericardial sac from its more stationary base to its moveable apex. Excessive fluid in the pericardial cavity can result in restricted expansion of the relaxed heart and compromise its ability to fill properly and propel adequate amounts of blood into the systemic circulation. This condition is called **cardiac tamponade** and may require the opening of the fibrous pericardium to remove this accumulated fluid in the pericardial sac.

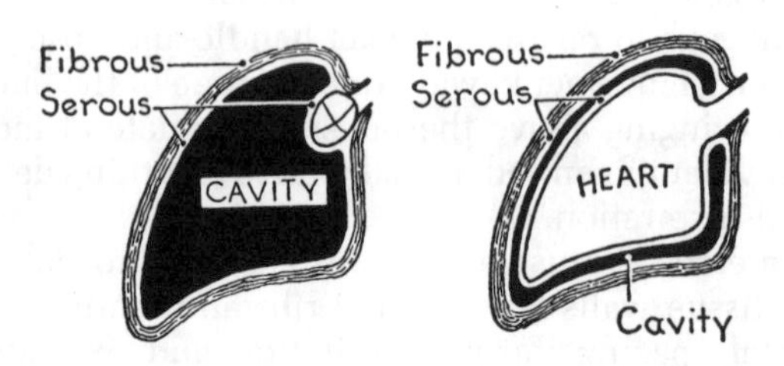

Figure 8.1. To explain the layers of the pericardium.

The fibrous pericardium blends inferiorly with the connective tissue of the central tendon of the diaphragm and superiorly with the connective tissue covering of the great vessels of the heart. It lies adjacent to the posterior aspect of the sternum and defines the middle mediastinum within the thorax.

Heart (Lat. = Cor; Gr. = Kardia)

This muscular pump is somewhat larger than a closed fist. It has 4 chambers, the **right** and **left atria** and the **right** and **left ventricles**. The atria are separated from the ventricles by a constriction that completely encircles the heart and is appropriately called the **coronary (atrioventricular) sulcus**. The ventricles are separated from each other by the **anterior and posterior interventricular (longitudinal) sulci**.

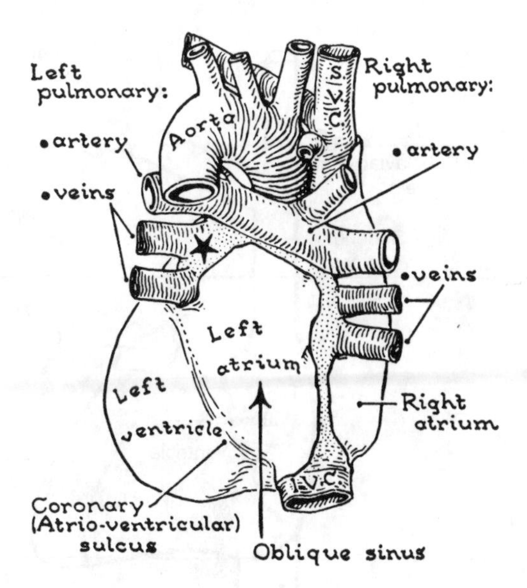

Figure 8.2. The posterior aspect of the heart. ★, site of contact of left bronchus with left atrium.

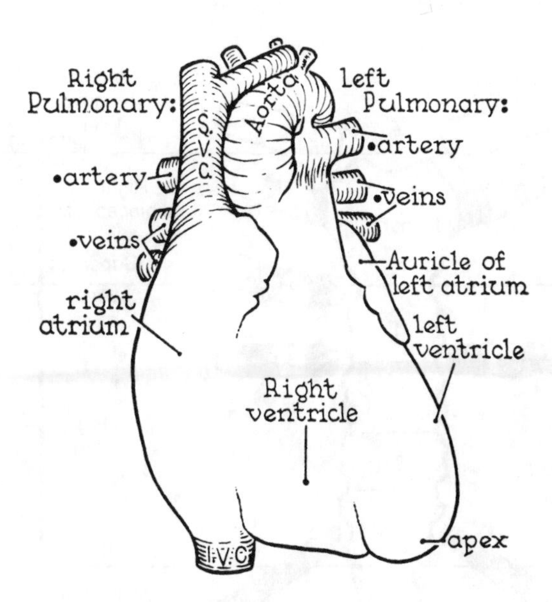

Figure 8.3. Anterior view of undissected heart.

The heart is a conical shaped organ with its base positioned posteriorly. This is the most stationary part of the heart and moves the least during contraction and relaxation. The apex is the most moveable part, and it is located in the *left* 5th intercostal space (approximately 10 cm from the midline) when the ventricles are relaxed and filling with blood. The surfaces of the heart are the **anterior** (right sternocostal) surface, the **left** (left sternocostal) surface and the **inferior** diaphragmatic) surface. The **anterior interventricular sulcus** is located on the left sternocostal surface and the **posterior interventricular sulcus** is located on the diaphragmatic surface. The **coronary blood vessels** are found in these sulci and in the **coronary sulci** which separate the atria from the ventricles on the surface of the heart.

The left atrium and the four **pulmonary veins** that enter the left atrium form the major part of the base of the heart. The serous pericardium on the base of the heart (the epicardium or visceral pericardium) reflects onto the inner surface of the fibrous pericardial sac as the parietal pericardium at this point. This reflection of the serous pericardium forms a blind ending sac (*fig. 8.2*) in the pericardial cavity over the base of the heart. This blind projection is termed the **oblique sinus**.

GREAT VESSELS AND SKIN-SURFACE ANATOMY

Students and physicians must be able to plot the heart and great vessels on the skin of a living person (refer to *figs. 8.3, 8.4, 8.19*).

1. A vertical line drawn from the neck to the abdomen one finger's breadth from the right margin of the sternum relates to the following (*fig. 8.4B*):
   a. The right subclavian vein lies under the right clavicle and joins the right internal jugular vein at the sternoclavicular joint to form the right brachiocephalic vein.
   b. The left brachiocephalic vein passes obliquely behind the body of the manubrium to join the right brachiocephalic vein posterior to the right side of the sternum superior to the sternal angle. The superior vena cava (SVC) descends posterior to the right side of the sternum to the level of the 3rd costal cartilage where it enters the superior aspect of the right atrium.
   c. The *inferior vena cava* (IVC) pierces the right dome of the diaphragm at the level of the xiphisternal joint (8th thoracic vertebral level) and after a 1-cm course enters the inferior aspect of the right atrium at the level of the 6th costal cartilage.
   d. The right atrium therefore bulges somewhat more than a finger's breadth from the right posterior aspect of the sternum between the 3rd and 6th intercostal cartilages.
2. A horizontal line through the xiphisternal joint from the right 6th costal cartilage to the 5th intercostal space at the left midclavicular line will define the inferior margin of the heart.
   a. The inferior vena cava (IVC) is located on the right side of the sternum at the level of the 6th costal cartilage.
   b. The right ventricle overlaps the IVC somewhat as it enters the more posteriorly situated right atrium inferiorly. The right ventricle then constitutes the majority of the anterior (sternocostal) surface of the heart posterior to the inferior part of the sternum and costal cartilages of the 4th and 5th ribs.

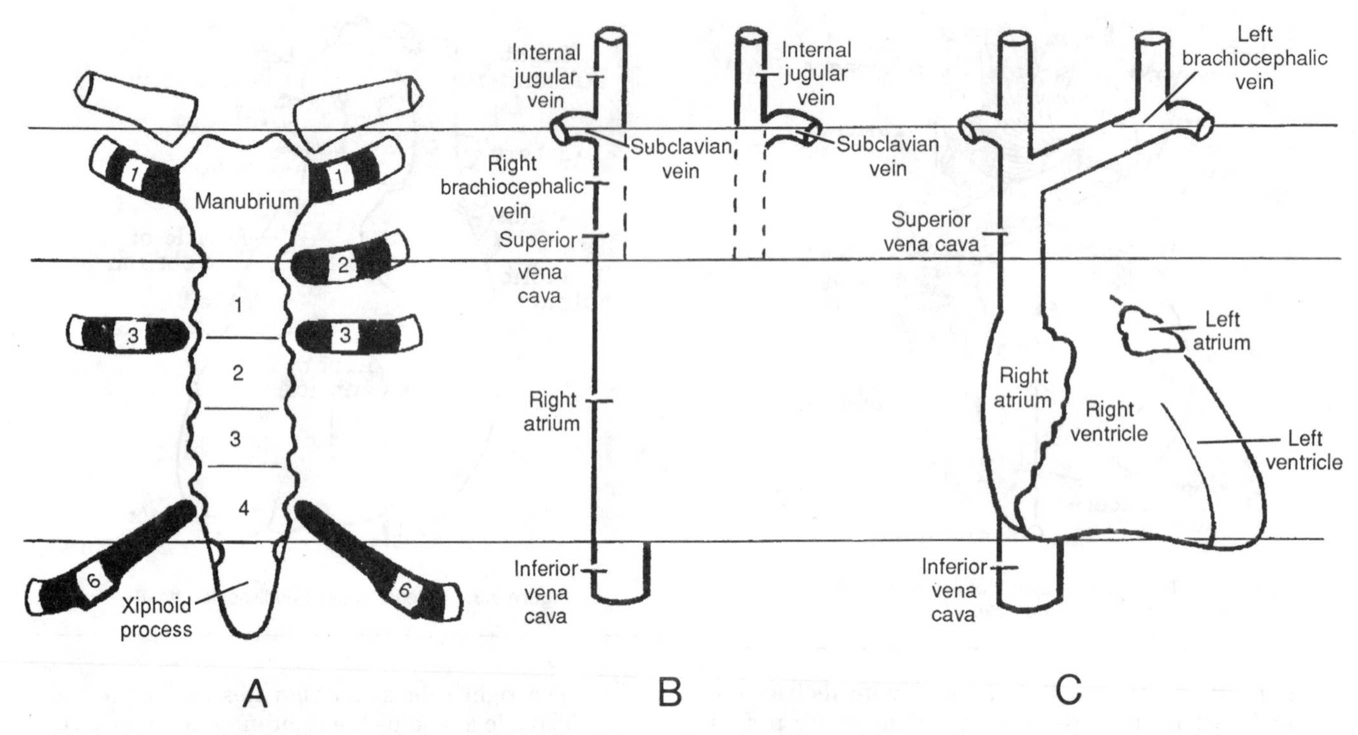

Figure 8.4. The sternocostal surface of the heart and great veins constructed on projection lines.

   c. The apex of the heart is the tip of the left ventricle and is related to the left 5th intercostal space, 10 cm from the midline of the sternum. The apex beat is palpable at this point.
3. The left margin of the heart parallels an oblique line drawn on the chest wall from the apex position in the left 5th intercostal space to the 2nd left intercostal space, a fingerbreadth from the sternum.
   a. The left ventricle forms the inferior left margin of the heart on this oblique line.
   b. The tip of the auricle of the left atrium is located on this margin posterior to the 3rd rib and the pulmonary trunk is superior to the left auricle in the 2nd intercostal space (*fig. 8.3*).
   c. The pulmonary trunk overlaps the more posteriorly positioned descending aorta in the left 2nd intercostal space (*fig. 8.3*). As the ascending aorta reaches the sternal angle, it becomes continuous with the arch of the aorta posterior to the sternum. The aortic arch projects posteriorly and to the left as it crosses the superior aspect of the root of the left lung. This image on a radiograph is continuous with the left margin of the heart and is called the "aortic knuckle" (Plate 8.1).
   d. The pulmonary trunk pierces the pericardium and divides into the right and left pulmonary arteries inferior to the arch of the aorta (*fig. 8.5*). The long right pulmonary artery passes posterior to the ascending aorta and the SVC to enter the hilum of the right lung. The shorter left pulmonary artery passes into the hilum of the left lung anterior to the descending aorta, which is the posterior continuation of the aortic arch. A fibrous connection between the proximal part of the left pulmonary artery and the undersurface of the arch of the aorta is the **ligamentum arteriosum**. In the embryo and fetus, this vestigial ligament was an arterial shunt (**the ductus arteriosus**) that allowed the arterial circulation to bypass the collapsed lungs. It normally becomes ineffective as a shunt after birth when the lungs expand. The ductus arteriosus postnatally fuses to form the ligamentum arteriosum.

ASCENDING AORTA AND PULMONARY TRUNK

The ascending aorta and the pulmonary artery in the embryo were initially a single vessel forming the arterial end of the developing heart, the **truncus arteriosus**. They became divided by an elaborate internal spiral septum to form the outlet vessels from the left and right ventricles.

The **pulmonary trunk**, 5 cm long, begins anterior to the aorta in the superior limits of the right ventricle (*fig. 8.5B*). It passes semispirally (superiorly, posteriorly, and to the left) around the ascending aorta until it reaches the undersurface of the aortic arch. Here it pierces the fibrous pericardium and divides into the **right** and **left pulmonary arteries**. The pulmonary arteries then trav-

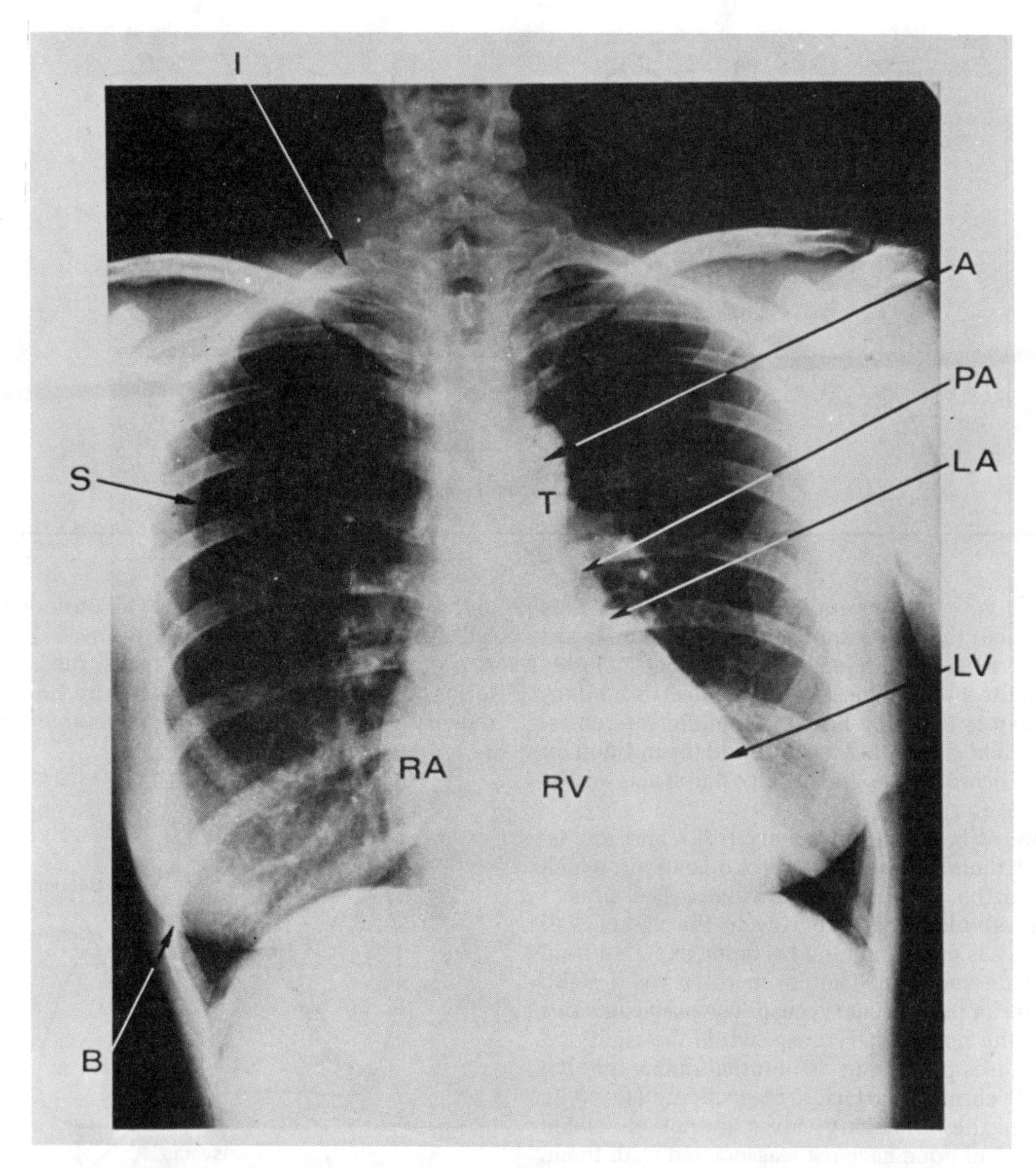

Plate A.1. Female chest (P-A view). *A*, aortic arch; *B*, breast shadows; *I*, first rib; *LA*, left auricle (auricular appendage); *LV*, left ventricle; *PA*, pulmonary artery; *RA*, right atrium; *RV*, right ventricle; *S*, vertebral border of scapula; *T*, descending thoracic aorta.

erse the middle mediastinum in a horizontal fashion between the main bronchi (posteriorly) and the pulmonary veins (anteriorly) to enter the hilum of their respective lung.

The **ascending aorta** begins at the superior aspect of the left ventricle posterior to the **pulmonary trunk** (*fig. 8.5A*). The **ascending aorta** passes obliquely (superiorly, anteriorly, and to the right) to reach the right side of the sternum in the pericardial sac. It pierces the pericardium at the level of the sternal angle to become the **arch of the aorta**.

Each of these great vessels have a valve within their lumen to prevent the reflux of blood back into the respective ventricles following systole (contraction of the ventricles) and during **diastole** (subsequent ventricular relaxation). Both valves are similarly constructed and have 3 semilunar valvules or cusps. The aortic valve is much stronger because it resists the much higher systemic blood pressure, which is nearly 5 times greater than the pressure of the pulmonary system.

Each **valvule** or **cusp** has a fibrous base covered on both surfaces with endothelium (*fig. 8.6*). A fibrous **nodule** is contained within the endothelium at the middle of the free edge of each nodule. The portion of the cusp that underlies the free concentric edge is the **lunule**. When the valve closes due to a higher pressure in the arterial lumen distal to the valve, the nodules meet in the center of the lumen and the ventricular surfaces of the lunules

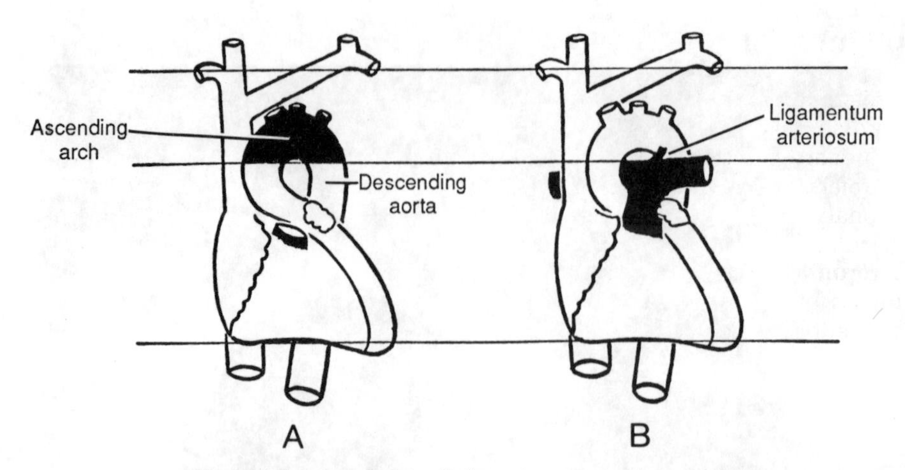

Figure 8.5. *A*, the aorta; *B*, the pulmonary trunk and arteries.

of contiguous valvules are applied to each other. This contact produces the valve sound that can best be heard at a point where the blood-filled great vessels most closely approximate the anterior chest wall. The heart sound for the aortic cusp is therefore heard at the 2nd intercostal space on the right side of the sternum, and the pulmonary valve is best heard at the 2nd intercostal space on the left side of the sternum.

At the root of both the pulmonary trunk and the ascending aorta, there are also 3 sinuses or dilatations, which correspond to the 3 leaflets of the valves. The sinuses prevent the valvules from adhering to the vessel wall when the valve is open and blood is being expelled from the appropriate ventricle. The aortic valve has a **right, left**, and **posterior (noncoronary) cusp**. The **posterior sinus** is related to the noncoronary cusp, while the right and left aortic sinuses possess an ostium that opens into the **right** and **left coronary arteries**, respectively (*fig. 8.7*). The sinuses of the pulmonary valves are not associated with arteries, and none have ostia associated with them. The cusps in the **pulmonary valve** are termed the **right, left**, and **anterior cusps**. This is easy to remember if you let the right and left coronary arteries define the right and left aortic cusps (*fig. 8.8*). The pulmonary right and left cusps are directly anterior to these. Since the aorta is posterior to the pulmonary trunk, the aorta has a posterior cusp. The third cusp in the anteriorly placed pulmonary valve is therefore called the anterior cusp.

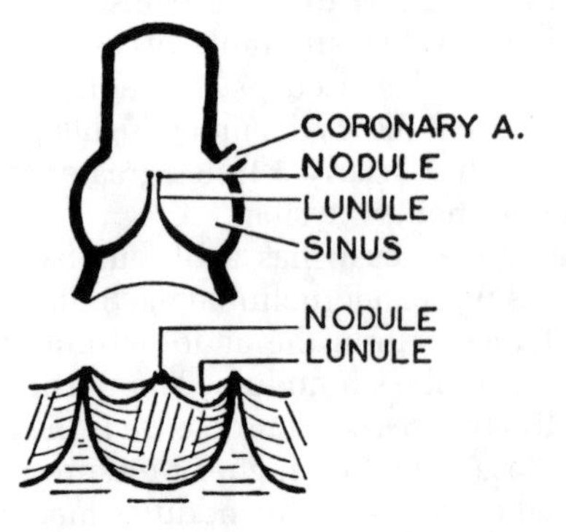

Figure 8.6. The aortic valve. On sagittal section, opened up.

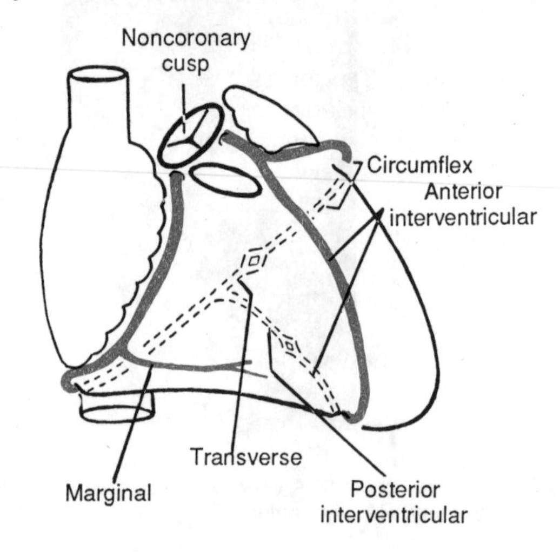

Figure 8.7. The coronary arteries. The left marginal is not shown.

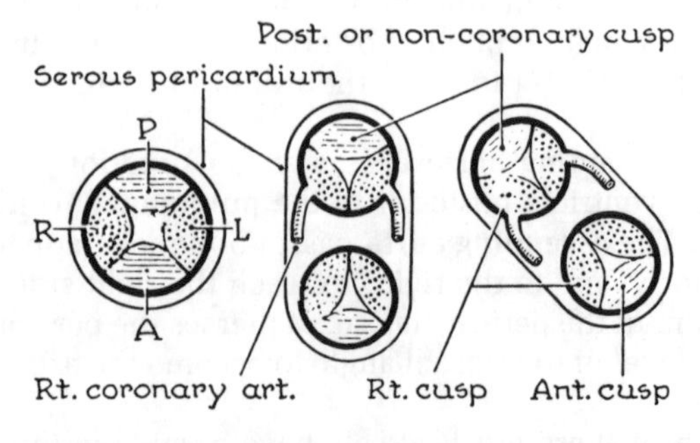

Figure 8.8. The 4-cusped valve of the truncus splits to form 2 valves, each with 3 cusps or valvules. Axial rotation occurs, but it need not affect the nomenclature.

BLOOD SUPPLY TO THE HEART

The term coronary refers to the arterial vessels of the heart, whereas the cardiac vessels are mainly venous structures. The term coronary is derived from the Latin term for heart. It also refers to the crown-like (corona) arrangement of the coronary arteries as they encircle the heart in the atrioventricular sulci.

The **right and left coronary arteries** arise from right and left coronary sinuses of the aortic valve. The **left coronary artery** passes anteriorly between the pulmonary trunk and the tip of the auricle of the left atrium (*fig. 8.7*). It divides into two major branches on the anterior aspect of the heart. The **anterior interventricular branch** descends in the anterior interventricular sulcus to the inferior margin of the heart and continues into the posterior interventricular sulcus on the diaphragmatic surface. This branch supplies the anterior aspects of the right and left ventricles and the anterior half of the interventricular septum. The second branch, the **circumflex branch** runs to the left in the atrioventricular sulcus between the left atrium and left ventricle. It gives off a large marginal branch (not shown in *fig. 8.7*) to serve the lateral margin of the thickened left ventricle and continues to run in the atrioventricular (coronary) sulcus onto the posterior surface of the heart. The circumflex branch of the left coronary artery forms an anastomosis (union) with the arterial vessels in the posterior interventricular sulcus which are usually derived from the right coronary artery.

The **right coronary artery** exits from the right coronary sinus to enter the atrioventricular sulcus, which separates the right atrium and right ventricle. Like its counterpart on the left, it gives off a marginal branch on the acute margin of the heart to supply part of the right ventricle. This right marginal branch is usually much smaller than the left marginal branch as the thickness of the right ventricle is approximately one-fifth of the thickness of the left ventricle. This is due to the difference in diastolic and systolic pressures between the pulmonary (4/25 mm Hg) and systemic (80/120 mm Hg) systems, which the right and left ventricle encounter during contraction. The right coronary artery terminates in the posterior interventricular sulcus as the **posterior interventricular artery**. It supplies mainly the posterior aspect of the right and left ventricles as well as the posterior half of the interventricular septum. The right coronary artery also frequently gives off a ***nodal*** artery in its first few centimeters. The nodal artery passes on the posterior aspect of the right atrium to supply the region of the **sinuatrial node** within the wall of the right atrium.

In general, the typical coronary arterial distribution as described here has the left coronary forming the anterior interventricular branch and the right coronary artery forming the posterior interventricular branch. In about 10% of the cases, the posterior interventricular branch is derived from the left coronary artery. In all cases, the two arteries are richly anastomotic and possess the potential for a rich collateral circulation.

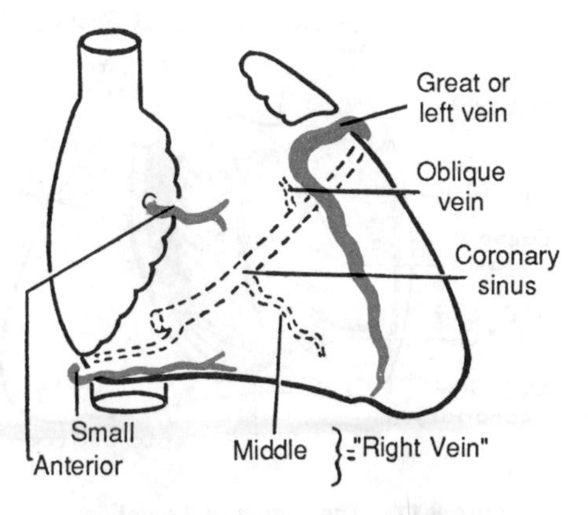

Figure 8.9. The cardiac veins.

The **cardiac veins** (*fig. 8.9*) accompany most of the coronary arteries and their branches. The veins tend to lie superficial to the larger arterial branches in the interventricular and atrioventricular sulci. Most of the larger cardiac veins (described below) drain into the coronary sinus on the posterior surface of the heart.

The **coronary sinus** is derived from the left horn of the primative receiving chamber in the developing heart, the sinus venosus. It is 3 to 4 cm long and lies in the coronary sulcus between the left margin of the heart and the posterior interventricular sulcus. The coronary sinus drains into the right atrium (*figs. 8.10* and *8.11*) through an opening at the left of the orifice for the inferior vena cava. The **great cardiac vein** forms in the anterior interventricular sulcus and joins the coronary sinus near the left margin of the heart. A middle cardiac vein occupies the posterior interventricular sulcus and enters the coronary sinus near its opening into the right atrium. A **small cardiac vein** parallels the course of the right marginal branch of the right coronary artery and also joins the coronary sinus near the site that the middle cardiac vein joins the coronary sinus. An **oblique vein**, draining the left atrium, and occasionally posterior ventricular veins, draining the diaphragmatic surface of the left ventricle, also enter the

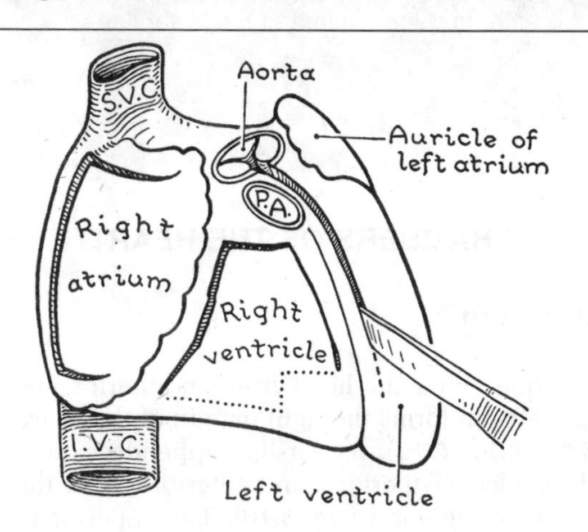

Figure 8.10. Incisions for opening chambers of heart.

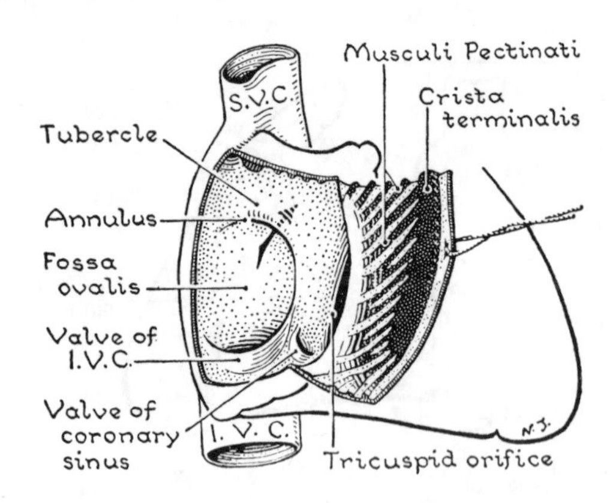

Figure 8.11. The interior of the right atrium.

coronary sinus. The veins draining the anterior surface of the right ventricle cross the atrioventricular sulcus as the **anterior cardiac vein(s)** and open directly into the right atrium. The mouth of the large veins sometimes have incompetent valves, and the coronary sinus opening and the inferior vena cava usually show prominent but nonfunctional valves (*fig. 8.11*). Additional small venous channels in the heart wall open directly into the chambers of the heart. These tiny vessels are called **venae cordis minimae** or **thebesian veins**.

Clinical Case 8.2

Patient Jane S. This 28-year-old mother of two children had rheumatic fever at age 7 and has had a loud heart murmur due to a damaged mitral valve. Having been fairly well until her first baby's birth, she has now developed many of the signs of inadequate heart function and is so severely handicapped that her case is presented at Cardiac Rounds (which you are attending) for a decision about intracardiac surgery. She is thin, listless, and short of breath, and has edema of the lower limbs.

CHAMBERS OF THE HEART

Right Atrium

When one views the heart from an anterior position, the right atrium forms the right margin of the heart (*figs. 8.4, 8.10, 8.11, 8.12*). Its ear-like appendage, the **right auricle**, projects superiorly and anteriorly over the base of the ascending aorta (*fig. 8.10*). The superior margin receives the SVC while the IVC and coronary sinus drain into the right atrium through openings on its posterior surface.

The interior walls of the right atrium show rough and smooth surfaces. The roughened surfaces are formed by raised bundles of cardiac muscle fibers called **musculi pectinati** (L. pecten = comb) (*fig. 8.11*). These musculi pectinati are found on the internal surface of the part of the atrial wall that was derived from the embryonic atrium (the auricle). They are prominent on the anterolateral wall of the right atrium, and they extend across this wall to merge with the **crista terminalis**. The crista terminalis is a prominent ridge of cardiac muscle that extends from the opening of the superior vena cava superiorly to the valve of the IVC near the floor of the right atrium. The crista terminalis underlies a depression on the exterior of the anterolateral wall of the right atrium, the **sulcus terminalis**. The **sinuatrial node** (pacemaker) is located in the floor of the sulcus terminalis near its junction with the superior vena cava. The sinuatrial node cannot be seen grossly but can be detected histologically in sections of the atrial wall.

The remaining interior of right atrial wall is smooth. This portion is derived from the large venous structures that formed the primitive receiving chamber of the developing heart, the **sinus venarum cavarum**. The openings of the SVC and IVC are readily seen in this part of the right atrium. The IVC opening is protected by a valve, which is usually fenestrated and nonfunctional in the adult. Between the valve of the IVC on the posterior aspect of the atrial wall and the tricuspid orifice medially is the opening for the coronary sinus (*fig. 8.11*). It also has a valve associated with it on the atrial wall. The **fossa ovalis** is a depression on the interatrial wall superior to the opening of the IVC. It demarcates the functional opening in the developing prenatal heart that permitted the blood to readily pass from the right atrium to the left atrium and bypass the pulmonary system. Superior to the fossa ovalis is the annulus. This crescent-shaped thickening is the inferior border of a rigid portion of the superior aspect of the interatrial wall. A right to left "flutter valve" is thus established by allowing the moveable inferior interatrial wall to open into the left atrium when blood flow increases the pressure in the right atrium. When the left atrial pressure is raised above that in the right atrium, the valve will close against the rigid superior interatrial wall above the annulus and seal the interatrial opening, the foramen ovale. The *arrow* in *Figures 8.11* and *8.12* illustrate the direction of blood flow from the right atrium to the left atrium in the fetus.

The **right atrioventricular or tricuspid orifice** takes the place of the anterior wall (*fig. 8.11*). It transmits the blood from the right atrium to the right ventricle when the latter is relaxed. It closes during ventricular contraction to prevent reflux of blood into the right atrium and thereby forces the emptying right ventricle to expel its blood into the pulmonary trunk.

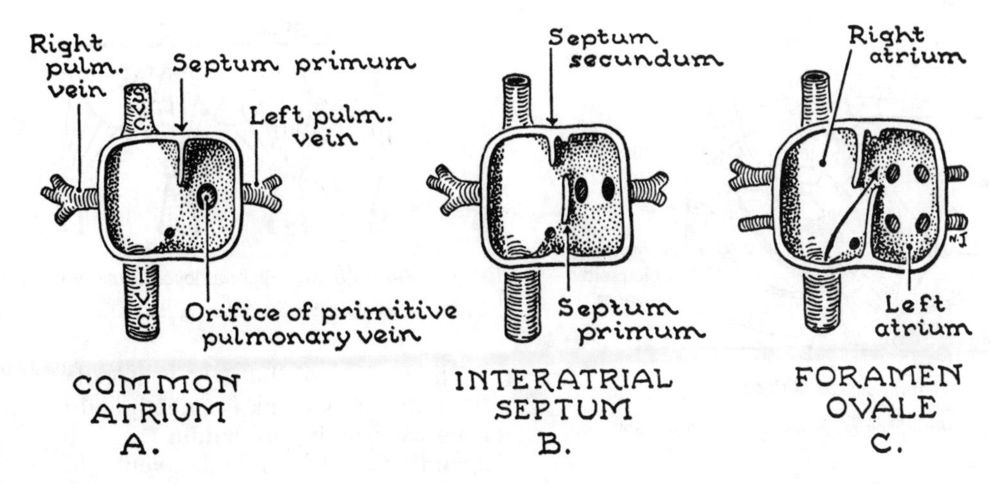

Figure 8.12. Development of the left atrium. Incorporation of the stem of the primitive "common pulmonary vein." The history of the foramen ovale.

Left Atrium

The left atrium forms two-thirds of the base of the heart (*fig. 8.2*), and its auricle is visible from the anterior aspect as it "peeks" around the left side of the pulmonary trunk (*fig. 8.10*). It is demarcated from the left ventricle below by the coronary sulcus. The right and left superior and inferior pulmonary veins open into the left atrium near its right and left margins, respectively (*figs. 8.2, 8.12*). Like the right atrium, the interior of the left atrium is characterized by smooth and roughened walls. The auricle has prominent **musculi pectinati** and is capable of contracting. The rest of the left atrium is derived from the **pulmonary veins** and is a noncontractible, smooth-walled chamber. The **left atrioventricular, bicuspid or mitral orifice** replaces the anterior wall of the left atrium and leads into the left ventricle. The interatrial system will show evidence of the embryonic **foramen ovale**. The superior edge of the septum primum, "flutter valve" is found on the left side of the interatrial septum. *Figure 8.12* graphically depicts the interatrial partitioning that occurs during development and permits the fetal circulation from right to left atria.

Ventricles

The **right** and **left ventricles** lie somewhat anterior to their respective atria (*fig. 8.13*). The **right ventricle** forms most of the anterior sternocostal part of the heart and the inferior margin behind the sternum (*fig. 8.4*). The **left ventricle** forms the left margin of the heart and most of the diaphragmatic base.

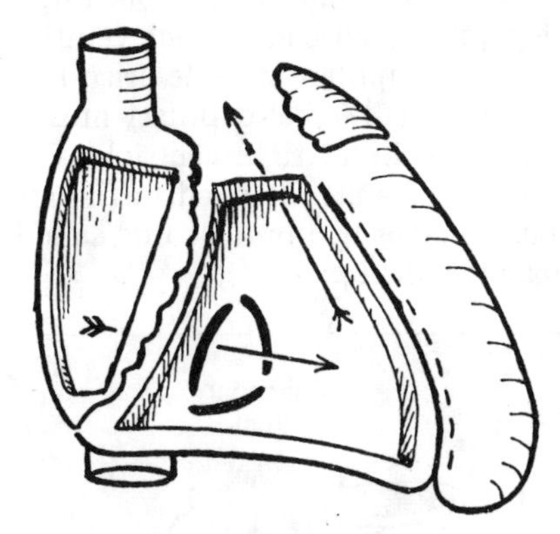

Figure 8.13. Interior of the right ventricle showing relative positions of orifices.

Walls

The thickness of the heart walls is proportional to the amount of work each chamber has to do to pump its blood into the next compartment of the circulatory pathway. While the atria are working in a very low pressure environment (approximately 4 mm Hg), the ventricles are working against the pressure of the pulmonary circulation (4/25 mm Hg) and the systemic circulation (80/120 mm Hg). The walls of the **left ventricle** are therefore approximately 5 times thicker than the **right ventricle** (*fig. 8.14*). Since the interventricular septum is a shared wall for both of the right and left ventricle, its thickness also approximates that of the left ventricular wall. The coronary blood supply is predominantly directed to the ventricles and proportionally distributed to the mass of cardiac muscle in the heart walls.

The ventricular chambers have their atrioventricular orifices on the posterior walls. The exits from the chambers are via the openings into the great vessels on the superior aspect toward the interventricular septum (*fig. 8.14*).

Except near the exits, the ventricular walls are lined with muscular bundles, **trabeculae carneae** (L. *Caro* =

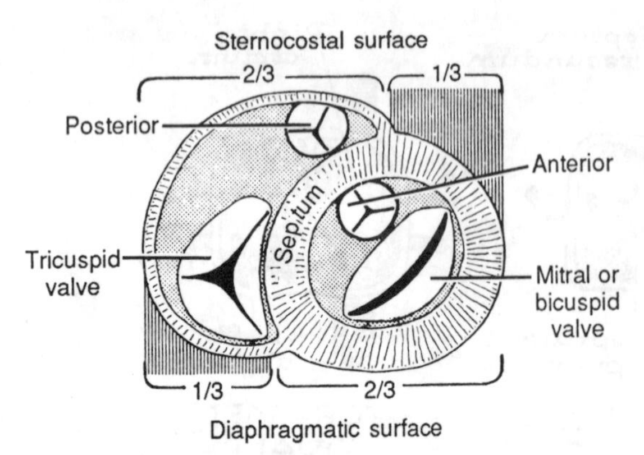

Figure 8.14. Ventricles on cross-section, front view.

flesh). Some of these cardiac muscle bundles are merely elevated ridges (*fig. 8.15*), while others are attached at both ends like bridges and others from finger-like projections, the papillary muscles. In each ventricle, an anterior and posterior papillary muscle arise from the corresponding wall. Small septal papillary muscles are also present particularly in the right ventricle. The apices of the papillary muscles are attached to the cusps of the atrioventricular valves by fibrous cords called the **chordae tendineae** (*fig. 8.16*).

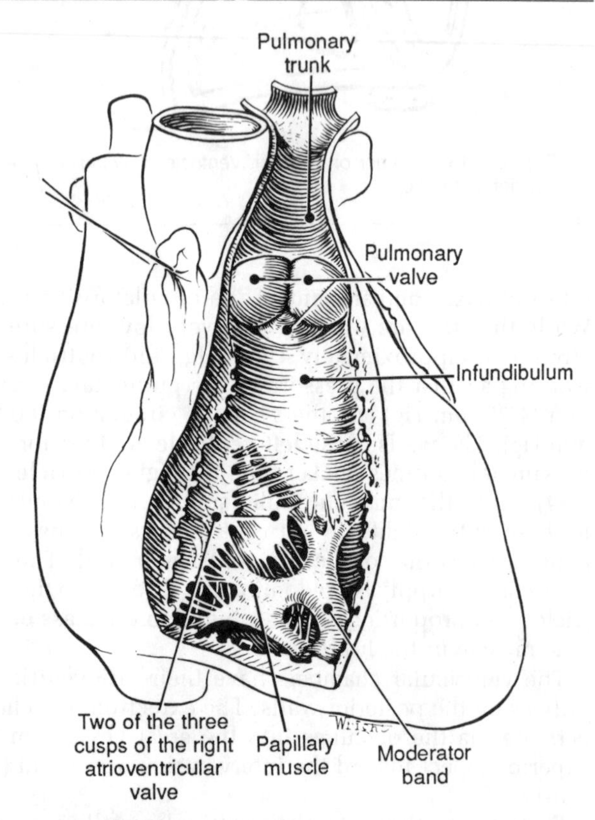

Figure 8.15. Interior of right ventricle and pulmonary trunk. (From Basmajian, J.V.: *Primary Anatomy*, 8th edition, Williams & Wilkins, Baltimore, 1982.)

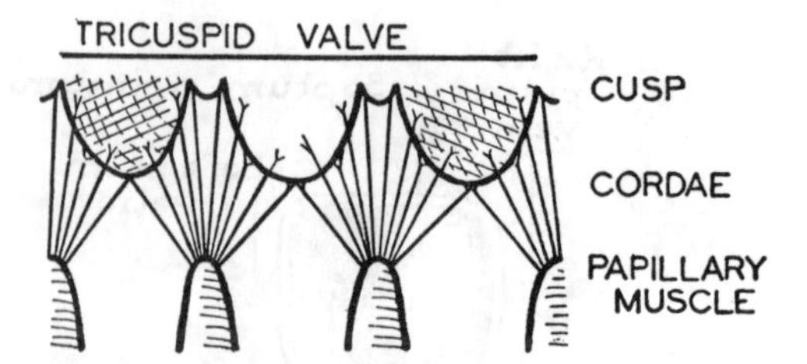

Figure 8.16. Right atrioventricular valve spread out.

The portion of the right ventricular cavity preceding the pulmonary trunk is smooth and is called the **conus arteriosus** or **infundibulum** (L. = funnel). The corresponding portion in the left ventricle also has a smooth wall and is called the **aortic vestibule**. These smooth areas of the ventricle are largely fibrous and noncontractible.

The interventricular septum is fleshy except at its uppermost part where a fibrous, membranous portion of the septum exists. This membrane portion is continuous with the membranous spiral septum that divided the primitive **truncus arteriosus** into a **pulmonary trunk** and an **ascending aorta** (*fig. 8.17*). A developmental defect occasionally occurs when this membranous septum fails to develop properly, and the ensuing interventricular (interseptal) defect permits venous blood from the right ventricle to mix with oxygenated arterial blood in the left ventricle.

The Atrioventricular Valves (A-V Valves)

The right A-V valve is tricuspid and the left is bicuspid. The bicuspid valve is shaped like a bishop's hat (the miter), and the left valve is also referred to as the **mitral valve** (*figs. 8.14 and 8.18*).

The **chordae tendineae** are attached to the edges and ventricular surfaces of the cusps. The chordae tendineae of each papillary muscle controls the contiguous margins of 2 cusps (*fig. 8.16*). These tendinous cords prevent the cusps from everting into the atria during ventricular contraction. The apical areas and the papillary muscles are the first regions of each ventricle to undergo contraction. The A-V valve leaflets are therefore initially drawn into the ventricle by the chordae tendineae on the shortening papillary muscles. The blood, which is being expelled

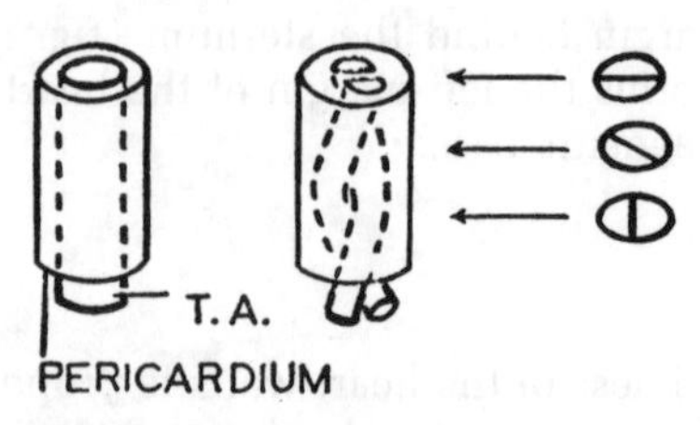

Figure 8.17. The spiral septum within the truncus arteriosus explains the twisted courses of the aorta and pulmonary trunk.

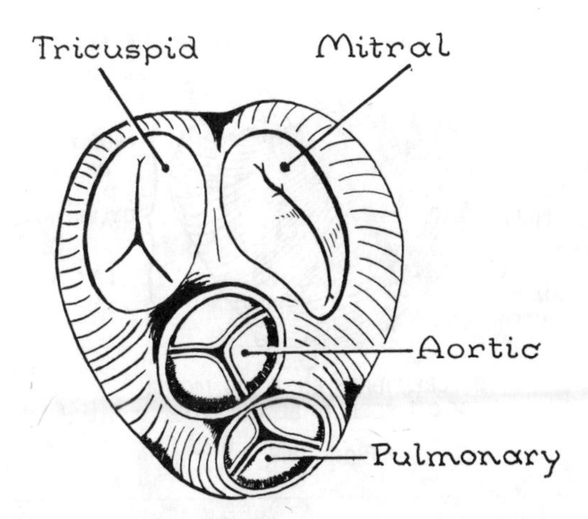

Figure 8.18. The 4 orifices guarded by valves, showing the cusps, also the superficial muscle layer of the ventricle.

superiorly, encounters the ventricular surface of the A-V valve leaflets and forces them toward the atria. As the leaflets resist the expelling blood, they divert the blood flow into the noncontractible area of the ventricles, the infundibulum of the right ventricle and the aortic vestibule of the left ventricle. Further contraction of the ventricular walls forces the blood through the semilunar cusps of the pulmonary artery and ascending aorta.

Cusps (figs. 8.14, 8.18)

The bases of the cusps unite to form a short cuff that is attached to the fibrous A-V orifice. The margins of the cusps are dentate (tooth-like) where the chordae tendineae are attached. The edges and atrial surfaces of the cusps must contact each other when the valve is closed, otherwise the valve will leak and allow the reflux of blood into the atrium. The 2 cusps of the mitral valve (the anterior and posterior cusps) parallel each other as well as the interventricular septum. The **anterior cusp** of the mitral valve diverts blood into the aortic vestibule. It is therefore called the **aortic cusp** by some clinicians. The cusps of the tricuspid valve are named anterior, posterior, and septal to correspond to their walls of origin and their relationship with similarly named papillary muscles.

Surface Anatomy of the Four Cardiac Orifices Guarded by Functioning Valves—Pulmonary, Aortic, Mitral and Tricuspid. These valves lie posterior to the sternum on an oblique line, joining the 3rd left sternocostal joint with the 6th right sternocostal joint (*fig. 8.19B*). The **orifice** of the **pulmonary trunk** is deep to the left 3rd sternocostal joint; the **aortic orifice** being slightly lower, more medial, and more posterior, is behind the sternum at the level of the 3rd intercostal space; the mitral orifice, is still lower and more medial at the level of the 4th costal cartilage; and the tricuspid orifice is on the right of the median plane at the level of the 4th and 5th spaces.

Heart sounds are produced when the chamber or vessel filled with blood causes the valves to close upon contraction of the appropriate ventricle or recoil of the appropriate elastic artery. Since the sound will travel more efficiently through fluid than through a gas, one can listen to the individual valves on the chest wall where the blood-filled chamber or vessel more closely approximates the anterior chest wall. Therefore one listens for the pulmonary valve heart sound at the 2nd intercostal space to the left of the sternum; the aortic valve is best heard in the right 2nd intercostal space next to the sternum; the right ventricle underlies the 5th intercostal space on the right side of the sternum and projects the sound on the tricuspid valve closure to this site when the right ventricle is filled with blood; likewise the blood-filled chamber of the left ventricle transmits the sound of the closing bicuspid (mitral) valve to the apex of the heart in the left 5th intercostal space 10 cms from the middle of the sternum.

Therefore, these important heart valves, which lie in a

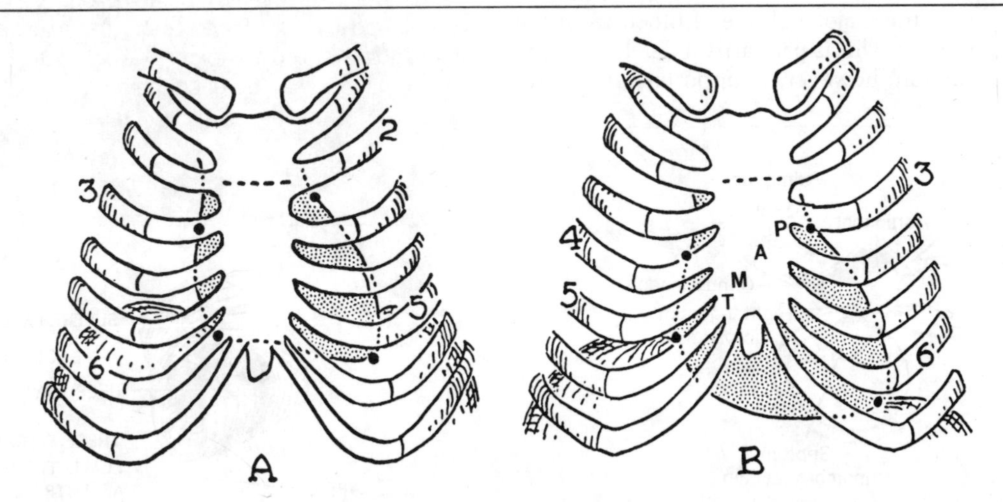

Figure 8.19. The surface anatomy of the heart; *A*, in the supine cadaver; *B*, in the erect living subject. (From Lachman, after Mainland and Gordon.) Modified to show the positions of the 4 valves of the heart. P, pulmonary valve; A, aortic valve; M, mitral valve; T, tricuspid valve.

close anatomical proximity to each other behind the thick sternum, can be separated and distinguished individually in the physical examination with a stethoscope. Incompetency and "murmurs" in defective valves can be detected readily by a physician who understands the relationships of the ventricles and the great vessels to the chest wall in the living patient (*fig. 8.19*).

STRUCTURES OF THE WALLS OF THE HEART

The connective tissue skeleton of the heart is the fibrous tissue to which the cardiac muscle and valves are firmly attached (*fig. 8.20*). The ventricles are emptied during systole by the blood being "wrung" from the cavities like water from a wet cloth. The blood is expelled first from the apices and then from the succeeding levels of the ventricle as it approaches the great vessel orifices. This requires a skeletal framework to allow the cardiac muscle to contract against a rigid base. A fibrous ring of connective tissue surrounds each of the 4 orifices of the heart (*fig. 8.20*) and provides the connective tissue skeleton necessary for the controlled contraction of the heart muscle. The aortic ring is the strongest and is like a cuff. The rings are joined to each other and the membranous part of the interventricular septum. In some animals, e.g., sheep, there is a central bone in this connective tissue of the heart.

The Musculature of the Heart (Myocardium)

Atrial Musculature

The walls of the atria are translucent. The superficial muscle fibers run transversely; the deep fibers arch over the atrium in an anteroposterior fashion and are attached to the skeleton by both ends; other fibers encircle the mouth of the great veins. Atrial contraction of the auricle does little to move the major volume of blood from the atrium to the ventricle. This is primarily aided by gravity. The atrial contraction, however, is important in the sequencing of the heart contraction cycle, and atrial arythmias can cause serious cardiac disorders. Likewise, the inadequate expulsion of blood from the atrium can cause stasis and coagulation of blood in the auricles. This may lead to clots that can be released into the circulation, causing occlusions in smaller blood vessels throughout the circulatory systems.

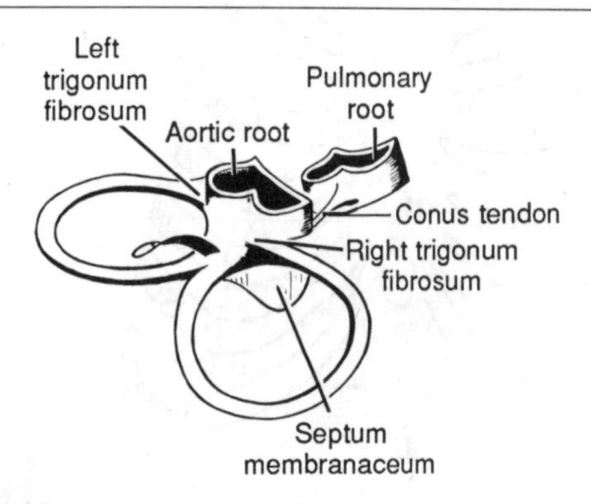

Figure 8.20. The fibrous skeleton of the heart. (From Walmsley, after Ungar).

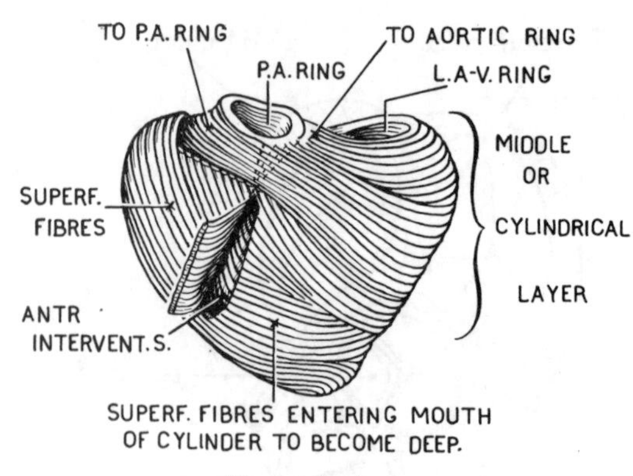

Figure 8.21.

Ventricular Musculature (figs. 8.21, 8.22, and 8.23

The ventricular musculature is composed of 3 layers: (1) superficial, (2) middle, and (3) deep.

All fibers of the ventricles arise from the skeleton of the heart, and eventually they return to be inserted into this connective tissue skeleton (*fig. 8.23*).

Clinical Case 8.3

Patient William J. This 61-year-old uncle of one of your college friends has recently had an artificial pacemaker inserted. As an "expert," you are questioned by your friend about the relevant anatomy and physiology (and general but not technical problems that may be encountered).

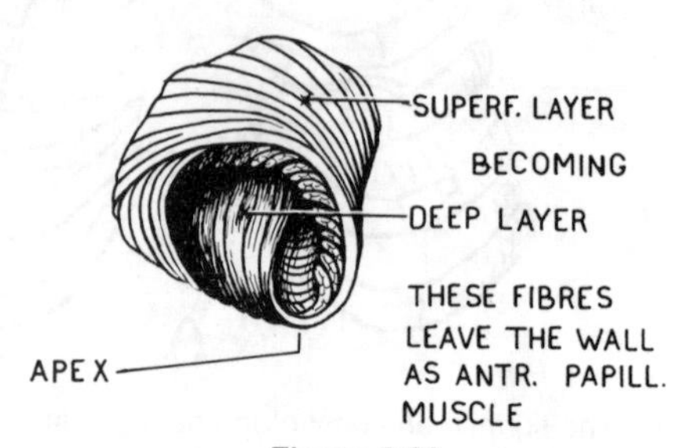

Figure 8.22.

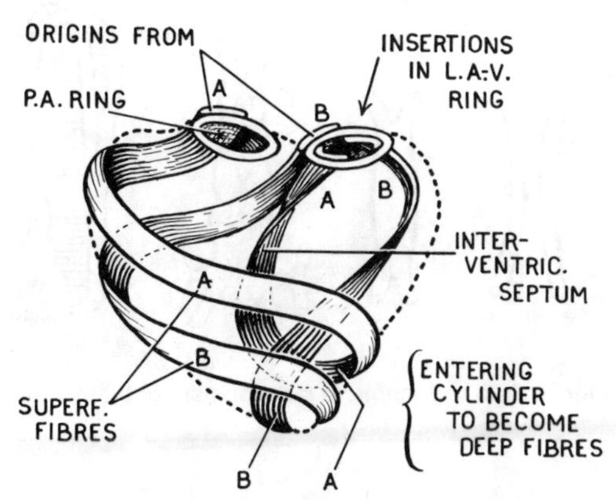

Figure 8.23. Dissections of the ventricles of the heart of the sheep. (By B. Leibel.)

Impulse Conducting System (*fig. 8.24*)

This is comprised of the **sinuatrial node**, **A-V node**, the **A-V bundle** and its **2 crura**, right and left.

1. The **sinuatrial node** (S-A) is the "pacemaker" and initiates the heart beat. It is situated in the atrial myocardium, which underlies the superior aspect of the **sulcus terminalis**. Its peculiar longitudinally striated cells receive their main blood supply from the nodal branch of the **right coronary artery**. The S-A node is innervated from the cardiac plexus by fibers from the parasympathetic vagus and the sympathetic chain ganglion.
2. The **atrioventricular node** (A-V) is composed of similar type cells and is located in the wall of the right atrium anterior to the mouth of the coronary sinus. It has an electrical continuity with the S-A node, but this connection is inapparent on a gross anatomical basis. A block in this system by vascular lesions in the atrial wall can cause serious arrhythmias. While the A-V node is normally a secondary "pacemaker," it may be used when an artificial pacemaker is implanted to establish the initiation of the cardiac cycle.
3. The **atrioventricular bundle (of His)** is a pale bundle of specialized myocardial cells. These specialized muscle fibers are about 2 mm thick and enveloped in a loose sheath beneath the endocardium. This slender bundle is the sole muscular connection between the musculature of the right atrium and the musculature of the ventricles. It extends from the A-V node, through the fibrous skeleton, to the interventricular septum. As it enters the superior aspect of the muscular part of the interventricular septum, it divides into a **right and left crus** (*fig. 8.24*).

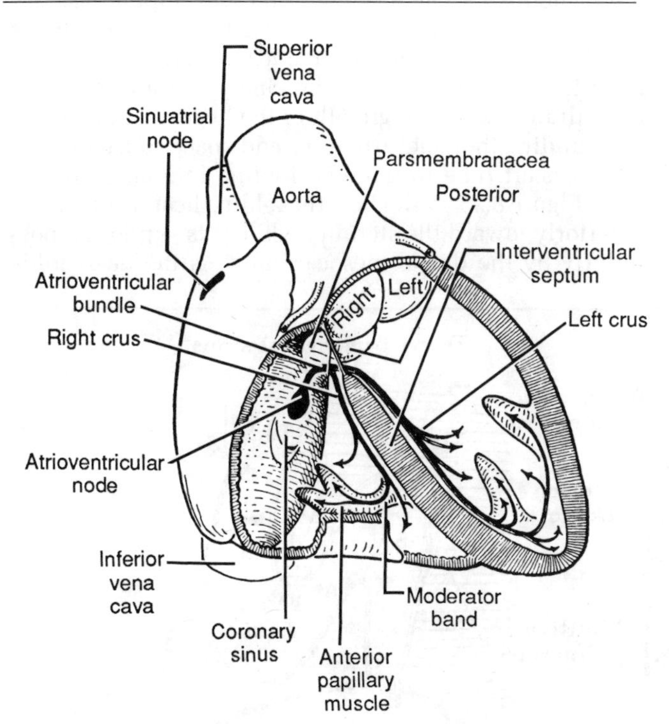

Figure 8.24. The conducting system of the heart.

The crura descend in the interventricular septum to reach the base of the papillary muscles and the apices of the ventricles. The most obvious route of this electrical pathway is visible in the right ventricle where the right crus leaves the septum and traverses the ventricular wall through the **septomarginal trabecula (moderate band)** to the anterior papillary muscle. The conducting fibers (Purkinje's fibers) project subendocardially to innervate the myocardium of the ventricles. They ensure that the wave of contractility in the ventricle begins at the apices of the ventricles and then spreads through the muscle wall to give the "wringing" action necessary to expel all the blood into the great vessels.

NERVE SUPPLY TO THE HEART

The heart is under the dual control of the autonomic nervous system and it receives both sympathetic and parasympathetic fibers through the cardiac plexus. (1) **Sympathetic**, via 1–3 cervical cardiac branches arising at variable levels from the cervical part of the sympathetic trunk; 2–3 cervicothoracic branches arising from the region of the cervicothoracic (stellate) ganglion; and 2–4 thoracic branches from the upper 4 thoracic levels of the sympathetic trunk (*fig. 9.8*). (2) **Vagus**, (Parasympathetic fibers) via a single cervical branch in its cervical course; 1 or 2 cervicothoracic cardiac branches from the main nerve at the inlet to the thorax (or its right recurrent laryngeal branch); and 2–4 thoracic cardiac branches from the thoracic part of the vagus nerve (and its left recurrent branch). Most of the cardiac nerves tend to fuse with each other in their descent to the **cardiac plexus** (Mizeres).

CARDIAC PLEXUS

Distribution

Branches of the **cardiac plexus** course around the right pulmonary artery to the posterior aspect of the atria (see p. 109 for details.) The S-A and A-V nodes are innervated by this plexus. The **coronary plexuses** are also derived from vagal and sympathetic fibers and distributed with the coronary arteries.

The vagal fibers are inhibitory to cardiac rate; the sympathetic are cardiac accelerators. Accompanying the sympathetic fibers on the coronary vessels are a group of visceral afferent (sensory) fibers. They possess chemoreceptors that are sensitive to chemical metabolites that accumulate in ischemic tissue (e.g., lactic acid). These sensory fibers also transmit painful stimuli from heart muscle that is deprived of an adequate blood supply. Since they follow the sympathetic pathways, they enter the spinal cord mainly at the level of the **T1 white rami** communicantes. As a result, referred pain to the T1 dermatome on the left arm is a common symptom of cardiac ischemia (**angina pectoralis**) in patients with heart disease.

Clinical Case 8.4

Development of the Heart

A knowledge of the development of the heart is necessary to understand and explain the basis for the congenital cardiac malformations that are encountered in clinical practice. Since embryology is usually much easier to understand and appreciate after one masters the relevant anatomical vocabulary, it is presented at this time as a review of the most important anatomy of the heart for beginning clinicians.

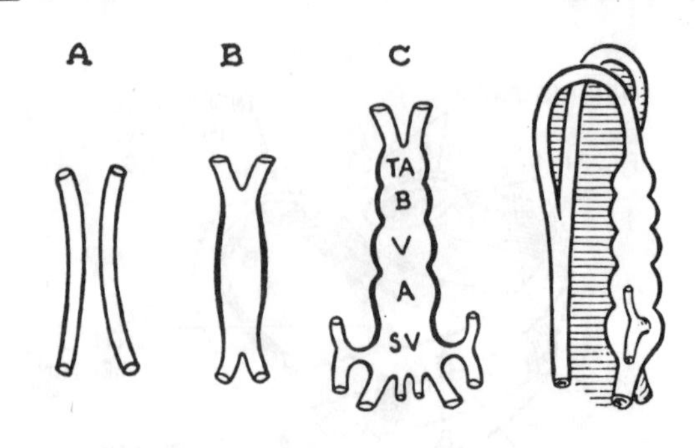

Figure 8.25. Scheme of development of tubular heart.

PRIMITIVE HEART TUBE

During the end of the 3rd week of embryonic development, the primitive heart is formed by the fusion of two endocardial tubes (*fig. 8.25*). The muscular wall of the heart tube undergoes thickening and contraction begins, propelling the blood from the caudal end to the cranial end and establishing a circulatory pathway. The primitive heart tube develops five sacculations as shown in *Figure 8.25C*. The **sinus venosus (SV)** and primitive single chambered **atrium (A)** are in the venous (caudal) region and the single-chambered primitive **ventricle (V)**, **bulbis cordis (B)** and **truncus arteriosus (TA)** constitute the arterial (cranial) region. The constriction between the primitive atrium and ventricle becomes the **atrioventricular** or **coronary sulcus**. The paired vessels that project cranially from the truncus arteriosus and arch posteriorly to form the dorsal aorta are the first aortic arches. They are evident in a lateral view of the developing heart as illustrated in *Figure 8.25D*. This lateral view also shows how the early developing heart and vessels are suspended from the posterior wall of the embryo in a **dorsal mesocardium**. Extensive growth of the heart relative to the surrounding thoracic structures and space cause the primitive heart tube to undergo folding into an S-shaped loop. *Figure 8.26* shows how the folding heart is projected anteriorly toward the sternum while it is supported posteriorly by the dorsal mesocardium. As the heart folds

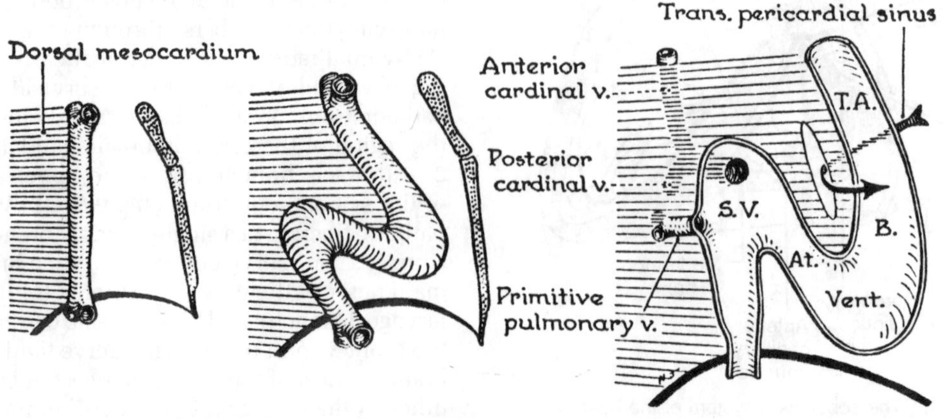

Figure 8.26. The elongated tubular heart becomes "S-shaped."

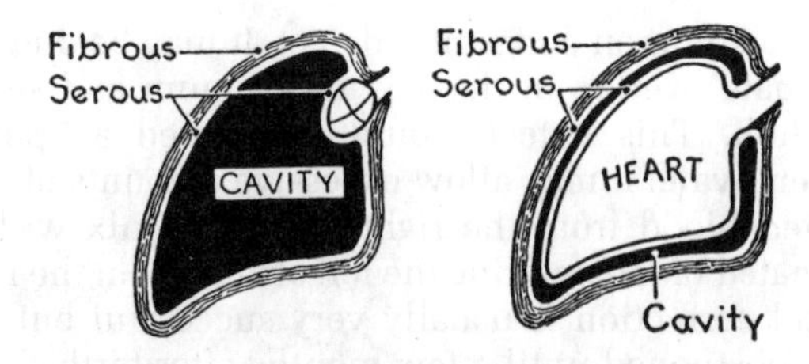

Figure 8.27. To explain the layers of the pericardium.

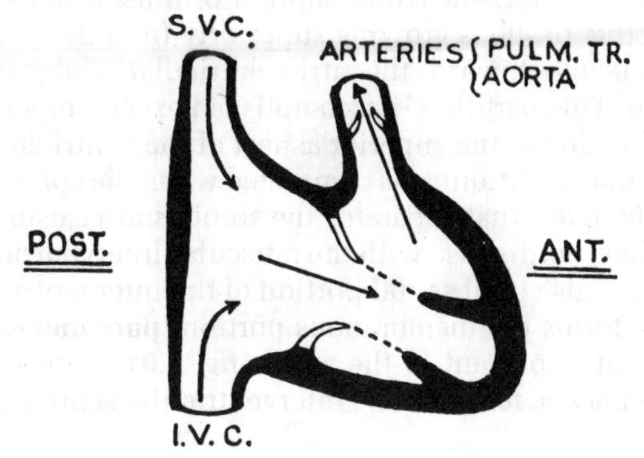

Figure 8.28. Diagram of heart in sagittal section to show that the atria and entering veins are posterior to the ventricles and emerging arteries.

and projects anteriorly, it invades the pericardial cavity and establishes its adult relationships to the pericardium (*fig. 8.27*). The result of the heart folding is that the venous portion of the heart (cardinal veins, pulmonary veins, sinus venosum, and the primitive atrium) are forming the base of the heart, abutting the structures within the dorsal mesocardium (*fig. 8.26*). The primitive ventricle projects anteriorly into the pericardial cavity and the ascending truncus arteriosus exits the pericardial sac superiorly. This is similar to the adult situation depicted in *Figure 8.28*.

DISAPPEARANCE OF THE DORSAL MESOCARDIUM

The breakdown of the dorsal mesocardium between the arterial end (bulbus cordis) and the venous end (primitive atrium) of the developing heart within the pericardial sac results in the formation of the **transverse pericardial sinus** (*fig. 8.26*). This sinus is demonstrable in the adult heart between the great arteries (ascending aorta and pulmonary trunk) and the great veins (pulmonary veins and SVC). In fact, the entire dorsal mesocardium disappears in the adult and probably contributes to the mesodermal connective tissue of the posterior mediastinum.

FORMATION OF A FOUR-CHAMBERED HEART FROM THE PRIMITIVE ATRIUM AND VENTRICLE

During the 4th week of embryological development, the partitioning of the primitive heart into 4 chambers begins. As already seen, the constriction between the primitive atrium and ventricle begins to define the heart tube into two distinct chambers. Each of these chambers then undergoes a sagittal partitioning to form right and left chambers.

The ingrowths of the heart wall that begin to separate the primitive atrium and ventricle are called **atrioventricular endocardial cushions**. Their fusion at the 5th week of development results in a partitioned 2-chamber heart but with two **atrioventricular canals** communicating between the atrium and the ventricle. The atrium is further divided by a vertically growing **septum primum** that arises from the superior aspect of the atrium (*fig. 8.29A*). This septum primum grows inferiorly toward the atrioventricular endocardial cushions that are separating the atrioventricular canals. Prior to fusing with the endocardial cushion inferiorly, the superior portion of the septum primum degenerates to form an opening in the superior aspect of this interatrial partition (*fig. 8.29B*).

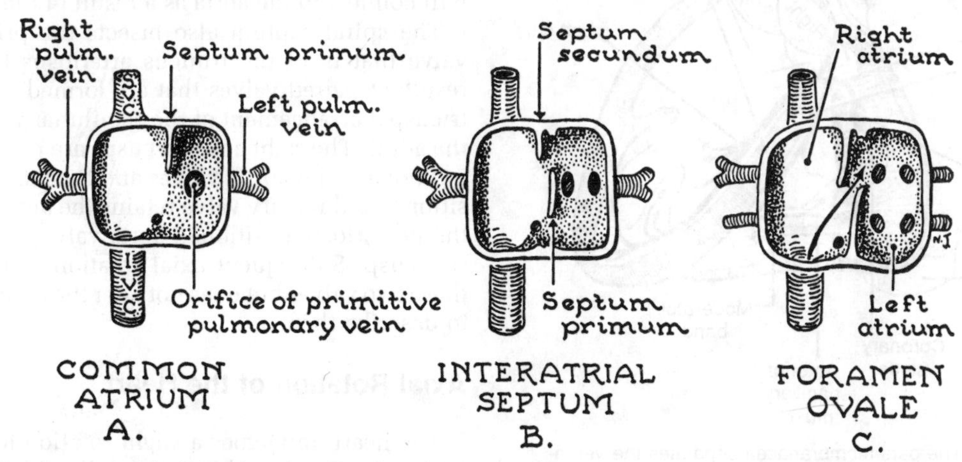

Figures 8.29. Development of the left atrium. Incorporation of the stem of the primitive "common pulmonary vein." The history of the foramen ovale.

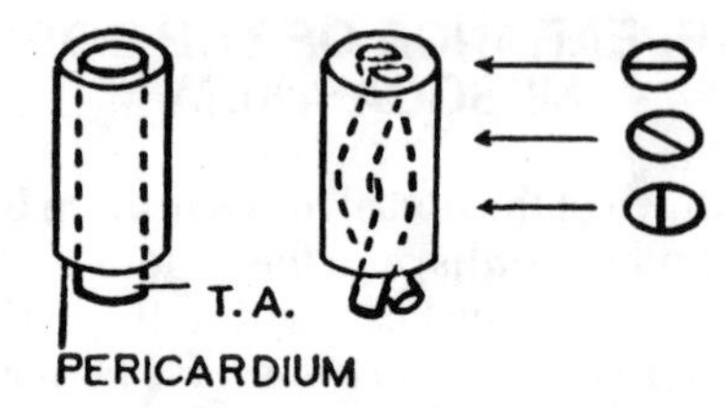

Figure 8.30. The spiral septum within the truncus arteriosus explains the twisted courses of the aorta and pulmonary trunk.

The preservation of an opening in the interatrial septum is required to maintain the embryonic circulation. If the atrium was completely separated by the interatrial septum, the embryo would not survive. As the septum primum finally fuses inferiorly with the endocardial cushions, a second, thicker and more rigid **septum secundum** descends from the roof of the atrium on the right side. *Figure 8.29C* shows how the final partitioning of the atrium is established by the septum primum inferiorly and the septum secundum superiorly. The intervening opening is the foramen ovale. The foramen ovale is opened as blood in the right atrium forces the flexible septum primum into the left atrium. Closure of the foramen ovale occurs when the septum is forced to the right and opposes the inferior left side of the septum secundum as the pressure in the left atrium becomes greater than the pressure in the right atrium. The septum primum will thus be seen as the **fossa ovalis** in the adult and the inferior lip of the septum secundum will be visible as the **annulus** in the right atrium. An opening in the interatrial septum after birth is a common congenital defect. It may be due to an inadequate overlap of the septum primum and septum secundum. This defect would be termed a **"patent" foramen ovale**. It may allow excessive amounts of deoxygenated blood from the right atrium to mix with the oxygenated blood entering the left atrium from the lungs. Surgical correction is usually very successful but commonly postponed until a few months after birth.

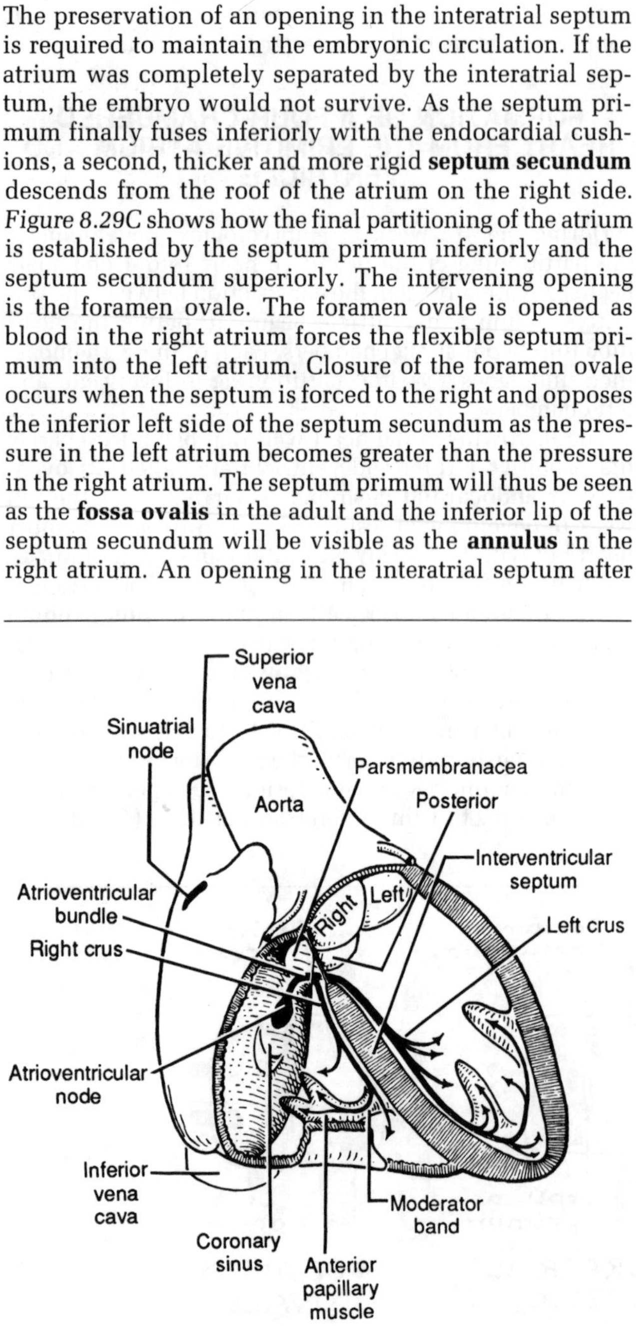

At the same time that the primitive atrium is being partitioned, the primitive ventricle is also being partitioned. An interventricular septum of muscle cells and connective tissue is growing superiorly from the floor of the ventricle toward the atrioventricular endocardial cushion. This partition is accomplished except for a small central region at the superior aspect of the ventricle. The ventricular partitioning is completed when the spiral septum (*fig. 8.30*) that separates the truncus arteriosus and the bulbus cordis fuse with the muscular interventricular septum. This spiral septal portion of the interventricular septum forms the membranous portion (**pars membranacea**) that is present in the adult (*fig. 8.31*). Congenital defects may be found in the interventricular septum also.

Figure 8.31. The pars membranacea separates the ventricles superior to the muscular interventricular septum.

FORMATION OF A FOUR-CHAMBERED HEART FROM THE PRIMITIVE ATRIUM AND VENTRICLE FORMATION OF THE ASCENDING AORTA AND PULMONARY TRUNK FROM THE TRUNCUS ARTERIOSUS

During the 5th week of development, a spiral septum forms in the truncus arteriosus and descends in a clockwise fashion to separate this tubular structure into the ascending aorta and the pulmonary trunk. *Figure 8.30* depicts this septal formation and shows how the formed pulmonary trunk spirals around the anterior base of the aorta to its posterior position where it divides to form the pulmonary arteries. The inferior attachment of this spiral septum will be onto the interventricular septum that is developing to separate the primitive ventricle into a right and left ventricle. The right ventricle will thus be emptied by the pulmonary trunk and the left ventricle will connect to the aorta as a result of the spiral septum.

The spiral septum also bisects the primitive 4 cusp valve that is in the truncus arteriosus (*fig. 8.32*). The resultant paired valves that are formed demonstrate the tricuspal arrangement of the semilunar valves present in the adult. The right and left cusps are present in both the pulmonary and aortic valves and the more anteriorly positioned pulmonary valve retains the anterior cusp while the posteriorly positioned aorta valve retains the posterior cusp. Subsequent axial rotation of the heart shifts these two valves but does not alter the nomenclature used to describe them.

Axial Rotation of the Heart

The heart undergoes a slight rotation to the left on its long axis. This results in the following adult relation-

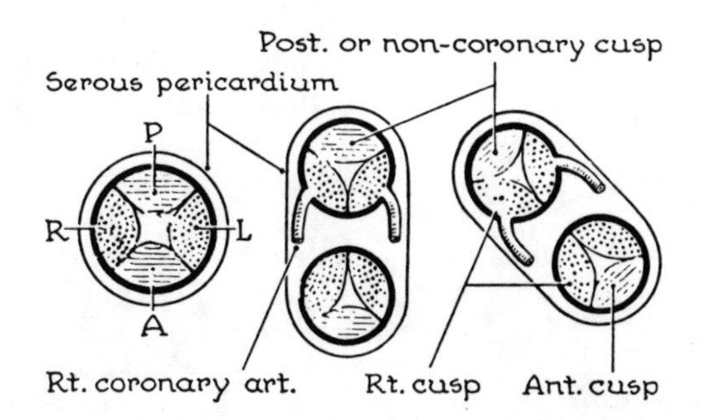

Figure 8.32. The 4-cusped valve of the truncus splits to form 2 valves, each with 3 cusps or valvules. Axial rotation occurs, but it need not affect the nomenclature.

ships: (1) the right atrium is conspicuous at the right margin of the heart and anteriorly, while the left atrium is most conspicuous at the posterior (base) aspect of the heart; (2) the right ventricle is largely anterior and inferior in the pericardial sac, and the left ventricle projects to the left and inferior onto the diaphragm; (3) the interatrial and interventricular septa come to face anteriorly and towards the right (and posteriorly toward the left); and (4) this rotation also shifts the orientation of the semilunar valves of the great vessels to parallel the interatrial and interventricular septa (*fig. 8.32*). Septal defects in newborns can therefore affect the formation, size, and position of the valves of the heart and create constricted (stenosis) or enlarged and leaky valves ("murmurs").

Clinical Mini-Problems

1. **Where would a physician place a stethoscope to listen to the competency of the aortic valve in a patient?**
2. **Where would one listen with the stethoscope to evaluate the competency of the pulmonary valve?**
3. **Cardiac ischemia (lack of blood) frequently induces a referred painful stimulus to the area of the left arm. Why does the patient sense this discomfort over the T1 dermatome?**
4. **Which two embryological structures could have been defective in a neonate that has a patent foramen ovale?**

(Answers to these questions can be found on p. 585.)

9

Superior and Posterior Mediastina

Clinical Case 9.1

Patient Maud W. This 56-year-old woman gives a history of "throat pain," difficulty in swallowing (dysphagia) over a 3-month period, along with some ill-defined changes in the quality of her voice. She is rushed to emergency department of the University Medical Center this evening after dinner, during which she suffered a severe coughing reaction and had blood in her expectorated sputum. You examine her neck and find some tenderness in the trachea at the level of the superior thoracic aperture and enlargement and tenderness in the supraclavicular lymph nodes. The physician-in-charge carefully has the patient drink some water, and coughing begins immediately following her first swallow. Radiographs are ordered and examined immediately. The anteroposterior views showed the radiolucent trachea deviating to the left side of the superior mediastinum. The tracheal bifurcation is also displaced somewhat superior to the level of the sternal angle. The superior aspect of the mediastinum (with all its contents of great vessels and nerves) is bulging to the right against the superior lobe of the right lung. The patient is premedicated with phenobarbital and given a barium "cocktail" immediately before taking the next set of chest films. The films demonstrate that there is a constriction of the esophagus at the level of the T2 vertebra and that some barium is also entering the trachea at this level. An esophageo-tracheal fistula is confirmed and the patient is scheduled for surgery. (This problem will continue as Clinical Case 9.2.)

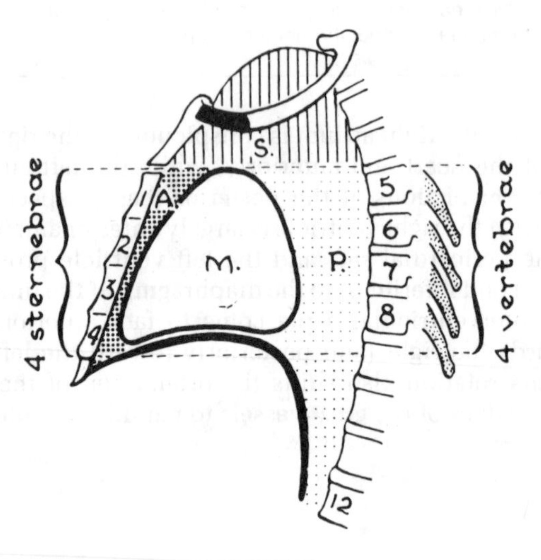

Figure 9.1. Four subdivisions of mediastinum.

The **mediastinum** is a septum that divides the thoracic cavity and contains all the thoracic viscera except the lungs. It extends from the posterior aspect of the sternum to the bodies of the thoracic vertebrae and from the thoracic inlet in the neck superiorly to the diaphragm inferiorly. It is subdivided into a **superior** and **inferior mediastinum** by an imaginary horizontal plane through the sternal angle and the intervertebral disc of T4/5 (*fig. 9.1*). The **inferior mediastinum** is then subdivided into an **anterior, middle**, and **posterior mediastinum**. The **middle mediastinum** is defined by the fibrous pericardium and contains the heart and great vessels. The **anterior mediastinum** may contain some fat and thymic tissue that is projecting below the limits of the superior mediastinum. The **posterior mediastinum** contains the major thoracic viscera which traverses the thorax from the neck to the abdomen.

Contents of the Superior Mediastinum

1. Retrosternal Structures (*fig. 9.2*)
   a. Thymus
   b. Great Veins

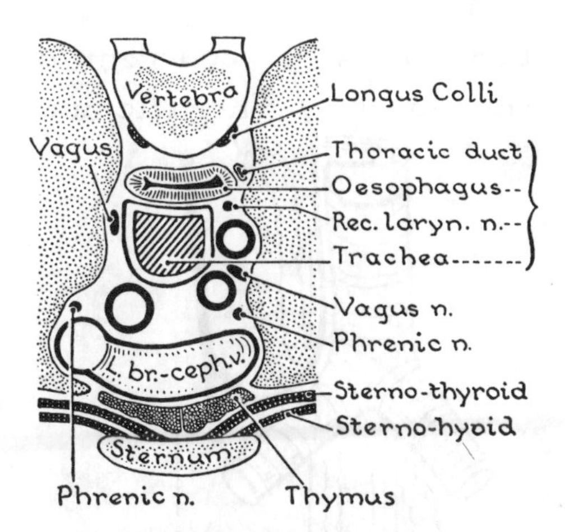

Figure 9.2. Cross-section of the superior mediastinum, showing the arrangement of the contents.

2. Prevertebral Structures
   a. Trachea
   b. Esophagus
   c. Left recurrent laryngeal nerve
   d. Thoracic duct
3. Intermediate structures
   a. Aortic arch and its 3 great branches
   b. Vagus nerves
   c. Phrenic nerves

RETROSTERNAL STRUCTURES

The **thymus** in the adult is an elongated, encapsulated, fatty, lymphoid mass, lying in the loose tissue posterior to the **manubrium sterni** (*fig.* 9.2). On each side of the thymus are the diverging borders of the lungs and pleura that constitute the costomediastinal recess. Posterior to the thymus is the left brachiocephalic vein and the aortic arch.

The thymus consists of 2 loosely joined, asymmetrical lobes. At birth, its superior extent reaches as high as the thyroid gland. It may even contain some inferior parathyroid tissue since the thymus and inferior parathyroid glands are both derived from the tissue of the 3rd pharyngeal pouch of the embryo. The inferior margin of the thymus may extend through the superior mediastinum into the anterior mediastinum and over the superior part of the pericardium (*fig.* 9.3).

The thymus is relatively largest and most extensive at birth. It attains its greatest size and weight at puberty and then begins to involute and gradually diminishes in size. It is extremely sensitive to steroids and may be depleted of its lymphoid content within 1 day after treatment with therapeutic doses of corticosteroids. The thymus may be prominent in adults who have not experienced debilitating or stressful diseases. Such illnesses would elevate the endogenous corticosteroids in affected patients. Lymphoid regeneration of a depleted thymus gland can take up to a month following stress or steroid therapy.

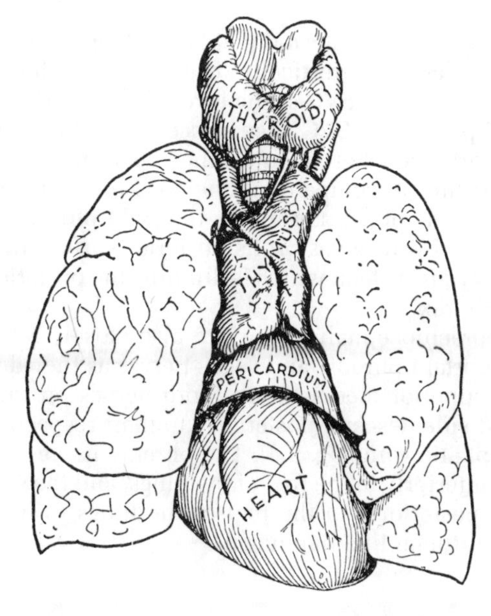

Figure 9.3. The thymus gland of a child.

Great Veins (*fig. 9.4*)

The **left brachiocephalic vein** is formed by the confluence of the **left jugular** and **left subclavian veins** posterior to the left sternoclavicular joint. It descends obliquely posterior to the superior half of the manubrium to unite with the **right brachiocephalic vein**. Halfway down the right margin of the manubrium this union of brachiocephalic veins forms the superior vena cava (SVC) (*fig.* 9.4). The left brachiocephalic vein passes anterior and superior to the 3 great branches of the aortic arch (*fig.* 9.4). In children, it may extend above the superior border of the manubrium and pose a surgical danger in neck operations.

The **right brachiocephalic vein** is formed in a similar fashion posterior to the right sternoclavicular joint by the

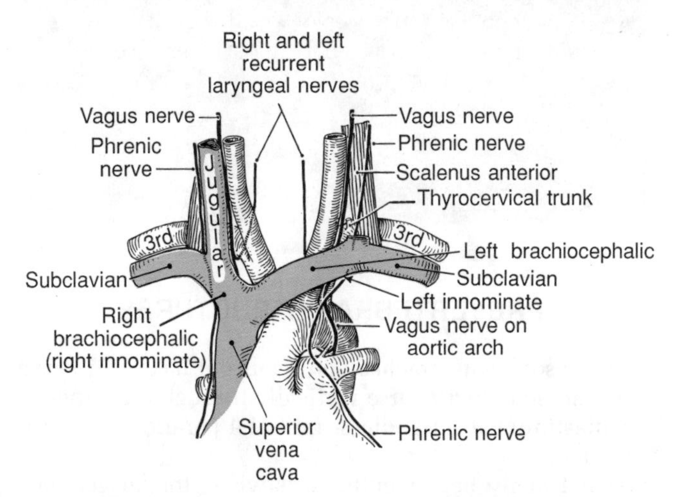

Figure 9.4. The courses of the phrenic and vagus nerves.

union of the **right internal jugular and right subclavian veins**. It descends vertically to join the left brachiocephalic vein as the SVC at the level of the 1st right intercostal space. The SVC then descends to the 3rd right costal cartilage where it enters the superior aspect of the right atrium. Accompanying the SVC in its descent on the right side of the sternum is the **right phrenic nerve**. The SVC also receives on its posterior surface the **arch of the azygos system**, which is draining the posterior wall of the thorax.

The brachiocephalic veins are also closely related to the important phrenic and vagus nerves as they descend in the superior mediastinum. Both nerves on the right and left side descend posteriorly to their respective brachiocephalic vein (*fig. 9.4*). The phrenic nerves will descend anteriorly to the root of the lungs, and the vagi will course posteriorly to the root of the lungs to leave the thorax through the diaphragm.

Clinical Case 9.2

Patient Maud W. (Cont'd from Clinical Problem 9.1 in which a serious condition was revealed involving the esophagus and trachea.) Now a thoracotomy is done to expose the structures in the superior mediastinum. A large mass found on the esophagus is displacing the trachea anteriorly and bulging onto the mediastinal surface of the superior lobe of the right lung. The mass has also fused with the posterior surface of the trachea. The mass is resected and the defects in the esophagus and trachea are closed. Biopsy of the resected material shows that it is an esophageal malignant tumor. Biopsies of the mediastinal lymph nodes also indicate that the tumor has spread into the surrounding lymphatic tissue. *Follow-up*: The patient is given postsurgical radiation therapy and followed closely. One year later, a chest radiograph reveals a mass in the right lung, biopsies of which show it to be a tumor derived from the esophageal tumor resected earlier. Lymphatic drainage is a vital topic in the mediastinum.

PREVERTEBRAL STRUCTURES

The esophagus, trachea, left recurrent laryngeal nerve and thoracic duct course vertically through the superior mediastinum as a bundle or unit of 4 parallel structures (*fig. 9.5*).

The **trachea** begins in the neck where the larynx ends at the level of the 6th cervical vertebra. Half of its 10 cm length is in the neck and the other half is in the superior mediastinum. It bifurcates into 2 main bronchi at the inferior limit of the superior mediastinum, i.e., the level of the sternal angle and the T4/5 intervertebral disc. The lumen of the trachea is maintained in a patent (opened) relationship by U-shaped cartilaginous rings. The **cricoid cartilage** forms a complete ring of cartilage at the superior end of the trachea, and the most inferior tracheal ring has a keel-like extension (**the carina**) that supports the "crotch" of the trachea. The deficiency of cartilage on the posterior aspect of the tracheal rings below the cricoid cartilage is replaced by distensible soft tissue. This allows the esophagus to expand against the trachea's posterior surface during swallowing.

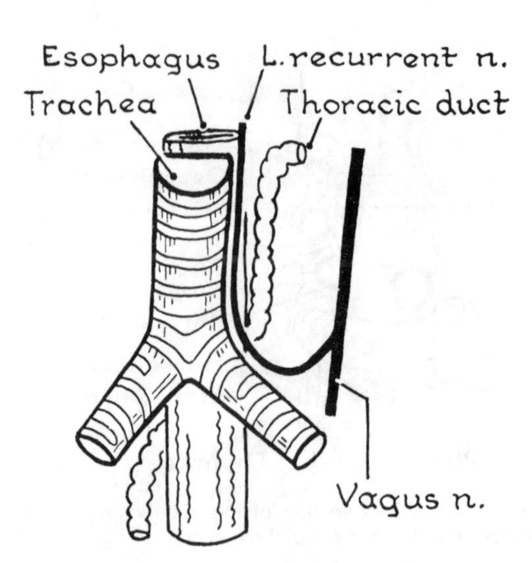

Figure 9.5. A unit of 4 parallel structures runs through the superior mediastinum.

The trachea occupies the median plane except at the inferior end, where the aortic arch deflects it to the right. This is the key structure to visualize in an anteroposterior (A-P) radiograph of the chest. Since it is filled with air, it appears radiolucent and can easily be seen from the neck to its bifurcation.

Figure 9.6 displays some of the important structures that are related to the trachea and interposed between it and the mediastinal parietal pleura. The **brachiocephalic trunk** (artery), **the right vagus**, and the **azygos arch** (venous) are related to the right side. The left common carotid and left subclavian arteries branching off of the arch of the aorta, the left vagus, and its left recurrent laryngeal branch are interposed between the left side of the trachea and the mediastinal pleura. The esophagus lies posterior to the trachea.

Extrapulmonary Bronchi: Right and Left (*fig. 9.7*)

Of these the right bronchus is much larger and more vertically oriented. Hence, foreign bodies are more commonly aspirated into the right lung than into the left lung.

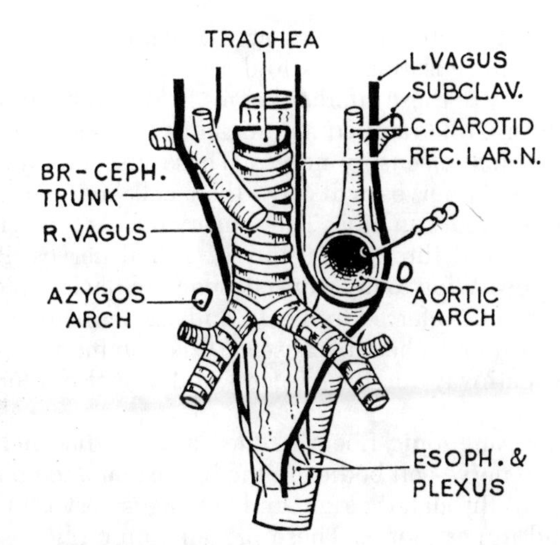

Figure 9.6. Trachea and extrapulmonary bronchi, and their lateral relations.

One must consider the position of patients at the time of aspiration. If lying on their left side in bed, the foreign object would likely go to the left lung due to gravity. If sitting, standing, or lying in the right side, gravity would favor localization of the aspirated material in the bronchial tree of the right lung.

Tracheobronchial Lymph Nodes (*fig. 9.7*)

The lymphatic drainage of the lungs is extensive, and lymph nodes are scattered throughout the connective tissue of the mediastinum. A consistent collection of lymph nodes is found in the 3 bifurcation angles of the trachea as shown in *Figure 9.7A*. These lymph nodes would drain into the right lymphatic duct and the thoracic duct to reach the venous system.

Sympathetic and Parasympathetic Nerves in the Mediastinum

The **cardiac plexus** is a plexus of sympathetic and vagal fibers situated anterior to the bifurcation of the trachea, inferior to the bifurcation of the pulmonary trunk, and therefore below the arch of the aorta (*fig. 9.8*). Subdivisions of the cardiac plexus are artificial (Mitchell, Mizeres) and should be disregarded, but extensions of it are the right and left pulmonary plexus and the plexus of the thoracic aorta.

The **right** and **left pulmonary plexuses** are formed by cervical, cervicothoracic, and thoracic cardiac branches of both sides. At the hilus of each lung, the pulmonary plexuses receive (both anteriorly and posteriorly) branches of the **vagus nerve** and **sympathetic trunk**.

Cardiac Nerves

On each side, the plexus receives slender cardiac nerves from the **sympathetic trunk** and the **vagus nerve**. The sympathetic fibers originate in large part as cardiac nerves from the cervical ganglia in the neck. Some vagal branches may arise in the neck, but most are branches as the vagi course from the inlet of the thorax to the root of the lung. The heart develops from cardiogenic cells that are in the head process just posterior to the mesoderm destined in part to become the diaphragmatic musculature (septum transversum). The innervation of these mesodermal derivatives of the heart and diaphragm is from nerves in the cervical region. Subsequent development and head folding in the embryo bring the heart and diaphragm into

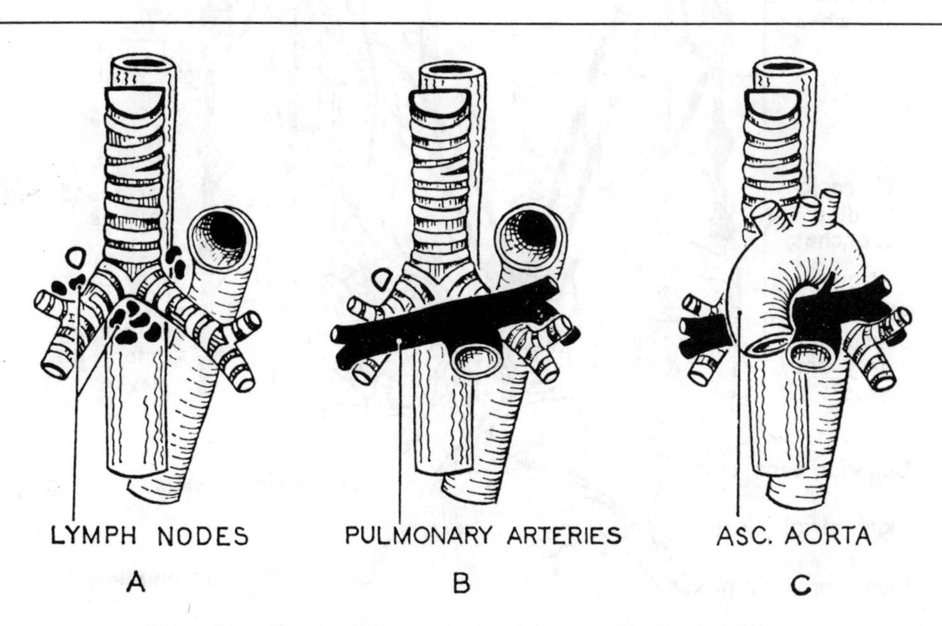

Figure 9.7. Relations at the bifurcation of the trachea, shown serially.

the thorax, and their cervical innervation is carried into the thorax as well.

Functions of the Cardiac Plexus

The preganglionic sympathetic nerves that contribute to the function of the cardiac plexus originate in cell bodies in the T1–T4 region of the intermediolateral column of the gray matter of the spinal cord. They enter the paravertebral sympathetic chain ganglion through the white rami communicantes of T1–T4. These preganglionic fibers may synapse in the upper thoracic ganglia or ascend in the sympathetic trunk to the cervical ganglia and synapse there. The postganglionic cardiac branches from the cervical and upper thoracic ganglia produce acceleration of the heart's contraction rate (tachycardia) and vasoconstriction of the coronary arteries. An entirely separate set of visceral afferent fibers parallel the course of the sympathetic fibers to carry visceral reflexes and visceral pain sensation back to the spinal cord. The typical pathway of these afferent nerves is to course through the highest white rami communicantes (T1 and T2) and join the somatic sensory nerves. These visceral and somatic sensory nerves will have their cell bodies in the same dorsal root ganglion. The sensory nerves then enter the spinal cord through T1 and T2 spinal nerves. Pain from the heart due to myocardial ischemia (lack of blood supply) can be referred to the T1 and T2 dermatomes of the left arm and chest because of this common central sensory pathway for the visceral and somatic afferent nerves.

The preganglionic fibers of the vagal cardiac nerves arise from nerve cell bodies in the brainstem and course to the heart through the right and left vagus nerves in the neck and upper thorax. These preganglionic fibers synapse with parasympathetic postganglionic cell bodies in the ganglia of the cardiac plexus. These synaptic ganglia

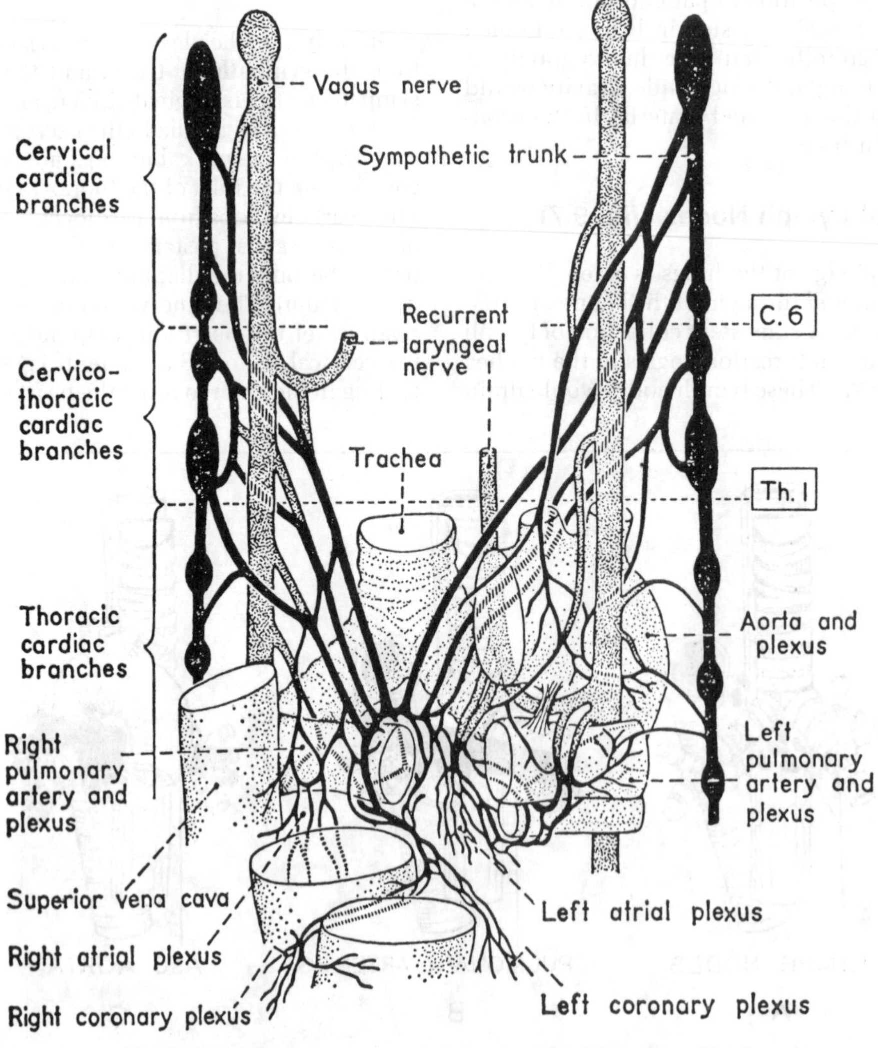

Figure 9.8. Cardiac nerves and plexuses. (After Mizeres, modified.)

are practically confined to the atria and interatrial septum area near the roots of the great vessels. Afferent vagal fibers from the heart, ascending aorta, and great veins are concerned with chemoreceptors and baroreceptors. These visceral vagal fibers have their cell bodies in the inferior ganglion of the vagus nerve. Activation of the vagal efferent system depresses the activity of the heart and produces bradycardia (slowing heart beat).

INTERMEDIATE STRUCTURES

Aortic Arch and Its Three Branches

The inferior aspect of the aortic arch lies on the transthoracic plane, which runs through the sternal angle and the T4/5 intervertebral disc (*fig. 9.9*). It runs from the anterior aspect of the thorax beginning with the ascending aorta posterior to the right margin of the sternum. It arches over the right pulmonary artery and left main bronchus as it courses posteriorly and to the left to become the descending aorta on the left side of the T5 vertebral body (*figs. 9.7C* and *9.9*). In its posterior course, it lies anterior to the terminal portion of the trachea, the left recurrent laryngeal nerve, the superior aspect of the esophagus, and the thoracic duct (*fig. 9.10*). It is crossed anteriorly by the left vagus and left phrenic nerves as well as the left superior intercostal vein.

Aortic branches are all on the superior aspect of the aorta in the adult, but a vitally important fetal branch existed on the inferior surface to shunt the blood from the pulmonary system (**ductus arteriosus**) in prenatal life (*fig. 9.11*). This shunt closes with expansion of the lungs and a decreasing resistance in the pulmonary arterial system after respiration commences postnatally. A connective tissue band connecting the pulmonary trunk of the left pulmonary artery to the inferior arch of the aorta (*figs. 9.7C* and *9.11*) is all that remains of this fetal shunt. This **ligamentum arteriosum** is intimately associated with the course of the left recurrent laryngeal nerve (*fig. 9.12*). The asymmetry of the course of the right and left recurrent laryngeal nerves in the thorax is due to the presence of this fetal shunt on the ***left side only***.

The three branches that arise from the superior aspect of the aortic arch are the brachiocephalic trunk, the left common carotid, and the left subclavian arteries. All of these vessels are crossed anteriorly by the left brachiocephalic vein (*fig. 9.12*). These vessels and the preceding two coronary arteries from the aortic sinuses of the ascending aorta transport a large proportion of the cardiac output to the heart, upper extremities, and head and neck.

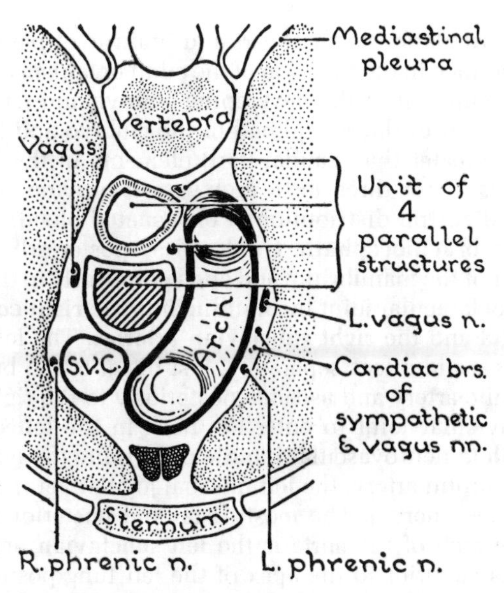

Figure 9.10. Cross-section of superior mediastinum showing the relations of the aortic arch.

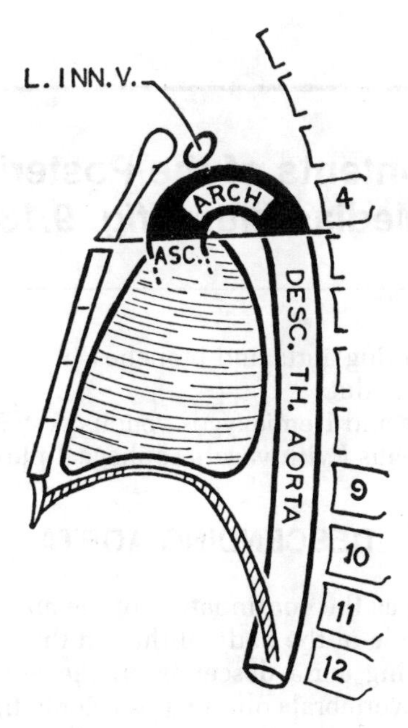

Figure 9.9. The thoracic aorta.

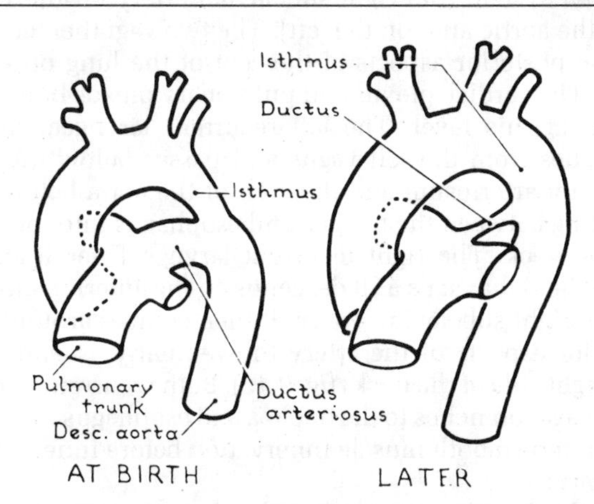

Figure 9.11. Aortic isthmus and ductus arteriosus, at birth and a few months later. (After Patten.)

This circulation to the heart and brain is of paramount importance in the function of the adult organism and was also dominant in the developing embryo and fetus. Development of the rostral end of the embryo and fetus is much greater than caudal (tail) elements in the human species. These three aortic arch vessels provide the basis for this rostral distribution of oxygenated blood.

The **brachiocephalic trunk** arises posteriorly to the center of the manubrium and ends posteriorly to the right sternoclavicular joint by dividing into the **right common carotid** and the **right subclavian arteries**. The left common carotid arises separately, just distal to the brachiocephalic artery and ascends posteriorly to the right sternoclavicular joint to enter the neck in the **left carotid sheath** (a neurovascular bundle containing the left common carotid artery, the left internal jugular vein, and the left vagus nerve). The most distal and posterior branch of the arch of the aorta is the **left subclavian artery**. It arches superior to the apex of the left lung posterior to the 1st costal cartilage to enter the neck. After a brief course in the root of the neck, the left subclavian artery crosses the 1st rib and then enters the axilla to supply the left upper extremity.

Some common anomalies in the arrangement of the great vessels arising from the aortic arch are that the left common carotid artery may arise from the brachiocephalic trunk, and the left vertebral artery (usually a branch of the subclavian artery) may arise from the aortic arch. These variations are commonly seen by radiologists doing angiography of the brain. A catheter introduced in the femoral artery in the leg is passed superiorly through the iliac arteries, abdominal aorta, and descending aorta to reach the aortic arch. Each branch of the aorta can then be selectively entered and visualized on a fluoroscope, as contract dye is introduced in the blood coursing through the given artery.

Each vagus nerve descends in the neck on the posterior aspect of the common carotid artery in the carotid sheath (*fig. 9.12*). Both nerves enter the thoracic inlet posterior to the 1st rib and then pass anterior to the underlying arterial system (the right subclavian artery on the right and the aortic arch on the left). The two vagi then course to the posterior aspects of the root of the lung on each side. The cardiac plexus and pulmonary plexus branches occur at this level. The left recurrent laryngeal nerve branches from the left vagus and passes below the ligmentum arteriosum and the arch of the aorta before ascending between the trachea and esophagus into the root of the neck. The right recurrent laryngeal nerve arises from the right vagus as it descends on the anterior surface of the right subclavian artery. It then courses around the inferior aspects of the artery to eventually ascend into the right side of the neck (*fig. 9.12*). Both recurrent nerves give vagal branches to the trachea and esophagus for glandular and smooth muscle innervation before innervating the larynx.

Each phrenic nerve enters the thoracic inlet like the vagus nerves (posterior to the 1st rib but anterior to the subclavian artery on the right and the arch of the aorta on the left). In the superior part of the thorax, both phrenic nerves course medially and the left phrenic nerve crosses anterior to the left vagus. The phrenic nerves then descend *anterior* to the root of the lungs between the lateral wall of the fibrous pericardium and the mediastinal pleura (*fig. 9.12*). The **right phrenic nerve** is associated with the superior vena cava in its upper thoracic course and pierces the right dome of the diaphragm with the inferior vena cava. The **left phrenic nerve** pierces the left dome of the diaphragm at the margin of the fibrous pericardial attachment to the central tendon. The phrenic nerves arise from the **anterior rami of cervical nerves 3, 4 and 5**. They carry both efferent and afferent fibers for motor and sensory supply to the diaphragm and parietal pleura, respectively.

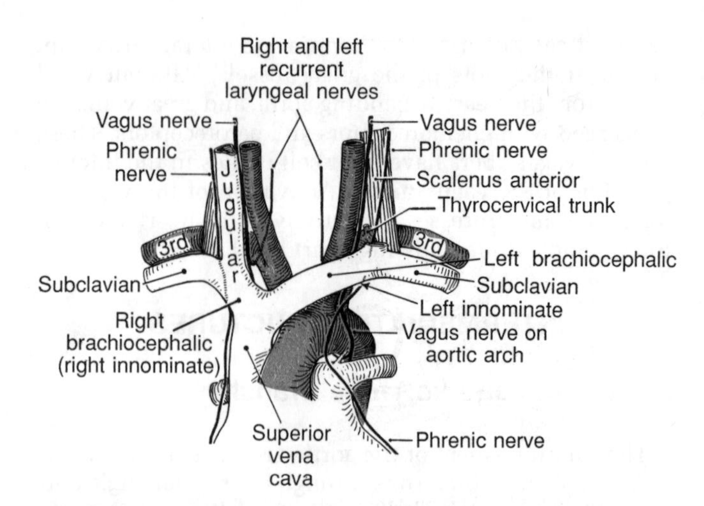

Figure 9.12. The courses of the phrenic and vagus nerves.

Contents of the Posterior Mediastinum (*fig. 9.13*)

A. 1. Descending aorta and branches
   2. Thoracic duct
   3. Azygos and Hemiazygos venous systems
   4. Esophagus (with vagal esophageal plexuses).

DESCENDING AORTA

Beginning as the continuation of the arch of the aorta on the left side of the body of the 5th thoracic vertebra, the descending aorta descends on the left side of the midthoracic vertebral column, posterior to the root of the left lung. It returns to a midline position at the level of

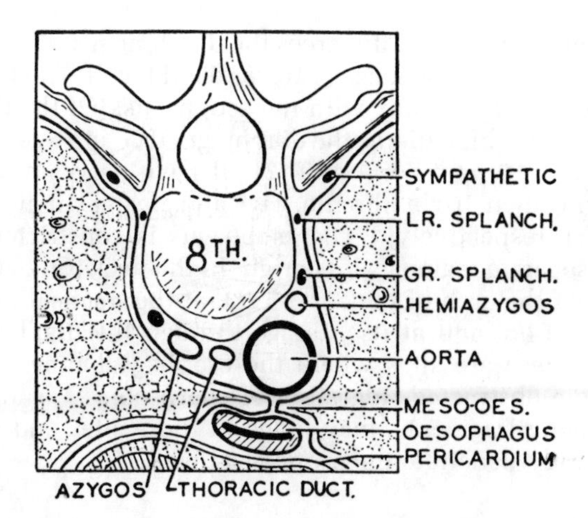

Figure 9.13. The posterior mediastinum (in transverse section).

T8. As it descends further to pass through the aortic hiatus of the diaphragm, the descending aorta is crossed anteriorly by the esophagus. The **greater splanchnic nerves** from the sympathetic trunk join the descending aorta in the aorta in the lower thorax and exit the thorax to enter the abdomen with the aorta at the level of T11/12.

The visceral branches (*fig. 9.14*) of the descending aorta are 1–3 **bronchial arteries**, 1–3 **esophageal arteries**, and twigs to the pericardium and diaphragm. The parietal (thoracic wall) branches are the **right** and **left posterior intercostal arteries** from the 3rd intercostal space to the subcostal arteries below the 12th ribs. The 1st and 2nd intercostal spaces are supplied from the **costocervical trunk** of the subclavian arteries. The **right posterior intercostal** arteries from T3 to T8 must cross the anterior surface of the vertebral column since the descending aorta at this point is on the left side of the vertebral column. The **posterior intercostal arteries** anastomose with the **anterior intercostal branches** of the **internal thoracic artery** at the midclavicular line on the thoracic wall.

THORACIC DUCT (*figs. 9.15, 9.16*)

The thoracic duct begins as the cisterna chyli at the posterior aspect of the abdominal aorta, just inferior to the diaphragm. It enters the thorax posterior to the descending aorta and stays on the right side of the anterior surface of the vertebral column in the lower and mid-thoracic level. When the descending aorta shifts to the left side of the thorax and the esophagus takes a midline position at the level of T8, the thoracic duct becomes related to the posterior aspect of the esophagus. It ascends in this relationship to the T5 level and then crosses the anterior aspect of the vertebral column and ascends into the root of the neck on the left side. The thoracic duct terminates in the junction of the left internal **jugular** and **left subclavian veins**. It receives and transmits all the lymphatic drainage of the body except the right thorax,

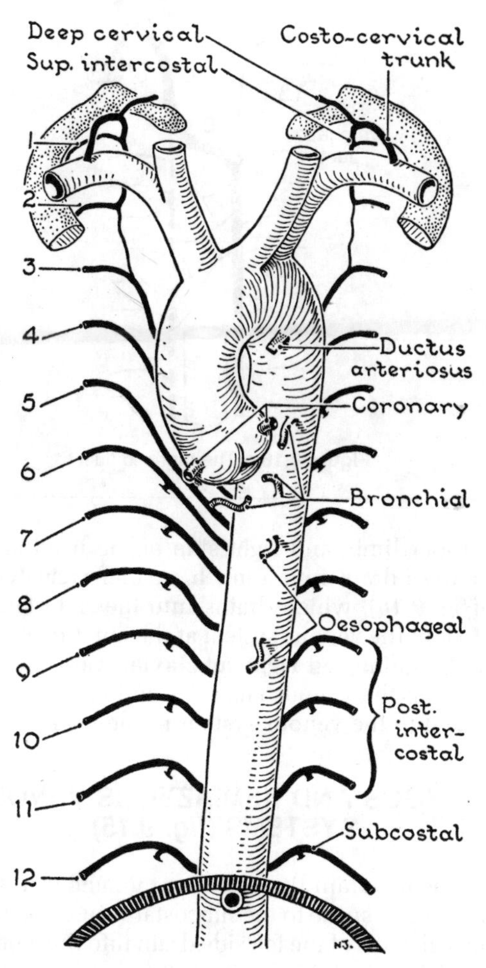

Figure 9.14. The branches of the thoracic aorta.

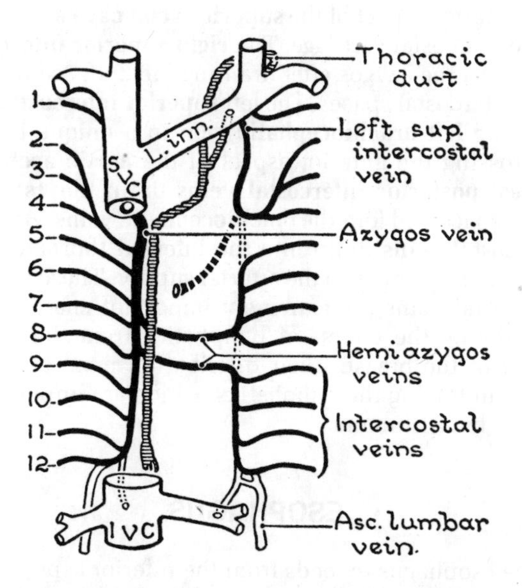

Figure 9.15. The thoracic duct and the intercostal veins. (Continued on *fig. 26.14*.)

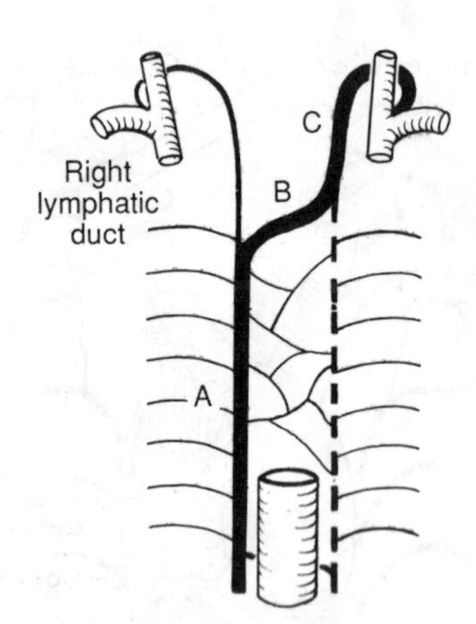

Figure 9.16. The thoracic duct.

right upper limb, and right side of the head and neck. These areas drain into a much smaller **right lymphatic duct** (*fig. 9.16*), which drains into the right side of the neck to join the venous system at the junction of the right **internal jugular** and **right subclavian vein**. Lymphatics evolve from the venous channels and are eventually reconnected to the venous system in the neck.

AZYGOS AND HEMIAZYGOS VENOUS SYSTEMS (*fig. 9.15*)

These veins drain the posterior thoracic wall from the 3rd intercostal space to the subcostal veins. The posterior intercostal veins of the left side drain into the **hemiazygos veins**. They join the azygos system in the midthoracic area by passing anterior to the vertebral column. The azygos system ascends on the right side of the vertebral column and arches over the right main bronchus to enter the posterior aspect of the **superior vena cava** at the level of the 3rd costal cartilage. The **right superior intercostal vein** joins the azygos after draining the 2nd, 3rd and 4th right intercostal spaces. The left **superior intercostal vein** joins the **left brachiocephalic vein** in a unique fashion by crossing the anterior aspect of the **aortic arch**. The **highest posterior intercostal veins** drain the 1st intercostal space and join the brachiocephalic veins. **Anterior intercostal veins** drain into the **internal thoracic veins** in a manner similar to the arterial supply. These anterior intercostal veins are extremely important channels, for they define the course of lymphatics from the medial aspect of the breast. Their distribution must be appreciated in tracing the lymphatics, which are important in certain breast tumors.

ESOPHAGUS

The esophagus extends from the inferior aspect of the pharynx (C6 level) to the stomach, below the left dome of the diaphragm. It therefore has cervical, thoracic, and abdominal portions. It pierces the diaphragm at the level of the 7th costal cartilage (T10 vertebral level). The vagus nerves exit the thorax with the esophagus. (Of the three structures which pierce the diaphragm [IVC at T8, esophagus at T10, and aorta at T12] all have nerves accompanying them [right phrenic, vagi, and greater splanchnic nerves, respectively].) The esophagus is constricted in four separate regions: at its origin in the neck (C6), at the level of the arch of the aorta T2/3, at the tracheal bifurcation T4/5, and at its passage through the diaphragm T10. These are sites where the esophageal lumen can become obstructed. Ingestion of sharp objects, such as fine bony fragments, may lodge in these sites and subsequently penetrate the esophageal wall.

Thoracic Relationships of the Esophagus (*fig. 9.17*)

The **esophagus** lies anterior to the vertebral bodies of C7–T8 and then "swings" to the left in the lower thorax in front of the descending aorta to pass through the left dome of the diaphragm at the T10 level. This course is readily evident in a patient that has had a barium swallow (*Plate 9.1*). It is also anterior to the thoracic duct, right

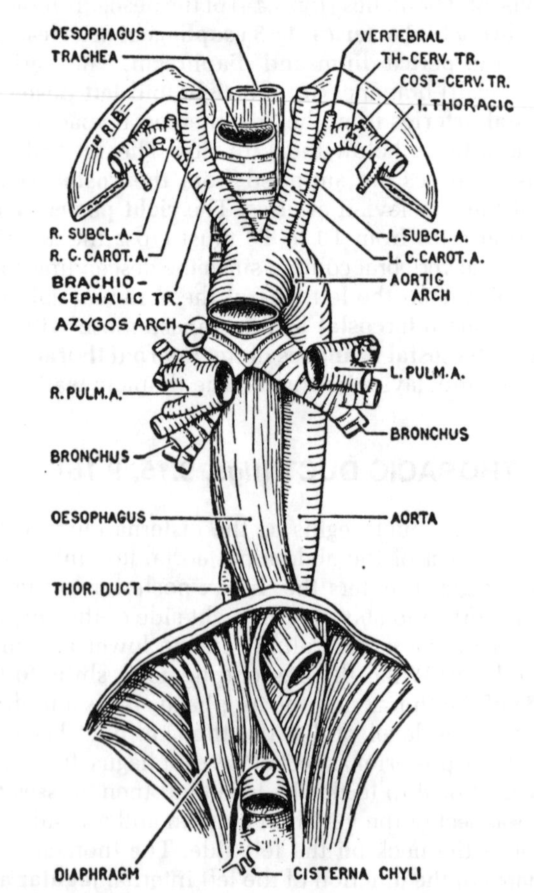

Figure 9.17. The esophagus, the aorta, and the 3 branches of the aortic arch.

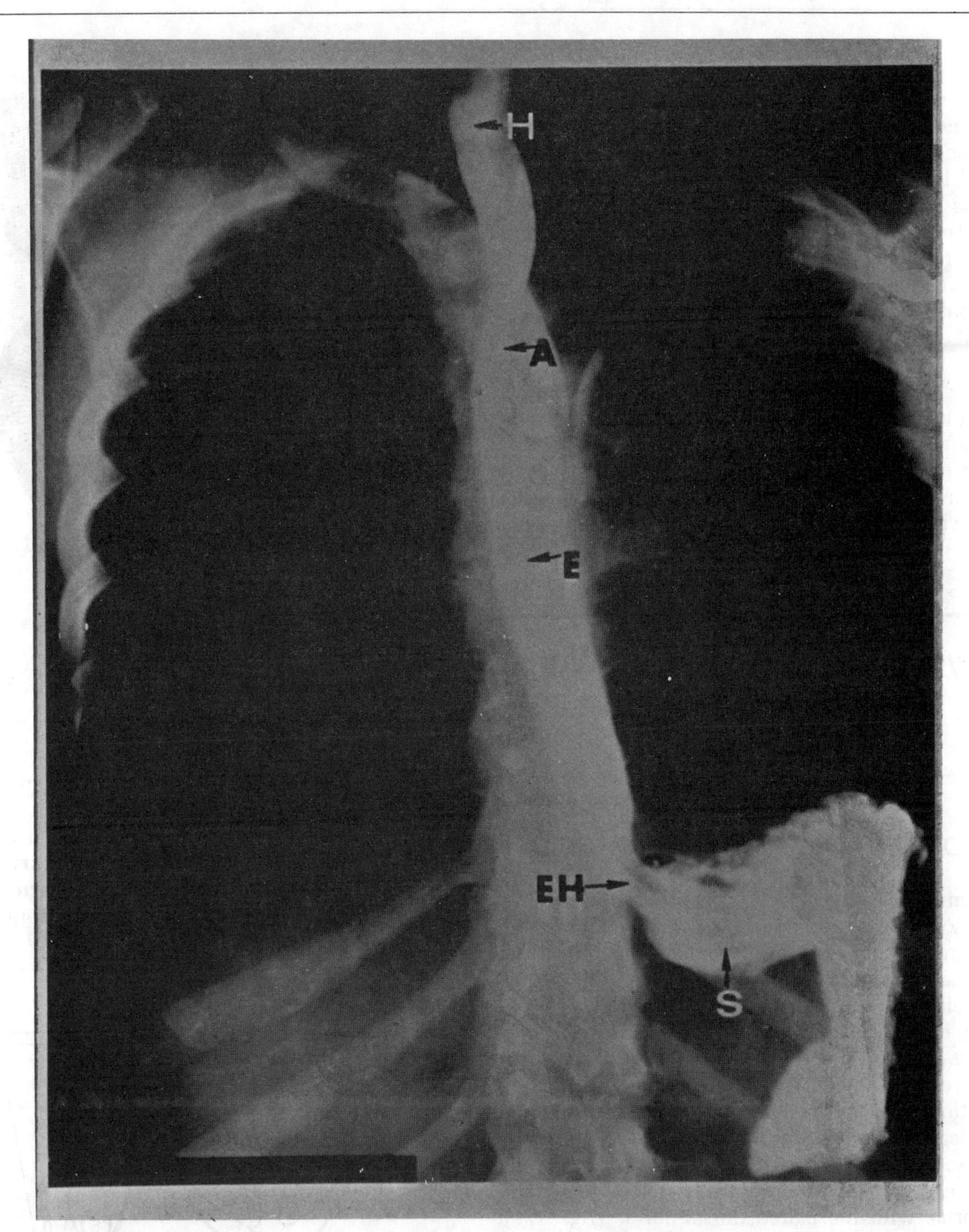

Plate 9.1. Esophagus, upper GI series (A-P view). Barium fills the lumen of the esophagus and has entered the stomach. *A*, aortic arch; *E*, heart shadow; *EH*, esophageal hiatus; *H*, hypopharynx (laryngopharynx); *S*, cardia of the stomach.

posterior intercostal arteries (T3–T7), and azygos and hemiazygos systems in the mid thoracic region. The anterior relationships of the esophagus are extremely important to the clinician (*fig. 9.17*). The trachea is anterior to the esophagus from C7–T4. As the esophagus descends below the bifurcation of the trachea, it is in contact with the base of the heart (the left atrium). Visualization of the left atrium can be appreciated much more in a lateral thoracic radiograph if a barium swallow is used to define the esophagus.

The **vagi** also are intimately related to the esophagus. In the cervical position, both **left** and **right recurrent laryngeal nerves** lie between the esophagus and the trachea, and give branches to the glandular and smooth muscle elements of the esophagus. Inferior to the root of the lung, the left vagus assumes an anterior relationship onto the esophagus and the right vagus disassociates and passes to the posterior surface of the esophagus. Branches of the two vagi contribute to form the **anterior** and **posterior esophageal plexuses**. The fibers of these plexuses recollect into single nerves on the inferior surfaces of the esophagus and pierce the diaphragm through the **esophageal hiatus**. Once they pass through the diaphragm they are renamed the **anterior** and **posterior gastric nerves**. The vagal branches to the esophagus and secretomotor to mucous glands, and they induce peristalsis by stimulating the smooth muscle to contract. Sympathetic branches from the sympathetic trunk are vasomotor to the blood vessels of the esophagus. The sympathetic and parasympathetic nerves also innervate the sphincter muscles at each end of the esophagus. The **inferior gastric** or **cardiac** (cardia of stomach, not heart) **sphincter** is a physiological rather that an anatomical one. It is supplied by the vagus nerve, which conducts the opening impulse (i.e., relaxes the musculature) and by the sympathetic fibers, which convey closing impulses (i.e., contract the musculature). The **superior sphincter** is the **cricopharyngeus**, which is described in the chapter on the pharynx.

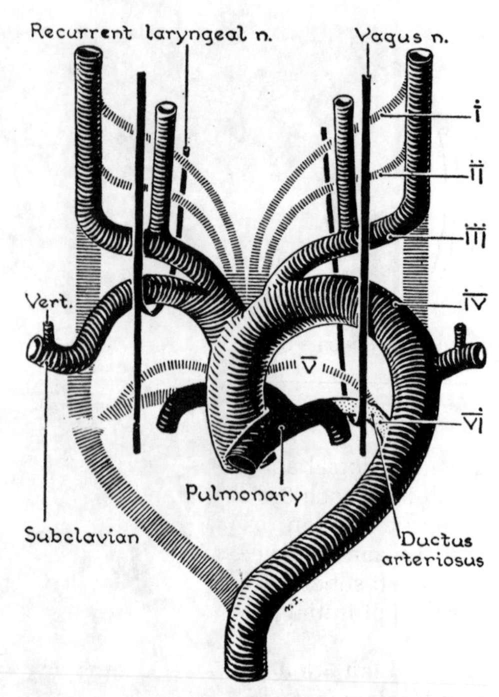

Figure 9.18. The 6 paired primitive aortic arches.

Embryological Development Within the Superior and Posterior Mediastina

DEVELOPMENT OF THE GREAT ARTERIES

The asymmetry of the great arteries in the adult thorax is easily explained by the embryological development of the aortic arches. Six paired aortic arches are formed from the aortic sac (rostral portion of the truncus arteriosus) of the developing heart during the 4th and 5th weeks of embryonic development. They appear successively during this 2-week period and are not all present at any one time. *Figure 9.18* depicts the symmetrical organization of the aortic arches in the embryo and their subsequent asymmetrical relationship in the adult.

The basic pattern (*fig. 9.18*) shows that the embryonic aortic arches connected the ventrally placed heart and ventral aortae with the paired dorsal aortae (*fig. 9.20*). The paired dorsal aortae fuse to form a single dorsal aorta in the caudal half of the embryo. As the aortic arches course between the ventral and dorsal aortae, they "embrace" the developing pharynx (foregut) and form the pharyngeal arches of the developing head and neck (*fig. 9.19*).

The fate of the individual aortic arches is such that the 1st aortic arch induces the formation of the **mandibular arches** (1st pharyngeal arches) in the head. Most of the arch disappears in development by contributing to the formation of the **maxillary artery** in the face. The **2nd aortic arches** are also transient and largely disappear during development. They induce the formation of the **hyoid arches** (2nd pharyngeal arches) in the head. The 2nd aortic arch vessels contribute to the very small **hyoid** and **stapedial arteries** in the head.

The **3rd aortic arches** are retained on both sides to form the **right** and **left common carotid arteries** and the proximal part of the **internal carotid arteries**. The remaining distal aspects of the internal carotid arteries are

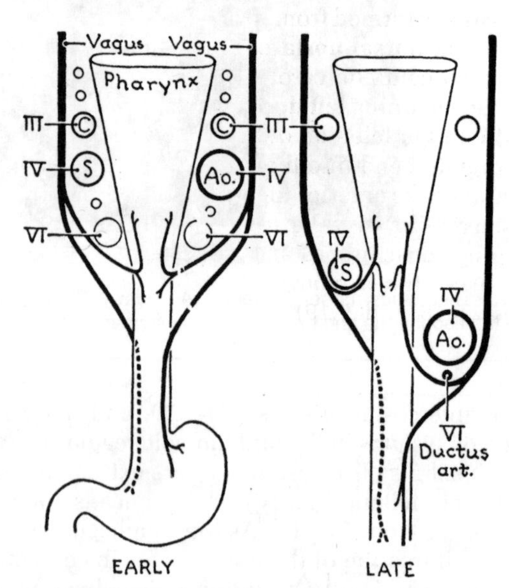

Figure 9.19. Coronal section of *fig. 9.1* showing that only the arches separate the vagus nerves from the digestive tract (*early*). A later stage of explaining the asymmetrical courses of the recurrent laryngeal nerves (*late*).

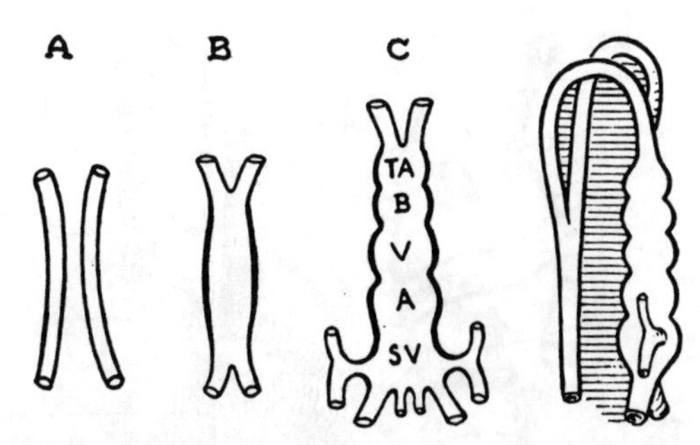

Figure 9.20. Scheme of development of tubular heart.

formed from the dorsal aortae. The **external carotid arteries** branch from each 3rd aortic arch.

The 4th aortic arch on the left side forms the part of the **arch of the aorta** between the left common carotid artery and the left subclavian artery. The **right 4th aortic arch** forms the left initial portion of the **right subclavian artery**.

The **right and left 5th aortic arches** both disappear in embryonic development. The proximal portions of the **6th aortic arches** form the **pulmonary arteries**. On the **left side**, the connection between the left pulmonary artery and the arch of the aorta, the **ductus arteriosus** is also derived from the **left 6th aortic arch**. The distal portion of the **right aortic arch** degenerates, as does the right dorsal aorta, which originally extended between the right common carotid artery and the fused dorsal aorta (*fig. 9.18*).

The **ascending aorta** and the **pulmonary trunk** are formed when the **truncus arteriosus** is divided by the spiral septum depicted in *Figure 9.21*. The **brachiocephalic trunk** is formed from the **ventral aorta on the right side**. The **left dorsal aorta** degenerates between the developing left common carotid artery and the arch of the aorta. The remaining **left dorsal aorta** forms the **descending aorta** on the left side of the thorax (*fig. 9.18*).

The **vagi** descend into the thorax through the superior thoracic aperture anterior to the developing aortic arches. The **recurrent laryngeal branches** "hook" around the six developing aortic arches and then ascend back into the neck between the esophagus, trachea, and the dorsal aortae (*figs. 9.18* and *9.19*). The **left recurrent laryngeal nerve** is retained in the thorax because the 6th aortic arch is present as the ductus arteriosus and persists in the adult as the **ligamentum arteriosus**. The distal portion of the right 6th aortic arch and the 5th aortic arch degenerates and therefore the **right recurrent laryngeal** nerve is "hooked around" the right 4th aortic arch or the right subclavian artery (*figs. 9.18, 9.19*). The counterclockwise rotation of the gastrointestinal system in the developing abdomen causes a 90° rotation of the vagi on the lower esophagus in the thorax. *Figure 9.19* shows the relationship of the **right vagus** to the posterior aspect of the esophagus, and the **left vagus** to the anterior aspect of the esophagus. In the adult, these relationships are retained in the formation of the esophageal plexuses.

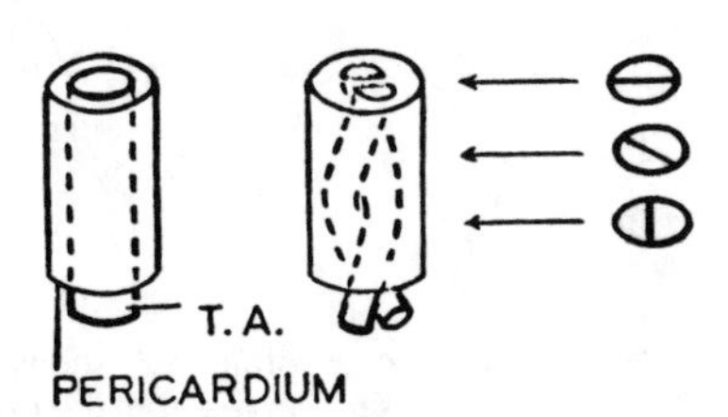

Figure 9.21. The spiral septum within the truncus arteriosus explains the twisted courses of the aorta and pulmonary trunk.

The position of the trachea anterior to the esophagus in the superior mediastinum reflects the origin of the endodermal diverticulum for the lungs from the ventral surface of the developing foregut. The enlarging arch of the aorta in the embryo displaces both the esophagus and the trachea to the right side of the superior mediastinum.

DEVELOPMENT OF THE VENOUS SYSTEM IN THE THORAX

The early arrangement of the developing veins in the embryo is also symmetrical, like the early arterial formation. Three major sets of paired veins (*fig. 9.22A*) are present at the 5th week of development: the **vitelline veins**, draining the primitive yolk sac; the **umbilical veins**, connecting the placenta to the venous end of the developing heart; and the **cardinal veins**, draining the body of the embryo. The cardinal veins are composed of **anterior cardinal veins**, which drain the cephalic part of the embryo, and **posterior cardinal veins**, which drain the caudal portion of the embryo. The anterior and posterior cardinal veins join to form the **common cardinal veins**, which enter the **horns of the sinus venosus**.

The right cardinal system dominates in the venous development over the next 2 weeks to form the major venous structures in the adult. The **superior vena cava** (SVC) is derived from the right anterior cardinal and right common cardinal veins (*fig. 9.22B, C*). The **azygos vein** and the **arch of the azygos** are derived from the right posterior cardinal system. The IVC is formed from the **right** vitelline veins and the right posterior cardinal veins in the developing abdomen. The developing IVC is surrounded by the developing septum transversum, and it penetrates the right dome of the diaphragm at the level of T8 in the adult.

The left posterior cardinal vein forms the **hemiazygos system** and the **left superior intercostal vein** that crosses the anterior surface of the arch of the aorta in the adult to drain into the left brachiocephalic vein. The **left brachiocephalic vein** is formed from an intercommunicating vein, which connects the two anterior cardinal veins. This arrangement brings the blood from the left side of the head, neck, thorax, and the left upper extremity to the right side of the developing heart.

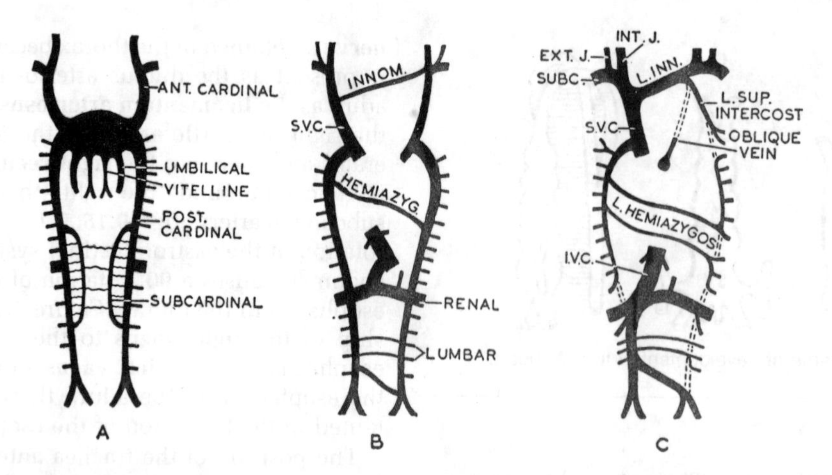

Figure 9.22. Veins. *A*, 6 veins join the sinus venarum. *B*, cross-communications appear. *C*, only 2 of the 6 veins survive. They become caval veins.

Clinical Mini-Problems

1. **At which vertebral level would one expect to visualize the bifurcation of the trachea in an A-P thoracic radiograph?**
2. **How could a bronchiogenic carcinoma in the left primary bronchus cause an alteration in the quality of a patient's voice?**
3. **Why would a quadriplegic patient (one with a transected spinal cord at the level of C5) demonstrate bradycardia (slowing of the heart rate)?**
4. **Why would inflammation of the parietal pleura on the right dome of the diaphragm refer a painful stimulus to the right shoulder?**
5. **Cannulation of the thoracic duct in the neck may be employed to reduce a lymphocytosis (increased lymphocyte population) in the circulating blood. Where does the thoracic duct enter the circulation in the neck?**

(Answers to these questions can be found on p. 585.)

10

Anatomical Features in the Physical Examination of Thorax

Walls

The thoracic wall is inspected for its symmetrical features. The clavicles that overlie the first ribs and the superior thoracic aperture are subcutaneous throughout their position from the shoulder to the superolateral border of the manubrium. The 2nd rib costal cartilage articulates at the sternal angle of the sternum, and this demarcates the level of the T4/5 intervertebral disc and the division between the superior and inferior mediastinum. This is also the level of bifurcation of the trachea and the inferior aspect of the arch of the aorta. The sternal angle is a definite and prominent landmark in surface anatomy and radiographic examination of the thorax. The second intercostal space is inferior to the second rib and its costal cartilage. The costal cartilage of the 6th rib articulates with the xiphisternal joint. The 5th intercostal space is superior to the 6th rib.

Lungs

The lungs are percussed (examined by tapping one's finger on the chest) and auscultated (listened to with a stethoscope) in the thoracic examination. The superior lobes are examined on the anterior surface above the 6th rib. The apex of each lung extends about 1 inch above the clavicle and extends into the root of the neck. On the right side, the horizontal fissure parallels the 4th rib from the midaxillary line to the right margin of the sternum. The middle lobe of the right lung is therefore examined between the 4th and 6th ribs. The left lung is displaced by the heart between the 4th and 6th ribs on the left side. The lingula of the left lung lies anterior to the heart at the midclavicular line in the 4th and 5th intercostal spaces.

The oblique fissures of both lungs parallel the 6th ribs on the posterior wall of the thorax. The inferior lobes of each lung are examined by percussion and auscultation below the 6th rib. If patients place their hand behind their head during this examination, the medial border of the rotated scapula will closely parallel this oblique fissure and serve as an additional guide to the examiner.

The extent of the inferior projection of the lung tissue in the pleural cavities during quiet breathing is somewhat above the floor of the pleural cavities. The medial margin of the right lung lies posterior to the midsternal line from the manubrium to the xiphisternal joint. The inferior margin of the right lung courses obliquely from the 6th rib at the midclavicular line to the 8th rib at the midaxillary line and to the level of the 10th rib on the back. Although the left lung is similar in its location, it is influenced by the position of the heart in the 4th and 5th intercostal spaces. The medial border is therefore displaced from the midsternal line above the 4th rib to the costochondral joints of the 5th and 6th ribs. The inferior margin of left lung will be at the 6th rib in the midclavicular line, the 8th rib in the midaxillary line and the 10th rib posteriorly on the back.

Pleura

Access to the pleural space for sampling pleural fluid or draining pleural exudates is done below these inferior levels of the lung but above the attachments of the diaphragm and the covering parietal pleural.

Mediastinum

TRACHEA

The trachea can be palpated in the anterior aspect of the root of the neck above the sternal notch. It is readily visible in radiographs in the midline behind the sternum. It ends at the level of the sternal angle by bifurcating into the primary bronchi.

HEART

Heart sounds are examined by auscultation with a stethoscope. The actual valve positions are posterior to the sternum in a line from the left 3rd costal cartilage level to the right 5th costal cartilage level. The sound of the closing valve is best heard where the blood-filled chamber that closes the valve lies near the chest wall, devoid of a bony covering. Therefore one listens to the aortic valve in the right 2nd intercostal space where the ascending aorta approaches the thoracic wall; the pulmonary valves are heard in the left 2nd intercostal space where the pulmonary trunk is projecting into the middle mediastinum; the tricuspid valve closes when the right ventricle is full of blood and is best heard in the 5th intercostal space at the right margin of the sternum; the bicuspid (mitral) valve is heard when the left ventricle is full at the apex, 10 cm from the left margin of the sternum in the 5th intercostal space. The apex beat in a thin individual can frequently be *seen* by observing the chest wall in the 5th intercostal space on the midclavicular line.

The position of the heart in the thorax can be outlined by percussion. The right margin of the heart (right atrium) extends from the 3rd costal cartilage on the right side of the sternum to the sternal attachment of the 7th costal cartilage at the xiphisternal joint. The heart lies approximately one finger's breadth lateral to the right sternal margin. The inferior border of the heart projects from the right side of the xiphisternal joint to the apex of the heart in the 5th or 6th intercostal space on the midclavicular line. The left margin of the heart projects obliquely from the apex region to the left 3rd costal cartilage attachment. The left atrial appendage is in the left 3rd intercostal space at the left margin of the sternum while the pulmonary trunk lies behind the left half of the sternum at the level of the second intercostal space. The aortic arch would rise to the level of the first intercostal space on the left side and form the "aortic knuckle" in the anterior-posterior (A-P) radiograph of the chest.

Female Breast

The breast should also be examined in the examination of the thorax in females. The level of the nipple is variable but is usually in the region of the 4th intercostal space. The lateral two-thirds of the breast is located between the midclavicular line and the midaxillary line. Vascular supply and lymphatic drainage of this component of the breast is related to the lateral thoracic vessels that are components of the axilla. The medial third of the breast is supplied by blood vessels from the internal thoracic arteries, the anterior intercostal vessels. The lymphatic drainage of the medial aspect of the breast follows the anterior intercostal veins to pass through the parasternal lymph nodes that lie posterior to the costal cartilages, which attach to the sternal margins.

Abdomen 3

11

Anterior Abdominal Wall and Scrotum

Clinical Case 11.1

Patient George G. This 29-year-old construction worker, who is the father of five, noticed a walnut-size growth just above his right pubic bone. A physical examination by your clinical tutor led to a diagnosis of an inguinal hernia through the superficial inguinal ring on the right side and a tendency for herniation on the left side when the patient coughed. In consultation with a surgeon, it was decided that a bilateral repair for direct inguinal hernias was required, and you were invited to assist. It is proposed that the surgeon should also perform bilateral vasectomies on the exposed spermatic cords. The patient is advised that he will be released from the hospital three days after the surgery and return to work after 4 weeks of convalescence.

Anterior Abdominal Wall

The anterior abdominal wall is composed of skin, fascia, and muscle. It extends between the anterior aspects of the rib cage and the pelvic girdle. This wall is inspected, and the underlying abdominal organs are palpated in the general physical examination. Surgical access to the abdomen frequently requires that the anterior abdominal wall be incised, and therefore a detailed knowledge of its composition is necessary for general surgery.

BOUNDARIES

The anterior abdominal wall is bounded superiorly by the inferior margin of the rib cage: the xiphoid process

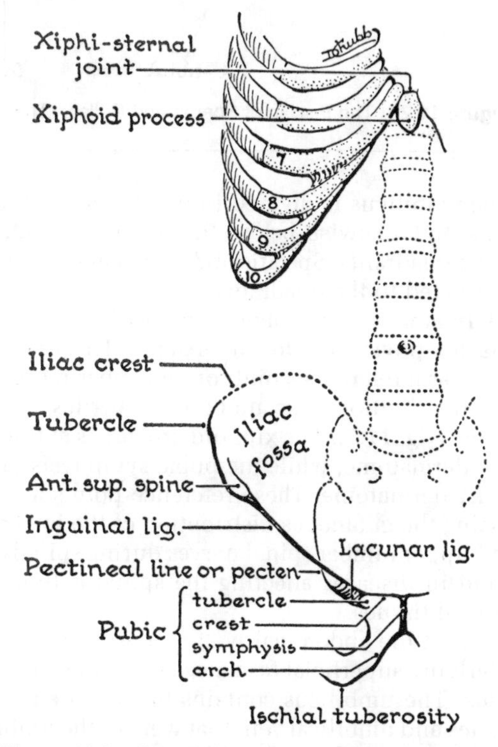

Figure 11.1. Boundaries of the anterior abdominal wall.

and the costal cartilages of ribs 7–10 (*fig. 11.1*). The inferior attachments of the anterior abdominal wall are the iliac crest, anterior superior iliac spine, inguinal ligament, pubic tubercle, pubic crest, and pubic symphysis.

SKIN

The most prominent feature of the skin of the abdominal wall is the **umbilicus**. In a well-conditioned patient, the umbilicus is usually in a midline position, midway between the xiphoid process and the pubic symphysis. The umbilicus therefore overlies the intervertebral disc between the 3rd and 4th lumbar vertebrae (*fig. 11.1*). This is not a reliable reference in an overweight patient due to the tendency for the anterior abdominal wall to bulge and become pendulous in obesity. The umbilicus, however, does serve a purpose for readily dividing the abdominal wall into quadrants. The anterior abdominal wall

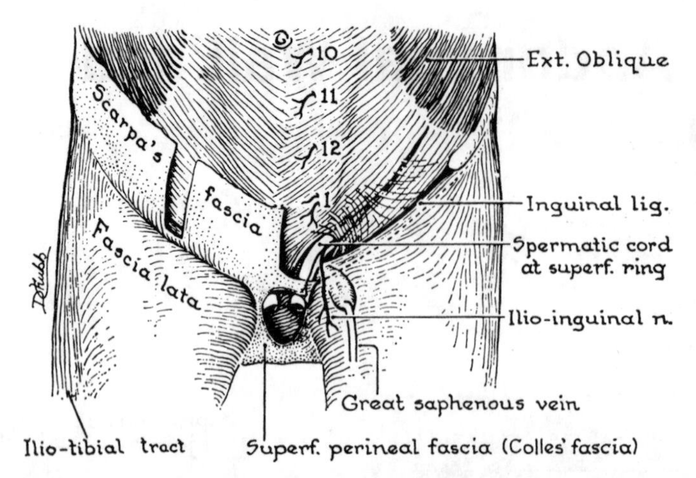

Figure 11.2. The fasciae of Scarpa and Colles. (The penis and scrotum are cut away and the spermatic cords are cut across.)

above the umbilicus is divided into left and right upper quadrants and likewise below the umbilicus, left and right lower quadrants. Specific abdominal and pelvis viscera are related to the quadrants.

The dermatomes of the abdominal wall are important concepts for physicians to understand. The umbilicus also serves as a reference point for remembering the dermatomal distribution. The umbilicus overlies the T10 dermatome (*fig. 11.2*). The xiphoid process is at the level of the T7 dermatome, while the pubic symphysis underlies the L1 dermatome. These reference points are used in assessing the cutaneous distribution of the lower thoracic and upper lumbar spinal nerves during spinal anesthesia and in diseases affecting the spinal column and neurological tissues.

The skin of the abdominal wall is loosely attached to the underlying superficial fascia except at the site of the umbilicus. The umbilicus contains the scarified umbilical arteries and umbilical vein that were in the umbilical cord in the fetus and embryo. These vessels penetrated the anterior abdominal wall and joined the arterial and venous channels of the abdominal cavity. We will see the vestiges of these vessels within the abdominal cavity and their relationship to the abdominal wall.

SUPERFICIAL FASCIA

The superficial fascia of the anterior abdominal wall is distinct and serves as a major site for fat deposits in overweight individuals. This subcutaneous tissue has two different patterns of distribution. Above the umbilicus, the subcutaneous fascia typically contains fat cells and connective tissue fibers as a single layer of tissue. Inferior to the umbilicus, this superficial fascia has two distinct layers. The most superficial layer (**Camper's fascia**) is typical subcutaneous fascia. The deeper membranous connective tissue layer (**Scarpa's fascia**) is a unique feature of the superficial fascia in this region (*fig. 11.2*).

The membranous (Scarpa's) fascia has two important clinical relations: (1) it serves as a firm unit for suturing the subcutaneous fascia in repairs of the anterior abdominal wall, following abdominal or pelvic surgery; and (2) as shown in *Figure 11.2*, Scarpa's fascia is continuous with the superficial perineal fascia (Colles' fascia) of the perineum and adheres to the fascia lata of the thigh slightly inferior to the inguinal ligament. In cases where the male urethra is ruptured in the perineum, urine can drain from the urethra into the space between the deep fascia of the abdominal wall musculature and the membranous layer of the superficial fascia and extravasate onto the lower abdominal wall inferior to the umbilicus. The presence of urine in this confined space where Scarpa's fascia is attached will lead the physician to suspect some urethral trauma and dysfunction.

ABDOMINAL WALL MUSCLES AND THEIR RELATIONSHIPS

The anterior abdominal wall is composed of 4 major muscles on each side of the midline:

1. Rectus Abdominis (and Pyramidalis) (midline muscles)
2. External Oblique
3. Internal Oblique (three flat muscles of the anterolateral wall)
4. Transversus Abdominis

The Rectus Abdominis and its Fascial Sheath

The rectus abdominis (*fig. 11.3*) is a long strap-like muscle that runs vertically from the rib cage to the pubic bone. The muscle is enclosed in a sheath derived from the fascia, which invests the 3 flat muscles of the abdominal wall and their aponeuroses. The rectus abdominis has **4 transverse tendinous intersections**, which subdivide the muscle into distinct segments. Each segment runs between the adjacent tendinous intersections. This

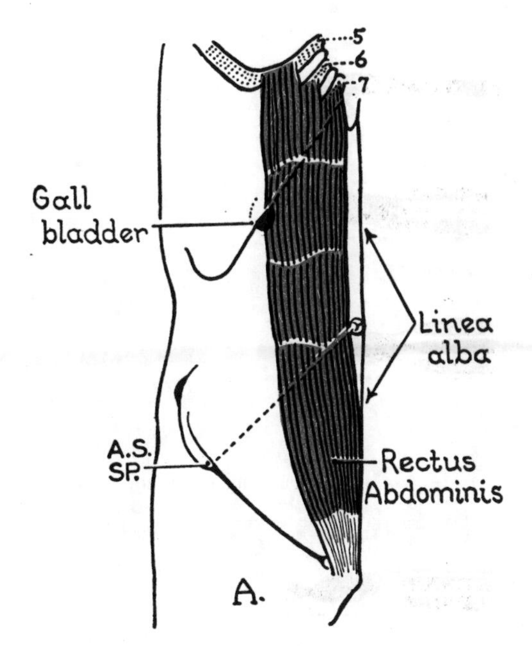

Figure 11.3. *A*, the rectus abdominis. ASSP, anterior superior (iliac) spine.

allows this long muscle to contract in shorter segments and assist in an efficient flexing of the trunk. The lower six thoracic nerves send motor fibers to the rectus abdominis, so that each segment of the transverse abdominis may have an innervation by one or more thoracic spinal nerves.

Attachments

The superior attachment of the rectus abdominis is anterior to the surface of the xiphoid process and the adjacent 5–7 costal cartilages (*fig. 11.3*). The inferior attachment of the rectus abdominis is to the pubic crest and symphysis.

Surface Anatomy (*fig. 11.3*). The lateral border of the rectus abdominis curves from the pubic tubercle, across the midpoint of a line between the anterior superior iliac spine and the umbilicus, and across the costal margin, about 8 cm from the midline. As it crosses the costal margin, the lateral margin of the right rectus abdominis overlies the gallbladder.

The medial border of the rectus abdominis is separated from the adjacent rectus abdominis by the midline **linea alba**. The linea alba is a relatively avascular insertion site of the aponeuroses of the 3 flat abdominal muscles. It extends the entire length of the anterior abdominal wall from the xiphoid process to the pubic symphysis. Gaps may occur in the linea alba due to congenital malformation of improper surgical repair. These gaps can be the site of herniation of abdominal contents through the abdominal wall.

Rectus Sheath (*fig. 11.4*)

The rectus sheath is an enveloping of the rectus muscle by the deep fascia, which invests the superficial and deep layers of the three flat abdominal muscles: external oblique, internal oblique, and transversus abdominis. Just superior to the umbilicus (level *b* in *fig. 11.4*), the three muscles converge onto the lateral margin of the rectus abdominis. The deep fascia of the external oblique then passes anterior to the rectus abdominis. The fascia covering the anterior surface of the internal oblique passes anteriorly to the rectus abdominis also. The fascia on the posterior surface of the internal oblique joins the fascia of the transversus abdominis to pass posteriorly to the rectus abdominis. This basic arrangement of the fascia forming the rectus sheath changes as one ascends or descends the abdominal wall. Below the umbilicus (level *a* in *fig. 11.4*), all of the deep fascia pass anterior to the rectus muscle, while superiorly, the fascial layers pass posteriorly. One therefore sees a strong anterior aspect to the rectus sheath inferiorly and a strong posterior aspect superiorly.

While the details of the rectus sheath are complex, one should generally be aware of the following relations of the sheath:

1. The principal blood supply to the rectus abdominis is the **inferior epigastric artery**. It gains direct access to the posterior surface of the muscle as it abuts the areolar tissue (**transversalis fascia**) above the pubic bone. Here the posterior aspect of the rectus sheath lacks the dense deep fascia. The inferior epigastric artery then ascends on the posterior surface of the rectus abdominis within the rectus sheath to anastomose with the **superior epigastric artery** near the level of the 8th costal cartilage (*fig. 11.5*).
2. The motor nerves to the rectus abdominis enter the rectus sheath from the lateral margin (*fig. 11.6*). Surgical approaches should therefore be made from the medial margin to avoid denervating the muscle.
3. Since the linea alba is relatively avascular, it will not heal as well as a fascial area with a more abundant blood supply. Surgeons therefore commonly elect to incise the anterior rectus sheath, which overlies the medial one half of the rectus abdominis muscle. They then retract the muscle laterally to avoid damaging the nerve supply and incise the posterior aspect of the rectus sheath to enter the abdominal cavity.

THE THREE FLAT MUSCLES OF THE ABDOMINAL WALL

These muscles arise from the posterior and lateral aspects of the abdominal wall and pass anteriorly to end in aponeurotic tendons at the lateral margin of the rectus abdominis. The aponeuroses blend with the rectus sheath

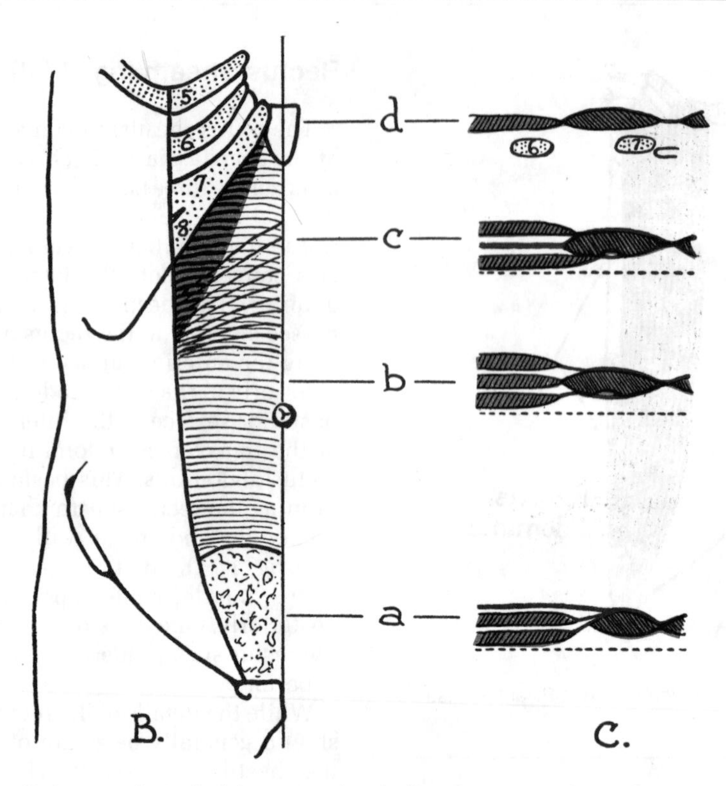

Figure 11.4. *B*, the posterior wall of the rectus sheath; *C*, transverse sections of the sheath at 4 levels.

and insert into the linea alba at the midline. The direction of the fibers of the three flat abdominal muscles are different, and this arrangement forms a strong "three ply" wall for protection of the abdominal viscera in an area where no bony elements are present.

The arrangement of the muscle layers is comparable to the intercostal musculature of the thoracic wall. The position of the neural and vascular elements which supply the abdominal wall are also arranged as they are in the thoracic wall. The motor nerves and arteries are found between the 2nd and 3rd layers of muscle (*figs. 11.5, 11.6*).

One major feature of the abdominal wall that exists is the passage created for the contents of the inguinal canal. This requires a series of defects in the muscular wall but in a fashion that does not compromise the strength of the wall and its ability to prevent the herniation of the contents within the abdominal cavity (*fig. 11.7*).

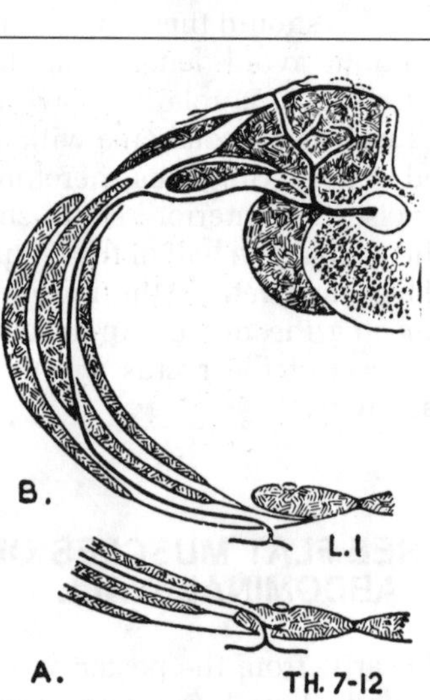

Figure 11.5. The nerves and arteries within the rectus sheath.

External Oblique (Obliquus Externus Abdominis)

The external oblique muscle arises as fleshy interdigitations from the external aspects of the lower eight ribs. The superior and lateral portions of the external oblique are muscular, while the inferior and medial aspects are aponeurotic (*fig. 11.2*). The majority of muscle fibers are oriented in a medial and inferior arrangement as the muscle projects to its multiple sites of insertion. The most superior fibers are nearly horizontal near the costal margin, and the most posterior fibers are nearly vertical as they descend from the 12th rib to the iliac crest.

Insertion

The fibers of the external oblique spread out like a fan as they protect in a medial inferior direction. (1) The

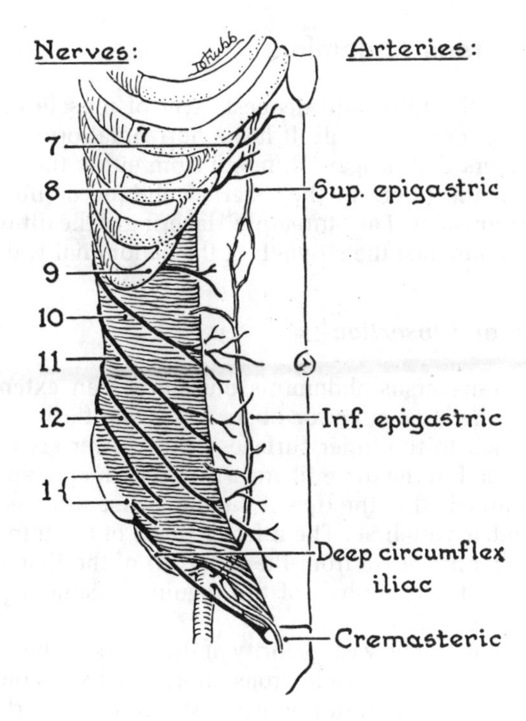

Figure 11.6. The course of a ventral nerve ramus in the abdominal wall. *A*, lower thoracic; *B*, 1st lumbar.

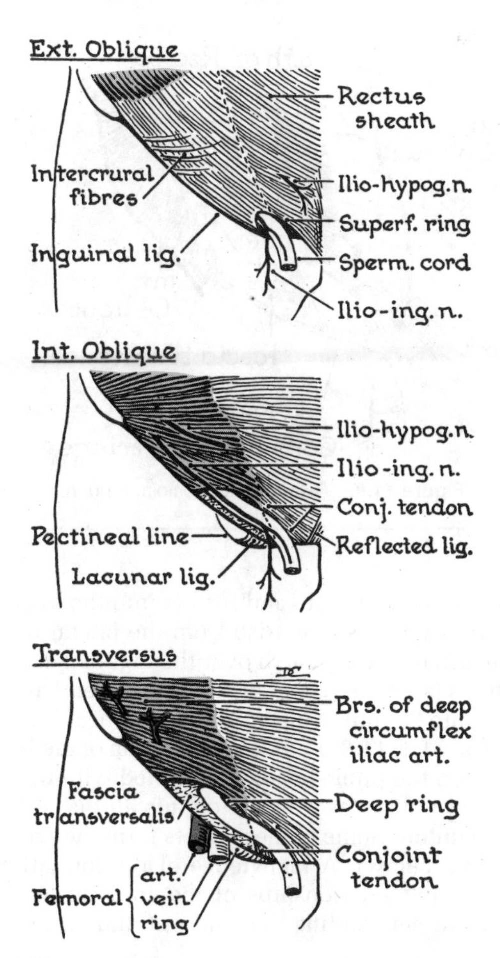

Figure 11.7. Three flat muscles below level of anterior superior spine. Walls of inguinal canal.

fibers of the superior half of the external oblique insert into the anterior aspect of the rectus sheath and then decussate (cross) the fibers of the opposite external oblique in the linea alba (*fig. 11.2*). (2) The fibers, which form the inferior half of the external oblique, insert onto elements of the pelvic girdle that are highlighted in *Figure 11.1* (iliac crest, tubercle, anterior superior iliac spine, inguinal ligament, and pubic bone). The lowest and most posterior fibers of the external oblique insert onto the external lip of the iliac crest, posteriorly, Successively higher fibers insert into the more anterior aspects of the iliac crest. As the insertion progresses to the anterior iliac spine, the external oblique is aponeurotic in its inferior and medial aspect. The free inferior border of the external oblique aponeurosis that extends from the anterior superior iliac spine to the pubic tubercle is termed the **inguinal ligament**. A reflection of the inguinal ligament onto the pectineal line of the pubic bone is termed the **lacunar ligament** (*fig. 11.1*). These specialized portions of the insertion of the external oblique onto the pelvic girdle are extremely important in understanding the anatomy relevant to physical examination and surgical repair in this region. An opening in the external oblique aponeurosis occurs superiorly to the pubic tubercle. This is the **superficial inguinal ring**, which allows the spermatic cord (round ligament in the female) to exit the inguinal canal and descend into the subcutaneous fascia and scrotum. The superficial inguinal ring is palpable when examining the abdominal wall.

Internal Oblique (Obliquus Internus Abdominis)

This muscle lies deep to the external oblique and has a fiber arrangement that is oriented at right angles to the external oblique. The fleshy component of the internal oblique is in its inferior and lateral aspect, and its aponeurotic component is in the superior and medial aspect of the muscle at the site of its insertion.

Origin and Insertion

The internal oblique **arises** principally from the **intermediate line of the iliac crest** and the lateral one-half of the **inguinal ligament**.

Insertion. (1) The most posterior fibers ascend to the cartilages of the lower 4 ribs (12, 11, 10, and 9). The fibers arising more anteriorly on the iliac crest, traverse superiorly and medially to form an aponeurotic tendon, which contributes to the formation of the anterior and posterior

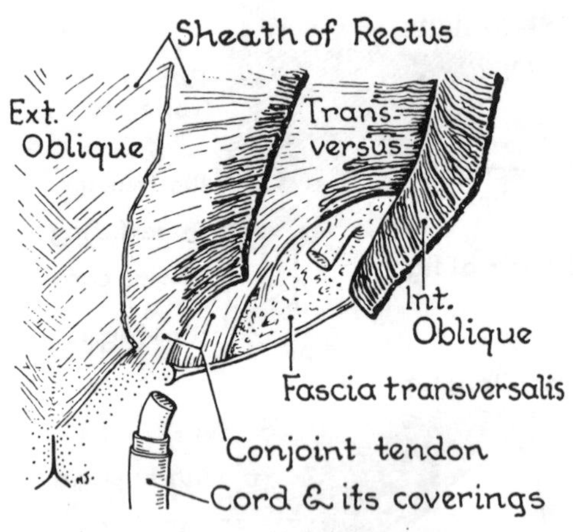

Figure 11.8. Dissection of conjoint tendon.

walls of the rectus sheath and inserts into the linea alba. (2) The fleshy fibers that arise from the lateral one-half of the inguinal ligament arch over the inguinal canal and its contents (spermatic cord or round ligament) and then descend onto the pubic tubercle and pectineal line of the pubis (*figs. 11.1, 11.8, 11.9*). This insertion of the internal oblique onto the pubic bone is combined with the pubic insertion of the transversus abdominis and is called the **conjoint tendon**. Some of these fibers form the cremaster muscle (see below). A knowledge of the formation, attachments, and relationships of the conjoint tendon is critical in understanding the cause of and repair of inguinal hernias.

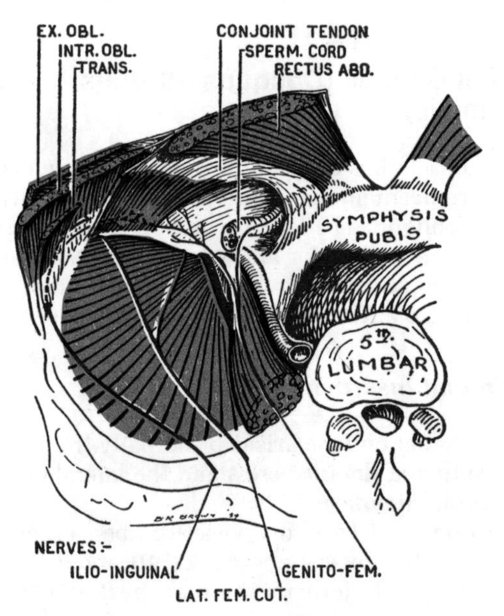

Figure 11.9. Branches of ilioinguinal and genitofemoral nerves to muscle fibers controlling conjoint tendon (dissection by Dr. R.G. MacKenzie).

Transversus Abdominis

This is the third and deepest layer of muscle in the anterior abdominal wall. It has a horizontal orientation to its fibers that is again different from either the orientation of the fibers in the external oblique or internal oblique muscles. The "three-ply" layering of the different muscles enhance the strength of the abdominal wall.

Origin and Insertion

The transversus abdominis muscle has an **extensive origin**. The most superior fibers arise from fleshy strips that attach to the inner surfaces of the lower six costal cartilages. Intermediate fibers arise from an aponeurosis that is attached to the tips of the transverse processes of the lumbar vertebrae. The inferior fibers of the transversus abdominis arise from the inner lip of the iliac crest and the lateral one-half of the inguinal ligament (*figs. 11.5* and *11.7*).

Insertion. The vast majority of the fibers of the transversus abdominis muscle project horizontally to contribute to the formation of the rectus sheath and insert into the **linea alba**. The fibers arising from the lateral one-half of the inguinal ligament arch superior to the inguinal canal then descend to insert onto the pubic tubercle and pectineal line with the internal oblique muscle as the **conjoint tendon** (*figs. 11.7* and *11.8*). The conjoint tendon (falx inguinalis) fuses and blends with the underlying **transversalis fascia** (*fig. 11.8*) to reinforce the abdominal wall behind the superficial inguinal ring, which is in the aponeurosis of the external oblique (*fig. 11.7*). The conjoint tendon is therefore an important anatomical element in preventing a direct herniation of abdominal contents and viscera through the superficial ring of the inguinal canal.

CREMASTER MUSCLE

This is a thin sheet of skeletal muscle that invests the spermatic cord within the inguinal canal. It is derived from the muscle fibers of the internal oblique that arch over the spermatic cord as they run from the inguinal ligament to the conjoint tendon. Contraction of the cremaster muscle causes the testes to be elevated in the scrotum and drawn toward the superficial inguinal ring. The cremaster muscle is innervated by the genital branch of the genitofemoral nerve (L1, 2). A cremasteric reflex is sometimes elicited in the pediatric neurological examination to test for normal development of the motor pathways in the spinal cord. The physician strokes the skin on the thigh (dermatomes L1, and 2), and this cutaneous stimulation evokes a contraction in the muscles innervated by L1, 2. The obvious sign in this reflex is an elevation of the testis and scrotum on the stimulated side of the body. A similar abdominal reflex can be elicited by stroking the skin over the lower quadrant of the abdominal wall and looking for a movement of the umbilicus toward the stimulus site as the abdominal wall muscles contract in an abdominal reflex. The latter reflex would be more suitable for use in assessing spinal cord development in the pediatric neurological examination in the female.

Functions of the 3 Flat Muscles

They are the chief lateral flexors of the torso. The esternal oblique fibers of one side and the internal oblique fibers of the opposite side are parallel and therefore act together in rotation movements of the torso.

All three flat muscles help to maintain the intra-abdominal pressure. This pressure is extremely important in maintaining the positions of the viscera within the abdominal cavity.

The three flat muscles of the abdominal wall function extensively with the diaphragm. When the abdominal muscle contracts and the diaphragm relaxes, forced expiration ensues. When the abdominal muscles and diaphragm both contract simultaneously, they aid in expelling the contents of the abdominal cavity through the pelvic orifices (in micturition, defecation, and parturition). Abdominal contraction also is involved in coughing, vomiting, and the mechanisms of venous blood return to the heart.

Layers of the Abdominal Wall and Their Clinical Significance

The anterior rami of the thoraco-abdominal nerves T7–12 innervate the anterior abdominal wall. These mixed (motor and sensory) nerves run between the 2nd (internal oblique) and 3rd (transversus abdominis) layers of the abdominal wall (*fig. 11.6*). Surgical care must be taken

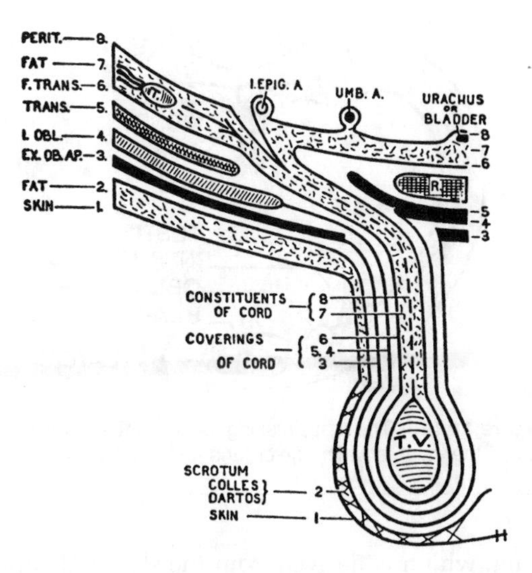

Figure 11.10. Scheme: The inguinal canal, and the layers of the anterior abdominal wall prolonged into the scrotum (*see Table 11.1*). (In this cross-section, the scrotum is raised to the horizontal position.)

not to sever these nerves as this causes motor paralysis in the segments of the abdominal muscle that they innervate, and subsequent weakness in the abdominal wall. *Figure 11.6* depicts the distribution of these nerves in this layer as well as the position of the inferior and superior epigastric arteries that lie posterior to the rectus abdominis within the rectus sheath.

Below the level of the anterior superior iliac spine, the abdominal wall consists of 8 layers (*Table 11.1*). These individual layers are important in surgical preparations and repairs. In addition, they have a corresponding relationship to the layers of the scrotum and the investing layers of the spermatic cord and testes (*Table 11.1; fig. 11.10*).

The testes develop from the intermediate mesoderm (gonadal ridge) in the extraperitoneal fat (layer 7) of the abdomen (*figs. 11.10, 11.11,* and *11.12; see also 11.14*). The testes descend in the extraperitoneal fat to enter the scrotum during the final month of fetal life. To reach the

Table 11.1. Corresponding Layers in the Walls of the Abdomen and Scrotum

The Layers of the Abdominal Wall	The Corresponding Layers in the Scrotum	
1. Skin	1. Skin	Scrotum
2. Superficial fascia: (a) fatty (Camper); (b) membranous (Scarpa)	2. Dartos muscle and fascia	Scrotum
3. External Oblique (aponeurotic)	3. External spermatic fascia	Coverings of the cord
4. Internal Oblique (fleshy)	4. Cremaster muscle	Coverings of the cord
5. Transversus (fleshy)	5. Cremaster muscle	Coverings of the cord
6. Fascia Transversalis (lining abdominal cavity in this region)	6. Internal spermatic fascia	Coverings of the cord
7. Extraperitoneal fatty tissue (layer inhabited by organs)	7. Areolar tissue with localized collections of fat	Two of the constituents of the cord
8. Peritoneum	8. Processus vaginalis	Two of the constituents of the cord

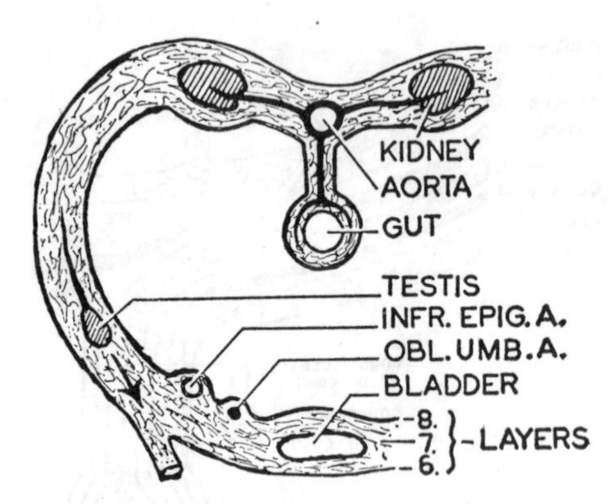

Figure 11.11. The testis making its descent in layer number 7, i.e., the layer in which the organs and great vessels reside.

scrotum, which is derived from the skin and superficial fascial layers of the abdominal wall (*Table 11.1; fig 11.10*), the testes must also traverse the muscular and associated deep fascial layers of the abdominal wall (*fig. 11.10*). The descending testes therefore first enter the deep inguinal ring in the transversalis fascia as it begins to traverse the anterior abdominal wall. It successively encounters the three fascial coverings of the 3 flat abdominal muscles: transversus abdominis, internal oblique, and external oblique. The testes approximate the scrotum in the fetus as they exit the superficial inguinal rings in the aponeurosis of the external oblique.

Contained within the coverings of the spermatic cord are the neurovascular elements that supply the testes and the cremaster muscle. In addition, a tube-like process of the peritoneum, the **processus vaginalis** lies adjacent to the testes and its neurovascular pedicle. The course that these structures take in penetrating the anterior abdominal wall is described as the **inguinal canal**.

The foregoing is a simplified description. In reality, the testis is not the active agent in producing the tubular prolongations within which it is later contained. The scrotum and the coverings of the cord are formed early, even before the abdominal muscles differentiate, and, as Curl and Tromly properly point out, there is a definite inguinal canal in the male fetus before the testes pass through the abdominal wall, and in the female it is just as definite. Originally, the inguinal canal is an almost straight anteroposterior passage through the abdominal muscles; subsequently, during the process of growth it becomes more oblique.

This almost straight anteroposterior inguinal canal in the newborn and postnatal infant is the basis for the "indirect hernias" in children, which pass through the 3 layers of the abdominal wall.

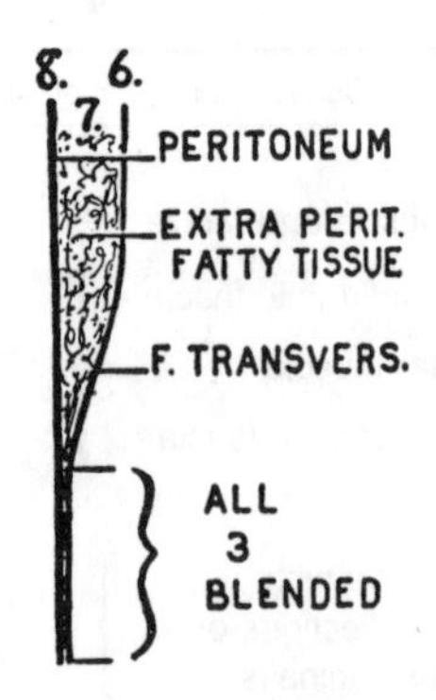

Figure 11.12. In the absence of fat the inner 3 layers of the abdominal wall blend to form a single layer.

INGUINAL CANAL

On each side of the abdomen, as the descending testis moves from the extraperitoneal fatty tissue and traverses the abdominal wall, it first encounters the abdominal wall layer formed by the **fascia transversalis** (*fig. 11.10*). This fascial lining is therefore the innermost fascial covering of the testis and spermatic cord. This tube-like fascial covering is called the **internal spermatic fascia**. The opening in the fascia transversalis that opens into the tube-like internal spermatic fascia is the **deep inguinal ring** and the entrance into the **inguinal canal**. The deep inguinal ring lies a fingerbreadth above the midpoint of the inguinal ligament (*fig. 11.7*). The deep inguinal ring also lies just lateral to the inferior epigastric artery, which is coursing superomedially in the extraperitoneal connective tissue. This relationship to the inferior epigastric artery is of considerable importance to the surgeon in classifying and repairing inguinal hernias.

The fascial coverings of the internal and external oblique muscles are also encountered by the descending testis as it traverses the abdominal wall. The fascia of the internal oblique muscle also contains muscle fibers that originate from the lateral aspect of the inguinal ligament. This muscle is the **cremaster muscle**, and this fascial investment of the spermatic cord is the **cremasteric fascia**. Contraction of the cremaster muscle will retract the testes back toward the inguinal canal.

External to the cremasteric fascia is a 3rd fascial covering of the testis and spermatic cord. This **external spermatic fascia** is derived from the fascia covering the aponeurosis of the external oblique muscle. The mouth-like opening of the external spermatic fascia is continuous with the **superficial inguinal** ring and the exit of the inguinal canal. The surface anatomy landmark for the superficial inguinal ring is the **pubic tubercle**. The superficial inguinal ring is immediately superior and medial to the pubic tubercle and can be palpated subcutaneously in the male by tracing the course of the spermatic cord from the scrotum to the abdominal wall.

The **inguinal canal** is therefore within the abdominal wall between the superficial and deep inguinal rings and is about 2 inches (4 cm) long. (1) The **anterior wall** of the inguinal canal is formed throughout by the aponeurosis of the external oblique. (2) The **floor** of the inguinal canal is formed by the inferior, "rolled-in" border of the aponeurosis of the external oblique. This floor is part of the **inguinal ligament** and **lacunar ligament** shown in *Figure 11.1.* (3) The **posterior wall** of the inguinal canal is formed throughout by the fascia transversalis. (4) The **roof** of the inguinal canal is formed by the fleshy fibers of the internal oblique and transversus abdominis that arch over the spermatic cord as they pass from their origin on the lateral one-half of the inguinal ligament to become the conjoint tendon, which inserts medially into the pubic crest, tubercle, and pectineal line.

The **cremaster muscle** is derived from the internal oblique fibers that arch over the spermatic cord. This shift in the position of the two deeper abdominal muscles as they relate to the lateral and medial aspects of the inguinal canal in the adult are important to the strength of the abdominal wall and the resistance to herniation when intra-abdominal pressures are increased. At the lateral aspect of the inguinal canal, the deep inguinal ring in the fascia transversalis is **reinforced anteriorly** by at least two of the abdominal muscles: internal oblique and aponeurosis of external oblique. The origin of the transversus abdominis is usually lateral to the position of the deep inguinal ring (*fig. 11.7*) and therefore not protective to the deep ring in the noncontracted state (Anson and McVay).

At the medial aspect of the inguinal canal, the superficial inguinal ring in the aponeurosis of the external oblique is **reinforced posteriorly** by the common tendinous insertion of the transversus abdominis and the internal oblique, the **conjoint tendon**. Breakdown in the structural continuity of these reinforcing anterolateral and postermedial walls of the inguinal canal underlie the susceptibility for inguinal hernia formation.

In the newborn and young child, the superficial inguinal ring is superimposed over the deep inguinal ring. The lack of a reinforcement by the abdominal muscles favors herniation, which could occur directly through both rings. With growth, the deep ring is shifted laterally and is not superimposed by the superficial ring.

Immediately posterior to the fascia transversalis that forms the posterior wall of the inguinal canal is the extraperitoneal tissue (*fig. 11.10*). This fatty loose connective tissue layer contains the important vascular elements of this region: the **inferior epigastric arteries** and the **obliterated umbilical arteries**. The inferior epigastric artery lies at the medial boundary of the deep inguinal ring and has the ductus (vas) deferens passing lateral to the artery to enter the deep ring and join the other neurovascular elements within the spermatic cord.

What Control Does One Have Over the Inguinal Canal: How Is It Closed?

Your conception of the inguinal canal may be helped by likening it to an arcade of 3 arches formed by the transversus abdominis, internal oblique, and external oblique muscles (*fig. 11.13*). During standing, there is continuous contraction of the internal oblique and transversus abdominis in the inguinal region. During coughing and straining, when the raised intra-abdominal pressure threatens to force a hernia through the canal, vigorous contraction of the arched fleshy fibers of the internal oblique and transversus abdominis "clamp down," without damaging the cord. The action is that of a half-sphincter.

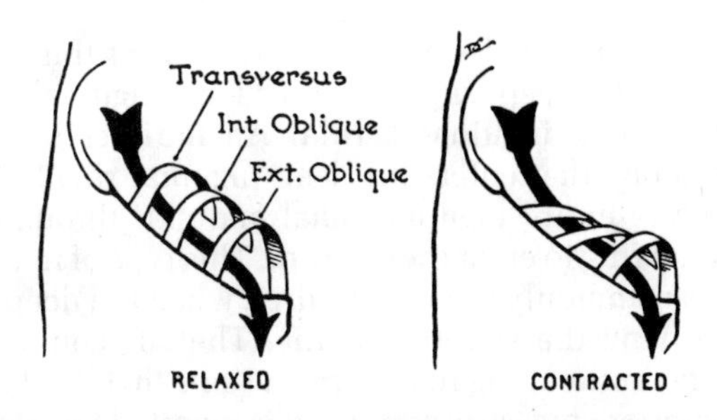

Figure 11.13. "The inguinal arcade." The inguinal canal likened to an arcade of 3 arches traversed by the spermatic cord.

Nerves

The nerves related to the inguinal canal are the **ilioinguinal, iliohypogastric**, and **genitofemoral nerves**. The ilioinguinal nerve (L1) pierces the internal oblique muscle and courses through the inguinal canal as a purely sensory nerve. It exits the inguinal canal through the superficial inguinal ring (*fig. 11.7*) and contributes to the innervation of the skin in the L1 dermatome. Its terminal branches in the male innervate the skin on the anterior aspect of the scrotum, and this nerve is anesthetized to make an incision in the anterior scrotum for vasectomy. The terminal branches of the ilioinguinal nerves in the female innervate the skin over the anterior aspect of the labium majus (pl., Labia majora). It is clinically important to note that this L1 nerve overlaps with the S2,3 dermatomes of the perineum. Surgical incisions in the skin of this region are done with care in operations utilizing local anesthetics.

The iliohypogastric nerve is also derived from the L1 spinal nerve and is sensory as it pierces the aponeurosis of the external oblique above the superficial inguinal ring (*fig. 11.7*). The **genital branch** of the **genitofemoral nerve** (L1,2) is the motor supply to the cremaster muscle in the inguinal canal (*fig. 11.9*). As noted before, eliciting the cremasteric reflex is done by stimulating the skin over the L1,2 dermatomes (medial aspect of the thigh) and observing the effect of the contraction of the cremaster muscle under the control of the L1,2 motor nerves in the genitofemoral nerve.

Inguinal Hernias

A hernia is the protrusion of a loop of organ or tissue through an anatomical opening. Since intraperitoneal fat, connective tissue and loops of bowel can abut against the lower abdominal wall, the protrusion of this tissue into either the deep, superficial, or both inguinal rings, constitutes one **inguinal hernia**. An **indirect inguinal hernia** is one that enters the deep inguinal ring. It may traverse the entire inguinal canal and exit through the superficial ring to enter the scrotum. This type of inguinal hernia is commonly seen in children when the deep ring is underlying the superficial ring. The surgeon would define the indirect inguinal hernia as one that lies lateral to the inferior epigastric artery and is invested by internal spermatic fascia.

The **direct inguinal hernia** only protrudes through the superficial inguinal ring. This occurs when the conjoint tendon no longer protects the superficial ring as the posterior wall of the inguinal canal. The indirect hernias are medial to the inferior epigastric artery and are covered by only external spermatic fascia. They occur most commonly in adult males.

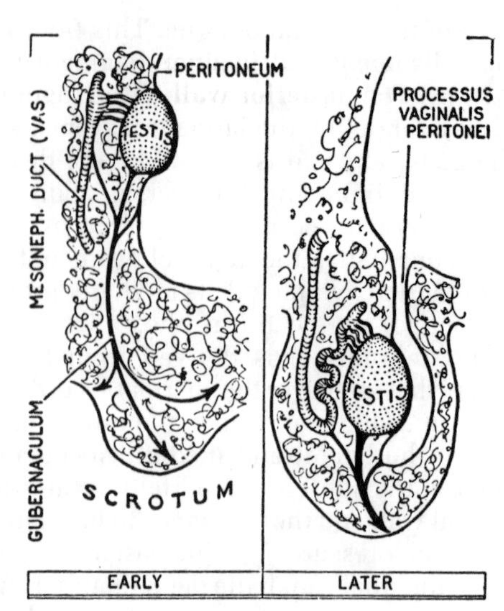

Figure 11.14. Descent of testis, processus vaginalis peritonei, and epididymis (diagrammatic).

Testis, Spermatic Cord, and Scrotum

DESCENT OF THE TESTIS

The testis (like the ovary) develops from mesodermal cells of the gonadal ridge covering the medial part of the primitive kidney (mesonephros). In early fetal life, a cord-like connective tissue band of cells, the **gubernaculum testis**, develops and connects the inferior pole of the developing gonad and surrounding mesodermal tissue to the skin that will become the scrotum (labium majora in the female) (*fig. 11.14*). Differential body growth in the embryo and fetus results in the descent of the gonads from the abdominal cavity to the pelvis (female) and scrotum (male). The gubernaculum serves to attach the inferior pole of the gonad to the skin.

In the male, the gubernaculum is also involved in drawing the processes vaginalis and the ductus (vas) deferens into the inguinal canal with the testis (*fig. 11.14*). This gonadal descent places the testis in the iliac fossa during the 3rd prenatal month; it traverses the inguinal canal during the 7th prenatal month and reaches the bottom of the scrotum after birth. The physical examination of the newborn male includes the assessment of the descent of the testis. Failure of the testis to descend into the scrotum requires a surgical correction in later postnatal life. Testes retained in the abdominal wall may produce nonviable sperm in adult life and may be more susceptible to testicular pathology.

In the female, the ovary remains in the pelvis but a **round ligament of the ovary** and its continuing **round ligament of the uterus** attaches the inferior pole of the ovary to the labium majus. These round ligaments are homologous structures to the gubernaculum testis. The **round ligament of the uterus** traverses the inguinal canal in a fashion similar to the spermatic cord. The smaller size of the inguinal canal and its contents in the female, result in far fewer inguinal hernias. The female is more susceptible to herniation in the femoral canal below the inguinal ligament.

The **scrotum** is the pouch of skin and subcutaneous tissue that contains the testes in adult life. The scrotum has a bilateral origin and is derived from the right and left labioscrotal folds (*fig. 11.15*). In the female, these folds remain individually distinct as the labia majora. In the male, the two labioscrotal folds fuse throughout in a midline raphe behind the penis. The scrotum contains two pouches separated by a midline partition and each testis occupies its own pouch.

The skin of the scrotum lacks a subcutaneous fascia but contains a sheet of smooth muscle, the **dartos muscle**, that is innervated by sympathetic nerves. This underlying smooth muscle is connected to the skin of the scrotum and causes wrinkling of the scrotum when a cold stimulus is applied to the skin. The function of the scrotum

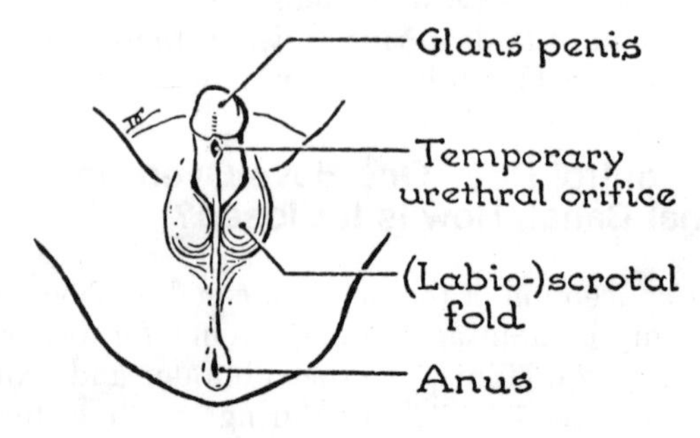

Figure 11.15. The scrotum has a bilateral origin.

is to "house" the testes in an environment that is below normal body temperature. Some forms of infertility in males are related to sustained elevated temperatures in the gonadal environment.

Nerves

The anterior scrotal (labial in female) region is innervated by the ilioinguinal nerve (L1). The posterior scrotal (labial in female) region is innervated by the 2nd, 3rd and 4th sacral nerves. While these nerves overlap to some degree in the skin, they are well separated in the spinal cord and body wall, and are therefore not frequently anesthetized in the same spinal or local anesthetic injection.

Vessels

The scrotum is supplied by the **external** and **internal pudendal arteries** and **veins**. The external pudendal arteries are derived from the femoral artery, and the internal pudendal arteries are derived from the internal iliac arteries. The corresponding veins accompany the arteries. It is clinically important to note that the lymphatic drainage of the scrotum to the superficial inguinal nodes is different than the lymphatic drainage of the testes, which follows the testicular arteries and veins up the spermatic cord to the abdominal lymph nodes.

The **spermatic cord** is made up of structures contained within its fascial coverings: external spermatic, cremasteric and internal spermatic fasciae. The constituents of the spermatic cord enter the deep inguinal ring lateral to the inferior epigastric artery. They pass through the inguinal canal and descends in the scrotum to the testis. As the spermatic cord emerges from the superficial inguinal ring, it rolls over the pubic tubercle and acquires its covering of external spermatic fascia.

The **constituents of the cord** are as follows.

1. Representatives of the inner 2 layers of the abdominal wall (usually fused and indistinct) (*fig. 11.10*):
   a. Processus vaginalis (derived from the parietal peritoneum)
   b. Areolar connective tissue that is continous with the extraperitoneal connection tissue.
2. Structures connected to the testis:
   a. Ductus (vas) deferens
   b. Testicular artery and accompanying sympathetic nerves
   c. Pampiniform plexus of veins, which forms the single testicular vein in the abdomen
   d. Lymphatics
   e. The artery of the ductus deferens and its accompanying veins and sympathetic nerves.
3. Other structures:
   a. The genital branch of the genitofemoral nerve, which innervates the cremasteric muscle.

The processus vaginalis is a tube of peritoneum from the fetal coelomic cavity that is pulled into the inguinal canal adjacent to the testis as the latter descends. The portion of the processus vaginalis that is contained in the inguinal canal normally becomes obliterated around the time of birth and becomes a fibrous thread of connective tissue. When this portion remains open, it potentiates the development of indirect inguinal hernia formation. The distal portion of the processus vaginalis remains patent and becomes the **tunica vaginalis**. The testis invaginates the balloon-like patent tunica vaginalis within the scrotum. The resultant effect is that the outer layer of the testis has a visceral layer of peritoneum covering its **tunica albuginea**, and a parietal layer of peritoneum lines the innermost fascial covering of the testis derived from the fascia transversalis. An intervening "mini" coelom, the **cavity of the tunica vaginalis**, exists between these two layers of peritoneum surrounding the testis (*fig. 11.16*). This cavity contains a small amount of serous fluid, and the accumulation of large amounts of fluid in this space is called a **hydrocele**.

The **ductus deferens** (vas deferens) (*fig. 11.17*) is a duct that conveys spermatozoa from the testis to the ejaculatory duct. The ductus deferens in a continuation of the epididymis. It is about 18 inches (45 cm) long and ascends from the testis, in the spermatic cord, through the inguinal canal, and then around the inferior epigastric artery to reach the back of the bladder and seminal vesicle in the floor of the pelvis. It may be ligated or transected within its course in the scrotum in a surgical vasectomy.

Testis (*fig. 11.16*)

The **testis** is an ovoid gland, measuring $4 \times 3 \times 2$ cm. Its normal size is quite variable (Farkas). It is enveloped in the **tunica vaginalis testis** except where the epidid-

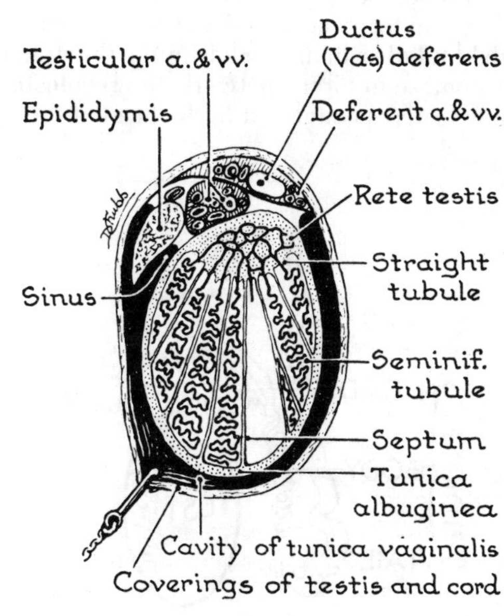

Figure 11.16. Right testis in transverse section.

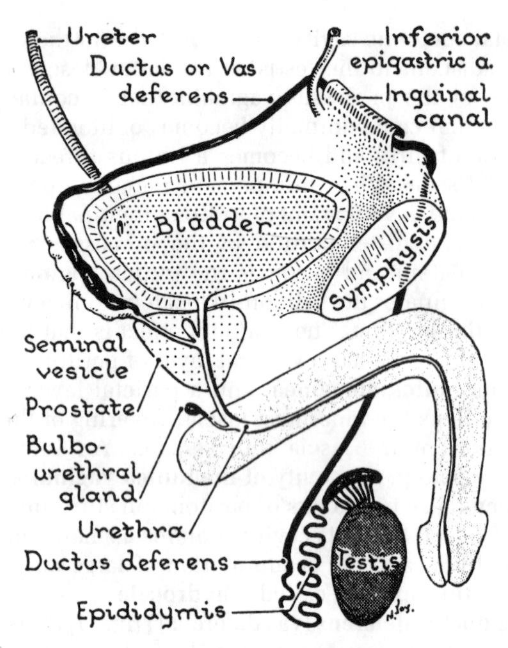

Figure 11.17. Diagram of ductus (vas) deferens.

ymis and the structures within the spermatic cord are attached to the upper pole and posterior surface. It has a tough white outer coat, the **tunica albuginea**. Posteriorly, the outer coat is thickened and less dense and is known as the **mediastinum testis** (*fig. 11.20*).

Areolar septa extend from the mediastinum testis to the tunica albuginea and divide the testis into about 250 elongated pyramidal compartments. Each compartment contains a lobule formed from 2 or more thread-like **seminiferous tubules**, which are each a half meter long (*fig. 11.16*). The seminiferous tubules are closely packed and convoluted, except at the apex of the compartment where they join together and take a short, straight course, the **straight tubules**. In the mediastinum, the straight tubules anastomose to form a network, the **rete testis**.

The rete testis is connected to the head of the epididymis by a series of 6–12 fine ducts, the **efferent ductules** (*fig. 11.18*).

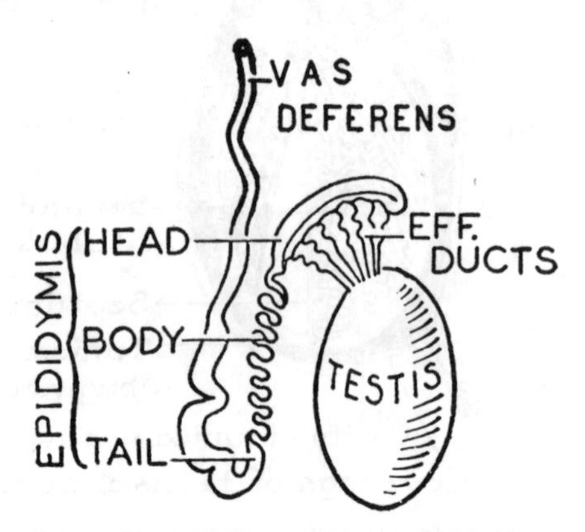

Figure 11.18. The testis and epididymis.

Epididymis

The epididymis is applied to the upper pole and posterior border of the testis. It is somewhat larger in diameter that a standard pencil. It tapers as it descends on the posterior border of the testis to connect to the vas deferens. The epididymis is subdivided into a **head** or upper part, the **body** or intermediate part, and a **tail** which turns upward as the vas deferens.

Structure

In the head of the epididymis, the efferent ductules become coiled to form lobules and empty into the duct of the epididymis. The duct of the epididymis forms the body and tail of the organ. The duct is greatly twisted and folded, and when unraveled, it is approximately 20 feet (about 7 meters) in length.

Vessels and Nerves of Testis, Epididymis, and Ductus

ARTERIES

The **testicular artery** is a branch of the abdominal aorta at the L2 vertebral level. It pierces the mediastinum testis in a variety of ways, branches, and anastomoses with the **artery of the ductus deferens** and the **cremasteric artery** (*figs. 11.19* and *11.20*). The artery of the ductus deferens is derived from the inferior vesical branch of the internal iliac artery and the cremasteric artery is a branch of the inferior epigastric artery.

Veins

Up to a dozen veins from the region form a anastomising plexus, the **pampiniform plexus**, which ascends, reducing to 2 or 3 veins and ultimately one testicular vein. The right testicular vein joins the inferior cava, and the left testicular vein drains into the left renal vein. The testicular veins may become dilated and tortuous, a clinical condition called **varicocele**. The venous channels of the testes are quite prominent and much more obvious than the arterial vessels. This prominence of veins is characteristic of all endocrine glands and probably ensures the ready transport of testosterone from the testis into the systemic venous circulation.

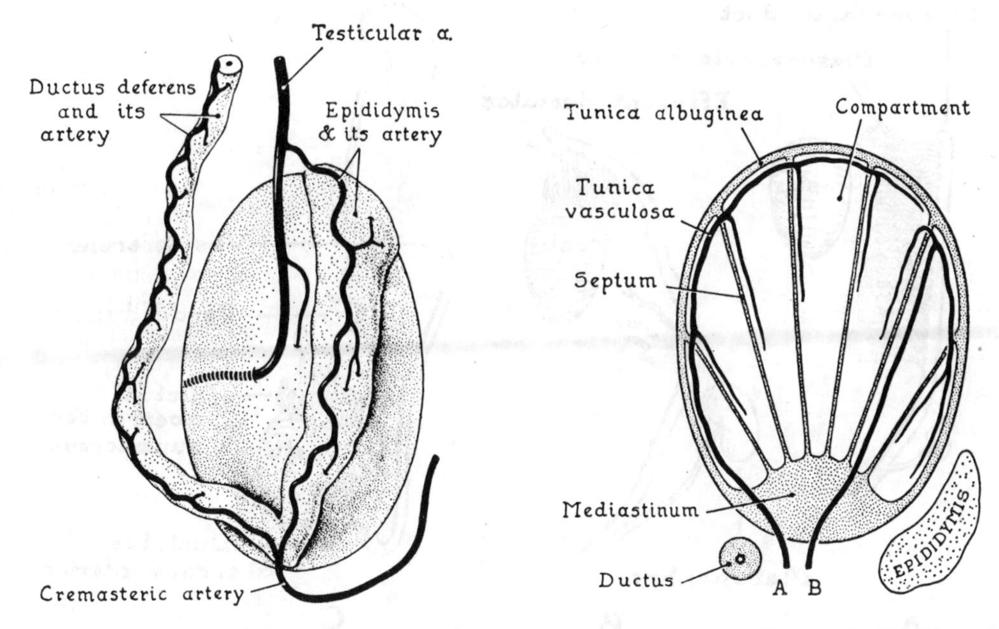

Figures 11.19 and 11.20. The testicular artery and its anastomoses (injection and dissection by Dr. Neil Watters).

LYMPHATICS

The lymphatics of the testis, like those of the ovary, end in lymph nodes situated between the common iliac and renal vessels (*fig. 11.21*).They are distinctively different from the lymph vessels of the scrotum and penis, which drain into the inguinal nodes. This anatomical difference is clinically important in the spread of carcinoma from these two closely related structures.

Nerves

The testis is innervated by sympathetic fibers derived from thoracic segments T6–10. These nerves are mainly vasomotor. Accompanying the sympathetic nerves are afferent fibers. Pain in the testis is referred to the "pit of the stomach" region or T8–10 dermatomes. This referral of pain is due to the entry of these visceral afferent fibers into the spinal cord at the same level as the cutaneous nerves that innervate the skin over the stomach region.

The ductus deferens is a tubular smooth muscle structure that is highly innervated by sympathetic fibers. These sympathetic fibers are derived from the inferior hypogastric plexus, which is made up of fibers from T11, 12 and L1 (Mitchell).

DEVELOPMENT (fig. 11.22)

The *duct system of the testis* is of 3-fold origin: (1) The seminiferous and straight tubules and the rete testis are developed from anatomosing cords of cells in the genital ridge between vertebral segments L4 and S2. (2) The efferent ductules and the lobules in the head of the epididymis are formed from the 6 or more mesonephric tubules that succeed in establishing connections between the rete testis and the duct of the epididymis. (3) The

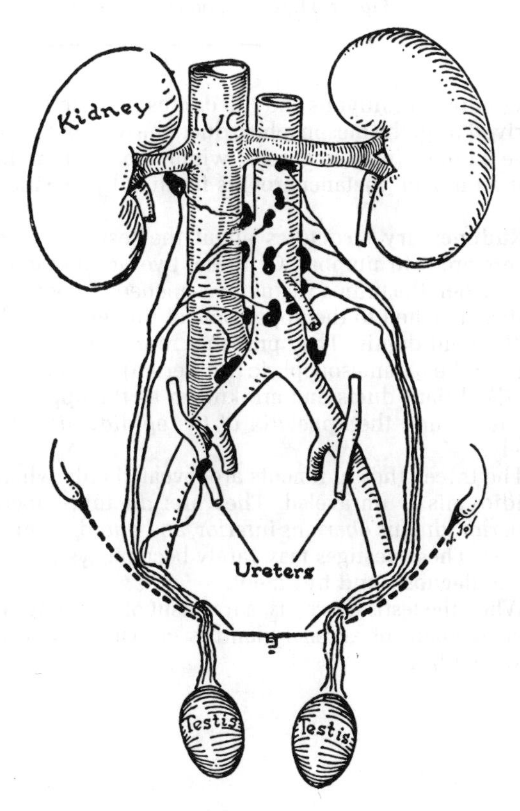

Figure 11.21. The lymph drainage of the testes (after Jamieson and Dobson).

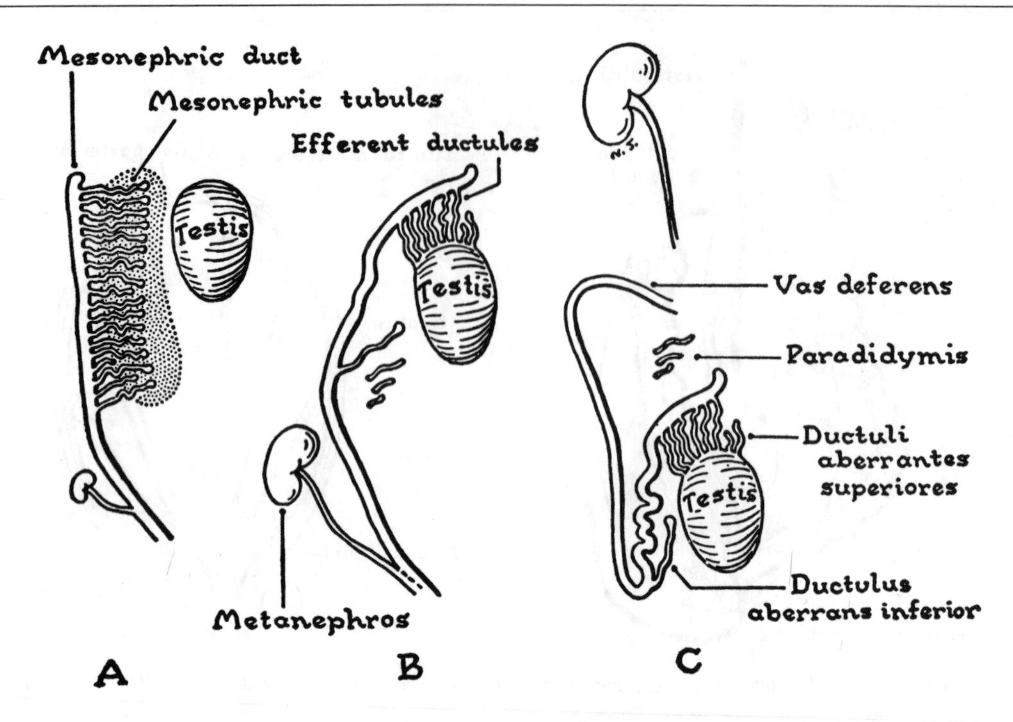

Figure 11.22. Some mesonephric tubules make connections; others persist as vestigial structures.

duct of the epididymis and the ductus (vas) deferens are derived from the mesonephric duct, the duct of the primitive kidney (mesonephros), while the modern mammalian model (metanephros) is forming by its side (*fig. 11.22*).

Rudimentary structures about the testis and epididymis are 5 in number. Of these, two are brought into view when the tunica vaginalis is opened. They are little bodies attached to the upper pole of the testis and head of the epididymis. They probably represent the cranial ends of the paramesonephric (Mullerian) and mesonephric (Wolffian) ducts and are known as the *appendix of the testis* and the *appendix of the epididymis*, respectively.

The three other rudiments are revealed only when the epididymis is unraveled. They are: *ductulus aberrans superior, ductus aberrans inferior,* and *paradidymis* (*fig. 11.22*). These vestiges may rarely become cystic and require diagnosis and treatment.

When the testis descends, a fragment of spleen is rarely, or a fragment of adrenal gland is occasionally, carried down with it.

Clinical Mini-Problems

1. **Which dermatome is associated with the skin (a) adjacent to the umbilicus? (b) overlying the pubic crest?**
2. **Which aortic branches are interconnected by the epigastric arteries in the rectus sheath?**
3. **Which nerve exits through the superficial inguinal ring? It can be anesthetized at this point in preparation for a vasectomy.**
4. **What is the insertion site of the conjoint tendon?**
5. **a. How many layers of spermatic fascia would cover the contents of an indirect hernia in a child? Would the herniation be medial or lateral to the inferior epigastric artery?**

(Answers to these questions can be found on p. 585.)

12

Abdominopelvic Cavity

Clinical Case 12.1

Patient Mary Ann T. This 14-year-old girl was rushed to emergency after she complained of acute pain in her lower abdomen. She is particularly sensitive to your palpation of the anterior abdominal wall between the umbilicus and the right anterior iliac spine. The resident on duty suspects appendicitis but has requested a consultation with a general surgeon. The surgeon includes Meckel's diverticulitis, a ruptured ovarian follicle, and bowel obstruction as unlikely but possible causes for the patient's signs and symptoms. Surgery is imperative to distinguish between these possible disorders.

Followup. The anterior abdominal wall and parietal peritoneum were opened in the lower right quadrant. The greater omentum was adherent to an inflamed and enlarged vermiform appendix, now the obvious diagnosis. The surgeon retracted the greater omentum; exposed, isolated, and ligated the appendicular artery within the mesoappendix; and ligated the appendix near its origin on the cecum. An appendectomy was performed by excising the appendix and appendicular artery distal to the ligatures. The paracecal area was irrigated with sterile saline, and all foreign debris was flushed from the peritoneal cavity. The abdominal wall defect was closed one layer at a time from the parietal peritoneum out to the skin. The patient was sent to the recovery room for observation and placed on antibiotic therapy. Five days later she was released from the hospital to continue her convalescence at home.

Landmarks

Surface anatomy features, such as the costal margin, the umbilicus, and the palpable components of the pelvic girdle, are important guides in orienting the clinician to the location of the viscera in the abdominopelvic cavity. Radiographic landmarks are even more consistent and constantly used as references for localizing the visceral structures in this region.

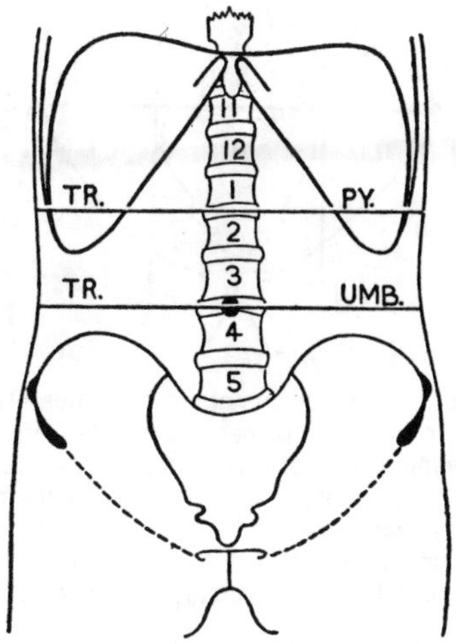

Figure 12.1. Horizontal planes with the vertebral column as a measuring rod. To remember 1:2 and 3:4 should not tax the memory. TR-PY, transpyloric; TR-UMB, transumbilical.

The **transpyloric plane** bisects the line joining the *top* of the sternum to the *top* of the pubic symphysis. It is a horizontal plane that passes through the intervertebral disc, which separates the 1st and 2nd lumbar vertebrae (*fig. 12.1*). This plane marks the level of the pyloric region of the stomach. The **transumbilical plane** passes through the umbilicus, or navel, and lies at the level of the intervertebral disc between the 3rd and 4th lumbar vertebrae. It is true that the position of the umbilicus varies somewhat with age, sex, obesity, and posture but still serves as a valuable landmark in clinical examinations.

The anterior abdominal wall can be simply divided into quadrants or further subdivided into the nine regions as depicted in *Figure 12.2*. Various abdominal viscera normally underlie these surface regions and are palpated or visualized by radiography in these locations.

Protection of the Viscera

The abdominal viscera are well protected by bony elements of the rib cage superiorly and the pelvic girdle

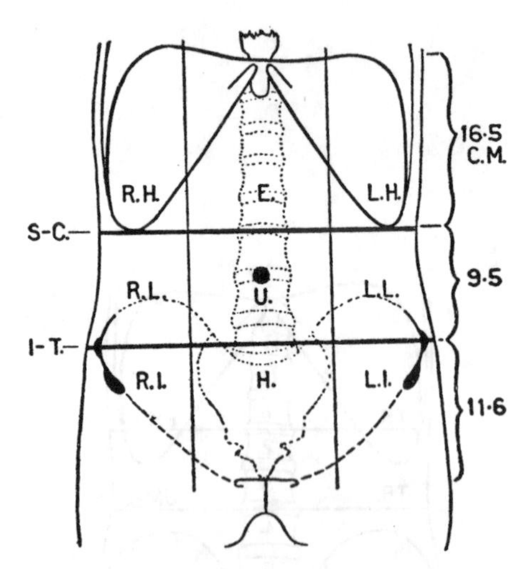

Figure 12.2. The 9 regions of the abdomen. The right and left sagittal or "vertical" planes are erected on the midpoint of the line joining the corresponding anterior superior spine of the ilium to the top of the symphysis pubis, the **midinguinal point**. *RH, LH,* right and left hypochondriac; *RI, LI,* right and left inguinal or iliac; *E,* epigastric; *U,* umbilical, *H,* suprapubic or hypogastric; *I-T,* intertubercular; *S-C,* subcostal.

inferiorly. Since the domes of the diaphragm rise to the level of the 5th rib in the midclavicular line, a large portion of the superior abdominal cavity is within the confines of the lower rib cage. The inferior aspects of the abdominal cavity are protected by the ilia of the hip bones. The 11th and 12th ribs protect the posterior and lateral aspects of the abdominal cavity and closely approximate the iliac crests at the level of the 4th lumbar vertebra. The midline and lower anterior abdominal wall lack bony elements but are reinforced by the rectus sheath and the aponeurotic components of the abdominal muscles, which afford a flexible but efficient protective wall for the abdominal viscera.

Definitions

The abdominopelvic cavity is the body space enclosed by the bones, muscles, and fasciae of the abdominal and pelvic walls. It is limited superiorly by the diaphragm and inferiorly by the pelvic floor (pelvic diaphragm). The cavity is divided into an abdominal cavity (proper) and a pelvic cavity as depicted in *Figure 12.3*. The two cavities are continuous but demarcated by the position of the pelvic inlet (pelvic brim).

The abdominal cavity and some of the organs and surfaces of the pelvic cavity are lined by a membrane, the **peritoneum**. This peritoneum is composed of a single layer of squamous cells supported by an external areolar connective tissue, which binds it to the walls of these cavities. It is analogous to the parietal pleural lining of the thoracic cavity and is sensitive to certain painful stimuli.

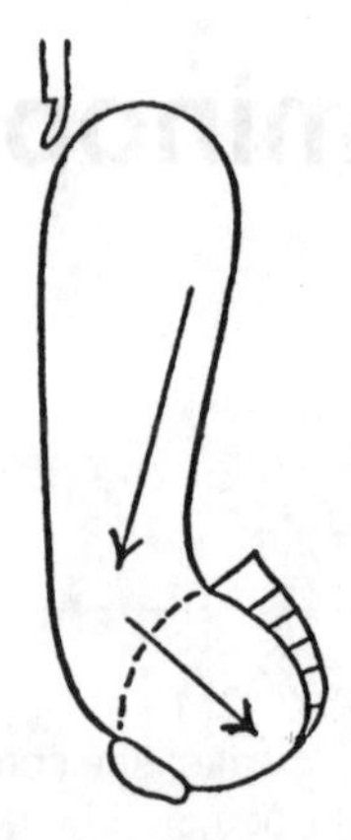

Figure 12.3. The abdominal and pelvic cavities are almost set at a right angle with each other.

The **peritoneal cavity** is a complicated sac, lined everywhere with peritoneum and moistened with a serous peritoneal fluid that is secreted into the cavity by the peritoneal squamous cells. The cavity is really a potential cavity. It contains a molecular thin layer of serous fluid, *in situ*, and is visualized as a space when air is admitted during an operation or when fluids collect in a number of conditions. Accordingly, each part of its free surface is in contact with, and rubs against, some other part of its free surface. The peritoneal fluid within the cavity lubricates the opposing surfaces and prevents fusion (peritoneal adhesions), which limit movement of the viscera and may produce painful stimuli when movement of the viscera is taking place. In the female, the peritoneal cavity communicates with the exterior through the uterine tubes, uterus, and vagina. Ovulation occurs into the abdominal cavity through the peritoneum lining the ovary, and the uterine tubes receive the ovulated egg intraperioneally and transport it to the uterine cavity, which is extraperitoneal (outside of the abdominal cavity). The abdominal cavity in the male is completely lined with peritoneum and has no normal communication with the exterior.

The peritoneum lining the walls of the peritoneal cavity is called the **parietal layer** of peritoneum or **parietal peritoneum**. The peritoneum lining the organs or viscera within the peritoneal cavity is called the **visceral layer** of peritoneum or **visceral peritoneum**. The two layers of peritoneum are continuous through certain folds and double layers formed when a layer of parietal peritoneum from the right and left side of the abdominal cavity approach the midline and then reflect onto the viscera within the cavity (see *fig. 12.39*). These double folds of peritoneum that exist between the parietal and visceral layers are the **mesenteries** and **ligaments** of the abdominal cavity. Use of the term "ligament" in this sense does not

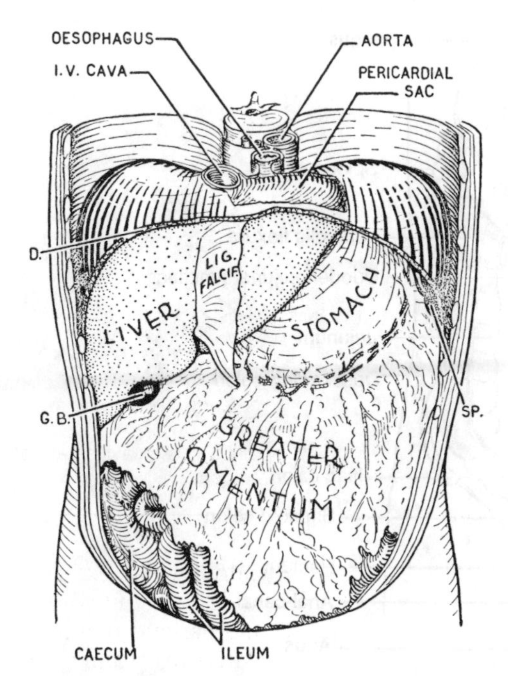

Figure 12.4. Abdominal contents undisturbed. *GB*, gallbladder; *SP*, spleen; *D*, cut edge of diaphragm; *IVC*, inferior vena cava.

imply a collagenous structure connecting two bony elements. An abdominal ligament is a special peritoneal fold, usually containing vessels and nerves that run from the body wall to an abdominal organ or structure.

A double layer of peritoneum connecting the stomach to another structure is called an **omentum**. The **greater omentum**, the most obvious (*fig. 12.4*), connects the greater curvature of the stomach to the posterior abdominal wall. The greater omentum serves to convey blood vessels and nerves from the posterior abdominal wall to the stomach and also functions as vital defensive organ within the abdominal cavity.

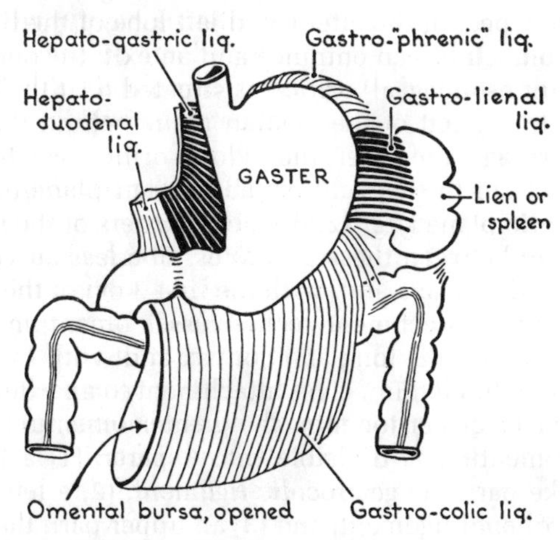

Figure 12.5. Diagram of the 2 parts of the lesser omentum and of the 3 parts of the greater omentum.

The **lesser omentum** arises from the lesser curvature of the stomach and connects to the liver (*fig. 12.5*). It is frequently described as the **hepatoduodenal ligament** and the **hepatogastric ligament**. Both of these structures provide pathways for blood vessels and nerves to pass from the posterior body wall, around the stomach, and to the liver and gallbladder.

Parts of the Gastrointestinal Canal

The digestive passage extends from the mouth to the anus. It is divisible into the following parts: mouth, pharynx, esophagus, stomach, small intestine, large intestine, rectum, and anus.

The stomach and intestines are collectively called the gastrointestinal canal or gut. The last 2 cm (1 inch) of the esophagus to the superior one-half of the rectum are situated within the abdominal linings of the abdominal cavity.

The esophagus pierces the left dome of the diaphragm at the level of the 10th thoracic vertebra and connects with the cardiac region of the **stomach**. The stomach lies to the left of the median plane in the upper left quadrant of the abdomen. It exits via the **pylorus** into the duodenum of the small intestine (*fig. 12.6*). The lower aspect of the stomach lies on the transpyloric plane as it crosses the midline to reach the duodenum. The small intestine is coiled within the central aspect of the abdominal cavity and is subdivided into 3 parts: **duodenum**, **jejunum**, and **ileum**.

The duodenum (approximately 12 fingerbreadths in length) is a horseshoe-shaped portion of the small intestine that has lost its primitive mesentery and is firmly attached to the posterior abdominal wall throughout most of its length. It begins at the pylorus and ends at the **duodeno-jejunal junction**, which is on the left side of the midline just below the transpyloric plane (*fig. 12.6*). Since it is, in large part, firmly attached to the posterior body wall, it serves as a significant reference organ for radiological examination of the abdominal viscera. The more mobile components of the small intestine are the **jejunum** and the **ileum**. This portion of the small intestine extends from the duodeno-jejunal junction to the right iliac fossa, where the ilium connects to the large intestine at the **ileocecal orifice** (ostium) (*fig. 12.6*).

The **large intestine** is subdivided into: vermiform appendix, cecum, ascending colon, right colic flexure, transverse colon, left colic flexure, descending colon, sigmoid colon, rectum, anal canal, and anus (*fig. 12.6*).

The cecum is the blind cul-de-sac situated below the ileocecal orifice. The vermiform (worm-shaped) **appendix** opens into the cecum about 2 or 3 cm (1 inch) below

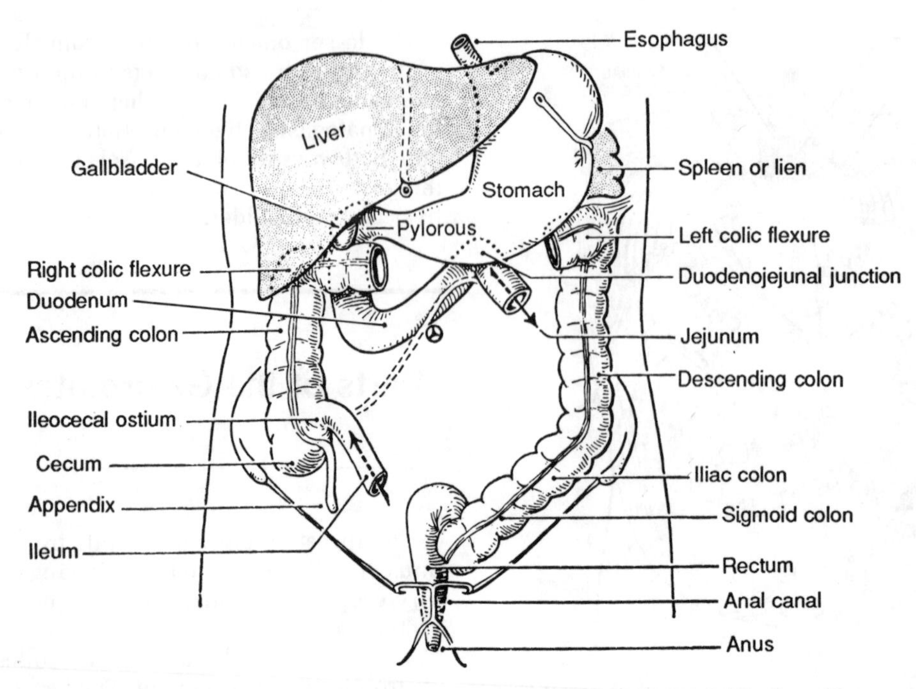

Figure 12.6. The names of the various parts of the digestive tract and their dispositions.

the ileocecal orifice (*fig. 12.6*). The **ascending colon** ascends from the right iliac fossa, across the iliac crest, to the undersurface of the right lobe of the liver. At that point the ascending colon, which is anterior to the right kidney (*fig. 12.7*), makes a bend (the **right colic or hepatic flexure**) and becomes the **transverse colon**. The transverse colon traverses the abdominal cavity just below the transpyloric plane. It extends from the right colic flexure, anterior to the duodenum, to the region of the spleen (*fig. 12.6*) and left kidney (*fig. 12.7*). Inferior to the spleen the colon bends again at the **splenic flexure** to descend on the left side of the abdomen. The **descending colon** descends into the left iliac fossa and over the pelvic brim

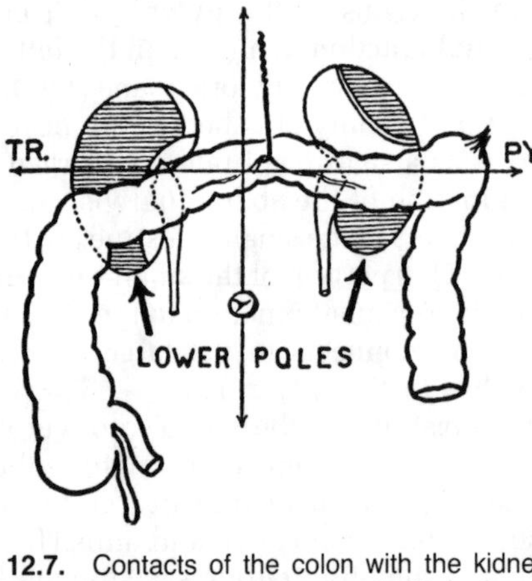

Figure 12.7. Contacts of the colon with the kidneys. The *shaded parts* of the kidneys are covered with peritoneum and are palpable when the peritoneal cavity is open. *Py*, pylorus.

where it becomes the **sigmoid colon**. The sigmoid colon continues into the pelvis to reach the midline anterior to the sacrum where the colon becomes the **rectum**. The rectum descends into the pelvic cavity to the level of the pelvic diaphragm (floor). The **anal canal** traverses the floor of the pelvis and the gastrointestinal tract opens through the **anus** onto the exterior surface of the perineum.

EXAMINATION OF GASTROINTESTINAL CANAL

By inspection and by handling, one may examine the abdominal portions of the alimentary canal.

The last part of the *esophagus* lies in a groove on the posterior aspect of the attenuated left lobe of the liver.

The **stomach** has an entrance and an exit: the one, the *cardiac* (or esophageal) *orifice*, is situated near the heart 2–3 cm to the left of the median plane behind the 7th costal cartilage; the other, the *pyloric orifice*, is situated about 2–3 cm to the right of the median plane on the transpyloric plane (*fig. 12.8*). The 2 borders of the stomach extend between these 2 orifices. The *less curvature* is short and concave and, with the first 3 cm of the duodenum, gives attachment to the *lesser omentum*. The *greater curvature* is long and convex and, with the first 3 cm of the duodenum, gives attachment to an extensive double layer of peritoneum, the *greater omentum*. The greater omentum is divisible into 3 parts: (1) a lower apron-like part, the *gastrocolic ligament*; (2) a left part, the *gastrolienal ligament*; and (3) an upper part, the *gastrophrenic ligament* (*fig. 12.5*). These are attached to the

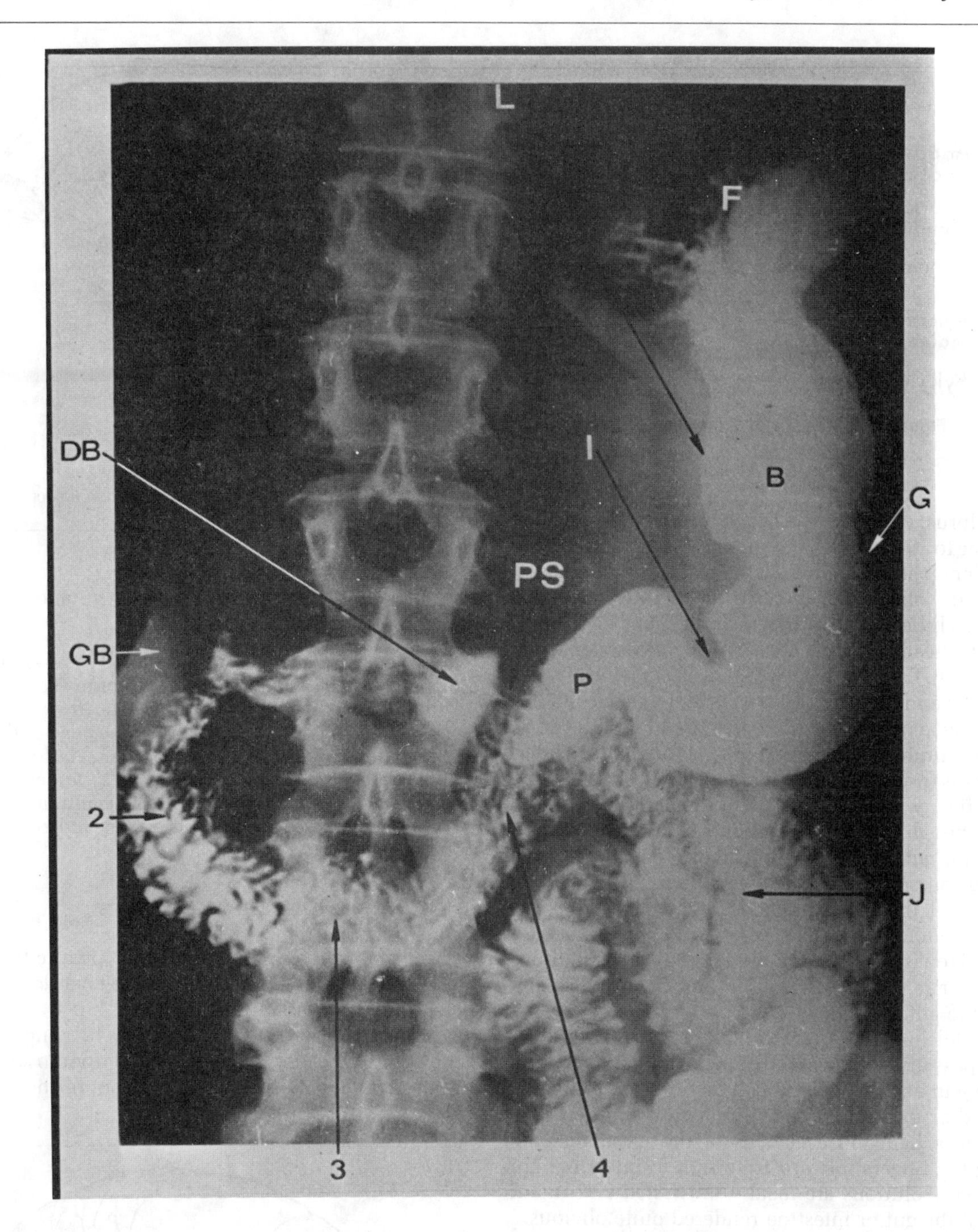

Plate 12.1. Stomach and duodenum in upper GI series (swallow of barium; A-V view, recumbent). *B*, body of stomach; *F*, fundus of stomach; *DB*, duodenal bulb; 2, 3, 4, second, third, and fourth parts of duodenum; *G*, greater curvature of stomach; *I*, incisura angularis; *J*, jejunum; *L*, lesser curvature of stomach; *P*, pyloric antrum; *PS*, pyloric sphincter; *GB*, gallbladder (partly opaque because of simultaneous intravenous cholecystography).

transverse colon, spleen (lien), and diaphragm, respectively.

The stomach is subdivided thus: a line drawn horizontally at the level of the cardiac orifice separates the *fundus* from the *body* (*fig. 12.8*). An oblique line joining an *incisura* (notch) on the lesser curvature (incisura angularis) to the greater curvature separates the body from the pyloric part of the stomach. Another oblique line passing from an indentation on the greater curvature (sulcus intermedius) to the lesser curvature subdivides the pyloric part into a large chamber, the *pyloric antrum*, and a more tubular portion, the *pyloric canal*, which ends at the *pylorus* (*Plate 12.1*).

The **duodenum** is the fixed part of the small gut or the part that has lost its primitive mesentery. Although 25 cm in length, its 2 ends are but 5 cm apart, for it begins

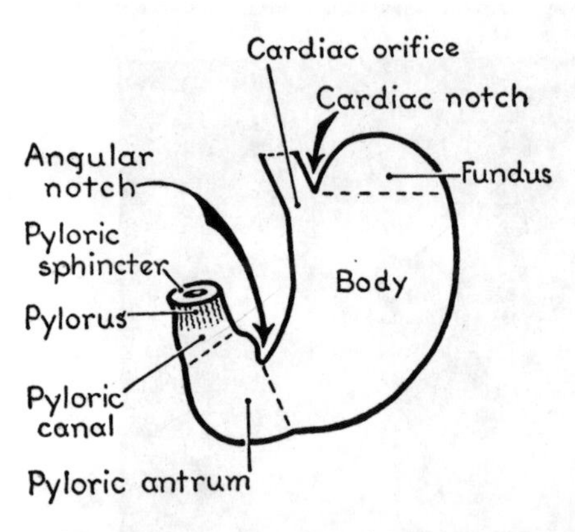

Figure 12.8. The parts of the stomach.

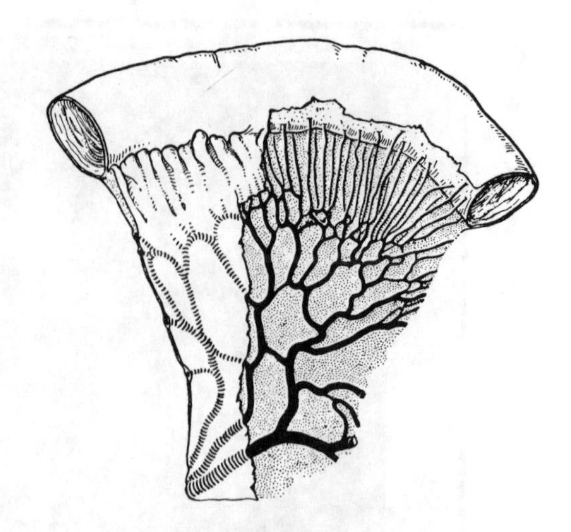

Figure 12.9. Arteries of jejunum injected.

at the pylorus, which lies on the transpyloric plane about 2 or 3 cm to the right of the median plane and ends at the duodenojejunal junction, where it is continuous with the jejunum, about 2 or 3 cm to the left of the median plane and a little below the transpyloric plane. It is molded around the head of the pancreas and is horseshoe-shaped (*see fig. 14.9*). Later (p. 174), we shall see that it is divided into 4 parts and that the 2nd part is largely concealed by the transverse colon which crosses it and adheres to it.

The **remainder of the small gut** retains its mesentery and extends from the duodenojejunal junction to the ileocecal orifice, which is situated in the right fossa (*Plate 12.2*). Although a distance of about 15 cm separates these 2 points, the gut, under dissecting room conditions, steers a varying course of 5–7 m between them. The root of the mesentery of the gut is attached diagonally across the posterior abdominal wall between the 2 points; accordingly, it likewise is about 15 cm long, whereas its intestinal border is elaborately ruffled and frilled to accommodate the gut.

The small intestine is so convoluted and mobile that you can pass many coils of it through your hands without being able to decide whether it is leading you to its duodenal end or to its cecal end. But, by the simple device of placing a hand on each side of the mesentery and drawing the fingers forward from root to intestinal border, the convolutions are locally untwisted and the direction of the gut or intestine rendered quite obvious.

Jejunum Contrasted with Ileum

The upper two-fifths of the free part of the small gut are called jejunum, and the lower three-fifths ileum. The jejunum has a greater digestive surface than the ileum, because (1) its *diameter* is greater; (2) its spirally arranged folds of mucous membrane, called *plicae circulares* (p. 174), are bigger and more closely packed; and (3) the minute finger-like projections of its mucous membrane, called *villi*, are larger and more numerous. Hence, its wall feels thick and velvety, whereas the wall of the smaller-calibered ileum with its fewer plicae and villi may be almost parchment-like in thinness.

The extraperitioneal *fat*, normally present in the mesentery, creeps along the vessels onto the ileal wall but fails to reach the jejunal wall; hence, there are translucent "windows" in the mesentery at the edge of the jejunum. Further, the disposition of the *vessels* from which the *vasa recta* spring (Arcades, p. 172) to become progressively more complex from the jejunum to the ileum (Michels). Also the vasa recta become progressively shorter (*figs. 12.9, 12.10*).

Features of the Large Intestine

The *large gut* forms 3½ sides of a "picture-frame" (*Plate 12.3*). Its outer longitudinal muscle coat does not form a complete coat, as in the small gut, but it is arranged in 3 narrow bands, the *teniae coli*, which, being shorter than the gut itself, cause it to be gathered up into *sacculations*. These teniae lead directly to the root of the appendix.

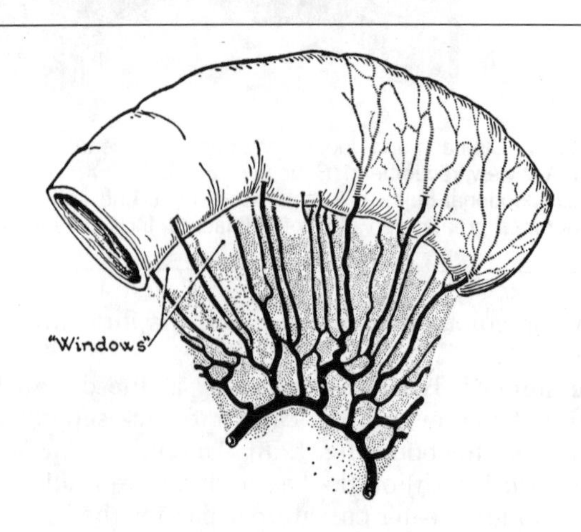

Figure 12.10. Arteries of ileum injected.

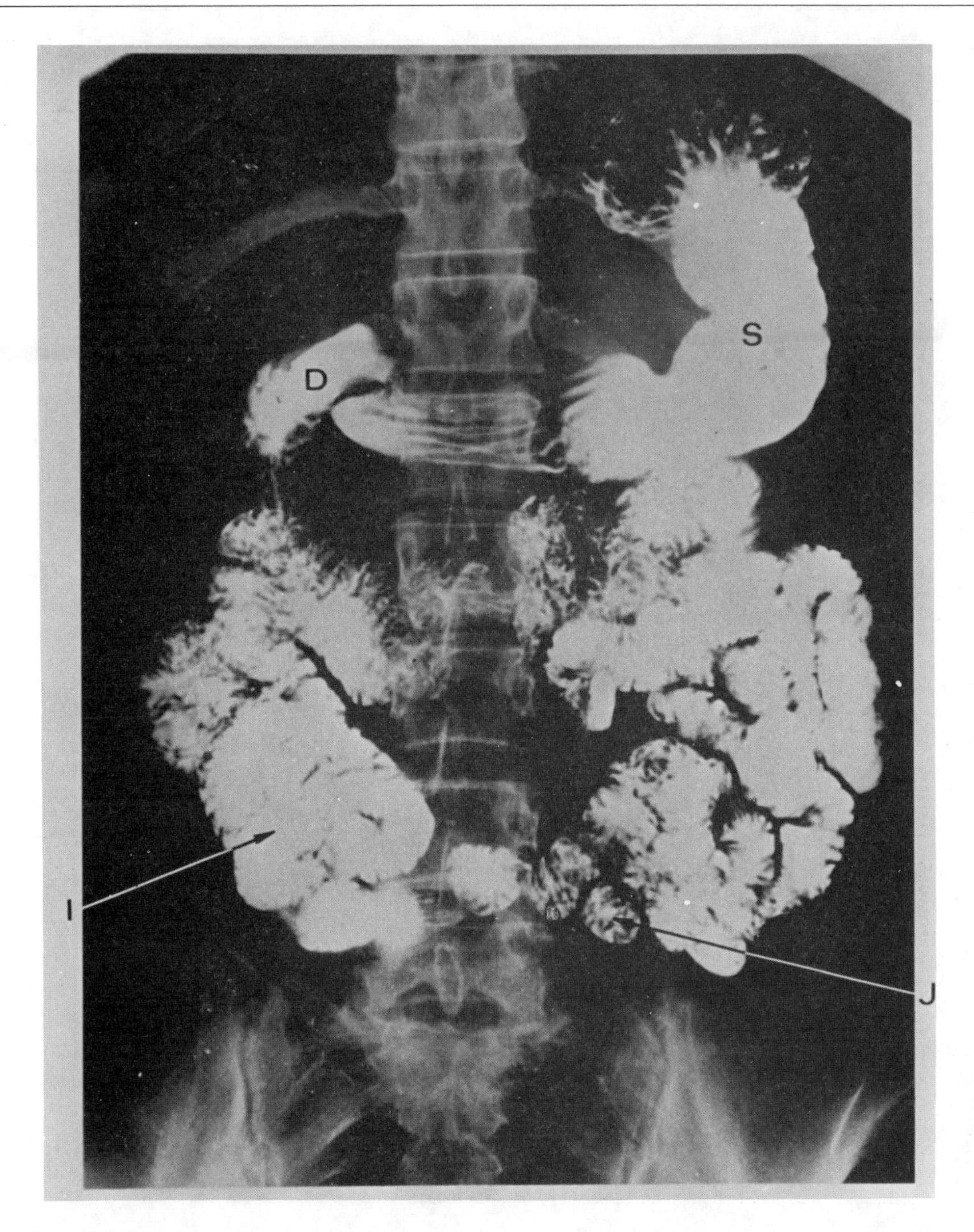

Plate 12.2. Jejunum and ileum outlined in barium (A-P view). *D*, duodenum; *I*, ileum; *J*, jejunum; *S*, stomach.

Peritoneal bags of fat, *appendices epiploicae*, hang from the large gut throughout its whole length (*fig. 12.11*). Those from the appendix, cecum, and rectum generally contain no fat.

Size alone does not necessarily distinguish large gut from small gut, the descending colon commonly having a *caliber* less than that of the small gut.

The *primitive mesentery*, possessed by the large gut during prenatal life, is constantly retained by the transverse and sigmoid colons, while the appendix acquires a mesentery and the cecum is free. The extent to which the ascending and descending colons lose their primitive mesenteries varies (*fig. 12.12*).

The **cecum** is free and lies in the right iliac fossa, commonly "hanging" over the pelvic brim. The cecum may have a short mesentery or even 2 mesenteries, with a *retrocecal fossa* extending upward between them, a favorite hiding place of the appendix. An extensive retrocecal fossa is, of course, a *retrocolic fossa*.

The **vermiform appendix** in fetal life opened into the

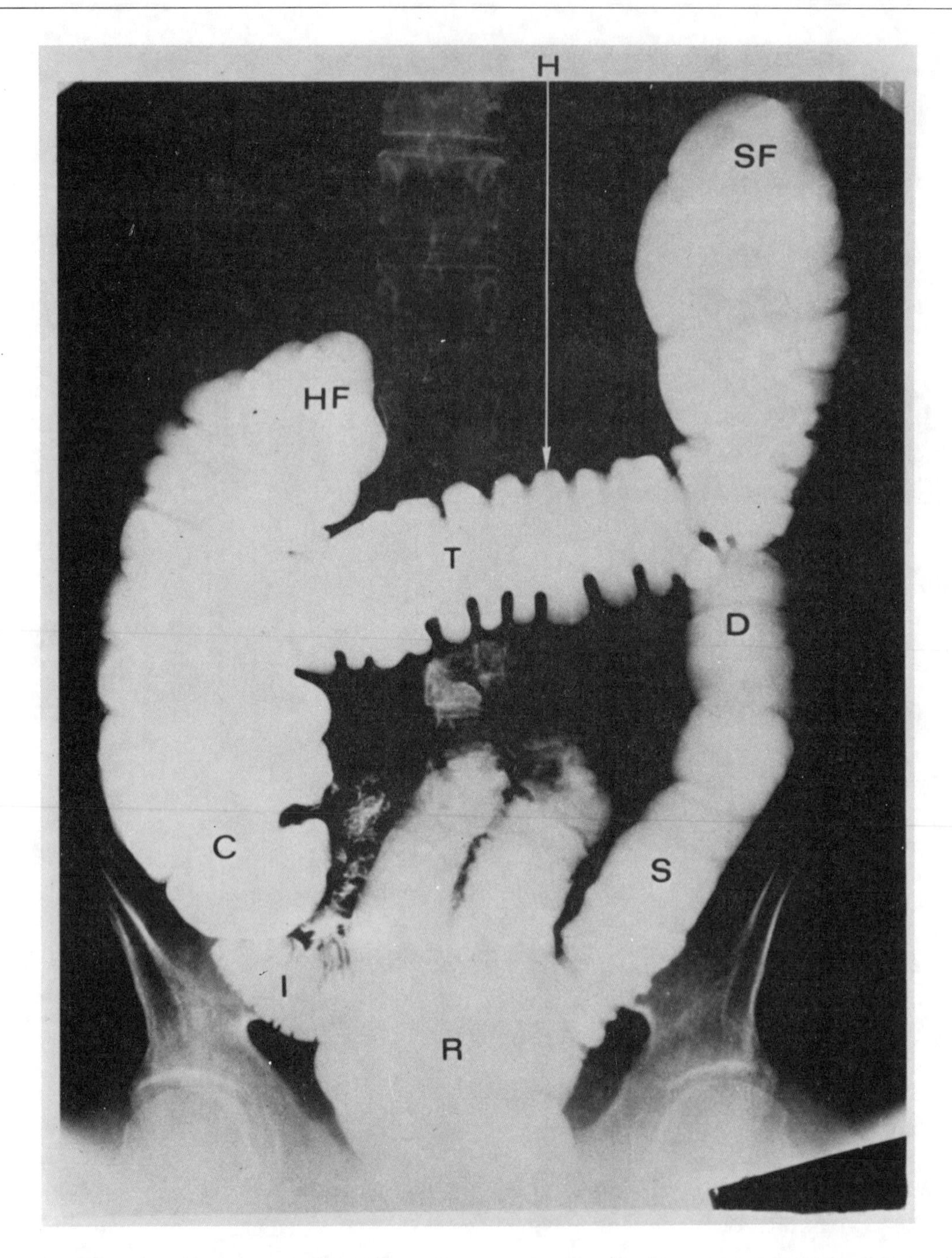

Plate 12.3. Colon, barium enema (A-P view). *C*, cecum; *D*, descending colon; *H*, haustral markings; *HF*, right colic (hepatic) flexure; *I*, ileum; *R*, rectum; *S*, sigmoid colon; *SF*, left colic (splenic) flexure; *T*, transverse colon.

apex of the cecum; now it opens into the cecum about 2 cm below the ileocecal junction. It may be long or short (4–20 cm in Solanke's series) and may occupy any position consistent with its length (*fig. 12.13*).

A triangular fold of peritoneum, known as the *mesoappendix*, attaches the appendix to the terminal part of the left (lower) layer of the mesentery of the ileum.

The appendix has a uniform external coat of longitudinal muscle fibers continuous with the teniae coli. On the cecum and colon, the teniae coli remain discrete until they reach the rectum when they form for it a nearly uniform coat again.

Colon

The *ascending colon* (*fig. 12.14*) crosses the iliac crest and ascends in front of the muscular posterior wall and

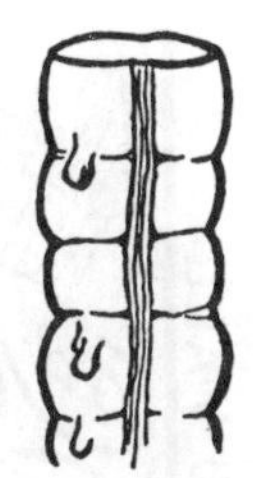

Figure 12.11. Segment of large intestine showing 1 of the 3 teniae, sacculations, and appendices.

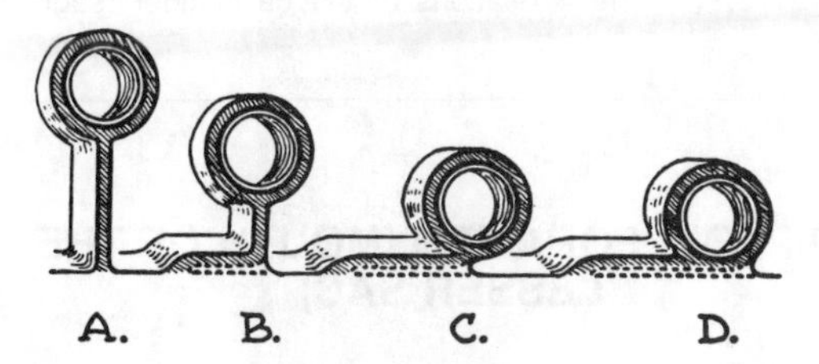

Figure 12.12. Primitive mesentery of the large gut in various stages of absorption.

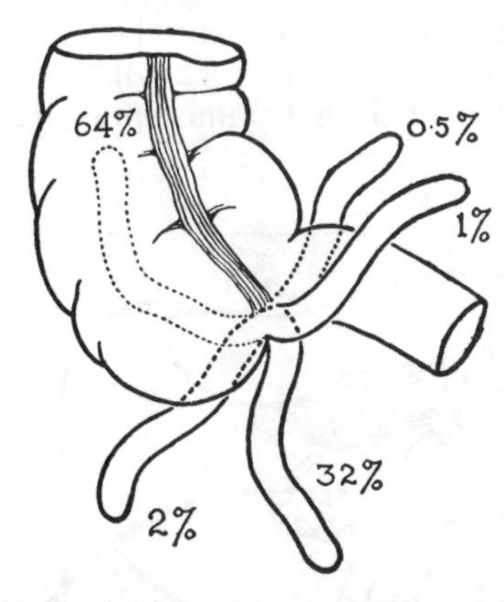

Figure 12.13. Various sites assumed by vermiform appendix and approximate frequencies. (After Wakeley.)

lower pole of the right kidney to the undersurface of the liver. Here, in front of the lower pole of the right kidney, it makes a right angle bend, the *right colic flexure*, and becomes the *transverse colon*.

Resting on the transverse colon are (*fig. 12.6*): the liver, gallbladder, and stomach. It is attached to the greater curvature of the stomach by the gastrocolic ligament (part of the greater omentum) (*fig. 12.21*) and to the pancreas by the *transverse mesocolon* (*fig. 12.15*).

The right extremity of the transverse colon crosses and adheres to the front of: the right kidney, the second part of the duodenum, and the head of the pancreas (*fig. 12.15*). The remainder hangs down for a varying distance but ascends again in front of the descending colon and makes with it an acute angle at the left colic flexure.

The *left colic flexure* is attached to the diaphragm be-

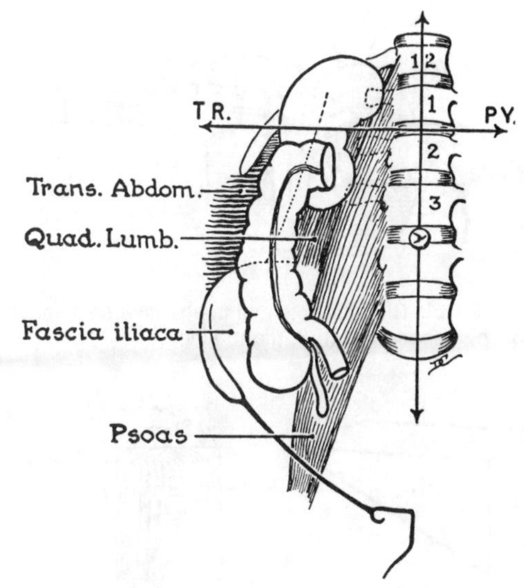

Figure 12.14. The ascending colon.

low the spleen (*fig. 12.6*) by a bloodless fold of peritoneum, the *phrenico-colic ligament*.

The *descending colon*, often much reduced in caliber, descends, crosses the iliac crest, and proceeds across the iliac fossa to the pelvic brim, where it becomes the sigmoid (pelvic) colon (*fig. 12.6*).

The *sigmoid colon* has a mesentery, the *sigmoid mesocolon*, whose root runs a /\-shaped course; upward and then downward in front of the sacrum as far as its middle or third piece. Its appendices epiploicae are very long fatty tags. The sigmoid colon may be of the short or long type—20–45 cm.

The long type forms quite a long loop, which may become twisted on itself, thereby causing an intestinal obstruction.

The *rectum* begins at the 3rd piece of the sacrum, and since it lies in the pelvis, it is described with the pelvic organs (Chapter 18).

Upper Abdominal Viscera and Their Connections

LIVER (L. = *hepar*)

The liver is a soft, pliable organ, molded by its surroundings and weighing almost 1½ kg. When hardened by embalming, its *inferior or visceral surface* (*fig. 12.16*) faces downward, to the left, and backward. It is coated with peritoneum.

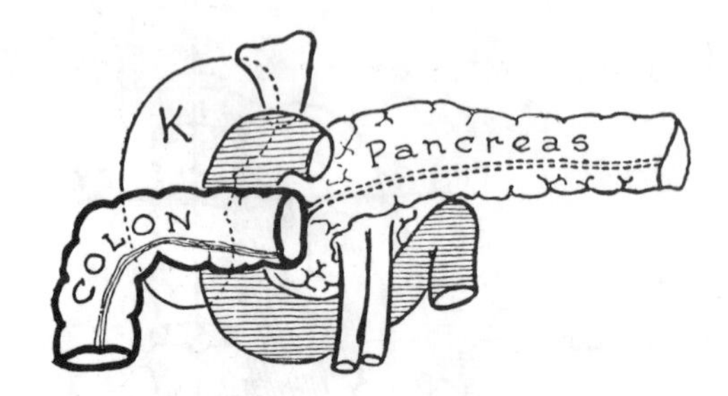

Figure 12.15. The attachment of the transverse mesocolon (*shown by dashed lines*). *K*, (right) kidney.

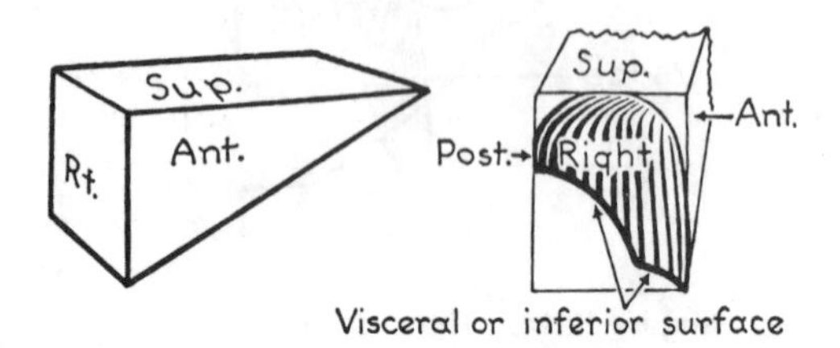

Figure 12.16. The general shape of the liver.

The remainder of the liver is in contact with the anterior abdominal wall and the diaphragm, which make it smooth and round. It is called the *diaphragmatic surface*. Most of it, too, is coated with peritoneum, except for a small "bare area" on the posterior aspect which is in contact with the diaphragm (p. 148).

The visceral and diaphragmatic surfaces are separated from each other (except behind) by the sharp, *inferior border*.

Far back on the visceral surface there is a deep transverse fissure, 5 cm long, the *porta hepatis*. It is the door through which vessels, nerves, and ducts enter and leave the liver. To its lips a part of the lesser omentum is attached. (cont'd on pp. 148 and 168).

GALLBLADDER

This pear-shaped vesicle, about 8 cm long and holding about 40–50 cc of bile, is divided indefinitely into *fundus, body*, and *neck*. The fundus of the gallbladder projects beyond the sharp, inferior border of the liver and comes into contact with the anterior abdominal wall, where the lateral border of the rectus abdominis crosses the costal margin (*fig. 11.3*). The body and neck adhere to the sloping inferior surface and run to the porta hepatis (*fig. 12.17*).

Relations (*fig. 12.17*). A gallstone could penetrate its way through the walls of the gallbladder (1) upward into the liver substance, (2) downward into the duodenum, (3) into the colon, or (4) forward through the anterior abdominal wall.

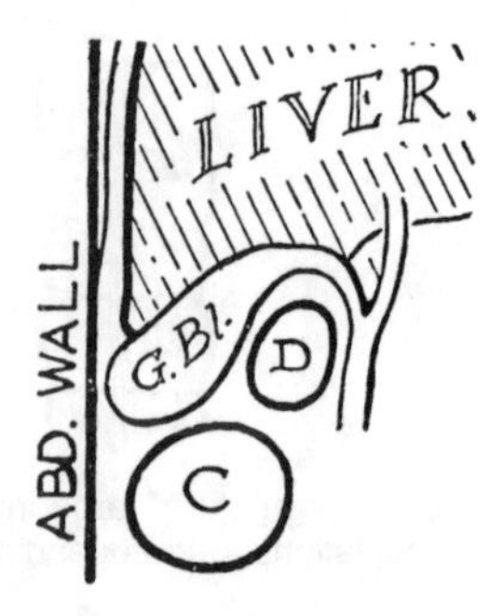

Figure 12.17. The 4 relations of the gallbladder (sagittal section).

EPIPLOIC FORAMEN (MOUTH OF THE LESSER SAC)

The gallbladder serves as a guide to the mouth of a diverticulum of the general peritoneal cavity, called the **omental bursa** (*lesser sac of peritoneum*). This is because the *cystic duct*, which drains the gallbladder, lies in the free edge of the lesser omentum and the mouth of the sac lies behind this free edge (*fig. 12.18*). This sac is not unlike an empty, rubber hot water bottle (*fig. 12.19*).

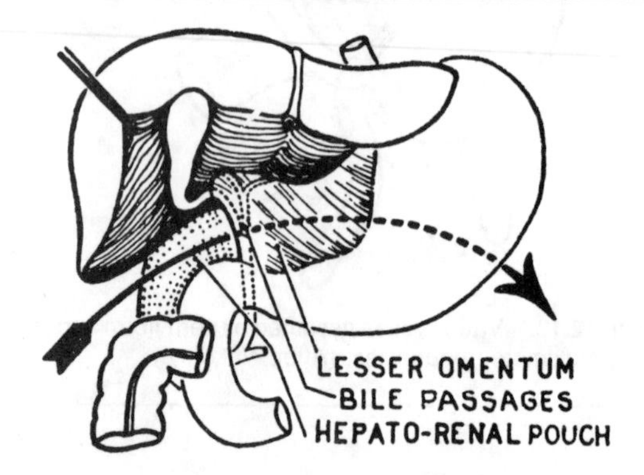

Figure 12.18. Showing why the gallbladder serves as a guide to the epiploic foramen.

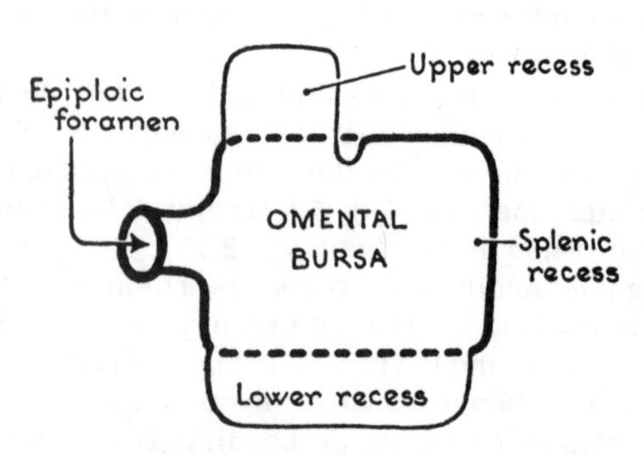

Figure 12.19. Scheme of omental bursa.

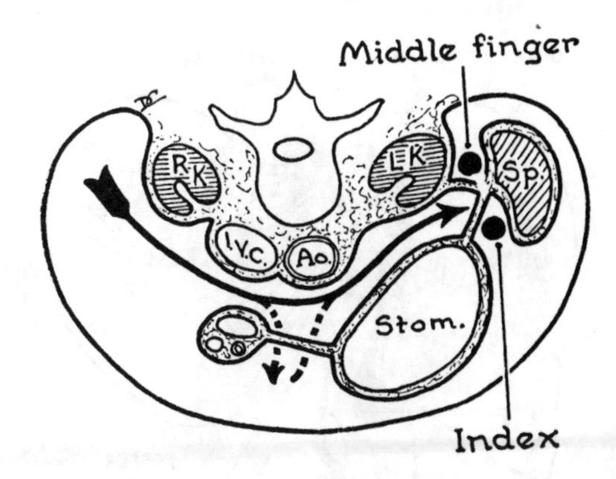

Figure 12.20. Palpating the hilus of the spleen while its pedicle is clamped between 2 fingers of the right hand. If you pass a finger to the left (as shown by the *arrow*), it will reach the hilus at the left limit of the omental bursa.

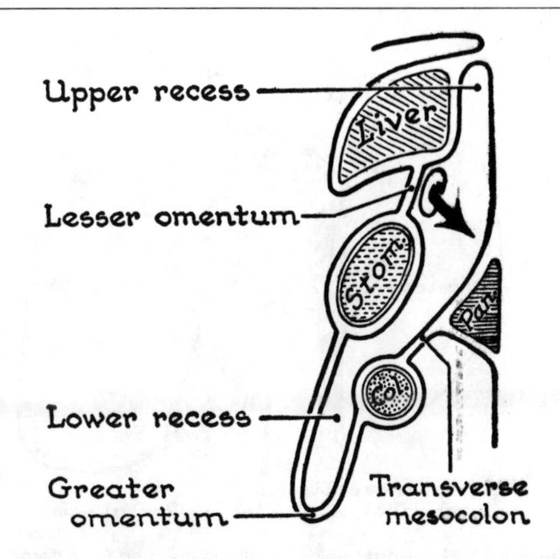

Figure 12.21. Showing the vertical extent of the omental bursa (lesser sac). The *arrow* passes through the epiploic foramen (mouth of the lesser sac).

PERITONEAL ATTACHMENTS OF SPLEEN (L. = LIEN)

A double layer of peritoneum, the *lienorenal ligament*, passes from spleen to kidney (*fig. 12.20*). A second double layer, continuous with the gastrocolic ligament, passes from the greater curvature of the stomach to the spleen; this is the *gastrolienal (gastrosplenic) ligament*. These 2 ligaments form a stalk or pedicle in which the vessels run to and from the spleen.

PEDICLE OF THE SPLEEN; OMENTAL BURSA (THE LESSER SAC OF PERITONEUM)

The "pedicle or stalk" of the spleen may be clamped between 2 fingers (*fig. 12.20*). It has a free lower border (and a free upper border), or you could not grasp it. Its linear site of attachment to the spleen is around the hilus. Clamped between your index and middle fingers are 4 layers of peritoneum (*fig. 12.20*).

The **omental bursa** is an expanded cul-de-sac of the peritoneal cavity that lies behind the stomach and anterior to the peritoneum, covering the pancreas (*fig. 12.20*). It has an upper recess that relates to the left lobe of the liver and a lower recess that is contained in the greater omentum (*fig. 12.21*). The only exit of the omental bursa (lesser sac) into the peritoneal cavity is the **epiploic foramen** (*figs. 12.19, and 12.21*). Since this is located in the superior aspect of the space, drainage of accumulated excess fluids (peritoneal exudates) can present problems. In cases of ulceration of the posterior wall of the stomach or 1st part of the duodenum, foreign material in the omental bursa may cause a **peritonitis**. The accumulation of an inflammatory exudate in the lesser sac may have to be drained by placing tubes in the sac and draining the peritoneal cavity through the anterior body wall. *Figure 12.21* shows the relationships that the surgeon would consider in placing an omental bursa drainage tube through the greater omentum or transverse mesocolon. Gravity would then assist the required drainage of the lesser sac.

Peritoneal Attachments of the Liver

The free edge of the **falciform ligament** contains the **round ligament**, which extends from the umbilicus to the sharp inferior border of the liver. Before birth, the round ligament was the *left umbilical vein*, that returned purified blood from the placenta of the mother to the liver of the fetus. After birth, it becomes a fibrous cord, the *round ligament of the liver* or *ligamentum teres* (see *figs. 12.22, 12.48*).

The falciform ligament is attached to the anterior abdominal wall and diaphragm in the median plane and thence to the convex surface of the liver (to the right of the median plane). On the inferior surface of the liver to the left end of the porta, the round ligament occupies the *fissure for the round ligament*.

Before birth, the left umbilical vein opened for a short time into the left portal vein and, so, poured its blood into the liver. But the *ductus venosus* develops to connect the left portal vein to the inferior vena cava (IVC) (*fig. 12.22*).

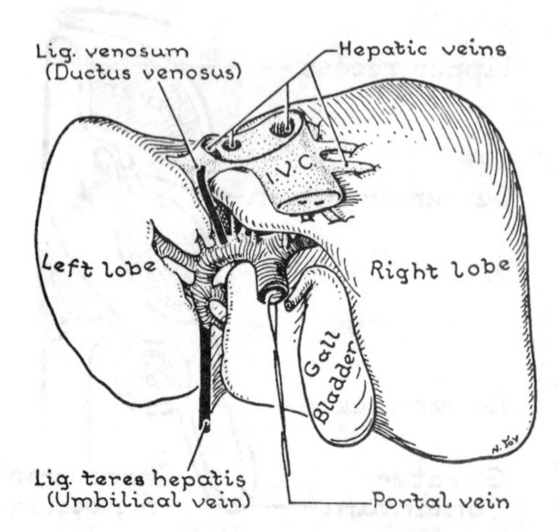

Figure 12.22. Portal vein, hepatic veins, and 2 obliterated veins, called ligaments. (Posteroinferior view.)

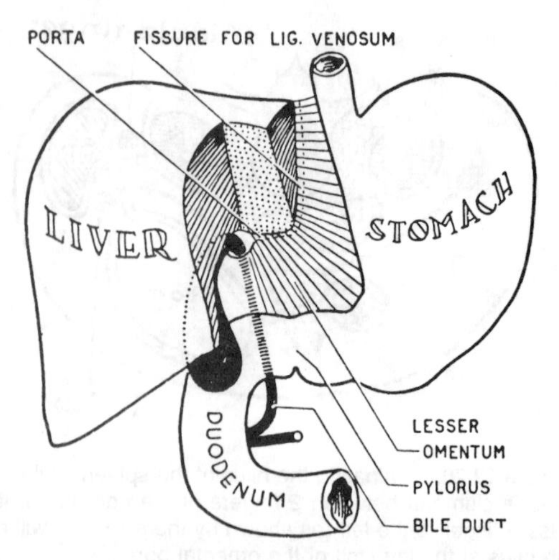

Figure 12.23. The attachments of the lesser omentum. (A section has been taken from the liver to show the fissure for the ligamentum venosum.)

The ductus venosus is obliterated after birth and becomes the **ligamentum venosum.** It continues the sagittal course of the umbilical vein, at the bottom of the *fissure for the ligamentum venosum*, on the posterior aspect of the liver.

For descriptive purposes, these 3 ligaments—falciform, round, and venosum—divide the liver into a right and left lobe.

The **lesser omentum** extends from the lesser curvature of the stomach and first 3 cm of the duodenum to the fissure for the ligamentum venosum and to the porta hepatis (*fig. 12.23*).

The **triangular ligaments** are the sharp, bloodless, peritoneal folds at the extreme right and left limits of the attachment of the liver to the diaphragm (*fig. 12.24*).

The **coronary ligament** is the peritoneal reflection from the edges of the bare area of the liver to those of the contracting "bare" area of the diaphragm. Generations of students have wasted many hours in fruitlessly learning its intricacies. A brief review of *Figure 12.24* is sufficient.

The IVC occupies the left or bursal part of the bare area of the liver (*fig. 12.24*). The coronary ligament and the right triangular ligament enclose the bare area of the liver.

The **caudate lobe** is the lobe with a tail, the tail or

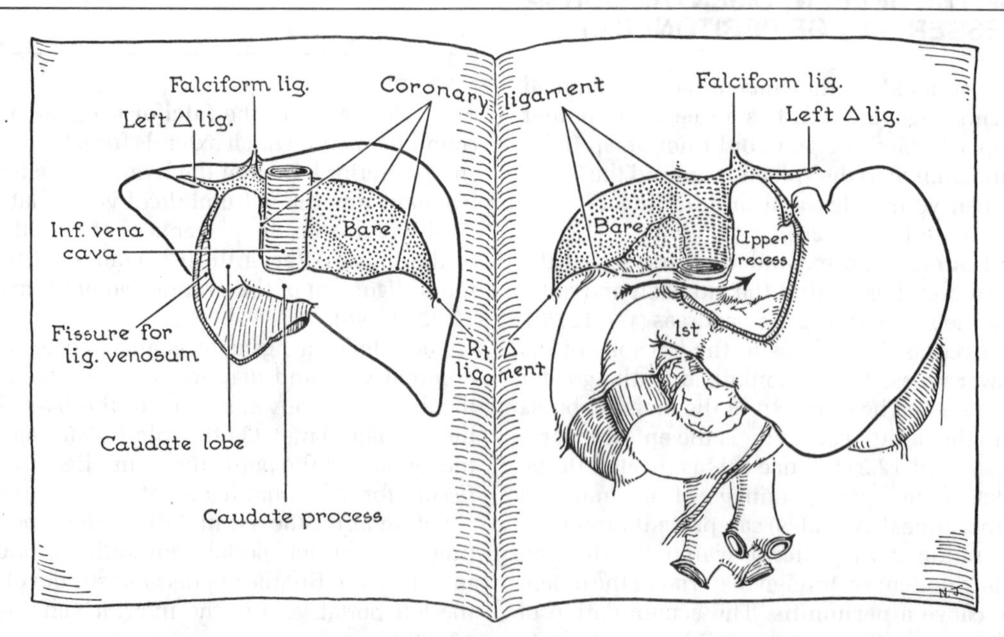

Figure 12.24. The coronary and triangular ligaments of the liver. (The attachments of the liver are cut through, and the liver is turned to the left, as you would turn the page of a book. Hence, the posterior aspect of the liver is revealed on the left page and its posterior relations on the right page. The *arrow* passes through the epiploic foramen.)

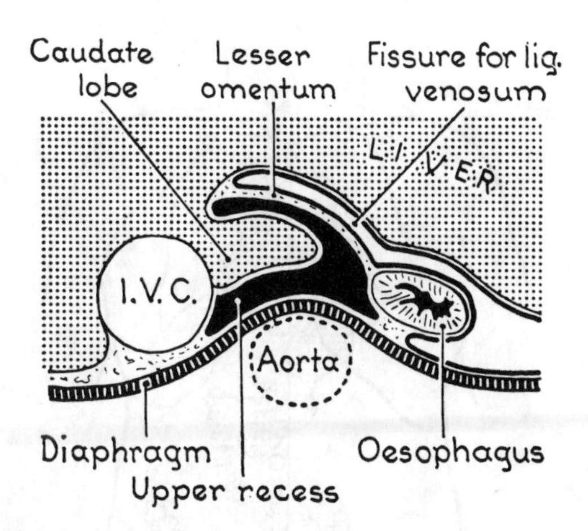

Figure 12.25. The boundaries of the upper recess, viewed from below.

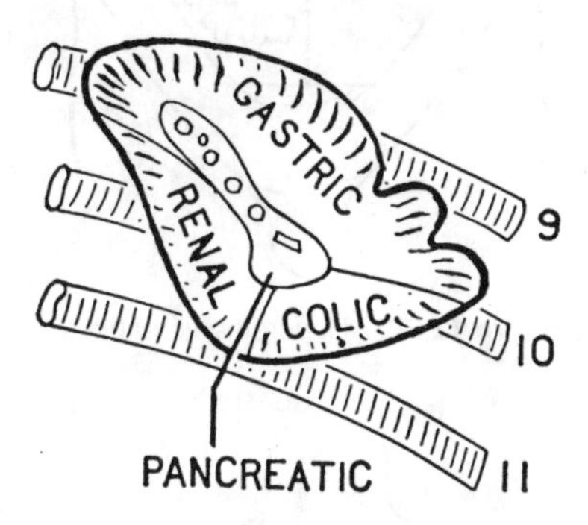

Figure 12.26. Visceral surface of spleen and its "circumferential" border.

caudate process being the narrow isthmus of liver that bounds the epiploic foramen above and connects the caudate lobe to the visceral surface of the right lobe (The liver is continued on p. 165) (*fig. 12.25*).

SPLEEN OR LIEN (cont'd. from p. 147)

A thin peritoneal covered and easily torn capsule encases the soft, vascular spleen, which is molded by the structures in contact with it. The size of a small fist after death, it is much larger in life. *Notches* on its superior border are a notable feature.

It has a convex diaphragmatic surface and a visceral surface shared unequally by *stomach, kidney,* and *colon* (*figs. 12.26 and 12.27*).

The *hilus* admits the vessels and nerves running in the pedicle along with the *tail of the pancreas,* which abuts against the hilus.

Development (see *figs. 12.48–12.50*)

Surface Anatomy

The spleen lies deep to the 9th, 10th, and 11th left ribs. Its long axis follows the 10th rib and extends from, or

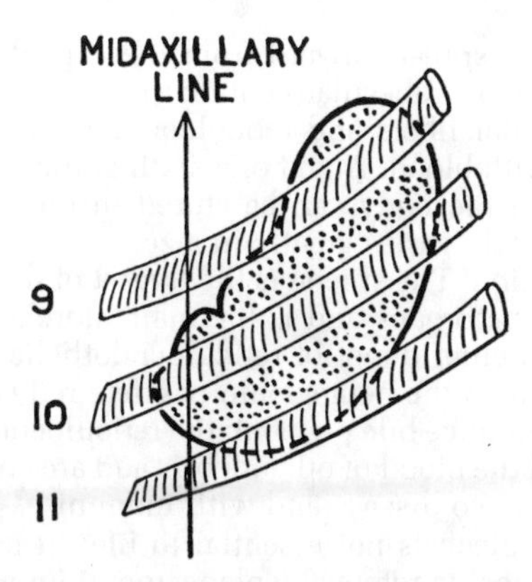

Figure 12.27. Parietal (diaphragmatic) surface of spleen.

almost from, the suprarenal gland to the midaxillary line. Separating it from the ribs are the peritoneal cavity, the diaphragm, and the pleural cavity; in its upper half the left lung also intervenes (*fig. 12.28*).

Physical Examination of the Spleen

The clinician palpates the internal surface of the left costal margin while the patient is lying in the supine position. While the patient inhales deeply, the clinician feels for the spleen as it descends along the intercostal surface as the diaphragm contracts and descends. Normally, the spleen will *not* be felt, but in cases of splenic enlargement (splenomegaly), it may be palpated to varying degrees. Greatly enlarged spleens following parasitic infections or in chronic alcoholics may extend far beyond the protective wall of the rib cage. These individuals are susceptible to internal bleeding if the spleen is damaged by abdominal or pelvic trauma.

Structure. The spleen has a capsule of white fibers, elastic fibers, and smooth muscle fibers that allow it to expand and contract. Supporting *trabeculae* of the same

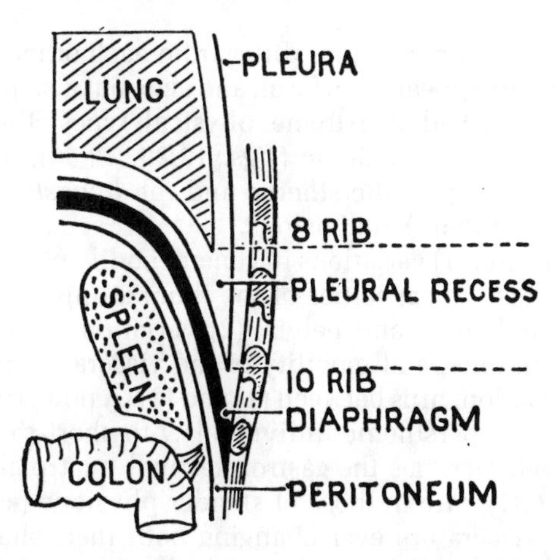

Figure 12.28. Coronal section in the midaxillary line, to show the parietal relation of the spleen.

materials spread inward from the capsule. The spaces between the trabeculae contain a supporting *spongework* of reticular fibers and reticuloendothelial cells and are filled with blood. About one-sixth of the total volume of blood in the body can be stored in the spleen, which accordingly varies greatly in size.

Function. The spleen is the largest of the lymphocyte-producing organs and is the main storehouse of blood. It is the chief depot of reticuloendothelial cells, which break down the hemoglobin of effete red cells and in so doing produce bile pigment; the reticuloendothelial cells also rid the blood of other debris and are concerned with resistance to disease and with immunity.

The spleen is not essential to life; in fact, in certain conditions its removal (splenectomy) improves health.

Rupture of the spleen, fortunately rare, can be caused by severe abdominal blows. Surgical removal of the organ seems to be easily compensated for by the body.

Accessory spleens, the size of large lymph nodes (and very small ones), may lie on the course of the splenic artery, its left gastroepiploic branch, and elsewhere. In performing splenectomy to relieve certain disorders of the blood, all accessory spleens must be removed if recurrence of the disorder is to be avoided (Curtis and Movitz; Halbert and Eaton).

Ever-Changing Positions of Viscera

The positions of the various abdominal viscera vary considerably from subject to subject, depending largely upon the body build—upon whether the subject is of the broad type (when characteristically, they are placed high in the abdomen) or of the intermediate and slender types (when they are placed lower).

BODY TYPES OR BODILY HABITUS

Healthy human beings differ from each other not only in outward appearance, form, and size, but also inwardly. Mills described 2 extreme physical types, the *hypersthenic* and the *asthenic* (*figs. 12.29, 12.30*), and 2 intermediate types, the *sthenic* and the *hyposthenic*, and these have their subdivisions.

There are: (1) variations in general bodily physique and in the relative capacities of the thorax, upper abdomen, lower abdomen, and pelvis; (2) variations in the form, position, tone, and mobility of the viscera; and (3) constant relationships between (1) and (2). A powerful, heavily built hypersthenic individual with short thorax and long abdomen has the gastrointestinal tract placed high (*fig. 12.31*), rare in those of slender physique (asthenic).

The viscera are ever changing both their shapes and their positions. They move with the movements of the diaphragm and of the anterior abdominal wall. They move when the posture alters, being highest when the subject is recumbent, lower when he stands, and still lower when he sits. They rise when the anterior abdominal wall is voluntarily retracted.

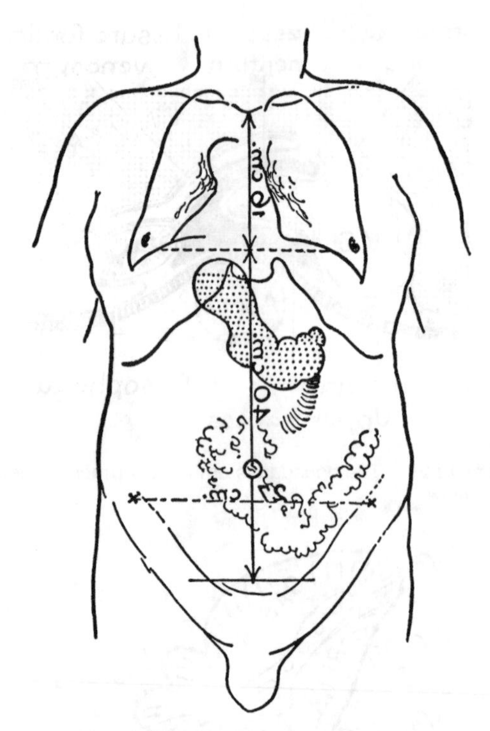

Figure 12.29. Hypersthenic habitus.

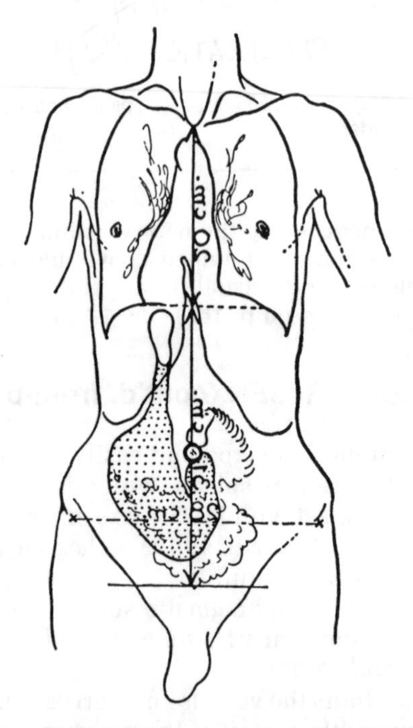

Figure 12.30. Asthenic habitus.

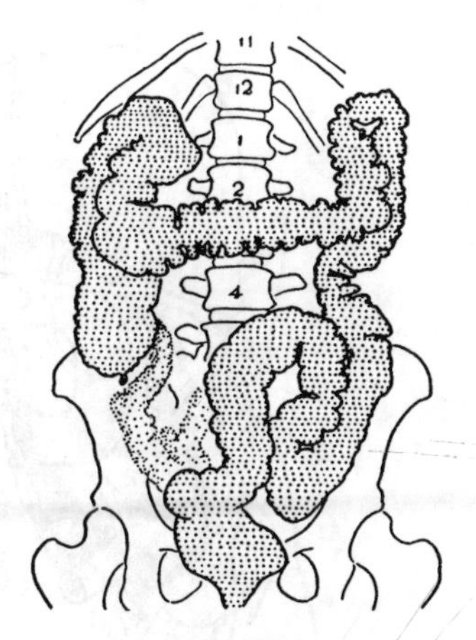
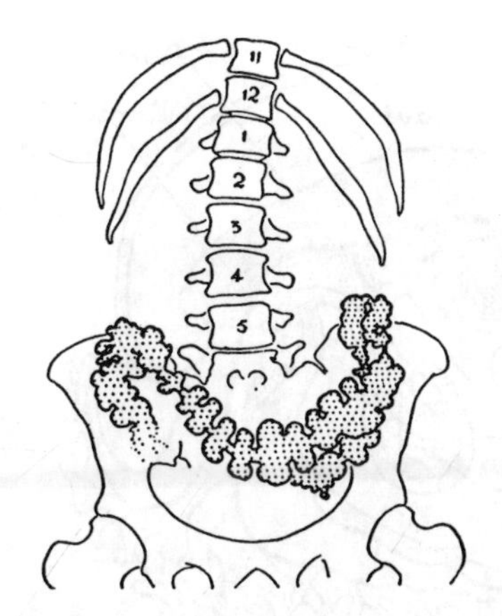

Figure 12.31. *Left*, hypersthenic type; *right*, asthenic type.

The hollow organs (e.g., stomach, intestines, bladder, and uterus) vary as they fill and empty, and they vary with the tone of their muscle coats (e.g., fear and other emotions result in relaxation of the stomach so that the greater curvature suddenly falls).

Features of the Hypersthenic Habitus

1. A powerful and massive *physique*, great body *weight*, and heavy *bony framework*.

2. The *thorax* is short, deep, and wide; the *abdomen* is long and of great capacity in its upper zone. The *subcostal angle* is very obtuse, and the xiphoid process is broad.

3. The *lungs* are wide at their bases, and contracted at their apices, which project but little above the clavicles.

4. The long axis of the *heart* is nearly transverse.

5. The *gastrointestinal tract* is high. The stomach is of the bullhorn type, the pylorus being the lowest, or nearly the lowest, part of the stomach. The entire colon is short; the cecum is well above the iliac basin, even when the subject is standing; the transverse colon is short, actually transverse, and high; consequently, the descending colon is long and straight. The relative proportions of the colon are characteristic; so are its fine, numerous haustrations. The gastric motility is fast; there is marked tone and rapid motility of the colon. Defecation takes place 2 to 3 times a day.

Features of the Asthenic Habitus

1. Frail and slender *physique*, light body *weight*, and delicate *bony structure*.

2. The *thorax* is long and narrow; the *abdomen* is short. There is disproportion between the *pelvic capacity* and that of the upper abdomen, the false pelvis being often as wide and capacious as that of a hypersthenic subject of twice the weight. The *subcostal angle* is narrow.

3. The *lungs* are widest above, and their apices reach well above the clavicles.

4. The long axis of the *heart* is approximately in the median plane.

5. The *gastrointestinal tract* is low. The stomach is atonic and largely pelvic when the subject is standing. The entire colon is long; the cecum is capacious and low in the pelvis; the transverse colon dips down toward or into the pelvis; the haustrations of the colon are coarse. The tone of the gastrointestinal tract is poor and its motility slow.

Umbilicus

When the umbilicus or navel is examined from its peritoneal aspect, 4 fibrous cords are seen radiating from it. They are the obliterated remains of 4 tubes, which in fetal life traversed the umbilical cord (*fig. 12.32*). The tubes are: the urachus, the right and left umbilical arteries, and the umbilical vein. Each of the 4 may produce for itself a peritoneal fold or mesentery, but whether oc-

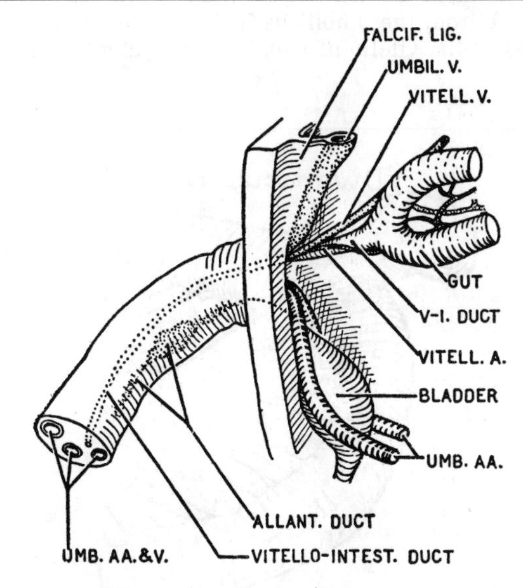

Figure 12.32. Structures in umbilical cord. (After Cullen.) *V* = vitellointestinal.

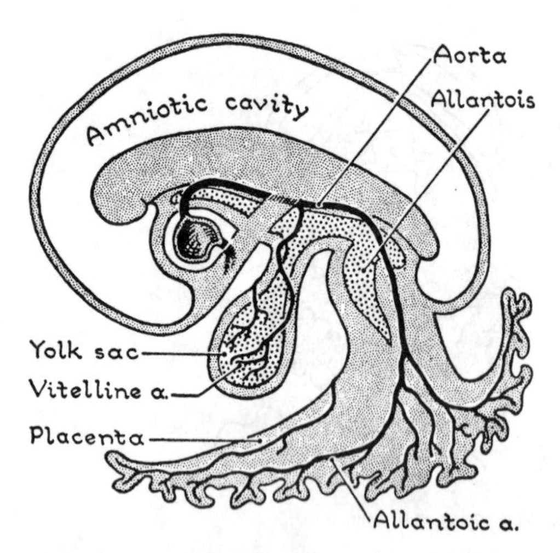

Figure 12.33. In the human embryo the allantois is superseded by the placenta. (After Paterson.)

cupying peritoneal folds or not, they are all situated in the extraperitoneal fatty-areolar layer (*figs. 11.10, 12.32*).

The *obliterated allantoic duct or urachus* ascends in the median plane from the apex of the urinary bladder to the umbilicus [*cf.* the *allantois* (*fig. 12.33*)].

On each side of the urachus an *obliterated umbilical artery* proceeds from the internal iliac artery to the umbilicus—equivalent of the allantoic arteries (*fig. 12.33*). After birth, the umbilical cord is cut, and the arteries thereafter become the obliterated umbilical arteries. (Each artery is lateral to the inferior epigastric artery that passes from the external iliac artery to the rectus sheath; occasionally, each creates a pronounced fold.)

The *obliterated umbilical vein* is, of course, the *round ligament of the liver.*

If at birth the cord is cut very short: (1) urine will escape from the umbilicus if the urachus is patent; (2) feces, if the vitellointestinal duct is patent; and (3) the peritoneal cavity will be opened if the extra-embryonic celom is patent.

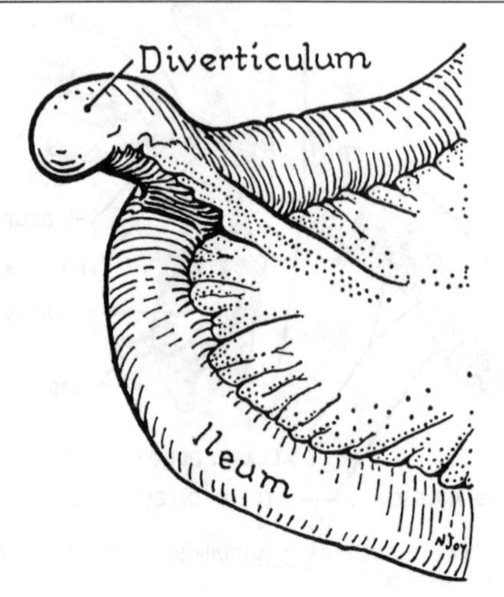

Figure 12.34. Diverticulum ilei (of Meckel).

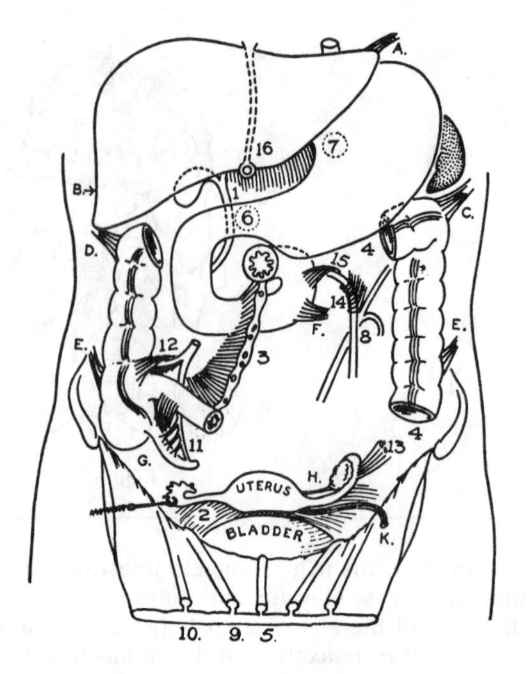

Figure 12.35. Peritoneal folds acting as mesenteries of tubes (see *table 12.1*).

MECKEL'S DIVERTICULUM (*fig. 12.34*).

The vitellointestinal duct persists in 2% of persons as a patent appendage of the gut, springing from the ileum within a meter or so of the cecum (Jay *et al.*).

Peritoneum

PERITONEAL FOLDS

Peritoneal folds are often commonly "*the mesenteries*" of tubes.

Sixteen folds containing tubes are represented in *Figure 12.35* and *Table 12.1*.

Folds not containing tubes are indicated alphabetically in *Figure 12.35*.

A. Left triangular ligament of the liver
B. Right triangular ligament of the liver
C. Phrenicocolic ligament
D. "Supporting ligament of liver"
E. Acquired folds lateral to the ascending and descending colons

Table 12.1. Peritoneal Folds Acting as "Mesenteries" of Tubes

Nature of Tube	Patent or Obliterated	Name of Fold	Name of Tube of Which the Fold is a Mesentery	Number on Fig. 12.38
Duct	Patent	The lesser omentum	The bile passages	1
		The broad ligament of uterus	The uterine tube	2
		The mesentery	The small intestine	3
		The mesocolon	The large intestine	4
	Obliterated	The median umbilical ligament	The urachus (allantoic duct)	5
Artery	Patent	The right gastropancreatic fold	The hepatic artery	6
		The left gastropancreatic fold		
		The fold of the paraduodenal fossa	The left gastric artery	7
			The ascending branch of left colic artery sometimes	8
	Obliterated	The medial umbilical ligament	Obliterated umbilical artery	9
Artery and Vein	Patent	The lateral umbilical ligament	Inferior epigastric vessels	10
		The mesentery of the appendix	Appendicular vessels	11
		The superior ileocecal fold		
		The suspensory ligament of ovary	Anterior cecal vessels	12
			Ovarian vessels	13
Vein	Patent	The fold of paraduodenal fossa	Inferior mesenteric vein	14
		The fold of superior duodenal fossa	Inferior mesenteric vein	15
	Obliterated	Falciform ligament of the liver	Ligamentum teres hepatis (Obliterated umbilical vein)	16

F. Fold guarding inferior duodenal fossa
G. Inferior ileocecal fold (bloodless fold) (*fig. 12.36*)
H. Ligament of the ovary

Peritoneal Fossae, Recesses, and Gutters occur as follows.

1. *Omental Bursa* (p. 146).

2. *Above the greater omentum*, the right and left *subphrenic spaces* lie between diaphragm and liver, one on each side of the falciform ligament. The *Hepatorenal Recess* or *Pouch* lies between the right lobe of liver, right kidney, and right colic flexure (*fig. 12.37*).

3. *Below the Greater Omentum* are the following fossae and gutters.

Duodenal Fossae. The (inconstant) superior duodenal, inferior duodenal, paraduodenal, and retroduodenal fossae are little pockets whose mouths face each other near the duodeno-jejunal junction.

Cecal Fossae. The superior ileocecal, inferior ileocecal, and retrocecal fossae are related to the cecum (*fig. 12.36*). An extensive retrocecal fossa is a retrocolic fossa.

An *Intersigmoid Fossa* is sometimes present. Its mouth opens at the apex of the $\wedge$-shaped root of the sigmoid mesocolon, where the left ureter crosses the common iliac vessels. A pencil can be pushed up the fossa for several centimeters in front of the ureter.

4. *Pelvic Fossae*. In the male the rectovesical fossa lies between rectum and bladder. In the female, the uterus and its broad ligaments divide the rectovesical fossa for several centimeters in front of the fossae.

5. *The "Retro-Omental" or Paracolic Gutters* (*fig. 12.38*).

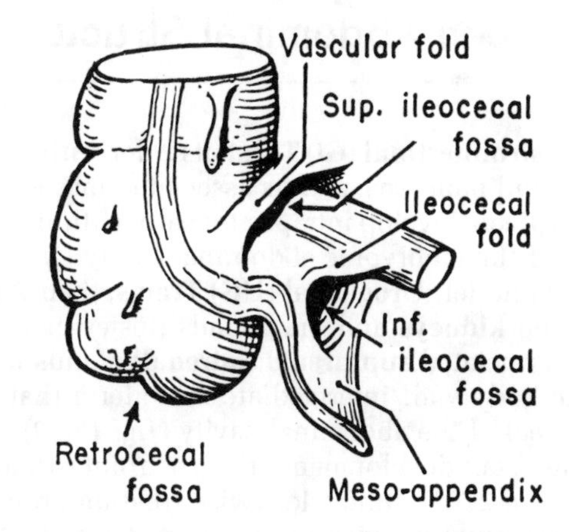

Figure 12.36. Two pouches which are the lowest parts of the peritoneal cavity when the subject is supine. Fluid (e.g., blood) from torn organs follow these paths.

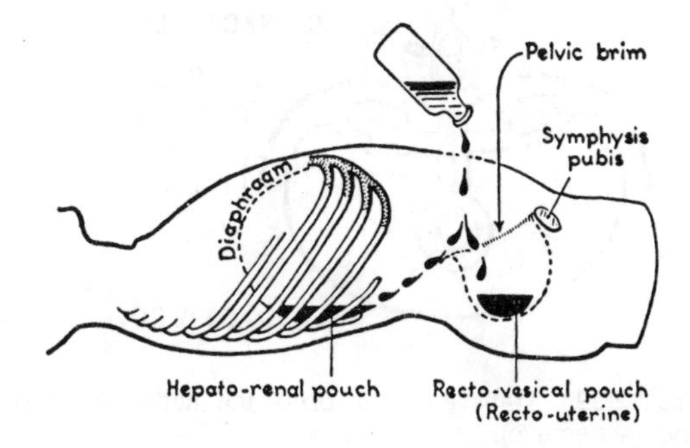

Figure 12.37. Peritoneal folds and fossae about the cecum.

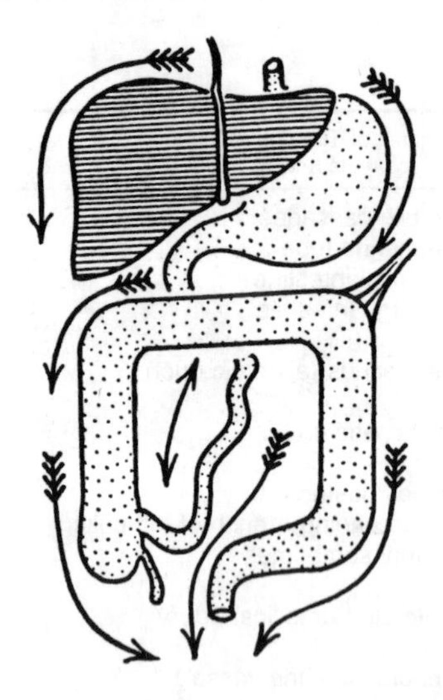

Figure 12.38. The 4 retro-omental or "paracolic" gutters and 3 "supra-omental spaces." The right lateral gutter is the only gutter open above. It would conduct fluid from the hepatorenal pouch and right subphrenic space past the appendix and into the pelvis—if the subject is sitting.

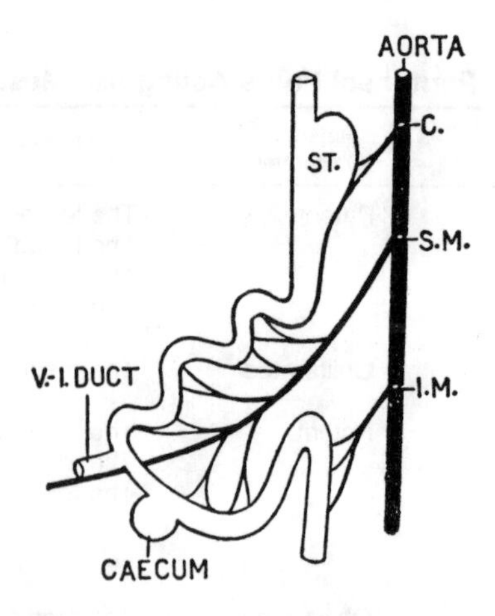

Figure 12.40. The 3 unpaired arteries of the "GI Tract & Co." in the primitive dorsal mesentery. *VI*, vitellointestinal; *ST*, stomach; *C*, celiac; *SM*, superior mesenteric; *IM*, inferior mesenteric.

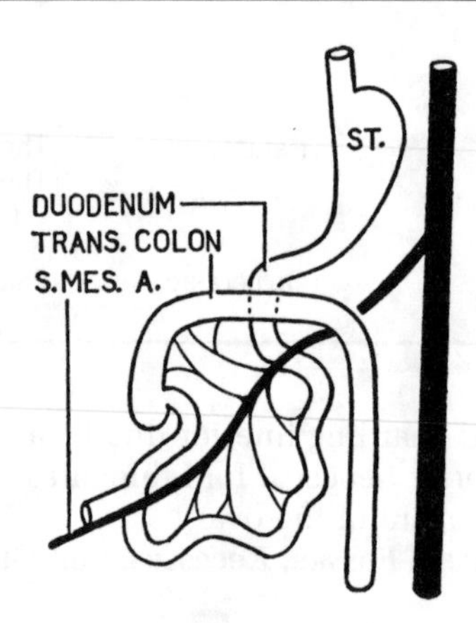

Figure 12.41. Rotation of gut around the superior mesenteric artery. *ST*, stomach.

Development: The Basis For Explaining Relationships Between Abdominal Structures

The **Gastrointestinal (GI) Tract**; its 2 derivative glands, the **liver** and **pancreas**; and its associated mesodermally derived **spleen** develop intraperitoneally within the mesenteries of the embryonic abdominal cavity (*fig. 12.39*). The right and left **Urogenital (UG) Tracts**; their 2 paired glands, the **kidneys** and the **gonads** (testes or ovaries); and the associated **suprarenal (adrenal) glands** develop from the body wall intermediate mesoderm that is retroperitoneal of the abdominal cavity (*fig. 12.39*).

During fetal development, the gastrointestinal tract undergoes a 270° counterclockwise rotation around the axis of the superior mesenteric artery (*figs. 12.40 and 12.41*). This causes some of the primitive mesenteries to fuse with the parietal peritoneum of the posterior body wall (*fig. 12.42*). This loss of mesentery support to certain elements of the gastrointestinal tract results in these organs becoming retroperitoneal or partially retroperitoneal and firmly fused to the posterior abdominal wall in the adult.

The adult intestinal canal may reach 6 or 7 m in length, but the abdominal cavity is less than ½ m long. Therefore, the gut ceases to be a straight tube confined to the median plane. The small intestine becomes convoluted, and a long loop of gut, involving the cecum and adjacent parts of the small and large intestine, taking the superior mesenteric artery as an axis, rotates counterclockwise around it. This brings the cecum temporarily to the undersurface of the liver. From there, it ultimately descends into the right iliac fossa (93% of adults) and so helps to encircle the small gut (*fig. 12.41*).

The blood supply to the developing gastrointestinal tract arises from 3 branches of the retroperitoneal abdom-

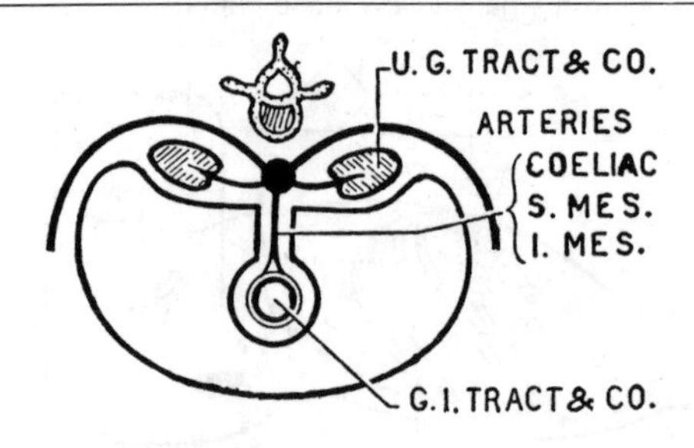

Figure 12.39. Transverse section of the abdomen of an embryo (schematic).

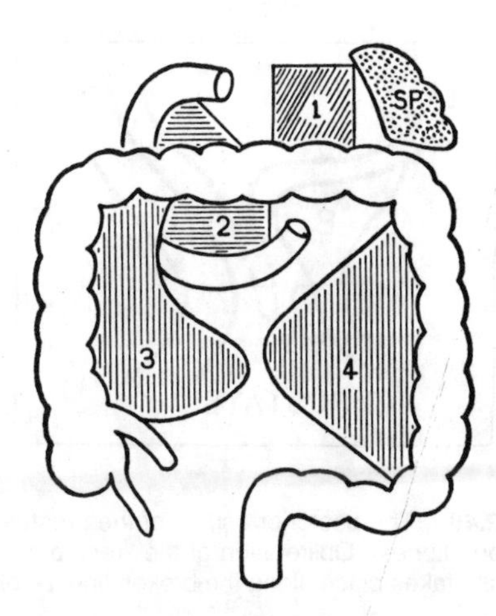

Figure 12.42. Sites where primitive mesenteries adhere to the posterior abdominal wall, obliterating the peritoneal cavity: (*1*) dorsal mesogastrium; (*2*) mesoduodenum; (*3*) mesocolon of ascending colon and right colic flexure; and (4) mesocolon of descending colon.

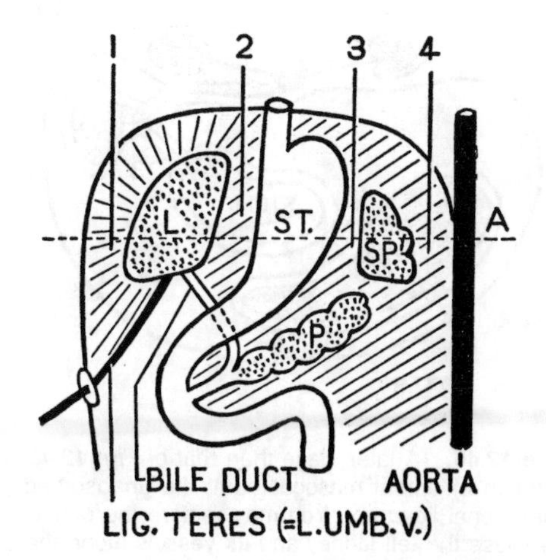

Figure 12.44. Primitive ventral and dorsal mesogastria give rise to: (*1*) falciform ligament; (*2*) lesser omentum; (*3*) gastrolienal ligament; and (*4*) lienorenal ligament. *ST*, stomach; *SP*, spleen; *L*, liver.

inal aorta. The **celiac, superior mesenteric**, and **inferior mesenteric** arteries utilize the dorsal mesentery (*fig. 12.39*) to reach the developing intraperitoneal gastrointestinal tract.

The **Celiac (Trunk) Artery** and its subsequent branches supply derivatives of the embryonic foregut and its associated derivatives: terminal esophagus, stomach, proximal duodenum, liver, pancreas, and spleen. The **Superior Mesenteric Artery** supplies the embryonic **midgut** and its adult derivatives: distal duodenum, jejunum, ileum, appendix, cecum, ascending colon, and transverse colon. The **Inferior Mesenteric Artery** supplies the embryonic **hindgut** and its adult derivatives: descending colon, sigmoid colón and rectum (*fig. 12.40*).

The venous drainage of the gastrointestinal tract and its associated structures is via the **portal vein** (*fig. 12.43*). The portal vein is formed by 3 unpaired veins: the **splenic, superior mesenteric**, and **inferior mesenteric veins**. The inferior mesenteric vein may join the splenic vein. The portal vein and its tributaries are mainly intraperitoneal as they form and ascend to enter the liver. The liver drains via the hepatic veins into the retroperitoneal **inferior vena cava**.

The counterclockwise rotation of the gut must bring some part of the large intestine in front of the superior mesenteric artery—the transverse colon—and some part of the small intestine behind it—the 3rd part of the duodenum (*fig. 12.41*).

The transverse colon would seem to have forced the entire duodenum (save its first 2 cm) and the pancreas, which lay in the mesoduodenum, against the posterior abdominal wall, for the duodenum has lost its mesentery.

A surgeon uses the above principles in order to free or mobilize a portion of the colon or duodenum without damage to vessels.

The **primitive ventral mesentery** exists only above the umbilicus and first 2 cm of duodenum, so it is called the **ventral mesogastrium**. The liver divides it into 2 portions, the falciform ligament and the lesser omentum (*fig. 12.44 and 12.45*).

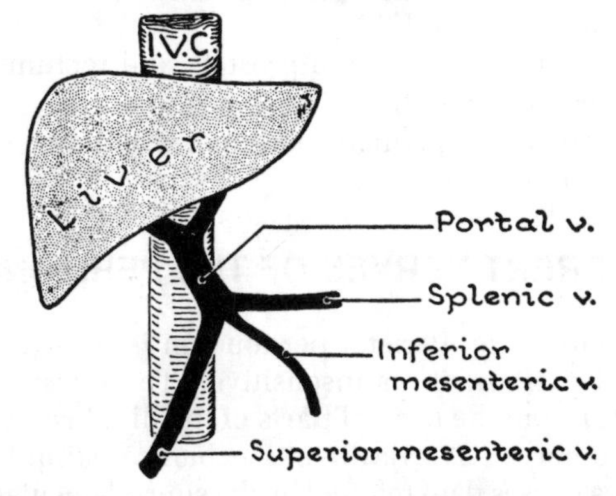

Figure 12.43. The 3 unparied veins of the "GI Tract & Co." form the portal vein. *IVC*, inferior vena cava.

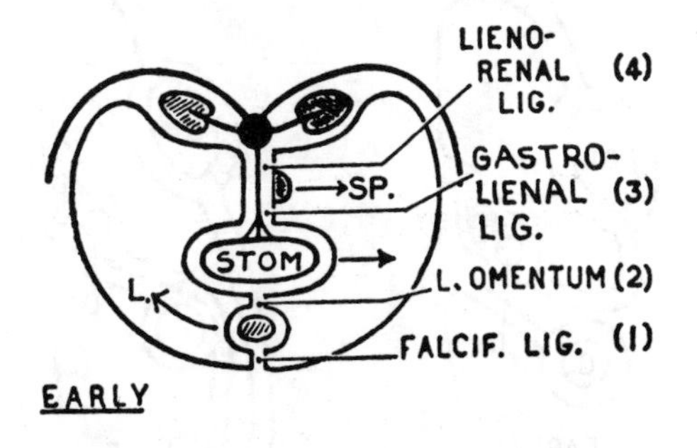

Figure 12.45. Transverse section of abdomen of embryo at level *A*, *Fig. 12.44*, indicating that the liver (*L*) moves to the right, the stomach and spleen (*SP*) to the left.

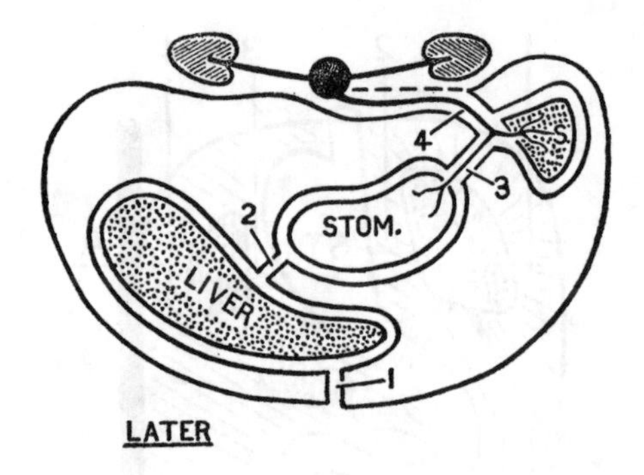

Figure 12.46. A later stage than that of *Fig. 12.45*. Partial absorption of dorsal mesogastrium; the unabsorbed part is the lienorenal ligament. You may restore the primitive state and expose the left kidney and its vessels along the broken line without damaging any structures. *S*, spleen.

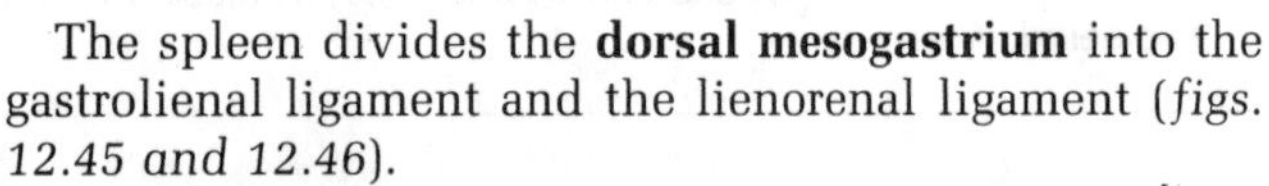

The spleen divides the **dorsal mesogastrium** into the gastrolienal ligament and the lienorenal ligament (*figs. 12.45 and 12.46*).

The liver comes to occupy especially the right side of the body and relegates the stomach and spleen to the left (*fig. 12.46*).

When the stomach moved to the left, it underwent a rotation on its long axis. As a result, its original left surface became the ventral surface, the original right surface, the dorsal surface.

The portion of the dorsal mesogastrium passing between the spleen and aorta is forced against the posterior abdominal wall and absorbed between the median plane and the front of the left kidney. The new attachment becomes the **lienorenal ligament** (*figs. 12.45 and 12.46*).

The duodenum, by losing its mesentery and adhering to the posterior abdominal wall, limited the epiploic foramen (mouth of the omental bursa) below; the liver, by enlarging, encroached on the mouth, narrowing it from above; and the bile passages, passing from liver to duodenum in the free edge of the lesser omentum, limited the mouth in front.

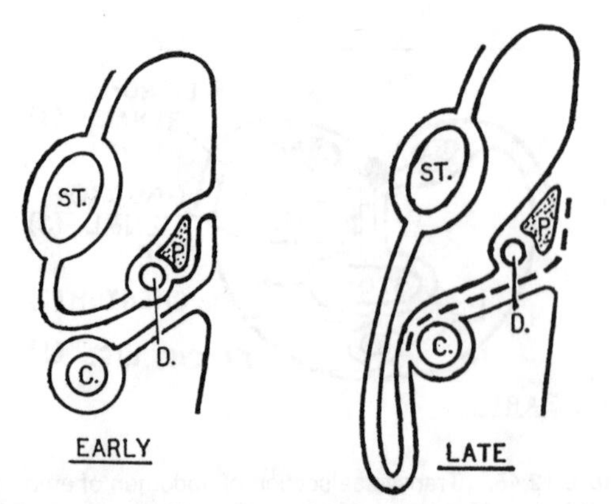

Figure 12.47. Development of the greater omentum. *ST*, stomach; *D*, duodenum; *P*, pancreas; *C*, colon.

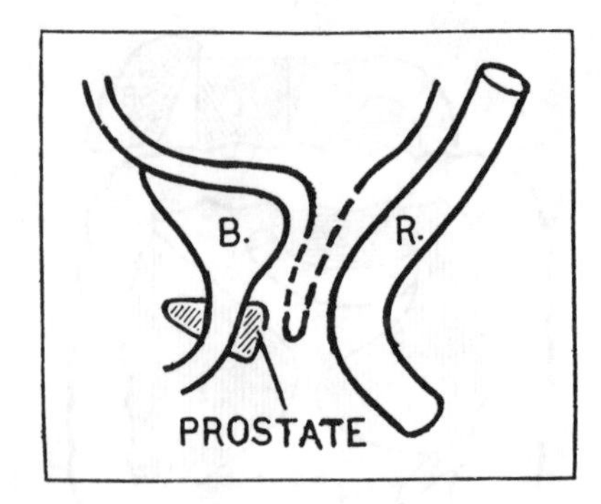

Figure 12.48. The posterior aspect of the prostate was formerly subperitoneal. Obliteration of the "rectoprostatic" peritoneal fossa takes place along the broken line. *B*, bladder; *R*, rectum.

The primitive omental bursa was at first limited below by the dorsal mesogastrium, but in time it bulged downward in front of the transverse mesocolon and transverse colon and adhered to them, thereby forming the gastrocolic portion of the greater omentum. Developmentally, therefore, the transverse mesocolon is 4 layers thick (*fig. 12.47*).

The portion of the peritoneal cavity between the base of the bladder and prostate anteriorly and the rectum posteriorly underwent obliteration. And, it can be safely opened (*fig.12.48*).

In summary, the chief sites of obliteration of the peritoneal cavity are as follows.

1. The portion of the dorsal mesogastrium between the aorta and the middle of the left kidney (*fig. 12.46*).
2. The mesoduodenum, including the part containing the pancreas (*fig. 12.47*).
3. The ascending and descending mesocolons (*fig. 12.42*).
4. The right portion of the transverse mesocolon adheres to the right kidney, 2nd part of the duodenum, and the head of the pancreas (*fig. 12.15*).
5. The greater omentum adheres to transverse colon and mesocolon (*fig. 12.47*).
6. The walls of the greater omentum commonly cohere, thereby obliterating the lower recess of the omental bursa.
7. The pouch between the prostate and rectum is obliterated (*fig. 12.48*).
8. The processus vaginalis peritonei is obliterated in part (*fig. 11.10*).

AFFERENT NERVES OF THE PERITONEUM

(*General Rule:* Parietal peritoneum is sensitive while visceral peritoneum is insensitive.) These travel as follows: (1) from the **central parts of the diaphragm** via the phrenic nerve (C3–5); direct mechanical stimulation of this area causes pain referred by the supraclavicular nerves (C3, 4) to the lower part of the anterior border of the trapezius; (2) from the **peripheral parts of the diaphragm** via the intercostal and subcostal nerves (Th7–12); here stimulation causes pain referred through these same nerves to the skin of the abdominal wall; (3) from the **parietal peritoneum** again via these same nerves (Th7–12) and L1 (here stimulation is correctly localized at the point stimulated); (4) the mesenteries of the small and large intestines are sensitive from their roots to near the intestine, whereas the **greater omentum** and the **visceral peritoneum** are insensitive to mechanical stimulation.

Clinical Mini-Problems

1. **Which specific part of the small intestine would first receive a gallstone that had passed through the common bile duct?**
2. **Which three structures are contained in the hepatoduodenal ligament?**
3. **Which abdominal quadrant usually becomes painful in acute appendicitis?**
4. **How could a peritoneal exudate in the omental bursa drain into the peritoneal cavity proper?**
5. **Which ribs are normally related to the position of the spleen?**

(Answers to these questions can be found on p. 585.)

13

Stomach, Liver, and Related Structures: Lesser Omentum, Bile Passages, and Celiac Trunk

Clinical Case 13.1

Patient Geoffrey L. This 41-year-old male accountant was admitted to emergency with severe abdominal pain above his umbilicus. You find that he has a fever and an elevated white blood cell count. The chief resident suspects an acute peritonitis, but no peritoneal exudate can be obtained by a needle aspiration of the peritoneal cavity. This is an emergency case requiring a laparotomy, and you will be an assistant. When the anterior abdominal wall is opened, a large accumulation of fluid in the lesser sac (omental bursa) is obvious. The greater omentum had occluded the epiploic foramen and confined the exudate to the lesser sac. When the latter is opened through the gastrocolic ligament, a bloody fluid can be drained. A perforated gastric ulcer is found on the posterior wall of the stomach. The lesser sac is flushed clean of the exudate and gastric debris. The stomach bed over the pancreas is examined and found to be intact. The gastric ulcer is repaired and closed, and abdominal drains are placed in the lesser sac. *Follow-up:* The tubes are removed after 5 days, and the patient released from the hospital 10 days after admission. He is placed on a modified preventive diet and gastric medication. He returns to a reduced work level 4 weeks later and has no recurrence of abdominal pain for more than a year.

Lesser Omentum and Bile Passages

The **lesser omentum** extends from the lesser curvature of the stomach and the first 2 cm of the duodenum (intraperitoneal portion) to the fissure for the ligamentum venosum and the porta hepatis (*fig. 13.1*). The bile passages run in the free edge of the lesser omentum as they descend to the posterior aspect of the first part of the duodenum.

The **cystic duct** (*fig. 13.2*) begins at the neck of the gallbladder and descends to join the **common hepatic duct** within the free edge of the lesser omentum (hepatoduodenal ligament). The **common bile duct** is therefore formed by the union of the cystic and common hepatic ducts above the duodenum. The common bile duct descends posterior to the first part of the duodenum and enters the head of the pancreas. It will unite with the main pancreatic duct within the pancreatic tissue and the two ducts then enter the duodenum on its posteromedial aspect to open at the **duodenal papilla** into the lumen of the second part of the duodenum.

The **common hepatic duct** is formed in the porta hepatis by the union of the **right** and **left hepatic ducts** that drain bile from the right and left halves of the liver. The common hepatic duct and cystic duct may be bound together by a tough areolar web of connective tissue. The bile passages lie lateral to the hepatic artery and its branches within the lesser omentum. Posterior to the bile passages and arteries is the large thin-walled portal vein

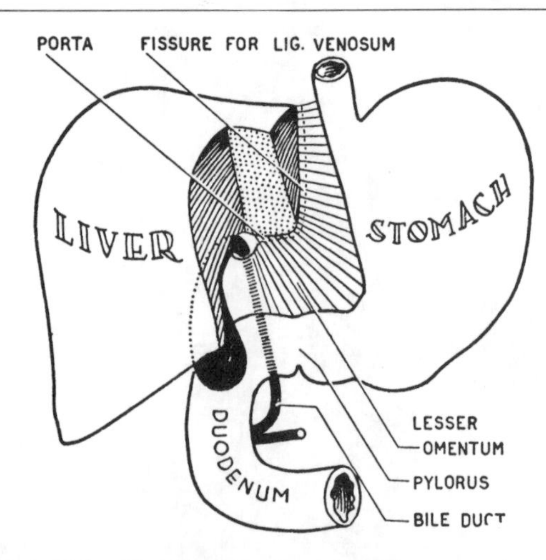

Figure 13.1. The attachments of the lesser omentum. (A section has been taken from the liver to show the fissure for the ligamentum venosum.)

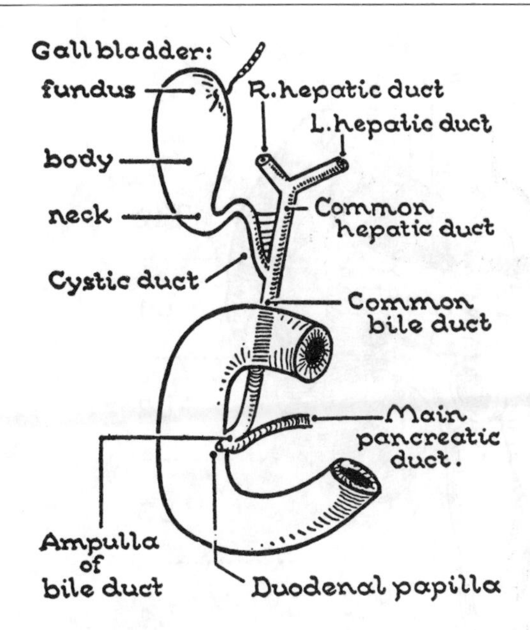

Figure 13.2. The bile passages. (The average lengths of the cystic, common hepatic, and (common) bile ducts are 3.4, 3.2, and 6.5 cm, respectively.)

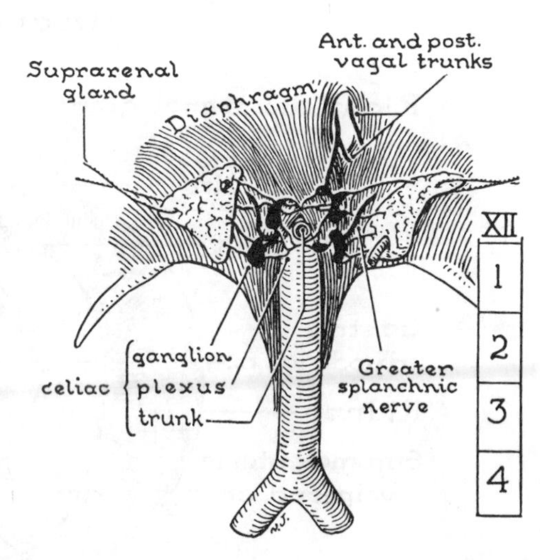

Figure 13.3. The relations of the celiac trunk and its vertebral level. (Glands are retracted.)

(see *fig. 13.4*). Accompanying these principal structures in the hepatoduodenal ligament of the lesser omentum are lymphatic vessels and autonomic nerve fibers that innervate the smooth muscles and some glands of the liver and gallbladder.

The **epiploic foramen** that opens into the lesser sac (omental bursa) lies between the free edge of the lesser omentum anteriorly and the inferior vena cava (IVC) posteriorly. *Figures 12.18* and *12.20* in the previous chapter clearly illustrate these relationships of the epiploic foramen.

Celiac Trunk

The celiac trunk is the first of the three unpaired abdominal aortic arteries that supply the gastrointestinal tract (*fig. 12.40*). It supplies all of the structures derived from the embryonic foregut below the diaphragm and the structures derived from the mesoderm supporting this portion of the foregut.

The celiac trunk, like the abdominal aorta, is retroperitoneal and arises from the abdominal aorta at the level of the 12th thoracic vertebra (*fig. 13.3*). Access to the celiac trunk is through the posterior wall of the omental bursa (lesser sac) as the trunk lies posterosuperior to the body of the pancreas (*fig. 13.4*).

SURFACE ANATOMY AND RELATIONS

The celiac trunk arises from the abdominal aorta between the **right** and **left crura** of the diaphragm, which straddle the trunk superiorly (*fig. 13.3*). The celiac trunk or artery is only about 1 cm long before it branches into its terminal branches. The celiac trunk is surrounded by the **celiac plexus** of nerves and the **celiac ganglia**. The celiac ganglia receives the **greater splanchnic nerves** (T5–9) from the sympathetic chain in the thorax. The greater splanchnic nerves penetrate the crura of the diaphragm as they exit the thoracic cavity and enter the abdominal cavity. The celiac plexus of nerves consists of both postganglionic sympathetic nerves from the celiac ganglia and preganglionic nerves from the vagus nerve (anterior and posterior gastric nerves). These autonomic fibers will course to the abdominal viscera that is derived from the embryonic foregut endoderm and its associated mesoderm on the arterial branches of the celiac trunk. The suprarenal glands are also situated laterally to the celiac trunk on the superior pole of each kidney (*fig. 13.3*).

BRANCHES AND DISTRIBUTION OF THE CELIAC TRUNK

The celiac trunk supplies the stomach, the adjacent parts of the esophagus and duodenum, and the three unpaired glands—liver, pancreas, and spleen (*figs. 13.4, 13.5*). It has three branches.

1. The **left gastric artery** branches into:
   a. esophageal branches
   b. gastric branches
   c. sometimes an aberrant left (accessory) hepatic branch

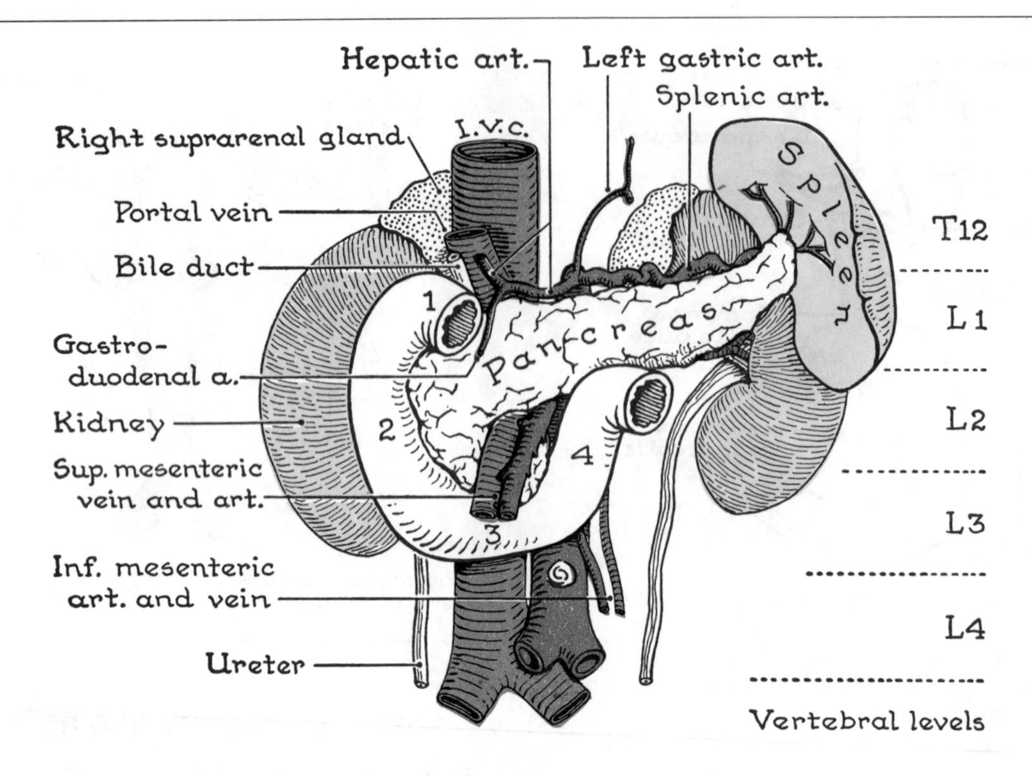

Figure 13.4. Abdominal viscera and vessels—the key picture. *IVC*, inferior vena cava.

2. The **splenic artery** branches into:
   a. pancreatic branches
   b. splenic branches
   c. short gastric arteries
   d. left gastroepiploic artery
3. The **common hepatic artery** branches into:
   a. gastroduodenal artery with branches:
      i. supra- and retroduodenal arteries
      ii. superior pancreaticoduodenal arteries
      iii. right gastroepiploic artery
   b. The **hepatic artery proper** branches into:
      i. right gastric artery
      ii. left hepatic artery
      iii. right hepatic and cystic artery

The Left Gastric Artery

The **left gastric artery** is the smallest of the 3 arteries of the celiac trunk but the largest of the five arteries that supply the stomach. It runs retroperitoneally in the posterior wall of the lesser sac to reach the superior margin of the lesser omentum. It then becomes intraperitoneal and courses along the lesser curvature of the stomach between the two layers of the lesser omentum. It sends off circumferential branches to both sides of the stomach and eventually anastomoses with the right gastric artery on the superior aspect of the pylorus (*fig. 13.5*).

Branches

As it enters the superior aspects of the lesser omentum, the left gastric artery gives off an esophageal branch that anastomoses with the esophageal branches from the thoracic aorta.

Variations

Aberrant left (accessory) hepatic arteries arise from the left gastric artery in 25% of patients. They course through the lesser omentum to enter the porta hepatis. These aberrant left hepatic arteries may completely replace the normal left hepatic artery that arises in the hepatoduodenal ligament from the hepatic artery proper (Michels et al.).

The Splenic Artery

The **splenic artery** is the largest branch of the celiac trunk. It takes a serpentine (tortuous) course to the left side of the abdomen along the superior border of the pancreas. This allows the artery to lengthen as the spleen moves in its intra-abdominal position during breathing. In its retroperitoneal course, the splenic artery crosses the left suprarenal and one-half the breadth of the left kidney. Anteriorly to the left kidney, the splenic artery enters the lienorenal ligament and passes anteriorly to

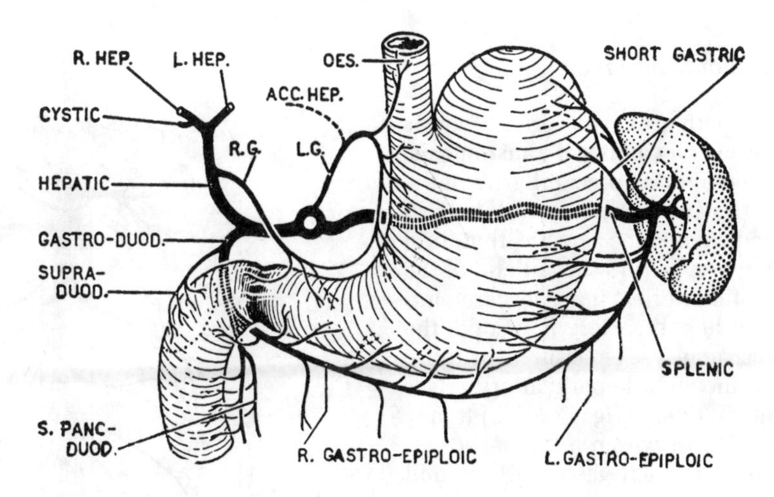

Figure 13.5. Distribution of the celiac trunk or artery. *RG*, right gastric artery; *LG*, left gastric artery.

the hilus of the spleen with the tail of the pancreas (*fig. 13.4*).

Branches

The splenic artery gives **pancreatic branches** to the body and tail of the pancreas while it is in its retroperitoneal position on the posterior abdominal wall. When it becomes intraperitoneal within the lienorenal ligament, it gives off **short gastric branches**, a **left gastroepiploic** and **splenic branches** (*fig. 13.5*). The short gastric arteries supply the fundus of the stomach and reach the greater curvature via the gastrolienal ligament of the greater omentum. The left gastroepiploic branch supplies a portion of the greater curvature of the stomach and anastomoses with the right gastroepiploic artery within the greater omentum. As the name implies, the gastroepiploic arteries supply both the stomach and the greater omentum.

The Common Hepatic Artery

The **common hepatic artery** courses retroperitoneally in the right side of the posterior abdominal wall as it leaves the celiac trunk. It divides into two terminal branches superior to the neck of the pancreas and posterior to the first part of the duodenum. The descending limb passes posteriorly to the duodenum and gives off the duodenal branches, the right gastroepiploic artery and the superior pancreaticoduodenal branches. The ascending limb becomes the **hepatic artery proper** and enters the hepatoduodenal ligament of the lesser omentum. The branches that arise from the hepatic artery proper are the intraperitoneal right gastric, the left hepatic, and the right hepatic arteries. The right hepatic artery gives off the important cystic artery, which is of great significance in surgical removal of the gallbladder.

Details Important Only to Surgeons

The cystic artery usually arises in the angle between the common hepatic duct and the cystic duct. It commonly has 2 branches that anastomose with each other. The branches supply the gallbladder and send twigs to the liver. In about 25% of cases, the 2 branches arise independently, i.e., there are 2 cystic arteries. The cystic artery may arise from any nearby artery. When it arises to the left of the common hepatic duct, it must cross that duct to reach the gallbladder, and the crossing usually takes place in front of the duct (Daseler et al.).

Collateral Branches of the Hepatic Artery

The **right gastric artery** (*fig. 13.5*) descends to the lesser curvature of the stomach. The **gastroduodenal artery** passes downward between the first part of the duodenum and the pancreas where, after a course of about 2 cm, it divides into the **right gastroepiploic** and the **superior pancreaticoduodenal artery** (*fig. 13.5*).

The gastroduodenal artery gives off an end-artery, the **"supraduodenal artery"** (Wilkie) (*fig. 13.5*), retroduodenal twigs to the back of the duodenum, and the **superior pancreaticoduodenal arteries** (**posterior** and **anterior**). These last effect a double arch with the inferior pancreaticoduodenal branches of the superior mesenteric artery (*fig. 14.18*), one arch lying in front of the head of the pancreas and the other behind it; both arches supply pancreatic and duodenal branches.

The right gastroepiploic artery runs between the 2 layers of the greater omentum, a fingerbreadth from the greater curvature of the stomach, and commonly anastomoses with the left gastroepiploic artery. Both vessels have **gastric branches** and long slender epiploic branches, which descend in the omentum.

Collateral Circulation about the Liver

To deprive the liver altogether of its arterial blood is usually fatal. There is however a **collateral anastomosis**. Thus, (1) the larger branches and the precapillaries of the right and left hepatic arteries anastomose so well both in the fissures of the liver and deep to the capsule, that fluid injected into the one artery flows from the cut end of the other. The deep intrahepatic arteries, however, are end-arteries. (2) If the common hepatic artery is ligated, the arterial supply to the liver may yet be assured in those 12% of persons in whom the right hepatic artery arises from the superior mesenteric artery (*fig. 13.5*). (3) It may also be assured in the 11.5% in whom a "replaced left hepatic artery" springs from the left gastric artery; and perhaps in some of those in whom an accessory left hepatic artery does so (p. 160). (4) If the hepatic artery is obstructed gradually on the aortic side of the origin of the right gastric artery, the circulation is maintained by the anastomosis the right gastric artery effects with the left gastric artery. (5) The inferior phrenic, the cystic, and the superior epigastric arteries send fine twigs to the liver, the last via the falciform ligament.

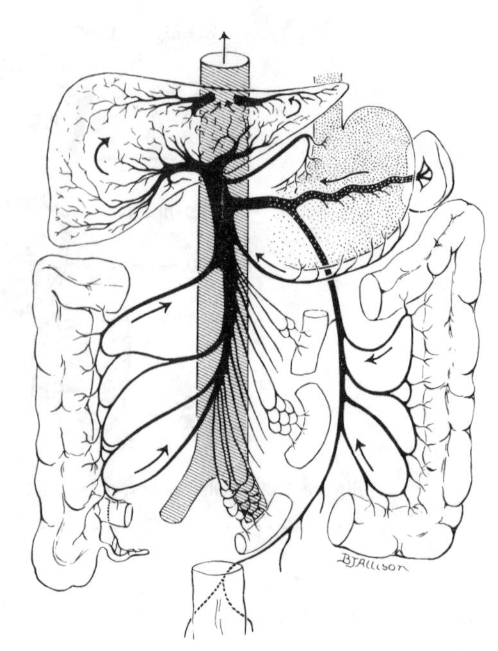

Figure 13.6. Scheme of portal circulation. (From *Primary Anatomy*.)

Portal Vein

The portal vein drains all, and drains only, the gastrointestinal tracts and its unpaired glands, with the exception of the liver. It returns to the liver the blood delivered by the celiac, superior mesenteric, and inferior mesenteric arteries to gastrointestinal organs (*fig 13.6*).

COURSE

The portal vein is formed between the head and neck of the pancreas by the union of the splenic (which drains organs supplied by the celiac artery), the superior mesenteric, and the inferior mesenteric veins. It ascends through the hepatoduodenal ligament to the porta hepatis and divides into the **right** and **left portal veins**. The right portal vein enters the right lobe of the liver and the left passes transversely through the porta hepatis to supply the caudate, quadrate, and left lobes of the liver (*fig. 13.6*).

There are no functioning valves in the portal system.

RELATIONSHIPS

The portal vein is intraperitoneal when it is formed in the root of the mesentary for the small bowel. It ascends retroperitoneally behind the neck of the pancreas, the first part of the duodenum, and the gastroduodenal artery (*fig. 13.4*). It then becomes intraperitoneal again as it enters the hepatoduodenal ligament of the lesser omentum. As it ascends in the free edge of the lesser omentum, the portal vein lies posterior to the hepatic artery proper and the common bile duct.

Posterior to the portal vein, is the inferior vena cava (IVC) (*fig. 13.7*). When the portal vein becomes intraperitoneal in the lesser omentum, it is separated from the IVC by the epiploic foramen.

The left portal vein that traverses the porta hepatis to drain into the left half of the liver has the **ligamentum teres** (from the umbilicus) and the **ligamentum venosum** (to the hepatic veins) attached to its outer walls (*fig. 13.8*).

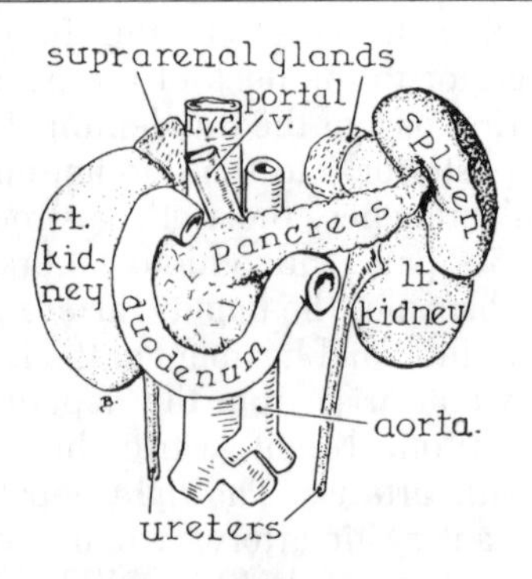

Figure 13.7. Relations of portal vein. *IVC*, inferior vena cava.

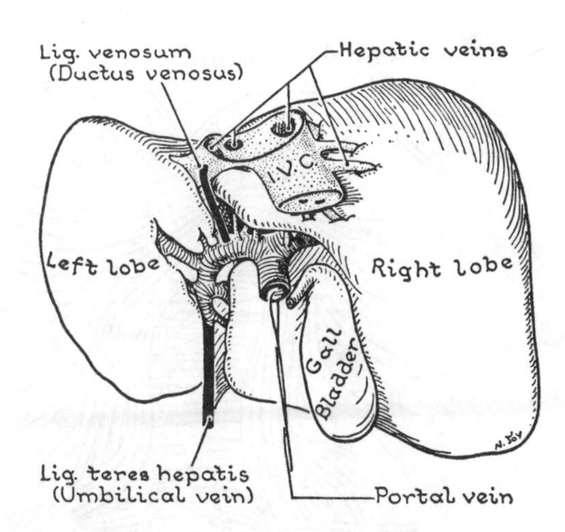

Figure 13.8. Portal vein, hepatic veins, and 2 obliterated veins, called ligaments. *IVC*, inferior vena cava. (Postero-inferior view.)

PORTACAVAL VENOUS ANASTOMOSES

When the portal vein is slowly obstructed, as a result of disease of the liver or from other causes, the portal blood may enter the IVC by way of certain anastomotic veins, which then become dilated and varicose. They may also burst.

They are as follows (*fig. 13.9*).

a. At the upper end of the gastrointestinal tract: the esophageal branches of the left gastric vein anastomose with esophageal branches of the azygos veins in the thorax (when varicose they are called esophageal varices).

b. At the lower end of the gastrointestinal tract: the superior rectal vein anastomoses with the middle and inferior rectal veins and, most important of all, with the **pelvic venous plexuses** (p. 223) (when varicose, the rectal or hemorrhoidal veins are called hemorrhoids).

c. Fine **paraumbilical veins** run with the round ligament of the liver from the left portal vein to the umbilicus, where they anastomose with the superficial and deep epigastric veins.

d. Twigs of the colic and splenic veins anastomose in the extraperitoneal fat with twigs of the renal vein and with veins of the body wall. Here may be included twigs from the bare area of the liver.

Since the anastomotic veins seldom possess valves, they can conduct blood equally readily in either direction, depending on whether the obstruction is in the portal vein or in the IVC (Edwards).

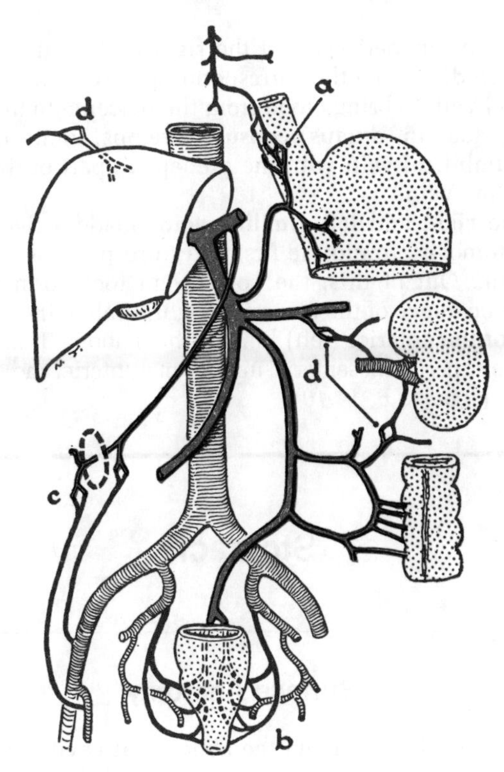

Figure 13.9. Diagram of the chief portacaval anastomoses (see text).

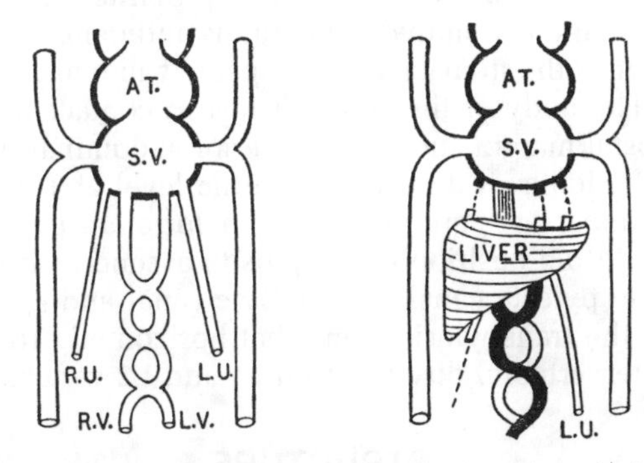

Figure 13.10. Development of portal vein and terminal part of inferior vena cava (see text). (See also *Fig. 16.5.*) *AT, atrium; RU*, right umbilical vein; *LU*, left umbilical vein; *RV*, right vitelline vein; *LV*, left vitelline vein; *SV*, sinus venarum.

DEVELOPMENT OF THE PORTAL SYSTEM

In prenatal life, the **right** and **left vitelline veins** from the yolk sac and the **right** and **left umbilical veins**, originally from the allantois but later from the placenta, opened independently into the common **sinus venosus** of the heart (*fig. 13.10*) until the developing liver intercepted and broke them up into the anastomosing sinusoids of the liver. Thereafter, the prehepatic parts of the right umbilical, left vitelline, and left umbilical veins disappeared, leaving only the prehepatic part of the right vitelline to conduct blood from the liver to the heart (prehepatic = cephalad to the liver, posthepatic = caudad to the liver). Definitively, this prehepatic part of the right vitelline vein becomes the terminal segment of the IVC (see *fig. 16.5*).

The posthepatic part of the right umbilical vein disappeared, leaving the corresponding part of the left umbilical vein to bring blood from the placenta to the liver. A shortcut, the **ductus venosus**, develops, connecting the left umbilical vein with the prehepatic part of the right vitelline vein.

The **right** and **left vitelline veins** made a figure-of-8 anastomosis around the first and third parts of the duodenum. Out of this, the portal vein took form by the disappearance of the posterior (right) limb of the 8 below and of the anterior (left) limb of the 8 above. It is joined by the superior mesenteric, inferior mesenteric, and splenic veins (*figs. 13.6, 14.10*).

Stomach

Surface Anatomy

The esophagus enters the abdominal cavity by penetrating the left dome of the diaphragm at the level of the 10th thoracic vertebra. The fundus of the stomach rises above this level to the left of the esophageal hiatus. Gas in the fundus of the stomach is a prominent feature in thoracic and abdominal radiographs and demarcates the position of the stomach in the upper left abdominal quadrant. The body of the stomach lies in contact with the left costal margin and upper anterior abdominal wall on the left side as it descends from the level of T10 to the mid-lumbar vertebral level. The pyloric antrum is in a near-midline position and begins to ascend as it blends into the pyloric canal. The pyloric canal and sphincter lie on the transpyloric plane that horizontally traverses the intervertebral disc between L1 and L2 vertebrae.

STRUCTURE

The coats of the stomach are: serous, subserous, muscular, submucous, and mucous.

Muscular Coat

It has 3 layers—an outer longitudinal, a middle circular, and an inner oblique (*fig. 13.11*). The **longitudinal fibers** are continuous with those of the esophagus; they are best marked along the curvatures. At the pylorus, they dip in to join the sphincter, and only a few are continuous with those of the duodenum.

The **circular fibers** are present everywhere except at the fundus, and they are greatly increased at the pylorus to form a sphincter. At the pylorus and anus, powerful sphincters are required to keep the contents from escaping from the stomach and rectum.

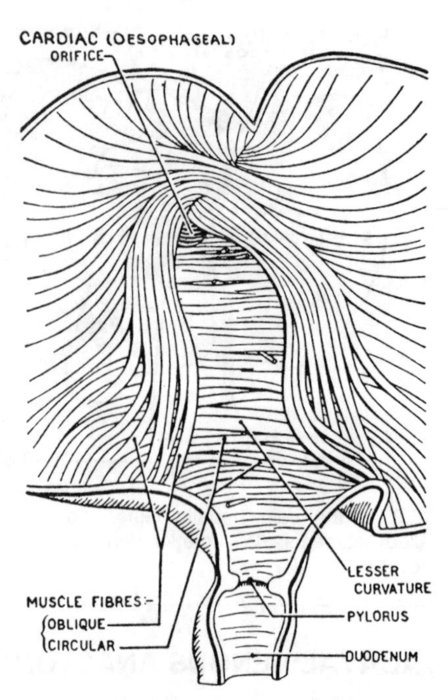

Figure 13.11. The muscular coat of the stomach seen from within. (The stomach was opened along the greater curvature and the mucous and submucous coats were removed.)

Anatomically, the presence of a sphincter at the esophageal or cardiac end of the stomach is disputed; perhaps the diaphragm suffices. Physiologically, however, there is a cardiac sphincter that opens on stimulation of the vagus and closes on stimulation of the sympathetic. Jackson demonstrated that spirally arranged muscle fibers in the esophagus can act as a sphincter.

The **oblique fibers** form ∩-shaped loops that extend over the fundus and down both surfaces of the stomach to the pyloric antrum, the cardiac notch forming their medial limit.

Mucous membrane is rugose (i.e., thrown into many folds). When the stomach is empty, 3 or 4 uninterrupted ridges lie along the lesser curvature from esophagus to pylorus, forming a gutter (*fig. 13.12*).

ARTERIES

The arteries of the stomach are branches of the celiac trunk. The right and left gastric arteries form an arch close to the lesser curvature; the right and left gastroepiploic arteries usually form a feeble arch at some distance from the greater curvature, the left artery receding from the stomach as it is traced to the left. Three or four short gastric arteries pass to the greater curvature at the fundus via the gastrolienal ligament (*fig. 13.5*).

All arterial branches supplying the stomach penetrate the muscular coats and enter the submucosa, where they form a very extensive network of comparatively large vessels. From this network in the submucosa, 2 systems of branches are given off. Of these, one turns back to the

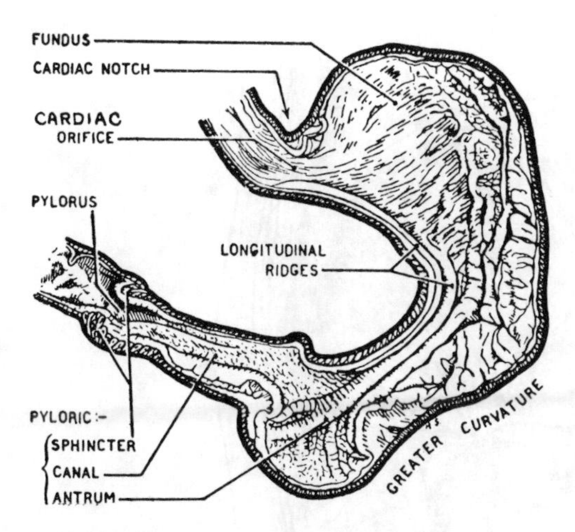

Figure 13.12. The mucous coat of the posterior half of the stomach.

muscular coats; the other continues to the mucosa. The branches to the mucosa usually divide twice, run spirally toward the muscularis mucosae, and pierce it to enter the mucosa, where they suddenly become smaller by giving off end-branches (i.e., vessels connected only by means of a capillary network). Each end-artery continues to run a spiral course, supplying an area of mucosa of about 2.5 mm in diameter. The submucous network on the lesser curvature is made up of long parallel vessels which are smaller, make fewer anastomoses, and run more than twice the distance of similarly sized vessels in other parts of the stomach (Reeves).

Lymphatics of the stomach (*fig. 16.23*) follow the arteries as do the lymphatics of most deep structures in the body.

NERVES OF THE STOMACH

Distribution of Vagal Trunks within the Abdomen

Due to the partial rotation undergone by the stomach during development (left becomes anterior), the anterior vagal trunk (left vagus) enters the abdomen in front of the esophagus and the posterior vagal trunk (right vagus) behind it. Either trunk is occasionally in 2 or 3 branches. Both trunks, each carrying fibers from both vagi, run close to the lesser curvature and send gastric branches to the respective anterior and posterior surfaces of the stomach as far as the pyloric antrum.

The **anterior vagal trunk** sends **hepatic branches** curving upward in the lower part of the lesser omentum to the porta hepatis. It is through **pyloric branches** descending from these hepatic branches that the pyloric antrum, pylorus, and first part of the duodenum are supplied.

From the **posterior vagal trunk**, a branch descends along the stem of the left gastric artery to the **celiac plexus**, where its fibers, in company with sympathetic fibers, are distributed along the branches of the aorta to the abdominal viscera, e.g., intestines (proximal to the left colic flexure), pancreas, and spleen. This is the only connection these organs have with parasympathetic nerves (*figs. 13.3, 13.4*).

Sympathetic fibers and visceral afferent fibers from cord segments 5–9 (and 10) via the splanchnic nerves relayed in the celiac ganglia, pass to the stomach along the blood vessels.

RELATIONS OF THE STOMACH

Anterosuperiorly are the left lobe of the liver, diaphragm, and anterior abdominal wall. The diaphragm separates the stomach from the pleura of the left lung and apex of the heart.

Posteroinferiorly, the omental bursa intervening, is **the stomach bed**, formed to the extent shown in *Figure 13.13*. Pathological processes in the bed of the stomach or lesser sac can erode the pancreas or splenic artery. Perforations of the posterior stomach wall due to ulcers are hazardous to this region when acid escapes the stomach and enters the lesser sac.

Liver

The liver occupies the vast majority of the space in the upper right abdominal quadrant (*fig. 13.13*). Its **superior, posterior**, and **anterior surfaces** contact the undersurface of the diaphragm. Since a visceral peritoneum covers most of the liver on these surfaces and an opposing parietal peritoneum covers most of the inferior surface of the diaphragm, a very thin space exists between these two squamous coverings. The **bare area** of the liver is the exception to this description of the liver's relationship to the diaphragm (as depicted in *Figure 13.15*).

The **inferior surface** of the liver is also covered by a visceral layer of peritoneum. This peritoneum is continuous with the layers of the lesser omentum and falciform ligament that continue onto the inferior surface of the liver (*figs. 13.14, 13.15*).

The fissures of the liver and fossae formed by the organs contacting the inferior surface of the liver serve to subdivide it into 4 distinct areas or lobes (*fig. 13.14*). The **right lobe** is the largest of these areas and approximately equals the volume of the remaining 3 lobes: the **quadrate, caudate** and **left lobes**. These areas can be demarcated

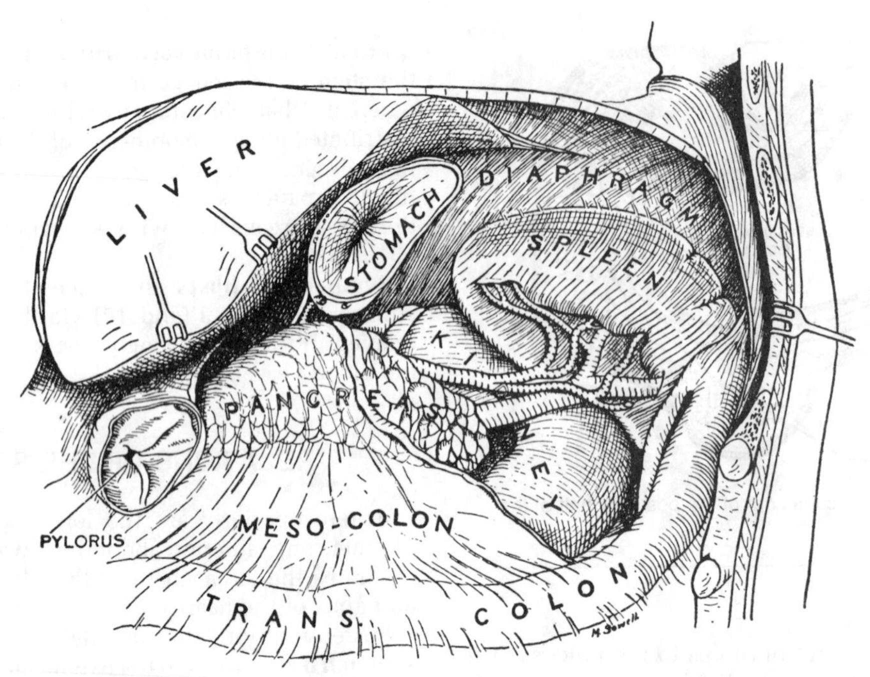

Figure 13.13. The stomach bed. (This pancreas is unusually short; the left suprarenal gland, the left gastric artery, and the branch of the posterior vagal trunk to the celiac plexus are not labeled.)

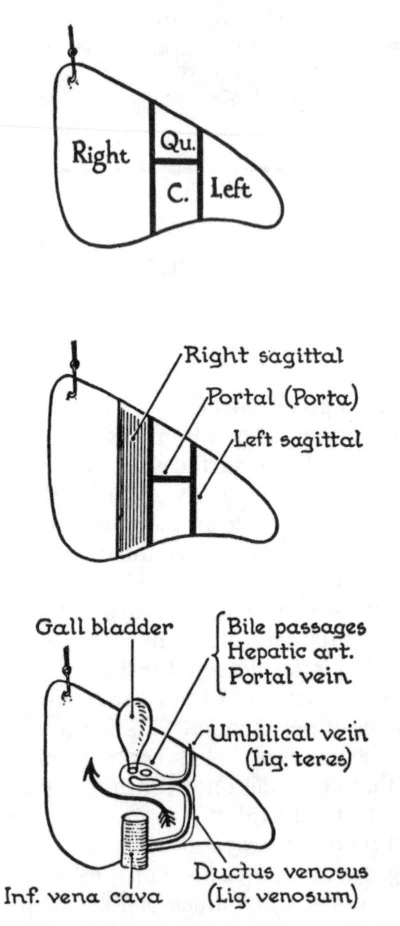

Figure 13.14. Diagrams of the liver, hooked upward to show (1) the H-shaped fissure on the inferior and posterior surfaces. (2) the subdivisions of this fissure, and (3) their contents. *Qu*, quadrate lobe; *C*, caudate lobe.

by an H-shaped visualization of the associated fissures and fossae (*fig. 13.14*). The crossbar of the H represents the porta hepatis, which contains the portal vein, hepatic arteries, and bile passages. The quadrate lobe is **anterior** to the porta hepatis, while the caudate lobe is **posterior** to it.

The left vertical limbs of the H are formed by the deep fissures containing the **ligamentum teres** that runs from the anterior abdominal wall (umbilicus) to the porta hepatis (left portal vein) and the **ligamentum venosum**, which is the obliterated venous pathway in the fetus (ductus venosus) that shunted the placental blood through the liver. At birth, when the umbilical cord is "tied off," the portal venous system becomes effective and establishes a venous return through the liver parenchyma.

The right vertical limbs of the H are representative of the fossae formed by the gallbladder anteriorly and the fossa formed by the inferior vena cava posteriorly. The posterior vertical limb of the right side of the H is incomplete. The caudate lobe is seen to continue onto the right lobe of the liver and the visceral peritoneum of these two surfaces is continuous (*figs. 13.14, 13.15*).

The **posterior surface** of the liver cannot be seen until the liver is removed from the abdomen. On the left, the posterior surface is covered with peritoneum and grooved by the esophagus.

On the right side, the posterior surface lies in contact with the diaphragm, the inferior vena cava, and the right suprarenal gland. This contact with retroperitoneal structures on the posterior abdominal wall constitutes the **bare area**, which is bounded by the **coronary ligaments** that reflect on the inferior surface of the diaphragm (*fig. 13.15*).

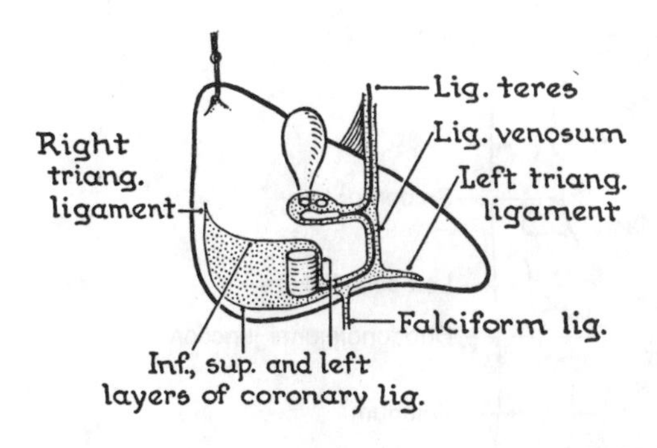

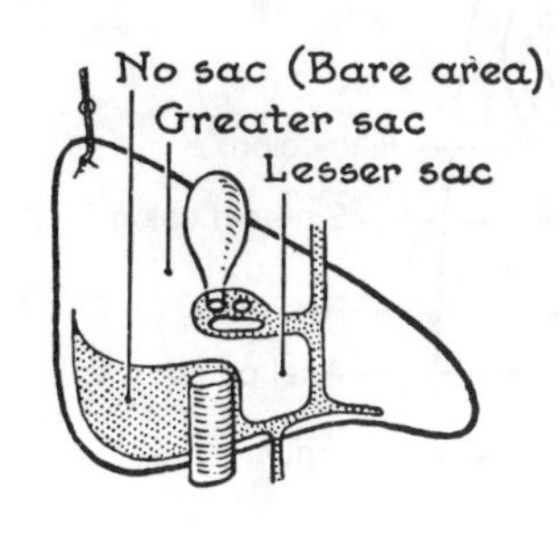

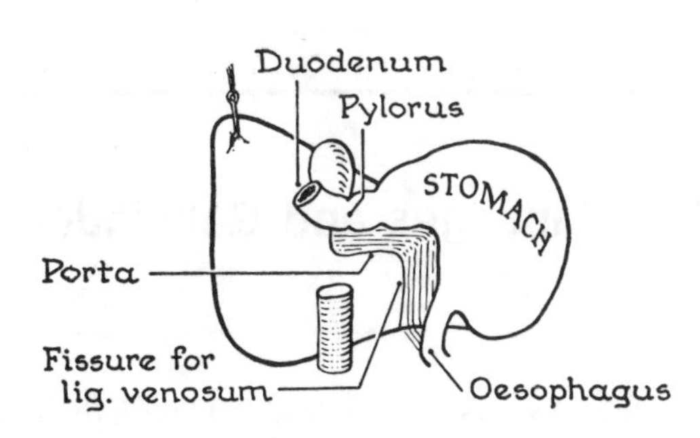

Figure 13.15. Peritoneal attachments of the inferior and posterior surfaces of the liver.

RELATIONSHIPS OF THE LIVER TO THE ABDOMINAL VISCERA

The esophagus grooves the posterior aspect of the left lobe of the liver and enters the stomach as it underlies the left lobe of the liver. The lesser curvature of the stomach is related to the porta hepatis through their common ventral mesentary, the lesser omentum (*fig. 13.1*). The pyloric canal and the duodenum underlie the gallbladder and the anterior aspect of the right lobe of the liver. The transverse colon contacts the inferior surface of the right lobe of the liver from the right colic flexure to the area anterior to the pyloric canal and duodenum (*fig. 13.16*). The right kidney and suprarenals underlie the posterior aspect of the visceral surface on the right lobe of the liver. Visceral peritoneum covers the inferior surface of the liver above the retroperitoneal right kidney and suprarenal gland. The intervening space in the peritoneal cavity is called the **hepatorenal pouch**. Intraperitoneal exudates and infections can collect in this space, particularly in patients that are lying on their back (supine position) (*fig. 13.17*).

SURFACE ANATOMY OF THE LIVER

1. The **base** or right lateral aspect of the liver extends from near the midlateral aspect of the right iliac crest in the midlateral line to cross the 11th, 10th, 9th, 8th and up to the 7th rib.
2. The superior limit is the diaphragm and the superior aspect of the right lobe of the liver is at the level of the 5th rib under the right dome of the diaphragm. The superior aspect of the liver is at the xiphisternal joint in the median plane and then contacts the medial aspect of the left dome of the diaphragm medial to the midclavicular line.
3. The sharp inferior border of the liver is formed by the junction of the inferior and anterior surfaces. This border crosses the transpyloric plane about 2–3 cm to the right of the midline. This area is below the right costal margin and may be susceptible to injury when the anterior abdominal wall is penetrated (*figs. 13.16, 13.18*).

The position of the gallbladder varies with that of the liver. The fundus of the gallbladder typically lies at the lateral border of the right rectus abdominis, somewhat below the costal margin (*figs. 11.3, 13.18*).

LOBES AND SEGMENTS OF LIVER

Descriptively, the liver is divided into unequal right and left lobes by the falciform ligament. **Structurally**, however, the right and left portal veins, hepatic arteries, and hepatic ducts serve approximately equal halves of the liver. The plane between the 2 halves (structural lobes) runs anteroposteriorly through the fossa for the gallbladder and the fossa for the IVC, i.e., through the right sagittal fossae, and between the right and left groups of vessels and ducts in the porta hepatis (*fig. 13.19*).

The entire quadrate lobe and half the caudate lobe are served by left vessels and ducts. And, according to the prevailing pattern, no branches of the portal vein, hepatic artery, or hepatic duct cross this right sagittal plane.

According to the biliary drainage, the left half of the liver is divided by the plane of the left sagittal fissures into a medial and a lateral segment, whereas the right half is divided by an oblique plane into an anterior and a posterior segment. Each of these 4 segments is subdivisible into an upper and a lower area, as in *fig. 13.19* (Healey et al.; Hjortsjo).

Clinical Case 13.2

Patient Arlene C. This 42-year-old female schoolteacher developed severe pain along the costal margin in the upper right abdominal quadrant fol-

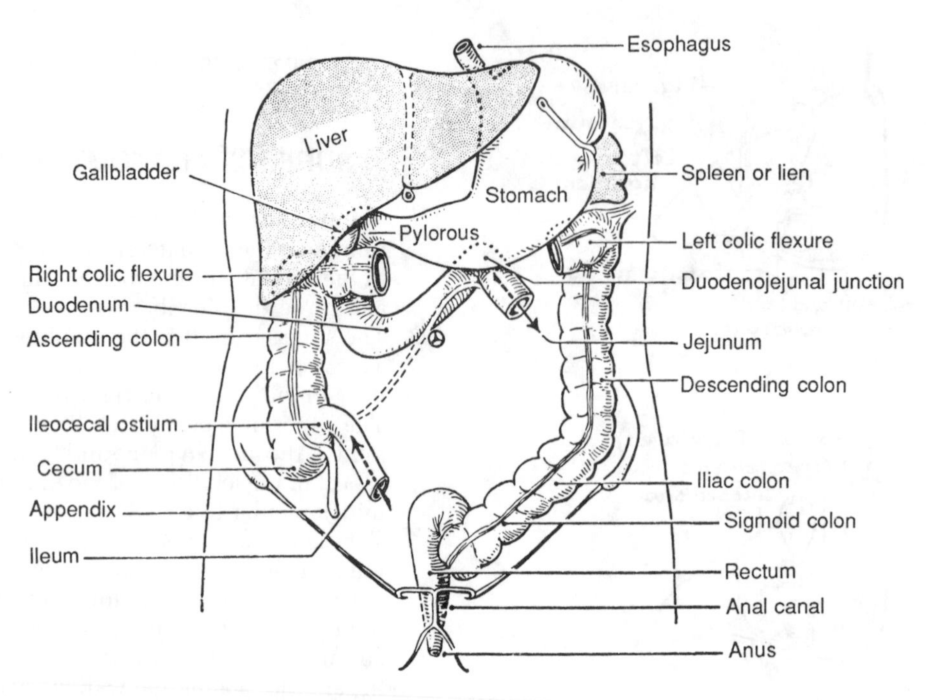

Figure 13.16. The names of the various parts of the digestive tract and their dispositions.

lowing a pre-Christmas eggnog party. She has been rushed to the emergency ward at a University hospital, where you are on the junior duty roster for the vacation period. The chief resident prescribes analgesics (painkillers) and injects phenothalein intravenously to prepare the biliary system for radiography the next morning, but he is too busy to explain his reasoning. The next morning's X-ray pictures reveal a gallstone lodged in the common bile duct at its junction with the main pancreatic duct. You attend the procedure when the radiologist removes the gallstone using a lithotriptor to disintegrate the stone within the common bile duct. The patient is released from the hospital and returns to teaching in the 1st week of January.

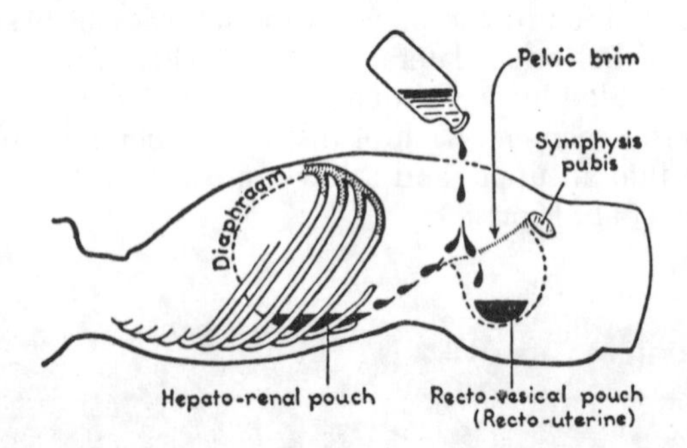

Figure 13.17. Two pouches which are the lowest parts of the peritoneal cavity when the subject is supine. Fluid (e.g., blood) from torn organs follow these paths.

Bile Passages and Gallbladder

These structures have been described with the liver (p. 152, *figs. 13.2 and 13.20*).

FUNCTION

The function of the gallbladder is to concentrate and store the bile brought to it from the liver via the cystic duct between meals and to discharge it into the intestine via the cystic duct during meals.

THE VESSELS AND NERVES OF THE GALLBLADDER

The **cystic artery** (see p. 161). The **cystic veins** mostly plunge through the fossa for the gallbladder into the liver substance and behave like branches of the portal vein. The **lymph vessels** pass to the cystic node at the neck of the bladder and then downward along the biliary chain. The parasympathetic and sympathetic **nerves** are derived from the anterior vagal trunk (p. 165), and from cord segments T(6), 7–9 (and 10) via the celiac plexus.

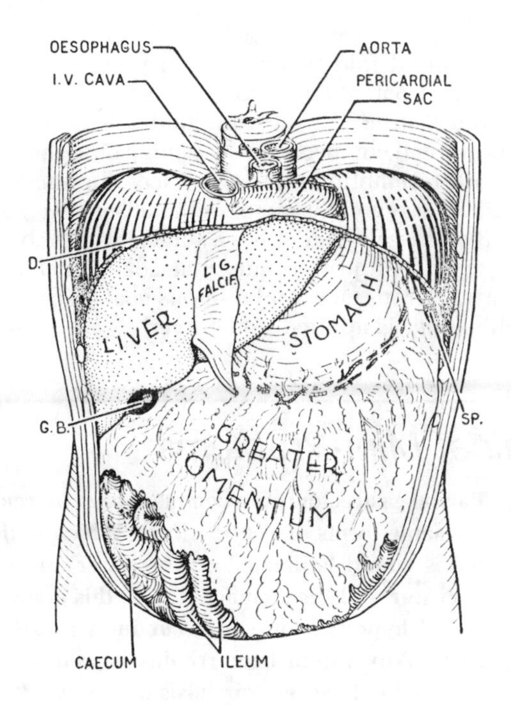

Figure 13.18. Abdominal contents undisturbed. *GB* = gallbladder; *Sp* = spleen; *D* = cut edge of diaphragm; *IV* = inferior vena (cava).

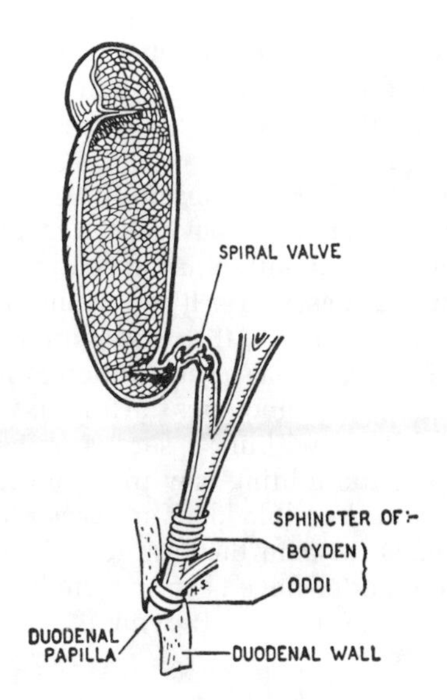

Figure 13.20. The mucous membrane of the gallbladder and extrahepatic bile passages. The 2 sphincters are shown diagrammatically. This bladder happened to have a folded fundus.

VARIATIONS

The bile duct develops as an outgrowth from the duodenum (*fig. 14.11*). It branches and rebranches more or less dichotomously. One of the main branches, the **cystic duct**, instead of branching gives rise to a blind vesicle, the **gallbladder**.

Irregular branching of the bile passages is common, and when the branch is in surgical danger, this fact is of importance. Thus, the cystic duct may end much lower than usual or much higher (*fig. 13.21A*), sometimes joining the right hepatic duct (*B*), and it may swerve across the common hepatic duct (C). When there are 2 right

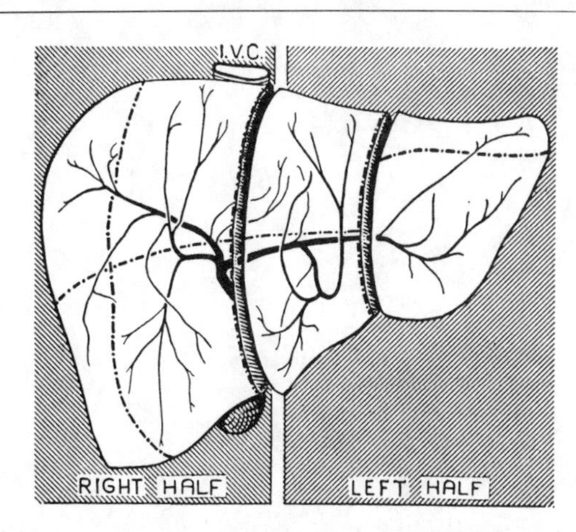

Figure 13.19. The segments of the liver, according to the biliary drainage. *IVC*, inferior vena cava. (After Healey et al.)

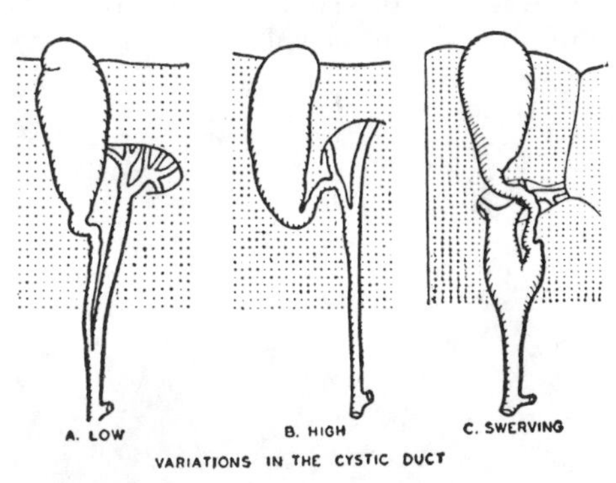

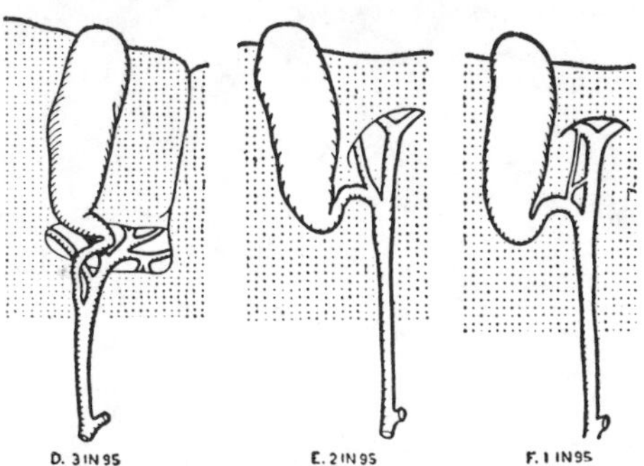

Figure 13.21. Variations of the bile passages. *A–C* = variations in length and course of cystic duct. *D–F* = accessory right hepatic ducts. (Of 95 gallbladders and bile passages injected *in situ* with melted wax and then dissected, seven had accessory hepatic ducts in positions of surgical danger. Of these, 4 joined the common hepatic duct near the cystic duct (*D*), 2 joined the cystic duct (*E*), and 1 was an anastomosing duct (*F*).

hepatic ducts, the lower is erroneously called an **accessory hepatic duct**—it is not an additional or supplemental duct, but merely one that arises unusually early (fig. 13.21 D–F).

The gallbladder is rarely **absent** (unless like an offending appendix, it has been removed). It is rarely **buried** in the liver, but occasionally it is suspended from the liver by an **acquired mesentery**. It is commonly attached to the transverse colon or to the duodenum by a **peritoneal fold**. Occasionally, it has a **sacculation** at its neck (Hartmann's pouch), and sometimes the fundus is **congenitally folded** upon itself within its serous or peritoneal coat (*fig. 13.20*), or the folding may include the serous coat. Although **bilobed** gallbladders are very common in domestic animals, in man they are rare. (Seventeen cases of a double bladder have been reported; however, each with a separate cystic duct [Boyden].)

STRUCTURE

The **intrahepatic ducts** are described on p. 45. The **extrahepatic ducts** (viz., right and left hepatic, common hepatic, cystic, and bile) are fibrous tubes that contain many elastic fibers. They are lined with columnar epithelium but do not have a muscular coat, except at the lower 5–6 mm of the bile duct, where there is a strong and effective submucous sphincter; the duodenal papilla (or the ampulla) also has a sphincter. In the cystic duct, there is a spiral fold (valve), which probably serves to keep the duct patent (*fig. 13.20*).

The inner surface of the **gallbladder** is covered with small polygonal compartments, opening onto the interior lumen and resembling the cut surface of a honeycomb. Like villi, these greatly increase the absorptive surface. The wall of the gallbladder has a single layer of columnar cells, a tunica propria, a muscular coat of decussating fibers, a subserous coat, and a serous coat (except where the gallbladder is applied to the liver).

Clinical Mini-Problems

1. **Partial gastrectomy is a common surgical treatment for some forms of peptic ulcers. Which *major* arteries would be ligated on the lesser and greater curvatures of the stomach during this operation?**
2. **Portal hypertension can occur in cirrhosis of the liver. Why would hemorrhoids be commonly associated with severe cirrhosis of the liver?**
3. **Structures such as the first part of the duodenum, the pylorus, and the pancreas are located on the transpyloric plane. Which vertebral level is represented by the transpyloric plane?**

(Answers to these questions can be found on p. 585.)

14

Mesenteric Vessels, Duodenum, and Pancreas

Clinical Case 14.1

Patient Lillian K. This 60-year-old musician developed a sudden, excruciating, central abdominal pain with abdominal wall rigidity while in the University Center for medical treatment of a connective tissue disease of her joints. You are a student on the ward and follow her case as she is rushed to the acute care unit and to the operating room even before X-ray reports are available. Upon opening the abdomen, the surgical team finds that she has a complete thrombosis of three major branches of the superior mesenteric artery to the small intestine, and already a local area of jejunum is obviously doomed. As you watch the surgeons prepare to perform a resection, you review all you can remember of the anastomotic blood supply of the small and large intestines.

Mesenteric Vessels

SUPERIOR MESENTERIC ARTERY

The superior mesenteric artery arises from the aorta just below the celiac trunk at the level of the L1 vertebra. It supplies the small and large intestines from the 2nd part of the duodenum to the left colic flexure. For embryological reasons (*see fig. 14.9*), it passes between the head and neck of the pancreas.

"Clamped" between it and the aorta like a nut in a nutcracker, are the left renal vein and the third part of the duodenum (*fig. 14.1*). Both—at least theoretically—may be obstructed by the "clamp." The superior mes-

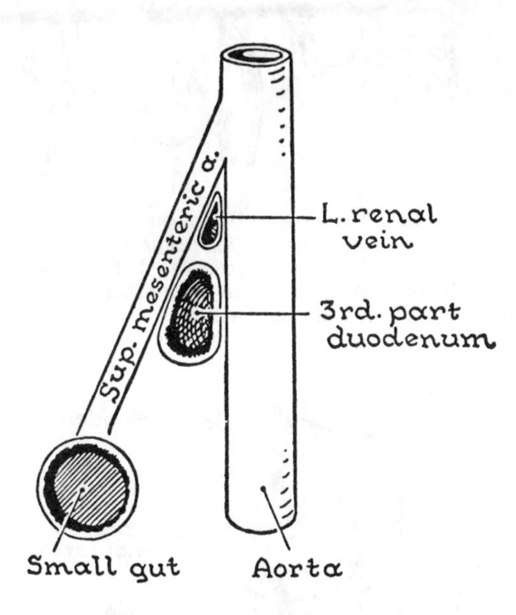

Figure 14.1. Compression of the left renal vein and the duodenum—as in a nutcracker.

enteric artery enters the root of the mesentery at the point where the mesentery is applied to the inferior vena cava (IVC). The artery then runs in the mesentery to the ileum 10 or 12 cm from the ileocecal junction. There it ends by forming an arch with one of its own branches (*fig. 14.2*).

Branches

1. The superior mesenteric artery's branches **to the small intestine** include the inferior pancreaticoduodenal, jejunal, and ileal.
2. Its branches **to the large intestine** include the ileocolic, right colic, and middle colic.

The **inferior pancreaticoduodenal arteries** (*fig. 14.3*), anterior and posterior, form arches that give branches to the head of the pancreas and the 2nd and 3rd parts of the duodenum. Their terminal branches anastomose with the anterior and posterior branches of the superior pancreaticoduodenal artery that arises from the gastroduodenal artery. In this way, the celiac trunk and the superior mesenteric artery are anastomotic and can share their blood supply if one of the major arteries is occluded.

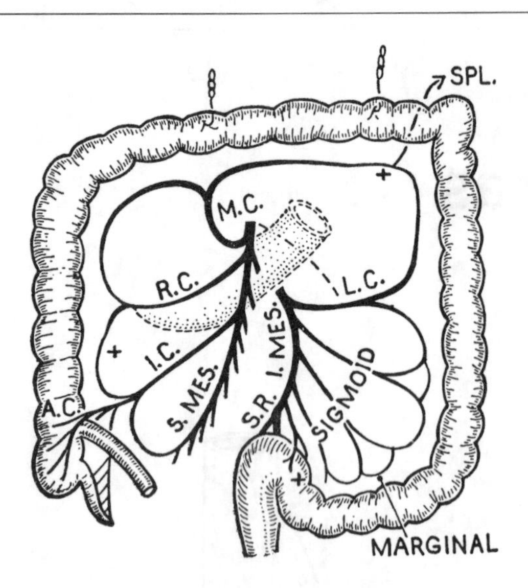

Figure 14.2. The superior and inferior mesenteric arteries. [+ denotes 3 weak points in the marginal anastomosis: between ileocolic (*IC*) and right colic (*RC*), between middle colic (*MC*) and left colic (*LC*), between lowest sigmoid and superior rectal (*SR*)] *AC* = anterior cecal.

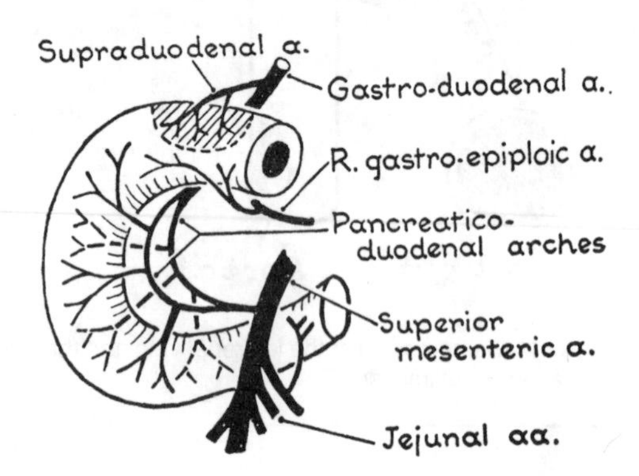

Figure 14.3. The blood supply of the duodenum.

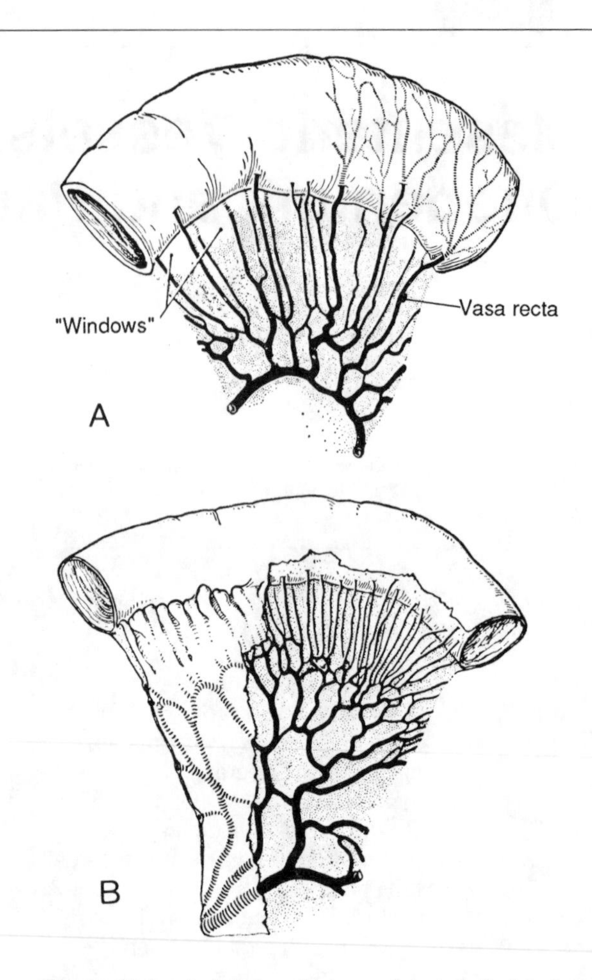

Figure 14.4. *A*, Arteries of jejunum injected. *B*, Arteries of ileum injected.

The **jejunal** and **ileal arteries**, 18 or so, fan out from the left side of the artery into the mesentery, where they unite to form loops or arcades from which the **vasa recta** arise.

Arcades (*fig. 14.4*)

The number of tiers of arcades varies from subject to subject and from area to area, the heaviest concentration generally being in the 2nd quarter of the length of the small intestine. In the 1st quarter there are 2–4 tiers (average 2); in the 2nd quarter, 3–5 (average 4); in the 3rd quarter, 2–4 (average 2); and in the last quarter, 0–4 (only 1 in the majority of cases) (Michels et al.).

The **vasa recta** do not anastomose themselves (*fig. 14.4*). In the submucosa, their twigs anastomose freely (Benjamin and Becker).

The **three colic branches** of the superior mesenteric artery arise from its right border (*fig. 14.2*), commonly from only 2 stems, one of which bifurcates. Each in turn bifurcates and forms loops or arches at a very variable distance (0.5–8 cm) from the colon. Similarly, the colic branches of the inferior mesenteric artery form loops. The result is a series of anastomosing links, "the **marginal artery**," which parallels the entire colon (*fig. 14.2*).

The **ileocolic artery** descends retroperitoneally toward the ileocecal region, crossing the IVC and ureter, and then bifurcates. A descending branch divides into: anterior cecal, posterior cecal, appendicular, and ileal branches (*fig. 14.5*).

The **appendicular branch** descends behind the end of the ileum and runs in the free edge of the mesentery of the appendix.

Variations. Commonly the appendix is supplied by 2 arteries (Shah and Shah; Solanke).

The **right colic artery** crosses the same structures as the ileocolic artery and divides into a descending and an ascending branch: the latter crosses the lower pole of the right kidney (*fig. 14.2*).

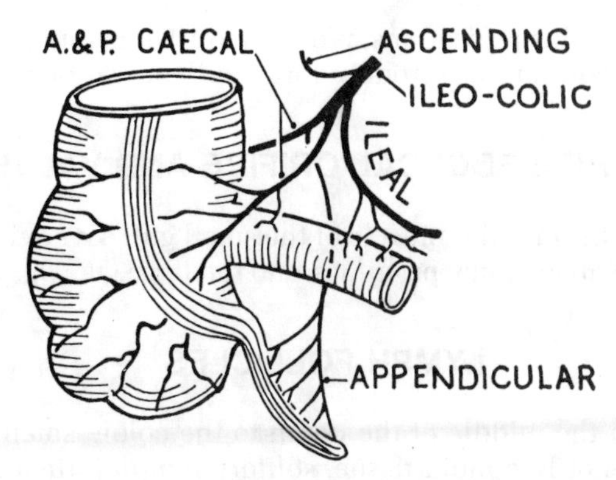

Figure 14.5. The branches of the ileocolic artery.

The **middle colic artery** arises at the lower border of the pancreas. It curves downwards in the right half of the transverse mesocolon and divides into a right and a left branch (*fig. 14.2*).

INFERIOR MESENTERIC ARTERY

The inferior mesenteric artery supplies the large gut from the left end of the transverse colon to the lower end of the rectum. It arises from the front of the aorta 4 cm above its bifurcation (therefore in front of the third lumbar vertebra), where the lower border of the duodenum overlaps it (*see fig. 14.9*). It descends retroperitoneally on the aorta and the psoas fascia, crosses the left common iliac artery, and enters the pelvis as the superior rectal (superior hemorrhoidal) artery (*fig. 14.6*).

Branches

The branches of the inferior mesenteric artery include the left colic (upper left colic), the sigmoid (lower left colic) branches, and the superior rectal artery.

The **left colic artery** passes retroperitoneally to the left across the inferior mesenteric vein, ureter, and gonadal vessels and divides into an ascending and a descending branch.

The **sigmoid arteries** are generally 2 to 4 in number (*fig. 14.6*). The upper branches cross the structures in front of the psoas muscle (viz., inferior mesenteric vein, ureter, and gonadal vessels). The lower branches cross the common iliac vessels and enter the sigmoid mesocolon to form 2 or 3 tiers of arches.

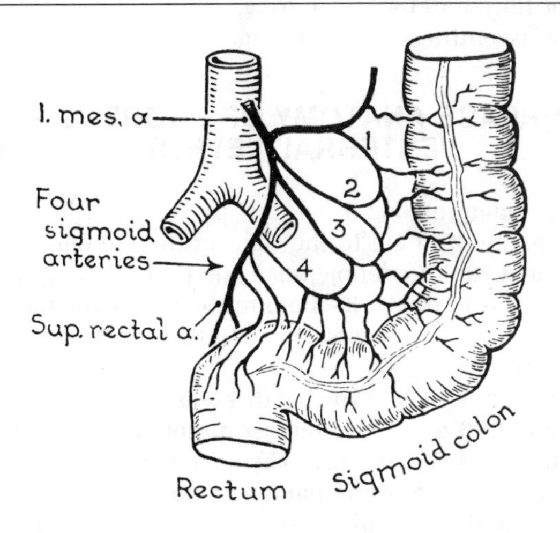

Figure 14.6. Branches of inferior mesenteric artery.

MARGINAL ARTERY (OF DRUMMOND)

The anastomosing loops of colic branches result in a continuous marginal artery situated from 0.5 to 8 cm from the wall of the large gut. It is closest along the descending and sigmoid colons.

Long and short terminal branches proceed from "the marginal artery" to the colon (*fig. 14.7*). The anastomoses they effect across the antimesenteric border are meager (Ross).

The short branches are 4–5 times as numerous as the long branches (*fig. 14.7*). Like the long branches, they pass to the submucous plexus after a short tortuous subserous course. The muscular coats are mainly supplied by recurrent branches from the submucous plexus.

Variations. (1) In about 5% of 100 specimens, the marginal artery is discontinuous, due to the ileocolic artery failing to anastomose with the right colic. (2) The right colic artery very commonly takes origin either from the middle colic or the ileocolic artery. (3) The middle colic artery commonly has an accessory left branch. (4) The left colic artery supplies the left end of the transverse colon in about two-thirds of cases, and in one-third it fails to reach the left colic flexure. (5) The middle colic and left colic arteries probably always anastomose, although the portion of the marginal artery so-formed is long and usually single. (6) A large branch (the arc of Riolan) not rarely connects the stem of the superior mesenteric artery with the left colic artery on the posterior abdominal wall. (7) Occasionally, a branch connects the left colic artery with the splenic artery. (8) The marginal

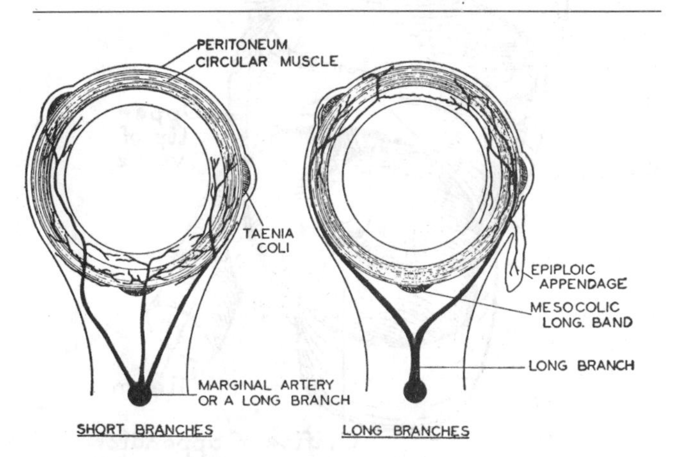

Figure 14.7. The blood supply of the colon. (After Steward and Rankin.)

artery does not link up the lowest sigmoid artery with a superior rectal artery—except occasionally and feebly. So, if the superior rectal artery is obstructed beyond the origin of the lowest sigmoid artery, there is little chance of an effective collateral circulation being established (*fig. 14.2*) (Steward and Rankin; Basmajian; Michels et al.).

MESENTERIC VEINS

The *superior mesenteric vein* lies on the right side of its artery and ends in the portal vein. The *inferior mesenteric vein* is the continuation of the superior rectal vein (superior hemorrhoidal vein). It ends behind the neck of the pancreas by joining the angle of union between the superior mesenteric and splenic veins as they form the portal vein (*fig. 13.6*).

Structure of Intestines

When the gut is opened, folds of mucous membrane, the *plicae circulares*, are seen running transversely for variable distances around the gut wall. They begin in the duodenum, 5 cm from the pylorus, and end beyond the middle of the ileum. In the duodenum and upper part of the jejunum they are high (about 6 mm) and closely set; lower down, they gradually become smaller and more widely separated.

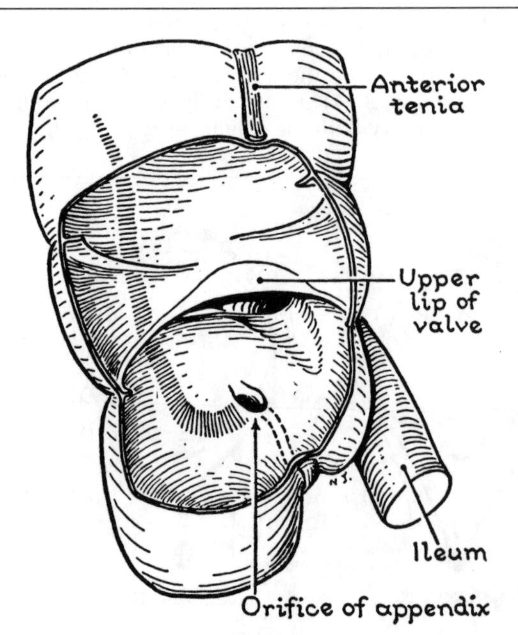

Figure 14.8. The ileocecal valve, as seen on opening a dried cecum. In life, the orifice is circular.

Finger-like projections, *villi* (0.5 to 1.5 mm long), cover the mucous surface from pylorus to ileocecal orifice.

THE ILEOCECAL ORIFICE AND VALVE

Here the circular muscle of the small gut, covered with mucous membrane, protrudes into the large gut (*fig. 14.8*).

LYMPH FOLLICLES

From the middle of the ileum to the colon, small collections of lymphoid tissue, *solitary lymph follicles*, 2–3 mm in diameter, are present. Also in the ileum, at the antimesenteric border, are 20 or more *aggregated lymph follicles* (Peyer's patches); they are oblong, 1 cm wide by 3 cm or more long, the long axis being in the long axis of the gut.

Duodenum and Pancreas

The word *duodenum* is a Latin corruption of the Greek word dodekadaktulos, meaning 12 fingers (cf. the 12 islands in the Levant called the Dodecanese Islands). About 300 B.C., Herophilus of Alexandria introduced the name dodekadaktulos from its being as long as 12 fingers are broad in certain animals (Finlayson).

In man, the duodenum is the part of the small intestine that has lost its dorsal mesentery. About 25 cm long, it is molded around the head of the pancreas in horseshoe fashion and is divided into 4 parts (*fig 14.9*):

1st or superior—5 cm long.
2nd or descending—7½ cm long.
3rd or horizontal—10 cm long.
4th or ascending—2½ cm long.

SURFACE ANATOMY OF DUODENUM: VERTEBRAL LEVELS

The duodenum begins at the pylorus in the transpyloric plane, 2–3 cm to the right of the median plane, and ends at the duodenojejunal junction slightly below the transpyloric plane, 2–3 cm to the left of the median plane. The ends of the horseshoe are, therefore, only about 5 cm apart (*fig. 14.9*).

The **pancreas** (Gk. pan = all, kreas = flesh) weighs only about 170 g and resembles the letter J, or a curve-handled cane, set obliquely. The curved part, or handle, known as the *head* of the pancreas, lies within the concavity of the duodenum, while the stem of the cane, divided indefinitely into *neck*, *body*, and *tail*, slants

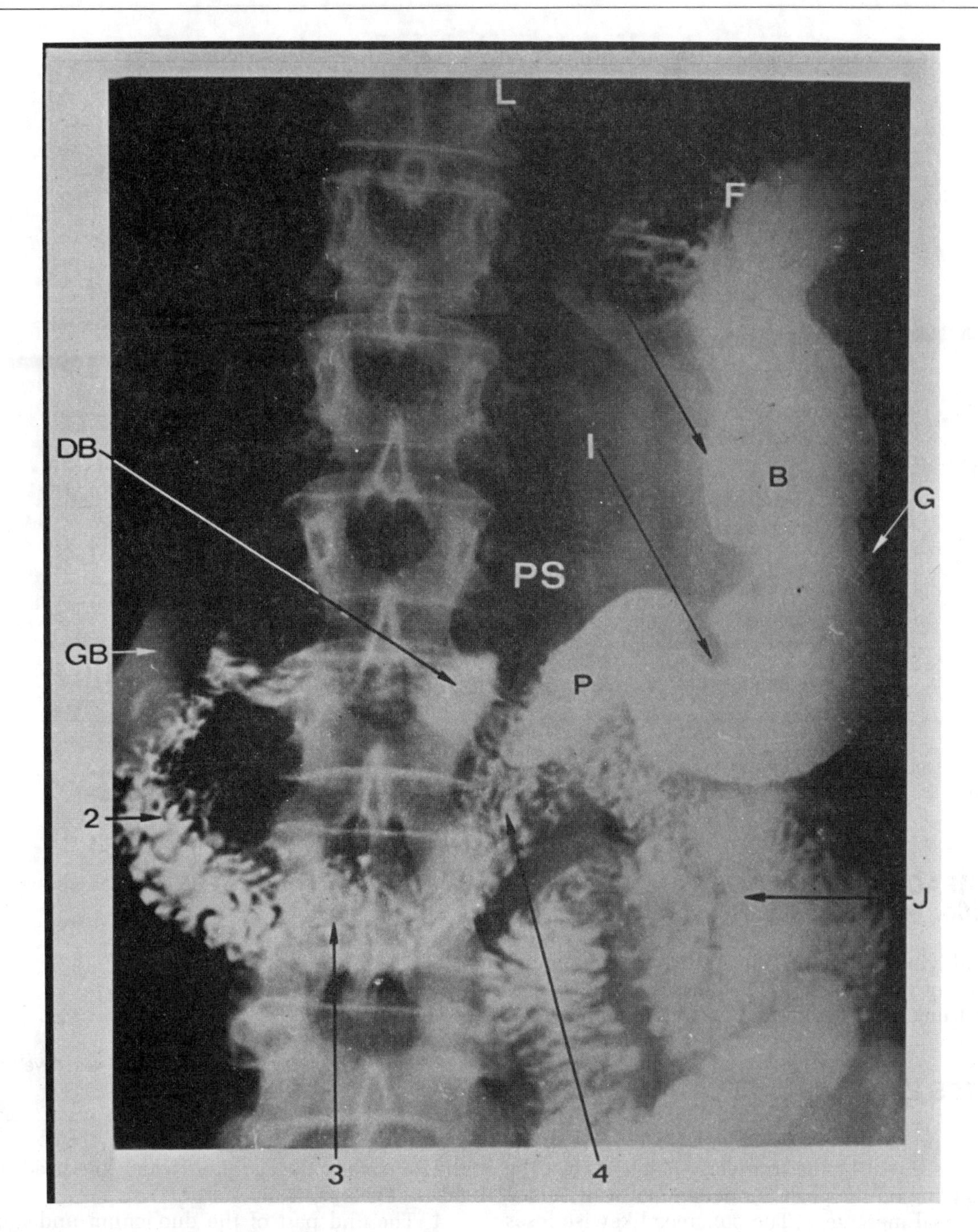

Plate 14.1 Stomach and duodenum in upper GI series (swallow of barium; A-V view, recumbent). *B*, body of stomach; *F*, fundus of stomach; *DB*, duodenal bulb; 2, 3, 4, second, third, and fourth parts of duodenum; *G*, greater curvature of stomach; *I*, incisura angularis; *J*, jejunum; *L*, lesser curvature of stomach; *P*, pyloric antrum; *PS*, pyloric sphincter; *GB*, gallbladder (partly opaque because of simultaneous intravenous cholecystography).

obliquely across the abdomen. The tail abuts against the spleen. The head projects medially, behind the superior mesenteric vessels, as the *uncinate process* (*fig. 14.9*).

The celiac trunk lies at the upper border of the pancreas and the splenic branch runs along the upper border of the body and tail.

The **main duct** of the pancreas resembles a herring bone, in that small ducts spring from the main duct, which is straight. This duct joins the common bile duct and empties the *exocrine secretion* of the pancreas into the second part of the duodenum.

The *endocrine secretion*, including glucagon and in-

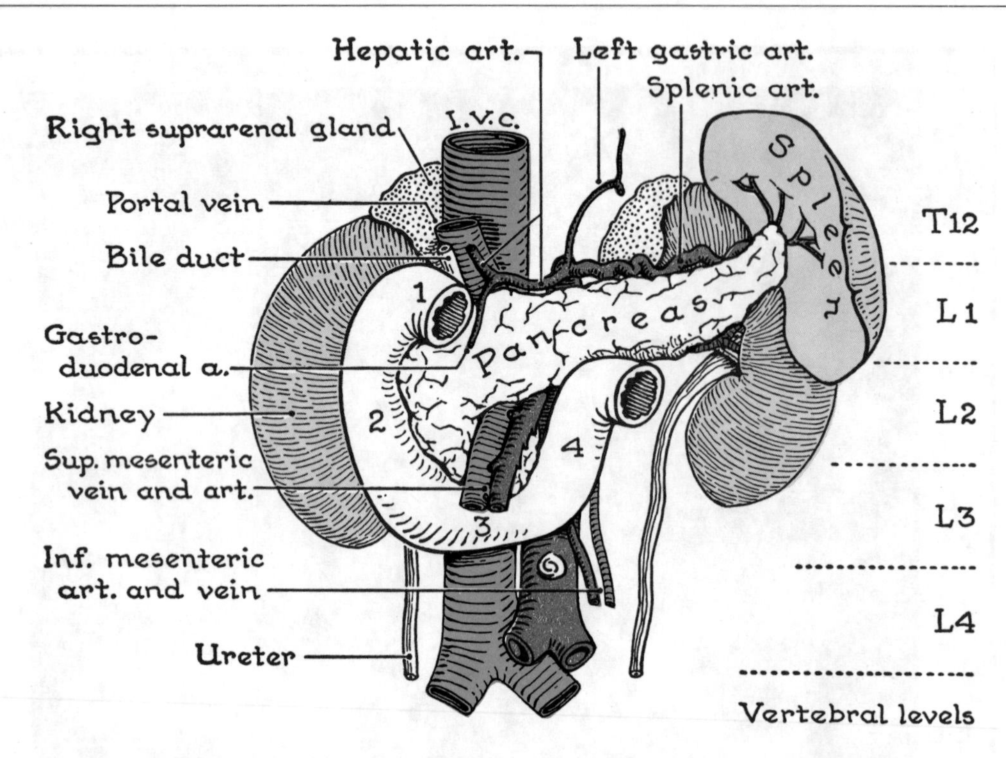

Figure 14.9. Abdominal viscera and vessels. *IVC* = inferior vena cava.

sulin, is formed by tiny clusters of cells, the *islets of Langerhans*. Its deficiency causes the disease diabetes mellitus.

SURFACE ANATOMY OF PANCREAS (*fig. 14.9*): VERTEBRAL LEVELS

The head lying within the concavity of the duodenum lies in front of vertebra L2; the body rises to the level of the celiac trunk and, therefore, is in front of vertebra L1.

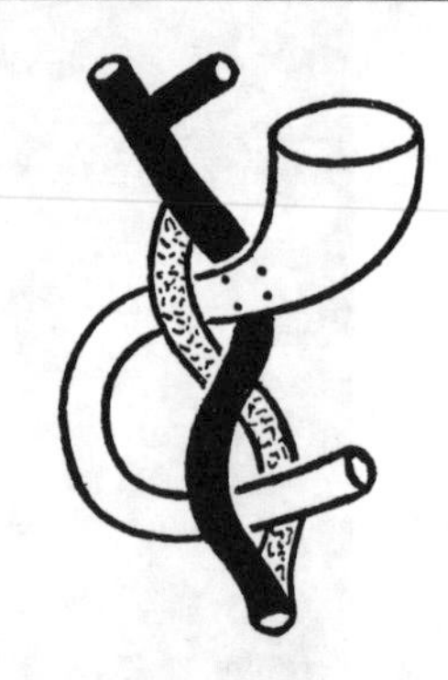

Figure 14.10. Showing how the portal vein develops from a figure-of-8 anastomosis around the duodenum.

NOTES EXPLAINING RELATIONSHIPS

1. During prenatal intestinal rotation (*fig. 12.41*), the *duodenum* is thrust by the transverse colon against the structures lying on the posterior abdominal wall, and it loses its dorsal mesentery. The *pancreas* likewise loses the layer of peritoneum that formerly clothed its right surface—now its posterior surface.

2. The right and left vitelline veins return blood from the yolk sac (*fig. 14.10*). The veins twine about the duodenum in figure-of-8 fashion. The portions anterior to the 1st part of the duodenum and posterior to the 3rd part disappear, with the result that the superior mesenteric and portal veins of adult anatomy take form.

3. The ducts of the liver and pancreas arise as 2 outpouchings or hollow buds, a dorsal and a ventral, of the endoderm of the duodenal wall (*fig. 14.11*). It is the rudiment of the bile passages and of the liver and also of the pancreatic duct around which the head of the pancreas develops.

4. The 2nd part of the duodenum undergoes partial rotation to the right on its own long axis (*fig. 14.12*). This explains why the (common) bile duct passes upward behind the accessory pancreatic duct and 1st part of the duodenum.

For descriptive purposes, the head of the pancreas may be regarded as swinging around behind the junction of the splenic, superior mesenteric, and portal veins, thereby causing them to occupy a position between it and the neck of the pancreas.

5. The primitive dorsal and ventral pancreatic ducts remain separate as one of several options (*fig. 14.13*). The common bile duct and the main pancreatic duct open

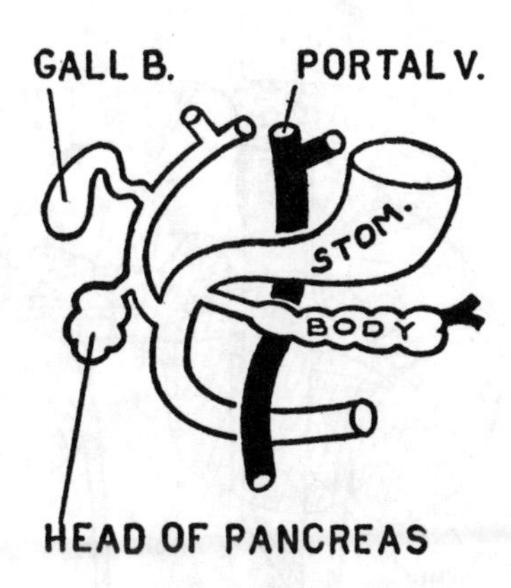

Figure 14.11. The rudiments of the liver and pancreas arise as outpouchings of the duodenum into the ventral and dorsal mesogastria.

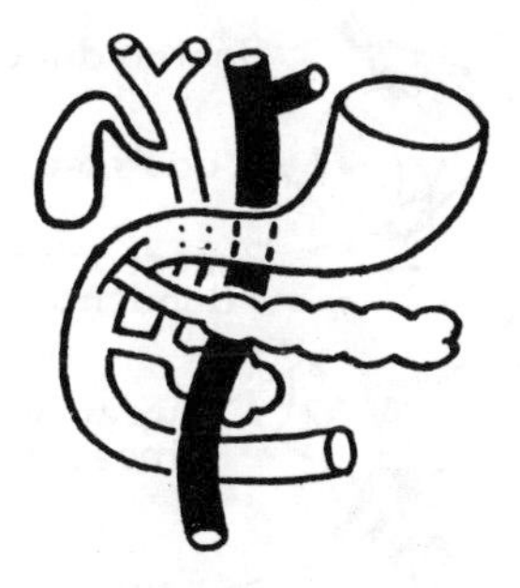

Figure 14.12. The 2 rudiments of the pancreas close on the portal vein (or superior mesenteric vessels) like a book on a bookmark.

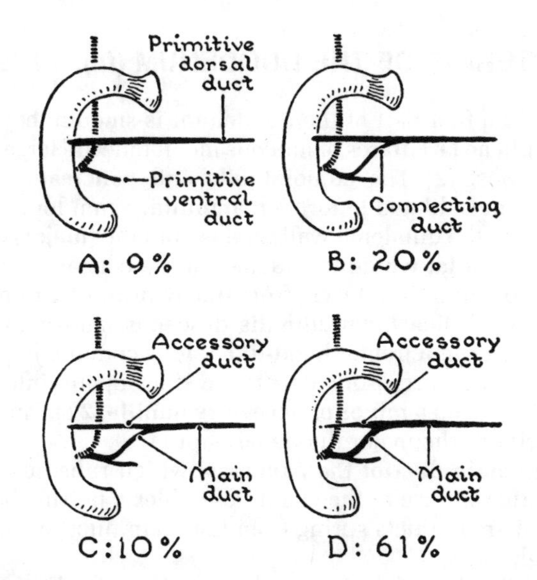

Figure 14.13. Varieties of pancreatic ducts in 200 specimens.

separately into the duodenum, one above the other in 5% of 200 cases (Millbourn).

Evidence of a 270° counterclockwise rotation of the embryonic gut are seen in the following adult relationships within the abdomen.

1. The jejunal and ileal branches of the superior mesenteric artery arise from its left side, and the large bowel branches arise from the right side of the superior mesenteric artery. In the embryo, the small intestine branches would have been anterior, and the large bowel vessels would have been posterior (*fig. 12.40*). This indicates that a 270° rotation has taken place around the superior mesenteric artery.
2. The common bile duct and the duct of the ventral pancreas were on the anterior border of the embryonic gut. The rotation of the gut causes a slightly greater than 180° rotation in the second part of the duodenum. The position of the duodenal papilla (opening of the bile and pancreatic ducts) is on the posteromedial aspect of the duodenum.
3. The vagi, which are on the right and left side of the esophagus in the superior aspect of the thorax, are anterior and posterior when they pierce the diaphragm. This reflects a 90° rotation of the esophagus and stomach.

ANTERIOR RELATIONS OF DUODENUM AND PANCREAS

The omental bursa lies anterior to the neck, body, and tail of the pancreas (*fig. 14.14*). The structures anterior to the C-shaped duodenum are: (a) gallbladder and quadrate and right lobes of the liver on each side of gallbladder; (b) transverse colon (direct anterior relation of the 2nd part of the duodenum) (*fig. 14.15*); (c) superior mesenteric vein and artery cross anterior to the 3rd part of the duodenum (*see fig. 14.18*); (d) coils of jejunum.

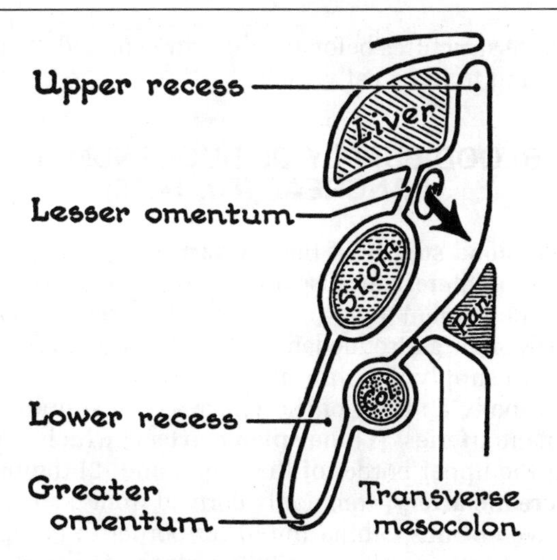

Figure 14.14. Diagram of the stomach, omental bursa, and stomach bed, on sagittal section.

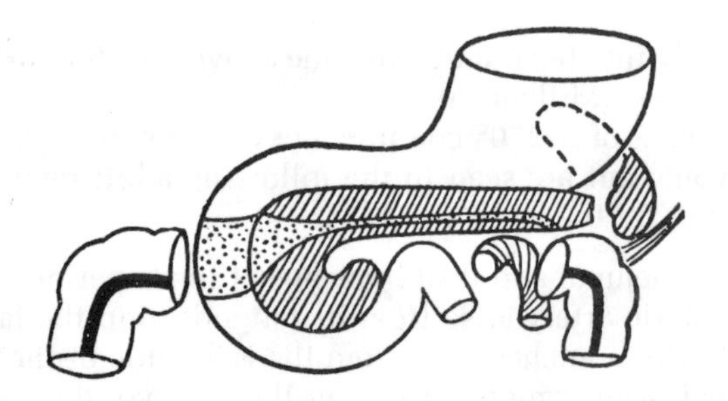

Figure 14.15. Relations of the gastrointestinal apparatus to the pancreas.

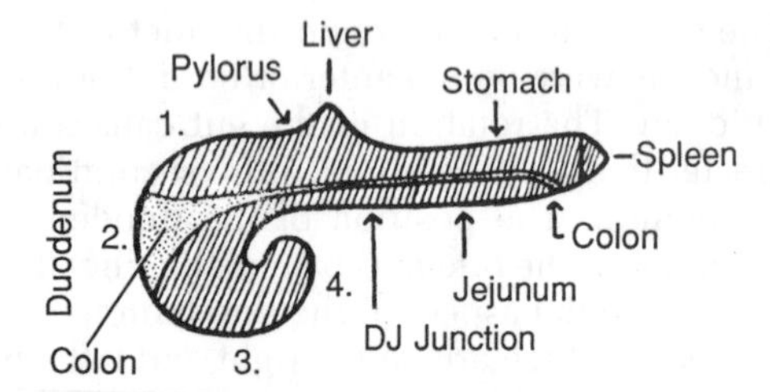

Figure 14.16. Key to Figure 14.15. *DJ* = duodenojejunal. Structures relating to the anterior surface of the duodenum.

The **pancreas** has 2 aspects, an **anterior** and a **posterior**. The transverse colon is attached to the anterior aspect of the body and tail (*fig. 14.14*) by the transverse mesocolon. Above and below the attachment, the pancreas is obviously covered with peritoneum. The tip of the tail extends into the lienorenal ligament and abuts against the spleen.

The pancreas is surrounded by various portions of the gastrointestinal tract (*figs. 14.15, 14.16*).

POSTERIOR RELATIONS OF DUODENUM AND PANCREAS (*fig. 14.17*)

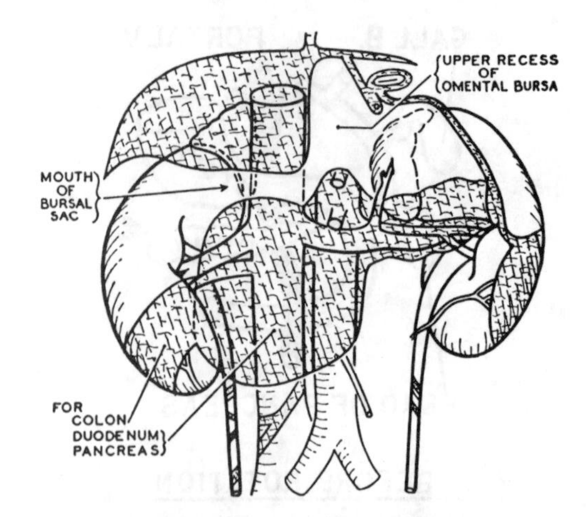

Figure 14.17. Especially to show the extent of the duodenum and pancreas.

These structures belong to the "three paired gland system" and to the great vessels.

BLOOD SUPPLY OF DUODENUM AND PANCREAS (*fig. 14.18*)

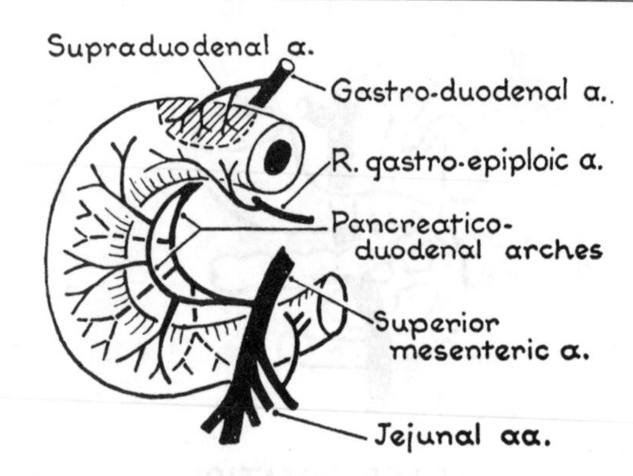

Figure 14.18. The blood supply of the duodenum.

The **blood supply of the 1st part** of the duodenum is of special interest because this is a common site of ulcers. An independent twig (or twigs), the **"retroduodenal" branch** of the gastroduodenal artery (not seen in *fig. 14.18*) helps to supply the posterior wall.

The body and tail of the pancreas are supplied by 2 constant arteries: (1) the **splenic artery**, which runs behind the upper border of the gland, and (2) the **inferior pancreatic artery**, commonly derived from the celiac artery which runs behind the lower border to the tail. Inconstant arteries also supply the gland. The various arteries anastomose freely to form a network around the lobules of the gland (Pierson; Wharton; Woodburne and Olsen; and Michels).

INTERIOR OF THE DUODENUM (*fig. 14.19*)

(1) The first part of the duodenum is smooth; beyond this, **plicae circulares** of mucous membrane are large and numerous. (2) The conjoint bile and pancreatic duct (sometimes dilated to form an **ampulla**, 5 mm long, as it traverses the duodenal wall) opens onto the **(major) duodenal papilla**, which is situated on the concave side of the duodenum about 8 cm from the pylorus. (3) From the papilla, a **plica longitudinalis** descends, and over the papilla, a semicircular **hood-like fold** is commonly present. (4) The accessory pancreatic duct opens into the duodenum on a **minor or accessory papilla**, 2 cm anterosuperior to the major duodenal papilla.

The **main duct of the pancreas**, which runs near the posterior surface of the gland, resembles a herring bone, in that small ducts spring from the main duct, which is straight.

Application of the embryology permits the **display of the (common) bile duct** (*fig. 14.20*). One has simply ro-

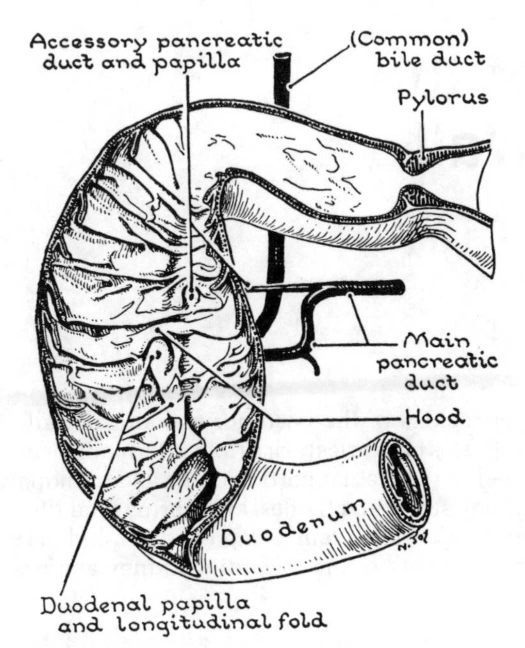

Figure 14.19. Interior of duodenum. Pancreatic ducts.

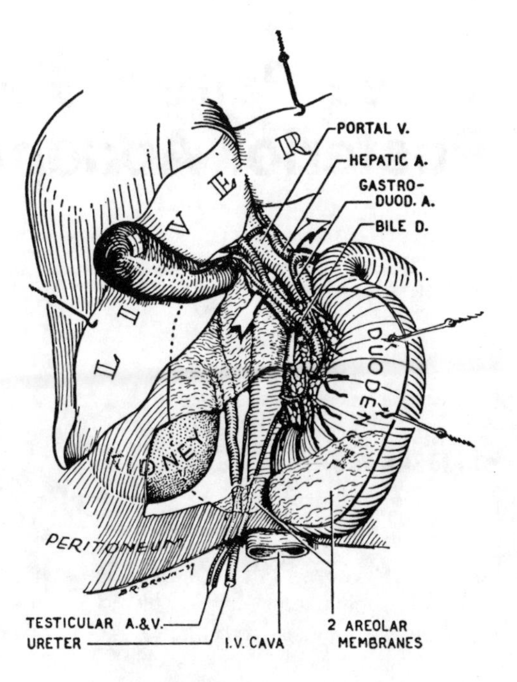

Figure 14.20. Display of bile duct by embryological approach. (*Arrow* passes through epiploic foramen.)

tated the duodenum back to a midline position that would be intraperitoneal.

Clinical Mini-Problems

1. **Since the parasympathetic fibers within the vagus follow the course of the arteries branching from the celiac trunk and superior mesenteric artery in the abdomen, where would the vagus nerve cease to influence peristalsis in the distal part of the gastrointestinal tract?**
2. **Which blood vessels would permit an anastomosis between the celiac trunk and the superior mesenteric artery?**
3. **In a patient with a transected spinal cord at the T12 level, how much of the large bowel loses its normal peristaltic movement?**
4. **How could a gallstone lodged in the bile duct orifice at the greater duodenal papilla result in damage to the pancreas?**

(Answers to these questions can be found on p. 585.)

15

Three Paired Glands of the Posterior Abdominal Wall

Clinical Case 15.1

Patient Raymond V. This 63-year-old painter developed severe pains in the area of his right iliac fossa. After your examination, which did not provide an obvious diagnosis, your tutor ordered an intravenous pyelogram (IVP) to examine the urogenital system. A sequence of radiographs showed a kidney stone "lodged" in the right ureter where it crosses the pelvic brim. The patient goes to surgery for removal of the stone and you assist. The surgeon opens the anterior abdominal wall and secures the ureter between the testicular vessels and the psoas muscle anterior to the right external iliac artery. She permits you to keep it lightly retracted away from the vessels. The stone is removed by creating a longitudinal incision in the ureter, which is then repaired, and the abdominal wall is closed. You are pleased that the operation was performed without entering the peritoneal cavity. (The patient convalesces for 4 weeks and returns to his work without subsequent episodes of "kidney stone attack."

Suprarenals, Kidneys and Gonads

MIGRATIONS (*fig. 15.1*)

All three glands develop from mesodermal tissues in the retroperitoneal tissues. The kidneys develop from more caudally placed intermediate mesoderm in the embryonic pelvis. The kidney then ascends into the abdominal cavity with its duct, the ureter, connected to the urinary bladder in the pelvis and the hilus of the kidney in the upper regions of the posterior abdominal wall. The gonads (testes or ovaries) arise from intermediate mesoderm located in the cranial portion of the abdominopelvic cavity. They subsequently descend during fetal life and drag their ducts, artery, vein, lymph vessels, and nerves anterior to the path of the ascending kidney and ureter.

TESTICULAR OR OVARIAN ARTERY

The gonadal arteries arise from the abdominal aorta just below the level of the renal arteries (L2 vertebra). The testicular artery descends in the retroperitoneal fascia covering the psoas muscle to enter the deep inguinal ring in the fascia transversalis where it joins the spermatic cord. The ovarian artery enters the pelvis by crossing the ureter at the pelvic brim. It reaches the ovary in the suspensory ligament that is attached to the superior pole of the ovary.

DUCTUS DEFERENS (VAS DEFERENS)

This duct has a short retroperitoneal course in the abdomen between the deep inguinal ring and brim of the pelvis as it passes into the pelvic fascia to join the seminal vesicles on the posterior aspect of the urinary bladder. The "vas" deferens crosses the external iliac vessels and the ureter as it courses over the iliac fossa and lateral wall of the pelvis.

SUPRARENAL GLANDS (ADRENAL GLANDS) (*fig. 15.2*)

These paired endocrine glands lie in contact with the superior pole of each kidney at the level of the 12th thoracic vertebra (*fig. 15.2*). The right suprarenal is triangular and is "wedged" between the superior pole of the right kidney and the IVC on the bare area of the liver (*fig. 15.3*). The left suprarenal gland is crescentic (half-moon-shaped) over the superomedial aspect of the left kidney (*fig. 15.3*).

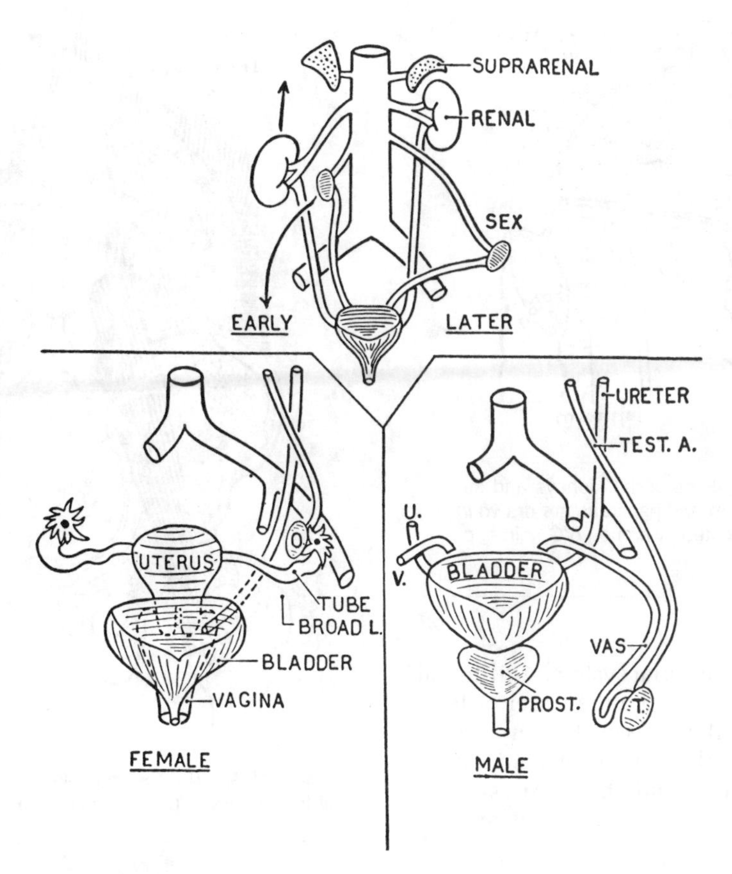

Figure 15.1. The migrations of the kidneys and sex glands. Definitively, the ovarian vessels and the uterine tubes in the female bear the same relationships to the ureter as do the testicular vessels and the ductus (vas) deferens in the male. *T* = testicle; *U* = ureter; *V* = vas.

Structure and Development

Each suprarenal gland is composed of 2 parts: an outer mesodermally derived cortex that is responsible for steroid production and an inner vascular medulla that is derived from neural crest ectoderm. The medulla produces adrenaline (epinephrine) and noradrenaline (norepinephrine) for sympathetic responses.

Vessels

The suprarenal glands are endocrine glands and therefore require a well-developed vascular supply. Numerous branches from 3 major arteries supply each suprarenal gland: (1) branches from the suprarenal artery that arises from the aorta; (2) branches from the inferior phrenic artery that branches from the aorta; and (3) branches from the renal arteries below each gland. A single large suprarenal vein leaves the anterior surface of each gland: the right suprarenal vein drains into the inferior vena cava and the left suprarenal vein drains inferiorly into the left renal vein.

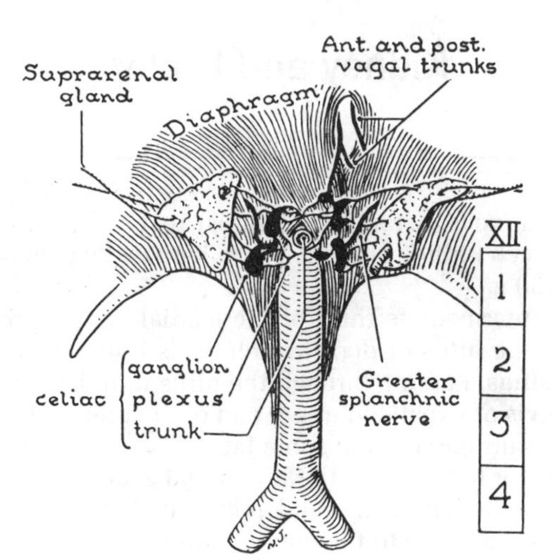

Figure 15.2. The celiac plexus and the suprarenal glands (retracted). Note vertebral levels.

Nerves

The greater and lesser splanchnic nerves pierce each crus of the diaphragm and synapse in the celiac and adjacent sympathetic ganglia. Postganglionic vasomotor sympathetic fibers enter the suprarenal cortex to control

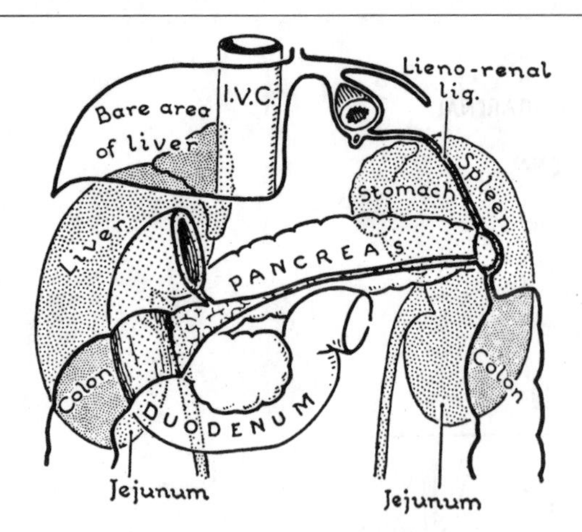

Figure 15.3. The anterior relations of the kidneys and suprarenal glands. (The duodenum and pancreas are drawn *in situ*; the other relations are indicated by name.) *IVC* = inferior vena cava.

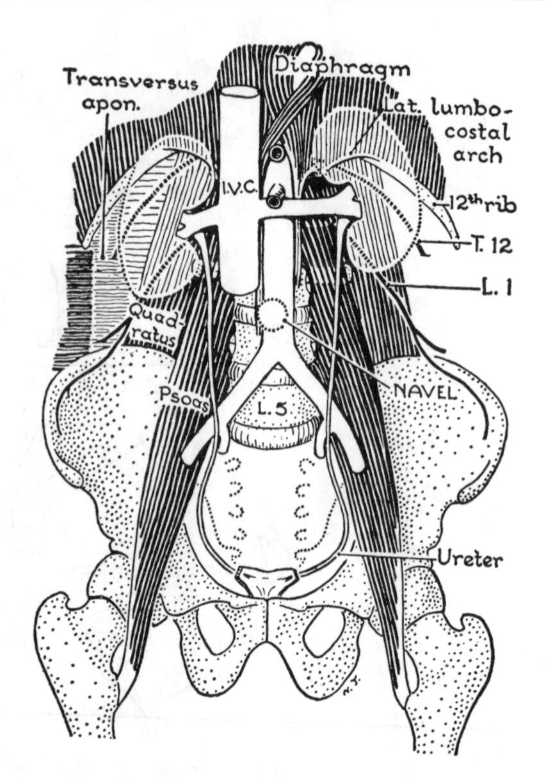

Figure 15.4. The posterior relations of the kidneys. The course of the ureters. *IVC* = inferior vena cava.

the blood flow throughout the arterial system. Additional preganglionic sympathetic fibers from the lesser or least splanchnic nerve innervate the cells of the suprarenal medulla. These medullary cells release catecholamines (adrenaline and noradrenaline) and therefore serve as "modified" postganglionic neuronal cells that are derived from neural crest cells.

Kidney and Ureter

The kidneys (L. *ren*, Gk. nephros) are paired (*figs. 15.3, 15.4*). Each kidney is about 11 cm long and weighs about 130–150 g.

The intermediate third of the medial border presents a cleft, the **hilus** or door, which leads into a cavity, the **renal sinus**. Passing through the hilus into the sinus are the **pelvis** or expanded upper end of the ureter, the renal vessels and nerves, and some fat.

Superior to the hilus, the suprarenal gland is in contact with the medial border and pole; inferior to the hilus, the ureter is close to the medial border.

Like other posterior abdominal organs, the kidney lies in the extraperitoneal fatty tissue. Within this fatty tissue, there is a tough areolar membrane, the **renal fascia**, which splits to enclose the kidney and a certain quantity of fat, the **perinephric fat or fatty capsule**.

The 2 layers of fascia do not blend below, so, the kidneys can move downward with the diaphragm during inspiration. Neither do they blend medially, but pass anteriorly to and posteriorly to the renal vessels, aorta, and IVC to unite indefinitely with the respective layers of the opposite side. Superiorly and laterally, the 2 layers of fascia blend.

SURFACE ANATOMY

The **inferior** poles of the kidneys lie about 2 cm above the transumbilical plane (level of 3–4 intervertebral disc); their **superior poles** reach the level of the 12th thoracic vertebra and, therefore, beyond the 12th rib (*fig. 15.4*).

Their **medial borders** are about 5 cm from the aorta. The hilus is at or near the transpyloric plane (*fig. 15.3*).

From behind, the kidney extends from a point one or two fingersbreadths superior to the highest part of the iliac crest (which is on a level with the spine of the 4th lumbar vertebra) (*fig. 15.5*), upward to or almost to the 11th rib (*Plate 15.1*).

ANTERIOR RELATIONS (*figs. 15.3, 15.5*)

The kidneys are in direct contact with some structures and indirectly in contact with others. These are given below for reference only.

The **direct contacts** are:

1. The 2nd part of the duodenum and the tail of the pancreas at the hili.
2. The right and left colic flexures.
3. Suprarenal glands.

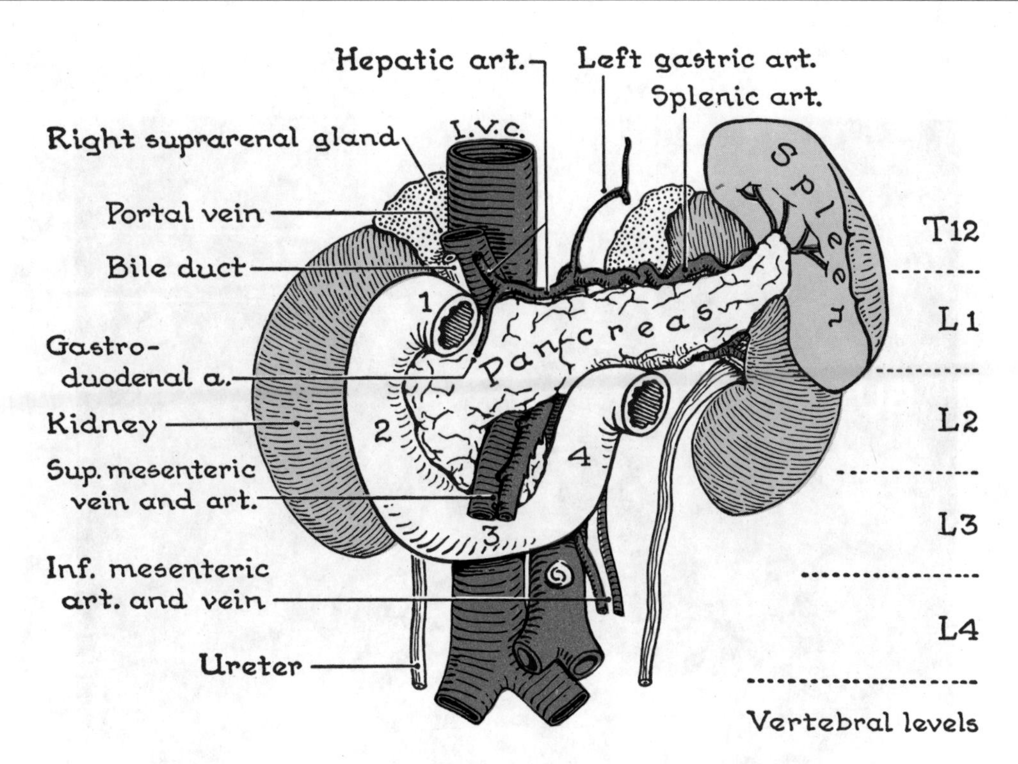

Figure 15.5. Abdominal viscera and vessels—the key picture. *IVC*, inferior vena cava.

4. Bare area of liver (on right side only).

The **indirect contacts** are:

1. Coils of jejunum.
2. Liver at the hepatorenal pouch on the right side.
3. Spleen on the left side.
4. Stomach on the left side.

POSTERIOR RELATIONS

Behind the kidney are parts of the roof and posterior wall of the abdomen (*fig. 15.4*). These relationships will be more evident after you have studied the structures of the posterior abdominal wall in Chapter 16. The following information is for reference and need not be memorized.

1. **Muscles and Bones**: The diaphragm, together with the medial and lateral arcuate ligaments (lumbocostal arches) from which it arises; the psoas and quadratus lumborum and, in the angle between them, the uncovered tips of the transverse processes of the 1st, 2nd, and (3rd) lumbar vertebrae; and the posterior aponeurosis of the transversus abdominis, which arises from these transverse processes (*fig. 16.16*).
2. **Nerves and Vessels**: T12 (subcostal nerve) and L1 (iliohypogastric nerves).

STRUCTURE OF THE KIDNEY

The kidney has a **fibrous capsule**, which is easily stripped off. When a kidney is divided with a sharp knife into anterior and posterior halves, its cut surface is seen to possess 6 or more smooth, darkish, longitudinally striated, triangular areas known as pyramids (*fig. 15.6*).

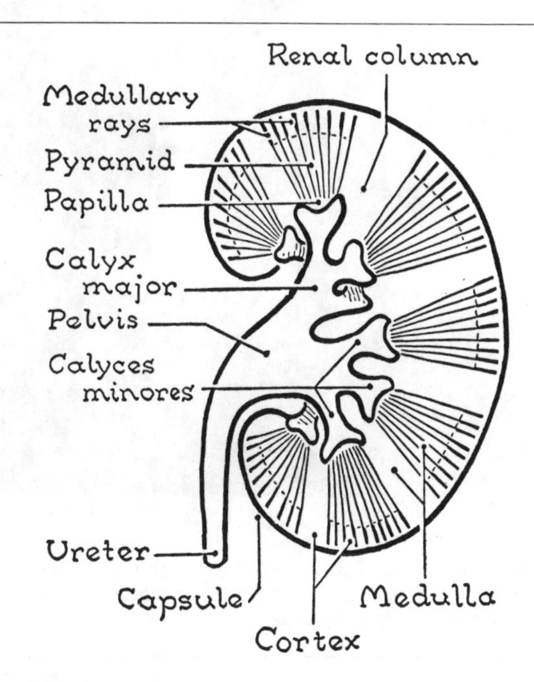

Figure 15.6. The macroscopic structure of the kidney (seen on longitudinal section).

Each **pyramid** is seen to have free, rounded apex of **papilla** (L. = nipple), which projects into the renal sinus and a **body** or main portion. From the **base** of the body,

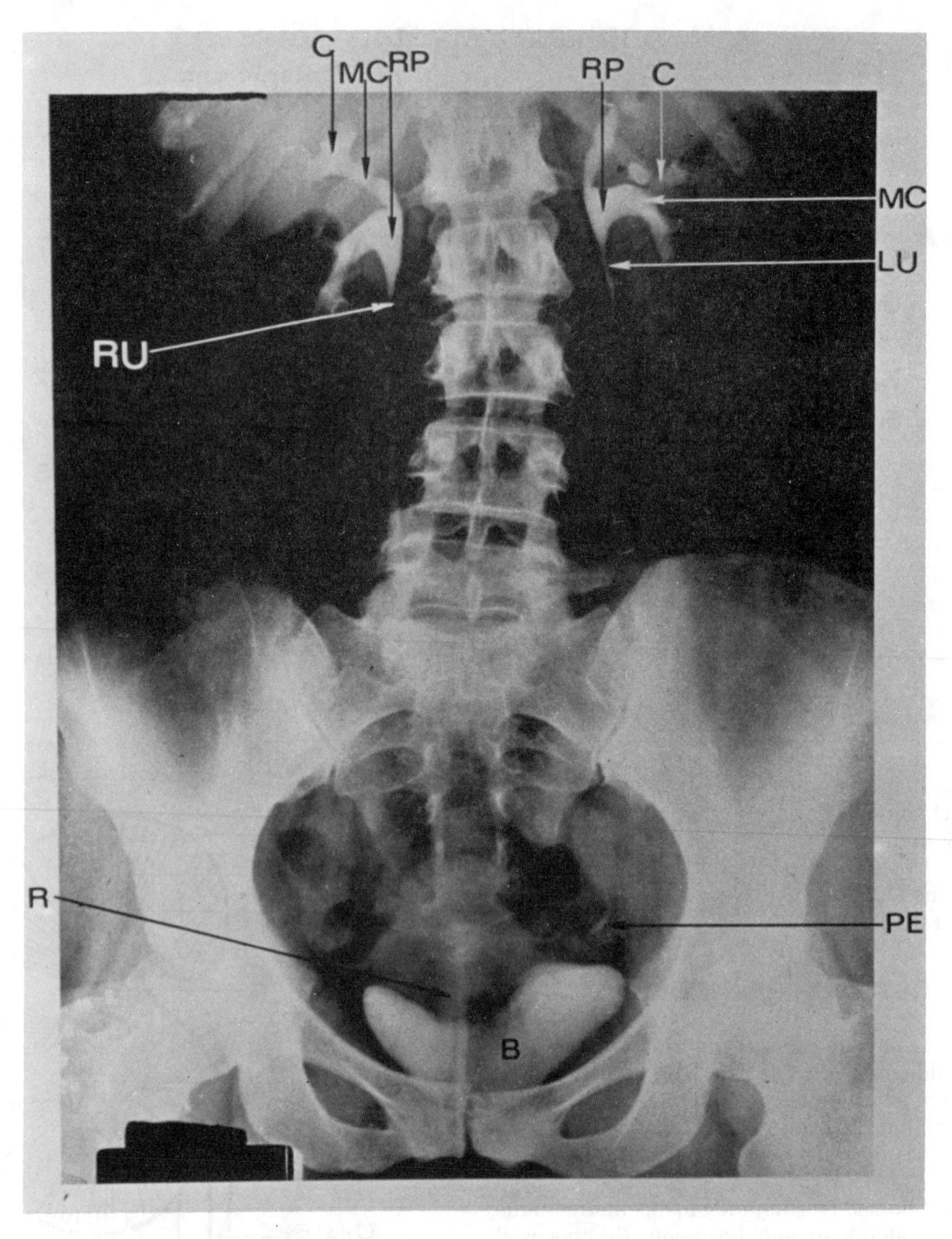

Plate 15.1. Intravenous urogram (A-P view). *B*, urinary bladder; *C*, minor calyx of kidney; *LU*, left ureter; *MC*, major calyx; *PE*, pelvic portion of left ureter; *R*, body of uterus; *RP*, renal pelvis; *RU*, right ureter (at junction with renal pelvis; remaining parts of the ureter are not visualized in this IVU).

radiations, the **medullary rays**, occupy the cortex of the kidney and extend to the surface. A papilla, a body, and a series of medullary rays constitute a complete pyramid.

The outer or surface layer of the kidney is the **cortex**. It is the part that lies superficial to the bases of the pyramids, and it comprises the entire outer one-third of the kidney substance. It looks granular. Cortical tissue also fills the areas between the pyramids, where it is known as **renal columns**. Details of fine structure are given for reference on p. 50.

RENAL VESSELS AND NERVES

The blood supply of the kidney is peculiar in that all, or almost all, of the blood passes through the glomerular

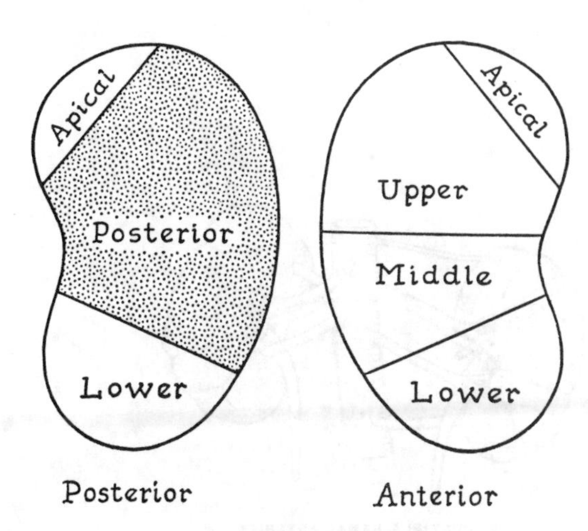

Figure 15.7. The 5 segments of the kidney, according to its arterial supply. (After Graves.)

capillaries. Here the blood is filtered and purified before it passes through a second set of capillaries from which it nourishes the kidney substance (see p. 51) and is further modified.

The **renal arteries**, one for each kidney, arise from the sides of the aorta 1 cm below the superior mesenteric artery (see also p. 51). Accessory renal arteries are commonly seen arising from the aorta or iliac arteries.

ARTERIAL SEGMENTS OF THE KIDNEY

According to the distribution of the branches of the renal artery, the kidney has 5 segments (*figs. 15.7* and *15.8*); these branches do not anastomose. The veins, however, anastomose freely (Graves).

The **renal nerves** include sympathetic vasomotor fibers and visceral afferent fibers derived from segments T12 and L1, 2 (for details see p. 52).

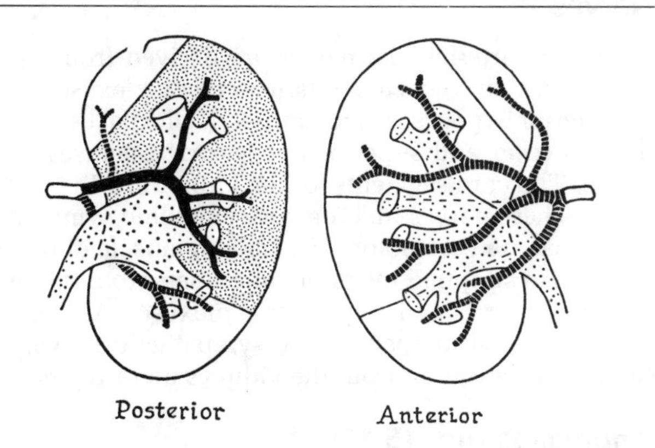

Figure 15.8. The 5 segmental branches of the renal artery. (After Graves.)

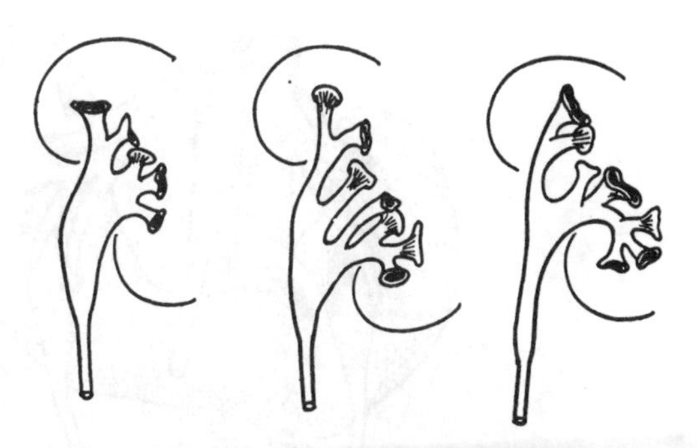

Figure 15.9. Sketches of casts of the renal pelvis.

PELVIS OF THE KIDNEY (PELVIS OF THE URETER) (*fig. 15.9* and *Plate 15.1*)

The pelvis is the expanded, funnel-shaped, upper end of the ureter. It lies partly within the renal sinus and partly outside it. Traced toward the kidney, it is seen to divide into 2 stalks, the **superior** and **inferior major calices**, with sometimes a third or middle calix. Each major calix divides into several goblet-shaped **minor calices**, into each of which one or more papillary ducts open onto each papilla.

To Explain Variations (Details)

The primitive ureter bifurcates to form a cranial and a caudal calix major. These continue to bud and divide progressively, the terminal branches being the collecting tubules in the pyramids of the renal medulla.

The **medulla** is composed of 7 pairs of pyramids (7 being ventral and 7 dorsal), of which 3 pairs are connected with the cranial calix and 4 with the caudal.

The **papillae** of the pyramids become crowded at the 2 poles of the kidney (especially the superior pole), with the result that groups of 2 or more papillae fuse to form compound papillae. Accordingly, the maximal number of papillae is 14 and the average is 9.

Commonly there is a third calix major, the **middle calix**, which receives the 4th and 5th pairs of papillae, leaving only the 6th and 7th for the caudal calix (Lofgren).

URETER (*fig. 15.4* and *Plate 15.1*)

The ureter or duct of the kidney is 25 cm long. Its upper half is in the abdomen; its lower half is in the pelvis; its terminal part pierces the posterolateral angle of the bladder.

Its abdominal part extends almost vertically from the lower part of the hilus of the kidney (less than 5 cm from the median plane) to the bifurcation of the common iliac artery (*fig. 15.4*).

The ureter lies in the retroperitoneal areolar tissue and adheres to the peritoneum. When the peritoneum is mo-

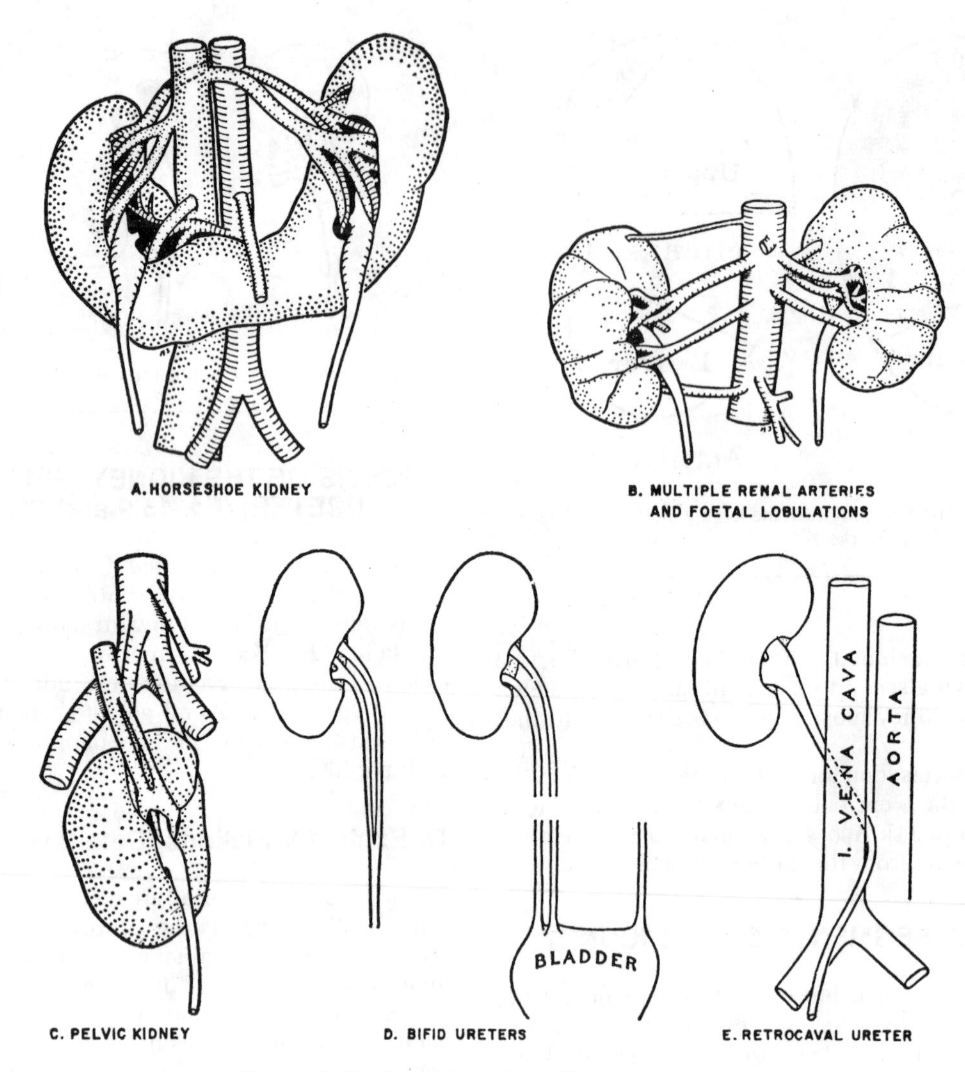

Figure 15.10. Anomalies of the kidneys and ureters.

bilized, the ureter is in danger of injury, for it moves with it.

The ureter descends on the psoas fascia and crosses the genitofemoral nerve. The IVC is close to the medial side of the right ureter; the inferior mesenteric vein is close to the medial side of the left ureter.

Otherwise, its **anterior relations** are the vessels of the testis or ovary, and parts of the gastrointestinal canal. (On the **right side**, they are the 2nd part of the duodenum, the right colic and ileocolic arteries, the root of the mesentery, and the end of the ileum. On the **left side**, they are the left colic and sigmoid arteries and the sigmoid mesocolon.)

Arteries

The pelvis and the ureter possess a longitudinal anastomosing network of arteries derived from the renal artery above and the vesical artery below. This is reinforced along its length by an aortic or testicular (ovarian) or common iliac branch.

Nerves

Like the arteries, the nerves are derived from nearby sources, i.e., the renal and intermesenteric plexuses above, the inferior hypogastric plexus (pelvic plexus) below, and the testicular and superior hypogastric plexuses in between. The cord segments for the kidney are T12 and L1 and 2. These nerves include postganglionic sympathetic fibers that are vasomotor. The smooth muscle of the ureters is also supplied by autonomic fibers from the renal, intermesenteric, and hypogastric plexuses. Visceral afferent fibers accompany these sympathetic nerves and carry pain sensation from the kidneys and ureters.

Anomalies (*fig. 15.10*)

The commonest gross anomaly of the urinary tract is a **bifid ureter** and pelvis, the result of premature division of the ureteric bud in the fetus. The condition is generally incomplete and unilateral, and the ureter is commonly constricted at the point of fission. When completely bifid, one of the ipsilateral ureters may open into other parts of the urogenital tract, e.g., the floor of the urethra, or roof of the vagina or seminal vesicle (rare). The **fetal lobulation** (*fig. 15.10B*) present at birth is commonly retained.

About 3% of kidneys have **2 renal arteries** arising from the aorta, of which one usually goes to the upper or lower pole. Fused kidneys: in 1 in 700 persons the right and left kidneys are fused, generally at their lower poles, forming a horseshoe kidney (*fig. 15.10A*). The ascent may be arrested by the inferior mesenteric artery crossing in front of the isthmus. This anomaly may present a problem at the time of delivery in a pregnant female.

Congenital absence of kidney and ureter or **rudimentary kidney with ureter** occur with the same frequency as horseshoe kidney.

Other Uncommon-to-Rare Anomalies. The kidney on one side may be **double**. One or the other kidney may migrate across the median plane. A **pelvic ectopic kidney** (*fig. 15.10C*) is one that, failing to ascend, remains lodged in the pelvis and is supplied by arteries of low origin. This may present a problem during childbirth and may be an indication for a Caesarean section. A **congenital cystic kidney** results when the secreting parts of a kidney and their ducts fail to unite.

A **retrocaval ureter** (Pick and Anson) is rare (*fig. 15.10E*).

Clinical Mini-Problems

1. **In surgical repair of an occluded right iliac artery, the surgeon must be careful not to damage the right ureter. Where would one expect to find the right ureter in relation to this vessel?**
2. **As the ureters approach the posterosuperior surface of the bladder, which major structure crosses the ureter superiorly in the (a) male? (b) female?**
3. **At which lumbar vertebral level would one expect to see the bifurcation of the (a) aorta? (b) the vena cava?**

(Answers to these questions can be found on p. 585.)

16

Posterior Abdominal Structures

Clinical Case 16.1

Patient Robert Q. This 29-year-old baggage-handler was injured by a terrorist bomb in the airport. He is rushed to the emergency department where you are a student assistant. He is suffering from acute shock and loss of blood and has a penetration wound in the right abdominal wall near the tip of the 11th rib. A piece of shrapnel had penetrated all eight layers of the anterior abdominal wall and entered the abdominal cavity. The ileum was penetrated in 3 separate regions, and the metal fragment had exited the abdomen by penetrating the parietal peritoneum of the posterior abdominal wall and severing the right colic artery. It also had damaged the right ureter, lacerated the lateral aspect of the IVC, and penetrated the psoas muscle before lodging into the right side of the body of the L2 vertebra. The right L2 spinal nerve is also severed. After briefly testing you about the anatomy, the emergency room surgeon repairs the IVC, right ureter, right colic artery, ileum, and peritoneum on the anterior and posterior abdominal walls, but does not attempt a nerve repair now. The patient is simultaneously treated for his shock and blood loss, and later placed on antibiotic therapy and moved to an intensive care unit. Five days later he is classified in fair condition and recovering well. Following his case, you find some weakness in his hip flexion and knee extension on the right side and "numbness" in the skin of his right thigh below his inguinal ligament. He continues to recover rapidly otherwise, except for the neurological symptoms. The question arises as to whether a nerve repair would be worthwhile.

Great Vessels of the Abdomen

The great vessels are the abdominal aorta, the inferior vena cava (IVC), and the common, external, and internal iliac arteries and veins.

ABDOMINAL AORTA AND ILIAC ARTERIES

The aorta enters the abdomen in the median plane at the disc between vertebrae T12 and L1. It ends in front of L4 by bifurcating into the right and left common iliac artery (*fig. 16.1*).

Each common iliac artery bifurcates into an internal iliac and an external iliac artery. The main trunk to the thigh curves and runs on the psoas to the midinguinal point where the name changes to femoral artery. Two branches arise from the external iliac artery just before it passes behind the inguinal ligament (*fig. 16.1*)—the *inferior epigastric* and *deep circumflex iliac arteries.*

The *veins* lie within the bifurcation of the arteries, as shown in *Figure 16.2.* The common iliac veins join to form the IVC behind the right common iliac artery, in front of vertebra L5.

Collateral Branches of the Abdominal Aorta

A. Celiac trunk, S. mesenteric a., I. mesenteric a. } To the GI tube and the 3 unpaired glands.

B. Suprarenal a., Renal a., Testicular a. } To the 3 paired glands.

C. (Inf.) Phrenic a., Lumbar arteries, Median sacral a. } To the roof and walls of the abdomen.

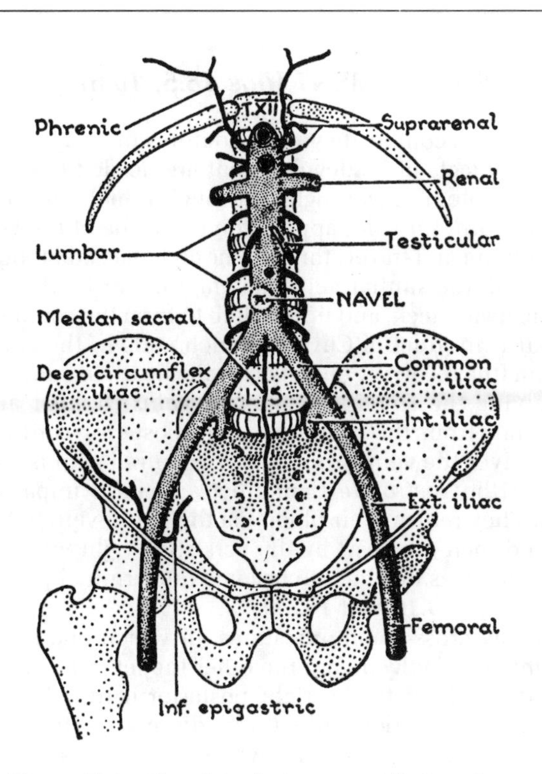

Figure 16.1. The abdominal aorta and its branches—collateral and terminal.

These arteries occupy 3 planes (*fig. 16.3*). The **3 unpaired arteries** to the gastrointestinal (GI) plane arise from the front of the aorta (*fig. 16.4, A*).

The Celiac Trunk or Artery

The median arcuate ligament which unites the 2 crura of the diaphragm, rests upon the celiac trunk (*fig. 16.4*). Its level is, therefore, the disc between the 12th thoracic and 1st lumbar vertebrae.

The **superior mesenteric artery** takes origin just below the celiac trunk, and therefore behind the neck of the pancreas, and behind the splenic vein, which embedded in the pancreas. It "clamps" and prevents ascent of the left renal vein, which crosses the aorta close below it *fig. 16.4, A*).

The **inferior mesenteric artery** takes origin 4 cm above the aortic bifurcation and 2 cm above the umbilicus, and therefore in front of the 3rd lumbar vertebra. It would arrest the ascent of a horseshoe kidney (*fig. 16.4, A*).

The **arteries to the 3 paired glands** arise close together (*fig. 16.4, B*).

From their inception the **phrenic, lumbar, and median sacral arteries** are part either of the roof or of the wall of the abdomen (*fig. 16.4, C*).

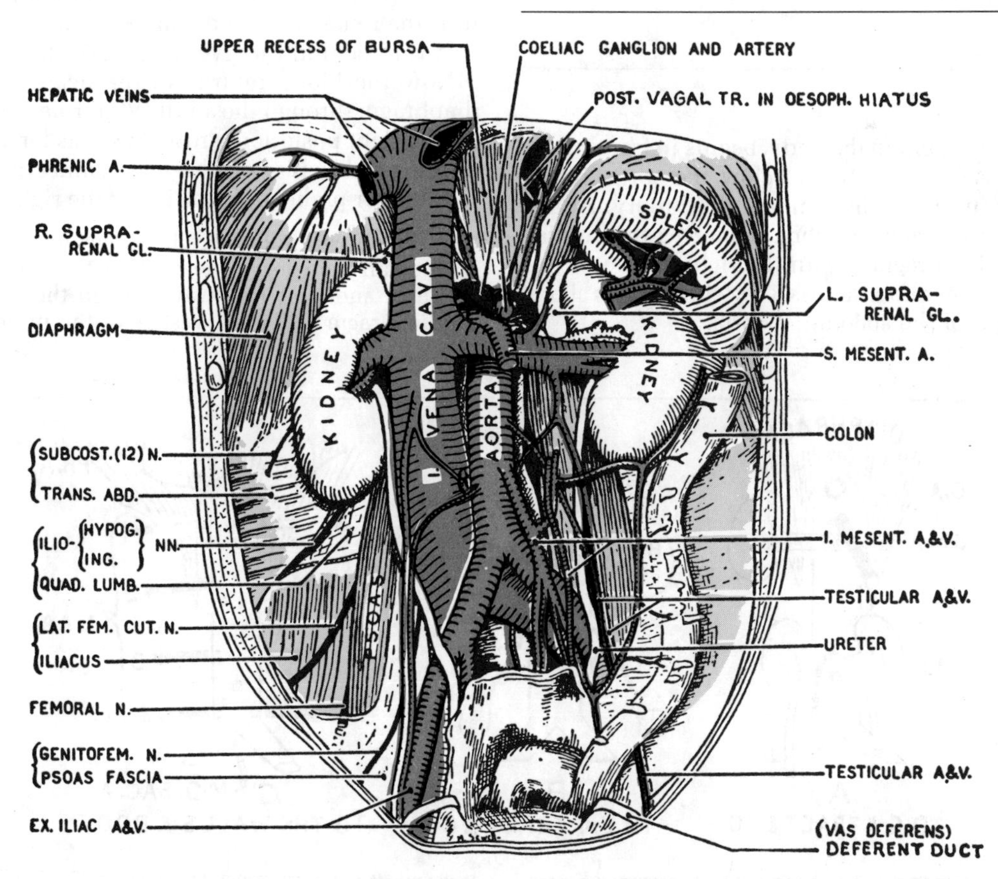

Figure 16.2. Primary retroperitoneal structures; also spleen and colon.

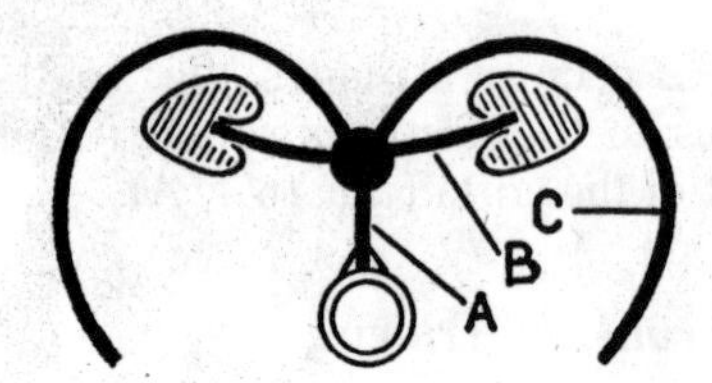

Figure 16.3. "The 3 vascular planes." (Cross-section, see *fig. 16.4*).

The **(inferior) phrenic artery** arises beside the celiac trunk and passes at once to the diaphragm.

Four pairs of **lumbar arteries** hug the bodies of the vertebrae 1–4 and pass dorsally medial to the sympathetic trunk and psoas. A 5th pair, the *iliolumbar arteries*, spring from internal iliac arteries.

The **median sacral artery** is the tiny "continuation" of the aorta (*fig. 16.1*).

Inferior Vena Cava

This, the largest vein in the body, begins in front of the 5th lumbar vertebra (*fig. 16.2*), below the aortic bifurcation and behind the right common iliac artery. At the level of the 8th thoracic vertebra, it pierces the central tendon of the diaphragm to join the right atrium of the heart. It extends, therefore, across 8 vertebrae and is about twice the length of the abdominal aorta.

DEVELOPMENT (*figs 16.5, 16.6*)

The IVC is a composite vein of complex origin. But the elementary facts of its development are needed to make its adult anatomy meaningful. A longitudinal vein, the *posterior cardinal vein*, appears on each side of the vertebral column and travels through the abdomen and thorax to join with the similar vein, the *anterior cardinal vein*, from the head, neck, and upper limb to form the *common cardinal vein* (duct of Cuvier), which ends in the sinus venarum (the receiving chamber) of the heart. In addition to receiving the *somatic segmental veins* (lumbar and intercostal) from the body wall, the posterior cardinal vein receives the *veins from the three paired glands*, but *not* from the *gastrointestinal tract* and its 3 unpaired glands. They return their blood via the portal vein to the liver, and then the heart by the persisting right vitelline vein, which was destined to become the terminal part of the IVC (*figs. 13.10 and 16.3*).

Two new cross-communications, of which one is the *left common iliac vein* and the other the *left renal vein*, divert their blood to the right posterior cardinal vein, while the left posterior cardinal vein disappears (*fig. 16.5, B*). (A left IVC is an anomaly which reminds us of this history.)

A new vessel sprouts from the posterior cardinal vein near the renal veins and connects it with the right vitelline vein behind the liver (*fig. 16.5, B, broken lines*).

Now the blood returns to the heart from below the diaphragm through the adult IVC formed (*fig. 16.6*) from: (1) the right posterior cardinal vein as far as the entrance of the renal veins, (2) the new connection, and (3) the terminal or prehepatic portion of the right vitelline vein.

> If you will pull the right kidney forward, you will see a small vein leave the back of the IVC near the renal veins and pass headway through the right crus of the diaphragm. This is a vestige of the prerenal portion of

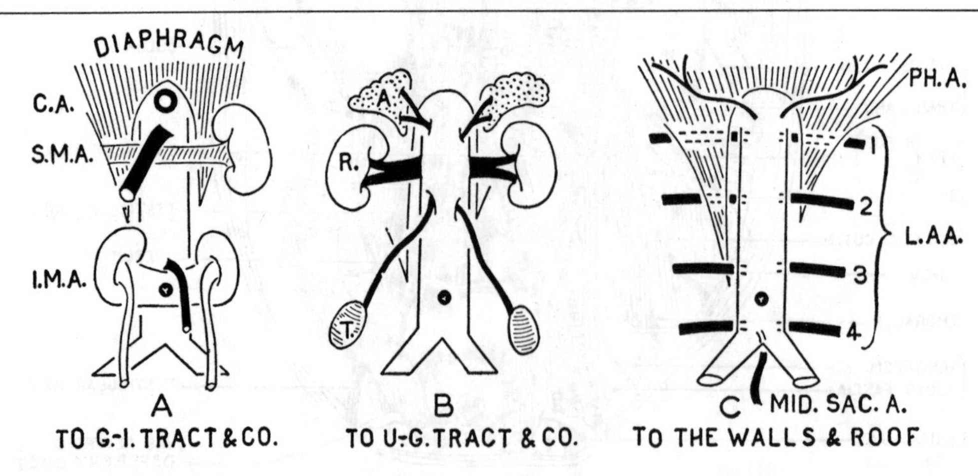

Figure 16.4. The branches of the abdominal aorta arranged according to the planes they occupy. *CA* = celiac artery; *SMA* = superior mesenteric artery; *IMA* = inferior mesenteric artery; *GI* = gastrointestinal; *UG* = urogenital; Mid SAC A = median sacral artery; *LAA* = lumbar arteries; *PHA* = phrenic artery; *R*; = renal artery; *T* = testicular artery.

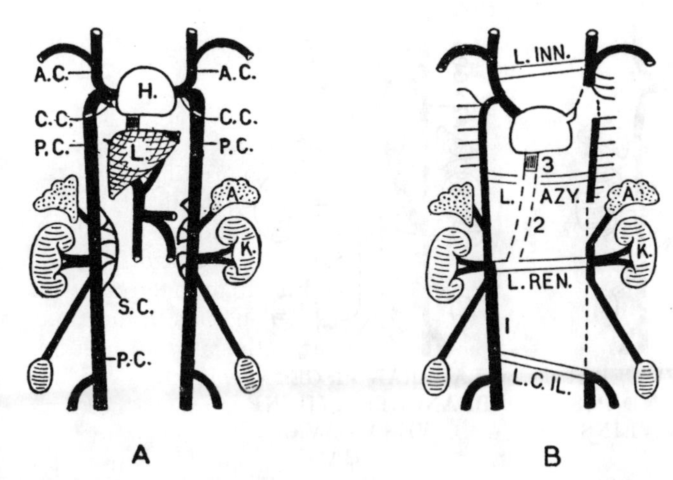

Figure 16.5. Development of the inferior vena cava. (*A*): *AC, PC, SC, CC* = anterior, posterior, sub-, and common cardinal veins; (*B*): 1 = postrenal segment of inferior vena cava; 2 = new connection; 3 = prehepatic segment. *A* = adrenal; *LAZY* = left azygous vein; *LCIL* = left common iliac; *LREN* = left renal vein; *K* = kidney; *LINN* = left innominate (brachiocephalic) vein.

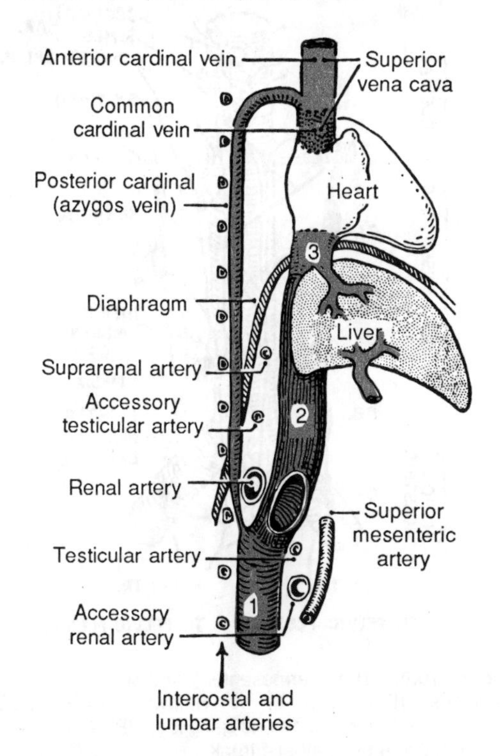

Figure 16.6. Scheme of the development of the inferior vena cava, explaining its relationship to the arteries that cross it (side view).

the right posterior cardinal vein. In the thorax, it joins intercostal veins which form the azygos vein.

Tributaries of the IVC

These fall into the same 3 groups as the branches of the abdominal aorta.

A. *The Blood from the Gastrointestinal Canal*, after circulating in the liver, leaves it via 3 large hepatic veins and 6 or more small hepatic veins to enter the last part of the IVC.
B. *Veins of the Three Paired Glands*. The right suprarenal, renal, and testicular (or ovarian) veins enter the IVC; the left renal vein drains all 3 left organs (*fig. 16.5, B*).
C. *Veins of the Body Wall:* (1) right and left (inferior) phrenic veins and (2) the lumbar veins.

The lumbar veins, on each side, are irregularly linked together by an **ascending lumbar vein**, which lies in front of the lumbar transverse processes (*fig. 26.14*). This vein begins caudally at the common iliac vein. The right and left ascending lumbar veins are (1) accessory to the IVC, and (2) they are to be classified with the vertebral venous system.

The right and left iliolumbar veins, which are the veins of the 5th lumbar segment, end in the right and left common iliac veins.

The median sacral vein ends in the left common iliac vein.

Relationships

Relationships of the branches of the aorta to the IVC are explained on a developmental basis in *Figures 16.6 and 16.7*.

1. The shifts from a posterior relationship to the aorta inferiorly to an anterior relationship superiorly. This is due in part to the aorta passing through the posterior aspect of the diaphragm at the level of the T12 vertebra and the IVC tranversing the central tendon of the right dome of the diaphragm inferior to the heart at the level of the T8 vertebra (see *fig. 16.18*).
2. The IVC is formed posterior to the right common iliac artery at the L5 vertebral level.
3. The IVC ascends the right abdominal wall on the right side of the midline aorta. The IVC is therefore on the lateral side of the lumbar vertebrae anterior to the attachment of the right psoas muscle. The midline aorta is anteriorly positioned due to the secondary (anterior) curvature of the lumbar spine, which is maximal at the L4 vertebra. Here the abdominal aorta attains its closest relationship to the anterior abdominal wall and may be compressed against the lumbar vertebra to reduce blood flow to the lower limbs.
4. At the level of the L4 vertebra, the IVC is crossed anteriorly by the right gonadal artery as it joins its accompanying vein (which drains into the IVC). The two right gonadal vessels also cross the right ureter at the lateral margin of the IVC (*fig. 16.2*).
5. The renal veins at the level of the L2 vertebra are both anterior to the renal arteries. This is identical to their relationship in the hilus of the kidneys. The IVC is also anterior to the proximal portion of the right renal artery.

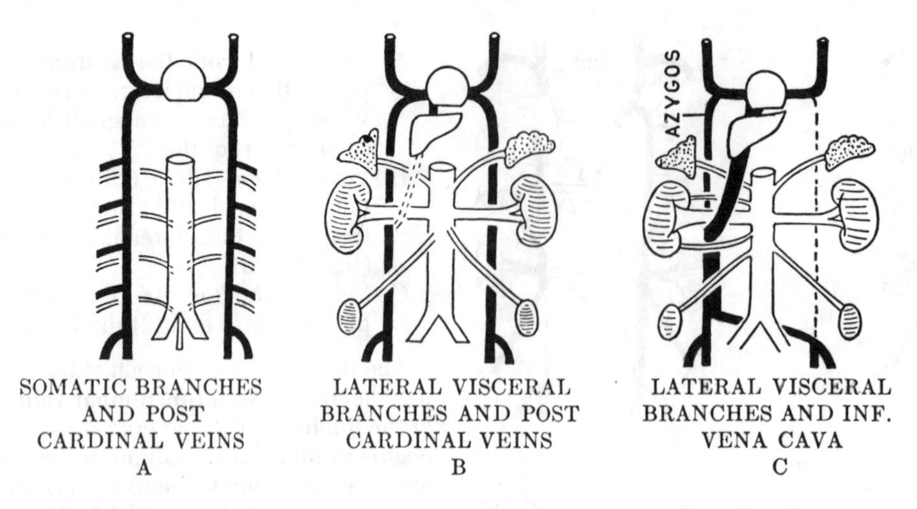

Figure 16.7. Explaining the relationships of the branches of the aorta to the cardinal veins and subsequently to the IVC. In **C** note the relations of the accessory renal arteries to the IVC.

6. The renal vein crosses the midline and the abdominal aorta to reach the IVC. Its major tributaries and relationships are important.
   a. It receives the left gonadal vein on its inferior surface medial to the left kidney.
   b. It receives the left suprarenal vein on its superior surface medial to the left kidney.
   c. It is anterior to the left renal artery and abdominal aorta.
   d. It is crossed in the midline by the superior mesenteric artery as it descends over the superior and anterior surfaces of the retroperitoneal left renal vein to enter the root of the mesentery.

Abdominal Autonomic Nerves

The autonomic nervous system within the abdomen is represented by:

1. The sympathetic trunks, white and gray rami communicantes, lumbar splanchnic nerves, and thoracic splanchnic nerves.
2. The prevertebral plexuses—celiac, intermesenteric, and superior hypogastric.
3. Parasympathetic nerves—vagus and pelvic splanchnic nerves.

SYMPATHETIC TRUNK (*fig. 16.8*)

In the abdomen, the sympathetic trunk follows faithfully the anterior border of the psoas muscle (*fig. 16.9*). It descends, therefore, on the bodies of the vertebrae and the intervertebral discs, the transversely running lumbar vessels alone intervening. It enters from the thorax with the psoas behind the medial arcuate ligament (lumbocostal arch), and it passes into the pelvis behind the common iliac vessels (*fig. 16.8*).

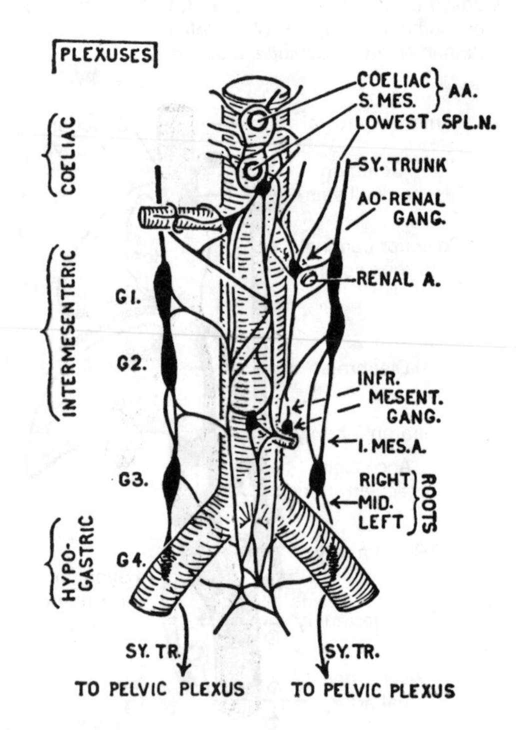

Figure 16.8. The intermesenteric and superior hypogastric plexuses. (Dissection by K. Baldwin.) INFR MESENT GANG = inferior mesenteric ganglion; SPLN = splanchnic nerve; SY TR, sympathetic trunk.

The **right trunk** is concealed by the IVC, and crosse[d] by the right renal artery. The **left trunk** is crossed by th[e] left renal vessels, the left testicular artery, and the inferi[or] mesenteric artery.

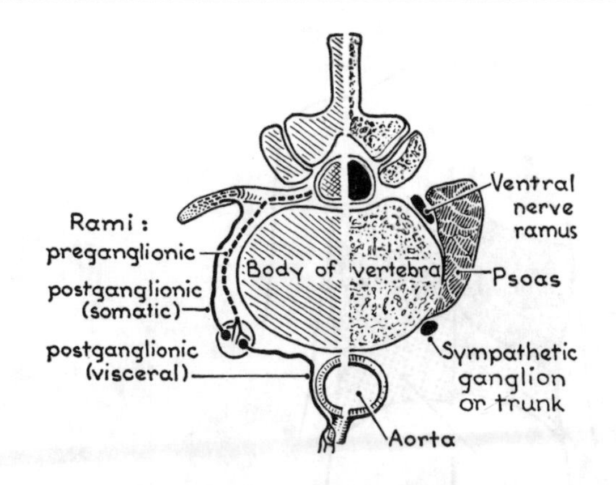

Figure 16.9. Pre- and postganglionic fibers of a sympathetic ganglion.

As a rule each trunk has 4 ganglia—not 5—with 2 probably having fused.

CONNECTIONS

Each trunk receives a **white ramus** from each of the upper 2 (or 3) lumbar nerves (L1, 2, [3]) and sends one or more **gray rami** to each of the 5 lumbar nerves to be distributed with nerves to somatic structures. These white and gray rami curve backward and laterally on the sides of the vertebrae either with the lumbar vessels or independently (*fig. 16.9*). Four postganglionic **lumbar splanchnic nerves** (visceral branches) run medially to the intermesenteric and superior hypogastric plexuses to be distributed largely with blood vessels to viscera.

CELIAC PLEXUS

The **celiac ganglia** (*fig. 16.10*) are tough nodular masses connected to each other by fibers that encircle the celiac trunk, the whole being known as the **celiac plexus**, or as the **solar plexus**, because its branches radiate like the rays of the sun.

Each ganglion lies behind the peritoneum, between the celiac trunk and the suprarenal gland and, therefore, on the crus of the diaphragm. The IVC largely conceals the right celiac ganglion, while the pancreas and splenic artery largely conceal the left one.

The celiac plexus extends down the front of the aorta and is reinforced by the lumbar splanchnic nerves to form the **intermesenteric and superior hypogastric plexuses**. The intermesenteric plexus lies in front of the aorta; the hypogastric plexus lies within the bifurcation of the aorta and therefore in front of the left common iliac vein and the 5th lumbar vertebra and disc (*fig. 16.8*).

From the celiac plexus, fine branches stream into the suprarenal gland (*fig. 16.10*).

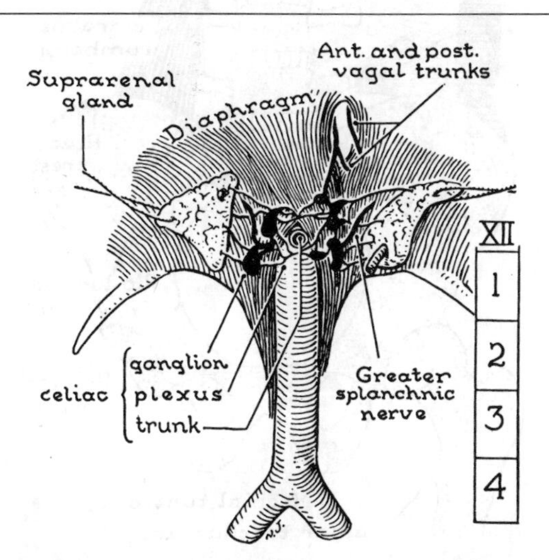

Figure 16.10. The greater splanchnic nerves, celiac ganglia, and plexus; also suprarenal glands and anterior and posterior vagal trunks.

SYMPATHETIC CONNECTION

The greater splanchnic nerve pierces the crus along the side of the celiac trunk and at once joins the celiac ganglion; the lesser splanchnic nerve pierces the crus just below and laterally, and also joins the ganglion; the lowest splanchnic nerve also pierces the crus and joins an offshoot of the celiac plexus, called the **renal plexus**. Some preganglionic sympathetic fibers enter the adrenal medulla to synapse on the medullary cells. All other preganglionic sympathetic neurons in the splanchnic nerves from the thorax must synapse in one of the abdominal prevertebral sympathetic ganglia. The postganglionic sympathetic fibers from these abdominal ganglia follow the arterial branches of aorta and the sympathetic plexuses to the appropriate smooth muscles and glands within the abdominopelvic cavity. Lumbar splanchnic postganglionic nerves from the sympathetic chain also join these preaortic plexuses of autonomic nerves.

PARASYMPATHETIC CONNECTIONS

Fibers of both **vagus nerves**, via a large branch of the posterior vagal trunk, pass through the celiac plexus (*fig. 16.10*) to be distributed with arteries (celiac, superior mesenteric, and renal) to the abdominal viscera.

Fibers of **pelvic splanchnic nerves** (S2–4) ascend from the pelvis in the hypogastric plexuses to be distributed with the inferior mesenteric artery to the gut.

The vagi control the gut as far as the left colic flexure, and the pelvic splanchnics innervate the descending colon, sigmoid colon, rectum, and internal anal sphincter.

SENSORY NERVES

Accompanying fibers of the autonomic nervous system are sensory (afferent) fibers, which ultimately enter the

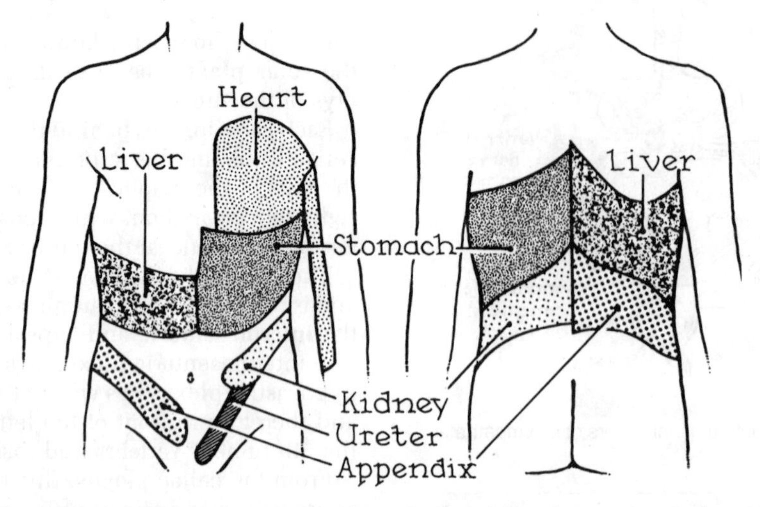

Figure 16.11. Cutaneous areas to which pains from internal organs are "referred."

spinal cord or brainstem along sensory roots of spinal or cranial nerves. Sensation from internal organs is diffuse and often "referred" at the level of consciousness to cutaneous areas (*fig. 16.11*) that are supplied by the same spinal cord level.

The key to understanding the areas of skin to which the abdominopelvic organs refer pain is to know the source of the sympathetic fibers to the given organ. These sympathetic fibers are associated with afferent (sensory) pathways that are principally sensitive to chemical stimuli. The lack of an adequate blood supply can cause ischemia to an organ, and the accumulation of waste substances, such as lactic acid, can activate the chemoreceptors. Since the greater splanchnic nerve contains T5–9 preganglionic sympathetic fibers that go to branches of the celiac trunk, pain from the spleen, stomach, liver, gallbladder and first half of duodenum is referred to T5–9 dermatomes. The least splanchnic (T12) goes to the renal plexus and therefore renal pain would be associated with the T12 dermatome. The gonads receive their arterial supply from their site of origin in the upper abdomen. These vessels are associated with sympathetic fibers derived from the T9–10 levels of the spinal cord. Therefore gonadal pain is felt in the T9 and T10 dermatome over the "pit of the stomach."

Posterior Wall of Abdomen Proper

BONY PARTS

In the median plane are the bodies, intervertebral discs, and transverse processes of the 5 lumbar vertebrae; laterally, the wall extends from the 12th rib above to the pelvic brim below, and it is divided into upper and lower parts by the iliac crest. The *bodies* of the lumbar vertebrae increase in height and width from 1st to 5th; so do the discs (*fig. 16.12*).

The *transverse processes* project laterally at the levels of the upper halves of the bodies. The 3rd projects farthest while those above and below it project progressively less. The 5th is stout and conical, and it projects upward,

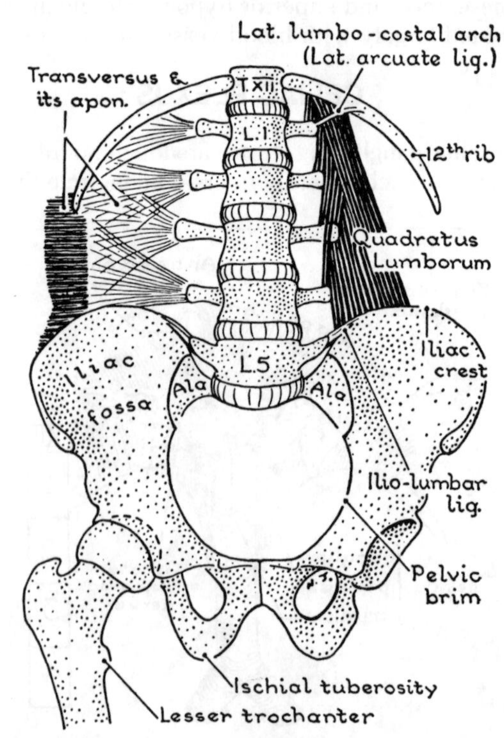

Figure 16.12. The skeleton of the posterior abdominal wall. Note that the discs bulge. The transversus abdominis and its aponeurosis.

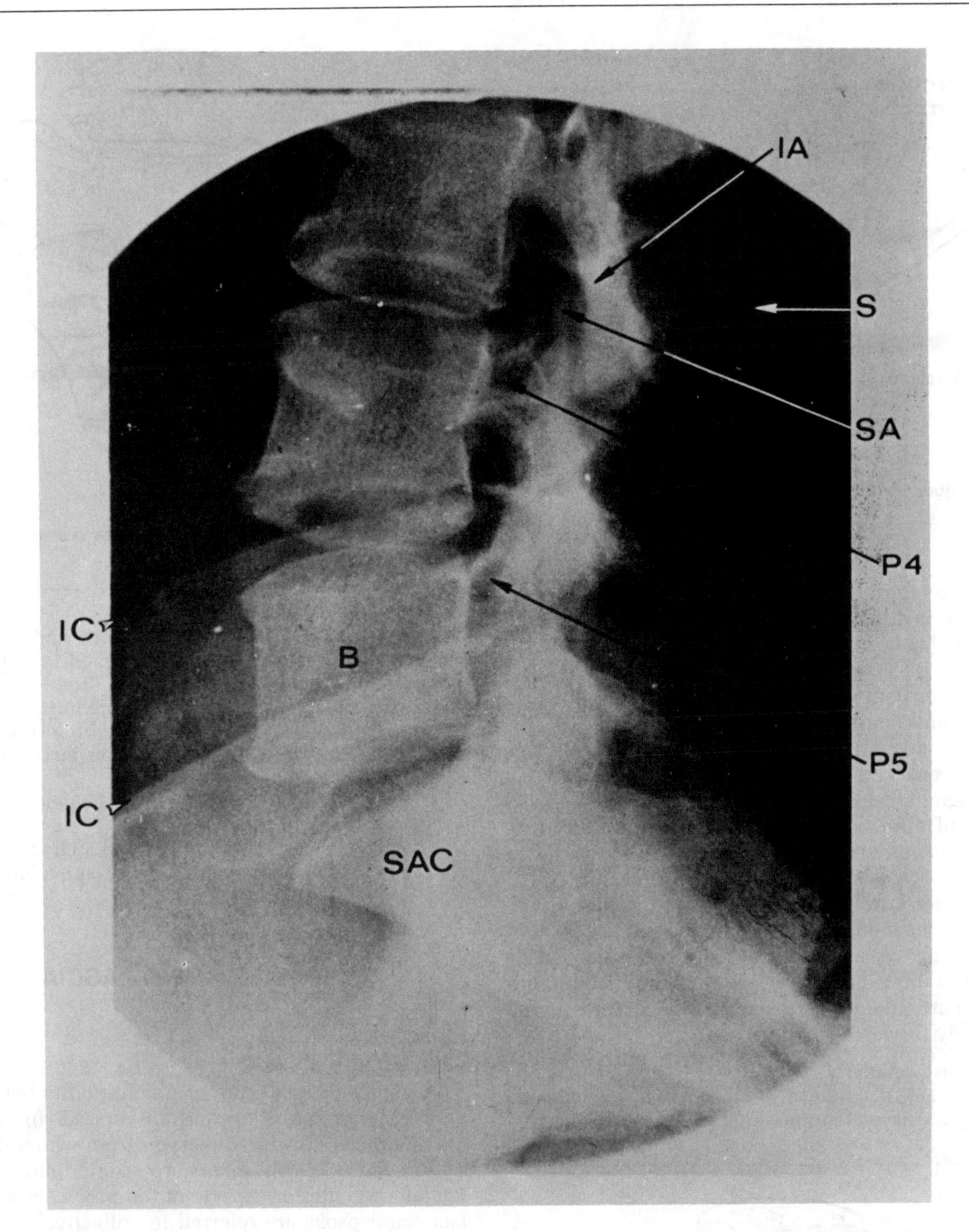

Plate 16.1. Lumbosacral region (lateral view). *B*, body of 5th lumbar vertebra; *IA*, inferior articular facet of L3; *IC*, iliac crest; *P4*, pedicle of L4; *P5*, pedicle of L5; *S*, spinous process of L3; *SA*, superior articular facet of L4; *SAC*, sacrum.

backward, and laterally. The *12th rib*, which can be variable in length, curves downward and laterally to about the level of the 2nd lumbar disc. The *iliac crest* curves upward and laterally to the level of the middle or lower part of the 4th lumbar vertebra (*fig. 16.12 and Plate 16.1*).

The region of the abdomen between the iliac crests and the pelvic brim is the *pelvis major*. It includes the alae of the sacrum and the iliac fossae described on p. 211.

The Lumbar Vertebrae (*fig. 16.13*)

The *body* is large, kidney-shaped, and flat above and below. In view of the lumbar curve, it is surprising that only L4 and 5 are deeper in front than behind.

The *pedicles* are directed backward and laterally. Above and below each pedicle, there is a small superior *intervertebral notch* and a large inferior one. The *laminae* are

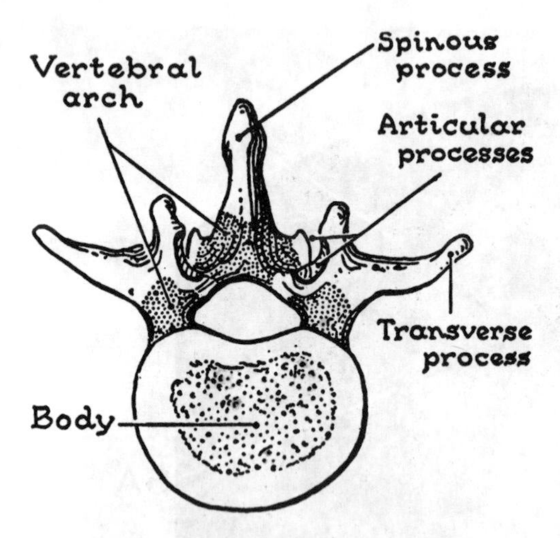

Figure 16.13. A lumbar vertebra from above.

thick and slope downward and backward, enclosing a triangular *vertebral foramen*.

The *spine* is a thick oblong plate that projects nearly horizontally backward and ends in a thickened posterior border.

The *transverse processes* each spring from the junction of pedicles and laminae.

Their ends give attachment to the transversus abdominis aponeurosis and to the quadratus lumborum. Owing to the topographical position of the 5th vertebra, its processes extend onto the body and are conical. The iliolumbar ligaments attach to the transverse processes of L5.

The superior *articular processes* spring from the pedicles and, facing medially, grasp the inferior processes of the vertebra above, which spring from laminae and face laterally. The directions, however, gradually change, with the inferior articular processes of the 5th lumbar vertebra facing nearly forward.

Variations (Details)

1. In 6–12% of adult males, one of the lower vertebrae (usually L5) is a *bipartite vertebra*, having a so-called "separate arch"—the spine, laminae, and inferior articular processes being detached (*fig. 16.14*). This is probably the result of an unhealed fracture.

The bodies, losing the restraining influence of the inferior articular processes, tend to slip forward (spondylolisthesis). This clinical condition increases with age and is uncommon in women.

2. The 5th lumbar vertebra is commonly partly sacralized (*fig. 16.15*).

3. The two sides of the vertebral arch of the lower (or of any) vertebrae may fail to meet (spina bifida).

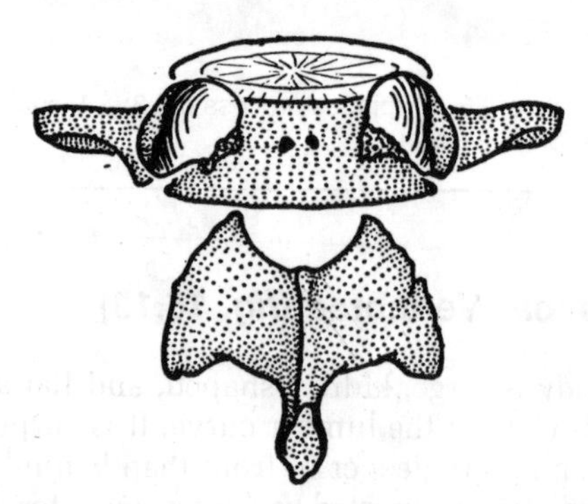

Figure 16.14. The 5th lumbar vertebra is in 2 pieces in 6% of individuals, rendering them liable to a deformity called spondylolisthesis.

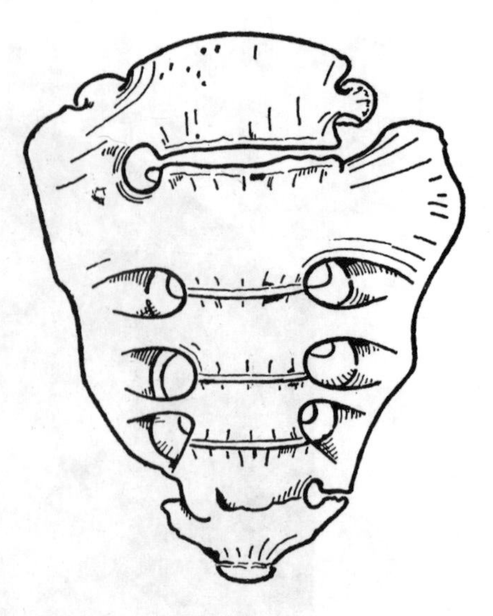

Figure 16.15. The 5th lumbar vertebra is commonly partly sacralized.

MUSCLES AND FASCIA

Muscles

Iliacus, psoas, psoas minor, quadratus lumborum, transversus abdominis, and intertransversarii (*fig. 16.16*).

The **iliacus**, like the iliac fossa from which it arises, is fan-shaped. Its fleshy fibers are inserted mostly into the lateral and anterior aspects of the psoas tendon. The iliacus and psoas are referred to collectively as the **iliopsoas**.

The **psoas** takes fleshy origin from the sides of the bodies and intervertebral discs of all the lumbar vertebrae and T12, as far forward as the sympathetic trunk.

The ventral rami of the upper 4 lumbar nerves plunge into the substance of the psoas when they emerge from the intervertebral foramina, but the 5th ventral ramus escapes under the medial margin of the muscle. On the sides of the vertebral bodies, which are constricted like a waist, the lumbar vessels and the rami communicantes of the sympathetic trunk run dorsally, protected by sheets of fascia that bridge them and afford the psoas an uninterrupted origin (*fig. 16.16*).

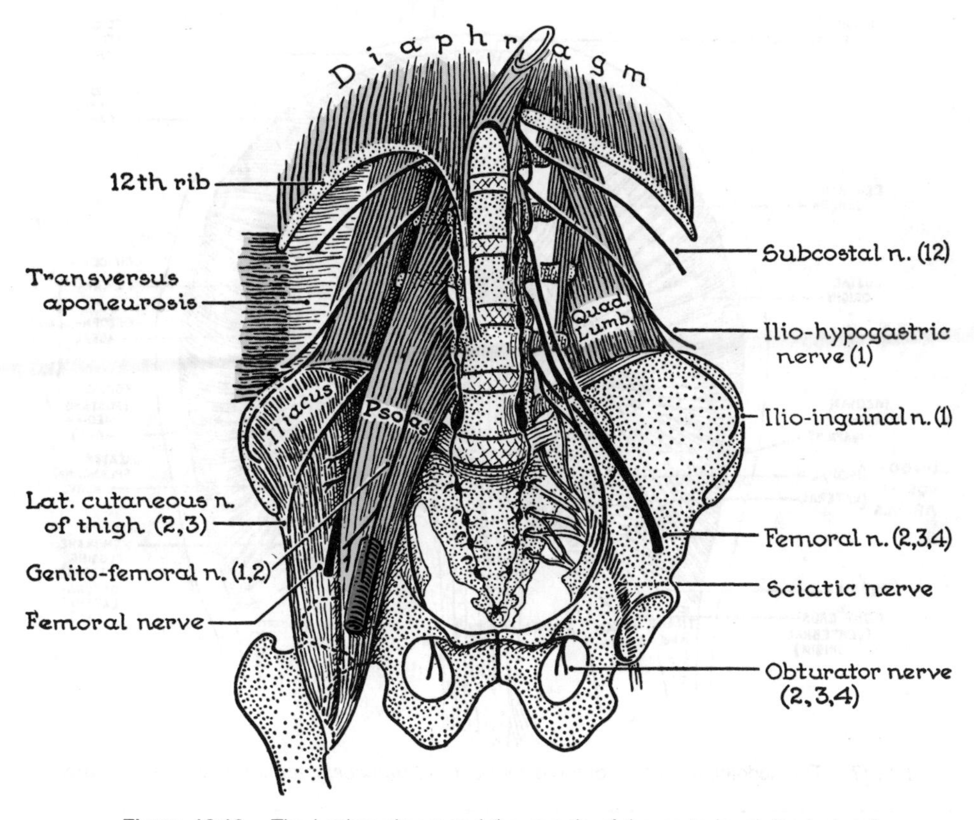

Figure 16.16. The lumbar plexus and the muscle of the posterior abdominal wall.

The psoas tendon inserts into the lesser trochanter of the femur. It lays in the groove between the anterior inferior iliac spine and the iliopubic eminence prior to crossing the middle of the front of the hip joint, of which it is a powerful flexor. The psoas has a minor effect on the lumbar segments of the column (increasing the lumbar curvature).

The **psoas minor** is either absent (50%) or insignificant. Its narrow tendon can be mistaken for the genitofemoral nerve (L12) on the anterior surface of psoas major.

The **quadratus lumborum** is quadrate but not rectangular, for its lateral border is oblique (*fig. 16.16*); this oblique border is a landmark when exposing the kidney from behind (*fig. 15.4*).

The quadratus attachments to the iliac crest, iliolumbar ligament, lumbar transverse processes, and 12th rib are well shown in *Figure 16.16*.

Its fascia is slightly thickened above to form the *lateral arcuate ligament* (lumbocostal arch)—which gives origin to the diaphragm—and is greatly thickened below to form the important *iliolumbar ligament* (*fig. 16.12*).

The **transverse abdominis** helps to form a sheath for the deep muscles of the back. Near the lateral border of the quadratus, it becomes aponeurotic. This posterior aponeurosis divides into 2 layers, the *anterior and posterior layers of the thoracolumbar fascia*. Of these, the posterior passes to the tips of the lumbar spines and supraspinous ligaments, enclosing back muscles; the anterior passes behind the quadratus and attaches itself to the tips of the lumbar transverse processes (*fig. 16.12*), the last rib above and the iliac crest below.

The **intertransversarii** are fleshy bands that pass between adjacent borders of transverse processes.

Fascia Iliaca and Psoas Fascia

These are part of the general fascia that lines the muscles that enclose the abdominal cavity. This strong fascial sheet covers the iliacus and the psoas. The psoas fascia is thickened above to form the medial arcuate ligament, which gives origin to the diaphragm (*fig. 16.17*).

It is carried downward in front of the iliacus and psoas into the thigh behind the inguinal ligament; it adheres to it and to the fascia transversalis; and in the thigh it lies deep to the fascia lata. But, the part covering the psoas (and also the pectineal fascia) is separated from the inguinal ligament by the femoral artery, femoral vein, and deep inguinal lymph vessels wrapped in extraperitoneal areolar tissue, the femoral sheath (*fig. 23.5*). Local anesthetics are injected into the psoas fascial compartment to produce "lumbar plexus blocks."

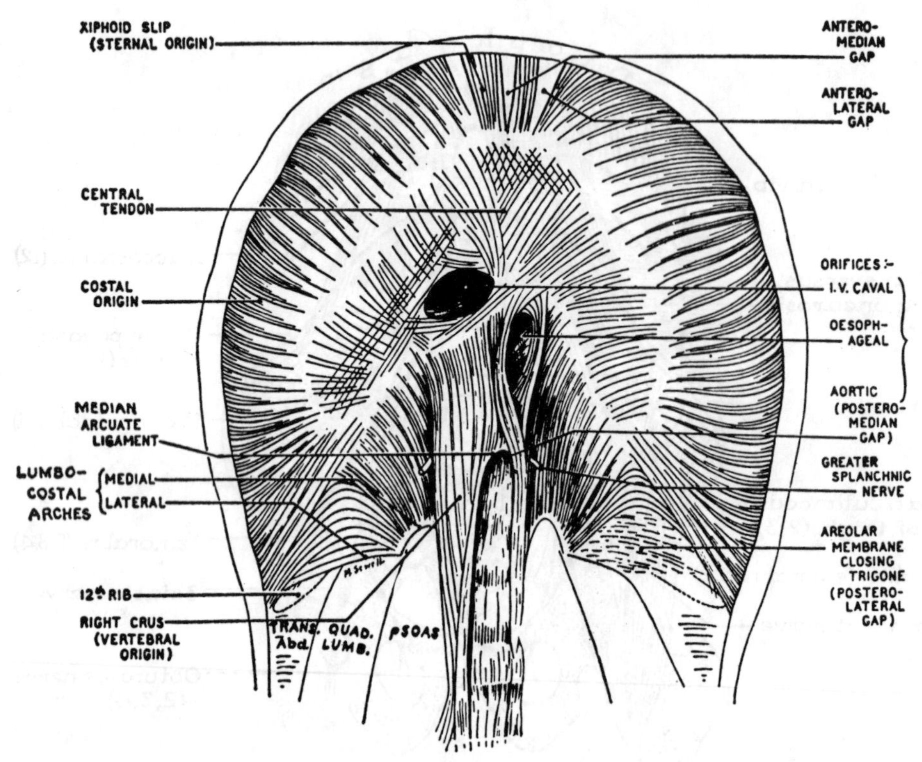

Figure 16.17. The abdominal surface of the diaphragm. (*Lumbocostal arches* = *Arcuate ligaments.*)

NERVES

The Lumbar Nerves

The *ventral rami* are large, and they increase in size from the 1st to the 5th. Each of the five receives 1 or 2 gray rami communicantes from the sympathetic trunk; each of the upper 2 (or 3) sends a white ramus communicantes to the sympathetic trunk; most of the lumbar nerves supply the psoas, quadratus lumborum, and intertransversarii. Each of the five continues downward across the front of the root of the transverse process of the vertebra directly below, the 5th crossing the ala of the sacrum (*fig. 16.16*). The ventral rami form the *lumbar plexus* (see below) and *lumbosacral trunk.*

LUMBAR PLEXUS

The lumbar plexus is formed by the ventral rami of the upper 3½ lumbar nerves (*fig. 16.16*). The first ramus is joined by a branch of the 12th thoracic ramus. [The lower half of the 4th ramus joins the 5th ramus near the anterior border of the ala of the sacrum to form the lumbosacral trunk.] The branches of the plexus run through the psoas. Its largest and most important branches are the *femoral* and *obturator nerves*, both of which spring from the segments L2, 3, and 4.

The **obturator nerve** courses to the upper part of the obturator foramen on the side wall of the pelvis. It appears from under cover of the medial border of the psoas muscle, pierces the psoas fascia, crosses the sacroiliac joint, passes lateral to the internal iliac vessels and ureter, and enters the pelvic cavity.

The **femoral nerve** lies lateral to the femoral sheath and enters the thigh behind the inguinal ligament. It appears at the lateral border of the psoas, runs downward in the angle between the psoas and iliacus. It does not enter the femoral sheath, which encloses the femoral vessels (*fig. 23.5*). In the false pelvis, it supplies the iliacus.

It is of interest to observe that although the femoral nerve supplies the muscles on the front of the thigh, its 3 roots arise behind the 3 roots of the obturator nerve, which supplies the muscles on the medial aspect of the thigh. The explanation is that during development the limb undergoes medial rotation whereby the femoral nerve region, originally behind, is brought to the front, and the obturator nerve region is carried from the front to the medial side.

In addition to the femoral nerve, 4 other nerves appear at the lateral border of the psoas. In ascending order they are as follows.

1. *The lateral (femoral) cutaneous nerve* (from L2 and 3 either directly or from the femoral nerve) enters the thigh behind the inguinal ligament lateral to the femoral nerve.

2. and 3. *The ilioinguinal and iliohypogastric nerves (from L1)* enter the abdomen behind the medial arcuate

ligament and cross in front of the quadratus and pierce the transversus near the anterior superior spine and then the internal oblique. Piercing the external oblique aponeurosis, they supply the skin of the suprapubic and inguinal regions.

The lateral cutaneous branch of the iliohypogastric nerve crosses the iliac crest and descends to the level of the greater trochanter of the femur (*fig. 24.21*).

4. *The subcostal nerve (ventral ramus of T12)* is not a branch of the lumbar plexus, but it enters the abdomen behind the lateral arcuate ligament, pierces the transversus aponeurosis, and then runs to the rectus sheath, which it enters (*fig. 11.6*). Its lateral cutaneous branch crosses the iliac crest and descends to the level of the greater trochanter.

The *genitofemoral nerve* (L1 and 2) comes to lie within the extraperitoneal fatty layer and, so, pierces the psoas muscle and the psoas fascia. It divides at a very variable level into 2 branches, *femoral* and *genital*, which descend in front of the psoas muscle.

The *femoral branch* supplies the skin of the femoral triangle, piercing the fascia lata. The *genital branch* supplies the cremaster muscle and traverses the inguinal canal to end in the skin of the scrotum. In its course it pierces the coverings of the spermatic cord.

Variations. Considerable interchange takes place between the above nerves, and so their territories are quite variable.

An *accessory obturator nerve* (from L3 and 4) is a small nerve, when present, along the medial border of the psoas. It crosses the superior ramus of the pubis to the pectineus.

Diaphragm

The diaphragm (Gk. dia = through, across; phragma = a partition) has a rounded dome on each side below the lungs and a depressed median portion on which the heart lies at the level of the xiphisternal joint (*figs. 16.18, 16.19*).

The right dome, occupied by the liver, rises to the 5th rib, just below the right nipple; the left dome falls short of the left nipple by 3 cm. (The nipples correspond to the inferior angles of the scapulae on the back.)

Attachments and Marginal Gaps

The diaphragm arises by fleshy digitations from the back of the xiphoid process, from the inner surfaces of the 7th to 12th costal cartilages (interdigitating with the transversus abdominis) and from the vertebrae L1–3 by the *crura* (*fig. 16.17*).

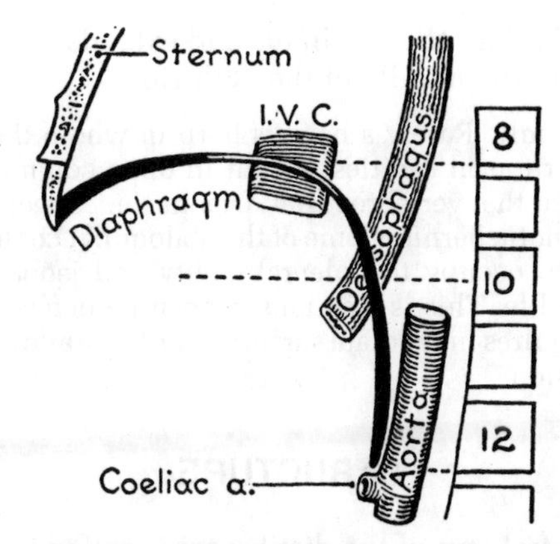

Figure 16.18. The higher the vertebral level, the more ventral is the hiatus in the diaphragm *IVC*, inferior vena cava.

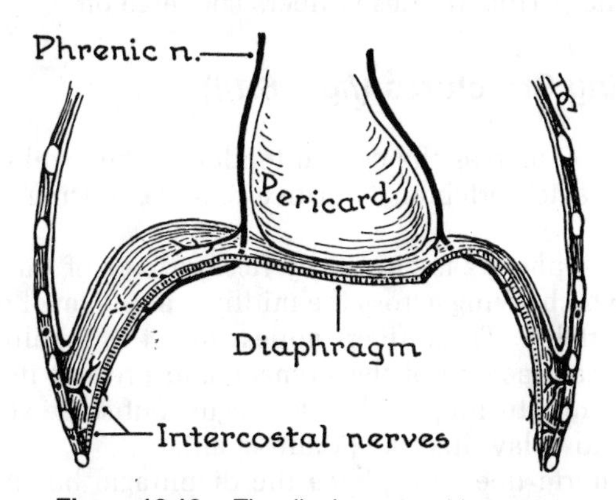

Figure 16.19. The diaphragm and its nerves.

Anteriorly, there are slight gaps both between the right and left xiphoid slips and lateral to them. The superior epigastric vessels pass into the rectus sheath through the lateral gaps.

Posteriorly, in the median plane, there is a large gap between the crura; through this the aorta passes. The medial parts of the crura are fibrous, and they join in front of the aorta immediately above the celiac trunk to form the fibrous *median arcuate ligament* (*fig. 16.17*).

Posteriorly, on each side, between the crus and the 12th rib, the pleural and peritoneal cavities are continuous with each other in prenatal life. Then crural fibers migrate across and generally separate the 2 cavities (*fig. 16.17*). Sometimes they fail, leaving the *vertebrocostal trigone* (lumbocostal trigone), above the 12th rib. The areolar and muscular fibers closing the space find attachment to the fascia covering the psoas and quadratus lumborum, which in response becomes thickened and strengthened to form the *medial* and *lateral arcuate ligaments* (lumbocostal arches). Of these, the medial ligament bridges the psoas and extends from the lateral border of the crus to the transverse process of the 1st lumbar vertebra; the lateral ligament bridges the quadratus and extends from the latter point to the middle of the 12th rib.

Anomaly. Rarely a baby is born in whom the pleural and peritoneal cavities remain in open communication through the vertebrocostal trigone, a *congenital diaphragmatic hernia.* Some of the abdominal contents may come to occupy the pleural cavity and jeopardize the child's life. This is a serious congenital defect and usually requires immediate surgery in order to allow the lung expansion.

STRUCTURE

The central part of the diaphragm is called the *central tendon.* It is composed of decussating and interwoven tendinous fibers, and it has the shape of a trefoil or clover leaf. The peripheral fleshy fibers converge on it.

Piercing Structures (*fig. 16.18*)

The *IVC* pierces the central tendon at the level of the 8th thoracic vertebra. Its orifice enlarges during inspiration.[1]

The *esophagus* is circled by fleshy fibers of the right crus (which swing across the midline) at the level of the 10th vertebra. These fibers appear to act as a sphincter for the cardiac end of the stomach and prevent its contents from returning to the esophagus, but some studies appear to delay this role (Mann et al.).

The *aorta* does not pierce the diaphragm but passes behind the median arcuate ligament at the level of the 12th vertebra.

The vertebral levels, then, are 8, 10, and 12.

Other Piercing Structures

Through the caval foramen pass some branches of the *right phrenic nerve.* The left phrenic and other branches of the right phrenic pierce the diaphragm independently to spread out on its abdominal surface (*fig. 16.19*).

Through the esophageal hiatus pass the *anterior and posterior vagal trunks* and the *esophageal branches of the left gastric artery and vein.* The vein is of special importance, because it anastomoses with esophageal branches of the azygos veins to connect the portal and systemic venous systems (p. 163, *fig. 13.9*).

Through the aortic hiatus passes the *thoracic duct* [also a vein connecting the right ascending lumbar vein to the azygos system] (*fig. 16.6*).

The *3 splanchnic nerves* (greater, lesser, and lowest) pierce the crura to end in the celiac (and aorticorenal) ganglia.

[1]This is not agreed upon for man; in some animals there is a certain degree of constriction of the IVC during strong contraction of the diaphragm (Franklin).

Nerve Supply (*fig. 16.19*)

The diaphragm is supplied by (1) the phrenic nerve (C3, 4, and 5), its only motor nerve—but it is also sensory—and by (2) the lower intercostal nerves, which are sensory to the peripheral parts.

Arteries

Pericardiacophrenic and musculophrenic arteries (from internal thoracic artery) and intercostal. superior phrenic, and inferior phrenic arteries (from the aorta).

Relations

The *abdominal relations* are the liver, stomach, and spleen and the celiac ganglion, suprarenal glands, and kidneys. The *thoracic relations* are the heart and pericardium, the lungs and pleurae, and pleural recesses, below the recesses are the lower intercostal spaces and ribs, the thoracic aorta, and the esophagus.

Development

The fully formed diaphragm is derived from the mesodermal partition of composite origin (*fig. 16.20*). The anteromedian part arose from the septum transversum; the posteromedian part, from the primitive dorsal mesentery; the lateral parts from the body wall. The *pleuroperitoneal canal* on each side is closed by a membrane, the *pleuroperitoneal membrane.*

In the young embryo, the hinder part of this composite partition lies at the level of vertebra C2. On passing vertebrae C3, 4, and 5, portions of the myotomes of these segments, supplied by the phrenic nerve, extend into it and pervade it, thereby forming the muscular diaphragm (Wells).

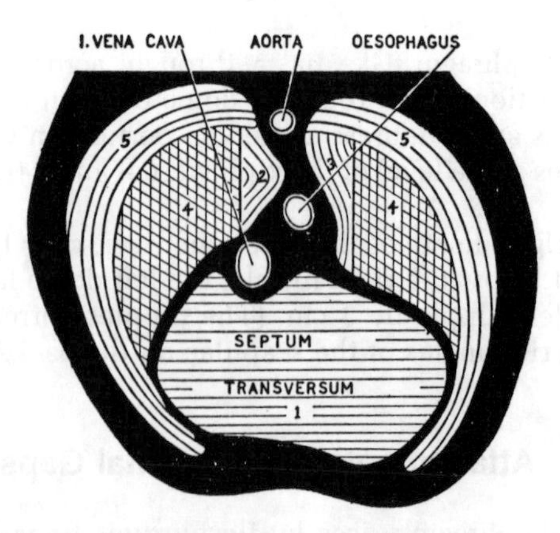

Figure 16.20. The elements from which the diaphragm is developed. *1* = septum transversum; *2, 3* = dorsal mesentery; *4* = pleuroperitoneal membrane; *5* = body wall.

The *septum transversum* is the thick mesodermal mass that surrounds the vitelline veins (also the common cardinal veins) prior to their entering the sinus venarum of the heart. At a certain period this septum projects horizontally backward from the anterior body wall to meet the dorsal mesentery at the level where the duct system of the liver (of endodermal origin) buds from the duodenum.

Abdominal Lymphatics

Lymph capillaries, lymph vessels, and lymph nodes occupy areolar and fascial planes, e.g., the deep fascia, and the submucous and subserous coats of viscera. Lymph capillaries form networks which drain by the nearest issuing lymph vessels. Retrograde flow is prohibited by numerous valves.

After a meal, the intestinal lymph vessels contain emulsified fat, and since this is white like milk, the vessels are called *lacteals*.

The external and common iliac chains of nodes (see *fig. 20.2*) continue upward along the sides of the aorta and around it as the *right* and *left aortic* or *lumbar chains of nodes*. They open by means of a *right* and a *left lumbar lymph trunk* into a thin-walled tubular sac, the *cisterna chyli*. The IVC runs through the right chain, making it less accessible than the left.

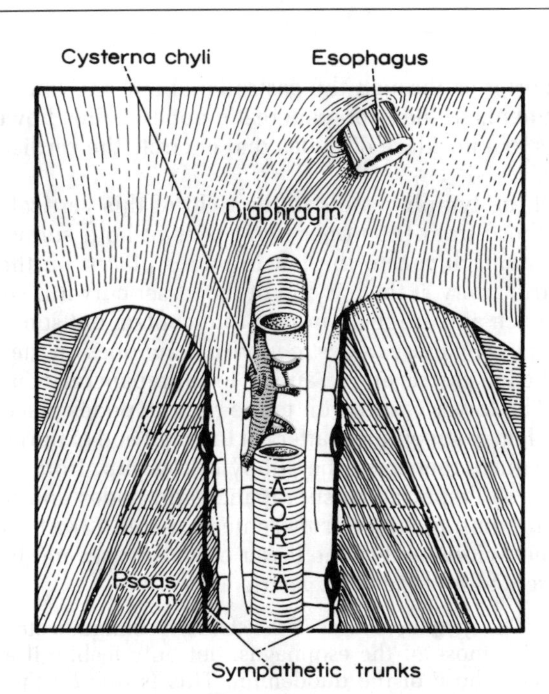

Figure 16.21. The cysterna chyli.

These chains receive (1) lymph already filtered through nodes and coming from the lower limbs, lower part of the anterior abdominal wall, external genitals, perineum, and pelvis; (2) lymph vessels that follow the lumbar arteries and drain the posterior abdominal wall; (3) the vessels from the 3 paired glands—suprarenal, kidney, and testis (the ovary, upper part of the uterus, and the uterine tube in the female); and (4) vessels from the part of the gastrointestinal tract supplied by the inferior mesenteric artery, draining into the left aortic chain. However, the parts supplied by the celiac and superior mesenteric arteries are drained by means of a gastrointestinal trunk into the cisterna chyli.

The **cisterna chyli** (*fig. 16.21*) resembles a segment of a vein about 5 cm long. Its diameter is less than that of a lead pencil, but it may be irregularly dilated. It lies between the aorta and the right crus of the diaphragm. It receives 5 or more trunks, namely, the *right* and *left lumbar trunks*, the *gastrointestinal trunk*, and a *pair of vessels* that descend from the lower intercostal spaces. On passing through the aortic hiatus, it becomes the thoracic duct.

LYMPHATIC DRAINAGE

Intestine

The nodes of the *large intestine* are numerous and are roughly divisible into 3 groups: (1) *paracolic nodes* on the marginal artery close to the gut wall; (2) *intermediate nodes* on the stems of the colic arteries; and (3) *main nodes* near the roots of the colic arteries; beside the aorta. Also, there are a few small *epicolic nodes* applied to the surface of the colon.

The lymph vessels from the intestine follow the arterial vessels fairly closely, and each vessel is interrupted by one or more groups of nodes (*fig. 16.22*). The lymph vessels from the segment of the large gut between the appendix and the left colic flexure follow the branches and stem of the superior mesenteric artery and, joining those of the small gut, form the *intestinal trunk*. Those from the left colic flexure and remainder of the large gut follow the branches of the inferior mesenteric artery and end among the nodes of the left lumbar (aortic) chain.

There may be an appendicular node in the mesentery of the appendix; there are many nodes clustered in the ileocolic angle; the vessels from the transverse colon have the longest distance to travel; those at the left colic flexure communicate with the splenic nodes, traveling probably along the route of the occasional artery described on p. 173.

The *small intestine* is drained through numerous nodes divisible, like those of the large intestine, into 3 groups. Ultimately a channel, the *intestinal trunk*, emerges and, by joining the *gastric trunk*, forms the *gastrointestinal trunk*, which opens into the cisterna chyli.

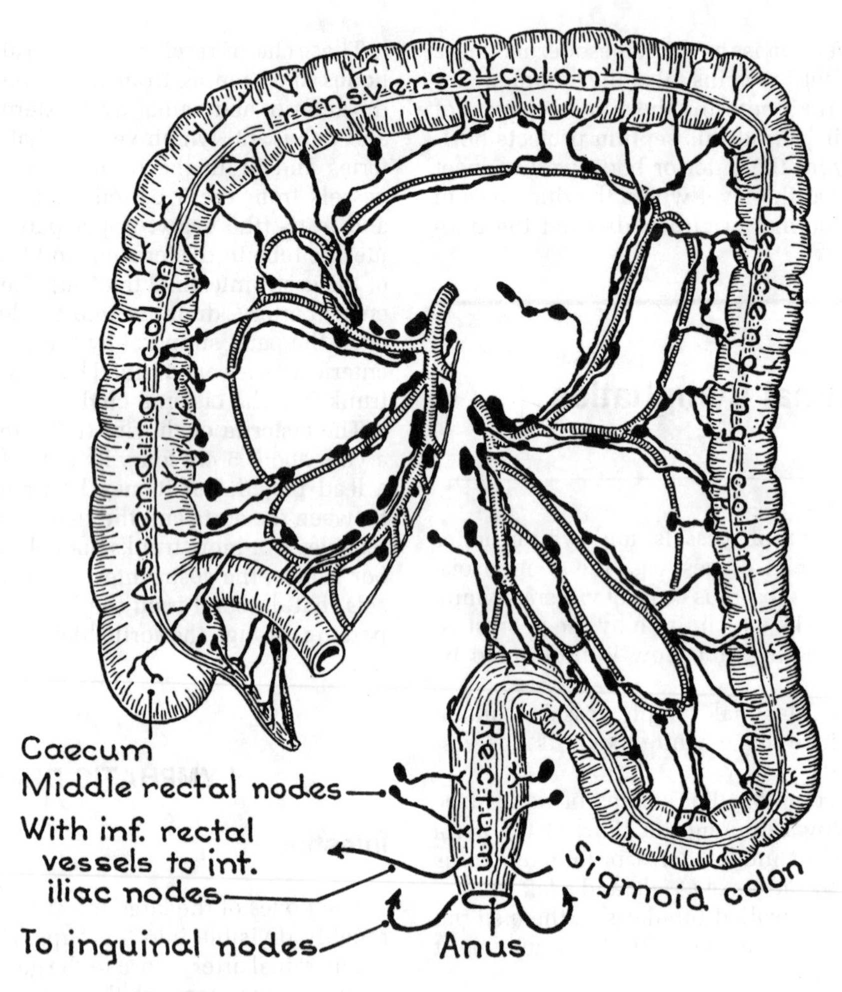

Figure 16.22. The lymphatics of the large intestine. (After Jamieson and Dobson.)

Pancreatic Groups of Nodes (Celiac and Superior Mesenteric Nodes, *fig. 16.23*)

Along the upper border of the pancreas there are *middle, right,* and *left suprapancreatic groups* of nodes related to the celiac artery and to its hepatic and splenic branches. A *subpyloric* group is applied to the front of the head of the pancreas below the pylorus. A *left gastric chain* lies on the course of the left gastric artery. A *biliary chain* extends along the bile passages from the porta hepatis above, through the lesser omentum, and behind the first part of the duodenum and head of the pancreas, to the second part of the duodenum below. And, there is the *main group of submesenteric nodes* at the root of the mesentery.

Stomach

The lymph vessels from the part of the stomach that lies to the left of a vertical line dropped through the esophagus pass with the left gastroepiploic and short gastric arteries through the gastrolienal and the leinorenal ligaments to the suprapancreatic nodes; some, however, pass to a necklace of nodes that encircle the cardiac orifice.

Of the lymph vessels to the right of this vertical line, (1) those from the upper two-thirds of both surfaces of the stomach run to the left to nodes placed on the left gastric artery at the left end of the lesser curvature of the stomach and are there in part intercepted, but some vessels pass by these nodes to more distant ones on the stem of the left gastric artery, and so to suprapancreatic nodes. (2) The lower vessels run to the right to nodes placed on the right gastroepiploic artery at the right end of the greater curvature, thence to the subpyloric group, from which they are dispersed to suprapancreatic nodes and to the main group of superior mesenteric nodes. (3) At the extreme pyloric end of the lesser curvature, several lymph vessels follow the right gastric artery.

The lymph plexuses of the stomach communicate freely with those of the esophagus, but only feebly, if at all, with those of the duodenum. This is due to: the connective tissue septum in the submucous coat at the py-

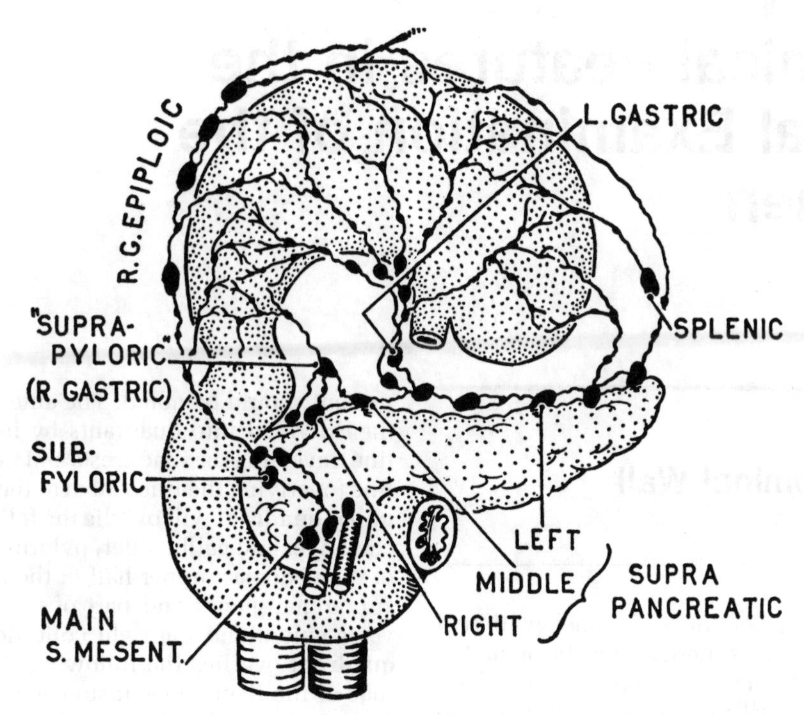

Figure 16.23. The lymphatics of the upper abdomen and stomach.

loric sphincter, discontinuity of the circular muscle fibers, and indipping of the longitudinal muscle fibers.

Liver

Lymph vessels from the upper surface of the liver pass through the falciform ligament to retrosternal nodes, which lie on the diaphragm and discharge into the parasternal (internal mammary) chain; some from the interior, following hepatic veins, pass with IVC through the diaphragm to diaphragmatic nodes, thence the thoracic duct; others, following the branches of the portal vein, emerge at the porta and travel down the biliary chain to be distributed to various pancreatic nodes. There is an intercepting gland, the *cystic node*, at the neck of the gall bladder, and there are other "hepatic nodes" in the porta hepatis.

Pancreas

The lymphatics of the pancreas drain into adjacent nodes.

Spleen

The lymphatics of the spleen drain into splenic nodes situated where the tip of the pancreas abuts against the spleen; these in turn drain into the suprapancreatic group.

Clinical Mini-Problems

1. **At which sites would one expect to find anastomoses of the portal and systemic venous systems?**
2. **(a) In the embryo, which vessel carries oxygenated blood from the placenta to the ductus venosus? (b) What becomes of that vessel when it is severed at birth?**
3. **Operations on the iliac arteries may produce hypotension (reduced blood pressure). What structures are affected to produce this condition?**

(Answers to these questions can be found on p. 585.)

17

Anatomical Features in the Physical Examination of the Abdomen

Abdominal Wall

While visual inspection of the abdominal wall will reveal abnormalities such as herniations through the muscle layers or lumps in the inguinal region, palpation is the primary approach. All routine physical examinations of males should include a test for inguinal hernia by gently invaginating loose scrotal skin with a fingertip into the external inguinal ring. When a man gives a cough, even a small hernia announces itself by pushing against the fingertip. This easily separates the diagnosis from enlarged lymph nodes and the uncommon obturator hernia (which, incidentally, is much commoner in women). Congenital hernias in little boys are often seen as quite obvious bulges into the scrotum. They are rather easily treated by surgical repair. Undescended testes are common and also obvious in male infants. They are rarely seen in teenagers, but when they are, they must be treated hormonally or surgically by a specialist.

Neurologists routinely test superficial reflexes of the abdominal wall. They also perform the cremasteric reflex test, which depends on the intact genitofemoral and ilioinguinal nerves (L1,2). Stroking the skin of the proximal thigh causes the testis of that side to be pulled vigorously upward toward the inguinal canal.

Inspection

The anterior abdominal wall is visually inspected for symmetry and obvious bony landmarks such as the costal margin and iliac crests. Any discoloration, scars, or lumps present should be noted. The anterior abdominal wall is then divided into quadrants by imagining a horizontal line traversing the iliac crests and intersecting the median line (overlying the linea alba) at the umbilicus. The **right upper quadrant** will overlie the following abdominal viscera: liver and gallbladder, pylorus and duodenum; head of the pancreas; upper half of the right kidney and right suprarenal gland; and part of the ascending and transverse colon; and the right colic flexure. The **left upper quadrant** overlies the following abdominal viscera: left lobe of the liver; spleen; stomach; body and tail of pancreas; upper half of the left kidney and the left suprarenal gland; and the left part of the transverse colon, its splenic flexure and the descending colon. The **lower left quadrant** overlies the following abdominopelvic viscera: lower pole of the left kidney and left ureter; lower descending colon and sigmoid colon; left overy and left uterine tube in the female and the left spermatic cord and inguinal canal in the male; the bladder and uterus may be palpated near the midline if they are distended or enlarged; the jejunum and ileum also occupy a midline position. The **right lower** quadrant overlies the following abdominopelvic viscera: the lower pole of the right kidney and the right ureter, the cecum, appendix, and lower portion of the ascending colon; the right ovary and right uterine tube in the female and the right spermatic cord and inguinal canal in the male; the ileum occupies a midline position above the uterus and bladder, which may also rise to the abdominal wall level if distended or enlarged. The bladder is consistently present below the umbilicus in the newborn and young child.

Auscultation

Bowel sounds can be heard with a stethoscope placed over the various components of the gastrointestinal system. Blood flow through the large vessels can also be heard with the stethoscope. The abdominal aorta lies on the anterior aspects of the bodies of the lumbar vertebrae.

Due to the lumbar curvature, the aorta approximates the anterior abdominal wall at the level of the umbilicus. Abnormal patterns of blood flow (bruits) may be heard in the abdominal aorta and external iliac arteries.

Percussion

Solid organs produce a dull sound when percussed, while hollow organs or cavities filled with air sound resonant. The upper border of liver can be detected from the lung around the 6th intercostal space in the right midclavicular line. Its lower border is just below the right intercostal margin in the midclavicular line.

The normal spleen can be percussed in the left midaxillary line in the 9th and 10th intercostal spaces.

Abdominal Contents

Palpation is rather a perfunctory procedure in a routine examination if no symptoms are reported. With the normal relaxed person in a recumbent position, the physician can make out very little specifically. One can barely feel the edge of the liver, inferior poles of the kidneys, and perhaps some loops of intestines if they are full. The pulsation of the aorta is easily felt in thin persons.

If symptoms are reported, however, one's palpation concentrates on the quadrants of the abdomen where one seeks to locate tenderness, enlargements, or masses associated with specific organs, e.g., liver, gallbladder, pancreas, stomach, spleen, kidneys, ureters, urinary bladder, etc. Radiology is resorted to rather early if any abnormality is detected. In some cases, direct visualization through optical devices introduced into the hollow organs or the peritoneal cavity is becoming common.

Palpation

The spleen may be felt by palpation when the examiner places their fingers under the left costal margin and asks the patient to inhale. The spleen descends in an inferoanterior fashion as the diaphragm contracts, and, if sufficiently enlarged, it will contact the examiner's fingers. Other abdominal organs are more difficult to palpate directly, but tenderness or enlargement of a given organ in the various quadrants can make them more detectable.

The superficial inguinal ring can be palpated in the male by placing the little finger in a fold of scrotum behind the spermatic cord and pressing upward and laterally. The patient is asked to cough to increase the intra-abdominal pressure. Any tendency for intra-abdominal tissue to herniate through the superficial ring can be felt next to the spermatic cord in the superficial ring. The superficial inguinal ring lies superior to the pubic tubercle.

The dermatomes of the abdominal wall can be tested with touch or a sharp pin. T6 is at the xiphoid level. T10 overlies the umbilicus, while T12 is in the suprapubic region. L1 dermatome overlies the inguinal canal and extends into the anterior scrotal (labial) region.

Pelvis and Perineum 4

18

Male Pelvis

Clinical Case 18.1

Patient Mannie Z. This 52-year-old truck driver visited his family physician with a complaint that he felt an increasing urge to urinate 2–3 times per hour and that the volume of urine at each micturation was very low. He felt a burning sensation during micturation and incomplete expelling of the urine from his urethra. The physician diagnosed a bladder infection (cystitis) and prescribed antibiotics. Two months later, the patient returned, stating that the antibiotics had helped somewhat, but he was again experiencing so much urinary discomfort and annoyance that he could not go to work. His physician examined him, tested a sample of urine and found no signs of infection. He referred him to the urologist to whom you are assigned as a student assistant. The urologist performed a transurethral cystoscopy to visualize the urethra and internal aspect of the bladder. A large growth was evident on the mucosa of the bladder over the lateral and posterior walls. A biopsy that included a number of tissue samples of the bladder mucosa was done. The pathologist's report indicated that 3 of the 6 biopsied specimens showed evidence of carcinoma.

Radical surgery was imperative to excise the entire bladder, including the prostatic urethra and prostate. The ureters were connected to the cecum. A biopsy of the iliac lymph nodes taken during the surgery revealed that the carcinoma had also spread beyond the confines of the pelvis. The patient was placed on postsurgical radiation therapy.

Follow-up. Postsurgical complications including impotency relate to the disruption of the sympathetic innervation of the vas deferens, ejaculatory ducts, and urethra. The ureters were eventually externalized by a second surgery through the abdominal wall, the urine having been collected in a prosthetic receptacle. A bone scan performed three months after surgery revealed a tumor in the 5th lumbar vertebra. The patient died within a year following his cystectomy and prostatectomy. Perhaps an earlier diagnosis would have saved his life, but many patients ignore urinary symptoms.

Male Pelvis Viewed from Above

The following *peritoneal fossae* are seen: on each side of the rectum, the *pararectal fossae*; on each side of the partly filled bladder, the *paravesical fossae*; and, between them, the *rectovesical pouch*.

Follow the Peritoneum in the Median Plane (fig. 18.1).

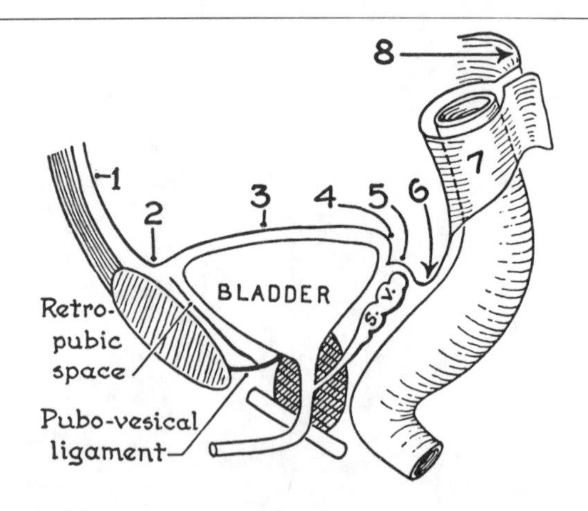

Figure 18.1. The peritoneum of the male pelvis in paramedian section (*see fig. 19.3*). *1*, anterior abdominal wall; *2*, back of the pubis; *3*, superior surface of the empty bladder; *4*, posterior surface for 1 cm; *5*, upper ends of the seminal vesicles on each side of the median plane; *6*, across the bottom of the rectovesical pouch; *7*, the rectum; *8*, the mesentery of the sigmoid colon.

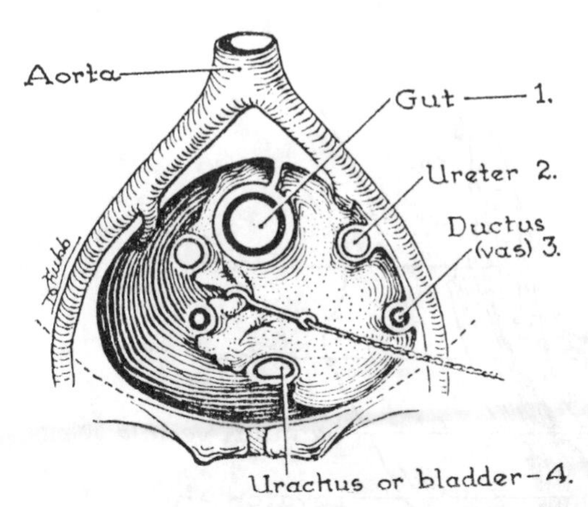

Figure 18.2. Four visceral tubes adhere to the peritoneum. (The peritoneum has been detached from the side wall of the pelvis.)

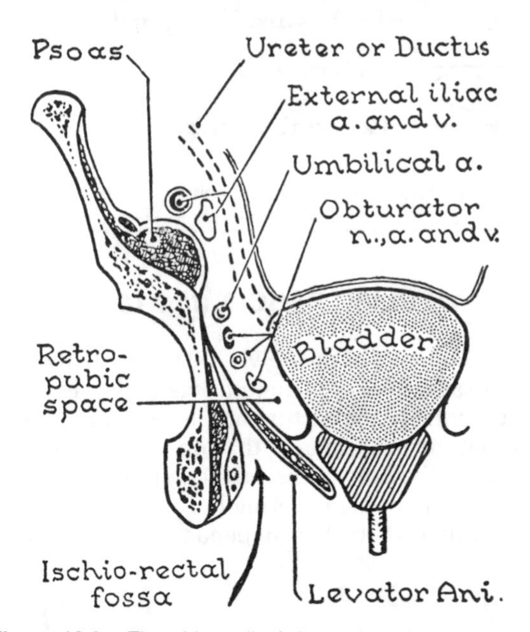

Figure 18.3. The side wall of the male pelvis on coronal section (diagrammatic).

Four "visceral tubes" adhere to the peritoneum, even when it is detached and mobilized (fig. 18.2).

The Retropubic Space (*fig. 18.1, 18.3*) is a continuous bursa-like cleft in the areolar tissue at the sides and front of the bladder which allows the bladder to fill and empty without hindrance. The space is limited *posteriorly* by a broad areolar sheet enclosing a leash of vessels that pass from the internal iliac artery and vein to the posterolateral border of the bladder (see *fig. 18.36*). Two taut cord-like thickenings, one on each side of the median plane, attach the neck of the bladder to the lower end of the symphysis; these are the *puboprostatic* or *pubovesical ligaments* (see *fig. 18.7*).

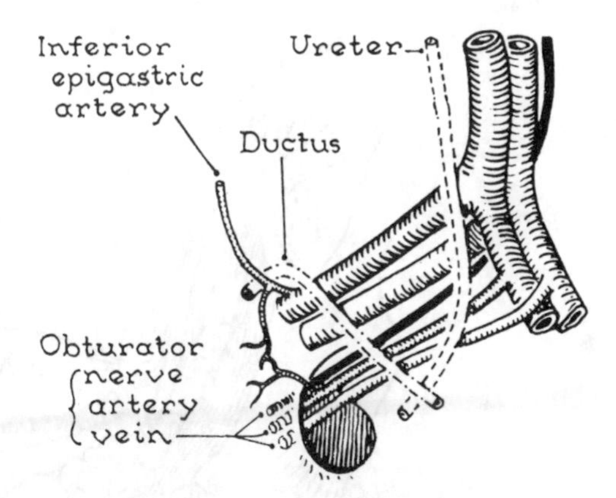

Figure 18.4. The structures on the side wall of the male pelvis. Note the medial positions of the ureter and vas or ductus deferens.

URETER AND DUCTUS (VAS) DEFERENS

The *ureter* crosses the external iliac artery, just in front of the bifurcation of the common iliac artery. It enters the lateral angle of the bladder, where its last 2–3 cm is enveloped in the leash of vessels just mentioned and is crossed by the ductus deferens (*fig. 18.4*).

The *ductus deferens* runs from the deep inguinal ring, where it turns around the inferior epigastric vessels.

The *seminal vesicles* are situated lateral to the deferent ducts behind the bladder (*fig. 18.5*).

Side Wall of the Pelvic Cavity

The side wall of the pelvic cavity is made up of bone and muscle (*fig. 18.6*). The pelvic brim constitutes the margins of the pelvic inlet and is devoid of muscle. Some neurovascular elements cross the pelvic brim in the posterolateral aspect.

The **obturator nerve** (L2,3,4) crosses the pelvic brim and traverses the lateral pelvic wall above the attachment of the obturator fascia (*fig. 18.6*). The nerve is joined by an accompanying artery and vein in its pelvic course. The neurovascular bundle leaves the pelvic cavity via the **obturator foramen** to enter the medial compartment of the thigh. The **obturator internus muscle** underlies the obturator fascia on the lateral wall of the pelvic cavity.

The *Levator Ani* forms the greater part of the pelvic floor or diaphragm. It arises from the inner surface of the body of the pubis, from the inner surface of the ischial spine and, between these 2 points, from the obturator

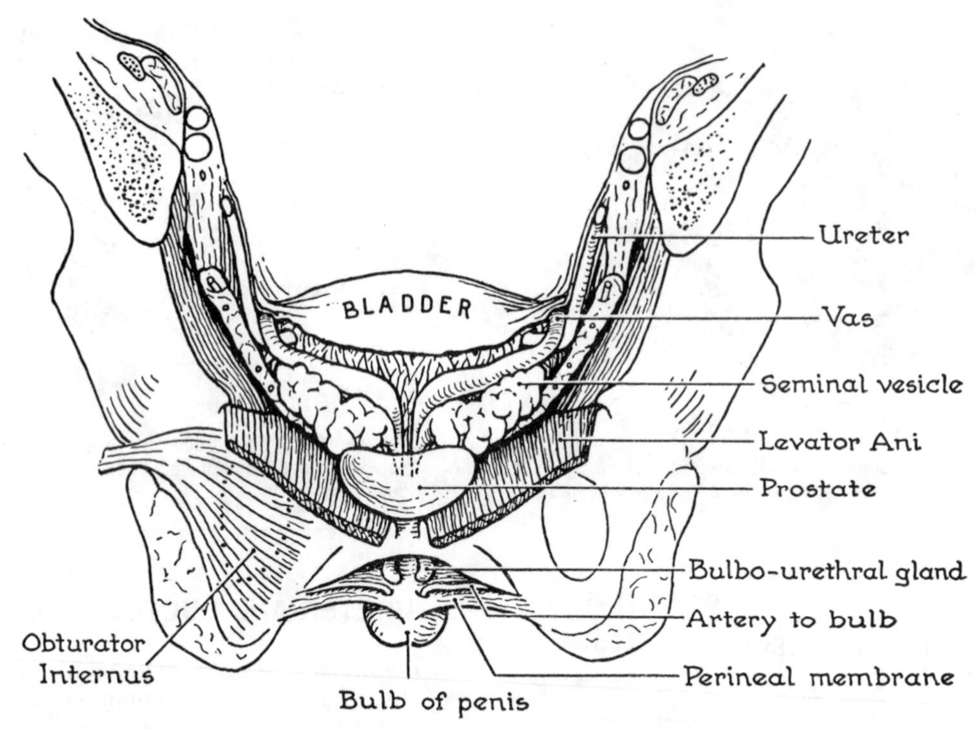

Figure 18.5. A coronal section of the pelvis to show the genitourinary organs from behind.

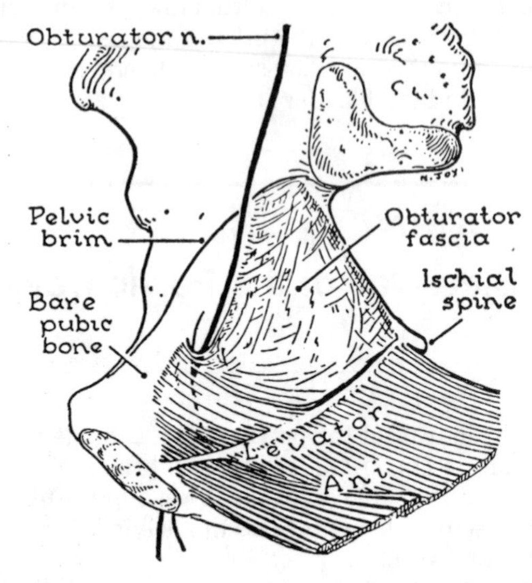

Figure 18.6. The side wall of the lesser pelvis divided into upper and lower (anterior and posterior) parts by the obturator nerve.

fascia, which is commonly thickened to form a *tendinous arch* of origin (*fig. 18.6*).

The structures encountered on the side wall are shown in *Figures 18.3* and *18.4*. They include the obturator nerve and vessels, the ureter, the vas deferens, and the internal iliac vessels. Also running forward from the internal iliac artery to the umbilicus is the (obliterated) umbilical artery (p. 152).

PELVIC FASCIA (*fig. 18.7*)

All the contents of the pelvis are covered all over with areolar tissue. The covering of organs that expand and contract, notably the rectum and bladder, is necessarily loose; that given to organs that do not expand may be dense, as in the case of the prostatic fascia, or thin as in the case of the fascia covering the seminal vesicles and deferent ducts. Its texture depends upon the strains put upon it.

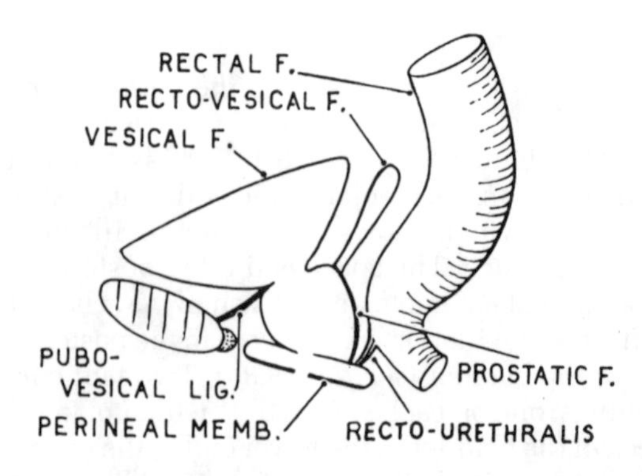

Figure 18.7. The pelvic fascia of the male in median sagittal section. Rectourethralis consists of strands of smooth muscles.

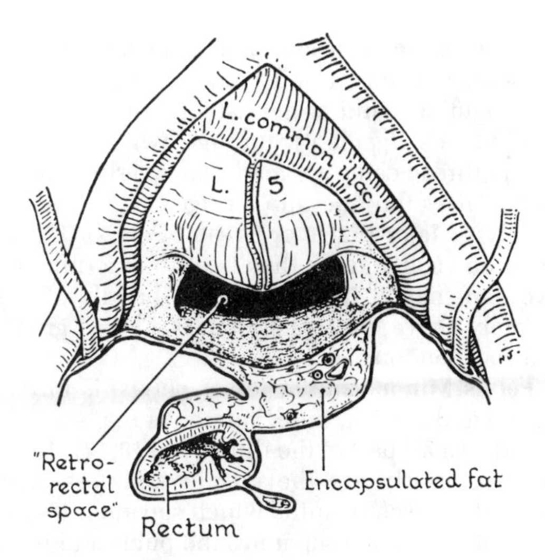

Figure 18.8. The "retrorectal space."

Subdivisions of the Pelvic Fascia

This fascia is divided into *parietal*, *diaphragmatic* (for the pelvic diaphragm or floor), and *visceral* layers. Note:

1. The puboprostatic or pubovesical ligaments, which anchor the prostate (and bladder) to the pubis, are thickenings of this fascia.
2. The fascia invests the numerous veins and few arteries that are passing from the internal iliac vessels to the base of the bladder and internal genital organs (within the rectovesical fascia) (see *fig. 18.36*). This limits the retropubic space posteriorly.
3. The rectovesical space and fascia separate the bladder from the rectum (*fig. 18.7*).
4. Dense fascia covers the back of the prostate (p. 219).
5. The potential "retrorectal space" is bounded on each side by an areolar fold containing the pelvic splanchnic nerves, which run from the 2nd, 3rd, and 4th sacral foramina to the side of the rectum (*fig. 18.8*).

Interior of Bony Pelvis

The pelvis (L. = *basin*) is formed by the right and left hip bones, the sacrum, and the coccyx. Each hip bone (L. = *os coxae*) articulates with the other anteriorly at the symphysis pubis; posteriorly, they articulate with the first 3 sacral vertebrae at the sacroiliac joints. The right and left hip bones constitute the pelvic girdle.

SACRUM AND COCCYX

The **Os Sacrum** is composed of 5 fused vertebrae, and the **os coccygis** of 3–5, though before birth it has 7–11 cartilaginous caudal rudiments. Both sacrum and coccyx are triangular with base above and apex below (Fig. 18.9).

The Base of the Sacrum (*fig. 18.9*) is divided into 3 parts. The median part is the oval upper surface of the body of the first sacral vertebra. Its anterior border is an important landmark named the *promontory* of the sacrum. Behind its posterior border is situated the somewhat compressed triangular entrance to the *sacral canal*, flanked by prominent superior articular processes. The right and left lateral parts, called the *alae*, are fan-shaped and represent fused costal and transverse elements (*fig. 1.16*).

Each ala is crossed by the constituents of the lumbosacral trunk, the iliolumbar artery, the obturator nerve, and the psoas. Only the 5th lumbar ventral ramus is so taut that it grooves the surface (*fig. 18.9*).

The Pelvic Surface of the sacrum is concave and is crossed by ridges at the sites of obliterated discs (*fig. 18.9*). Lateral to them are the 4 pelvic (or *anterior*) *sacral foramina*. Their margins are shaped by the emerging ventral rami of the upper 4 sacral nerves which pass laterally. The mass of bone lateral to the foramina is the *pars lateralis* (lateral mass).

The **sides** of the upper 3 pieces of the sacrum form a large *auricular* or *ear-shaped facet* that articulates with a corresponding auricular facet on the ilium (*fig. 18.10*); the sides of the lower 2 pieces, as well as the sides of the coccyx, are thin for the attachments of the sacrotuberous and sacrospinous ligaments (see *fig. 18.18*).

The **Posterior Surface** is related to the gluteal structures (p. 274).

The first piece of the coccyx possesses a pair of *transverse processes*, each of which is joined by a ligament to the sacrum (inferolateral angle), thereby making a foramen through which the ventral ramus of the 5th sacral nerve enters the pelvis (see *fig. 18.37*). The remaining

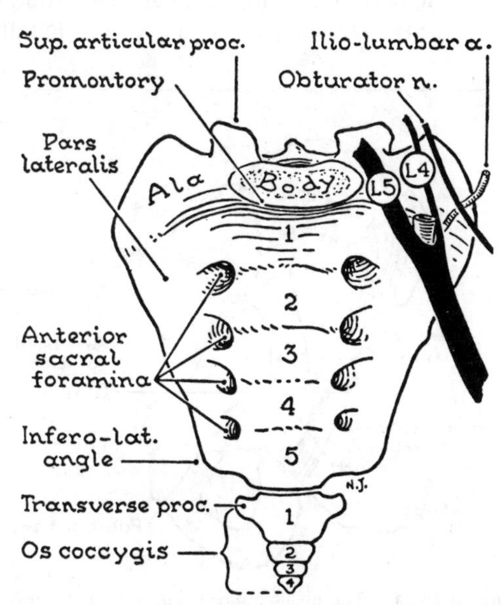

Figure 18.9. Pelvic (anterior) aspect of sacrum and coccyx.

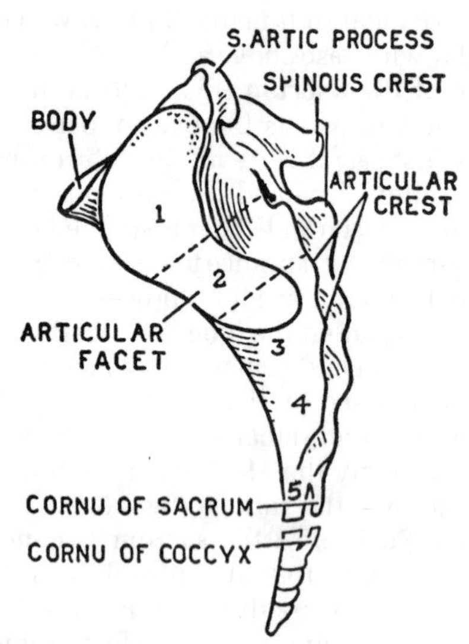

Figure 18.10. The lateral aspect of sacrum and coccyx.

pieces of the coccyx are nodular. The 1st piece commonly fuses with the sacrum.

Ossification. The 5th, 4th, and 3rd pieces of the sacrum are always completely fused together by the 23rd year, the 3rd and 2nd by the 24th year, and the 2nd and 1st by the 25th year or even later (McKern and Stewart).

The Pelvic Brim is the boundary line between the greater or *major pelvis* above and the lesser or *minor pelvis* below. Its component parts are the promontory of the sacrum, the anterior border of the ala of the sacrum, the iliopectineal line, the pubic crest, and the upper end of the symphysis pubis (*fig. 18.11*). The *iliopubic eminence* (iliopectineal eminence) marks the site of union of ilium and pubis.

The pubic part of the iliopectineal line, called the **pecten pubis**, gives attachment in its whole length to the fascia covering the pectineus, a muscle of the thigh, and to the lacunar ligament and conjoint tendon. A strong fibrocartilaginous band of periosteum, the *pectineal ligament* (Cooper's ligament), through which a surgeon's needle and thread can get a good grip, overlies the pecten pubis and raises the line into a ridge.

The Pelvis Major (False or Greater Pelvis) (*fig. 18.11*) is formed on each side by the ala of the sacrum and the concave, fan-shaped iliac fossa. The handle of the fan immediately above the acetabulum is a broad groove for the iliopsoas tendon.

The **Pelvis Minor** (*true* or *lesser pelvis*) is formed by the sacrum and coccyx, inner surface of the ischium and pubis, and a small part of the ilium (*fig. 18.12*). The main features posteriorly are the *greater* and *lesser sciatic notches* and the *ischial spine* which separates them. The oval *obturator foramen* separates the pubis and ischium.

Foramina in the Walls of the Pelvis Minor

The posterior wall is perforated by the 4 *anterior* or *pelvic sacral foramina*. Two ligaments, the *sacrotuberous* and *sacrospinous*, so unite the posterior wall to the side wall as to leave 2 gaps, the *greater* and *lesser sciatic foramina* (see *figs. 18.18, 18.21*). The large *obturator foramen* is closed by the obturator membrane, except above, where a gap transmits the obturator nerve and vessels and may be the site of a rare type of hernia.

Muscles of the Walls of the Pelvis Minor

The *Obturator Internus* arises by fleshy fibers from almost the entire inner surface of the side wall of the pelvis minor (*fig. 18.13*). It leaves the pelvis through the lesser sciatic foramen to be inserted into the greater trochanter of the femur.

The *Piriformis* arises by fleshy fibers from the anterior surface of the sacrum and enters the gluteal region through

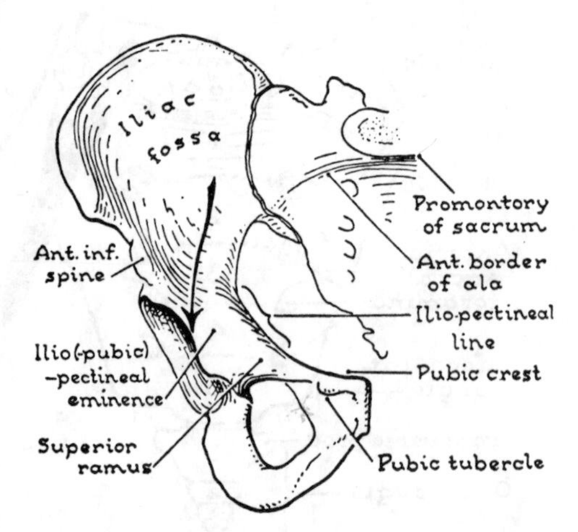

Figure 18.11. The pelvis major (false pelvis) and the pelvic brim. *Arrow* denotes the pathway of the iliopsoas tendon.

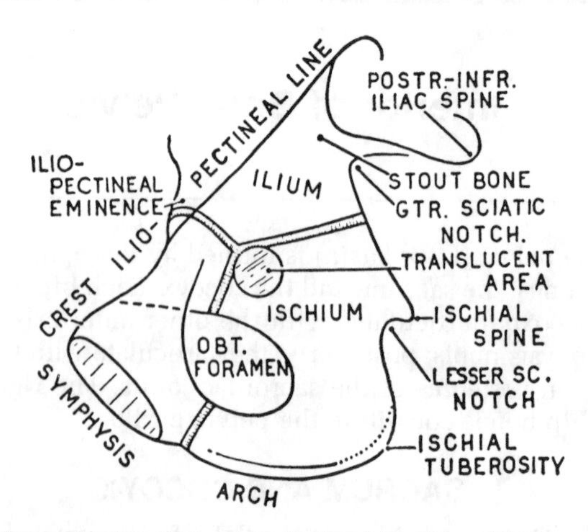

Figure 18.12. Lateral wall of pelvis minor. (*Eminence: iliopectineal = iliopubic.*)

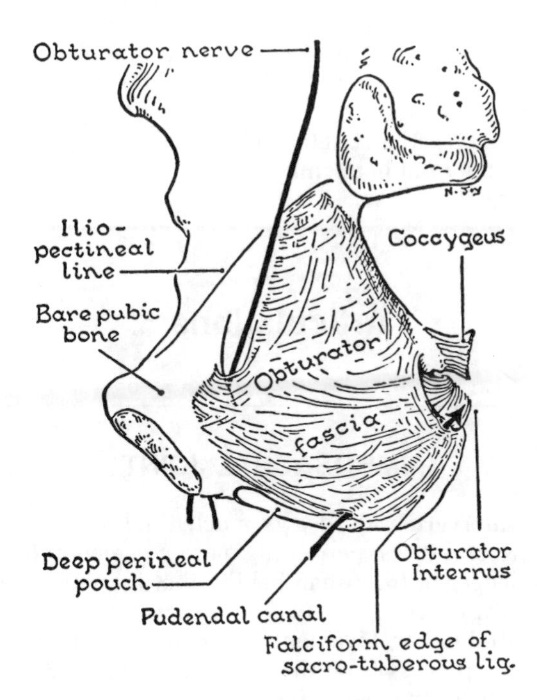

Figure 18.13. The obturator internus fascia forms a pocket. In it runs the pudendal canal and from it arises the levator ani (*cf. fig. 21.5*).

the greater sciatic foramen (see *fig. 18.36*) and is inserted into the greater trochanter (*fig. 24.5*). For the *obturator fascia*, see Figure 18.13.

THE PELVIC DIAPHRAGM

The *Levator Ani* (*and Coccygeus*) of opposite sides stretch across the pelvis like a hammock, separating the pelvic cavity from the perineum. This pelvic "floor" is perforated by the urethra (*fig. 18.5*) and the anal canal and, in the female, by the vagina also. More specifically, the levator ani has a free anterior border, which is separated from its fellow by the urethra, vagina, and anal canal. The structure of this muscle is quite complex (Ayoub), suggesting a complexity of function.

The pelvic diaphragm (levator ani) arises anteriorly from the body of the pubis, behind from the spine of the ischium, and between these 2 points from an arched thickening of the obturator fascia, called the *tendinous arch* (*fig. 18.6*).

Functions

The pelvic diaphragm supports the pelvic viscera. The integrity of the pelvic floor depends especially on the levator ani, including the part known as the puborectal sling, since these hold forward the lower part of the rectum, which in turn helps to support the bladder, prostate and seminal vesicles in the male and (which is more important) the bladder and vagina in the female (*fig. 18.14*).

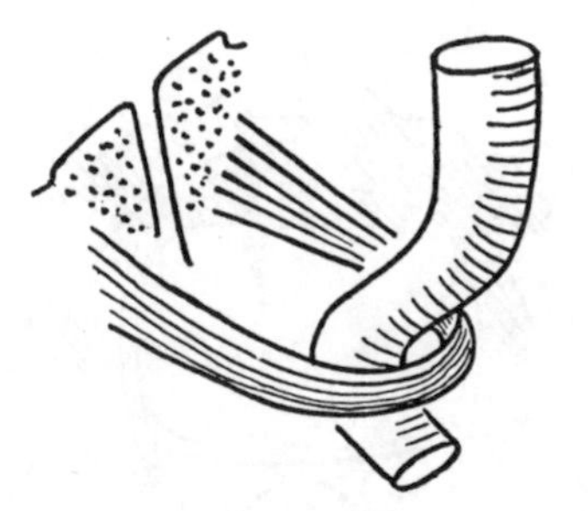

Figure 18.14. "The puborectal sling."

The puborectal sling keeps the anorectal angle closed, but during defecation it relaxes and allows the anorectal junction to straighten, while other fibers draw the anal canal over the feces that are being expelled.

Nerve Supply

Branches of S (2), 3, and 4 supply the muscle on its pelvic surface; twigs of the perineal nerve (S 2, 3, and 4) supply it on its perineal surface.

The Coccygeus (*figs. 18.13, 18.15*) stretches like a fan from the ischial spine to the sacrum and coccyx, lining the sacrospinous ligament.

The *Pubococcygeus*, which is the thickest and most important part of the levator ani, runs downward and backward from the pubis and meets its fellow (a) in the perineal body in front of the anus and (b) in the anococcygeal raphe behind the anus; between the two (c) it is carried down by the rectum and anal canal, which perforate it and with which it blends. On the pelvic surface of the muscle many fibers pass backward from pubis to coccyx; others meet those of the opposite side in the angle between the rectum and anal canal and so form a U-shaped puborectal sling that maintains this angle (*fig. 18.14*). The different parts of the diaphragm, though overlapping somewhat, blend to form a single sheet.

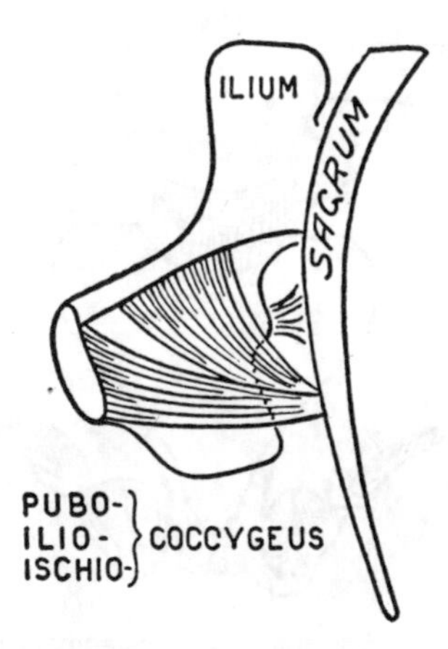

Figure 18.15. Pelvic diaphragm of monkey. (After Keith.)

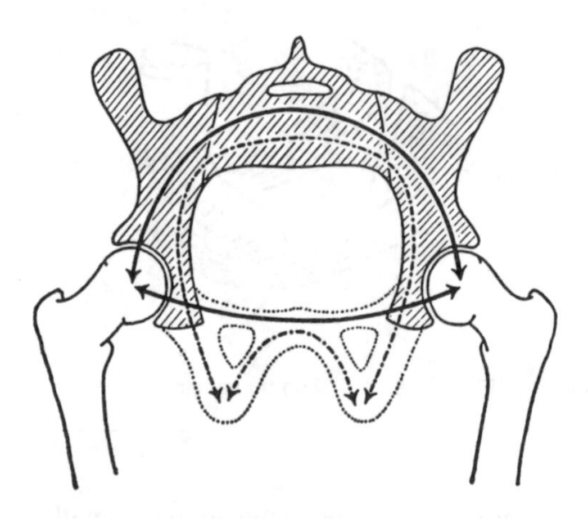

Figure 18.16. The mechanism of the pelvis. *Solid lines* = the standing arch and its tie beam or counter arch which, like the clavicle, resists compressive forces; *broken lines* = the sitting arch and its tie beam or counter arch which resists spreading forces. (After Braus.)

ORIENTATION OF THE PELVIS

In the "neutral" position of the pelvis, the anterior superior iliac spines and the top of the symphysis are in a vertical plane. This is approximated by applying the pelvis to a wall.

In the female, the anterior superior spines overstep the vertical slightly. In compensation for the forward tilting of the pelvis in the female, the lumbar curvature of the spine is increased. Hence, the lower part of the back is more concave in the female and the buttocks more prominent.

MECHANISM OF THE PELVIS (*fig. 18.16*)

The weight of the body superimposed on the 5th lumbar vertebra is transferred to the base of the sacrum, thence to the upper 3 pieces of the sacrum, across the sacroiliac joints to the ilia, and thence (1) when standing, to the acetabula and so to the femora (plural of femur), or (2) when sitting down, to the ischial tuberosities. Along these lines, the bony parts are thickened. In the standing posture, the acetabula and the side walls of the pelvis tend to be forced together, but the pubic bones, acting as struts, prevent this from happening (*fig. 18.17*).

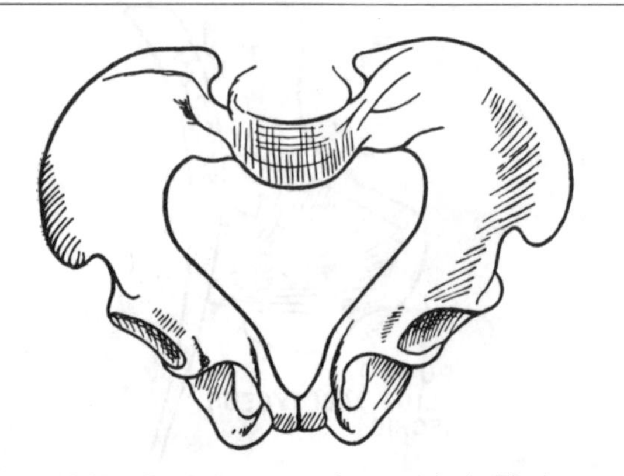

Figure 18.17. Pelvis from case of osteomalacia. The femora have driven in the softened bones, narrowing the pelvis.

Articulations

LUMBOSACRAL JOINT

The Intervertebral Disc is much thicker than other intervertebral discs, permitting more movement. It is so much deeper in front than behind that it contributes to the lumbar curve.

The Iliolumbar Ligament suspends the 5th lumbar transverse process from the iliac crest (*fig. 18.21*). The iliolumbar ligaments are important, for they limit axial rotation of the 5th vertebra on the sacrum, and they assist the articular processes in preventing forward gliding of the 5th vertebra on the sacrum.

JOINTS OF THE PELVIS

The joints of the pelvis are: (1) sacrococcygeal joint; (2) symphysis pubis; and (3) right and left sacroiliac joints.

Sacrococcygeal Joint

This is an atypical intervertebral joint. Movement backward of the coccyx takes place on defecation and on parturition.

The bodies of the last sacral and 1st coccygeal vertebrae are united by an intervertebral disc, and the transverse processes and the cornua are united by ligamentous bands. And a very tough membrane, which is a downward prolongation of the supraspinous and interspinous ligaments, closes the sacral canal posteriorly and extends to the posterior surface of the coccyx.

Symphysis Pubis

Here, as between the bodies of 2 vertebrae, the opposed bony surfaces are coated with hyaline cartilage and are united by fibrocartilage. Dense anterior decussating fibers and a strong arcuate ligament (inferior pubic ligament) unite the pubic bones.

Aging

The surface and margins of the pubic symphysis undergo small changes especially between the 20th and 40th years, which serve as reliable criteria of age (Todd). This symphysis normally never fuses.

SACROILIAC JOINT

The Bony Surfaces of this synovial joint are:

(1) *The Internal Surface of the Ilium* behind the iliac fossa. This part is bounded above by the posterior one-third of the iliac crest, below by the greater sciatic notch, and behind by the posterior superior and posterior inferior iliac spines and the slight notch between them (*fig. 18.18*).

It is subdivided into 2 parts: a lower, the *auricular surface*, and an upper, the *tuberosity*. The auricular or ear-shaped part articulates with the sacrum, is covered with cartilage, and is traversed by a longitudinal sinuous ridge. The iliopectineal line begins at its most anterior part. The tuberosity is rough and tubercular for the numerous short fibers of the strong interosseous sacroiliac ligament.

(2) *The Sacrum* possesses the counterpart: thus, on the side of the pars lateralis there is an *auricular surface* with a sinuous furrow. Posterosuperior to the auricular surface is the *sacral tuberosity* (*fig. 18.10*).

Structural Requirements

Weight on the sacrum by the superimposed vertebral column will tend to cause its upper end to rotate forward and its lower end with the coccyx to rotate backward (*fig. 18.19*). Ligaments are so disposed as to resist this tendency. Furthermore, the articular surfaces of the sacrum are farther apart in front than behind so the sacrum behaves not as a keystone, but as the reverse of a keystone, and tends therefore to sink forward into the pelvis. As it does so, the posterior ligaments become taut and draw the ilia closer together, with the result that the interlocking ridge and furrow engage more closely. Here is an **automatic locking device** (*fig. 18.20*).

Ligaments Resisting Forward Rotation of Upper End of Sacrum are in 3 pairs.

1. *The Interosseous Sacroiliac Ligament* (*fig. 18.20*), which is a ligament of great strength, unites the iliac and sacral tuberosities. It lies posterosuperior to the joint.
2. *The Posterior Sacroiliac Ligaments*, very strong bands, unite the transverse tubercles of the sacrum to the posterior superior iliac spine (*fig. 18.21*).
3. *The Iliolumbar Ligament* (p. 213) (*fig. 18.21*).

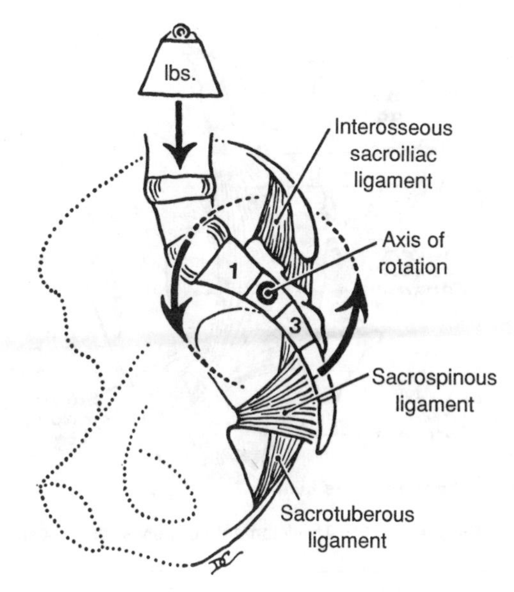

Figure 18.19. Ligaments resisting rotation of sacrum.

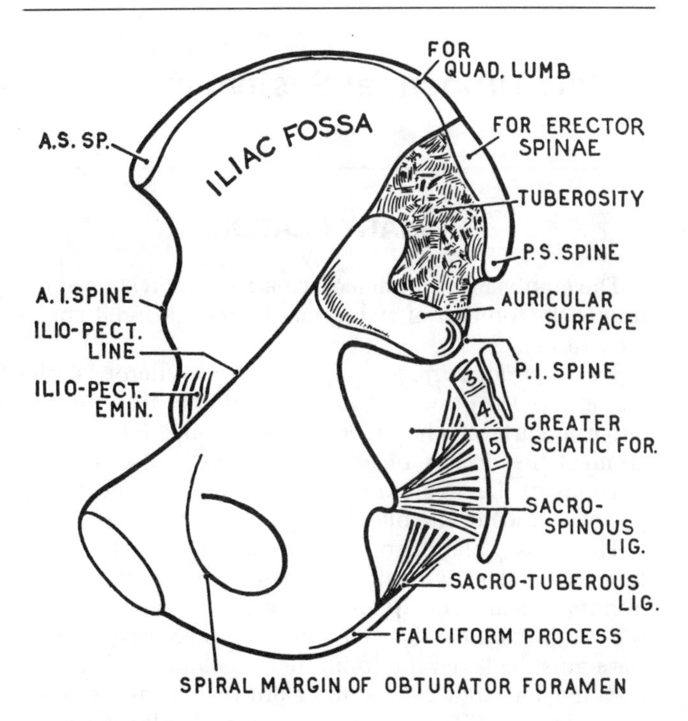

Figure 18.18. Inner surface of hip bone. *AS* = anterior superior; *AI* = anterior inferior; *PS* = posterior superior; *PI* = posterior inferior.

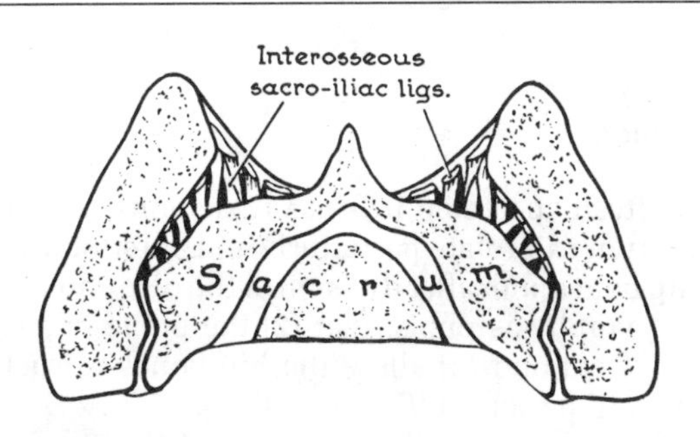

Figure 18.20. The sacroiliac joint on transverse section. Note the locking device.

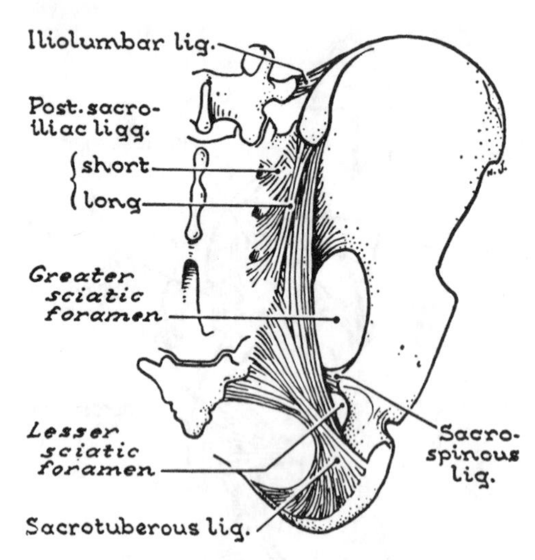

Figure 18.21. The ligaments of pelvis (posterior).

Ligaments Resisting Backward Rotation of Lower End of Sacrum include the *sacrotuberous* and *sacrospinous ligaments* (*figs. 18.19* and *18.21*). The one passes from the tuberosity of the ischium, the other from the spine of the ischium, to the available parts of the side of the sacrum and coccyx, i.e., to the lateral border of the sacrum and coccyx below the articular facet.

The *sacrotuberous ligament* is a broad band from the medial part of the tuber ischii to the sides of the sacrum and coccyx and adjacent dorsal surfaces.

[The lower end of the ligament curves forward as a falciform process of which one edge creates a line (or ridge) on the ischium, while the other is continuous with the obturator internus fascia. The lateral part of the ligament arches from ischial tuberosity to posterior iliac spines (inferior and superior); its only function is to afford origin to the gluteus maximus.]

The *sacrospinous ligament* is a triangular sheet, coextensive with the coccygeus; it extends from the ischial spine to the side and dorsum of the sacrum and coccyx.

The sacrotuberous and the sacrospinous ligaments convert the greater and lesser sciatic notches into foramina (*fig. 18.21*). They are no doubt responsible for the curvature of the sacrum, which is a characteristic of man, who alone walks erect.

Movements

In a fresh specimen it is easily demonstrated (1) that the sacrum can rotate posteriorly and anteriorly between the hip bones and, after division of the disc and ligaments of the symphysis pubis, that (2) the pubic bones easily spread 1 cm and (3) allow the hip bones to rock fairly freely on the sacrum (Frigerio *et al.*).

During pregnancy, the ligaments of the pelvis are relaxed and movements are more free.

Anterior Relations (fig. 18.37)

The lumbosacral trunk, the superior gluteal artery and the 1st sacral nerve cross the pelvic surface of the capsule.

Surface Anatomy

The posterior superior iliac spine, which lies at the level of the 2nd sacral vertebra, is readily palpated. It lies in a prominent skin dimple. The joint lies lateral to this.

Development and Variations. The sacroiliac joint and the symphysis pubis do not develop, like other joints, as clefts in a continuous rod of condensed mesenchyme but by the coming into apposition of the ilium and sacrum posteriorly and of the pubic bones of opposite sides anteriorly. The auricular surface of the sacrum is usually covered with hyaline cartilage, that of the ilium with fibrocartilage, and between them there is a joint cavity which is present before birth.

The anterior part of the joint cavity is closed by a strong fibrous capsule called the *anterior sacroiliac ligament.* After middle life, particularly in males, this ligament may ossify, forming a crust of bone that prevents movement of the joint (synostosis).

Accessory articular facets, 1–2 cm in diameter, are commonly found between the opposed tuberosities of the sacrum and ilium.

Male Urogenital System in Pelvis

URINARY BLADDER

The empty and contracted urinary bladder, shaped not unlike the forepart of a ship has 4 surfaces and 4 angles (*fig. 18.22*).

The *superior surface* and 1 cm of the inferoposterior surface are the only parts covered with peritoneum. The superior surface is bounded by the rounded borders that connect the ureters of each other and to the urachus. Portions of the sigmoid colon and ileum lie on the superior surface of the bladder.

The urinary bladder owes its shape largely to the structures in contact with it. At birth the bladder is an abdominal organ (*fig. 18.23*), fusiform in shape, lying in the extraperitoneal tissue of the anterior abdominal wall. Care must be taken not to injure it in abdominal surgery in young children. By about the 6th year, a child's pelvis has enlarged sufficiently to allow the bladder to sink to its permanent pelvic position and assume a tetrahedral shape.

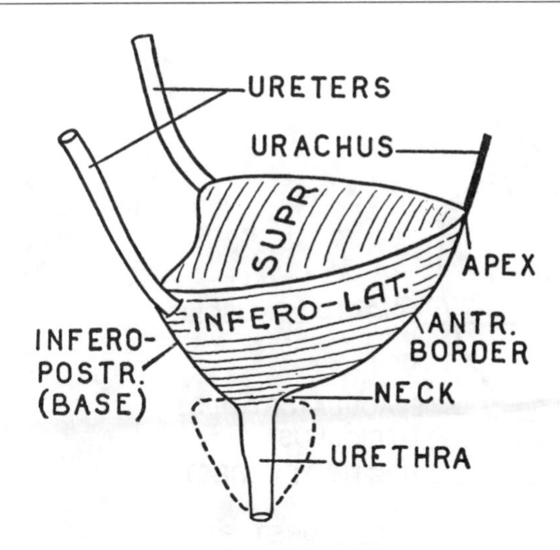

Figure 18.22. The 4 surfaces, 4 angles, and 4 ducts of the urinary bladder.

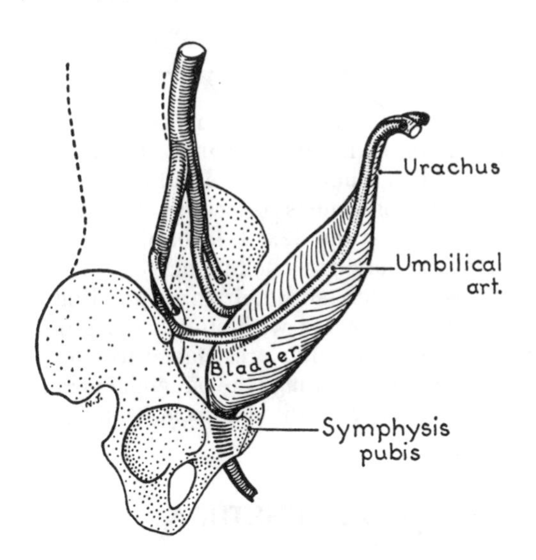

Figure 18.23. The bladder is abdominal at birth.

Bed of Bladder. The bed or mould in which the bladder lies is formed on each side by bare pubic bone, the obturator internus, and the levator ani, and behind by the rectum, hence the inferolateral and inferoposterior surfaces (*figs. 18.5* and *18.24*).

The entire organ is enveloped in areolar tissue, the *vesical fascia*. A dense plexus of veins, the *vesical plexus*, lies in this fascia on the side of the bladder, clings to the bladder, and separates it from the retropubic space.

The base or *inferoposterior surface* of the bladder is separated from the rectum by the seminal vesicles and ampullated ends of the deferent ducts, which are enclosed between the 2 layers of retrovesical fascia; above these the peritoneum of the rectovesical pouch separates the upper 1 cm of the bladder from the rectum (*figs. 18.1* and *18.7*).

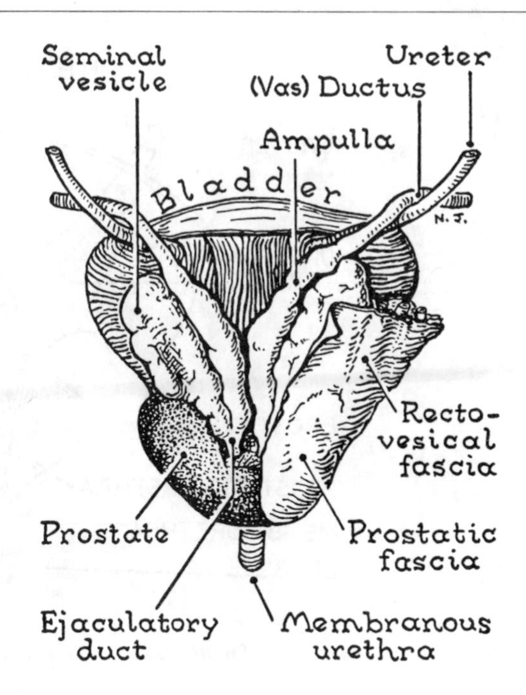

Figure 18.24. Posterior relations of the bladder.

PELVIC PORTION OF THE URETER

Each ureter enters the pelvis at the bifurcation of the common iliac artery and descends immediately in front of the internal iliac artery (*figs. 18.4* and *18.8*). It lies subperitoneally, is crossed by the vas deferens near the bladder, and it crosses the apex of the seminal vesicle (*fig. 18.24*).

The Interior of the Bladder

The mucous membrane lies in folds, except over the trigone where it is smooth. The *Trigone* (Gr. = triangle) is a relatively fixed part of the base of the bladder (*fig 18.25*). The 2 ureters and the urethra open at its angles about 2 cm apart. The internal urethral orifice lies at the lowest point of the bladder and, therefore, is situated advantageously for drainage.

The orifice of each ureter is guarded by a flap of mucous membrane and is collapsed. A ridge, the *interureteric fold*, connects the 2 ureters at the upper border of the trigone.

An elevation, the *uvula*, overlying the middle lobe of the prostate, is sometimes to be seen at the apex of the trigone behind the internal urethral orifice.

The *Trigonal Muscle* is a submucous sheet distinct from the muscular wall proper. It is continuous with the muscular wall of the ureters above; its apex descends in the posterior wall of the urethra to the utricle. This trigonal area of the bladder and urethra is derived from the Wolffian or mesonephric ducts. It is supplied by the hypogastric plexus (L1, 2 sympathetic nerves).

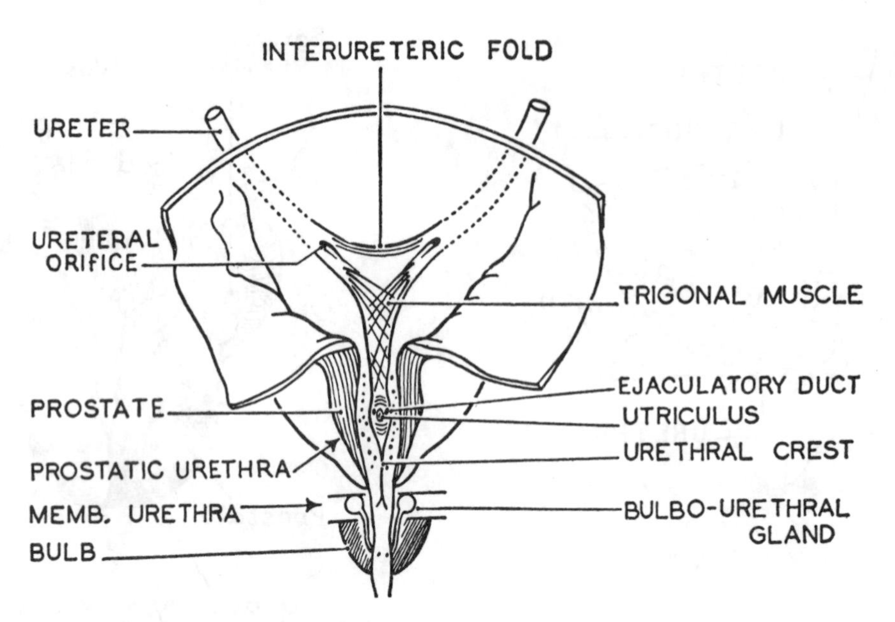

Figure 18.25. The trigone of the bladder and the prostatic urethra.

Though 500 cc can be injected into the bladder without causing discomfort, urine is usually voided when little more than half this volume has accumulated, the exact volume depending largely on acquired habit.

As the bladder fills, the trigone enlarges but little, so the 3 orifices are only slightly displaced, but the rest of the bladder stretches. When the fundus reaches the level of the umbilicus, the peritoneum is stripped from the anterior abdominal wall for several centimeters above the symphysis pubis.

Vessels and Nerves of the Bladder

Arteries

The arteries are the superior and inferior vesical. *Veins* pass to the vesical plexus and thence to the internal iliac veins. For *Lymph Vessels*, see p. 240.

Nerves

Pelvic splanchnic nerves and hypogastric plexus are the nerves. The sympathetic postganglionic neurons from the hypogastric plexus are vasomotor to the vessels of the bladder and also motor to the trigone region and urethral smooth muscle around the internal urethral orifice. Contraction of these fibers during ejaculation prevents the reflux of semen into the bladder and directs it into the prostatic and penile urethra. The parasympathetic neurons are preganglionic and synapse in ganglia in the wall of the bladder. The intravesical postganglionic neurons innervate the smooth muscle of the bladder wall (except the trigone). This **detrusor muscle** contracts and expels the urine through the internal urethral orifice into the urethra. Visceral afferent (sensory) fibers take a corresponding course to the parasympathetic fibers. Expansion of the bladder induces a sensation of fullness and an urge to void. A bladder reflex is stimulated when an adequate volume of urine is collected within the bladder. This reflex occurs at the S2,3,4, level of the spinal cord and involves the pelvic splanchnic nerves that are motor to the detrusor muscle and their corresponding sensory nerves that are associated with stretch receptors in the bladder wall. The voluntary resistance to override the "full bladder reflex" is through the pudendal nerve (S2,3,4 also) to the sphincter urethrae muscle in the urogenital diaphragm.

MALE URETHRA

The male urethra is a fibroelastic tube, 20 cm long (*figs. 18.25* and *18.26*). It is divided by the superior and the inferior fascia of the urogenital diaphragm into 3 parts: *prostatic*, *membranous*, and *spongy*. The part above the superior fascia traverses the prostate; the part between the 2 fasciae has no covering; the part beyond the inferior fascia or perineal membrane traverses the corpus spongiosum penis.

The *Prostatic Urethra* (*fig. 18.25*) is the widest and most dilatable part of the urethra. It runs almost vertically and is the length of the prostate—2 cm or more. In transverse section it is crescentic, owing to a ridge on the posterior wall called the *urethral crest*. The gutter on each side of the crest is called the *prostatic sinus*. The urethral crest rises to a summit, the *colliculus*. Opening on it is a diverticulum within the prostate, the *prostatic utricle* (= little uterus), up which a probe can be passed about 1 cm. Onto the lips of the utricle open the pinpoint orifices of the right and left *ejaculatory duct*. Into the prostatic sinus, several dozen *prostatic ducts* open (*fig. 18.25*).

The *Membranous Urethra* passes through the middle

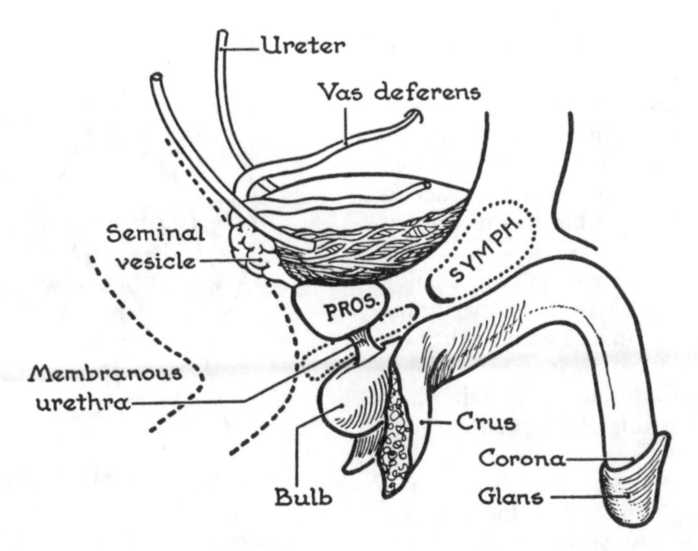

Figure 18.26. The lower parts of the genital and urinary tracts and their relations.

of the urogenital diaphragm and its 2 fasciae and is only 1 cm long. Behind it on each side lies a bulbourethral gland. The *Sphincter Urethrae* part of the diaphragm forms its sphincter of voluntary muscle, supplied by the perineal branch of the pudendal nerve. Like the sphincter ani externus, also supplied by the pudendal nerve, the sphincter urethrae constantly maintains its muscular tone and relaxes only during micturition.

The Spongy or Penile Urethra traverses the bulb, body, and glans of the corpus spongiosum penis. It enters the bulb on its upper surface and ends near the lower part of the apex of the glans at the external urethral orifice. Its lumen is dilated both in the bulb and in the glans. The dilatation in the glans is known as the navicular (terminal) fossa (*fig. 21.14*).

The external urethral orifice—like the opening of most ducts—is the narrowest part of the urethra. A small kidney stone may stick at the vesical orifice of the ureter or, having passed, it may yet stick at the external urethral orifice.

Palpation

A catheter passed into the bladder can be palpated in the *spongy urethra* from the under surface of the penis, in the *membranous urethra* from the perineum, and in the *prostatic urethra* by a finger in the rectum.

Nerve Supply (*fig. 20.1*; p. 239).

Structure of the Urinary Tract

Although the membranous and spongy urethrae are lined with stratified columnar epithelium, the navicular fossa is lined with stratified squamous epithelium, as are other orifices opening onto the skin surface, such as the nostrils, mouth, mammary ducts, sebaceous ducts (but not the sweat ducts), anal canal, and vagina.

The *Tunica Propria* is thick and loose in the ureter and bladder, and it falls into folds when they are empty.

Muscle. The urinary tract has both an inner longitudinal and an outer circular coat of muscle. The lower part of the ureter and the bladder have an additional outer longitudinal coat; the layers are interwoven in the bladder. Fibers from the longitudinal and circular muscle coats loop from behind forward around the front and sides of the upper part of the prostatic urethra. These slings are optimistically called the *sphincter vesicae*.

The smooth muscle fibers in the urethra are not numerous.

The intramural part of the ureter has longitudinal fibers, but no circular ones. Of these, some join their fellow in the interureteric fold; others radiate into the trigonal muscle (*fig. 18.25*).

Urethral Glands. Mucous cells, singly and in clusters lining recesses (lacunae urethrales), as well as in branching outpouchings (urethral glands) in the tunica propria, occur especially on the dorsum of the anterior two-thirds of the spongy urethra. They may extend to the neck of the bladder.

Common Direction. The mouths of all ducts and tubes opening into the urethra open forward, in the direction in which the urine flows. Hence, urine does not enter them during micturition, but fine instruments passed in the reverse direction may do so. *Catheterization of the urethra for bladder management is a common medical procedure.*

Internal Parts of the Male Genital System

PROSTATE (*fig. 18.26*)

The prostate surrounds the urethra between the bladder and the urogenital (UG) diaphragm. It occupies the same bed as the bladder, and it resembles the bladder in shape, its surfaces being superior, inferolateral, and inferoposterior. Of these, the superior blends with the bladder; the 2 inferolateral ones lie on the levatores ani; the inferoposterior lies on the rectum. Its apex, from which the urethra emerges, abuts against the superior fascia of the diaphragm between the anterior borders of the levatores ani (*fig. 18.5*).

The prostate is encased in a strong envelope of pelvic fascia, which is continuous below with the superior fascia of the UG diaphragm and which is anchored to the pubes by the puboprostatic (-vesical) ligaments. The *prostatic fascia* is distinct from the outermost part of the gland proper, which is called the *capsule of the prostate* (*figs. 18.7* and *18.24*).

On each side of the prostate the capsule is separated from the fascia by a venous plexus, called the *prostatic plexus of veins*. This plexus receives the deep dorsal vein of the penis in front, communicates with the vesical plexus superiorly, and drains into the internal iliac veins posteriorly. The prostatic plexus of veins also anastomose with the vertebrae plexus of veins. Prostatic cancer frequently spreads to the central nervous system (CNS), lower lumbar vertebrae, pelvic bones, and femora.

The posterior part of the prostatic fascia forms a broad strong sheet, called by the surgeon the *fascia of Denonvilliers*. It is easily separated from the loose rectal fascia behind it.

The prostate can be palpated per rectum and exposed readily from the perineum (p. 249).

The urethra passes vertically through the forepart of the prostate; the prostatic utricle projects into the posterior part, and the ejaculatory ducts pierce its upper surface and open onto the lips of the utricle.

Structure. The prostate is a modified portion of the urethral wall. In composition it is one-half glandular, one-fourth involuntary muscle, and one-fourth fibrous tissue. The glands are arranged in 3 concentric groups (*fig. 18.27*). The innermost or *mucosal* are short and simple. They open all around the urethra above the level of the colliculus. "All hypertrophies of the prostate arise from these mucosal or suburethral glands" (Young). The intermediate or *submucosal* glands open into the prostatic sinus at the level of the colliculus. The outermost or *prostatic glands proper* are long and branching. They envelop the other 2 groups, except in front where those of opposite sides are joined by a nonglandular isthmus. Their ducts open into the prostatic sinus.

The Ductus (Vas) Deferens is about 45 cm long. It has a course in the scrotum, in the inguinal canal, on the side wall of the pelvis, and between the bladder and rectum, where it rounds the ureter. Medial to the seminal vesicle, it is ampullated and thin-walled (*fig. 18.28*).

Each **Seminal Vesicle** is a tortuous, branching diverticulum developed from the ampullated end of the ductus deferens (*fig. 18.29*). Below, it joins the ductus deferens to form the ejaculatory duct.

The vesicles, prostate, and bulbourethral glands each add a distinctive secretion to the semen, but seminal vesicles are absent in dogs and other Carnivora.

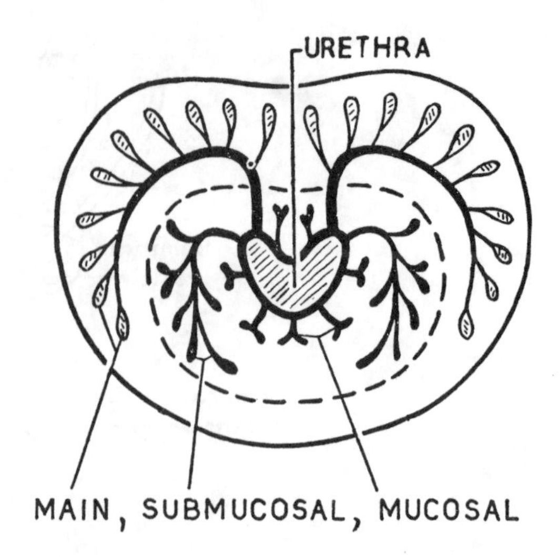

Figure 18.27. Prostate in transverse section showing 3 concentric groups of tubules (schematic).

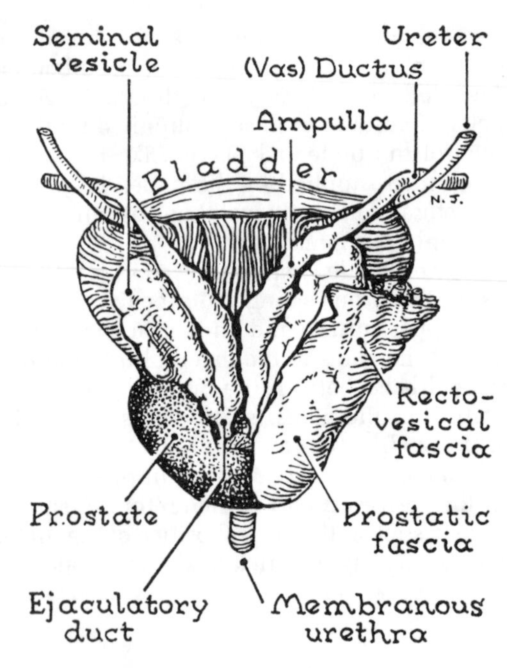

Figure 18.28. The seminal vesicles and the deferent ducts (vasa deferentia).

The **Ejaculatory Duct (paired)** is common to the ductus deferens and the seminal vesicle. It has the diameter of the lead of a pencil. It is easily torn away from the prostate, the upper half of which it pierces obliquely to open beside the prostatic utricle.

Vessels. The *inferior vesical arteries*, with a little help from the middle *rectal* arteries are the **vessels** that supply all the structures in between the bladder and the rectum, namely prostate, seminal vesicles, ampullae of the de-

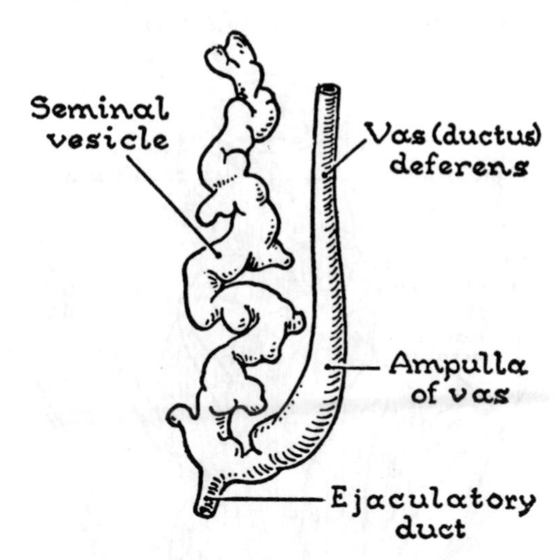

Figure 18.29. The seminal vesicle unraveled.

ferent ducts, and the ends of the ureters. The *deferent artery* (artery of the vas deferens) itself springs from an inferior vesical artery.

Nerves (see p. 238).

Rectum and Anal Canal

The rectum and anal canal are the terminal parts of the large intestine. The rectum begins where the sigmoid colon ceases to have a mesentery, which is in front of the 3rd piece of the sacrum, and it is about 12 cm long. The rectum continues the curvature of the sacrum and coccyx downward and forward for 4 cm beyond the coccyx and there, at the apex of the prostate, makes a right-angled bend and becomes the anal canal (*fig. 18.26*). Maintenance of this angle by the levator ani is an important element in the voluntary retention of bowel contents (Shafik) (*fig. 18.14*). The anal canal passes downward and backward for about 3 cm or more to its orifice, the *anus*.

RECTUM (*fig. 18.30*)

The upper part of the rectum is covered with peritoneum in front and at the sides; the middle part is covered in front only; and the lower part, which is commonly dilated to form the *ampulla of the rectum*, lies below the level of the rectovesical pouch and, therefore, is devoid of peritoneal covering. The distance from the skin surface to the peritoneal cavity (i.e., to the rectovesical pouch), measured in front of the anal canal and rectum, is 8–10 cm (*fig. 18.30*).

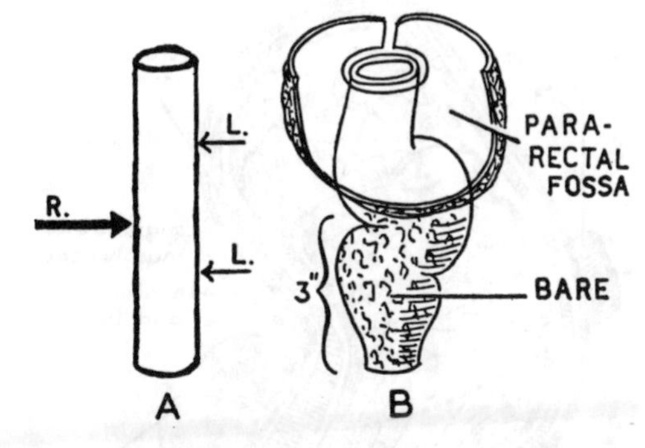

Figure 18.30. The peritoneal coverings and lateral flexures of the rectum, front view. (3″ = 8 cm.)

The rectum in man is not straight, as its name would imply; it has the anterposterior curve and bend just described; it has also 3 lateral curvatures. Thus, at the bottom of the rectovesical pouch, the right wall is indented with the result that a transverse fold or shelf projects into the lumen, and similar but less pronounced indentations from the left occur about 2–3 cm below and above this. The shelves within the gut are the *plicae transversales* (*fig. 18.30*). They consist of mucous membrane and circular muscle. Their form is maintained by the prolongations of the 3 teniae coli, which spread out to form a continuous outer longitudinal muscle coat, thick ventrally and dorsally but thin at the sides (*see fig. 18.33*).

The plicae must be avoided by tubes and instruments introduced into the rectum.

ANAL CANAL (*figs. 18.31 and 18.32*)

The anal canal extends from the anorectal junction, which lies above the level of the puborectal sling (*fig. 18.31*) and the sphincters, to the anus below. It remains closed, except during defecation. Accordingly, it is surrounded throughout by 2 sphincters: (1) an *external* (voluntary), described on p. 244 and (2) an *internal* (involuntary), which is merely the much thickened lower end of the circular muscle coat of the gut (*fig. 18.31*). Of the two, the voluntary sphincter is the more important (Hardy; Stephens; Varma).

The upper part of the anal canal possesses 5–10 permanent longitudinal folds of mucous membrane, the *anal columns* whose lower ends are united by semilunar folds, the *anal valves* (*fig. 18.32*).

Between the 2 sphincters, the longitudinal muscle coat of the rectum, reinforced by fibers of the levator ani and its fasciae, descends and splits into a number of *fibroelastic septa* (*fig. 18.31*).

One of these, the *anal intermuscular septum* (*fig. 18.32*), passes below the internal sphincter to reach the mucous membrane of the canal, whereas the others swing through the subcutaneous external sphincter to reach the skin

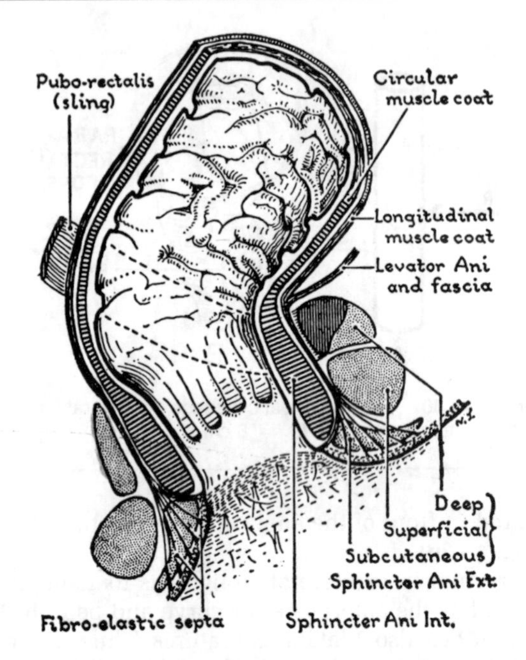

Figure 18.31. The sphincters of the anus and the puborectal sling.

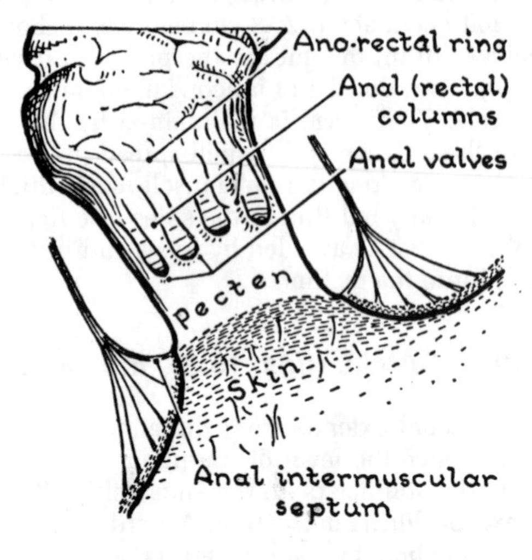

Figure 18.32. Interior of anal canal.

(*fig. 18.31*). Not only are there differences in detail between the 2 sexes, but even different parts of the circumference of the sphincter in individuals show significant differences in detail (Oh and Kark).

Above the anal valves the canal has columnar epithelium containing goblet cells. The 1 cm of canal below the valves is smooth, is lined with stratified squamous epithelium, and is known to the surgeon as the "pecten." The remainder is lined with skin.

Anteroposterior versus side-to-side flattening. When the rectum is empty, its anterior and posterior walls are in apposition; when the anal canal is empty, its lateral walls are in apposition. The same is true of the vagina and the vaginal orifice, and of the male spongy urethra and the external urethral orifice.

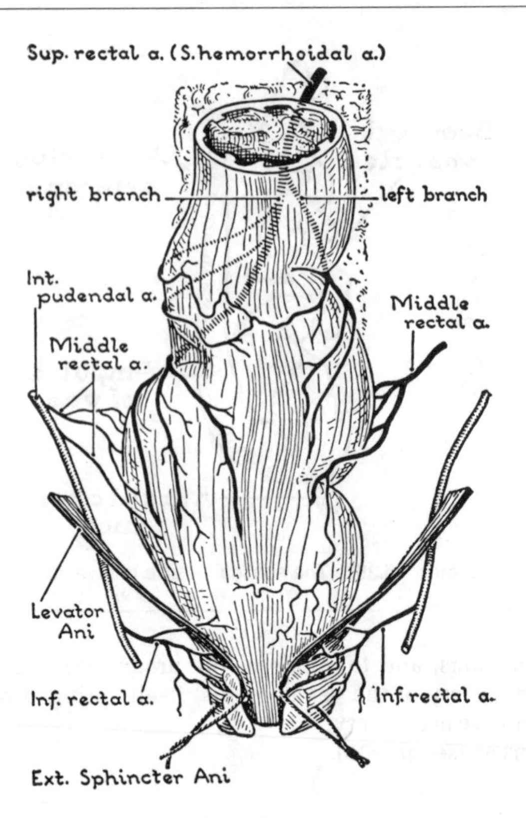

Figure 18.33. The arterial supply of the rectum and anal canal, front view.

Arteries (*fig. 18.33*)

The superior, middle, and inferior rectal (*hemorrhoidal*) arteries supply the rectum and anal canal, making 5 arteries in all—for the superior artery is unpaired.

The *Superior Rectal Artery* is the continuation of the inferior mesenteric artery. It divides into a right and a left branch.

Terminal branches, given off irregularly from these, pierce the muscle coat, ramify in the submucosa, and descend in the anal columns (*fig. 18.32*). They anastomose with each other and with the middle and inferior rectal arteries lower down.

Other Arteries (e.g., median sacral, inferior gluteal, internal pudendal) send twigs to the lower part of the rectum.

Veins

The superior, middle, and inferior rectal or hemorrhoidal veins accompany their arteries and drain corresponding parts of the rectum and anal canal. The superior vein becomes the inferior mesenteric vein and, therefore, belongs to the portal system. It is valveless. The middle and inferior veins are paired and belong to the caval system. They have valves.

The *Superior Vein* begins in the anal columns. It has

extensive mucous and submucous plexuses, and it receives branches from the circumrectal tissues.

The *Middle Rectal Vein* is a much more important vessel than the corresponding artery. It drains the rectum above the internal sphincter and communicates both submucously and in the rectal fascia with the inferior rectal vein, and it makes free anastomoses submucously with the superior rectal vein. Its branches communicate with the prostatic (vaginal and uterine) plexus. It is the chief link between the portal and caval systems—a point of clinical significance—and it ends in the internal iliac vein. (See Portal Anastomoses, Chapter 13, *fig. 13.9*.)

The *Inferior Vein* begins in the sphincter ani externus, the walls of the anal canal, and the subcutaneous veins begin at the anal margin. These subcutaneous veins are continuous above with the submucous veins of the anal canal and rectum.

Lymph Vessels. See p. 000.

Nerves. See p. 000.

Relations. Posterior to the rectum (*figs. 18.34* and *18.37*): 3 pieces of the sacrum, the coccyx, and sacrospinous ligament; piriformis, coccygeus, and levator ani; the median sacral artery and vein; the sympathetic trunks and ganglion impar, parts of the last 3 sacral and the coccygeal nerves, the lateral sacral vessels and the rectal fascia, which is usually thick and fatty posteriorly and encloses branches of the superior rectal vessels and lymph nodes.

Laterally are: the pararectal fossa containing sigmoid colon or ileum, the middle rectal vessels, the pelvic splanchnic nerves and the inferior hypogastric plexus, the levator ani, and the ischiorectal fossa.

Anteriorly are: the bladder separated by rectovesical pouch, seminal vesicles, and ampullae of the deferent ducts; below these is the prostate (*fig. 18.26*).

At the anorectal junction are: the puborectal sling, the base of the UG diaphragm, and the membranous urethra.

Anterior to the anal canal are the perineal body and the bulb of the uretha.

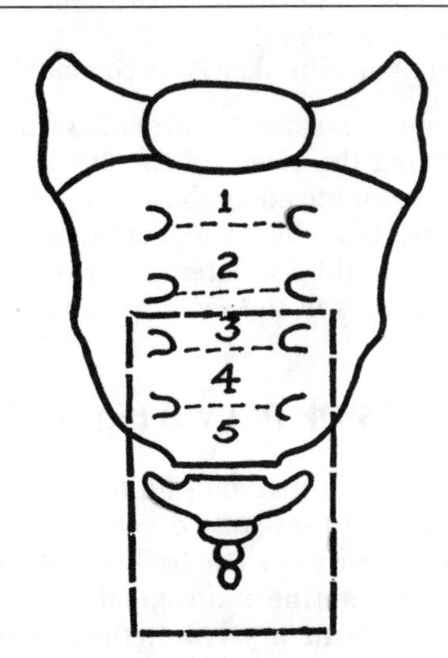

Figure 18.34. The parts of the sacrum and coccyx covered by the rectum (see *fig. 18.37*).

Vessels and Nerves of Pelvis

ARTERIES OF THE PELVIS

The arteries of the pelvis are:

internal iliac artery— paired.
median sacral artery unpaired (previously
superior rectal artery described).

The Internal Iliac (Hypogastric) Artery (*fig. 18.35*) takes origin from the common iliac artery one-third of the way, i.e., 5 cm, along the line joining the aortic bifurcation to the midinguinal point. It descends subperitoneally and crosses medial to the external iliac vein, the psoas, and the obturator nerve. In front of it runs the ureter; behind it lies its vein.

Branches of Internal Iliac Artery arise erratically. They may be grouped thus:

1. Visceral branches
   Umbilical *Superior vesical*
   Inferior vesical *Middle rectal*
   (also, *Uterine* and *Vaginal* in the female)
2. Branches to the limb and perineum
   Superior gluteal *Inferior gluteal*
   Obturator *Internal pudendal*
3. Somatic segmental branches
   Iliolumbar *Lateral sacral*

1. *The Visceral Branches.* (a) The *umbilical artery* (*fig. 12.32*; p. 151) is obliterated as far back as the branches to the bladder, i.e., the *superior vesical arteries.* (b) The

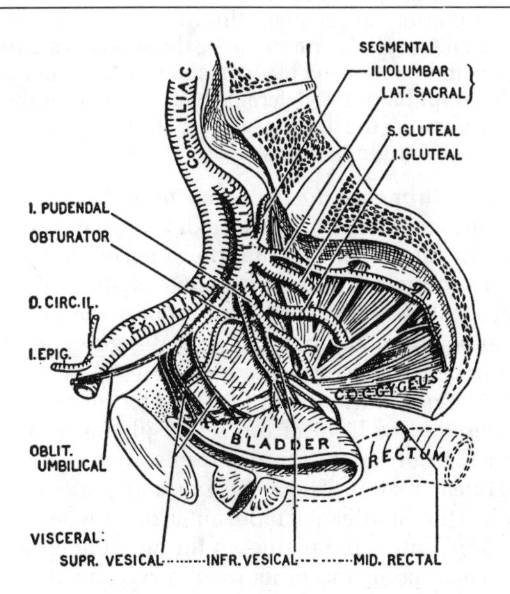

Figure 18.35. Internal iliac artery and branches.

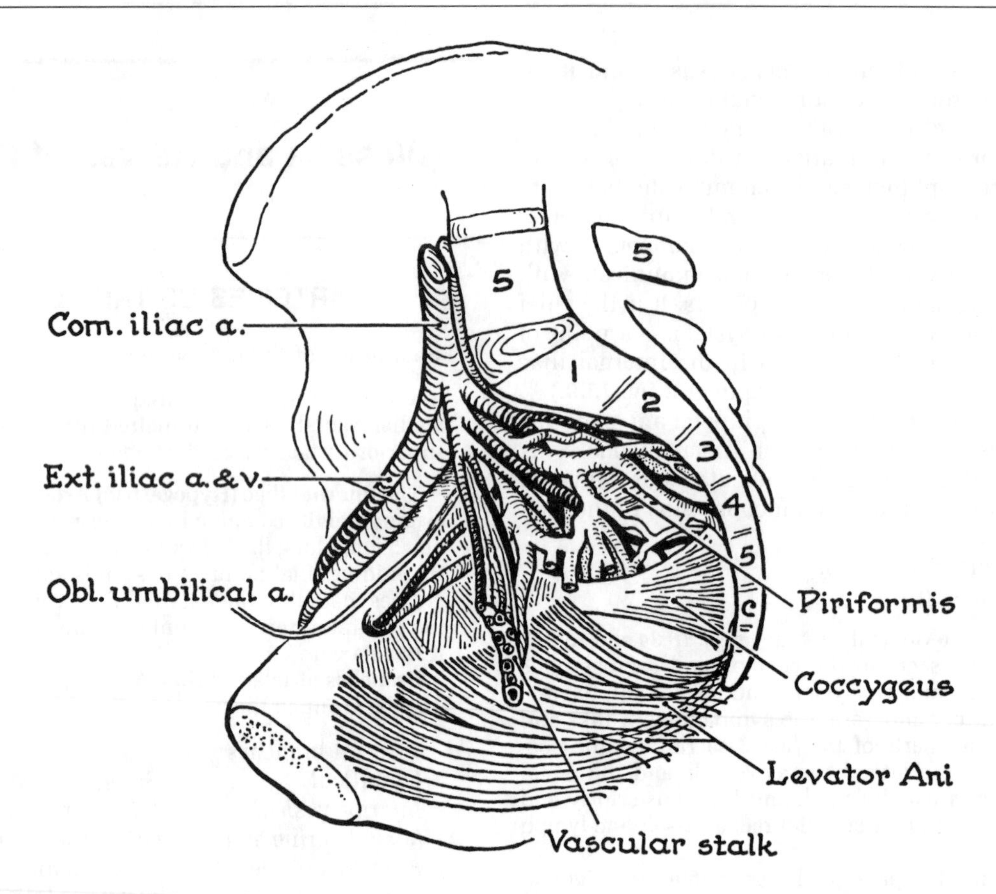

Figure 18.36. Blood vessels of pelvis. The vascular stalk of pelvic viscera limits retropubic space posteriorly.

inferior vesical arteries and (c) the *middle rectal artery* run in the leash of veins that form the posterior limit of the retropubic space.

Of the two inferior vesical arteries one (the *vesiculo-deferential artery*) is constant in arising from the umbilical artery at its origin (Braithwaite). The other has a variable origin. On each side, these arteries supply the side and base of the bladder, the prostate, seminal vesicle, ampulla of the deferent duct, and end of the ureter. One gives off the deferent artery. The *middle rectal artery* varies in origin and in size (*fig. 18.33*).

2. *The Branches to the Limb and Perineum.* These branches leave the pelvis through the greater sciatic and obturator foramina. Thus: (a) the *superior gluteal artery*, which is much the largest branch of the internal iliac artery, makes a U-shaped turn around the angle of the greater sciatic notch into the gluteal region. Its vein and the superior gluteal nerve accompany it (*fig. 24.14*). (b) The *inferior gluteal artery* and (c) the *internal pudendal artery* descend in front of the sacral plexus and pass between the borders of the piriformis and coccygeus into the gluteal region. There the internal pudendal artery, which is the smaller and more anterior, crosses the ischial spine and enters the perineum through the lesser sciatic foramen in company of its own nerve and the nerve of the obturator internus. (d) The *obturator artery* runs forward on the side wall of the pelvis to the obturator foramen between its nerve and vein.

The obturator and inferior epigastric arteries both supply branches to the posterior aspect of the pubis. These pubic branches anastomose, and the anastomotic channel is commonly (33%) so large that the obturator artery derives its blood from the epigastric artery. This is known as an *accessory obturator artery*.

The obturator vein likewise is commonly "abnormal."

3. *The Somatic Segmental Branches.* (a) The *iliolumbar artery*, being the artery of the 5th lumbar segment, ascends in front of the ala of the sacrum and divides into iliac and lumbar branches. The iliac branches anastomose in the iliac fossa. (b) The *lateral sacral artery* descends in front of the roots of the sacral plexus.

VEINS OF PELVIS (*fig. 18.36*)

The pelvic viscera may be said to lie within a basket woven out of large thin-walled veins among which the arteries thread their way. The basket is divided into vesical, prostatic (or uterine and vaginal), and rectal venous plexuses, which drain largely into the internal iliac vein, but partly via the superior rectal (hemorrhoidal) vein into the inferior mesenteric vein and so to the portal vein.

Table 18.1. Ventral Rami or Lumbar Plexus, Sacral Plexus and Coccygeal Plexus

Ventral Rami of	Th	L.	S.	C.
Lumbar plexus	12	1, 2, 3, 4		
Sacral plexus		4, 5,	1, 2, 3, 4	
Coccygeal plexus			4, 5,	1

The *middle rectal vein* is relatively large. It emerges from the lower part of the side of the rectum and passes to the internal iliac vein. It anastomoses with the superior and inferior rectal veins and also with the other plexuses of pelvic veins, thereby taking a prominent part in the portacaval anastomoses (p. 163).

NERVES OF PELVIS

Sacral and Coccygeal Nerves

The lumbar, sacral, and coccygeal plexuses are derived from the ventral rami of spinal nerves T12–Co1, as in *Table 18.1*.

Note that rami L4 and S4 both contribute to 2 plexuses.

The ventral rami of the sacral and coccygeal nerves, like all other ventral rami, receive gray sympathetic rami communicantes, which they conduct to the blood vessels, sweat and sebaceous glands, and arrectores pilorum in their territory. No white sympathetic rami communicantes arise caudal to L3, but parasympathetic fibers, called the *pelvic splanchnic nerves*, arise from the ventral rami of S(2), 3, and 4 (*fig. 18.38*).

Sacral Plexus (*figs. 18.37 and 18.38*)

Of the **six roots** of the sacral plexus, L4 and L5 cross and groove the ala of the sacrum as the **lumbosacral trunk**, which joins S1 to pass in front of the piriformis muscle to form the sacral plexus. The plexus has many collateral branches, and it ends as 2 terminal branches, the *sciatic* and *pudendal nerves*.

Arteries Piercing the Plexus (fig. 18.37). Four branches of the internal iliac artery pierce the plexus: the *iliolumbar*, the *superior gluteal*, the *inferior gluteal*, and the *internal pudendal arteries*.

The *Branches of the Sacral Plexus (Table 18.2)* may be grouped thus:

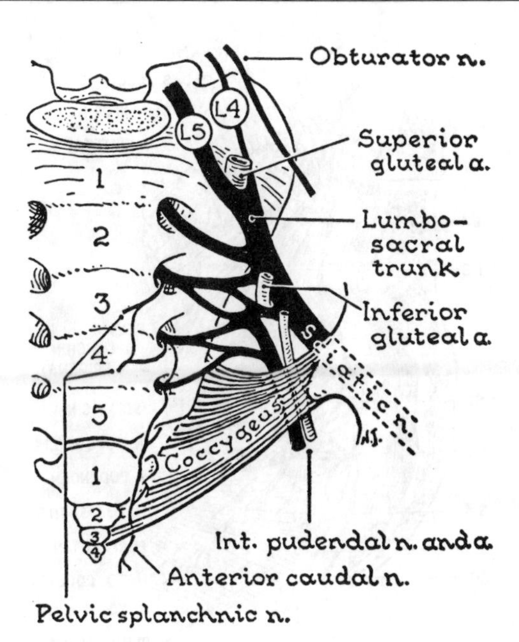

Figure 18.37. The (left) sacral plexus.

1. Branches from roots of plexus:
   a. *Muscular* (to piriformis, levator ani, and coccygeus)
   b. The *Pelvic splanchnic nerves.*
2. Branches that pass through the greater sciatic foramen:
   (a) *Two terminal* (sciatic and pudendal)
   (b) *Five Collateral.*
3. Branches that emulate the coccygeal plexus in piercing the structures attached to the side of the coccyx in order to become cutaneous:
   a. *Perforating cutaneous of S2, 3*
   b. *Perineal of S4.*

1. *The branches arising from the roots of the sacral plexus* are short twigs to piriformis (S. 1, 2), long branches to levator ani and coccygeus (S3, 4), and the pelvic splanchnic nerves (S2, 3, 4).

The *pelvic splanchnic nerves*, like the pudendal nerve, arise from S2, 3, 4, but are "mixed" parasympathetic nerves. They supply the involuntary muscles derived from the cloaca that cause the involuntary sphincters guarding the rectum and bladder (sphincter ani internus and sphincter vesicae) to relax while the organs contract, and they are sensory to them. They also cause dilatation of the arteries of the erectile tissue of the penis or clitoris and thereby produce erection, hence the alternative name *nervi erigentes*.

Table 18.2. Branches of the Sacral Plexus

	Collateral		
Terminal	From the Back	From the Front	From Front and Back
1. Sciatic 2. Pudendal	3. Superior Gluteal 4. Inferior Gluteal	5. I. to Quadratus Femoris 6. N. to Obturator Internus	7. Posterior cutaneous of the thigh

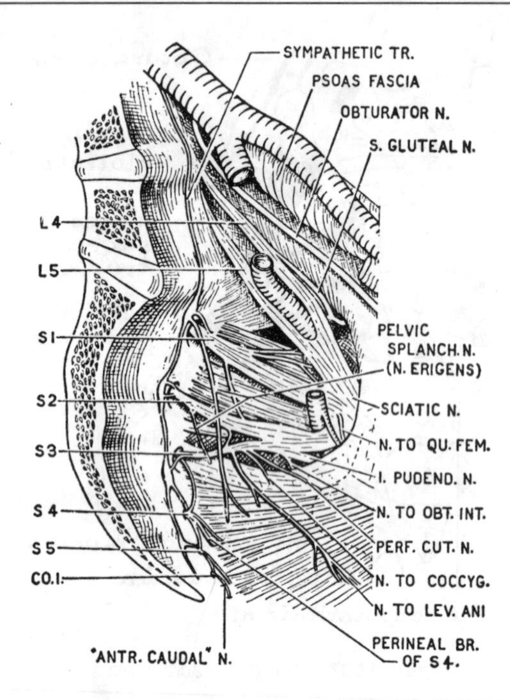

Figure 18.38. The (left) sacral and coccygeal plexuses.

2. *The branches that pass through the greater sciatic foramen* enter the gluteal region, and they should be studied with it. They are 7 in number, 2 being terminal and 5 collateral. The exact segments of origin of the collateral branches are unimportant.

The *sciatic nerve* is the largest nerve in the body, and it forms the greatest part of the sacral plexus. It arises by 5 roots (L4, 5, S1, 2, and 3) and leaves the pelvis between the piriformis and the ischial border of the greater sciatic notch.

The *pudendal nerve* (S2, 3, 4), already seen in the perineum, escapes between the piriformis and coccygeus just medial to the sciatic nerve.

The *superior gluteal nerve* (L4, 5, S1) and the *inferior gluteal nerve* (L5, S1, 2) arise from the back of the plexus. The nerves to *quadratus femoris* (L4, 5, S1) and *obturator internus* (L5, S1, 2) arise from the front of the plexus; the former descends in front of the sciatic nerve, the latter on the medial side of the sciatic nerve. The *posterior cutaneous nerve of the thigh* (S1, 2, 3) arises from the back and front of the plexus.

The superior gluteal nerve and its companion artery and vein escape from the pelvis above the piriformis at the angle of the greater sciatic foramen. All other structures passing through the foramen pass below the piriformis. The inferior gluteal nerve and the posterior cutaneous nerve of the thigh escape below the piriformis and behind the sciatic nerve.

3. *The 2 branches that emulate the terminal branch of the coccygeal plexus* are the *perforating cutaneous branch* of S2 and 3 and the *perineal branch* of S4. These descend in front of the coccygeus and then pass through it, reaching the skin.

The Coccygeal Plexus is formed by the ventral rami of S5 and Co1, which emerge from the sacral hiatus onto the dorsum of the sacrum and supply skin as the *anococcygeal nerve*, which corresponds to the ventral *caudal nerve* of tailed mammals. A similarly formed nerve, derived from the dorsal rami of S4, 5, and C1, corresponds to the dorsal *caudal nerve* of tailed mammals. In man, they supply a circular area of skin around the coccyx—the area on which one sits down (*fig. 24.22*).

Pelvic Sympathetic Nerves

Postganglionic sympathetic nerves enter the pelvis by three primary routes. (1) Sympathetic neurons on the iliac blood vessels carry vasomotor fibers to the lower limb for blood pressure and temperature control as well as sympathetic fibers to sweat glands and **arrectores pilorum** in the dermatomes of the skin for temperature control. (2) Postganglionic sympathetic fibers in the pelvic fascia derived from the superior and inferior hypogastric plexuses. These fibers are vasomotor to the vessels of the pelvic viscera and are also motor to some smooth muscles mainly in the male reproductive system. The vas deferens, ejaculatory ducts and prostatic urethra smooth muscle are heavily innervated by sympathetic fibers. Spinal cord injuries that interrupt the descending neuronal pathways to L1 cord level frequently present major problems with the function of the male reproductive tract. (3) The lumbar plexus and sacral plexus receive gray rami communicantes and carry postganglionic sympathetic fibers to the dermatomes of the perineum and lower limb through their cutaneous nerves. These sympathetic fibers are mainly concerned with body temperature regulation.

Clinical Mini-Problems

1. **What fibrocartilaginous structure on the pelvic brim is used by surgeons as an "anchor" in suturing the conjoint tendon to the pelvic brim in hernial repair?**
2. **Which components of the spinal nerves enter the pelvis on the anterior surface of the sacrum?**
3. **Which nerves carry the afferent and efferent limbs of the bladder reflex from the detrusor muscle?**
4. **Where does the male urethra bend sharply in its course and require some added caution when catheterizing the bladder?**
5. **Which route of spread would permit a primary cancer in the prostate to metastasize (form a new focus) in the brain?**
6. **Which spinal cord segments are being tested during a rectal examination when the external anal sphincter reflexively contracts during digital palpation?**

(Answers to these questions can be found on p. 585.)

19

Female Pelvis

Clinical Case 19.1

Patient Millie D. This 56-year-old shopkeeper was diagnosed by your tutor as having an adenocarcinoma of the endometrium, and he scheduled her for a hysterectomy at which you assist. The surgeon makes a suprapubic incision in the anterior abdominal wall and reflects the peritoneum from the superior surface of the bladder and fundus and body of the uterus. The uterine tubes are tied off at the isthmus and the round ligament of the ovary severed from its inferior pole. The ascending branches of the uterine arteries supplying the body and fundus of the uterus are ligated. A supravaginal hysterectomy is done by amputating the uterus at its isthmus, 1.5–2.0 cm above the cervix. The ovaries and distal ends of the ligated uterine tubes are left in the broad ligament. Lymph nodes from the common iliac nodes are biopsied and found to be free of carcinoma. *Follow-up:* The patient spends 5 days in the hospital after surgery and is sent home for 3 weeks of convalescence. She returns to work and remains free of any pelvic disease or discomfort. She is on a regular postsurgical follow-up schedule, with an examination every 6 months at a local cancer clinic.

Female Pelvis in Median Section

Because of its obstetrical importance you should be able to sketch the skeletal parts, then insert the viscera and, lastly, put on the peritoneal covering.

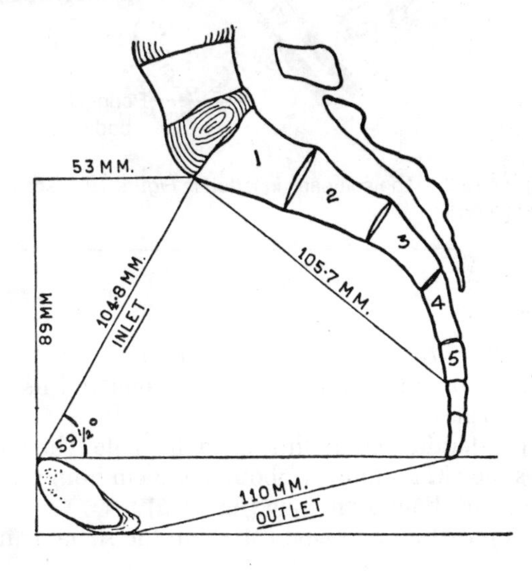

Figure 19.1. Outlines of female pelvis in median section (drawn to scale).

1. The symphysis pubis is 3–4 cm (or about 1½ inches) long. Its posterior surface is flat and faces more upward than backward (*fig. 19.1*).

2. Erect a vertical line 9 cm (or about 3½ inches) high on the symphysis. Draw 2 parallel horizontal lines, one at the level of the top of the symphysis, the other at the top of the vertical line.

3. The sacral promontory lies on the upper horizontal line, 3 times the length of the symphysis (about 11 cm) from the upper end of the symphysis.

4. The tip of the coccyx lies on the lower horizontal line, 3 times the length of the symphysis more or less (about 11 cm) from the lower end of the symphysis.

5. The upper part of the sacrum is straight and parallel to the symphysis; its lower part and the coccyx are often much curved.

6. A line joining the upper end of the symphysis to the promontory of the sacrum indicates the plane of the pelvic brim, which bounds the pelvic inlet. The inlet makes an angle of 60° with the horizontal.

7. A line joining the lower end of the symphysis to the tip of the coccyx is the plane of the outlet of the pelvis. It makes an angle of 15° with the horizontal.

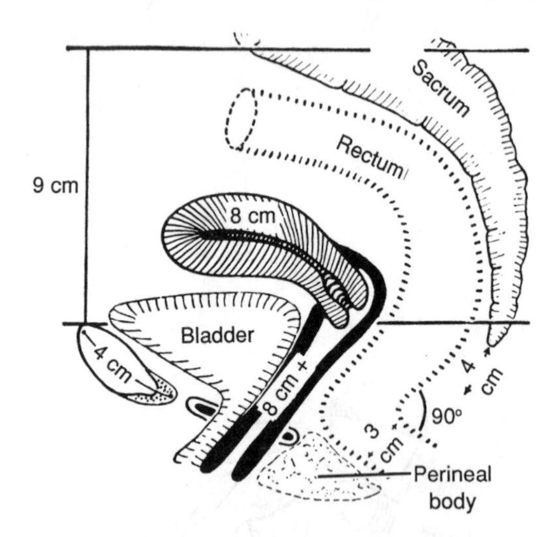

Figure 19.2. The soft parts inserted in *Figure 19.1* (scale in centimeters).

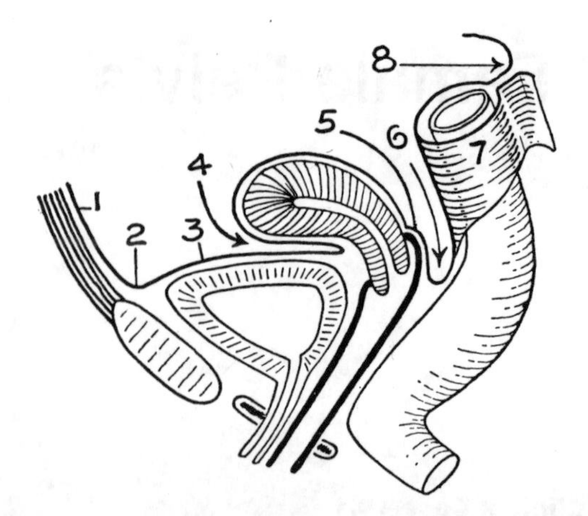

Figure 19.3. The peritoneum of the female pelvis in median section (see text and *fig. 18.1*). *1*, anterior abdominal wall; *2*, back of the pubis; *3*, superior surface of the empty bladder; *4*, posterior surface for 1 cm; *5*, upper ends of the seminal vesicles on each side of the median plane; *6*, across the bottom of the rectovesical pouch; *7*, the rectum; *8*, the mesentery of the sigmoid colon.

The soft parts are now to be inserted:

8. Insert the urogenital (UG) diaphragm and its upper and lower fasciae.

9. The bladder and urethra, as in the male: The urethra pierces the UG diaphragm about 2 or 3 cm from the symphysis. It is about 4 cm long (*fig. 19.2*).

10. The rectum and anal canal are the same as in the male.

The uterus and vagina occupy the positions taken by the seminal vesicles, ampullae of the deferent ducts, and prostate in the male.

11. The vagina is more than 9–10 cm long. It lies nearly parallel to the pelvic brim. Its anterior wall is in structural continuity with the urethra in its lower part, it is in contact with the bladder in its middle part, and it is pierced by the cervix of the uterus in its upper part. Like the urethra, it pierces the UG diaphragm.

12. The posterior wall of the vagina is separated from the rectum by the rectal and the vaginal layer of fascia (rectovaginal septum) and from the anal canal by the triangular mass of fibromuscular tissue called the *perineal body*, much larger than in the male.

13. The uterus is 8 cm long. It lies nearly at right angles to the pelvic brim and to the vagina. Its upper 5 cm (or fundus and body) are 2–3 cm thick; its lower 3 cm (or cervix) is 2 cm thick. The external orifice of the uterus lies on (or below) the lower of the 2 parallel horizontal lines. The fundus does not reach to the pelvic inlet. The body and cervix meet at a single angle, so the uterus is said to be anteflexed.

14. The anterior and posterior fornices (fornix L. = an arch) of the vagina are the shallow depression in front of the anterior lip of the cervix and the deeper depression behind the posterior lip. The depression runs like a gutter all round the cervix, so there are also a right and a left lateral fornix, but they are not seen in sagittal section.

15. The peritoneum passes from the symphysis onto the upper surface of the bladder and from the third piece of sacrum onto the rectum, as in the male. It falls at least 2 cm short of the anterior fornix of the vagina in the vesicouterine pouch (item 4 in *fig 19.3*), but it clothes 1 cm or more of the posterior fornix (item 5 in *fig. 19.3*).

Female Pelvis Seen from Above (*fig. 19.4*)

The bladder and rectum have the same disposition as in the male. However: (1) the uterus and vagina situated medianly replace the seminal vesicles, ampullae of the deferent ducts, and prostate of the male; (2) the ovaries and their ducts (the uterine tubes) lie on either side, with the *broad ligaments of the uterus*, which pass from the lateral margins of the uterus to the side walls of the pelvis.

The anterior subdivision is the *vesicouterine pouch*; the posterior is the *rectouterine pouch*. The uterus rests on the empty bladder and the shallow vesicouterine pouch is empty; the deeper *rectouterine pouch* (item 6 in *fig. 19.3*) is occupied by sigmoid (pelvic) colon and ileum.

The ovary is attached to the back of the broad ligament. The uterine tube occupies the upper border of the broad ligament, except at its lateral end (*fig. 19.4*). There the broad ligament is continued as a fold (the *suspensory ligament of the ovary*) across the external iliac vessels. A cord of fibromuscular tissue curves backward and upward from the junction of the body and cervix of the

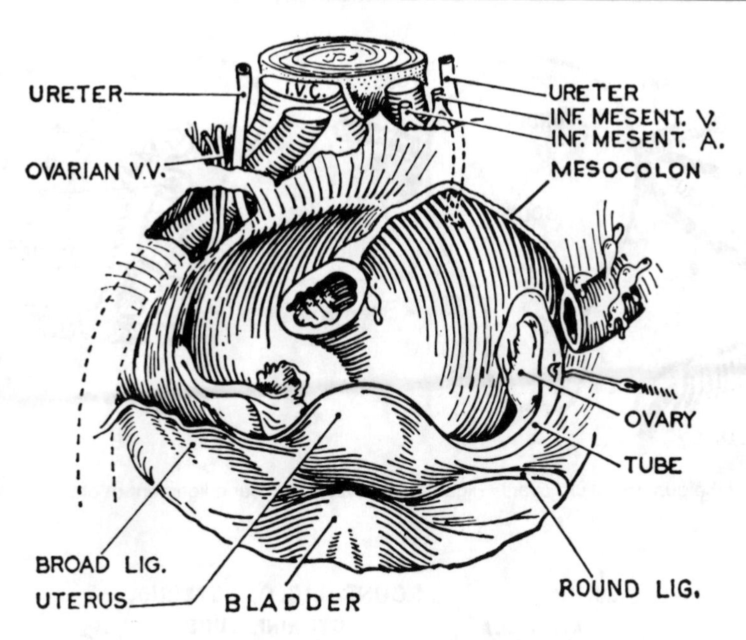

Figure 19.4. Female pelvis (from above).

uterus, past the rectum, to the sacrum. It helps to suspend the uterus and is called the *uterosacral ligament*. The overlying crescentic fold of peritoneum is the *rectouterine fold* (items 5 and 6 in *fig. 19.3*).

A round fibromuscular cord, the *ligament of the ovary*, stands out in relief from the back of the broad ligament. It joins the lower pole of the ovary to the angle between the side of the uterus and the uterine tube. A similar cord, the *round ligament of the uterus* (ligamentum teres uteri), stands out from the front of the broad ligament. It passes from the angle between uterus and tube across the pelvic brim to the deep inguinal ring (*fig. 19.5*).

Female Internal Genital Organs

These comprise: ovaries, uterine tubes, uterus, and vagina.

OVARY (*fig. 19.6*)

The ovary has the shape of a testis but is only half the size. It is covered with cuboidal epithelium. The pits and scars on its surface mark the sites of the absorbed corpora lutea (sing., *corpus luteum* L. = yellow body). These result monthly from the shedding of ova. The *mesovarium* attaches its anterior border to the back of the broad ligament. The *suspensory ligament of the ovary* suspends the tubal (upper) pole of the ovary from the external iliac vessels and conducts the ovarian artery, vein, lymph vessels, and nerves to the ovary. The uterine (lower) pole of the ovary is attached to the lateral margin of the uterus by the *ligament of the ovary*.

The typical position of the ovary is on the side wall of the pelvis, behind the broad ligament, and in the angle between the external iliac vein and the ureter. It is hidden by the uterine tube which falls over it medially. But its position is variable.

UTERINE TUBE (OF FALLOPIUS) (*fig. 19.6*)

The uterus is 5 cm across at its widest part, which is where the tubes enter it. Each tube is about 11 cm long. The spread, therefore, of the uterus and its 2 tubes greatly exceeds the diameter of the pelvic inlet (13 cm). Evidently the tubes cannot lie in a straight line. Each tube occupies the free edge of the broad ligament, its "mesentery," and runs upward, laterally, and backward to the side wall of the pelvis and there curves backward over the ovary.

Each tube has the following parts: *infundibulum, abdominal orifice, ampulla, isthmus, uterine part*, and *uterine orifice*.

The *abdominal orifice*, 2–3 mm in diameter, lies at the bottom of a funnel-shaped depression, the *infundibulum*. Fringes or fimbriae lined with ciliated epithelium, project from the infundibulum and encourage ova, when shed, into the tube. One fimbria is attached to the ovary.

The main part of the tube is long, irregular, and dilated and is called the *ampulla*. The succeeding shorter part is straight and narrow and is called the *isthmus*. The *uterine part* passes through the uterine wall, which is at least 1 cm thick. Four longitudinal folds, bearing sec-

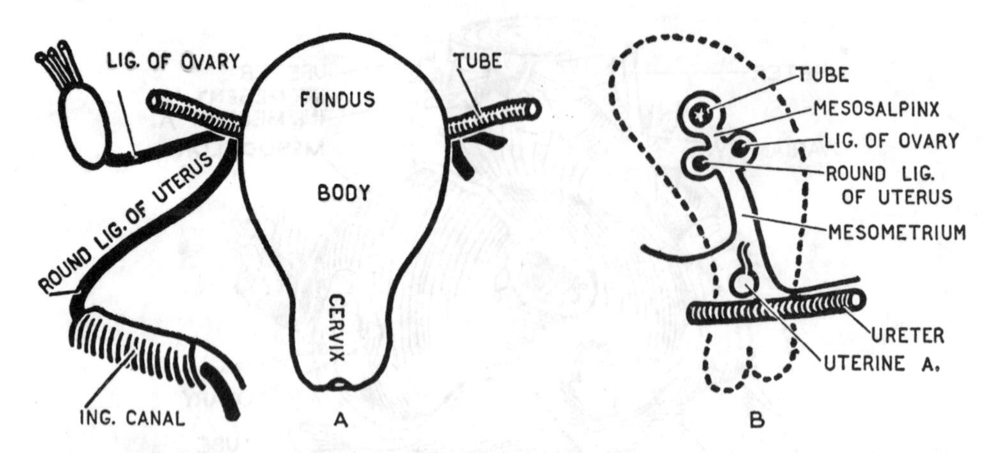

Figure 19.5. *A*, derivatives of the gubernaculum ovarii—ligament of ovary and round ligament of uterus; *B*, sagittal section through broad ligament of uterus.

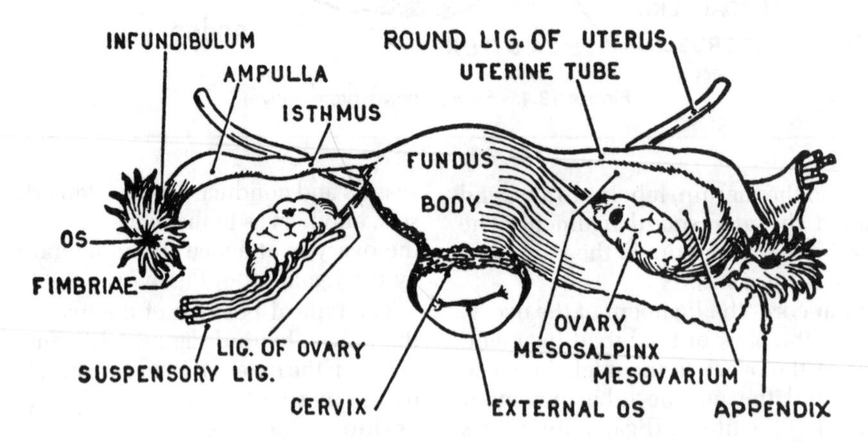

Figure 19.6. The uterus and appendages (from behind).

ondary folds, project into the lumen of the tube, which is lined with columnar ciliated epithelium.

UTERUS

The uterus is thick-walled and muscular. It is pear-shaped, 8 cm long, 5 cm at its widest part, and 2 or 3 cm at its thickest part. It is flattened in front where it rests on the bladder and is convex behind. The uterine tubes enter it at its widest part. The broad ligaments are attached to its margins; the ligament of the ovary and the round ligament of the uterus are attached just below the tube (*figs 19.5* and *19.7*).

The uterus is divisible into 3 parts—*fundus, body,* and *cervix.* The fundus and body form the upper 5 cm, the cervix the lower 3 cm. The fundus is the part that rises above the tubes. The uterine artery runs tortuously up the side of the uterus between the layers of the broad ligament.

The *external os* of the uterus is round until the birth of the first child; thereafter, it becomes a transverse slit guarded by an anterior and a posterior lip. There is a slight angle at the junction of the body and cervix, so the uterus is said to be anteflexed. The long axis of the uterus seldom lies in the median plane, but is deflected to one side or the other.

The potential *cavity of the body* of the uterus is triangular; the uterine orifices of the tubes, which are about 1 mm in diameter, open at the upper lateral angles, the *internal os of the uterus* at the lower angle. The anterior and posterior walls are applied to each other. The *cervical canal* extends from the internal os to the external os of the uterus. It is spindle-shaped and 3 cm long.

Structure: The uterus has 3 coats—serous, muscular, and mucous. *In the fundus and body,* the muscular coat (myometrium) is more than 1 cm thick and consists of interlacing bundles of smooth muscle, an arrangement which, after the birth of a child, brings about the natural arrest of hemorrhage by constricting the penetrating vessels. The serous (peritoneal) coat is adherent to the muscular coat except at the sides, where it passes onto the broad ligaments. The mucous coat (endometrium) is thick and is lined with columnar cells, many of which are ciliated, and it possesses numerous tubular glands which extend to, or even into, the muscle coat.

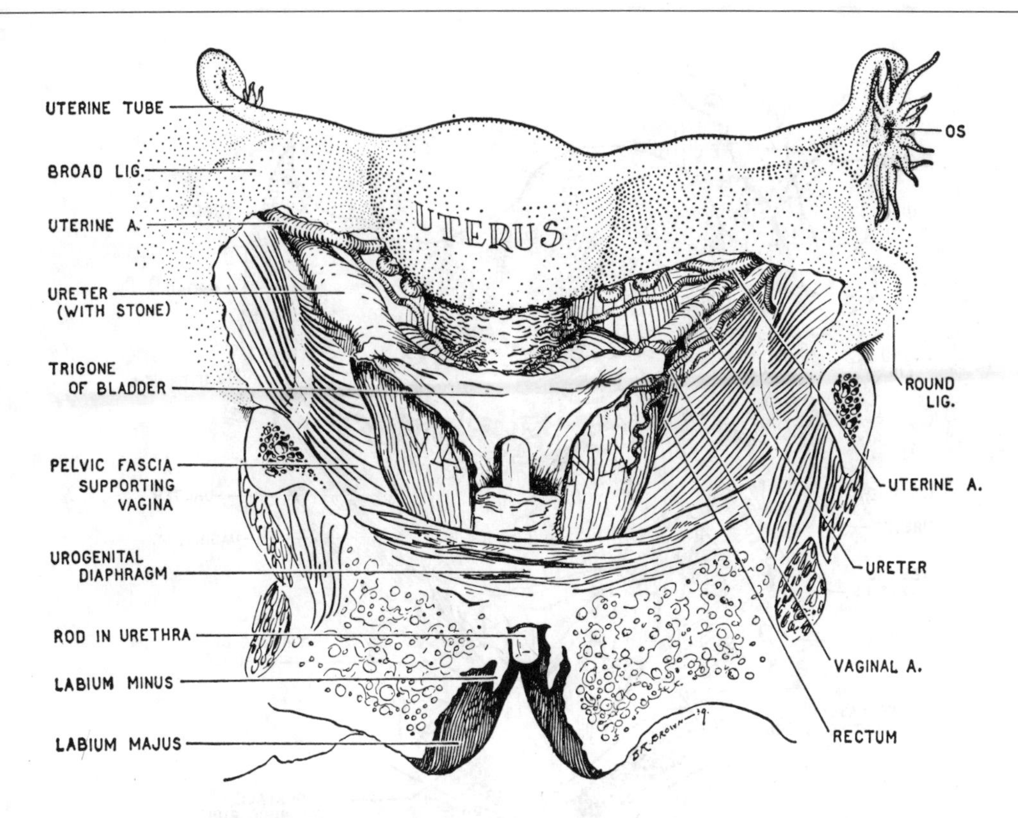

Figure 19.7. The female internal genitalia. (Parts of the pubic bones and all of the bladder, the trigone excepted have been removed.) The uterus is here symmetrically placed; hence, the ureters are nearly equidistant from the cervix. (Note the stone in the right ureter. Dissection by Dr. B. L. Guyatt.)

In the cervix, the muscular bundles are largely circularly arranged, as in a sphincter. The mucous membrane is thrown into branching folds, is lined with columnar mucus-secreting cells, and possesses both simple and branched tubular mucus-secreting glands.

VAGINA

The vagina (L. = a sheath or scabbard) is about 7–9 cm long and is approximately parallel with the pelvic brim. It extends from the vestibule of the vagina, which is guarded by the labia minora and, where its orifice opens, through the UG diaphragm to reach the recto-uterine peritoneal pouch (item 6 in *fig. 19.3*). Its anterior and posterior walls are applied to each other.

Its anterior wall is structurally continuous with the urethra in its lower third, is in contact with the bladder in its intermediate third, and is pierced by the cervix in its upper third. Around the cervix there is a circular gutter, described as the anterior, posterior, and lateral **fornices of the vagina**. Since the anterior aspect of the cervix is not covered with peritoneum, the anterior fornix is 2 cm from the vesicouterine pouch of peritoneum, but the posterior fornix is clothed with peritoneum of the recto-uterine pouch.

Note that an instrument, forced upward through the posterior fornix, would enter the rectoterine pouch, which is the lowest part of the peritoneal cavity and, therefore, well placed for surgical drainage.

Each lateral fornix is crossed by the base of the broad ligament, the uterine artery, and the ureter. Because the uterus is seldom median in position, the ureter is almost apposed to the cervix on one side (generally the left) and is 1–2 cm away on the other (*figs. 19.7* and *19.8*). Surgical approaches to the uterus via the vagina require extreme caution in the region of the lateral fornices. Severing the ureter or the uterine artery in the broad ligament (cardinal ligament portion) above the lateral fornix is a hazard.

Relations

The posterior wall of the vagina is separated from the rectum by the vaginal and the rectal layer of pelvic fascia and from the anal canal by the perineal body.

In the pelvic cavity, the side walls of the vagina rest on the levatores ani. On passing between the free anterior borders of the levatores ani, from which it receives fibers, the vagina enters the perineum and at once encounters, carries before it, and perforates, the UG diaphragm and its fasciae. These fasciae and the levator ani fascia and the superficial perineal fascia all fuse and blend with the laminated outer surface of the vaginal wall. To a great extent these form the support of the female pelvic viscera

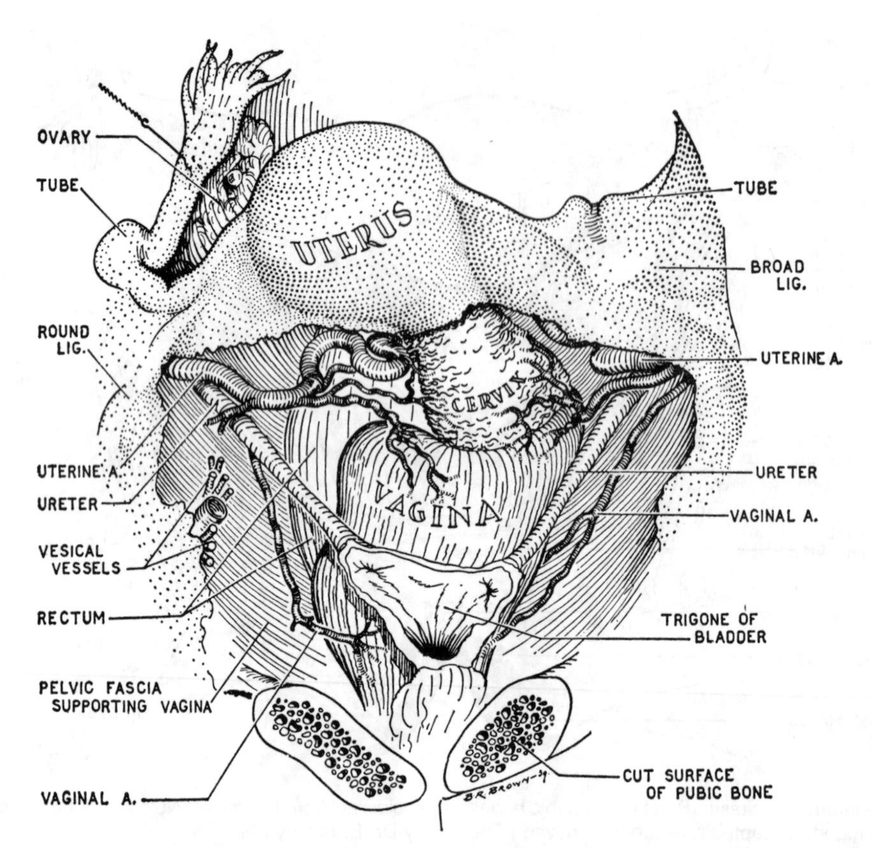

Figure 19.8. The female internal genitalia. (Parts of the pubic bones and all of the bladder, the trigone excepted, have been removed.) The uterus is asymmetrically placed; hence, one ureter is close to the cervix and the other is far removed from it.

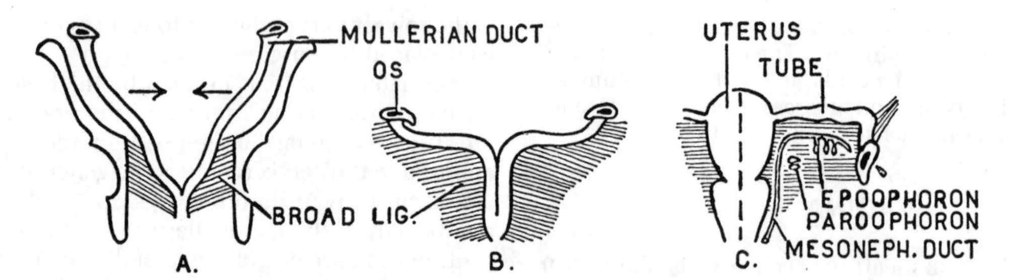

Figure 19.9. Scheme of the development of the broad ligament of uterus as the "mesentery of the paramesonephric or Mullerian duct."

(*fig. 19.9*). Applied to this wall are the bulbs of the vestibule, the bulbourethral glands, and the bulbospongiosus muscle (*fig. 21.15*).

Structure: The vagina is lined with stratified squamous epithelium. This epithelium lines a *tunica propria* which presents numerous transverse folds (rugae). Outside the tunica propria there is a thin *muscular coat* of longitudinal fibers and some interlacing circular ones, and a thick fibroareolar *adventitious coat.*

The stratified squamous epithelium is reflected from the vagina onto the cervix, clothing it as far as the external os. Though the vagina possesses no glands, its epithelial surface is moist from the secretion received from the cervical canal.

Development

The ovaries like the testes arise from the intermediate mesoderm (gonadal ridges) in the upper abdomen. The ovaries descend into the pelvis as shown in *Figure 19.10*. The inferior pole of each ovary is attached to the **gubernaculum ovarii**. This connective tissue band is attached to the junction of the body and fundus of the uterus and becomes the **round ligament of the ovary** in the adult. This is continuous with the **round ligament of the uterus** that exits the pelvis and traverses the body wall through the inguinal canal and attaches to the skin of the labium majus (*fig. 19.11*).

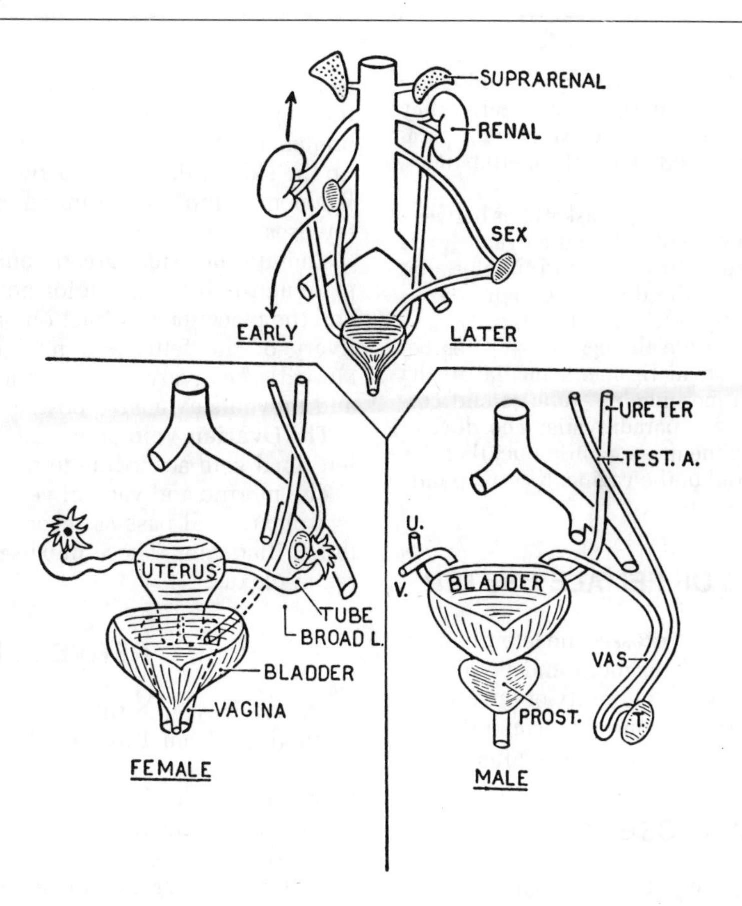

Figure 19.10. The migrations of the kidneys and sex glands. Definitively, the ovarian vessels and the uterine tube in the female bear the same relationships to the ureter as do the testicular vessels and the ductus (vas) deferens in the male. *T* = testicle; *U* = ureter; *V* = vas.

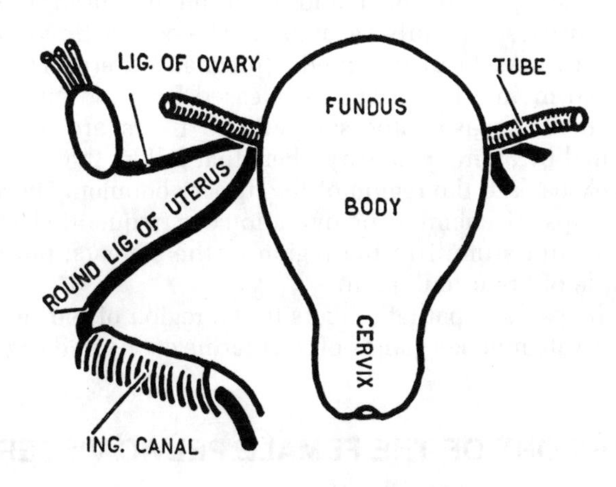

Figure 19.11. Derivatives of the gubernaculum ovarii—ligament of ovary and round ligament of uterus.

The round ligament in the female practically repeats the course taken by the vas deferens in the male, i.e., it is subperitoneal; it crosses the side wall of the pelvis and the external iliac vessels, and it turns around the inferior epigastric artery, passes through the inguinal canal, and ends in the labium majus, the homologue of the scrotum.

Very rarely, the ovary does follow the gubernaculum into the labium majus.

Derivatives of the Mullerian or Paramesonephric Duct. Before the sex of the embryo is apparent, 2 vertical ducts appear on each side of the posterior abdominal wall. They run to (but do not empty into) the ventral subdivision of the cloaca, the *urogenital sinus.* They are the Wolffian or *mesonephric ducts* (which predominate in the male) and the Mullerian or paramesonephric ducts (which predominate in the female).

The mesonephric ducts in the male serve as sperm ducts and on each side become the duct of the epididymis, vas deferens, and ejaculatory duct. The paramesonephric ducts in the female serve as ducts for ova (*fig. 19.9*), their cranial parts becoming the uterine tubes, their intermediate parts fusing to form the uterus (drawing their mesenteries, the broad ligaments, off of the side walls of the pelvis), and their caudal ends fusing to form the upper part of the vagina. As growth proceeds, a solid plate of cells, the *vaginal plate,* extends downward between the UG sinus and the rectum. Eventually this plate becomes canalized and joined by a right and a left upwardly growing diverticulum of the sinus to form the lower part of the vagina.

Remnants of the (Wolffian) Mesonephric Duct and Tubules (*fig. 19.9*) persist in the female as the *epoophoron, paroophoron,* and *duct of Gartner (of the epoopho-*

ron). Because they commonly become cystic and cause trouble, they are of clinical importance. Naturally, they are to be sought for at the sides of the (Mullerian) paramesonephric ducts, now converted into the uterine tubes, uterus and vagina.

The *epoophoron* (= above the egg basket) lies between the layers of the broad ligament, above the ovary. It is a vestigial part of the mesonephric duct and tubules, and corresponds in the male to the duct of the epididymis and efferent ductules of the testis (*fig. 11.22*, p. 135).

The *paroophoron* (= beside the egg basket) lies between the layers of the broad ligament, medial to the ovary. It is formed from mesonephric tubules and corresponds in the male to the paradidymis. *The duct of Gartner* is the segment of the mesonephric duct that lies in front of the anterior wall of the vagina. It corresponds to the end of the vas deferens.

UNIQUE RELATIONS OF FEMALE URETER

The ureter (*figs. 19.7* and *19.8*) crosses the lateral fornix of the vagina below the broad ligament and below the uterine artery, and because of the obliquity of the uterine axis, it lies closer to the cervix on one side (generally the left) and enters the bladder in front of the vagina.

SPECIAL VESSELS

The Ovarian Artery (*fig. 19.12*) arises from the aorta, like the testicular artery, but differs from it in crossing the external iliac vessels, about 1 cm in front of the ureter (*fig. 19.4*), to enter the suspensory ligament of the ovary. It anastomoses freely with the uterine artery, which may practically replace it.

The **Uterine and Vaginal Arteries** spring from the internal iliac artery (*figs. 19.7, 19.8,* and *19.12*). The uterine artery is large. It descends in front of the ureter to the base of the broad ligament and, at the lateral fornix of the vagina, it crosses above the ureter. After sending branches to the vagina and cervix, it continues tortuously up the side of the uterus between the layers of the broad ligament, supplies the uterus and the tubes, and anastomoses freely.

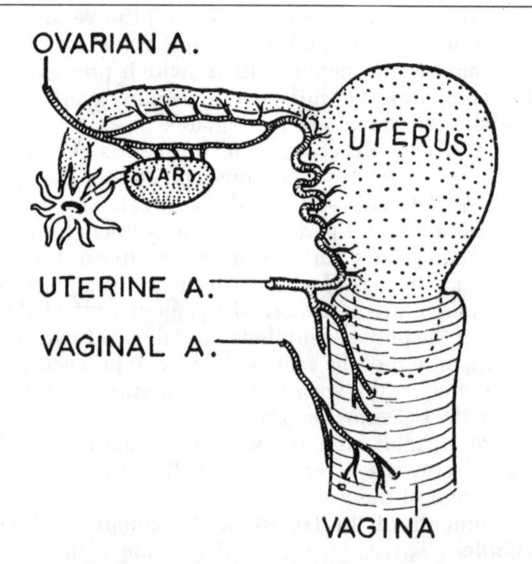

Figure 19.12. The uterine vessels.

The uterine artery greatly enlarges in the pregnant female and is responsible for nourishing the uterine wall and the placenta. Contraction of the uterus following delivery of the fetus and fetal membrane (placenta included) effectively occludes the enlarged uterine vessels and prevents extensive postpartum hemorrhage.

The Ovarian Vein opens into the inferior vena cava or left renal vein according to the side.

The uterine and vaginal *venous plexuses* join the vesical plexus and pass as several large branches to the internal iliac vein. These plexuses communicate with the rectal plexus.

NERVE SUPPLY

The ovaries, like the testes, are supplied by sympathetic fibers from T10; the fibers travel with the ovarian vessels and some of them supply the *uterine tubes*. The *uterus* is supplied by the hypogastric plexus and the pelvic splanchnic nerves. The afferent fibers from the fundus and body pass through the hypogastric plexus to T11, 12; those from the cervix and vagina via the pelvic splanchnic (parasympathetic) nerves (S2, 3, 4); but the lowest part of the vagina is supplied by the (somatic) pudendal nerve (also S2, 3, 4).

The sympathetic innervation of the uterus, uterine tubes, and ovaries is mainly vasomotor. The contraction of the smooth muscle in the gravid (pregnant) uterus is controlled mainly by hormones released from the pituitary gland. The vasomotor sympathetic fibers are accompanied by afferent sensory fibers that follow the ovarian artery back to the region of the upper abdomen. Uterine "cramps" (ischemia) and ovulation can frequently be felt as painful stimuli in the region of the stomach (an example of "referred" pain).

The parasympathetic fibers to the region of the cervix innervate mucous glands of the uterine cervix and vagina wall.

SUPPORT OF THE FEMALE PELVIC VISCERA

Muscular

As in the male, so in the female, the thick pubic parts of the levatores ani form a puborectal sling for the rectum, drawing it forward until it forms a sloping shelf (*fig. 18.14*). Upon this shelf the vagina rests, and on the vagina rests the bladder. This is the essential support.

The pubic parts of the levatores ani are also inserted into the perineal body, and thereby they act as a sling for the lower part of the posterior wall of the vagina.

The UG diaphragm and its fasciae blend with the lower third of the vagina and assist the levatores ani to support it.

Fascia

The bladder and the rectum are clothed with vesical and rectal fasciae, as in the male (*fig. 18.7*). The uterus and the vagina likewise have their own fasciae, which are thick and tough around the cervix and vagina.

The apposed layers of rectal and vaginal fasciae are so loosely connected that a rectovaginal areolar space may be said to exist between them, whereas the apposed layers of the vesical and vaginal fasciae, being more closely blended, constitute a vesicovaginal fascial septum.

There are 3 paired suspensory structures:

1. *The perivascular stalk* is the first and most important of these. In effect, it includes (1) those branches of the internal iliac vessels that have been referred to previously as the leash of vessels (uterine, vaginal, vesical) that limit the retropubic space posteriorly (*fig. 18.36*) and (2) the fascia in which these vessels are imbedded. This fascia is conducted by its contained vessels to the side of the junction of the cervix and vagina and to the side of the bladder, and there it blends firmly with the fasciae of these organs, and with that of the rectum as well. This mass of fascia and vessels plus an accession of fibers from the region of the ischial spine is commonly referred to as the *lateral cervical ligament* or the *cardinal ligament*.

2. *The uterosacral ligament* lies in the rectouterine fold of peritoneum and may be regarded as the posterior, free, curved margin of the lateral cervical ligament. It extends from the middle of the sacrum to the junction of the body and cervix of the uterus, where it meets its fellow. It supports the cervix posterosuperiorly.

Slips of involuntary muscle everywhere pervade the pelvic fascia and give it a supporting value during life.

3. *The round ligament* of the uterus, by being attached along the side of the uterus as far caudally as the cervix, perhaps helps to stabilize the uterus and vagina anterosuperiorly.

Differences Between the Male and Female Pelvis

The female pelvis may be contrasted with the male pelvis under the following headings:

1. Features dependent on the fact that women are generally smaller.
2. Features related to the unique function of the female pelvis.
3. Markings on the pubic arch for the crura clitoridis.
4. Other features.

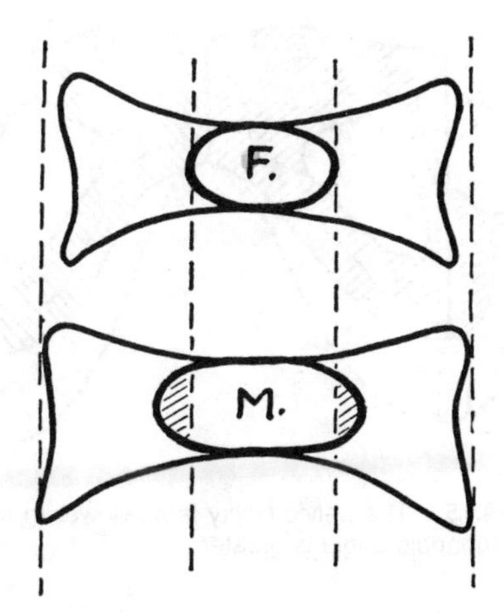

Figure 19.13. Base of sacrum: male and female compared.

Women are generally less muscular than man; they are shorter and lighter. For these and other reasons, their bones, including the bones of the pelvis, are lighter, and the ridges and markings for tendons, aponeuroses, and fasciae are less pronounced.

JOINTS

The articular surfaces are relatively and absolutely smaller in the female. Thus, the oval articular facet on the base of the sacrum and the acetabulum are both strikingly small (*fig. 19.13*). The symphysis pubis is short. The auricular facet of the sacrum barely encroaches on the 3rd vertebra.

FUNCTIONS RELATED TO CHILDBIRTH

1. *The cavity* (*figs. 19.14–19.16*). **All diameters** are absolutely greater in the female than in the male (*fig. 19.17*), each of the 3 diameters of the inlet being about ½ cm

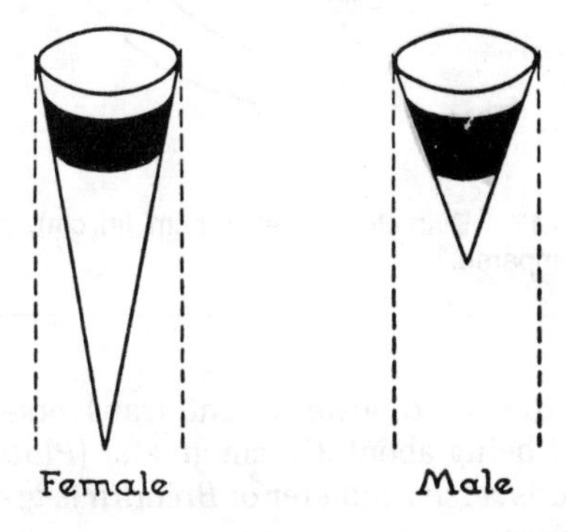

Figure 19.14. The true pelves (colored *black*) as segments of cones. Female—short segment of long cone; male—the reverse.

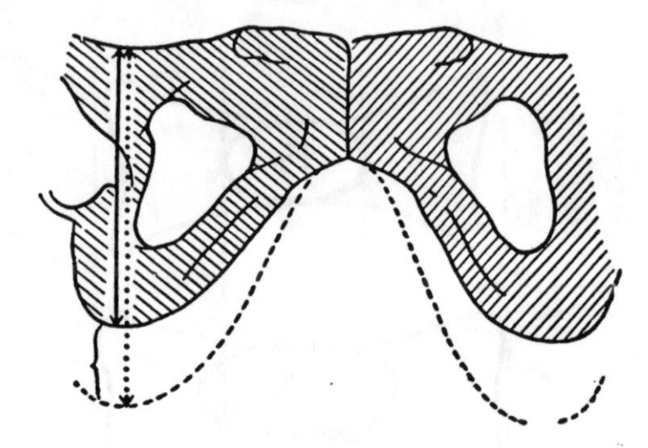

Figure 19.15. The pelvic cavity is shallower in the female, and the subpubic angle is greater.

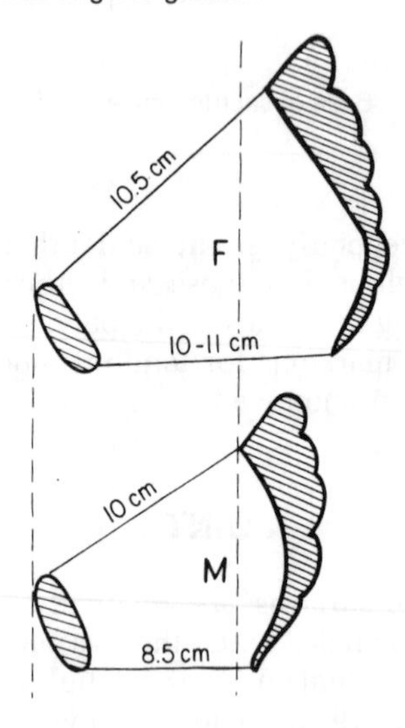

Figure 19.16. Pelvic cavities on median section: male and female compared.

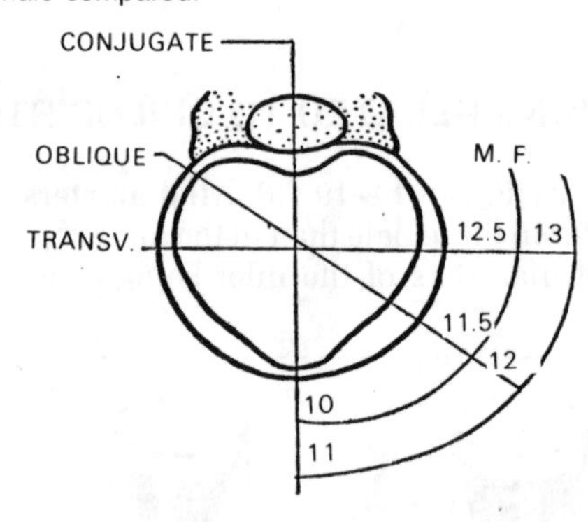

Figure 19.17. Diameter of pelvic brim (in cm), male and female compared.

greater and the anteroposterior and transverse diameters of the outlet being about 2½ cm greater (*Plate 19.1*).

2. *The Transverse Diameter or Breadth* is greater in the female because:

a. The body and crest of the pubic bone are wider in the female.

b. The acetabulum is approximately its own diameter distant from the symphysis in the male, but in the female it is relatively much more.

c. The base of the female sacrum is relatively wide.

d. The female pubic arch is almost a right angle; it equals the angle between the outstretched thumb and the index finger (*fig. 19.15*). In the male, it is an acute angle, equal to the angle between the index and middle fingers when spread.

e. The female ischial tuberosities are everted.

3. *The Anteroposterior Diameter* is greater in the female than in the male:

a. *At the inlet*, the promontory is less prominent, so the female inlet tends to be round, the male to be heart-shaped (*fig. 19.17*).

b. *At the outlet* the coccyx is carried backward in the female by increasing the angle of the greater sciatic notch to approximately a right angle in the female (*fig. 19.18*).

Other Features. In the female the area for the crus clitoridis is narrow and the arch is thin. The pelvis major is shallow and the anterior superior spines rather point forward (*fig. 19.19*).

Skirting the anterior margin of the auricular facet on

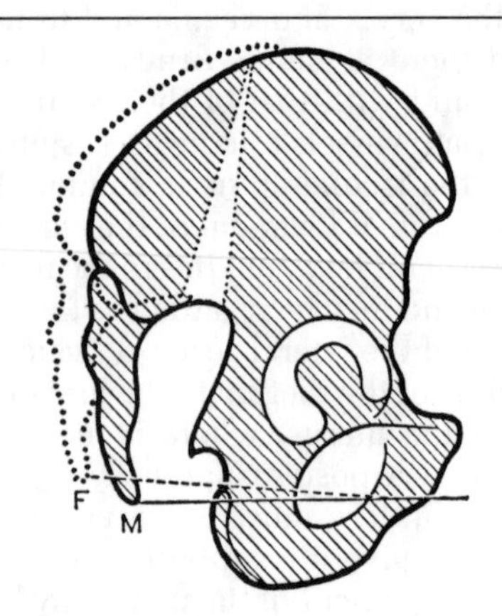

Figure 19.18. Conversion of the male outlet into the female enlarges the angle of the greater sciatic notch.

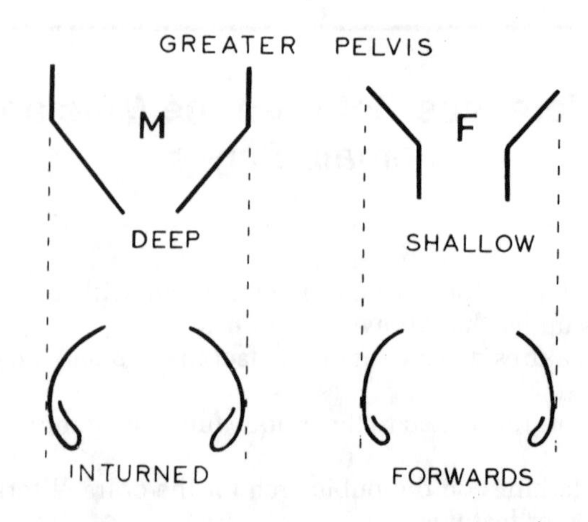

Figure 19.19. Iliac crests: male and female compared.

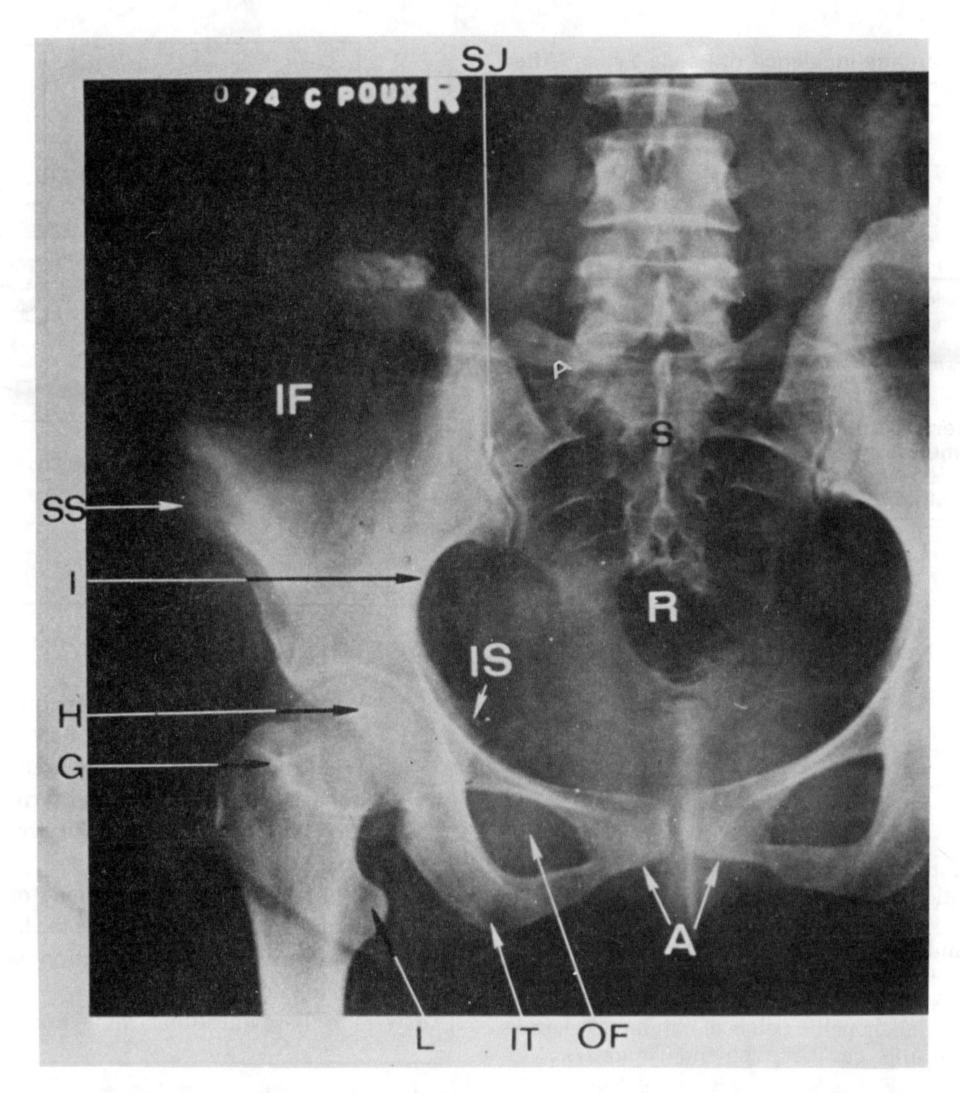

Plate 19.1. Female pelvis (A-P view). *A*, subpubic angle; *G*, greater trochanter of femur; *H*, head of femur in acetabulum; *I*, iliopectineal portion of linea terminalis; *IF*, iliac fossa; *IS*, ischial spines; *IT*, ischial tuberosity; *L*, lesser trochanter of femur; *OF*, obturator foramen; *R*, gas in rectum; *S*, sacrum; *SJ*, sacroiliac joint; *SS*, anterior superior iliac spine.

the ilium there may be an identifying deep groove, the *preauricular sulcus* (Derry).

Identification. In cases where it is difficult to arrive at a decision as to the sex of a given pelvis, greatest weight should be placed upon: (1) the area for the attachment of the crus penis or clitoridis; (2) the angle of the pubic arch; (3) the size of the acetabulum; (4) the size of the facet on the base of the sacrum relative to the alae; (5) the distance of the acetabulum from the symphysis pubis; and (6) the size of the greater sciatic notch. This last measure was found to be useless for sexing purposes by Singh and Potturi—unless the posterior angle was depended on (the last suggesting that the widening of the notch in females occurs primarily in its posterior part).

The medicolegal expert also uses the *Ischium-Pubis Index* (*fig. 19.20*), which reveals the sex of an adult pelvis in more than 90% of instances; that is because the pubic bone is relatively and absolutely longer in the female and the ischium relatively and absolutely longer in the male. The index is less than 90 in adult males and more than 90 in adult females (Washburn). Other features of the pubis are valuable in "sexing" adult pelves (Phenice).

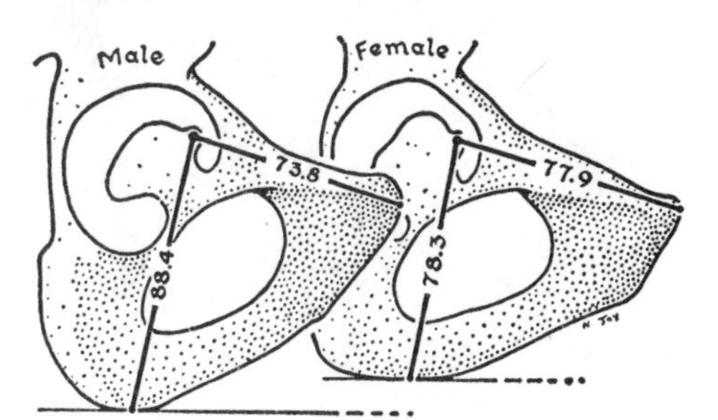

Figure 19.20. The ischium-pubis index.

$$\frac{\text{length of pubis} \times 100}{\text{length of ischium}}$$

Types of Female Pelvis. Female pelves by no means all conform to the description given above; and since for the obstetrician it is of practical importance to know the

Table 19.1. Percentage Incidence of Pelvic Types (After Greulich and Thoms)

Type	300 Clinic Patients	100 Nurses	107 Girls Age 5–15
Dolichopellic*	16.3	37	57.9
Mesatipellic†	44.0	46	33.6
Brachypellic‡	36.3	17	8.3
Platypellic§	3.3		

*Dolichopellic (long): anteroposterior (AP) diameter > transverse diameter.
†Mesatipellic (round): transverse > AP diameter by 0–1.0 cm.
‡Brachypellic (oval): transverse > AP diameter by 1.1–2.9 cm.
§Platypellic (flat): transverse > AP diameter by 3 cm or more.

Table 19.2. Average Anteroposterior (Conjugate) and Transverse Diameters of the Pelvic Inlet in White Females, Classified According to the Four Pelvic Types of Caldwell and Moloy.*

Pelvic Type	No. of Cases	Conjugate Diameter (cm)	Transverse Diameter (cm)
Gynecoid	26	10.86	13.76
Android	25	10.59	13.56
Anthropoid	19	11.75	12.94
Platypelloid	3	8.55	14.45
Mixed average for females	73	10.90	13.51
Average for males	43	10.10	13.00

*After T. W. Todd

dimensions and shapes of the pelves of his patients, various classifications have been proposed (*Tables 19.1, 19.2; fig. 19.21*).

The size and shape of the pelvis are influenced by various factors (hormonal, environmental, hereditary, and mechanical). At puberty, over a period of 1½ years, the predominantly dolichopellic pelvis of childhood changes to one of the adult types (Greulich and Thomas).

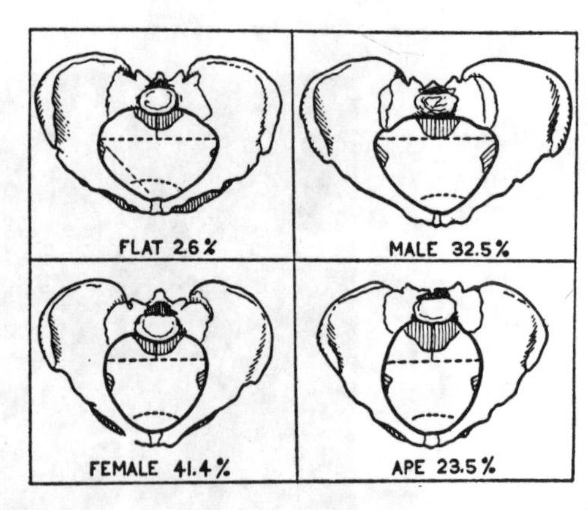

Figure 19.21. The inlets of the 4 major types of female pelvis (after Caldwell and Moloy).

Clinical Mini-Problems

1. **Which compressible hollow structures pass over the pelvic brim of the female?**
2. **What relationship does the ureter have to the uterine artery above the lateral fornix of the vagina?**
3. **Why would pain in the ovary refer to the stomach area of the patient?**
4. **Where is the greatest diameter of the female pelvis usually found? This will be the best position for the anterior-posterior orientation of the fetus' head during delivery.**

(Answers to these questions can be found on p. 585.)

20

Pelvic Autonomic Nerves and Lymphatics

Clinical Case 20.1

Patient Doug. S. This 24-year-old taxi driver suffered a broken back in a car accident. The fracture of the T12 vertebral lamina produced an underlying spinal cord lesion that destroyed the sensory and motor neurons in the L3–5 segment. The spinal column fractures were stabilized and the patient was placed in a body cast for four months. He never regained the lost sensory or voluntary motor function in his lower limbs during his convalescence and was rendered permanently paraplegic by this accident.

When the body cast was removed, he was sent to a special center for rehabilitation. The motor deficits in his lower limbs were treated by fitting him with special leg braces and teaching him to walk with crutches. He also relied heavily on the use of a wheelchair. Initially he had lost voluntary control of his external anal sphincter and urethral sphincter, so that control of defecation and micturition were unstable. Gradually he learned to control the "emptying reflexes" for bowel and bladder by stimulating the afferent fibers in the walls of these organs through increased intraorgan pressure. These afferent impulses activated the pelvic parasympathetic fibers which initiated the contraction of the smooth muscle in the walls of the bladder and bowel and relaxed the internal sphincters. Now that he is able to control his bladder and bowel function, he is being prepared to go home. You attend the discharge conference at the rehabilitation facility at the invitation of your tutor.

His major concern on leaving hospital was whether he would ever regain normal sexual functions, and he discusses this intelligently. He has lost all sensation to the skin of his penis but still has sensation over the anterior aspect of his scrotum. He is able to ejaculate semen into his urethra due to sympathetic fibers from the intact L1, 2 segments. He is incapable of initiating tumescence (erection) in concert with seminal emission. Therefore he is scheduled for further therapy in a sexual medicine clinic as an outpatient, and he will also be seen at regular outpatient clinics in rehabilitation medicine.

Pelvic Autonomic Nerves

SYMPATHETIC TRUNK

Within the abdomen the sympathetic trunk descends on the bodies of the lumbar vertebrae, crosses behind the common iliac vessels and enters the pelvis.

Within the pelvis, the trunks of the two sides, each having four ganglia, converge to the front of the coccyx to end in the unpaired unimportant *ganglion impar*. They lie medial to the pelvic sacral foramina (*figs. 16.16, and 18.38*) and send gray rami laterally to each of the sacral and coccygeal nerves and a few visceral twigs to the inferior hypogastric plexus.

SUPERIOR HYPOGASTRIC PLEXUS [PRESACRAL NERVE]

This plexus lies below the bifurcation of the aorta (*fig. 16.8*). It is a downward prolongation of the pre-aortic (intermesenteric) plexus, joined by the 3rd and 4th lumbar splanchnic nerves (*fig. 20.1*).

Branches enter the pelvis and descend in front of the sacrum as *right and left hypogastric nerves*. As these descend, they become plexiform and are joined by *twigs from the sacral sympathetic ganglia* and the pelvic splanchnic nerves to form the inferior hypogastric plexus [pelvic plexus].

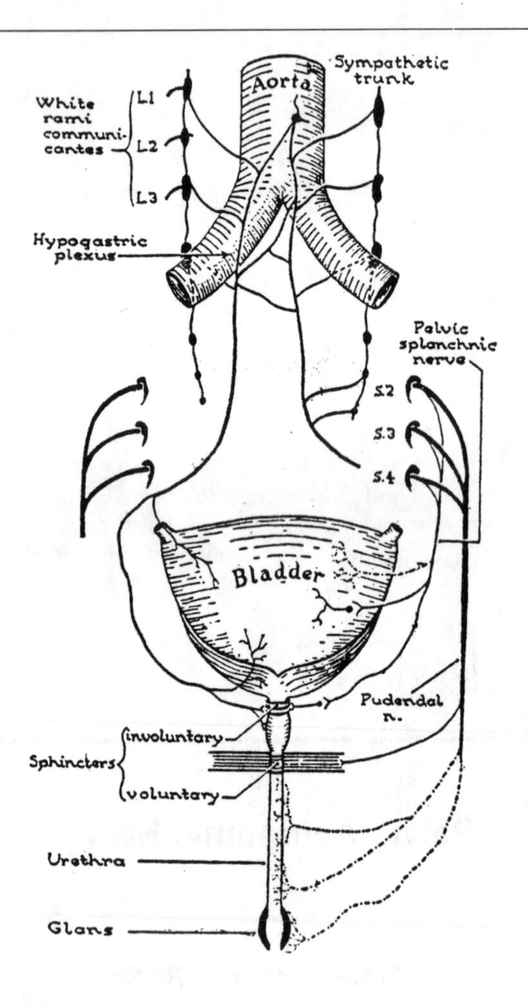

Figure 20.1. Diagram of the nerve supply of the bladder and urethra.

PELVIC SPLANCHNIC NERVES [NERVI ERIGENTES]

These threads run forward on each side from the sacrum to the rectum (*fig. 20.1*). Parasympathetic nerves, they spring from S(2), 3, 4. They join the corresponding inferior hypogastric plexus, which then becomes mixed sympathetic and parasympathetic.

Mixed branches are distributed with the vessels to the various pelvic viscera.

The pelvic splanchnic nerves (both right and left) send ascending fibers across the left common iliac artery to the inferior mesenteric artery and are distributed with it to the descending and sigmoid colons (Stopford).

Functions

It is generally believed that the parasympathetic has exclusive control on the muscular walls of the bladder, membranous and penile urethra, and rectum (Sheehan). The pelvic splanchnics also cause relaxation of the arteries to the erectile tissue, producing erection of the penis (or clitoris), hence the alternative name—*nervi erigentes* (*fig. 20.1*).

The hypogastrics (sympathetics), when stimulated, cause the epididymis, vas deferens, seminal vesicle, and prostate to contract and empty their contents; at the same time the region of the neck of the bladder is shut off (Learmonth). Hence, on ejaculation, the seminal fluid is hindered from entering the bladder.

The pelvic splanchnic nerves also carry many afferent fibers from the pelvis (e.g., pain, distension). But, the visceral afferents that accompany the sympathetics are sensory to the body and fundus of the uterus (Mitchell).

Pelvic Lymphatics

Many lymphatics leave the superficial and deep inguinal nodes (p. 258), pass behind the inguinal ligament, and end in iliac nodes (*fig. 20.2*). These lymph vessels surround the female artery and vein. The medial vessels pass through the femoral canal and may be interrupted by a node (of Cloquet) that lies in the canal.

The pelvic nodes are in 2 groups: (1) *nodes near the*

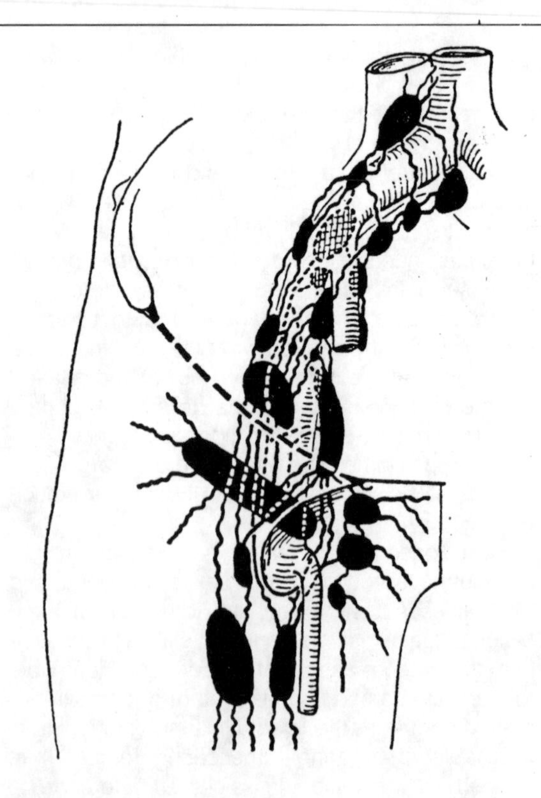

Figure 20.2. Dissection of the inguinal lymphatics and those at the pelvic brim.

pelvic brim: (a) the *external and common iliac nodes*, arranged as several intercommunicating chains around the respective blood vessels, and (b) the nodes above the *sacral promontory*; and (2) *nodes within the cavity*: (a) *the internal iliac, lateral sacral,* and *median sacral nodes* arranged on the respective blood vessels and (b) others in the *vesical fascia*, in the *rectal fascia* mainly behind the rectum (pararectal nodes) and on the course of the superior rectal artery, *in the broad ligament* near the cervix uteri, and between the prostate and rectum.

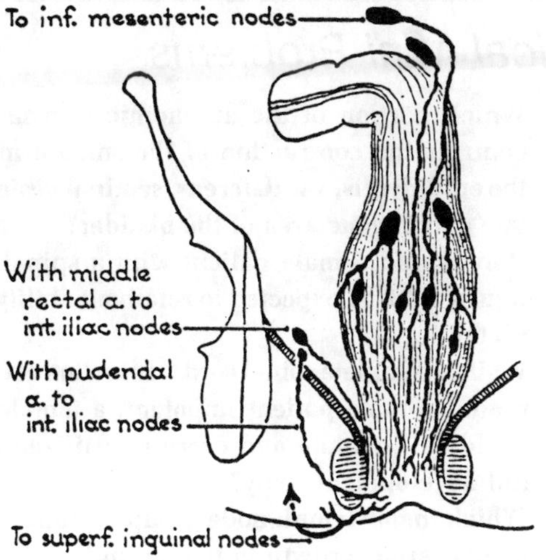

Figure 20.3. Lymphatics of rectum (After Rouviere).

STRUCTURES DRAINED

The following structures are drained by lymph vessels traveling to lymph nodes, thus:

The Skin of the Penis and the Prepuce → with the superficial dorsal vein in the subcutaneous tissues to the superficial inguinal nodes of both sides.

The Glans Penis and the Penile Urethra → accompany the deep dorsal vein deep to the fascia of the penis to the deep inguinal nodes of both sides, then to the external iliac nodes. Some vessels pass without interruption through the femoral and inguinal canals to the external iliac nodes. The lymph vessels of the glans and of the prepuce anastomose.

Bulbar Urethra → follows the internal pudendal artery to internal iliac nodes and the deep dorsal vein below the infrapubic ligament, then to external iliac nodes.

Membranous and Prostatic Urethra (or Whole Female Urethra) → drain into the internal iliac nodes.

Bladder (superior and inferolateral surfaces) → follows the general course of the branches of the superior vesical artery, deferent duct, and ureter to external iliac nodes

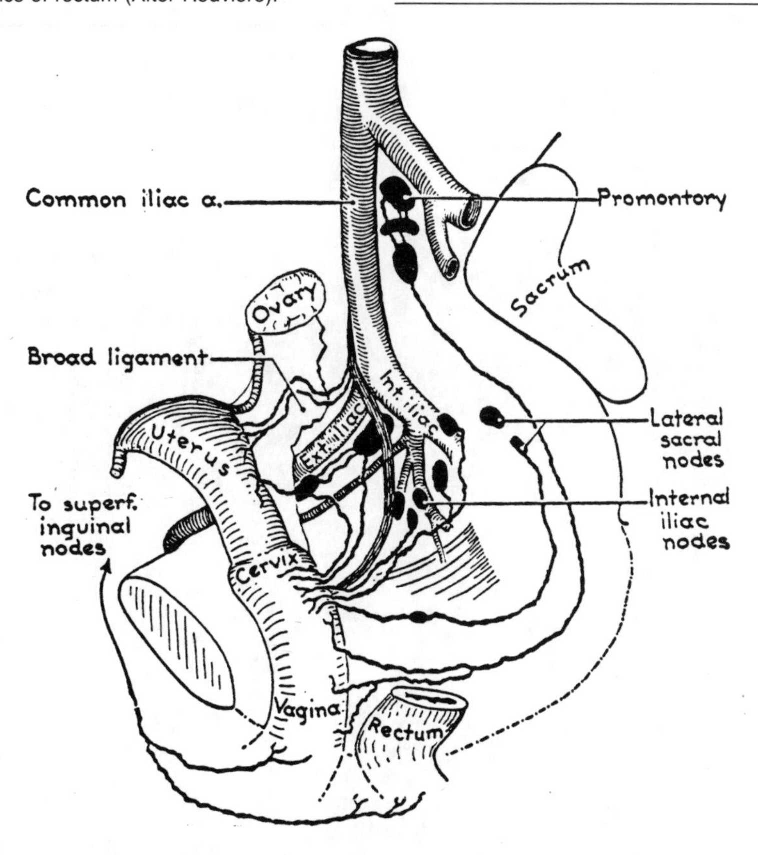

Figure 20.4. The lymphatics of the uterus and vagina.

lying along the medial side of the external iliac vein, some being interrupted by anterior and lateral vesical nodes.

Base of the Bladder and the Male Internal Genital Organs (prostate, seminal vesicles, ampullae of the deferent ducts) → internal iliac nodes, also to the external iliac nodes, and on the levator ani on the sacral nodes.

Anus and Lowest Part of the Anal Canal → by cutaneous vessels to the superficial inguinal nodes.

Anal Canal → internal iliac nodes by vessels piercing the levator ani (others crossing the ischiorectal fossa with branches of the pudendal vessels) and by others following the middle rectal vessels.

Rectum, above the anal valves (*fig. 20.3*), → (1) pararectal nodes, which lie in the fatty tissue behind the rectum enclosed within the rectal fascia, then to the superior rectal and inferior mesenteric nodes; (2) lateral and median sacral nodes; and (3) with the middle rectal artery to internal iliac nodes.

Ovary, like the testis (*fig. 11.21*), → lateral aortic and preaortic nodes, between the levels of the common iliac vessels below and the renal vessels above.

Uterine Tube and Fundus of the Uterus → with those of the ovary, ending in the lower preaortic and lateral aortic nodes. They traverse the broad ligament and cross the pelvic brim.

Body of Uterus (*Fig. 20.4*) → via broad ligament to the external and common iliac nodes and by a single vessel running in the round ligament to the superomedial group of superficial inguinal nodes.

Cervix of Uterus and Upper End of Vagina → external iliac nodes, with uterine and vaginal arteries to internal iliac nodes, and, by vessels passing close to the rectum in the uterosacral fold, to lateral sacral nodes and nodes of the promontory.

Lower End of Vagina and labium majus, like the scrotum → to superficial inguinal nodes.

Clinical Mini-Problems

1. **Which portion of the autonomic nervous system controls the contraction of the smooth muscle in the epididymis, vas deferens, seminal vesicle, prostate and trigone area of the bladder?**
2. **Why would a male patient with a spinal column injury at L2 be expected to retain an ability to ejaculate?**
3. **While a complete spinal cord transection at C6 would render a male patient impotent, a female patient could still ovulate and carry a fertilized ovum to full development. Why?**
4. **Which major lymph node group is stimulated first by cancerous growth in the uterus?**

(Answers to these questions can be found on p. 585.)

21

Perineum

Clinical Case 21.1

Patient Phil S. This 48-year-old man (in the clinic for a complete insurance assessment) is the first patient on whom you have performed a digital examination of the anus. Under the supervision of your clinical supervisor and with the consent of the patient, you proceed without any chance to review your books. What should one be able to palpate in all directions?

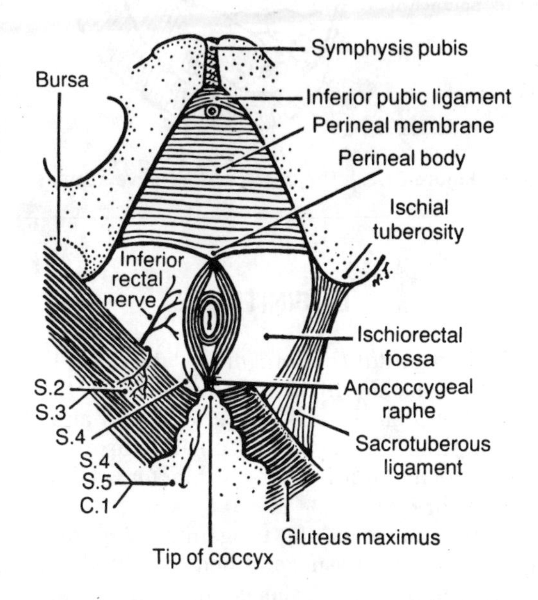

Figure 21.1. Boundaries and subdivisions of the perineum (*Inf pubic lig = arcuate ligament*).

The perineum is the diamond-shaped region at whose angles are the arcuate (inferior) pubic ligament, the tip of the coccyx, and the ischial tuberosities (*fig. 21.1*). The pubic arch and the sacrotuberous ligaments form its sides. The sacrotuberous ligament, however, is hidden by the lower border of the gluteus maximus.

The anterior half of this diamond is the *urogenital region (triangle)*; the posterior half is the *anal region (triangle)*.

The adult differences in the form of the external genitalia disguise the fact that developmentally they are homologous. This theme will be emphasized in this chapter.

Developmental Considerations

In the embryo the endodermal-lined alimentary canal ends in a blind receptacle, the *cloaca*, shaped somewhat like a coffee pot, the spout being the *allantoic diverticulum*, definitively the *urachus* (*figs. 21.2, 21.3*).

The *mesonephric duct* (definitively the *ductus deferens* in the male) grows caudally and opens into the anterior part of the cloaca. The *ureter* develops as an outgrowth from the mesonephric duct, and the two have for a period a common terminal duct. This common duct is absorbed subsequently into the posterior wall of the bladder and prostatic urethra with the result that the ureter and ductus deferens come to have independent openings.

In reptiles and birds, the cloaca opens on the skin surface through an orifice, guarded by a sphincter of striated muscle, the *cloacal sphincter*. In mammals, including man, a septum of mesoderm, the *urorectal septum*, divides the cloaca into (1) an anterior or urogenital part, and (2) a posterior or intestinal part (*fig 21.3*). The cloacal sphincter also divides into anterior and posterior parts: the posterior part becomes the sphincter ani externus; the anterior part becomes the other perineal muscles (p. 247).

From these considerations you will understand why a single nerve, the *pudendal nerve* (S2, 3, 4), serves the muscles as well as the skin of the region, and why its companion artery, the *internal pudendal artery*, nourishes the entire territory; you will also see why the bladder and rectum have a common nerve supply (pelvic splanchnic nerves and hypogastric plexus).

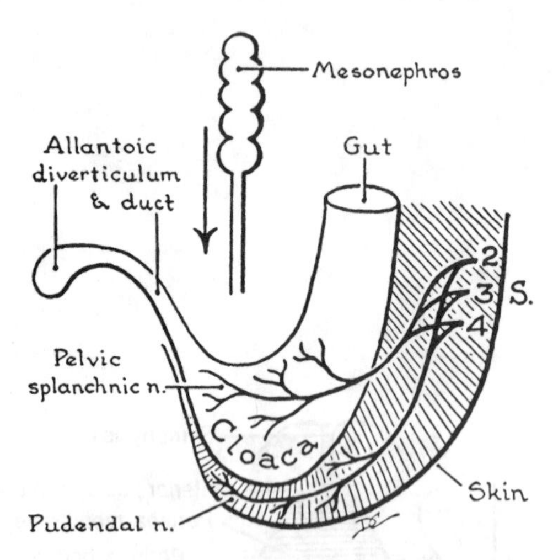

Figure 21.2. The cloaca and its nerve supply.

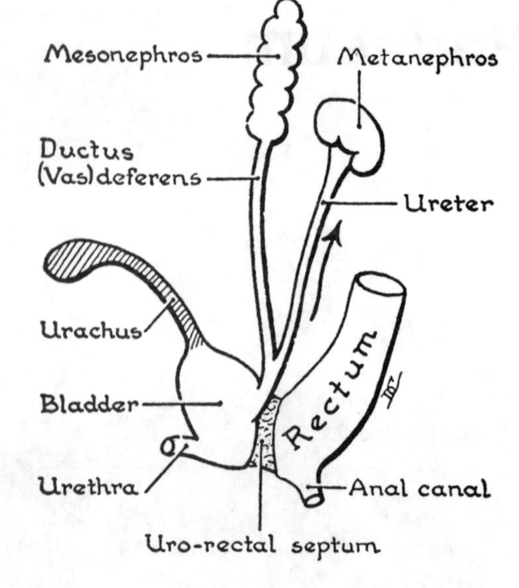

Figure 21.3. Connections and subdivisions of cloaca.

DEFINITIONS

1. *The Urogenital Diaphragm* (*fig. 21.4*) is a thin sheet of striated muscle which stretches between the 2 sides of the pubic arch. Its most anterior fibers and its most posterior fibers (transversus perinei profundus) run transversely. Its middle fibers (sphincter urethrae) encircle the urethra.

Like other muscles, the urogenital diaphragm is enveloped in areolar tissue, and because the diaphragm is flat, its envelope forms 2 sheets, the *inferior and superior fasciae of the urogenital diaphragm. Perineal membrane* is another name for the inferior fascia.

2. *The Superficial Perineal Fascia* (of Colles) is continuous with the membranous layer of the superficial fascia of the abdominal wall (Scarpa's Fascia). This superficial fascia is firmly attached to the fascia lata of the thigh, the pubic arch, and the base of the perineal membrane (*fig. 21.4*). Anteriorly, this superficial fascia is prolonged over the penis and scrotum.

3. *The Superficial Perineal Pouch* is the fascial space deep to the superficial perineal fascia. Should the urethra rupture into this space, the attachments of the fascia will determine the direction of flow of the extravasated urine—not to the anal triangle or the thigh, but into the scrotum, around the penis, and upward into the abdominal wall.

4. *The Deep Perineal Pouch* is the space enclosed by the superior and inferior fasciae of the urogenital (UG) diaphragm. Among its contents are the membranous urethra and the sphincter urethrae (*fig. 21.4*).

5. *The Perineal Body* is a small fibrous mass at the center of the perineum (*figs. 21.1, 21.4*). Attached here are the base of the perineal membrane and several muscles that converge from 5 directions: sphincter ani externus, transversus perinei superficialis (r. and l.), bulbospongiosus, and in part the levator ani (*see fig. 21.10*).

6. *The Anococcygeal Raphe* is a fibrous band between the anus and coccyx.

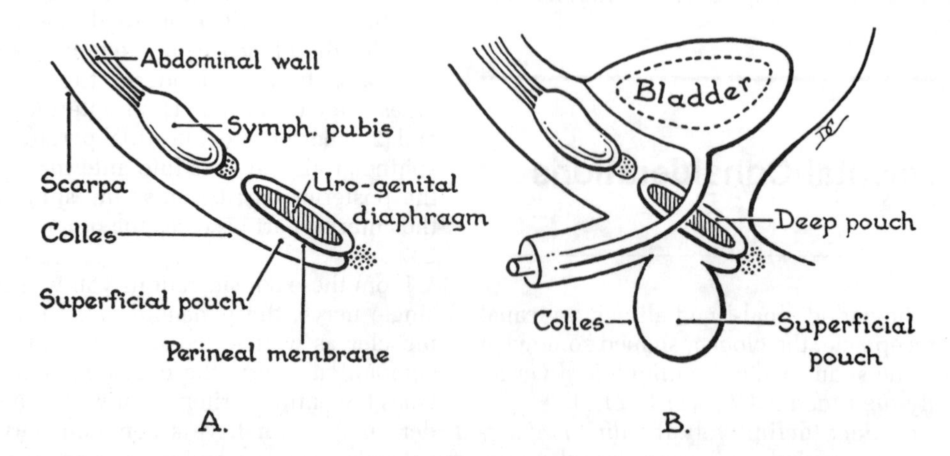

Figure 21.4. To explain the urogenital diaphragm and the perineal pouches (schematic).

Anal Region (Triangle)

SPHINCTER ANI EXTERNUS

This is a sphincter of voluntary muscle, 2–3 cm deep, placed around the anal canal. It has 3 parts—subcutaneous, superficial, and deep (*see figs. 21.10, 18.31*). (L. *anus* = a ring.)

The Subcutaneous Part is slender and encircles the anal orifice. *The Superficial Part* is elliptical and extends from the tip of the coccyx and anococcygeal raphe to the perineal body (*see fig. 21.10*). It supports the circular anus in the median plane between the tip of the coccyx and the perineal body.

The Deep Part encircles the anal canal like a collar; in front, however, some fibers decussate and join the opposite superficial transverse perineal muscle. Above, it blends with the levator ani. It is supplied by many branches of the inferior rectal (hemorrhoidal) vessels and nerve (*see fig. 18.33*). Its nerve is the inferior rectal branch of the pudendal nerve, chiefly from S4 (*see fig. 21.10*).

ISCHIORECTAL FOSSA

The ischiorectal fossae are the fascia-lined, wedge-shaped spaces, one on each side of the anal canal and rectum. Filled with fat, they allow the rectum to become distended and to empty. Each fossa is bounded laterally by the ischium, from which the obturator internus arises (*fig. 21.5*); medially, by the rectum and anal canal, to which the levator ani and external sphincter are applied; posteriorly, by the sacrotuberous ligament and the overlying gluteus maximus; anteriorly, by the base of the urogenital diaphragm and its fasciae.

The fascia covering the obturator internus is fairly strong and it extends upwards beyond the fossa in the pelvis minor, reaching the pelvic brim (*fig. 21.5*). The fascia covering the levator ani is weak. The apex or roof of this fascia-lined, wedge-shaped fossa is formed by the levator ani where it arises from the obturator internus fascia less than the length of an index finger above the ischial tuberosity.

A fingertip passed forward enters a short cul-de-sac above the base of the UG diaphragm. Here the superior fascia of the diaphragm and the fasciae of the medial and lateral walls and of the roof adhere.

The fascial linings of the right and left fossae blend posteriorly, between anal canal and coccyx, to form a weak, median, areolar partition whose lower free border is the **anococcygeal raphe.**

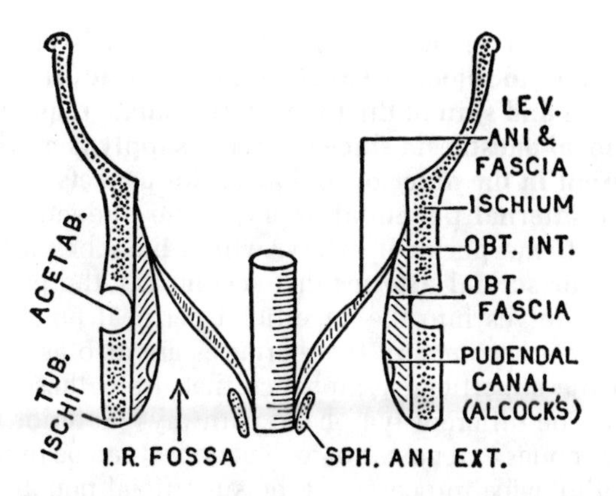

Figure 21.5. The pelvis in coronal sections to show the ischiorectal fossa (schematic). *OBT INT* = obturator internus; *SPH ANI EXT* = Sphincter ani externus; *IR* = ischiorectal.

Contents of the Fossa

1. Fat.
2. Internal pudendal vessels and pudendal nerve (L. *pudere* = to be ashamed; cf. impudent) run forward in an areolar sheath, the *pudendal canal*, in the obturator internus fascia, 2–3 cm above the tuberosity. Far back they give off the inferior rectal (hemorrhoidal) vessels and nerve, which pass forward and medially through the ischiorectal fossa toward the surface to supply: external sphincter ani, skin around the anus, and mucous membrane of the lower canal.
3. Two other cutaneous nerves, the *perforating cutaneous* branch of the 2nd and 3rd sacral and the *perineal branch* of the 4th sacral, are shown in *Figure 21.1*.

Urogenital Region in the Male

SUPERFICIAL PERINEAL POUCH

The superficial perineal pouch has been defined on p. 243).

The nerve and vascular supply to the structures in the superficial pouch comes from two different sources. The **anterior scrotal nerves** are branches of the **ilioinguinal nerve** (L1) and the **anterior scrotal arteries** are from the **superficial** and **deep pudendal arteries** that arise from the **femoral artery**. The **posterior scrotal nerves** are branches of the **perineal nerve**, which arises from the **pudendal nerve** (S2,3,4) in the pudendal canal (*see fig. 21.10*). These nerves enter the superficial pouch by piercing the posterior aspect of the perineal membrane. They are usually accompanied by a **perineal branch** of the

posterior cutaneous nerve of the thigh. **It is clinically important to remember that surgical procedures on the scrotum and skin of the urogenital triangle require precise local anesthesia since the nerve supply is markedly different in the anterior and posterior aspects.**

The **internal pudendal artery** in the pudendal canal gives off the **perineal artery**, which branches into the **posterior scrotal arteries** that accompany the posterior scrotal nerves into the superficial perineal pouch. The veins that accompany these arteries are also associated with the superficial lymphatics that drain these structures. The different lymphatic pathways (anterior to inguinal nodes and posterior to iliac nodes) can be involved in pathological processes of the superficial pouch structures. These pathways are independent of the lymphatic pathways that drain the testis along the testicular artery to the abdominal lymph nodes.

The Contents of the Superficial Perineal Pouch

1. The root of the penis.
2. Three superficial nerves (paired).
3. Three superficial arteries (paired).
4. Three superficial muscles (paired).

PENIS

The penis is composed of 3 fibroelastic cylinders, the right and left *corpora cavernosa penis* and the *corpus spongiosum penis* (*fig. 21.6*), which are filled with erectile tissue and are enveloped in fasciae and skin (L. *penis* = tail *and* of tail; *corpus* = body; plural, *corpora*).

The corpora cavernosa fuse with each other in the median plane, except behind where, as 2 diverging *crura*, they separate to find attachment on each side to 2 cm of the pubic arch. They support the corpus spongiosum, which lies below and between them. (L. *crus* = leg; plural, *crura*).

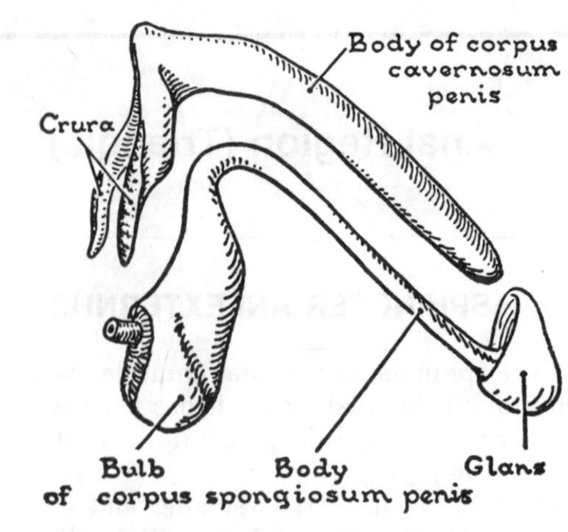

Figure 21.6. The parts of the penis.

The corpus spongiosum is traversed by the urethra; it is enlarged anteriorly where, as the *glans penis*, it fits onto the blunt end of the united corpora cavernosa and expanded posteriorly where, as the *bulb of the penis*, it is fixed to the perineal membrane. This membrane moors the bulb between the pubic arches.

The crura and the bulb are the attached parts or *root of the penis* (*fig. 21.7*).

Skin and Fasciae (*fig. 21.8*)

The skin and fasciae of the penis are prolonged loosely over the glans as the *foreskin* or *prepuce*. The skin is devoid of hairs and the fascia of fat. Deep to these, a closed tube of denser and more tightly fitting fascia, *the*

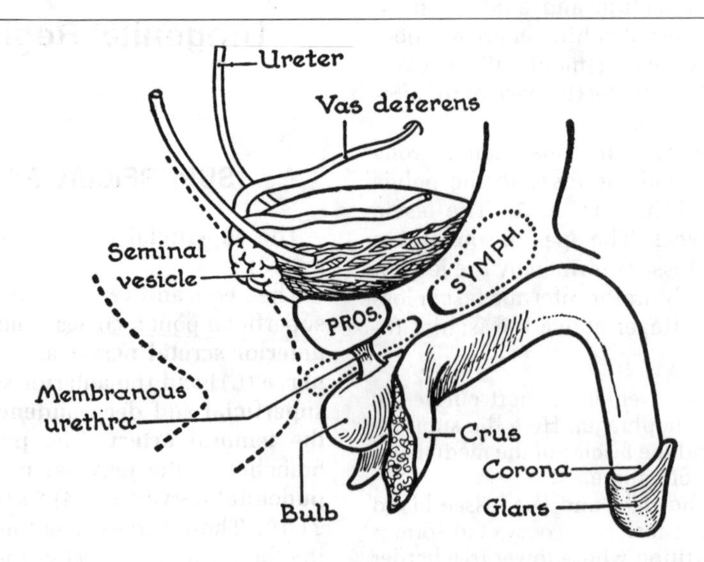

Figure 21.7. The lower parts of the genital and urinary tracts and their relations.

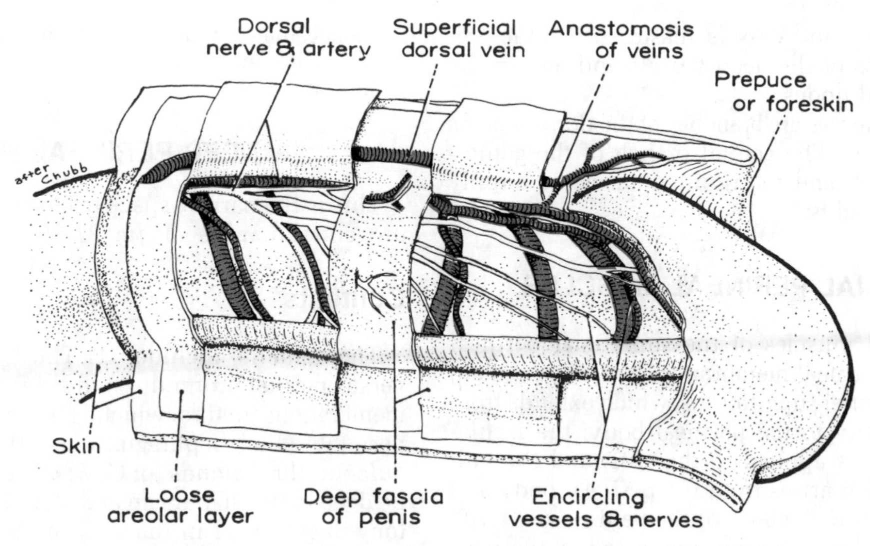

Figure 21.8. The skin, fascia, vessels, and nerve of the penis. (Modified after *Grant's Atlas*.)

deep fascia of the penis, envelops the body of the penis from its root to the corona of the glans (*fig. 21.9*).

The Suspensory Ligament of the Penis is a thick, triangular fibroelastic band. Above, it is fixed to the lower part of the linea alba and upper part of the symphysis pubis; below, it splits to form a sling for the penis at the junction of its fixed and mobile parts (i.e., where the organ is bent), and there it blends with the fascia penis (deep fascia of the penis [*fig. 21.9*]).

Vessels and Nerves of the Penis (*fig. 21.9*)

There is a superficial and a deep dorsal vein of the penis. Each is single and occupies the median plane. The superficial dorsal vein is accompanied by lymph vessels, but it has no companion artery. It drains a cutaneous plexus that bulges through the thin penile skin. The deep dorsal vein, on the other hand, has a companion dorsal artery and nerve on each side and lymph vessels. They run along the dorsum penis deep to the deep fascia of the penis. Branches of the arteries conduct blood to the corpora cavernosa and corpus spongiosum, to be returned by branches of the vein (*fig. 21.9*). The sensory nerves end mainly in the very sensitive glans.

The Deep Dorsal Vein (unpaired) passes below the symphysis pubis to end in the prostatic plexus of veins.

The *Lymph Vessels* end in the deep inguinal nodes (p. 240).

The *Erectile Tissue* of the penis is supplied by 3 paired arteries that branch from the internal pudendal arteries—artery to the bulb, artery to the crus (deep artery, p. 248), and *dorsal artery. The last named artery sends encircling twigs to assist the former 2 arteries* (*fig. 21.9*).

The Vasomotor Nerves are derived from the pelvic splanchnics (*fig. 20.1*; p. 239).

The *Coverings of the Penis* (skin and fasciae) are supplied by the dorsal arteries of the penis and by the external pudendal branches of the femoral arteries. They are drained by the *superficial dorsal vein*, which divides the right and left branches, then passes via the external pudendal veins to the great saphenous veins of the thigh.

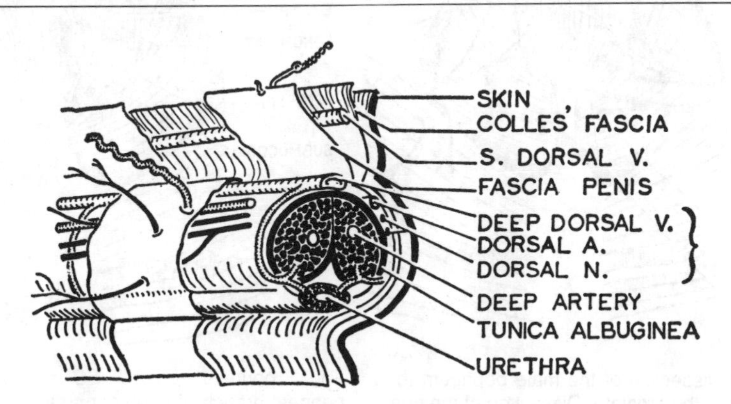

Figure 21.9. The penis on cross-section: its coverings and its vessels.

The *Superficial Lymph Vessels* anastomose in the prepuce with branches of the deep vessels and end in the superficial inguinal nodes.

The *Cutaneous Nerves* are branches of the dorsal nerves of the penis (S2,3,4). The genital branch of the genitofemoral nerve (L1,2) and the ilioinguinal nerve supply the parts near the pubis.

SUPERFICIAL PERINEAL MUSCLES

On each side, 3 muscles of the superficial perineal pouch cover the perineal membrane. They are: the slender *transversus perinei superficialis*, which extends from the ischial tuberosity to the perineal body; the *ischiocavernosus*, which is applied to the crus; and the **bulbospongiosus**, which arises from the perineal body and a median raphe below the bulb of the penis (*fig. 21.10*). All of these skeletal muscles are innervated by the perineal nerve.

The most posterior fibers of each bulbospongiosus pass to the perineal membrane; the intermediate fibers of the 2 sides meet on the dorsum of the corpus spongiosum; and the most anterior fibers meet on the dorsum of the penis, where they blend with the fascia penis. The bulbospongiosus is a compressor that empties the bulb and the posterior part of the spongy urethra. There is evidence that it is essential in maintaining erection, and involved in circulation of semen through the penile urethra.

DEEP PERINEAL POUCH

This is the narrow space enclosed by the fasciae of the urogenital diaphragm (*fig. 21.4*).

Contents

The contents include: (1) The UG diaphragm (transversus perinei profundus and sphincter urethrae); (2) the membranous urethra, which is thin-walled and 1 cm long, perforates the diaphragm; and (3) 2 small glands, the *bulbourethral glands* (of Cowper), each the size of a pea, lie deep to the diaphragm and alongside the urethra. Their long ducts travel in the wall of the urethra for 2–3 cm before opening into the spongy urethra (*figs. 21.4, 21.5*); and (4) vessels and nerves to be described now.

INTERNAL PUDENDAL VESSELS AND PUDENDAL NERVE

The artery, which is a branch of the internal iliac artery, and the nerve, which arises from sacral segments 2, 3,

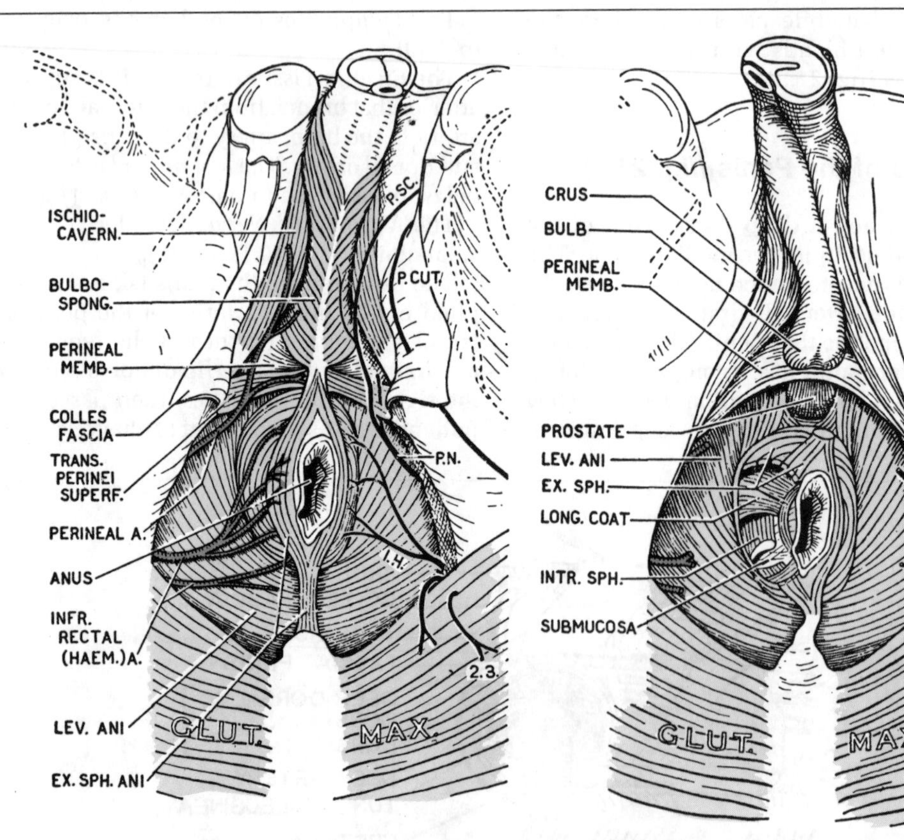

Figure 21.10. *Left*, superficial dissection of the male perineum (by Dr. H. C. Hair). *Right*, exposure of the prostate. Dissection of the anal canal (by Dr. V. P. Collins). *IH* = inferior rectal (hemorrhoidal) nerve; *PN* = perineal nerve; *P SC* = posterior scrotal nerve; *P CUT* = perineal branch of posterior cutaneous nerve of thigh.

and 4, together leave the pelvis through the greater sciatic foramen and flit through the gluteal region (*see fig. 21.12*), crossing the ischial spine and passing through the lesser sciatic foramen to enter the pudendal canal on the side of the ischiorectal fossa. They have 3 territories to supply: (1) anal triangle; (2) urogenital triangle and scrotum (labium majus in female); and (3) penis (clitoris in the female).

Dermatomes (*see fig. 21.11*)

The nerve divides while entering the pudendal canal into 3 terminal branches (*figs. 21.12, 21.13*)—(1) inferior rectal; (2) perineal; and (3) dorsal nerve of penis. *Details* are as follows.

1. The *inferior rectal (hemorrhoidal) nerve* supplies the external sphincter ani, the skin about the anus, and the lining of the canal below the anal valves. It may enter the region independently by piercing the sacrospinous ligament (Roberts and Taylor). This nerve can be tested on rectal examination.
2. The *perineal nerve* runs through the pudendal canal to the perineal membrane, where its 2 cutaneous branches, the *posterior scrotal nerves*, become superficial and continue to the scrotum. Its motor branch, the *deep perineal nerve*, innervates the skeletal muscles within the urogenital diaphragm and the 3 superficial muscles, and sends twigs to levator ani and external sphincter ani.
3. The *dorsal nerve of the penis* burrows through to the dorsum of the penis, where it lies deep to the fascia penis. It supplies the glans, the prepuce, the skin of the penis, and the spongy urethra (*fig. 21.9*).

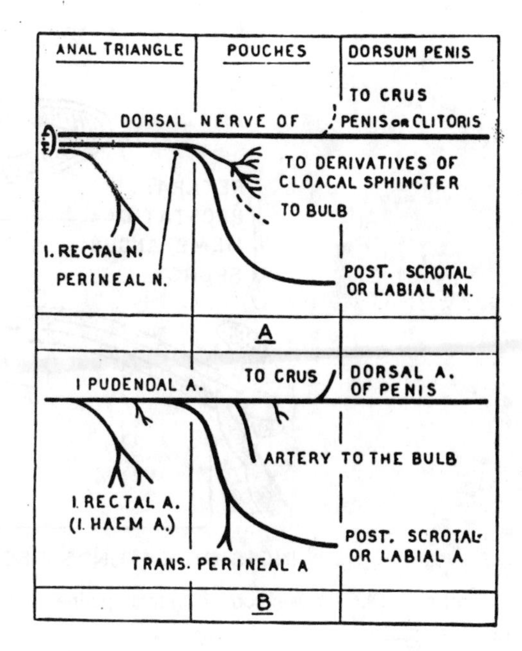

Figure 21.13. **A**, the 3 *divisions* of the pudendal nerve: each for a region; **B**, the 3 *branches* of the pudendal artery. A necessary difference in terminology, but an essential similarity in anatomy.

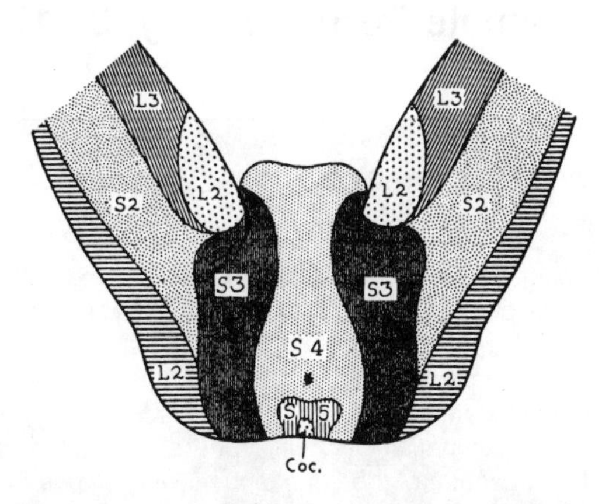

Figure 21.11. The dermatomes of the perineum.

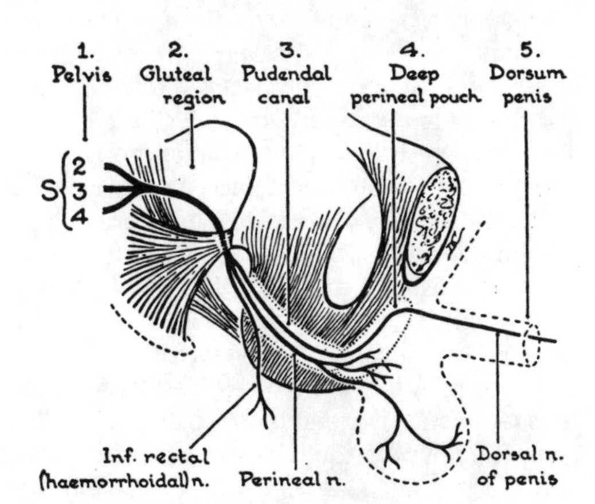

Figure 21.12. The course of the pudendal nerve: its 3 divisions and 5 regions traversed.

The **Internal Pudendal Artery** travels through the pudendal canal and deep perineal pouch to become the *dorsal artery of the penis*. Deep to the fascia penis, on the dorsum of the penis, it lies between its companion vein and nerve.

Its *branches* are: (1) The *inferior rectal artery* (inferior hemorrhoidal artery) which supplies the anal triangle; (2) The *perineal artery*, which gives off the slender *transverse perineal artery* and *posterior scrotal arteries*; (3) The *artery to the bulb*, a large vessel to the bulb; and (4) The *artery to the crus* (deep artery of the penis), which arises deep to the crus, and at once plunges into it.

The Vasomotor Nerves to the cavernous tissue are derived from the pelvic splanchnic (parasympathetic) nerves and the hypogastric plexus (sympathetic). They pass below the symphysis pubis to the erectile tissue.

Erection

Stimulation of the pelvic splanchnic nerves produces erection of the penis by causing dilatation of the arteries

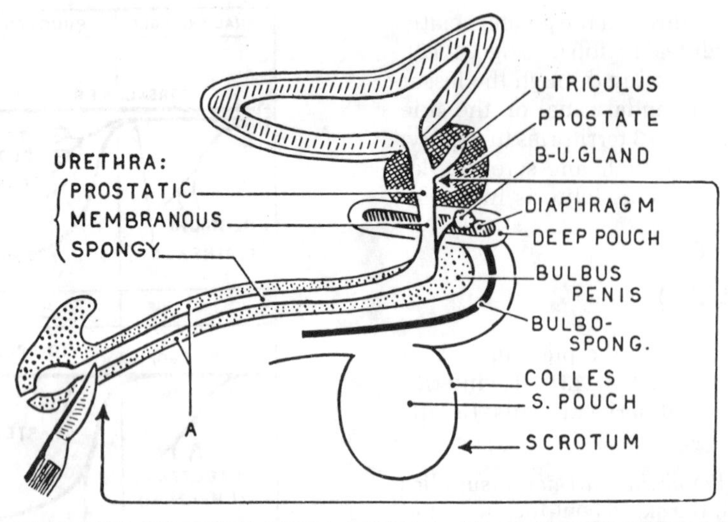

Figure 21.14. Incision converting male perineum into female. *B-U* = bulbourethral gland; *S POUCH*, superficial pouch.

and cavernous tissue. In the normal reflex act, the afferent limb is the pudendal nerve (S2, 3, 4); the efferent is the pelvic splanchnic (S2, 3, 4) (*see fig. 20.1*). By causing the bulbospongiosus and ischiocavernosus to contract, the pudendal nerve is responsible for increasing the erection. These two muscles compress the veins draining the erectile tissue.

Ejaculation

The impulse-spreading, sympathetic nerves are stimulated to cause closure of the internal urethral orifice, to set up peristaltic waves, which empty the epididymis and propel its contents through the vas to the urethra, and to empty the seminal vesicles and prostate of their secretions. The cord segment is L1. The path is probably via the intermesenteric and hypogastric plexuses, because the usual operation for removal of the lumbar sympathetic chain does not impair ejaculation, but removal of the hypogastric plexus does so permanently (Learmonth).

Exposure of the Prostate from the Perineum (*fig. 21.10*). Between the borders of the 2 levatores ani, pushing the anal canal and rectum backward exposes the tough fascia covering the posterior surface of the prostate.

An incision into the male urethra, entering it on the undersurface just behind the glans penis and carried back to the prostatic urethra, restores it to its fetal condition, which is similar in both sexes (*fig. 21.16*). The incision (*fig. 21.14*) divides everything encountered, including the urethra, scrotum, bulbospongiosus, bulb of the penis, and UG diaphragm. However, the female penis, called the clitoris, is diminutive and is not traversed by the urethra. It comprises 2 *corpora cavernosa clitoridis* (L. = *of clitoris*) and a *glans clitoridis*, which caps the conjoint corpora cavernosa (*fig. 21.15*).

Female Perineum (*fig. 21.15*)

Clinical Case 21.2

Patient Rose T. This 26-year-old woman in the 36th week of her first pregnancy is admitted to the obstetrics ward of a community hospital where you are doing a clinical elective. She has been experiencing labor pains at intermittent intervals. A perineal examination by your supervisor has revealed the presence of the fetus in the lower region of the birth canal. Some dilation of the vaginal orifice is evident. You accompany her and her husband to the delivery room and observe the ever-magic event of parturition three hours later. The delivery proceeds normally, but the physician decides that an episiotomy is necessary to prevent the tearing of the perineum through the perineal body. A local anesthetic is injected bilaterally into a specific part of the ischiorectal fossae to anesthetize the pudendal nerves. A painless incision in the perineal skin and fascia (at a 45° angle to the midline from the posterior vaginal orifice toward the right ischial tuberosity) relieves the perineal tissue. A normal birth occurs within 30 minutes, and the newborn's respiratory tract is cleared as regular breathing begins. The physician sutures the episiotomy incision and the mother is moved to the maternity ward, where she reassures you that the perineum (dermatomes S2, 3 and 4) have stayed anesthetized for two hours; but then some discomfort is experienced in the area of the episiotomy when sensation returns to this area. This is managed with some mild analgesics and facilitated by her elation. Mother and child go home after four days.

Each *labium minus* is a thin cutaneous fold, devoid of fat and lying alongside the orifice of the vagina. The posterior end is free. The anterior end divides into 2 lesser folds that unite with their fellows across the median plane, the upper folds forming a hood, the *prepuce of the clitoris*, over the glans, the lower joining to form a band, the *frenulum of the clitoris*, which is attached to the undersurface of the glans.

Each *labium majus* is a broad, rounded, cutaneous ridge lying lateral to the labium minus and covering a long finger-like process of fat. This process extends backward from a median skin-covered mound of fat, the *mons pubis*, situated in front of the pubis. Entering the fat of the labium posteriorly and running forward are the posterior labial (cf. scrotal) nerves, arteries, and veins and also the perineal branch of the posterior cutaneous nerve of the thigh. Entering it anteriorly are branches of the ilio-inguinal nerve and external pudendal artery and vein and a fibrous band called the *round ligament of the uterus* (*fig. 19.5*).

Sebaceous glands open onto both surfaces of the labia minus and majus; hair covers the mons and the lateral surface of the labium majus.

The superficial perineal fascia and pouch are split into right and left parts and contain the bulbospongiosus, which is split into right and left parts (sphincter vaginae).

The right and left *bulbs of the vestibule* underlie the bulbospongiosus.

Each *bulb* is a loosely encapsuled mass of erectile tissue, shaped like a half pear. The convex surface is lateral and is covered with bulbospongiosus; the flat or concave surface is medial and is applied to the perineal membrane which intervenes between it and the wall of the vagina; the enlarged end of the pear is posterior; the narrow stalk is anterior and, after joining its fellow, ends as the *glans clitoridis*.

To allow for the vaginal canal, the urogenital diaphragm, its inferior fascia (perineal membrane), and its superior fascia are split into right and left parts. These blend with the vaginal wall.

The *vestibule* of the vagina is the cleft between the labia minora. The *hymen* is a thin membranous fold of

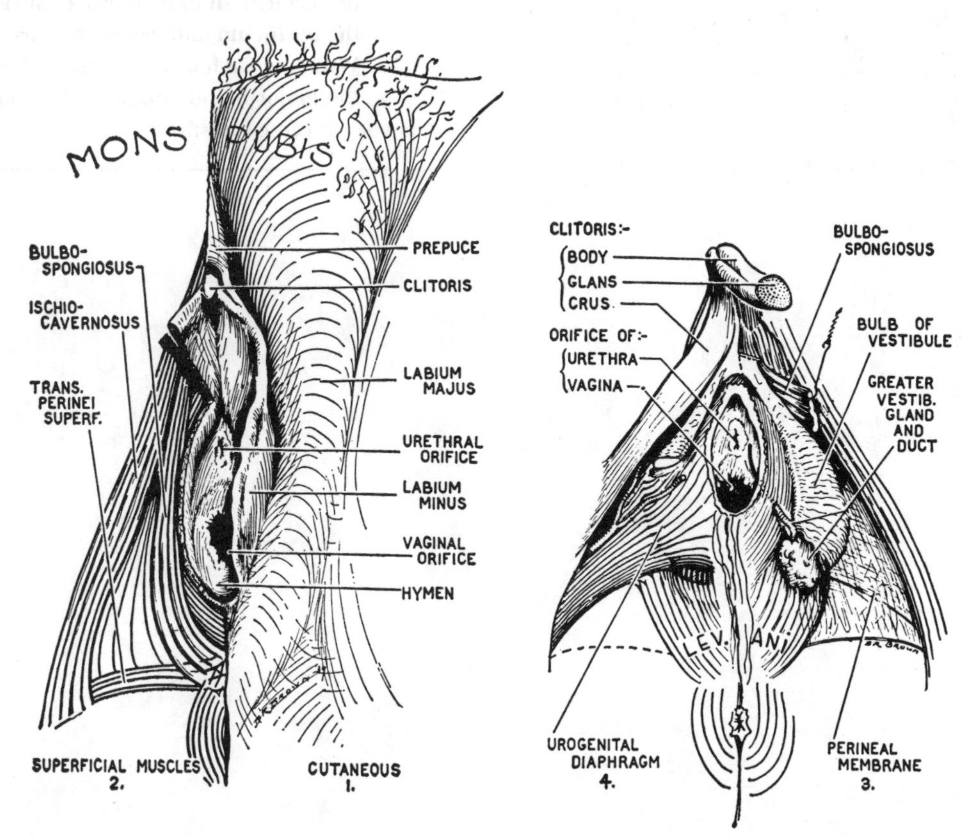

Figure 21.15. Successive dissections (1, 2, 3, and 4) of the female urogenital triangle.

irregular outline that surrounds the vaginal orifice like the ruptured membrane of a drum. Opening into the vestibule are the urethra, vagina, paraurethral glands, and greater and lesser vestibular glands. (The lesser glands are the many small mucous glands that open here.)

The *greater vestibular glands* (Bartholin's glands) open by a 2-cm long duct onto the tissue between the hymen and the labium minus in the vestibule.

Each *gland* is larger than a pea (but smaller in the aged) and is situated at the posterior end of the bulb.

The female urethra lies immediately in front of the anterior wall of the vagina and is intimately adherent to it (*fig. 19.2*).

The *urethral orifice* opens just in front of the vaginal orifice and is 2–3 cm behind the glans.

Though a prostate is not found in the female, the *paraurethral glands*, whose ducts open one on each side of the female urethra, are probably homologous with prostatic glands.

The ejaculatory ducts, which in the male open onto the lips of the prostatic utricle (*fig. 18.25*), generally disappear in the female, but, as the ducts of Gartner (*fig. 4.6*, p. 53), they may persist as blind tubes on the anterior wall of the vagina and become cystic; rarely, they open onto the skin surface.

In the male, the primitive urethral orifice opened in the perineum. Later, when the lips of the genital folds met and fused, a secondary orifice was formed behind the glans penis (*fig. 21.16*); subsequently, the urethra traversed the glans; so, a third and permanent orifice opens at the end of the glans. In the female, and as a rare anomaly in the male, the primary perineal orifice is the permanent one. The glans clitoridis is not canalized.

In the male, the gubernaculum testis passed to the scrotum, and the processus vaginalis peritonei and the testis followed it. In the female, the gubernaculum ovarii and the processus vaginalis (of Nuck) entered the labium majus but, owing to a side attachment that the gubernaculum makes with the uterus, the *ovary* enters the pelvis; only rarely does it descend into the labium (*fig. 19.5*).

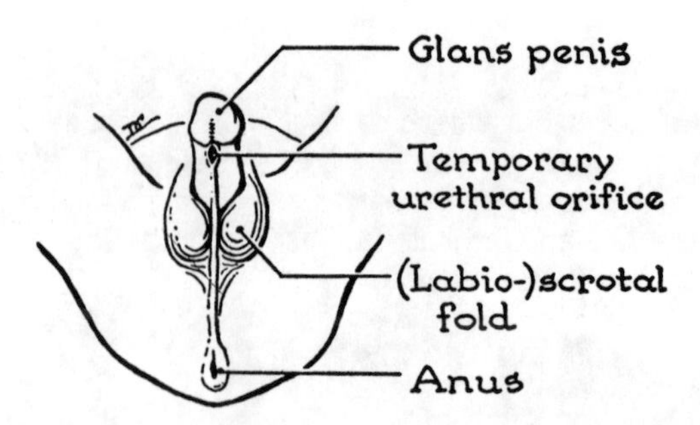

Figure 21.16. Showing (1) the paired right and left labioscrotal (genital) folds about to unite below the penis to form the scrotum and (2) the urethral orifice moved forward.

Clinical Mini-Problems

1. **Which dermatomes are represented on the scrotum and the labia majora?**
2. **Which nerve is anesthetized by a local anesthetic to perform an episiotomy (a surgical incision into the perineum and vagina for obstetrical purposes)?**
3. **Which branch of the internal iliac artery provides the major blood supply to the penis?**

(Answers to these questions can be found on p. 586.)

22

Anatomical Features in the Physical Examination of the Male and Female Pelvis

Rectal Examination

The rectal examination is considered essential for a complete examination. In the male, it will reveal some of the possible abnormalities of the prostate and of the rectum itself. During childbirth, a rectal examination provides an approach to the deepest part of the vagina and external os of the uterus that avoids possible contamination of the birth canal in the early stages.

Vaginal examination with the examining physician's other hand palpating through the abdominal wall ("bimanual") can reveal to the expert the existence or character of abnormalities of all the female internal genitalia. Direct visual examination of the external os and "Pap smear" are routine in many physicians' full physical examination of adult women.

Pelvic Examination

FEMALE

With the patient in the lithotomy position (supine with thighs abducted and knees flexed), the external genitalia (labia majora, labia minora, clitoris) and urethral opening are inspected and palpated. The perineum is also inspected and palpated. An intravaginal examination to examine the cervix of the uterus is done by placing a speculum in the vagina. Epithelial specimens should be obtained for Papanicolaou smears during this examination. An anteflex (anteverted) uterus can be palpated manually by inserting the index and middle finger of one hand into the vagina and anterior fornix and placing the palm of the other hand on the anterior wall of the abdomen between the pubis and the umbilicus. Moving the two hands gently toward each other, one will feel the fundus of the uterus interposed between the hands just above the level of the pubis.

MALE

Examine the penis and retract the foreskin from the glans if the patient is uncircumcised. The external meatus of the urethra is visualized by pressing the glans between the thumb and index finger. Inspect the scrotum and test for herniations through the superficial inguinal ring. Palpate the testes, epididymis and vas deferens.

The rectal examination may be required in perineal examination of both sexes, but it is of critical importance in the male. With the patient in a knee-chest position, the anus is inspected for tissue integrity and normal vascularity (color and venous structures). The examiner's gloved index finger is lubricated and inserted into the anal canal. This action will evoke an anal reflex, causing the external anal sphincter to tighten around the examiner's finger. The integrity of the S2,3,4, levels of the spinal cord can be assessed in this examination. One can make assumptions about the integrity of the pelvic splanchnic nerves from this examination result. All four walls of the rectum can be palpated. The prostate can be palpated from the anterior wall of the rectum. A normal prostate protrudes slightly into the rectum. Hypertrophied or neoplastic prostates will project more prominently into the rectum and will have a different consistency than a normal prostate when palpated.

Lower Limb 5

23

Femur and Front of Thigh

Front of Thigh

BONY LANDMARKS

The bony parts labeled in *Figure 23.1* should be located on the skeleton and palpated on the living model, which will likely be yourself or a fellow student; also feel the ischial tuberosity by sitting on your fingertips.

The greater trochanter lies about 10 cm below the tubercle on the iliac crest. Grasp it. It is subaponeurotic and, so, difficult to feel unless the parts are relaxed (e.g., by lying down).

The *body of the femur* is buried in muscles and runs obliquely to its lower end at the knee.

Clinical Case 23.1

Patient Ann A. For the past two days, this woman, age 48, has complained of a tender lump in the right anteromedial aspect of the right thigh near the external genitalia. This focuses attention on various structures of the front of the thigh and femoral triangle, including vessels, and associated nerves, lymph nodes, muscles, fasciae, the femur, hip joint, and the passageways from the abdomen into the thigh. Any of these may account for the lump, which clinically could be very important or quite innocuous. If surgery is indicated, the relationships are very important. Almost all of this chapter bears upon the differential diagnosis and treatment of this problem.

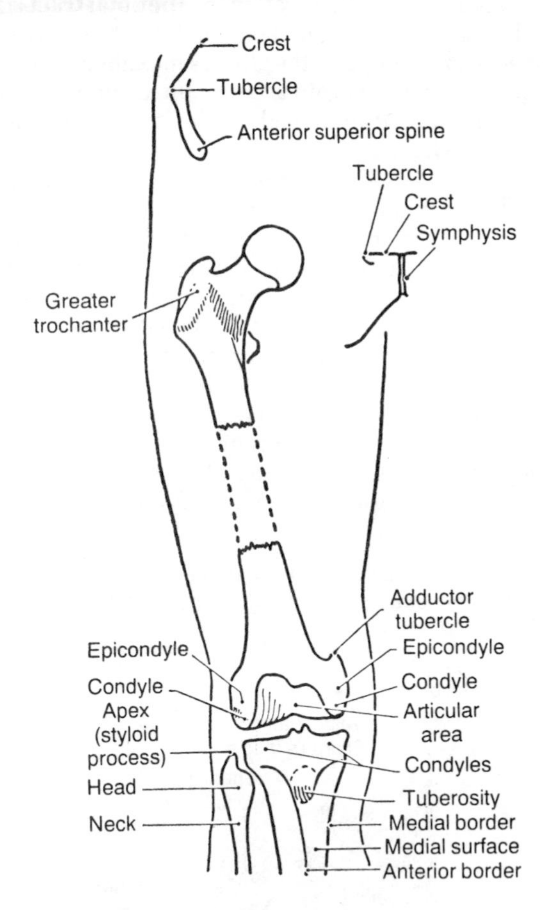

Figure 23.1. Bony landmarks of the thigh that you should palpate.

Superficial Structures

The skin, both dermis or true skin and epidermis, is thick and tough where it needs to be: on the back, buttocks, lateral aspects of both limbs, palm, and sole.

The superficial fascia over bony points, such as the patella, and that under skin creases is generally reduced; superficial to the ischial tuberosity and the heel, the fat is imprisoned in much fibrous tissue to form cushions. On the buttocks, large quantities of fat may be deposited.

The deep fascia is especially strong; in the thigh it is called the **fascia lata**. According to rule, it is attached to bone at those areas where the deep fascia and bone come into direct contact.

It is extremely strong laterally because in between 2 thin layers of circularly disposed fibers there runs a broad band of tough vertical fibers called the **iliotibial tract** (*fig. 23.3*). This band of fascia is the conjoint aponeurosis of the tensor fasciae latae and the gluteus maximus muscles.

The great saphenous vein ascends unaccompanied by an artery throughout the length of the limb, in the subcutaneous fat (*fig. 23.2*).

It begins at the medial end of the *dorsal venous arch* of the foot, passes anterior to the medial malleolus, crosses the lower third of the medial surface of the tibia, and follows 1 cm posterior to its medial border as far as the knee. At the knee, it is found by incising a handbreadth (or slightly more) posterior to the medial border of the patella. From there it courses up the thigh to join the femoral vein 4 cm inferolateral to the pubic tubercle.

It anastomoses freely with the small saphenous vein inferior to the knee and communicates along intermuscular septa with the deep veins. It receives numerous tributaries, including 3 superficial veins near the inguinal ligament.

These 3 veins accompany branches of the femoral artery: (1) the *external pudendal artery*, which crosses anterior to the spermatic cord to supply the scrotum or labium majus; (2) the *superficial epigastric artery*, which passes toward the navel; and (3) the *superficial circumflex iliac artery*, which passes laterally inferior to the inguinal ligament.

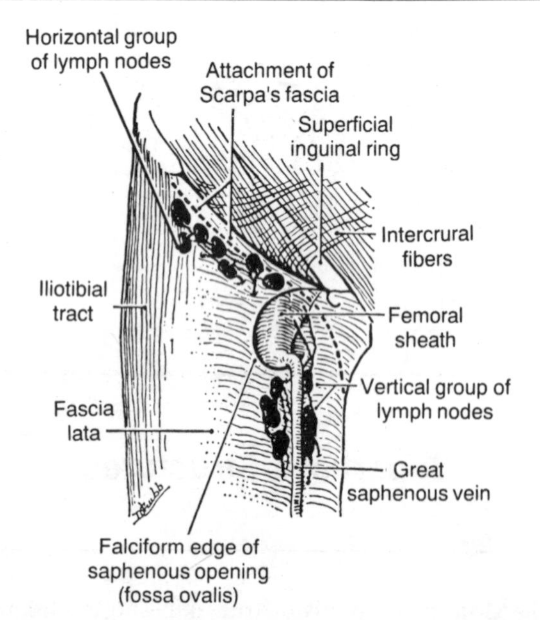

Figure 23.2. Saphenous opening; great saphenous vein joining the femoral vein within the femoral sheath; superficial lymph nodes; and the attachment of "Scarpa's" fascia (*broken line*).

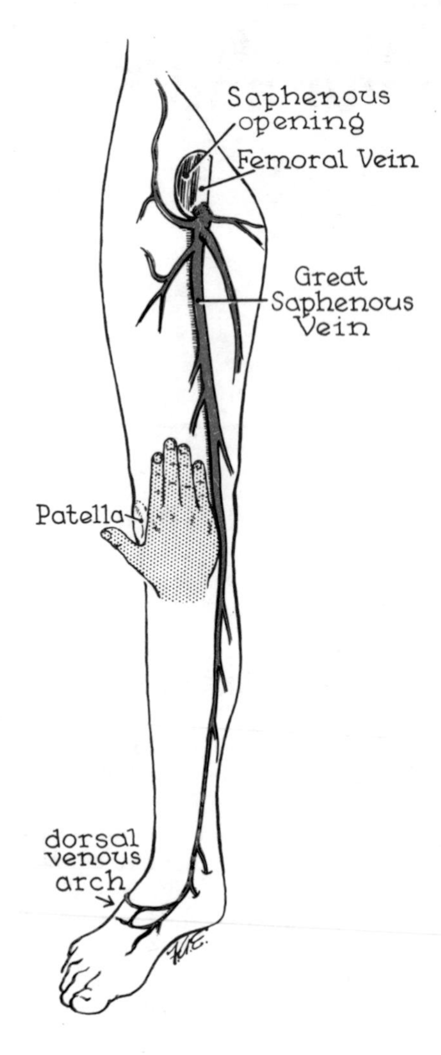

Figure 23.3. The great saphenous vein.

Saphenous Opening or Hiatus (*fig. 23.2*)

The great saphenous vein has to pass through the fascia lata in order to reach the femoral vein 4 cm inferior to the pubic tubercle; it passes through a large hole called the *saphenous opening*.

From the pubic tubercle, the lateral border of the saphenous opening is crescentic (falciform margin) and blends with some areolar tissue (cribriform fascia) that bridges the opening.

Lymph Nodes

In the lower limb, the lymph nodes are situated at the knee and the groin, those at the knee (popliteal) being deep and impalpable but those at the groin (inguinal) being superficial and palpable.

The Inguinal Nodes (*fig. 23.3*) are in 2 groups, a superficial and a deep. The *superficial nodes* are subdivided into (1) a superior horizontal group and (2) an inferior vertical group distributed along the great saphenous vein.

The inferior superficial inguinal nodes receive all the superficial lymph vessels of the lower limb except a few that, following the small saphenous vein, end in the popliteal nodes. *The superior superficial inguinal nodes* drain the regions supplied by the 3 superficial inguinal blood vessels, namely, the subcutaneous tissues of the anterior abdominal wall inferior to the navel, of the penis and scrotum and the vulva and distal part of the vagina, and of the gluteal region, perineum, and distal part of the anal canal, **but not of the testis or ovary**.

The *deep nodes*, 1–3 in number, lie deep to the fascia lata on the medial side of the femoral vein, in and inferior to the femoral canal (soon to be described). They receive the deep lymph vessels of the limb and of the glans penis or glans clitoridis and spongy urethra.

Cutaneous Nerves (*fig. 23.4*)

The cutaneous nerves of the front of the thigh are derived from the ventral rami of L1, 2, 3, 4. The *lateral, intermediate*, and *medial cutaneous nerves of the thigh* and the *saphenous nerve* are branches of the femoral nerve. They pierce the deep fascia along an oblique line that roughly marks the sartorius muscle (*see fig. 23.10*), but their exact disposition is unimportant. Details are given below for specialists.

The *lateral cutaneous nerve* commonly arises independently from the lumbar plexus and enters the thigh close to the anterior superior spine. When it springs from the femoral nerve, it enters at some distance medial to the spine. A posterior branch passes to the gluteal region.

A posterior branch of the medial cutaneous nerve extends to the calf, and the saphenous nerve extends halfway along the medial border of the foot. The other branches remain proximal to the knee.

The *saphenous nerve* comes to the surface between the sartorius and gracilis muscles, posterior to the adductor tubercle (*fig. 23.1*). It gives off a *patellar branch* inferior to the patella.

The skin just inferior to the inguinal ligament (*fig. 23.4*) is supplied by (1) the *ilioinguinal nerve* and (2) the femoral branch of the *genitofemoral nerve*.

A branch of the *obturator nerve* becomes cutaneous at the middle of the thigh and may extend to the calf.

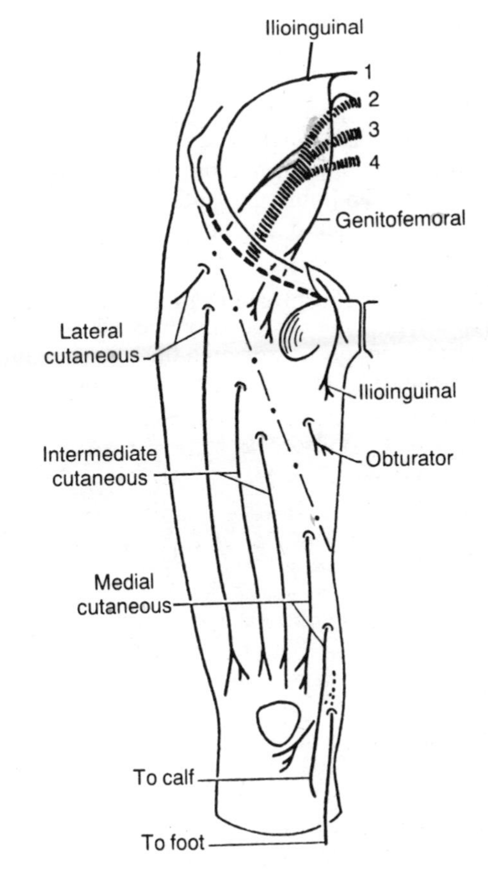

Figure 23.4. Cutaneous nerves of the front of the thigh, above and below the line of the sartorius.

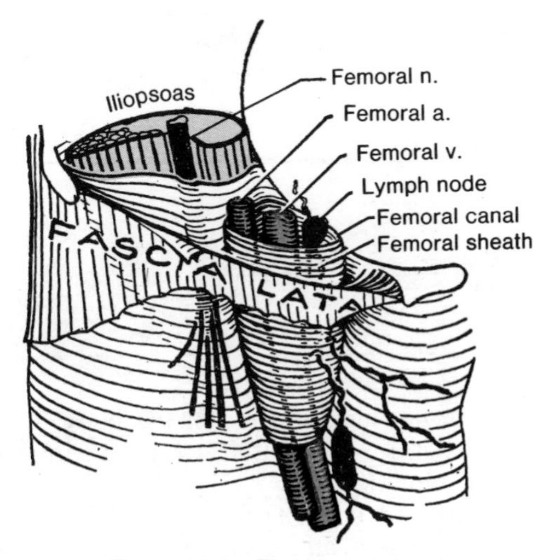

Figure 23.5. The femoral sheath.

Femoral Sheath (*fig. 23.5*)

Deep to the plane of the fascia lata, the femoral artery, vein, and some lymph vessels are wrapped up in a prolongation of the extraperitoneal areolar tissue that envelops the external iliac vessels in the abdomen. This wrapping, the femoral sheath, is funnel-shaped; it has 3 compartments—a lateral one for the artery, a middle one for the vein, and a medial one, the **femoral canal**, partly occupied by lymph vessels and deep nodes. It is into the femoral canal that a femoral hernia may bulge from the abdominal cavity above.

The femoral ring is the mouth of the femoral canal. It is bounded *laterally* by the femoral vein; *posteriorly* by the superior ramus of the pubic bone covered with a coating of the unimportant pectineus muscle and its fascia; *medially* at a variable distance (Doyle) by the lacunar ligament and the conjoint tendon, both of which are attached to the pecten pubis; and anteriorly by the inguinal ligament and the round ligament of the uterus or the spermatic cord.

To enlarge the ring in an operation to release a femoral hernia, you dare not cut laterally into the vein; it would be useless to cut posteriorly onto the bone; you must cut either medially or anteriorly (see *fig. 23.14*).

Clinical Case 23.2

Patient Brian B. This 10-year-old boy has been complaining of aching pain deep in the upper part of his left thigh for several weeks. He has a slight fever and no definite history of trauma. Your clinical supervisor in pediatrics suspects the possibility of a serious inflammatory or a neoplastic disease of the femur, and even after careful physical examination, she requires radiographs of the region. A clear knowledge of femoral anatomy and relationships is crucial to the diagnosis and treatment of this child's condition, especially if emergency surgical treatment (e.g., amputation) is necessary.

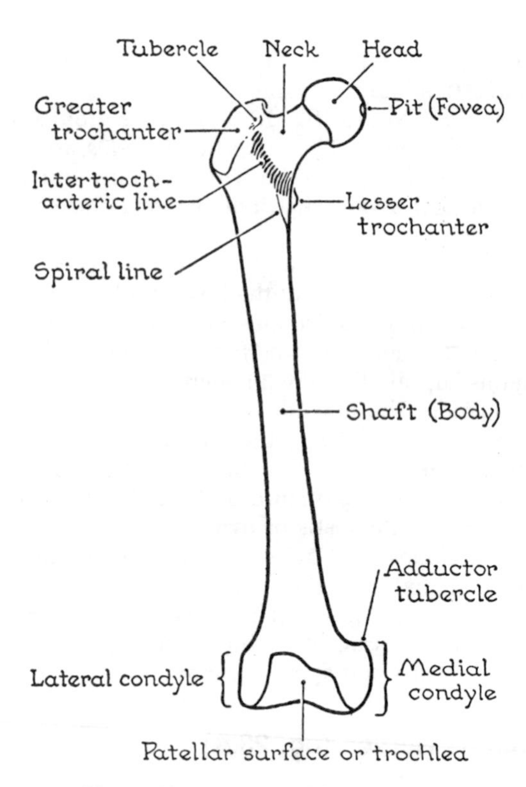

Figure 23.6. The femur (anterior view).

Femur

The femur is the longest bone in the body, being a quarter of the stature, or about 45 cm (or 18 inches) in an average individual.

The proximal end presents for examination a head, neck, greater trochanter, and lesser trochanter (*figs. 23.6, 23.7*).

The *Head* forms two-thirds of a sphere, and is directed medially and anterosuperiorly. It is much more secure in its socket than the head of the humerus.

The *Neck* is obliquely placed. It is limited laterally by the greater trochanter and really is the medially curved proximal end of the shaft (*fig. 23.8*). A broad, very rough, oblique line, the *intertrochanteric line*, is due to the attachment of the massive iliofemoral ligament.

The posterior aspect of the neck is separated from the shaft by a prominent, rounded ridge, the *intertrochanteric crest.*

In the child, the pelvis is narrow before the bladder descends into the pelvis; so, the neck and shaft of the femur are nearly in line with each other. As the pelvis widens, the neck becomes more horizontal and the angle between neck and shaft becomes smaller (125°, male).

The *Lesser Trochanter* is the cone-shaped traction epiphysis of the iliopsoas muscle. It projects from the posterior surface of the bone, but it points medially (*fig. 23.7*).

The *Greater Trochanter* is a traction epiphysis, especially for the gluteus medius, which draws it superomedially and posteriorly.

Some Details. The *quadrate tubercle* marks the site where the epiphyseal line crosses the intertrochanteric crest. On the medial side of the greater trochanter, there is a depression, or the *intertrochanteric fossa*, in which the obturator externus muscle is inserted. From the fossa, a shallow groove for the tendon of the obturator externus runs horizontally across the posterior surface of the neck (see *fig. 25.12*).

The body is circular on cross-section; from its posterior surface a broad rough line, the **linea aspera**, stands out. Many muscles and 3 intermuscular septa crowd onto this line. The linea aspera bifurcates superiorly and inferiorly (*fig. 23.7*).

The anterior and lateral aspects of the body are smooth because they give origin to the fleshy fibers of the vastus

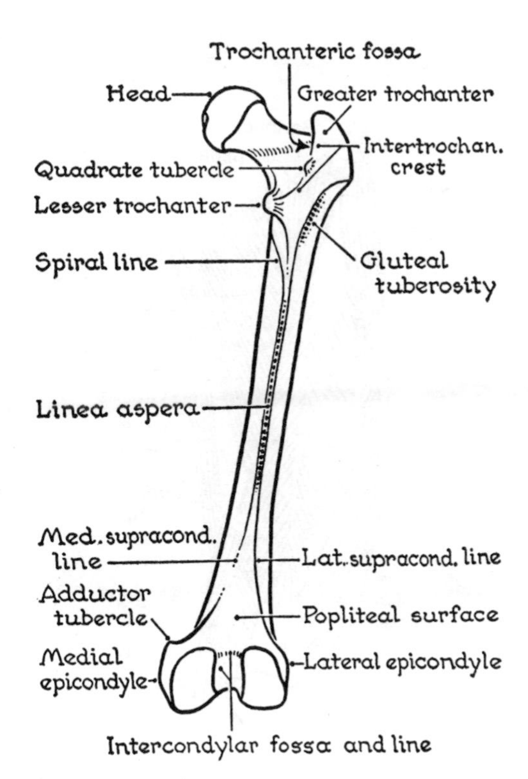

Figure 23.7. The femur (posterior view).

Figure 23.8. The neck is the curved shaft to which the trochanters are added.

intermedius; the medial aspect is bare, and therefore, also smooth.

The distal end of the femur is divided into 2 large knuckles, the *medial* and *lateral condyles*. The condyles project posteriorly like thick discs beyond the popliteal surface (*fig. 23.9*). Between the opposed surfaces of these discs is a U-shaped notch, the *intercondylar fossa*, the width and depth of a thumb. At the center or hub of the nonopposed surface of each condyle, there is a fullness, the *epicondyle*, to which the medial and lateral ligaments of the knee joint are attached (*see fig. 26.9*). The femoral condyles are covered with cartilage both inferiorly and posteriorly for articulation with the tibia, and the cartilage of the 2 sides meets in front in a V-shaped trough, the *patellar surface* in which the patella plays. The lateral

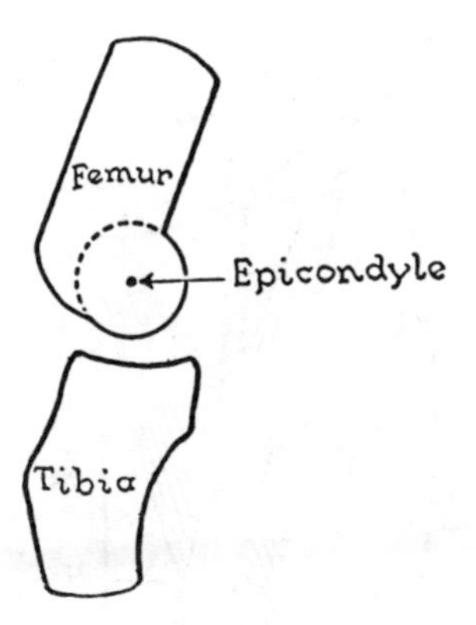

Figure 23.9. A disc-shaped condyle and an epicondyle.

lip of this pulley projects further anteriorly and superiorly than the medial lip.

The **epiphyseal line** of the *distal end* of the femur runs irregularly through the adductor tubercle and along the intercondylar line (*fig. 23.7*). At *birth* the distal epiphysis is an ossific nodule about ½ cm in diameter.

Variations. A *third trochanter* is an enlarged gluteal tubercle; it is constant in certain rodents and the horse, and may appear in any mammal, including man. *Platymeria* or marked anteroposterior flattening of the proximal part of the shaft of the femur occurs in some races, but is not common in Europeans.

Clinical Case 23.3

Patient Charles C. This 42-year-old man has symptoms and signs of cardiovascular problems and requires the threading of a fine catheter through his arteries into the aorta to inject radiographic contrast media for X-ray studies. The femoral artery route is chosen, and you will be assisting the physician with the procedure. Obviously, you must know the anatomy of the vessels in the femoral triangle and adductor canal as well as all the neighboring structures. The rest depends on the professional and technical skill of the clinician.

Femoral Triangle

The base of the femoral triangle (*fig. 23.10*) is formed by the inguinal ligament. Its apex lies about 10 cm below

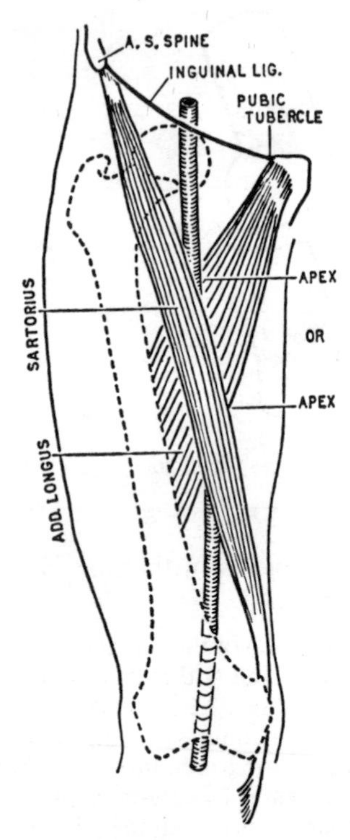

Figure 23.10. The sides of the femoral triangle. The course of the femoral artery.

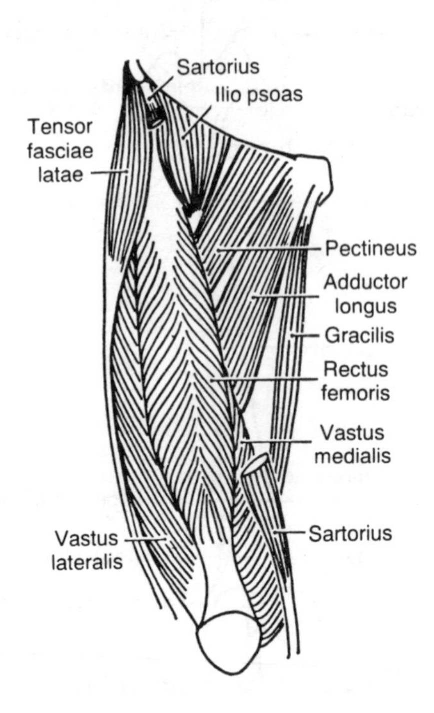

Figure 23.11. The floor of the femoral triangle. The walls of the adductor canal.

the inguinal ligament, where the sartorius crosses the lateral border of the adductor longus.

The central and dominant structure within this triangular frame is the *femoral artery* (*Plate 23.1*). It begins at the *midinguinal point*, immediately anterior to the femoral head. It leaves the triangle at its apex and enters the adductor canal (p. 264), and so only its proximal 10 cm lie in the triangle; its remaining 15 cm travel through the adductor canal.

The floor of the triangle is muscular (see *fig. 23.11*).

CONTENTS OF THE TRIANGLE (*figs. 23.12* and *23.13*)

These are the femoral artery, vein, and nerve, certain of their branches, and the deep inguinal lymph nodes and profunda femoris artery and vein and their circumflex branches (p. 264).

The *profunda femoris artery* usually takes origin several centimeters inferior to the ligament. It is only slightly smaller than the continuation of the femoral artery itself because it supplies many muscles of the thigh.

The *femoral vein* lies medial to the femoral artery in the *femoral sheath*, but at the apex of the triangle it has slipped posterior to the femoral artery; along with the profunda artery and vein, the 4 vessels are almost inseparably united in a tough areolar sheath. A stab or bullet wound at the apex of the triangle may penetrate in succession the 4 great vessels of the limb (*fig. 23.12*).

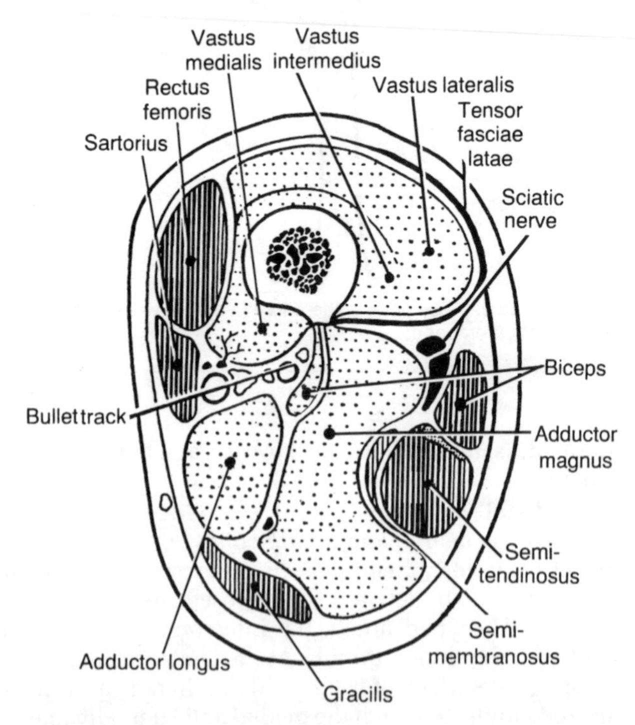

Figure 23.12. Cross-section of thigh near apex of the femoral triangle showing (1) attachments of muscles to linea aspera, (2) 6 muscles (hatched) that span the femur, and (3) that a bullet might penetrate the femoral and profunda femoris vessels.

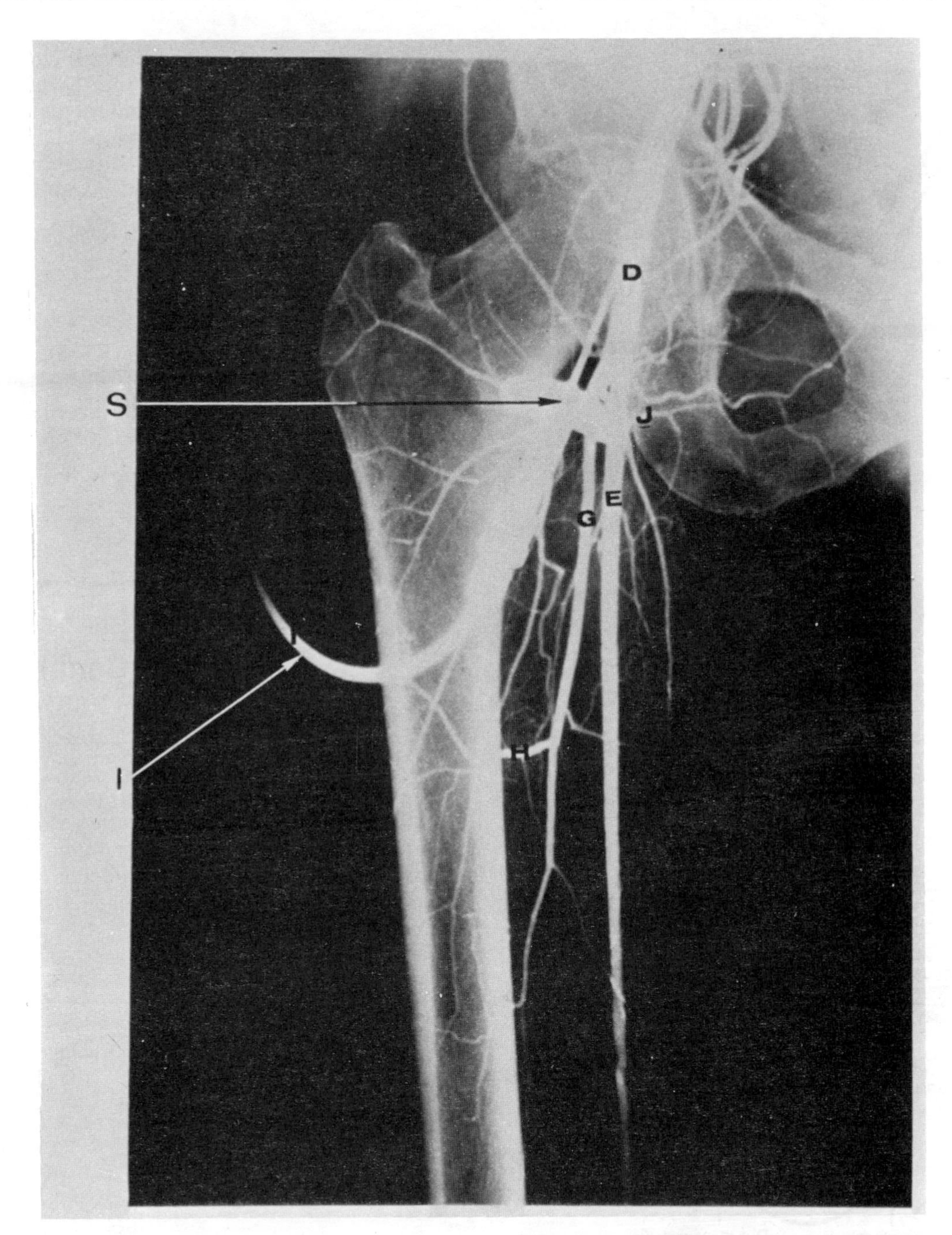

Plate 23.1. Femoral arteriogram, right. *D*, femoral artery; *E*, femoral artery distal to origin of profunda femoris; *G*, profunda femoris artery; *H*, first perforating artery; *I*, the catheter; *J*, superficial external pudendal arteries; *S*, syringe and needle for injecting contrast medium through the catheter into femoral artery.

The great *saphenous vein* is the last tributary to join the femoral vein.

Venous Valves

The saphenous and femoral veins have valves throughout their course in the lower limb. The inferior vena cava and the common iliac vein have no valves.

Indeed, in 24% of 200 limbs no valve was found between the mouth of the saphenous vein and the heart. This may be important in relation to varicose veins. In 76%, one valve was found above the saphenous vein, usually in the femoral vein but sometimes in the external iliac, and sometimes in both. Usually (90%), a valve occurs just below the opening of the profunda vein, and there may be 2 or 3 others (Basmajian). The saphenous vein is well supplied with valves.

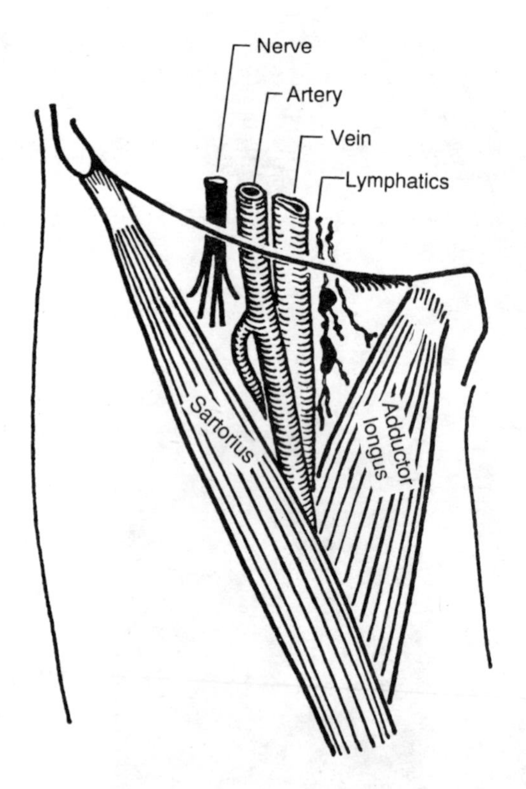

Figure 23.13. The contents of the femoral triangle.

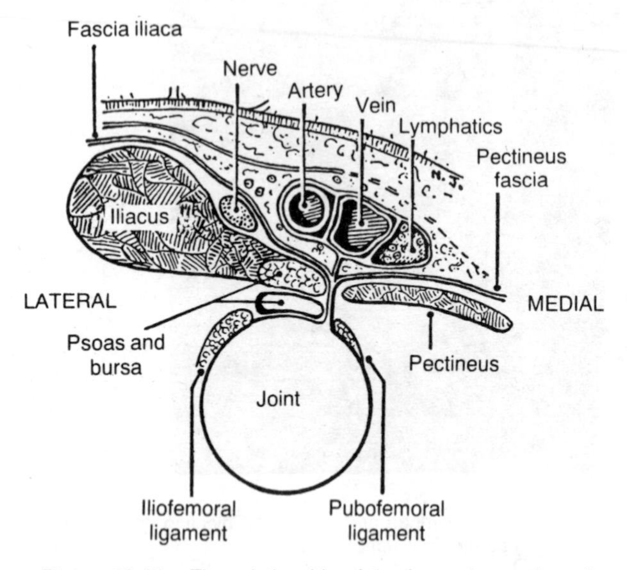

Figure 23.14. The relationship of the femoral vessels and nerve to the hip joint.

The *femoral nerve* enters the thigh slightly lateral to the artery, deep to the fascia lata. Soon after crossing deep to the inguinal ligament, it breaks up into numerous motor and sensory branches. Only two of these follow the artery, applied to its lateral side, into the adductor canal: one, the *nerve to the vastus medialis* is motor; the *saphenous nerve*, is sensory.

The *medial femoral circumflex* branch of the profunda artery passes posteriorly (leaving the triangle between pectineus and iliopsoas). The *lateral femoral circumflex* branch passes laterally through or behind the branches of the femoral nerve, disappears under cover of the sartorius and rectus femoris muscles and breaks up into 3 terminal branches.

Relationships

The femoral artery lies in front of the **psoas tendon**; the femoral nerve, in front of the iliacus; and the femoral vein and most lymph vessels, in front of the pectineus (*fig. 23.14*).

No motor nerve, except a *twig to the pectineus*, crosses the femoral artery.

Adductor Canal

The adductor canal (of Hunter) is the narrow outlet of the femoral triangle, enclosed by the sartorius and thereby converted into an intermuscular tunnel, 15 cm long. It ends 10 cm above the adductor tubercle (*fig. 23.10*). Only the femoral vessels and the saphenous nerve run right through it (*fig. 23.15*).

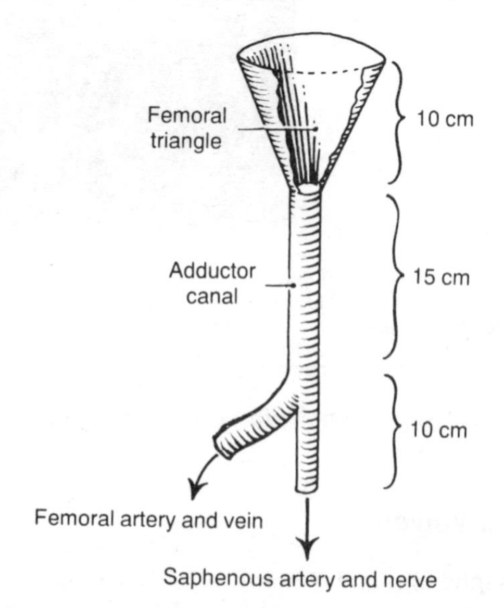

Figure 23.15. The femoral triangle and adductor canal (subsartorial "tunnel") and its 2 outlets.

SARTORIUS

This muscle arises from the anterior superior iliac spine and is inserted into the medial surface of the tibia, inferior to the level of its tuberosity (*fig. 23.10*). It is commonly pierced by the femoral cutaneous nerves in transit (*fig. 23.4*).

Nerve Supply—Femoral Nerve.

When origin and insertion are approximated, the limbs are brought into the position that the *tailor* (L. *sartor* = a tailor) traditionally assumes when at work—hence, its name, sartorius.

WALLS OF THE ADDUCTOR CANAL

The adductor canal (*see fig. 25.14*) is enclosed by the sartorius anteriorly and the iliopsoas and vastus medialis laterally. The posteromedial wall is composed of the 5 **adductors**, which radiate from the pubis and ischium to the linea aspera (*fig. 23.16*). They spread out fanwise as they descend to their aponeurotic insertions. *Adductor brevis* appears in the interval between *pectineus* and *adductor longus*; the *adductor magnus* appears between *adductor longus* and *gracilis* (*fig. 23.16*). The gracilis is inserted into the medial surface of the tibia deep to the sartorius, and it lies immediately posterior to the sartorius as the two cross the medial femoral condyle. The magnus descends as far as the adductor tubercle.

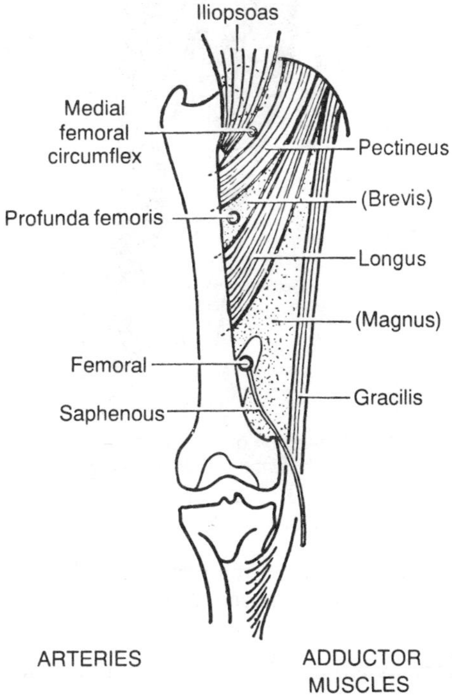

Iliopsoas
Medial femoral circumflex
Profunda femoris
Femoral
Saphenous
Pectineus
(Brevis)
Longus
(Magnus)
Gracilis
ARTERIES
ADDUCTOR MUSCLES

Figure 23.16. The muscles posterior to the femoral and saphenous arteries. Adductor muscles spread fanwise to their insertions. Arteries course between them, the femoral artery piercing the adductor magnus.

CONTENTS OF THE ADDUCTOR CANAL

The Femoral Artery and Its (Small) Branch, the Saphenous Artery

Their course is mapped out by a line joining the mid-inguinal point to the adductor tubercle. Its proximal two-thirds maps out the course of the femoral artery, the distal third that of the saphenous artery. The femoral artery becomes the popliteal artery by passing through a gap or hiatus in the insertion of the adductor magnus 10 cm superior to the adductor tubercle. However, the saphenous artery becomes cutaneous (*fig. 23.16*).

The femoral artery and vein are so firmly bound together in a common sleeve that one cannot move without the other.

The *nerve to vastus medialis* accompanies the femoral artery far into the canal, clinging to its muscle (*fig. 23.12*).

The *saphenous nerve* accompanies the femoral artery through the femoral triangle and adductor canal and then it accompanies the saphenous artery to the surface; here it joins and accompanies the great saphenous vein through the leg, ending halfway along the medial border of the foot.

Regions of the Thigh (Details)

Theoretically, the thigh may be regarded as divided into 4 muscular regions—flexors, extensors, adductors and abductors. Each is separated from its neighbor by an intermuscular septum and each contains a group of muscles and their nerve (*fig. 23.17*).

Actually, this scheme is departed from in the following main respects. Of the 3 abductors, all of which are supplied by the superior gluteal nerve, two (glutei medius and minimus) do not descend beyond the greater trochanter, while the third, the tensor fasciae latae, spans the femur and gains attachment to the anterior aspect of the lateral condyle of the tibia and to the side of the patella by means of the **iliotibial tract**. It falls to the 3 remaining groups to surround the femur.

The anterior muscles envelop and monopolize the shaft of the femur, except along its posterior border, which alone is free and available to the medial and posterior muscles and to all intermuscular septa. The attachments of the muscles to this restricted border are necessarily aponeurotic; hence, the roughness of the border and its name—the *linea aspera*.

Clinical Case 23.4

Patient Douglas D. This male middle-aged truck driver had a fractured right femur two months ago. Now, after his cast has been removed, the muscles appear obviously shrunken (atrophied), and he

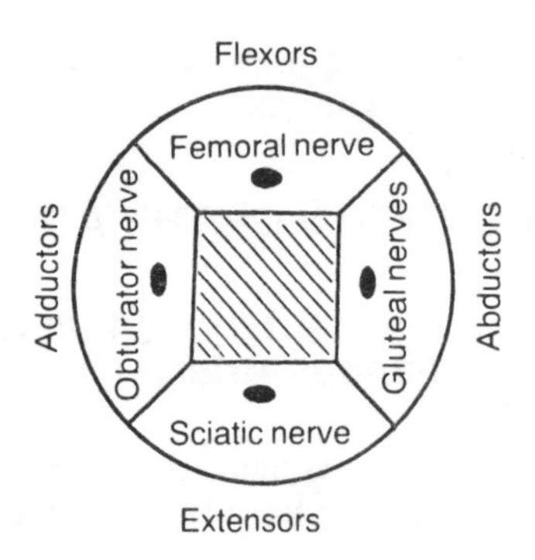

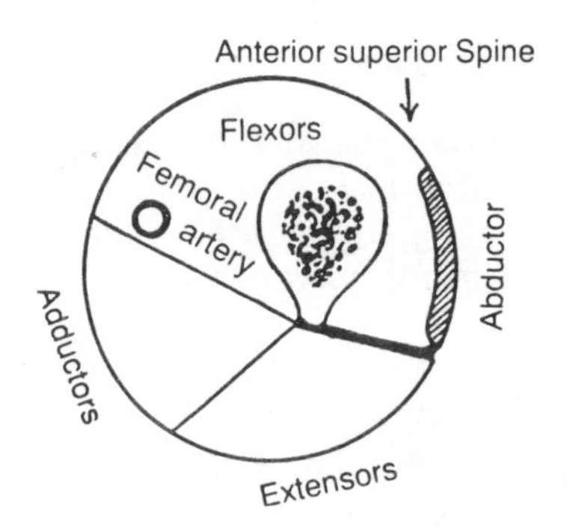

Figure 23.17. The regions of the thigh; theoretical (left) and actual (right). *AS* = anterior superior.

complains of serious weakness with knee extension, which makes thrusting movements of his right foot difficult. You can see that the muscles of the front of the thigh present the chief problem, and you rapidly review your knowledge of them before embarking on an appropriate course of exercise therapy.

Quadriceps Femoris

The 4 heads of this very powerful knee extensor are the rectus femoris and the 3 vasti (medialis, lateralis, and intermedius) (*fig. 23.18*). Of these the rectus arises from the ilium and so also flexes the hip joint; the 3 vasti arise from the body of the femur (*fig. 23.12*).

This enormous muscle is inserted into a small area of bone, namely, to the tuberosity of the tibia, so the insertion must be tendinous. Where the tendon plays across the anterior aspect of the distal end of the femur, a sesamoid bone, the *patella*, is developed. The portion of the tendon distal to the patella is called the *ligamentum patellae*.

The rectus femoris (*fig. 23.18*) has a tendinous origin from the anterior inferior iliac spine (*straight head*) and to the acetabular margin for about 3 cm behind it (*reflected head*).

Internervous Line between the Motor Territories of the Femoral and Gluteal Nerves. If a surgeon makes an incision vertically downward from the anterior superior spine, the sartorius may be pulled medially, the tensor fasciae latae laterally, and the anterior inferior spine exposed; thereafter, the glutei medius and minimus may be pulled posteriorly, the rectus and iliacus anteriorly, and the hip joint exposed. No motor nerve crosses this line.

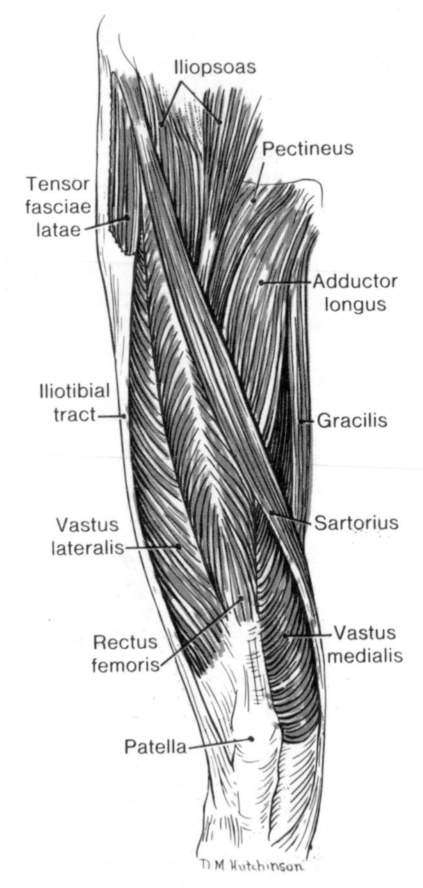

Figure 23.18. Dissection of muscles of front of (right) thigh. (From Basmajian JV: *Primary Anatomy*, ed 8. Williams & Wilkins, Baltimore, 1982.)

The **three vasti** clothe the shaft of the femur.

ORIGINS

The vastus intermedius arises by fleshy fibers from the anterior and lateral aspects of the femur. No muscle arises from the medial aspect, but the vastus medialis overlies it. The vasti lateralis and medialis arise largely by aponeuroses from the lateral and medial lips, respectively, of the linea aspera, as well as from the proximal and distal continuations of these lips (*see fig. 25.14*).

The most distal fibers of origin of the intermedius, called the **Articularis Genu**, are attached to the synovial capsule of the knee joint to retract the capsule during extension of the knee.

Insertions on Patella (Detail). The vasti medialis and lateralis are continuous at their insertions, and they occupy a plane between the rectus femoris and vastus intermedius; all 4 are inserted into the superior border and sides of the patella (*see fig. 26.3*).

DISTRIBUTION OF THE FEMORAL NERVE (L2, 3, 4)

Motor Distribution (*see fig. 28.42* on p. 333)

The psoas is supplied by the roots of the femoral nerve. The *iliacus* is supplied while in the iliac fossa. The *pectineus* is supplied near its lateral border by a twig that passes posterior to the femoral sheath. The *sartorius* and *rectus femoris* are both supplied by one or two branches that enter them from 8 to 15 cm from the anterior superior spine. The *three vasti* receive short, stout branches at their proximal ends. The vasti medialis and lateralis both receive 1 long branch also, which enters their respective anterior borders near their middles.

Cutaneous Distribution (*fig. 23.4*; p. 259).

Articular Distribution. The femoral nerve is distributed to the hip joint and to the knee joint, thus:

To the capsule of the *hip joint* via the nerve to pectineus and either the nerve to rectus femoris or the nerve to vastus lateralis or directly from the femoral nerve itself.

To the capsule of the *knee joint* via the nerves to the 3 vasti and the saphenous nerve (Gardner).

Vascular Distribution. According to rule, sympathetic fibers run to all companion vessels.

Anatomical Features of the Physical Examination

By itself, visual inspection of the front of the thigh usually yields very little useful clinical information except for any substantial quadriceps atrophy (Clinical Case 23.4). In the presence of gross architectural deformities of the femur, distortions are often obvious. Even palpation has limited usefulness (Clinical Case 23.2) because of the relative depth of the most significant structures (femur, great saphenous vein during its course in the thigh).

The inguinal ligament cannot be palpated, but the femoral artery pulse is easily felt by pressing on it posteriorly (against the femoral head) just inferior to the middle of the ligament (Clinical Case 23.3). In that same area, lumps may be felt, e.g., swollen lymph nodes, femoral and/or inguinal hernias, etc. (Clinical Case 23.1). Sensory and motor testing of the femoral and obturator nerve territories almost coincides with the dermatome area for L2,3,4. Clinical Case 12.1 obviously can only be solved and treated by clinicians who know the details summarized above along with extensive clinical training based in great part on this knowledge. The description of the femoral triangle that followed Clinical Case 23.3 in the text also bears on Clinical Case 23.1, of course.

A clear knowledge of the femur is necessary for the clinician who suspects osteosarcoma in the patient of Clinical Case 23.2. If that is what the tests confirm, the orthopedic surgeon who is called in will certainly expect the student assistants and interns to know the anatomy of this region very well because major surgery, probably even high amputation, is indicated.

Clinical Case 23.4 requires clear knowledge of the great knee extensor for the student to be able to help manage this and similar cases. The three vasti (lateralis, intermedius, and medialis) arise from much of the length of the femur but the companion rectus femoris arises from the hip bone—and so flexes the hip, too. All are supplied by the femoral nerve (L2,3,4). Immobilization of any joint will produce atrophy of its main muscles; this being a notorious consequence for the quadriceps femoris, aggressive retraining is often necessary.

Clinical Mini-Problems

1. **When repairing a femoral hernia, what important structures lie lateral to the hernia in the femoral sheath?**
2. **Which major thigh structure is located in the femoral triangle at the midinguinal point?**
3. **Which *palpable* bony landmarks are attachments for the inguinal ligament that forms the base of the femoral triangle?**
4. **What structures are contained in the adductor canal?**
5. **Which spinal cord segments are tested when the patellar tendon is "tapped" with a reflex hammer to elicit the knee jerk?**

(Answers to these questions can be found on p. 586.)

24

Hip Bone, Hip Joint, and Gluteal Region

Clinical Case 24.1

Patient Arnold D. This male construction worker, aged 38, was pinned to the side of a wall by a heavy truck that crushed his pelvis. Radiographs reveal two fractures of his pubis, two of his ischium, and one of his ilium that extends into the superior part of the acetabulum. Knowledge of the anatomy described in the following pages is essential for an appreciation of the present and future risks and complications, the general prognosis, and the best course of therapy.

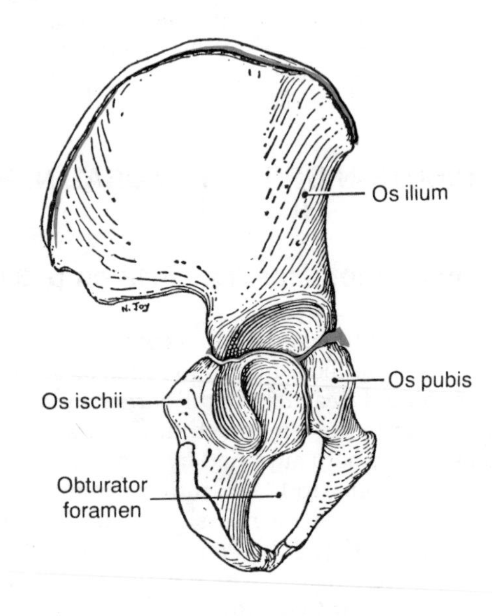

Figure 24.1. A young hip bone (cartilage in blue).

Hip Bone (Os Coxae)

COMPONENT PARTS

The hip bone is composed of 3 elements—ilium, ischium, and pubis. These three meet at the cup-shaped cavity for the head of the femur, the *acetabulum*. Here they are united by a Y-shaped cartilage (quite obvious in radiographs) until the 16th year, when fusion takes place (*fig. 24.1*).

The *ilium* is a flat bone; the *ischium* and the *pubis* are irregular V-shaped bones. The ilium lies superior to the acetabulum, the ischium posteroinferior, the pubis anteroinferior. The pubis and ischium surround the *obturator foramen*.

Gluteal Aspect (Posterolateral Aspect) of the Hip Bone (*fig. 24.2*)

This aspect is wide and thin superiorly and is there bounded by the whole length of the iliac crest. Below, it narrows and thickens to form the ischial tuberosity on which one sits. There are 3 obviously different areas: (1) superiorly, the fan-shaped *dorsum ilii*, (2) the rough, oval *ischial tuberosity*, and (3) inferiorly, between these, the *dorsum ischii*, the large quadrate area posterior to the acetabulum.

Adductor Aspect (Anterior Aspect) of the Hip Bone (*fig. 24.3*)

It provides origin to the adductor muscles.

The **Acetabulum** is described with the hip joint on p. 270.

Ilium

The fan-shaped ilium (*figs. 24.1 and 24.2*) provides areas of origin for the gluteal muscles (gluteus minimus, medius, and maximus) on its **dorsum ilii**.

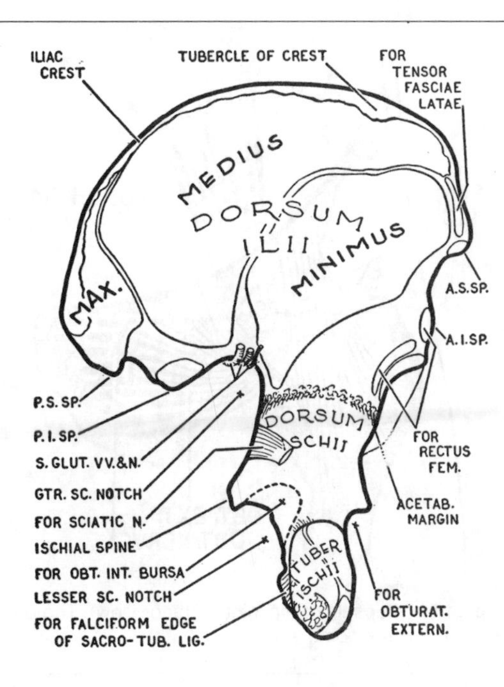

Figure 24.2. The gluteal aspect of the hip bone.

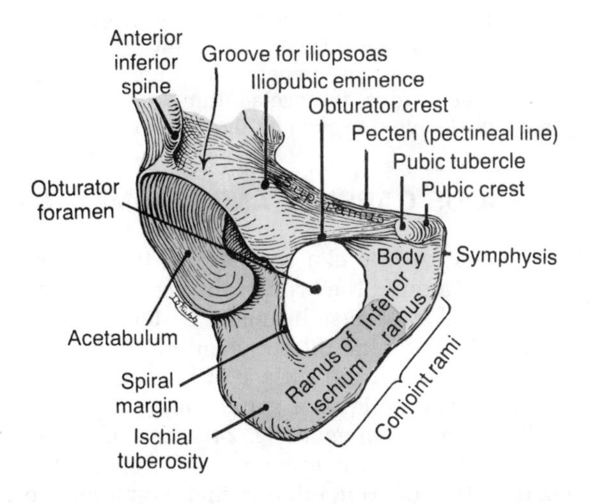

Figure 24.3. The adductor aspect of the hip bone.

The superior border, the *iliac crest*, is palpable throughout. Feel it on yourself or a companion. Anteriorly, it ends in the rounded *anterior superior iliac spine* (easily felt), and posteriorly in the sharp *posterior superior iliac spine* (difficult to feel). From its most lateral point a wide *tubercle* projects inferiorly. The pull of the abdominal muscles creates a traction epiphysis along the whole length of the crest (*fig. 24.1*).

Not palpable in the living are the *anterior inferior spine* for the straight head of the rectus femoris and the *posterior inferior spine* (at the posterior limit of the sacroiliac joint) below their respective superior spines (*fig. 24.2*). The V-shaped *greater sciatic notch* completes the outline.

The **sciatic notches**, greater and lesser, are separated from each other by a beak-shaped process, the *ischial spine*. This gives attachment to the sacrospinous ligament which, with the sacrotuberous ligament, converts the sciatic notches into sciatic foramina (*fig. 18.21*).

Ischium

The ischium (os ischii) has 3 parts—a *body* adjoining the ilium, a *tuberosity* running inferiorly from the body, and a *ramus* passing from the tuberosity anterosuperiorly inferior to the obturator foramen.

The **Ramus of the Ischium** (*fig. 24.3*) is a flattened bar that unites with the inferior ramus of the pubis, fusing in the 8th year to form the *conjoint ramus* of the ischium and pubis. The joined rami of the 2 sides constitute the *pubic arch*.

Pubis

The pubis (os pubis) (*fig. 24.3*) has 3 parts—a body lying medially, a *superior ramus* passing superiorly and laterally from the body, and an *inferior ramus*, descending from the body and forming part of the pubic arch.

Body. Its *symphyseal surface* is elliptical, covered with cartilage, and 4 cm long. The pubic crest ends laterally at the *pubic tubercle*.

The **Superior Ramus** is a 3-sided pyramid whose base forms a fifth of the articular part of the acetabulum.

The 3 borders are formed by 2 lines that diverge from the pubic tubercle: (1) the *pecten pubis*, a part of the brim of the pelvis minor; (2) another line, the spiral *margin of the obturator foramen*. The obturator vessels and nerve emerge from the superior part of the obturator foramen in contact with the inferior surface of the superior ramus of the pubis.

The *pectineal surface* faces anterosuperiorly and gives origin to the pectineus.

Clinical Case 24.2

Patient Betty E. This 67-year-old woman, who has had painful "arthritic" hips for several years, slipped and fell. Radiographs reveal both an acute fracture of the right femoral neck and osteoarthritis of the joint. The surgical team is debating whether a total hip replacement or "pinning" the neck of the femur with a metal nail should be done. A clear knowledge of the region is obviously important.

The Hip Joint

THE BALL AND SOCKET

The *head of the femur* forms two-thirds of a sphere that is slightly flattened superiorly where its socket, the *acetabulum* rests most heavily upon it. It faces medially and anterosuperiorly so its anterior part is not within the socket when the limb is in the anatomical position.

The *acetabulum* (L. = vinegar cup) is deeply notched inferiorly at the *acetabular notch*, which expands centrally as the nonarticular *acetabular fossa*. The remainder is horseshoe-shaped and covered with cartilage (*fig. 24.4*).

The acetabular notch is converted into a foramen by the *transverse acetabular ligament* (*fig. 24.4*). The acetabulum is deepened and its rim completed by a complete ring of pliable fibrocartilage, the *acetabular labrum*, which is attached to its brim and to the transverse ligament. The labrum grasps the head of the femur just beyond its "equator."

Epiphyses. The epiphysis of the head of the femur fits like a cap on a spike, and the epiphyseal line encircles the articular margin. It lies entirely within the joint space as does most of the neck (*fig. 24.8*). *Ossification* of the head, which starts during the 1st year, is always complete before the 20th year (McKern and Stewart).

The epiphyseal line, where the ilium, ischium and pubis meet to form the acetabulum, is triradiate; synostosis is always complete by the 17th year.

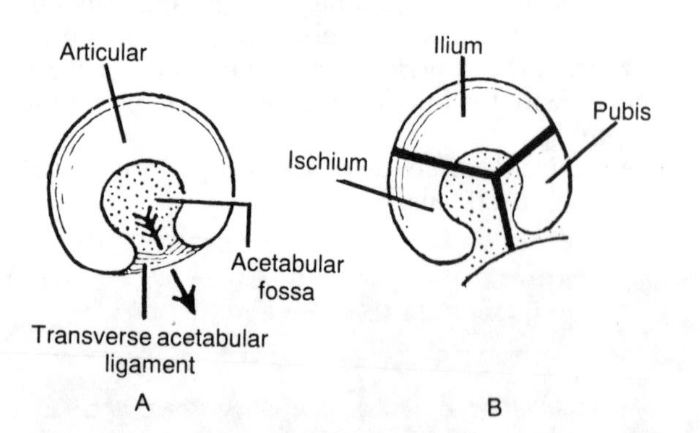

Figure 24.4. *A*, the acetabulum shown in correct orientation and the transverse acetabular ligament. The *arrow* passes through the acetabular foramen. *B*, triradiate "epiphyseal" cartilage.

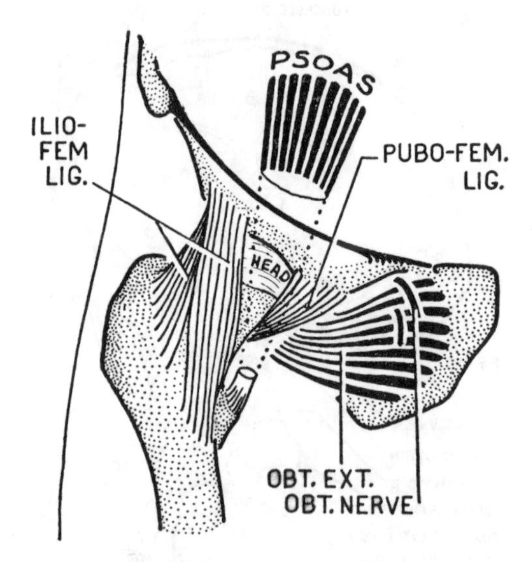

Figure 24.5. Capsule of hip joint (anterior view). The psoas "guards" the weak point.

FIBROUS CAPSULE AND LIGAMENTS

The fibrous capsule is a very strong, thick sleeve attached firmly around the brim of the acetabulum and to the labrum and transverse ligament. Distally, its femoral attachment is along the intertrochanteric line in particular and to a line on the surface of the neck inferiorly. The fibers run an oblique or spiral course (*fig. 24.5*).

The *iliofemoral ligament* (*fig. 24.5*) is a broad, strong band, shaped like an inverted Y or V. Superiorly, it is attached to the anterior inferior iliac spine and to the acetabular margin, deep to the 2 heads of origin of the rectus femoris. Inferiorly, it creates the broad, rough intertrochanteric line of the femur.

The *pubofemoral and ischiofemoral ligaments* are attached to the pubic and ischial parts, respectively, of the acetabular margin, and they pass across the back of the neck of the femur to the superior end of the intertrochanteric line (*fig. 24.6*) "avoiding" superior and posterior attachments to the femoral neck.

An *orbicular zone* is a collection of deep fibers that run circularly and cause an hourglass constriction within the capsule (*see fig. 24.8*).

SEQUENCES

When you stand erect, your line of gravity passes just behind the hip joints. Your trunk tends to rotate poste-

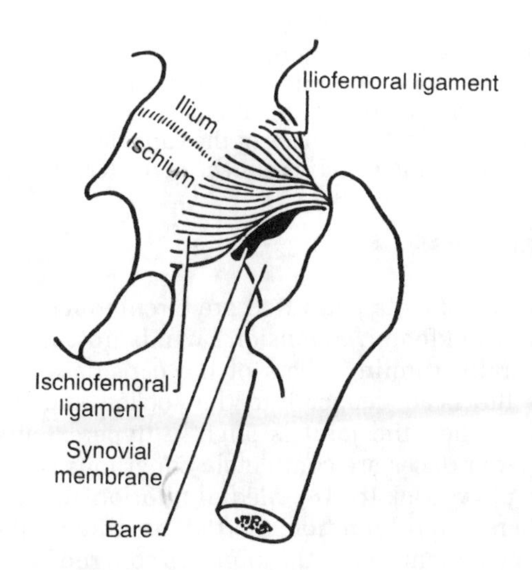

Figure 24.6. Capsule of hip joint (posterior view).

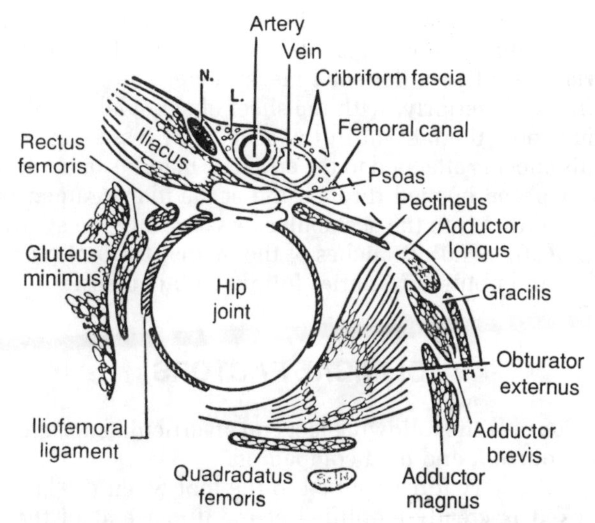

Figure 24.7. Sequences: the hip joint and its relations; coronal section, semidiagrammatic. Start at the middle and work outward in each of the four directions. *A* = femoral artery; *L* = lymphatics; *N* = femoral nerve; *ScN* = sciatic nerve.

riorly. This rotation is checked by the iliofemoral ligament.

Between the iliofemoral and pubofemoral ligaments, there is no ligament and the capsule is commonly perforated. However, guarding this area is the **tendon of the psoas**. Its *bursa* communicates with the joint through the perforation (*fig. 24.7*)—rarely in the young, but commonly in the adult (20% of 478 limbs, aged 20–92 years, in our series).

Clinical Case 24.3

Patient Charles F. This 11-year-old very active but overweight boy began limping because of pain in the right hip region three weeks ago and complains still of a pain deep in his right groin and buttock. There is mild spasm of the muscles that move the hip joint. Radiographs of the femoral head have been ordered with a provisional diagnosis of either an aseptic necrosis of the head or a slipping of its epiphyseal plate. Both are serious problems requiring clear knowledge of the anatomy that follows.

Synovial membrane lines all parts of the interior of the joint, except where there is cartilage. It lines the fibrous capsule, the neck of the femur completely in front, and less completely posteriorly, and roofs the acetabular fossa.

The synovial membrane "droops" posteroinferior to the obturator externus tendon, which runs laterally on the neck (*fig. 24.6*). The synovial membrane surrounding the neck is raised into several loose longitudinal folds, in which important arteries ascend to the head.

The fat in the obturator region extends (with vessels and nerves) deep to the transverse ligament into the acetabular fossa where it forms a pliant extrasynovial fat pad (*fig. 24.8*).

Ligament of the Head

Within the joint, there is a hollow cone of synovial membrane that transmits blood vessels to the head of the

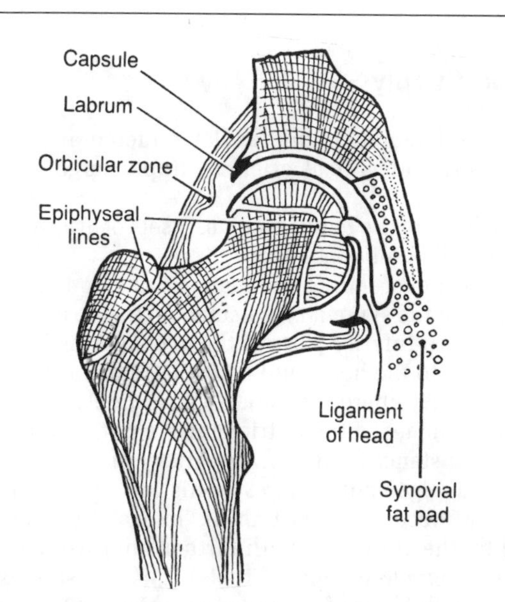

Figure 24.8. The hip joint in coronal section. Note the lines of strain and stress.

femur. This lies between the pit or *fovea* on the head of the femur and the acetabular fossa, so it is flattened and triangular like a flattened megaphone. The base is continuous inferiorly, with the sheet of synovial membrane that roofs the fat pad in the acetabular fossa (*fig. 24.8*); this sheet is attached to the articular margins of the fossa.

A probe pushed through the acetabular foramen will pass either into the acetabular fossa, or into the synovial cone (*fig. 24.8*). Branches of the medial femoral circumflex and obturator arteries follow both courses.

CLINICAL FACTORS

Confirm the following facts on an articulated skeleton, on yourself, and on a companion.

1. A rotary force applied to the foot when the knee is locked is greatly amplified at the upper end of the femur—as at the end of a long screwdriver—and may fracture the neck of the femur.

2. A pillow placed under your knee, when you lie down, automatically results in flexion of the hip, knee, and ankle and a general relaxation of the entire limb.

3. The posture conducive to **dislocation of the hip joint** is one in which the joint is fully flexed and medially rotated, as on bending forward with toes turned in. Flexion brings the shallow part of the acetabulum to rest on the femur; medial rotation brings the head of the femur to the back. A weight (such as a sack of potatoes) then falling on the back can dislocate the head of the femur onto the dorsum ilii. Or conversely, a massive thrust to the soles of the feet in a seated person not wearing a seatbelt can produce the same results.

Relations are shown in *figure 24.7*. The relationship of the sciatic nerve is appreciated only when the obliquity of the acetabular margin is kept in mind (*see figs. 24.13* and *24.14*).

Blood Supply

Medial femoral circumflex, lateral femoral circumflex, obturator, and gluteal arteries supply large twigs (*fig. 24.9.*).

The head of the femur receives 3 sets of arteries. Details of clinical significance are:

(1) The never failing and main set of 3 or 4 arteries that ascends in the synovial retinacula on the posterosuperior and postero-inferior parts of the neck, perforate the neck just distal to the head, and bend at 45° toward the center of the head, where they anastomose freely with (2) terminal branches of the nutrient artery of the shaft, and in 80% of instances with (3) the artery of the ligament of the head (ligamentum teres). This last artery has established a precarious anastomosis in the epiphysis of the head by the 10th year, with further improvement occurring during adolescence (Trueta). This anastomosis persists even in those of advanced age, but in 20% it is never established (Wolcott).

Nerve Supply. *Femoral nerve* (via nerve to rectus femoris), p. 267. *Obturator nerve* (via 1 or 2 branches that pass laterally), p. 288. *Sciatic nerve* (via nerve to quadratus femoris) to the back of the capsule, and sometimes twigs from the *superior gluteal nerve* (Gardner).

Movements

The movements permitted are circumduction and rotation of the femur. *Extension* "winds up" and tightens the spirally running fibers of the capsule and thereby forces the head deeper into the socket, and is self-arresting. When the joint is slightly hyperextended, the articular surfaces are completely congruous (Walmsley), i.e., in close-pack (p. 18). *Medial rotation* also winds up the fibers; *lateral rotation* and *flexion* unwind them and these movements make the joint less congruous and more free (loose-pack).

To understand the *Actions* of the abductors and medial rotators, the pelvis and femur must be held in correct orientation. Thus, (1) the anterior superior spines and the upper end of the symphysis lie on the same coronal plane, and (2) the head of the femur is directed anteriorly as well as medially and superiorly, i.e., the greater trochanter lies posterior to the plane of the head. The *function of the abductors* is to prevent the pelvis and the body from falling to the unsupported side when one foot is off the ground, as in walking (*fig. 24.10*). The *function of the medial rotators* is to rotate the unsupported, or opposite, side of the pelvis forward and thereby to increase the stride (*see fig. 24.11; Table 21.1*).

The gluteus maximus (p. 274) is necessary for rising from the sitting position, for climbing stairs, and for running and jumping, but it is hardly required in walking which essentially involves movement of the flexors and, there, the hamstring muscles suffice as extensors. Details on these muscles will follow in this chapter.

Clinical Case 24.4

Patient Doris H. When 3-years-old, this 21-year-old woman had an injection of penicillin into her left buttock area, which resulted in a persistent partial paralysis of a number of muscles in her lower limb and a slight limp. She now seeks advice about the diagnosis and prognosis. These require a thorough assessment of muscle power of all the muscles of the area and may require electromyographic studies with intramuscular electrodes. Only a good knowledge of the region (described in the following section) will permit a good assessment.

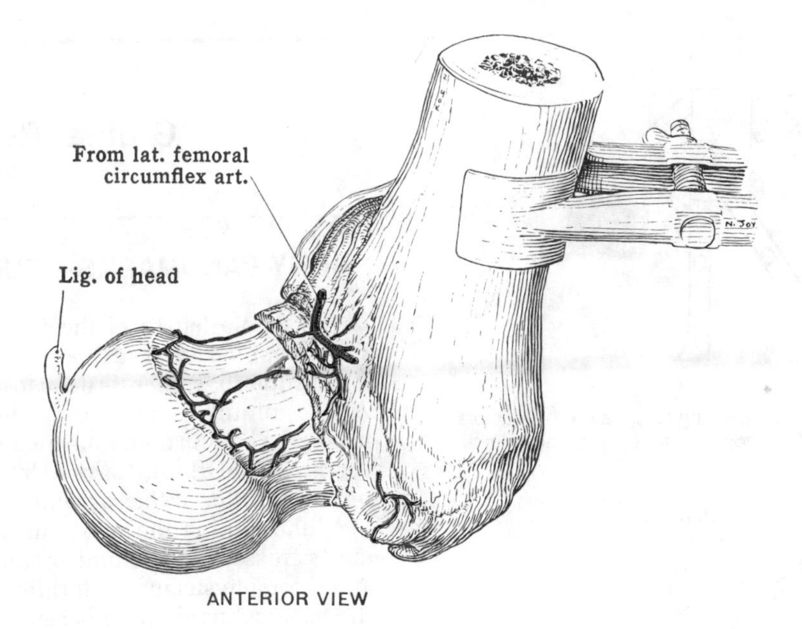

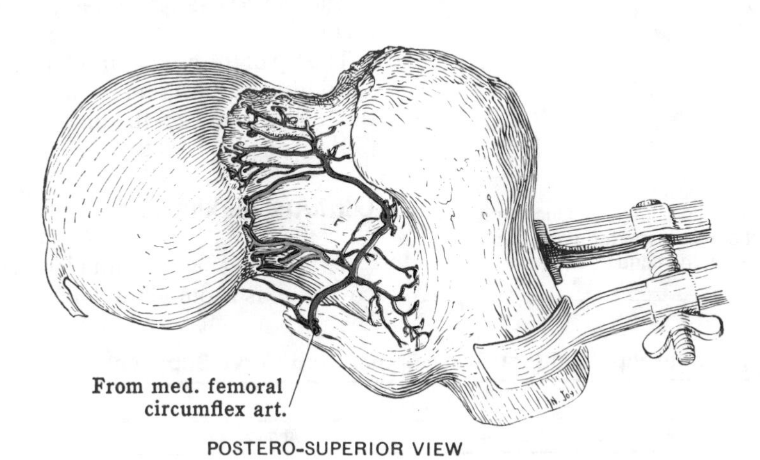

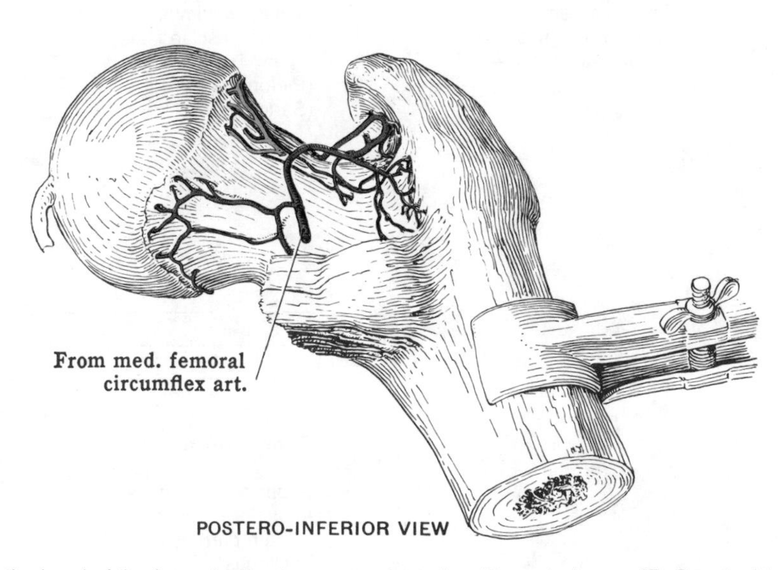

Figure 24.9. Blood supply to the head of the femur by the three sets of arteries. (From Anderson JE: *Grant's Atlas*, ed. 8. Williams & Wilkins, Baltimore, 1983.)

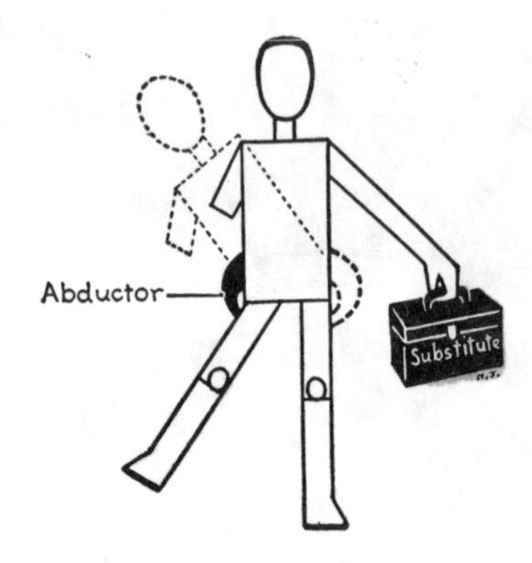

Figure 24.10. Weight, substituting for paralyzed abductors, demonstrates the chief function of gluteus medius and gluteus minimus.

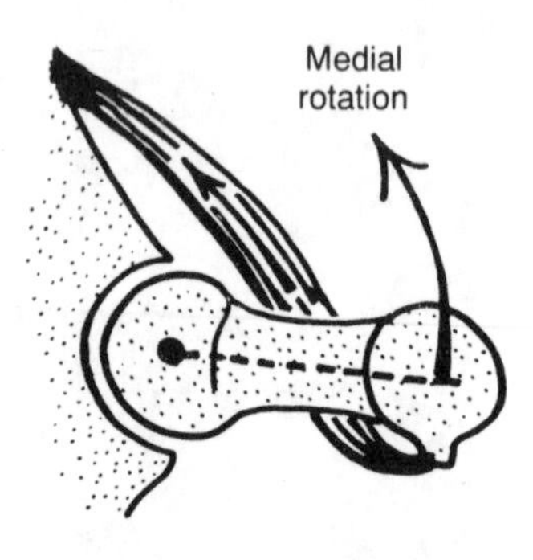

Figure 24.11. Explaining why the superior adductor muscles, although inserting on the linea aspera (on the back of the femur), are medial rotators of the femur.

Gluteal Region

BONY LANDMARKS—SURFACE ANATOMY

The whole length of the *iliac crest* and the *anterior superior spine* are easily palpated. The right and left *posterior superior spines* lie deep to a pair of visible dimples. A line joining them crosses the *2nd sacral spine* and marks the bottom of the subarachnoid space that contains cerebrospinal fluid. Carried laterally, this line crosses near the *center of the sacroiliac articulation (fig. 24.12).*

A line joining the most superior points on the iliac crests crosses the *4th lumbar spine.* It is the guide to that spine for physicians performing "spinal punctures" of the subarachnoid space because the inferior end of the spinal cord is safely superior to this at the L2 vertebral level.

The tip of the *greater trochanter* is palpable at the greatest width of the "hips"—as is the tip of the coccyx in the furrow between the buttocks about 3–4 cm posterior to the anus.

Gluteus Maximus

This very coarse-grained muscle, with a thin fascial covering, is rhomboidal (*fig. 24.13*). It **arises** from avail-

Table 24.1. Muscles Acting on the Hip Joint (and Their Spinal Cord Nerve Supplies)

Circumductors			
Flexors (L2,3,4)	Extensors (L5,S1,2)	Abductors (L5,S1)	Adductors (L2,3,4;S1,2)
Iliopsoas	Gluteus maximus	Gluteus medius	Adductor magnus
Tensor fasciae latae	The three hams	Gluteus minimus	Adductor brevis
Sartorius	Adductor magnus	(Tensor fasciae latae)	Adductor longus
Pectineus	(ham part) (L2,3,4)		
		(Piriformis)	Pectineus
		(Sartorius)	
Rectus femoris			Gluteus maximus
Adductor longus			
Adductor brevis			Short muscles
Adductor magnus			Obturator internus
(Obt. part)			Gemelli
			Obturator externus
			Quadratus femoris

Rotators	
Medial (L2,3,4,5;S1)	Lateral (L2,3,4,5;S1,2)
Gluteus medius	Gluteus maximus
Gluteus minimus	Short muscles
(Tensor fasciae latae)	Piriformis
	Obturator internus
Upper adductor mass	Gemelli
Adductor magnus	Obturator externus
(ham part)	
Pectineus	Quadratus femoris
	Iliopsoas

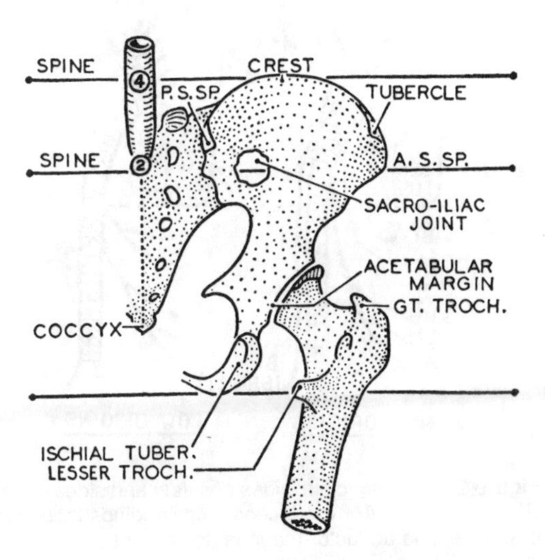

Figure 24.12. Bony landmarks of the gluteal region.

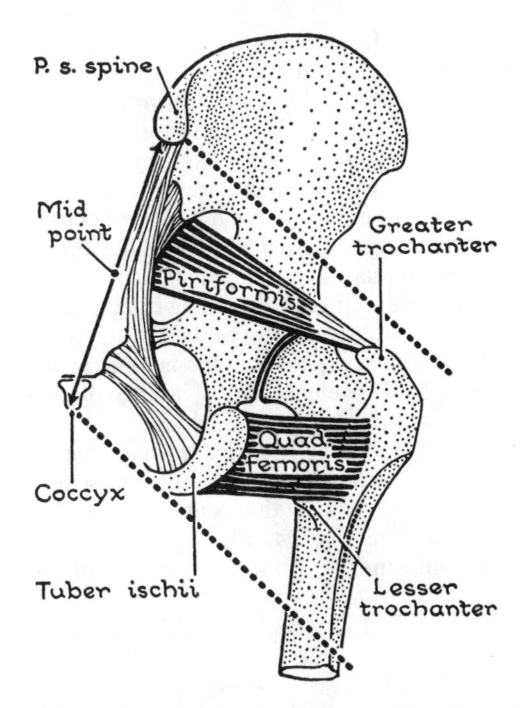

Figure 24.13. To map gluteus maximus, piriformis, and quadratus femoris. The lower border of the piriformis is the key line of the gluteal region.

able surfaces of bones and ligaments along a strip between the posterior superior spine and the tip of the coccyx.

The inferior border of the gluteus maximus extends from the tip of the coccyx across the ischial tuberosity and onward to the shaft of the femur, but the tuberosity is uncovered when you sit down.

The superior parallel border runs from the posterior superior iliac spine to pass superior to the greater trochanter.

Insertion

One-quarter of the gluteus maximus is inserted into the gluteal tuberosity which runs distally from the greater trochanter to the linea aspera of the femur; the rest joins the tensor fasciae latae to form the **iliotibial tract**. The tract descends between the circularly disposed layers of the fascia lata to be attached to the lateral condyle of the tibia anterior to the axis of the knee joint. Owing to this attachment (*fig. 26.9*), it assists in maintaining the extended knee joint in the extended position. You can feel it become prominent and taut on the side of the knee.

The gluteus maximus also attaches to the whole length of the linea aspera via the iliotibial tract and its continuation as the lateral intermuscular septum. Where gluteus maximus slides across the greater trochanter, there must be a bursa (*see fig. 25.1*).

Nerve and Vessels (fig. 24.14)

The inferior gluteal nerve and vessels enter the middle of its deep surface, and other large neighboring arteries give it branches.

Functions

The gluteus maximus is the great extensor of the hip joint, acting in rising from the sitting position, straight-

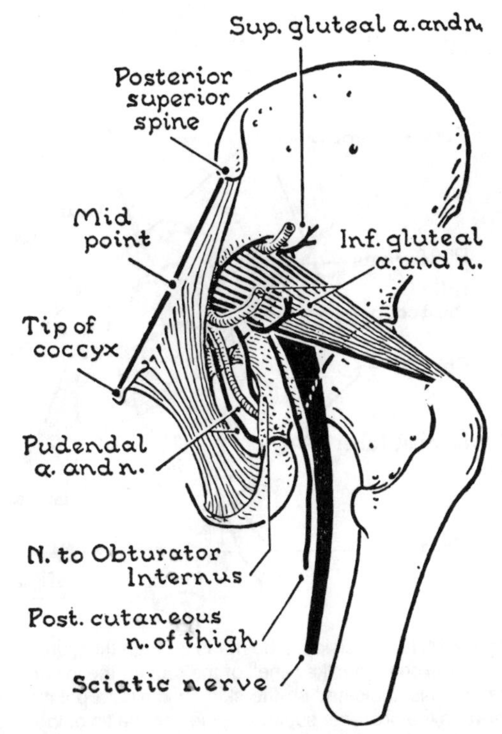

Figure 24.14. Structures passing through the "door" to the gluteal region.

ening, walking upstairs, and running, but it is used very little in gentle walking. It is also a lateral rotator against resistance.

Structures Deep to Gluteus Maximus

The main "door" for the passage of structures from the pelvis to the gluteal region is the greater sciatic foramen.

The **piriformis** itself emerges through the greater sciatic foramen here and occupies a *key position*. Only the superior gluteal vessels and nerve emerge *above* the muscle (*fig. 24.14*).

The *Obturator Internus* is the only structure that enters the gluteal region by the lesser sciatic foramen.

At the Lower Border of Piriformis two groups of nerves and vessels emerge (*fig. 24.14*):

5 {
- Sciatic nerve, which hides nerve to quadratus femoris.
- Inferior gluteal nerve.
- Inferior gluteal vessels.
- Posterior cutaneous nerve of thigh.

3 {
- Nerve to obturator internus.
- Internal pudendal vessels.
- Pudendal nerve.

The group of 3 exit the greater sciatic foramen and swing forward through the lesser sciatic foramen into the ischiorectal fossa. The internal pudendal vessels exit the greater sciatic foramen by crossing the ischial spine, but the accompanying pudendal nerve lies medial to them on the sacrospinous ligament. Clinicians take advantage of the protection afforded the vessels by the bony spine when they perform a transvaginal perineal block. This procedure involves first palpating the ischial spine through the wall of the vagina. Next a hypodermic needle is passed through the vagina medial to the ischial spine and through the sacrospinous ligament. The sensory nerve impulses from the perineum are blocked by injecting an anesthetic agent in the vicinity of the pudendal nerve, as it crosses the sacrospinous ligament. This is an excellent example of how a physician may take advantage of his knowledge of precise anatomical relationships.

Various surgical approaches through the gluteus maximus to the region of the hip joint and sciatic nerve depend on a knowledge of the location of the inferior border of the piriformis (*fig. 24.15*).

The **sciatic nerve**, from the lower border of the piri-

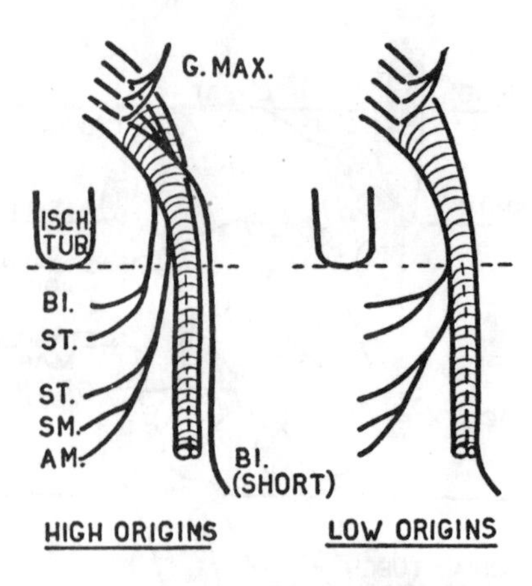

Figure 24.16. Principle: sides of safety and sides of danger. *BI*, *ST*, *SM* and *AM* are biceps, semitendinosus, semimembranosus, and adductor magnus (lower part).

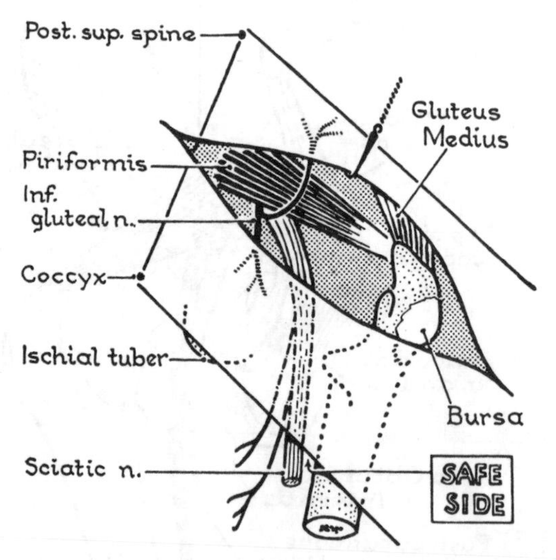

Figure 24.15. Incision to be made through the gluteus maximus to expose the "key line" of the region, the lower border of piriformis, indicated on the skin by joining a point midway between the posterior superior spine and the tip of the coccyx to the top of greater trochanter (cf. *fig. 24.13*).

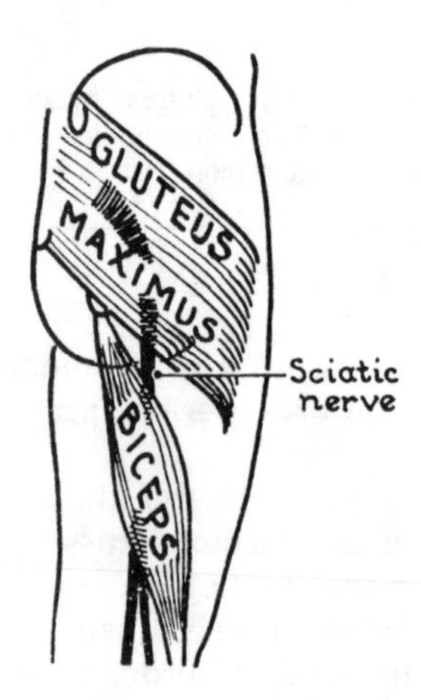

Figure 24.17. Posterior relations of sciatic nerve.

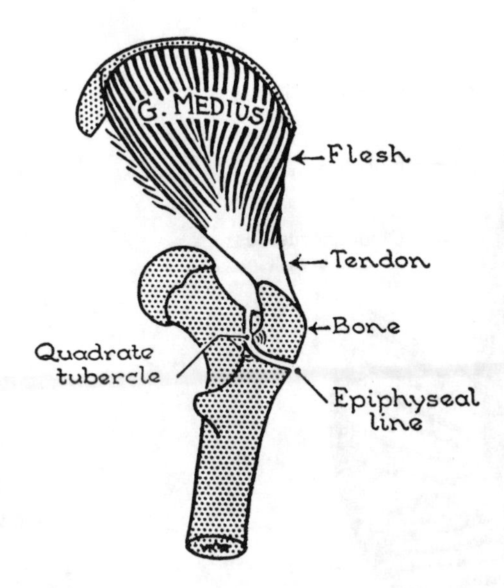

Figure 24.18. The fleshy, tendinous, and bony portions of the gluteus medius.

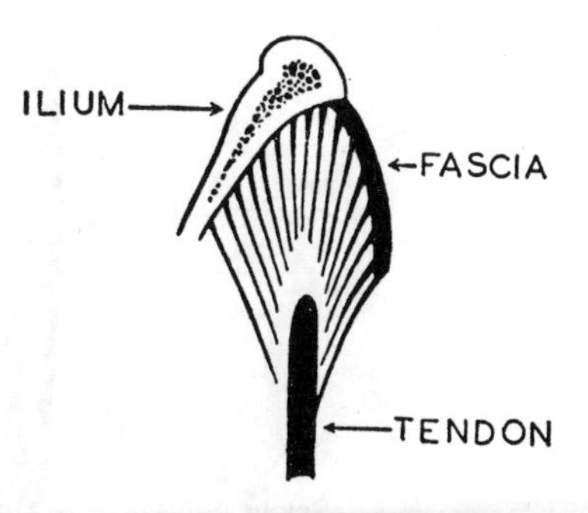

Figure 24.19. The anterior part of the gluteus medius arises largely from the fascia covering it, hence, the thickness of this fascia.

The sciatic nerve, from the lower border of the piriformis, curves inferiorly midway between the ischial tuberosity and the greater trochanter, covered by the gluteus maximus (*figs. 24.15, 24.17*). It has a side of danger and a side of safety—its lateral side, which has no branches (*fig. 24.16*).

Greater Trochanter

The tendons of 5 muscles are inserted into this great lever, and of these the two greatest are (1) the gluteus minimus and (2) the gluteus medius (*fig. 24.18*); the others are (3) the piriformis, (4) the obturator internus and its gemelli, and (5) the obturator externus (to a pit on the medial surface). These muscles will be described below.

The **superior gluteal nerve** (*fig. 24.14*), accompanied by the superior gluteal vessels, passes through the greater sciatic foramen, superior to the piriformis, in contact with the bony angle of the foramen, and it runs laterally across the surface of the ilium between the gluteus medius and gluteus minimus. It supplies the 3 abductors of the hip joint, viz., the gluteus medius, gluteus minimus, and tensor fasciae latae.

Glutei Medius and Minimus

The anterior borders of these two large and essential abductor muscles are commonly fused. The more superficial gluteus medius receives many fibers from the overlying fascia (*fig. 24.19*). Their tendons of insertion to the greater trochanter dominate all the other muscles there.

The **Tensor Fascia Latae** arises from the anterior superior iliac spine and adjacent outer lip of the iliac crest. The iliotibial tract is its aponeurosis of insertion (p. 275).

The 6 Small Lateral Rotators

(1) The Quadratus Femoris (*Figs. 24.13, 25.1*). This relatively unimportant muscle is part of the "red carpet" for the sciatic nerve as it runs down into the thigh. Its nerve arises from the anterior aspect of the sciatic nerve, and it penetrates the muscle on its deep surface. The nerve also supplies a twig to the hip joint. **(2) The Obturator Internus and (3 and 4)** its twin **Gemelli** form the carpet above the quadratus (*fig. 24.20*). The *Obturator Internus* arises from the pelvic surface of the obturator membrane and most of its periphery of bone. It makes a sharp turn as it passes through the lesser sciatic foramen (*fig. 18.13*) lubricated on the bone by a *bursa*. Its tendon inserts on the upper border of the greater trochanter.

The *Superior Gemellus* arises from the upper margin of the lesser sciatic foramen (ischial spine) and the *Inferior Gemellus* from the lower margin of the lesser foramen (ischial tuberosity), and their fleshy fibers join the tendon of the obturator internus.

(5) Piriformis also is a lateral rotator of the hip joint, as is the deeply placed **(6) Obturator Externus**, which spirals from its origin on the lateral surface of the obturator membrane to the back of the femoral neck and is deeply situated.

The **Gluteal Arteries**, superior and inferior branches of the internal iliac artery, accompany their companion nerves and the piriformis muscle out of the pelvis; they are responsible for supplying the entire region, their largest branches being *muscular*.

The **Posterior Cutaneous Nerve of the Thigh** runs vertically, duplicating, at a subcutaneous depth, the line of the sciatic nerve (*fig. 24.21*) and has 2 large branches, the *gluteal* and *perineal*. It ends on the calf.

For *dermatomes*, see *Figure 24.22*.

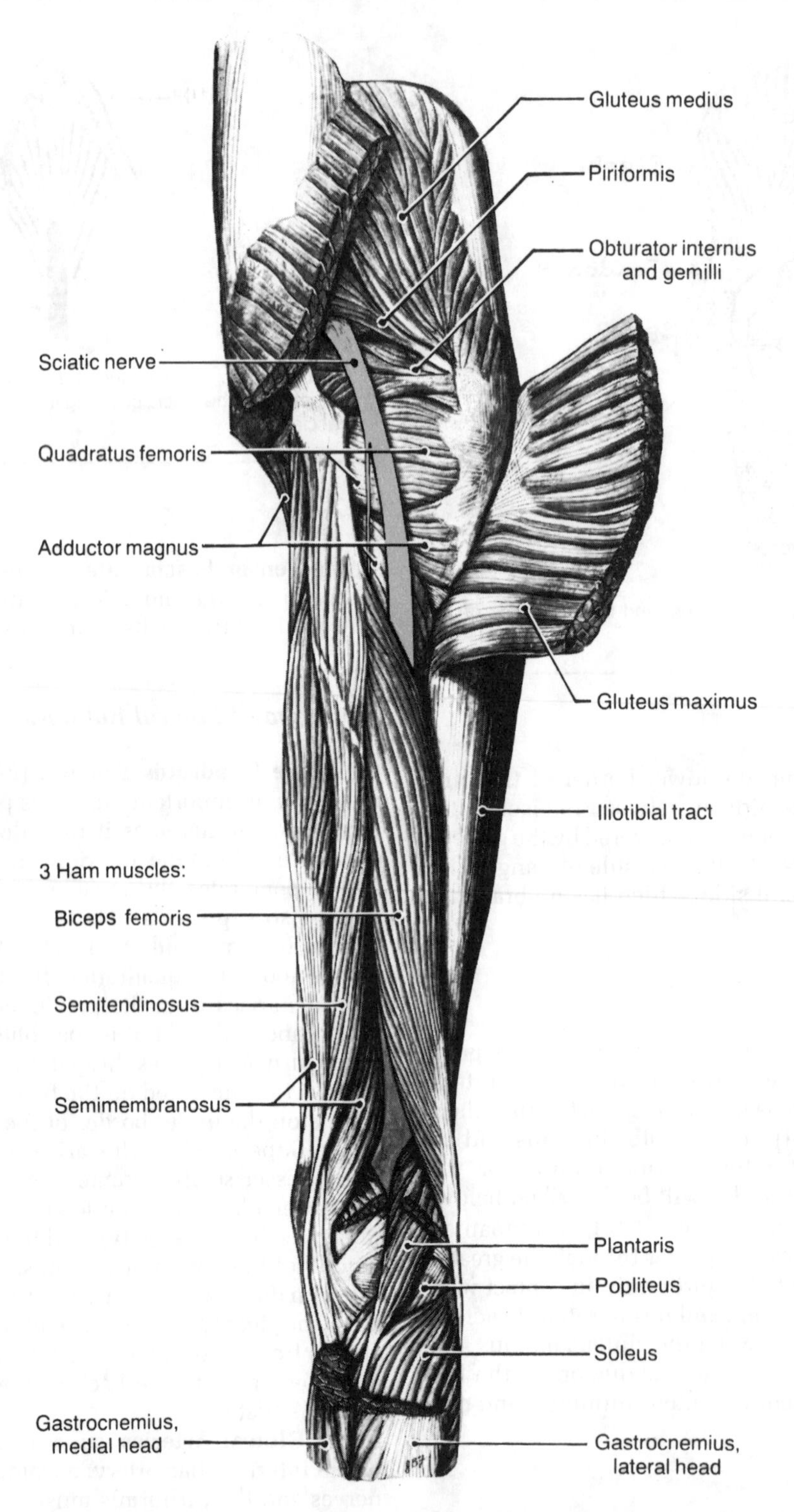

Figure 24.20. Muscles of the gluteal region and back of the thigh and the "muscular red carpet bed" for the sciatic nerve in the gluteal region. (From Anderson, JE: *Grant's Atlas*, ed. 8. Williams & Wilkins, Baltimore, 1983.)

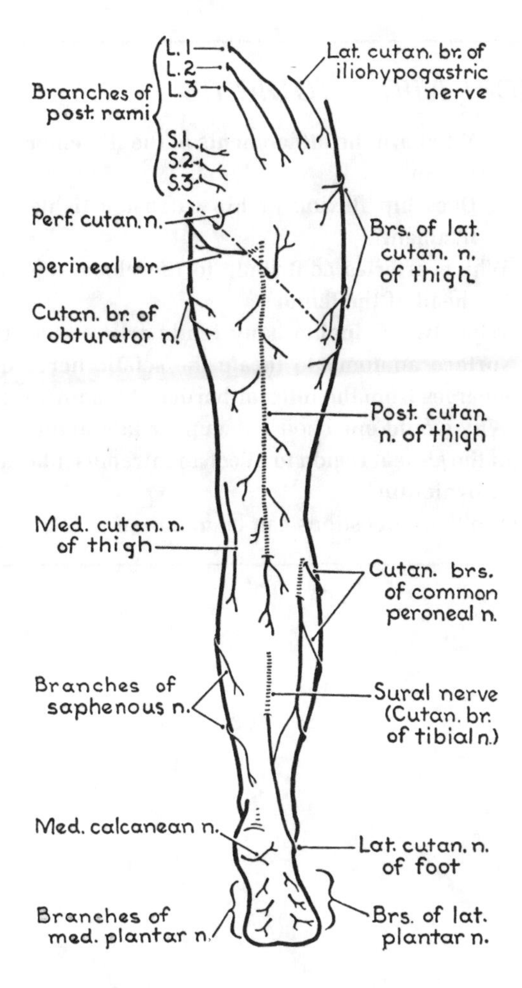

Figure 24.21. Cutaneous nerves of back of lower limb.

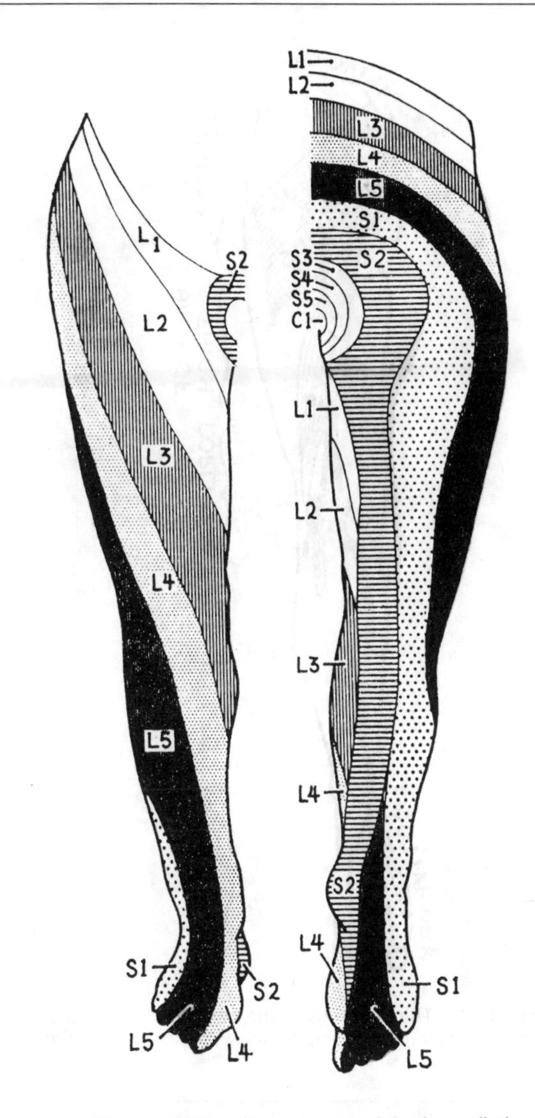

Figure 24.22. Dermatomes of the lower limb.

Five Regions of the Back of the Limb (fig. 24.23)

Four muscles delineate 5 regions; gluteus maximus, biceps femoris, superior border of soleus, and abductor hallucis. Passing deep to all four is the tibial division of the sciatic nerve.

Anatomical Features of the Physical Examination

Both visual inspection and palpation provide the experienced clinician with many clues in interpreting abnormalities of the ilium, sacrum, and coccyx, ischial tuberosity, gluteal muscles, and greater trochanter (Clinical Case 24.1). The deeper bony structures are inaccessible (Clinical Cases 24.2 and 24.3) except to radiography. Aseptic avascular necrosis of the femoral head and/or a slipped epiphysis must be considered with Clinical Case 24.3. The sciatic nerve lying deep to the gluteus maximus (Clinical Case 24.4) is not palpable but it is at risk when intramuscular injections go too deep. Tests and force-ratings of the gluteus medius and minimus depend on their being almost exclusively the great abductor of the hip joint. Observing the patient while standing on one foot reveals their weakness (Trendelenburg sign: the opposite hip "drops downward"). General motor strength ratings of the moving muscles of the hip joint and sensory testing of the dermatomes will be essential whenever any form of nerve or spinal cord damage is suspected.

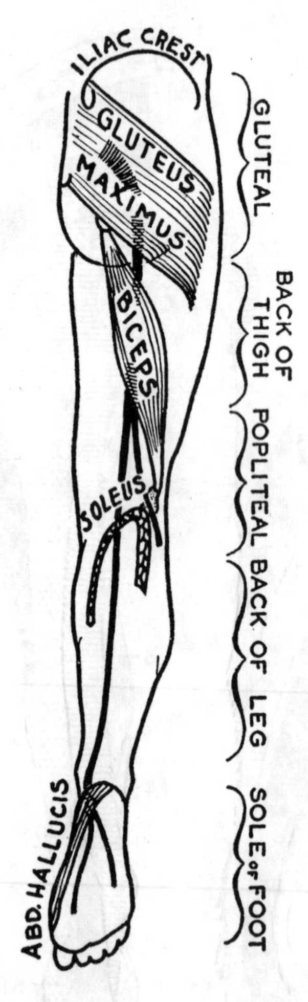

Figure 24.23. The 5 regions of the back of the limb, the 4 muscles separating them, and the tibial division of the sciatic nerve.

Clinical Mini-Problems

1. a. **What are the attachments of the iliofemoral ligament?**
   b. **Does hip flexion or hip extension tighten this ligament?**
2. **Which arteries contribute to the blood supply to the head of the femur?**
3. **What two palpable bony landmarks are used in surface anatomy to locate the sciatic nerve as it emerges from the inferior border of the piriformis?**
4. **Why would one choose the upper lateral quadrant of the gluteal region to inject an intramuscular dose of penicillin?**

(Answers to these questions can be found on p. 586.)

25

Posterior and Medial Regions of Thigh

Clinical Case 25.1

Patient Andrew I. This 50-year-old diabetic man, who has smoked heavily since age 19, is at risk of gangrene in the right foot and leg. The surgical team believes a high amputation is necessary, i.e., through the midthigh, because the arteries are inadequate. You will assist at the operation and need to know the major structures that will be encountered (described below and in the previous two chapters).

Posterior Region of Thigh

The "floor" of the region (actually its *anterior wall*) is flat (*fig. 25.1*). It is formed by the adductor magnus and vastus lateralis muscles, continued inferiorly to the posterior surface of the knee joint.

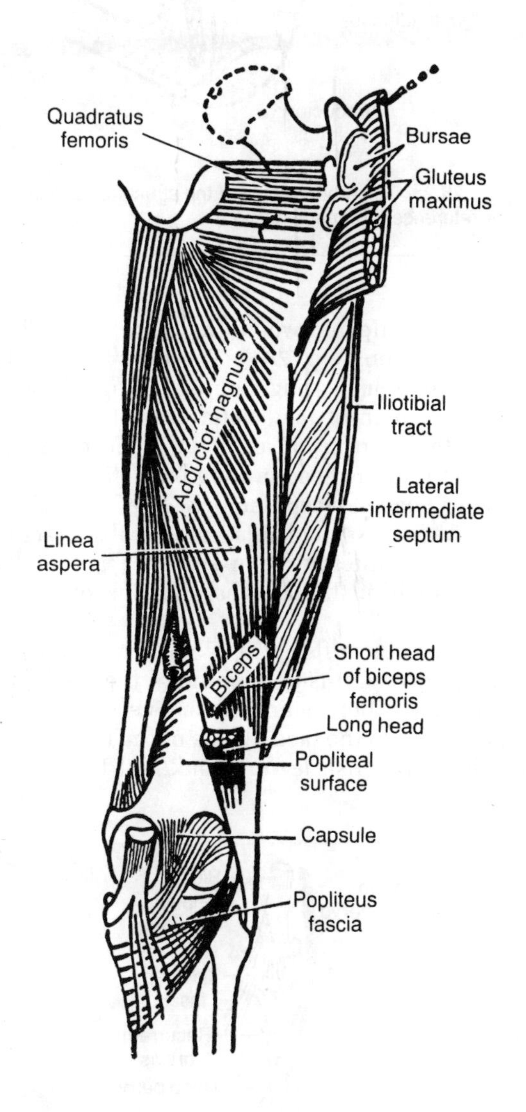

Figure 25.1. The "floor" of the posterior region of the thigh and popliteal fossa.

CONTENTS

1. The hamstring muscles
   Long head of biceps femoris
   Semitendinosus
   Semimembranosus
2. The short head of biceps femoris
3. The sciatic nerve
4. The posterior cutaneous nerve
5. Certain vessels

The Hamstring Muscles

A hamstring, by definition: (1) arises from the ischial tuberosity (*fig. 25.2*); (2) is inserted into the tibia or fibula and (3) is supplied by the tibial division of the sciatic nerve. They extend the hip joint and flex the knee joint, but you cannot do both fully at the same time—try it!

The bulk of the fleshy belly of **semitendinosus** covers the broad aponeurosis of origin of **semimembranosus** (see Cross-section, *fig. 23.12*). Both of these "medial hamstrings" narrow down to rounded tendons of insertion

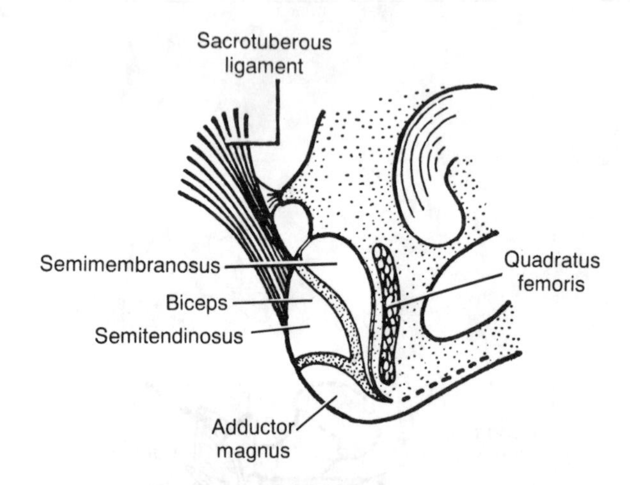

Figure 25.2. The tuber ischii and the structures attached to it (for reference, not memorization).

which can be palpated where they run to the tibia (see "Palpable Tendons," p. 284). The **long head of biceps femoris** ("lateral hamstring") in its sloping course covers most of the length of the sciatic nerve (*fig. 24.17*) and then joins the short head (see below) to form the tendon of insertion into the head of the fibula (*fig. 25.1*).

The short head of biceps femoris arises by fleshy fibers from the linea aspera and adjacent lateral supracondylar line and lateral intermuscular septum (*fig. 25.1*). It receives a branch of the peroneal division of the sciatic nerve.

When the knee is flexed it is not locked and so the biceps, attached as it is to the head of the fibula, rotates the leg laterally; the semimembranosus and semitendinosus, attached to the medial side of the tibia, rotate the leg medially. (English *leg* or shank and L. = *crus*, are

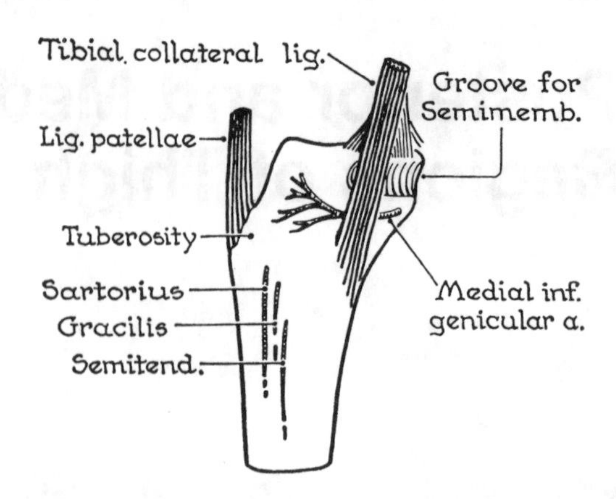

Figure 25.4. Insertion of 4 tendons into tibia (medial view).

specific terms for the limb from knee to ankle, i.e., the leg or shank.)

Origins—see *Figure 25.2.*

Insertions

See *Figures 25.3–25.5.* In addition to the direct bony attachments, the tendons send aponeurotic expansions to the fascia around and below the knee.

Hybrid Muscles. The biceps clearly is an amalgamation of 2 muscles. The long head belongs developmentally to the front of the limb, the short head to the back.

The *Adductor Magnus* also is hybrid. The part arising from the pubic arch is supplied by the obturator nerve, whereas the part arising from the tuber ischii is supplied by the tibial division of the sciatic nerve (*fig. 25.5*).

The *Pectineus* is supplied by both femoral and obturator nerves.

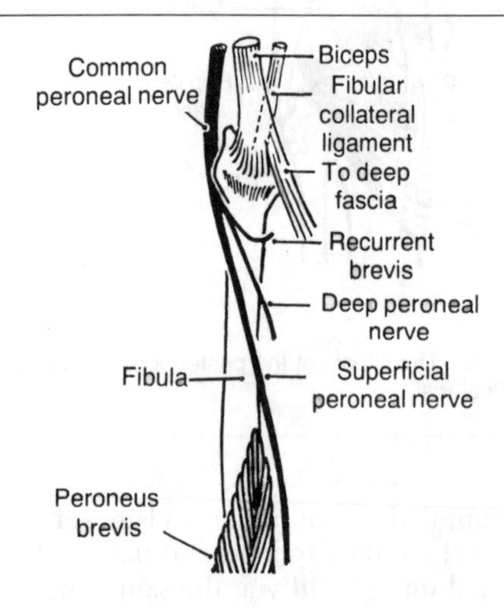

Figure 25.3. Insertion of biceps femoris (with peroneus brevis removed from lateral aspect of fibula superior to peroneus brevis).

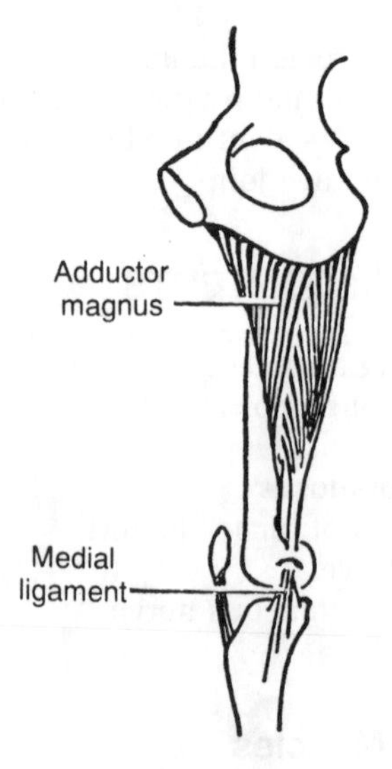

Figure 25.5. The ischial part of the adductor magnus almost "made it" as a hamstring.

NERVES AND ARTERIES

The **sciatic nerve** (*Nervus Ischiadicus*) runs down vertically in the midline of the thigh; it is only accessible deep in the angle between the gluteus maximus and the long head of the biceps (*fig. 24.23*).

Posterior Relations

Superiorly, the gluteus maximus covers it; inferiorly, the biceps (long head).

Termination

Deep to the biceps, the sciatic nerve divides into its 2 terminal branches—*tibial* and *common peroneal nerves* (Gk. perone = L. *fibula* = a pin or skewer). They are merely bound together by loose areolar tissue right up to the sacral plexus (*fig. 25.6*).

Only the nerve to the short head of the biceps springs from the lateral side of the sciatic nerve (peroneal division). The branches to the hamstrings spring from its medial side (*fig. 24.16*).

AXIOM. The side from which a motor nerve leaves its parent stem is constant. It leaves from the side nearest the muscle it supplies, but the level at which it leaves is variable.

Posterior Cutaneous Nerve of the Thigh (*fig. 24.21*)

The posterior cutaneous nerve (S1, 2), a branch of the sacral plexus, emerges inferior to the middle of the inferior border of the gluteus maximus. It gives cutaneous twigs to right and to left as it descends to end on the calf.

Arteries to back of thigh (see p. 289).

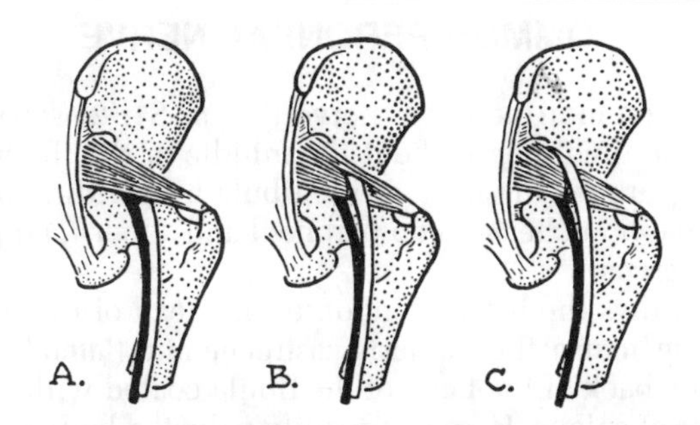

Figure 25.6. The relationship of the sciatic nerve to the piriformis. *A*, in 87.5% of 420 limbs; *B*, in 12% the peroneal division passed through the piriformis; *C*, in 0.5% (i.e., in both limbs of one subject) it passed above.

Clinical Case 25.2

Patient Barry J. This 34-year-old policeman was shot through the left lower thigh while apprehending a burglar 4 weeks ago and now has severe motor, sensory, and circulatory problems in the leg and foot. A loud "roaring" noise with each heartbeat is heard through a stethoscope over the popliteal fossa (back of the knee).

Popliteal Fossa

The *"Floor" of the Popliteal Fossa* (*fig. 25.1*) is formed by: (1) the popliteal surface of the femur; (2) the capsule of the knee joint; and (3) the fascia covering popliteus muscle.

The fossa is a potential one because the circular fibers in the deep fascia bandage the structures together. Dissection converts it into a diamond-shaped recess (*fig. 25.7*).

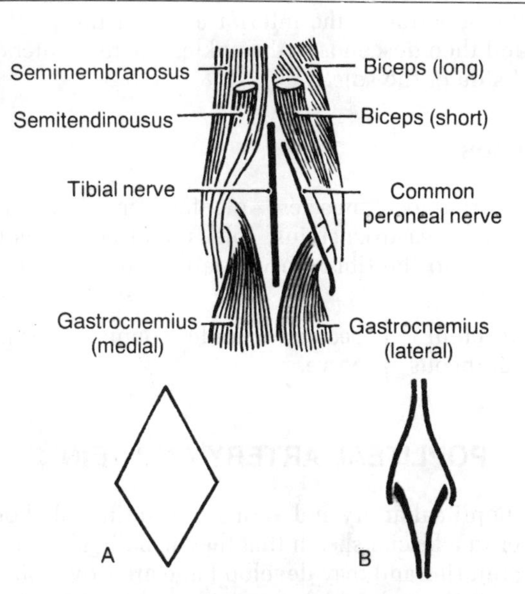

Figure 25.7. Boundaries of the popliteal fossa.

PALPABLE TENDONS

While sitting down and using both hands to feel, intermittently press the heel backward against a chair leg, noting that laterally the *biceps tendon* becomes taut and is easily followed to the head of the fibula; do not mistake it for the prominent posterior border of the much broader *iliotibial tract*, which runs a full fingerbreadth in front of it. The *biceps tendon* also becomes taut on lateral rotation of the leg (knee being flexed); it is the only lateral rotator. Medially, the *semitendinosus tendon* can be made to spring, like a bowstring, backward from the rounder *semimembranosus tendon*. You should then be able to feel them both.

CONTENTS

The fossa contains 4 important structures:

1. tibial nerve.
2. and 3. popliteal artery and vein.
4. common peroneal nerve.

RELATIONSHIPS

As the tibial nerve and the popliteal vessels are passing through the narrow ravine bounded on each side by a femoral condyle and a head of the gastrocnemius, they are crowded together one behind the other—nerve, vein, artery. The customary rule—nerve, artery, vein—is broken here.

TIBIAL NERVE

The tibial nerve (L4, 5 and S1, 2, 3), which is the larger of the 2 terminal branches of the sciatic nerve, passes from the superior to the inferior angle of the popliteal fossa and then descends to the ankle, where it enters the medial side of the sole.

Branches

1. *Motor*: to local muscles—plantaris, medial, and lateral heads of gastrocnemius, soleus, and popliteus (*fig. 25.8*). Deep to the tibial nerve are the popliteal vessels (*fig. 25.8*).

2. *Articular* }
3. *Cutaneous* } See below under common peroneal nerve.

POPLITEAL ARTERY AND VEIN

The popliteal artery and vein are so intimately bound together in a fascial sheath that they usually share penetrating injuries and may develop large arteriovenous anastomoses, through which arterial blood is pumped into the vein. This explains the loud roaring noise in Clinical Case 25.2. The artery divides into its 2 terminal branches, the *anterior* and *posterior tibial arteries* behind the tibia (*fig. 25.8*); soon after, the latter gives off the peroneal artery. The arteries below the knee are accompanied by venae comitantes; so, the *popliteal vein* begins as an assembly of veins.

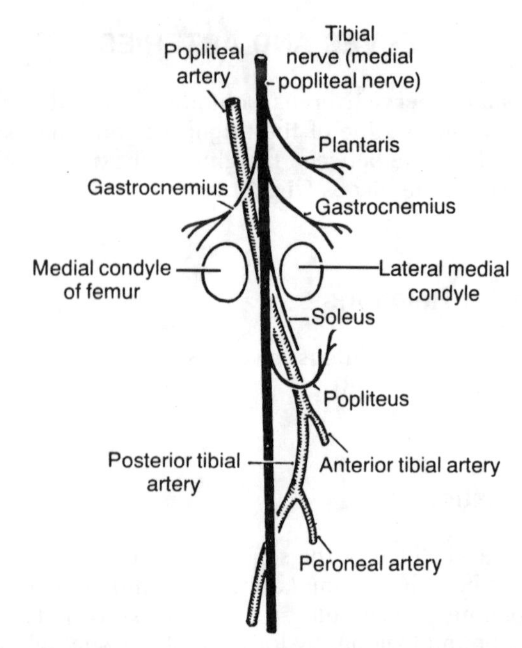

Figure 25.8. The tibial nerve (posterior views): its side of safety and side of danger and its relation to the arterial stem.

Anterior Relations (*fig. 25.9*)

Branches

Terminal: Anterior and posterior tibial arteries. *Collateral*: (1) cutaneous, (2) muscular, and (3) articular (*see fig. 25.15*).

COMMON PERONEAL NERVE

The common peroneal nerve (L4, 5, S1, 2) separates from the tibial nerve about the middle of the thigh and ends lateral to the neck of the fibula by dividing into 2 terminal branches—*deep peroneal* and *superficial peroneal nerves*.

It follows the biceps tendon to the head of the fibula, crossing in turn: the plantaris, gastrocnemius (lateral head), and the back of the head of the fibula coated with a thin veneer of soleus. Here it is readily palpated by fingertips drawn horizontally across it. If appropriately "flicked," it will reveal its sensory distribution to you through referred lightning pains. Try it.

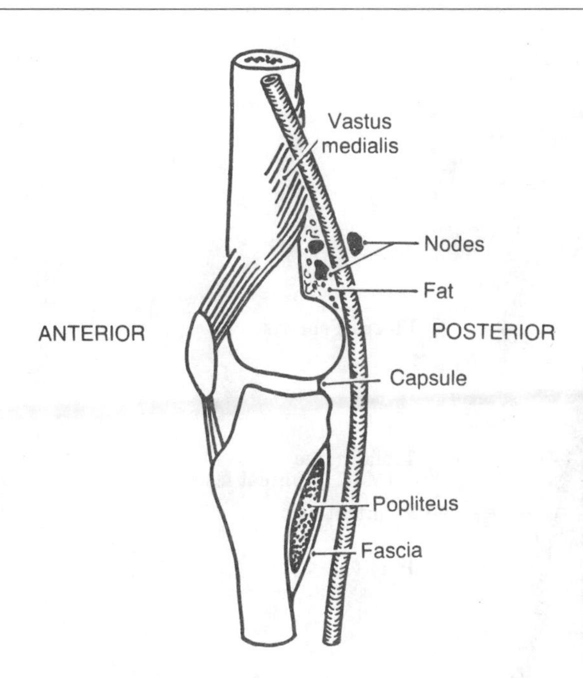

Figure 25.9. Anterior relations of popliteal artery (diagrammatic, medial view).

Branches

Terminal: Deep and superficial peroneal nerves (*fig. 25.3*).

Collateral

Articular and *cutaneous branches* spring from both tibial and peroneal nerves.

The cutaneous branch of the tibial nerve, called the *sural nerve*, runs inferiorly with the small saphenous vein in the furrow between the 2 bellies of the gastrocnemius (*fig. 25.10*). It is joined by a branch of the common peroneal nerve (called the *communicating peroneal nerve*) to travel distally on the gastrocnemius.

THE SMALL SAPHENOUS VEIN

The small saphenous vein starts on the foot as the lateral continuation of the dorsal venous arch. It passes below and then behind the lateral malleolus and is accompanied in its course by the sural nerve. It turns deep through the popliteal fascia and, after dividing, ends in the popliteal and profunda femoris veins.

A large superficial branch connects it to the upper end of the great saphenous vein.

LYMPH NODES

The most distal nodes drain into the popliteal nodes (*fig. 25.9*) and have little clinical significance. They lie in the fat around the popliteal vessels and are impalpable. Their efferents follow the femoral vessels to the deep inguinal nodes.

Clinical Case 25.3

Patient Carl K. This 6-year-old boy has had cerebral palsy since birth and has serious locomotor problems, due to spastic "scissoring" gait. You must know which muscles and nerves are involved in order to "neutralize" them by appropriate surgery or physical therapy.

Medial Region of the Thigh

MUSCLES

The muscles of the medial or adductor region of the thigh arise collectively from the anterior aspect of the hip bone and the obturator membrane (*fig. 25.11*).

From this compact area of origin (the details of which have no interest for the clinician and may be safely ignored), they spread out fanwise to a linear insertion that extends the length of the femur and beyond to the tibia (*fig. 25.12*). They share their chief actions—adduction and medial rotation of the hip joint, and their nerve, the obturator. The profunda femoris and obturator arteries nourish them (see *fig. 25.15*).

The large muscle mass consists of the following 6 individual muscles: pectineus, adductor longus, gracilis, adductor brevis, adductor magnus, and obturator externus. Read but do not memorize the following details.

Origins (*fig. 25.11*). The *pectineus, adductor longus,* and *gracilis* have a continuous curvilinear origin that extends from the iliopubic eminence to the ramus of the ischium. The origin of the pectineus, which is fleshy, meets that of the adductor longus, which is tendinous, at the pubic tubercle; the gracilis has an aponeurotic origin. If the thigh is abducted, the tendon of the longus becomes prominent and palpable and acts as a guide to the pubic tubercle. The *adductor brevis, adductor magnus,* and *obturator externus* arise by fleshy fibers in successively deeper, overlapping planes. (Some of the upper fibers of adductor magnus occasionally form a separate muscle, *adductor minimus*.)

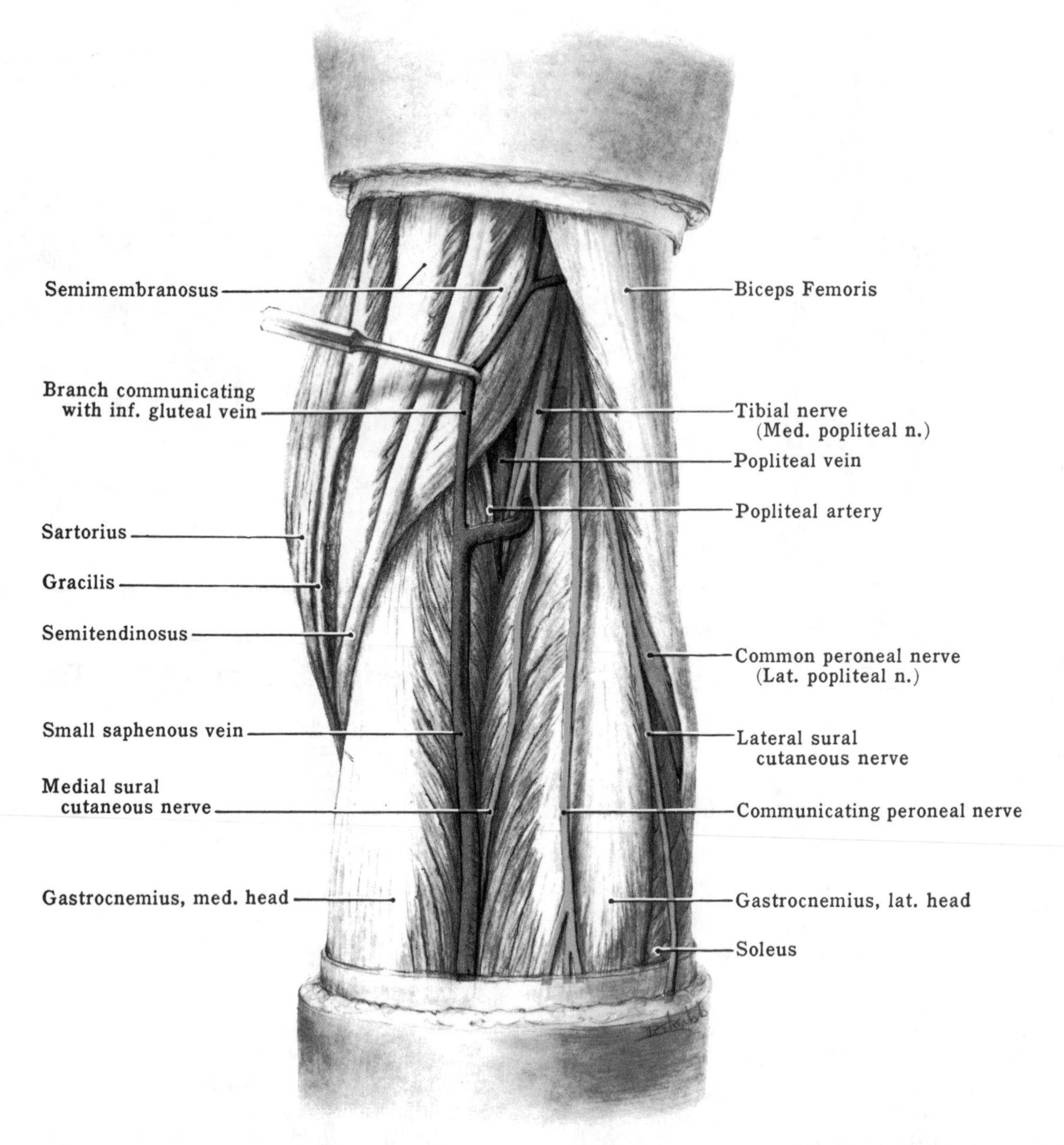

Figure 25.10. Superficial dissection of the popliteal fossa. Observe:

1. The two heads of Gastrocnemius, embraced on the medial side by Semimembranosus, which is overlaid by Semitendinosus, and on the lateral side by Biceps. The result is the lozenge-shaped popliteal fossa.
2. The The small saphenous vein running between the two heads of Gastrocnemius. Deep to this vein is the medial sural cutaneous nerve which, followed proximally, leads to the tibial nerve. The tibial nerve is superficial to the popliteal vein, which in turn is superficial to the popliteal artery.
3. The common peroneal nerve following the posterior border of Biceps, and here giving off two cutaneous branches. (From Anderson JE: *Grant's Atlas*, ed. 8. Williams & Wilkins, Baltimore, 1983.)

Insertions. The restricted and therefore fibrous insertions of these 6 muscles are (*fig. 25.12*) as follows.

The *Pectineus:* to the pectineal line running from the lesser trochanter toward the linea aspera.

The *Adductor Longus:* to almost the whole length of the linea aspera in line with the pectineus. (Between the pectineus and longus the brevis can be seen from the anterior view.)

The *Gracilis:* to the medial surface of the tibia inferior to the level of the tuberosity and between the insertions of the sartorius and semitendinosus. The gracilis is the only muscle of the adductor group to cross the knee joint. (Between the longus and gracilis the magnus can be seen.) (*fig. 23.16*).

The *Adductor Brevis:* to the inferior part of the pectineal line and upper part of the linea aspera. Its superior part is overlapped by the pectineus; its lower part by the adductor longus.

The *Adductor Magnus:* to the linea aspera, extending upward onto its lateral continuation (the gluteal tuber-

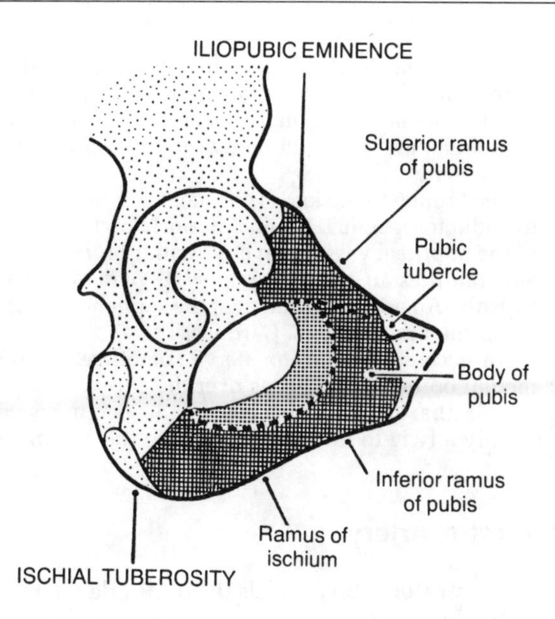

Figure 25.11. Compact area of origin in adductors.

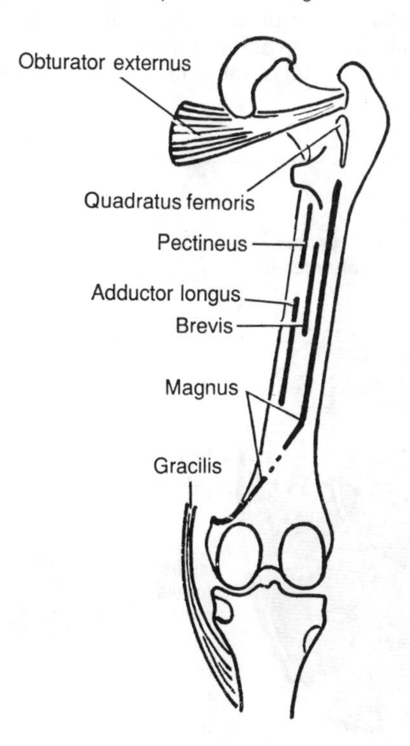

Figure 25.12. The linear insertion of the adductors.

osity) and downward onto its medial continuation (the medial supracondylar line). In fact, it extends from the level of the lesser trochanter, where it is continuous with the quadratus femoris, to the adductor tubercle distally. The portion of the adductor magnus that arises from the tuber ischii does so by tendon, belongs developmentally to the hamstring muscles, is supplied by the tibial nerve, and is inserted mainly by a palpable tendon into the adductor tubercle and into the supracondylar line just superior to it.

The *Obturator Externus*: to the trochanteric fossa. It passes inferior to the head of the femur and grooves the posterior surface of the neck. It is better seen with the hip joint (p. 271).

THREE HERNIAL SITES

Three hernial sites of clinical significance are illustrated in *Figure 25.13*. Only the inguinal and femoral hernias are common and rather easy to palpate.

NERVES AND VESSELS

Obturator Nerve

The **obturator nerve** (see *fig. 28.42* on p. 333), like the femoral nerve, is derived from L2, 3, and 4 and has motor, cutaneous, articular, and vascular distribution.

On passing through the obturator foramen, the nerve divides into anterior and posterior divisions, which supply the 6 adductors (*fig. 25.14*; *Table 25.1*).

The sole *cutaneous branch* is the continuation of the nerve to the gracilis. It reaches the surface at about the middle of the thigh, where it supplies a restricted area, but it may extend to the calf.

Articular branches supply both the hip joint and the knee joint. The following details are for reference only.

Table 25.1. Segmental Innervation of Muscles of Hip and Thigh*

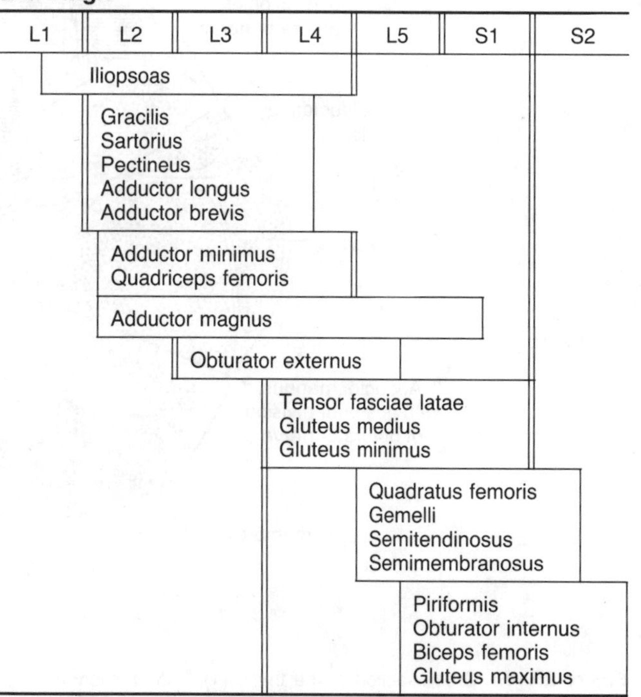

L1	L2	L3	L4	L5	S1	S2
Iliopsoas						
	Gracilis Sartorius Pectineus Adductor longus Adductor brevis					
	Adductor minimus Quadriceps femoris					
	Adductor magnus					
		Obturator externus				
			Tensor fasciae latae Gluteus medius Gluteus minimus			
				Quadratus femoris Gemelli Semitendinosus Semimembranosus		
				Piriformis Obturator internus Biceps femoris Gluteus maximus		

*Modified after Bing, and Haymaker and Woodhall (For leg and foot, see Table 28.1 on p. 331).

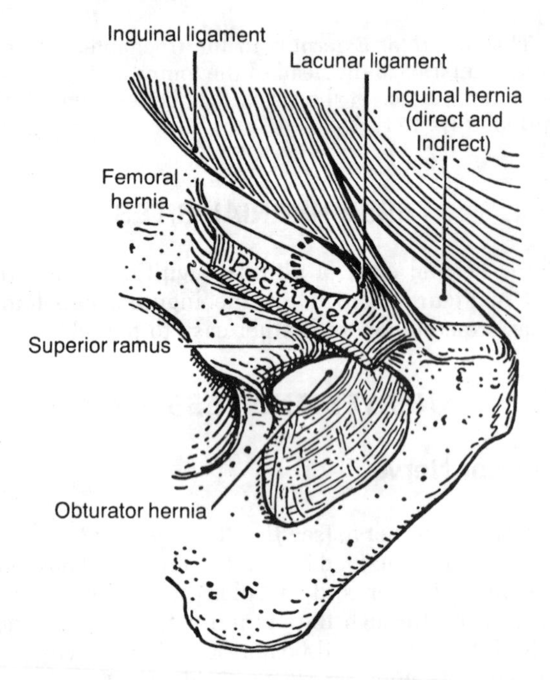

Figure 25.13. Three hernial sites and the structures separating them.

The branch to the hip springs from the main nerve within the obturator canal. Its numerous twigs ramify in the pubofemoral ligament. Some reach the synovial membrane and also run along the ligament of the head of the femur.

The branch to the knee is the continuation of the nerve to adductor magnus. It runs inferiorly on the medial side of the popliteal vessels, gives *vascular branches* to them, and ramifies in the posteromedial part of the fibrous capsule. An articular branch of the saphenous nerve may reach the knee joint (E. Gardner).

An *accessory obturator nerve* sometimes follows the medial border of the psoas over the superior ramus and rejoins the main nerve deep to the pectineus. It may supply a twig to the hip joint and to the pectineus.

Obturator Artery

The **obturator artery** assists the profunda artery to supply the adductors.

Articular twigs run through the acetabular foramen to the acetabular fossa and often a branch traverses the ligament of the head of the femur (ligamentum teres; *fig. 24.8* and p. 271).

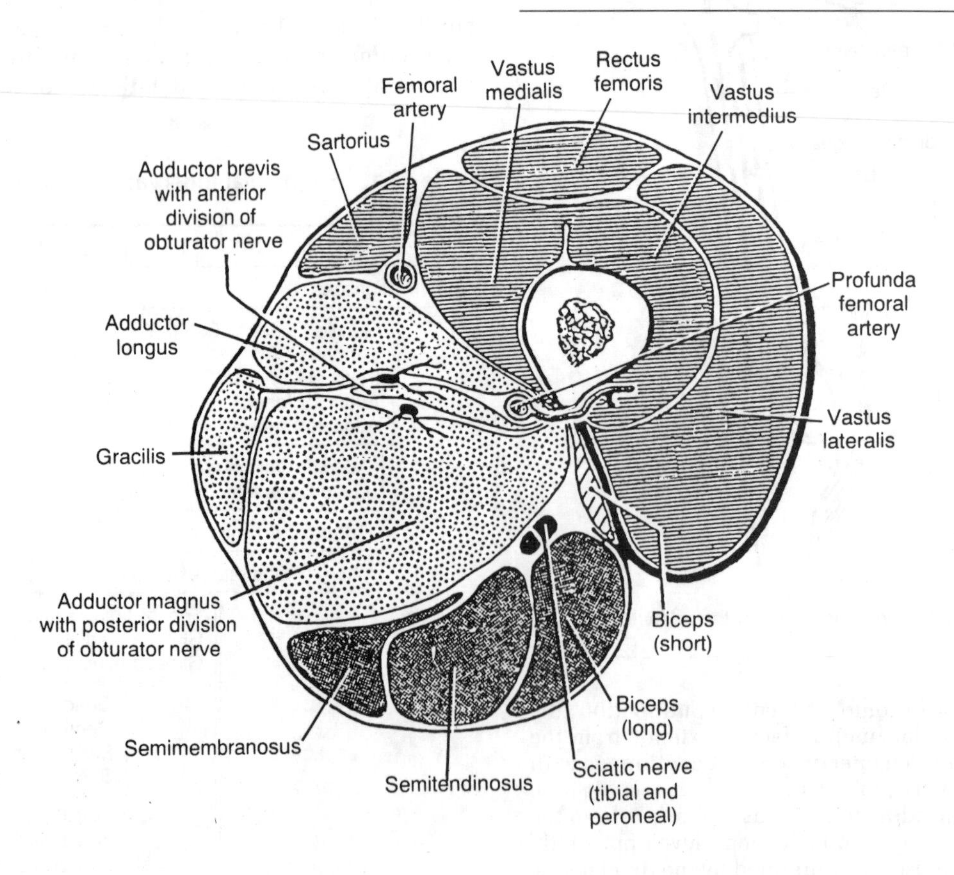

Figure 25.14. Cross-section of the thigh. (1) Adductor longus separates the femoral and profunda femoris arteries, (2) Adductor brevis separates the 2 divisions of the obturator nerve, and (3) the perforating arteries hug the bone.

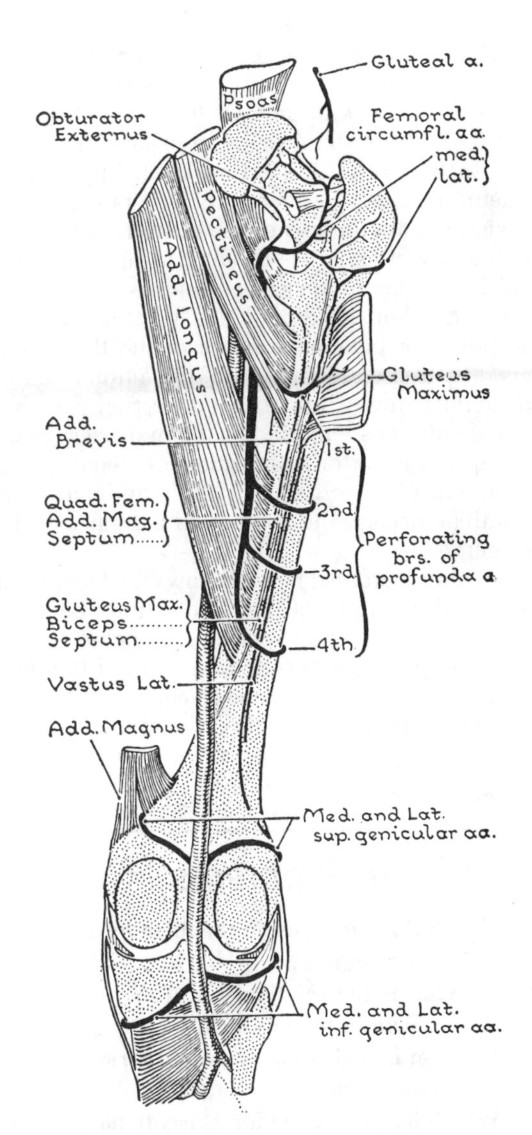

Figure 25.15. Arteries of thigh and knee from behind.

Profunda Femoris Artery

The **profunda femoris artery** usually arises from the lateral side of the femoral artery about 4 cm below the inguinal ligament. At the apex of the femoral triangle it lies posterior to the femoral vessels with its own vein. It descends among the adductors (*figs. 25.14 and 25.15*).

Distribution

Its various branches supply most of the *muscles* of the thigh, *articular branches* to the hip and knee joints, and a *nutrient* branch (or two) to the femur, and they effect numerous *anastomoses*. The named branches of the profunda femoris artery are:

1. Lateral femoral circumflex
2. Medial femoral circumflex
3. 1st, 2nd, 3rd, and 4th perforating
4. Muscular (unnamed branches)

The *lateral femoral circumflex artery* is a large artery that runs laterally between the branches of the femoral nerve and divides into 3 branches.

The *ascending* and *transverse* branches anastomose in the gluteal region, and they send branches along the anterior surface of the neck of the femur to the head. The large *descending* branch follows the anterior border of the vastus lateralis and anastomoses at the knee.

The chief duty of the *medial circumflex artery* (*fig. 25.15*) is to supply the neck and head of the femur. An *articular* branch passes through the acetabular foramen with a branch of the obturator artery (p. 288). Other branches are *muscular* and *anastomotic*.

The medial circumflex artery divides into (1) a transverse branch, and (2) an ascending branch, which anastomoses posterior to the hip.

The Four Perforating Arteries

The *four perforating arteries* encircle the shaft of the femur, hugging it so closely that they must surely be torn if the shaft is fractured. They perforate the muscles they encounter (*figs. 25.14–25.16*).

Distribution

The perforating arteries are essentially muscular in this most muscular of regions, but one also supplies the nutrient artery to the femur.

Anastomoses

The term *cruciate anastomosis* is applied to the union of the medial and lateral femoral circumflex arteries with the inferior gluteal artery superiorly and the 1st perforating artery inferiorly (*fig. 25.15*).

Primary Route of Arteries. In the embryo, the primary arterial trunk arises as a branch of the internal iliac artery and passes inferiorly on the back of the limb, accompanying the sciatic nerve. A vessel, which becomes the external iliac and femoral arteries, grows down the anterior aspect of the thigh and joins the primary trunk proximal to the knee (*fig. 25.16*). The proximal part of the primary trunk is resorbed, except in the gluteal region. In the adult, main vascular trunks cross joints on their flexor aspects—for reasons that should be obvious.

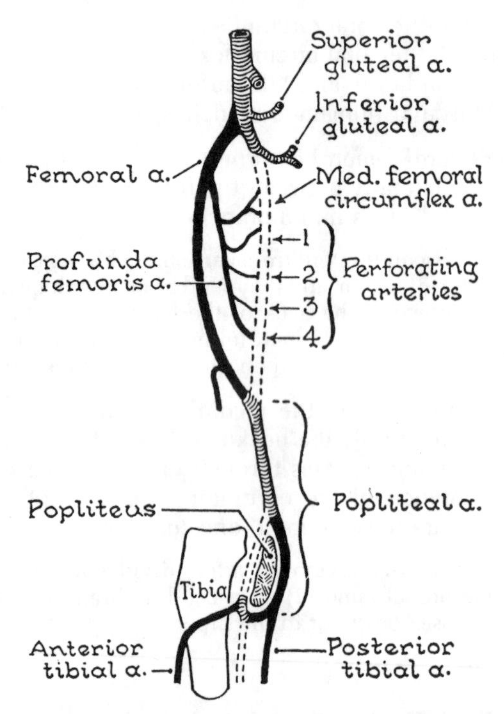

Figure 25.16. Development of the main arterial trunk: (*hatched segments* are primary, *solid black* are secondary, *clear* disappear). (After Senior.)

Anatomical Features of the Physical Examination

Palpation is the prominent technique for examining the posterior and medial regions of the thigh, along with testing of motor and sensory functions. The adductor muscles (Clinical Case 25.3) are tested as a group, without any attempt at discriminating specific functions. The emphasis for the hamstring muscles is on the knee region, where all three tendons are easily palpated. Between them in the popliteal fossa, the popliteal artery is felt faintly because of a dense wrapping of deep fascia, and one of the rare uses of the stethoscope in examining the lower limb is illustrated in Clinical Case 25.2. Palpation of the infrequent obturator hernia is probably rarely possible.

The significant contents of the posterior region of the thigh all run vertically: (1) the hamstring muscles (long head of biceps femoris, semitendinosus and semimembranosus), (2) short head of biceps, (3) the sciatic nerve, (4) the posterior cutaneous nerve of the thigh, and (5) various deep vessels. In addition, the amputation at the middle of the femur (Clinical Case 25.1) transects the four heads of quadriceps femoris, the femoral artery and vein lying deep to the sartorius muscle, cutaneous and some motor branches of the femoral nerve, adductor magnus and gracilis muscles, and some final branches of the obturator nerve.

The common peroneal nerve follows the biceps tendon to the posterior surface of the head of the fibula, where it splits into its deep and superficial branches. The tibial and common peroneal nerves supply all of the muscles and most of the skin below the knee, and the popliteal artery and vein (which are closely wrapped together) essentially supply everything below this level.

Clinical Mini-Problems

1. **Which three muscles in the thigh have an unusual dual nerve supply?**
2. **a. What is the relationship of the femoral artery and vein in the popliteal fossa?**
   b. What is their relationship superiorly in the adductor canal?
3. **Which hamstring tendon is easily palpable on the lateral aspect of the popliteal fossa?**
4. **Which gluteal muscles form the "bed" for the sciatic nerve as it descends deep to the gluteus maximus into the posterior compartment of the thigh?**

(Answers to these questions can be found on p. 586.)

26

The Knee Joint

Clinical Case 26.1

Patient A. This 21-year-old female college athlete temporarily dislocated her patella during a vigorous tennis match yesterday. Now the only finding is the presence of easily felt fluid in the joint. Still in considerable discomfort at the Sports Medicine Clinic, she seeks advice on the reason for the dislocation and how to prevent its recurrence. This requires a clear knowledge of the "tracking" of the patella in its groove and muscular lines of pull. The surgeon comtemplates draining the excess fluid, which demands precise knowledge of the various safe approaches for the needle to penetrate the interior of the joint cavity.

General Observations

It is apparent that the chief movements occurring at the knee joint are *flexion* and some *extension*. The knee joint is therefore classified as a hinge joint. However, check the degree of axial rotation (medial and lateral) permitted while your knee is in the position of flexion and semiflexion. Later you will see that an imperceptible medial rotation of the femur always occurs at the completion of knee extension.

When standing up and leaning forward, as when washing your face, you can, using your hands, *move your patella from side to side* because the quadriceps femoris is then relaxed. Remember how the unexpected blow on the back of the knee may cause you to fall. The quadriceps is not required to be in action when you stand erect, because the *line of gravity* passes in front of the axis of the knee joint (*fig. 1.15*). A considerable economy in muscular effort occurs in human bipedal stance compared with the bent knee stance of most other mammals.

BURSA

A bursa, the *prepatellar bursa*, lies between the skin and patella to permit free movement of the skin.

Anterior to the ligamentum patellae, there develops another bursa, the *superficial infrapatellar bursa*.

The precise depth of the *prepatellar bursa* varies: it may be either in the subcutaneous areolar tissue, or deep to the fascia lata, or actually in the substance of those fibers of the quadriceps tendon that pass across the front of the patella. Commonly, 2 communicating bursae are present, one in front of the other.

DIVISIONS OF THE JOINT

The bones taking part in the knee joint are: femur, tibia, and patella (*fig. 26.1* and *Plate 26.1*); the fibula is only indirectly associated forming a small joint with the tibia. Primitively, there were 3 *joint cavities* now merged into one. One is situated between the medial condyles of the femur and tibia, one between the lateral condyles of the femur and tibia, and one between the patella and the femur. They may be referred to as the *medial and lateral condylar articulation* and the *patellar articulation*. The condylar articulations are partly subdivided by medial and lateral menisci into upper and lower parts.

THE PATELLA AND THE PATELLAR ARTICULATION

Because of the obliquity of the long axis of the femur, the femur and tibia form an open angle at the lateral side of the knee, toward which the patella would tend to dislocate. Dislocations are not common, largely because of 2 factors.

1. The forward projection of the lateral condyle of the femur (*fig. 26.2*).

2. The low attachment to the patella of the lowest part of the vastus medialis (*fig. 26.3*) draws the patella medially.

The *posterior surface* of the patella (*fig. 26.4*) has a series of paired facets that articulate in turn during extension, slight flexion, flexion, and full flexion (*fig. 26.5*).

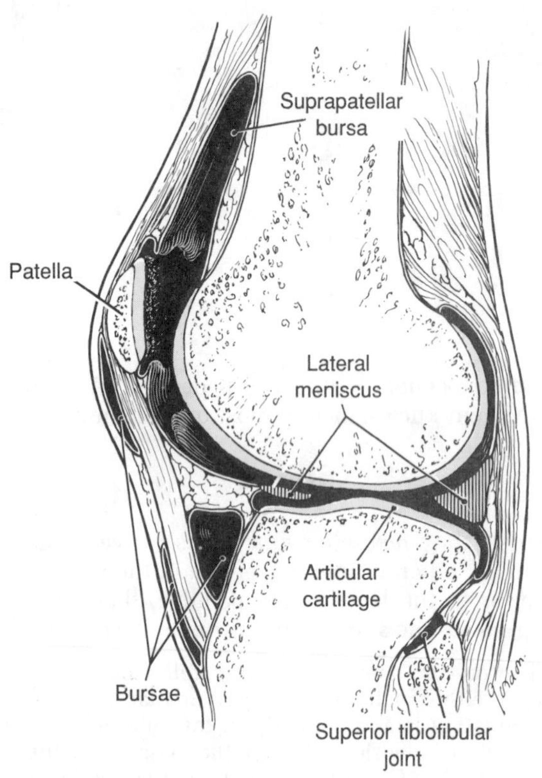

Figure 26.1. Dissection of knee joint; sagittal section through lateral condyles of femur and tibia. (From Basmajian JV: *Primary Anatomy*, ed. 8. Williams & Wilkins, Baltimore, 1983.)

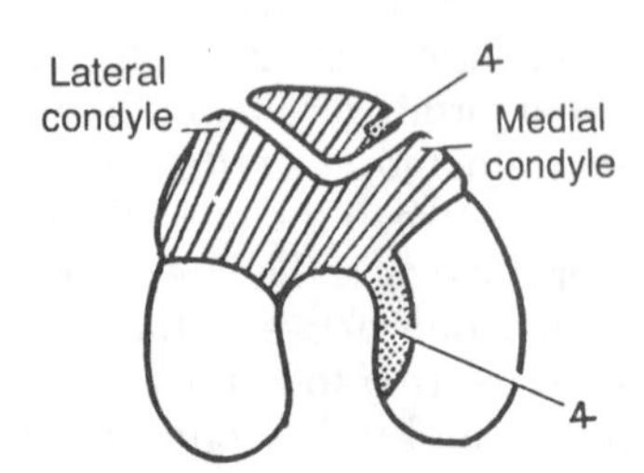

Figure 26.2. The patella in its "trochlea."

Figure 26.3. The patellar attachments of the vasti medialis and lateralis.

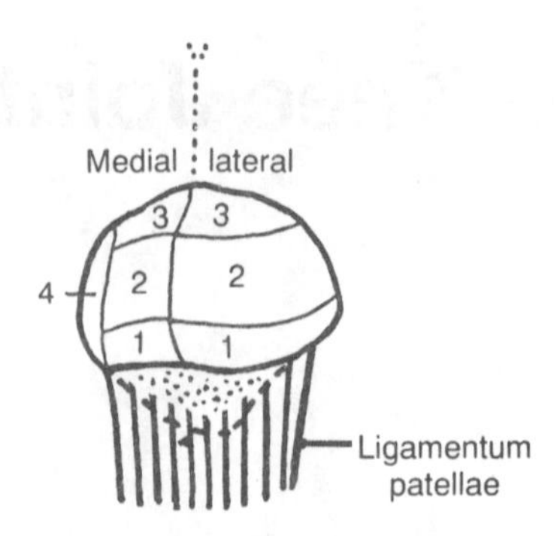

Figure 26.4. The 3 paired facets on the posterior surface of the patella articulate with the femur as shown in *Figure 26.5.* The medial vertical facet (4) articulates along the margin of the intercondylar notch during full flexion (see *fig. 26.2*).

The human patella is subject to fracturing into an upper and lower fragment due to trauma. Would it be less liable to fracture if it possessed a single concave facet as in lower mammals (*fig. 26.6*)?

Palpation

A considerable portion of the articular surfaces of the knee joint can be palpated, but the muscles must be relaxed.

Functions of the Patella. Experimental work has shown that when the knee joint is flexed, the patella is a mechanical hindrance to extension. Indeed, the removal or excision of the patella results in increased extensor efficiency. But in the later degrees of extension (say 150–180°), the patella improves efficiency by holding the patellar tendon away from the axis and thereby increasing the extending momentum of the quadriceps pull (Haxton) (*fig. 26.7*).

Clinical Case 26.2

Patient Blair McM. During an amateur ice-hockey game last night, this 28-year-old motor mechanic was driven against the boards by a defenseman. He suffered extreme pain immediately and now has localized tenderness and moderate swelling on the lateral side of the knee just superior to the head of the fibula, but no significant swelling of the knee joint itself. Cautious palpation and stress-testing for ligamentous tears convinces you that probably the lateral (fibular) collateral ligament was the only important structure damaged. Radiographs rule out bony lesions. Further management depends on preventing stress to the ligament until it heals.

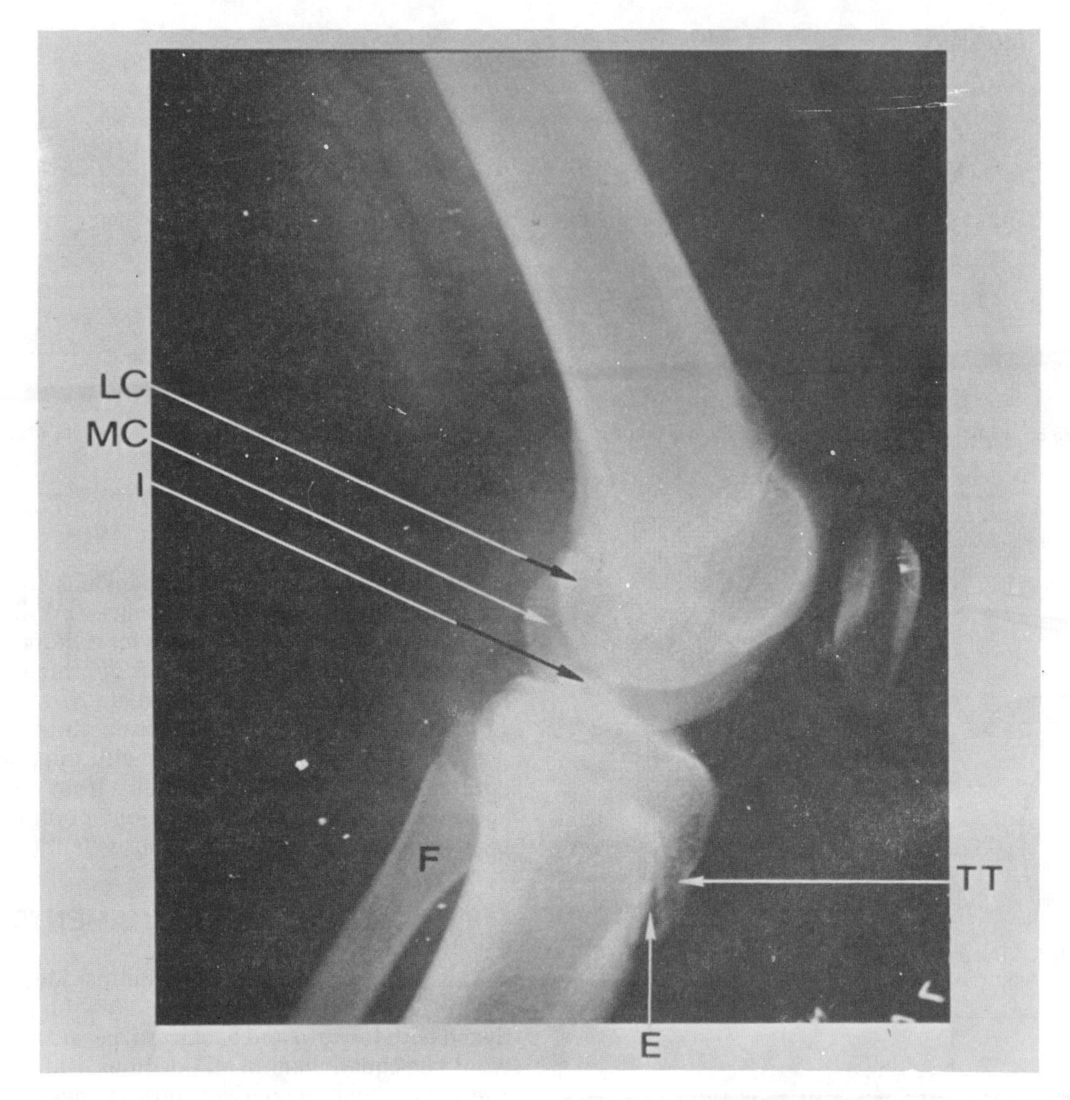

Plate 26.1. Knee (lateral view). *E*, epiphyseal line of tibia; *F*, fibula; *I*, intercondylar eminence; *LC*, lateral condyle of femur; *MC*, medial condyle of femur; *TT*, tibial tuberosity.

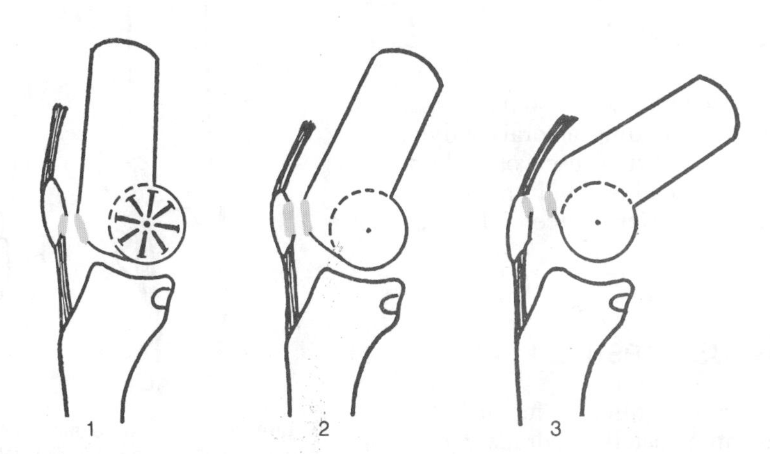

Figure 26.5. The knee joint during (1) extension, (2) slight flexion, (3) flexion. At the hub of the wheel, or center of the disc, lies the epicondyle.

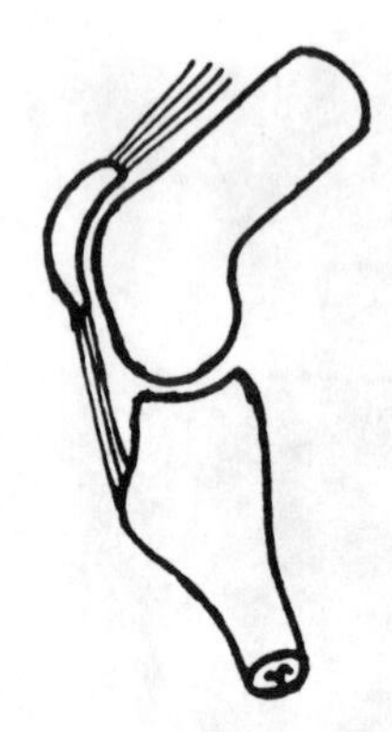

Figure 26.6. Knee joint of a sheep with concave patellar facet.

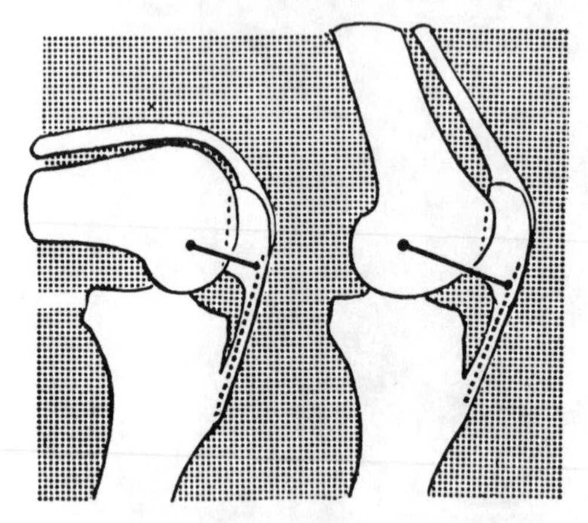

Figure 26.7. The patella hindering (*left*) and assisting (*right*) extension. (After Haxton.)

Tibiofemoral or "Condylar" Articulations (*Plate 26.2*)

On the upper surface of each tibial condyle there is an oval articular area for the corresponding femoral condyle. The articular areas are separated from each other by a narrow nonarticular area, which widens in front and behind into an *anterior* and a *posterior intercondylar area* (see *fig. 26.11*).

THE FIBROUS CAPSULE

This "tube" of fibrous tissue joining the femur to the tibia includes the ligamentum patellae, coronary ligaments (p. 298) and medial ligament, but not the lateral ligament. The fibrous capsule is reinforced in front and at the sides by aponeurotic expansions from the vasti, sartorius, semimembranosus, biceps, and iliotibial tract. (For *details*, which are important for orthopedic surgeons who must repair disrupted knees of athletes, see Harty; Hughston and Eilers; Marshall et al.; Noyes and Sonstegard; Robichon and Romero.) Behind, the capsule is composed of fibers that run parallel with the popliteus, i.e., obliquely downward and medially from femur to tibia; and one band, known as the *oblique popliteal ligament*, is attached to the semimembranosus tendon.

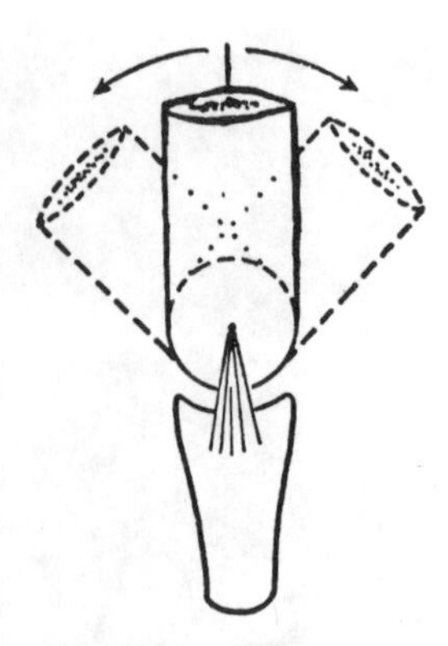

Figure 26.8. The hypothetical perfect hinge, which the knee certainly is not.

COLLATERAL LIGAMENTS

If the femoral condyles were round and their collateral ligaments arranged as in *Figure 26.8*, the joint could be flexed both forward and backward, because the ligaments would be equally taut in all positions. But: (1) the medial and lateral femoral condyles project backward like discs

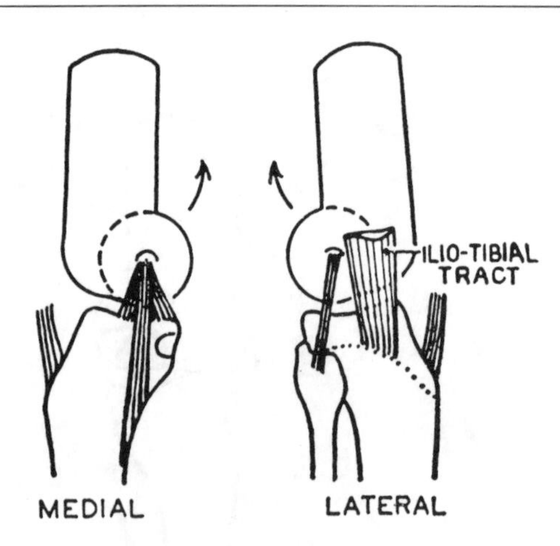

Figure 26.9. The eccentric attachments of collateral ligaments. The insertion of the iliotibial tract in front of the transverse axis of the joint helps to keep the extended joint extended.

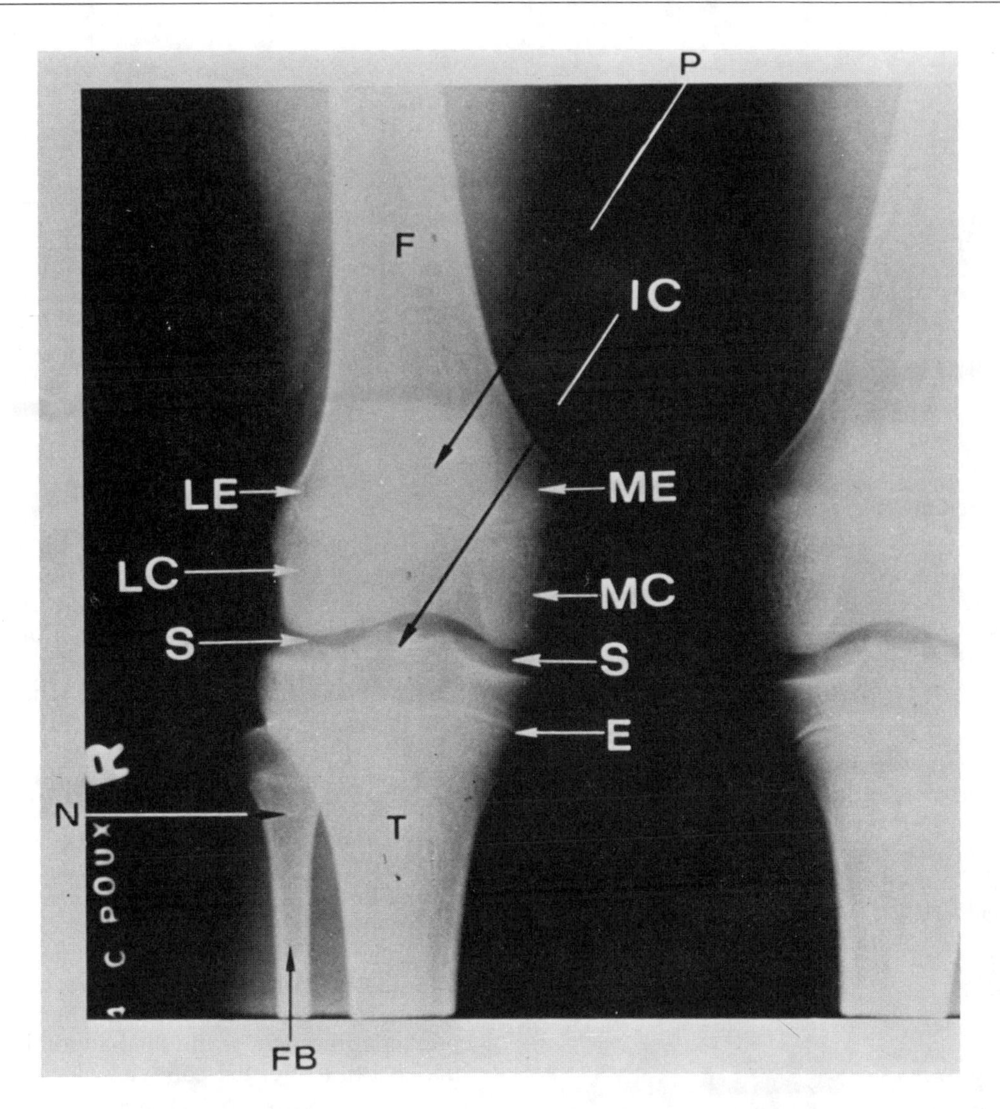

Plate 26.2. Knees, right and left (A-P view). *E*, epiphyseal cartilage of right tibia; *F*, femur; *FB*, fibula; *IC*, intercondylar eminence; *LC*, lateral condyle of femur; *LE*, lateral epicondyle of femur; *MC*, medial condyle of femur; *ME*, medial epicondyle of femur; *N*, neck of fibula; *P*, patella; *S*, menisci (semilunar cartilages); *T*, tibia.

or wheels; (2) the collateral ligaments are attached above to the epicondyles at the centers of the superadded wheels; and (3) below they are attached far back (*figs. 26.9 and 26.10*). As a result, when the knee is flexed, the collateral ligaments are slack and permit medial and lateral rotation. When the knee is extended, they become taut.

The *tibial collateral ligament (Medial Collateral Ligament)* has a superficial and a deep part. The superficial part is a long band that bridges the tibial condyle and the hollow below (*fig. 26.9*). The deep part is deltoid in shape (and name) and is attached to the margin of the tibial condyle. One or more *bursae* lie deep to the long band.

The band is crossed by 3 tendons—sartorius, gracilis, and semitendinosus, tendinous expansions of which reinforce the capsule of the joint (details below).

The *tendinous expansions* of the sartorius, gracilis, and semitendinosus find attachment to the medial surface of the tibia below the level of the tuberosity (*see fig. 27.2*). Collectively, they are called the "*pes anserinus*" (goose's foot) by clinicians. Each represents a different region of the thigh—a curious but insignificant detail; each is supplied by a different nerve (femoral, obturator, or sciatic); each passes across the medial ligament of the knee to reach its insertion, a *bursa* intervening. Much more important to surgeons is the spread of fibers from these tendons which reinforce the medial side of the knee capsule as part of the *medial patellar retinaculum*.

The *fibular collateral ligament (lateral collateral ligament)* is a cord that extends from the lateral epicondyle downward and backward to the head of the fibula (*fig. 26.9, see also fig. 26.19*). This cord is partly overlapped by the biceps tendon—with a bursa intervening (*see fig. 25.3*).

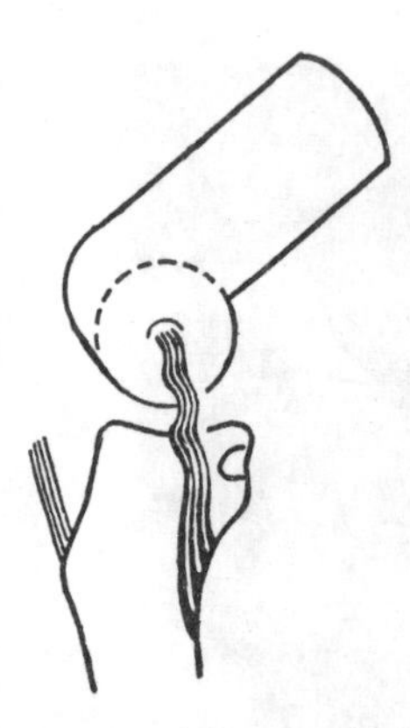

Figure 26.10. The collateral ligaments are slack during the flexion and permit rotation.

Clinical Case 26.3

Patient Casey N. As this 21-year-old college football quarterback was about to throw a forward pass, he was "sacked." When he attempted to rise, he found his left knee was very painful and locked in the flexed position. As a volunteer assistant to the team doctor, you are expected to help with the acute care. Your anatomical knowledge should be adequate to explain the prognosis and later management of the case.

Menisci, Cruciate Ligaments, Condyles, and Movements

CRUCIATE LIGAMENTS

Within a midline synovial septum (to be discussed later), the important *anterior* and *posterior cruciate ligaments* cross each other like the limbs of a St. Andrew's cross or X. The limb attached to the tibia anteroinferiorly (anterior cruciate) is attached in the intercondylar notch of the femur posterosuperiorly, and vice versa. It is from the tibial attachments that the cruciates are designated anterior and posterior (*figs. 26.12–26.15*). They will be discussed further on p. 298 from developmental and clinical points of view.

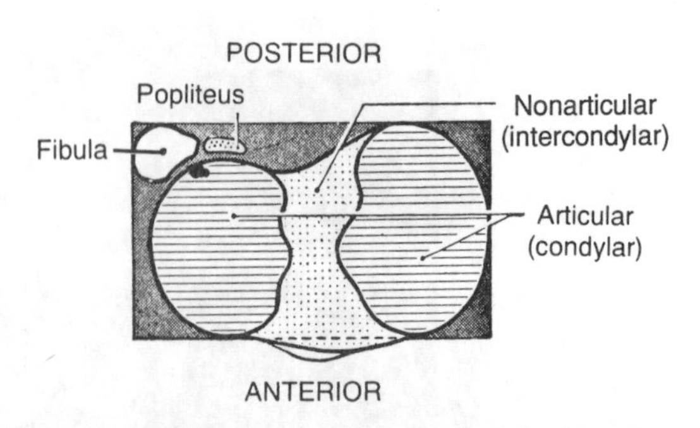

Figure 26.11. Upper ends of tibia and fibula in oblong frame.

SEQUENCES

In drawing the upper ends of the tibia and fibula from above, you may arbitrarily fit them into an *oblong frame*, in which the head of the fibula occupies the posterolateral angle (*fig. 26.11*). From this fact the following sequence of facts might be surmised.

1. The lateral condyle of the tibia must be shorter from front to back.

2. *Therefore*, of the 2 **menisci** that rest on and fit the tibial condyles, the lateral is shorter than the medial (*fig. 26.12*). The shorter lateral meniscus is shaped like a small "o," the longer medial meniscus like a capital "C." The ligamentous horns of the "C" are attached far apart, embracing those of the "o" on the nonarticular part of the tibial plateau (*fig. 26.12*).

3. *Therefore*, the portion of the lateral condyle of the femur that articulates with the lateral condyle of the tibia (and its meniscus) is shorter anteroposteriorly than the corresponding part of the medial condyle of the femur (*fig. 26.12*). (Note: the condyles of the femur also have

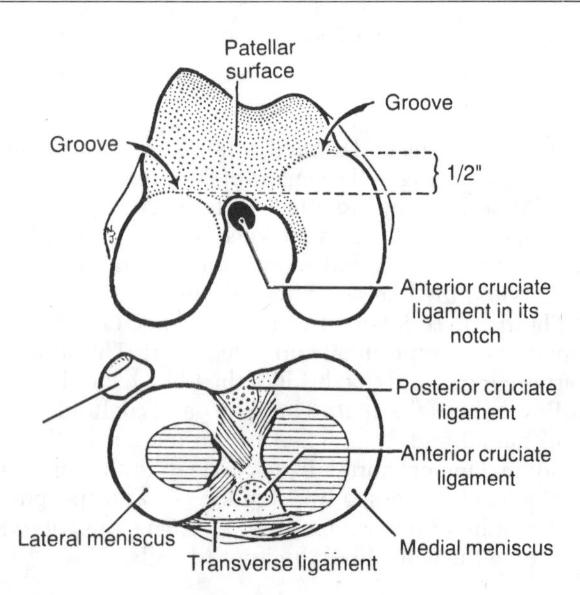

Figure 26.12. The articular surface of the knee joint.

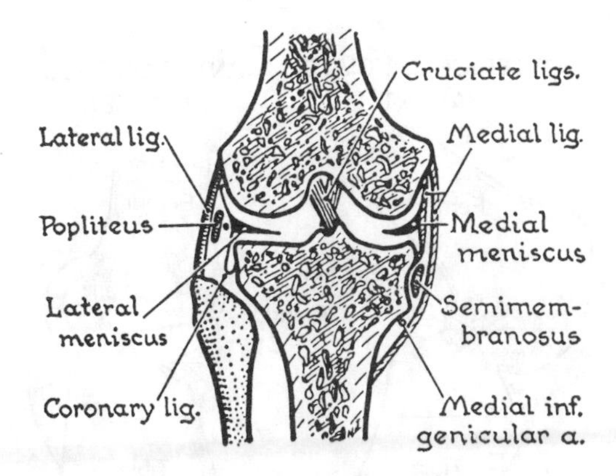

Figure 26.13. The knee joint, in coronal section.

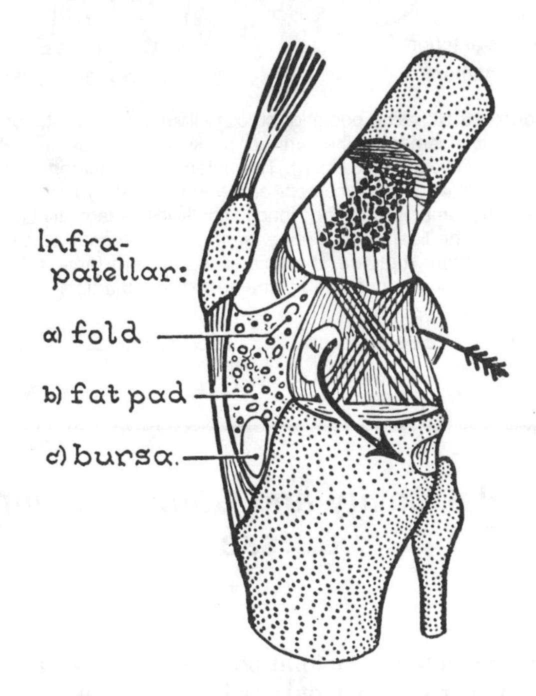

Figure 26.14. Scheme of the intercondylar septum.

patellar areas. Just now reference is to the tibiomeniscal areas.)

4. *Therefore*, as the joint passes from full flexion to full extension, the medial condyle of the femur has room to roll farther than the lateral condyle.

5. During extension, the 2 femoral condyles also revolve or "spin" on the tibia and its menisci. The *posterior cruciate ligament*, acting as a drag, greatly restricts the forward roll of the condyles and causes them to spin. In effect, the movement is a *hinge movement* with the addition of some true *forward roll*.

When the lateral femoral condyle can revolve no more, the longer medial condyle still has surface available (1 cm) to revolve (*fig. 26.12*).

6. During extension, the turning of the lateral femoral condyle is arrested by 2 obstacles: (1) the anterior margin of the lateral meniscus, which fits into a *curved groove*

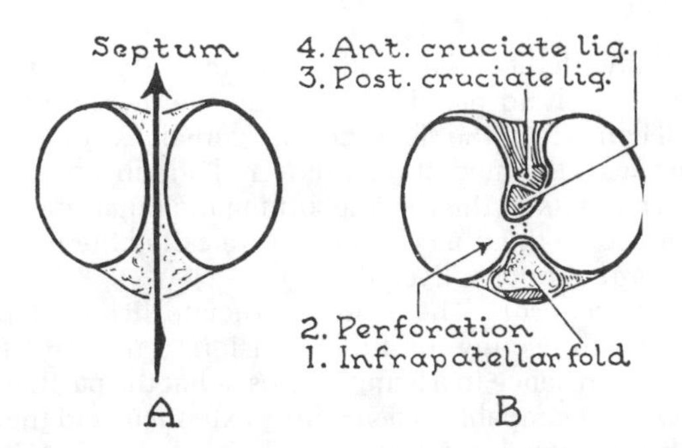

Figure 26.15. Intercondylar septum as seen from above on the tibia. *A*, early; *B*, later. Four derivatives are seen.

on the femur; and (2) the taut *anterior cruciate ligament*, fits into its own secondary notch in the intercondylar notch (*fig. 26.13*). But extension is not complete.

7. *Therefore*, while the medial femoral condyle is completing its final spin and roll, the femur must rotate medially on its long axis to permit it. The pivot around which it rotates is the anterior cruciate ligament.

8. At the same time, the lateral femoral condyle, which is no longer rolling, and the lateral meniscus, whose sharp anterior margin is locked to it by its groove, slide forward together on the tibia, moving as one structure.

It is a matter of interest and significance that the posterior end of the lateral meniscus is attached to the femur by an oblique band that passes either in front of or behind the posterior cruciate ligament. In many mammals, this greatly enlarged oblique band is almost the sole posterior attachment of the lateral meniscus, indicating thereby that its allegiance is to the femur rather than to the tibia.

Part of the popliteus is attached to the posterior end of the lateral meniscus and may serve as a retractor muscle (Basmajian and Lovejoy; Last; Lovejoy and Harden).

9. The medial femoral condyle is, then, completing the process of extension at the same time that the whole femur is undergoing axial rotation. No special rotator muscles are provided. When these movements are completed, the anterior border of the medial meniscus also fits into its *curved groove* on its femoral condyle (*fig. 26.12*).

And finally some *details for students interested in orthopedics and similar specialties*:

10. It follows that the upper surface of the lateral tibial condyle is flat (and not concave) to allow forward gliding, and the lateral meniscus is broad and expansive in order to act as a carriage or toboggan for the lateral femoral condyle (*fig. 26.13*).

11. The upper aspect of the medial tibial condyle is also flat but allows flexion, extension, and rotation. And its meniscus, having only the restricted sliding action, is narrow; indeed, it tapers anteriorly.

12. The fibular collateral ligament passes to the lateral margin of the head of the fibula. Therefore, it lies wide of the lateral tibial condyle and lateral meniscus. In fact, the gap is wide enough to afford a passage for the popliteus tendon (*fig. 26.13*)

13. But with the tibial collateral ligament it is different: its deeper deltoid part is attached to the margin of the medial condyle of the tibia and therefore does come into contact with the medial meniscus and adheres to it.

14. The parts of the capsule joining the outer edges of the menisci to the tibia (*fig. 26.17*) are called the *medial* and *lateral coronary ligaments*.

Prehistoric Man. The anterior cruciate ligament has been seen to occupy, on full extension, a notch of the femur. Its presence in a femur suggests that the particular knee joint was capable of being fully extended, and therefore that its owner walked erect. The femora of certain prehistoric men present such well-marked notches for the anterior cruciate ligament that they unquestionably walked erect, but Neanderthal Man, whose femora have no suggestion of the notch, probably walked with a crouching gait.

THE INTERCONDYLAR SEPTUM (*figs. 26.14 and 26.15*)

In prenatal life, a vertical septum separates the medial and lateral condylar joints from each other. Its lower border remains attached to the intercondylar area of the tibial plateau. The posterior half of its upper border is attached to the intercondylar notch of the femur; the anterior half is free and extends from the intercondylar notch of the femur to the patella just below its articular surface.

A *perforation* appears in the septum and extends backward to the anterior cruciate ligament (*fig. 26.14*). It divides the septum into an anterior part, the *infrapatellar synovial fold*, and a posterior part in which the *anterior* and *posterior cruciate ligaments* develop.

The cruciate ligaments develop in the posterior part of the septum and cross each other obliquely. It is from their tibial attachments that they take their names. *Figure 26.16* makes their functions clear and explains the *drawer sign* for diagnosing which cruciate ligament is torn—pulling forward or pushing backward on the tibia (with the knee bent).

The articular surfaces of the tibial condyles rise gently to 2 peaks, the *intercondylar eminence*, in the nonarticular area (*fig. 26.13 and Plate 26.2*). A sideways slide of the femur would cause one or other condyle to mount an incline. This the cruciate and collateral ligaments resist.

The infrapatellar synovial fold is a flattened and fat-filled cone of synovial membrane. Its apex remains attached to the most anterior point of the intercondylar notch of the femur (*fig. 26.14*).

Its open base extends from just below the articular cartilage of the patella to the anterior intercondylar area of the tibia; its 2 sides are free, and from each a small wing, the alar fold, projects.

An *infrapatellar pad of fat* (i.e., the fat behind the ligamentum patellae) is continued upward into the infrapatellar fold and, in stout subjects, into the alar folds also.

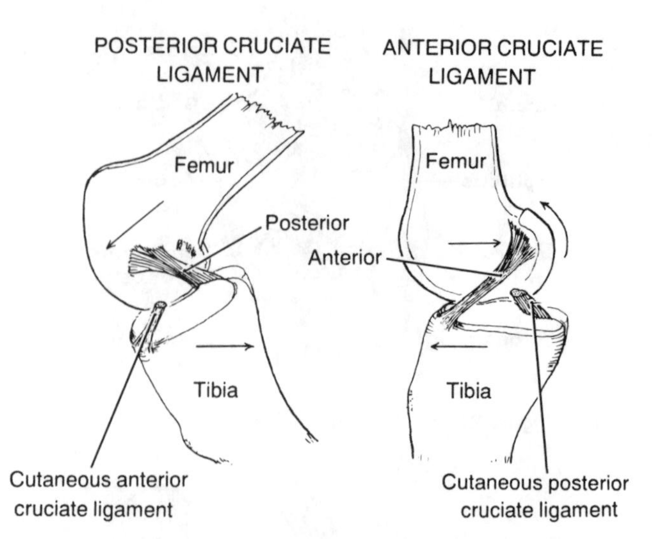

Figure 26.16. The posterior cruciate ligament prevents forward displacement of the femur or backward displacement of the tibia indicated by *arrows*. The anterior "X" ligament prevents backward displacement of the femur and hyperextension. The femur is sectioned longitudinally in the sagittal plane to reveal the ligaments. *Drawer signs:* with the anterior cruciate ligament ruptured, the tibia slides forward (anteriorly); with the posterior cruciate ligament ruptured, the tibia slides posteriorly.

Synovial Cavity and Communicating Bursae

Developmentally, the joint possesses 3 *synovial cavities*: a patellar and 2 condylar, which later communicate freely.

Each condylar cavity is divided into an *upper* and a *lower part* by the meniscus (*fig. 26.13*). The coronary ligaments—a part of the capsule—attach the convex margins of the menisci to the upper end of the tibia just below the articular margin (*fig. 26.17*).

Three bursae communicate with the knee joint (*figs. 26.17 and 26.18*). These bursae lie deep to the tendons of quadriceps femoris, popliteus, and gastrocnemius. The bursa deep to the **quadriceps femoris** tendon, known as the **suprapatellar bursa**, almost always opens into the patellar cavity (*figs. 26.18 and 26.19*). During extension, the *articularis genu* retracts it.

The *popliteus bursa* opens into the lateral condylar cavity below the meniscus (*figs. 26.17 and 26.19*).

Sometimes the partition between bursa and tibiofibular joint dissolves, bringing these joints into communication. Abnormal fluid in the knee joint can fill the bursa, which can produce a sausage-shaped, soft mass deep to the skin. The *gastrocnemius bursa* (deep to the medial head) com-

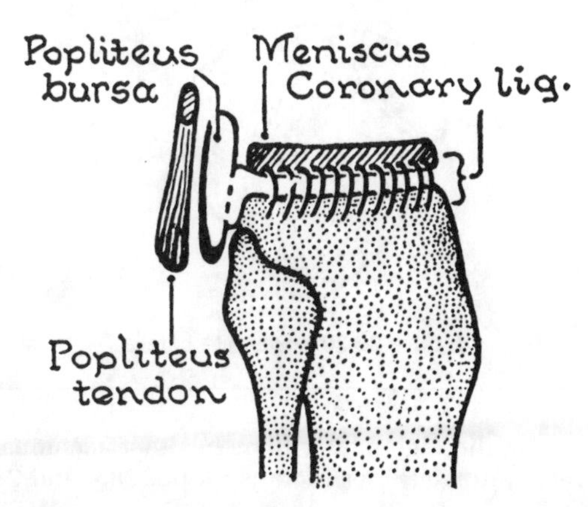

Figure 26.17. Popliteus bursa communicating with the joint cavity and coronary ligament attaching convex border of meniscus to tibia.

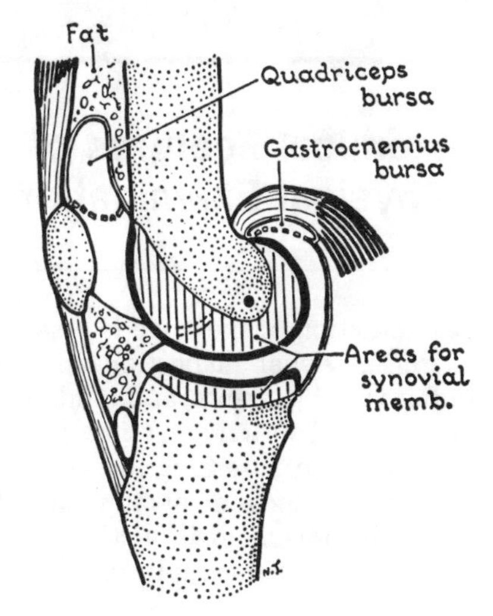

Figure 26.18. The quadriceps and gastrocnemius bursae communicating with the synovial cavity.

monly communicates. This bursa may also communicate with a bursa deep to the semimembranosus, which may be palpated if filled with fluid from the knee.

Extent of the Synovial Cavity (*figs. 26.18 and 26.19*). How far may an incision be carried downward on the femur without invading the joint cavity? In front, to the top of the suprapatellar bursa, that is 2 fingerbreadths above the patella; at the sides, the level must fall below the epicondyles, for they give attachment to the collateral ligaments.

Details of Bursae about the Knee, for reference. There are 11 or more:

3 communicate with the joint—quadriceps (suprapatellar), popliteus, and medial gastrocnemius.

3 related to the patella and ligamentum patellae—prepatellar, superficial infrapatellar, and deep infrapatellar (*fig. 26.19*).

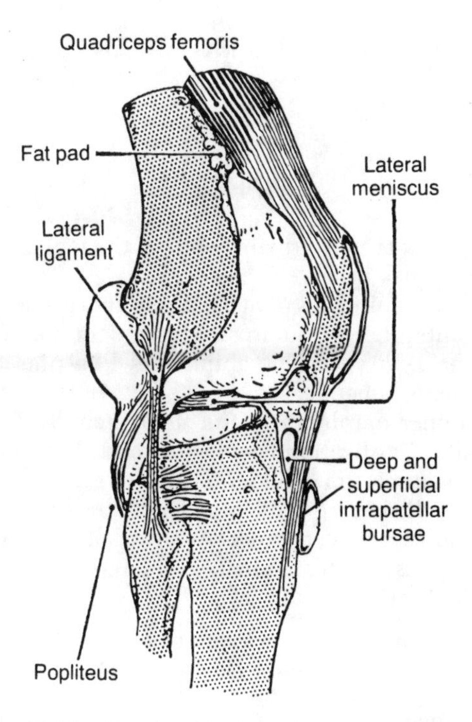

Figure 26.19. Knee joint filled with latex to show the extent of the synovial cavity (lateral view).

2 semimembranosus bursae—the one between semimembranosus and gastrocnemius tendons may communicate with the gastrocnemius bursa and so may communicate indirectly with the knee joint; the other lies between semimembranosus tendon and the tibial condyle.

2 superficial to the collateral ligaments—the one between the fibular collateral ligaments and the overlying biceps tendon, the other between the tibial collateral ligaments and the 3 overlying tendons (sartorius, gracilis, and semitendinosus). The latter is commonly continuous with a bursa between sartorius superficially and gracilis and semitendinosus deeply.

1 bursa between the superficial and deep parts of the tibial collateral ligaments.

RELATIONS

All muscles crossing the joint are relations of the joint. The **iliotibial tract** should be noted especially, on account of its great protective value to the exposed lateral side of the knee. About 2–3 cm (or more) wide and placed between the ligamentum patellae and the biceps tendon, it alone separates the skin from the synovial membrane. The common peroneal nerve follows the posterior border of the biceps. The tibial nerve is behind the popliteal vein, which in turn is behind the popliteal artery (*figs. 25.8, 25.10, and 25.15*).

MUSCLES AND MOVEMENTS

See Table 26.1.

EPIPHYSES

The more actively growing ends of the femur and tibia are at the knee (p. 6).

The *lower epiphysis of the femur* begins to ossify about the 9th intrauterine month and fuses with the diaphysis about the 19th year. The epiphyseal line runs through the adductor tubercle.

The *upper epiphysis of the tibia* includes the tibial tuberosity. Ossification begins before birth or soon after birth; fusion occurs at about the 19th year.

The *patella* begins to ossify at about the 3rd year, probably from several centers. The superolateral angle may remain unossified (emarginate patella) or may ossify independently (*fig. 26.20*).

Blood Supply. Of the 5 articular branches of the popliteal artery, the middle genicular artery passes forward and supplies the structure in the intercondylar septum. The 2 medial and 2 lateral genicular arteries course deep to all muscles and ligaments they encounter and embrace either the femur or the tibia. The lateral inferior genicular artery is an exception because it passes (1) behind the popliteus tendon (*fig. 25.15*), and then (2) runs along the margin of the lateral meniscus.

These arteries anastomose with each other, with the lateral circumflex (descending branch), the descending genicular, and the anterior tibial recurrent artery.

Nerve Supply. This is derived from the *femoral nerve* via branches to the vasti and the saphenous nerve; the *obturator nerve* via the branch to the adductor magnus, and the tibial and common peroneal branches of the *sciatic nerve* via the 6 genicular branches that accompany the corresponding arteries (details on pp. 285, and 288).

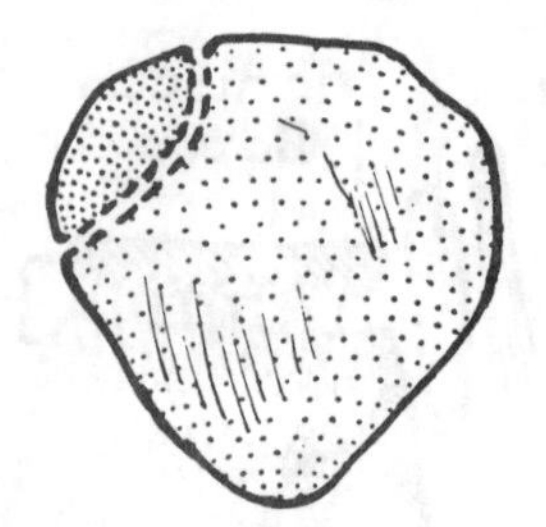

Figure 26.20. Bipartite patella.

TIBIOFIBULAR JOINTS

See p. 309.

Anatomical Features of the Physical Examination

Clinical examination of the knee is essentially an anatomical exercise. Visual inspection and palpation of bones, joint borders, ligaments, and tendons and testing for tenderness depend entirely on anatomical knowledge and guided experience. Stressing the knee joint, by the examiner's attempting movements in various obvious directions in which movement is not normally possible, reveals problems in the collateral ligaments (Clinical Case

Table 26.1. Muscles Acting upon the Knee Joint (All the Muscles That Cross It)

Nerve supply	Muscles	Accessory actions	Main actions
	Iliotibial tract	Retain knee in the extended position	
Sup. gluteal	T. fasciae latae		
Inf. gluteal	Gluteus max. (part)		Extensors
	Quadriceps femoris		
	Rectus femoris		
Femoral	V. intermedius		
	V. lateralis		
	V. medialis		
	Sartorius		Flexors
Obturator	Add. gracilis	Rotate leg medially	
	Semitendinosus		
	Semimembranosus		
	Popliteus		
Tibial division of sciatic	Gastrocnemius		
	Plantaris		
	Biceps (long)		
Peroneal division of sciatic	Biceps (short)	Rotate leg laterally	

26.2) and the anterior and posterior cruciate ligaments. The latter are tested by pulling the flexed tibia anteriorly or pushing it posteriorly, remembering the obliquely running cruciate ligaments are named for their tibial attachments.

With experience, one can easily feel the hamstring tendons bordering the popliteal fossa, the iliotibial tract and the common peroneal nerve lying on the head and neck of the fibula.

Fluid in the knee joint (Clinical Case 26.1) is palpable if there is more than about 50 ml. It can be demonstrated also by forcing all the fluid proximally, which causes a visible suprapatellar bulging. Also a quick forceful posterior tap on the patella will reveal that it is "afloat" on excess fluid separating it from its femoral trough. Occasionally, fluid may form a sausage-like mass when it has filled the communicating bursae posterior to the knee (e.g., popliteus bursa).

The menisci are normally inaccessible to palpation except for the anterior part of the lateral meniscus, which can be felt as a vague bulging during full extension. Access to the interior of the knee joint by arthroscopy permits extensive viewing and direct minisurgical operations on the menisci and ligaments.

The patella is anchored inferiorly by its large, palpable ligament to the prominent tibial tubercle; this is really the insertion of the quadriceps muscle, the great extensor. Side extensions of the tendon of this and all other muscles crossing the knee reinforce the rather loose fibrous capsule, which is especially loose above the patella (deep to the quadriceps) and quite redundant upward for two to three fingerbreadths—the suprapatellar bursa. Other bursae, several of which communicate directly or indirectly with the knee joint lie between tendons and the joint capsule.

The capsule is reinforced by collateral ligaments running from inconspicuous femoral epicondyles. The lateral ligament is palpable where it runs to the head of the fibula.

The medial aspect of the fibrous capsule is thickened in complex layers to form the tibial (medial) collateral ligament but the fibular (lateral) ligament is a separate rounded cord. Both ligaments are vulnerable to severe traumatic stresses, especially in athletes, as illustrated in Clinical Case 26.2.

Clinical Mini-Problems

1. a. **Which cruciate ligament prevents hyperextension (tendency for tibia to slide posteriorly) of the knee joint?**
   b. **Which cruciate ligament prevents hyperflexion of the knee joint?**
2. **Which two palpable structures are attached to the head of the fibula?**
3. **What features of the patella and its attachments prevent its lateral displacement during the contraction of the quadriceps muscles?**

(Answers to these questions can be found on p. 586.)

27

Leg and Dorsum of Foot

The leg or **crus** is the segment of the lower limb between the knee and ankle. Its main regions (fascial compartments) are (*figs. 27.1, A* and *B*):

1. Anterior crural
2. Posterior crural (much larger)
3. Lateral or "fibular" or "peroneal."

Clinical Case 27.1

Patient Andy O. During a parade, this 17-year-old male suddenly darted out to cross the road. He was struck on the lateral side of his right leg by the bumper of an official's car. Even to ordinary spectators, a fracture of the leg bones was obvious. He has just been brought into the emergency department where you recently started a clinical clerkship. The following pages are directly related to this type of case, but, of course the problem includes the sections on the soft parts, not just the bones.

Medial Surface and Body of Tibia

Run your fingers over the medial surface of your tibia to confirm that it is subcutaneous and smooth from the medial malleolus, the prominent lump at the ankle, to the medial condyle; note that it is bounded in front by the subcutaneous *anterior border*, the "shin," throughout, and that it is bounded posteriorly by the subcutaneous *medial border* of the tibia (*fig. 27.2*).

At the medial side of the knee, you have already felt 2 of the 3 tendons. (What were they? —p. 284.) Sartorius tendon is much more difficult to feel and not worth the effort.

The tibial collateral ligament (*medial ligament of the knee*) and the three tendons of the "pes anserinus" (p. 295) attach to the tibia as shown in *Figures 27.2 and 27.3*.

The **great saphenous vein** crosses the lowest third of the medial surface of the tibia obliquely and continues

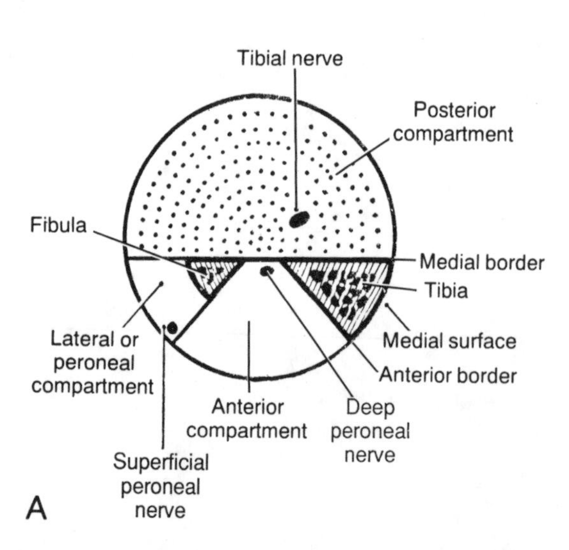

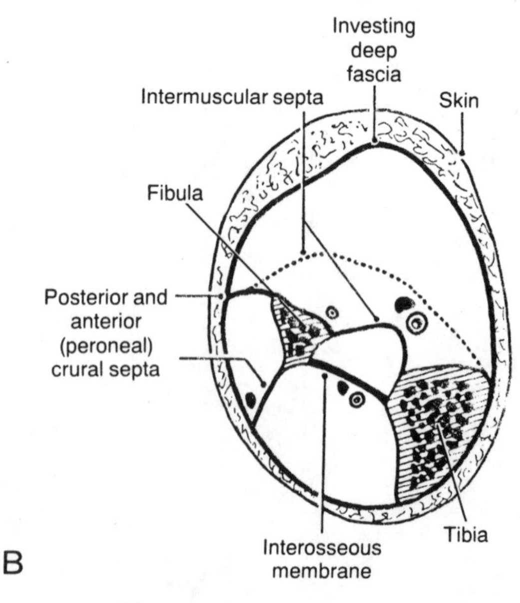

Figure 27.1. *A*, Scheme of regions of the leg on transverse section, showing their relative sizes and nerves. *B*, The leg on cross-section, showing the interosseous membrane and the various intermuscular fascial septa enclosing the fascial compartment.

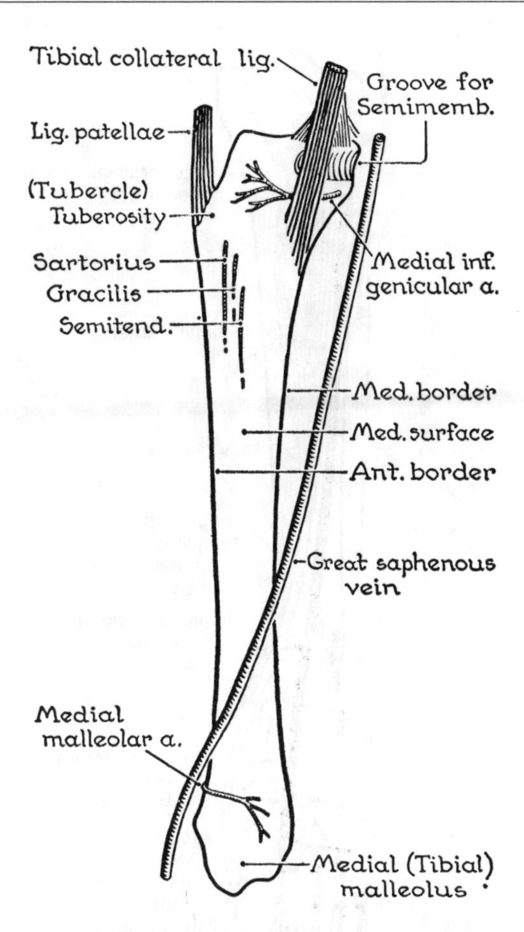

Figure 27.2. The subcutaneous or medial aspect of the tibia.

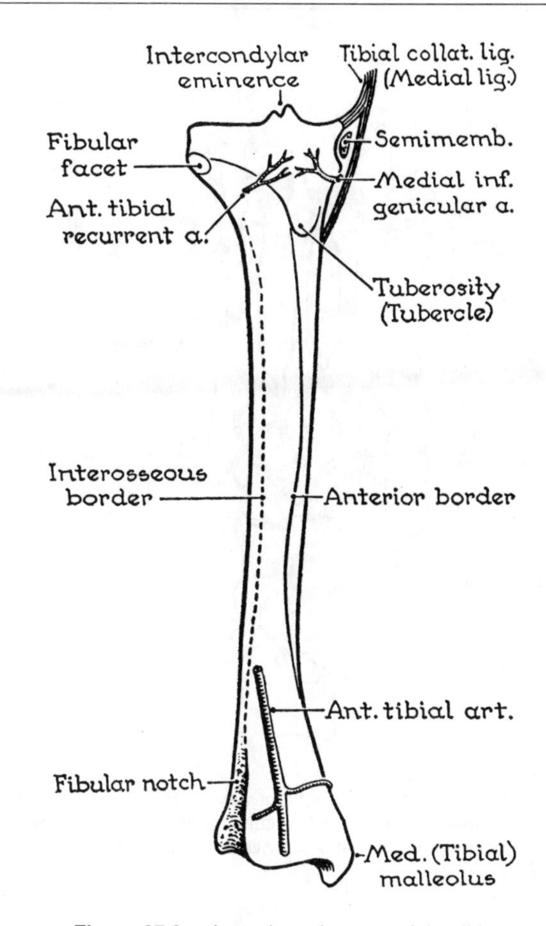

Figure 27.3. Anterolateral aspect of the tibia.

proximally. At the knee, it lies a hand's-breadth behind the medial border of the patella (*figs. 23.2 and 27.2*).

The terminology of surfaces and border is clarified in *Figure 27.4*. The details are for reference only.

INTEROSSEOUS BORDER OF TIBIA

Because the *lateral border* gives attachment to an interosseous membrane, which unites the fibula to the tibia, it is sharp and, quite appropriately, called the *interosseous* border (*fig. 27.4*). It runs to the apex of the deep, rough triangular notch that accommodates the distal end of the fibula. As you would expect, the fibula needs to be firmly united to the tibia at its lower end by an interosseous ligament. So the *fibular notch* is filled with strong interosseous ligaments and therefore is rough.

The distal end of the tibia must be expanded to offer a large quadrangular bearing surface to sit on the talus (the "ankle bone").

The lateral surface lies deep to and is devoted to tibialis anterior.

The posterior surface (see p. 310).

BONES OF FOOT

Although the bones of the foot are considered in detail in the next chapter, a preliminary knowledge of their general disposition may be gained from *Figure 27.5*.

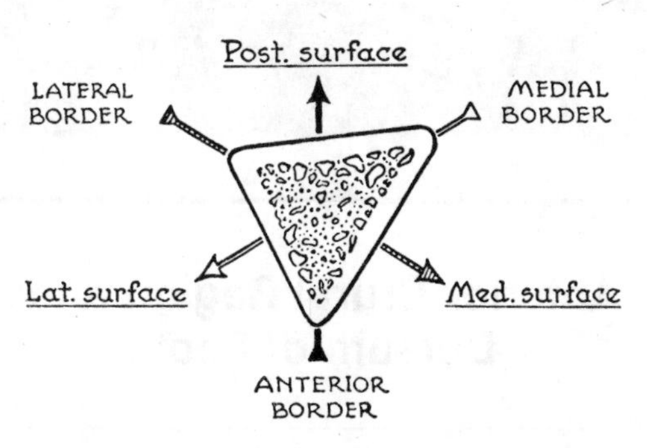

Figure 27.4. Terminology: surfaces and borders are named by opposites. (From Basmajian JV: *Primary Anatomy*, ed. 8. Williams & Wilkins, Baltimore, 1982.)

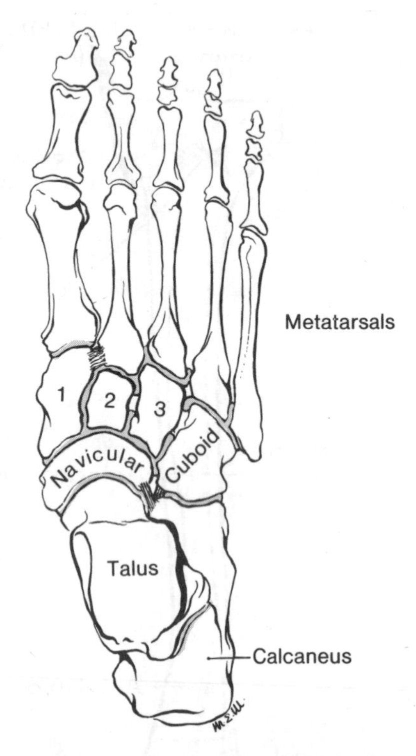

Figure 27.5. Bones of right foot viewed from above (cf., *fig. 2.72*). Joints in *blue; 1, 2, 3* = cuneiforms. (From Basmajian JV: *Primary Anatomy*, ed. 8. Williams & Wilkins, Baltimore, 1982.)

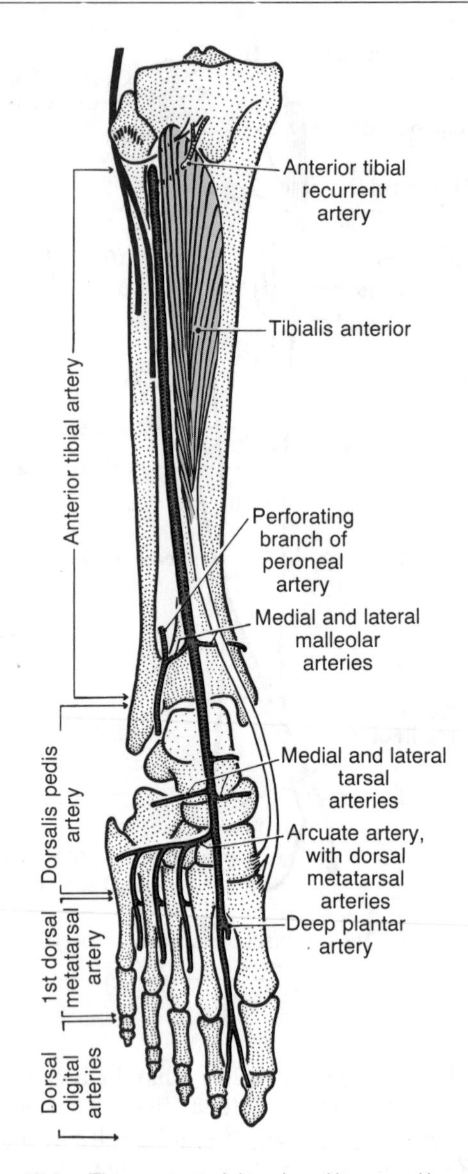

Figure 27.6. The great arterial trunk and its named branches lie on the skeletal plane.

Clinical Case 27.2

Patient Brian P. Yesterday, this 29-year-old man, an ardent amateur rugby football player, was "pulled" out of a game when he began to suffer severe and continuous pain in the muscles of the anterior aspect of the left leg. He does not recall being kicked there. This evening, surgeons are planning an emergency operation to release the pressure in the swollen, hardened and painful leg.

Anterior Crural Region and Dorsum of Foot

The **arterial trunk** of the anterior compartment of the leg and dorsum of the foot, with its venae comitantes, enters the anterior osseofascial compartment as the **anterior tibial artery**. In contrast with the nerve of the region (check this on p. 306), it lies in contact with the medial side of the neck of the fibula. After its vertical descent, it emerges from the depths to enter the foot, and changing its name to **dorsalis pedis artery**, it ends near the web between the great and 2nd toes by dividing into dorsal digital branches for them. A line joining these 2 points and crossing the middle of the ankle gives the course of the great arterial trunk (*fig. 27.6*). Feel the pulse of the artery in the midline of the ankle.

The arteries hug the "skeletal plane," crossing in turn the interosseous membrane, the lowest third of the lateral surface of the tibia, the ankle joint, the 7 tarsal bones of the posterior ½ of the foot, and the fascia covering the first dorsal interosseous muscle in the sole.

At the proximal end of the first intermetatarsal space, a *deep plantar branch* plunges into the sole of the foot.

Branches (*fig. 27.6*) should be easily "guessed" at later by the student and surely need not clutter the memory now. They are:

anterior tibial recurrent artery
branch accompanying the superficial peroneal nerve
medial and lateral malleolar arteries
medial and lateral tarsal arteries
arcuate artery
dorsal metatarsal arteries
dorsal digital arteries

VARIATION

Clinicians feeling for its pulse at the ankle should know that the anterior tibial artery may fail to grow more than a short way down the leg, in which case the dorsalis pedis artery springs from the perforating branch of the peroneal artery (*figs. 27.6, 27.7*).

The following details are for reference but not memorization at this time:

The *Anterior Tibial Recurrent Artery* (*fig. 27.6*) is one you would *not* guess. With a companion nerve (anterior tibial recurrent) it runs to the lateral condyle to take part in the genicular anastomoses. Another is the *lateral malleolar artery*, which joins the perforating branch of the peroneal artery. The common stem thus formed runs distally, anterior to the inferior tibiofibular joint to take part in the anastomoses on the lateral side of the ankle. It may become the chief source of the dorsalis pedis artery, as noted above.

The *lateral tarsal artery* and the *arcuate artery* run laterally on the dorsum of the foot. The *dorsal metatarsal branches* to the 2nd, 3rd, and 4th spaces arise from the arcuate artery; each is joined by a perforating branch of the deep plantar arch, and each in turn divides into 2 *dorsal digital arteries*.

Figure 27.7. Arteries are formed from a network of primitive channels, accounting for the common occurrence of variations through chance or genetics.

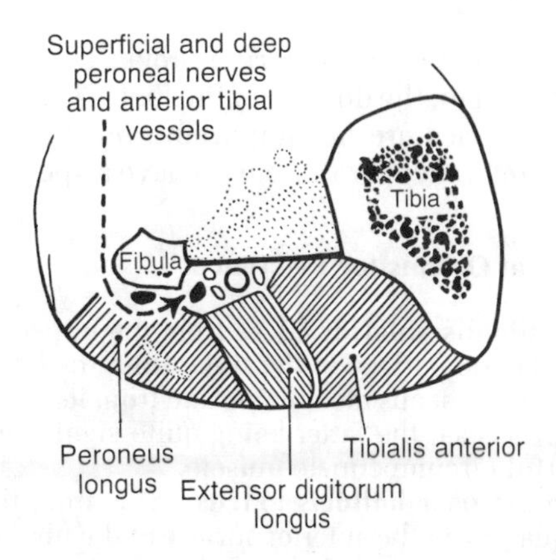

Figure 27.8. Section through the proximal third of the anterior surface of the leg. There are two muscles in the anterior region; one in the lateral. The vessels and nerves cling to the skeletal plane.

The medial side of the 1st toe and the lateral side of the 5th toe receive digital branches from the 1st and 4th metatarsal arteries, respectively. Hence, the arrangement of vessels on the dorsum of the foot is almost identical with that on the dorsum of the hand.

MUSCLES OF ANTERIOR CRURAL REGION

As may be seen in the cross sections (*figs. 27.8, 27.9*), there are 2 fleshy muscles: *tibialis anterior* and *extensor digitorum longus*, in the upper part of the region; plus two more, *extensor hallucis longus* and *peroneus tertius*, in the lower part. At the ankle, these are represented by their 4 tendons.

The tibialis anterior takes origin from the anterior surface of the tibia, as its name proclaims, forcing the other three to crowd onto a narrow strip of the fibula. Only a

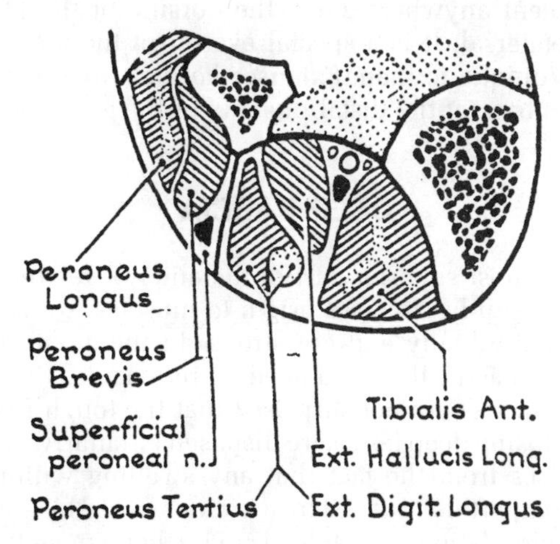

Figure 27.9. Section through the distal third of the anterior and lateral (peroneal) compartments of the leg. There are 4 muscles in the anterior region, 2 in the lateral.

fifth muscle, *extensor digitorum brevis*, arises from, and confines itself to, the dorsum of the foot. The exact origins of these muscles are not important, but they are given below for reference for those who have a special interest.

Details of Origins for Reference

The tibialis anterior arises from the upper two-thirds of the lateral surface of the tibia, from the adjacent part of the interosseous membrane, and from its own covering of deep fascia, the latter being quite significant for this powerful circumpennate muscle.

The **extensor hallucis longus** arises from the middle two-quarters of the anterior surface of the fibula and from the interosseous membrane.

The **extensor digitorum longus** is a thin, unipennate muscle. It arises from the entire length of the narrow anterior surface of the fibula and from adjacent parts of the interosseous membrane, anterior crural septum, and deep fascia.

The **extensor digitorum brevis** arises from the anterior part of the calcaneus and extensor retinaculum. Its 4 tendons pass to the medial 4 toes—i.e., it has an *extensor hallucis brevis* portion, whose tendon is inserted into the base of its proximal phalanx; the 3 other tendons join the dorsal expansions of the extensor digitorum longus to the 2nd, 3rd, and 4th toes.

The fleshy belly of the extensor digitorum brevis is responsible for the soft swelling seen in life on the dorsum of the foot anterior to the fibular malleolus. Novice clinicians sometimes "diagnose" it as a contusion.

Insertions

The stout tendon of the *tibialis anterior* inserts into the medial surface of the 1st metatarsal and 1st cuneiform, a *bursa* intervening. *Extensor digitorum longus* inserts by means of dorsal expansions into the distal 2 phalanges of the lateral 4 toes. Its distal quarter, known as the *peroneus tertius*, fails to reach the toes, but it gains attachment anywhere along the dorsum of the (4th or) 5th metatarsal. It is a special evertor of the foot and is almost unique to man. *Extensor hallucis longus* inserts into the base of the distal phalanx of the great toe.

DEEP FASCIA

In the most superior part of the anterior region of the leg, the deep fascia gives origin to muscles, so its fibers run longitudinally and are strong. In the ankle region, you can imagine there is a need for retinacula (*fig. 27.11*; p. 60), so, you are not surprised that the tough fibers of the enclosing deep fascia are disposed circularly. A problem arises from the fact that any swelling within this tight compartment can start a vicious cycle by shutting off the circulation, which leads to further worsening and the serious condition called *anterior compartment syndrome*.

Clinical Case 27.3

Patient Connie Q. A year ago, this 14-year-old girl broke her left tibia just above the ankle while skiing. At a small hospital near the slopes, a doctor put a cast on her leg. When the cast was removed at the medical center, it was obvious that she had a severe foot-drop due to muscle paralysis, possibly due to the treatment. The condition remains unchanged and now may be the cause of a suit for damages.

Peroneal Nerves

COMMON PERONEAL NERVE

This subfascial nerve, the lateral division of the sciatic, follows the posterior border of the biceps tendon. Crossing the head of the fibula, from which it is separated by a film of soleus, it comes finally into direct contact with the lateral side of the neck of the fibula; there it divides into its 2 clinically important terminal branches (*see fig. 27.12*):

1. Deep peroneal nerve.
2. Superficial peroneal nerve.

These 2 nerves begin on the skeletal plane and, except for their terminal cutaneous branches, literally never leave the skeletal plane (see *fig. 27.12*); therefore they burrow through various structures.

These are: the posterior crural septum, peroneus longus, anterior crural septum, and extensor digitorum longus.

The *anterior tibial recurrent nerve* (really only a twig) joins an artery of the same name to the knee joint and a bit of the tibialis anterior.

The **deep peroneal nerve** approaches the anterior tibial artery from the lateral side and accompanies it through the leg; thereafter, it accompanies the dorsalis pedis artery onto the foot and, supplying the joints it crosses, it becomes cutaneous and divides into 2 *dorsal digital nerves* (*fig. 27.10*) that supply the opposed surfaces of the great toe and its neighbor. Otherwise, it is entirely motor.

It supplies the 4 muscles in the anterior compartment of the leg, and on the dorsum of the foot it sends a lateral branch to supply the extensor digitorum brevis.

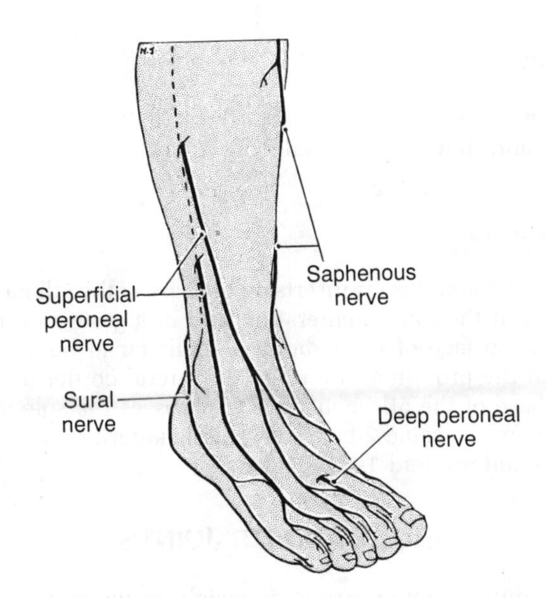

Figure 27.10. Cutaneous nerves of dorsum of foot.

In 22% of limbs, an accessory deep peroneal nerve runs posterior to the lateral malleolus and is motor to the extensor digitorum brevis (Lambert).

The **superficial peroneal nerve** runs inferiorly and anteriorly in contact with the shaft of the fibula. It is covered by and is motor to the peroneus muscles (longus and brevis) (*see fig. 27.12*). It becomes superficial, and entirely sensory, along the line of the anterior crural septum, a variable distance above the ankle (*fig. 27.10*).

As noted, it supplies peroneus longus and peroneus brevis and provides dorsal digital branches to all the toes, except adjacent sides of the 1st and 2nd (*deep peroneal nerve*) and the lateral side of the 5th (*sural nerve*). To these, it sends communicating twigs (*fig. 27.10*).

It has a rare distinction: one of its branches on the dorsum of the foot is easily felt and may even be seen bowstringing like an extensor tendon (Lemont).

Retinacula for Tendons

The **extensor retinaculum** is in 2 parts—superior and inferior (*fig. 27.11*).

The *inferior part* anterior to the ankle has the appearance of a Y-shaped band. The stem of the Y is attached to the anterior part of the superior surface of the calcaneus, the largest tarsal bone that forms the heel. The fibers of the Y form compartments within loops or slings—especially for peroneus tertius, extensor digitorum longus, and extensor hallucis longus—that efficiently prevent the tendons from bowstringing anteriorly and also medially.

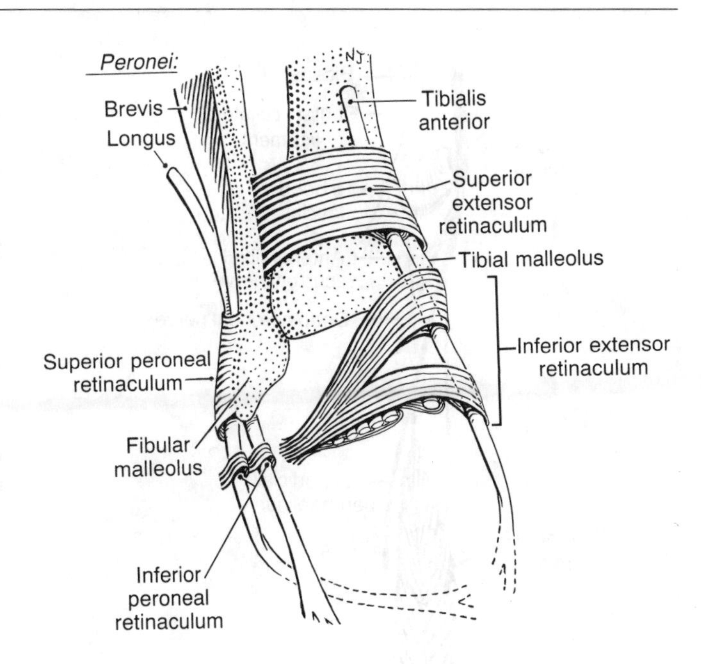

Figure 27.11. Extensor and peroneal retinacula. The attachment of the loops of inferior retinacula to the calcaneus is omitted for clarity.

The **inferior peroneal retinaculum** is attached to the lateral surface of the calcaneus; it consists of 2 loops for the peroneus brevis and the peroneus longus. (For the *superior peroneal retinaculum*, see Figure *27.11*.)

Synovial sheaths that extend about 2 or 3 cm proximal and distal to the points of friction envelop the tendons. They are subject to inflammation, with athletes and fitness addicts particularly prone to *tenosynovitis* here.

Peroneal (Fibular) Region

By **palpation** of your own limb, determine the following points: only the proximal and distal ends of the fibula are subcutaneous; the *head* is rounded and the *common peroneal nerve* can be rolled behind it; the *malleolus* is triangular; its anterior and posterior borders are conspicuous and palpable; its lateral surface is a subcutaneous isosceles triangle (*fig. 27.12*).

The cord-like *fibular collateral ligament* of the knee joint can be felt (when the knee is flexed) where it runs obliquely posteroinferiorly to be attached to the head of the fibula, just anterior to its apex. Indeed, when it is acutely injured by an abnormal stress to the opposite (medial) side of the knee, an exquisitely tender localized focus can be located superior to the head of the fibula.

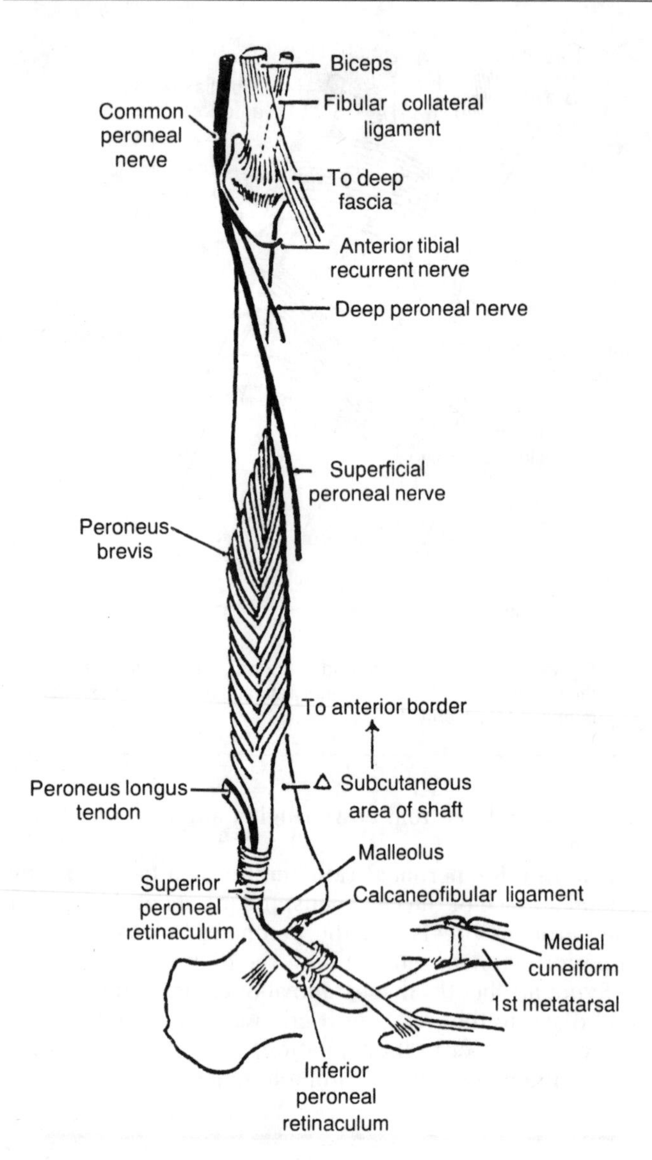

Figure 27.12. Peroneal or lateral crural region. Fleshy fibers of peroneus longus removed.

At the ankle the cord-like *calcaneofibular ligament* runs posteroinferiorly from the fibular malleolus just anterior to its tip. It cannot be felt because the peroneal tendons cross it (*fig. 27.12*).

Boundaries and Contents. (See *figs. 27.8, 27.9,* and *27.12*).

MUSCLES

Peroneus longus and **Peroneus brevis** fill the peroneal (fibular) compartment.

The 2 tendons use the posterior aspect of the fibular malleolus as a pulley, and both cross the calcaneofibular ligament. They are bound down by the *superior* and *inferior peroneal retinacula* (*fig. 27.12*) and "use" a single synovial sheath, which bifurcates distally.

Origins

They arise from the lateral aspect of the fibula and septa, and they are bipennate (*fig. 27.12*).

Insertions

The peroneus brevis inserts on the base of the 5th metatarsal, but the longus enters the sole in a groove on the inferior surface of the cuboid, a small tarsal bone just behind the prominent lump on the lateral border of the foot (base of the 5th metatarsal). It crosses the sole and inserts on the same 2 bones as tibialis anterior (i.e., medial cuneiform and 1st metatarsal).

FIBULA AND ITS JOINTS

The human fibula carries no weight to the ground although it holds the talus in its socket, as described under

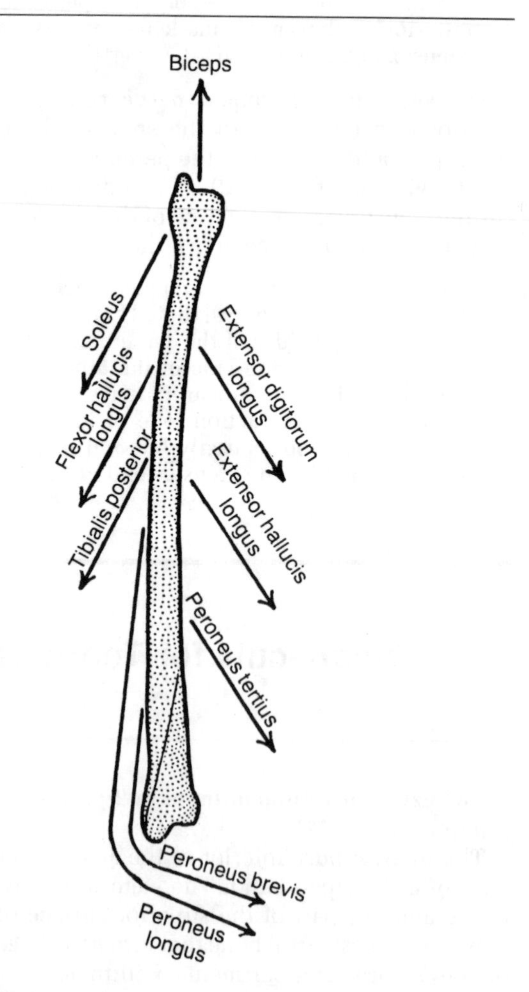

Figure 27.13. The muscles attached to the fibula pull inferiorly, except the biceps.

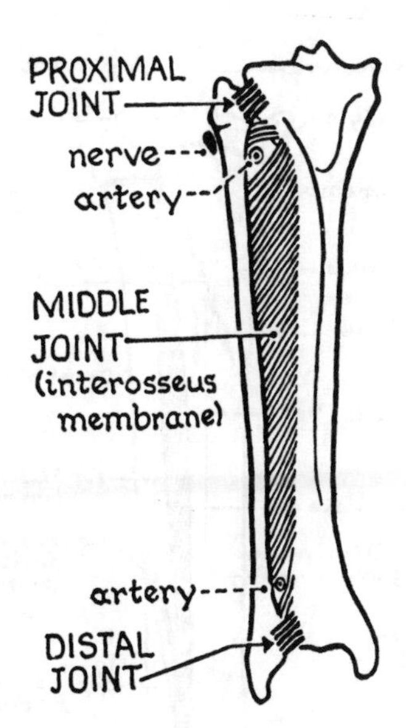

Figure 27.14. The tibiofibular articulations. Note unity of direction of ligamentous fibers.

the ankle joint in the next chapter. A main function of the bone is to provide attachments for 9 muscles, 8 of which pull inferiorly (*fig. 27.13*). Weinert *et al.* have shown that, contrary to expectations, the fibula moves inferiorly to deepen the ankle socket at the strike phase of running.

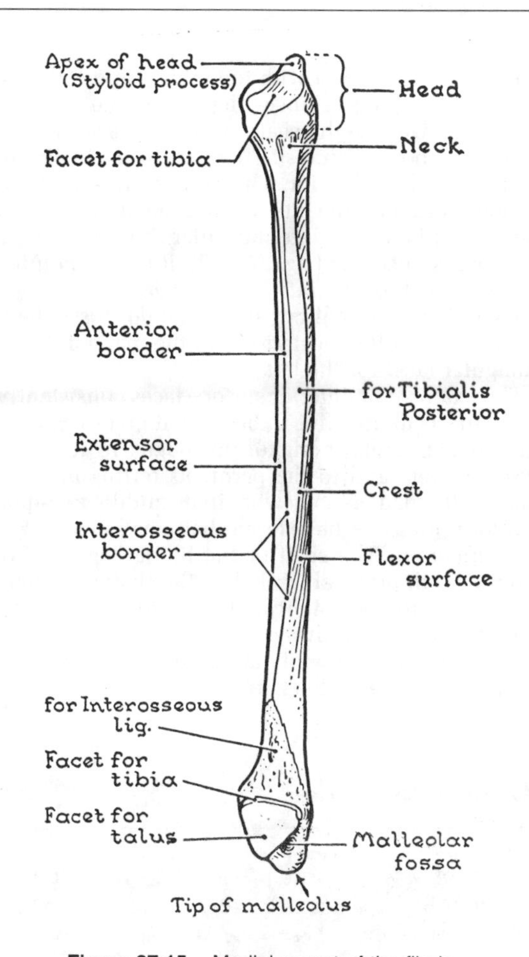

Figure 27.15. Medial aspect of the fibula.

Tibiofibular Joints

The fibula is moored to the tibia at its proximal end, along its shaft, and at its distal end—at proximal, middle, and distal joints (*fig. 27.14*).

The *proximal tibiofibular joint* is a synovial plane joint between the head of the fibula and a facet on the posterolateral part of the lateral condyle of the tibia. Posterior to the joint is the popliteus tendon, separated by the popliteus *bursa*, which may communicate with the joint space (*fig. 26.17*; p. 298).

The *middle* and *distal tibiofibular joints* are syndesmoses (i.e., ligamentous). The *interosseous membrane* (middle joint) stretches between the tibia and fibula, producing a sharp line on each. Distally, each line expands into a large, rough, triangular area (*fig. 27.15*) for a strong interosseous ligament that binds the opposed areas together just above the ankle joint.

Body of the Fibula (*fig. 27.15*)

The shape of the body depends largely on the muscles and septa attached to it. In general terms, the peroneal surface is dedicated to the peronei; the flexor (posterior) surface to the soleus and flexor hallucis longus, with a special strip just posterior to the interosseous membrane for the tibialis posterior; the narrow anterior surface gives origin to the extensors of the toes.

Details. If you must learn details for a special purpose, begin with the important peroneal surface. What follows should be appreciated but *not memorized* by most students.

The *peroneal surface* is found by placing a finger posterior to the malleolus, which is the pulley for the peronei, and letting it run up the shaft to the head of the bone. This surface is broad and spiral, facing posteriorly at first and laterally in the superior part. It is bounded by lines (the *anterior* and *posterior borders*) which give attachment to the anterior and posterior crural septa. To make doubly sure of the anterior border, which separates the peroneal from the extensor surface, place a finger on the subcutaneous isosceles triangle (*fig. 27.12*) superior to the malleolus and run the finger straight up to the head of the bone.

The *flexor* or *posterior surface* is broad and it gives origin to the soleus (p. 311) in its proximal third and to the flexor hallucis longus (p. 312) in its distal two-thirds. It is spiral, like the peroneal surface next to it.

The *surface* for the *tibialis posterior* (p. 311) is the enigma; it is fusiform and is to be found thus: put a finger on the rough area for the interosseous ligament—i.e., the area superior to the smooth, triangular facet on the malleolus for articulation with the talus—and follow it superiorly. It becomes a line which splits, one-third to one-half of the way up the shaft, into an *anterior line* and a *prominent posterior crest.* These enclose a fusiform area for the tibialis posterior. The anterior line is the *interosseous border* for the interosseous membrane; the prominent crest is for the intermuscular septum proximal to the tibialis posterior (*fig. 27.1, B*). *It is common to mistake the crest for the interosseous border.*

There is a deep hollow, the *malleolar fossa,* between the grooved pulley posteriorly to the malleolus and the triangular facet for the talus.

The *extensor* or *anterior surface* faces consistently forward and is almost linear, because it gives origin to the unipennate extensor digitorum longus in its proximal three-quarters and to the peroneus tertius in its distal quarter. It broadens somewhat in its middle two-quarters to afford origin to the external hallucis longus.

Ossification. The shaft of the fibula begins to ossify about the 8th prenatal week, like the shafts of other long bones. The upper epiphysis begins to ossify about the 5th year; fusion is always complete by the 22nd. The distal end begins to ossify about the 2nd year; fusion is always complete by the 20th.

Clinical Case 27.4

Patient Doyle R. While playing volleyball with his children 3 hours ago, this 48-year-old man felt a sudden pain above his heel while jumping upward. Now he cannot plantarflex his ankle and is suffering great pain, although he can bear weight on the foot.

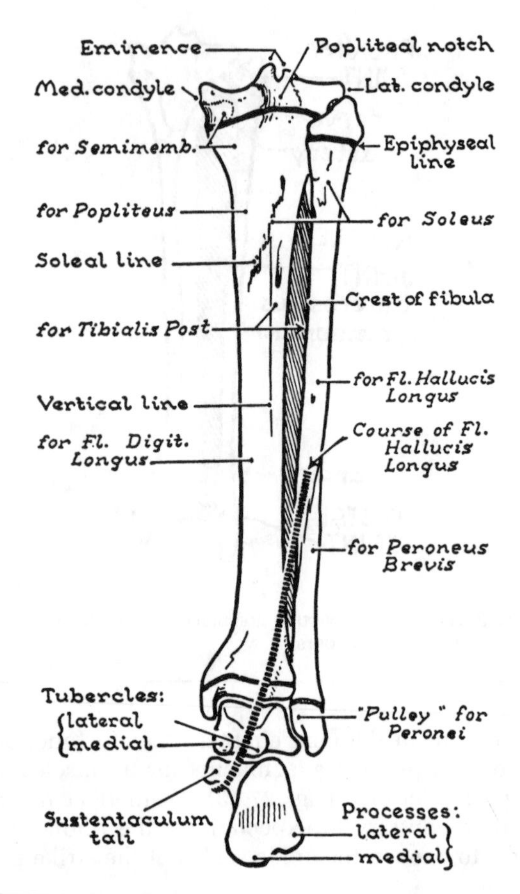

Figure 27.16. Posterior aspect of the bones of the leg and foot, and the epiphyseal lines of an adolescent.

Posterior Crural Region (The Back of the Leg)

BONY FRAMEWORK

Posterior view (*fig. 27.16*).

The *tibial condyles* overhang the shaft of the tibia posteriorly and at the sides. The rounded *head of the fibula,* supported by its neck, is quite inferior to the knee joint.

The posterior surface of the body of the tibia is crossed obliquely by the *soleal line.* The *tibial malleolus* is 2 cm shorter (higher from the ground) than the fibular malleolus (*fig. 27.16*).

In view are: the posterior aspect of the talus and the *posterior third of the calcaneus* projecting posteriorly to form the heel. It ends in a large medial process which rests on the ground (*fig. 27.16*).

MUSCLES OF POSTERIOR CRURAL REGION (POSTERIOR COMPARTMENT OF THE LEG)

Superficial group (*fig. 27.17*):
1. Gastrocnemius,
2. Soleus, and
3. Plantaris, between them.

Deep group:
1. Popliteus, a rotator of the knee,
2. Flexor hallucis longus, and
3. Flexor digitorum longus to the terminal phalanges of the toes.
4. Tibialis posterior to every small tarsal and most metatarsals.

The **gastrocnemius** has 2 heads of origin from just superior to the lateral and medial epicondyles of the femur (*fig. 27.18*). The 2 resulting bellies unite to form the fullness of the calf, and then end at the middle of the leg in a broad tendon that blends with that of the soleus to form the **tendo calcaneus (tendon of Achilles)** (*fig. 27.19*). The gastrocnemius and soleus are often referred to as the *tri-*

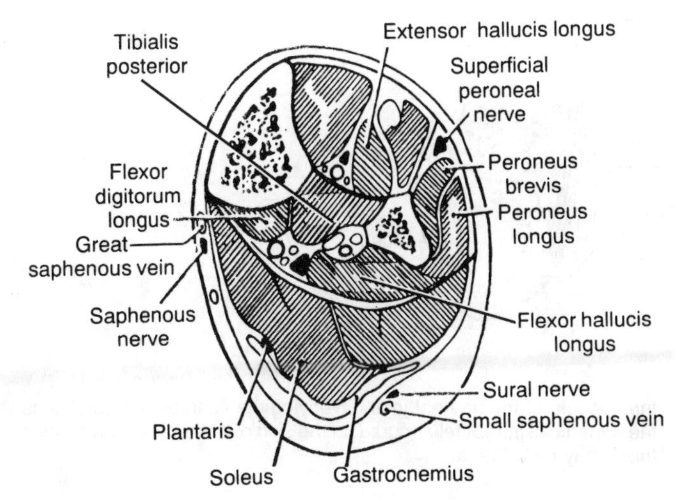

Figure 27.17. Transverse section through middle of leg.

ceps surae (L. *sura* = the calf) (*fig. 27.18*). The *insertion* is into the middle of the posterior surface of the calcaneus.

A constant *bursa* between the tendon of the semimembranosus and the medial head of the gastrocnemius may communicate with a *bursa* between the latter and the capsule of the knee joint. This in turn may communicate with the knee joint (in 42.4% of 528 dissecting-room limbs). So, fluid in the joint, resulting say from a sprain, can pass from the knee joint via the gastrocnemius bursa into the semimembranosus bursa and cause a swelling at the back of the knee. A *sesamoid bone* is commonly found in the lateral head of the gastrocnemius and less commonly in the medial head.

The **soleus** is shaped both like the sole of a boot and like a flatfish. It arises from the proximal part of the posterior surface of the fibula and the soleal line of the tibia, and even more distally on the tibia (*fig. 27.20*). So, its *origin is horse-shoe shaped* (*fig. 27.21*). Short fleshy fibers of this powerful muscle join the tendo calcaneus.

The **plantaris** is between the gastrocnemius and the soleus. It has a small fleshy belly and a very long ribbon-like tendon (*fig. 27.18*). It arises near the lateral head of the gastrocnemius and it is inserted by joining the tendo calcaneus. Some cases of sudden pain in the calf during severe exertion, called a "charley-horse," are due to its rupture.

Popliteus, attached distally to the posterior surface of the tibia by fleshy fibers (*figs. 27.22 and 27.23*), arises by tendon just inferior to the lateral epicondyle of the femur. It runs posterior to the lateral meniscus and the proximal tibiofibular joint.

A variable bundle of fleshy fibers from its superior border gains a fibrous attachment to the posterior part of the lateral meniscus, thereby forming an *articular muscle* (Last; Lovejoy *et al.*) that may pull the lateral meniscus posteriorly.

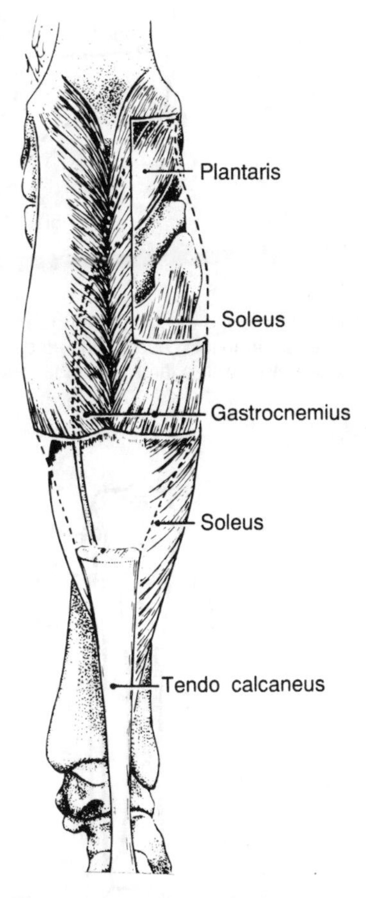

Figure 27.18. The gastrocnemius, soleus, and plantaris muscles form the tendo calcaneus.

DEEP MUSCLES

Inferior to *the horseshoe-shaped origin* of the soleus there are 3 bipennate muscles—tibialis posterior, flexor digitorum longus, and flexor hallucis longus (*figs. 27.21* and *27.22*). Their tendons are the quills of the feathers. They pass inferiorly and medially to enter the sole of the foot.

Origins and Insertions

The **tibialis posterior** is the deepest, *arising* from the interosseous membrane and adjacent bones. Its tendon clings faithfully to the skeletal plane, using the medial malleolus as its pulley. For *insertion* see *Figure 28.35* on p. 330.

The **flexor digitorum longus** arises from the posterior surface of the tibia and the fascia covering the tibialis posterior.

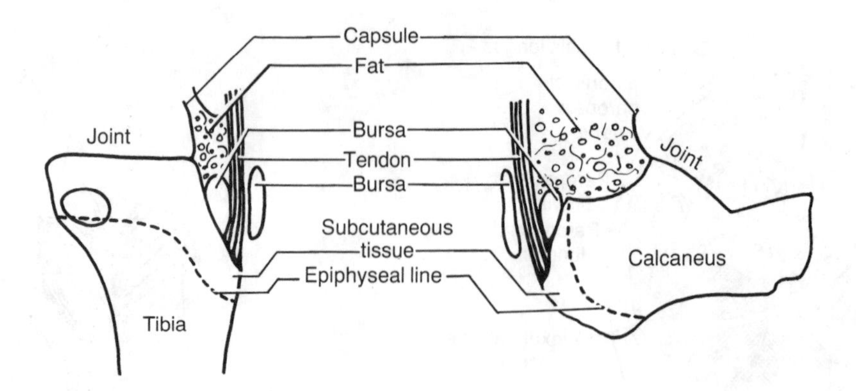

Figure 27.19. The tendo calcaneus and ligamentum patellae compared. About equal in length, each is inserted into an epiphysis and has a bursa deep to it with a large pad of fat between the bursa and the capsule of the nearest joint. Overlying the ligamentum patellae is the superficial infrapatellar bursa: in the skin over the tendo calcaneus there may be a bursa.

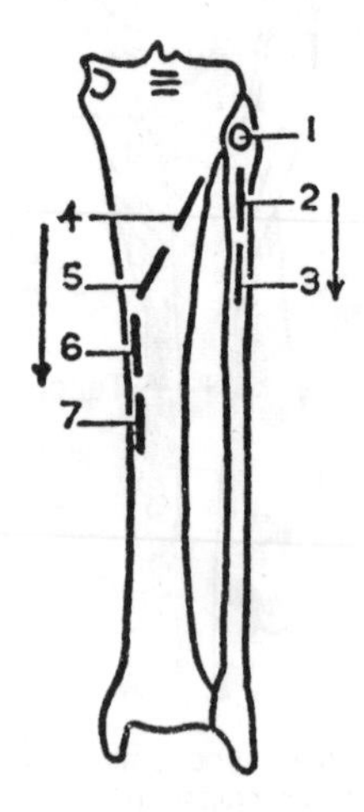

Figure 27.20. The increasing origin of the soleus is horseshoe-shaped.

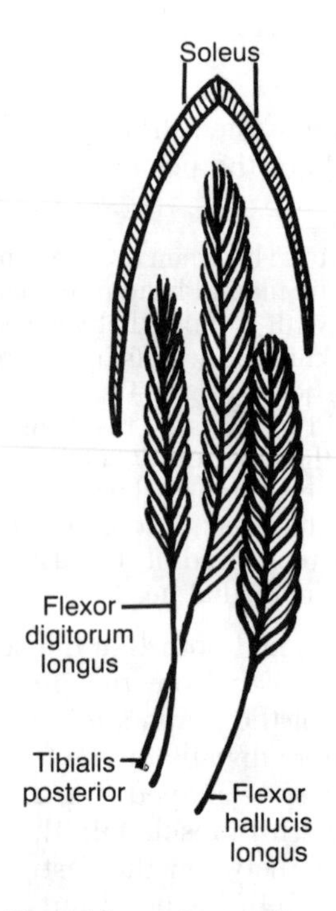

Figure 27.21. Three bipennate muscles resembling three feathers.

The **flexor hallucis longus** is the largest of the three. Its origin from the fibula distal to the soleus overflows to the posterior crural septum, the fascia covering the tibialis posterior, and even the interosseous membrane and tibia inferiorly (*fig. 27.22*). (Continued into the foot on p. 330.)

The **tendons posterior to the ankle** (*fig. 27.24*) are the tendo calcaneus and 5 others. Of these, two, the peronei brevis and longus, groove the posterior surface of the fibular malleolus; two, the tibialis posterior and flexor digitorum longus, groove the posterior surface of the tibial malleolus, and one, the flexor hallucis longus, grooves the distal end of the tibia midway between the malleoli (*fig. 27.22*).

Table 27.1 summarizes the muscles that act on the ankle joint.

NERVES, ARTERIES, AND VEINS

There are 2 large arteries accompanied by deep veins, one nerve, and an important subcutaneous vein in this region (*fig. 27.23*), namely:

Tibial nerve
Posterior tibial artery (and veins)
Peroneal artery (and veins)
Small saphenous vein

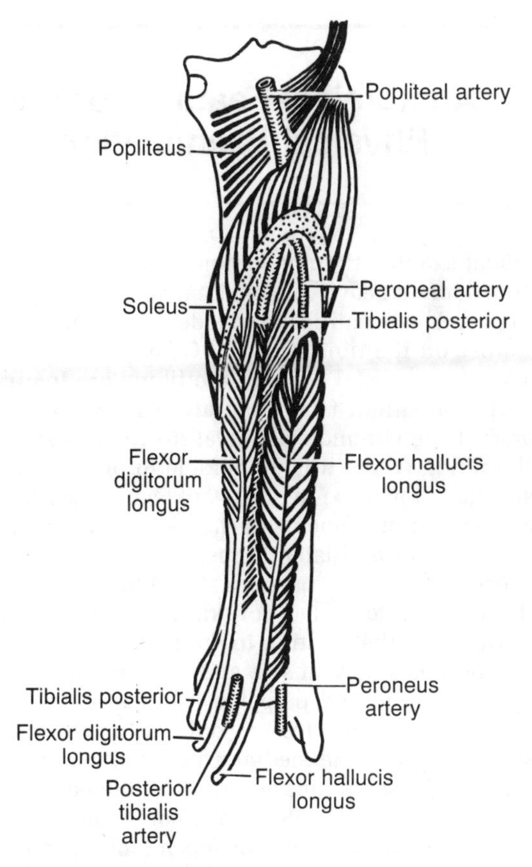

Figure 27.22. The deep muscles of the back of the leg—1 proximal and three distal to soleus.

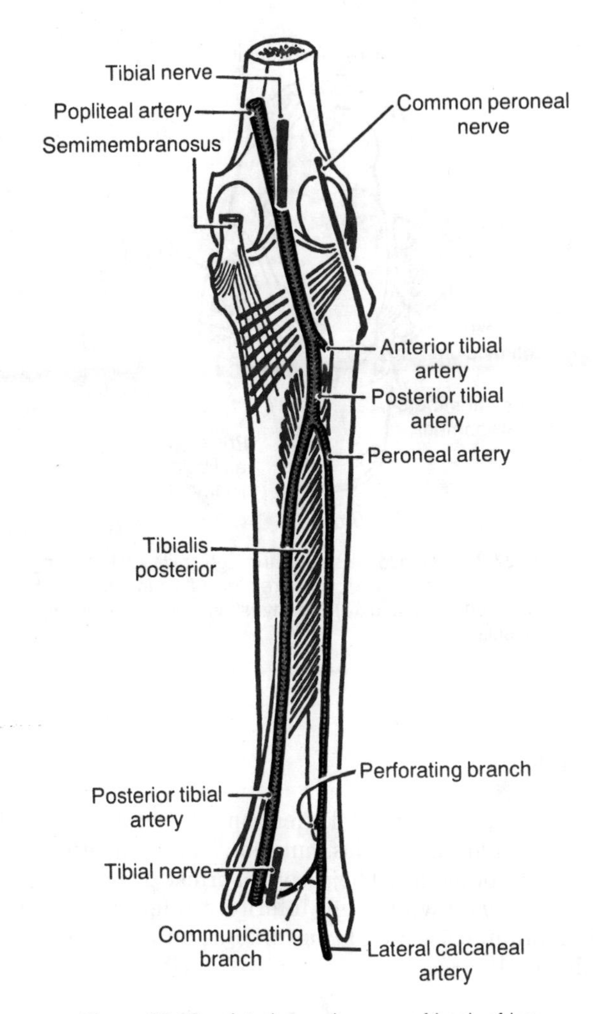

Figure 27.23. Arteries and nerves of back of leg.

Tibial Nerve

The tibial nerve takes a straight course on the posterior surface of the fascia, covering the deep-lying tibialis posterior and then the tibia. It reaches the distal end of the tibia, where it lies deep to the flexor retinaculum, and divides into the *medial* and *lateral plantar nerves* (p. 331). It accompanies the posterior tibial artery (*fig. 27.23*).

Side Branches

Branches include: (1) *muscular;* (2) *cutaneous,* to the heel (medial calcanean nerves) (*fig. 28.40*); (3) *articular,* to the ankle joint.

Table 27.1. Muscles Acting on the Ankle Joint (and Their Spinal Cord Nerve Supplies)

Plantarflexion (L5;S1,2)	Dorsiflexion (L4,5)
Soleus	Tibialis anterior
Gastrocnemius	Extensor hallucis longus
Peroneus longus	(Extensor digitorum longus and peroneus tertius)
(Plantaris)	
(Peroneus brevis)	
(Tibialis posterior)	

Cutaneous nerves are summarized in *Figures 24.21* and *27.10*. The patterns formed by these nerves are numerous.

Posterior Tibial Artery

This artery begins at the superior border of the soleus and ends deep to the flexor retinaculum by dividing into the *medial* and *lateral plantar artery*. Near the ankle, 2 layers of fascia cover it (*fig. 27.24*). When this fascia is relaxed by inverting the foot, the pulsations of the artery can be felt posterior to the medial malleolus.

Peroneal Artery

This artery arises from the posterior tibial artery and descends posterior to the fibula, distal tibiofibular joint, and ankle joint. It ends on the lateral surface of the calcaneus as the *lateral calcanean artery* (*fig. 27.23*).

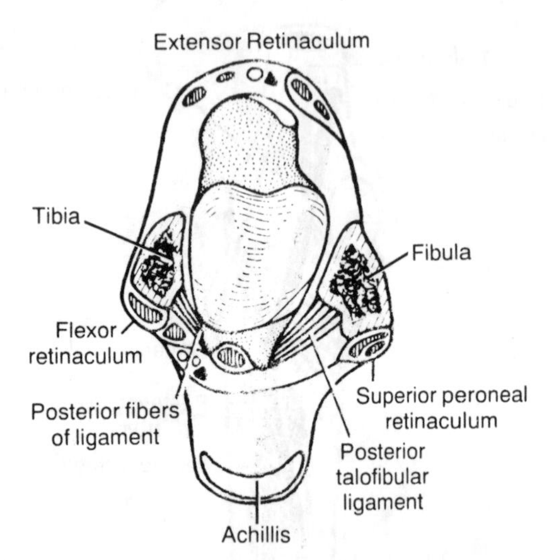

Figure 27.24. Horizontal section through ankle joint showing (1) wedge-shaped socket, (2) direction of ligaments, (3) 5 posterior tendons, and (4) the investing and intermuscular deep fascia.

Branches

Both the posterior tibial artery and its peroneal branch have *muscular, cutaneous, nutrient, communicating,* and *calcanean* branches. The peroneal artery gives off a *perforating branch* which runs inferiorly anterior to the distal tibiofibular joint and anastomoses on the lateral side of the ankle.

Small Saphenous Vein

This cutaneous vein drains the lateral end of the dorsal venous arch of the foot. It runs from inferior to the lateral malleolus to the groove between the 2 heads of the gastrocnemius and, after piercing the popliteal fascia, ends partly in the popliteal vein and partly in the profunda femoris vein.

Communicating Veins

The cutaneous veins form an open network between the great and small saphenous veins, and they communicate along the intermuscular septa with the many deep veins. Superior to the ankle the valves in the communicating veins are so directed that the blood flows from the superficial veins into the deep veins and from the small saphenous vein into the great saphenous vein. Excessive inactivity (as in the case of hospitalization) can lead to inflammation (thrombophlebitis) and deep venous thrombosis in this system.

Anatomical Features of the Physical Examination

Visual examination, palpation, and motor and sensory testing depend heavily on anatomical knowledge and experience. Abnormalities can be detected from one end to the other on the tibia (Clinical Case 27.1), but only the proximal and distal parts of the fibula are freely accessible to palpation. One must always bear in mind the course of the common peroneal nerve against the head of the fibula where it is vulnerable to pressure (Clinical Case 27.3). Dorsiflexion of the ankle and toes and inversion-eversion movements of the foot are obvious tests of all the anterior and lateral muscles (Clinical Case 27.2). Sensory testing may be indicated and is quite easy to perform and interpret if you know the dermatome and specific nerve distribution to the leg and foot.

The triceps surae muscle and its tendo calcaneus are easily visualized and palpated (Clinical Case 27.4), but the deeper calf muscles are only palpated as tendons posteroinferior to the medial aspect of the ankle region. Bursitis of these tendons and those of the peroneus muscles posterior to the lateral malleolus can be quite an obvious diagnostic exercise dependent on anatomical knowledge. Quite prominent when it becomes distended and especially when varicose, the great saphenous vein courses obliquely from the dorsum of the foot obliquely proximally and posteriorly, to be found a handsbreadth behind the medial border of the patella.

Enclosed by the two bones of the leg, the tough interosseous membrane that joins them deep in the leg, and the tough enclosing deep fascia, the anterior compartment muscles have little extra room to swell, helping to explain what happened in Clinical Case 27.2. The muscles are supplied by the anterior tibial artery, which enters from superior to the interosseous membrane; it runs distally in the depths to the middle of the anterior surface of the ankle to become the dorsalis pedis artery, whose pulsations should be felt in clinical examinations. The anterior crural compartment muscles include: the tibialis anterior (which is a powerful invertor and dorsiflexor of the foot), the extensor hallucis longus, and extensor digitorum longus (a split-off part of which is the peroneus tertius). The belly of the extensor digitorum brevis on the dorsum of the foot is palpable anterior to the lateral malleolus.

Peroneal Nerves and Fibula

The common peroneal motor nerve of the anterior crural and lateral (peroneal) compartments (peroneus longus and peroneus brevis) enters the leg on the posterolateral side of the fibular head. Here it is vulnerable to compression, which was caused by the cast as illustrated in Clinical Case 27.3. The common nerve divides here into the superficial branch (for the peroneal muscles and the skin of most of the toes) and the deep branch (for all extensor muscles and the great dorsiflexor-invertor—hence the footdrop suffered by the patient—and also the skin between toes 1 and 2). The slender fibula is only palpable at its two ends (head and lateral malleolus), and you should be quite familiar with these important landmarks among essential structures. The fibular body is surrounded by muscles that arise from its surfaces.

In the calf, the tibialis posterior, flexor hallucis longus and flexor digitorum longus are too deep to be easily palpated. Covering them is triceps surae consisting of the soleus (from both bones), and the two bulging bellies of gastrocnemius. The triceps surae narrows to the tendo calcaneus and provides powerful plantarflexion of the ankle, as in leaping. Obviously, Clinical Case 27.4 illustrates a case of ruptured tendon, which will require surgical repair. The nerve supply of all these muscles is the tibial nerve. It accompanies the posterior tibial on the skeletal plane to enter the sole of the foot (as its great motor and sensory supply) posterior to the medial malleolus. Its pulsations should be felt.

Clinical Mini-Problems

1. **Where is the location of the great saphenous vein in the ankle and leg region?**
2. **How could inflammation of the right tibialis anterior muscle result in a right foot drop?**
3. **Which tendons pass posterior to the lateral malleolus?**
4. **Between which palpable tendons at the medial malleolus would one take a posterior tibial artery pulse?**

(Answers to these questions can be found on p. 586.)

28

Foot and Ankle

Clinical Case 28.1

Patient Ann S. This 13-year-old girl was born with a severe right club foot. She has had a series of major operations over the years to fuse the ankle joint, subtalar joint, and transverse tarsal joint (triple arthrodesis) in the most useful position of the ankle and foot. Her parents seek advice on further therapy and prognosis at the special pediatric orthopaedic clinic where you are a student.

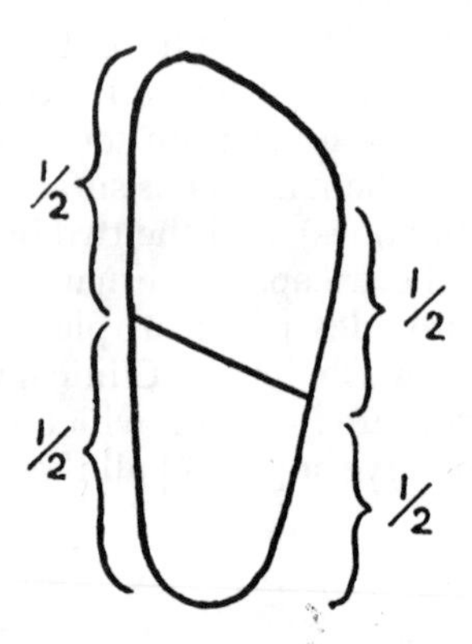

Figure 28.1. The outline of the foot. All of the long bones occupy the anterior ½ (cf. *fig. 28.2*).

Bones of Articulated Foot

Study of the bones of the foot need not be as forbidding as most books make it. If you follow the remarks describing the accompanying diagrams, verify them by referring to an articulated foot and your own foot, you will understand and learn the essential facts and the more important details in a practical way.

Examine the dorsum of an articulated foot and make the following observations.

1. **Outline.** The medial border of the foot is almost straight; the most projecting toe is usually the big toe or *hallux* (*figs. 28.1 and 28.2*).

2. Draw or imagine a line joining the midpoints of the medial and lateral borders. In front of it lie only long bones—metatarsal bones and phalanges; behind it lie all 7 tarsal bones.

The middle and distal phalanges are quite nodular and rudimentary. The hallux has 2 stout phalanges.

The 1st *metatarsal* is the stoutest and strongest; the 2nd is the longest; the 5th has an easily palpated tuberosity at its base (*fig. 28.2*).

3. **The Transverse Tarsal Joint.** You can only flex and extend your ankle joint. To obtain a view of the sole, you invert your foot by rotating joints other than the ankle. The *head of the talus*, which is globular, fits into the posterior surface of the *navicular*, which is cup-shaped; and the anterior surface of the *calcaneus*, which is sinuous, articulates with the reciprocal posterior surface of the cuboid (*Plate 28.1, fig. 28.2*). At these 2 joints, collectively known as the transverse tarsal joint, part of the movement of *inversion* and *eversion* takes place; it augments the more important movements at the joints between the talus and the supporting calcaneus (see *fig. 28.4*).

4. The *transverse tarsal joint* further divides the foot so that there are 3 units: anterior, middle, and posterior (*fig. 28.3*). The "middle unit" comprises 5 small tarsal bones; the "posterior unit," 2 large tarsal bones—talus and calcaneus (*figs. 28.2, 28.4, and 28.5*).

The **talus** rests on the anterior two-thirds of the calcaneus and projects slightly anterior to it (*fig. 28.5*). It is shaped like a tortoise with a body, neck, and head. The superior surface of the body supports the tibia, is entirely articular, and is saddle-shaped; its inferior surface forms the large and important *subtalar joint*. The sides of the body are grasped by the malleoli, and the facets that result correspond in length and shape with those of the malleoli. The posterior surface of the body tapers to 2 tubercles separated by a groove for the flexor hallucis lon-

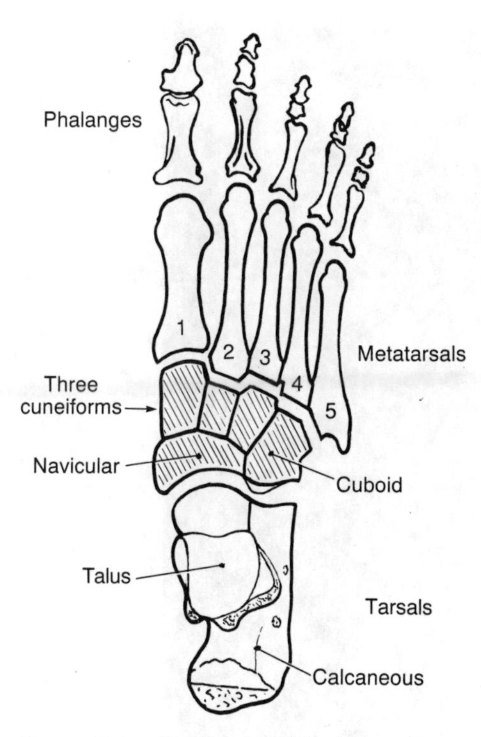

Figure 28.2. The dorsum of the foot and toes.

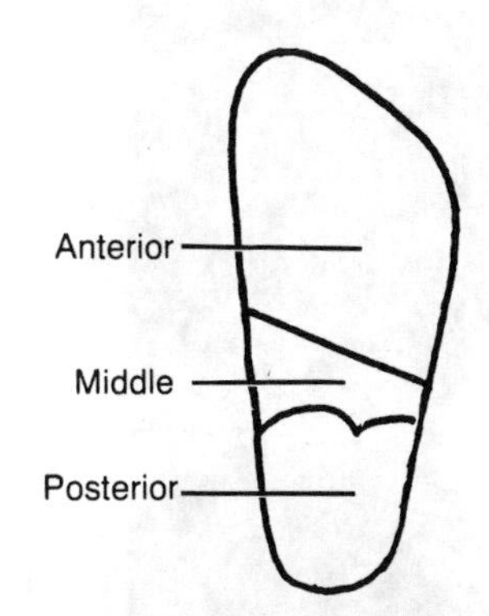

Figure 28.3. The thee "units" of the foot.

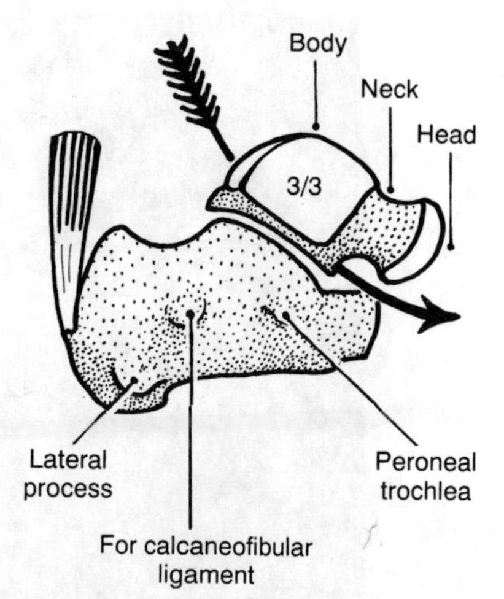

Figure 28.4. Lateral aspects of talus and calcaneus. (A probe may be passed, like the *arrow*, traversing the tarsal tunnel, which is filled with an interosseous ligament.)

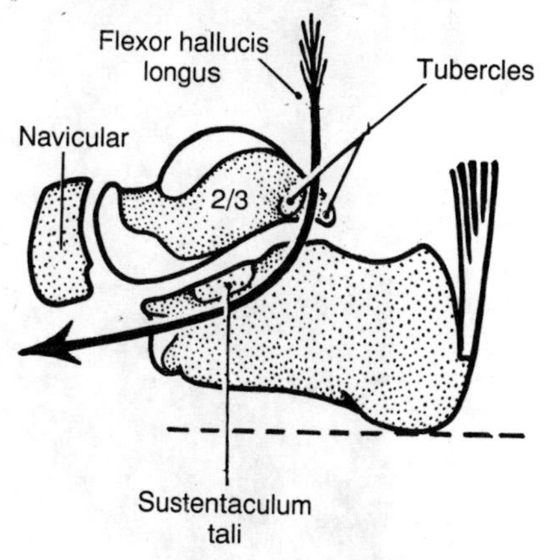

Figure 28.5. The medial aspects of the talus and calcaneus.

Figure 28.6. Pattern of tarsal bones as an aid for beginners to the names and relative positions of the 7 tarsal bones—calcaneus, cuboid, and 3 cuneiforms, navicular, and talus. This very simple design has been found useful. (Right foot, dorsal aspect, after Paff.)

gus tendon on its way into the sole (*fig. 28.5*). The head is rounded anteriorly to articulate with the navicular bone. Inferiorly, it partly rests on the calcaneus.

The **calcaneus** is by far the largest tarsal bone. It is divided into 3 thirds. The anterior two-thirds supports the talus; the posterior one-third forms the prominence of the heel, which rests on the ground (*fig. 28.5*). The anterior surface articulates with the *cuboid* (*fig. 28.6*). The posterior third of the superior surface is saddle-shaped and nonarticular; the intermediate third supports the body of the talus and is convex, articular, and oblique; the anterior third is nonarticular laterally and forms a projecting shelf, the *sustentaculum tali*, medially; two small facets (which may be continuous) support the head of the talus (*see fig. 28.7*).

The *lateral surface* (*fig. 28.4*) is palpable in the living foot and is crossed by peroneal tendons.

The posterior surface [tuber calcanei] is wider inferiorly than superiorly. It is continuous, on the plantar surface, with a large medial process or "tubercle" (weight-bearing) and a small lateral process (*fig. 28.4*).

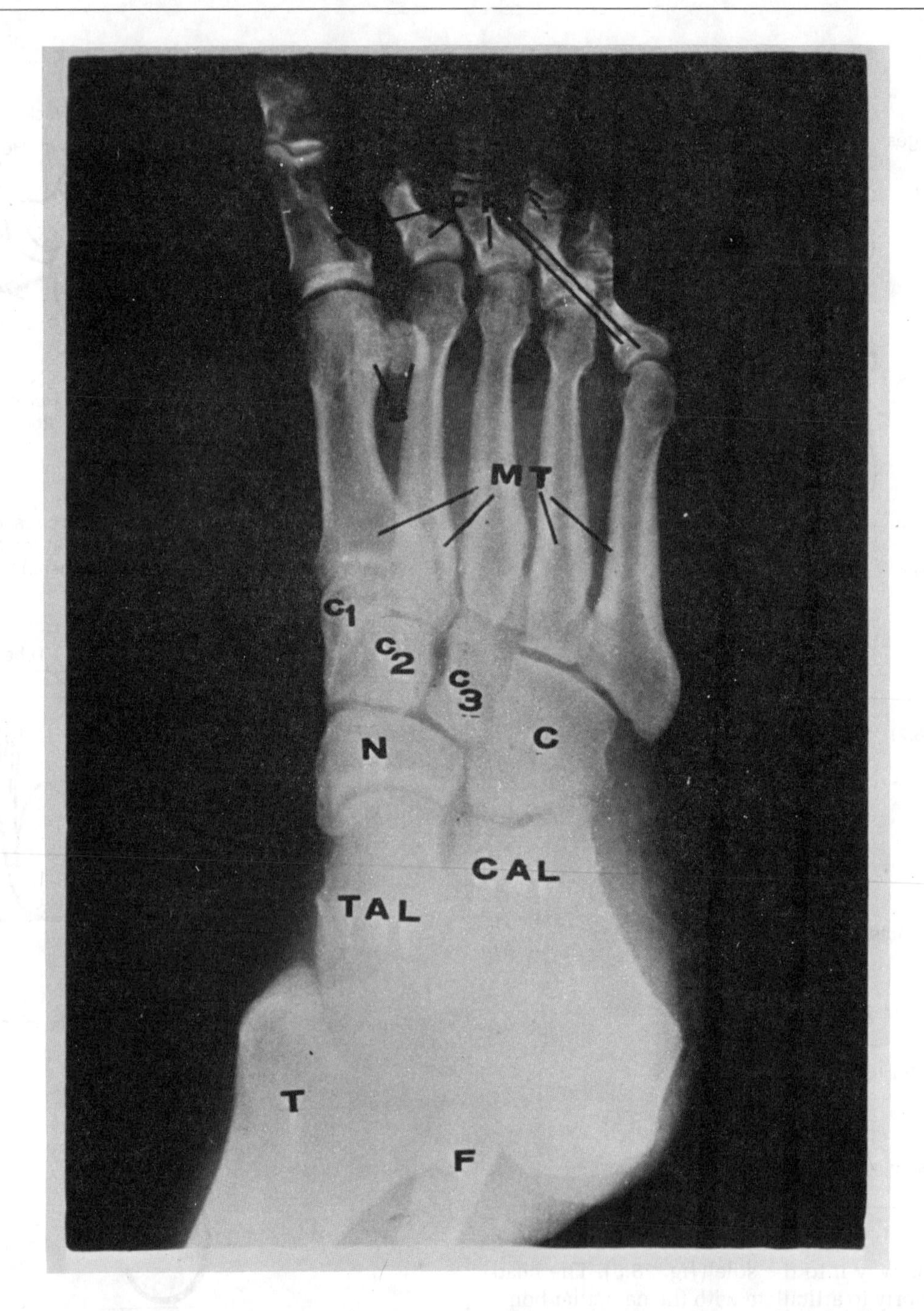

Plate 28.1. Foot (oblique view). *C*, cuboid bone; C1, C2, C3, the three cuneiform bones; *CAL*, calcaneus; *MT*, metatarsals numbered from 1–5 beginning on the medial aspect; *N*, navicular; *PP*, proximal phalanges; *S*, sesamoid bones (2 at the first metatarsal phalangeal articulation); *TAL*, talus; *T*, tibia; *F*, fibula.

The *middle unit* comprises the 5 small tarsal bones (*figs. 28.6–28.8*). The anterior side is indented between the 1st and 3rd cuneiforms to receive the base of the 2nd metatarsal, which is morticed between them, preventing side to side shifting.

The lateral surface of the cuboid is grooved for the tendon of the peroneus longus. On the medial side, the navicular presents a tuberosity, which is easily felt in your own foot.

Two continuous articular facets never, or hardly ever, lie in precisely the same plane; even when facets on one bone are continuous with one another, they always, or almost always, meet at an angle, forming inclined planes. See if this is not so (*fig. 28.7*).

The *contiguous sides of the bases of most metatarsals have articular facets* (*fig. 28.9*). But there are no such facets between the bases of the 1st and 2nd metatarsals.

The 2nd metatarsal is morticed between the medial and lateral cuneiforms. The interlocking of the various

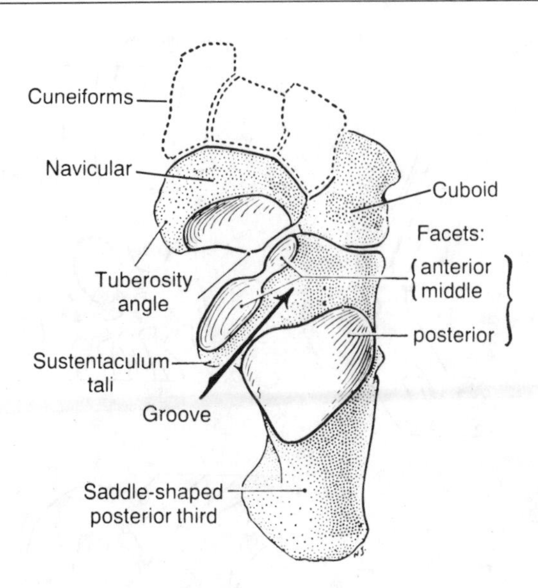

Figure 28.7. The upper surface of the calcaneus.

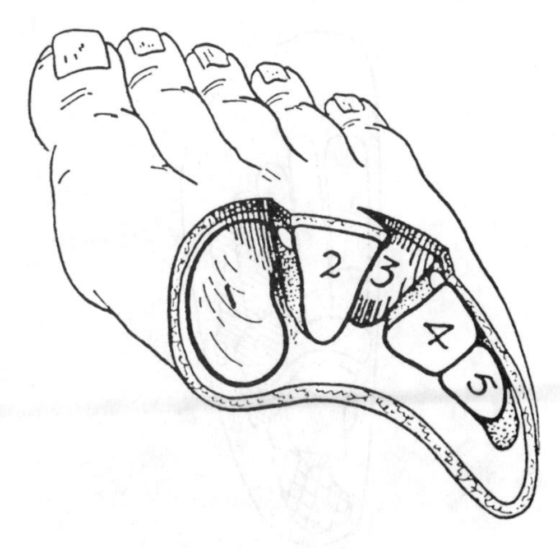

Figure 28.10. The half-arch across the bases of the metatarsals.

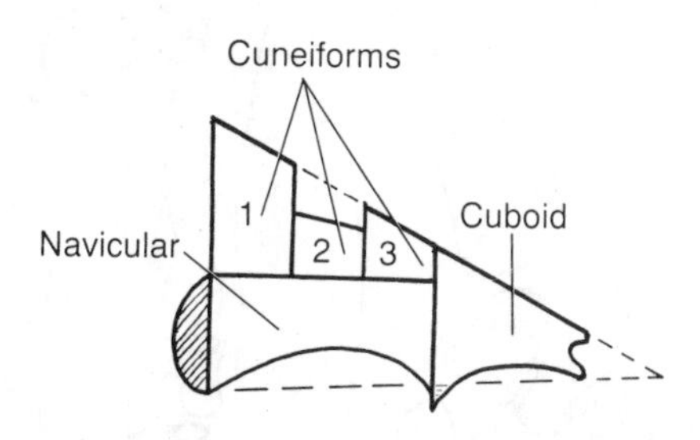

Figure 28.8. The middle "unit," comprising the 5 small tarsal bones, is a modified wedge.

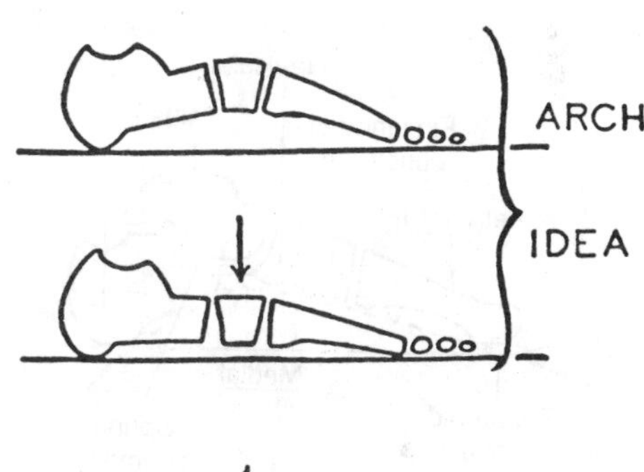

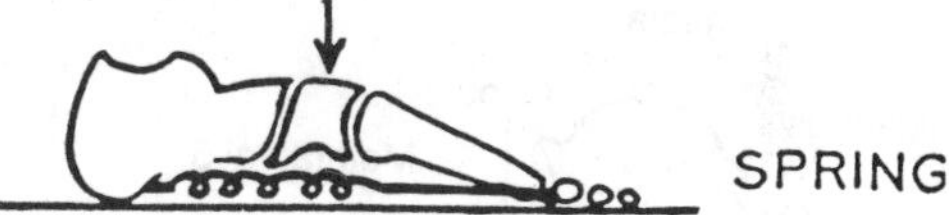

Figure 28.11. The foot is not an inert masonry arch; it is a spring.

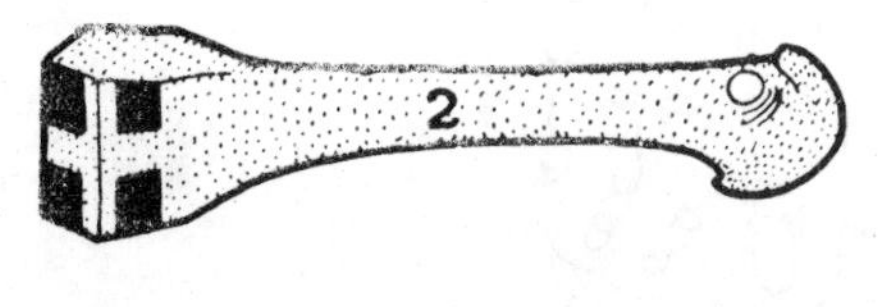

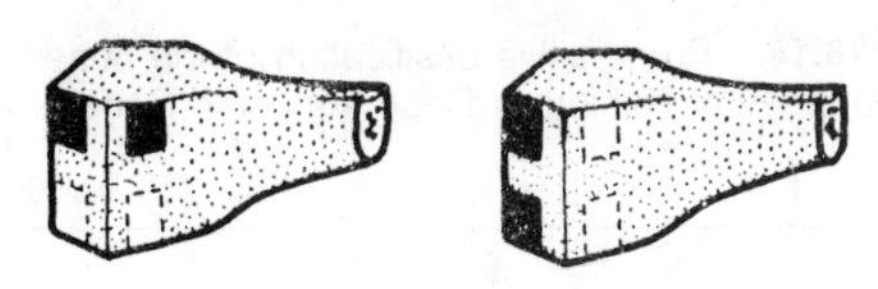

Figure 28.9. Variations in metatarsal facets.

bones prevents sideshifting of the bones (*figs. 28.2 and 28.10*).

ARCHES

The foot has 2 longitudinal arches but is not a piece of masonry. It is a spring rather than an arch (*fig. 28.11*).

The 3 medial digits, their metatarsals, and cuneiforms, the navicular and talus (supported by calcaneus) are collectively known as the *medial longitudinal arch*—it is the vital one. The 2 lateral digits, their metatarsals, the cuboid and the calcaneus, collectively known as the *lateral longitudinal arch* (*fig. 28.12*), represent a very low arch and not much of a spring (*fig. 28.13*).

The foot also forms a transverse half-arch at its midpoint because the medial arch is so much higher than the lateral one (*fig. 28.10*). To call this an arch as some authors do is poor biomechanics, and the theory will stand up no better than a half-arch.

Viewed from the lateral side, the foot is a low, flat arch, which flattens under your weight.

Viewed from the medial side, the foot is highly arched. The medial process of the calcaneus forms the posterior pillar; the heads of the 1st, 2nd, and 3rd metatarsals and the sesamoid bones under the head of the 1st form the anterior pillar (*fig. 28.13*).

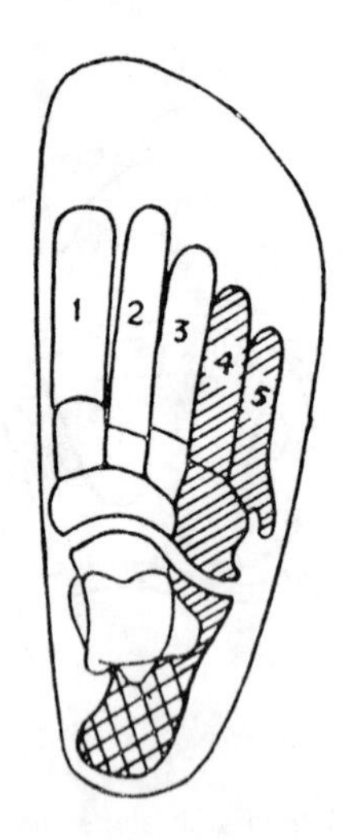

Figure 28.12. The medial and lateral longitudinal arches. The calcaneus is common to both.

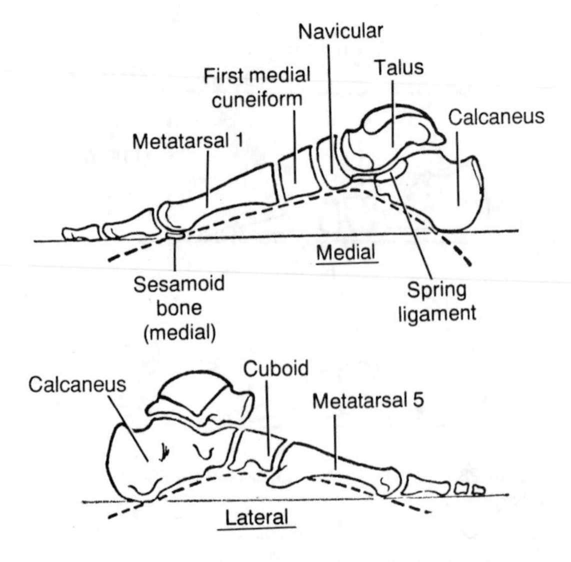

Figure 28.13. Medial and lateral longitudinal arches.

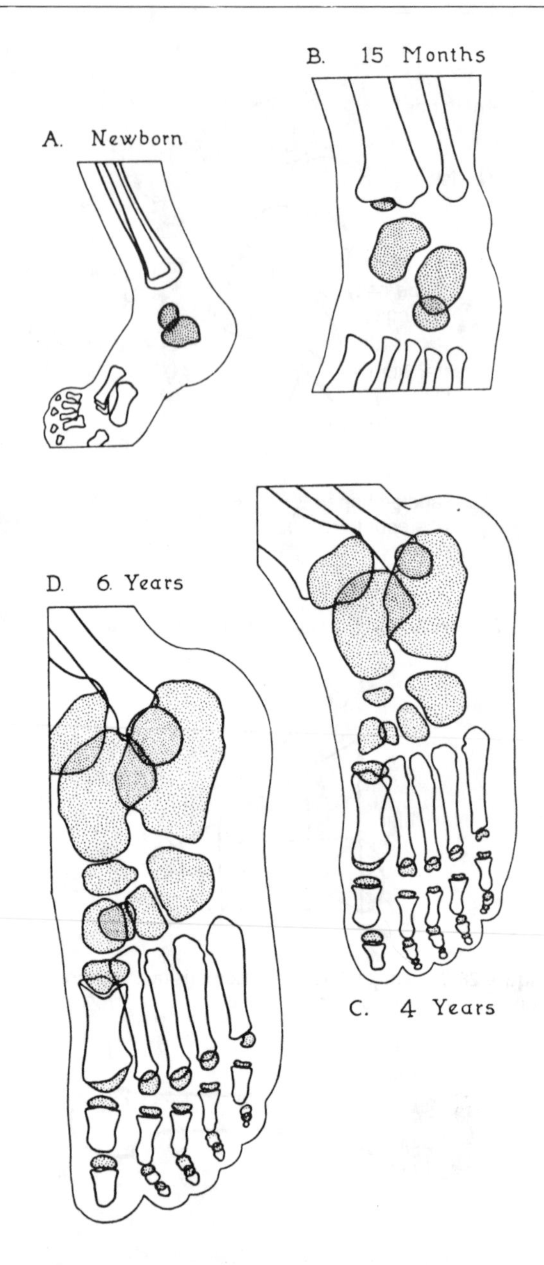

Figure 28.14. Progressive ossification of the bones of the foot. (Courtesy of Radiology Dept., Kingston General Hospital).

The round head of the talus is at the summit of the arch between the sustentaculum tali and the tuberosity of the navicular, and there receives no bony support. But the *spring ligament* (plantar calcaneonavicular), which extends between these 2 bony processes, lies below the head and supports it (*fig. 28.13; see also figs. 28.46–28.48*).

Ossification of the Bones of the Foot. X-ray photographs (*fig. 28.14*) show that at the time of birth the calcaneus and talus are well ossified and the cuboid is starting to ossify. The sequence is calcaneus, talus, cuboid, 3rd, 1st, and 2nd cuneiform, navicular, epiphysis of calcaneus (*see Table 28.2* on p. 339).

The calcaneus is the only short bone (either tarsal or carpal) that regularly has an epiphysis. This epiphysis gives attachment to the tendo calcaneus and the plantar aponeurosis; hence, it includes the posterior surface and the processes of the calcaneus (*see fig. 27.19*). It appears about the 11th year and fuses about the 17th year.

The metatarsals and phalanges have each a primary center for the body, which appears about the 3rd prenatal month, and secondary centers for the heads of the 2nd–5th metatarsals and for the bases of the 1st metatarsal and all the phalanges. These appear about the 3rd year and fuse about the 18th year.

Clinical Case 28.2

Patients Brenda T. and Brian U. On a warm early summer's evening when you are a clinical clerk in the emergency department, two patients are brought in almost together. The first is a woman apparently in her sixties who stepped down off a curb and twisted her left ankle painfully. There is extreme and very localized tenderness immediately anterior to the lateral malleolus. Even before she is sent to radiology, your anatomical knowledge permits you to tentatively diagnose her condition and predict her treatment. The other patient, a young man, was playing softball and after hitting a "two-bagger" was forced to slide into second base. As his left foot met the firmly planted foot of the second baseman, all who were near heard a snap followed by the runner's howl of pain. In great pain, he is brought to the desk by his supporting friends. Your quick investigation suggests he has a (Pott's) fracture of the medial malleolus and the lower part of the fibula. Fortunately, your anatomical training helps you to understand the pathomechanics and to initiate the series of actions that will be needed.

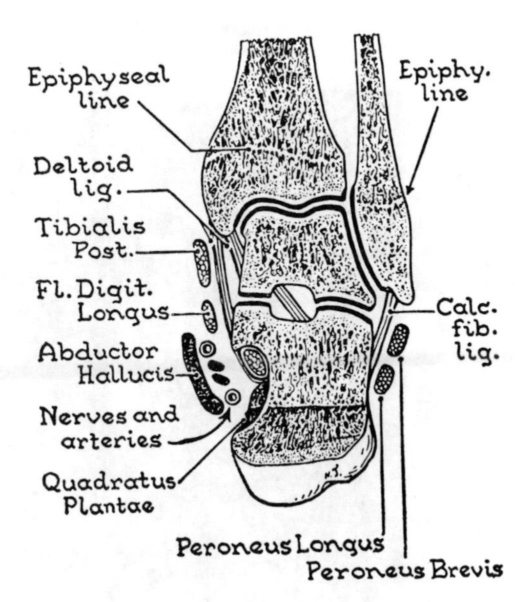

Figure 28.15. Coronal section of ankle region.

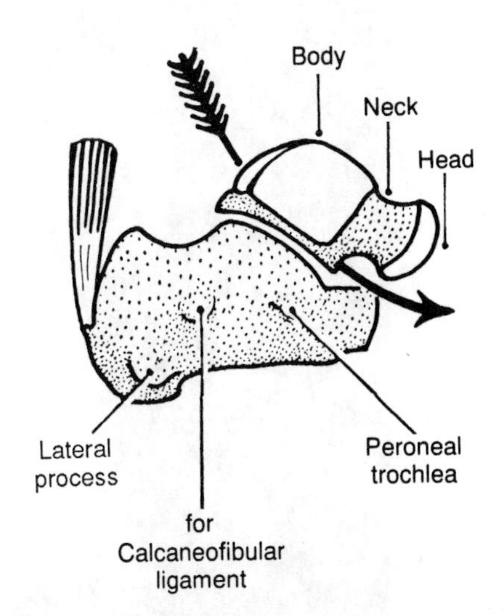

Figure 28.16. Lateral aspect of calcaneus and talus. The *arrow* traverses the tarsal tunnel (tarsal sinus).

The Ankle Joint (Talocrural Joint)

NINE BASIC OBSERVATIONS

1. In walking, the triceps surae (i.e., 2 heads of the gastrocnemius and the soleus) raises the heel from the ground—it causes *plantarflexion* of the ankle joint at the end of the "stance phase." During the act of advancing the limb ("swing phase") anterior crural muscles cause *dorsiflexion* of the foot to clear the ground.

2. The weight of the body is transmitted to the talus through the tibia. The malleoli grasp the sides of the talus (*fig. 28.15 and Plate 28.2*)

3. The sharp tip of the fibular (lateral) malleolus is felt 2 cm distal to the level of the blunt end of the tibial (medial) malleolus. The reciprocal articular facets on the two sides of the talus correspondingly articulate with the malleoli (*figs. 28.16 and 28.17*).

4. The sides of the malleoli and of the shafts of the bones proximal to them are subcutaneous—there are no muscles at the sides of the ankle, for this is an ideal hinge joint.

5. Because of the hinge movements, the superior surface of the body of the talus is articular and convex anteroposteriorly.

Comparative Anatomy. In quadrupeds, a pronounced anteroposterior flange projects from the tibia into a slot in the talus (*fig. 28.18*), but this is greatly reduced in the human ankle (*fig. 28.15*).

6. With the exception of the tendo calcaneus, all tendons cross the ankle and transverse tarsal joints, acting on both.

7. While the integrity of a joint depends upon 4 factors—bones, ligaments, muscles, and gravity—here muscles and gravity will tend toward forward displace-

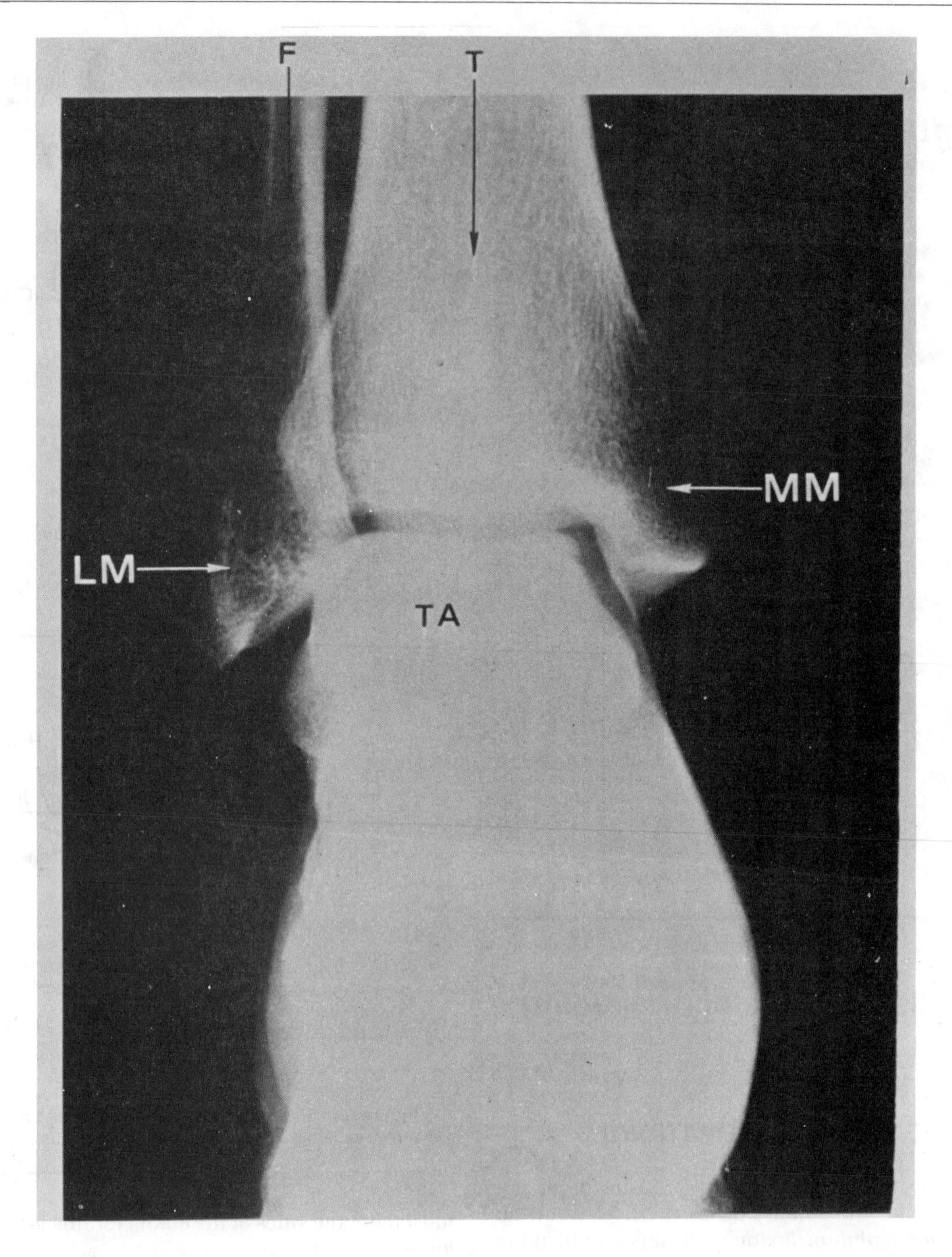

Plate 28.2. Ankle (A-P view). *F*, fibula; *LM*, lateral malleolus; *MM*, medial malleolus; *T*, tibia; *TA*, talus.

ment of the leg bones, so it falls to the bony parts and the ligaments to resist (*fig. 28.19*).

8. This demands that the socket shall be so fashioned that it cannot slide forward on the talus.

9. It also demands that the ligaments shall pass downward and backward.

BONY PARTS OF THE ANKLE JOINT

These include (1) parts of the talus and (2) its socket.

The Talus

The whole of the superior surface of the body is articular, and it continues down both sides with the large facet for the fibular malleolus and restricted facet for the tibial malleolus (*figs. 28.16, 28.17*). The body is wedge shaped (*fig. 28.20*), the broad end being anterior.

The *socket* comprises the malleoli, the inferior articular surface of the tibia, and a transverse tibiofibular ligament that deepens the socket posteriorly.

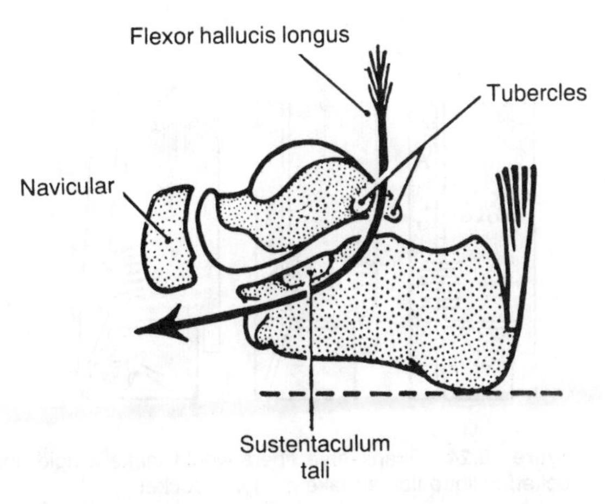

Figure 28.17. Medial aspect of calcaneus and talus.

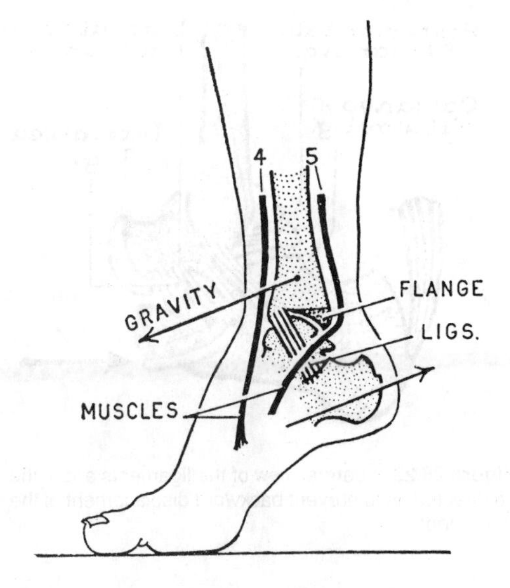

Figure 28.19. At the ankle joint the ligaments and bones resist the forces of the muscles and of gravity.

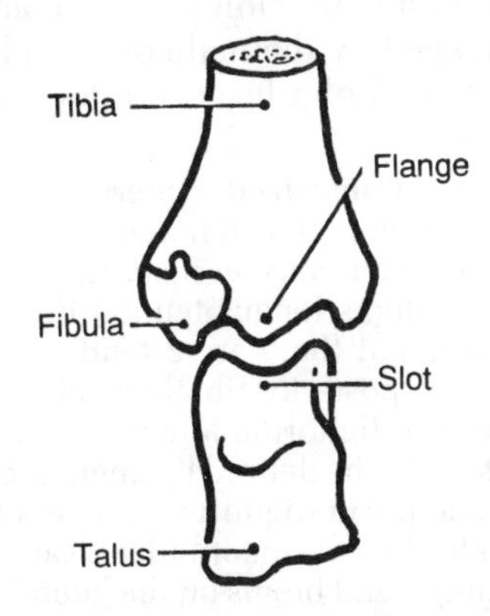

Figure 28.18. The ankle joint of the sheep, showing pronounced flange and slot.

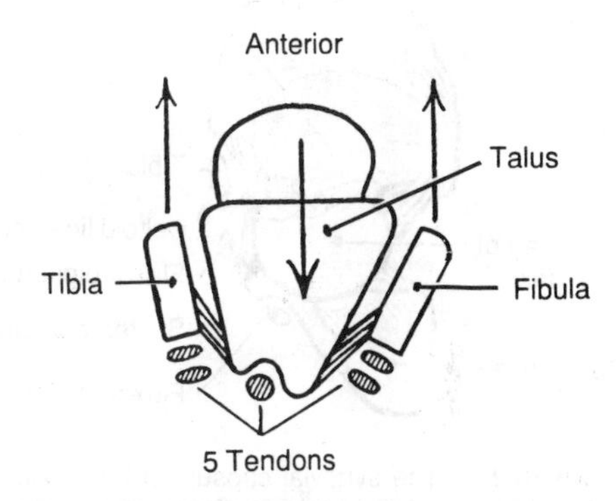

Figure 28.20. Diagram of the ankle joint in horizontal section to show that the direction of the ligaments and the converging malleoli prevent posterior displacement of the talus.

Collateral Ligaments. The **medial** or **deltoid ligament** (*fig. 28.21*) arises from the blunt end of the tibial malleolus and has 2 parts, a *superficial* and a *deep*, like the medial ligament of the knee. The superficial band passes posteroinferiorly to the sustentaculum tali.

The deep or *deltoid part* fans out to the nonarticular part of the medial aspect of the talus. It even reaches the navicular bone. Between the sustentaculum and the navicular it is continuous with the "spring" ligament, forming part of the socket for the head of the talus (*fig. 28.21*).

The **lateral ligament** (*fig. 28.22*) has 3 parts: (1) The *calcaneofibular ligament* is a cord that runs posteroinferiorly. (2) The *posterior talofibular ligament* lies deep (*fig. 28.22*) passing horizontally (from the *malleolar fossa* of the fibula, *fig. 27.15*) to the lateral tubercle of the talus. (3) The *anterior talofibular ligament* is a thin, weak band in the anterior part of the capsule—it is the most easily "sprained" ligament in the ankle region—as it was in Clinical Case 25.2.

The **synovial membrane** extends quite far onto the neck of the talus (*fig. 28.23*).

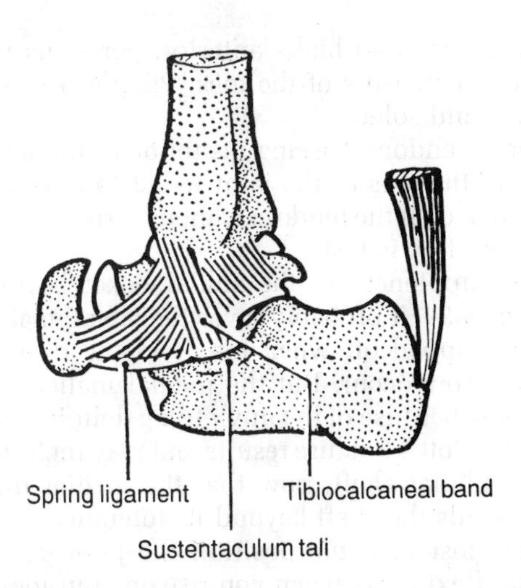

Figure 28.21. Medial or deltoid ligament of ankle joint.

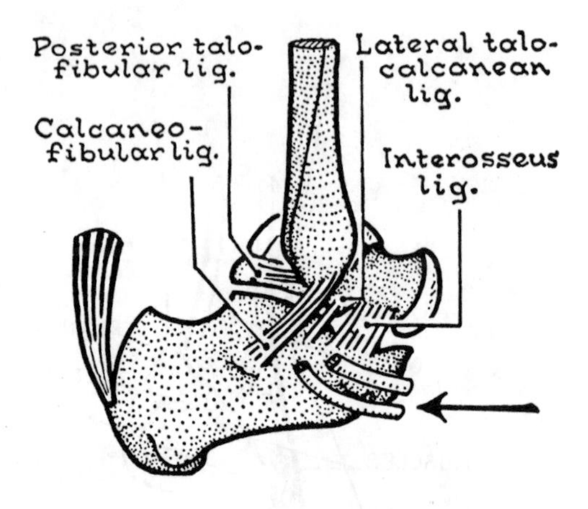

Figure 28.22. Lateral view of the ligaments about the ankle, so directed as to prevent backward displacement of the bones of the foot.

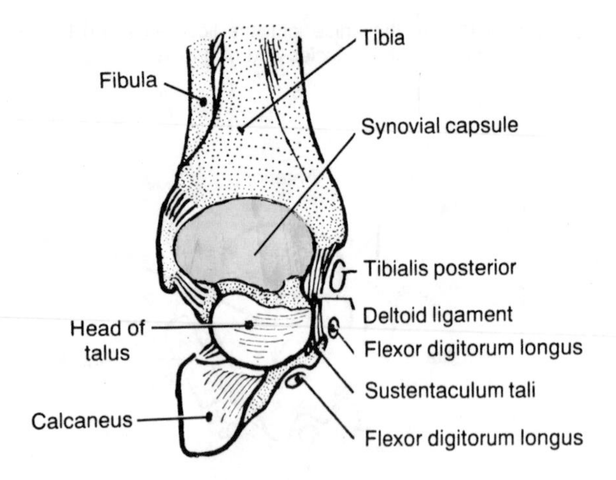

Figure 28.23. The synovial capsule of the ankle joint, distended. The anterior articular surfaces of the talus and calcaneus: sustentaculum tali and related tendons.

MOVEMENTS

Dorsiflexion—tibialis anterior, peroneus tertius, and the long extensors of the toes. *Plantarflexion*—gastrocnemius and soleus.

The 5 tendons passing posterior to the ankle are too close to the axis of the joint to act to advantage on the joint. In fact, if the tendo calcaneus is ruptured, the power to plantarflex is lost.

The suppleness of the ankle socket is illustrated in *Figure 28.24*. It probably saves the lateral malleolus from being snapped off when forces in the direction of the white arrow occur. But the medial malleolus is not so fortunate because of the unyielding deltoid ligament (*fig. 28.21*)—Pott's fracture results and may include snapping of the fibular shaft, now that the medial thrust of the talus bends the shaft beyond its tolerance.

The most unstable position the joint can assume is plantar flexion, as when you rise on your toes. On going downhill you instinctively dig your heels in because in this position of dorsiflexion the broad edge of the wedge is closely grasped by the malleoli, and because the heel is at the short end of a lever, the toes being at the long end.

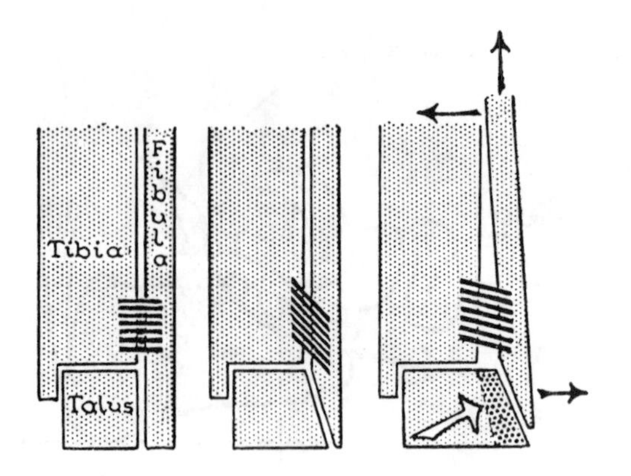

Figure 28.24. Transverse fibers would make a rigid ankle socket; oblique fibers make a supple socket.

Relations Summarized. Anteriorly, the anterior tibial vessels and deep peroneal nerve lie midway between the malleoli with 2 tendons on each side of them. Posteriorly, the flexor hallucis longus tendon lies midway between the malleoli, and there are 2 tendons posterior to each malleolus; the posterior tibial vessels and tibial nerve lie between flexor digitorum longus and flexor hallucis longus (*fig. 28.25*). The deltoid ligament is crossed by tibialis posterior and flexor digitorum longus, on the medial side of the ankle the calcaneofibular ligament is crossed by peronei longus and brevis on the lateral side of the ankle.

Epiphyses. The distal epiphyseal plate of the fibula is at the level of the superior surface of the joint; that of the tibia is a centimeter superior to the joint (*fig. 28.15*). The distal fibular and tibial epiphyses are always completely fused to their diaphyses by the 20th year.

Blood Supply. From the anastomoses of all the vessels around the joint.

Nerve Supply. From the deep peroneal and tibial nerves, i.e., from the major nerves that cross it.

Identifiable Parts in Living Foot

Now you can and should review with your eyes and your fingertips:

1. The blunt end of the medial malleolus.
2. The sustentaculum tali, more than a fingerbreadth below the malleolus.
3. The tuberosity of the navicular, 2 fingerbreadths in front of the sustentaculum.
4. The head of the talus, occupying the space between these two (*see Plate 28.3*). All four are easily palpated,

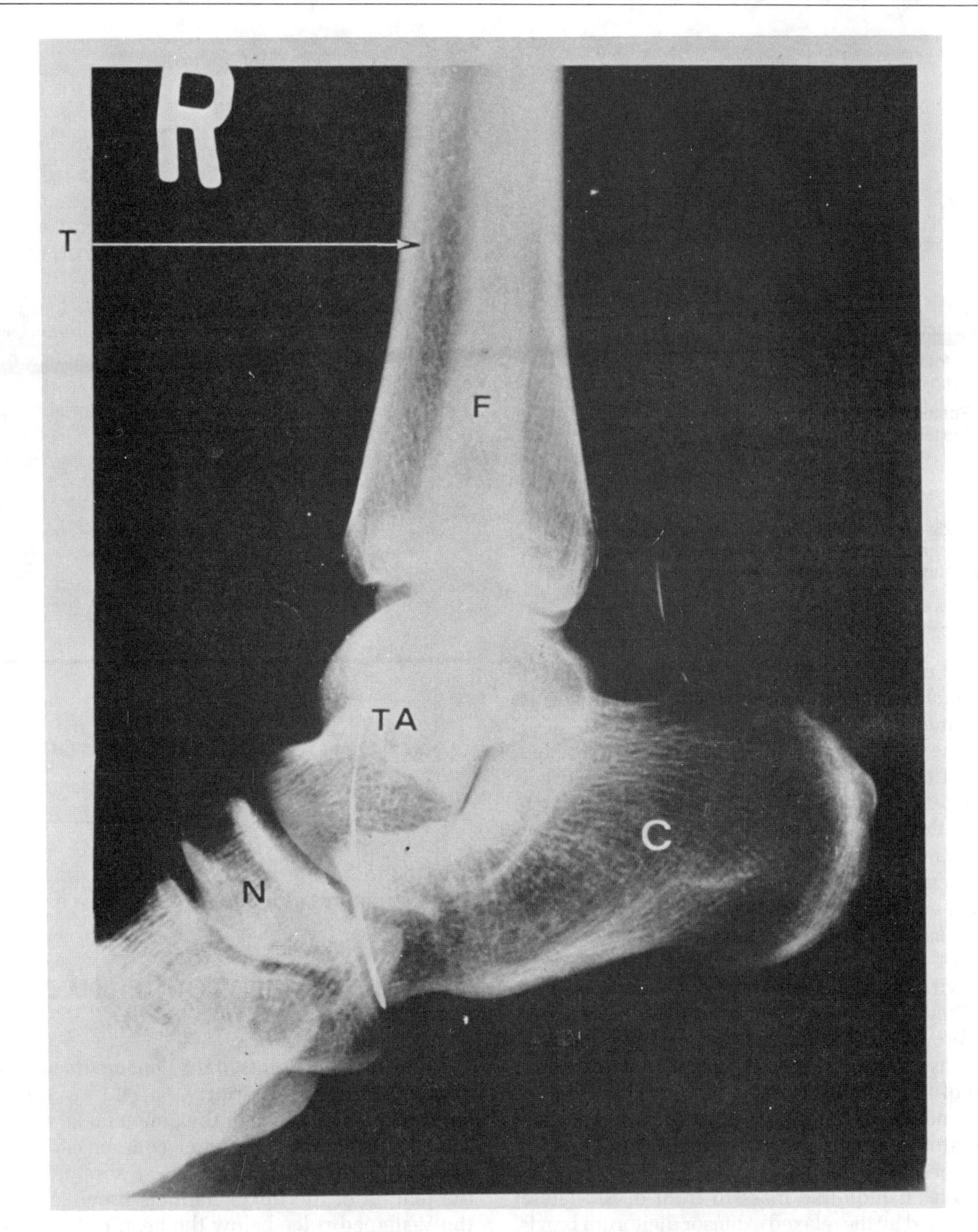

Plate 28.3. Ankle (lateral view). *C*, calcaneus; *F*, fibula; *N*, navicular; *T*, tibia; *TA*, talus.

and you can make them visible through the skin. To really feel the tip of the malleolus, the sustentaculum, and the tuberosity, *you must approach them inferiorly.*

5. The sesamoid bones, which play on the inferior surface of the head of the 1st metatarsal, can and should be felt under the ball of the big toe.

6. The lateral malleolus is visible; its pointed end is 2 cm more inferior (longer) than the medial malleolus and more posterior.

7. The projecting base of the 5th metatarsal is easily felt about halfway along the lateral border of the foot where shoes must bulge to accommodate it.

8. The calcaneocuboid joint lies more than midway between the tip of the lateral malleolus and the projecting base of the 5th metatarsal. Verify this by inverting the foot and palpating the superolateral portion of the anterior surface of the calcaneus, which is thereby uncovered. This is easily done.

9. With the foot still inverted, palpate the superior part of the round head of the talus, just superomedial.

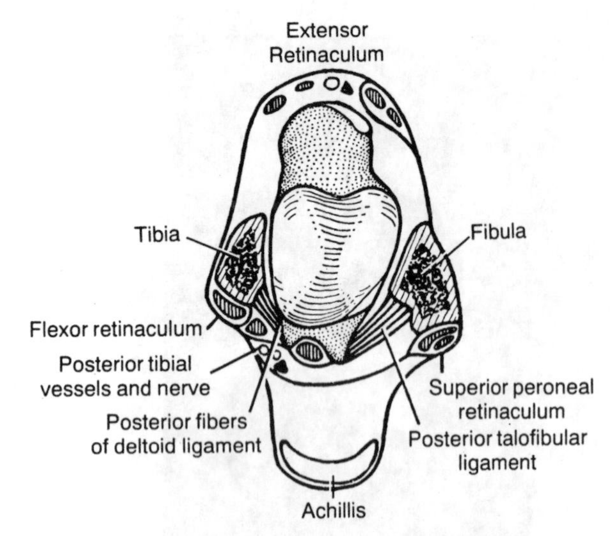

Figure 28.25. Horizontal section through ankle joint showing (1) wedge-shaped socket, (2) direction of ligaments, (3) 5 posterior tendons, and (4) the investing and intermuscular deep fascia.

Clinical Case 28.3

Patient Charles V. This 27-year-old male graduate student and marathoner had been increasing his practice runs until he began having severe pains in his right ankle and foot. Preliminary diagnoses being mentioned in the Sports Injury Clinic include: "tendinitis or bursitis," "fatigue fracture(s)," and "fasciitis," etc. Tenderness behind the medial side of the ankle continues deep into the sole of his foot. To reach a correct diagnosis, one must have a clear knowledge of the anatomy in this chapter in general and the structures that run from the leg into the medial aspect of the sole (e.g., see *figs. 28.39 and 28.40*).

10. Is it not obvious by now that the muscles of inversion and eversion, the 2 *tibiales* and the 3 *peronei*, must be inserted anterior to the transverse tarsal joint?

(a) With the foot inverted and the ankle dorsiflexed, the tendon of the tibialis anterior may be seen easily as it juts out anterior to the medial malleolus, on its way to its insertion (the medial surface of the 1st cuneiform and adjacent part of the 1st metatarsal) (*see fig. 28.35*).

(b) The tibialis posterior is not so obvious. With the foot inverted and ankle plantarflexed, its tendon may be felt as it passes from its pulley (the medial malleolus) to its insertion—the tuberosity of the navicular.

(c) Now, with the foot everted, trace the tendons of the peronei as they turn anteriorly inferior to the lateral malleolus; the brevis turns to the base of the 5th metatarsal, the longus passes into the sole of the foot just posterior to the base of the 5th metatarsal.

11. The tendons of the extensor hallucis longus and extensor digitorum longus can generally be made to stand out on the dorsum of the foot.

12. The soft cushion-like mass in front of the lateral malleolus, caused by the relaxed extensor digitorum brevis, should not be mistaken for a swelling due to a sprain. Extend your toes and feel it swell and harden.

13. The pulsation of the anterior tibial artery where it becomes the dorsalis pedis artery should and must be felt midway between the 2 malleoli by all physicians and their students because the blood supply of the foot is critically important. Here 2 tendons (tibialis anterior and extensor hallucis longus) lie medial to it and two tendons (extensor digitorum longus and its peroneus tertius) lie lateral.

14. The posterior tibial artery must be felt to pulsate behind the medial malleolus, but the foot should be slightly inverted (passively) in order to relax the fascia, and you should palpate firmly forward and laterally (*see fig. 28.40*).

The Sole of the Foot

In the following pages on the foot, you will recognize many structures as homologues of those met in the hand.

ARTICULATED FOOT FROM BELOW (*fig. 28.26*)

1. *The Bearing Points of the Foot are the medial process of the calcaneus* posteriorly, and 6 (not 5) equal pillars anteriorly—the heads of the 5 metatarsal bones with the 2 sesamoid bones. The latter compensate for the shortness of the first metatarsal. These sesamoid bones lie in the tendon of the flexor hallucis brevis on each side of the V-shaped ridge below the head of the 1st metatarsal to raise that head off the ground (*figs. 28.26 and 28.27*). *There is no transverse arch!*

2. The *flexor hallucis longus* grooves the inferior surface of the sustentaculum tali (*fig. 28.5*) and subsequently passes between the 2 sesamoid bones. The *peroneus longus* grooves the cuboid as it passes obliquely across the sole to the adjacent parts of the 1st metatarsal and 1st cuneiform.

3. The *Transverse Tarsal Joint—Some Details for Specialists*. The inferior surface of the cuboid is large and triangular. It ends posteroinferiorly in an *angle*, inferior to the calcaneus, which is cut away to receive it. The inferior surface of the navicular also has an *angle* opposite its *tuberosity*. The head of the talus is without

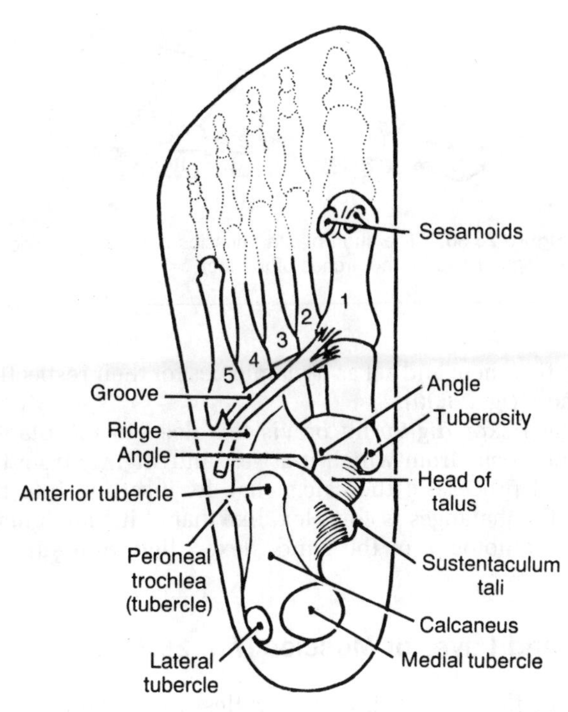

Figure 28.26. The chief features of the plantar aspect of the articulated foot.

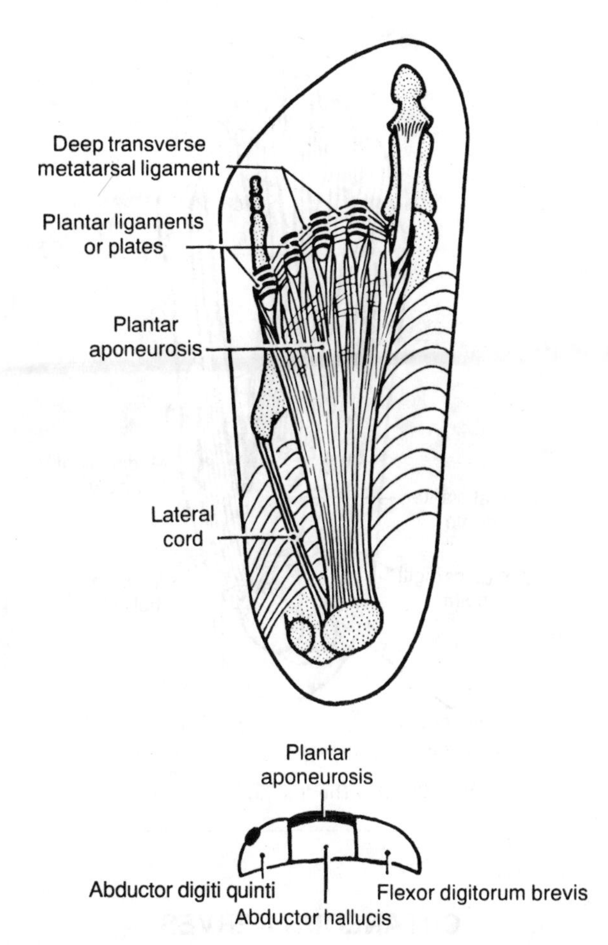

Figure 28.28. The plantar fascia (inferior view and cross-section) consisting of the plantar aponeurosis centrally and thinner fascia enclosing the abductor hallucis and abductor digiti quinti.

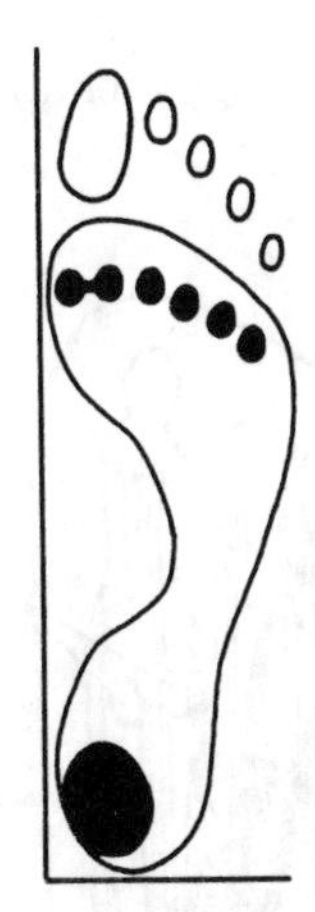

Figure 28.27. The bearing points of the foot.

bony support between the sustentaculum tali posteriorly and the tuberosity and angle of the navicular anteriorly.

COVERINGS

The Plantar Fascia (*fig. 28.28*).

Its *medial* and *lateral portions* cover the abductors of the great and little toes and are thin. The central portions, the **plantar aponeurosis** (developmentally a detached part of the tendon of the plantaris) is a strong tie beam for the longitudinal arches of the foot. Posteriorly, it is attached to the medial process of the calcaneus; anteriorly, it splits into 5 bands, which attach to the connective tissue around the metatarsal heads (*fig. 28.28*).

Observe and feel how dorsiflexion of the toes renders the plantar aponeurosis taut and increases the arches. This happens during walking too, preventing collapse of the springy arches.

The **epidermis** and **dermis** are both much thicker on the palms and soles than elsewhere—even at birth. With pressure and friction the thickness increases. The subcutaneous fat is bound down within small fibrous compartments that serve as cushions.

FUNCTION OF THE TOES

They press into the ground and form a friction surface during the act of walking and provide the purchase for the forward thrust of the body during running and pushing—though not so much in walking. (Look at wet footprints.)

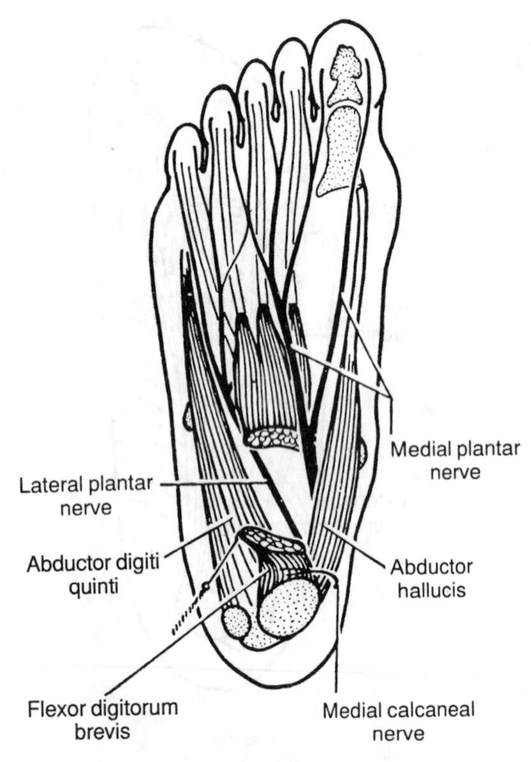

Figure 28.29. The first layer of muscles.

CUTANEOUS NERVES

The medial plantar nerve supplies the 3½ digits on the hallux side of the foot, just as the median nerve supplies the 3½ digits on the pollex side of the hand, leaving the lateral plantar nerve, which corresponds to the ulnar nerve in the hand, to supply 1½ digits (*fig. 28.29*).

The skin under the heel is supplied by the medial calcaneal branches of the tibial nerve.

PLANTAR MUSCLES

These are arranged in 4 layers.

First or Superficial Layer

These arise from the calcaneus (*fig. 28.29*).

1. Abductor Hallucis and } inserted into proximal phalanges
2. Abductor Digiti Quinti } inserted into proximal phalanges
3. Flexor Digitorum Brevis—inserted into middle phalanges.

The abductors of the 1st and 5th digits have a continuous fleshy origin across the heel. The insertions are into the bases of the proximal phalanges. Though these 2 muscles are called "abductors," they seldom abduct the toes; they flex them and act as dynamic ties for their respective arches (*fig. 28.30*).

Figure 28.30. The short plantar muscles can act as elastic springs or ties for the arches of the foot.

The *flexor digitorum brevis* lies deep to the plantar aponeurosis, from which it arises (and slightly from the medial process of the calcaneus). Its insertion into the middle phalanges is complex, like that of its much more vital homologue in the hand, flexor digitorum superficialis.

Second Layer of Muscles (*fig. 28.31*)

Long flexor tendons and short fleshy muscles attached to them form this layer.

1. Flexor Digitorum Longus tendon.
2. Quadratus Plantae (or Flexor Accessorius).
3. Lumbricals.
4. Flexor Hallucis Longus tendon.

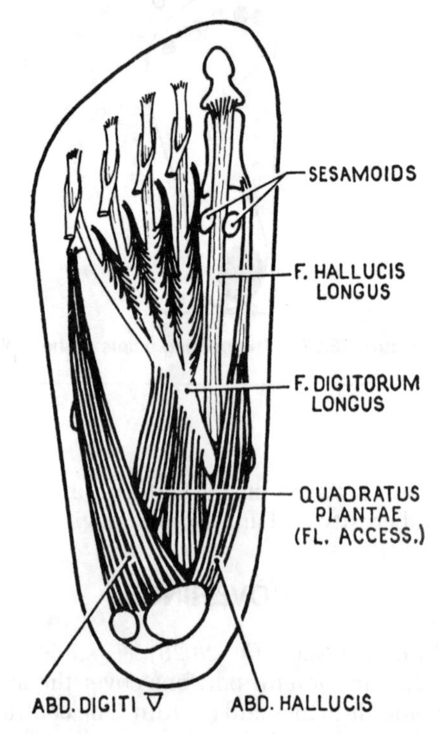

Figure 28.31. The second layer of muscles displayed by removal of the flexor digitorum brevis.

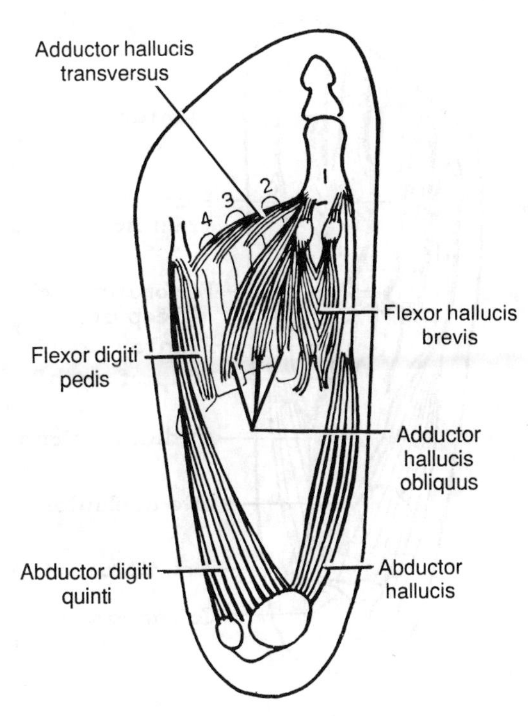

Figure 28.32. The 3rd layer of muscles displayed by excision of the 2nd layer but leaving the abductors of the 1st layer.

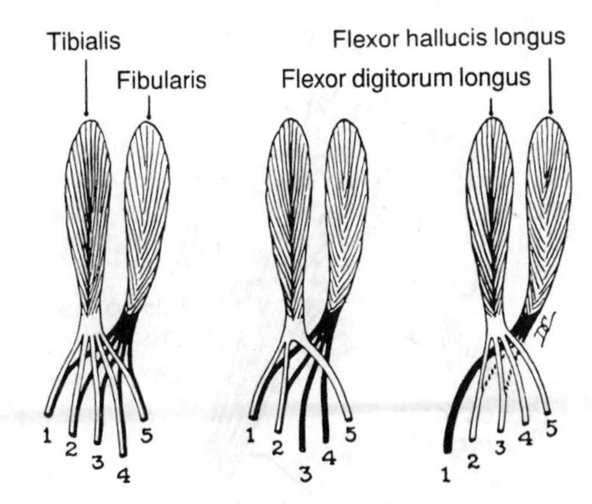

Figure 28.33. Scheme explaining the crossing and variable union of long digital flexors in various mammals and man. (After Straus.)

The tendon of the *flexor digitorum longus* appears from under cover of the abductor hallucis, crossing inferior to the tendon of the flexor hallucis longus. Its *insertion* (as for flexor digitorum profundus in the hand) is into the bases of distal phalanges.

The *quadratus plantae* [flexor accessorius] arises from the medial side of the calcaneus, and it is inserted into the tendon of the flexor digitorum longus apparently to redirect the line of pull of the oblique tendons (*fig. 28.31*). Most physicians will never hear of it again.

The *four lumbricals* arise from the tendons of the flexor digitorum longus and are inserted into the dorsal digital expansions.

The *flexor hallucis longus* tendon passes onward between the sesamoid bones developed in the tendons of insertion of flexor hallucis brevis (*fig. 28.32*) and finally is inserted into the distal phalanx of the hallux. *Figure 28.33* illustrates an interesting phylogenetic explanation of its variations.

Third Layer of Muscles

All are short (*fig. 28.32*).

1. Flexor Hallucis Brevis.
2. Flexor Digiti Quinti and
3. Adductor Hallucis Transversus (form 3 sides of a square open posteriorly).
4. Adductor Hallucis Obliquus (largely fills the square).

The muscles of both 3rd and 4th layers practically confine themselves to the anterior half of the foot. Students need only be aware of their existence. Specialists who need details of the 3rd layer of small muscles, *see Figure 28.32*, and note that the *abductor hallucis* is inserted in conjunction with the medial head of flexor hallucis brevis; the *adductores hallucis obliquus* and *transversus* with the lateral head. Two sesamoid bones develop in the tendons of insertion (*figs. 28.32 and 28.34*).

Fourth Layer of Muscles

1. Seven Interossei.
2. Two tendons on the skeletal plane.

Interossei, 4 Dorsal and *3 Plantar*. These 7 muscles are like their more important homologues in the hand, but their real *function* emerges during walking. The interossei flex the metatarsophalangeal joints and thereby draw the heads of the metatarsals together (i.e., they keep the foot from spreading), and they extend the interphalangeal joints and thereby keep the toes from curling up.

Two Long Tendons on Skeletal Plane

The *tibialis posterior* and the *peroneus longus* belong to the ligamentous or deepest layer of the posterior half

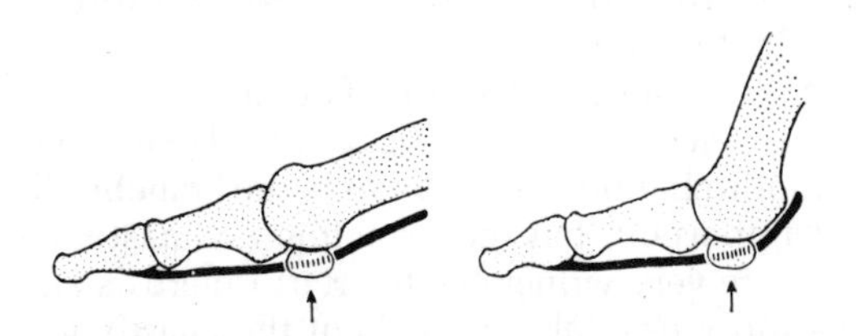

Figure 28.34. The sesamoids are always the bearing points for the head of the 1st metatarsal, allowing free play for the long flexor tendon.

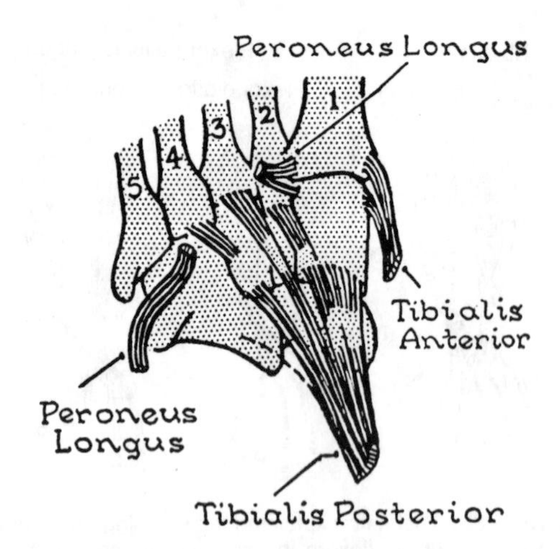

Figure 28.35. The grasp of the tibialis posterior extends to the small tarsals and to the metatarsals. Peroneus longus and tibialis anterior insert on the same two bones.

of the foot. One being an invertor, the other an evertor, they must find attachment in front of the transverse tarsal articulation (*fig. 28.35*).

1. The **tibialis posterior** (*fig. 28.35*) sends two-thirds of its tendon to insert into the tuberosity of the navicular. One-third divides into a number of finger-like bands that pass inferior to the spring ligament and diverge to reach the cuboid, cuneiforms, and 2nd, 3rd, and 4th metatarsal bases.
2. The **peroneus longus** enters the sole deep to the abductor digiti V and passes obliquely across the sole in the groove on the cuboid to its insertion into the lateral side of the base of the first metatarsal and adjacent part of the first cuneiform (*fig. 28.26*). The **Tibialis anterior** is inserted into the same 2 bones as the peroneus longus, but on their medial side (*fig. 28.35*)

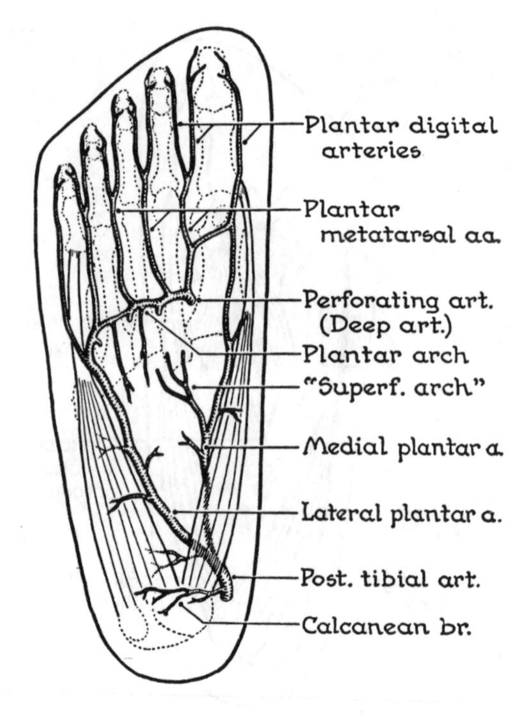

Figure 28.36. The plantar arteries (for reference).

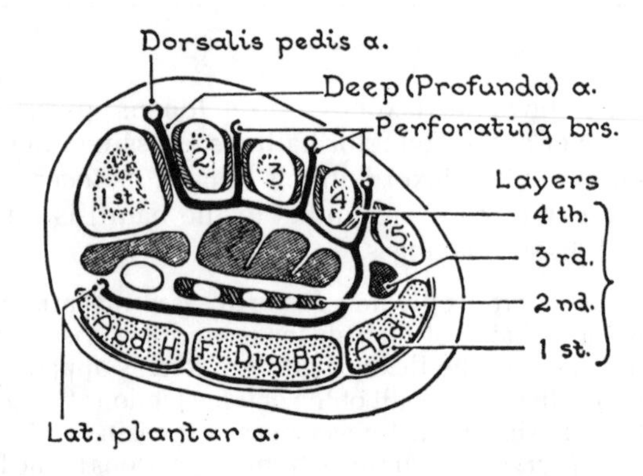

Figure 28.37. Scheme to show the lateral plantar artery coursing first between the 1st and 2nd layers of muscles, then between the 3rd and 4th layers.

PLANTAR VESSELS AND NERVES

The **lateral plantar artery** (*fig. 28.36*) is much larger than the *medial plantar artery* and is, in fact, the continuation of the posterior tibial artery. It enters the sole deep to the abductor hallucis and, with its companion nerve, runs anterolaterally between the 1st and 2nd layers of muscles. Then, dipping deeper, it runs medially between the 3rd and 4th layers (*fig. 28.37*), forming the (deep) **plantar arch**.

The arteries of the anterior half of the foot are arranged like the arteries of the hand (*fig. 28.38*). There are similar anastomotic channels. The names of the branches (below) are not important, but the principle is: anastomoses in this area are very vulnerable to arteriosclerosis and poor blood supply from above. Without the anastomoses, almost everyone would have serious trouble with their feet after the age of 60.

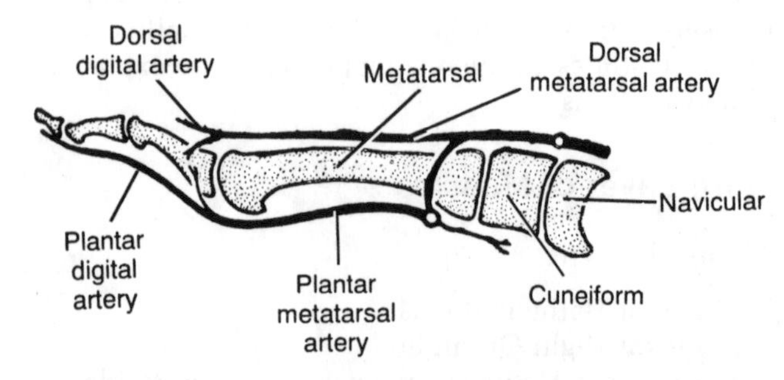

Figure 28.38. Scheme of arteries in the anterior part of the foot resemble that of the hand.

Branches

The *deep branch* of the dorsalis pedis artery, is just an enlarged (1st) perforating artery (*fig. 28.37*). From the deep plantar arch spring: (1) the 3 other *perforating arteries*, which join the dorsal metatarsal arteries; (2) 4 *plantar metatarsal arteries* which bifurcate into *plantar digital arteries;* and (3) a branch to the medial side of the big toe and one to the lateral side of the little toe (*fig. 28.36*).

The **lateral plantar nerve** follows its artery closely and sends cutaneous branches to the lateral 1½ digits and motor branches to all the muscles of the sole not supplied by the medial plantar nerve. This includes quadratus plantae, abductor and flexor digiti V, several lumbricals, all seven interossei, and the adductor hallucis (*cf.* ulnar nerve in the hand).

The **medial plantar nerve**, like the median nerve of the hand, sends:

1. Cutaneous branches to 3½ digits which fan out subjacent to the plantar fascia.
2. Motor branches to the 4 muscles between which the nerve runs, namely:
   a. Abductor Hallucis
   b. Flexor Digitorum Brevis
   c. Flexor Hallucis Brevis
   d. Lumbricalis I.

Table 28.1. Segmental Innervation of Muscles of Leg and Foot*

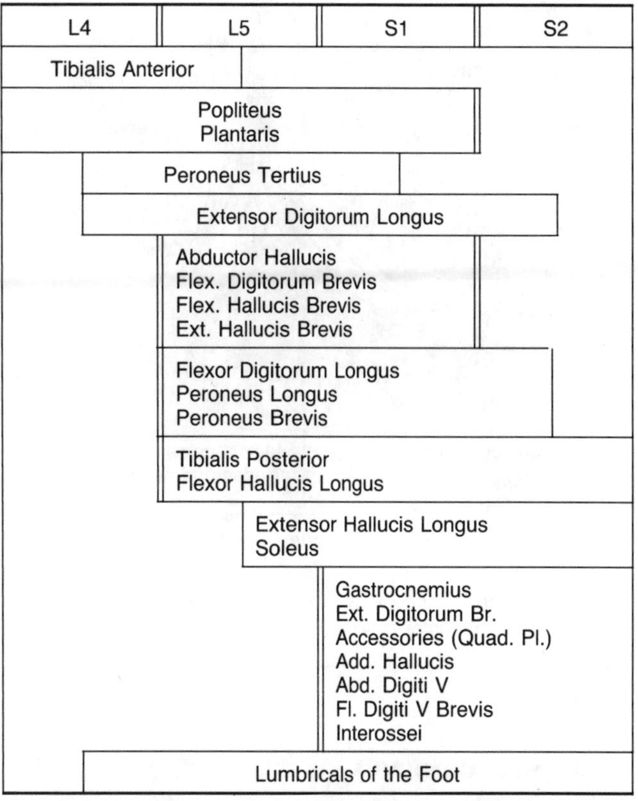

Muscles	L4	L5	S1	S2
Tibialis Anterior	X	X		
Popliteus, Plantaris	X	X	X	
Peroneus Tertius	X	X	X	
Extensor Digitorum Longus	X	X	X	X
Abductor Hallucis, Flex. Digitorum Brevis, Flex. Hallucis Brevis, Ext. Hallucis Brevis		X	X	
Flexor Digitorum Longus, Peroneus Longus, Peroneus Brevis		X	X	X
Tibialis Posterior, Flexor Hallucis Longus		X	X	X
Extensor Hallucis Longus, Soleus		X	X	X
Gastrocnemius, Ext. Digitorum Br., Accessories (Quad. Pl.), Add. Hallucis, Abd. Digiti V, Fl. Digiti V Brevis, Interossei			X	X
Lumbricals of the Foot	X	X	X	X

* Modified after Bing, and Haymaker and Woodhall (for Hip and Thigh, p. 287).

"THE DOOR OF THE FOOT"

Since the lateral and medial plantar vessels and nerves and 3 tendons enter the sole on the medial side by passing deep to the abductor hallucis, this entrance may be thought of as the door of the sole of the foot. The peroneus longus enters the lateral side deep to the abductor digiti V, through the back door, so to speak. The perforating vessels pass through the intermetatarsal spaces, as it were, through windows. The abductor hallucis then, guards the door, where *quadratus plantae* provides a soft doormat on the bone (*figs. 28.39 and 28.40*).

COMMENT ON VESSELS AND NERVES

Recall the rule that *an artery accompanies its companion nerve on the side from which it approaches it.* But in the lower limb, the rule is tested by several exceptions, which, happily, have little other importance (*fig. 28.41*).

Review of Nerve Supplies

See Tables 25.1 (on p. 287) and 28.1 and *Figure 28.42*.

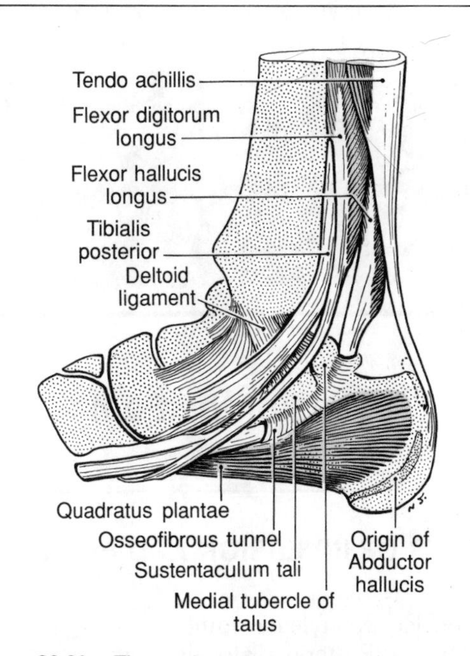

Figure 28.39. The tendons of the 3 deep muscles, at the ankle.

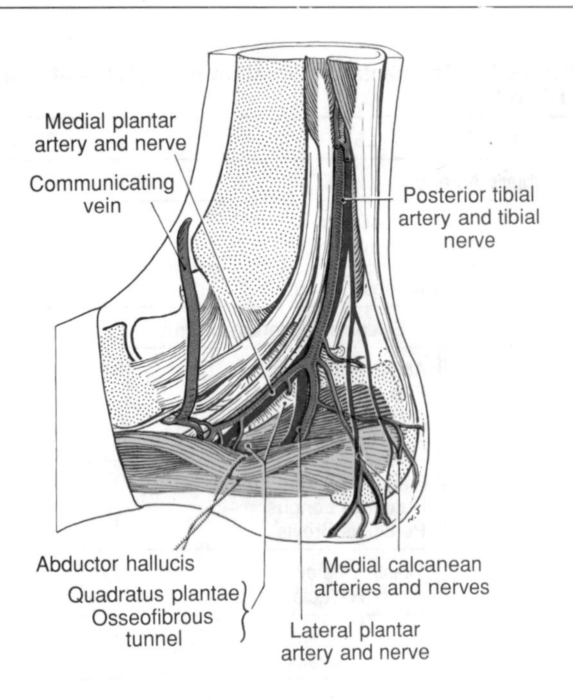

Figure 28.40. "At the door of the foot"—structures passing deep to the abductor hallucis.

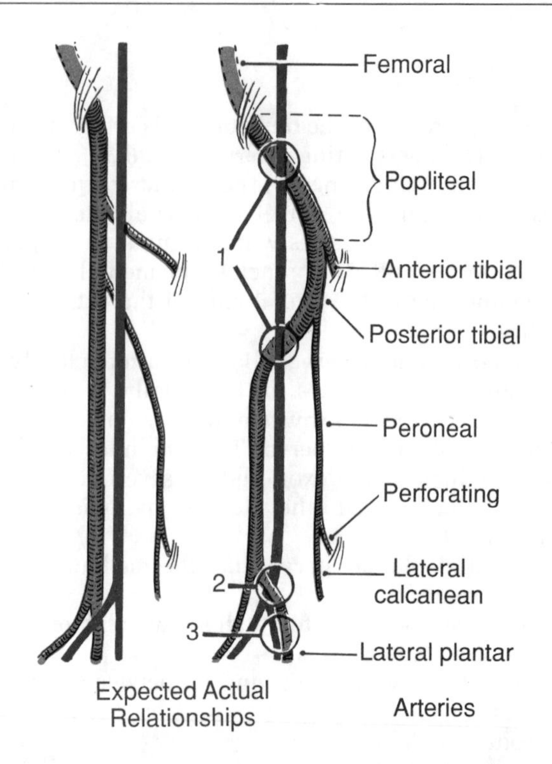

Figure 28.41. Unexpected relationships of arteries to nerves.

Clinical Case 28.4

Patient Dianne W. This infant girl seen today with your tutor in a well-baby clinic is like the patient in Clinical Case 28.1 had been 13 years ago. She has a grossly twisted right foot in marked plantarflexion. Again, a knowledge of the anatomy that follows will help you to understand the various treatments (mechanical, manipulative, surgical) that might be recommended by your tutor.

Joints of the Foot

JOINTS OF INVERSION AND EVERSION

1. Talocalcaneonavicular Joint
   Subtalar ("Posterior Talocalcanean") Joint.
   Talonavicular Joint (and "Anterior Talocalcanean") Joint(s).
2. Calcaneocuboid Joint, which is auxiliary to the talonavicular joint, providing it extra freedom.

Definitions of Movements

Most of the **inversion**, the movement of turning the sole of the foot to face the opposite sole, is a combination of *supination* and *adduction*. Conversely, **eversion** is a combination of *pronation* and *abduction*, and it results in the sole facing slightly laterally. But the movements are complex and the terms are often used loosely, e.g., inversion as synonymous with supination, and eversion with pronation.

Inversion almost automatically recruits plantarflexion of the ankle joint, and eversion is uncomfortable without dorsiflexion. Try it!

Inversion and eversion are movements of 2 composite joints: *the talus*; here the articular surfaces involved are all the facets inferior and anterior to the talus and (as seen in *fig. 28.43*) all 3 facets above the calcaneus and the concave facet behind the navicular; and (2) *of the anterior part of the foot on the posterior part* at the *transverse tarsal joint*.

The axes of rotation of these 2 joints are so similar that they may be thought of as a single "oblique hinge," of which the axis runs obliquely posteroinferiorly, and laterally from the superomedial aspect of the head of the talus to the inferolateral aspect of the heel (*fig. 28.44*).

Function of These Two Joints. This oblique hinge allows side-to-side swinging of the body, walking on slop-

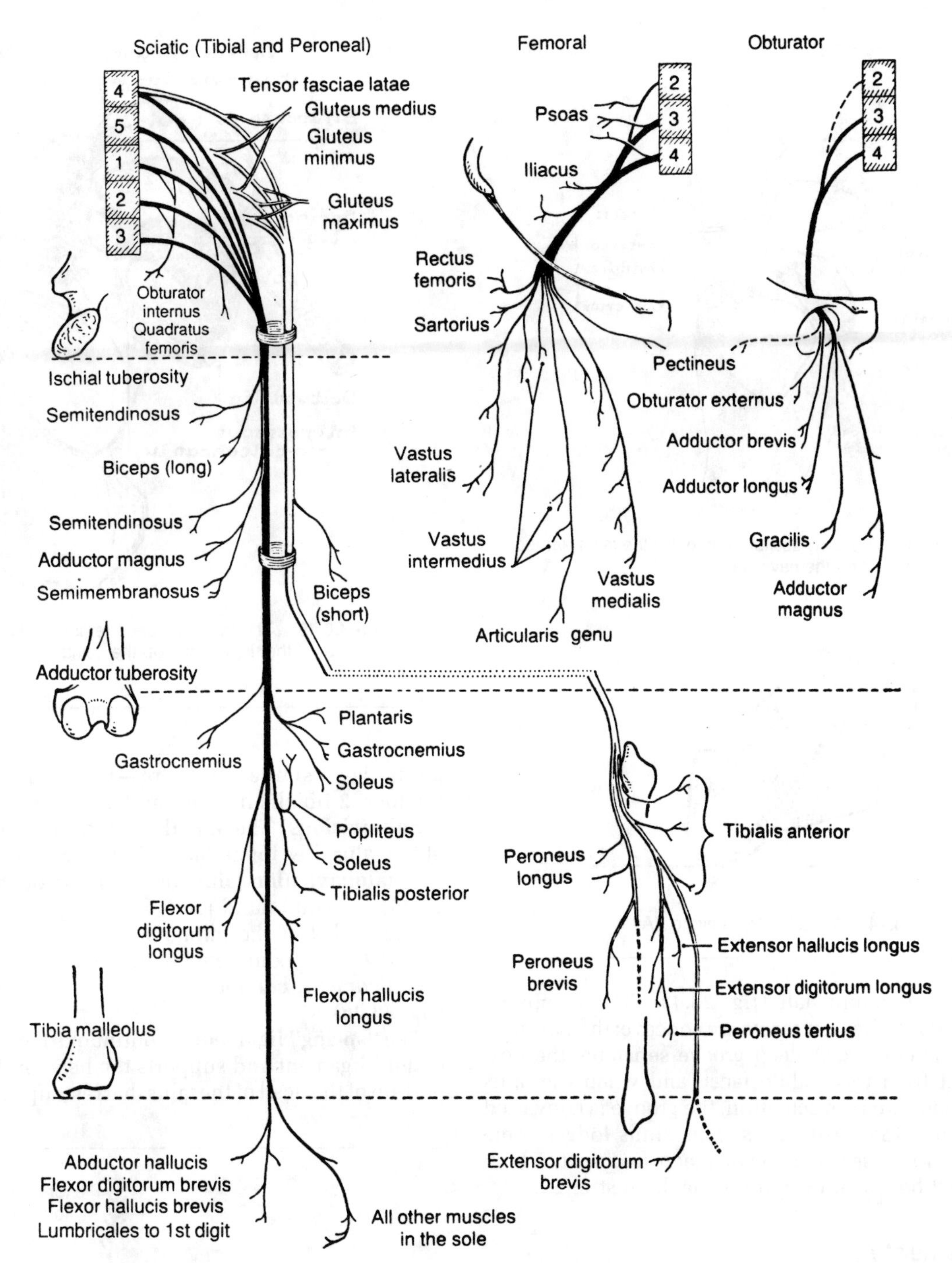

Figure 28.42. The motor distribution of the nerves of the lower limb. (You may find it profitable to compare the levels of origin of the branches in the limb you may be dissecting with these average levels.)

ing surfaces and, above all, control of side-to-side balance. Indeed, without this hinge you cannot maintain side-to-side balance while standing on 1 leg.

TALOCALCANEONAVICULAR JOINTS

Much of the movement of inversion and eversion occurs between the talus and calcaneus (*fig. 28.45*).

Bony Parts Reviewed

The entire body and part of the head of the talus rest upon the anterior two-thirds of the calcaneus and project slightly in front of it (*fig. 28.26*). The *upper surface of the calcaneus* presents a very large posterior talar facet and a small anterior facet, which is usually continuous posteromedially with a larger facet (middle talar facet)

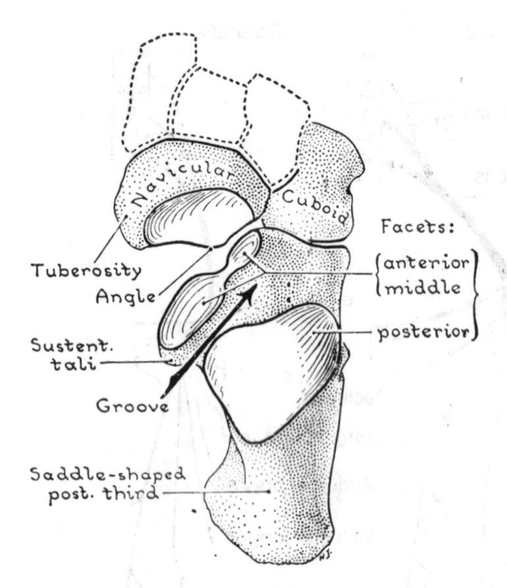

Figure 28.43. The bony socket, or bed, for the talus formed by the calcaneus and the navicular.

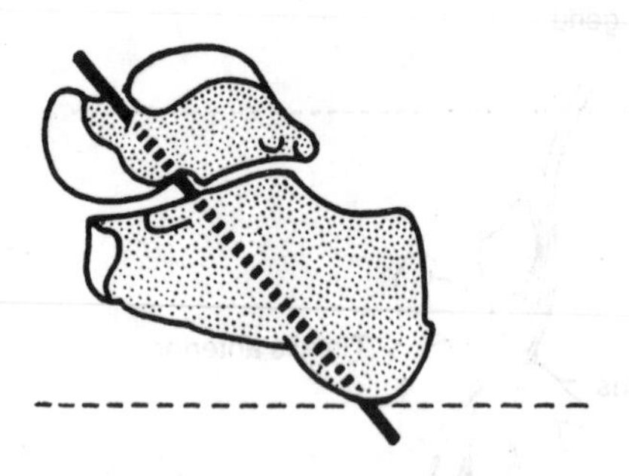

Figure 28.44. Axis of oblique hinge. (After Hicks.)

on the sustentaculum tali (*fig. 28.43*). The combined anterior and middle talar facets are concave; the posterior talar facet is convex. A deep groove separates the posterior facet from the middle facet, and when the talus and calcaneus are in articulation, the groove is converted into a tunnel, the *tarsal sinus*. This sinus lodges some fat and the *interosseous talocalcanean ligament*, which unites the 2 bones and is quite variable in strength.

The Ligaments

The 2 ligaments uniting the calcaneus to the talus are:

1. The interosseous talocalcanean—usually not very strong (*figs. 28.15 and 28.45*).
2. The lateral talocalcanean—a mere slip (*fig. 28.22*).

The 2 ligaments uniting the calcaneus to the bones of the leg—and therefore spanning the talus—are:

3. The calcaneofibular portion of the lateral ligament of the ankle (*fig. 28.22*).
4. The calcaneotibial portion of the medial or deltoid ligament of the ankle (*fig. 28.21*).

The **tendons**, so important as stabilizers of the ankle joint, play a similar role here—flexors and extensors of the toes, 2 tibialis muscles and 2 peronei.

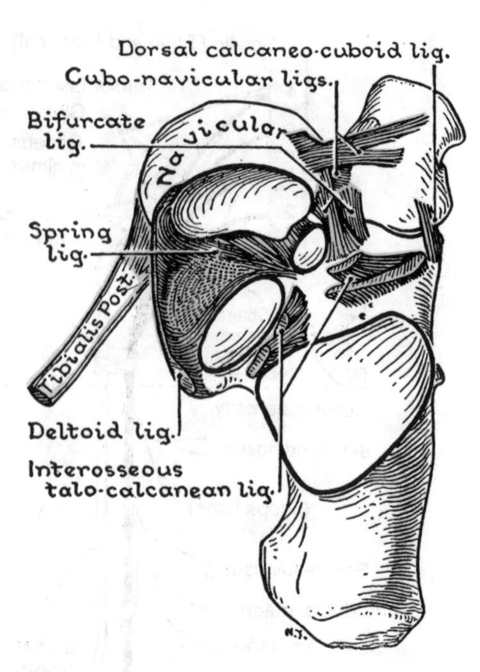

Figure 28.45. The talus has been removed in order to show its bed and the ligaments of the joints of inversion and eversion.

Subdivisions. The **subtalar joint**, inferior to the body of the talus, lies immediately in line with the ankle joint. The **talonavicular joint** for the head of the talus (*fig. 28.45*) is a compound joint.

The *socket of the talonavicular joint* is well seen in *fig. 28.45*. It is completed by the *plantar calcaneonavicular ligament*, commonly called *"the spring ligament"* (*fig. 28.46*).

The **"spring" ligament** is continuous medially with the deltoid ligament and supports the head of the talus. The portion of the head of the talus that rests upon the "spring"

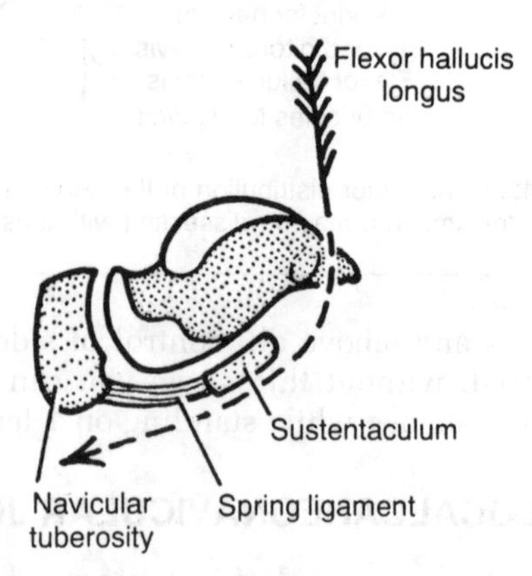

Figure 28.46. The "spring" ligament. The *arrow* indicates the position of the flexor hallucis longus tendon.

ligament is at the summit of the medial arch of the foot. If the tibialis posterior, which is a dynamic assistant to the ligament, is paralyzed, the spring ligament may stretch. This is one type of flat foot (paralytic), but much commoner is the type resulting from primary weakness of the ligaments, which become overstrained and let the head of the talus sag to the ground.

The **bifurcate ligament** (*fig. 28.45*) is really the collateral ligaments for the talonavicular and calcaneocuboid joints.

TRANSVERSE TARSAL JOINT

The *transverse tarsal joint* has 2 component parts; together they account for less than half of all inversion and eversion of the foot:

1. Talonavicular Joint.
2. Calcaneocuboid Joint.

The talonavicular joint has been described above with the talocalcaneonavicular joint, but it must be considered here also. Following surgical fusion (arthrodesis) of the calcaneus and talus (and tibia) into one mass, inversion and eversion are restricted, but not abolished. The 2 muscles of inversion and 3 of eversion are of necessity inserted anterior to the transverse tarsal joint. Various ligaments unite the cuboid to the navicular and cuneiforms, so when the 2 tibial muscles cause inversion, the cuboid must follow.

CALCANEOCUBOID JOINT

The opposed surfaces of the calcaneus and cuboid are sinuous, permitting inversion and eversion also.

The **plantar calcaneocuboid ligament** (or *"short plantar" ligament*) stretches from the inferior surface of the calcaneus to a large area inferior to a ridge on the cuboid (*fig. 28.47*). It belongs solely to the calcaneocuboid joint and is on the same plane as the spring ligament (*fig. 28.48*).

The **long plantar ligament** (*fig. 28.47*) bridges the "short" one and converts the groove on the cuboid into a tunnel for the peroneus longus tendon.

The "short plantar" and spring ligaments form an almost continuous sheet of thick parallel fibers that run anteromedially from calcaneus to cuboid and navicular (*fig. 28.48*).

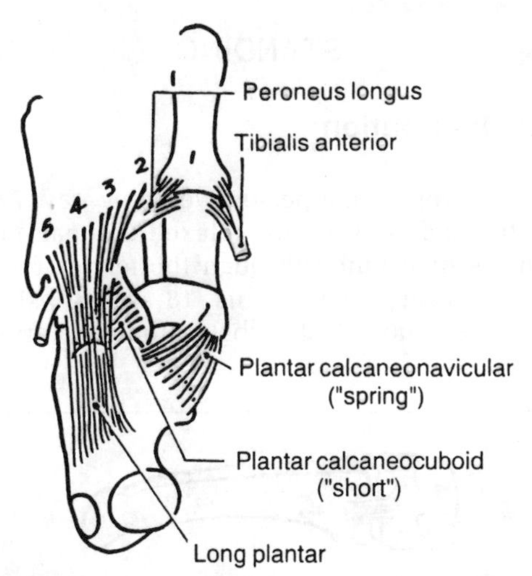

Figure 28.47. Three plantar ligaments and the 2 tendons that form a stirrup.

Clinical Case 28.5

Patient Ella Z. This 48-year-old saleswoman who has worked at that job for 20 years now finds that standing for more than a few minutes produces considerable foot discomfort and that longer periods can be agonizing. Even at a glance she appears to have "flat feet," but a more definitive diagnosis and treatment plan are needed, and you are preparing to discuss the case with a specialist consultant.

Arches of the Foot

The **lateral longitudinal arch** is low (*fig. 28.49*). During walking the lateral border of the foot receives and bears the weight of the body before the medial arch comes into play. It yields or flattens at the hinge surfaces between the cuboid and metatarsals IV and V.

The **medial longitudinal arch** is formed by the calcaneus, "spring" ligament, navicular, 3 cuneiforms, 3 medial metartarsals, and the 2 sesamoid bones. It is a high arch. At its summit lies the head of the talus resting on the spring ligament between the sustentaculum tali and the navicular. By inverting the foot, tibialis posterior (inserted into the tuberosity of the navicular bone) elevates the arch. At the hinge surfaces between the talus and the navicular and also between the other tarsal bones, the spring (or arch) can flatten and recoil.

A so-called "transverse arch" does not exist because the half-arch of the cuboid and cuneiforms (*fig. 28.50*) cannot act as an arch; it has no medial pillar.

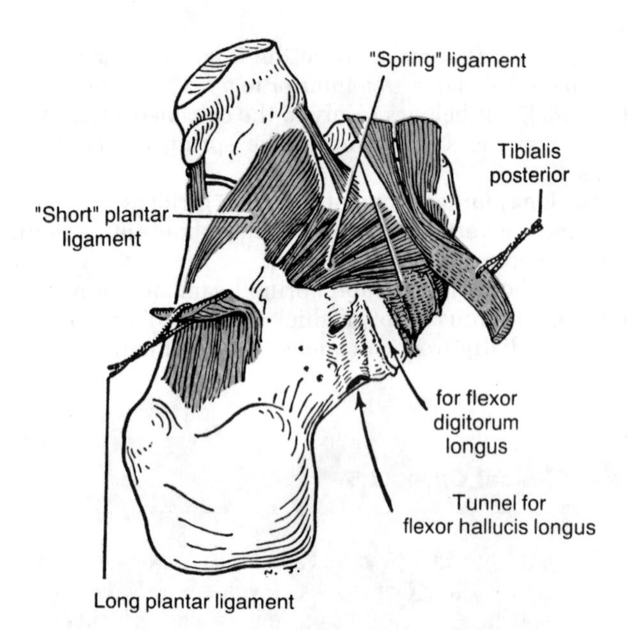

Figure 28.48. Calcaneonavicular and cancaneocuboid ligaments (i.e., spring ligament and short plantar ligament).

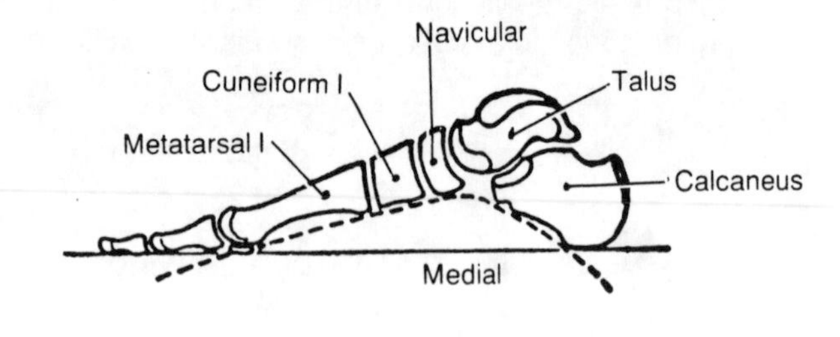

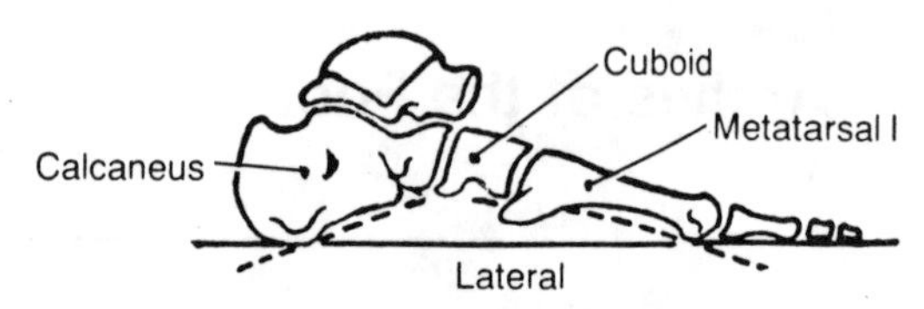

Figure 28.49. The longitudinal arches.

JOINTS DISTAL TO TRANSVERSE TARSAL JOINT

These include: small intertarsal, tarsometatarsal, intermetatarsal, metatarsophalangeal, and interphalangeal joints.

The *metatarsophalangeal* and *interphalangeal joints* are fashioned and supplied with ligaments like the corresponding joints of the hand. However, the *deep transverse metatarsal ligaments* (ligaments of metatarsal heads) extend to the hallux, but in the hand the corresponding ligaments do not, permitting the pollex to move extensively.

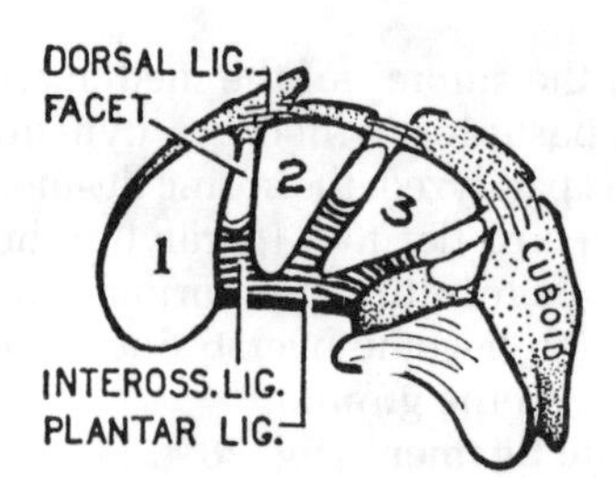

Figure 28.50. The ligaments or ties of the "side-to-side" joints. Note that the cuboid supports the cuneiforms and the navicular.

The *small intertarsal joints* (i.e., between cuboid, navicular, and cuneiforms), the *tarsometatarsal joints*, and the joints between the *bases of metatarsals* may be considered together under the headings: (1) *side-to-side joints* and (2) *end-to-end joints.*

The **side-to-side joints** have strong *plantar ligaments* which act as transverse ties (*fig. 28.50*). These ties are not limited to the plantar aspect but, as *interosseous ligaments*, partly replace the *articular facets.*

The **end-to-end joints** have *strong plantar ligaments, weak dorsal ligaments*, and in some instances, *collateral ligaments* (*figs. 28.50 and 28.51*). The side-to-side facets are commonly continuous with the end-to-end facets.

Arrangement of Ligaments. The ligaments of the foot are so disposed as to resist certain forces, (as in walking and kicking). They share in conducting the thrust via the cuneiforms and navicular to the talus and so to the tibia (*fig. 28.52*).

Dynamics of Foot

STANDING

Weight Distribution

The body weight of a person weighing, say, 72 kg (approx. 160 lbs) and "standing relaxed in a naturally held position" is distributed through the feet as follows: 36 kg (80 lbs) to each foot [of this, 18 kg (40 lbs) is to the heel (calcaneus) and 18 (40 lbs) to the metatarsals].

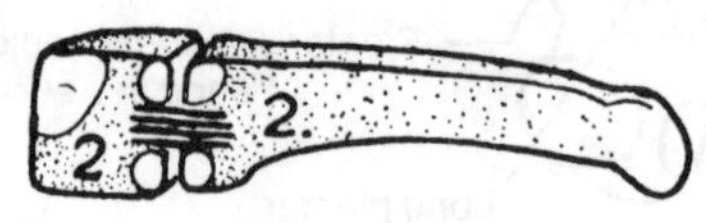

Figure 28.51. The collateral ligament of an "end-to-end" joint.

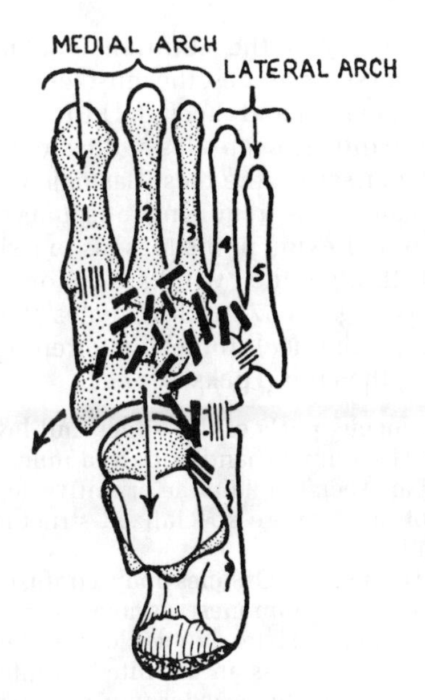

Figure 28.52. The dorsal ligaments of the foot are not scattered in a haphazard way. They collect forces from the toes toward the central axis of the foot to the ankle.

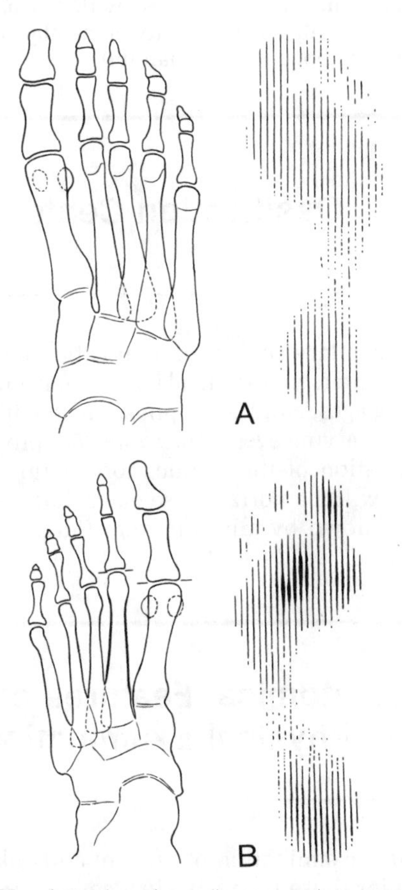

Figure 28.53. *A*, tracing of a radiogram and an accompanying print of a normal right foot made by walking on a corrugated mat; *B*, tracing, and a print of a left foot with a short 1st metatarsal bone. (After Morton.)

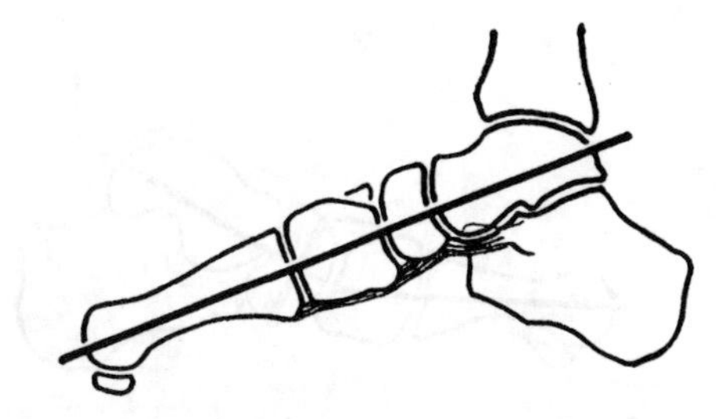

Figure 28.54. Tracing of radiogram of a normal foot, standing. (After Jack.)

The metatarsals have 6 points of equal contact with the ground, namely, the 2 sesamoids inferior to the head of the 1st metatarsal and the heads of the lateral 4 metatarsals, each supporting approximately 3 kg; the 1st metatarsal through its sesamoids supports a double load (Morton). (See *figs. 28.27 and 28.53*).

ARCH SUPPORT

Jones calculated that the plantar ligaments and aponeurosis bear the greatest stress. The chief function of the muscles is to preserve a relative constancy in the ratio of the weight distribution among the heads of the metatarsals.

The height of the arch varies with your posture. Babies have no arch at all; their arches develop with maturity. Indeed, there is great controversy as to whether treatment is either needed or effective—probably the truth lies between.

If the ligaments become lax, the metatarsals being mobile, will largely cease to be weight-distributing (*figs. 28.54, 28.55, and 28.56*).

Any failure of the arch is related to the type and duration of the stress to which it is subjected rather than to the severity of the stress; e.g., athletes and joggers who subject their arches to great stress intermittently will suffer injuries to the bones, joints and tendons, whereas those who stand immobile often develop trouble with their arches.

Electromyographic studies show that tibialis anterior, peroneus longus, and the intrinsic muscles of the foot play no important role in the normal static support of the long arches of the foot. They are commonly completely

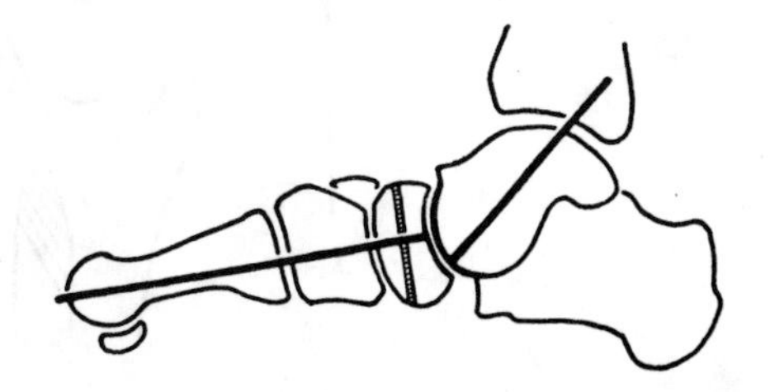

Figure 28.55. One form of flatfoot; talonavicular joint, standing. (After Jack.)

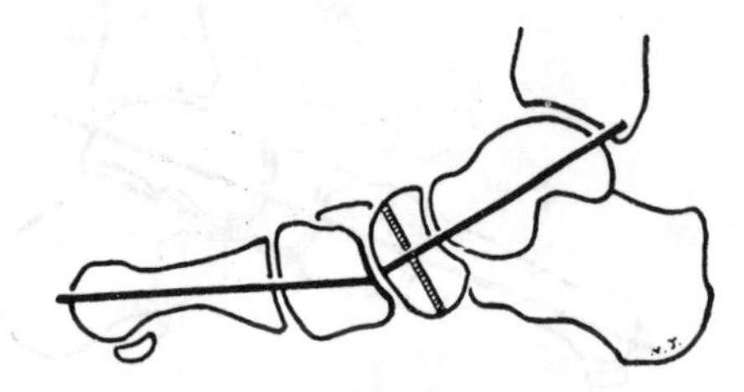

Figure 28.56. Another form of flatfoot; naviculocuneiform joint, standing. (After Jack.)

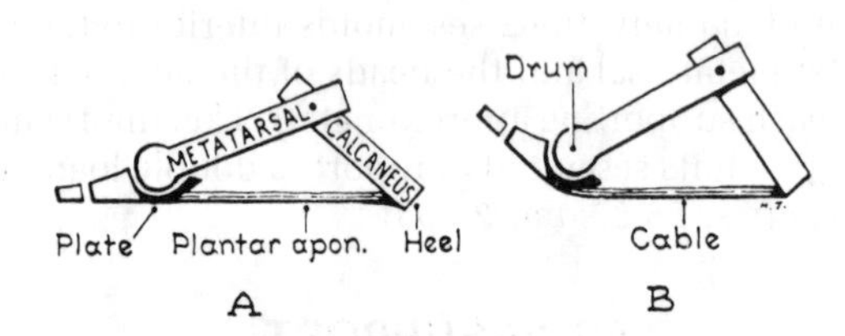

Figure 28.57. Traction through the windlass effect on the plantar aponeurosis caused by dorsiflexing (extending) the toes (either actively or passively) causes heightening of the arch. (After Hicks.)

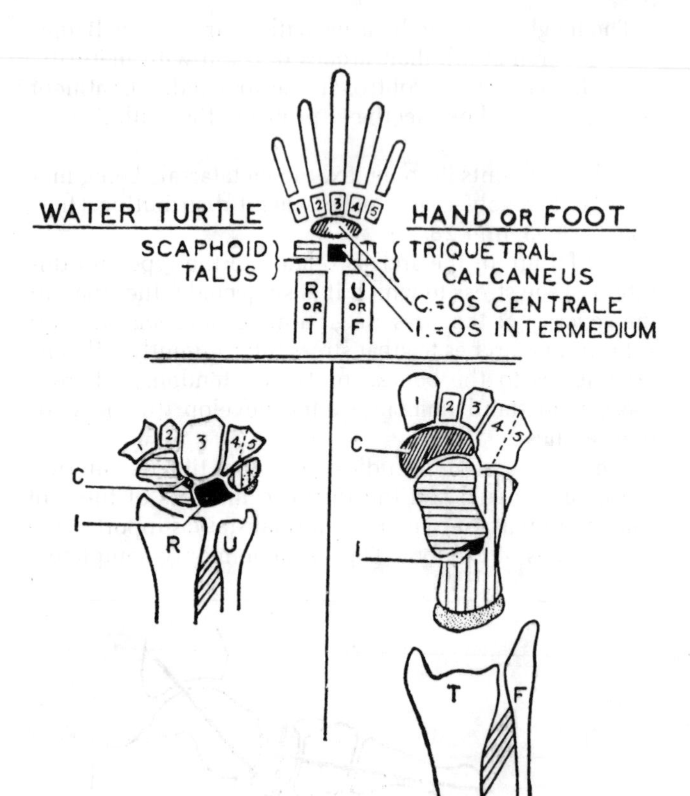

Figure 28.58. Homologies in the bones of the hand and foot compared with the primitive model.

inactive. Even with the addition of abnormally great weights to the static arches, the muscles remain relatively inactive (Basmajian and Stecko). During locomotion it is different, the intrinsic muscles of the foot then being very active as one rises on the toes (Basmajian and Bentzon); this is a greater stress requiring muscular reinforcement.

Passive dorsiflexion of the toes (as in rising on the ball of the foot) through the "windlass action" of the plantar aponeurosis (*fig. 28.57*) approximates the heads of the metatarsals to the heel, thereby shortening the foot and heightening the arch (Hicks).

Homologous parts of the upper and lower limbs (*fig. 28.58*). The human hand and, to a much lesser degree, the foot are seen to retain the primitive generalized plan. The foot is the more specialized structure, contrary to expectations.

Supernumerary Ossicles may confuse you in X-ray pictures. The commonest ossicles are: (1) The *os trigonum* or separate lateral tubercle of the talus (*fig. 28.59*). Phylogenetically it is an ununited os intermedium (*fig. 28.58*). (2) The *tibiale externum* or separate navicular tuberosity. (3) A *bipartite medial cuneiform.* (4) A fibrocartilaginous nodule in the peroneus longus tendon lateral to the cuboid is common, and commonly it is ossified. (5) A sesamoid bone in the tibialis posterior tendon—not to be confused with a separate navicular tuberosity. (6) The tuberosity of metatarsal V, existing as a separate bone (os vesalianum), is rare.

Ossification Centers

The secondary centers of ossification (epiphyses) and primary centers for the tarsal bones appear in radiographs at varying ages, and the epiphyses fuse with the primary centers at varying ages also. *Table 28.2* provides a useful approximation of these times for reference. Only a radiologist will memorize these dates, but the general rule must be known by all: *important (large) epiphyses appear "early" and fuse "late."*

Anatomical Features of the Physical Examination

Physical examination of the foot and ankle becomes a travesty for those who have inadequate training in anatomy. For the well trained, visual inspection of the resting and standing foot and careful palpation of the resting foot

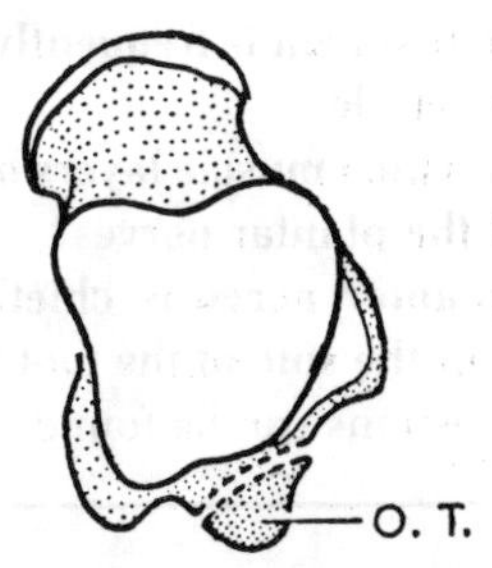

Figure 28.59. Os trigonum. [In 558 paired lower limbs (i.e., 279 adult cadavers) 7.7% had one. It occurred twice as commonly unilaterally as bilaterally (Storton).]

can be almost as informative as multiple radiographs (Clinical Cases 28.1, 28.2, and 28.4). Even with radiographs, a clear knowledge of anatomy is essential, and most radiologists are excellent anatomists of the skeletal systems. However, there are also tendons to be seen and felt (Clinical Cases 28.3 and 28.5), movements to be tested, the pulses of two arteries to be assessed, and the skin's condition and sensory supply to be judged. None of these activities is trivial, and all depend on anatomy covered in Chapter 28.

A line joining the middles of the medial and lateral edges of the foot finds all 7 tarsal bones posterior to it and all long bones anterior. The transverse tarsal joint crossing anterior to the talus and calcaneus leaves the 5 smaller tarsal bones (cuboid, navicular and 3 cuneiforms) in the middle of the foot. Only flexion and extension of the ankle occur between the talus and the bones of the leg, inversion and eversion occurring at the subtalar joint inferior to the body of the talus and the transverse tarsal joint. The calcaneus carries weight to the ground at the heel (through a cushion of encapsulated fat), and it articulates with the cuboid near the middle of the lateral edge of the foot. All clinicians should have this basic information whenever they look at and feel a patient's foot or view a radiograph of the region.

Essential plantar flexor muscles are the gastrocnemius and the soleus, which attach not to the talus but to the projecting calcaneus through the tendo achillis; the essential dorsiflexor is tibialis anterior. With the stronger invertor, tibialis posterior, tibialis anterior also inverts the foot at tarsal joints for inversion and eversion. The strongest ligament is the medial collateral connecting the tibial malleolus to both the talus and calcaneus. Three parts of the lateral ligament run to these bones, the anterior talofibular being the most vulnerable (cf., the first patient in Clinical Case 28.2).

You cannot consider yourself a good clinician if you cannot easily locate the following by palpation: *On the medial side*—(1) tibial malleolus; (2) posterior tibial artery pulse; (3) sustentaculum tali; (4) tuberosity of the navicular; (5) head of the talus; (6) paired sesamoid bones under the head of metatarsal 1; (7) tibiales posterior and

Table 28.2. Table of Appearance and Fusion Times of Epiphyses in Lower Limb including Primary Centers for Tarsus (in Years Unless Otherwise Stated)*

	Appears	Fuses
Hip bone, 3 primary parts	Before birth	16†
Ischium and pubis	Before birth	9
Ischial tuberosity	19	20
Iliac crest	16	22
Femur, head	1	18†
Greater trochanter	−5	18†
Lesser trochanter	−14	18†
Lower end	Birth	19†
Tibia—upper end	Birth	19
Lower end	2	18
Fibula—Upper end	5	19
Lower end	2	18
Calcaneus—before birth	−30 weeks‡	
Tuberosity	11	17
Talus—before birth	−30 weeks	
Navicular	3	
Cuboid—before birth	+30 weeks	
1st Cuneiform	3	
2nd Cuneiform	−4	
3rd Cuneiform	1	
Great toe metatarsal base	−3	18
2nd–5th metatarsal heads	−5	18
Great toe, proximal phalanx base	−4	19
Distal phalanx base	2	19
2nd–5th toes—proximal phalanx base	−5	19
middle phalanx base	−5	19
distal phalanx	−5	19
Patella	−5	
Sesamoids of great toe	−12	22†

*Since you will be concerned with the age periods at which you may reasonably be sure that fusion is complete, the latest times of fusion are given. Female bones fuse distinctly earlier.

†Times that are found to be quite constant. The table is a compilation from several authorities and is based on X-ray findings. Definite ages are given rather than the actual spread. Figures denote years unless otherwise stated.

‡+ = a later tendency; − = earlier tendency.

anterior tendons; (8) extensor hallucis longus tendon; and (9) abductor hallucis muscle. *On the lateral side*—(1) lateral fibular malleolus; (2) peronei tendons; (3) tuberosity of metatarsal 5; (4) extensor digitorum longus tendon and the belly of extensor digitorum brevis muscle. In the *midline of the dorsum* identify the dorsalis pedis artery pulse, and medial to it the tendons of tibialis anterior and extensor hallucis longus.

Inversion and eversion occur at (a) the subtalar joint (between the body of the talus and the posterior articular facet of the calcaneus) and (b) the associated talonavicular and calcaneocuboid joints that sinuously cross the foot and so form the transverse tarsal joint. Inversion and eversion are produced by muscles inserted anterior to the transverse tarsal joint: inversion—tibiales posterior and anterior; eversion by three peronei (longus, brevis and tertius). Orthopedic surgeons who treat club feet (as in Clinical Cases 28.1 and 28.4) and related conditions and all who deal with fractures depend heavily on clear anatomical knowledge of this area.

Clinical Mini-Problems

1. **Which major ligament supports the head of the talus as it attaches to the bones forming the medial longitudinal arch of the foot?**
2. **Stress ("march") fractures are common in the longest metatarsal. Which metatarsal is that?**
3. **Which ankle ligament is frequently traumatized in a "sprained" ankle?**
4. **(a) Between which muscle layers of the sole would one find the plantar nerves?**
   (b) Which plantar nerve is chiefly motor to the muscles in the sole of the foot?

(Answers to these questions can be found on p. 586.)

Section

Upper Limb and Back 6

29

Back and Scapular Regions

Clinical Case 29.1

Patient Arnold R. This 38-year-old chartered accountant has come into the hospital emergency room at midnight complaining of severe stiffness, soreness, and pain between his shoulder blades whenever he turns his head and neck or lifts his right arm overhead. The problem began slowly during a tennis match this afternoon but became so intense tonight that he sought emergency care. He had never had a similar problem before. Before calling the surgical resident who has gone to bed, you must assess the anatomical parts involved.

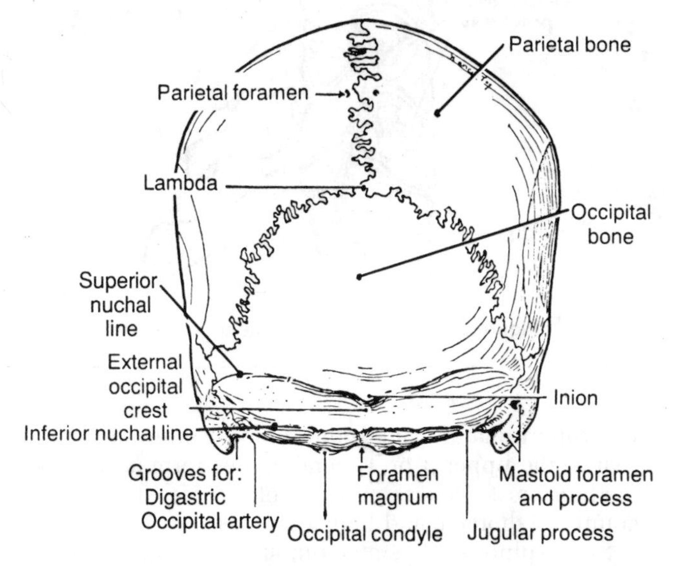

Figure 29.1. The skull, from behind. (Norma occipitalis of anthropologists.)

So much dissection of the muscles of the upper limb is required to approach the intrinsic muscles of the vertebral column that they are referred to "officially" as the first 2 layers of the muscles of the back. Because the intrinsic muscles attach to the back of the skull, that area is described here too. However, the *suboccipital region* rightly belongs with the head and neck (Chapter 42).

Skeletal Landmarks

SKULL FROM BEHIND

The **outline** is horseshoe-shaped and extends from one mastoid process over the vault to the other (*fig. 29.1*). On each side, it crosses the mastoid and parietal bones. Across the base of the skull, the outline is nearly horizontal, crossing the occipital condyles and the foramen magnum.

The **nuchal area** (nucha = neck) is convex and includes parts of the parietal, occipital, and temporal bones. At the center is the *lambda*, from which the sagittal (interparietal) suture runs up and over the vertex, and the lambdoid (parieto-occipital) suture down to the postero-inferior corner of the parietal bone on each side. (Hence the name *lambda*, which refers to the Greek letter shaped like an inverted Y.)

Palpate in the midline the *inion* or *external occipital protuberance*. From it the *superior nuchal line* curves to the *mastoid process* and the *external occipital crest* for the *ligamentum nuchae* runs to the foramen magnum.

VERTEBRAL COLUMN FROM BEHIND

Spinous Processes

C1 (the atlas) has only a posterior tubercle; C2–C6 spines are usually bifid (*fig. 29.2*). They are not easily felt. C7 is prominent but less so than T1; T5–8 are almost perpendicular, and are markedly overlapping; T10 is often

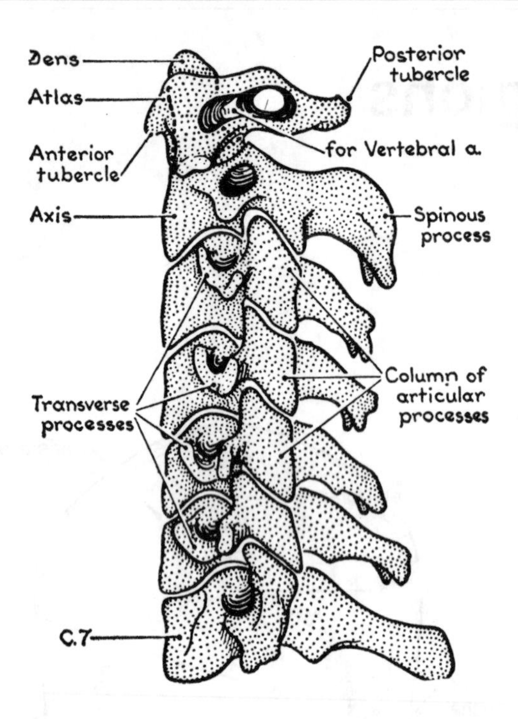

Figure 29.2. Cervical vertebrae on side view.

disproportionately small; the lumbar spines are oblong plates; the upper 3 or 4 sacral spines form an irregular *median crest.* You should be able to feel the tips and count the thoracic and lumbar spines.

Supraspinous and interspinous ligaments unite the spines. In the neck, the supraspinous ligament forms the **ligamentum nuchae**, which attaches to the cervical spines. Superiorly, it is attached to the inion and external occipital crest; and posteriorly, it is subcutaneous and palpable when stretched by strong flexion of the neck anteriorly.

The **laminae** tend to overlap like shingles and are united to each other by *ligamenta flava.* Their inferolateral angles project down as the inferior articular processes.

Superior and inferior to the slender posterior arch of the atlas there is a wide interlaminar space closed by membranes (posterior atlanto-occipital and atlantoaxial). The lumbar interlaminar spaces also are wide. Flexion of the spine, of course, enlarges all of them. The upper sacral laminae are fused while the lower ones and the coccygeal laminae are absent (*fig. 29.3*).

The **articular processes** for cervical vertebrae, except for the atlanto-occipital and atlantoaxial joints, look like segments of a round vertical column cut obliquely (*fig. 29.2*).

On the back of the rim of each lumbar superior articular process there is a large tubercle, the *mamillary process.* It gives origin to the multifidus.

The **transverse processes** of C1 and C7 project far beyond those of C2–6 (*fig. 42.4*). T1–12 diminish progressively in projection. L1–5 project farther than C1, L3 being the most projecting of all transverse processes.

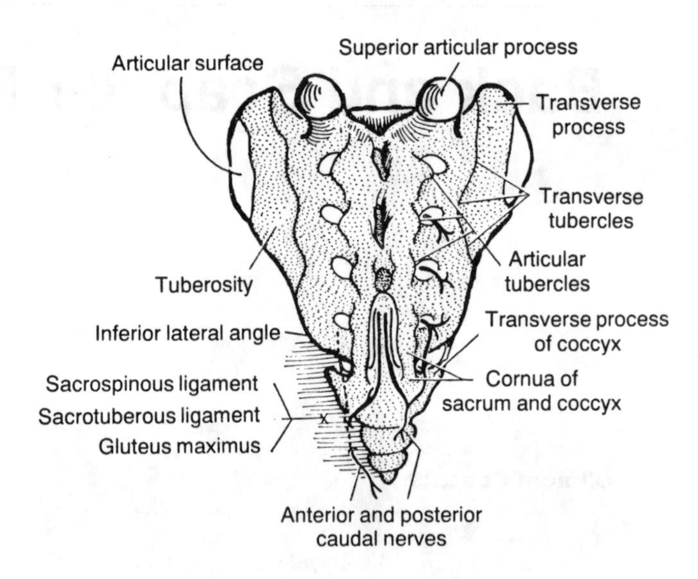

Figure 29.3. Dorsal aspect of sacrum and coccyx.

The sacral transverse and articular processes fuse to form irregular crests.

The **posterior sacral foramina** (*fig. 29.3*) communicate with the sacral canal.

RIBS AND ILIAC CREST

The angles are marked by the attachment of the iliocostalis.

Iliac Crest

The posterior 8 cm of the iliac crest gives part origin to the erector spinae.

Scapula

The scapula is a flat bone with 2 surfaces, a *costal* (or *anterior*) and a *dorsal* (or *posterior*). It is triangular in shape, with 3 borders and 3 angles, and it has 2 processes, the *coracoid process* and the *spine*, which ends as the *acromion* (*fig. 29.4*).

The *spine* of the scapula crosses most of the dorsal surface, dividing it into a smaller *supraspinatus* and a larger *infraspinatus* fossa. These 2 fossae communicate with each other at the "*spinoglenoid notch*," which lies between the lateral border of the spine and the *glenoid cavity*. The free border or *crest of the spine* ends superior

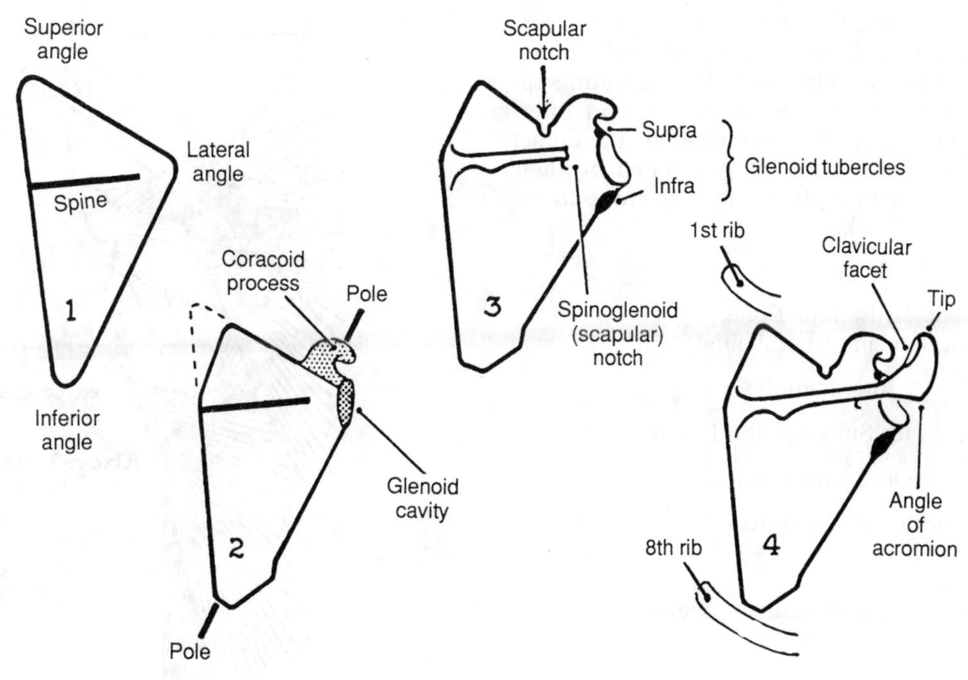

Figure 29.4. The progressive stages in sketching a scapula, from behind.

to the shoulder joint in a free, flattened, and expanded piece of bone, the *acromion*. On its medial border near its tip, the acromion has a small oval beveled facet for articulation with the flat acromial end of the clavicle.

PALPABLE PARTS

The crest of the spine, the acromion, and the clavicle form the limbs of a V which you can and must learn to palpate easily in patients from end to end.

The *acromioclavicular joint* is likewise subcutaneous and can be easily located. Sometimes it is a visible bump.

The *medial border* of the scapula is covered in its superior half by the trapezius muscle, but you can easily palpate it from the *inferior angle* to the *superior angle*. The *lateral border* you can trace with some difficulty. The inferior angle and the tip of the coracoid are, so to speak, at opposite "*poles*" of the scapula. By grasping the inferior angle with one hand and palpating the coracoid process with 2 fingers of the other, you can, with a well relaxed subject, manipulate the scapula in the same manner as a surgeon does to reveal a fracture between the "*two poles*."

The *medial border* is almost parallel to the spines of the vertebrae and is about 5 cm distant from them. It crosses ribs 2–7.

TO MEASURE THE LIMB

The *angle of the acromion* is the usual point from which to measure the length of the limb, since you can readily palpate it between your thumb and index finger.

Cutaneous Nerves of the Back

Each spinal nerve divides into ventral (anterior) and dorsal (posterior) rami. A typical dorsal ramus divides into a medial and lateral branch (see *fig. 30.20*). Both branches supply muscles, and then one or the other of the branches becomes cutaneous. Above the midthoracic region, the medial branches become cutaneous close to the median plane; below, the lateral branches become cutaneous at some distance from the median plane. The cutaneous branch of T2 extends to the shoulder area and is the longest of the dorsal rami.

The area of skin supplied by the sensory terminal branches of both ventral and dorsal rami of a single dorsal (i.e., sensory) nerve root is called a *dermatome*. On the trunk, the dermatomes form obliquely encircling bands (*fig. 2.15*).

Muscle Layers of the Back

The muscles of the back are arranged in superficial, intermediate, and deep groups. The *superficial group* acts upon the upper limb and the *intermediate group* is respiratory in action. Both have spread posteriorly across the *deep group* which are intrinsic to the back and supplied by dorsal nerve rami. The superficial and intermediate group muscles are not innervated by the dorsal rami of the spinal nerves, but are penetrated by their cutaneous branches that pass from the deep group to the overlying skin on the back.

Superficial

1. Trapezius and Latissimus Dorsi
2. Levator Scapulae and Rhomboidei

Intermediate

3. Serratus Posterior Superior and Inferior

Deep (Intrinsic)

4. Splenius (Cervicis and Capitis)
5. *Longitudinal muscles*
   Erector Spinae:
   Iliocostalis, Longissimus, Spinalis
6. *Oblique muscles*
   Transversospinalis:
   Semispinalis, Multifidus, Rotatores
7. *Remaining deep muscles*:
   Interspinales, Intertransversarii, Levatores Costarum, Suboccipital Muscles.

It is **profitless to memorize** even for a short period the reference book description of the attachments of the muscles of the back because you can make *no use of that particular information*. The back is, however, a region of such great importance that you must acquire a good grasp of the scheme of things.

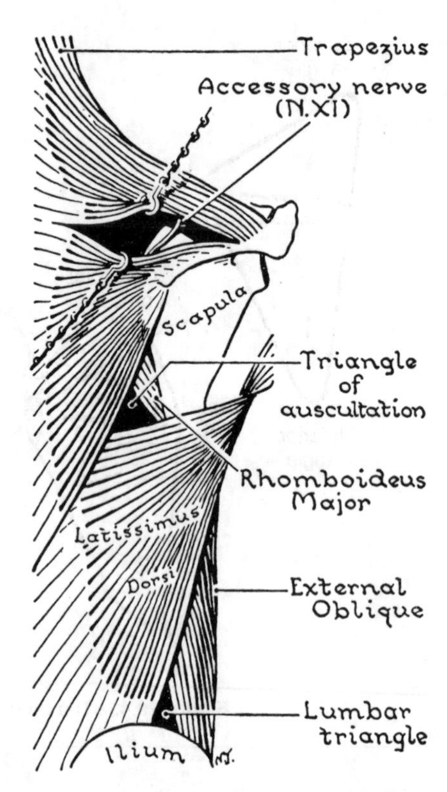

Figure 29.5. Exposure of superior border of scapula. Two triangles.

First Layer of Muscles of the Back

LATISSIMUS DORSI AND TRAPEZIUS

Latissimus Dorsi (*fig. 29.5*)

Origin

This fan-shaped muscle arises by an aponeurosis from the inferior 6 thoracic, all lumbar, and sacral spines, and the outer lip of the iliac crest. This aponeurosis is the *thoracolumbar fascia*. The latissimus dorsi also arises from ribs 9–12.

Insertion and Nerve Supply

See pages 358 and 361 (*Table 30.2*).

Actions and Functions

It extends, medially rotates, and adducts the humerus. It brings the outstretched arm from above the head to behind the back, as in swimming, raising the body when one hangs from a horizontal bar, rowing, pulling, or elbowing one's way through a crowd.

Two triangles with some minor clinical interest are associated with latissimus dorsi (*Fig. 29.5*). They are the *triangle of auscultation* superiorly and the *lumbar triangle* inferiorly.

Trapezius

This, too, is a triangular muscle. The trapezii of the 2 sides together form a trapezoid shape, hence the name.

The *origin* of the trapezius is from the skull and all the spinous processes of the cervical vertebrae via the ligamentum nuchae and all 12 thoracic spines.

Insertion

The superior fibers of the muscle run down to the lateral third of the clavicle; the intermediate fibers pass horizontally to the acromion and the crest of the spine of the scapula; the inferior fibers converge to form an aponeurosis which is inserted into the well-marked tubercle on the inferior lip of the crest of the spine.

Functions

The trapezii are the suspensory muscles of the shoulder girdles (*see fig. 30.4*). The inferior fibers assist the superior fibers to rotate the scapula on the chest wall by pulling on their tubercle of insertion to depress the medial border. In health they square the shoulders and elevate the shoulder girdles. Their function can be tested by asking the patient to "shrug" the shoulders.

Nerve Supply

The trapezius is supplied on its deep surface by (1) the accessory nerve (external branch), which arises from spinal cord segments C1–5 and takes a devious course (*fig. 43.8*), and (2) by the ventral rami of C3 and 4, which take a direct course and are not purely sensory in man as they are in some species (Corbin and Harrison; Wookey).

The **upper border of the scapula** lies deep to the trapezius and is the key to the suprascapular region. It extends from the superior angle, where the levator scapulae is inserted, to the superior part of the glenoid cavity, where the long head of the biceps arises from the *supraglenoid tubercle* (*fig. 29.4*). Laterally, it is abruptly deeper—the *scapular notch* (suprascapular notch). This notch is bridged by a sharp, taut band, the *transverse scapular ligament* (suprascapular ligament). Between the notch and the supraglenoid tubercle, the border gives rise to the stout *coracoid process*.

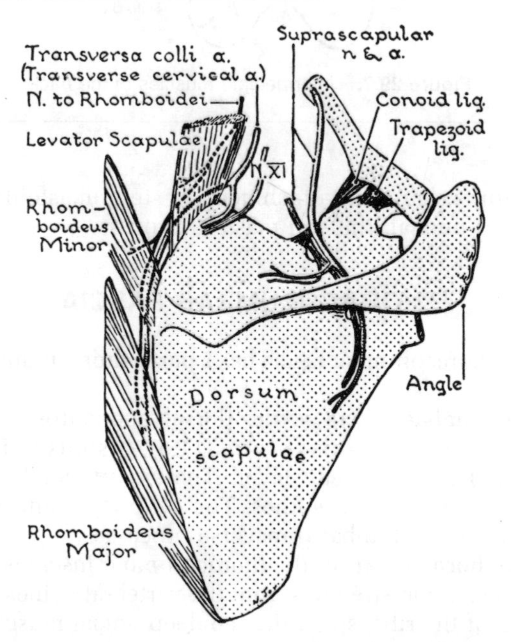

Figure 29.6. Superior and medial borders of scapula.

Suprascapular Vessels and Nerve

The vessels cross superior to the ligament, the nerve, inferior to it. On the dorsum of the scapula, they lie in contact with the bone, in the supraspinatus fossa, through the spinoglenoid notch, and into the infraspinatus fossa (*fig. 29.6*).

Distribution

The nerve supplies the supraspinatus and infraspinatus, and it sends important sensory twigs to the shoulder joint. Shoulder joint pain is a common and important clinical problem.

Second Layer of Muscles of the Back

LEVATOR SCAPULAE AND RHOMBOIDEI

When the trapezius is reflected, the **levator scapulae** and the **two rhomboids** are exposed, occupying the whole length of the medial border of the scapula (*fig. 29.6*).

Origins

The *levator scapulae* arises from the transverse processes of the cervical vertebrate 1–4 (their posterior tubercles). The 2 *rhomboids* arise from the ligamentum nuchae and the spines of cervical and first 4 thoracic vertebrae.

Insertions

From the superior angle and along the whole length of the medial border, as seen in *Figure 29.6*.

Nerve Supply

Levator: C3, 4 and twigs from 5. Rhomboids: C5 (as dorsal scapular nerve) (*fig. 29.6*).

TRANSVERSA COLLI ARTERY (*fig. 29.6*)

This artery (transverse cervical) runs posteriorly in the neck to the anterior border of the levator scapulae, where it divides. A *superficial branch* accompanies the *accessory nerve* while the *deep branch* accompanies the *nerve to the rhomboids.*

Clinical Case 29.2

Patient Barbara S. While you are an observer in an orthopedic clinic, this school girl, aged 14, is referred for assessment of an increasingly obvious S-shaped lateral deformity of the spine. This is easily diagnosed by the surgeons as idiopathic scoliosis. After Barbara and her parents leave for the Radiology Department, you must take an active part in a discussion of the anatomy and possible pathomechanics of the disturbance. Not only are the bones important but also the muscles and joints, and the nerves that supply them.

Intermediate Muscles and Splenius

The deep muscles are bridged by the *serrati* in the thoracic region, by the *lumbar fascia* in the lumbosacral region, and by the *splenius* in the neck.

The **serratus posterior** (*fig. 30.7*) is divided into the *serratus posterior superior* and the *serratus posterior inferior.* The latter blends with the thoracolumbar fascia and through it gains attachment to the spines T12–L2 (*fig. 29.7*)

Both parts of this muscle elongate the thoracic cavity and may act as muscles of inspiration. They are migrants, still supplied by ventral nerve rami.

The **splenius** (*fig. 29.7*) is wrapped around the other deep muscles in the neck, as its name implies (L. splenius = a bandage). It arises from the inferior half of the ligamentum nuchae and from thoracic spines T1–5. It separates into 2 parts—splenius cervicis and splenius capitis.

The *splenius cervicis* joins the levator scapulae to share its attachments to transverse processes C1–4. The *splenius capitis* shares the attachments of sternomastoid to the superior nuchal line and mastoid process.

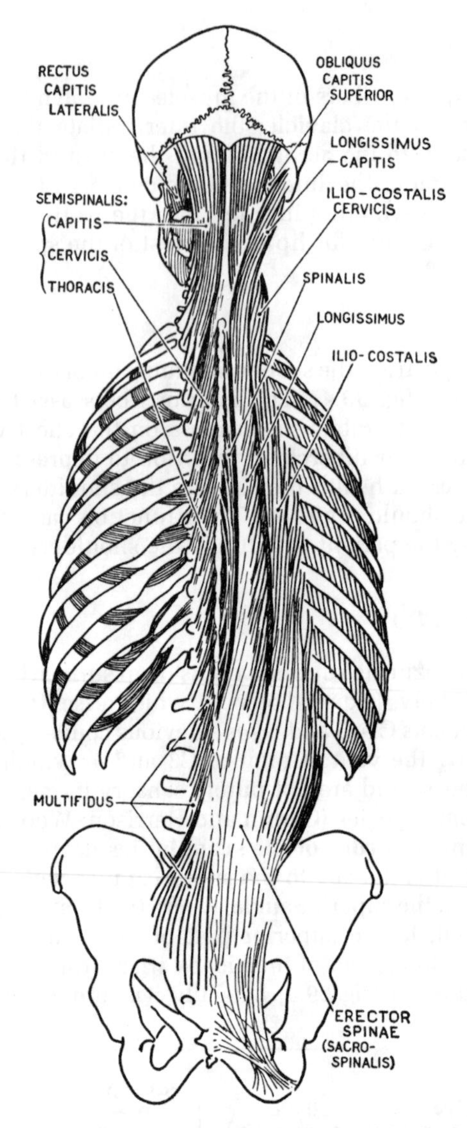

Figure 29.7. Intermediate muscles of the back.

THORACOLUMBAR FASCIA

The thoracolumbar fascia has 2 parts: thoracic and lumbar.

The **lumbar fascia** is the dorsal aponeurosis of the transversus abdominis (*see fig. 11.5*); it splits to form a superficial and a deep layer. The *superficial layer* attaches to the lumbar spines. The *deep layer* attaches to the tips of the lumbar transverse processes.

The **thoracic part** of the *thoracolumbar fascia* is a delicate sheet that stretches from the vertebral spines to the angles of the ribs, spanning the deep dorsal muscles.

Deep or Intrinsic Muscles

ERECTOR SPINAE (SACROSPINALIS) (*fig. 29.8*)

This muscle extends from the pelvis to the skull. It has a dense aponeurotic origin, which covers the fleshy origin of the multifidus and arises as shown in *Figure 29.9*.

Near the last rib, the erector spinae splits into 3 columns: iliocostalis, longissimus, and spinalis (*figs. 29.9 and 29.10*).

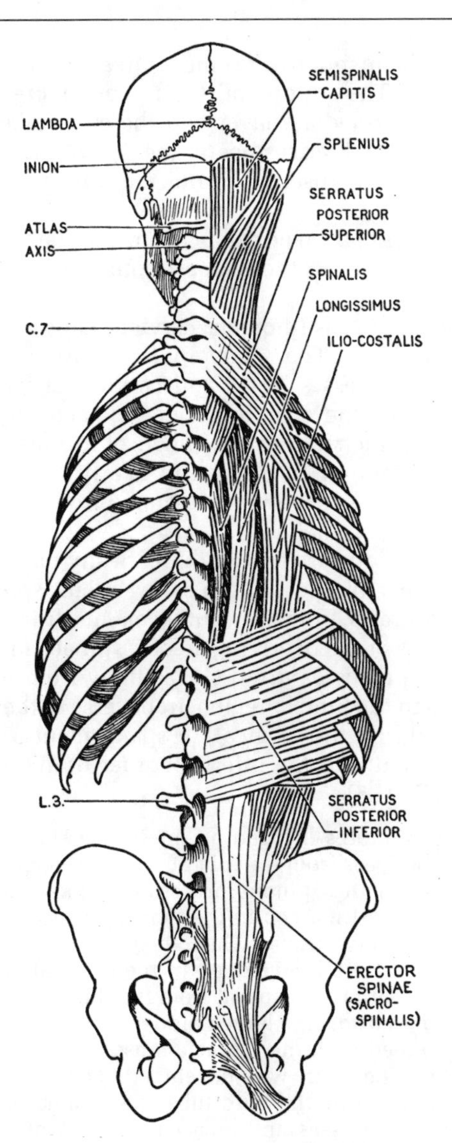

Figure 29.8. Deep muscles of the back.

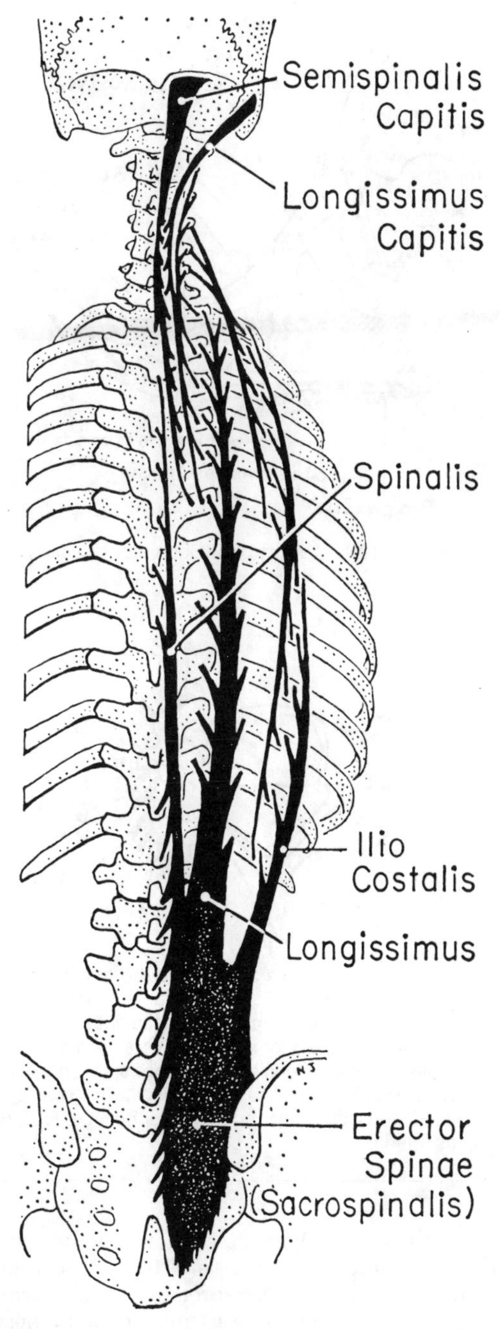

Figure 29.9. Plan of the erector spinae and semispinalis capitis.

Details of the crowded muscular attachments to cervical transverse and articular processes are shown for special interest groups in *Figures 29.11* and *29.12*.

The *iliocostalis* is inserted into the angles of the ribs and into the cervical transverse processes (C4–6) by a series of relayed bundles that extend over about 6 segments—where one slip is inserted another slip arises on its medial side.

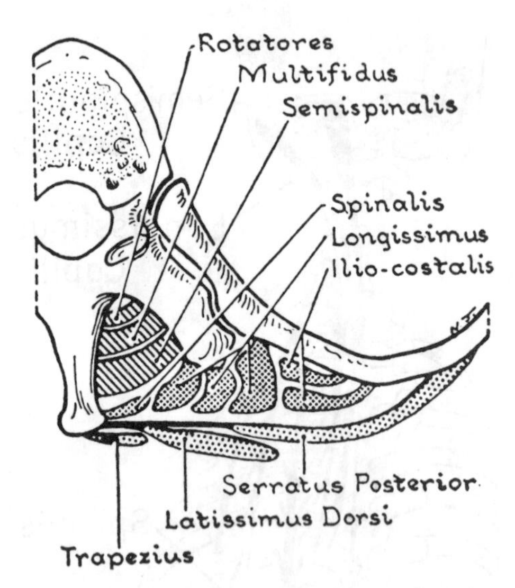

Figure 29.10. Muscles of back on cross-section.

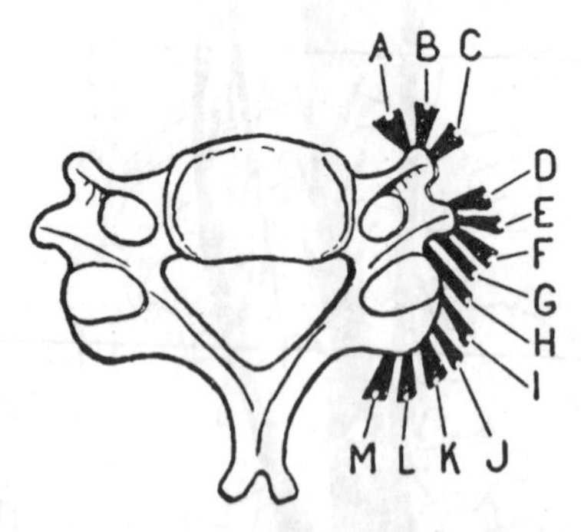

Figure 29.11. The muscles attached to the transverse and articular processes of a cervical vertebra. *A*, longus colli; *B*, longus capitis; *C*, scalenus anterior; *D*, scalenus medius; *E*, scalenus posterior; *F*, levator scapulae; *G*, splenius cervicis; *H*, iliocostalis cervicis; *I*, longissimus cervicis; *J*, longissimus capitis; *K*, semispinalis capitis; *L*, semispinalis cervicis; *M*, multifidus. IO and SO, inferior and superior oblique; RCI, rectus capitis inferior.

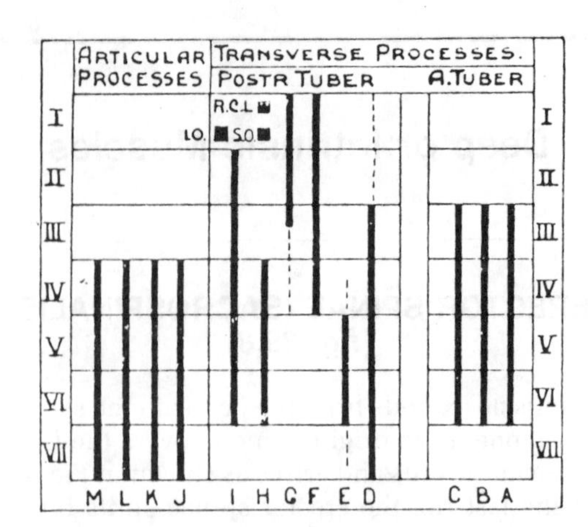

Figure 29.12. Graphic representation of *Figure 29.11*.

The *longissimus* (thoracis, cervicis, and capitis) is inserted into lumbar accessory and transverse processes, and into thoracic transverse processes and nearby parts of the ribs (T2–12). Bundles arising medial to these (T1–4) are relayed to the cervical transverse processes (C2–6). Other bundles arising medial to these again (thoracic transverse and cervical articular processes) extend to the mastoid process deep to the splenius capitis and sternomastoid.

The *spinalis* is largely ligamentous and narrow. It is found only in the *thoracic region*.

TRANSVERSOSPINALIS

This oblique group of muscles is concealed by the erector spinae. Its fibers pass obliquely superomedially from transverse processes to spines. It is disposed in three layers: (1) *semispinalis*, (2) *multifidus*, and (3) *rotatores* (*fig. 29.10*). The superficial layer spans more segments than the intermediate layer and therefore takes a more vertical course. It arises nearer the tips of the transverse processes and is inserted nearer the tips of the spinous processes.

The superficial layer spans about 5 segments; the intermediate about 3; the deepest connects adjacent segments.

The *semispinalis* (thoracis, cervicis, capitis) forms the superficial layer. The *semispinalis capitis* passes from transverse processes T1–5 and cervical articular processes C4–7 to the occipital bone between the superior and inferior nuchal lines. The fibers of this massive muscle in the nape of the neck run nearly vertically. The medial border is free and is separated from its fellow by the ligamentum nuchae.

The *multifidus* arises as a thick, fleshy mass from the dorsal aspect of the sacrum between its spinous and transverse crests, from adjacent ligaments, from the dense aponeurosis of the overlying erector spinae, and from all transverse processes up to C4 (actually from mamillary processes in the lumbar region, transverse in the thoracic, and articular in the cervical). It spans about 3 segments to be inserted into the inferior border of every spinous process (C2–L5).

The *rotatores* bridge one interspace. They are small slips that pass from the root of one transverse process to the root of the spinous process or lamina next above. They are best marked in the thoracic region.

Other Deep Muscles. The *interspinales* and *intertransversarii* are well developed in the cervical and lumbar regions, but mostly absent from the thoracic region. The *levatores costarum*, though hidden by the erector spinae, are grouped with the thoracic muscles.

Interspinales are well-developed, median-paired muscles. They unite the bifid tubercles of adjacent cervical spinous processes and adjacent borders of the oblong lumbar spinous processes.

Intertransversarii (Anteriores and Posteriores) unite

adjacent anterior tubercles and adjacent posterior tubercles of the cervical transverse processes, the highest posterior intertransverse muscle being the rectus capitis lateralis. In the lumbar region, they are well marked.

Suboccipital Muscles. See Chapter 42.

ACTIONS OF DEEP MUSCLES OF BACK

The deep dorsal muscles acting together *extend* the vertebral joints. They also prevent, or regulate, *flexion* of these parts. Much, however, of the apparent extensor and flexor movement of the vertebral column takes place actually at the hip joints. When the deep muscles of one side act, *lateral bending* and *rotation* occur, along with the effects of the more powerful oblique muscles of the abdomen (see Functions, etc., p. 129).

Nerves of the Back (Dorsal Rami of Spinal Nerves)

GENERAL CONSIDERATIONS

The dorsal rami of each of the 31 pairs of spinal nerves are similar in size, course, relations, and distribution, with the exception of the first 2 and last 3.

A *typical dorsal ramus* takes origin just beyond a spinal ganglion (*see fig. 30.20*) and passes dorsally on the side of a superior articular process. The dorsal sacral rami emerge from the dorsal sacral foramina.

The dorsal rami supply a serially segmented territory. They supply the skin and the "native" deep muscles of the back medial to the angles of the ribs. Hence, they are much smaller than the corresponding ventral rami (except for C1 and 2). The prevalence of back pain problems has focussed attention on the zygoapophyseal ("facet") joints between articular processes. Being synovial joints, they are subject to arthritis and are one of the possible causes of pain. Articular nerves (at least one from each neighboring dorsal ramus) supply them.

Details.

Certain cutaneous branches trespass beyond the angles of the ribs and a few wander afar. Thus: C2 (greater occipital nerve) ascends to the scalp; T2 extends toward the acromion; L1, 2, and 3 descend to the buttock.

Each dorsal ramus divides into a medial and a lateral branch (except C1; S4, 5; and Co1), and one or another ends as a cutaneous nerve (except C1; C6, 7, 8; and L4, 5). Above the midthoracic region, the medial branches become cutaneous; below it, the lateral branches.

Courses of Nerve Trunks C1 and 2. *Explanatory:* These are peculiar because the "true" articular processes of the atlas and superior articular processes of the axis disappear; new joints appear anterior to the nerve trunks (*fig. 36.5*). So now the anterior rami wind anteriorly around the 2 joints.

Sacral and Coccygeal Nerves. The nerve trunks of S5 and Co1 lie exposed at the inferior end of the sacral canal (*fig. 29.3*). The dorsal rami of S4, 5, and Co1 unite to form a small descending cutaneous nerve that supplies the skin over the coccyx. This nerve is the homologue of the dorsal nerve of the quadruped's tail.

Clinical Case 29.3

Patient Charles T. This 69-year-old pensioner is in for his 6-month checkup in the Veterans' Administration Urology Clinic where he has been under care for 3 years for prostatic cancer. His only new complaint is a severe aching pain in the lumbar spine area. After sending the patient off for X-rays, the surgeon turns to you with questions on the routes by which the cancer may possibly spread to the vertebrae.

Veins of the Back

These veins are large and they form liberal plexuses that tend to follow the arteries. In the neck they mainly descend around the vertebral and profunda cervicis (deep cervical) arteries.

VEINS OF THE VERTEBRAL COLUMN (VERTEBRAL VENOUS SYSTEM)

Because they are difficult to dissect and to demonstrate, these veins are not treated in classrooms with the great respect they deserve. They must not be ignored! The vertebral canal contains a dense plexus of thin-walled, valveless veins that surrounds the spinal dura mater like a basket. Superiorly, this plexus communicates through the foramen magnum with the occipital and basilar sinuses of the cranium. Anterior and posterior longitudinal channels (venous sinuses) can be discerned in this internal *vertebral venous plexus* (*fig. 29.13*).

At several spinal segments (but not all), the plexus receives a vein from the spinal cord and at each segment it receives a vein, the basivertebral vein, from the body of a vertebra. In turn, it is drained by intervertebral veins,

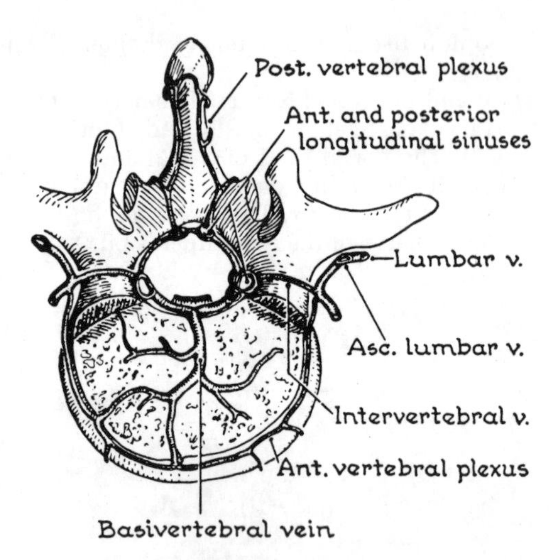

Figure 29.13. The veins of the vertebral column.

which pass through the intervertebral (and sacral) foramina to the vertebral, intercostal, lumbar, and lateral sacral veins.

Through the body of each vertebra come veins which form a meager *anterior vertebral plexus*, and through the ligamenta flava pass veins that form a well-marked *posterior vertebral plexus*.

OTHER LONGITUDINAL CHANNELS

In the cervical region, these plexuses communicate freely with the occipital, vertebral, and profunda cervicis veins, and in the thoracic, lumbar, and pelvic regions, each segment is linked to another by the azygos (or hemiazygos), ascending lumbar, and lateral sacral veins.

The *ascending lumbar vein* (*fig 29.14*) is an anastomotic vein that ascends anterior to the lumbar transverse processes, linking one lumbar vein to another and connecting the common iliac vein with the azygos (or hemiazygos) vein.

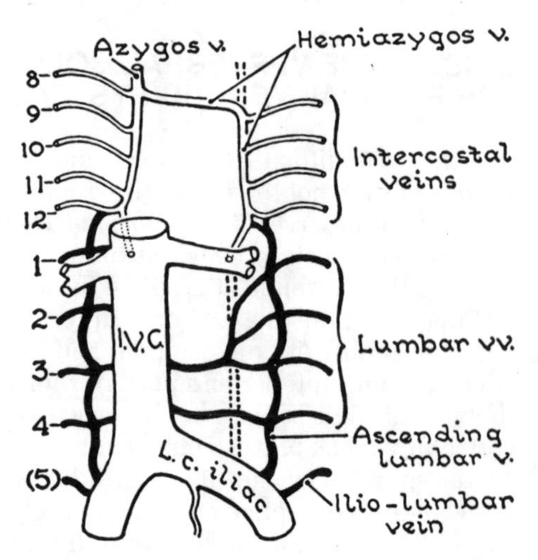

Figure 29.14. Ascending lumbar vein. (*Cont'd. from fig. 9.16.*)

SIGNIFICANCE

Batson's ingenious experiments forced us to add to the recognized pulmonary, portal, and caval venous systems a fourth or **vertebral venous system**. This system may be considered a separate, although overlapping, system of veins. It comprises the veins of the brain, skull, neck, viscera, vertebral column (and their valveless connections in the limb girdles), and the veins of the body wall (*fig. 29.15*).

The Batson experiments indicate (1) that compression of the thorax and abdomen with the larynx and other sphincters closed, as occurs in straining, coughing, and lifting with the upper limbs, not only prevents blood from entering the thoracoabdominal veins but also squeezes it out of them into the vertebral system; (2) that the increase in the intraspinal and intracranial pressure that occurs during coughing, sneezing, and straining is active—not passive; (3) that tumors and abscesses having connection with this venous system may spread anywhere along this system without involving the portal, pulmonary, or caval

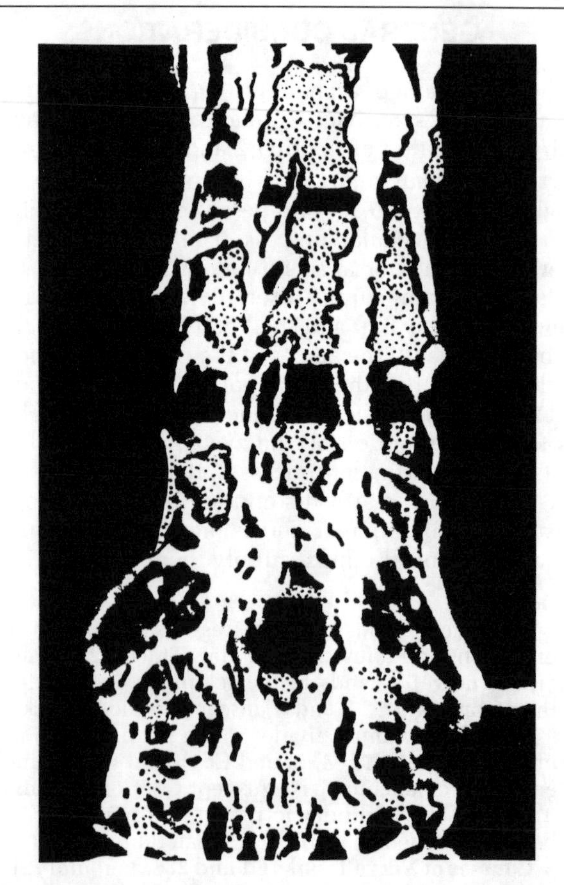

Figure 29.15. Venogram in lumbar region of the vertebral venous system (traced from an X-ray photograph; *dotted* areas are the vertebrae).

systems; and (4) that the cranial and spinal parts of the system, as well as being pathways, are blood depots or storage lakes of blood; furthermore, (5) that they reveal the channels through which blood from the lower limbs and pelvis may in favorable circumstances return to the heart after the inferior vena cava has been obstructed below the renal veins. In addition, Eckenhoff shows that the CSF pressure is regulated by the vertebral venous pressure.

Anatomical Features of the Physical Examination

The significant features of the posterior aspect of the neck and torso described above are viewed and palpated fairly easily by the experienced examiner, but skill is acquired by careful, repeated testing. In addition, mobility of the scapula and of the back is assessed by having the subject perform a full range of motion of the shoulder region and of the back and head and neck.

Measurements are possible with various devices. The simplest is a tape measure, but experience is needed to interpret the results. Certainly every beginning clinician should soon learn to recognize deformities (such as scoliosis in Clinical Case 29.2), winging of the scapula due to weakness of serratus anterior (p. 359), and limited mobility and tenderness of specific structures of the region. Counting the spinous processes should become a routine skill very early.

Because two large and several smaller muscles of the upper limb and also the scapula cover the muscles of the back, they are classified with it. Any one of these may be the site of the inflammatory problem suffered by the patient in Clinical Case 29.1. Included are the two muscles of the superficial or first layer: the trapezius and the latissimus dorsi.

The intermediate (or second) layer includes two thin, transverse muscles devoted to minor respiratory movements—serratus posterior superior and inferior and they cannot be felt. The deep or intrinsic muscles of the back—devoted to posture primarily—are complicated, and variable; but they have descriptive names and follow a general plan, being disposed in three layers, none of which can be felt clearly. Idiopathic scoliosis (Clinical Case 29.2) almost certainly is due to an imbalance of muscular functions rather than a primary bone defect.

The veins of the vertebral column are important because they form profuse ascending and interlocking valveless plexuses that can carry most of the blood as a bypass of the inferior vena cava from the pelvis up and through the bones. They even communicate with veins in the cranium and may be a path for the spread of infections or cancer from the pelvis to the vertebrae or the cranial contents (Clinical Case 29.3).

Clinical Mini-Problems

1. **Which vertebrae have the most prominent and palpable spinous processes?**
2. **(a) Trauma to the accessory nerve as it passes through the cervical fascia in the posterior triangle would cause weakness or paralysis in which muscle?**
   (b) How would you test this?
3. **The triangle of auscultation is a region of the back where a minimum of muscle tissue overlies the lung. What lobe of the lung do physicians examine when they place their stethoscope over this triangle and ask the patient to breathe deeply?**

(Answers to these questions can be found on p. 586.)

30

Pectoral Region and Axilla

Clinical Case 30.1

Patient Arthur U. In a rough and tumble ice-hockey game, this teenager was driven against the boards and immediately suffered severe pain and disability in the region of his left clavicle. Now in the emergency department, you are carefully removing his shirt while noting any visible abnormalities you will need to report to the physician-in-charge. In addition, you are reviewing in your mind the anatomy of the clavicular, pectoral, and axillary regions because you know you will be asked.

Introduction

Human upper limbs move freely, being adapted to purposes of prehension. Each articulates with the trunk at one small joint, the *sternoclavicular joint*, unlike the lower limbs which are united to the vertebral column at the sacroiliac joints and to each other at the symphysis pubis anteriorly (*Table 30.1*).

In prenatal life, the thumb and the radius, like the big toe and the tibia, are situated on the cranial or head side of the central axis of the limb and are said to be *preaxial*. However, the 5th finger and ulna are *postaxial*.

On assuming the **anatomical posture** (*fig. I.1* on p. 3), the upper and lower limbs undergo rotation but in opposite directions; the thumb by rotating laterally brings the palm of the hand to face anteriorly, while the big toe, by rotating medially, brings the sole of the foot to the ground.

MOORING MUSCLES (*fig. 30.1*)

The upper limb is moored to the head, neck, and trunk by muscles, which may be likened to dynamic guy ropes.

Lines of Force Transmission

The following fundamental points should be verified by reference to the skeleton and *Figure 30.2*.

The *clavicle* is a strut that, through the medium of a strong ligament, the *coracoclavicular*, holds the scapula away from the median plane. The clavicle's enlarged medial end articulates with a shallow socket formed by the sternum and 1st rib cartilage. A strong ligament, in the form of an *articular disc*, prevents the clavicle from being driven medially onto the sternum. The lateral end of the *clavicle* is not enlarged, doing little more than making contact at a small synovial joint with the *acromion*.

At the shoulder joint, the rounded head of the humerus

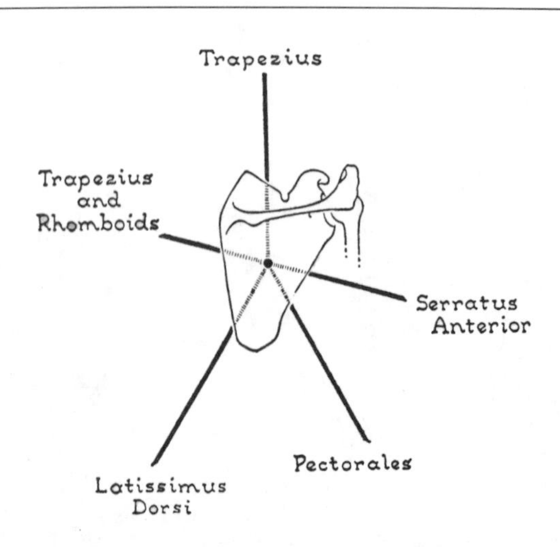

Figure 30.1. Upper limb is moored to body by muscles.

Table 30.1. Bones of Upper Limb

Region	Bones
Shoulder or pectoral girdle	Clavicle Scapula
Arm or brachium	Humerus
Forearm or antebrachium	Radius Ulna
Hand or manus	Carpal bones Metacarpal bones Phalanges 1st or Proximal 2nd or Middle 3rd or Distal

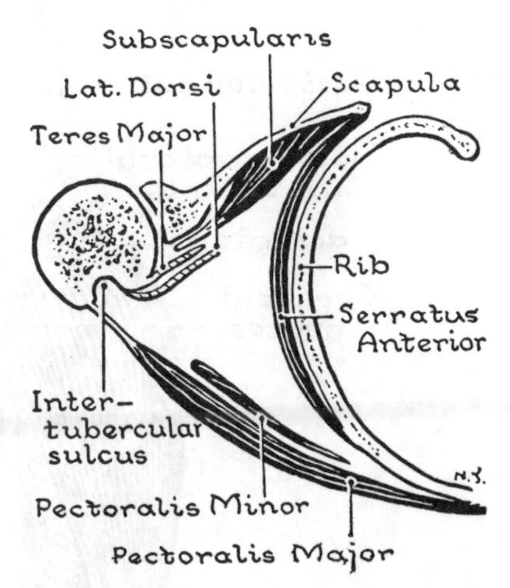

Figure 30.3. Walls of axilla (on cross-section).

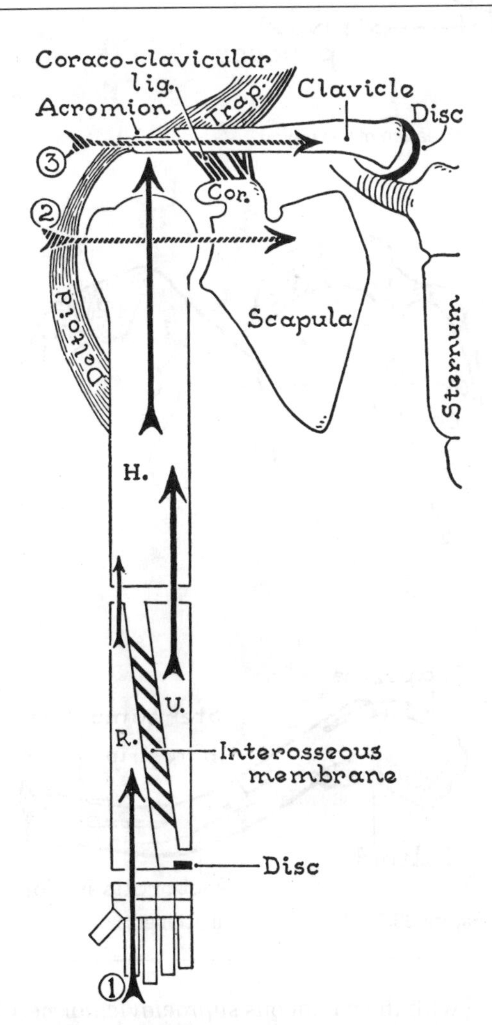

Figure 30.2. Scheme of skeleton of limb showing lines of force transmission and ligaments that may transfer force as the result of a fall:
1. on the hand—interosseous membrane.
2. on the shoulder—coracoclavicular ligament.
3. on the acromion—articular disc. *H*, humerus; *R*, radius; *U*, ulna.

fits into the shallow glenoid cavity of the scapula. The acromion and related *coracoacromial ligament* form a hood-like roof or arch, which prevents upward dislocation of the humerus. The lower end of the humerus articulates with the ulna and radius.

The ulna and radius are united by a strong *interosseous membrane.* At a glance, you can see that at the elbow the ulna is more important than the radius; at the wrist, the contrary is true. The enlarged lower end of the radius articulates with the carpal bones, but the ulna does not. *Figure 30.2* summarizes the lines of force transmission.

DEFINITION AND BOUNDARIES OF AXILLA

The **axilla** is the pyramidal space above the arm pit. It has four walls, an apex, and a base (*fig. 30.3*). The *anterior wall* is fleshy and is formed by the pectoralis major and 2 muscles that lie behind it enclosed in a sheet of fascia, the *clavipectoral fascia.* (This wall is practically synonymous with the pectoral region, which includes the breast or *mamma.*)

The *posterior wall* is formed by the scapula overlaid by the subcapularis and below this by the latissimus dorsi and teres major. Grasp the thick fleshy lower borders of the anterior and posterior walls between the fingers and the thumb and estimate how much you have enclosed.

The *medial wall* is formed by the upper ribs (2nd–6th) covered with serratus anterior. The very narrow *lateral wall* is the intertubercular sulcus (bicipital groove) of the humerus. It lodges the long tendon of the biceps, and its lips give attachment to muscles of the anterior and posterior walls.

The *base* is the skin and fascia of the armpit. The *truncated apex* is the triangular space bounded by 3 bones—the clavicle, the upper border of the scapula, and the 1st rib.

The *contents of the axilla* (*see fig. 30.13*) will be described on p. 359. The great vessels and nerves of the upper limb pass through the axilla on their way to the rest of the limb—they are the chief contents. The other

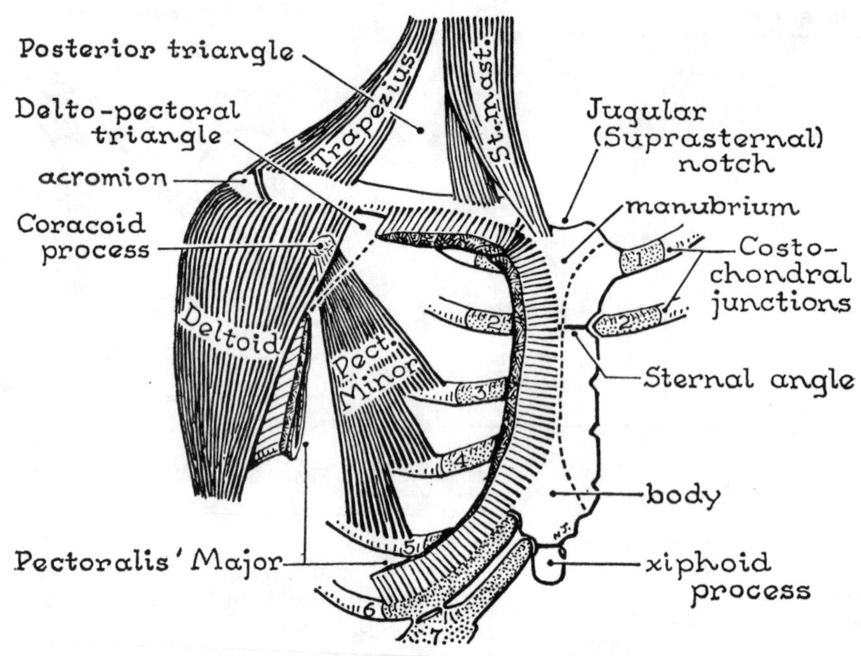

Figure 30.4. The pectoralis minor is the central feature of this region.

contents are the 2 heads of the biceps, the coracobrachialis, and lymph nodes.

LANDMARKS: BONY AND MUSCULAR (*fig. 30.4*)

The *sternum*, or breast bone, is a flat bone, shaped like a short flat sword. It consists of 3 parts: manubrium, body, and xiphoid process. The manubrium, or handpiece, meets the body at a slight angle, forming a ridge called the *sternal angle*. On each side, the sternum articulates with a clavicle and 7 costal (rib) cartilages. (For details, see Chapter 5).

The *coracoid process* of the scapula (Gk. korax = a crow) is misnamed (*fig. 30.5*). Its *tip* lies inferior to the clavicle under shelter of the anterior edge of the deltoid muscle (*fig. 30.4*). To palpate this essential landmark in the living person, press firmly and laterally.

The **clavicle**, or collar bone (*figs. 30.4 and 30.6*) is nearly horizontal and has a double curve. The part medial to the underlying coracoid process (medial three-fourths) is triangular on cross-section like long bones in general, but the part lateral to the process is flattened and is rough below for the ligaments that bind the scapula, via its coracoid process, to the clavicle (*fig. 30.7*). When the clavicle is fractured ("broken collar bone"), the shoulder region falls in a typical direction; will you accept this to be medially, forward and downward? How can it be otherwise!

The clavicle, although almost visible through the skin, is not merely subcutaneous; a thin broad sheet of muscle, the *Platysma*, descends from the neck to the 2nd or 3rd rib along with the cutaneous *supraclavicular nerves* that lie superficial to the clavicle (*fig. 30.8*).

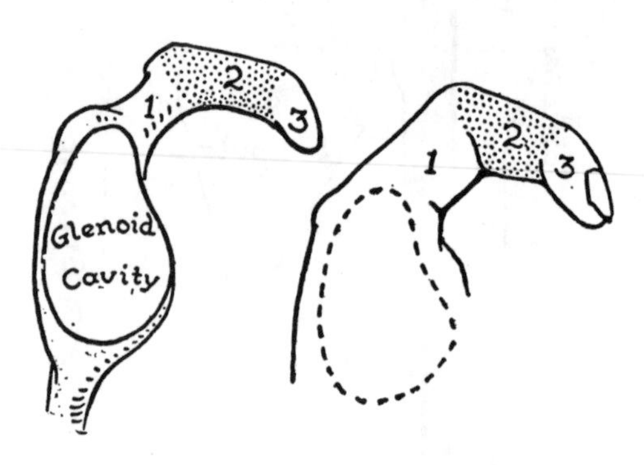

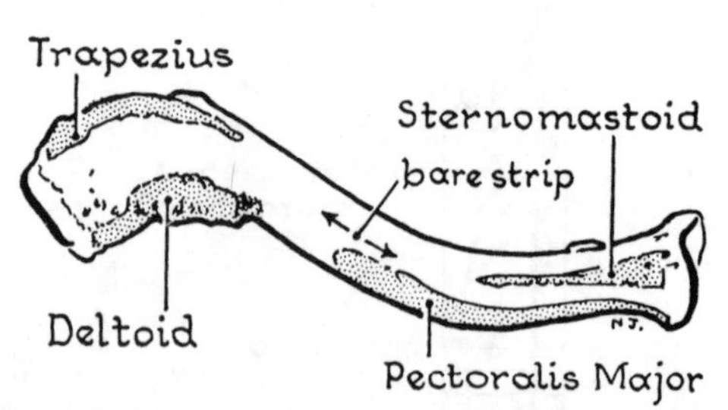

Figure 30.6. Clavicle and attachments of muscles.

The **pectoralis minor** (*fig. 30.4*), completely concealed by the pectoralis major, is the central landmark for surgeons. It arises from the (2nd), 3rd, 4th, 5th, (and sometimes 6th) ribs, where bone and cartilage join, and it is

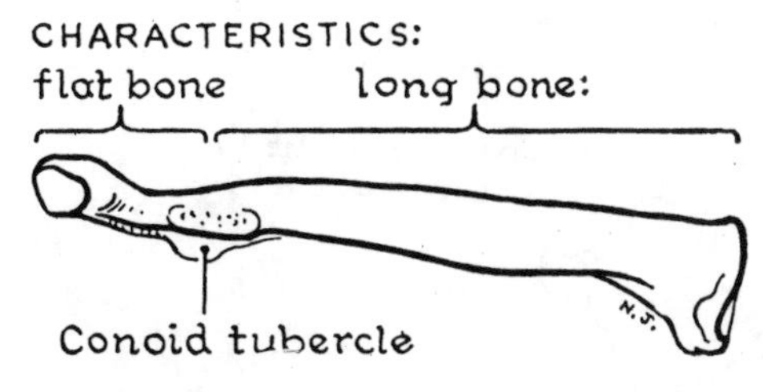

Figure 30.7. The 2 parts of the clavicle.

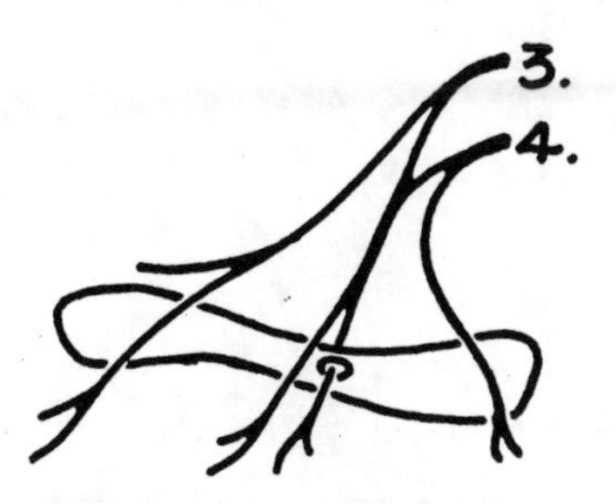

Figure 30.8. The clavicle is crossed and occasionally pierced by a branch of the supraclavicular nerves from C3, 4.

inserted into the medial border of the coracoid process near its tip.

SURFACE ANATOMY

The only certain way of identifying any rib lies in reckoning from the *sternal angle*, which is at the level of the 2nd costal cartilages (*fig. 30.9*). About 5 cm inferior to the *jugular notch* (suprasternal notch), you encounter the sternal angle, the transverse ridge of which indicates where the manubrium and body of the sternum articulate. Follow the ridge laterally to the 2nd costal cartilage; you may then readily palpate the 3rd, 4th, and 5th cartilages through the substance of the pectoralis major. By joining a point on the 3rd rib about 5–7 cm from the midline to the coracoid process, you define the superior border of the pectoralis minor; similarly, by joining the 5th rib near its cartilage, say 10 cm from the median plane, to the coracoid process you map out the inferior border of this triangular muscle (*fig. 30.10*).

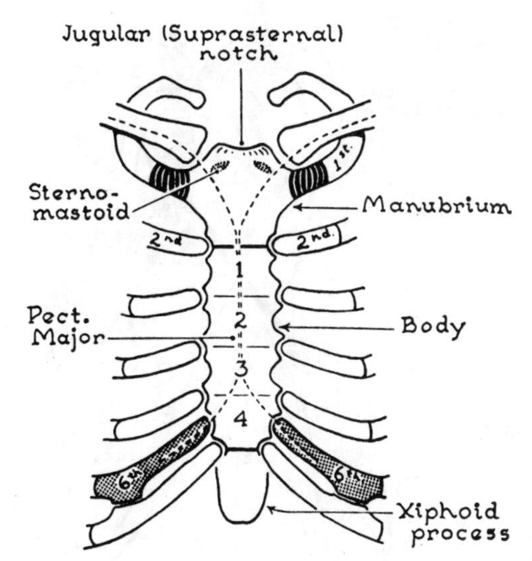

Figure 30.9. Palpable landmarks for sternum and costal cartilages.

Clinical Case 30.2

Patient Brent V. Three days ago while riding his motorcycle on wet pavement, this 28-year-old man skidded, was flung off, and landed on his right shoulder. His various injuries have been attended to, but now, before he is discharged from hospital, you (the student on the ward) discover that he has no power to abduct his right arm. You must review in your mind (or this book) the appropriate anatomical possibilities and suggest the necessary tests.

Walls of Axilla

ANTERIOR WALL OF AXILLA

The anterior wall of the axilla is formed throughout by pectoralis major, and behind this by pectoralis minor and subclavius within their sheaths of clavipectoral fascia (*fig. 30.11*).

Pectoralis Major

The *clavicular head* arises in line with the deltoid from the anterior aspect of the clavicle. The *sternal head* arises from the sternum: its origin curves inferolaterally along the 5th (or 6th) costal cartilage (*fig. 30.9*).

The pectoralis major is inserted by means of a folded (3-layer) aponeurosis (*figs. 30.3 and 30.10*) into the crest of the greater tubercle of the humerus (lateral lip of the bicipital groove) (*fig. 30.12*).

Actions

Obviously both parts of the pectoralis major adduct the humerus and rotate it medially. If, while palpating your

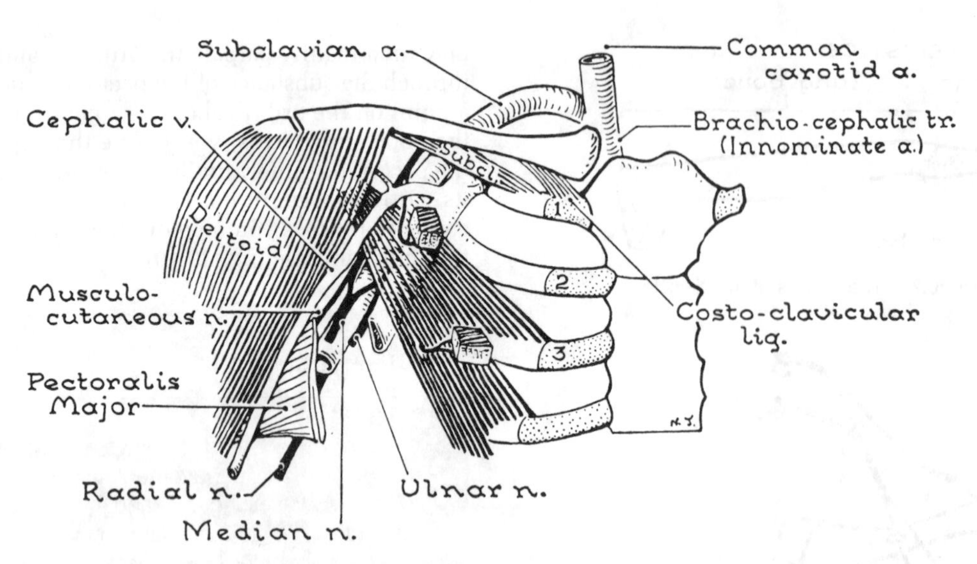

Figure 30.10. Cephalic vein. Two buttons of pectoralis major hang from its nerves.

pectoralis major, you put your fist under the edge of a heavy table and try to raise it, you can feel the clavicular part of the muscle come into action. If you press downward against resistance, you find the sternal part comes into action.

The pectoralis major and deltoid diverge slightly (1 cm) from each other and with the clavicle form the *deltopectoral triangle* (*fig. 30.4*), which allows the passage of the cephalic vein.

The **cephalic vein** (*fig. 30.10*), which occupies the furrow between the deltoid and the pectoralis major, plunges through the deltopectoral triangle and costocoracoid membrane to join the axillary vein.

> *The practical significance* of the cephalic vein is that an expert can pass a fine, pliable tube through it and the axillary and subclavian veins and onward into the heart.

The **costocoracoid membrane** stretches from the 1st and 2nd costal cartilages to the coracoid process. It is part of a larger fascial sheet, the **clavipectoral fascia** (*fig. 30.11*), which splits to enclose the pectoralis minor and subclavius muscles.

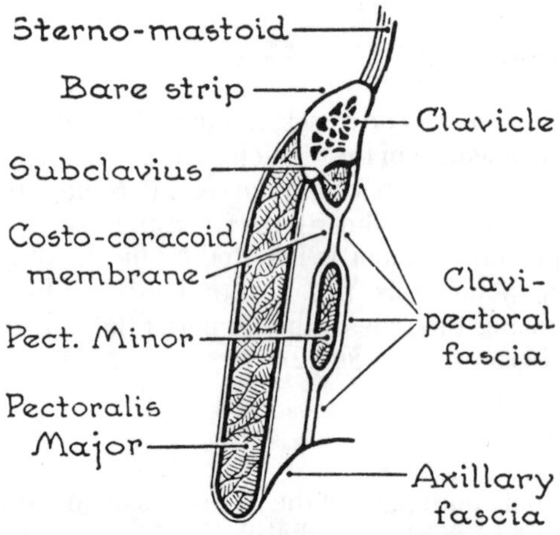

Figure 30.11. Anterior wall of axilla (sagittal section).

POSTERIOR WALL OF AXILLA

The posterior wall is formed by the subscapularis, teres major, and latissimus dorsi, the scapula being the background (*fig. 30.12*). Their tendons insert on the humerus on the lesser tubercle and medial lip of the intertubercular sulcus.

(For the origin of the latissimus dorsi from the vertebral spines and iliac crest, see p. 346.)

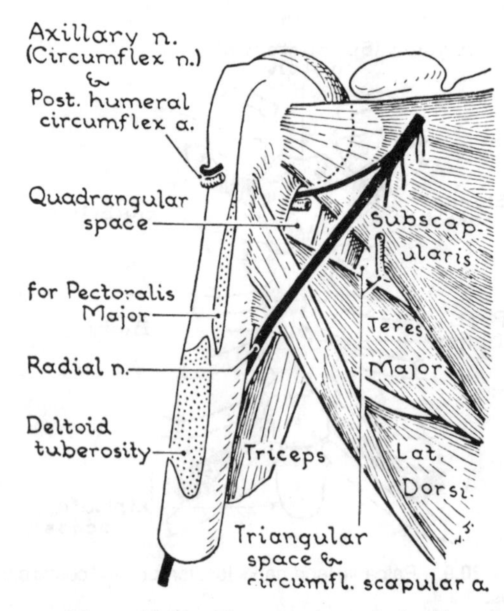

Figure 30.12. The posterior wall of axilla.

If the subscapularis is pulled away from the teres major, a long triangular space is opened up. Its lateral border is the surgical neck of the humerus. In its depths is the long head of the triceps brachii, subdividing it into a small, unimportant, medial *triangular space* and the important, lateral *quadrangular space* (*fig. 30.12* and p. 371).

LATERAL AND MEDIAL WALLS OF AXILLA

The **lateral wall** (*figs. 30.12 and 30.13*) is the narrow interval between the 2 crests of the humerus on which the anterior and posterior walls attach—the *intertubercular sulcus*.

The **medial wall** is formed by ribs and intercostal muscles covered with serratus anterior.

Serratus Anterior

Serratus anterior is the chief muscle to protract or pull the scapula anteriorly, as when pushing (*fig. 30.14*). If it is paralyzed, attempts to raise the arm in front of the body largely result in causing the inferior angle of the scapula to project from the back ("winged scapula"). Its nerve (long thoracic nerve) runs vertically near the mid-axillary line (*fig. 30.13*).

Origin

The serratus anterior arises from the outer surfaces of the first 8 ribs by a series of fleshy digitations (*fig. 30.14*).

Insertion

The serratus is inserted (*fig. 30.15*) into the medial border of the scapula with concentration on the inferior angle. The usefulness of this is apparent; the fibers are concentrated where they act to best advantage—at the end of a lever (*fig. 30.14*).

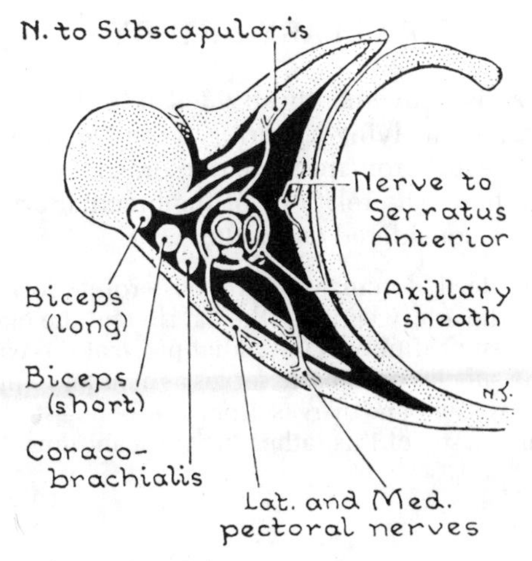

Figure 30.13. Contents of axilla (cross-section).

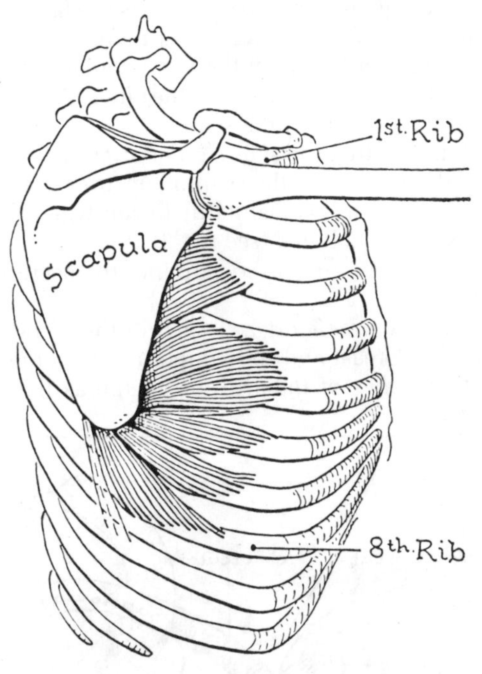

Figure 30.14. The serratus anterior.

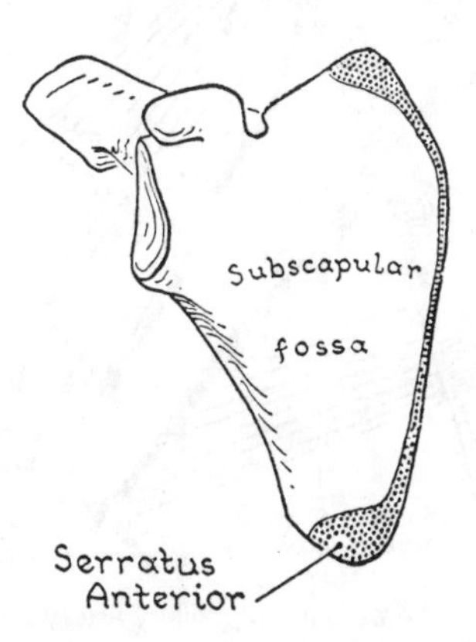

Figure 30.15. The insertion of serratus anterior.

Vessels of Axilla

The great axillary vessels and nerves are enveloped in the **axillary sheath**, continuous with prevertebral fascia in the neck. They enter the axilla at its apex through the triangle formed by the clavicle, 1st rib, and superior border of the scapula. They curve downward to the humerus.

The **Great Arterial Trunk** of the limb (*fig. 30.16*) is called the *subclavian artery* until it reaches the inferior border of the 1st rib. In the axilla, it is known as the *axillary artery*. In the arm or brachium, it becomes the *brachial artery* (p. 390).

THE AXILLARY ARTERY

This artery is conveniently divided into 3 parts (*fig. 30.16*), the 2nd part lying posterior to the pectoralis minor, a fingerbreadth from the tip of the coracoid process—this is important. The relations of the brachial artery to the nerves are considered on p. 390.

On voluntarily forcing your arm posteroinferiorly, you compress the artery between the 1st rib and the clavicle and so arrest or diminish the radial pulse at the wrist—try it! The subclavius muscle serves as a buffer to protect the great vessels and nerves from the bone—the only significant "use" of this rather insignificant muscle.

Branches

One arises from the first part, two from the second, and three from the third. Details are shown in *Figure 30.16*. Of these 6 branches, the *subscapular* is the largest by far; it is the artery of the posterior wall (*fig. 30.17*), and it follows the lower border of the subscapularis to the medial wall. It sends a large branch, the *circumflex scapular artery*, to the dorsum of the scapula.

Anastomoses that the subscapular artery makes on the chest wall and around the scapula ensures that the circulation in the limb is maintained even when the main trunk has been occluded (*see fig. 31.22*).

The *thoracoacromial artery* supplies the pectoral muscles.

The *lateral thoracic artery* is part of the blood supply of the female breast and so is important in surgery.

The *3 branches of the 3rd part* of the axillary artery arise near the surgical neck of the humerus, anterior to the quadrangular space (*fig. 30.12*).

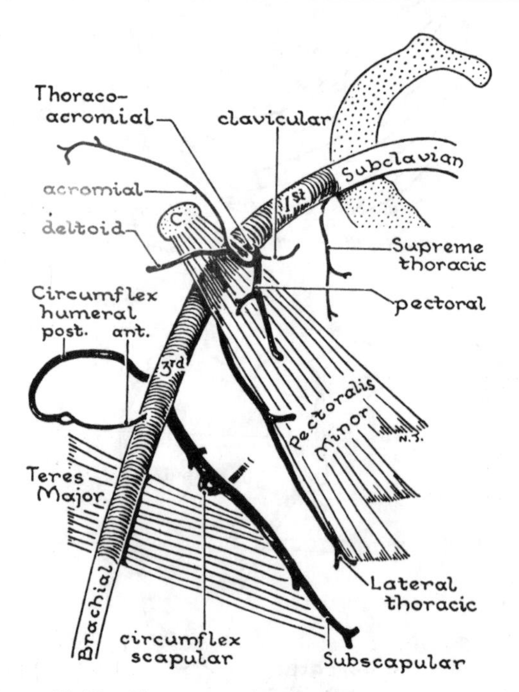

Figure 30.16. Three parts of axillary artery and its branches.

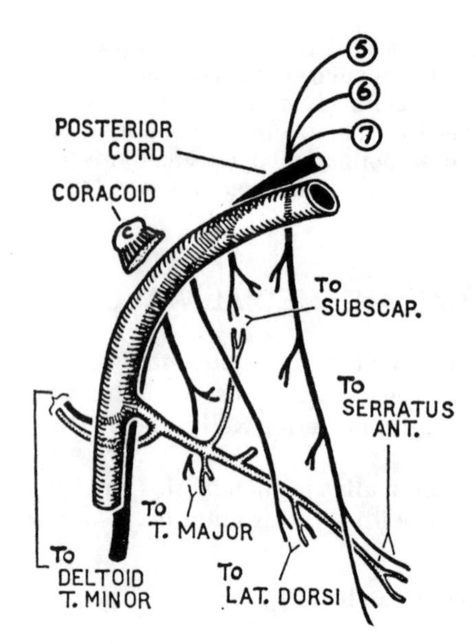

Figure 30.17. Branches of the subscapular artery accompanying 5 motor nerves.

Emboli or blood clots expelled from the heart are apt to lodge at this site, where the main vessel is much reduced in caliber beyond all these branches.

The circumflex humeral arteries encircle the surgical neck of the humerus. The posterior artery accompanies the axillary nerve and is large.

THE AXILLARY VEIN

This vein lies on the medial or concave side of its artery, but it overlaps and conceals the artery when the arm is abducted. It is the continuation of the basilic vein, and at the 1st rib becomes the **subclavian vein**; it has 2 or 3 bicuspid valves.

Tributaries

In addition to receiving tributaries corresponding to the 6 branches of the axillary artery, it receives the 2 **venae comitantes** of the brachial artery and the **cephalic vein**. (For venae comitantes, *fig. 2.3* and p. 26.)

Brachial Plexus

The brachial plexus is formed by the alternate union and bifurcation of nerves. Thus: 5 ventral nerve *rami*

Table 30.2. Derivatives of the Brachial Plexus within the Limits of the Axilla

Cords	Terminal branches	Collateral branches	
	Mixed (motor and sensory)	Motor	Cutaneous
Lateral	1. Musculocutaneous 2. Lateral root of median	1. Clavicular head of pectoralis major (and upper part of sternocostal head)	
Medial	1. Medial root of median 2. Ulnar	1. Sternocostal head of pectoralis major 2. Pectoralis minor	1. Med. cutan. n. of arm 2. Med. cutan. n. of forearm
Posterior	1. Axillary 2. Radial	1. Subscapularis 2. Latissimus dorsi 3. Teres major	
From the musculocutaneous		1. Coracobrachialis	
From the radial		1. Triceps (long head) and	Post. cutan. n. of arm
From the 5, 6, and 7 roots of the plexus		1. Serratus anterior	

unite to form 3 *trunks*, which bifurcate to form 6 *divisions*, which unite to form 3 *cords*, which bifurcate to form 6 *terminal branches* (*fig. 30.18*). Of these 6 terminal branches, two soon unite to form the median nerve; hence, the plexus may be said to begin as 5 big nerves and to end as 5 big nerves.

The rami and the trunks of the plexus lie in the neck, the divisions posterior to the clavicle, the cords above and posterior to the pectoralis minor, and the terminal branches distal to it.

Each of the 3 trunks divides into an anterior and a posterior division. The 3 posterior divisions unite together to form a single *posterior cord*. Of the anterior divisions, the lateral and intermediate unite to form the *lateral cord*, while the medial continues its course as the *medial cord*.

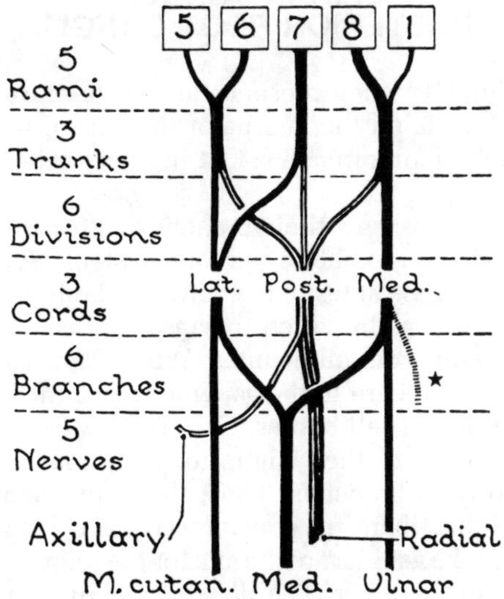

Figure 30.18. Scheme of the brachial plexus. ★ Indicates the medial cutaneous nerve of the forearm.

Each of the 3 cords gives off one or more *collateral branches* and ends by dividing into 2 *terminal branches* (*Table 30.2*), the lateral cord dividing into the musculocutaneous nerve and the lateral root of the median nerve, the medial cord into the ulnar nerve and the medial root of the median nerve, and the posterior cord into the radial and the axillary nerve (circumflex nerve).

From a glance at *Figure 30.18* it should be clear that the musculocutaneous nerve and the lateral root of the median nerve can derive fibers from the 5th, 6th, and 7th cervical nerve segments; the ulnar nerve and medial root of the median nerve from the 8th cervical and 1st thoracic segment; the median nerve itself and the posterior cord from each of the 5 segments.

In point of fact their origins are: musculocutaneous 5, 6, 7; ulnar (7) 8, 1; median (5) 6, 7, 8, 1; radial 5, 6, 7, 8 (1); axillary 5, 6.

The medial and lateral cords might better have been called the "anteromedial" and "anterolateral" cords. The posterior cord is well named because it is destined to supply all the muscles on the posterior or extensor aspect of the limb.

Relation of Plexus to Artery

The brachial plexus (*fig. 30.19*) emerges from the cervical portion of the vertebral column; thus in the neck, it lies posterosuperior and lateral to the artery.

Identification

The medial and lateral cords and their terminal branches have the form of a capital "M" (*fig. 30.19*). This fact should be used in identifying them.

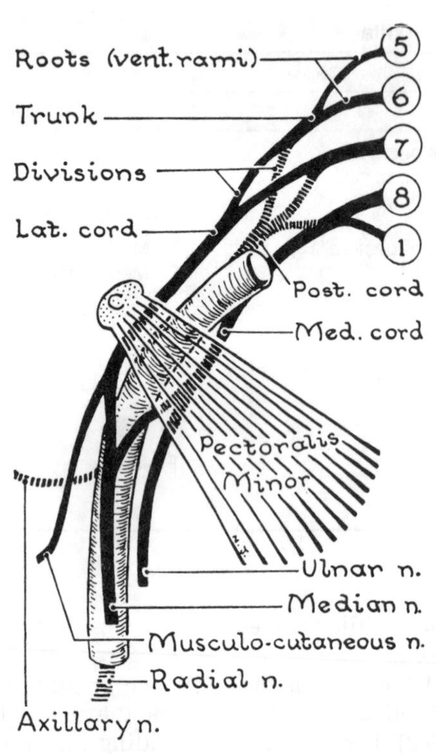

Figure 30.19. The brachial plexus.

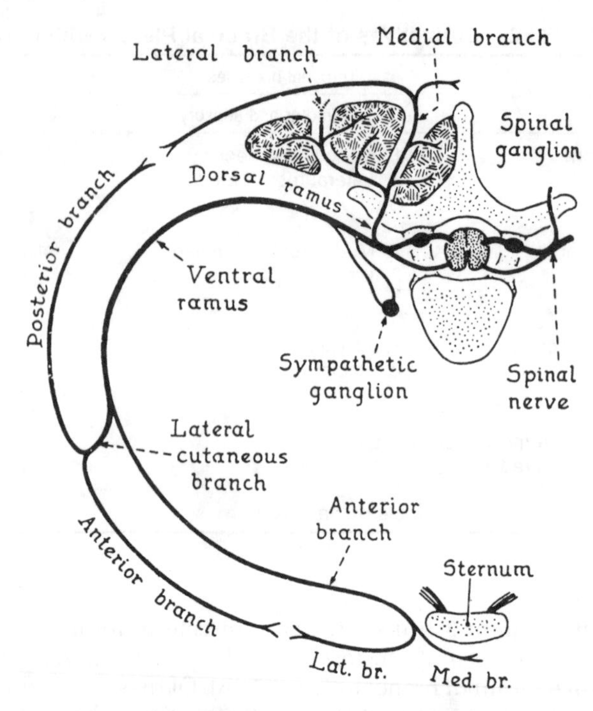

Figure 30.20. A typical segmental nerve.

Branches

Two cutaneous branches also can be found. The larger of these is the *medial cutaneous nerve of the forearm.* It and the much smaller *medial cutaneous nerve of the arm* arise from the medial cord just before it bifurcates.

If the derivatives of the lateral and medial cords are held aside, the only remaining large nerve, namely, the *radial nerve*, will be identified by this process of exclusion.

Unless the foregoing procedure is followed, the radial nerve may easily be mistaken for the ulnar nerve.

The *radial nerve* and the *axillary nerve* will be seen to be the 2 terminal branches of the posterior cord. Of the 5 terminal nerves of the plexus, the axillary nerve alone does not run longitudinally, but disappears posteriorly into the quadrangular space.

Motor Nerves of the Axilla in Review. See *Figures 37.18 and 37.19.*

A **typical spinal nerve** (*fig. 30.20*) is a serially segmental structure which is attached to the spinal cord by *2 roots*—a ventral (anterior) that is motor or efferent, and a dorsal (posterior) that is sensory or afferent. The 2 roots leave the vertebral canal through an intervertebral foramen and join immediately beyond it to form a *spinal nerve.* Having both motor and sensory fibers, the nerve is said to be mixed. After the course of a few millimeters, the spinal nerve divides into a *ventral* and a *dorsal ramus.* Roughly, the dorsal rami supply the muscles of the back that act on the vertebral column and the skin covering them, whereas ventral rami supply the muscles and skin of the anterior three-quarters of the body wall.

There is an enlargement on the dorsal root as it lies in the intervertebral foramen. This enlargement, known as a *spinal ganglion* (posterior or dorsal root ganglion), consists of the cell bodies of all the afferent (sensory) fibers in that spinal nerve.

SPINAL GANGLIA (POSTERIOR ROOT GANGLIA)

Ganglion C1, when present, lies on the posterior arch of the atlas; C2 on the lamina of the axis; C3–L4 in the intervertebral foramina; L5–Co1 in the vertebral and sacral canals.

Spinal nerves and their branches also carry efferent fibers of the sympathetic nervous system. Sympathetic ganglia lie on the sides of the vertebral bodies in front of the spinal nerves to which they are connected by white and gray rami communicantes. White rami carry fibers from the spinal cord to the ganglia, where they may synapse with nerve cell bodies. Gray rami carry fibers from the cell bodies in the ganglia to the spinal nerves and thence to their branches. Thus, the white rami contain preganglionic fibers, the gray rami contain postganglionic fibers, and the ganglia are the site for the relay of impulses from one to the other. The fibers traveling along the peripheral nerves supply only involuntary (smooth) muscle and glands (see *fig. 2.10*).

THE CUTANEOUS NERVES

The ventral rami of the thoracic nerves are usually called *intercostal nerves*. A typical intercostal nerve gives off a *lateral cutaneous branch* in the midaxillary line and ends, after piercing the pectoralis major at the side of the sternum, as an *anterior cutaneous branch* (*fig. 30.20*).

The *first 3 intercostal nerves* are atypical: the 1st has neither a lateral nor an anterior cutaneous branch; the 2nd sends its lateral branch across the dome of the axilla in the laminated fascia and then descends on the posteromedial aspect of the arm as the *intercostobrachial nerve* (*see fig. 32.5*); the 3rd sends a small lateral branch to the medial side of the arm.

Development

In prenatal life, the ventral rami of nerves C5, 6, 7, 8, and T1 are drawn out from the trunk into the developing limb to supply it with both motor and sensory fibers. They form the brachial plexus and their simple segmental arrangement becomes obscured. Since these 5 rami (5, 6, 7, 8, and 1) send no cutaneous branches to the pectoral region, the supraclavicular branches of C3 and 4 descend in front of the clavicle to the level of T2, thereby closing the gap or hiatus thus occasioned. In consequence, a person whose spinal cord is injured due to a fracture of the 5th cervical vertebra might retain sensation to pinprick as low as the 2nd intercostal space (*see fig. 32.4*).

The cutaneous supply to the limb draws upon more segments than the motor supply, because branches from C3 and 4 descend to the deltoid region; branches from T1 and 3 descend to the medial side of the arm.

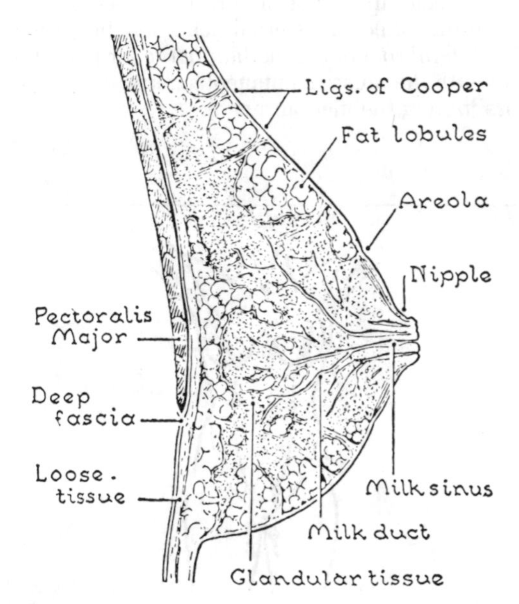

Figure 30.21. The mamma on vertical section. (After Testut.)

Clinical Case 30.3

Patient Carol W. A mammogram has strongly indicated that this 48-year-old mother of 3 children has a probable cancer of the left breast, which will require surgery. As a junior member of the surgical team preparing for the operation, you must recall as best you can the anatomical landmarks, blood and lymph supply, muscles, and fascia of the region.

Fascia, Mammary Gland, Lymph Nodes, and Anomalies

FASCIA

The large quantities of very loose, moist, areolar tissue found between the various muscles of this region are necessary to allow the limb a wide range of movements. Where it forms the dome-shaped floor of the axilla, it is irregularly laminated.

The *clavipectoral fascia* (*fig. 30.11*) which is the sheath of pectoralis minor and subclavius, acts as a suspensory ligament for the dome-shaped floor of the axilla.

MAMMARY GLAND

The *mamma* or breast (*fig. 30.21*) lies in the superficial fascia. it is made up of 15–20 units of glandular tissue, whose lobules, enclosed in a fibroareolar stroma, radiate from the nipple into the surrounding superficial fat, a major component of the adult woman's breast. The periphery of the gland extends from the 2nd to the 6th rib in the vertical plane and from the side of the sternum to near the midaxillary line in the horizontal plane.

About two-thirds of the gland overlies the pectoralis major, one-third the serratus anterior. Although easily separated from the fascia covering these muscles, the gland is firmly connected to the true skin by *suspensory ligaments (of Cooper)* that pass from its stroma between lobules of fat.

In the pigmented *areola* surrounding the nipple, there are a number of nodular rudimentary milk glands, *the areolar glands*, and deep to the areola are some involuntary muscles, a lymph plexus, and an absence of fat.

Vessels and Nerves

Nerves

Intercostal nerves (2nd–6th), via lateral and anterior cutaneous branches. These or the vessels convey sympathetic fibers.

Arteries

Perforating branches (especially the 2nd and 3rd) of the *internal thoracic artery* (internal mammary artery) and 2 branches of the *lateral artery* approach the gland from the sides, ramify on its superficial surface, send branches into it, and anastomose around the nipple. Twigs from the *intercostal arteries* may enter the deep surface of the gland.

Lymphatics

The main lymph vessels of the mamma, like the ducts, converge on the nipple. Deep to the areola, they form a subareolar lymph plexus (*fig. 30.22*). (1) From this plexus, 2 or 3 distinct vessels course superficially to the pectoral lymph nodes. (2) From the medial border of the mamma, lymph vessels pass to the parasternal lymph nodes, which lie along the internal thoracic artery. (3) From the deep surface of the mamma, lymph vessels pass through the pectoralis major to the interpectoral nodes, which lie superficial to the pectoralis minor and costocoracoid membrane or, passing through these, end in the apical nodes.

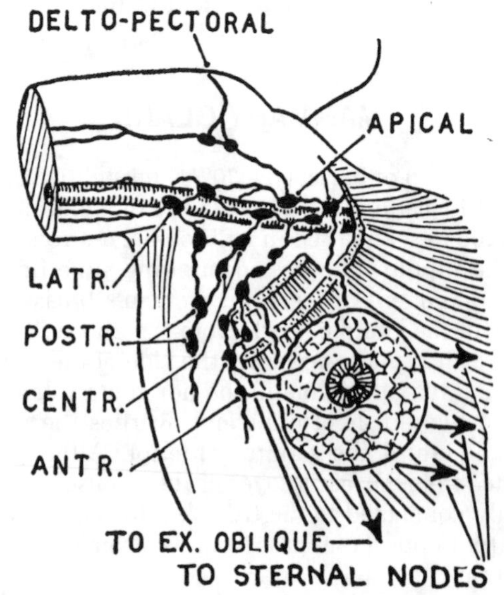

Figure 30.22. Lymphatics of breast and axilla (*sternal* nodes = *parasternal*).

AXILLARY LYMPH NODES

The axillary lymph nodes (*fig. 30.22*) are arranged in several main groups: (1) the *lateral* nodes lie along the inferior parts of the axillary vein. They receive the lymph vessels that ascend along the medial side of the arm and they empty into (2) the *apical* (*infraclavicular*) nodes that lie along the superior part of the axillary vein between the pectoralis minor and the clavicle. All the vessels of the limb, including those following the cephalic vein, drain either directly or indirectly into this group, and it in turn drains into the *subclavian lymph trunk*, which ends in the right lymph duct or (on the left) the *thoracic duct*. (3) The *pectoral* (*anterior*) nodes lie along the inferior border of the pectoralis minor with the lateral thoracic vein. (4) The *subscapular* (*posterior*) nodes lie along the subscapular veins. (5) The *central* nodes lie between the layers of fascia at the base of the axilla or in the fat deep to it. (6) Occasionally, 1 or 2 small nodes occur along the cephalic vein in the *deltopectoral triangle*.

ANOMALIES YOU MAY ENCOUNTER

Accessory Nipples or even *Mammae* are occasionally found along the "milk line" (*fig. 30.23*).

Sternalis Muscle. A narrow band of muscle, anterior to pectoralis major and in line with sternomastoid and rectus abdominis, is commonly found.

Axillary Arch. A variable band of voluntary muscle that stretches across the base of the axilla from latissimus dorsi to pectoralis major or coracoid process.

Pectoralis Major. Its sternal head may be absent.

The *lateral root of the median nerve* (or part of it) may travel with the musculocutaneous nerve far into the arm before joining the median nerve.

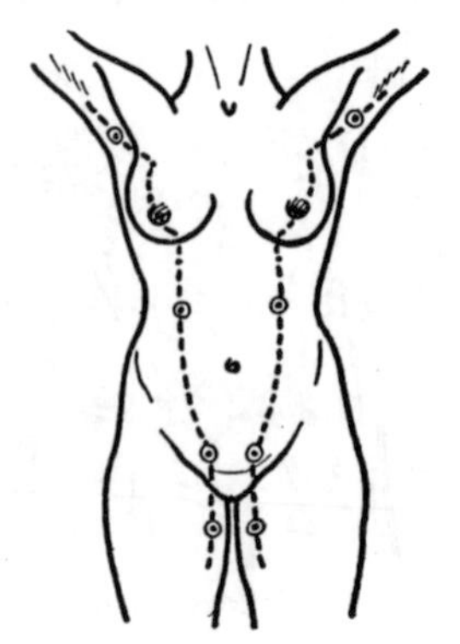

Figure 30.23. Accessory nipples may appear on the milk line.

Anatomical Features of the Physical Examination

Visual examination and palpation of the clavicle, scapula, proximal half of the humerus both at rest and during the range of possible movements quickly reveal problems to the experienced clinician. The beginner must perform these examinations frequently and carefully. Soon the angle of the acromion, most prominent parts of the scapula and clavicle, and the coracoid process will be easily located as landmarks. The anterior and posterior walls may be grasped while their enclosed muscles are contracting, and with the palmar aspect of the fingers, limited palpation of the contents of the axilla may be possible. The chief clinical feature that interests all physicians is a search for palpable lumps in the female breast and associated palpable lymph nodes in the axilla.

Behind the huge pectoralis major muscle, which arises from a sweeping C-shaped origin from the clavicle, sternum, and ribs, lies the axilla whose posterior wall is the scapula and its muscles. The floor is the skin and fascia of the armpit, and the medial wall is formed by the ribs coated by serratus anterior. This thin muscle runs from the ribs anteriorly to the medial border of the scapula posteriorly. Hence it pulls the scapula anteriorly unless its nerve (C5, 6, 7) which descends vertically on its lateral surface becomes incompetent ("winged scapula" resulting). Pectoralis minor muscle, ensheathed in clavipectoral fascia just deep to the pectoralis major, runs from several ribs obliquely superolaterally to the coracoid process, which is palable below the lateral part of the clavicle. The tendon of insertion is a surgical landmark because the brachial plexus and vessels cross posterior to it. The axillary artery and veins give off large and small branches to the area, the largest being the subscapular, beyond which the main artery gets definitely smaller [accounting for circulating "foreign" bodies (emboli, bullets) being arrested here].

The brachial plexus is so clinically important that you should review *Table 30.2* on p. 361, and *Figure 30.19* and *Figure 31.20*.

The mammary gland, its fascial planes and lymphatic drainage, the lymph nodes in the axilla, and the muscles that may need to be excised for carcinoma of the breast are summarized in *Figure 30.22*.

Clinical Mini-Problems

1. **Which bone in the shoulder "girdle" is most commonly fractured?**
2. **At which point of articulation would one expect to palpate a shoulder separation?**
3. **In radical mastectomies (breast and anterior thoracic soft tissue removal), the clavicular head of pectoralis major is frequently preserved to facilitate shoulder flexion after recovery. Which specific nerve innervates the clavicular head of the pectoralis major?**
4. **Most nerves lie on the deep surface of the muscles that they innervate. Which nerve in the axilla lies on the *superficial* surface of the muscle that forms the medial wall of the axilla?**

(Answers to these questions can be found on p. 586.)

31

Shoulder Region and its Joints

Clinical Case 31.1

Patient Amy Y. Several days after a radical mastectomy, this 52-year-old woman was told by her surgeon that she is ready to begin rehabilitation training of the shoulder region. She makes you think by asking you, a clinical assistant on the ward, searching questions about the movements possible and the muscles of the region. Will you be ready to explain?

Shoulder Girdle and Its Joints

As we noted in Chapters 29 and 30, the clavicle is both a strut and a mooring post that pivots on its sternal at-

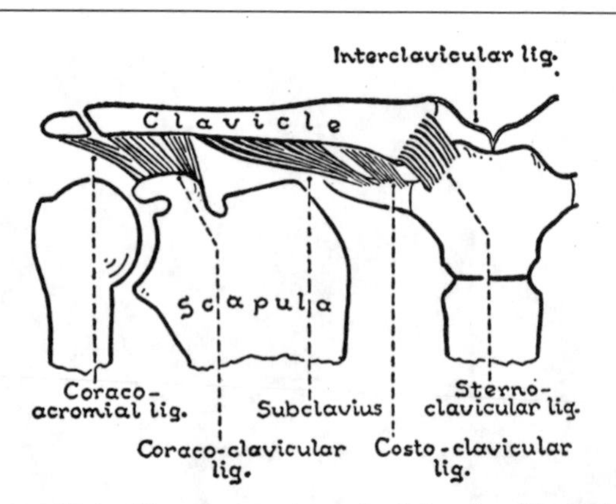

Figure 31.1. Structures having unity of direction. The feeble interclavicular ligament may be homologous with the wishbone of birds.

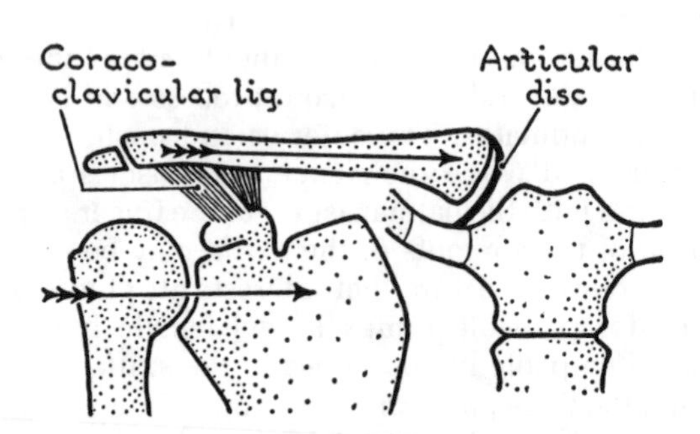

Figure 31.2. Structures having unity of function.

tachment, permitting the scapula to move extensively on the chest wall where it is kept moored by its large muscles, which act like dynamic mooring ropes.

Joints in which the clavicle takes part:

1. Sternoclavicular joint
2. Coracoclavicular ligament
3. Acromioclavicular joint

The chief duty of the clavicle is to thrust the scapula, and with it the arm, laterally and to prevent it from being driven medially. This function is performed by 2 tough ligaments, the *coracoclavicular ligament* laterally and the *articular disc* medially (*figs. 31.1 and 31.2*). A fall on the side of the shoulder must put a strain on them (*fig. 31.2*).

STERNOCLAVICULAR JOINT

Structure

The enlarge sternal end of the clavicle articulates in the shallow socket at the superolateral angle of the manubrium and adjacent part of the 1st costal cartilage. The end of the clavicle rises higher than the manubrium, the two making a poor fit, but the strong thick *articular disc* of fibrocartilage divides the joint cavity into two and prevents medial displacement of the clavicle. Its attachments must be to the clavicle superiorly and to the 1st costal cartilage inferiorly. Even in older persons when the disc may be perforated centrally, its hold is not lost. Strong *anterior* and *posterior ligaments*, to which the margins of the disc are attached, strengthen the joint.

Movements and Function

You can prove by palpation that movements at this joint allow the scapula considerable mobility. Much less obvious, the sternoclavicular joint allows the clavicle to undergo axial rotation during elevation (p. 379).

The *costoclavicular ligament* passes from the 1st costal cartilage to the sternal end of the clavicle (*fig. 31.1*).

CORACOCLAVICULAR LIGAMENT

This ligamentous joint has no articular surfaces—the bones are joined by a powerful ligament—i.e., a syndesmosis. The coracoclavicular ligament is in 2 parts, a *conoid* and a *trapezoid*, of interest to orthopedists and other specialists (*fig. 31.2*).

Functions

(1) Owing to their medial (and downward) direction, the ligament prevents the scapula from being driven medially. (2) It is the mainstay of the acromioclavicular joint and, so long as it is intact, that joint may, indeed, undergo subluxation (partial dislocation), but the acromion cannot be driven inferior to the clavicle (*fig. 31.3*). (3) It suspends the scapula.

ACROMIOCLAVICULAR JOINT

The medial border of the acromion has near its tip a small oval facet that forms a synovial joint with a similar small facet on the end of the clavicle (*fig. 31.3*). Strong parallel fibers form a complete capsule and a small *articular disc* hangs into the cavity.

Function

(1) This joint, while not vital to force transmission, does permit the scapula to move vertically on the chest wall when the pectoral girdle rises and falls (e.g., as when shrugging the shoulders). (2) It also permits the scapula (and with it the glenoid cavity) to glide forward and backward on the clavicle and so to face directions convenient to the head of the humerus (e.g., forward when striking a blow). (3) Also, its freedom is essential to free elevation of the limb (p. 379). It is vulnerable to injury in contact sports.

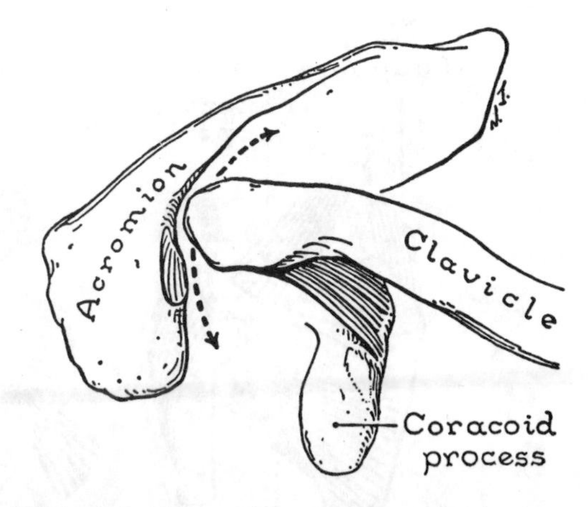

Figure 31.3. The acromion may swing forward or backward but, so long as the coracoclavicular ligament is intact, it cannot be driven inferior to the clavicle.

Relations

The joint is subcutaneous and is easily felt on pressing medially.

Movements of the Shoulder Girdle

1. *Simple elevation of scapula, i.e., the scapula moves superiorly.* A low level of activity in trapezius (upper), levator scapulae, and serratus anterior (upper) is sufficient to suspend the girdle, but when a weight is either supported on the shoulder or carried in the hand, these muscles contract vigorously (*fig. 31.4*). (See *Table 31.1.*)

Table 31.1. Muscles Acting upon the Shoulder Girdle*

Simple elevation	Simple depression	Elevation with upward rotation of glenoid cavity	Depression with downward rotation of glenoid cavity	Protraction or forward movement	Retraction or backward movement
Trapezius (upper) Lev. Scapulae Serratus anterior (upper)	Pect. minor Subclavius Pect. major and Lat. dorsi	Trapezius (upper) Trapezius (lower) Serratus anterior	Lev. scapulae Rhomboids Pect. minor Trapezius (mid.) Pect. major and Lat. dorsi	Pect. minor Lev. scapulae Serratus anterior and Pect. major	Trapezius (mid.) Rhomboids and Lat. dorsi

* *Note:* Pectoralis major and latissimus dorsi act on girdle indirectly, through the humerus.

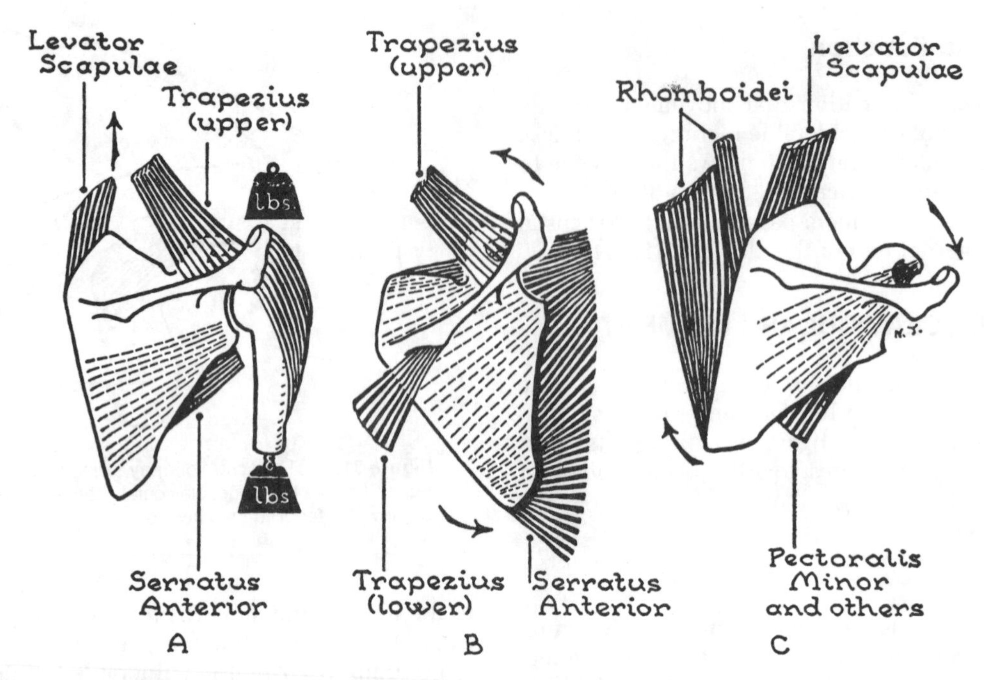

Figure 31.4. *A*, simple elevation of the shoulder girdle. The suspensory muscles of the girdle; *B*, elevation of the girdle with superior rotation of the glenoid cavity; *C*, depression of the girdle with inferior rotation of the glenoid cavity.

2. *Simple depression is brought about by the weight of the limb.* As an active movement, e.g., pressing downward or resting on parallel bars, depression of the girdle calls into action the pectoralis minor, which acts on the girdle, and the pectoralis major and latissimus dorsi, which act on the humerus (*fig. 31.5*). The timely contraction of these muscles saves the clavicle from fracture, when one falls on one's outstretched hand (p. 354).

3. *Elevation with upward rotation of glenoid cavity.* In this movement, the acromion rises, the superior angle of the scapula descends, and the inferior angle swings laterally. The trapezius (superior fibers), trapezius (inferior fibers), and the serratus anterior combine in this rotation. This movement is almost always part of a larger movement involving either abduction or flexion of the shoulder joint, as when the hand reaches for some object above the head, i.e., the entire limb is elevated (see p. 379).

4. *Depression with inferior rotation of glenoid cavity, that is, recovering from the last movement or overstepping the recovery, e.g., chopping wood (fig. 31.4).* The pectoralis minor, rhomboids, levator scapulae, and trapezius (especially the middle portion) are called into play, and the pectoralis major and latissimus dorsi, which act indirectly through the humerus, give them powerful assistance.

5. *Protraction of the scapula or forward movement, e.g., pushing.* The serratus anterior, pectoralis minor, and levator scapulae act together with the pectoralis major.

6. *Retraction of the scapula or backward movement, that is, recovering from the last movement or overstepping the recovery, e.g., pulling.* The trapezius (middle portion) and the rhomboids act with the latissimus dorsi.

Clinical Case 31.2

Patient Barry Z. This 60-year-old man sustained multiple injuries including fractures of the right scapula and upper part of the humerus in a serious car accident 2 weeks ago. Your clinical supervisor begins to quiz you about possible axillary nerve involvement, the general anatomy of the region, and possible surgical approaches to the fracture site and the "circumflex" (axillary) and radial nerves nearby.

Proximal Half of Humerus and the Shoulder Region

PROXIMAL HALF OF THE HUMERUS (*Plate 31.1*)

This half of the humerus (*figs. 31.6 and 31.7*), being devoted to the shoulder is quite different from the distal half, described with the arm on p. 387.

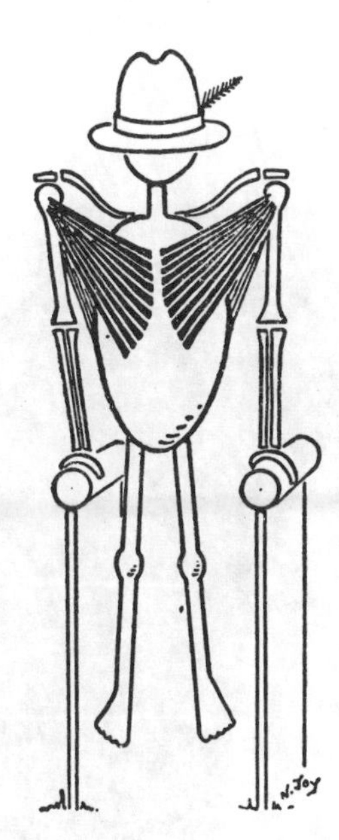

Figure 31.5. Pressing inferiorly or resting on parallel bars calls into action the pectoralis major and the latissimus dorsi.

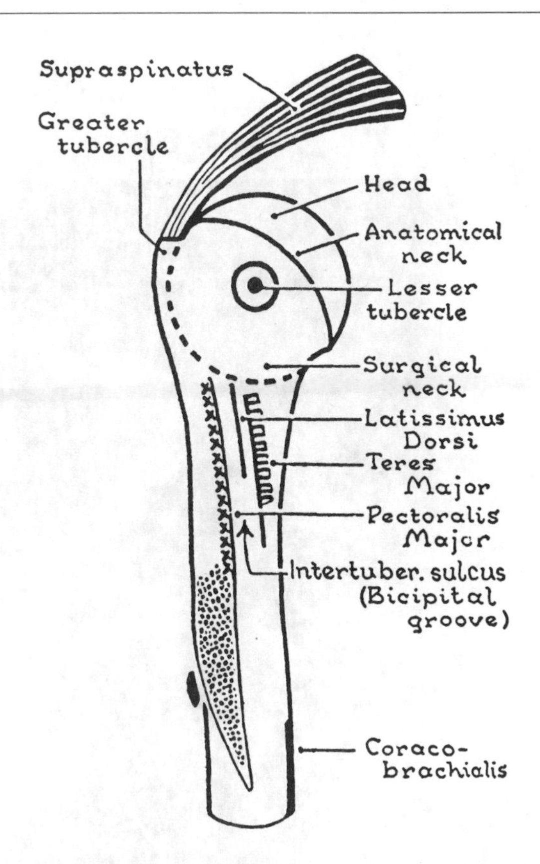

Figure 31.6. Proximal half of humerus (anterior view, schematic).

The **upper end** includes: (a) an articular portion, the *head*, which is covered with cartilage, directed superomedially; (b) the *anatomical neck* (to which the fibrous capsule of the joint is attached) and (c) and (d) 2 tubercles, the *lesser and greater tubercles* (tuberosities), separated by (e) the *intertubercular sulcus* (bicipital groove) for the long tendon of the biceps (*fig. 31.6*).

The prominent **lesser tubercle**—the insertion for subscapularis—points anteriorly; palpate it (through the deltoid) about 4 cm inferolateral to the coracoid process—but make sure the limb is in the anatomical position.

The **greater tubercle** bulges laterally beyond the acromion, giving the shoulder its roundness. Feel it through the deltoid. It possesses 3 flat contiguous facets (*fig. 31.7*) for the insertions of supraspinatus, infraspinatus, and teres minor.

The **surgical neck** is the zone between the head and tubercles proximally and the body distally. It is completely encircled by the circumflex humeral vessels and partly encircled by the axillary nerve (see *fig. 31.12*).

The cylindrical **proximal half of the body** features 2 *crests* and a *sulcus* on its anterior aspect and a *tuberosity* for the insertion of the deltoid laterally.

Into the *crest of the greater tubercle* (lateral lip of the bicipital groove) the pectoralis major is inserted; into the *crest of the lesser tubercle* (medial lip of the bicipital groove) the teres major is inserted.

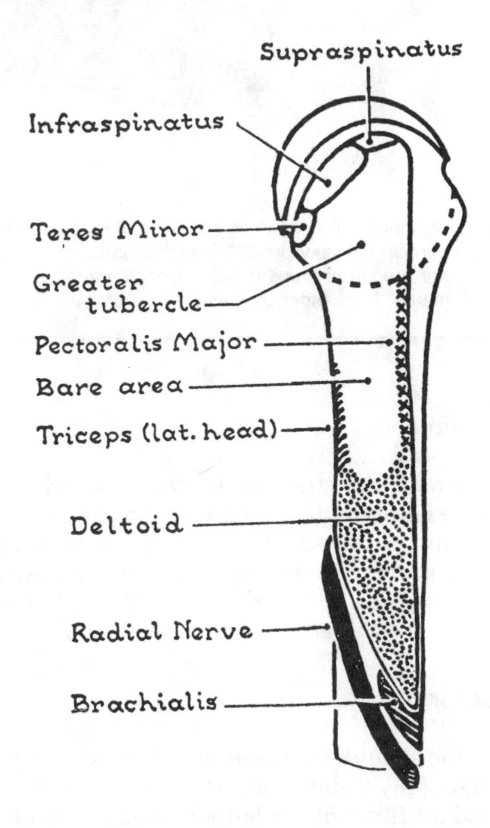

Figure 31.7. Proximal half of humerus (lateral view, schematic).

Deltoideus (*fig. 31.8*)

This muscle is shaped like an inverted Greek letter delta.

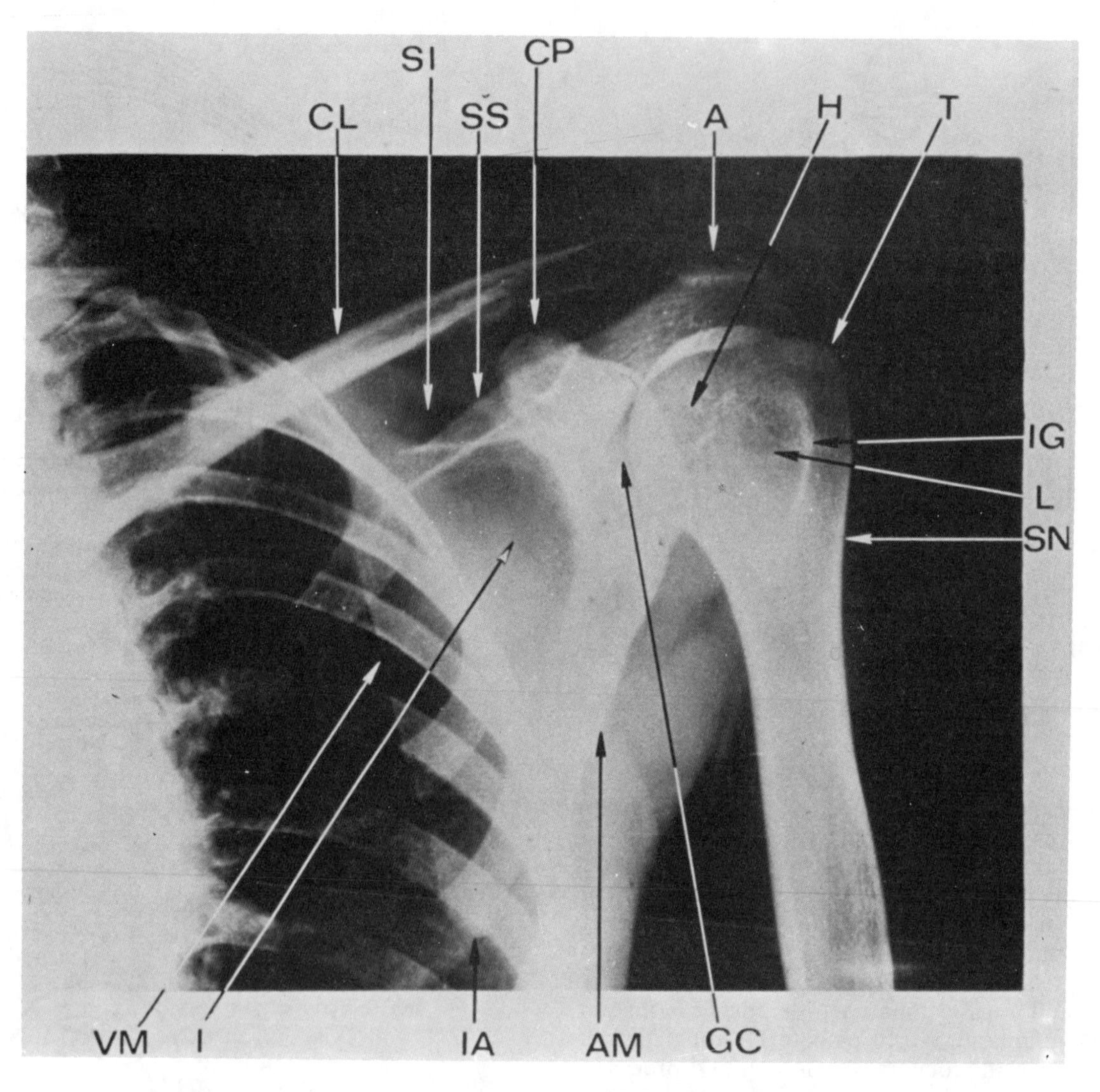

Plate 31.1 Shoulder (A-P view). *A*, acromion of scapula; *AM*, axillary margin of scapula; *CL*, clavicle; *CP*, coracoid process of scapula; *IG*, intertubercular groove of humerus; *GC*, glenoid cavity of scapula; *H*, head of humerus; *I*, infraspinous fossa; *IA*, inferior angle of scapula; *L*, lesser tubercle of humerus; *SI*, supraspinous fossa of scapula; *SN*, surgical neck of humerus; *SS*, scapular spine; *T*, greater tubercle of humerus; *VM*, vertebral (medial) margin of scapula.

Attachments

The deltoid arises from the lateral third of the clavicle, the lateral border of the acromion, and the whole length of the spine of the scapula. From this long origin, the muscle descends to the *deltoid tuberosity* halfway down the humerus (*figs. 31.7 and 31.8*). Posterior to the tuberosity is the *(spiral) groove* for the radial nerve.

Abductors

Since the shoulder joint has only 2 muscles superiorly, it can have only 2 *abductors*—the supraspinatus and the intermediate fibers of the deltoid (*fig. 31.9*). Each is supplied by the 5th and 6th cervical nerve segments, the one via the suprascapular nerve, the other via the axillary nerve.

Internal Structure of Deltoid (*fig. 31.10*). The intermediate part of the deltoid is multipennate, the muscle fibers being very numerous but very short. Thus it is very powerful, but its range of action is very short. The anterior and posterior parts have a different internal structure; they are composed of long parallel fibers because they take part in the more extensive movements of flexion and extension of the shoulder.

Deep to Deltoid. If you detach its origin and turn it downward you expose the surgical neck of the humerus and axillary nerve (circumflex nerve) and posterior humeral circumflex artery, which emerge from the quadrangular space, adhere to the deep surface of the deltoid and supply it (*see fig. 31.12*).

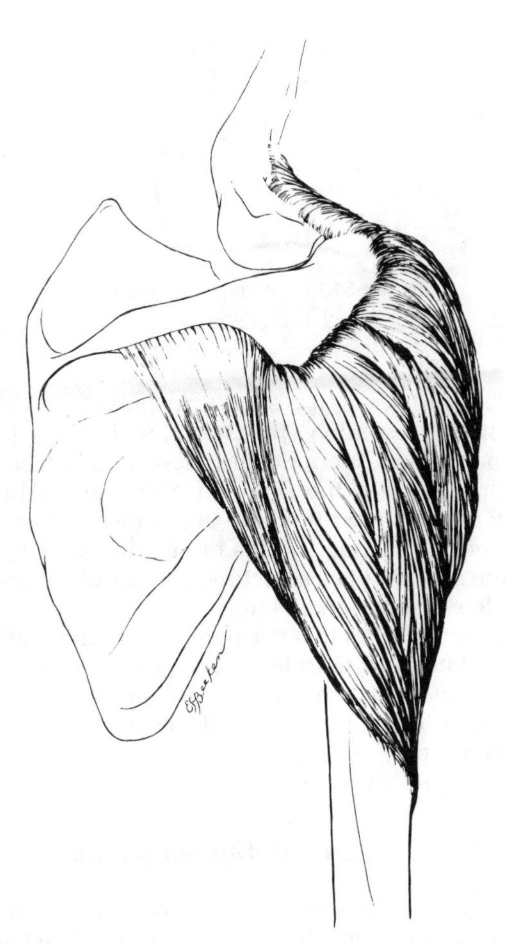

Figure 31.8. The right deltoid—posterior view (From Basmajian JV: *Primary Anatomy*, ed. 8. Williams & Wilkins, Baltimore, 1982.)

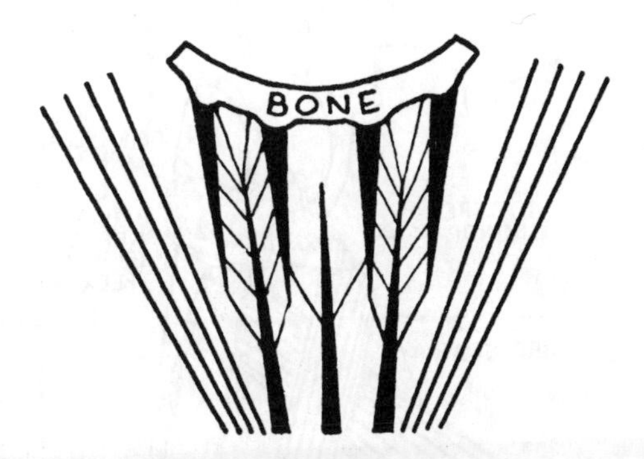

Figure 31.10. Architecture or internal structure of deltoid.

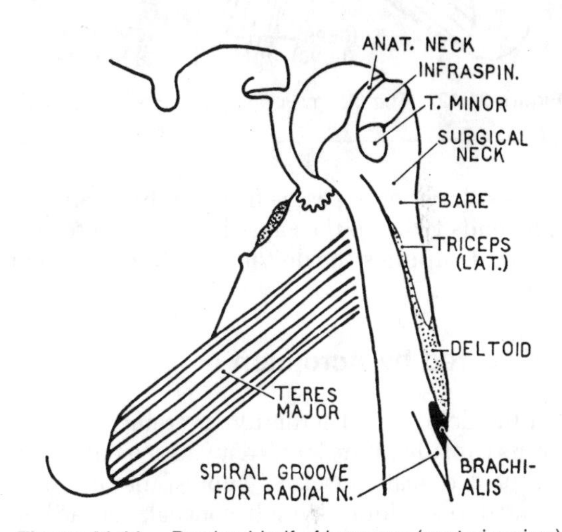

Figure 31.11. Proximal half of humerus (posterior view).

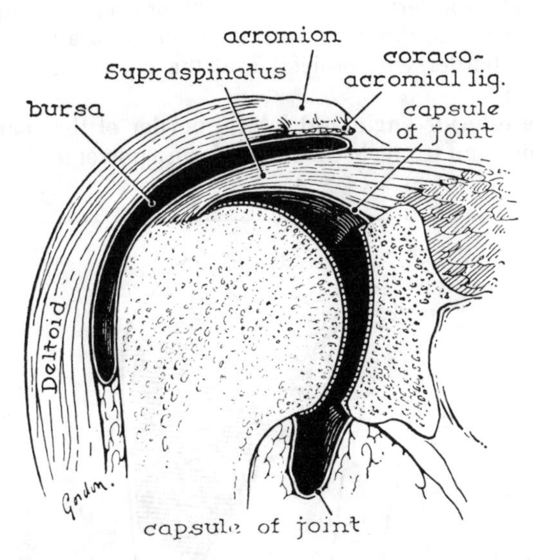

Figure 31.9. Deltoid and supraspinatus in coronal section. (From Basmajian JV: *Primary Anatomy*, ed. 8. Williams & Wilkins, Baltimore, 1982.)

The Quadrangular Space

The quadrangular space is bounded superiorly by the lateral border of the scapula and the capsule of the shoulder joint; *laterally*, by the surgical neck of the humerus; inferiorly, by the teres major (*figs. 31.11 and 31.12*); and *medially*, by the *long or scapular head of the triceps* (*fig. 31.12*). The last arises from the infraglenoid tubercle of the scapula. Passing through the quadrangular space are the *axillary nerve* and the *posterior humeral circumflex artery* and veins.

Medial to the long head of triceps (in an unimportant *triangular space*), a large anastomotic artery, the *circumflex scapular branch* of the subscapular artery which, on its way to the infraspinous fossa, grooves the axillary border of the scapula (*fig. 31.12*).

Axillary Nerve (Circumflex N.) (C5 and 6).

This nerve has 2 important relations:

1. The capsule of the joint superiorly.
2. The surgical neck of the humerus laterally.

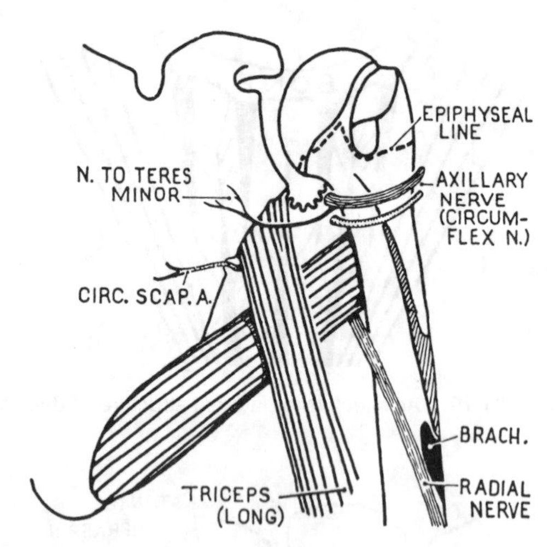

Figure 31.12. The triangular and quadrangular spaces.

As the axillary nerve passes through the quadrangular space, it sends twigs to the capsule and the teres minor. Thereafter, it supplies the deltoid and the skin covering it (*see fig. 32.5*).

Parts Covered by Acromion

When the deltoid and acromion are removed, it is not the fibrous capsule of the joint that you see but the tough tendons of 3 dorsal muscles—supraspinatus, infraspinatus, and teres minor—which conceal and adhere to the capsule.

The **subacromial bursa** lies between the acromion and the supraspinatus tendon. It plays the part of an accessory joint cavity and is large, extending beyond the acromion laterally deep to the deltoid as the subdeltoid bursa (*fig. 31.9*). When the arm is abducted, the greater tubercle passes completely under cover of the acromion, hence, the necessity for an extensive subacromial (plus subdeltoid) bursa.

Ossification of Scapula

The primary center appears at the 8th fetal week. Secondary centers: for the coracoid in the 1st year, and for the subcoracoid, including the superior end of the glenoid cavity, in the 10th year; these fuse with the scapula

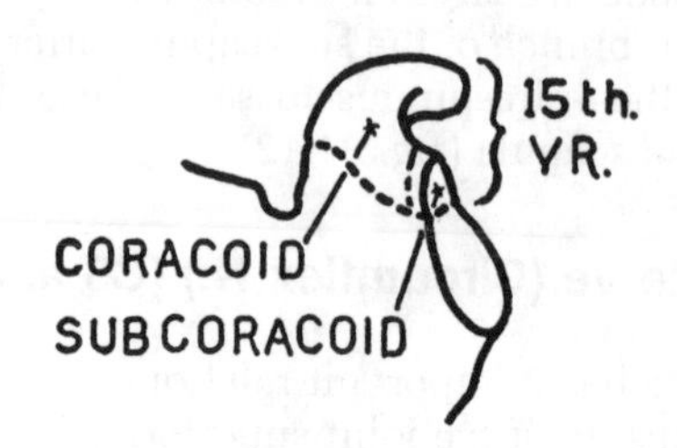

Figure 31.13. Coracoid epiphyses.

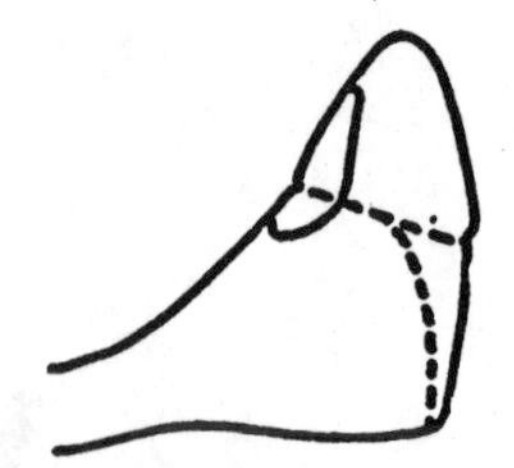

Figure 31.14. Acromial epiphyses.

during the 15th year (*fig. 31.13*). At puberty, 2 centers appear in the acromion (*fig. 31.14*), and others in the medial border and inferior angle; these usually fuse with the scapula between the 18th and 20th years and in all cases by the 23rd year (McKern and Stewart). The acromial epiphysis commonly fails to fuse and on X-ray examination may simulate a fracture. (For age of maturation of complete fusion, see p. 416.)

Four short muscles that surround the shoulder joint—called "**rotator cuff muscles**" by surgeons—along with the **teres major** are of great surgical importance in the cause and treatment of shoulder dislocations. The 4 "rotator cuff muscles" are the *subscapularis, supra-* and *infraspinatus,* and *teres minor.*

Subscapular Fossa and Subscapularis

The concave *subscapular fossa* provides for the **subscapularis** a multipennate origin (*fig. 31.15*). The tendon of insertion (to the lesser tubercle) grooves the anterior margin of the glenoid cavity and helps to give it a pear-shaped appearance (*fig. 31.16*).

Serratus anterior (p. 359) is interposed between the subscapularis and the chest wall (*fig. 31.17*). The rhomboids and the serratus anterior together keep the scapula applied to the thoracic wall. If either is paralyzed, the medial border will project from the back—a "winged scapula."

The inferior angle and lateral border of the scapula are thick and strong because they form a power lever (*figs.*

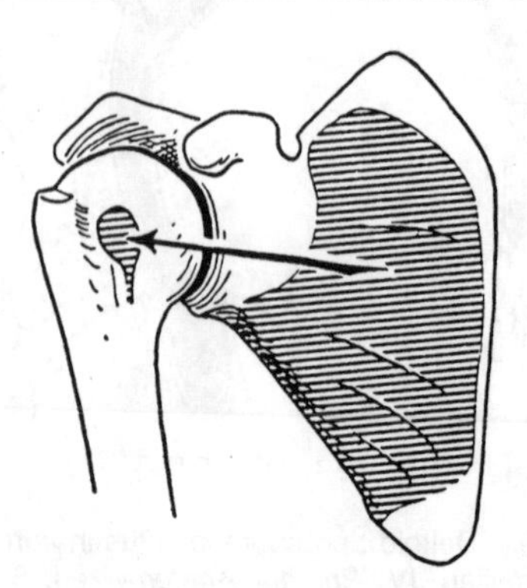

Figure 31.15. The attachments of subscapularis.

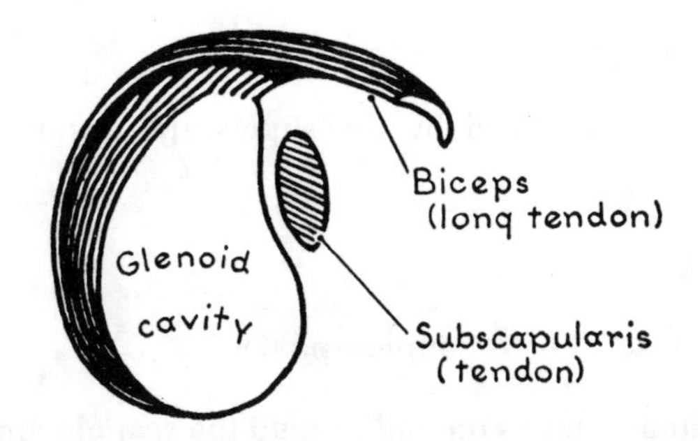

Figure 31.16. Glenoid cavity, biceps, and subscapularis.

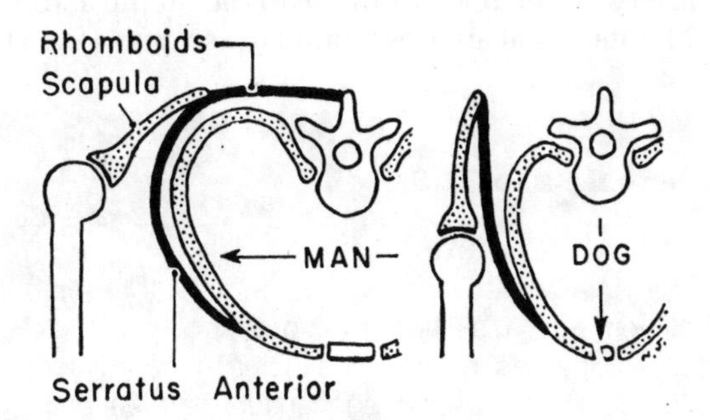

Figure 31.17. To show that the rhomboids and the serratus together hold the medial border of the scapula applied to the thoracic wall. In the quadruped, the thorax is suspended by the serrati.

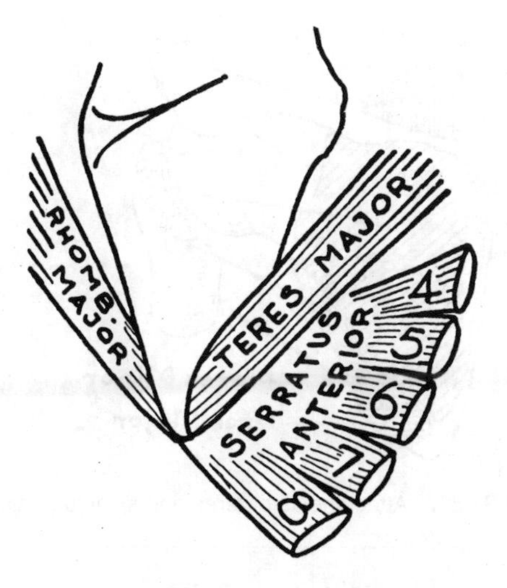

Figure 31.18. The 3 muscles attached to the end of the lever.

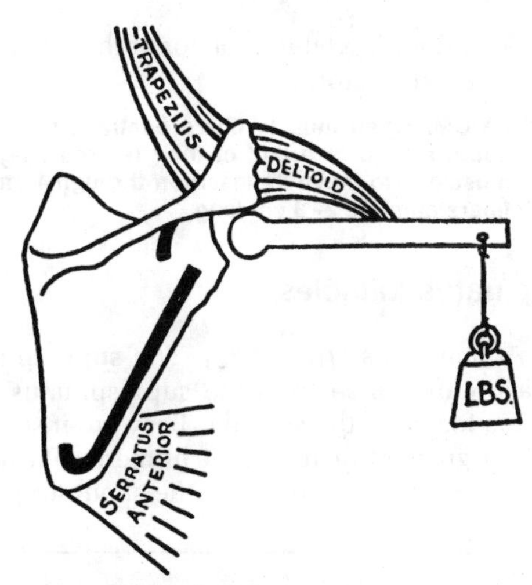

Figure 31.19. Three chief muscles concerned in elevation of limb; 2 strengthening bars on scapula.

31.18–31.20). The infraglenoid tubercle gives origin to the long head of the triceps.

Teres Muscles (L. teres = round) (*see fig. 31.21*)

Origins

The *teres minor* arises from the lateral border of the scapula. The *teres major*, a very large muscle, arises chiefly from a large impression on the dorsum of the inferior angle of the scapula.

Insertions

Both muscles are inserted into the humerus by tendon—the minor into the greater tubercle posteriorly, the major into the crest of the lesser tubercle, anteriorly.

Nerve Supply

Both muscles are supplied by nerve segments C5, 6, but via different nerves. The axillary nerve innervates teres minor and the lower subscapular nerve innervates teres major.

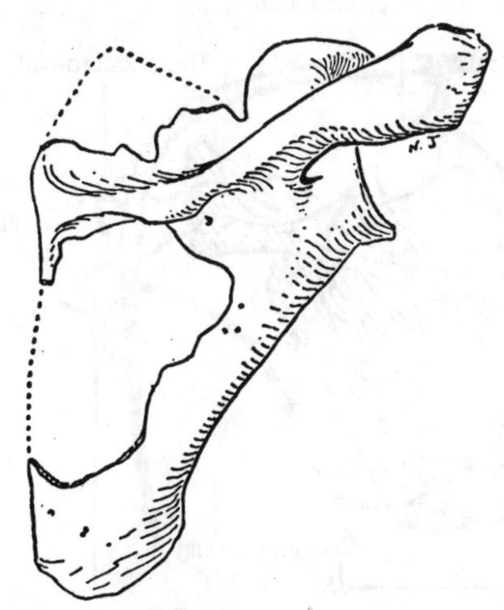

Figure 31.20. The lateral or axillary border is the strong border; the others can be broken away with the fingers, as seen here.

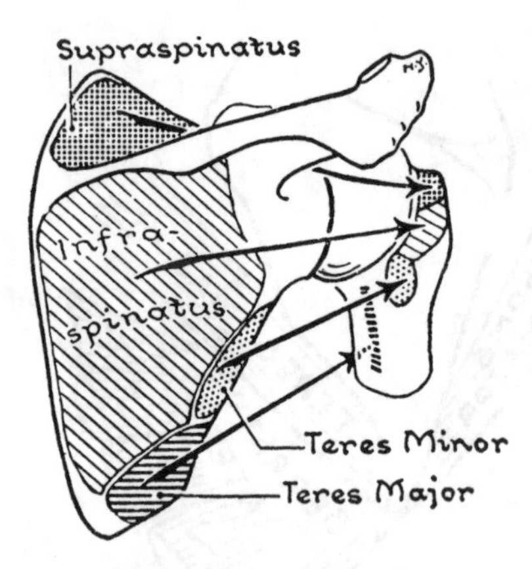

Figure 31.21. Attachments of teres and spinatus muscles.

Actions

The minor is a lateral rotator of the humerus; the major is a medial rotator.

AXIOM. When muscle fibers contract, they shorten by about a third or a half of their relaxed length, i.e., if a muscle is to move its insertion through 1 cm, its fleshy fibers must be 2–3 cm long.

Spinatus Muscles

Attachments (*fig. 31.21*). The **supraspinatus** and **infraspinatus** *arise* from the supraspinatus and infraspinatus fossae of the scapula. They are *inserted* by tendon on the greater tubercle of the humerus, the tendons fusing with the fibrous capsule of the shoulder joint.

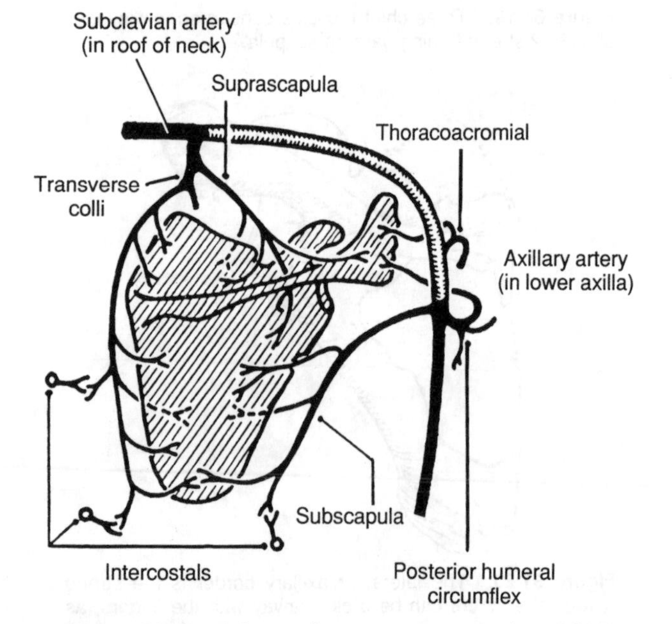

Figure 31.22. Scheme of anastomoses around scapula.

Nerve Supply

They are supplied by the suprascapular nerve (C5 and 6).

Actions

See Muscle Force Couples, p. 379.

The anastomoses on and around the scapula are very free. They bring notably the 1st part of the subclavian artery into communication with the 3rd part of the axillary artery on the body of the scapula, on the acromion, and with intercostal arteries around the scapula (*fig. 31.22*).

Clinical Case 31.3

Patient Carl A. While playing touch football just a half hour ago, this 21-year-old pre-med college student dislocated his shoulder joint, his third such episode. He has been rushed in for treatment at the adjacent student health clinic where you are helping. You will need to have enough anatomical knowledge to understand and to assist the qualified staff and also to explain the anatomy, pathomechanics, and treatment procedures both to the patient and his classmates who brought him in.

Shoulder Joint

This multiaxial ball-and-socket joint is also known by its Latin name, **articulatio humeri**, and by its unofficial but descriptive name, *glenohumeral joint*.

BONES

The **ball** is the head of the humerus. It forms one-third of a sphere and faces medially and posterosuperiorly.

The **socket** is the shallow pear-shaped glenoid cavity of the scapula. The tendon of the subscapularis is responsible for the anterior concavity that results in the pear shape. At the superior end is the supraglenoid tubercle for the long head of the biceps. Inferiorly, on the lateral border of the scapula is the infraglenoid tubercle for the long head of the triceps. A strip of fibrocartilage, the *glenoid labrum*, runs around the rim of the socket and makes a pliable elastic cushion (*fig. 31.24*).

MOVEMENTS

It is obvious that there is more freedom at the shoulder joint than at any other joint in the body. The movements are flexion and extension, abduction and adduction—and, therefore, circumduction, which combines them. The humerus can rotate medially and laterally on its own long axis. (To demonstrate these movements of axial rotation, bend your elbow to a right angle to eliminate confusion with pronation-supination movements of the forearm.)

FIBROUS CAPSULE AND LIGAMENTS

To allow such free movement, the capsule of the joint is very loose. In fact, when the shoulder muscles are removed, leaving the humerus attached to the scapula only by the capsule and ligaments, the head of the humerus can be pulled 1 cm away from its socket. This also happens clinically when C5, 6 nerve segments are damaged. However, in the normal adducted position of the hanging arm, the capsule becomes very tight superiorly, especially a part called the *coracohumeral ligaments* (*fig. 31.23*), effectively preventing downward displacement of the humerus.

The *fibrous capsule* stretches as a loose tube from just around the margin of the glenoid cavity of the scapula to the anatomical neck of the humerus. Inferiorly, however, it passes well down onto the surgical neck. When the arm is adducted, this inferior part lies in folds (*see fig. 31.28*).

The *synovial membrane* lines the fibrous capsule and is reflected along the anatomical (and surgical) neck as far as the hyaline cartilage of the head of the humerus.

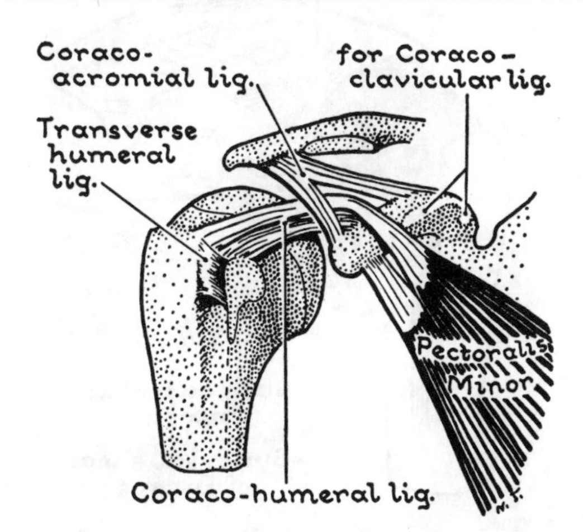

Figure 31.23. *Variation*: part of tendon of pectoralis minor augmenting the coracohumeral ligament.

The **coracohumeral ligament** extends from the lateral border of the coracoid process to the front part of the anatomical neck of the humerus (*fig. 31.23*). Its posterior fibers so blend with the capsule that the ligament can only be distinguished in an anterior view. The ligament prevents inferior dislocation of the adducted humerus and also limits lateral rotation (*fig. 31.23*).

Inferior dislocation of the shoulder joint is prevented by a **locking mechanism** dependent on 3 factors: (1) the slope of the glenoid fossa, leading to (2) the tightening of the superior part of the capsule (including the coracohumeral ligament), and (3) the activity of the supraspinatus muscle (Basmajian and Bazant).

INTERIOR OF THE SHOULDER JOINT

Figure 31.24 shows 3 unimportant bands, the superior, middle, and inferior glenohumeral ligaments; these are slight thickenings of the fibrous capsule, which may stand out in relief when viewed from within the joint. The middle band often stands out distinctly due to perforation of the capsule above and below it into the subscapularis bursa. Hence the *subscapularis tendon* enters into this picture (*figs. 31.24 and 31.25*).

The **long head of the biceps** may be followed from the intertubercular sulcus (which faces anteriorly, *fig. 31.26*) across the head of the humerus to its origin from the supraglenoid tubercle and posterior lip of the glenoid cavity, where it continues as the glenoid labrum (*figs. 31.16 and 31.24*).

The long tendon is retained in the intertubercular sulcus by the *transverse humeral ligament* (*fig. 31.26*); its synovial bursa obviously communicates with the joint space (*see fig. 31.29*).

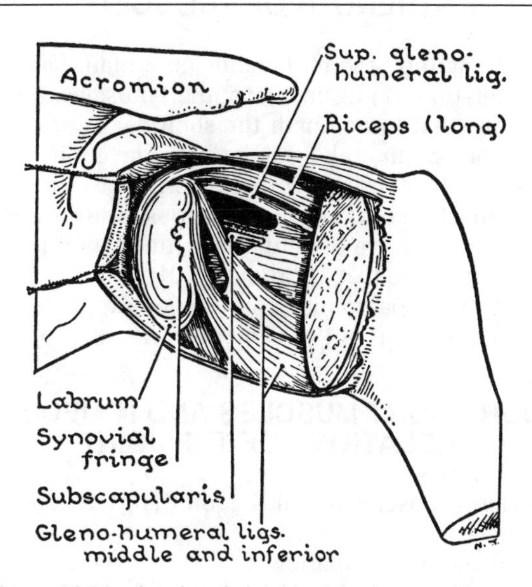

Figure 31.24. Interior of shoulder joint (exposed from posterior approach).

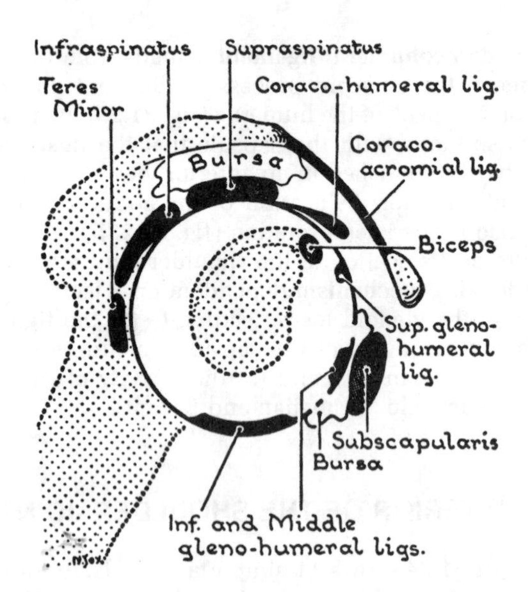

Figure 31.25. Scheme of shoulder joint on sagittal section (lateral view).

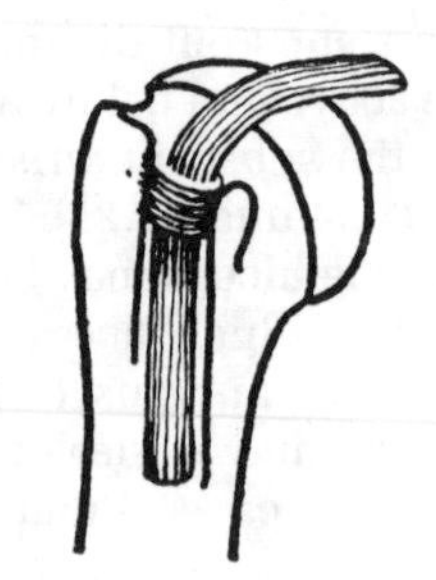

Figure 31.26. Proximal end of humerus (anterior view). Long tendon of biceps and transverse ligament.

STRENGTH OF THE JOINT

The strength of a joint depends on 3 main factors: (1) bony formation; (2) ligaments; and (3) muscles. It should be obvious that for strength the shoulder joint depends on the coracohumeral ligament at some times and on muscles at other times. While the long muscles perform movements, the short muscles—disposed closely around the head—also have the important function of retaining the head in its socket as "dynamic ligaments." The overhanging coracoacromial arch obviously prevents upward displacement of the humerus.

FOUR SHORT MUSCLES AND IMMEDIATE RELATIONS OF THE JOINT

The short muscles round the joint (*figs. 31.25, 31.27*) are:

Supraspinatus—superior
Infraspinatus and Teres Minor—posterior
Subscapularis—anterior

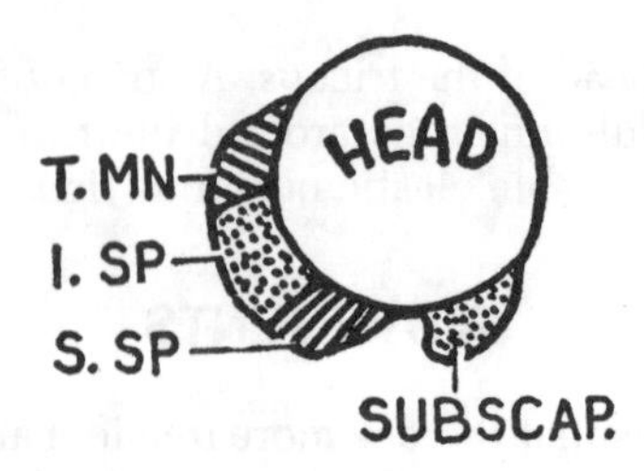

Figure 31.27. Insertions of the 4 short muscles that act as "accessory ligaments" of the shoulder joint viewed from above.

The tendon of supraspinatus fuses with the underlying fibrous capsule, as one would expect. The 4 short muscles are often called the "rotator cuff" but a better descriptor is either "guardians" of the joint or its *accessory dynamic ligaments*.

Inferior to the loose unsupported capsule lies the quadrangular space. Through it pass the axillary nerve (which supplies deltoid and teres minor) and its companion artery.

Lying side by side on the insertion of the subscapularis (*see fig. 30.12*), close to the joint, are (1) the brachial plexus and great vessels, (2) the coracobrachialis, and (3) the short head of the biceps, which in turn lies side by side with the long head.

BURSAE ASSOCIATED WITH THE SHOULDER JOINT

The **subscapularis bursa** lies between the subscapularis tendon and its groove on the anterior border of the glenoid cavity (*fig. 31.28*). Usually, bursal wall and joint capsule break down so that their cavities come into communication (*fig. 31.24*).

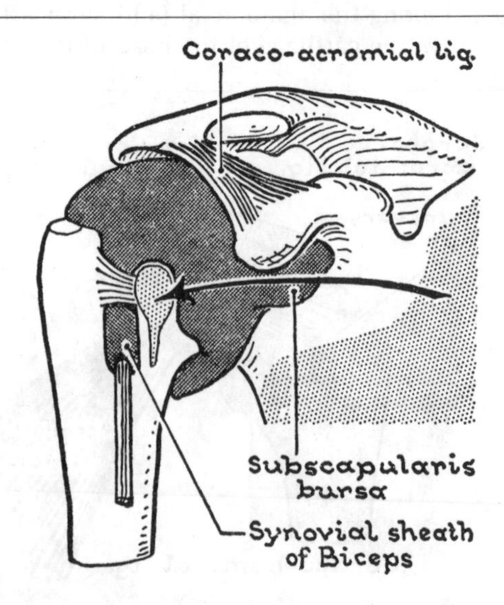

Figure 31.28. Synovial capsule of shoulder joint and communicating bursae (distended).

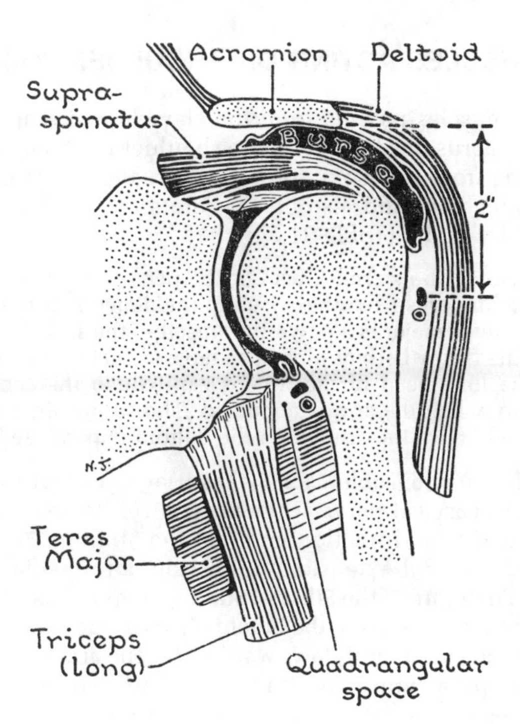

Figure 31.29. Shoulder joint (on coronal section). Observe: (1) the upper epiphyseal plate of the humerus; (2) the axillary (circumflex) nerve in the quadrangular space and 5 cm below the acromion; (3) the capsule in folds during adduction; and (4) the bursa, which is both subacromial and subdeltoid.

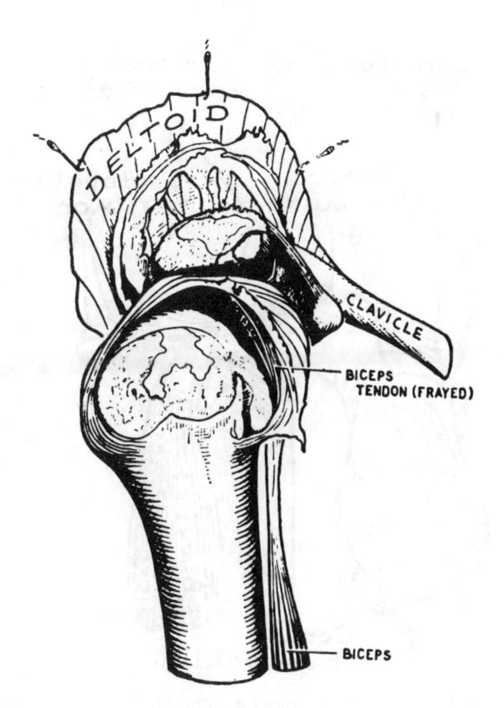

Figure 31.30. The supraspinatus tendon and the underlying capsule are commonly worn through. This is an advanced stage with eburnation (ivory-like polishing) of the contact surfaces of humerus and acromion.

The *biceps synovial sheath* in the intertubercular sulcus (*fig. 31.28*) has been mentioned above.

The **subacromial bursa** (*fig. 31.29*) lies between the acromion and the supraspinatus tendon, and it extends distally between the deltoid and the greater tubercle. How far down? As far as is necessary, i.e., to cover the part of the greater tubercle that passes inferior to the acromion during abduction of the humerus. *Bursitis*—probably beginning with degeneration and inflammation of an irritated supraspinatus tendon—is a common affliction of middle age.

As the result of breakdown of the supraspinatus tendon and underlying capsule, it is quite common in older people for the subacromial bursa to be in wide open communication with the synovial cavity of the shoulder joint (*fig. 31.30*). In our series, none (of 16) under 50 years old was perforated, but 3 (of 17) between 50 and 60 years old and 16 of the 46 over 60 years old were perforated.

The **coracoacromial arch** is formed by the coracoid, the coracoacromial ligament, and the acromion (*figs. 31.28 and 31.31*). The ligament joins the tip of the acromion to the lateral border of the coracoid process (*fig. 31.32*).

The arch, with the subjacent subacromial bursa, forms a resilient **secondary socket** for the head of the humerus, preventing its upward displacement. When the arm is abducted, an automatic lateral rotation of the humerus permits the greater tubercle to slip under the acromion.

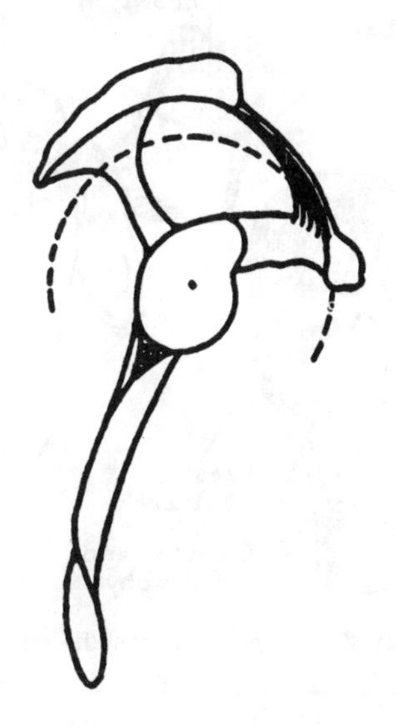

Figure 31.31. The coracoacromial arch.

EPIPHYSES NEAR SHOULDER (*fig. 31.33*)

The proximal epiphysis of the **humerus** rests on the spike-like end of the diaphysis. The epiphyseal line lies at the superior limit of the surgical neck. The proximal epiphysis of the humerus is an amalgamation of 3 smaller epiphyses: a pressure epiphysis for the head, which appears during the 1st year, and 2 traction epiphyses, one for the greater tubercle (3rd year), the other for the lesser

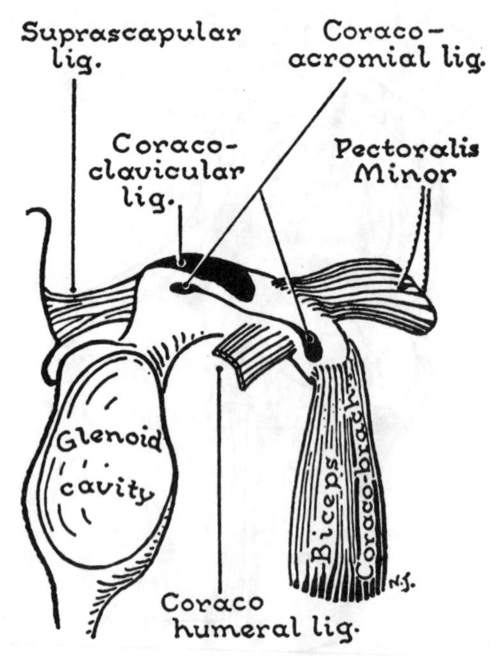

Figure 31.32. The structures attached to the finger-shaped coracoid process. (See *fig. 27.5*.)

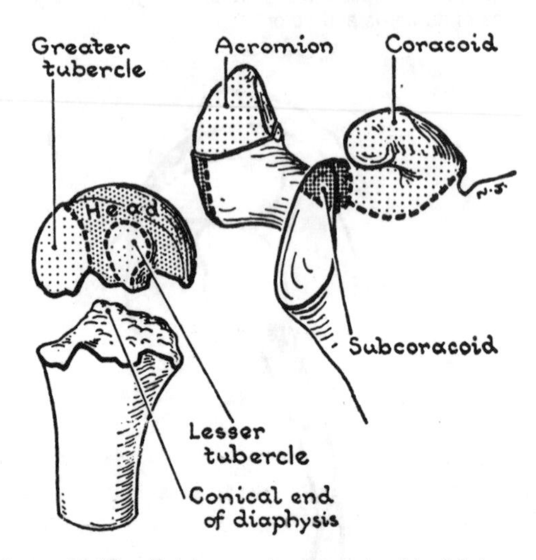

Figure 31.33. Epiphyses about the shoulder joint.

tubercle (5th year). All three fuse together before the 7th year, and the resulting single mass fuses with the shaft completely by the 18th year in some cases and by the 24th year in all cases (McKern and Stewart).

The epiphyseal line of the **coracoid process** crosses the superior part of the glenoid cavity. Union occurs during the 15th year. The epiphyseal line of the acromion crosses the clavicular facet. If the epiphysis has not fused by the 23rd year, it will remain separate (os acromiale).

Table 35.2 (on p. 416) summarizes the times of appearance and fusion of all epiphyses in the upper limb. It is intended for reference, not memorization.

MUSCLES ACTING ON SHOULDER JOINT

All muscles passing from the clavicle and scapula to the humerus must act upon the shoulder joint, and those passing from the trunk to the humerus must act on the shoulder joint and also on the joints of the clavicle or shoulder girdle (*Table 31.2*).

Note: The sternal fibers of the pectoralis major bring the humerus from the raised to the dependent position; the teres major is a powerful muscle—the sectional area of its fleshy belly is almost as large as that of the biceps. The long head of the triceps can extend the shoulder only when the arm is abducted. The supraspinatus can barely raise the arm when the deltoid is paralyzed.

Reference to *Table 31.2* makes it evident that the 5th and 6th cervical nerve segments control flexion, abduction, and lateral rotation and that the 5th, 6th, 7th, 8th, and T1 control extension, adduction, and medial rotation. Therefore, if the 5th and 6th segments are paralyzed, as when these roots of the brachial plexus are excessively stretched by forceful downward pulls, the arm will come to occupy a typical posture—the extended, adducted, and medially rotated position—described as the "position of a waiter taking a tip" (*fig. 31.34*). C5–6 root damage can result from being thrown forcefully on the shoulder or being born shoulder first.

NERVE SUPPLY OF THE SHOULDER JOINT

Sensory nerves come from C5 and 6 via the suprascapular, axillary, and lateral pectoral nerves, and posterior cord. Perhaps also some pain fibers come from the surface of the axillary artery (Gardner).

Elevation of the Upper Limb

The raised position can be attained either through forward flexion or through abduction. In this movement the sternoclavicular, acromioclavicular, and shoulder joints take part simultaneously. During the elevation through 180°, the humerus and scapula move in the ratio of 2:1 from almost the beginning of the movement to the termination. Thus, in general, for every 15° of elevation, 10° occur at the shoulder joint and 5° are due to movement of the scapula (Inman *et al.*).

As elevation progresses beyond shoulder level (90°), the clavicle rotates posteriorly on its own long axis and its lateral end rises. Without this clavicular movement, elevation above shoulder level is greatly restricted. If the scapula is held fixed, the humerus can move only through a right angle, and the power of movement is greatly di-

Table 31.2. Muscles Acting on the Shoulder Joint with Their Approximate Spinal Nerve Segments

Flexion		Extension		Abduction	
Deltoid (clav.)	5, 6	Deltoid (postr.)	5, 6	Deltoid (mid.)	5, 6
Supraspinatus	5, 6	Pect. major (st.)	7, 8, 1	Supraspinatus	5, 6
Pect. major (clav.)	5, 6	Teres major	5, 6		
Biceps	5, 6	Lat. dorsi	7, 8		
Coracobrachialis	7	Triceps (long)	7, 8		
5, 6 (7)		5, 6, 7, 8, 1		5, 6	
Adduction		**Med. rotate**		**Lat. rotate**	
Deltoid (postr.)	5, 6	Deltoid (clav.)	5, 6	Deltoid (postr.)	5, 6
Pect. major (cl.)	5, 6	Pect. major (clav.)	5, 6	Infraspinatus	5, 6
Pect. major (st.)	7, 8, 1	Pect. major (st.)	7, 8, 1	Teres minor	5, 6
Coracobrachialis	7	Subscapularis	5, 6		
Teres major	5, 6	Teres major	5, 6		
Lat. dorsi	7, 8	Lat. dorsi	7, 8		
Triceps (long)	7, 8				
5, 6, 7, 8, 1		5, 6, 7, 8, 1		5, 6	

minished. Therefore, in elevation of the limb the scapula must be free to rotate, and it does so with the permission of the clavicular joints (Inman *et al.*).

During the early phases of elevation, the sternoclavicular joint passes through its greatest range of movement, and in the terminal phase the acromioclavicular joint does so.

MUSCLE FORCE COUPLES

(1) *Movements of the humerus.* The supraspinatus and the deltoid act together and progressively in elevating the arm. The supraspinatus alone is unable to initiate abduction.

Electromyography reveals that both muscles act together throughout the entire range of the movement. These elevator muscles are assisted by the 3 short depressors—subscapularis, infraspinatus, and teres minor. The 2 groups act as a force couple, the one elevating and the other depressing (*cf.* prime mover and synergists).

(2) *Movements of the scapula.* The superior parts of the trapezius and serratus anterior (and the levator scapulae) constitute the superior component of the force couple necessary to the rotation "up" of the scapula, whereas the lower parts of the trapezius and serratus anterior constitute the inferior component.

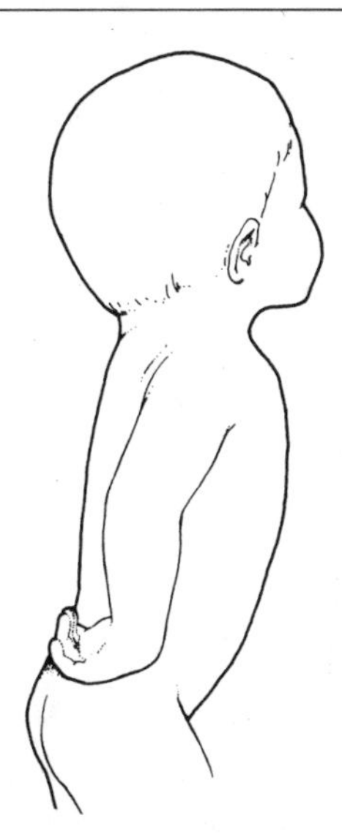

Figure 31.34. The "tip-taking" position of the upper limb of a baby with a damaged superior trunk of the brachial plexus.

Anatomical Features of the Physical Examination

General inspection and palpation of the region have been covered in Chapters 29 and 30. Restrictions of movements of the shoulder joint and pain and tenderness over the tendon of the long head of biceps are added considerations. Through an arthroscope, orthopedic surgeons now are able to scrutinize the interior of the shoulder joint and so identify and correct abnormalities of the capsule and biceps tendon with special instruments.

Force transmission and mobility of the pectoral (or upper-limb) girdle (clavicle and scapula) depend on the mooring of the clavicle by the very strong fibrocartilaginous disc at the sternoclavicular joint and the strong coracoclavicular ligament. Movements of the scapula on the chest wall are essential in raising of the arm, and several large muscles contribute to the rotation of the scapula "up" (trapezius and serratus anterior) and "down" (pectorals and latissimus dorsi). On the scapula, the *acromion*, especially its tip and its angle, is easily palpated and so are important landmarks.

Because the supraspinatus is squeezed between the head of the humerus and the inferior surface of the coraco-acromial arch the large subacromial bursa may become inflamed. The accumulation of fluid in it may be drained through a needle that perforates the origin of the deltoid muscle.

Clinical Mini-Problems

1. **Which important neurovascular structures are jeopardized in fractures of the surgical neck of the humerus?**
2. **"Rotator cuff" injuries may cause the supraspinatus muscle to be evulsed from its attachment to the superior facet on the greater tubercle of the humerus. How would one differentiate between an evulsed supraspinatus or paralysis of the supraspinatus due to damage of the suprascapular nerve?**
3. **What collateral route of circulation could be established to allow a sufficient blood flow to the right hand in a patient that has a gradual narrowing of the right axillary artery?**
4. **(a) Weakness in which muscles might potentiate posterior dislocation of the head of the humerus in the glenohumeral joint?**
   (b) Which nerve(s) supply these muscles?

(Answers to these questions can be found on p. 586.)

32

Cutaneous Nerves and Superficial Veins

Clinical Case 32.1

Patient Anne L. Four months ago, this 23-year-old woman broke her neck at level C6–7 of the spinal cord in a diving accident. She is now undergoing rehabilitation for severe paralysis of all four limbs (quadraplegia) with major sensory losses. You are asked two questions by the resident physician to whom you are assigned. Why is sensation excellent from the chest down to the level of the sternal angle but absent bilaterally down parts of the arm, forearm and hand? And which parts should have normal sensation? (Lucky for you, a more advanced student gets the questions on existing motor functions.)

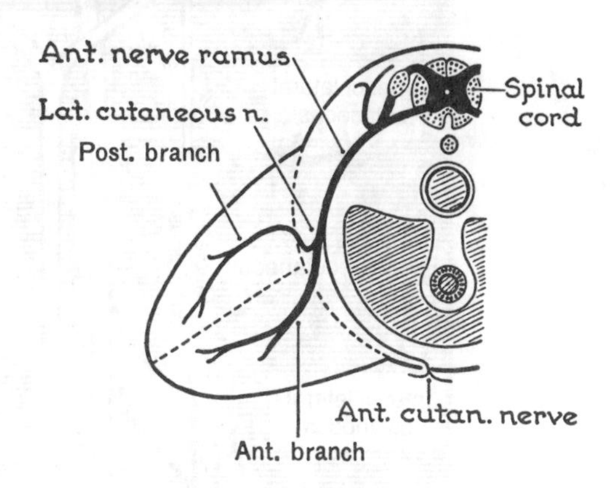

Figure 32.1. Scheme of the developing limb bud.

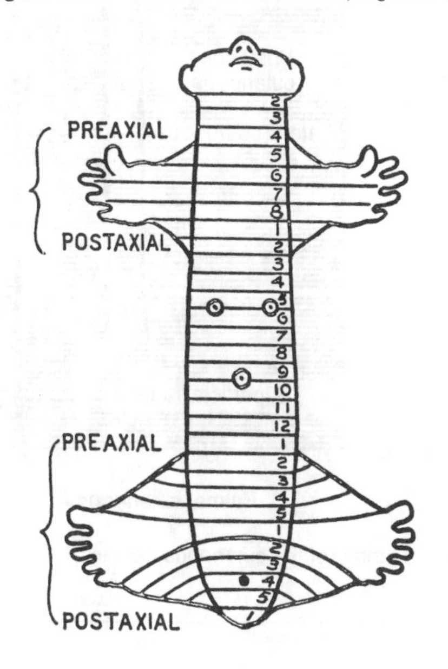

Figure 32.2. Scheme or primitive segmental nerve distribution.

Cutaneous Nerves

GENERAL

The muscles of the upper limb are supplied by nerve segments C5, 6, 7, 8 and T1 by way of the brachial plexus. The cutaneous supply is more extensive: it depends not only on the plexus, but also on 2 additional higher and 2 additional lower segments. It comes from segments C3 to T3. The limbs sprout and grow from the ventral half of the body; so, their nerve supply is derived from ventral nerve rami (*fig. 32.1*).

Dermatomes is the name given to the areas of skin supplied by individual spinal nerves in numerical sequence from the shoulder along the preaxial border of the limb to the thumb, from the thumb to the little finger, and from the little finger proximally along the postaxial border to the axilla (*figs. 32.2 and 32.3*). *Figure 32.4* summarizes the arrangement of the nerves themselves.

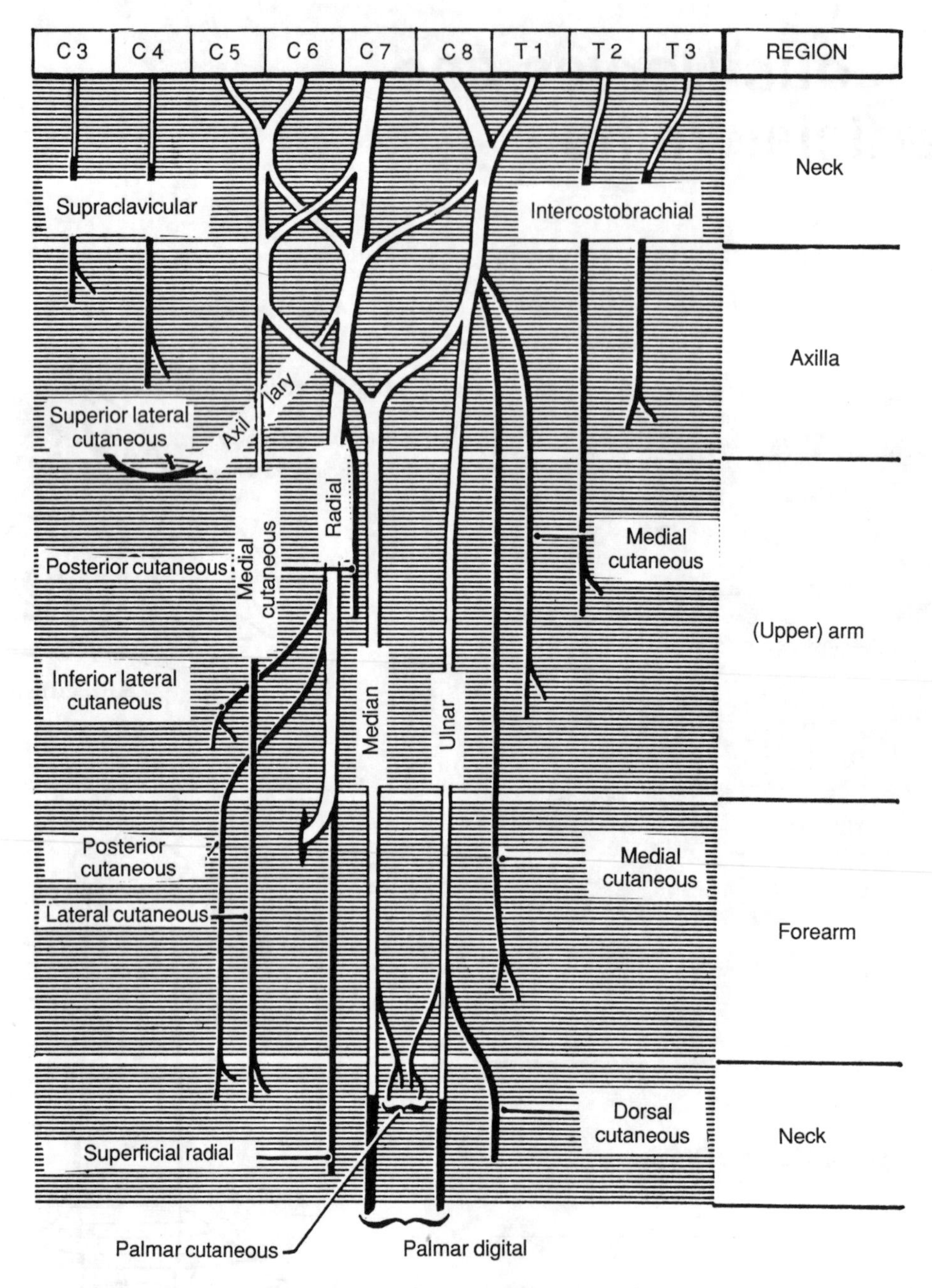

Figure 32.3. Scheme of the cutaneous nerves of the upper limb (in *black*). It shows 3 features—their source, level of origin, and level of termination.

AXILLA

The *intercostobrachial* nerve (T2) and small branches of T3 supply the skin of the armpit.

PECTORAL REGION

Superior to the 2nd rib, this region is supplied by the *supraclavicular nerves* (C3, 4) and inferior to the 2nd rib by the *intercostal nerves* T2–6 (*fig. 32.5*).

SHOULDER

Supraclavicular nerves (C3, 4) also supply the upper half of the deltoid region. The cutaneous branch of the axillary nerve (C5, 6), the *superior lateral cutaneous nerve of the arm*, appears at the posterior border of the deltoid and spreads anteriorly over the inferior half of the deltoid. Cutaneous branches of the dorsal rami of spinal nerves, especially of T2, reach the posterior aspect of the shoulder (*fig. 32.5*).

BRACHIUM OR ARM

The *posterior cutaneous nerve of the arm* (C5, 6, 7, 8) arises from the radial nerve while it is still in the axilla along with the motor nerve to the long head of the triceps. The *intercostobrachial nerve* (T2), accompanied by a branch of the 3rd intercostal nerve, pierces the fascial floor of the axilla and becomes cutaneous. The *medial cutaneous nerve of the arm* (C8 and T1) springs from the medial cord of the plexus and becomes cutaneous in the proximal third of the arm, medial to the brachial artery. The inferior *lateral cutaneous nerve of the arm* (C5, 6), a branch of the radial nerve, becomes cutaneous along the lateral intermuscular septum.

ANTEBRACHIUM OR FOREARM

It is supplied by *medial, lateral,* and *posterior cutaneous nerves of the forearm.* Each comes directly or indirectly from the similarly named cords of the brachial plexus.

The *medial cutaneous nerve of the forearm,* a direct branch of the medial cord ([7] 8, 1), becomes cutaneous halfway down the arm along with the basilic vein. Its 2 terminal branches pass anterior to the elbow and then descend anterior and posterior to the medial side of the forearm as far as the wrist.

The *lateral cutaneous nerve of the forearm* is the end branch of the musculocutaneous nerve (C5,6 [7]), and therefore of the lateral cord. It emerges at the lateral border of the biceps and divides into anterior and posterior branches. These both pass anterior to the elbow before descending on the front and back of the lateral side of

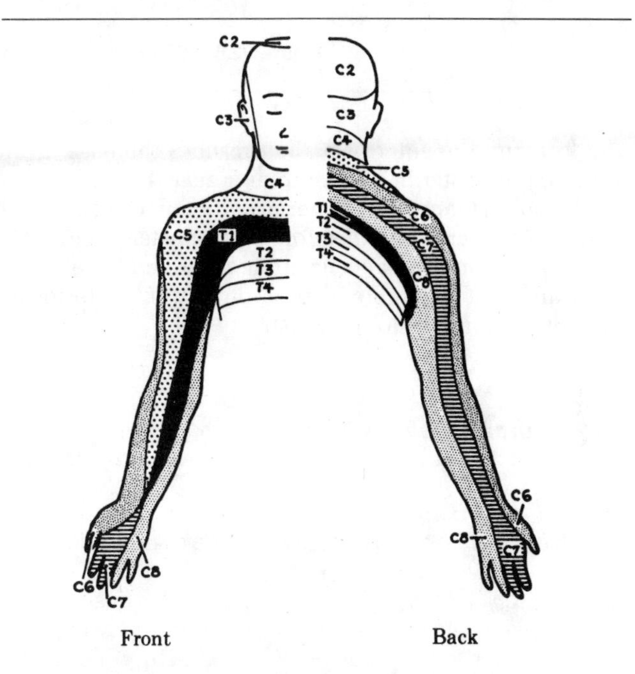

Figure 32.4. The cutaneous distribution of the spinal nerves—dermatomes. (After Keegan and Garrett.)

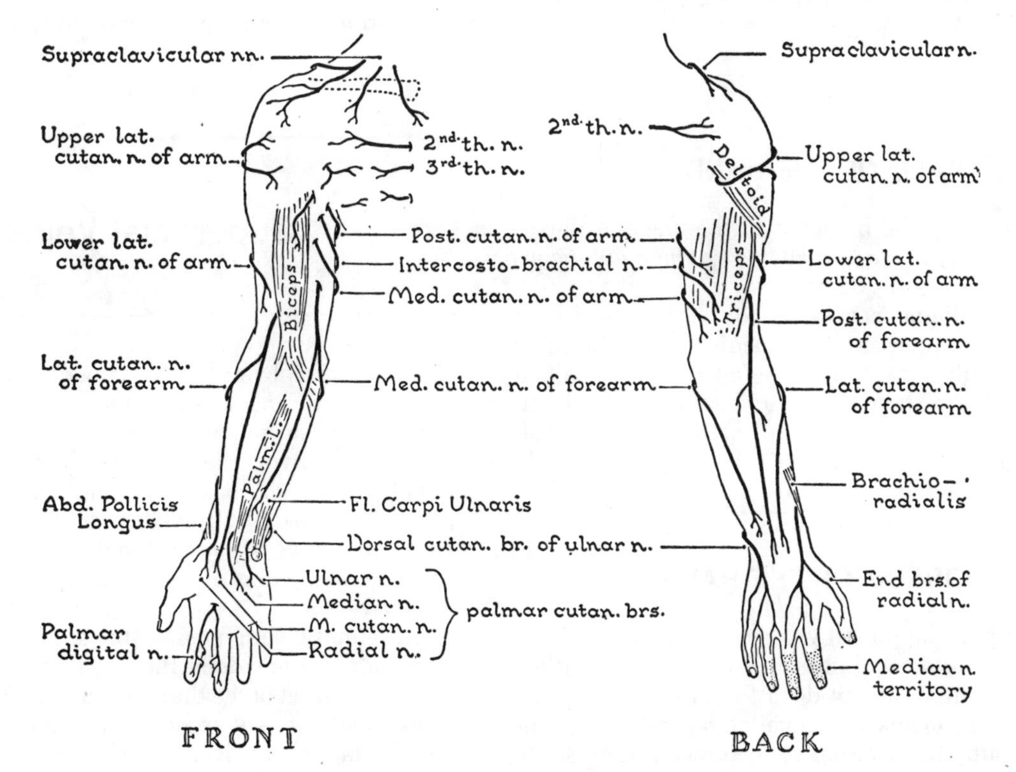

Figure 32.5. The cutaneous nerves of the upper limb.

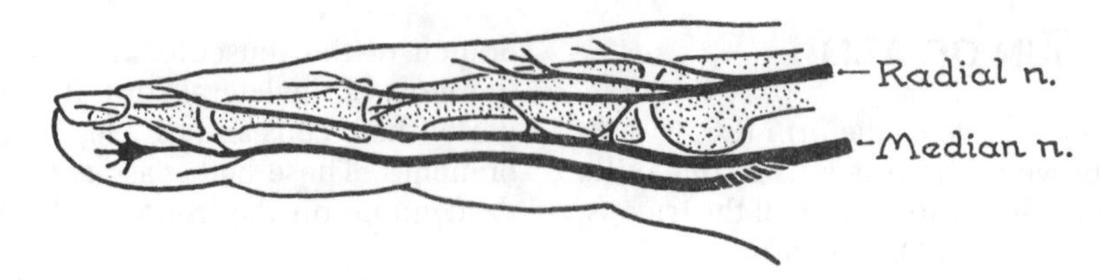

Figure 32.6. Digital nerves of the index. (From a dissection by M. Wellman.)

the forearm; the anterior branch reaches the base of the thumb, the posterior branch to end near the wrist.

The *posterior cutaneous nerve of the forearm*, a branch of the radial nerve and therefore of the posterior cord (C5, 6, 7, 8), becomes cutaneous along the line of the lateral intermuscular septum and runs down the posterior aspect of the forearm to the wrist.

Clinical Case 33.2

Patient Bernard M. Two weeks ago, this 35-year-old construction worker had a steel plate slash across his left ulnar nerve at the elbow. "What areas of sensory loss should you expect on the basis of simple anatomy?" is the question thrown at you in the first clinical ward rounds you attend. (A fellow student gets asked about the motor losses that are covered in later chapters.)

PALM OF THE HAND

Each of the terminal branches of the brachial plexus except the axillary nerve contributes a *palmar cutaneous branch* to the hand (*fig. 32.5*)

The **palmar digital nerves**—1 for each side of each digit—are very important. The median nerve supplied 3½ digits, the ulnar nerve 1½ (*see fig. 32.7*). The palmar digital nerves furnish branches to the entire palmar surfaces of the digits, to the distal parts of the dorsal surfaces including the subungual regions (i.e., deep to the nail), and to the local joints (*fig. 32.6*)

DORSUM OF THE HAND

The dorsum is supplied by the radial, ulnar, and median nerves. The *dorsal cutaneous branch of the ulnar nerve* (C7, 8), after passing deep to the flexor carpi ulnaris, crosses the medial (ulnar collateral) ligament of the wrist. The superficial radial nerve crosses the "snuff-box." The ulnar nerve supplies the dorsum of 1½ digits

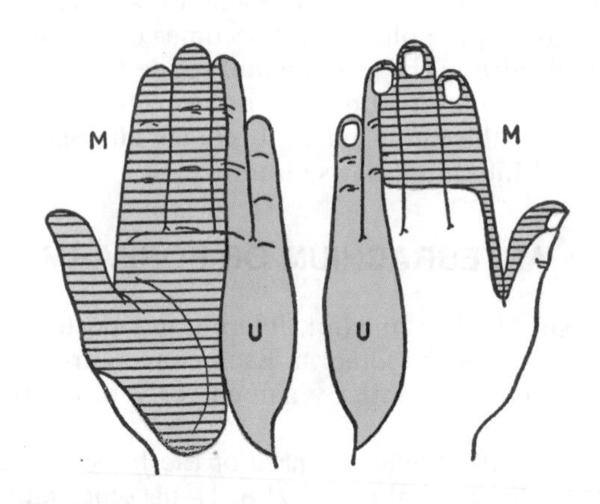

Figure 32.7. Distribution of median and ulnar nerves in the hand. (After Stopford.)

(*fig. 32.7*). The radial supplies the remainder, except for the distal parts of the lateral 3½ digits, which the median nerve supplies.

Variations in pattern in the hand (*fig. 32.8*) are common.

Superficial Veins

The superficial veins of the upper limb (*fig. 32.9*) form many patterns. The *dorsal venous arch*, seen as a plexus on the dorsum of your own hand, receives *digital branches*. It ends medially as the basilic vein and laterally as the cephalic vein. The **basilic vein** ascends on the medial side of the forearm to the elbow and then to the middle of the arm where it pierces the deep fascia. It then joins the companion veins of the brachial artery to become the axillary vein.

The **cephalic vein** crosses "the snuff-box" (that is, the depression at the side of the wrist proximal to the base of the metacarpal of the thumb) superficial to the branches of the radial nerve. It ascends on the radial border of the forearm, lateral to biceps in the arm, and in the cleft between the deltoid and pectoralis major at the shoulder.

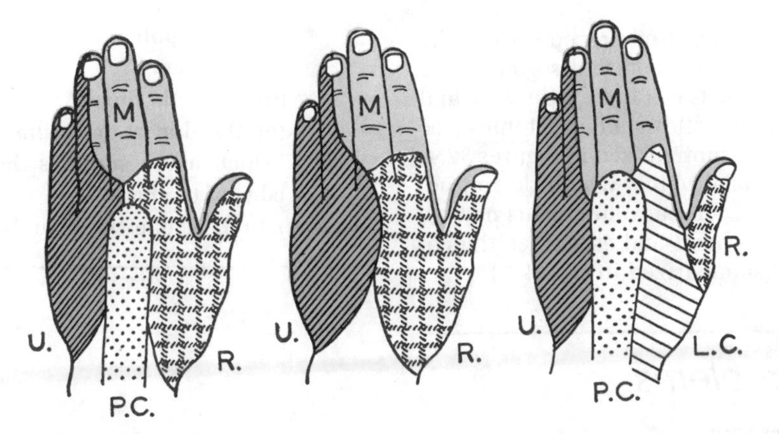

Figure 32.8. Patterns of distribution of cutaneous nerves on dorsum of hand. (After Stopford.)

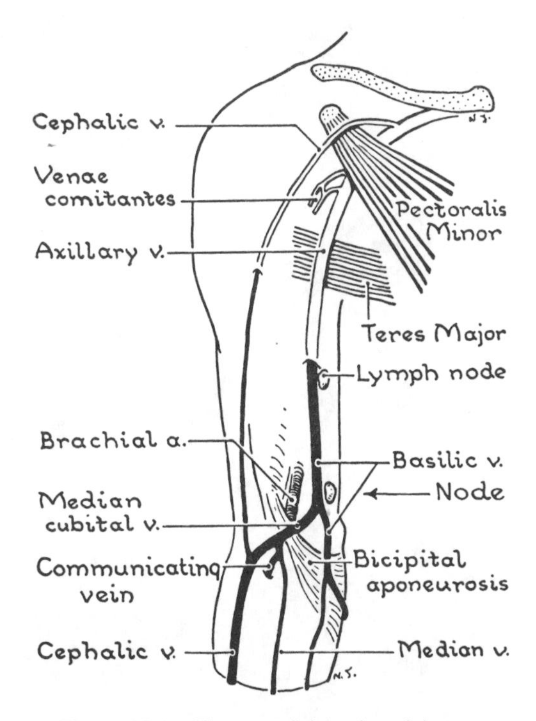

Figure 32.9. The superficial veins of the arm.

It pierces the deltopectoral triangle and ends in the axillary vein (*fig. 32.9*).

A small vein, the *median vein*, runs up the anterior aspect of the forearm and, after communicating with a *deep vein*, bifurcates into a medial and a lateral branch, which join the basilic and cephalic veins, respectively. This is a simple M-like pattern, but more commonly, a large oblique vein, the **median cubital vein**, lying anterior to the elbow, joins the cephalic vein to the basilic vein. The median cubital vein passes superficial to the bicipital aponeurosis (medial to the tendon). That alone separates it from the brachial artery and median nerve, and it passes superficial to (or between) the branches of the medial cutaneous nerve of the forearm. Beware of confusing an anomalous *superficial ulnar artery* for a vein. This can be disastrous in clinical practice.

Lymph Nodes

The most distal superficial node in the upper limb is the *cubital (supratrochlear) node*, placed 5 cm above the medial epicondyle. There may be a node where the basilic vein pierces the deep fascia (*fig. 32.9*). Though not palpable, these become tender when inflamed.

Anatomical Features of the Physical Examination

Sensory testing of the skin must be done whenever nerve damage of the distribution of C5–T3, or the individual parts of the brachial plexus and all its branches are suspected. Light touch and pinprick are the main sensations tested, but temperature and vibration (with a tuning fork) also may be tested in special cases. To permit a localizing diagnosis, areas of abnormality (anesthesia or paresthesia) are mapped out and matched with the known distribution of dermatomes, of parts of the plexus, and of the individual nerves.

Cutaneous nerves of the upper limb traced back to the spinal cord origin of their fibers represent segments C2 to T3 with the emphasis on C5–T1, i.e., the brachial plexus, following the orderly pattern of dermatomes (*see fig. 32.3*). Specific nerves are summarized in *Figures 32.4, 32.5 and 32.6*. The superficial veins, which do not accompany arteries but may accompany cutaneous nerves, begin on the dorsum of the hand and ascend as the cephalic and basilic veins (summarized in *fig. 32.9*).

Clinical Mini-Problems

1. **What dermatomes supply—**
   (a) the tip of the shoulder?
   (b) the nipple?
   (c) the thumb?
   (d) the little finger?
   (e) the skin of the axilla?
2. **Which nerve supplies the dorsum of the medial edge of the hand?**

(Answers to these questions can be found on p. 586.)

33

Arm (Brachium) and Elbow

Clinical Case 31.1

Patient Arthur S. During a violent street fight several hours ago, this 23-year-old man was slashed across the middle of the front of his right arm and was saved from rapidly bleeding to death by a policeman who applied a tourniquet. You are asked by the Emergency Room surgeon to loosen the tourniquet as she explores the depths of the wound. Which muscles, nerves, and vessels will be encountered that may need repair?

Distal Half of Humerus

The **distal end** of the humerus has articular and nonarticular parts. The former is divided into the rounded *capitulum* for the head of the radius and the spool-shaped *trochlea* for the trochlear (semilunar) notch of the ulna (*fig. 33.1*). Immediately above and wide of these 2 are 2 projections, the *medial* and *lateral epicondyles*. The medial epicondyle can be easily grasped through the skin; the posterior surface of the less prominent lateral epicondyle is smooth, subcutaneous, and palpable.

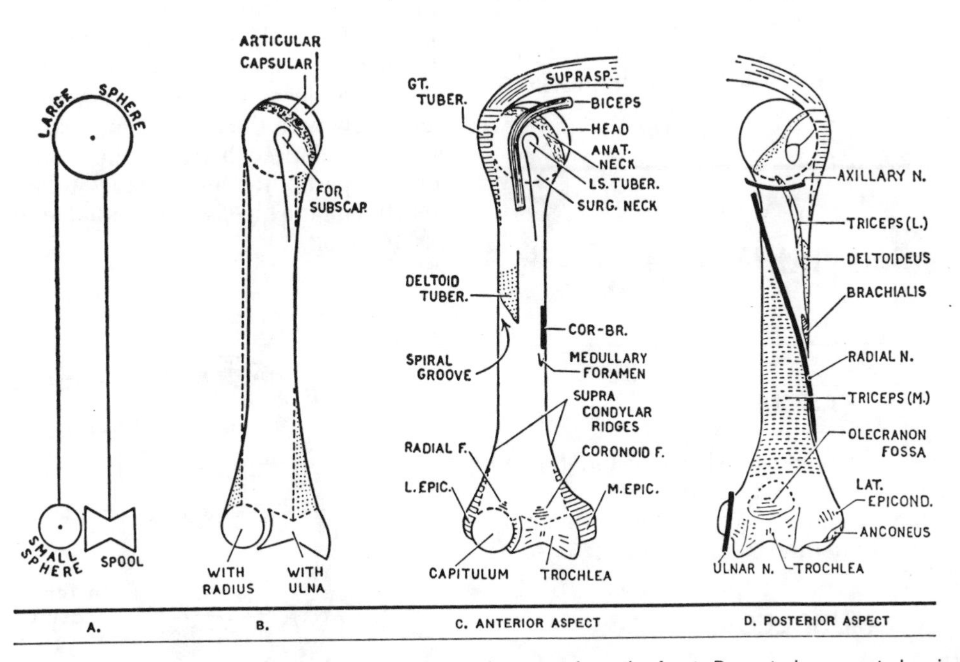

Figure 33.1. *A*, *B*, and *C* represent progressive stages in sketching a humerus from the front. *D*, posterior aspect showing attachments of muscles and contacts with nerves.

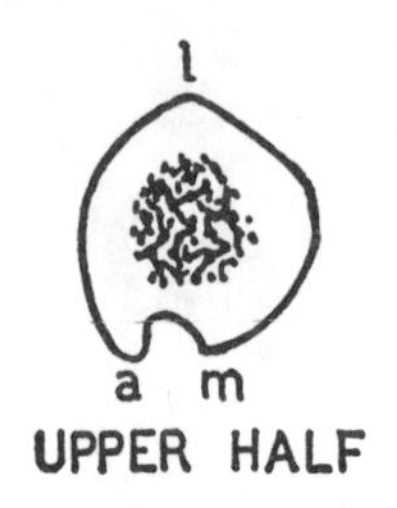

Figure 33.2. Cross-sections of humerus: borders are indicated by letters.

There are also 3 fossae (*fig. 33.1*); the one posteriorly, the *olecranon fossa*, receives the olecranon of the ulna when the elbow is extended.

The **lower half of the body of the humerus** is flattened and is divided by the medial and lateral supracondylar ridges into *anterior* and *posterior surfaces* devoted to fleshy fibers of the brachialis and medial head of the triceps.

The *supracondylar ridges* give attachment to the intermuscular septa for additional attachment of muscles. The lateral and more prominent ridge ascends to a broad shallow groove, the *groove for the radial nerve* (spiral groove), which intervenes between the ridge and the *deltoid tuberosity*. A rounded strengthening bar of bone (*figs. 33.2 and 33.3*) makes the distal half of the humerus triangular on cross-section.

Fasciae and Muscles

FASCIA

The muscles are enveloped in a sleeve of tough deep fascia, whose fibers take a circular course. This sleeve is divided into an anterior and a posterior compartment by the medial and lateral intermuscular septa (*fig. 33.4*).

MUSCLES

In the anterior compartment there are 3 muscles—coracobrachialis, biceps brachii, and brachialis. In the posterior compartment there is 1 muscle with 3 heads—triceps brachii.

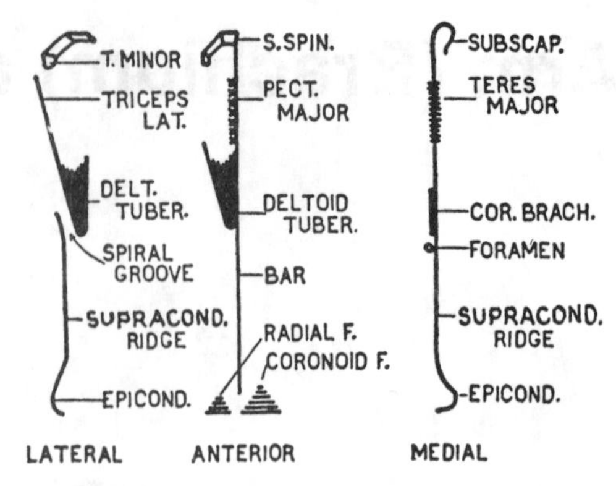

Figure 33.3. The borders of the humerus.

The Coracobrachialis

This muscle shares a common tendon from the tip of the coracoid process with the **short head of the biceps**. Its humeral insertion and function are of little significance.

The Long Head of the Biceps

This muscle descends in the intertubercular sulcus, encased in a *synovial sheath* continuous with the synovial cavity of the shoulder joint.

Anterior to the brachialis, the 2 heads of the biceps unite. Their common tendon is inserted into a posterior rough ridge on the tuberosity of the radius, located below the medial side of the neck of the radius (*fig. 33.5*); so, when the biceps contracts, as when driving a screw or corkscrew with the right hand, it supinates the radius. A *bursa* is required between the tendon and the smooth anterior part of the olive-shaped tuberosity. A secondary insertion of the biceps, the *bicipital aponeurosis*, is shown in *Figure 33.6*.

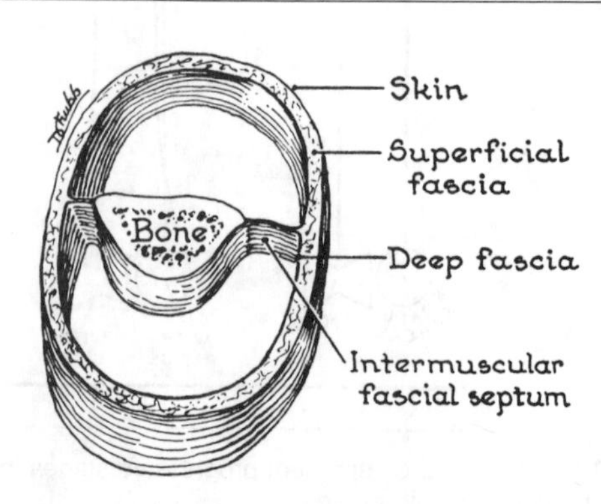

Figure 33.4. The 2 compartments of the arm.

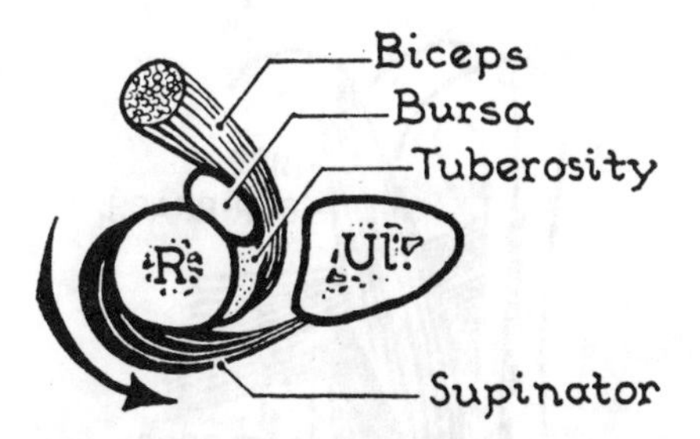

Figure 33.5. The insertion of the biceps. The biceps and the supinator supinate in direction of the arrow.

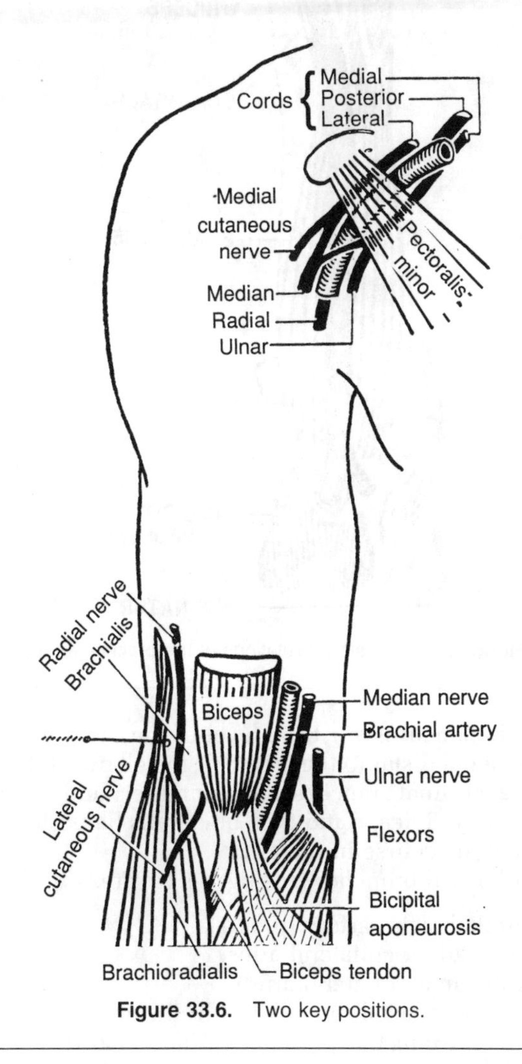

Figure 33.6. Two key positions.

The Brachialis

This muscle arises from the anterior aspect of the distal half of the humerus and from the intermuscular septa. Its fibrous insertion produces a rough elevation (tuberosity of the ulna) on the anterior aspect of the coronoid process of the ulnar.

Actions

Brachialis is the main flexor of the elbow joint. The biceps is both a powerful flexor and a powerful supinator of the forearm. But at the shoulder, its actions (in the expected directions) are much less powerful. Coracobrachialis is unimportant.

Nerve Supply

These 3 muscles are exclusively supplied by the musculocutaneous nerve (C5, 6) (*fig. 33.6*).

The Triceps Brachii

Its *long* head springs from the *infraglenoid tubercle* of the scapula. It has 2 humeral heads, *medial* and *lateral*; the medial head arises by fleshy fibers from the posterior surface of the humerus below the level of the (spiral) groove for the radial nerve and from the intermuscular septa (*fig. 33.1*). The tendinous lateral head is responsible for the line that ascends from the posterior margin of the deltoid tuberosity.

The common tendon of insertion is attached to the olecranon process of the ulna.

Action and Function

The triceps is the extensor of the elbow joint. Its chief function or use is to keep the elbow extended when one is pushing an object.

Nerve Supply

All 3 heads are supplied by the radial nerve.

Palpable Structures Around Elbow

Before studying the vessels and nerves of the arm, become familiar with the disposition of the structures around the elbow, for they assume a key position to both the anterior aspect of the arm and the forearm (*figs. 33.6 and 33.7*). You can, therefore, at any time roll up your sleeve, palpate the structures, and thereby refresh your memory.

Note that when the elbow is flexed to a right angle and the forearm forcibly supinated (the palm of the hand then faces upward), the *biceps tendon* and the *bicipital aponeurosis* stand out at the middle of the elbow as a prominent central landmark, 2 cm wide. They can be grasped between the index finger and the thumb. When the biceps

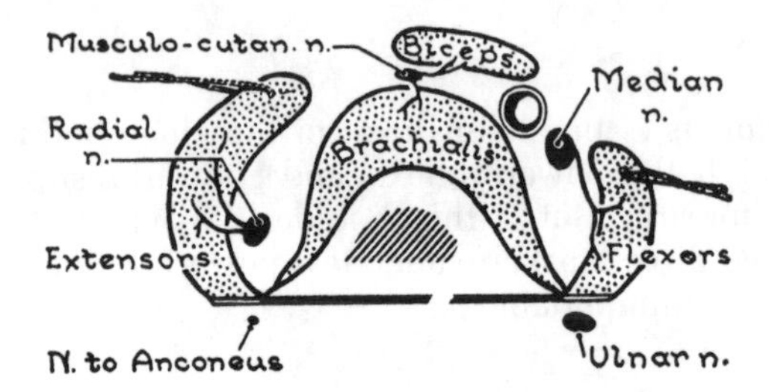

Figure 33.7. Section showing nerves just proximal to the elbow. Note their "sides of safety."

is relaxed (rest forearm on the table), the pulsations of the *brachial artery*, can be felt just medial to the biceps tendon; and with the tips of the fingers the *median nerve*, which lies just medial to the artery, can be rolled on the brachialis.

The *ulnar nerve*, lying behind the medial epicondyle, can be felt to slip from under the finger tips as they are drawn across it, and tingling feelings in the little finger may be felt. It can be traced halfway up the arm. Brachioradialis is seen and felt along the lateral edge of the elbow.

Cubital Fossa

The cubital fossa is the triangular space anterior to the elbow, bounded laterally by brachioradialis and medially by pronator teres. It is covered by deep fascia reinforced with bicipital aponeurosis, superficial to which are cutaneous nerves and the clinically important superficial veins (*see figs. 32.5 and 32.9*).

The *contents* are biceps tendon, brachial artery and its 2 terminal branches, median nerve, and (on retracting brachioradialis) radial nerve (*figs. 33.6; see also 33.11*).

The *floor* is chiefly brachialis.

Arteries and Nerves of the Arm

BRACHIAL ARTERY (*fig. 33.8*)

This is the largest artery whose pulsation and whose walls can be felt satisfactorily in a living person. Feel it along the medial bicipital furrow throughout the length of the arm to the point where it disappears deep to the

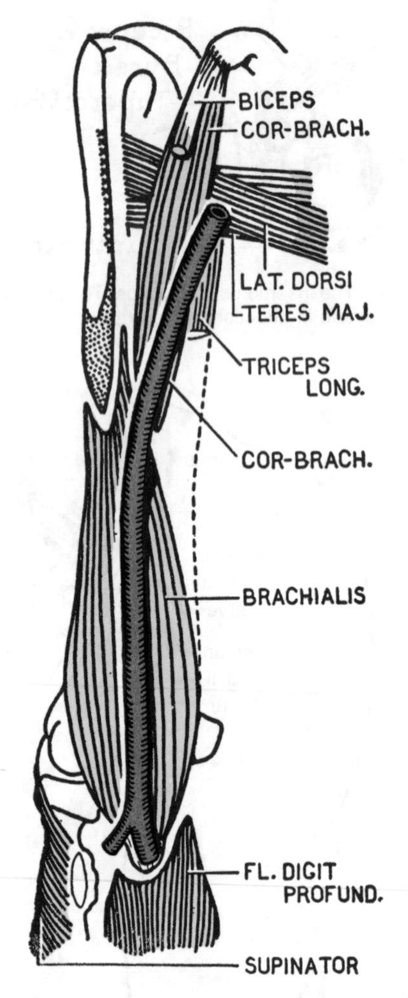

Figure 33.8. Posterior relations of the brachial artery.

bicipital aponeurosis. At the level of the neck of the radius, 2–3 cm distal to the crease of the elbow, it divides into its 2 terminal branches, the *ulnar artery* and (smaller) *radial artery*. Triceps and then brachialis lie posterior to it through its course in the arm.

Its *collateral branches* (*see fig. 34.14*) are:

Profunda brachii artery (p. 391),
Superior ulnar collateral artery
Inferior ulnar collateral artery,
whose special interest lies in the anastomoses they effect at the shoulder and elbow.
Muscular branches
Nutrient artery to the humerus.

Venae comitantes accompany the brachial artery and make a very open network around it. They join the axillary vein.

MAJOR NERVES

The **musculocutaneous nerve** arises from the lateral cord of the brachial plexus (C5, 6, 7) and must course distally and laterally. It pierces the coracobrachialis and continues between the biceps and brachialis, and it supplies these 3 muscles. It emerges at the lateral margin of the biceps, and the *lateral cutaneous nerve of the forearm* (*fig. 33.6*).

The **ulnar nerve** arises from the medial cord of the brachial plexus C(7), 8; T1) and, without any branching, extends to the posterior surface of the medial epicondyle (*fig. 33.6*) and onward into the forearm. It is applied to the medial side of the brachial artery down to the middle of the arm. There it leaves the artery and passes posterior to the medial intermuscular septum into the posterior compartment of the arm, where it lies subfascially, applied to the medial head of the triceps.

[Accompanying it in the elbow region are the superior ulnar collateral artery and a nerve, called the *ulnar collateral nerve*, sent by the radial nerve to the medial head of the triceps.]

The 2 roots of the **median nerve** (C(5), 6, 7, 8, and T1) unite in the axilla on the lateral side of the artery. It slowly slips across the brachial artery, lying medial to it at the elbow (*fig. 34.9*). The median nerve gives off no branches in the arm. It commonly receives a large communication from the musculocutaneous nerve and occasionally it gives one to it.

The **radial nerve**, being the "continuation" of the posterior cord (C5, 6, 7, 8(1)), lies posterior to the axillary artery and then the brachial artery anterior to the long head of the triceps, which conducts it into the posterior compartment of the arm. The nerve enters the spiral groove which is converted into a tunnel by the lateral head of the triceps. Within the tunnel, the radial nerve descends almost vertically along the origin of the medial head of the triceps (*fig. 33.10*).

On escaping from the tunnel, the radial nerve enters

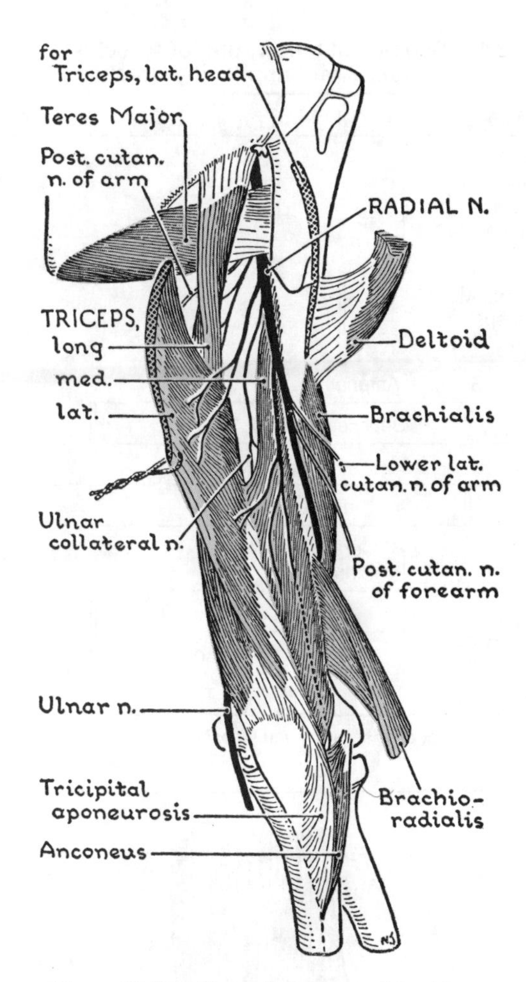

Figure 33.10. The radial nerve and the triceps.

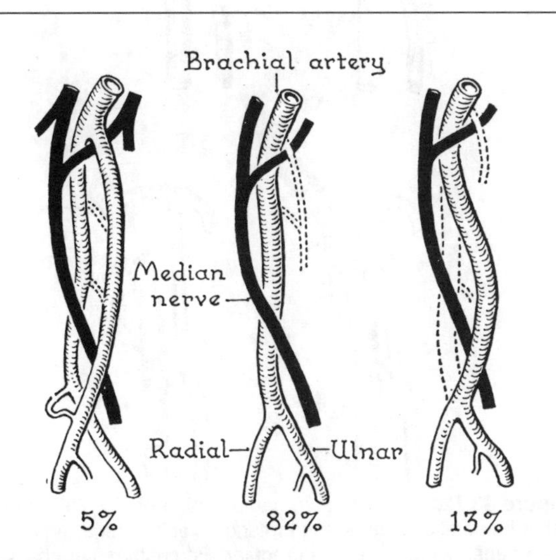

Figure 33.9. Developmental explanation of the variable relationship of median nerve to brachial artery. (Based on 307 limbs.)

the anterior compartment to lie between the brachioradialis and the brachialis (*fig. 33.11*). Next it lies on the capsule of the elbow joint and on the supinator. At a variable point it divides into 2 terminal branches: (1) a sensory branch, the *superficial radial nerve*, and (2) a branch that is essentially motor, the *posterior interosseous nerve* (deep radial nerve). The latter disappears into the substance of the supinator just below the elbow (*fig. 33.11*).

Collateral branches in the axilla and arm are shown in *Figures 33.10 and 33.11*. *At the lateral side of the arm and at the elbow*, branches pass from the radial and posterior interosseous nerves to 4 muscles—brachioradialis, extensor carpi radialis longus, extensor carpi radialis brevis, and supinator. The radial nerve usually supplies the proximal 2 muscles and the posterior interosseous nerve the distal 2 muscles.

Table 33.1 summarizes the motor distribution (to the elbow level) of the roots of the brachial plexus.

The **profunda artery** arises from the brachial artery just distal to the teres major. It is the companion artery of the radial nerve. By its anastomoses, it brings the axillary

Table 33.1. Segmental Innervation of Muscles of Shoulder and Upper Arm Supplied by Brachial Plexus*

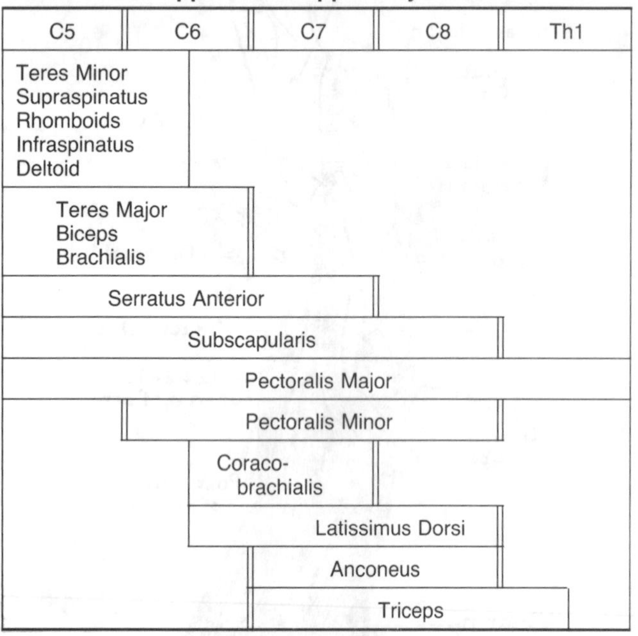

Muscle	C5	C6	C7	C8	Th1
Teres Minor, Supraspinatus, Rhomboids, Infraspinatus, Deltoid	X	X			
Teres Major, Biceps, Brachialis	X	X			
Serratus Anterior	X	X	X		
Subscapularis	X	X	X	X	
Pectoralis Major	X	X	X	X	X
Pectoralis Minor		X	X	X	
Coraco-brachialis		X	X		
Latissimus Dorsi		X	X	X	
Anconeus			X	X	
Triceps			X	X	X

* Modified after Bing, and Haymaker and Woodhall.

artery above into communication with the radial and ulnar arteries below.

Thus (for those who need these details):

A *recurrent branch* (commonly present) anastomoses with the posterior humeral circumflex artery (*fig. 33.12*).

Two terminal branches, an anterior and a posterior, which form anastomoses at the elbow (*see fig. 34.13*).

A *nutrient branch* is commonly given off.

Clinical Case 33.2

Patient Carla N. Two hours ago this 9-year-old farm girl fell off a tractor onto her outstretched hands and severely injured her left elbow. On preliminary examination in the Emergency Room, where you are an assigned student clerk, the surgical resident tells you "She has an obvious fracture-dislocation of the elbow." All the anatomy of the region you can remember must come to your aid since you are expected to help in the treatment.

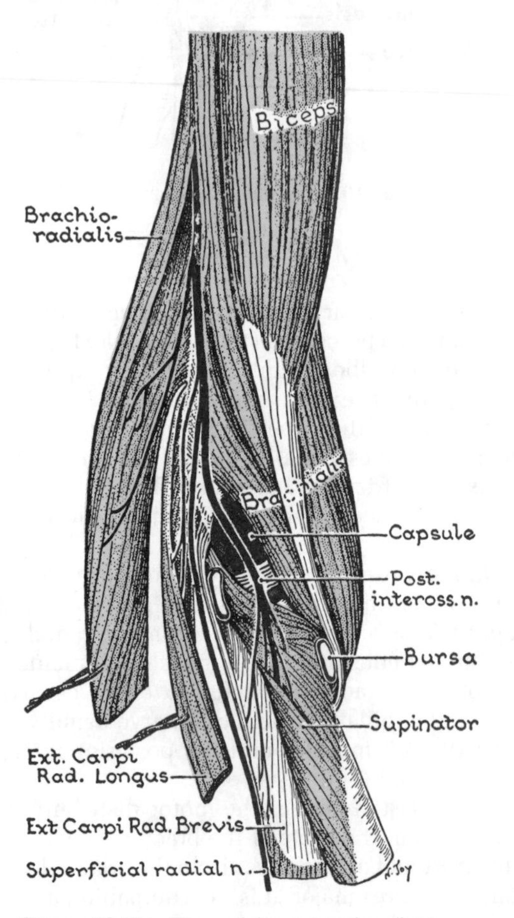

Figure 33.11. The radial nerve at the elbow.

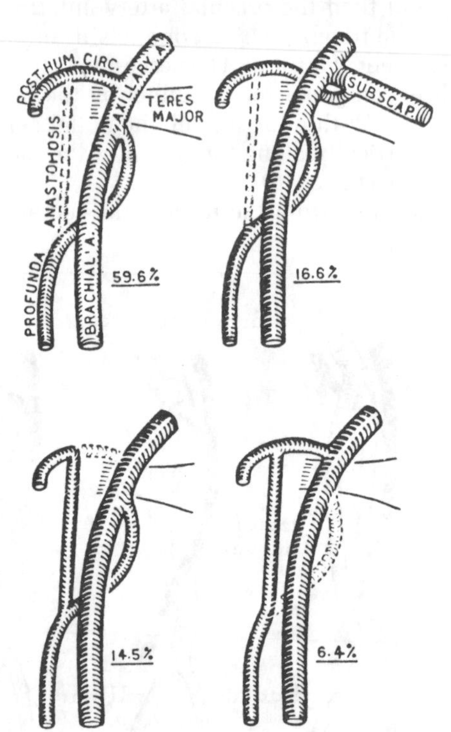

Figure 33.12. Four types of variations in origin of the posterior humeral circumflex and profunda brachii arteries; in 2.9% the arteries were otherwise irregular. Percentages are based on 235 specimens.

In 7.3% of 123 specimens the profunda artery arose from the stem of the subscapular artery.

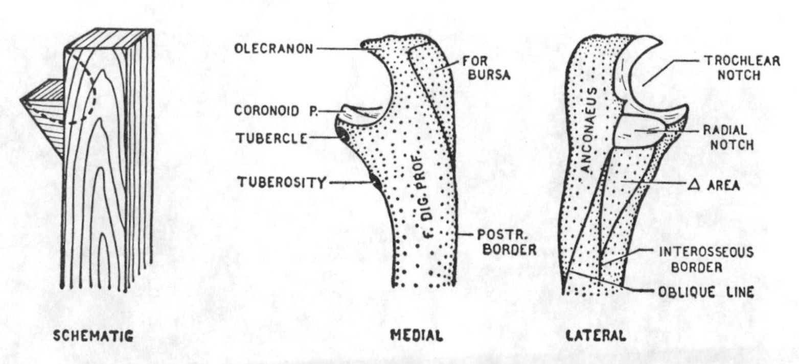

Figure 33.13. Upper end of ulna.

Elbow Joint

The correct Latin name for this joint is **articulatio cubiti**; it is a hinge or *ginglymus* joint, continuous with the proximal radioulnar joint. A hinge demands that: (1) the capsule will be loose anteriorly and posteriorly in order to permit flexion and extension; and (2) collateral ligaments will be required to prevent medial and lateral movements. *Plate 33.1* should be referred to while reading the next few pages.

BONES

The Humeroulnar Parts

Distal end of Humerus. This part has the spool-shaped pulley, the *trochlea*, which leads anteriorly to a depression, the *coronoid fossa*, and posteriorly to a broad triangular hollow, the *olecranon fossa*. Medially the epicondyle is very prominent (*fig. 33.15 and Plate 33.1*).

Upper End of Ulna

A triangular bracket, the *coronoid process*, projects anteriorly (*fig. 33.13*). The cubical portion of the ulna continued upward beyond the level of the coronoid process is the *olecranon*. Together they embrace a concavity, the *trochlear notch* (semilunar notch) which is reciprocally saddle-shaped to fit the trochlea of the humerus (*see fig. 33.16*).

Posteriorly, the olecranon presents a triangular, subcutaneous (and easily palpated) surface, overlaid by the subcutaneous **olecranon bursa** (*fig. 33.14*).

The lateral surface of the coronoid process is largely a concave facet, the *radial notch* of the ulna (*fig. 33.13*).

The brachialis tendon is inserted into the anterior surface of the coronoid process, where it produces a rough area, the *tuberosity of the ulna*.

The Humeroradial Parts

The upper concave surface of the *disc-shaped* **head of the radius** rotates on the inferior and anterior aspects of the rounded articular **capitulum** (L. = little head). Laterally, the capitulum merges with the lateral epicondyle (*fig. 33.15*).

THE PROXIMAL RADIOULNAR JOINT

The head of the radius is held in position by the **anular** (*annular*) **ligament**, which is attached to the ends of the radial notch of the ulna. The notch forms one-fourth of a circle, the ligament three-fourths.

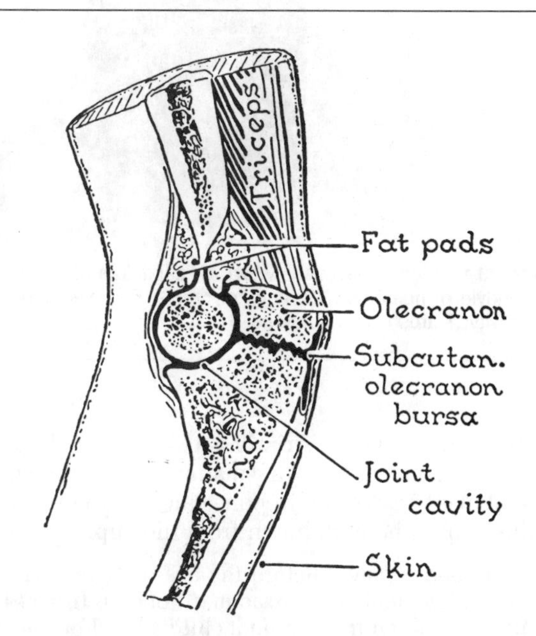

Figure 33.14. The joint cavity and the olecranon bursa united by a fracture (uncommon).

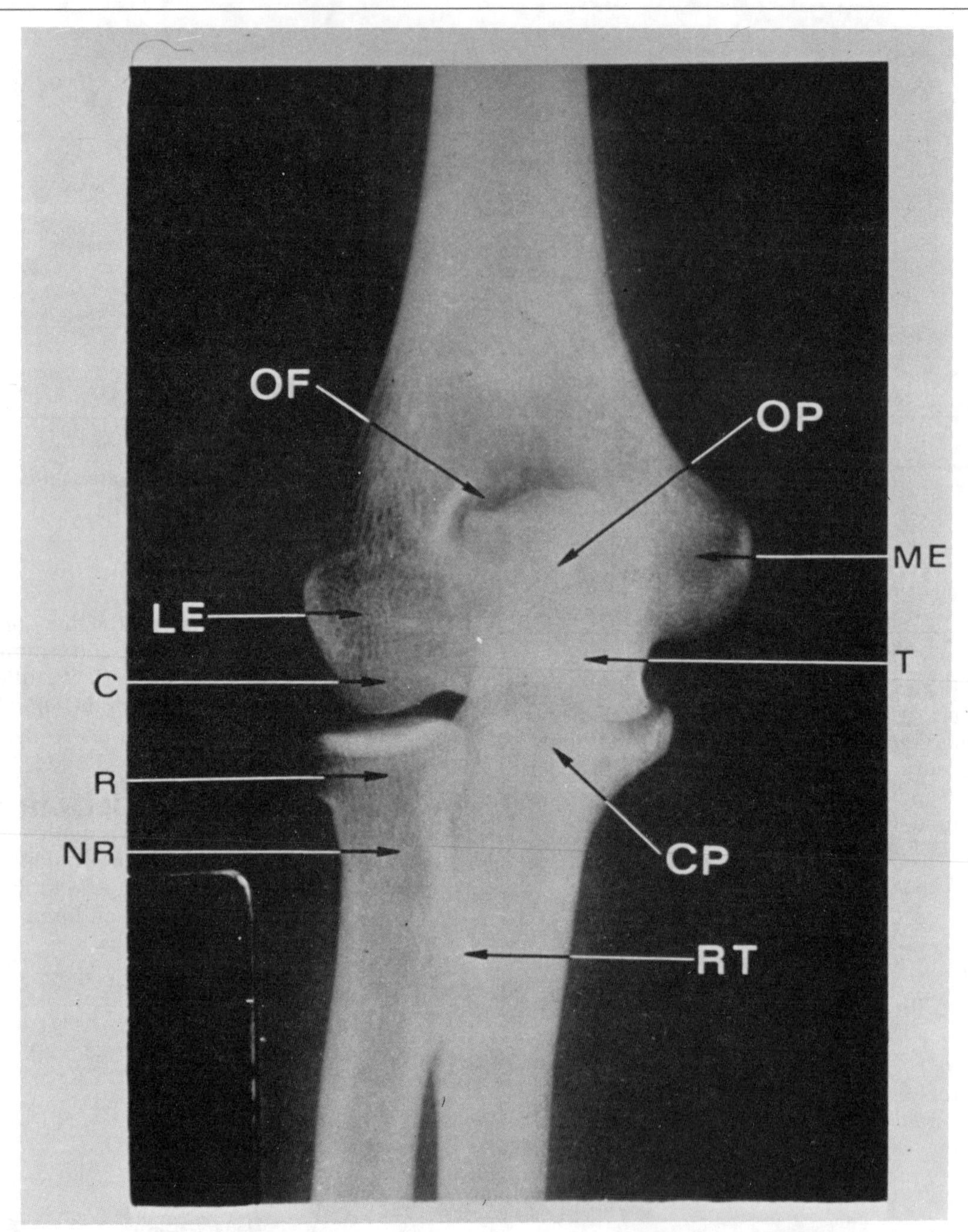

Plate 33.1 Right elbow (A-P view). *C*, capitulum of humerus; *CP*, coronoid process of ulna; *LE*, lateral epicondyle of humerus; *ME*, medial epicondyle of humerus; *NR*, neck of radius; *OF*, olecranon fossa of humerus; *OP*, olecranon process of ulna: *R*, head of radius; *RT*, radial tuberosity; *T*, trochlea of humerus.

The ligament is not strictly speaking ring-shaped, but rather cup-shaped, being of smaller circumference below than above (*fig. 33.16*). In consequence, the head of the radius cannot be withdrawn from the cup.

"Pulled Elbow." Before the age of 7, the grip of the anular ligament on the head may not be as firm as in later life, so sudden traction on a child's hand or forearm, as when pulling it out of the way of a passing motor car, may result in partial dislocation of the radius downward.

LIGAMENTS AND CAPSULE

The *radial collateral ligament* (lateral ligament) of the elbow joint is fan-shaped. It extends from the lateral epicondyle to the side of the anular ligament (*fig. 34.16*); so, indirectly it helps to retain the head of the radius in position. The supinator and the extensor carpi radialis brevis in part arise from it.

The *ulnar collateral ligament* (medial ligament) of the

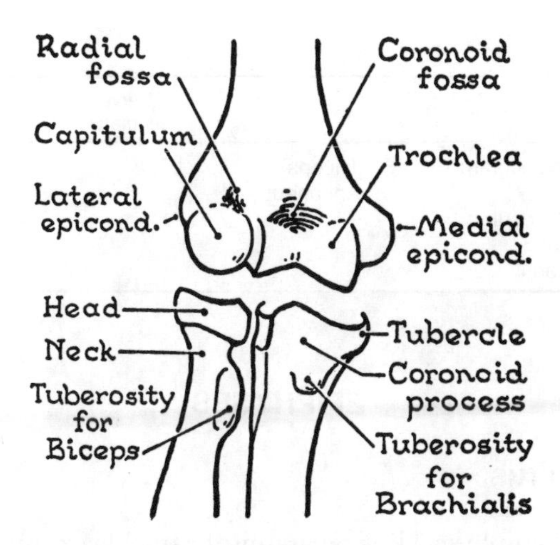

Figure 33.15. The bony parts concerned in the elbow joint (front view).

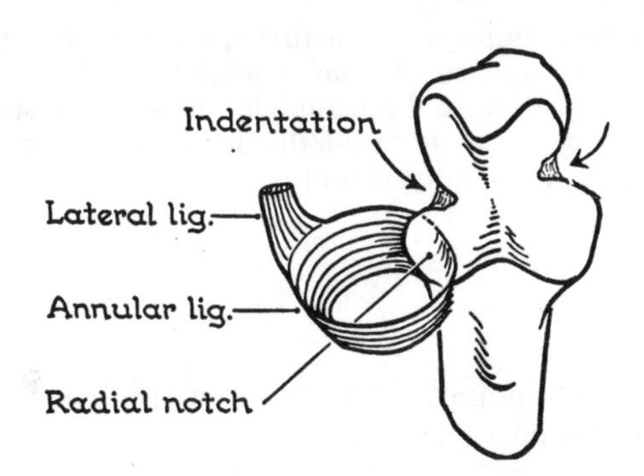

Figure 33.16. Socket for head of radius (anterosuperior view). The indentations are occupied by fat pads.

elbow joint fans out from the medial epicondyle to the medial margin of the trochlear notch (*fig. 33.17*).

The anterior fibers form a thick cord that is attached to a tubercle on the medial side of the coronoid process. This cord gives partial origin to the flexor digitorum superficialis.

The *fibrous capsule* extends to the upper margins of the coronoid and radial fossae anteriorly, but not quite as far on the olecranon fossa posteriorly. The distal attachment is to the margins of the trochlear notch and anular ligament.

The *synovial capsule* does not reach so high in the radial, coronoid, and olecranon fossae as the fibrous capsule. The intervals between the 2 capsules in these fossae are occupied by fat-filled synovial fat pads ("haversian glands" (*fig. 33.14*). The synovial capsule droops a short distant (½ cm) below the lower free margin of the anular ligament and surrounds the neck of the radius in a saclike manner (*fig. 33.18*). This device obviously allows the radius to rotate without tearing the synovial membrane. Redundant *folds of synovial membrane* project

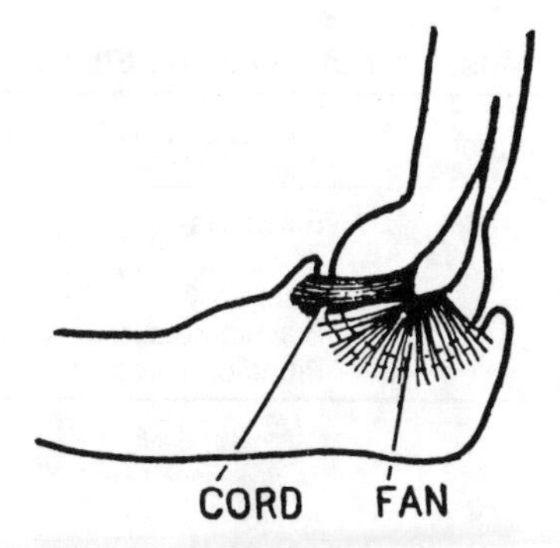

Figure 33.17. Medial or ulnar collateral ligament of elbow joint.

into this joint—as they do into other joints—and assist the fat pads to fill the unoccupied spaces.

"Carrying Angle." The medial lip of the trochlea of the humerus descends 5–6 mm lower than the lateral lip, for the lower end of the humerus is oblique. On account of this obliquity you will find that when your elbow is extended, the humerus and ulna do not lie in line with one another but form an angle, especially obvious in women.

The medial lip of the trochlea also projects forward about 5 mm in advance of the lateral lip; therefore, though the humeroulnar joint may be called a hinge joint, the ulna does not move on the trochlea like a door on a hinge, but rather it revolves on a cone. As a result the distal end of the ulna remains lateral to the axial line of the humerus in the flexed as well as in the extended position.

MUSCLES

Flexion is more powerful than extension in the ratio of 14:9. (Is this true for you?)

The chief flexor of the elbow joint is the *brachialis*. The biceps, as a flexor of the elbow, acts best when the forearm is supinated, less well when semipronated, and does not flex when pronated. As a supinator of the fore-

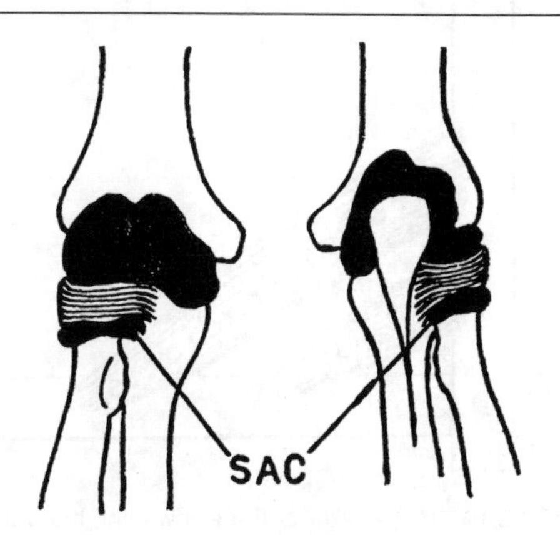

Figure 33.18. Synovial capsule of elbow joint (distended).

Table 33.2. Muscles Acting on the Elbow Joint

Subdivisions of the joint	Flexors	Nerve segments	Nerve	Extensors	Nerve segments	Nerve
Humeroulnar	Brachialis	5, 6	Musculocutan.	Triceps Anconeus	7, 8	Radial
Humeroradial	Biceps Brachioradialis Pronator Teres	5, 6 (5), 6 6	Musculocutan. Radial Median			

arm, it does not act when the elbow is extended, unless against resistance. (Basmajian and Latif.) (See *Table 33.2.*)

The brachioradialis creeps up the humerus (*fig. 33.19*), and after the brachialis and biceps are paralyzed, the brachioradialis may prove to be a useful flexor.

The common use of the *triceps* is not so much to extend the elbow as to prevent flexion of the elbow, or to regulate flexion, as in pushing. It is also an aggressive muscle, as in boxing.

RELATIONS

The brachialis (*fig. 33.20*), though thick in the middle, is thin and attenuated at its edges. In consequence, the musculocutaneous nerve, which lies anterior to the middle of the muscle, is far removed from the joint, but the median nerve is separated from the capsule of the joint merely by the thinness of the medial edge of the brachialis; the radial nerve (or its 2 terminal branches) is in direct contact with the capsule (*fig. 33.11*). The ulnar nerve is always in immediate contact with the ulnar collateral ligament, and it is covered by flexor carpi ulnaris. The nerve to the anconeus crosses the capsule posterior to the lateral epicondyle.

Anastomoses around the joint are shown in *Figure 34.14*.

Figure 33.19. As a flexor of the elbow joint, the brachioradialis is more advantageously situated than the pronator teres, though neither is a strong flexor.

EPIPHYSES

Humerus

The epiphyseal line separating the trochlea, capitulum, and lateral epicondyle from the diaphysis runs transversely just proximal to the articular cartilage (*fig. 33.21*).

This distal epiphysis is the first of all long bone epiphyses to fuse, synostosis being complete by the 17th or 18th year. The medial epicondyle fuses with a spur of bone that descends from the diaphysis and separates it from the distal epiphysis proper.

Ulna

The proximal epiphysis is a traction epiphysis of the triceps. It may be a mere scale.

Radius

The proximal epiphyseal line lies just distal to the head.

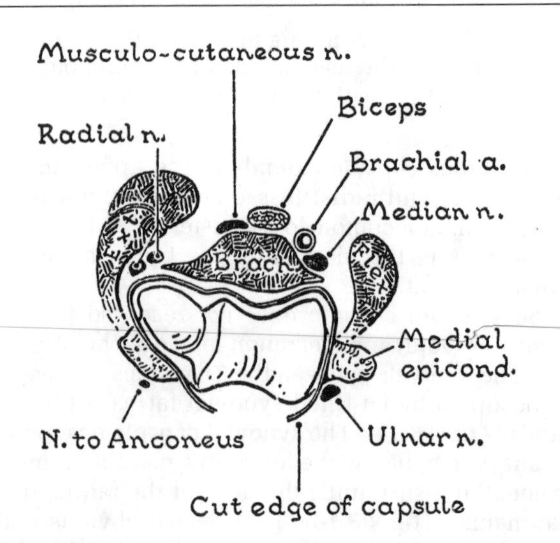

Figure 33.20. Cross-section through elbow joint showing important relations.

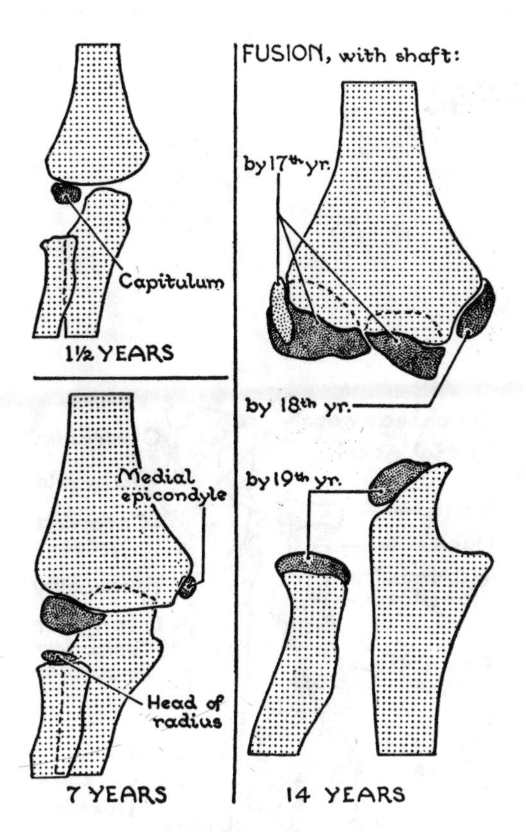

Figure 33.21. Epiphyses about the elbow joint. (Courtesy of Dr. J. D. Munn, The Hospital for Sick Children, Toronto.)

NERVE SUPPLY

Each of the 5 nerves shown in *Figure 33.20*, including the nerve to the anconeus, sends a twig or twigs to the joints.

Anatomical Features of the Physical Examination

General inspection and palpation of the arm and elbow is relatively easy with localization of exact landmarks—a simple exercise in surface anatomy, e.g., the pulse of the brachial artery at the elbow is felt by pressing deeply medial to the biceps tendon. With a syringe and needle excess elbow joint fluid can be withdrawn from several sites around the circumference, with careful attention to avoiding the important nerves and vessels in this region.

To take a systemic blood pressure reading requires that an encircling cuff be placed and inflated at mid-arm; then one listens below its edges for the sounds of the pulsations of blood during the slow release of the air in the cuff. The systolic pressure is the reading (normally about 120 millimeters of mercury) when sounds are first heard from the artery because some blood is getting through with each systole and causing a knocking sound. This sound increases as the cuff is deflated further. Then it disappears, normally when the cuff pressure is about 80 mm Hg (diastolic blood pressure).

The elbow flexors, principally brachialis and biceps brachii, are supplied by the musculocutaneous nerve (C5, 6, 7) which runs between the flexors and becomes the lateral cutaneous nerve of the forearm. Its mirror image, the radial nerve (C7, 8; T1) supplies triceps while running vertically in the spiral groove. At the elbow the radial nerve divides into motor nerves for the extensor muscles in the posterior compartment of the forearm and the posterior cutaneous nerve of the forearm, and the superficial radial nerve which are sensory.

The brachial artery descends first on the long head of triceps and then till it divides at the elbow joint on the surface of the brachialis medial to the biceps. Accompanying it are the median nerve (C(5), 6, 7, 8; T1) throughout the arm and the ulnar nerve (C7, 8; T1) to the middle. The ulnar nerve passes backward in contact with the medial head of triceps and then passes posterior to the medial epicondyle.

Clinical Mini-Problems

1. **Which neurovascular elements are jeopardized (a) in a midshaft fracture of the humerus? (b) In a fracture of the medial epicondyle?**
2. **(a) Which segments of the spinal cord are being tested when the bicep tendon is tapped with a tendon hammer in the cubital fossa?**
   (b) Which nerve carries the nerve fibers from these spinal cord segments to the biceps brachii?
3. **With reference to the very palpable bicep tendon, where would one palpate for a brachial artery pulse in the cubital fossa?**
4. **Why would damage to the radial nerve in the spiral groove of the humerus only cause a weakness in extension of the forearm and not a total inability to extend at the elbow?**

(Answers to these questions can be found on p. 586.)

34

Flexor Region of Forearm

Clinical Case 34.1

Patient Andrew L. This 48-year-old man has consulted the rehabilitation specialist in the neurodiagnostic unit (on which you are doing a clinical elective) complaining of weakness of his left forearm and hand muscles and numbness of some of his fingers. The specialist decides that the ulnar nerve is entrapped at the elbow and should be moved from the "back to the front," and he begins quizzing you about the relevant anatomy of the flexor region.

Radius and Ulna

The proximal end of the **radius** consists of: *head, neck, and radial tuberosity* for the biceps (*fig. 34.1*). Its distal end consists of: *inferior articular surface, ulnar notch, styloid process,* anterior aspect and posterior aspect. Also on the posterior surface are the *dorsal radial tubercle (of Lister)* and several grooves.

The proximal end of the **ulna** consists of: *olecranon, coronoid process, trochlear notch, radial notch,* and *ulnar tuberosity* for the brachialis. Its distal end consists of: *head, styloid process, pit* for the attachment of the articular disc, and *groove* for extensor carpi ulnaris.

The **bodies of the radius** and **ulna** are cylindrical. The expanded lower two-thirds of the radius and the expanded upper two-thirds of the ulna are triangular on cross-section (*fig. 34.2*).

The facing interosseous borders of the 2 bones are united by an **interosseous membrane** believed by many to transfer from radius to ulna the force of an impact received by the hand (*fig. 34.3*)—though this is highly improbable (Halls and Travill).

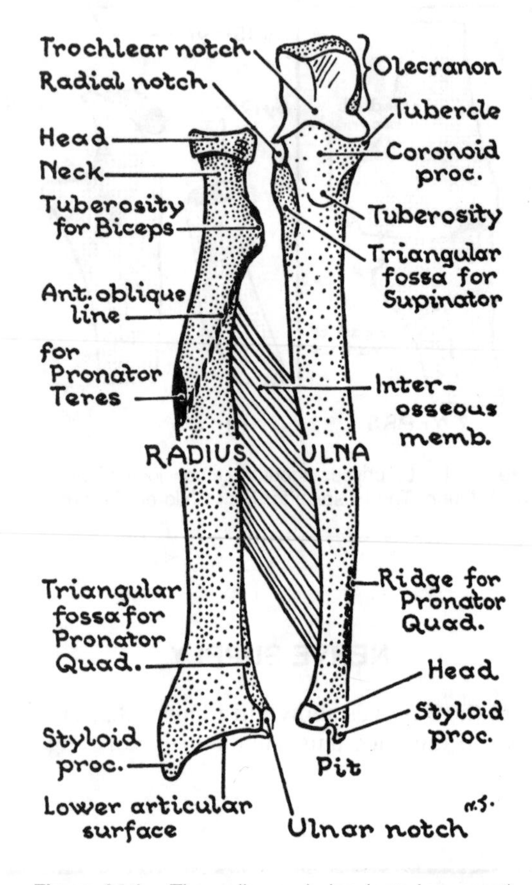

Figure 34.1. The radius and ulna (anterior aspect).

At the summit of the convexity of the radius there is a rough *area for the pronator teres.* From it the *anterior* and *posterior oblique lines* ascend on the front and back of the bone to the radial tuberosity. The area between them is the insertion of the *supinator.*

The posterior border of the ulna extends from the apex of the posterior *subcutaneous surface of the olecranon* to the back of the styloid process. This border is sharp, subcutaneous, and always palpable from end to end, and when the forearm is raised in self-protection, it is liable to be struck. The anterior border is rounded and covered with muscles (*fig. 34.4*).

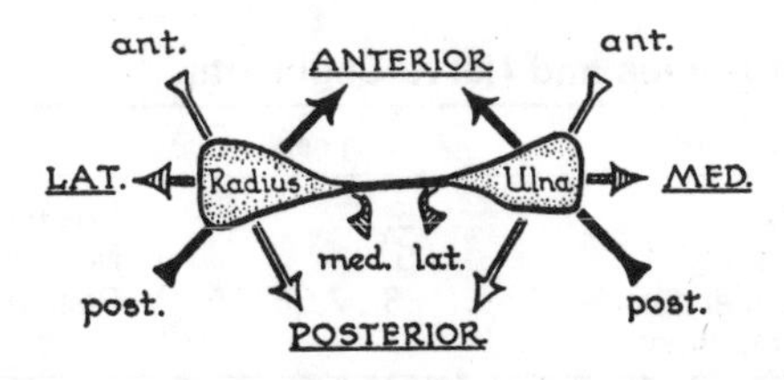

Figure 34.2. Surfaces and borders illustrated on cross-section. Surfaces in large type, borders in small type.

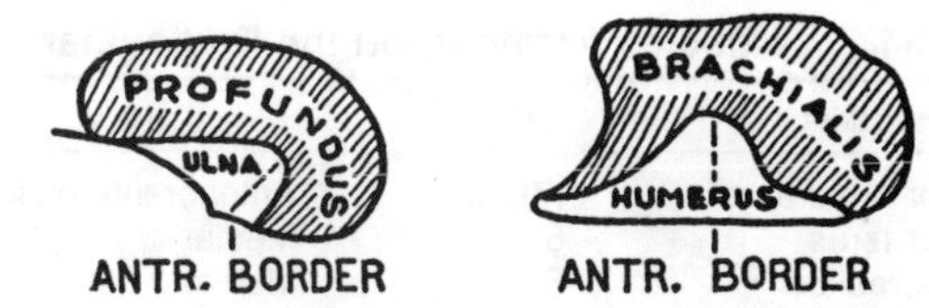

Figure 34.4. Rounded borders are usually strengthening bars: sharp and rough borders give attachment to fibrous tissue.

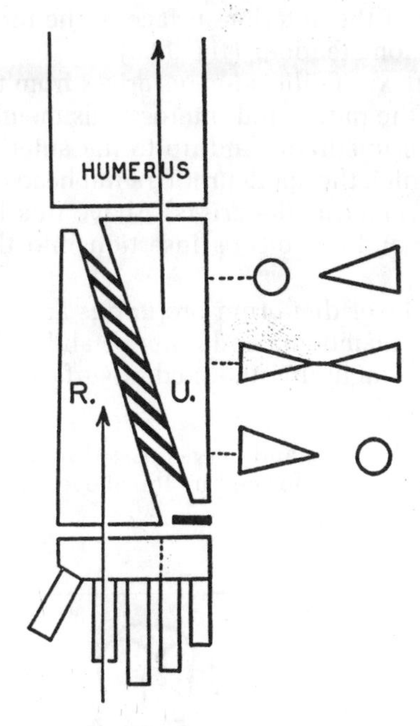

Figure 34.3. Scheme of the radius and ulna.

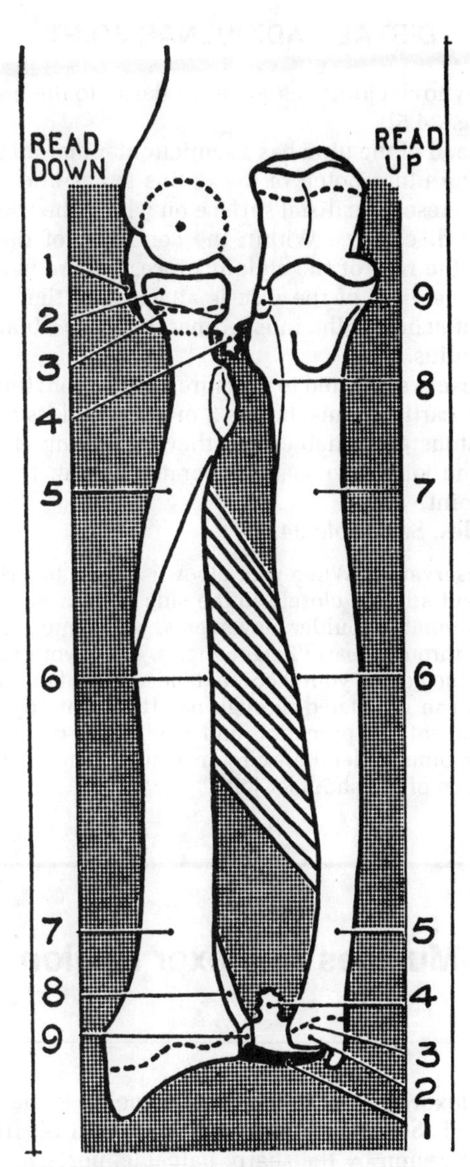

Figure 34.5. Points of similarity between the proximal and distal radioulnar joints.

1. Anular ligament———articular disc.
2. Circular head———semicircular head.
3. Epiphyseal line enters the joint cavity.
4. Sacciform recess of synovial membrane.
5. The bone here circular on cross-section.
6. The interosseous border is here sharp.
7. The bone here triangular on cross-section.
8. Triangular fossa between border and notch.
9. Ulnar notch of radius———radial notch of ulna.

Radioulnar Joints

The 2 bones of the forearm are united at the proximal, "intermediate," and distal radioulnar joints.

PROXIMAL RADIOULNAR JOINT

Considered with the elbow joint (p. 393).

INTERMEDIATE RADIOULNAR JOINT

The bodies of the radius and ulna are united to each other by the *interosseous membrane*, which is best marked at the middle two-fourths of the forearm (*figs. 34.1 and 34.5*). It is about equally taut (or lax) in all positions—pronation, semipronation, and supination.

Table 34.1. Muscles Acting upon the Radioulnar Joints and Their Nerves and Nerve Segments

Pronators	Segment	Nerve	Supinators	Segment	Nerve
Pronator quadratus	(7), 8, 1	Anterior interosseous	Supinator	5, 6	Posterior interosseous
Pronator teres	6	Median	Biceps	5, 6	Musculocutaneous
Fl. carpi radialis	6	Median	2 * dorsal muscles of the thumb	(6), 7, (8)	Posterior interosseous

* The extensor pollicis brevis arises from the radius and therefore cannot assist in rotation.

DISTAL RADIOULNAR JOINT

This synovial joint has some similarity to the proximal joint (*fig. 34.5*).

The **head of the ulna** has a semicircular margin around which the ulnar notch of the radius slides and a semilunar or crescentic distal surface on which the triangular articular disc plays. Within the concavity of the semilune, at the root of the styloid process, there is a *fovea or pit*. The apex of the disc is attached to the pit. The base is attached to the inferior margin of the ulnar notch of the radius.

The **disc** is subjected to pressure and friction; therefore, it is fibrocartilaginous, but its 2 margins and its apex are ligamentous and pliable, and they are strong. The disc closes the joint cavity and separates it from the radiocarpal joint.

Muscles. See *Table 34.1*.

Observation. When your elbow is flexed to a right angle and applied closely to the side of your body, so as to eliminate shoulder movements, your hand can be rotated through nearly 2 right angles. When your shoulder is abducted and your elbow extended, as in fencing, your hand can be rotated through nearly 3 right angles. The movements of pronation and supination of the radioulnar joints under these circumstances augment the axial rotation of the shoulder joint.

Muscles of Flexor Region

The flexor region of the forearm includes the medial surface of the ulna. Medially, it is marked off from the extensor region by the sharp, palpable, posterior border of the ulna. Laterally, it is marked off from the extensor region by the impalpable anterior border of the radius (*fig. 34.6*), for which the course of the radial artery is a practical guide.

Since neither of these internervous lines is crossed by motor nerves, they are relatively safe for surgical explorations.

The *pronator quadratus* arises from the pronator ridge of the ulna (*fig. 34.7*) and inserted into the smooth, distal quarter of the anterior surface of the radius. Bridging it are the long tendons (*fig. 34.8*).

The *flexor pollicis longus* arises from the anterior surface of the radius and interosseous membrane above the pronator quadratus and up to the anterior oblique line, from which the aponeurotic radial head of the *flexor digitorum superficialis* arises. Above this level lie the *supinator* and the *biceps* insertion into the radial tuberosity.

The *flexor digitorum profundus* arises from the interosseous membrane and anterior and medial surfaces of the ulna including the medial surface of the olecranon (*fig. 34.7*).

The vessels and nerves pass through the flexor region on the carpet formed by the above 7 muscles.

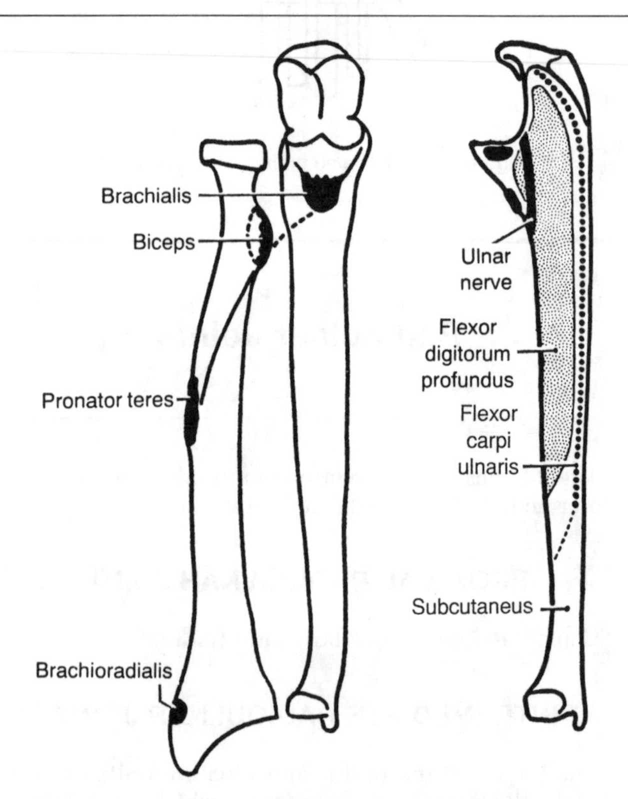

Figure 34.6. Flexor aspect of the bones of the forearm. The 4 flexors of the elbow (brachialis, biceps, pronator teres, and brachioradialis) like sentinels guard the boundary between flexor and extensor territories.

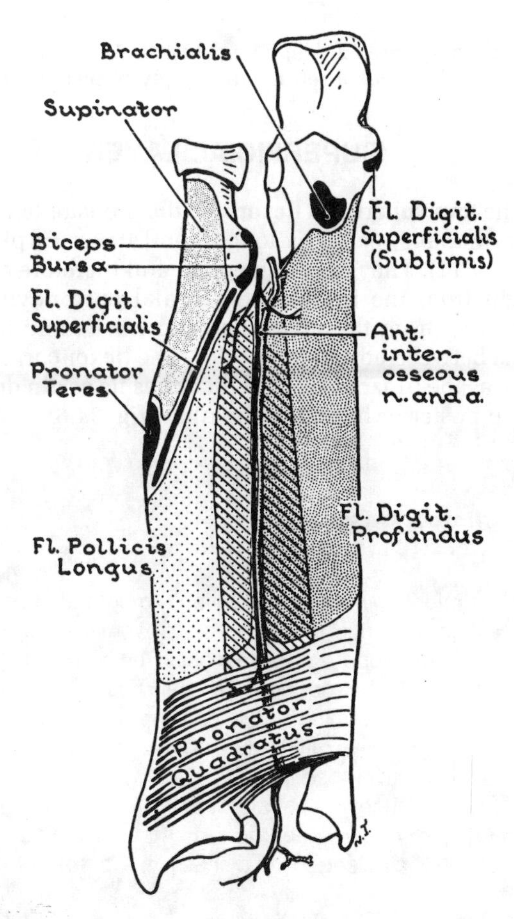

Figure 34.7. Radius, ulna, and interosseous membrane, showing attachment of muscles (front view).

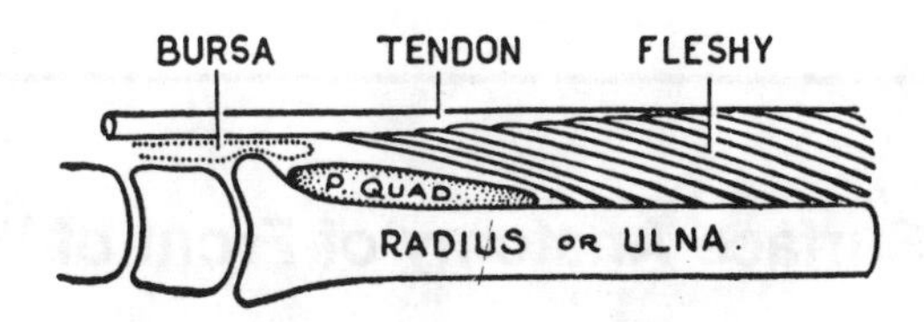

Figure 34.8. Tendon of deep digital flexor.

Tendons

The 5 tendons of the deep digital flexors converge on the midline of the limb in order to enter the carpal tunnel side by side (*fig. 34.9*), allowing the radial artery to come to lie on the pronator quadratus.

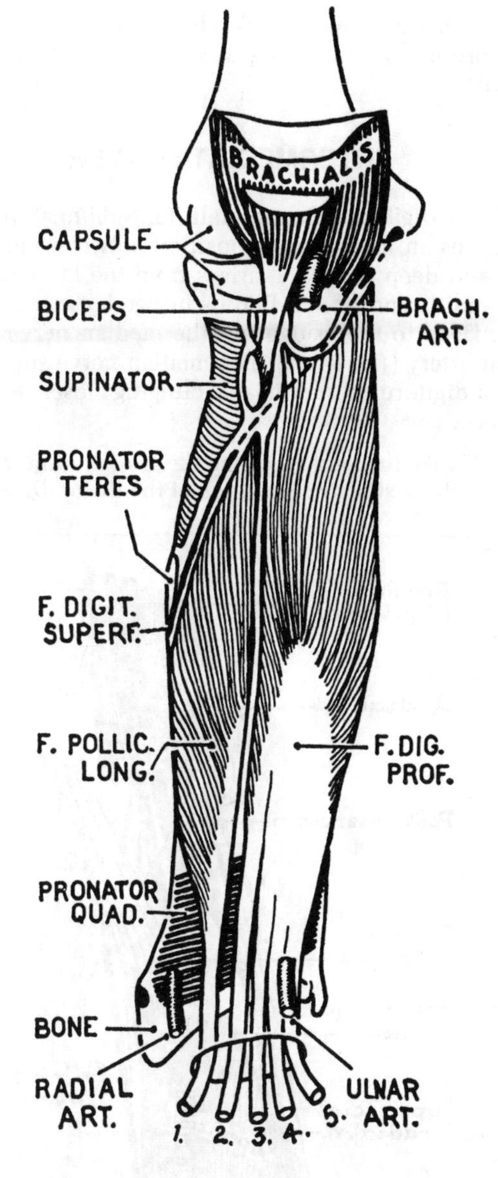

Figure 34.9. The 7 muscles clothing the flexor aspect of the radius and ulna.

Anterior Interosseous Nerve and Artery

This nerve and artery arise from the median nerve and common interosseous artery 2 or 3 cm distal to the elbow. They cling to the interosseous membrane (*fig. 35.7*). The *nerve* supplies the 3 deep muscles (flexor pollicis longus, distal ½ of flexor digitorum profundus (for digits 2 and

3) and pronator quadratus). The remainder of the flexor digitorum profundus (medial ½) is supplied by the ulnar nerve.

INTERMEDIATE LAYER

Flexor Digitorum Superficialis (or Sublimis) (*fig. 34.10*) occupies an intermediate position between the superficial and deep flexors. It arises from the humerus, ulna, and radius and from a fibrous bridge between the latter two. Deep to the bridge run the median nerve and the ulnar artery (*fig. 34.10*). The median nerve supplies the flexor digitorum superficialis, clinging closely to its deep surface.

Unlike the tendons of the flexor digitorum profundus which lie side by side, those of the flexor digitorum superficialis are "2 deep" (*fig. 34.10*). As the tendons pass behind the flexor retinaculum, they come to lie 4 abreast.

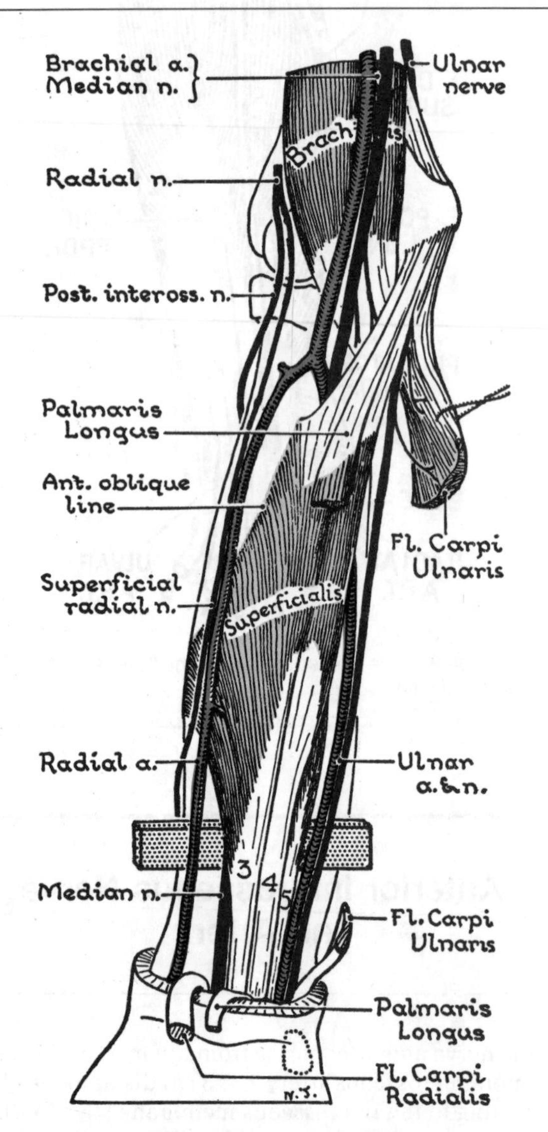

Figure 34.10. The key position of the flexor digitorum superficialis (sublimis) with reference to the nerves and arteries.

SUPERFICIAL LAYER

The **four superficial flexors** are the pronator teres, flexor carpi radialis, palmaris longus, and flexor carpi ulnaris (*fig. 34.11*). They have a common and therefore a fibrous origin from the front of the medial epicondyle of the humerus, from the investing deep fascia, and from the septa between adjacent muscles. The *flexor carpi ulnaris* bridges the ulnar nerve and increases its origin down the sharp posterior border to the ulna (*fig. 34.6*).

Clinical Case 34.2

Patient Billie M. In the seventh month of her first pregnancy, this pleasant young woman who is a typist and hopes to work almost to the end of "term," finds lately that her right hand and most of her fingers often become numb and that she has some substantial difficulty in moving her thumb. The obstetrician to whose clinic you are assigned seems to know much more anatomy of the wrist region than you ever expected her to remember, and you find yourself challenged to discuss this *carpal tunnel syndrome* intelligently.

Surface Anatomy of Front of Wrist

The *distal skin crease* at the wrist (*fig. 34.12*), slightly convex toward the palm, corresponds to the upper border of the **flexor retinaculum** (transverse carpal ligament) which creates the carpal tunnel. It crosses the 2 prominent proximal bony pillars to which the retinaculum is attached, namely, the *tubercle of the scaphoid bone* laterally, the *pisiform bone* medially.

Where the prominent tendon of the *flexor carpi radialis* crosses this skin crease at the junction of its lateral one-third with its medial two-thirds, it passes in front of the scaphoid tubercle and so serves as a guide to it. The flexor carpi radialis is rendered very prominent when the clenched fist is fully flexed against resistance. The tendon of the *flexor carpi ulnaris*, the most medial of all the

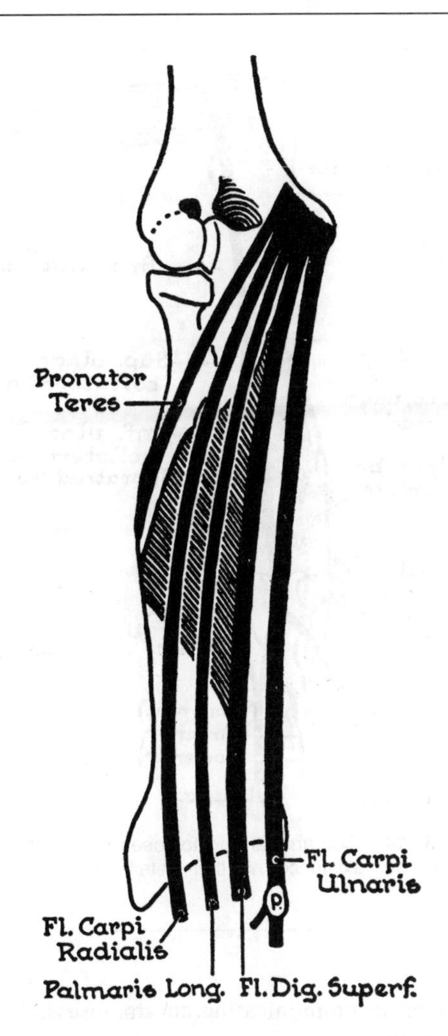

Figure 34.11. The 4 superficial flexors and the flexor digitorum superficialis.

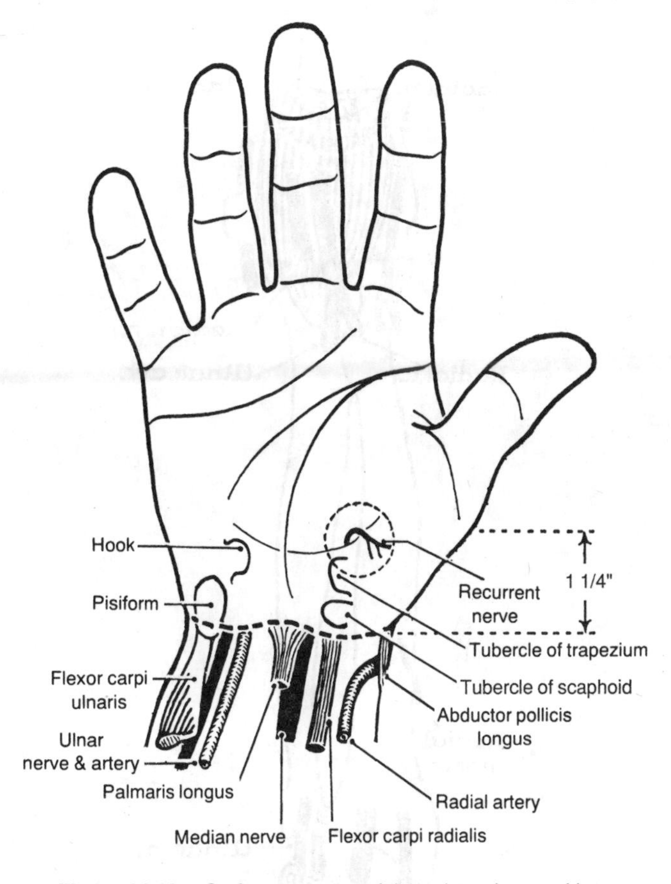

Figure 34.12. Surface anatomy of the wrist: a key position.

structures at the wrist, can be traced to the pisiform at the medial end of the crease.

At the middle (or *median*) of the distal skin crease of the wrist is the most important of all the structures—the *median nerve*. It cannot be palpated, for layers of fascia and the *palmaris longus* tendon covers it. The latter is sometimes absent.

Palmaris longus is absent on both sides in 7%, absent on the right in 7%, and absent on the left in 6% (George).

The *radial artery* is easily felt pulsating lateral to the flexor carpi radialis. The *ulnar artery* and *nerve* pass into the palm immediately lateral to the pisiform (*fig. 34.13*). A strong fascia covers them, and so the artery's pulse is not easily felt.

Flexor digitorum superficialis bulges between the flexor carpi ulnaris and the palmaris longus when the wrist is extended. The *lower end of the radius* can be grasped between the fingers and thumb. The anterior margin of its lower end is surprisingly prominent.

Arteries and Nerves of Flexor Region of Forearm

Lines drawn from the bifurcation of the brachial artery distal to the skin crease at the elbow to the points where the ulnar and radial arteries were identified at the wrist, indicate the general directions of the 2 arteries (*fig. 34.13*).

ULNAR ARTERY

In the forearm, the ulnar artery passes deep to the fibrous arch of the flexor digitorum superficialis, and crosses posterior to the median nerve (*fig. 34.10*). Behind them are the muscles clothing the ulna (*fig. 34.9*).

Its pulse is not readily felt at the wrist because it is bridged by deep fascia (the *volar carpal ligament*).

AXIOM. Most branches of arteries are muscular branches and they are usually unnamed. Certain muscular and named branches anastomose, especially about joints.

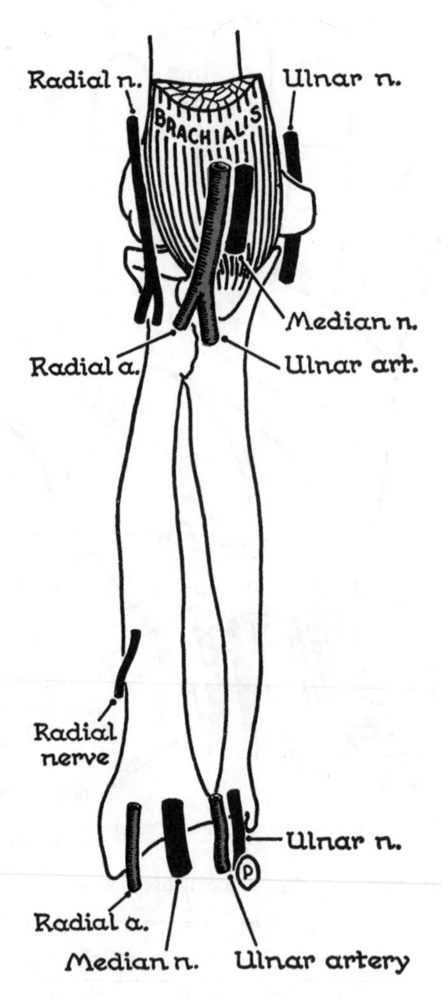

Figure 34.13. The vessels and nerves of the forearm at key positions. Complete the picture by filling in the missing segments.

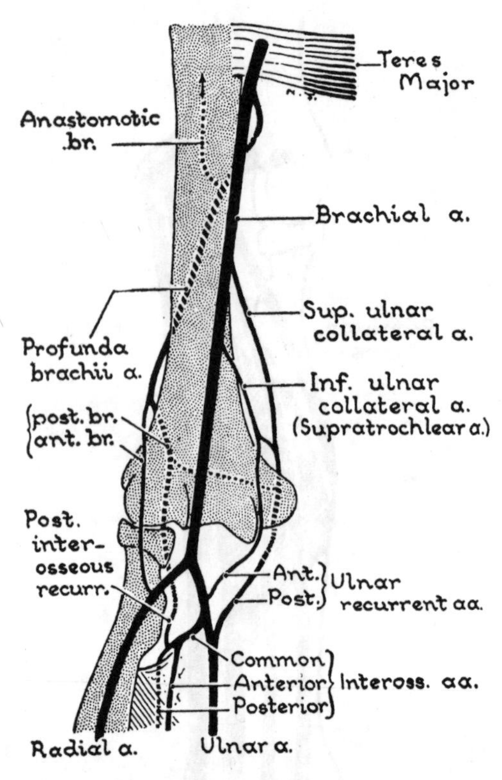

Figure 34.14. The arterial anastomoses around the elbow (details). Their existence, not their details, must be remembered.

Branches of ulnar artery in the forearm include the *recurrent* arteries that anastomose around the medial epicondyle; the *common interosseous* artery (*fig. 34.14*), which supplies the bulk of the forearm muscles and bones and has anastomotic branches around the elbow and wrist; the *muscular arteries*, and *carpal arteries* at the wrist.

Details of Branches for advanced specialists and reference:

1 and 2. *Anterior and posterior ulnar recurrent arteries* take part in the anastomoses around the medial epicondyle (*fig. 34.14*).

3. *Common interosseous artery* is a short stem that passes to the superior border of the interosseous membrane and divides there into the anterior and posterior interosseous arteries. Their branches are:

Anterior interosseous artery
- Median, companion to median nerve.
- Nutrient, to radius and ulna
- Muscular
- Anterior communicating, anastomoses in front of the wrist
- Terminal, anastomoses behind the wrist (*see figs. 36.16 and 36.18*).

Posterior interosseous artery
- Recurrent, anastomoses behind the lateral epicondyle
- Muscular
- Terminal, anastomoses behind wrist

4. *Muscular branches*

5, 6. *Palmar and dorsal ulnar* carpal arteries take part in the anastomoses around the wrist (*fig. 36.16*).

The *median artery* is of historical interest. For a brief period in embryonic life it accompanies the median nerve.

RADIAL ARTERY

This trunk is not crossed by any muscle, but the brachioradialis is lateral and overlaps it; the flexor carpi radialis is medial in its distal two-thirds.

Immediately behind it are 6 muscles that clothe the anterior aspect of the radius (*fig. 34.9*), and beyond these are the distal end of the radius and the capsule of the wrist joint. The artery gains the dorsum of the wrist by passing below the styloid process and deep to the abductor pollicis longus.

The superficial radial nerve accompanies the radial artery only briefly in the middle third of the forearm.

Branches of the radial artery given off in the forearm, are *muscular*, recurrent (*fig. 34.14*), *palmar (anterior) carpal* (p. 426), and *superficial palmar* (*see fig. 36.9*).

ULNAR NERVE

The ulnar nerve runs straight from the medial epicondyle to the lateral side of the pisiform on a carpet of the flexor digitorum profundus. The flexor carpi ulnaris covers it proximally and overlaps it distally. Its artery approaches it, obviously from the lateral side, and runs in contact with it through most of its course.

Branches

At the elbow the ulnar nerve supplies 1½ muscles—the flexor carpi ulnaris and the medial half of the flexor digitorum profundus—and it sends twigs to the elbow joint. Somewhat above the wrist its palmar and dorsal cutaneous branches arise (*see fig. 32.5*).

Plastic surgeons take advantage of the fact that if you sever the humeral origin of the flexor carpi ulnaris, you can bring the ulnar nerve to the front of the elbow, shortening its course by several centimeters (see Clinical Case 34.1).

SUPERFICIAL RADIAL NERVE

This is the sensory continuation of the radial nerve distal to the origin of the motor (posterior interosseous) nerve. To expose it, one must peel the brachioradialis laterally (*see fig. 33.11*).

After crossing the capsule of the elbow joint, the nerve descends with the brachioradialis, to which it passes deep to become cutaneous about 5 cm superior to the styloid process of the radius.

The radial artery joins and accompanies it through the middle third of the forearm.

MEDIAN NERVE

It plunges with the ulnar artery deep to the "superficialis bridge," running a straight course to the midpoint of the wrist (*fig. 34.10*), where it passes deep to the flexor retinaculum. Deep to it are the muscles clothing the front of the ulna, especially the flexor digitorum profundus.

Branches

The median nerve supplies all the flexor muscles of the forearm, except the 1½ supplied by the ulnar nerve. It adheres to the deep surface of the flexor digitorum superficialis, giving it branches in its course.

It also sends articular twigs to the elbow and wrist joints.

INTERNERVOUS LINES

The flexor digitorum superficialis is supplied by the median nerve and the flexor carpi ulnaris by the ulnar nerve; so, the septum between them marks an internervous line. This may safely be opened up, the 2 muscles pulled apart, and access gained to the deeper parts of the forearm without fear of damaging a motor nerve—until a point is reached about 6 cm below the medial epicondyle, for at this level motor branches of the ulnar nerve are encountered.

Other Internervous Lines

The posterior border of the ulna marks the internervous line between the motor territories of the ulnar and radial nerves; similarly, the course of the radial artery marks the internervous line between the motor territories of the median and radial nerves.

Anatomical Features of the Physical Examination

When symptoms are reported implicating structures of the forearm, one's knowledge of its surface anatomy becomes essential. Flexor carpi ulnaris, flexor digitorum superficialis, palmaris longus (sometimes absent) and flexor carpi radialis are fairly easy to trace en route to the distal skin crease of the wrist. Beyond the elbow, the median and ulnar nerves and the radial and ulnar arteries do not emerge from the depths of the forearm until just proximal to the wrist. Then they are accessible to careful palpation.

From their common tendon of origin from the medial epicondyle and ligament, the superficial layer of four flexor muscles can be traced as they radiate to their insertions—pronator teres to the radius halfway along its lateral aspect at its greatest convexity; flexor carpi radialis, which crosses the distal skin crease of the radius in front of the tubercle of the scaphoid; palmaris longus to its area of adherence on the front of the flexor retinaculum; flexor carpi ulnaris (which also arises by aponeurosis from the sharp palpable, posterior border of the ulna), running to the pisiform carpal bone, also palpable on the distal skin crease.

The median nerve supplies branches to all the flexors except 1½ muscles supplied by the ulnar nerve; it passes deep to the superficial flexor muscles and reappears at the wrist in the midline. Then it passes deep to the flexor retinaculum into the palm where it supplies five small muscles and most digital sensory nerves.

From the posterior surface of the medial epicondyle, the ulnar nerve tunnels through the origin of the flexor carpi ulnaris to the flexor region. There it descends between the flexor carpi ulnaris and the medial part of the flexor digitorum profundus, both of which it supplies (i.e., 1½ forearm flexor muscles). It enters the palm lateral to the pisiform and supplies most of the muscles of the hand and the skin of the medial 1½ fingers (described in later chapters).

Clinical Mini-Problems

1. **Which nerve is being tested when a patient is asked to pronate the forearm against resistance?**
2. **Why is the lateral side of the anterior (volar) aspect of the forearm considered the "safe side" by surgeons?**
3. **The radial pulse can be taken at the wrist on a point between the tendons of the flexor carpi radialis and the brachioradialis. Which bone is the radial artery adjacent to at this point?**

(Answers to these questions can be found on p. 587.)

35

Bones and Joints of Wrist and Hand

Clinical Case 35.1

Patient Anthony E. Several hours ago, this 36-year-old father of three teenage boys "jammed" his left thumb during an exuberant free-for-all wrestling match. Severe pain and tenderness at the base of the thumb (in the anatomical snuffbox) suggests a fracture of the scaphoid. Your supervisor in the orthopedic clinic asks you for a brief review of the anatomy of the carpal and first metacarpal bones.

While reading the following remarks, it is highly desirable to have beside you the bones of the hand, preferably strung on catgut. Also make the suggested observations on your own hand. To assist you, the drawings in this chapter conform to the way in which you look at your own right hand.

The skeleton of the hand consists of:

1. The carpal bones, or bones of the wrist, or carpus.
2. The metacarpal bones, or bones of the palm, or metacarpus.
3. The phalanges or bones of the digits.

Carpal Bones and Joints

The carpal bones are short or cubical bones (p. 5). Two surfaces, the *anterior* and *posterior*, are rough for the attachment of ligaments, while 4 surfaces articulate with adjacent bones and are, therefore, covered with cartilage. Like all short bones, they retain red marrow for many years after it has disappeared from the bodies of long bones. Carpal bones do not start to ossify until shortly before birth or within a few years after it. The sequence provides the radiologist with an index of "bone age" in children.

The carpal bones, 8 in number, are arranged in a proximal and a distal row, each of 4 bones. Their names express their general appearance. From radial to ulnar side they are:

Proximal row—*scaphoid, lunate, triquetrum,* and *pisiform.*
Distal row—*trapezium, trapezoid, capitate,* and *hamate.* (See *Plate 35.1.*)

It may be helpful first to picture these as 8 uniform cubes arranged as in *Figure 35.1,* but a glance at an articulated skeleton shows a more complex arrangement. The series of diagrams in *Figure 35.2* illustrates the plan.

Observe that:

1. the pisiform articulates with only the triquetrum and lies entirely anterior to it; therefore,
2. to the wrist joint the remaining 3 proximal bones present a common articular surface, which is a convex ovoid that allows movements in all directions (except rotation);
3. the capitate is the largest bone, it is centrally located and around it the others are organized, and its caput or head fits into a concavity provided by the proximal row;
4. on the lateral side, the scaphoid is elongated distally while the trapezium and trapezoid are relatively small;
5. the hamate reaches the lunate between the capitate and triquetrum, and so

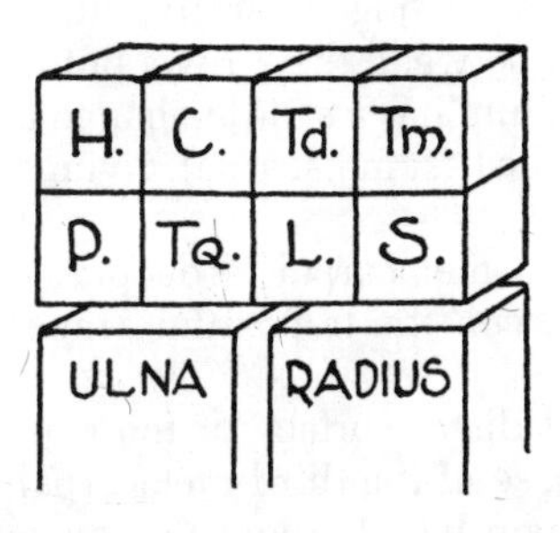

Figure 35.1. The carpal bones as cubes. *H,* hamate; *C,* capitate; *Td,* trapezoid; *Tm,* trapezium; *P,* pisiform; *TQ,* triquetrum; *L,* lunate; *S,* scaphoid.

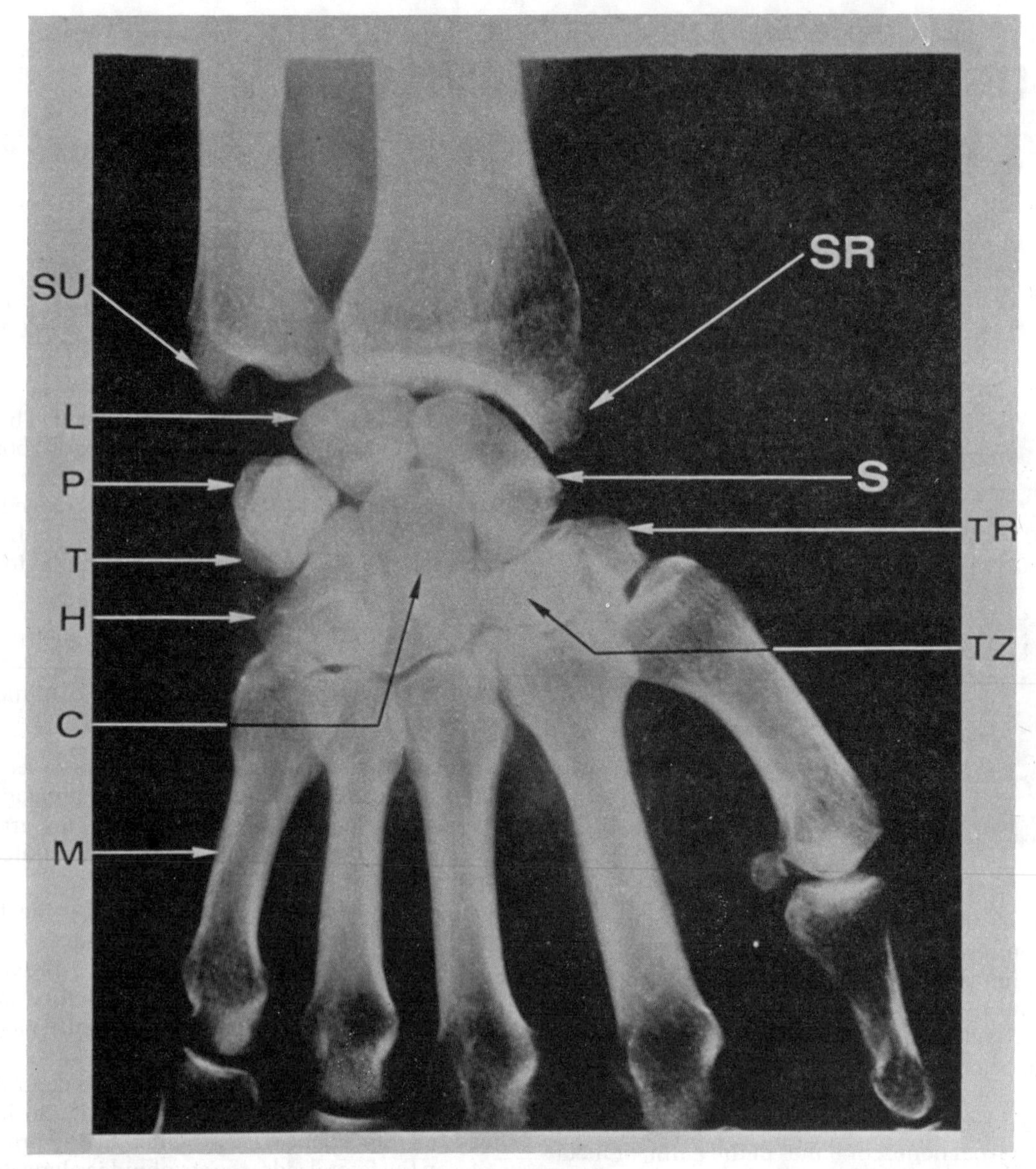

Plate 35.1. Wrist (P-A view). *C*, capitate; *H*, hamate; *L*, lunate; *M*, metacarpal 5; *P*, pisiform; *S*, scaphoid; *SR*, styloid process of radius; *SU*, styloid process of ulna; *T*, triquetrum; *TR*, trapezium; *TZ*, trapezoid.

6. the line of articulation between proximal and distal rows—called the **midcarpal joint**—is seen to be sinuous, having a very sharp "jog" laterally; this limits abduction and adduction between the 2 rows but allows flexion;

7. the trapezium has a saddle-shaped articular surface for the base of the first metacarpal, giving it great mobility (*fig. 35.3*);

8. the second metacarpal is deeply wedged and immobilized between the trapezium, trapezoid, and capitate;

9. the broad distal surface of the capitate articulates with the flat base of the third metacarpal;

10. the hamate has 2 adjoining concave surfaces for the convex base of the 4th and 5th metacarpals, and these permit a slight degree of flexion; and

11. the sides of the bases of the metacarpals 2, 3, 4, and 5 articulate with each other.

DISTAL SURFACE OF CARPUS AND BASES OF METACARPALS

Here we do not find plane gliding surfaces, such as would allow the metacarpals to shift from side to side and backward and forward. They are very irregular (*fig. 35.3*). Although the metacarpal of the thumb or pollex can be flexed and extended, abducted and adducted, and rotated, it does not possess the spherical base you naturally look for, but a saddle-shaped surface which fits onto the distal surface of the trapezium, which is reciprocally saddle-shaped.

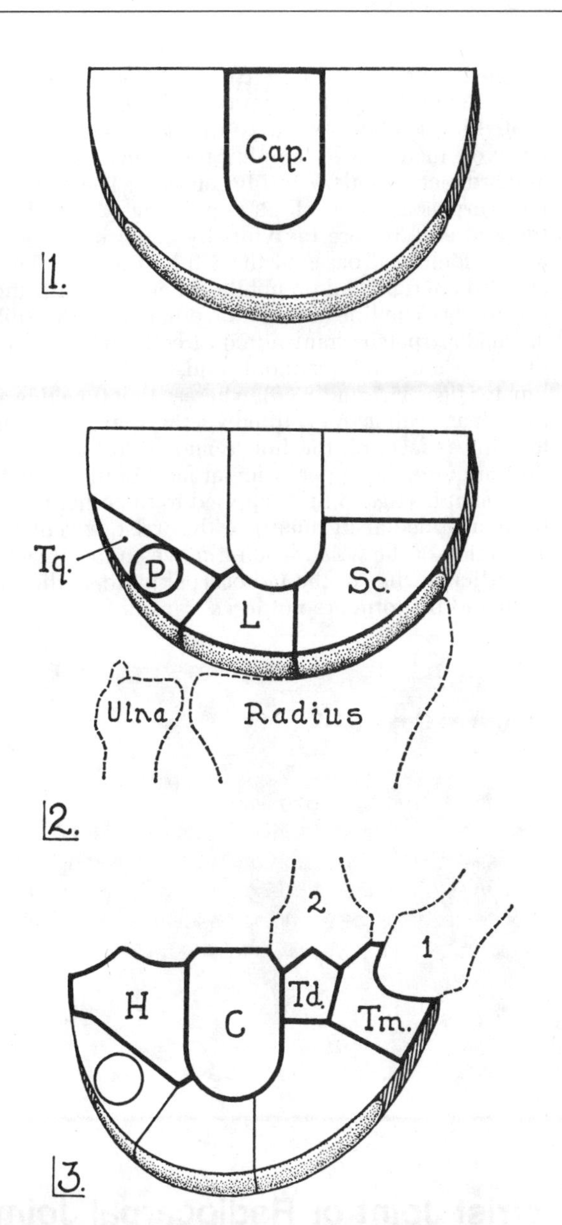

Figure 35.2. Steps in drawing carpal bones (see text). *Tq*, triquetrum; *Cap*, capitate; *P*, pisiform; *L*, lunate; *Sc*, scaphoid; *H*, hamate; *C*, capitate; *Td*, trapezoid; *Tm*, trapezium.

Details. The articulated hand reveals that the bases of the 4th and 5th metacarpals articulate with the hamate. Grasp in turn the knuckles or heads of the metacarpals of your own fingers, noting that the 5th metacarpal can be moved freely backward and forward, that the 4th metacarpal can be moved to a less degree, and that the 3rd and 2nd metacarpals are almost immobile. An examination of the carpus gives the reason: the 5th and 4th carpometacarpal joints are clearly hinge joints.

The capitate has an expansive plane surface for the base of the 3rd metacarpal. The trapezoid does not project so far distally as the carpals on each side of it, and in consequence, the base of the 2nd metacarpal is mortised between the capitate and trapezium and in part articulates with them. The trapezoid, moreover, possesses an

Figure 35.3. The carpal bones and the bases of the metacarpal bones.

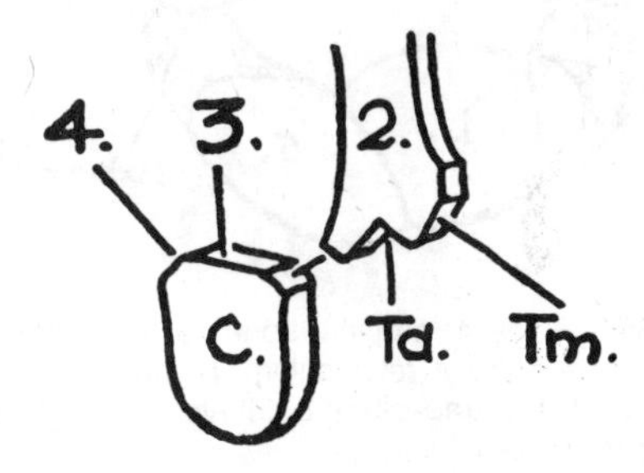

Figure 35.4. The capitate (C.) articulates with 3 metacarpals. The 2nd metacarpal articulates with three carpals. *Td*, trapezoid; *Tm*, trapezium.

anteroposteriorly placed ridge which fits like a wedge into the guttered base of the 2nd metacarpal (*fig. 35.4*).

DORSAL AND PALMAR ASPECTS OF CARPUS

Observe that the dorsum of your wrist or an articulated carpus is transversely arched or convex. The arched condition is maintained by a tie beam, the **flexor retinaculum** (transverse carpal ligament), which unites the marginal bones of the carpus. Its proximal part extends between 2 rounded prominences, the *pisiform* and the *tubercle of the scaphoid* (*fig. 35.5*), whereas its distal part stretches between 2 crests, namely, the *hook of the hamate* and the *tubercle of the trapezium.*

The tubercle of the trapezium is also the lateral lip of a vertical groove in which the tendon of the flexor carpi radialis runs on its way to the base of the 2nd (and 3rd) metacarpal, where it is inserted.

MIDCARPAL JOINT (TRANSVERSE CARPAL JOINT)

This composite joint lies between the proximal and distal rows of carpal bones (*fig. 35.3*) and is sinuous. The greater part of the flexion that appears to take place at

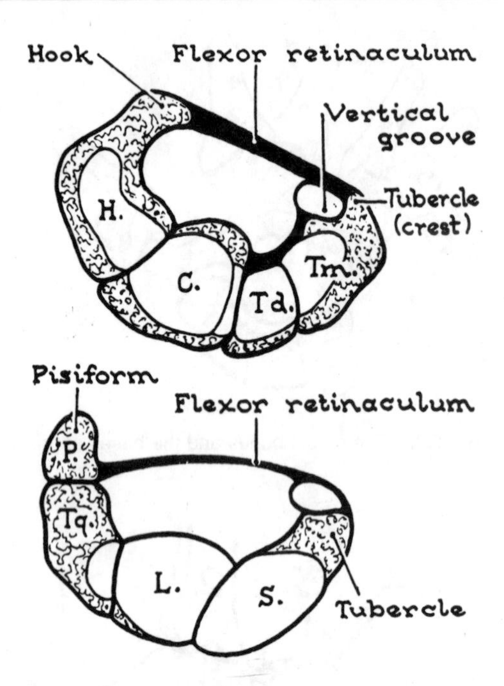

Figure 35.5. The 4 marginal bones of the carpus give attachment to the flexor retinaculum. *H*, hamate; *C*, capitate *Td*, trapezoid; *Tm*, trapezium; *L*, lunate; *S*, scaphoid; *Tq*, triquetrum; *P*, pisiform.

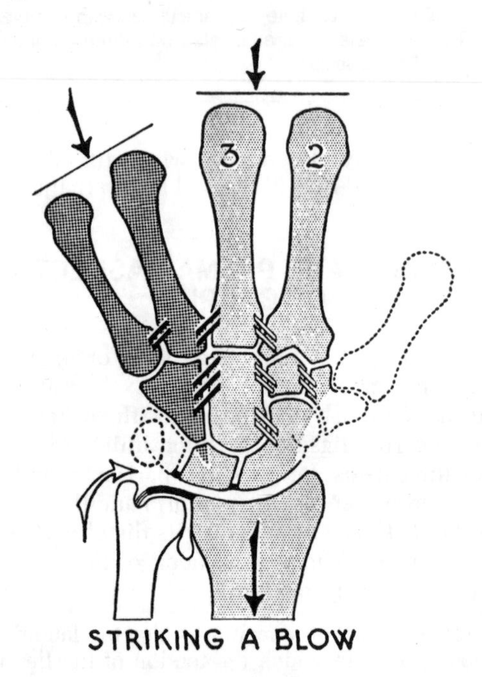

Figure 35.6. When striking a blow, use the 2nd and 3rd knuckles. Observe that the ligaments take the direction of usefulness.

the radiocarpal joint actually takes place at the midcarpal joint.

Intercarpal Ligaments. The various carpal bones are bound to each other by dorsal and palmar bands, which largely radiate from the capitate, and by interosseous ligaments. Between the individual bones of each row, a slight amount of gliding can take place.

Striking a Blow. When you strike a blow with your fist, you instinctively employ the knuckles of the 2nd and 3rd metacarpals. Why? Because: (1) The 2nd and 3rd metacarpals are long, stout, and strong. (2) The bases of the 2nd and 3rd are individually equal in surface area to the combined bases of the 4th and 5th. (3) The 2nd and 3rd are rigid and immobile, whereas the 4th and 5th are not. (4) The line of force traveling along the 2nd and 3rd metacarpals is transmitted directly to the radius via the trapezoid and scaphoid, and via the capitate and lunate (*figs. 35.3 and 35.6*), whereas that traveling along the 4th and 5th is transmitted via the apex of the hamate to a linear facet on the lunate and so to the radius. (5) Furthermore, the upper articular facet of the triquetrum, such as it is (*fig. 35.5*), is applied to the ulnar collateral ligament (medial ligament) of the wrist, except during adduction of the wrist, when it moves into contact with the articular disc of the radiocarpal joint; so the triquetrum and pisiform are not force transmitters.

Clinical Case 35.2

Patient Brenda F. This woman, age 38, has had chronic rheumatoid arthritis for 12 years. It now is causing severe pain, deformity and swelling of both her wrist joints. The professor of rheumatology asks you to describe the main anatomical features of this joint to the rest of your tutorial group.

Wrist Joint or Radiocarpal Joint

OBSERVATIONS

You can easily observe on your own wrist that the movements permitted are flexion, extension, adduction, and abduction. These together comprise the movement of circumduction. This, then, is a biaxial or ellipsoid articulation (*fig. 35.7*). It is at the midcarpal joint that the greater part of the flexion often attributed to the wrist joint actually takes place.

JOINT SURFACES

The socket is formed by the inferior articular surface of the radius plus the articular disc. The egg-shaped or ellipsoidal convex surface is formed by the proximal ar-

Figure 35.7. Circle *versus* ellipse.

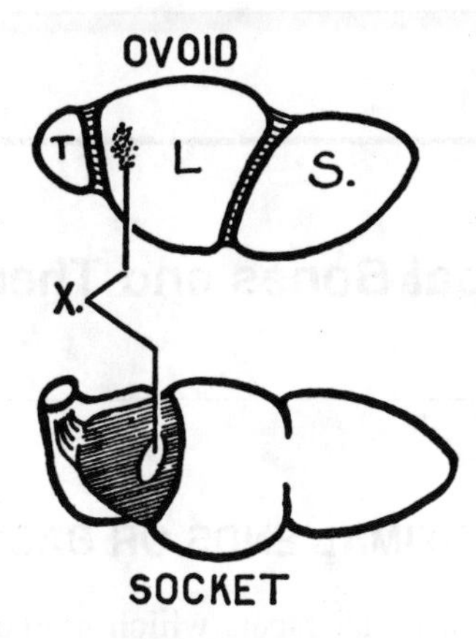

Figure 35.8. "Ovoid" and socket of wrist joint. "X" marks the sites at which the disc is sometimes perforated and the lunate softened. *L*, lunate; *T*, triquetrum; *S*, scaphoid.

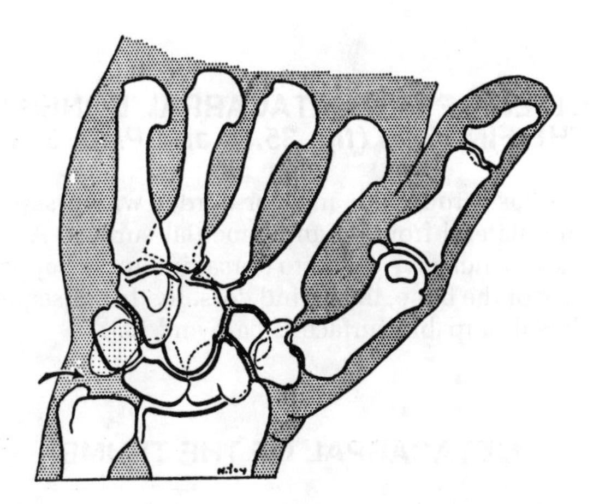

Figure 35.9. Tracing of an X-ray photograph of the wrist in abduction. (*Arrow* indicates a space, proximal to the triquetrum, present in abduction.)

ticular surfaces of the scaphoid, lunate, and triquetrum, plus the interosseous ligaments binding these 3 bones together (*fig. 35.8*).

Details Revealed by X-rays. In *abduction* (radial deviation, *fig. 35.9*), the scaphoid rotates forward, the lunate moves onto the articular disc, and the space between the triquetrum and the ulna is increased.

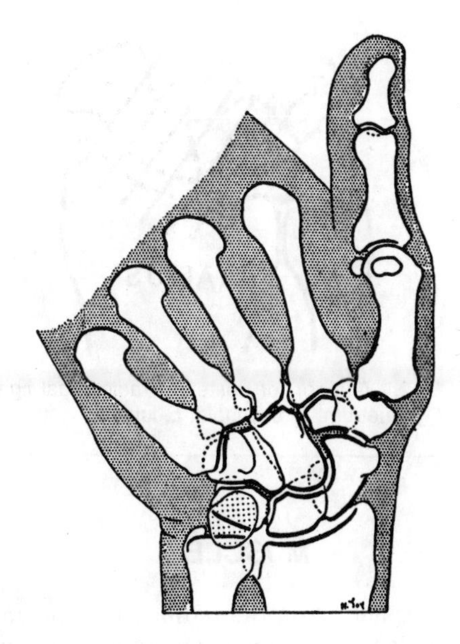

Figure 35.10. Tracing of an X-ray photograph of the wrist in adduction.

In *adduction* (ulnar deviation, *fig. 35.10*), the scaphoid, lunate, and triquetrum move laterally, the lunate coming entirely onto the radius and the triquetrum making contact with the articular disc (*fig. 35.11*). Marked changes in orientation between neighbors occurs as they glide on each other.

LIGAMENTS

The *palmar and dorsal radiocarpal ligaments* pass obliquely inferomedially from the front and back of the lower end of the radius to the front and back of the proximal row of carpal bones and to the capitate (*fig. 35.12*)

The *radial and ulnar collateral ligaments* (lateral and medial ligaments) extend from the styloid processes of the radius and ulna to the scaphoid and triquetrum, respectively.

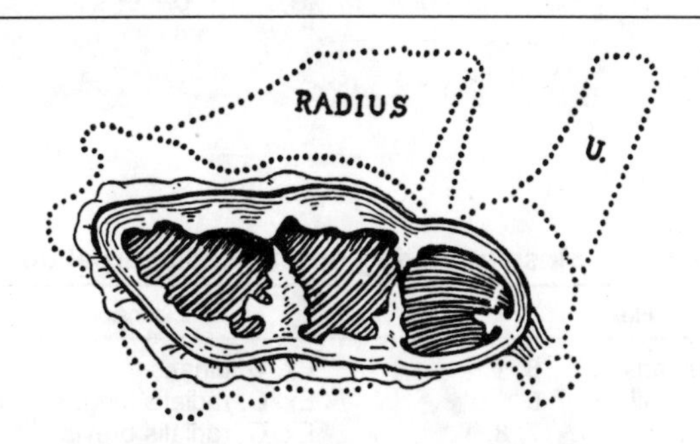

Figure 35.11. Socket for proximal row of carpals. Note the transparent synovial fringes.

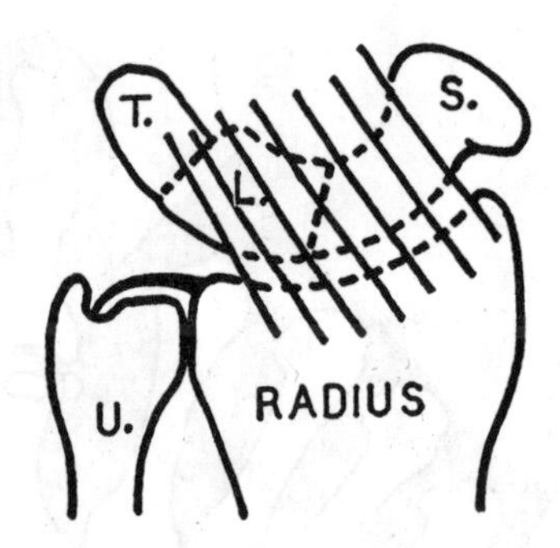

Figure 35.12. Anterior (or posterior) radiocarpal ligament. *U*, ulna; *T*, triquetrum; *S*, scaphoid; *L*, lunate.

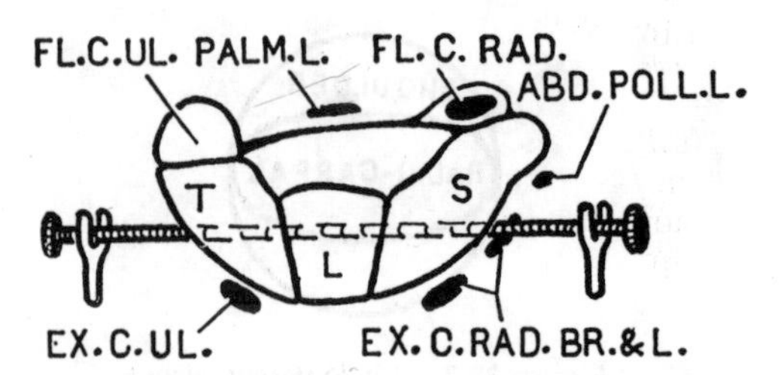

Figure 35.13. The flexors and extensors of the wrist (on transverse section). *T*, triquetrum; *L*, lunate; *S*, scaphoid.

MUSCLES

The muscles acting on the radiocarpal and midcarpal joints and their nerve segments are given in *Table 35.1*.

Actions. The flexors and extensors of these joints are indicated in *Figure 35.13*. The 3 important flexors are more advantageously situated than the 3 extensors. Flexion is more powerful than extension in the ratio of 13:5 (Fick).

NERVE SUPPLY

The sensory innervation of the wrist joint follows Hilton's rule: *Those nerves that pass a joint will innervate it*. Therefore the sensory branches are derived from the anterior interosseous branch of the median, posterior interosseous branch of the radial, and deep palmar and dorsal cutaneous branches of the ulnar.

Clinical Case 35.3

Patient Connie G. Although you do not see this young girl herself, the professor of pediatric radiology does put her radiographs on display and begins to quiz your group of students on the features of the metacarpals, ossification stages, and the probable chronologic age of the patient.

Metacarpal Bones and Their Joints

PROXIMAL ENDS OR BASES

These have articular facets which are counterparts of the distal surfaces of the carpals (*fig. 35.3*). Furthermore, the apposed surfaces of the bases of the 2nd, 3rd, 4th, and 5th metacarpals articulate with each other and, therefore, carry articular facets.

BODIES OF THE METACARPAL BONES OF THE FINGERS (*fig. 35.14 and Plate 35.2*)

Each has a rounded, anterior border which separates an anterolateral from an anteromedial surface. As these surfaces, which give origin to dorsal interossei, approach the base of the bone, they wind dorsally. In consequence, the dorsal palpable surface is bare and flat.

METACARPAL OF THE THUMB

This is short and stout. It is rounded posteriorly and is easily palpated, but anteriorly it is covered by the fleshy attachment of muscles (*fig. 35.15*). The base is saddle-shaped. On the palmar aspect of the head, lie 2 sesamoid bones.

Table 35.1. Muscles Acting on the Radiocarpal Joint and Their Nerve Segments

Flexion		Extension		Abduction		Adduction	
Fl. c. ulnaris	8, 1	Ex. C. ulnaris	7	Ex. c. radialis longus	6, 7	Ex. c. ulnaris	7
Fl. c. radialis	6	Ex. C. radialis longus	6, 7	Ex. c. radialis brevis	6, 7, 8	Fl. c. ulnaris	8, 1
Palm. longus	7, 8, 1	Ex. C. radialis brevis	7	Fl. c. radialis	6		
Abd. poll. long.	7			Abd. poll. long.	7		

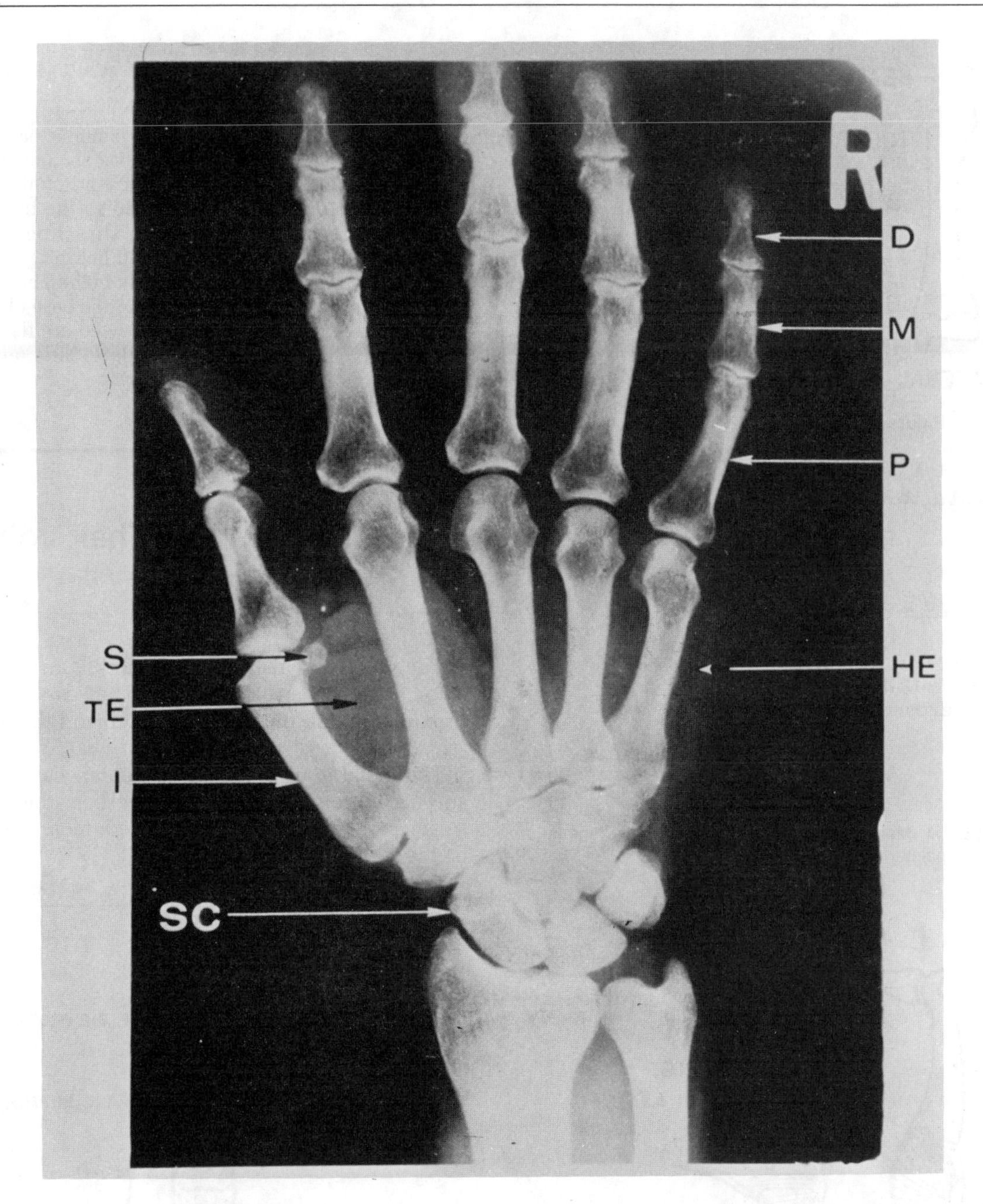

Plate 35.2. Hand (P-A view). *D*, distal phalanx; *HE*, hypothenar eminence; *I*, thumb metacarpal; *M*, middle phalanx of digit; *P*, proximal phalanx; *S*, sesamoid bone at metacarpophalangeal joint of thumb; *SC*, scaphoid bone; *TE*, thenar eminence.

THE METACARPOPHALANGEAL JOINT

Observe on your knuckles that the heads of the metacarpals of the fingers are spheroidal. The bases of the proximal phalanges fit the metacarpal heads and, so, are concave.

The metacarpophalangeal articulations of the 4 fingers can be flexed and extended, and, when the hand is open, they can also be abducted and adducted, that is to say, they can perform the 4 component movements of circumduction. These movements define condyloid joints (Gk. kondulos, = a knuckle).

When the joints are flexed, neither abduction nor adduction is possible because (1) the heads of the metacarpals are flattened in front; and (2) the **collateral ligaments**, though slack on extension, are taut on flexion, due to their eccentric attachments to the sides of the heads of the metacarpals (*fig. 35.16*).

Posteriorly, there are no ligaments, but the extensor (dorsal) expansions of the extensor muscles serve the

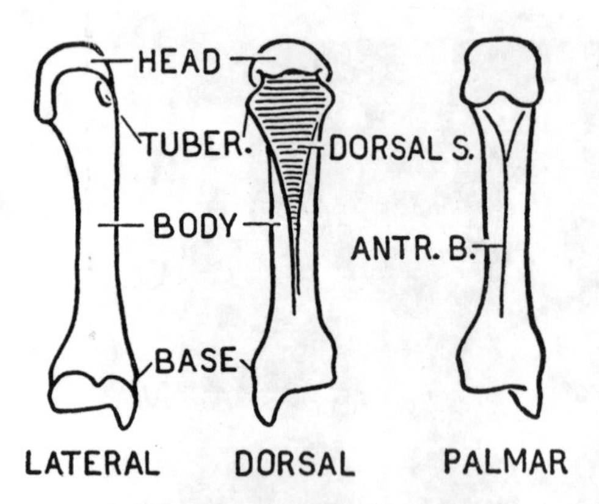

Figure 35.14. A metacarpal bone.

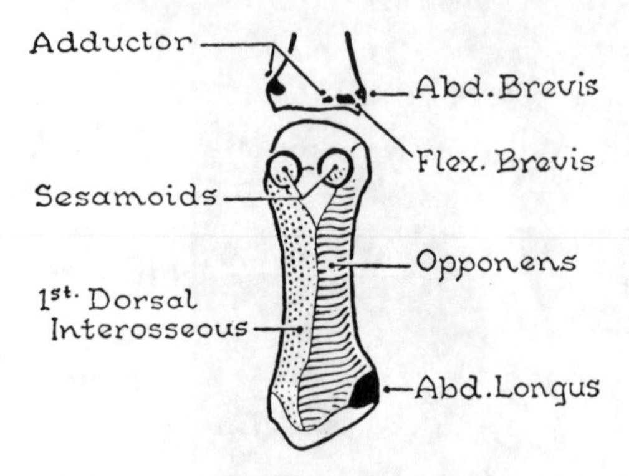

Figure 35.15. Metacarpal of thumb, showing attachments of muscles, palmar aspect.

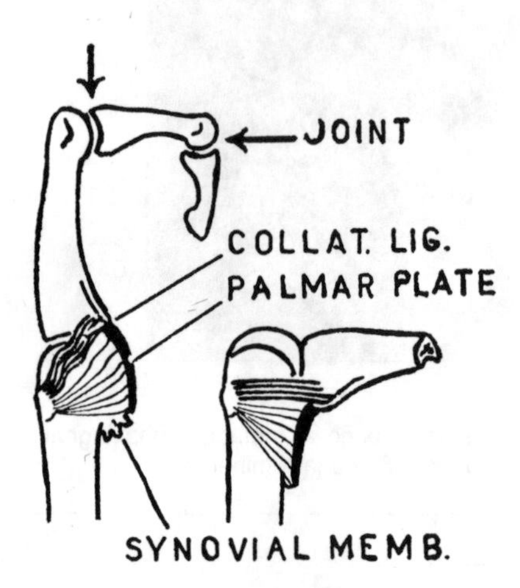

Figure 35.16. Metacarpophalangeal and interphalangeal joints.

part. Anteriorly, the capsule is thickened to form the **palmar ligament or plate** (volar accessory ligament). Fibers of the collateral ligaments radiate to the sides of this plate and keep it firmly applied to the front of the head of its metacarpal like a visor (*fig. 35.16*).

The sides of the palmar ligaments of the fingers are united to each other by 3 ligamentous bands, the **deep transverse metacarpal ligaments**, which help to prevent the metacarpals from spreading.

Anterior to the transverse ligaments pass the digital vessels and nerves and the lumbricals; posterior to them pass the interossei. Slips from the 4 digitals bands of the palmar aponeurosis are attached to the transverse ligaments (*see fig. 36.10*) in front; slips from the extensor expansions are attached to them behind.

The Metacarpophalangeal Joint of the Thumb. The head of the metacarpal is rounded and the base of the phalanx is concave. The movements allowed are flexion and extension, some rocking from side to side, and some rotation.

Phalanges and Their Joints

The phalanges also are long bones, 2 for the thumb and 3 for each of the other digits. They are known as the proximal or 1st, the middle or 2nd, and the distal or 3rd (*fig. 35.17*).

The distal ends of the distal phalanges have a smooth surface which the fingernail overlies; the area for the finger-pad is rough, owing to the attachment of fibrous bands that bind the skin to it.

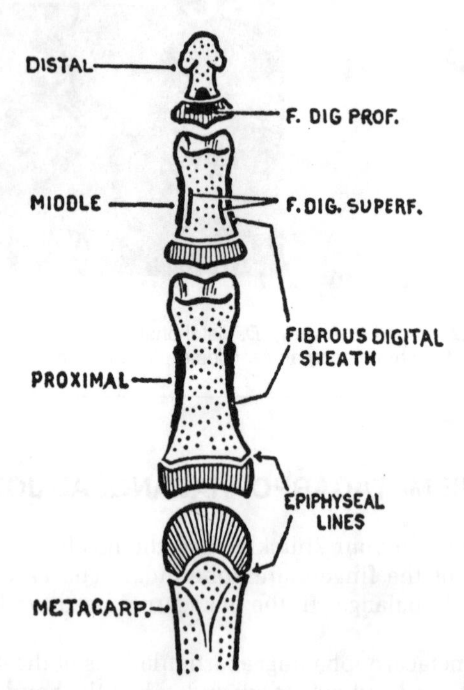

Figure 35.17. Phalanges (palmar aspect) showing epiphyseal lines and attachments of fibrous sheath and flexor tendons.

The dorsal aspects of the proximal and middle phalanges are smooth, rounded, and covered with the extensor expansion. The palmar surfaces take part in the floor of the osseofibrous tunnels (*see fig. 36.11*) in which the flexor tendons glide. They are smooth and flat.

Each phalanx has 2 borders. The borders of the proximal and middle phalanges provide attachment for the fibers of the fibrous digital sheath. Those of the middle phalanx also receive the slips of insertion of the flexor digitorum superficialis.

When your hand is closed, the heads of the middle and proximal phalanges are uncovered. They have 2 little condyles. The bases of the distal and middle phalanges have 2 little depressions.

The interphalangeal joints are hinge joints with their movements restricted (more or less) to flexion and extension.

Carpometacarpal and Intermetacarpal Joints

These are described with the carpal bones on pp. 408–409. Note again that your 4th and 5th metacarpal bones have slight hinge movements (*fig. 35.18*).

Each metacarpal head is separated from its neighbor by adjacent collateral ligaments, interosseous tendons areolar tissue, and often a **bursa**.

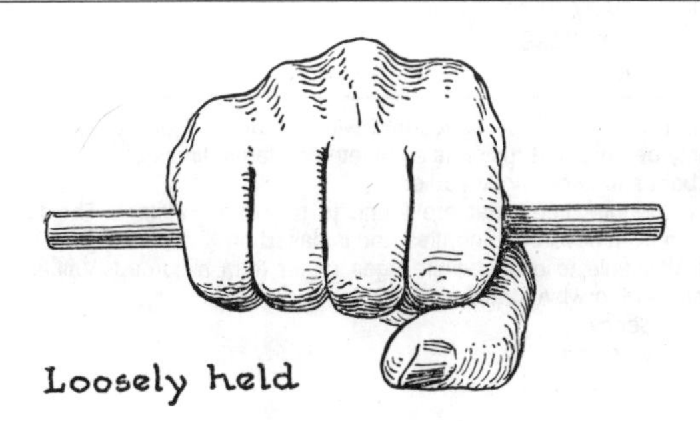

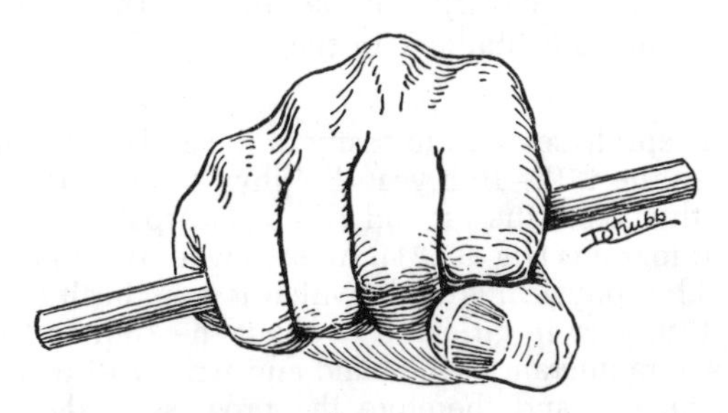

Figure 35.18. Because the 4th and 5th carpometacarpal joints are hinge joints, the grip on a rod is more secure.

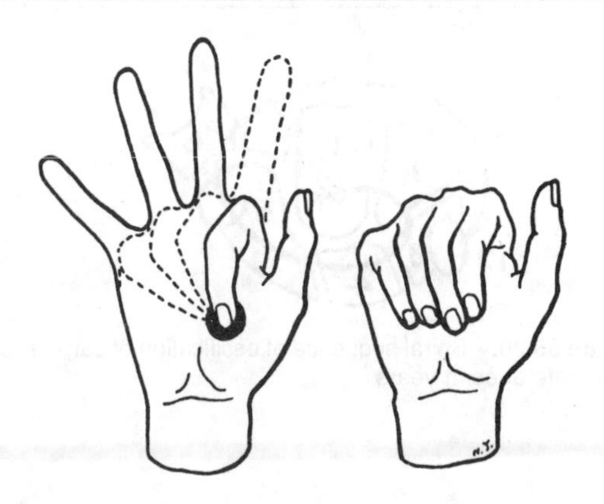

Figure 35.19. To demonstrate that flexed fingers buttress each other.

Metacarpophalangeal and Interphalangeal Joints

Described with the metacarpal bones and phalanges on pp. 413–415.

Buttressing of Fingers. The flexed fingers are adducted so that they steady each other (*fig. 35.19*).

Ossification of Bones of Hand and Wrist

PRIMARY CENTERS

The bodies of the metacarpals and phalanges start to ossify during the 3rd prenatal month. The carpal bones, unlike the tarsal bones, have not started to ossify at the time of birth, although, in the female, centers may have appeared in the capitate and hamate. The carpals proceed to ossify in orderly spiral sequence, approximately in the following years; capitate and hamate 1st, triquetrum 3rd, lunate 4th, scaphoid 5th, trapezoid and trapezium 6th—and the pisiform 12th (*fig. 35.20*).

EPIPHYSES

Each of the long bones has 1 epiphysis. In the metacarpals these occur at the heads and in the phalanges at

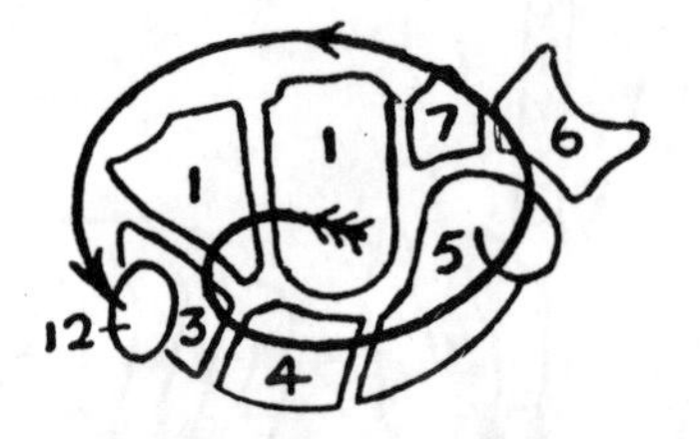

Figure 35.20. Spiral sequence of ossification of carpals; approximate ages in years.

Table 35.2. Table of Appearance and Fusion Times of Epiphyses in Upper Limb Including Primary Centers for Carpus (in Years Unless Otherwise Stated)

	Appears	Fuses*
Scapula, acromion	15	19
Coracoid	1 }	15†
Subcoracoid	10 }	
Inf. angle	15	22
Vertebral border	15	22
Clavicle sternal end	18‡	22‡
Humerus, head	Birth }	
Greater tubercle	3 }	6} 19†
Lesser tubercle	5 }	
Medial epicondyle	6	18
Lateral epicondyle	12 }	
Capitulum	1 }	15} 17
Trochlea	10 }	
Radius		
proximal end	7	19
distal end	1	19†
Ulna		
proximal end	11	19
distal end	6	19†
Scaphoid	6	
Lunate	5	
Triquetrum	4	
Hamate	½	
Capitate	½	
Trapezoid	6	
Trapezium	6	
Pisiform	12	
1st Metacarpal base	3§	19
2nd–5th Metacarpal heads	3§	19
Thumb		
proximal phalanx base	2	18
distal phalanx base	3§	18
2nd–5th Fingers		
proximal phalanx base	2	18
middle phalanx base	3§	18
distal phalanx base	3§	18§
Sesamoids of thumb	13§	

* Since the student will be concerned with the age periods at which he may reasonably be sure that fusion is complete, the latest times of fusion are given. Female bones fuse distinctly earlier.

† Daggers denote times that are found to be quite constant. The table is a compilation from several authorities and is based on X-ray findings. It has been deemed advisable to give definite ages rather than a spread. Values denote years unless otherwise stated.

‡ A later tendency.

§ An earlier tendency.

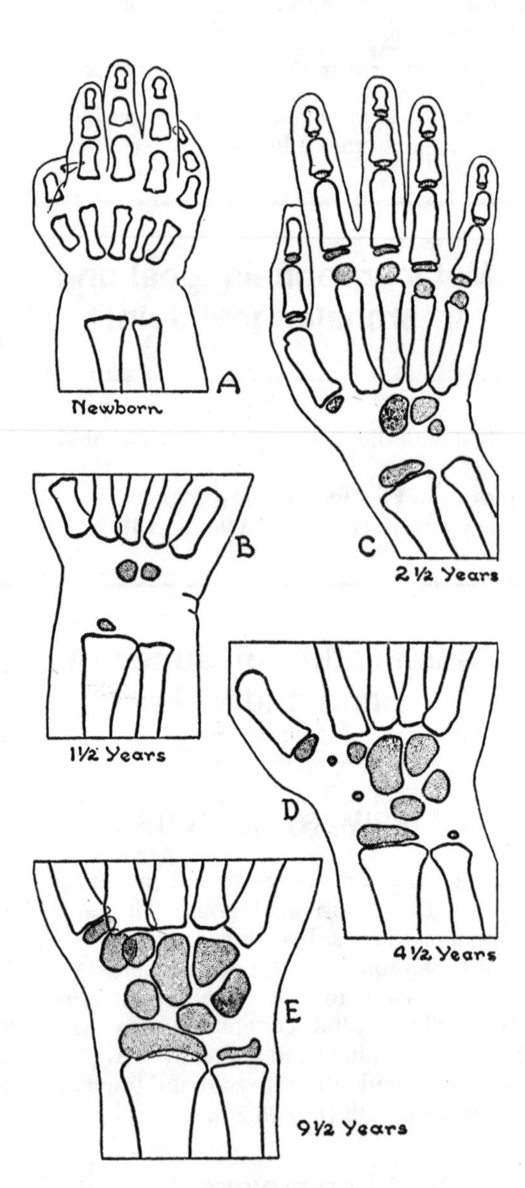

Figure 35.21. Progressive ossification of the bones of the hand. (Courtesy of Dr. J. D. Munn, Hospital for Sick Children, Toronto.)

the bases—the metacarpal of the thumb is the exception, for it resembles a phalanx in that its epiphysis is at its base.

The epiphyses start to ossify in the 2nd–3rd year and fuse in the 17th–19th year. Epiphyses have been found at both ends of the 1st and 2nd metacarpals.

Radiograms (*fig. 35.21*). At any given time during the period of physical growth, which is from birth to about the 17th year in girls and to about the 18th or 19th in boys, a radiogram of the hand and wrist will reveal the *skeletal age* and therefore the progress made toward physical maturity—which may be either accelerated or retarded—and it may reveal much else. The skeletal state of a girl aged 13½ years is not achieved by a boy until he is 15½ years old (Greulich and Pyle).

Garn *et al.* find that the hand alone cannot be used as a standard of skeletal development for the whole body, that all ossific centers do not have équal predictive value, and that a small number of centers of high predictive value in the hand and foot provide more useful information than does the entire number. (For details on the reliability of the data, see Johnson *et al.*)

Radius and Ulna. (*fig. 35.21*). The distal **radial epiphysis** appears before that of the **ulna** (*fig. 35.21*). Radiographically, these 2 epiphyses unite with the diaphyses about the 16th or 17th year in the female and the 18th or 19th in the male (see Greulich and Pyle); it may be delayed until the 23rd year (McKern and Stewart).

Table 35.2 summarizes the time of appearance and fusion of the secondary centers of ossification as well as the time of appearance of the primary centers for the carpal bones. These data are approximations of the mean dates and are offered for reference, not memorization. Memorize the rule, however: the important centers appear "early" and fuse late.

Anatomical Features of the Physical Examination

Injuries and arthritic conditions of the wrist and hand are so common that a clear knowledge of the visible and palpable landmarks is essential for the clinician. Although the flexor retinaculum is not palpable (except as a resistance to pressure), identifying its four pillars of attachment should be routine. When there is atrophy of the thenar muscles, it is a very important clue to median nerve problems—often a carpal tunnel syndrome (entrapment deep to the flexor retinaculum).

The intercarpal joints are covered by tendons, reducing access to the clinician's palpating fingers, but the metacarpophalangeal and interphalangeal joints are easily viewed and palpated posteriorly. Their movements and any deformities are easily compared with the normal.

Forces from the thumb pass through the scaphoid, which can be fractured across its narrow middle if the force is excessive. Tenderness in the "snuff-box" may be one of the clinical signs of this.

Ossification of both primary and secondary centers is of great interest to radiologists and pediatricians. The "spiral sequence" of ages at which the carpal bones start to ossify is best seen in *Figures 35.20* and *35.21*. The secondary centers are summarized in *Figure 35.21*.

Clinical Mini-Problems

1. **Which carpal bone is most frequently (a) broken? (b) dislocated?**
2. **When the lunate is dislocated it shifts anteriorly into the carpal tunnel. Which structure in the carpal tunnel is most susceptible to compression forces?**
3. **Wrist X-rays are frequently utilized by pediatricians to access the endocrinological development of a child. When would one expect to find 8 fully ossified carpal bones in a child?**

(Answers to these questions can be found on p. 587.)

36

Hand and Fingers

Clinical Case 36.1

Patient Arnold C. This electrician who is employed by the power company suffered severe electrical burns to the palm of his right hand 3 months ago. This morning, you are a student member of the plastic-surgery team preparing that scarred region (extending from the wrist into the fingers) for the reparative operation that your tutor will perform. You feel certain he will question you about the flexor retinaculum, palmar aponeurosis, thenar muscles, and nerves and tendons.

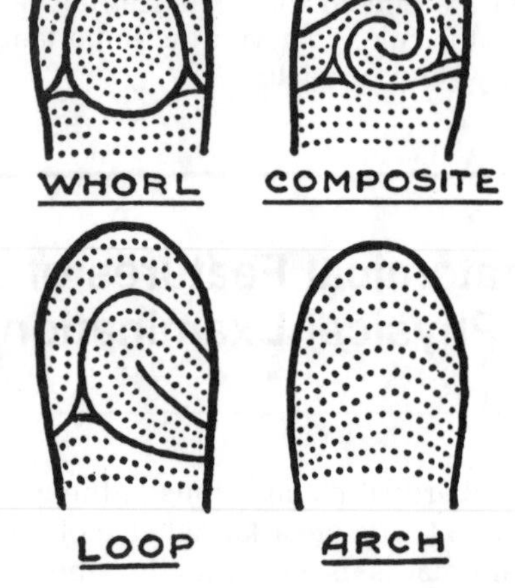

Figure 36.1. Types of finger prints. (After Wilder.)

Palm of the Hand

The skin of the grasping or palmar surface of the hand is very thick. It rests on a protective pliable layer of loculated fat that is anchored to the underlying tissues by fibrous bands, most dense at the pads of the fingers and in front of the palmar aponeurosis.

To improve the grip, the skin is ridged and furrowed, with a convenient absence of greasy, sebaceous glands. On the summits of the ridges, the mouths of numerous sweat glands open. The disposition of the ridges in arches, loops, and whorls differs in detail from person to person; so does the spacing, the shape, and the size of the mouths of the sweat glands. The impressions left by the ridges and gland mouths are known as **finger prints** (*figs. 36.1, 36.2*).

Permanent **skin creases** occur in the hand in response to its movements. On the digits they are transverse; in the palm they have the form of the letter M. At the creases, fat is absent. The relation of the creases to the joints is shown in *Figure 37.3*.

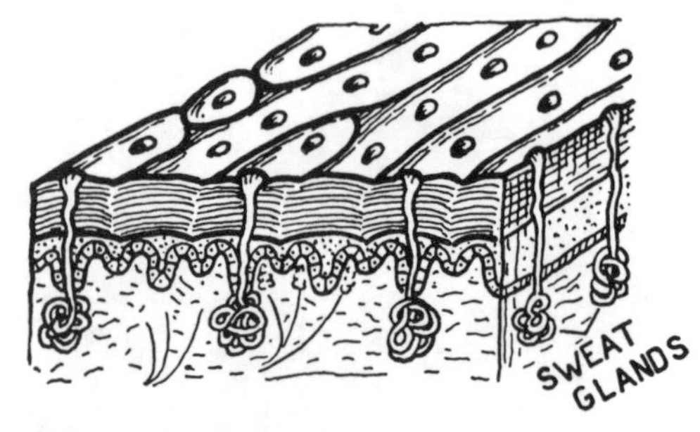

Figure 36.2. Friction ridges and orifices of ducts of sweat glands on the fingers. (After Wilder.)

Digital Formula. The most projecting digit in most people is the middle finger; next in order come the ring finger, index, little finger, and thumb, so the digital formula reads $3 > 4 > 2 > 5 > 1$ (Jones). This is the primitive arrangement, common to apes and man, but in one in 5 the ring projects equally ($3 > 4 = 2 > 5 > 1$) and in one-third the index exceeds the ring finger ($3 > 2 > 4 > 5 > 1$).

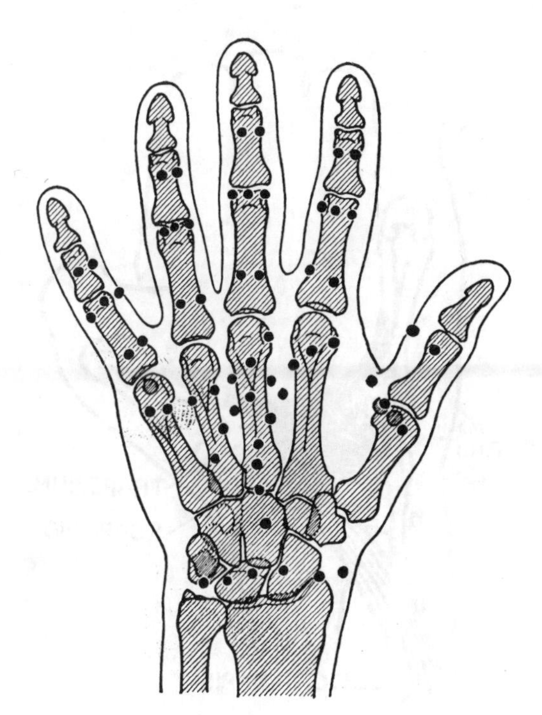

Figure 36.3. Tracing of X-ray of hand. Shot was placed on the skin creases to show their relations to the joints (see *fig. 34.12*).

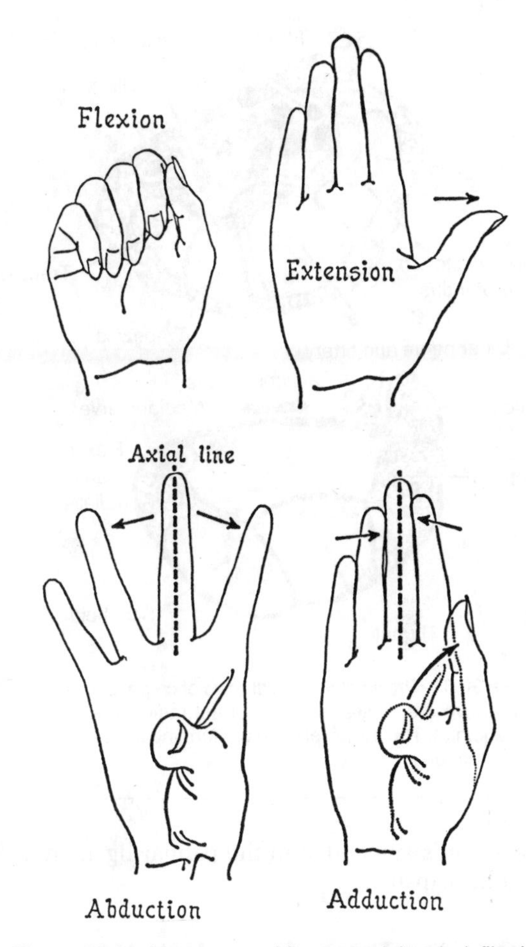

Figure 36.4. Movements of fingers and thumb defined.

MOVEMENT OF DIGITS

A line drawn through the middle finger, middle metacarpal, and capitate is called the *axial line of the hand* (*fig. 36.4*). Movement of a finger away from this axial line is called abduction and movement toward it, adduction, and the movements take place at the metacarpophalangeal joints. The thumb is set at right angles to the other digits. It also can be abducted and adducted, but those movements take place, at its carpometacarpal joint; the movement anteriorly being abduction, and posteriorly adduction; that is, in the side-to-side (coronal) plane you abduct and adduct your fingers and flex and extend your thumb.

CARPOMETACARPAL JOINT OF THUMB

It is to movements of medial and lateral rotation that the thumb owes its peculiar value, a value that exalts it above all the fingers. The base of its metacarpal sits on the trapezium as though astride a saddle, enjoying the movements of a multiaxial joint of the saddle variety (p. 17).

FLEXOR RETINACULUM

A tie-beam, the *flexor retinaculum* (*fig. 36.5*), forms with the carpal bones an osseofibrous tunnel, the *carpal tunnel*. The retinaculum stretches between 2 pillars in the proximal row—the *tubercle of the scaphoid* and the *pisiform*, and 2 in the distal row—the *tubercle of the trapezium* and the *hook of the hamate* (*fig. 36.6*).

The flexor retinaculum not only prevents the long flexors of the digits from "bow-stringing" (moving anteriorly during flexion) but also affords chief origin to the thenar and hypothenar muscles.

THE THREE THENAR MUSCLES

These **muscles of the base of the thumb** arise from the flexor retinaculum and its 2 lateral bony pillars (tubercles of scaphoid and of trapezium). Most superficial is the *abductor pollicis brevis* (*fig. 36.7*). Obviously, it draws the thumb anteriorly (i.e., abducts it), and the movement takes place at the saddle-shaped carpometacarpal joint.

Reflection of the abductor uncovers a fleshy sheet (*fig. 36.7*) that pronates or medially rotates the whole thumb and flexes its metacarpophalangeal joint. The portion inserted into the metacarpal is named the *opponens pollicis*, the portion into the phalanx, the *flexor pollicis brevis* (*fig. 36.7*). The latter insertion is largely via the lateral

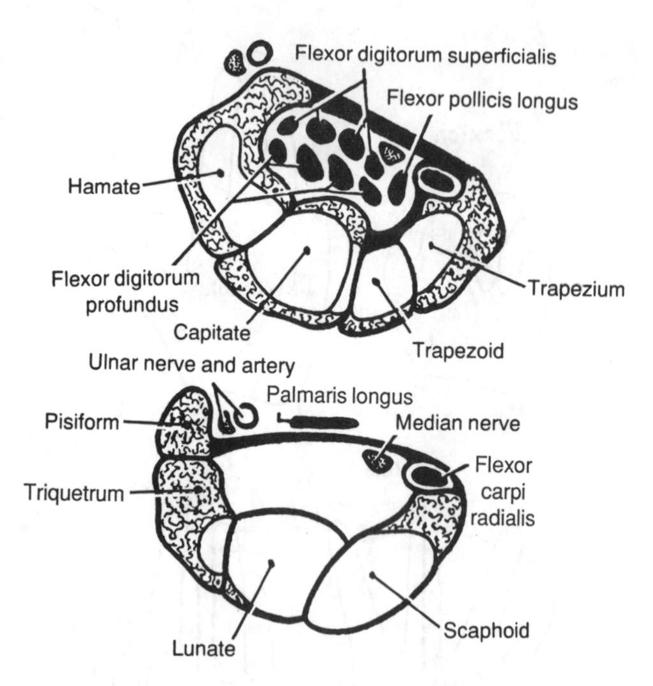

Figure 36.5. Proximal and distal rows of carpal bones viewed from above (i.e., distal surfaces of the radiocarpal and midcarpal joints). The flexor retinaculum forming the osseofibrous carpal tunnel.

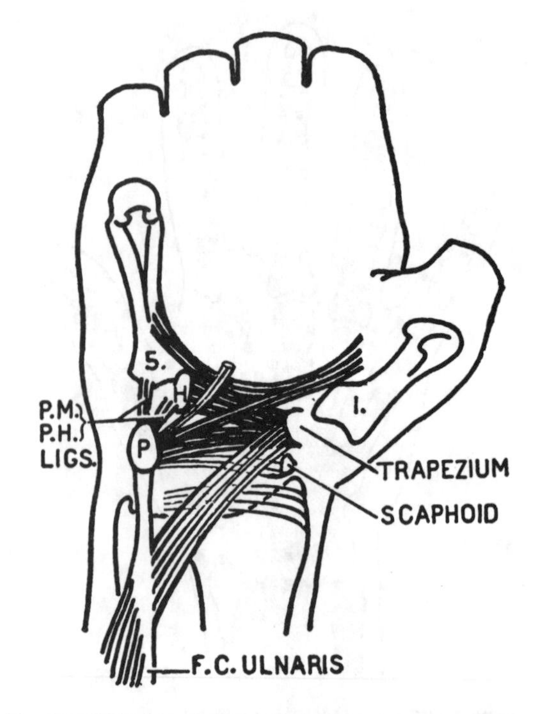

Figure 36.6. The flexor retinaculum (details). *PM*, pisometacarpal; *PH*, pisohamate; *P*, pisiform.

of 2 sesamoids embedded in the palmar ligament (plate) of the joint capsule.

Nerves to the Three Thenar Muscles

A **recurrent branch of the median** bends around the lower border of the retinaculum 3 cm below the palpable tubercle of the scaphoid (*fig. 36.8*). A small coin will cover the nerve, whose importance is bound up with the importance of the thumb and its own vulnerable position. Deep branches of the ulnar nerve in the palm supply variable amounts of the thenar muscles, uncommonly replacing the recurrent branch of the median completely (p. 422), and confusing many clinicians. Also anomalies of the median nerve and its recurrent branch are surprisingly common, including occasional burrowing of the main nerve or its recurrent branch(es) through the flexor retinaculum—a source of disaster for the surgeon performing a carpal tunnel release operation. (See Clinical Case 34.2 on p. 402).

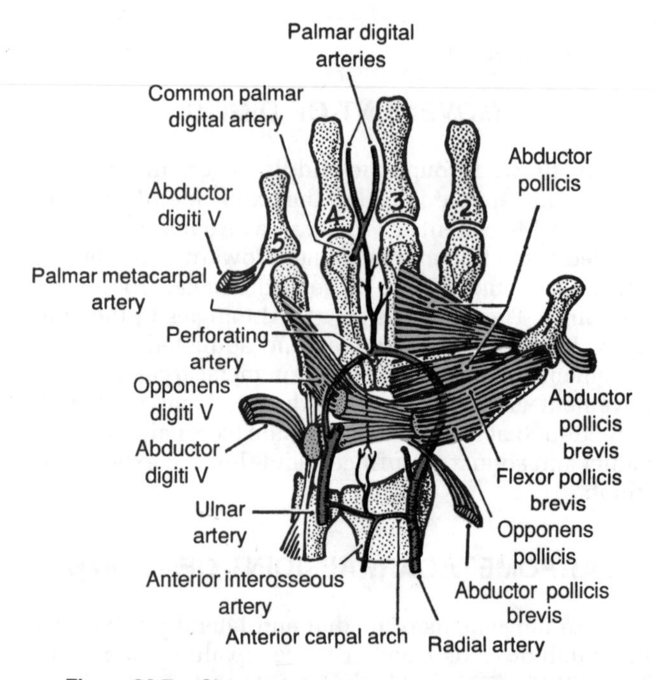

Figure 36.7. Short muscles of thumb and 5th digit, and deep arteries of hand (semischematic).

THE THREE HYPOTHENAR MUSCLES (*Muscles of the Base of the Little Finger*)

These muscles are in most respects mirror images of the thenar muscles. The common origin is from the flexor retinaculum and its 2 medial pillars (pisiform and hook of hamate). They act on a metacarpophalangeal joint, but, of course, opposition is impossible.

Nerve Supply

The hypothenar muscles are supplied by the ulnar nerve (deep branch) as it runs between the pisiform and the hook of the hamate.

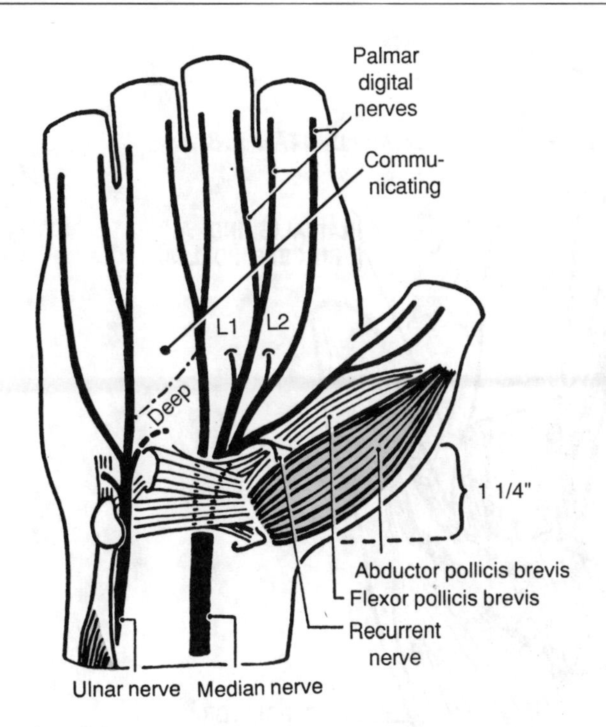

Figure 36.8. Median and ulnar nerves and thenar muscles. (L1 and L2, Lumbrical branches of median nerve.)

ADDUCTOR POLLICIS

The thumb possesses a fourth short muscle, the adductor pollicis. It lies in the depths of the palm. A transverse head arises from the palmar border of the middle metacarpal and an *oblique head* from the corresponding carpal bone (the capitate). It is inserted into the base of the first phalanx of the thumb (*fig. 36.7*). Obviously the adductor draws the thumb posteriorly to the palm.

Nerve Supply

Contrary perhaps to expectation, it is supplied by the ulnar nerve.

Opposition is bringing of the pad of the thumb to the pad of a finger and holding it there, as in pinching, writing, holding a cup by its handle, or fastening a button. Most delicate actions performed by the hand involve opposition.

As regards the thumb, the movements executed almost simultaneously are circumduction, rotation, and flexion. Rotation takes place principally at the metacarpophalangeal joint but also at the carpometacarpal joint, while the trapezium moves on the scaphoid and the scaphoid angulates forward (Bunnell).

The 3 joints of the opposing finger (or fingers) are flexed by the profundus, the superficialis, a lumbrical, and 2 interossei.

Note. Without 2 muscles supplied almost always by the median nerve (opponens and flexor pollicis brevis), there would be no rotation. Without 2 muscles supplied almost always by the ulnar nerve (adductor pollicis and an interosseus), the grip would be weak. An interosseus (usually the 1st dorsal to the index) is required to prevent ulnar deviation of the finger against which the adductor pollicis is exerting pressure. Hence, ulnar paralysis results in weak opposition.

ULNAR NERVE

It enters the hand by passing vertically between the pisiform and the hook of the hamate (*figs. 36.8 and 36.9*) in front of the flexor retinaculum and the pisohamate ligament where it divides into a *deep* and a *superficial branch*. It is covered first by a slip of deep fascia and then by a slip of cutaneous muscle, the **palmaris brevis** (*fig. 36.9*).

Distribution

The superficial branch supplies cutaneous branches to the medial 1½ fingers and the motor branch to the palmaris brevis. The superficial branch communicates with the median nerve.

The deep branch supplies 3 muscles of the hypothenar eminence and then curves around the lower edge of the hook of the hamate into the depths of the palm, where it supplies all the short muscles of the hand, except the 5 usually supplied by the median nerve.

MEDIAN NERVE

Crossing the midpoint of the skin crease of the wrist, it enters the palm through the carpal tunnel adhering to the deep surface of the flexor retinaculum (*fig. 36.5*). It appears in the palm deep only to the palmar aponeurosis, where it breaks up into "recurrent" and digital branches. These are distributed to 5 muscles and to the skin of the lateral 3½ digits, to the joints of these digits, and to the local vessels.

Motor Branches

The 5 muscles usually supplied by the median nerve are the 3 thenar muscles and the 2 lateral lumbrical muscles:

Abductor Pollicis Brevis
Opponens Pollicis } by the recurrent branch
Flexor Pollicis Brevis
and
1st Lumbrical
2nd Lumbrical } by the palmar digital branches

These are shown in *Figure 36.8*, and the "recurrent" branch was described above. Variations are common.

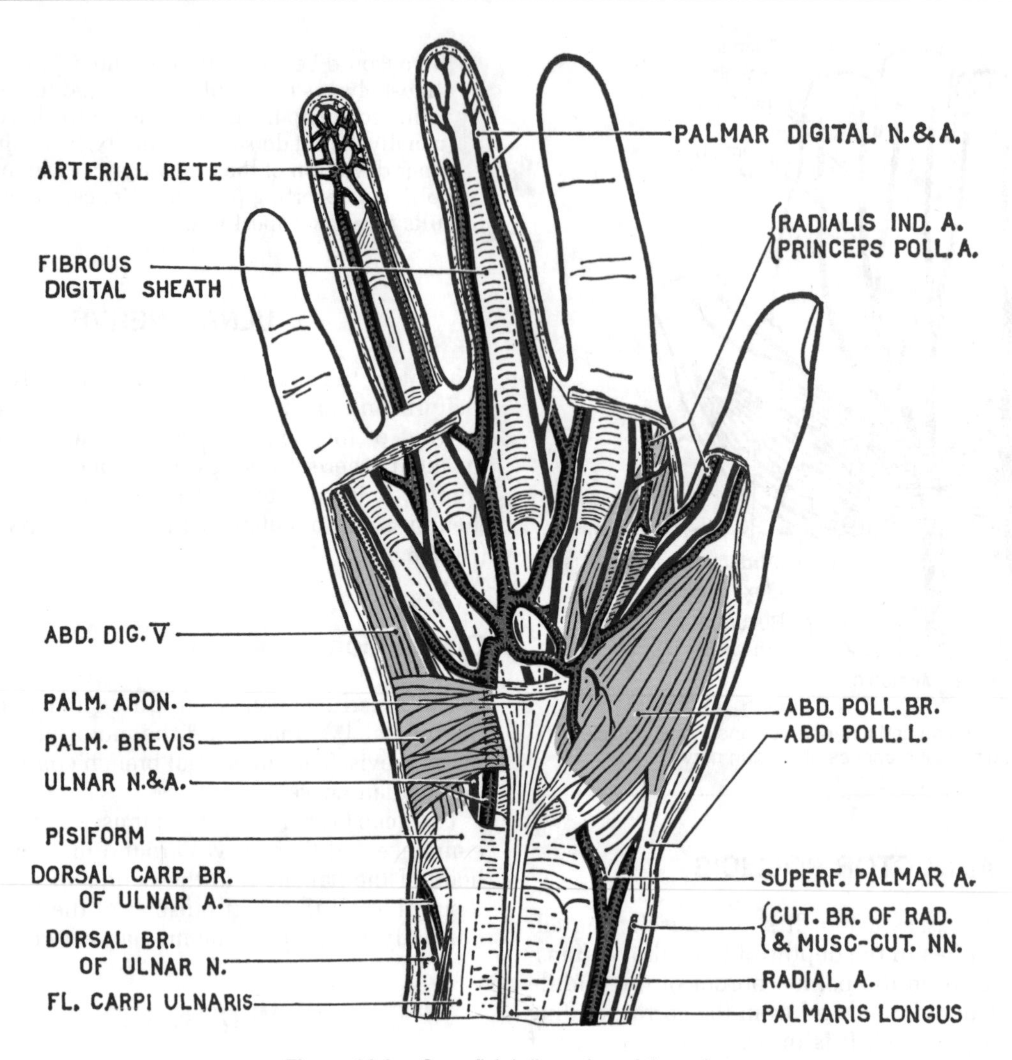

Figure 36.9. Superficial dissection of the palm.

Details of Variations in Motor Distribution. The standard, or textbook, pattern of motor innervation to the hand just described is commonly departed from either by the *ulnar nerve* encroaching on median nerve territory, or vice versa. Thus, in 226 hands the ulnar nerve supplied flexor pollicis brevis partially in 15% and completely in 32%, abductor pollicis brevis in 3%, and all 3 thenar muscles in 2%. Conversely, the *median nerve* supplied adductor pollicis in 3% and adductor pollicis + first dorsal interosseus in 1%. Neither the musculocutaneous nor the radial nerve has been proved to supply a thenar muscle (T. Rowntree, clinical observations). Forrest confirmed the variability of thenar nerve supply, but using electrical stimulation and electromyography revealed that 85% of flexor pollicis brevis muscles in 25 hands were supplied by both median and ulnar nerves. Morphological studies by Harness and Sekeles provide overwhelming confirmation.

The palmar digital branches of the median and ulnar nerves accompany the digital arteries on the sides of the fibrous digital sheaths (*fig. 36.9*).

Details for Surgeons. The 3 common palmar digital nerves descend to the 3 interdigital clefts (*fig. 36.8*), protected by the tough palmar aponeurosis and crossed by the superficial palmar arch. The 2 digital branches to the thumb accompany the flexor pollicis longus tendon (*fig. 36.9*). The ulnar nerve commonly extends its influence to median nerve territory and vice versa, through communicating branches to the palm.

PALMAR APONEUROSIS

The palmaris longus tendon adheres to the superficial surface of the flexor retinaculum, enters the palm, and divides into 4 broad, diverging bands, which enter the 4 fingers to blend with subcutaneous tissues (*fig. 36.10*).

In the distal half of the palm, it sends fibrous septa dorsally to the palmar ligaments (plates) and to the deep fascia (*see fig. 36.15*).

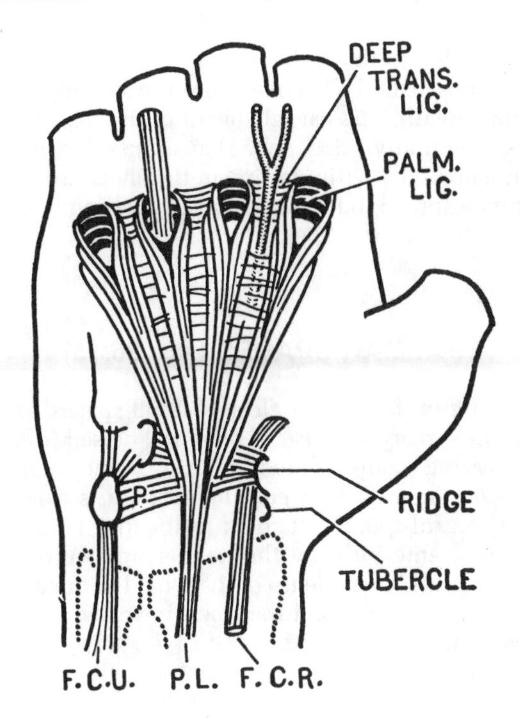

Figure 36.10. The 3 flexors of the wrist. Palmar aponeurosis. Flexor retinaculum. *Ridge of trapezium, tubercle*. *FCU*, flexor carpi ulnaris; *PL*, palmaris longus; *FCR*, flexor carpi radialis; *P*, pisiform.

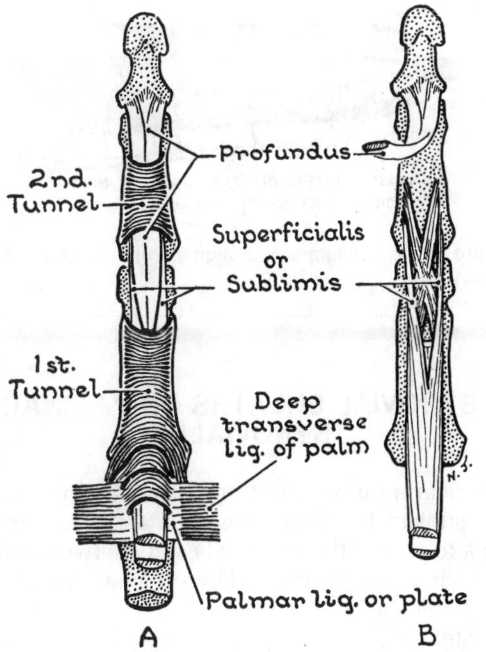

Figure 36.11. *A*, a fibrous digital flexor sheath showing the 2 osseofibrous tunnels. *B*, mode of insertion of the long digital flexors.

Clinical Case 36.2

Patient Barry D. This 52-year-old professor of anatomy walks into your classroom to present himself as a patient with a clinical problem. He tells you he has had pseudogout for several years, which flares up as tendonitis and synovitis in several different parts of his body. For several days, he has had soreness and swelling in the right wrist. Now the inflammation is spreading into his little finger, thumb and index but not the ring and middle fingers. Grimacing with pain, he asks the class to explain "what's going on" anatomically in his hand.

FIBROUS SHEATHS OF THE DIGITS

To each finger, a pair of tendons, a *superficialis* and a *profundus*, descend deep to the palmar aponeurosis. In front of the head of a metacarpal bone each pair enters a *fibrous digital sheath* (*fig. 36.11*). Each sheath extends from the palmar ligament (plate) of a metacarpophalangeal joint to the insertion of a profundus tendon into the base of a distal phalanx. Each sheath, therefore, crosses 3 joints. In front of the joints the sheath, for mechanical reasons, must be pliable and thin.

INSERTIONS OF LONG DIGITAL FLEXORS

Each profundus tendon is inserted into the anterior aspect of the base of a distal phalanx (*fig. 36.11*). Each superficialis tendon splits in front of a proximal phalanx into medial and lateral halves (*fig. 36.11, B*) which ultimately insert on the margins of a middle phalanx. Each profundus tendon passes through a perforation in a superficialis. The intricate perforation remains so neatly open that you can easily slip a cut profundus tendon through it again.

THE LUMBRICALS (*see fig. 36.14*)

These 4 palmar muscles resemble earthworms—hence their name. They arise from the profundus tendons. They lie dorsal to the digital vessels and nerves, accompany them to the radial side of the fingers, and join the extensor (dorsal) expansions beyond the interossei (*see figs. 37.9* and *37.14*). The medial 2 are supplied by the ulnar nerve, the others by the median.

Actions are described on p. 434.

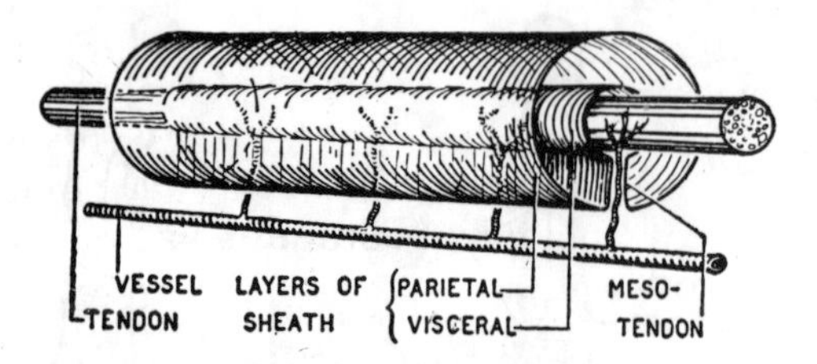

Figure 36.12. Diagram of a segment of a synovial sheath (distended).

SYNOVIAL SHEATHS OR VAGINAE SYNOVIALES

A lubricating device (a tubular bursa) ensheaths any tendon subject to friction inside a tunnel. It is a tube within a tube, and the potential cavity between the inner and the outer tube (lubricated by synovial fluid) is closed at both ends (*fig. 36.12*).

Every tendon within a synovial sheath has, or once had a **mesotendon** (*fig. 36.13*)—a double layer of synovial membrane that conveys vessels, as do the mesenteries of the gut.

A synovial sheath may extend 2 or 3 cm proximal to and distal to the tunnel. But this depends upon the excursion the tendon makes. The very end portions of the original mesotendons of the long flexors remain as little triangular membranes, the *vincula brevia*, and several threads persist in front of the proximal phalanges as *vincula longa*. The vincula convey tiny blood vessels required by the tendons. (The singular of vincula is *vinculum*, L. = *fetter*.)

Infections as a result of perforations entering the sheaths are serious because synovial fluid is an excellent culture medium for bacteria—and death of the poorly vascularized tendons is a real possibility. Before the days of antibiotics, heroic surgical opening and drainage of the sheaths often became necessary.

The long flexor tendons require synovial sheaths for both the carpal tunnel and the digital tunnels, i.e., *carpal synovial sheaths* and *digital synovial sheaths*. The great range of movement and the shortness of the metacarpals of the thumb and of digit V result in their carpal and digital sheaths being continuous (*fig. 36.14*). Those of the thumb probably always unite; those of the little finger fail to unite in about 10% of persons.

The carpal sheaths of the 4 superficialis and 4 profundus tendons usually become one, the **common carpal synovial sheath**. The carpal sheath of the flexor pollicis longus commonly joins, too. Hence an infection or an inflammatory synovitis starting in the sheath of the little finger may spread to the wrist and to the thumb (Clinical Case 36.2).

PALMAR SPACES (*fig. 36.15*)

There are in the palm 4 closed fascial spaces, also important in surgery because of potential closed infections from piercing wounds. Muscles occupy the *thenar* and *hypothenar spaces*. Between them, there is a large triangular *central space* that contains the long tendons and is enclosed anteriorly by the palmar aponeurosis. The posterior "wall" is made up of the 3 medial metacarpals, associated ligaments, and the fascia covering the medial interossei and the adductor pollicis. Its sides are fascial fusions of the palmar aponeurosis with the thenar and hypothenar fasciae. The 4th space lies posterior to the adductor pollicis.

The central space is divided distally into 8 subdivisions or tunnels (*fig. 36.15*) by septa from the palmar aponeurosis. The tendons of the lumbricals prolong the spaces downward onto the dorsum of the digits. By this route infection in the central palmar space may spread to the dorsum of the fingers and hand.

Clinical Case 36.3

Patient Connie C. In a fit of depression last night, this 23-year-old dancer attempted suicide by slashing her left wrist and only succeeded in severing the radial artery. Emergency treatment included tying off that large artery. The next morning at rounds, you are quite surprised to find the circulation in her hand and fingers appears to be quite normal. Why? What are the anastomoses?

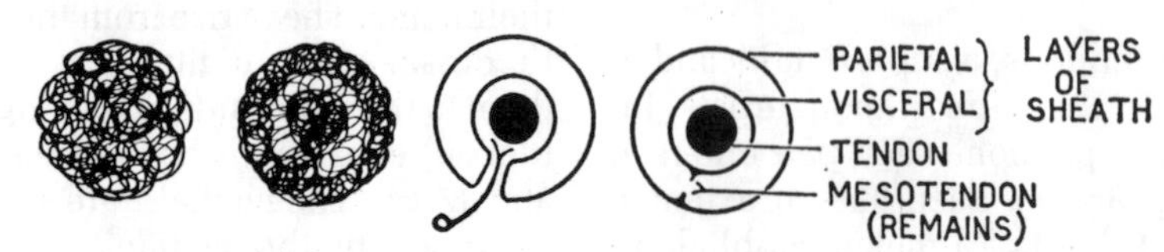

Figures 36.13. Stages in the development of tendon, synovial sheath, and mesotendon.

Arteries of the Hand

The blood supply from the ulnar and radial arteries is good and cross anastomoses are excellent through 4 arterial arches (*fig. 36.16*):

1. The superficial palmar arch lying deep to the palmar aponeurosis (*fig. 36.9*).
2. The deep palmar arch } on
3. The dorsal carpal arch } skeletal
4. The palmar carpal arch } plane

THE SUPERFICIAL PALMAR ARCH

This arch, the largest and most distal, is the continuation of the ulnar artery (*fig. 36.9*) completed by a (variable) branch of the radial artery. (The deep branch of the ulnar artery accompanies the deep branch of the ulnar nerve and completes the deep arch.)

Palmar Digital Arteries

The superficial palmar arch supplies the medial 3½ digits, leaving the lateral 1½ to the care of the deep palmar arch.

Palmar digital arteries and nerves run on the sides of the flexor tendons in their fibrous sheaths, the nerve being anteromedial to the artery (*fig. 36.17*). To expose them, the dissector or surgeon should feel for the edge of a phalanx and make a longitudinal incision in front of it.

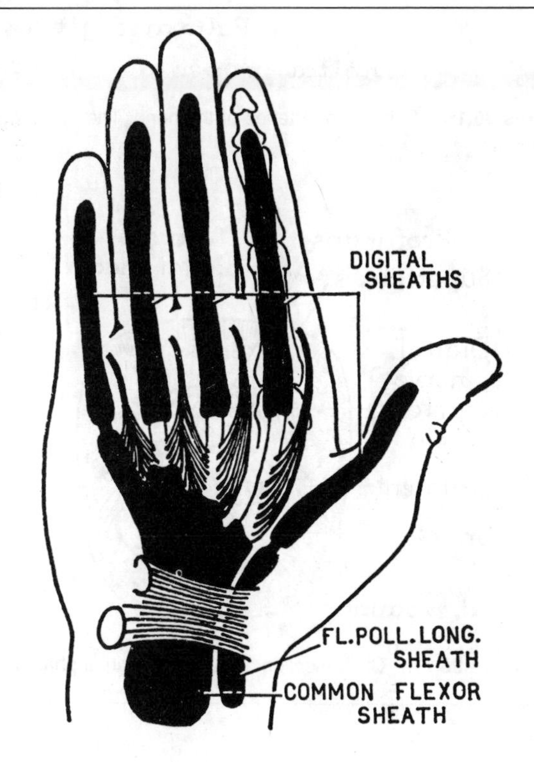

Figure 36.14. Synovial sheaths or tubular bursae are required for the long digital flexor tendons at the osseofibrous carpal tunnel and at the osseofibrous digital tunnels. The lumbricals are shown.

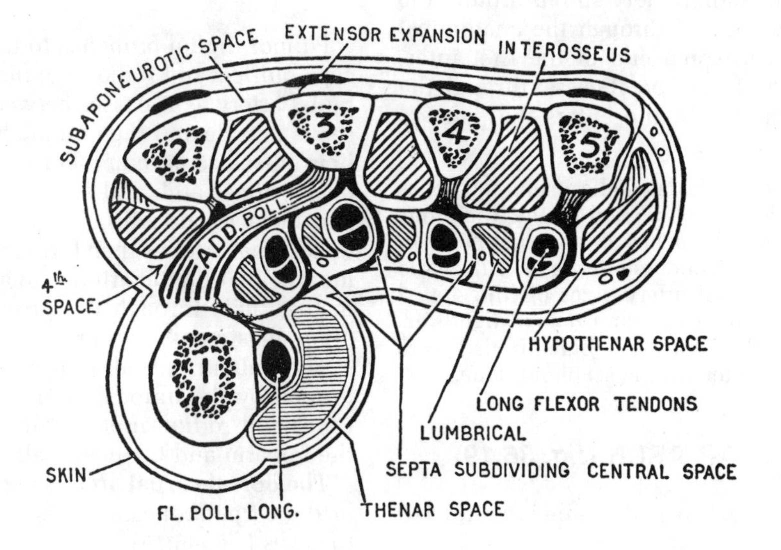

Figure 36.15. The palmar spaces in cross-section.

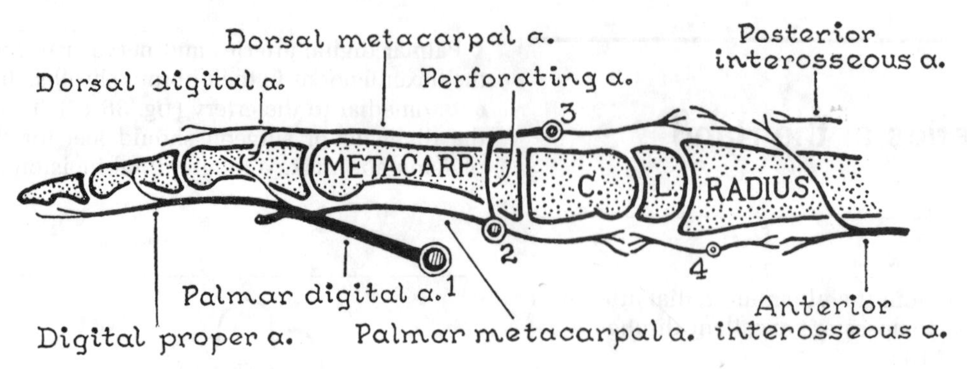

Figures 36.16. Scheme of arteries of the hand. The 4 arches are numbered in order of size—3 cling to the skeletal plane. *C*, capitate; *L*, lunate.

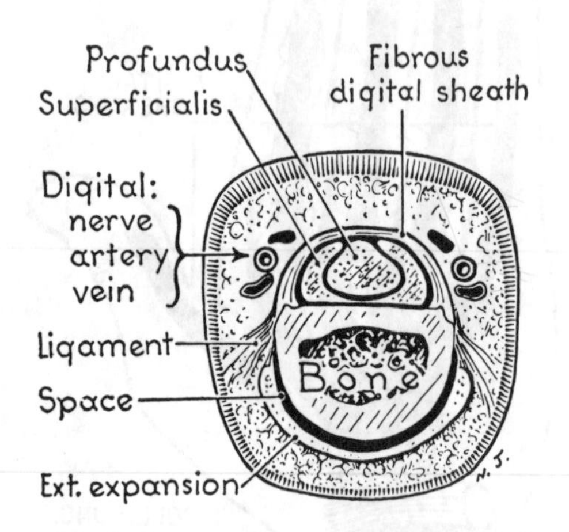

Figure 36.17. Cross-section through a proximal phalanx.

Figure 36.18. The arteries of the dorsum of the hand.

RADIAL ARTERY

After giving off the *palmar radial carpal* and *superficial palmar arteries*, the radial artery spirals around the lateral border of the wrist and through the anatomical snuffbox to reach the proximal end of the first intermetacarpal space (*fig. 36.18*), where it passes through the first dorsal interosseus muscle and enters the palm to become the *deep palmar arch*.

The radial artery crosses the lateral ligaments and bones of the carpus and, in turn, is crossed by the 3 tendons that bound the snuffbox, branches of the radial nerve to the thumb, and the dorsal venous arch (*fig. 36.18*). While in the snuffbox, the radial artery gives off the *dorsal radial carpal artery* and sends small *dorsal digital arteries* to the sides of the lateral 1½ digits. It also gives an important nutrient artery to the scaphoid bone.

THE DEEP PALMAR ARCH (*fig. 36.19*)

The radial artery continuation in the palm is completed by the deep branch of the ulnar artery.

Branches

Palmar digital branches to the lateral 1½ digits (*princeps pollicis* and *radialis indicis arteries*) arise from the radial artery as it passes between the 1st dorsal interosseous muscle and the oblique head of the adductor pollicis muscle; 3 *palmar metacarpal arteries* arise from the deep palmar arch and supply the interosseous muscles and metacarpals adjacent to these arteries. The palmar metacarpal arteries then form anastomoses with the common palmar digital arteries to assist in the blood supply to the digits through the proper digital arteries. Many other anastomoses occur.

The palmar carpal arch is actually a **rete** or network formed by the union of *palmar carpal branches of the ulnar* and *radial arteries* and anastomosing twigs from the forearm and the deep palmar arch (*fig. 36.19*).

The dorsal carpal arch is applied to the dorsal surface of the carpal bones, and its branches have many anastomoses (*fig. 36.18*).

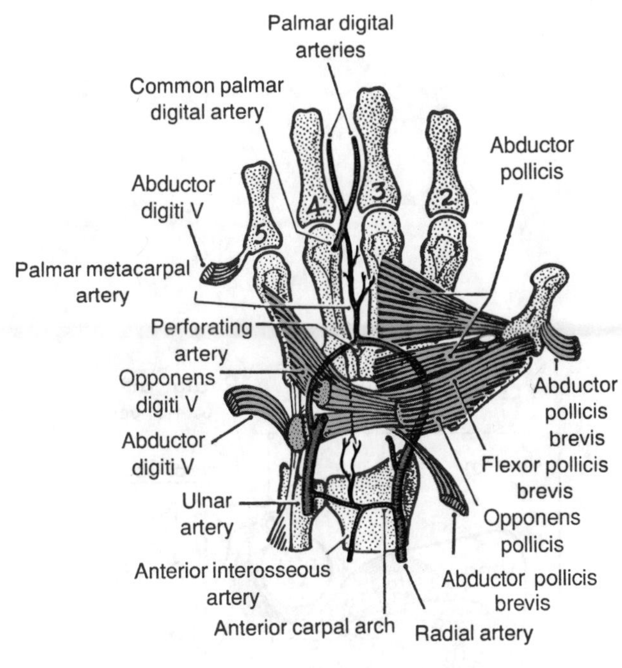

Figure 36.19. Deep arteries of the hand.

Anatomical Features of the Physical Examination

Deformities, motor and sensory disturbances, and pain are important clinical signs sought by the clinician on examining the hand. Atrophies of the thenar eminence (median nerve) and of the adductor pollicis and interossei (ulnar nerve) viewed from the dorsum are important. Testing for the power of the long muscles and intrinsic muscles can quickly establish the level of disturbance in these nerves.

Great mobility at the saddle-shaped carpometacarpal joint of the thumb permits opposition to the other digits. This is carried out by the thenar muscles (flexor brevis, abductor brevis and opponens), *usually* supplied by the recurrent branch of the median nerve, which comes off the median nerve as it emerges in the midline from the carpal tunnel (Clinical Case 36.1).

The ulnar nerve does not pass through the tunnel but enters the palm lateral to the pisiform bone. It divides into digital sensory branches to digits 5 and (half of) 4, and motor branches, particularly to the hypothenar muscles (named like the thenars), all the 7 interossei, the medial two lumbricals and the fan-shaped adductor pollicis deep in the palm. The adductor pollicis forcefully pulls the thumb posteriorly toward and into the palm for a firm grasp. The digital sensory branches of the median nerve run to the lateral 3½ digits.

The 7 interossei and 4 lumbricals contribute to powerful flexion of the MP joints and extension of the IP joints of their fingers. In addition the interossei abduct and adduct the fingers at the MP joints; these actions are easily tested during physical examination as are movements of the thumb.

Crowding through the carpal tunnel deep to the flexor retinaculum are the median nerve and the long tendons of flexor tendons ensheathed in their common carpal synovial sheath. The tendons enter tunnels that start in front of the metacarpal heads—the fibrous digital sheaths. The carpal synovial sheath continues on the tendons of the thumb and little finger as they enter their tunnels, but the other digital synovial sheaths do not usually communicate (Clinical Case 36.2).

Four potential areolar spaces are of surgical interest: a central space for the long tendons and related structures deep to the palmar aponeurosis; the thenar and hypothenar space occupied by those muscles and enclosed by their fascia; and a fourth in the interval between the thumb and index metacarpals and separating the adductor pollicis and the first dorsal interosseus muscles (Clinical Case 36.1).

Excellent arterial anastomoses are provided by (1) the superficial palmar arch, the continuation of the ulnar artery just deep to the aponeurosis and completed by a branch of the radial artery; (2) the deep palmar arch on the skeletal plane, a continuation of the radial artery as it plunges from the dorsum of the hand through the first dorsal interosseus and is completed by the deep palmar branch of the ulnar artery; (3) and (4) anterior and posterior carpal anastomotic plexuses (Clinical Case 36.3).

Clinical Mini-Problems

1. **Abduction and adduction of the fingers is a test for which nerve?**
2. **In a carpal tunnel syndrome (compression of the median nerve at the wrist), why is flexion of the thumb still possible?**
3. **In a patient with numbness over the palmar surface of the right, thumb, index, and middle fingers, how would you differentiate between a lesion in the medial nerve at the wrist or the elbow?**
4. **Compression of the radial artery can control the hemorrhage resulting in laceration of the thumb. Which carpal bone can the radial artery be compressed against in the wrist?**

(Answers to these questions can be found on p. 587.)

37

Extensor Regions of Forearm, Hand, and Fingers

Clinical Case 37.1

Patient Anna F. This 68-year-old grandmother of one of your classmates had a (Colles') fracture across the lower end of the right radius last winter and apparently recovered. Suddenly one morning 5 weeks ago, she discovered she could flex but not actively extend the distal joint of her thumb and she "consulted" your classmate for advice. Not being as good an anatomy student as you are, he turns to you for an explanation.

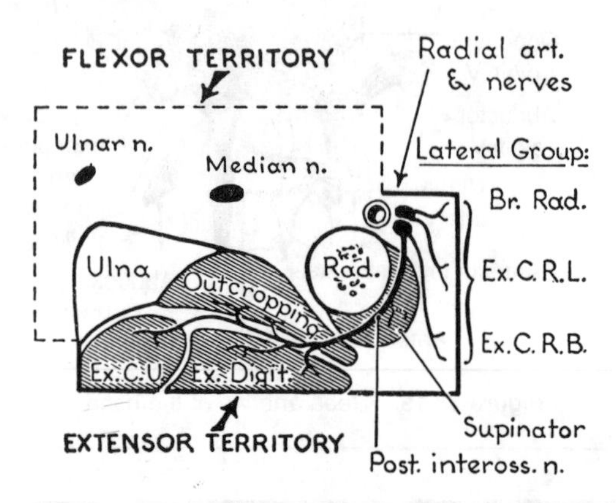

Figure 37.1. The extensor region of the forearm (on cross-section).

Forearm

BOUNDARIES

No motor nerve crosses the boundary lines, which are: *medially*, the subcutaneous border of the ulna and the ulnar border of the hand and, *laterally*, the course of the radial artery in the forearm and the radial border of the hand (*fig. 37.1*).

PALPABLE PARTS OF ULNA AND RADIUS (*fig. 37.2*)

While you can and should palpate the whole length of the ulna, only the upper and lower ends of the radius are easily felt. The head of the radius lies immediately distal to the smooth posterior aspect of the lateral epicondyle. When the elbow is extended, the head lies in the depth of a visible hollow in which its superior margin is easily felt on firmly pressing inferiorly. When the elbow is flexed and the forearm alternately pronated and supinated, the head can be felt moving deep to the palpating fingers. The *styloid process* of the radius lies in the anatomical snuffbox and is more than 1 cm distal to the level of the ulnar styloid process—a fact of importance to surgeons restoring the fractured bones of this region to their original shape. In order to palpate the tips of these processes, grasp the sides of the wrist between your thumb and index and press proximally.

TENDONS (*fig. 37.3*)

All tendons passing from the forearm to the dorsum of the hand span the carpus and reach the bases of metacarpals or phalanges. There are no fleshy muscles and, so, no motor nerves.

Snuffbox and the Three Dorsal Tendons of the Thumb

When the thumb is fully extended, a hollow, called the "**snuffbox**," can be seen on the dorsum of the wrist,

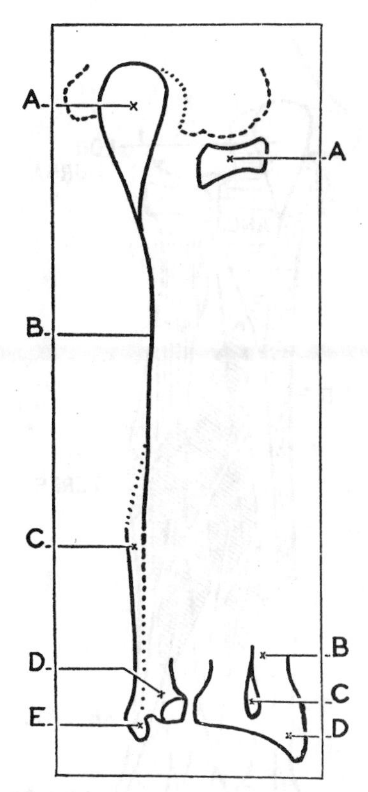

Figure 37.2. Palpable parts of ulna and radius (posterior view). *Ulna: A*, olecranon; *B*, posterior border; *C*, area between flexor and extensor tendons; *D*, head; and *E*, styloid process. *Radius: A*, head; *B*, distal end dorsally; *C*, dorsal tubercle; and *D*, styloid process.

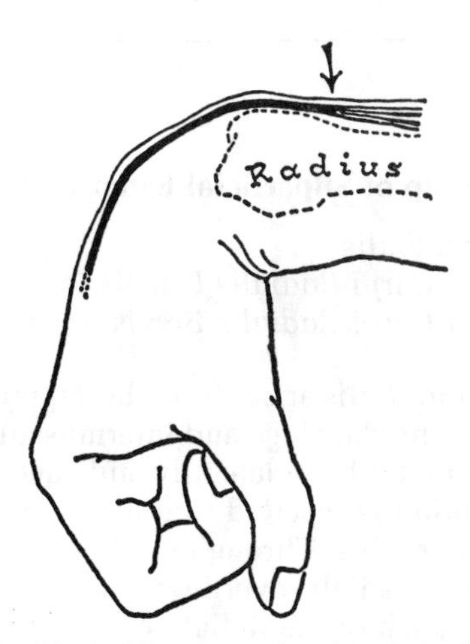

Figure 37.3. Fleshy fibers must give place to tendon beyond the *arrow* at the distal end of the radius.

at the root of the thumb. Bounding the snuffbox are the tendons running to the dorsum of the thumb (*fig. 37.4*) and easily observed beneath the overlying skin.

Crossing the snuffbox superficially are the *cephalic vein* and *superficial radial nerve*. The *radial artery* crosses on the skeletal plane.

THE THREE DORSAL TENDONS OF THUMB TRACED PROXIMALLY (*fig. 37.4*)

The line of the "*3 outcropping muscles*" of the thumb divides the superficial muscles of the extensor region into a lateral and a posterior group, each with its own nerve supply. It is an **internervous line**, and so the safest approach to the deeper parts including the supinator, which is wrapped around the proximal third of the radius.

Clinical Case 37.2

Patient Billy G. This morning, this 7-year-old boy fell off a coin-operated mechanical horse at the shopping plaza. He stopped the fall with his outstretched left hand and now has severe pain in the area of the head of the radius. The orthopedic chief resident is assigned to operate on him. He plans to approach the diagnosed fractured neck of the radius from behind. You are the surgical assistant. To your dismay the surgeon, apparently forgetting the anatomy, cuts boldly across the fibers of the supinator down to the bone. You foresee the consequent disaster. How bad will it be?

Muscles, Vessels, and Nerves

SUPINATOR

This, the primary supinator—the other being the biceps—arises from the lateral ligaments of the elbow (*fig. 37.4*) and an adjacent area of the ulna inferior to its radial notch (*see figs. 33.5, 33.13*). It inserts into the body of the radius between the anterior and posterior oblique lines (*fig. 37.5*).

POSTERIOR INTEROSSEOUS NERVE

This motor nerve (better called *deep radial nerve*) is one of the two terminal branches of the radial nerve—the other branch being the superficial radial nerve. The deep radial nerve starts anterior to the capsule of the elbow joint under cover of the brachioradialis and spirals round the radius while tunneling through the supinator,

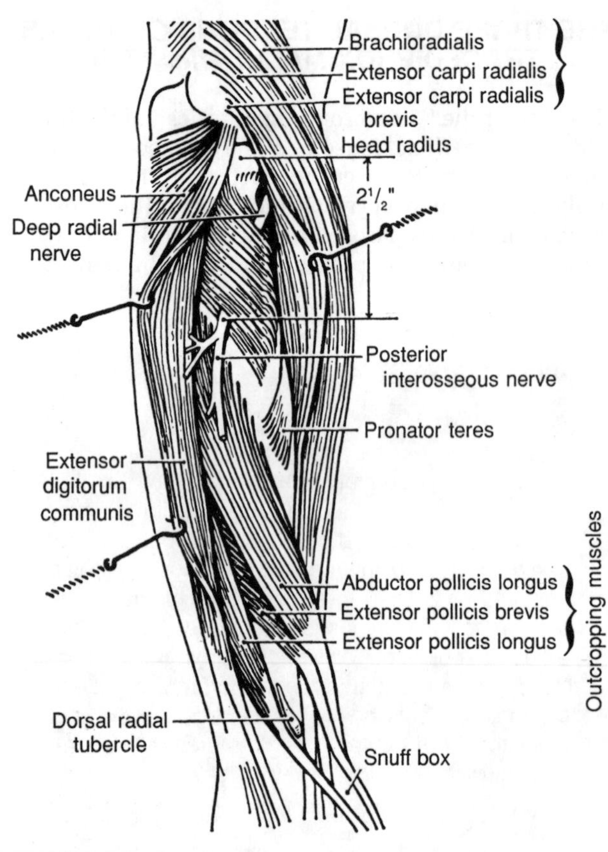

Figure 37.4. The furrow of "the 3 outcropping muscles of thumb" opened up—line of relative safety.

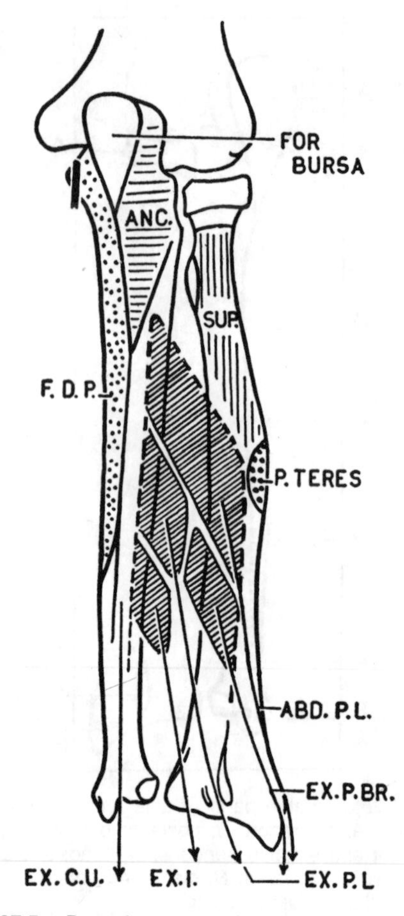

Figure 37.5. Posterior aspect of ulna and radius (clothed).

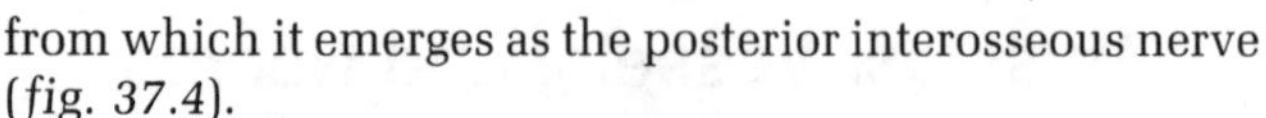

from which it emerges as the posterior interosseous nerve (*fig. 37.4*).

The radial and posterior interosseous nerves together supply the 3 muscles of the lateral group (brachioradialis, extensor carpi radialis longus, extensor carpi radialis brevis) and the supinator before the latter nerve enters the supinator. On emerging from the supinator, it supplies all the remaining muscles on the dorsum of the forearm. Damage to the radial nerve or the posterior interosseous nerve may result in a "wrist drop" similar to that depicted in *Figure 37.3*. This may be evident in the patient in Clinical Case 37.2 following his surgery.

MUSCLES OF EXTENSOR REGION OF FOREARM

The greatest origin of the superficial muscles is from a *common extensor tendon* attached to the front of the lateral epicondyle, neighboring fascia and supracondylar ridge.

Lateral Group of Superficial Extensors:

1. *Brachioradialis*
2. *Extensor Carpi Radialis Longus*
3. *Extensor Carpi Radialis Brevis*

The *brachioradialis* arises from the superior part of the lateral supracondylar ridge and intermuscular septum. It bounds the cubital fossa laterally and shelters the radial nerve. Its tendon is inserted into the base of the styloid process of the radius. Throughout its course in the forearm, the brachioradialis overlies the radial artery and the superficial branch of the radial nerve.

Clearly, the brachioradialis can act as a flexor, though it is supplied by an "extensor" nerve. Functionally, it acts as a flexor against resistance when the forearm is semiprone and flexed and during both fast extension and fast flexion, apparently as a "shunt muscle" (p. 00).

The *extensor carpi radialis longus* arises from the inferior third of the supracondylar ridge and lateral intermuscular septum. *Extensor carpi radialis brevis* arises from the common tendon, adjacent capsule and septa.

Their tendons cross the snuffbox and pass to the bases of the 2nd and 3rd metacarpals, where you should feel them while dorsiflexing the wrist.

A *bursa* exists at each end of the extensor carpi radialis brevis. One cause of "tennis elbow" is believed to be bursitis at the origin from the lateral epicondyle.

Posterior Group of Superficial Extensors:

1. *Extensor Digitorum*
2. *Extensor Digiti Minimi*
3. *Extensor Carpi Ulnaris*
4. *Anconeus*

The 3 *extensors* arise from the common tendon of origin, the deep fascia that covers them, and the intermuscular septa. Their *grooves* on the distal end of the radius and ulna and *insertions* are given on p. 432.

The *anconeus* (a semidivorced part of the triceps) arises by tendon from the back of the lateral epicondyle. It is inserted by fleshy fibers into the lateral surface of the olecranon and adjacent part of the ulna (*figs. 37.4* and *37.5*).

Deep Group of Extensor Muscles of Forearm:

1. *Abductor Pollicis Longus.*
2. *Extensor Pollicis Brevis.*
3. *Extensor Pollicis Longus.*
   (1–3: *"Outcropping Muscles"*)
4. *Extensor Indicus.*
5. *Supinator* (considered on p. 429).

Although the exact origin of the 3 outcropping tendons to the thumb (*fig. 37.5*) is unimportant, after outcropping (*fig. 37.4*) they run to the bases of the 3 long bones of the thumb; abductor longus inserts on the metacarpal and extensor brevis and extensor longus on the proximal and distal phalanges (*see fig. 36.18*). The extensor pollicis longus angles around a pulley (dorsal radial tubercle) and so is separated from the other two—hence creating the snuffbox (*fig. 37.4*).

Extensor indicis joins the extensor expansion to the index finger.

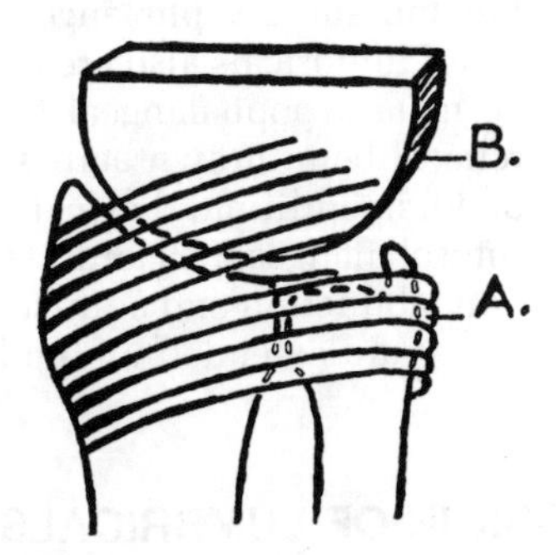

Figure 37.6. The extensor retinaculum encircles the ulnar head (*A*) but attaches to the dorsum of the carpus (*B*).

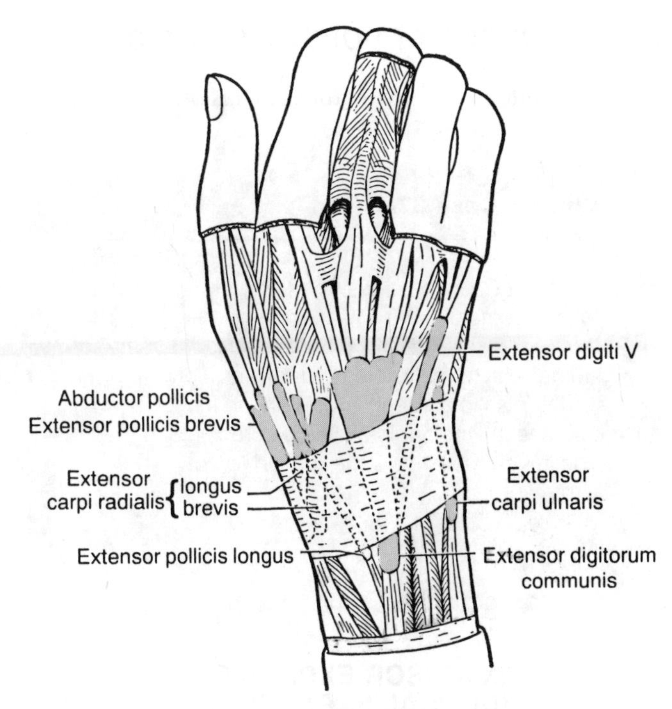

Figure 37.7. The 6 synovial sheaths on the dorsum of the hand and a dorsal expansion—see also *Figure 37.8.* (Dissection by C. P. Rance and J. W. Rogers.)

ARTERIES

The *posterior interosseous artery*, the smaller of the 2 terminal branches of the common interosseous artery (see *fig. 34.14*), does not travel with its nerve, but passes over the upper border of the interosseous membrane. It descends between the superficial and deep muscles, supplies them, and takes part in the anastomoses at the elbow and wrist.

CUTANEOUS NERVES

These are illustrated in *Figures 32.5* and *32.8*.

ENCLOSING DEEP FASCIA

For some distance *below the elbow* the deep fascia gives origin to the extensor muscles and, so, its fibers are strong and must run vertically for reasons that should be obvious.

At the distal end of the forearm, the fibers are required to retain the extensor tendons in place, so you will surmise that they form an **extensor retinaculum** (*fig. 37.6*). However, you may fail to guess that septa attach the retinaculum to the radius and medial carpals to form tunnels for the tendons. There are 6 of these, each lubricated by a synovial sheath (*fig. 37.7*).

TENDONS AT DORSUM OF WRIST

The 9 tendons occupy 6 grooves (*figs. 37.7* and *37.8*).

Clinical Case 37.3

Patient Carlotta H. Assigned as a clinical student to a rheumatoid arthritis team, you are asked to thoroughly test the movements of this 48-year-old woman's extremely deformed fingers in preparation for the plastic surgeon to operate on the joints and tendons. All the anatomy you can recall (and probably much more) must be rushed to your aid.

EXTENSOR EXPANSIONS (DORSAL EXPANSIONS)

The 4 flat tendons of the extensor digitorum traverse the most medial tunnel on the distal end of the radius and, diverging, pass to the 4 fingers. The tendons of the index and little fingers are joined near the knuckles by their respective extensor indicis and extensor digiti minimi (*figs. 37.7* and *37.8*).

Three oblique bands unite the 4 tendons proximal to the knuckles (*fig. 37.7*). Hence, independent action of the fingers is restricted.

Structure and Attachments (*fig. 37.9*). On the distal ends of the metacarpals and on the digits, the extensor tendons become flattened to membrane thinness as the *extensor* or *dorsal expansions.*

Each expansion wrap around the dorsum and sides of a metacarpal head and proximal phalanx. The *visor-like hood* around the metacarpal head is anchored on each side to the palmar ligament or plate and a broad *fibroareolar ribbon* attaches the hood to the base of the proximal phalanx.

On the proximal phalanx, the expansion divides into a median band, which passes to the base of the middle phalanx, and into 2 side bands, which pass to the base of the distal phalanx.

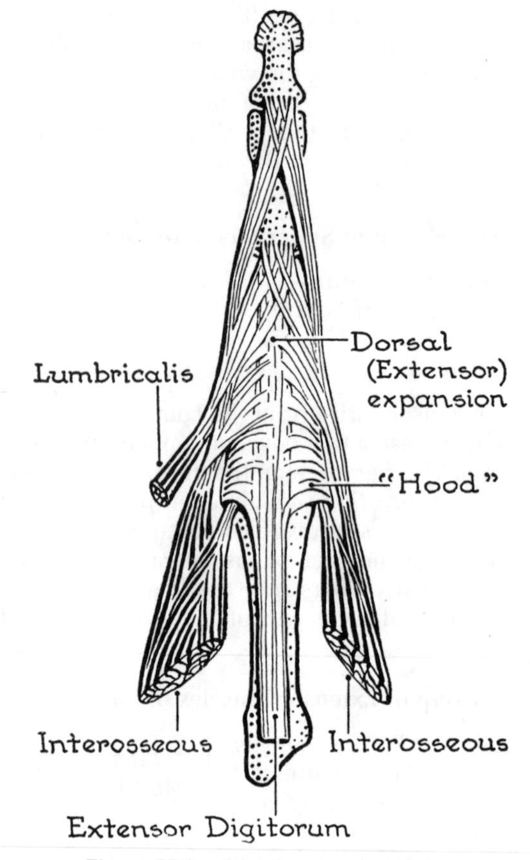

Figure 37.9. An extensor expansion.

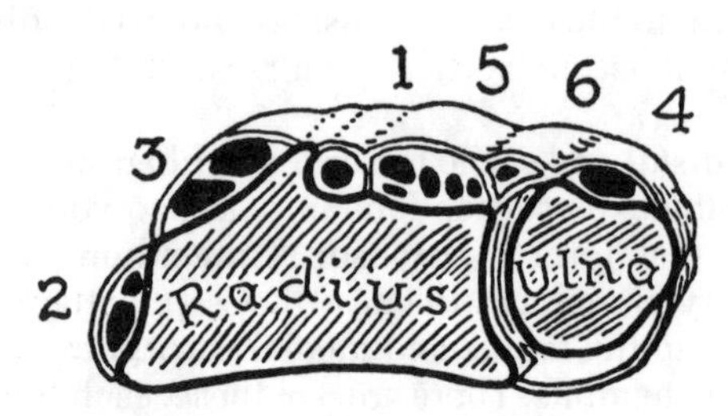

Figure 37.8. *Figure 33.7* on cross-section.

The extensor tendon pulls mostly via the median band and its attachment to the proximal phalanx and pulls only slightly via the side bands, so its effect on the distal joint is slight. However, each side band is joined by half an interosseus tendon and, more distally, on the radial side, by an entire lumbrical tendon, and these provide the extensor power to the distal joint. Their tendons are united across the dorsum of the proximal phalanx by a sling of transverse fibers, and via the side bands they run to the bases of the 2nd and 3rd phalanges (*fig. 37.10*).

The interossei and lumbricals also are in excellent position to flex the metacarpophalangeal (MP) joints by means of the arched fibers thrown across the proximal phalanges (*fig. 37.9*). In addition, they extend both proximal and distal interphalangeal (IP) joints (*fig. 37.11*) and impart side motion (abduction and adduction) at the MP joints, provided they are extended by the long extensor tendons).

ORIGIN OF LUMBRICALS

See p. 423.

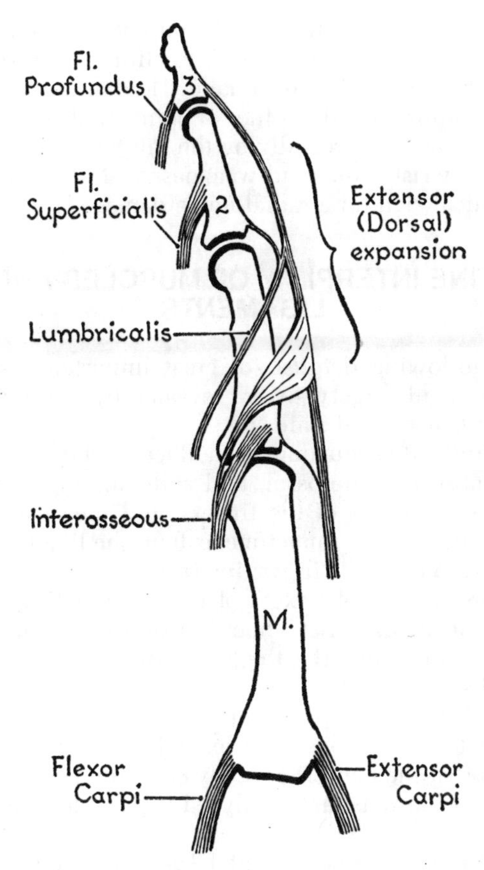

Figure 37.10. The insertions of the tendons of a finger (side view). *M*, metacarpal.

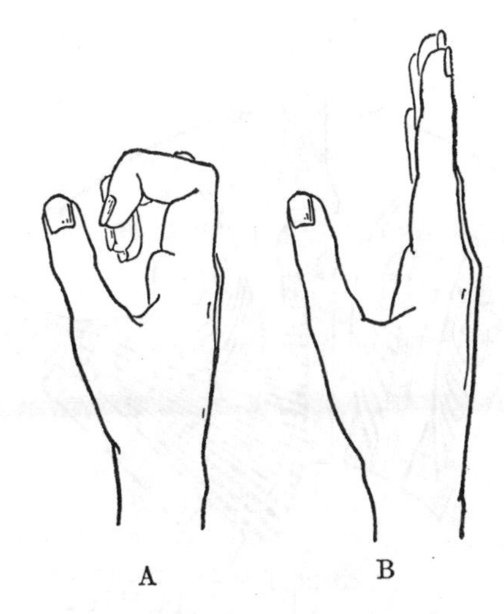

Figure 37.11. *A*, long digital extensors extend metacarpophalangeal joints; *B*, lumbricals and interossei extend interphalangeal joints.

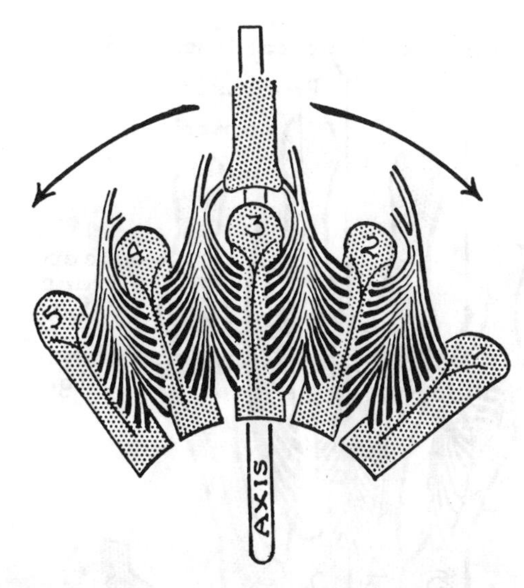

Figure 37.12. Four dorsal interossei abduct the 2nd, 3rd and 4th fingers.

INTEROSSEI

Each of the 5 digits can be abducted and adducted, that is, moved away from and moved toward a line passing through the middle finger, the *axial line of the hand*. The adductors and the 2 abductors (longus and brevis) of the thumb attend to its requirements, and the abductor digiti minimi abducts the little finger. The remaining 7 movements are performed by 7 interossei (*figs. 37.12* and *37.13*).

Four dorsal interossei (*fig. 37.12*) are conspicuous from the dorsum of the hand (especially the 1st), filling the 4 intermetacarpal spaces and arising by double heads from the facing sides of the bodies of the 5 metacarpals. They are abductors.

The 3 palmar interossei arise by single heads from the anterior borders of those finger metacarpals with available borders, namely, 2nd, 4th, and 5th (*fig. 38.13*)—the 3rd anterior border giving origin to the transverse head of the adductor pollicis. Each palmar interosseus adducts its own MP joint (2nd, 4th or 5th).

Course

The interossei pass behind the deep transverse metacarpal ligaments (*fig. 37.14*); but the lumbricals and the palmar digital vessels and nerves pass in front.

Insertions

The sides of the fingers to which the interossei must pass to perform these movements are apparent. Their insertions are partly into the bases of the proximal phalanges and partly into the extensor expansions.

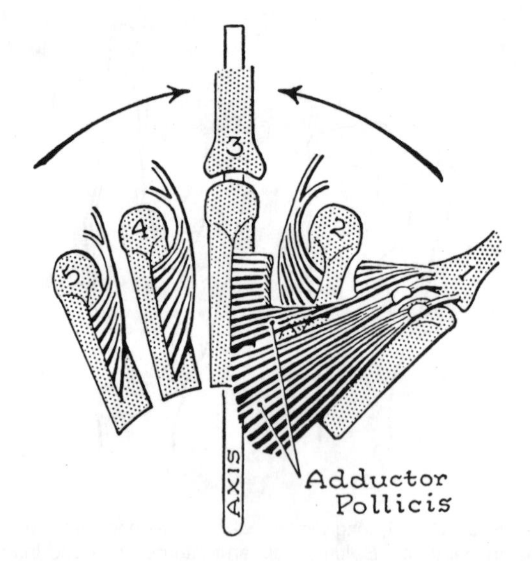

Figure 37.13. Three palmar interossei adduct the fingers.

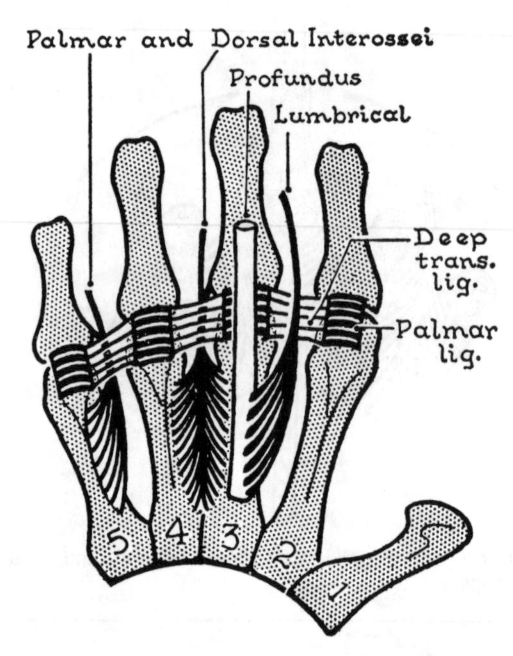

Figure 37.14. The different relationships of lumbricals and interossei to the deep transverse metacarpal ligaments (palmar view).

Nerve Supply

All interossei are supplied by the deep ulnar nerve.

Nerve Lesions and Extension of the Fingers

Since the ulnar nerve supplies the interossei and at least 2 of the lumbricals, its loss produces inability to extend the IP joints. There is also a reactive hyperextension of the MP joints through the pull of the radial nerve-supplied extensor digitorum. The clinical sign of ulnar nerve damage is a "clawed hand." Radial nerve loss results in failure to extend the MP joints, but the IP joints can be extended forcefully by the interossei and lumbricals. A "wrist drop" and weakness in MP extension is the clinical sign for a radial nerve damage (*fig. 37.3*).

FINE INTERPLAY OF MUSCLES AND LIGAMENTS

The following *details*—of great importance in specialized hand surgery—were revealed by Braithwaite *et al.*, Landsmeer, and Haines.

In addition to abduction and adduction of the MP joints, the lumbricals, interossei, and abductor digiti minimi flex the MP joints of the fingers, and they can, either separately or at the same time, extend the IP joints, as in making the upstroke in writing (*fig. 37.15*).

Synergists. The 4 flexors of the wrist, acting as synergists, steady the wrist when the extensors of the fingers are in action; similarly, the 3 extensors of the wrist act synergically when the flexors of the fingers are in action (*fig. 37.16*). Indeed, you cannot clench your fist or grasp an object firmly unless you extend your wrist. And, it follows that the hand is severely compromised if the extensors of the wrist are paralyzed (e.g., in lesions of the radial nerve).

Shortness of the Long Digital Tendons. See if you can fully flex your wrist while the hand is closed. The fingers must be extended owing to the relative shortness of the extensor tendons. Nor is it easy to extend your wrist fully while the fingers are extended due to the relative shortness of the flexor tendons.

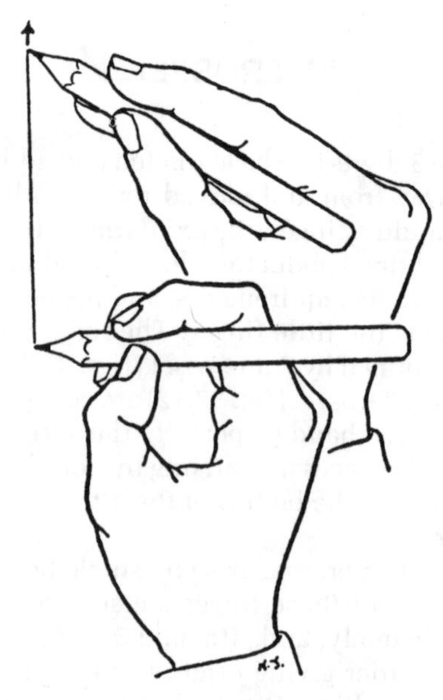

Figure 37.15. The interossei and lumbricals making the upstroke in writing.

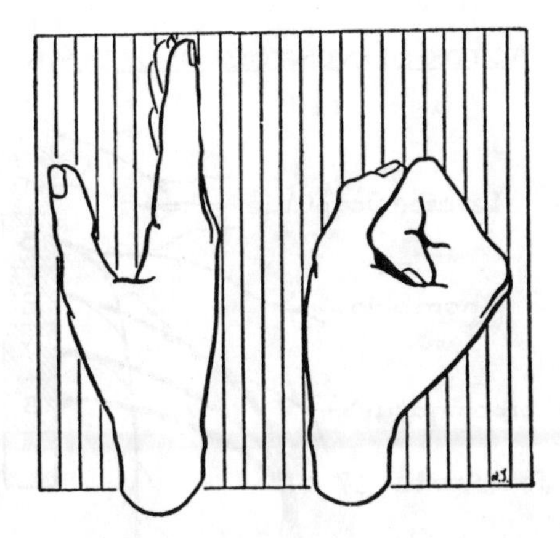

Figure 37.16. The extensors of the wrist act synergically with the flexors of the fingers.

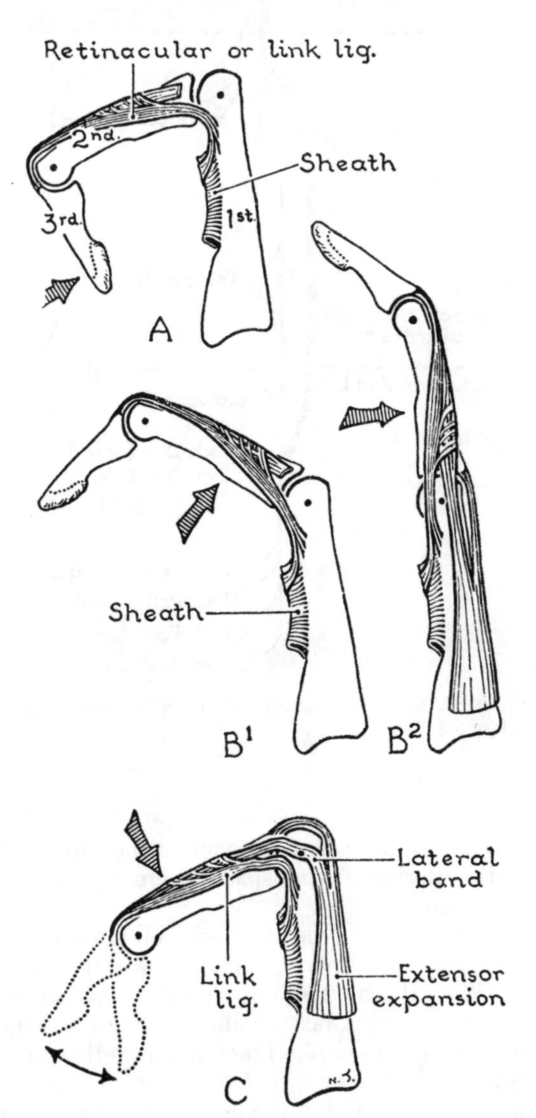

Figure 37.17. The functions of the retinacular or link ligaments demonstrated.

Ligamentous Functions. A lateral component of the extensor expansion is a ligamentous band (*fig. 37.17*), called the **retinacular ligament** (Landsmeer) or the *link ligament* (Haines). It spans the 2nd phalanx, crossing dorsal to the axis of the distal IP joint and palmar to the axis of the proximal joint.

Observations. 1. On flexing the distal IP joint either passively or voluntarily in your own hand (*fig. 37.17, A*) the retinacular ligament becomes taut and pulls the proximal IP joint into flexion.

2. Similarly, on extending the proximal IP joint (either passively or voluntarily), the distal IP joint is pulled by the retinacular ligament into nearly complete extension (*fig. 37.17, B*).

3. While the proximal IP joint is fully flexed passively (e.g., by pressing on the 2nd phalanx) the 3rd phalanx cannot be extended voluntarily, though it offers no resistance to passive extension. Indeed, it is flail because the lateral bands of the extensor expansion slip forward and hence are slack and ineffective (*fig. 37.17, C*).

You should have reached the correct conclusion by now that the MP joint of a finger can move independently of the IP joints but the IP joints are compelled by the retinacular or link ligaments to move together.

Clinical Case 37.4

Patient Anne L. This is the same patient who was in Clinical Case 32.1—the woman whose diving accident crushed her spinal cord at level C6–7. Now the resident physician is quizzing you about the expected motor loss and retention of power in the upper limbs. Will she be able to operate a wheelchair with her own power? What about her ability to control the joystick of an electric wheelchair?

Review of Nerve Supplies. The student will find Figures 37.18 and 37.19 and Table 37.1 useful for review.

Anatomical Features of the Physical Examination

The dorsum of the forearm is limited laterally by the easily felt brachioradialis and medially by the equally palpable sharp posterior border of the ulna. Between them is the mass of fleshy musculature, which few clinicians attempt to identify until these muscles reach the wrist region. Then the outcropping tendons of the deep exten-

Table 37.1. Segmental Innervation of Muscles of Forearm and Hand*

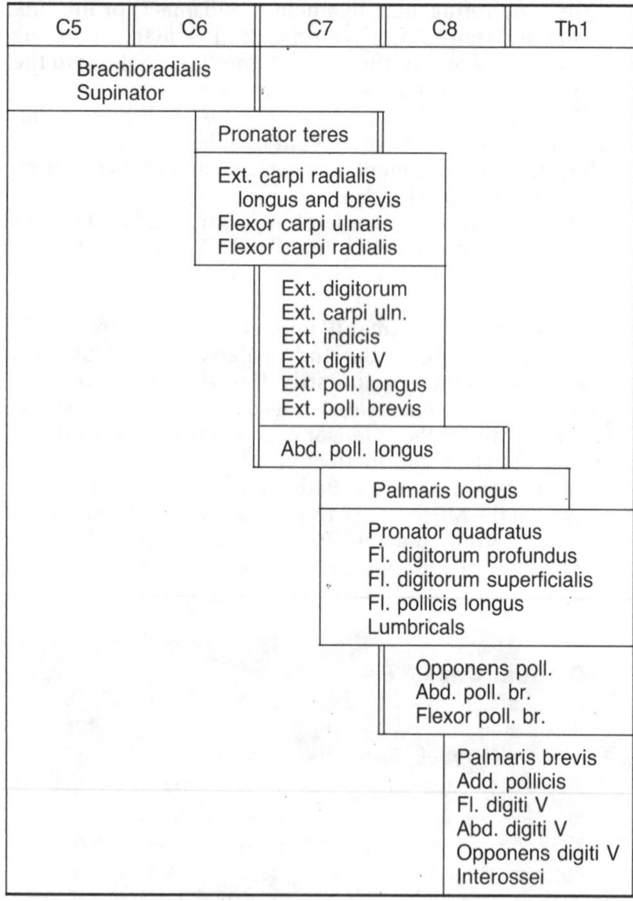

C5	C6	C7	C8	Th1
Brachioradialis Supinator				
	Pronator teres			
	Ext. carpi radialis longus and brevis Flexor carpi ulnaris Flexor carpi radialis			
		Ext. digitorum Ext. carpi uln. Ext. indicis Ext. digiti V Ext. poll. longus Ext. poll. brevis		
		Abd. poll. longus		
		Palmaris longus		
		Pronator quadratus Fl. digitorum profundus Fl. digitorum superficialis Fl. pollicis longus Lumbricals		
			Opponens poll. Abd. poll. br. Flexor poll. br.	
			Palmaris brevis Add. pollicis Fl. digiti V Abd. digiti V Opponens digiti V Interossei	

* Modified after Bing, and Haymaker and Woodhall.

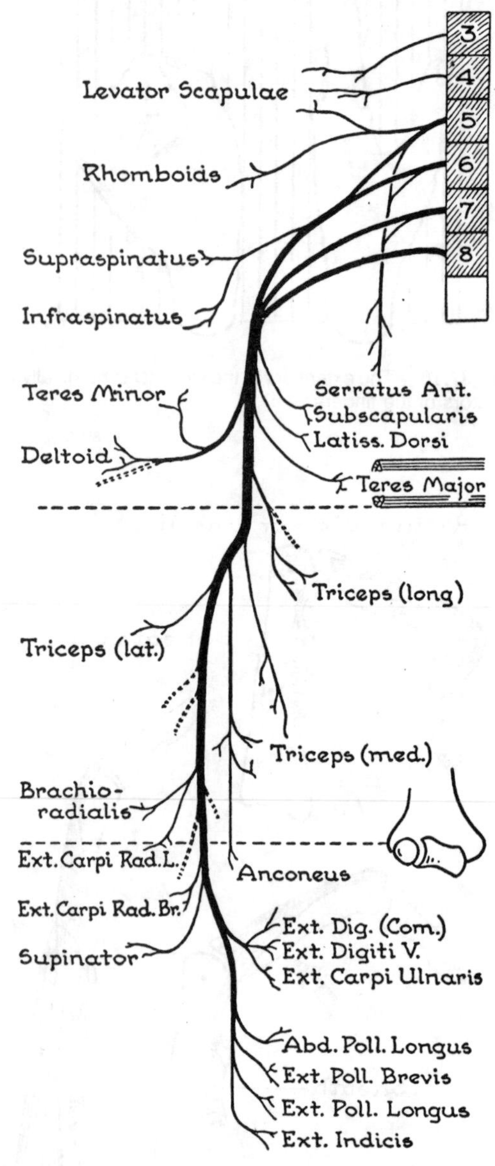

Figure 37.18. The motor distribution of the nerves of the back of the limb.

sors, the anatomical snuffbox they define and the other extensor tendons hold some interest for specialists.

The extensor muscle region is sharply limited medially by the posterior border of the ulna from the olecranon to the styloid process which you can easily palpate from end to end. The lateral boundary corresponds to the course of the radial artery. Thus, the closely grouped fleshy bellies of the three lateral superficial extensor muscles—brachioradialis and the extensor carpi radialis longus and brevis—form the obvious bulge on the lateral border of the forearm. The insertion of the brachioradialis is into the distal end of the radius but the other two muscles (which you can prove by testing to be true wrist extensors) run through the anatomical snuffbox to insert into metacarpal bases 2 and 3. You should feel the boundaries of the snuffbox on the lateral aspect of the carpus. They are the three thumb tendons—the abductor pollicis longus and extensor pollicis brevis laterally and the extensor pollicis longus medially. The last takes the long way around the dorsal radial tubercle and can be severed by sharp edges of a Colles' fracture there (as in Clinical Case 37.1). Follow the pulse of the radial artery *en route* through the snuffbox on the scaphoid bone to the proximal end of the first intermetacarpal space where it plunges anteriorly into the hand.

Traced proximally, your three thumb muscles plunge deep to the medial superficial extensors of the wrist and fingers. When the medial and lateral groups are pulled apart at an operation (as in Clinical Case 37.2) the supinator muscle is exposed. Emerging from the supinator (through which it has tunnelled across the direction of the muscle fibers), is the motor terminal branch of the radial nerve. Although it has already supplied the lateral extensor muscles, it now supplies all the others.

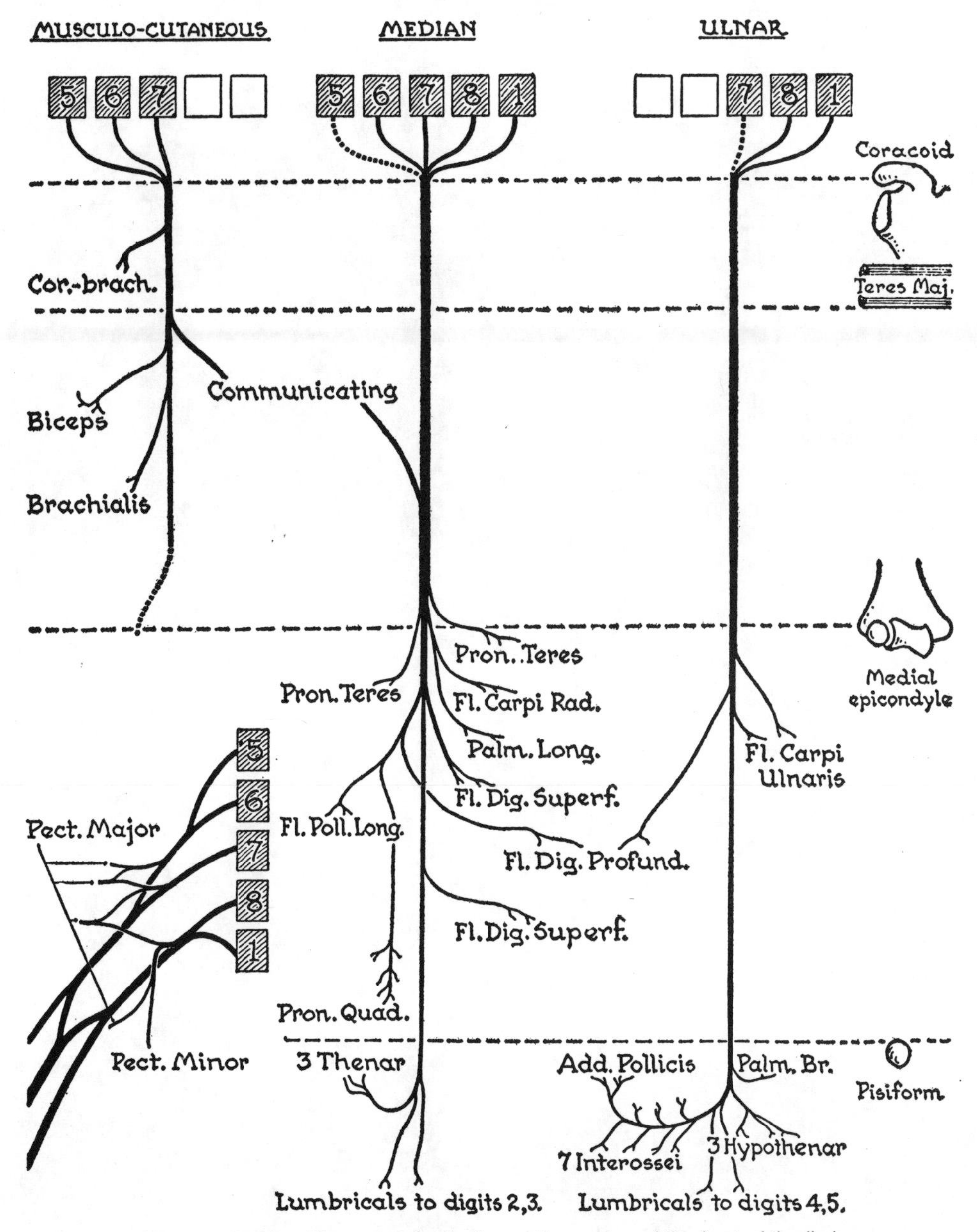

Figures 37.19. The motor distribution of the nerves of the front of the limb.

The extensor tendons descend through six discreet tunnels of the extensor retinaculum on the dorsum of the radius and ulna where they can be followed to their separate insertions. The digital tendons turn into broad, thin dorsal (extensor) expansions which are impalpable on the dorsum of the fingers.

Clinical Mini-Problems

1. **Which specific nerve is jeopardized in a fracture of the neck of the radius?**
2. **How would one differentiate between a lesion in the radial nerve at the level of the lateral epicondyle and a lesion in the deep radial nerve as it passes through the supinator?**
3. **(a) Which muscles supinate the forearm?**
   (b) What is their nerve supply?
4. **When the palmar interossei muscles and dorsal interossei both contract simultaneously,**
   (a) What movement is produced in the hand?
   (b) Which nerve is responsible for this movement?

(Answers to these questions can be found on p. 587.)

Head and Neck 7

38

Face, Scalp, and Skull

Clinical Case 38.1

Patient Mark M. This 16-year old motorcyclist was brought to the emergency room following a collision with a car. Among other injuries, he has a skull fracture of his left parietal bone, a depressed fracture of the left zygoma ("cheek bone"), a broken nose, and severe lacerations on his scalp and face on the left. You accompany your clinical tutor, a plastic surgery specialist, who is treating this case. You assist in the repositioning of the zygoma and nose and suturing of the skin. Three days later when you visit him, he has bilateral blackened eyes and is unable to completely close his left eye. Sensation to the skin over the scalp and face is found to be normal 10 days after the accident, but the orbicularis oculi muscle encircling the left eye remains weak and the lower eyelid droops.

Skin of the Face

The skin of the face is an important feature to consider when initially examining a patient. Since the face of a patient is usually exposed to the clinician on initial presentation, the expression, color, texture, and temperature can give important clinical information about the patient. The skin of the face is highly mobile, particularly around the orifices of the head (mouth, nose, ears, and eyes). Through the sphincter and dilative action of muscles that underlie the skin, these orifices can be opened or closed to varying degrees by the skin that approximates them.

In man, these facial muscles have evolved to have an additional function in both nonverbal and verbal communication. As a group, they are called the **muscles of facial expression**. They were derived from a common embryological origin, the mesoderm of the **2nd pharyngeal arch (hyoid arch)**, and they are all innervated by branches of the seventh cranial nerve, the **facial nerve**. As this mesoderm migrates over the face, the branches of the facial nerve are drawn superficially from the trunk of the nerve in the posterior aspect of the **parotid gland** to the deep surface of the individual facial muscles. The muscles are unique in that (1) they arise from bone but insert into the skin, (2) they are located in the subcutaneous tissue of the face, (3) they lack a demonstrable deep fascial covering (except for the buccinator muscle) and (4) the ratio of muscle fibers to axons in the facial nerve is markedly low. These features allow these muscles to move small areas of facial skin in an adept fashion and produce a number of "expressions" that connote emotion.

Figure 38.1 illustrates the individual muscles of facial expression found on the superficial and deeper layers of the face. Additional facial muscles, the **platysma** muscle of the superficial neck, the **occipitalis** muscle of the scalp, and the **auricular** muscles associated with the ear, are not shown in this illustration.

Muscles of the Rima Oris (Aperture of the Mouth)

The **orbicularis oris** lies within the lips and encircles the oral aperture. A number of smaller muscles interlace with the fibers of the orbicularis oris to make up the muscles of the oral aperture.

Five muscles converge on the angles of the mouth on each side. The **levator anguli oris** arises from the **maxilla** just below the infraorbital foramen; the **zygomaticus major,** or smiling muscle, arises more laterally from the body

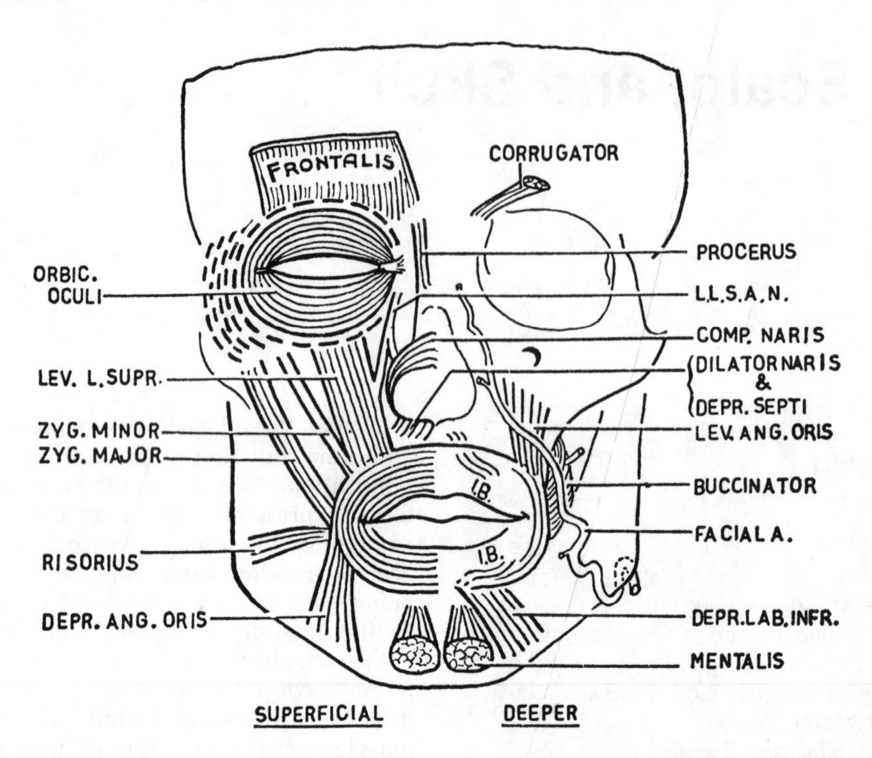

Figure 38.1. The muscles of the face. *LLSAN*, levator labii superioris alaeque nasi.

of the **zygoma** (the cheek bone); the **risorius,** or grinning muscle, arises from deep fascia of the face over the **mandible** and is joined by the fibers of the **platysma** that pass inferiorly over the mandible and into the skin of the neck and upper chest; the **depressor anguli oris** (triangularis) arises inferiorly from the oblique line of the mandible and inserts into the corner of the mouth.

Muscles of the Lips (L = *labium*, sing. *labia*, pl., *labii* = of the lip; *labiorum* = of the lips)

Inserting into the upper lip are three bands of muscles that arise from the medial and inferior borders of the orbital margin. They are the **levator labii superioris alaeque nasi, levator labii superioris,** and **zygomaticus minor** (L. *que* = and; *alae* = of the wing; *nasi* = of the nose).

Inserting into the lower lip is the **depressor labii inferioris** (quadratus), which arises from the oblique line of the mandible deep to the origin of the **depressor anguli oris.** The paired **mentales** originate from the mandible between the right and left depressor labii inferioris muscles. Each **mentalis** muscle inserts into the skin overlying the chin (L = *mentum*).

Muscles of the Cheek (L = *bucca*)

The **buccinator** is a flat muscle that exists between the mucous membrane lining of the oral cavity and the skin over the cheek. The buccinator possesses a definitive deep fascia that is continuous with the fascia of the **superior constrictor muscle** of the pharynx. Both the buccinator and the superior constrictor utilize the **pterygomandibular raphe,** the maxilla and the mandible for their sites of origin. While the superior constrictor passes posteriorly into the wall of the pharynx, the buccinator muscle passes anteriorly and laterally around the upper and lower molar teeth to fill the cheek and blend with the muscle fibers of the upper and lower lips. Its fibers contribute to the **orbicularis oris** musculature. The buccinator is pierced by the parotid duct and the long buccal branch of nerve V^3, which carries sensory innervation from the buccal mucosa.

ACTIONS

The buccinator aids in mastication by pressing the cheeks against the teeth and forcing food onto the **occlu-**

sal surfaces of the molar and premolar teeth. It prevents food from being forced into the **vestibule** of the mouth (*fig. 38.2*). The buccinator also aids in blowing and sucking actions.

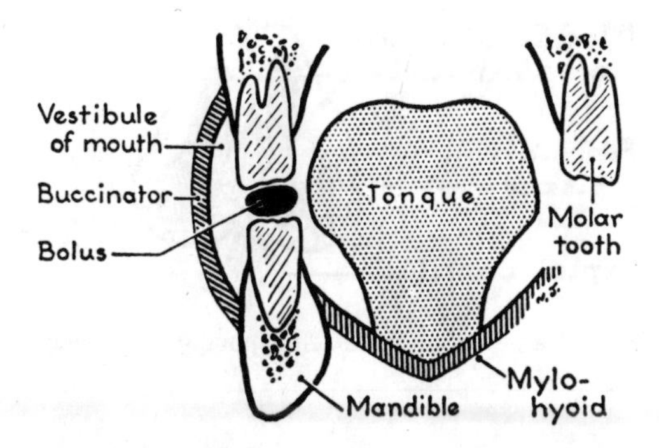

Figure 38.2. The buccinator and the tongue hold the food between the teeth.

NERVE SUPPLY

The **facial nerve** (cranial nerve VII) innervates *all* muscles of facial expression.

Lips

If you run the tip of your tongue across the back of your lower or upper lip, you will feel the small nodular **labial glands**. Grasp the margin of either lip between your fingers and thumb to feel the pulsations of the **superior** and **inferior labial arteries** between the labial musculature and the glandular epithelium of the lip mucosa. Profuse bleeding of the lip can be controlled by compressing these labial arteries between the point of hemorrhage and the corner of the mouth.

The lip margins are red partly because the skin is translucent and partly because the vascular papillae or **thelia** are unusually long. Historically, the term "epithelium" was first used to describe the cells covering the thelia of the lip. The close approximation of blood vessels to epithelial surfaces allows one to qualitatively assess the oxygen content in the circulating blood. Bluish tones to the color of the lip or fingernail "beds" are a clinical sign of **cyanosis** (decreased O_2 content and increased CO_2 content in the hemoglobin).

External Nose (*figs. 38.3 and 38.4*)

The framework of the nose is made up of the paired nasal bones, hyaline cartilage, and fibroareolar tissue. The cartilages are **septal, lateral** and **alar.** The right and left **lateral cartilages** are not separate entities but are the wing-like expansions of the midline septal cartilage that projects posteriorly between the right and left nasal cavities. The septal cartilage is firmly attached to the maxilla and ethmoid bones as it assists in the partition of the two nasal cavities. The **alar** cartilages are loosely connected to the septal cartilage and underlie the skin of the anterior and medial aspect of the **nares** (nasal openings).

Auricle

The framework of the **auricle** is made of a single piece of elastic cartilage, except at its most dependent part, the **lobule,** which is fibrocartilage. The cartilage is continuous with the cartilage of the **external acoustic meatus** (L. *meatus* = canal). *Figures 38.5* and *38.6* provide the names of the elevations and depressions of the auricle.

Three extrinsic auricular muscles—**auricularis posterior, auricularis superior,** and **auricularis anterior**—act on the auricle to cause movement. These are all muscles of facial expression and they are innervated by branches of the facial nerve (Cranial Nerve VII).

Sensory supply to the skin of the auricle is important in neurological diagnosis. The upper half of the auricle is innervated by the **auriculotemporal branch** of the V^3. The lower half of the auricle is innervated by the **greater auricular nerve** and **lesser occipital nerve** (C2, 3) from the cervical plexus. The epithelium of the external auditory meatus and the opening onto the auricle, the **concha,** is innervated by the **vagus** nerve (Cranial Nerve X)

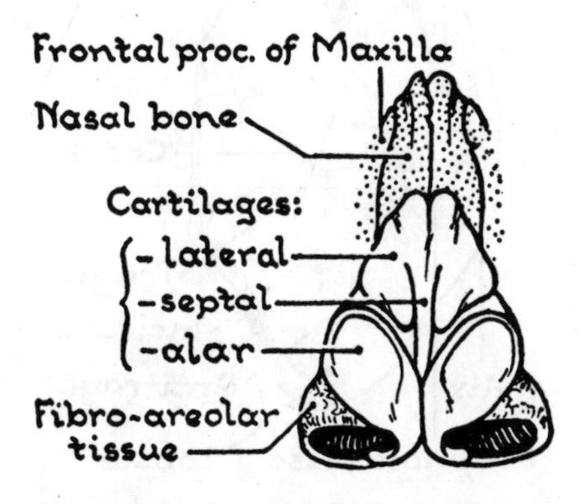

Figure 38.3. Framework of external nose.

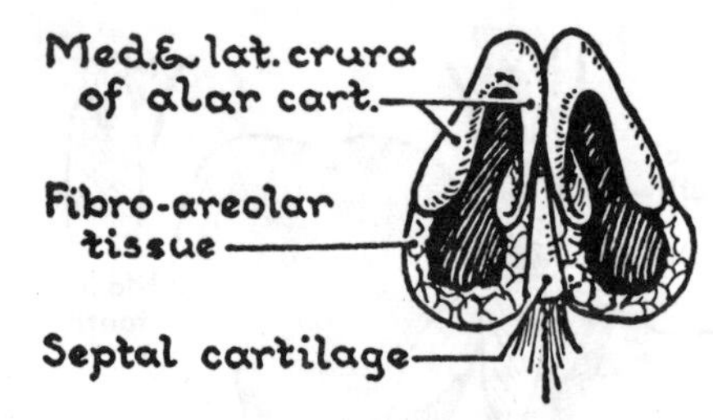

Figure 38.4. Framework of the external nose (from below).

for pain and perhaps the facial nerve (VII) for touch sensation.

Eyelids, Conjunctival Sac, and Tear Apparatus

DEFINITIONS

The upper and lower eyelids, or palpebrae, are united at the medial and lateral angles by the corresponding palpebral commissures. The eyeball is situated behind the palpebrae. The posterior five-sixths of the eyeball has an outermost coat of white, tough fibrous tissue called the **sclera.** The anterior one-sixth of the eyeball is transparent and called the **cornea.** Through the cornea you can see the varicolored **iris** with a central aperture, the **pupil.**

The potential space between the eyeball and the eyelids is the **conjunctival sac.** The membrane lining the sac is the **conjunctiva** (see *fig. 41.3*). At the upper and lower limits of the conjunctival sac, the conjunctiva reflects off of the eyeball onto the undersurface of the eyelid to form the **fornices.**

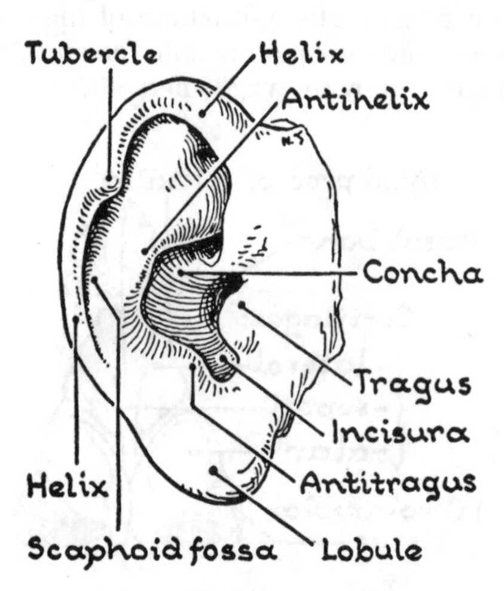

Figure 38.5. The auricle.

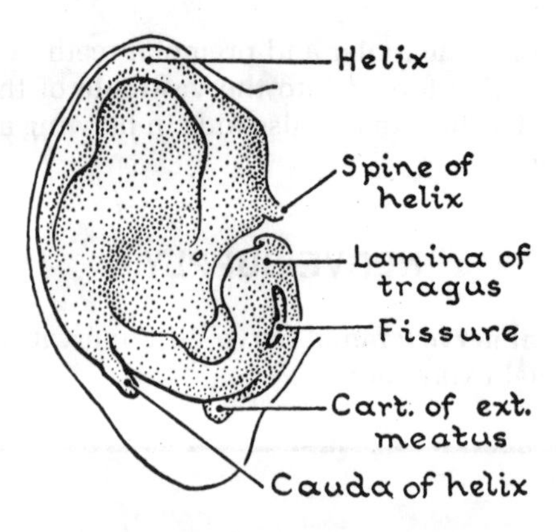

Figure 38.6. The cartilage of the auricle.

INSPECTION

Examine your eye in a mirror:

1. The margin of the lower eyelid crosses the lower limit of the cornea; the margin of the upper eyelid encroaches on the cornea (*fig. 38.7*).
2. The lateral five-sixths of the margins of the lids are flat and carry eyelashes or **cilia.** The medial one-sixth is devoid of hairs and rounded. This rounded area contains the **canaliculus** that drains away the tears.
3. At the medial angle, there is a triangular area, the **lacus lacrimalis,** bounded laterally by a free crescentic fold of conjunctiva, the **plica semilunaris.** In the lacus, there is a reddish area, the **caruncle.**

Gently pull down the lower lid to note:

4. A **papilla** on which the **punctum**, or entrance to the inferior lacrimal canaliculus, can easily be seen.

Evert the upper lid.

5. The hairs or cilia projecting from the lid margin are in 2 to 3 irregular rows.
6. Hairs imply the presence of sebaceous glands, and these open into each hair follicle. Sweat glands likewise open into or beside the hair follicles (*fig. 38.8*).
7. The **tarsal glands**, which waterproof the lids are embedded in the **tarsus**, a tough fibrous plate. The glands are visible as yellow streaks through the conjunctiva.

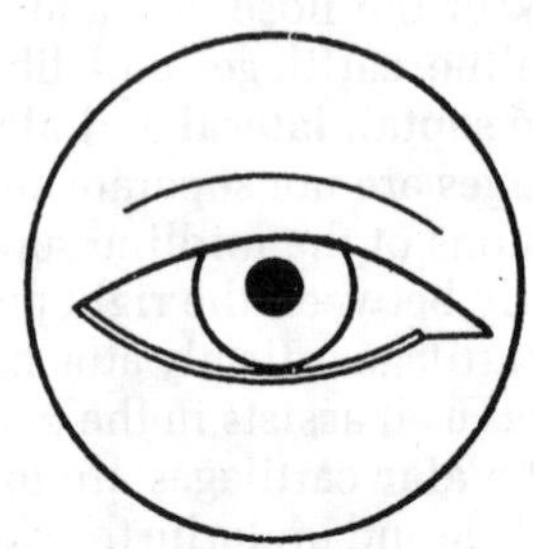

Figure 38.7. The margins of the eyelids.

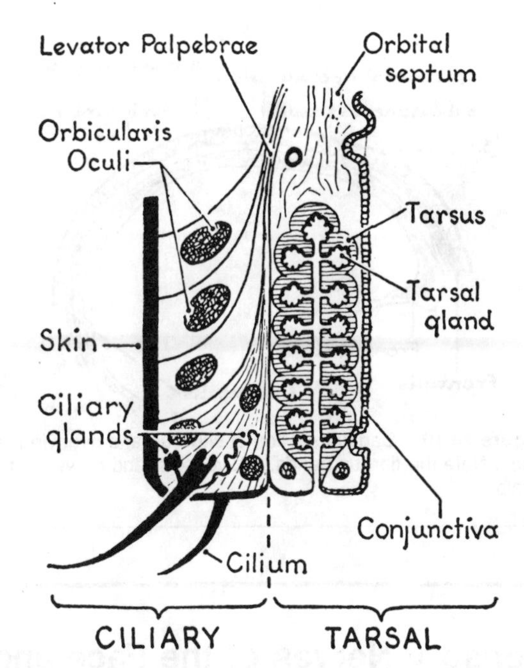

Figure 38.8. Section through the upper eyelid. (After Whitnall.)

An obstructed ciliary gland or inflamed hair follicle, a **stye** will project on the front of the eyelid while an obstructed tarsal gland projects internally onto the globe of the eye.

ORBITAL SEPTUM AND TARSI (*fig. 38.9*)

The eyelids develop as folds of skin that come together and adhere along their edges during the middle 3 months of intrauterine life. When they become free again the palpebral fissure is re-established.

While the eyelids are closed, the orbital septum, which runs from the orbital margin into the eyelids, forms a complete diaphragm for the orbital cavity. The medial aspect of the orbital septum passes behind the lacrimal sac to gain attachment to the lacrimal bone. This attachment creates a sharp bony crest, the **posterior lacrimal crest**. Operations on the lacrimal sac are therefore anterior to the orbital cavity proper.

Condensations and thickening of the septum takes place in the lids, forming the tarsal plates. These plates are anchored to the orbital margins by the **medial** and **lateral palpebral ligaments**. The medial palpebral ligament is a strong band of connective tissue that crosses in front of the lacrimal sac (*fig. 38.9*).

MUSCLES OF THE EYELIDS

The **orbicularis oculi** is the sphincter of the palpebral fissure. The fibers within the lids are the **palpebral portion**. The **orbital portion** makes a complete circle from the medial palpebral ligament, having no lateral attachment (*fig. 38.1*). It is responsible for the production of the "crow's feet" wrinkles around the eyes.

Closing the eyes against the resistance of an examiner's fingers is an excellent test for the functional status of the superior division of the **facial** nerve (VII).

The eye is opened by the **levator palpebrae superioris.** This is not a muscle of facial expression. It is a voluntary muscle of the orbit and is innervated principally by the **oculomotor** nerve (Cranial Nerve III). It also contains some smooth muscle fibers within its muscle mass that are innervated by sympathetic fibers. Damage to nerve III or the sympathetic fibers in the head and neck can produce a drooping of the eyelid. This clinical sign is called **ptosis.**

Muscles of the Forehead and Eyebrows (*fig. 38.1*)

The **frontalis** inserts into the eyelids as it descends over the frontal bone. It produces the transverse wrinkles on

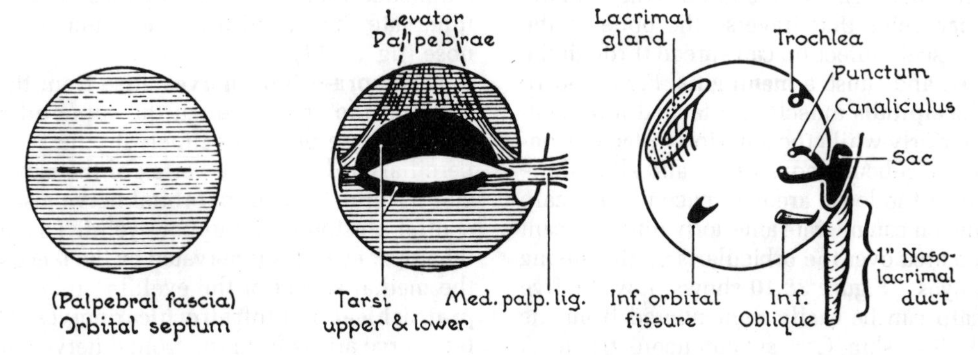

Figure 38.9. *A*, the orbital septum; *B*, the tarsi, ligaments, and levator palpebrae; *C*, features at the 4 corners of the orbital margin, and the tear apparatus (schematic).

the forehead that are produced when you raise your eyebrows.

The **corrugator supercilii** causes the short vertical wrinkles in front of the gabella above the root of the nose.

Lacrimal or Tear Apparatus (*fig. 38.9*)

The upper and lower **lacrimal canaliculi** are about 10 mm long and run near the free margin of the lid from the **lacrimal punctum** to the **lacrimal sac**. The lacrimal sac is the blind upper end of the **nasolacrimal duct** that lies just posterior to the medial palpebral ligament. The duct runs inferiorly in a bony canal of the lacrimal bone and empties into the nasal cavity in the inferior meatus. Thus, crying will produce tearing of the eyes as well as excessive nasal drainage.

Scalp

The scalp is a continuation of the skin above the face from the forehead to the posterior occipital area of the head. The scalp is regarded as a unit composed of (1) skin, (2) dense subcutaneous tissue, and (3) an epicranius muscle composed of **frontalis** and **occipitalis muscles** interconnected by an aponeurotic fascia. All three of these layers are firmly bound together and separated from the periosteum of the outer skull by a very loose areolar space. This space allows the upper three layers to move relatively freely over the underlying bone. It is, however, a "dangerous area" if it becomes infected. This space communicates with the meningeal venous sinuses around the brain via emissary veins that traverse the bone of the skull. Therefore, a scalp infection can spread through the bone via the veins and cause a **meningitis**. *Figure 38.10* shows how the **occipitalis** muscle is attached to the occipital bone posteriorly while the anteriorly placed **frontalis** inserts into the subcutaneous tissue and skin of the eyebrows. Blood in the loose areolar space of the scalp following a contusion can migrate anteriorly into the scant connective tissue area over the orbicularis oculi causing bilateral "black eyes." *Figure 38.10* shows how the five layers of the scalp can be easily remembered from the acronym SCALP (S = skin, C ≐ subcutaneous tissue; A = aponeurosis; L = loose areolar tissue; and P = periosteum or pericranium).

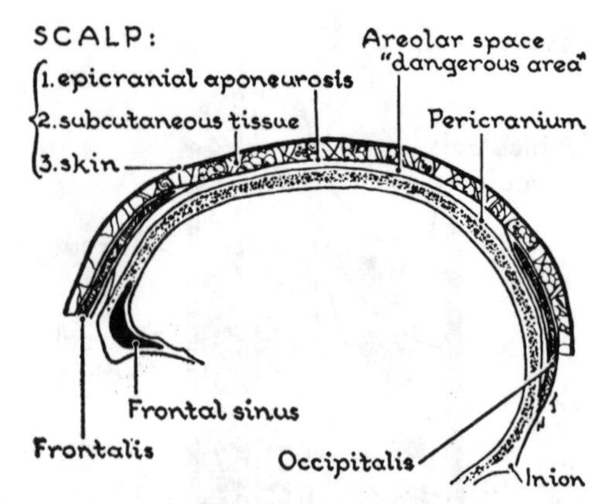

Figure 38.10. Sagittal section of skullcap and overlying tissues. Note the fibrous bands, fat, vessels, and nerves in the scalp.

Sensory Nerves of the Face and Scalp and Companion Arteries

The face develops from 3 rudiments, the **unpaired frontonasal process** and the bilateral **maxillary** and **mandibular** processes. Each process is supplied by one of the three divisions of the **trigeminal** nerve (Cranial Nerve V). The frontonasal process is innervated bilaterally by the ophthalmic divisions (V^1) while the maxillary divisions (V^2) and the mandibular divisions (V^3) are associated with the skin overlying their respective processes. Additional sensory innervation to the skin of the neck, parotid region, ear and angle of the mandible is provided by spinal nerves from the cervical plexus.

THE OPHTHALMIC DIVISION (V^1) (*fig. 38.11*)

Within the orbit the three major branches of V^1, (nasociliary, frontal and lacrimal nerves) give rise to the five cutaneous branches. Four of these cutaneous branches innervate the eyelid and one innervates the tip of the nose (*fig. 38.11*).

The **supra-orbital nerves** arise from the frontal nerve and exit the orbit through the supra-orbital foramen or notch. These branches innervate the scalp from the superciliary ridges to the vertex of the skull (occipitoparietal fissure). The supra-orbital nerve also innervates the central portion of the upper eyelid. The lateral aspect of the upper eyelid is innervated by the **lacrimal nerve** while the medial aspect of the eyelid is innervated by the **supratrochlear** and **infratrochlear** nerves. The supratrochlear nerve arises from the frontal nerve and passes above the superior oblique muscle to innervate the upper eyelid and forehead above the root of the nose. The infratro-

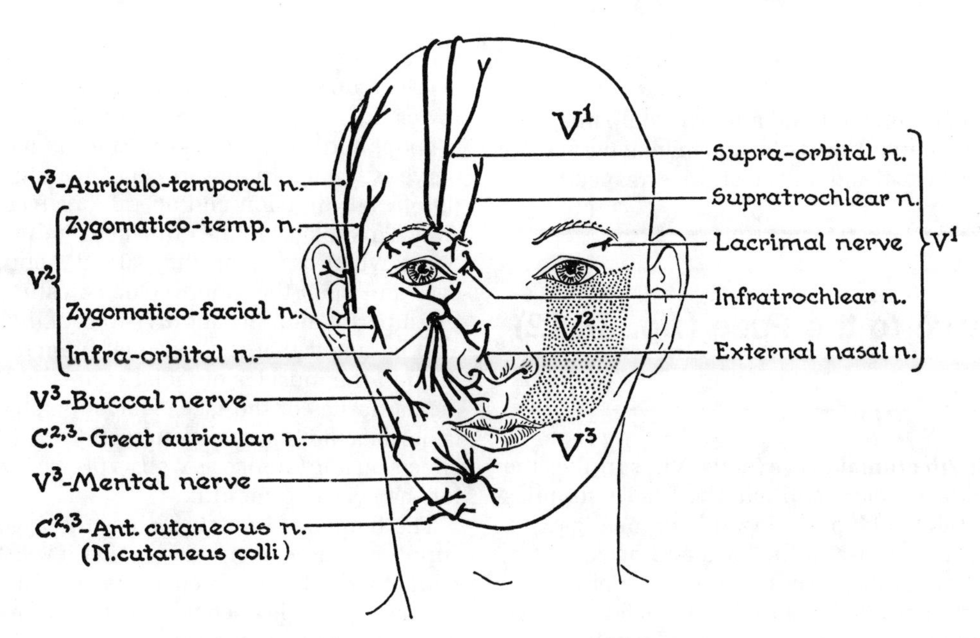

Figure 38.11. The sensory nerves of the face and front of the scalp.

chlear nerve is a branch of the nasociliary nerve within the orbit, and it emerges below the superior oblique muscle to innervate the skin of the upper eyelid and lateral aspect of the nose. The **external nasal branch** is also derived from the nasociliary nerve as the terminal component of the **anterior ethmoidal nerve**. It emerges at the lower border of the nasal bones and descends on the nasal cartilages to the tip of the nose.

Arteries

All of the cutaneous branches of V^1, are accompanied by a branch from the **ophthalmic artery** within the orbit. The arteries are named in accordance to the nerves they accompany.

Veins

The nerves and arteries are also accompanied by similarly named veins. These veins are important communications between the vascular system of the face and the dural sinuses of the brain. Since these veins lack valves, they may transmit infectious material from the face and scalp to intra-orbital and intracranial sites.

THE MAXILLARY NERVE V^2 (*fig. 38.11*)

Three branches of V^2 reach the cutaneous tissues of the face and scalp. The **infra-orbital branch** is the largest and emerges through the infra-orbital foramen. Its cutaneous branches innervate the lower eyelid and conjunctiva; the lateral aspect of the nose and the inside of the nostril; the upper lip, mucous membrane of the upper aspect of the cheek, the gingiva of the upper teeth and the anterior upper teeth as well.

Two smaller V^2 branches arise in the temporal region. They are the **zygomaticotemporal** and **zygomaticofacial nerves**. They innervate the skin over the **zygoma** (cheek bone).

All V^2 cutaneous nerves are accompanied by arteries and veins of the same name.

THE MANDIBULAR NERVE V^3 (*fig. 38.11*)

This nerve has several motor and sensory branches. Only 3 branches are cutaneous and discussed in this section.

The **mental nerve** emerges from the mental foramen of the mandible to innervate the skin of the chin, lower lips, and mucous membrane and gingiva adjacent to the lower lip.

The **buccal branch** of V^3 (the long buccal nerve) lies on the external aspect of the buccinator muscle. It runs from the depths of the cheek to the angle of the mouth. It supplies the skin overlying the buccinator and then pierces the muscle to supply the mucous membrane that lines the inner surface of the cheek and lower gums adjacent to the buccinator muscle.

The auriculotemporal branch emerges from the superior aspect of the parotid gland with the **superficial temporal** vessels and crosses the zygoma just in front of the ear. Its terminal distribution is to the auricle (superior 1/2) and the temporal region.

Arteries

While the superficial temporal artery accompanies the auriculotemporal nerve, the buccal and mental nerves are accompanied by buccal and mental arteries, respectively.

Motor Nerve to the Face (*fig. 38.12*)

The **facial** or **7th cranial nerve** (nerve VII) supplies the muscles of the face, scalp, and auricle. It also supplies the platysma, stylohyoid, posterior belly of the digastric, and the stapedius muscles in the head and neck.

Its terminal branches appear at the margins of the parotid gland and "fan out" to supply the facial muscles on their deep surfaces (*fig. 38.12*). Cervical branches cross the angle of the mandible into the neck to supply the platysma. These are vulnerable to damage during parotid, upper neck, and mandibular surgery. Paralysis of the platysma causes severe cosmetic effects when the facial tissue is allowed to "sag" under the jaw. Mandibular branches also extend over the jaw to supply the muscles of the lower lip.

Loops and communications are common, and the facial nerve is frequently described as having a superior (temporofacial) division and an inferior (cervicofacial) division. This relates to the clinical examination of the facial nerve when assessing the patient's ability to close the eyes and open the mouth. One is testing the integrity of the superior and inferior divisions of the nerve. It should also be noted that the terminal branches of VII are just deep to the muscles of facial expression in the subcutaneous tissues of the face. They are subject to trauma in facial lacerations, and denervation can have disastrous effects on appearance as well as protective reflex function for the eyes and mouth.

The **temporal branches** cross the zygomatic arch and supply all the muscles above that level (tested when the patient wrinkles the forehead and elevates the eyebrows). The **zygomatic branches** pass forward above the parotid duct to supply the muscles of the infra-orbital region (tested with temporal branches by having the patient close the eyes while the physician tries to open them with his fingers). The **buccal branch** of VII supplies the buccinator and other muscles of the cheek (ask the patient to "purse"

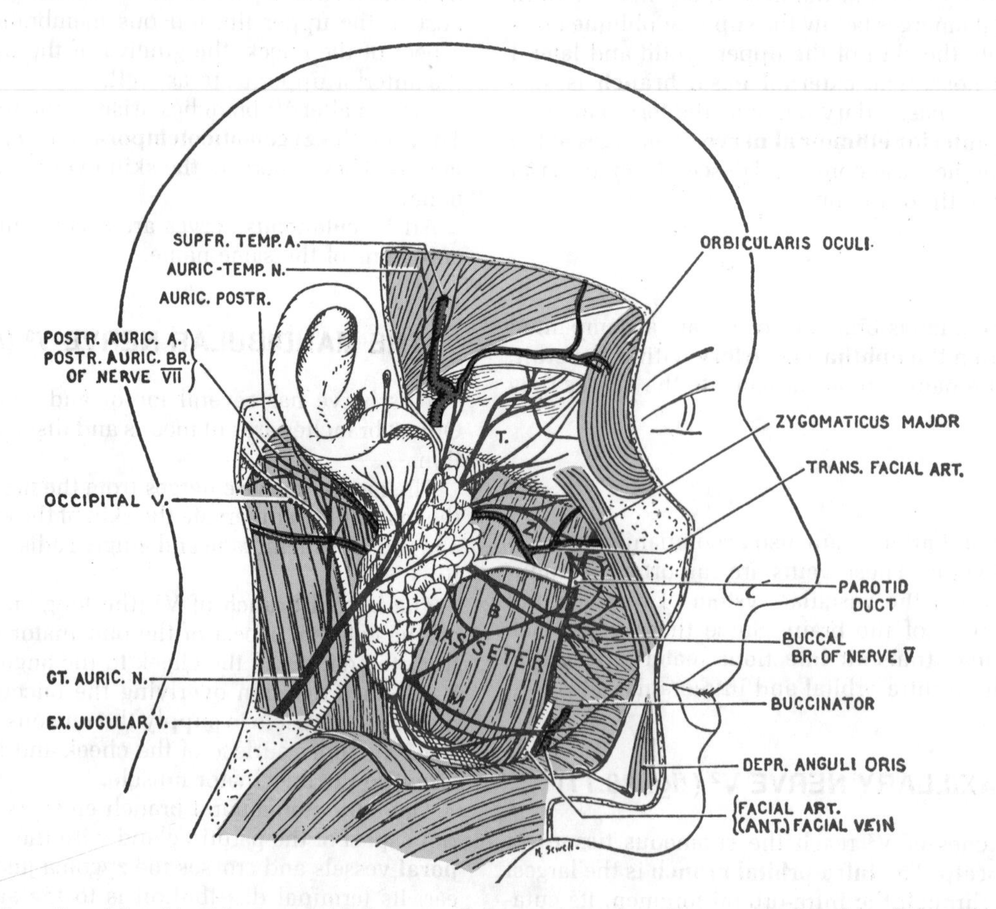

Figure 38.12. Dissection of the side of the face and facial nerve: *T*, temporal branches; *Z*, zygomatic branches; *B*, buccal branches; *M*, mandibular branch.

the lips and whistle to test these branches). The **mandibular branch** supplies the muscles of the lower lip and chin. The **auricular branches** arise behind the ear and innervate the muscles that move the auricle.

Blood Supply to the Face

The two principal arteries that supply the face are the **facial artery** and the **superficial temporal artery.** In addition, many branches of the ophthalmic and maxillary arteries that accompany cutaneous branches of the trigeminal nerve add to the rich vascular supply to the face.

FACIAL ARTERY (*figs. 38.1 and 38.12*)

The facial artery appears on the front of the face at the base of the jaw, immediately anterior to the insertion of the masseter muscle. The facial artery pulse can be taken as it courses over the mandible at this point. It rises on the face in a sinuous course to pass the corner of the mouth, the sides of the nose and terminates as the **angular artery** at the medial border of the eye. It lies deep to the muscles of facial expression and its sinuous nature allows for elongation when the central face is moved excessively. The **facial vein** is more posterior in its position and is not subjected to extensive elongation with movements of the mouth. The vein is therefore much straighter in its course from the medial aspect of the eye to the external jugular vein in the neck.

The important branches of the facial artery and vein are the **inferior** and **superior labial vessels**, a **lateral nasal vessel** that passes above the alar cartilage, and the **angular vessels** at the medial angle of the eye. The angular vessels allow for a communication between the facial artery of the face and the ophthalmic artery of the orbit. Venous connections of the facial vein with the cavernous sinus via the superior ophthalmic veins is an important clinical consideration for assessing routes of infection between the face and dural sinuses.

THE SUPERFICIAL TEMPORAL ARTERY (*fig. 38.12*)

The superficial temporal artery arises in the parotid gland as one of the two terminal branches of the **external carotid artery**. It emerges from the superior aspect of the parotid gland in front of the ear. It is accompanied in its course by the auriculotemporal branch of V^3. The artery divides into a frontal and parietal branch above the zygoma and is distributed to the scalp overlying the frontal and parietal bones, respectively. A **transverse facial artery** may arise from the superficial temporal artery within the parotid. The transverse facial artery courses between the parotid duct and the zygomatic arch to supply the facial tissue over the body of the zygoma. A **temporal pulse** is frequently taken by compressing the superficial temporal artery against the squamous part of the temporal bone above the ear.

A summary illustration of the cutaneous nerves and vessels of the scalp is shown in *Figure 38.13*.

Skull Related to Scalp and Face

SKULL VIEWED FROM ABOVE (*fig. 38.14*)

Three major skull bones are evident from a superior view. The **frontal bone** is usually a singular bone anteriorly. It was originally a right and left frontal bone separated by a suture (*fig. 38.15*). Fusion of the two frontal bones occurs in early childhood but a persisting **metopic** suture may occur in approximately 8% of adults. This retained frontal suture can be confused in radiology with a skull fracture. The frontal bone joins the **parietal** bones at the **coronal (frontoparietal) suture**. The two parietal bones are separated from each other in the midline by the **sagittal (interparietal) suture**. *Figure 38.14* depicts the sagittal suture in a newborn skull. It is shaped like an arrow (L. *sagitta*) with the arrow point (bregma) di-

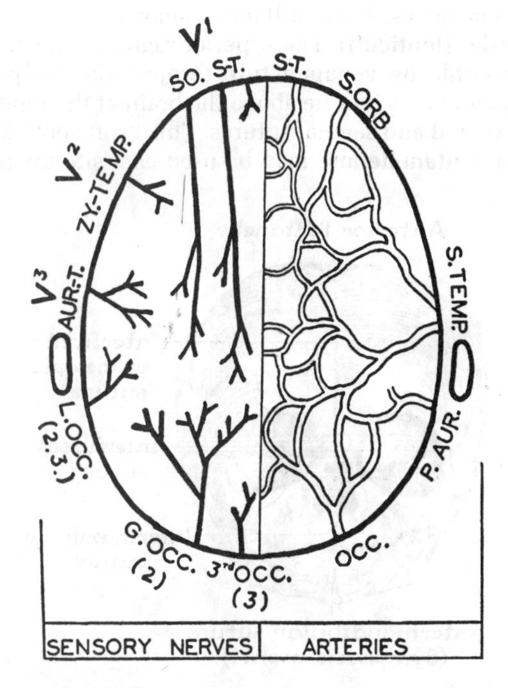

Figure 38.13. Sensory nerves and arteries of scalp.

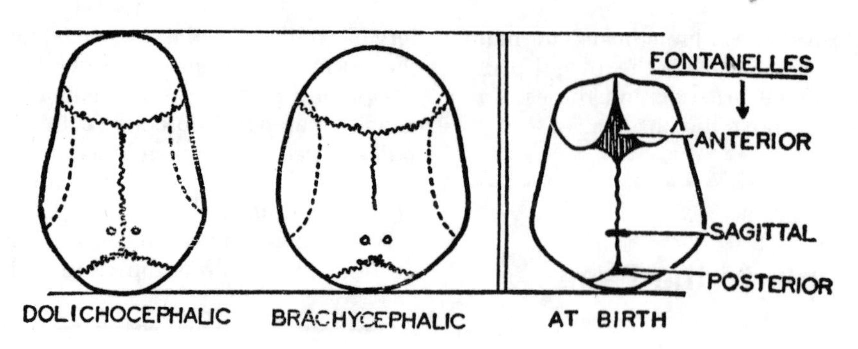

Figure 38.14. Skulls viewed from above. Fonticuli or fontanelles.

rected anteriorly. This can be palpated in the skull of the fetus during its descent in the birth canal. It is a valuable clinical sign to tell the physician the position of the head prior to delivery. A single **occipital** bone is seen posterior to the right and left parietal bones. The bones are separated by the **lambdoid (occipitoparietal) suture. Lambda** (the Greek letter λ) is the point where the lambdoid suture meets the sagittal suture. The parietal bones may show foramina for the parietal emissary veins just anterior to the lambdoid suture. These emissary veins traverse the parietal bones and communicate with the venous sinus within the dura of the cranium. They provide a means for blood (and infections) of the scalp to enter the meningeal venous sinuses.

The bones of the roof of the skull (calvaria) develop in membrane. Ossification of the frontal and parietal bones begins during the 2nd fetal month at their points of greatest fullness, the **frontal** and **parietal eminences**.

At birth, ossification has not reached the 4 corners of the parietal bones. These still membranous sites are called **fontanelles (fonticuli)**. The superior sagittal sinus is easily accessible by venapuncture through the scalp and dural membranes that lie above the brain at the junction of the coronal and sagittal sutures. This "soft spot" is the **anterior fontanelle** and may be used as a site for intravenous cannulas in hospitalized newborns. The anterior fontanelle is present as a soft membranous site on the child's head until 1–1½ years. It is 3–5 cm in length and shaped like a flat kite (*fig. 38.14*) at birth. It is normally obliterated before the end of the second year. The bregma marks its site in the completely ossified calvaria.

SKULL ON FRONTAL VIEW

The **zygomatic arches** lie at the widest parts of the face (*fig. 38.16*). Above them, the outline of the skull is rounded because it is formed by the **cranium** or brain case. The cranium bulges slightly to extend a few millimeters beyond the width of the zygomatic arches. Below the zygomatic arches, the skull is angular and is outlined by the prominent **mastoid processes** of the temporal bones and the posterior borders of the **ramus, angle,** and **base** or lower border of the mandible.

At birth (*fig. 38.15*), a median suture line bisects the skull vertically, separating the **parietal, frontal, nasal, maxillary** and **mandibular** bones of the opposite sides.

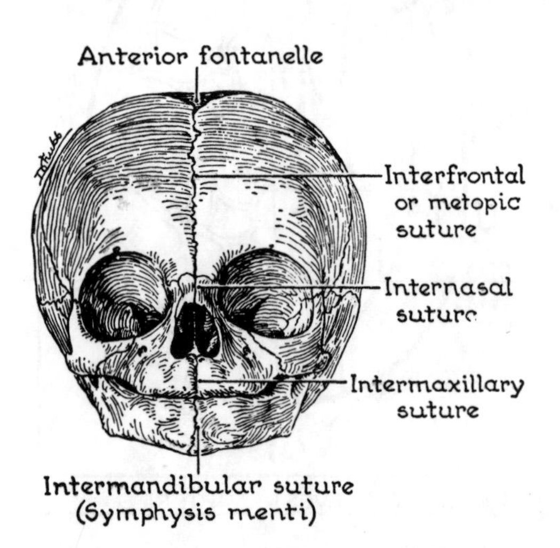

Figure 38.15. The skull at birth (front view).

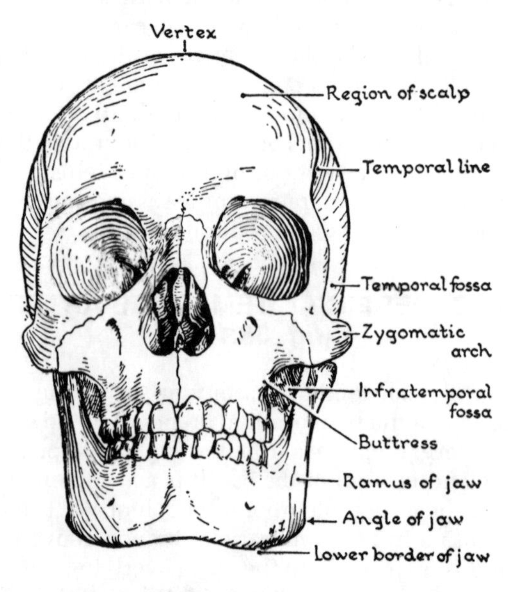

Figure 38.16. The skull (on front view).

During the 2nd year, the two halves of the mandible fuse at the **symphysis menti**.

The two halves of the frontal bone also fuse about the 2nd year. In some skulls, an interfrontal or **metopic** suture may persist into adulthood. The interparietal or sagittal suture usually persists but forms a **synostosis** (bony joint) between the two parietal bones. Connective tissue from the periosteum of the outer cranial bony table traverses the sagittal suture to join the connective tissue of the periosteum covering the inner cranial bony table.

The highest point of the skull is termed the **vertex** (*fig. 38.16*). This point marks the posterior extension of the V^1 dermatome on the scalp. In this vicinity, the branches of the supra-orbital nerve (V^1) overlap with the branches of the greater occipital nerve (C2) (*fig. 38.13*). The **nasion** is the point at the root of the nose where the frontal nasal suture crosses the median plane (*fig. 38.17*).

ENTRANCE TO THE ORBIT (ADITUS ORBITAE)

The frontal, maxillary, and zygomatic bones contribute to the formation of the orbital margin (*fig. 38.17*).

The fullness of the medial part of the supra-orbital margin of the frontal bone is the **superciliary arch**. This is well marked in the male and is a feature used to sex-type osteological remains in forensic pathology and anthropology. The elevation between the superciliary arches is the **glabella** because the overlying skin is glabrous or bald.

The **anterior nasal aperture (piriform aperture—L. = pear-shaped)** is formed by the nasal bones superiorly and the maxilla laterally and inferiorly. A median spine of bone, the **anterior nasal spine** projects forward from the maxilla and helps in supporting the septal cartilage of the nose.

Lateral to the orbit, the **zygomatic arch** (cheek bone) extends posteriorly toward the ear and fuses with the temporal bone above the **temporomandibular joint** where the mandible articulates with the base of the skull.

FORAMINA

Three osseous foramina are apparent from the frontal view of the skull. The **supra-orbital, infra-orbital** and **mental** foramina all open onto the face in a vertical line that passes sagitally between the premolar teeth. They penetrate the frontal bone, maxilla and mandible, and transmit sensory branches of V^1, V^2, and V^3, respectively. These foramina also transmit accompanying vascular elements with each nerve branch.

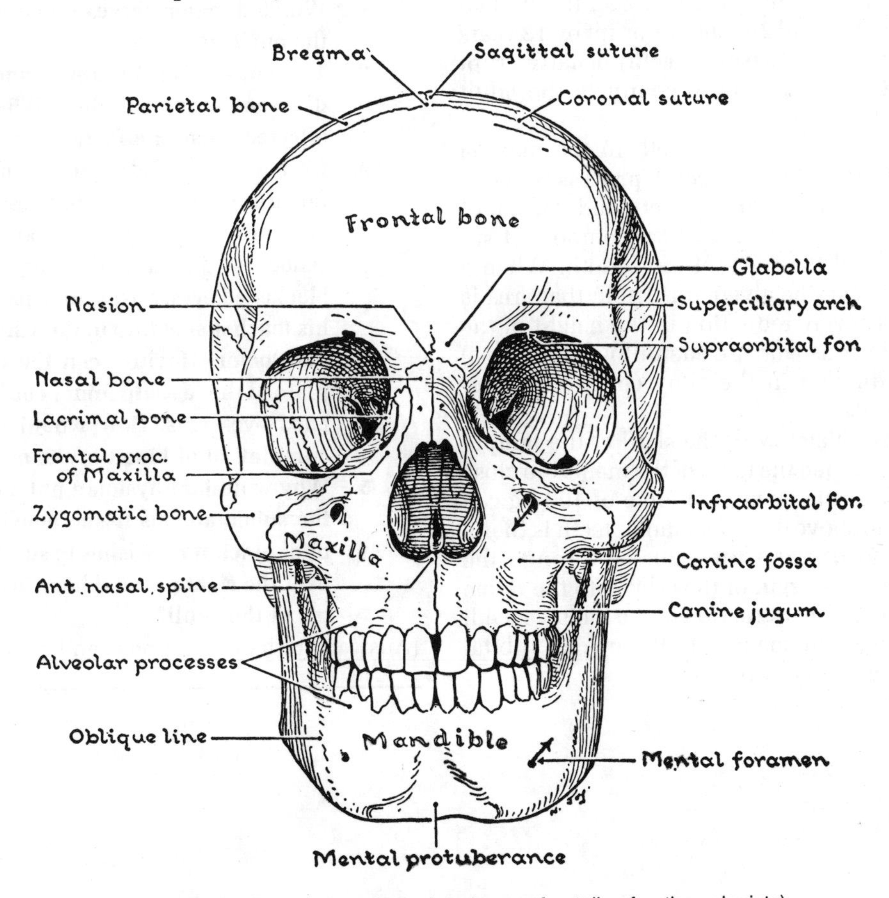

Figure 38.17. The skull on front view (norma frontalis of anthropologists)

Teeth

There are 32 teeth in the fully dentulous adult. Sixteen teeth are embedded in the bone of the maxilla, and 16 are embedded in the mandible. The teeth are further grouped into upper and lower quadrants of 8 teeth. Each quadrant from the midline to posterior aspect of the dental arch contains 2 **incisors,** 1 **cuspid,** 2 **premolars** and 3 **molars**. The incisors and cuspids are for cutting and tearing while the premolars and molars are for crushing food upon the **occlusal surfaces** of the premolar and molar teeth in the opposing arches.

Children under six years of age have a **deciduous** (L., to fall off) dentition of 20 teeth. Each quadrant contains 2 incisors, 1 cuspid and 2 **primary molars**. This first dentition erupts from the bone of the maxillary and mandibular arches between 6 months and 2 years of age with the incisors preceding the cuspids and primary molars in appearance. These teeth are then "shed" or exfoliated between 6 and 12 years of age to be replaced by the permanent adult dentition. The incisors and first permanent molars erupt around 6 years of age, the permanent cuspids, premolars and 2nd molars erupt by 12 years of age and the 3rd molars **(wisdom teeth)** usually erupt between 17 and 21 years of life to complete the adult dentition of 32 teeth.

The bony ridges supporting the teeth in the maxilla and mandible are termed the **alveolar processes**. They are produced by the eruption of the teeth and consist of mainly spongy bone with a thin cortical component superficially and around the roots of each tooth. When a permanent tooth is lost, the alveolar ridge in that area is also lost, causing a severe reduction in the height of the alveolar process. Persons who are edentulous (toothless) have a marked reduction in the bony content of their maxilla and mandible.

The mandible articulates with the skull at the **temporomandibular joint**. When the teeth of the mandible close on the upper teeth in the maxilla, a force of 100 kg can be exerted. The bone above the upper molar teeth is therefore strengthened to transmit these forces through the zygoma and the lateral margin of the orbit into the dome of the cranium. Stress from the action of the incisive teeth are dispersed through the maxilla to the medial orbital margin and into the frontal process.

The "point of the jaw" is formed by a triangular area of raised bone, the **mental protuberance**. From its lateral angle, an **oblique line** passes posteriorly and superiorly over the **body of the mandible** to reach the **ramus of the mandible**. The body and the ramus are oriented to form an obtuse **angle of the mandible**. The body is oriented in the horizontal plane, while the ramus is oriented in a vertical plane.

Clinical Mini-Problems

1. **The buccinator has two nerves entering its external surface. They are the buccal branch of the facial (VII) and the long buccal nerve of V^3. Only one of these is motor. (a) Damage to which nerve (VII or V^3) causes paralysis of the buccinator? (b) Which branchial arch mesoderm gives rise to the buccinator muscle?**
2. **Examination of the external ear (auricle) can reveal the integrity of the brainstem and spinal cord from the level of the C2 vertebra to the midpons region of the brainstem within the posterior cranial fossa. Which 3 major nerves innervate the skin covering the auricle?**
3. **The muscles of the upper and lower eyelid have different nerve supplies. Which eyelid would be affected most in a facial nerve (VII) palsy?**
4. **On a Sunday afternoon, a young boy had a severe fall from his bicycle that resulted in a scalp laceration over the parietal bone. When he went to school on Tuesday morning, he had two severely blackened eyes. Suspecting that he had been abused, his teacher sent him to the school nurse who called his doctor. Having seen the child earlier, he explained that a scalp injury could produce the blackened eye signs. How would you explain this accumulation of blood in the orbital region?**
5. **Why would a physician put a local anesthetic into the subcutaneous tissue above the right supra-orbital notch if he wishes to suture a scalp laceration over the right parietal bone just anterior to the vertex of the skull?**

(Answers to these questions can be found on p. 587.)

39

Interior of Cranium

Clinical Case 39.1

Patient Andrew W. This young driver suffered a closed head injury following a collision with another car. He arrives in the emergency room in a semicomatose state. An x-ray reveals a skull fracture in the temporal area and an underlying epidural hematoma. The neurosurgeon, whom you assist, removes a core of cortical bone from the calvaria overlying the hematoma and flushes the extravasated blood from the epidural site. He also ligates a lacerated middle meningeal artery to prevent further bleeding. The next day, the patient is recovering normal consciousness and is expected to be discharged in a few days.

Skull Cap or Calvaria

The **calvaria** is made up of superior portions of the frontal, parietal, and occipital bones. It forms the vault or "skullcap" of the cranium.

Structure. The bones of the roof of the cranial vault consist of an **outer** and **inner plate** or **lamina** of compact bone with an intervening layer of spongy bone called the **diploë**. The cortical bony plates give strength to the cranial vault, while the diploë serves to lighten the weight of the cranium and provide a site for blood marrow production. Blows to the head may cause fractures of either or both cortical plates. When the inner is fractured, it has a tendency to shatter and can lacerate the underlying vascular elements within the dura.

The bone in the cranial vault of a newborn consists of a single compact layer. This bone is eventually invaded by veins, which branch and rebranch, and unite with adjacent veins.

Bone marrow cells are "seeded" into the spaces around the venous channels to form the diploë. The four principal diploic veins are illustrated in *Figure 39.1*. These venous channels drain mostly into the dural venous channels within the cranial cavity. They lack valves and may communicate with veins that drain the scalp overlying the cranial vault. Scalp infections may spread via these veins to the diploë or the meninges of the brain that contain the intracranial venous sinuses.

The spongy bone of the diploic layer may also be "invaded" by mucous membranes from the nasal cavity and mastoid antrum to form **air sinuses**. The air sinuses are lined by respiratory epithelium and drain their secretions into the nasal cavity. They should not be confused with the venous sinuses of the cranium.

Diploë does not form in the cranial bones that are covered with thick, fleshy muscle, i.e., the squamous portion of the temporal bone and the nuchal part (base) of the occipital bone. The bone in these areas remains thin, translucent and cortical.

Meninges

Beneath the calvaria, three meningeal or membranous coverings envelop the brain and give it further protection within the cranial cavity. The three meninges—**the dura mater, arachnoid mater** and **pia mater**—are illustrated with their relationship to the calvaria and the brain in *Figure 39.2*.

The **dura mater (or dura)** consists of two closely adherent fibrous layers. The **outer layer of dura** is the periosteum for the inner plate of the calvaria and the cortical bone of the entire cranial cavity. This endosteal layer is also called the **endocranium**, and like other periostea, has a rich vascular supply. **Anterior, middle,** and **posterior meningeal arteries** supply the endocranium in the cranial cavity.

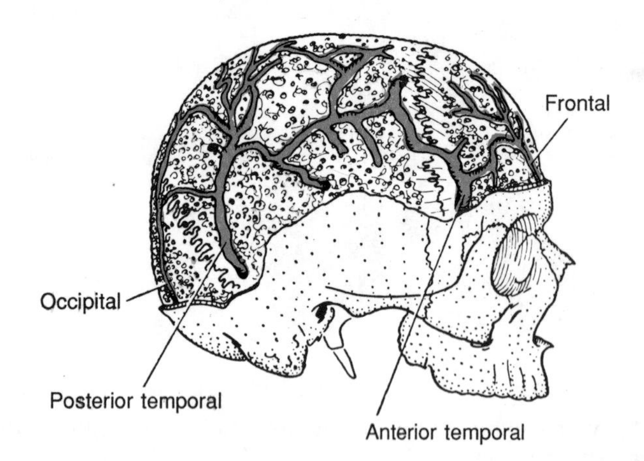

Figure 39.1. The 4 diploic veins.

The **inner layer of the dura** is a smooth serous membrane that opposes the arachnoid mater. It is also termed the **meningeal layer of dura**. While the periosteal dura is consistently attached to the bone of the cranial cavity, the meningeal dura maintains a close association to the underlying brain. When it approaches the **longitudinal fissure** of the brain, which separates the right and left cerebral hemispheres in the sagittal plane, the meningeal layer of the dura separates from the periosteal dura and approximates the meningeal dura from the opposite side to form the **falx cerebri** (*fig. 39.2*). The meningeal dura also forms partitions between the occipital lobes of the brain and the cerebellum to produce the **tentorium cerebelli** (*figs. 39.3 and 39.4*). These two major dural partitions subdivide the cranial cavity into compartments that support the brain and prevent gross shifting of the brain within the cranium.

When the periosteal dura and the meningeal dura become separated during the formation of these dural partitions, they create a cavity within the dura (*fig. 39.2*). These cavities become lined by endothelium and form the **venous sinuses** of the cranial cavity (*fig. 39.3*). Venous blood draining the brain, the meninges, the diploë and even some of the scalp will collect within these meningeal venous sinuses and eventually drain into the **internal jugular vein** at the base of the posterior cranial fossa (*fig. 39.3*).

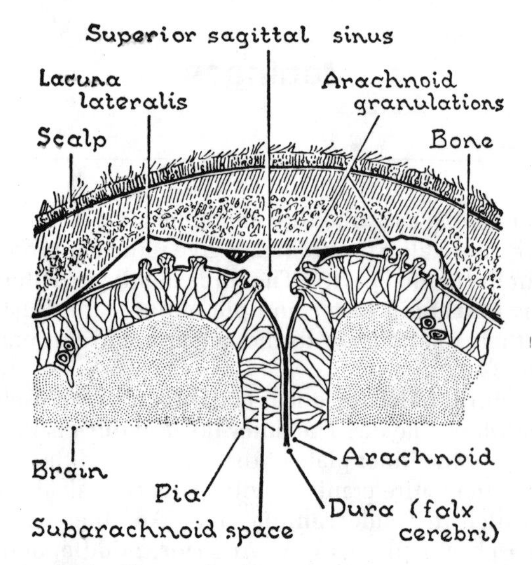

Figure 39.2. The arachnoid granulations return the cerebrospinal fluid.

Two **potential spaces** are related to the dural lining of the cranium. The **epidural space** is between the periosteal dura and the bone of the cranial cavity. If the meningeal arteries are damaged and bleed into this space, an **epidural hematoma** will occur between the bone and dura. Since this is arterial and a high pressure system, a considerable amount of blood may accumulate in this space following a head injury. The **subdural space** exists between the meningeal dura and arachnoid. Bleeding by the cerebral veins that traverse this space as they enter the **superior sagittal sinus** may expand this potential space. This is a low-pressure system on the venous side and much smaller amounts of blood would be present in this type of intracranial "bleed." Both situations, however, pose major problems for the patient and require different treatments.

The **arachnoid mater (or arachnoid)** is a delicate membrane that exists between the dura and the pia mater on the surface of the brain. The arachnoid is actually a split layer with intervening trabeculae of arachnoid tissue connecting the layers associated with the overlying dura and the underlying pia. This "cobwebbed-like" membrane is the basis for naming this layer arachnoid **(L. = spider)**.

Two essential features of the arachnoid are depicted in *Figure 39.2*. First, the **subarachnoid space** is between the arachnoid and pia. This space contains the **cerebrospinal fluid** that surrounds the entire brain and virtually "floats" the brain in the cranial cavity. The cerebrospinal fluid serves to protect as well as nourish the brain. Also within the subarachnoid space are the major cerebral arteries and veins that supply and drain the blood supply to the brain. Bleeding into the subarachnoid space is a common sign following a **cerebrovascular hemorrhage** or **"stroke."**

The second essential feature is that the subarachnoid space is extended into the **superior sagittal sinus** by the **arachnoid granulations** (*fig. 39.2*). These arachnoid granulations allow the cerebrospinal fluid to be drained into the venous system. Cerebrospinal fluid is continually produced within the brain by the arterial **choroid plexuses** and must therefore be removed by the venous system to maintain a balance in the volume of cerebrospinal fluid that exists in the subarachnoid space. The arachnoid granulations may be extensive and cause the periosteal dura to cavitate the inner surface of the parietal bones. They may be calcified to some degree in older adults and become visible in radiographs of the head.

The **pia mater (or pia)** is closely adherent to the outer surface of the brain. It follows all of the contours of the brain and actually adheres to the cerebral vessels as they

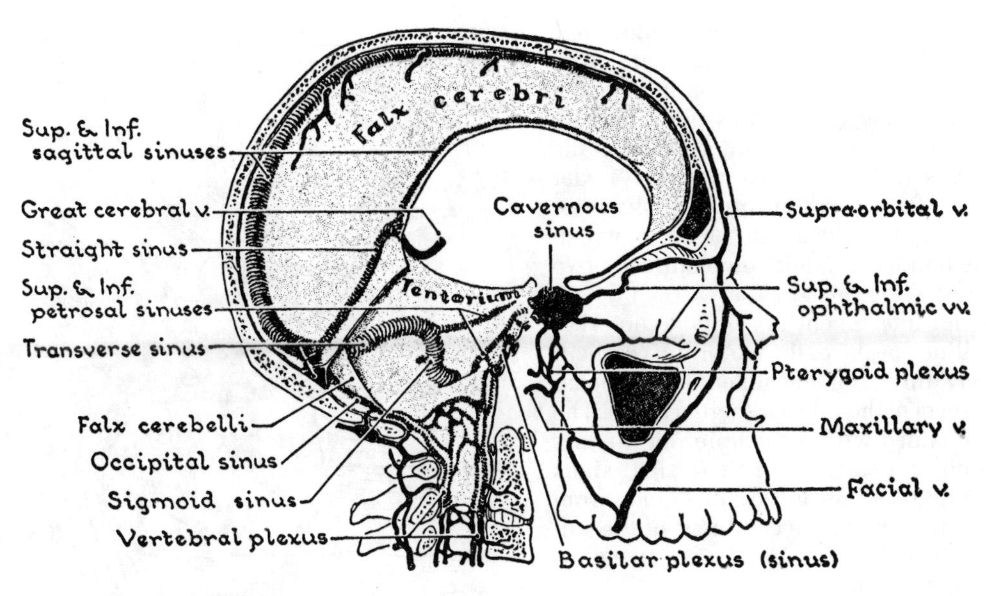

Figure 39.3. The folds of the dura mater. Venous sinuses. Vertebral venous plexus.

penetrate the brain tissue. One cannot separate the pia from the brain in a gross manner. It can be readily seen on the surface of the brain, but it is best appreciated in histologic preparations as a single cell layer external to the brain tissue.

DURAL FOLDS

There are 4 major double-layered meningeal dural folds in the cranial cavity. They are the "sickled-shaped" **falx cerebri** and the **falx cerebelli** in the midline sagittal plane and the horizontally oriented **tentorium cerebelli** and the **diaphragma sellae**, which forms a roof above the cerebellum and the pituitary gland, respectively. The margins of these dural folds contain some of the major venous sinuses of the cranial cavity.

The **falx cerebri** is suspended from the inner surface of the skull and projects inferiorly between the two cerebral hemispheres (*figs. 39.2 and 39.3*). It is attached anteriorly within the anterior cranial fossa to the **crista**

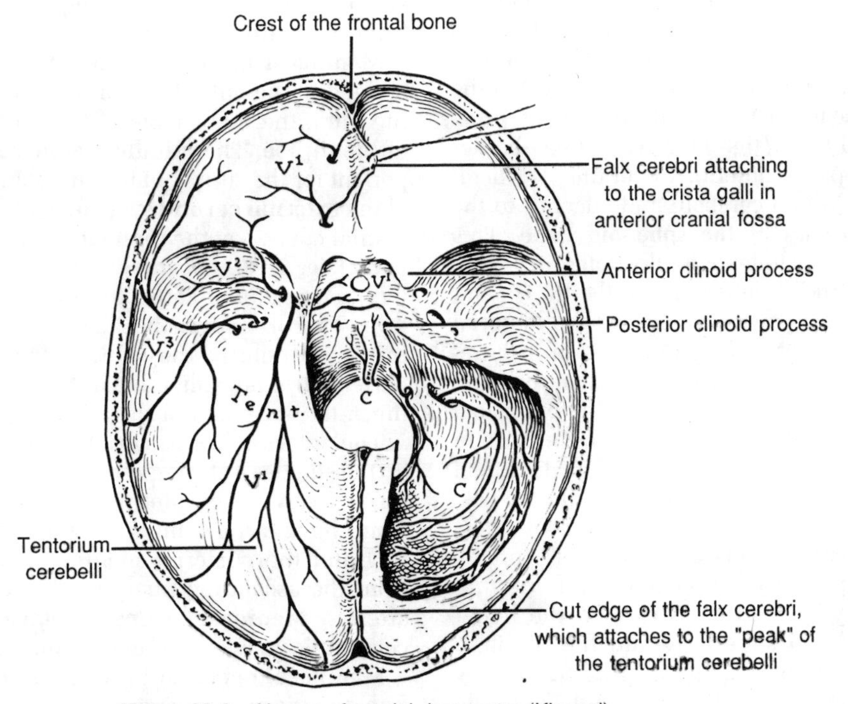

Figure 39.4. Nerves of cranial dura mater (Kimmel).

galli of the ethmoid bone and **crest of the frontal bone** (*fig. 39.4; see also fig. 39.15*). Its convexed upper border attaches to the lips of the sagittal sulcus on the frontal, parietal, and occipital bones as it passes posteriorly in the cranium. The attached superior border contains the **superior sagittal sinus** while the "free" inferior border contains the smaller **inferior sagittal sinus**. This inferior margin overlies the corpus callosum of the brain, which is a large transverse bundle of axons connecting the two cerebral hemispheres.

The broad posterior attachment of the falx cerebri is mainly into the midline "peak" of the tentorium cerebelli (*fig. 39.4*). Anteriorly, this basal attachment is associated with the terminal aspect of the inferior sagittal sinus. The **straight sinus** is contained within the union of the falx cerebri and the tentorium cerebelli. This straight sinus drains the inferior sagittal sinus and venous blood from the underlying brain tissue into the **confluens of the sinuses** on the inner aspect of the occipital bone at the **internal occipital protuberance**.

The **falx cerebelli** is a slight fold attached inferior to the tentorium cerebelli on the **crest** of the occipital bone. An **occipital sinus** is found in its posterior border as it attaches to the occipital bone.

The **tentorium cerebelli** attaches medially to the base of the falx cerebri and the falx cerebelli and forms a roof over the posterior one-half of the posterior cranial fossa. The anterior one-half of the posterior cranial fossa is opened by a "notch" in the tentorium cerebelli, which allows the brainstem to connect to the forebrain in the middle cranial fossa (*figs. 39.4 and 39.5*). The peripheral attachments of the tentorium cerebelli are well shown in *Figure 39.4*. From the internal occipital protuberance, the tentorium cerebelli attaches to the occipital bone and the temporal bone and encloses the **transverse sinus**. At the junction of the posterior and middle cranial fossae, the tentorial attachment continues in an anteromedial direction on the **crest** of the **petrous temporal bone** to reach the **posterior clinoid processes of the sphenoid bone** (*fig. 39.4*). The "free" border of the tentorial notch passes anteriorly over the peripheral attachment of the tentorium cerebelli and attaches to the **anterior clinoid processes of the sphenoid bone**. This "overlap" of tentorial attachments is illustrated in *Figure 39.5* and has an important relationship to the III, IV, and V cranial nerves and the **cavernous sinus**. The tentorial notch is also a site for "herniation" of the brain in intracranial disorders. Since the tentorium separates the cerebral cortices and forebrain superiorly from the brainstem and cerebellum in the posterior cranial fossa, pressure in the "supratentorial space" can force parts of the temporal lobes of the brain through the tentorial notch into the posterior cranial fossa.

The **diaphragma sellae** forms a "tentorium" for the hypophysis cerebri (pituitary). It has a central aperture to allow the pituitary stalk to connect with the hypothalamus of the forebrain superiorly and the pituitary gland inferiorly. The pituitary gland is situated in bony depression within the body of the sphenoid bone, the **sella turcica** (Turkish Saddle). Because the diaphragma sellae separates the pituitary from the base of the forebrain, care must be taken in elevating the brain from the middle cranial fossa. Disconnecting the pituitary gland from the hypothalamus constitutes a **hypophysectomy**. Surgical procedures on the pituitary gland are usually done from the nasal cavity by removal of the body of the sphenoid bone due to this unique separation of the gland from the brain by the diaphragma sellae (*fig. 39.5*).

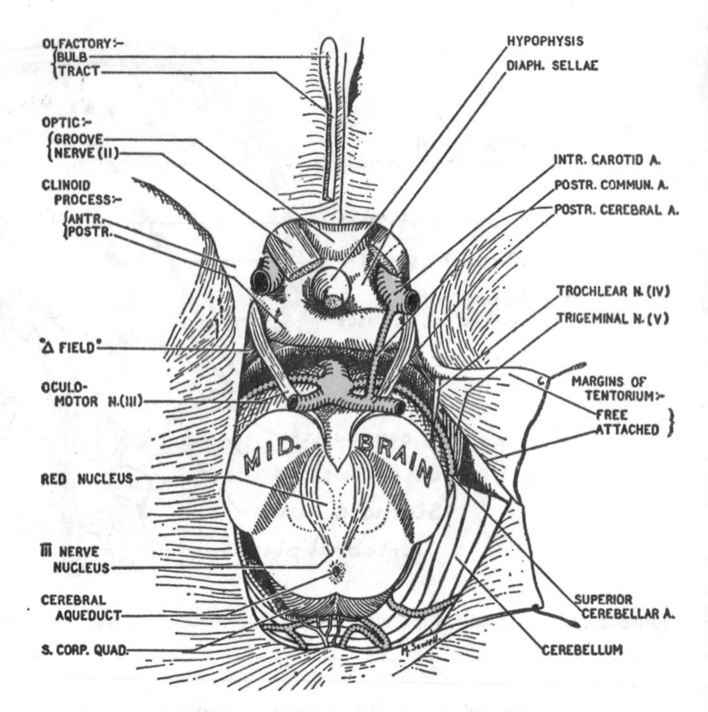

Figure 39.5. A stage in the removal of the brain.

Meningeal arteries, being periosteal arteries, lie embedded in the outer layer of the dura mater. They supply the dura, the inner table of the skull, and bone marrow of the diploë. The **middle meningeal artery** is most important for the supply of the supratentorial dura. A branch of the **external carotid system**, it gains access to the inner cranial cavity via the foramen spinosum in the sphenoid bone (*fig. 39.6*). Laceration of the middle meningeal artery is a common cause for **epidural hematoma** formation. The artery can be ligated to control bleeding by drilling a hole in the skull at the **pterion** (junction of parietal, frontal, sphenoid and temporal bones) to expose the anterior branch of the artery (*fig. 39.6*).

The anterior and posterior cranial fossae are supplied by the smaller **anterior** and **posterior meningeal arteries**, respectively. The anterior meningeal artery is derived from the anterior ethmoid branch of the ophthalmic artery and the posterior meningeal arteries are branches from the ascending pharyngeal, occipital and vertebral arteries. **Meningeal veins** accompany the arteries and communicate with the venous sinus and the diploic veins.

Meningeal nerves supply the cranial dura by branches from the trigeminal nerve (V^1, V^2, V^3) and cervical nerve(s) C2(3) in a pattern comparable to the overlying skin area

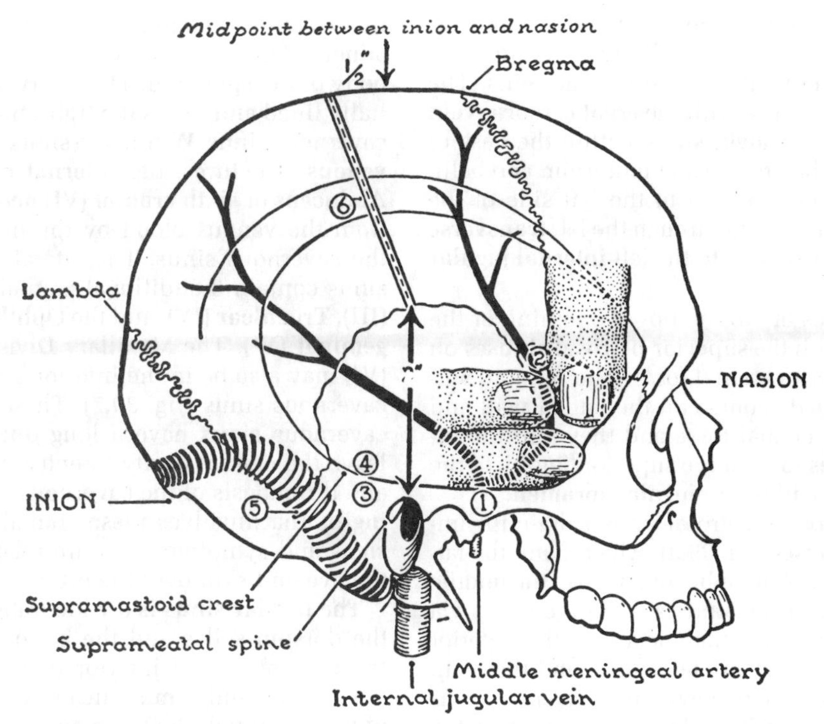

Figure 39.6. Surface anatomy of the skull. *1*, the stem of the *middle meningeal artery*, passing through the foramen spinosum, deep to the head of the mandible, which is readily located by palpation on opening and closing the mouth; *2*, the *anterior branch* of the middle meningeal artery crossing the pterion. The **pterion** is the point where 4 bones meet (parietal, frontal, greater wing of sphenoid and temporal squama). To locate it on the skin, place the thumb behind the frontal process of the zygomatic bone and 2 fingers above the zygomatic arch, and mark the angle so formed (Stiles). This great landmark overlies anterior branch of the middle meningeal art, and the lateral cerebral sulcus. *3*, the *Suprameatal Triangle* lies below the supramastoid crest and behind the suprameatal spine; a hole drilled here enters the *mastoid antrum*; *4*, a hole drilled above the supramastoid crest enters the *middle cranial fossa*. *5*, the "*lateral sinus*," i.e., transverse and sigmoid sinuses, passing from the inion to a point 2 cm or less behind the external acoustic meatus to become the internal jugular vein deep to the anterior border of the mastoid process. *6*, the *central sulcus* of the cerebrum running from a point 1 cm (½") behind the mid inion-nasion point to a point 5 cm above the external acoustic meatus. *Application* of key in *Figure 39.16*. The *crescent of foramina* on surface projection runs from the head of the mandible to the pterion.

(*figs. 39.4 and 38.13*). The dura is sensitive to pain and may be the source for "headache" when tensed by vasodilatation of the meningeal vessels, intracranial inflammation and swelling, or extensive shifts in brain tissue. The arachnoid, pia and brain tissue are insensitive to pain. Neurosurgery is frequently done under local anesthesia to relieve pain in the scalp, skull and dura only. The patient can be awake and feel no discomfort from the surgical procedures within the brain tissues that underlie the dura.

Clinical Case 39.2

Patient John R. This 20-year-old farmer's son, home from college for the summer, began feeling severely dizzy and developed headaches during the day while working on a tractor. He told his father he thought he was going blind. The headaches became so severe that he was admitted to the neurology clinic of the University Medical Center. Radiology quickly revealed the presence of an abnormal mass in the hypophyseal (pituitary) fossa. The neurosurgeon called in on the case plans to explore and remove a pituitary tumor believed by all to be the diagnosis. As your clinical tutor, he quizzes you on the anatomy of the middle cranial fossa, relationship of all the nerves and venous sinuses surrounding the hypophysis (pituitary gland), and the possible surgical approaches to the area through the nasal cavity, orbit, forehead, or temporal region.

VENOUS SINUSES OF THE DURA MATER (*figs. 39.3 and 39.7*).

The **superior sagittal sinus** within the superior aspect of the falx cerebri usually joins the right **transverse sinus** within the right peripheral margin of the tentorium cerebelli. Blood from these sinuses will drain via the right **sigmoid sinus** into the right internal jugular veins. The **inferior sagittal sinus** unites with the **great cerebral vein** (of **Galen**) to form the **straight sinus** within the base of the falx cerebri and the crest of the tentorium cerebelli. The straight sinus tends to flow to the left side of the **confluens of sinuses** and drain through the left **transverse sinus** and **left sigmoid** sinus into the left **internal jugular vein**.

The **sigmoid sinuses** are formed by the joining of the **transverse sinuses** with the **superior petrosal sinuses** on each side of the posterior cranial fossa. The sigmoid sinuses have an "S-shaped" course on the lateral wall and floor of the posterior cranial fossa and they join the **inferior petrosal sinuses** to form the superior "bulb" of the internal jugular vein within the **jugular foramen**.

The inferior and superior petrosal sinuses drain the important **cavernous sinuses** (*fig. 39.3*), which flank the lateral borders of the body of the sphenoid bone in the middle cranial fossa. The cavernous sinuses drain the orbits via the superior and inferior ophthalmic veins, the anterior cranial fossa via the sphenoparietal sinuses and the brain via the middle cerebral vein (most other cerebral veins drain into the superior sagittal sinus in the roof of the cranial cavity). The right and left cavernous sinuses are also joined by an **intercavernous sinus** that extends posteriorly and inferiorly around the pituitary and body of the sphenoid bone.

Figure 39.7 shows how the meningeal dura forms the lateral wall of the cavernous sinus as it is drawn laterally in the middle cranial fossa. The periosteal dura layer adheres faithfully to the bone on the lateral side of the body of the sphenoid. The intervening space is endothelially lined and forms the trabeculated (honeycomb-like) cavernous sinus. Within the sinus are two important **non-venous** structures; the **internal carotid artery** and the **Abducens** or **sixth cranial (VI) nerve**. They are separated from the venous blood by the internal endothelium of the cavernous sinus. The lateral wall of the cavernous sinus contain 3 additional cranial nerves: **Occulomotor (III), Trochlear (IV)**, and the **Ophthalmic Division of Trigeminal (V^1)**. The Maxillary Division of the Trigeminal (V^2) may also be in the inferior border of the wall of the cavernous sinus (*fig. 39.7*). These structures within the cavernous sinus have a long dural course before they leave the cranial cavity to enter the orbit. Inflammation and thrombosis of the cavernous sinus can cause a **meningitis** that involves these cranial nerves and produces signs and symptoms that are related to the function of the eye and skin over the orbit.

The **basilar sinus** is a wide trabeculated space behind the dorsum sellae and the basioccipital bone. It unites the cavernous and inferior petrosal sinuses of the opposite sides and communicates with the vertebral plexus of veins through the foramen magnum.

The dural sinuses, like most venous channels above the heart (including the vertebral plexus of veins), are drained by gravity and contain **no valves**. Therefore, infections in the face, scalp, skull, vertebral column, meninges or brain can spread to the cranial cavity and cause meningitis. Since the cavernous, inferior petrosal and

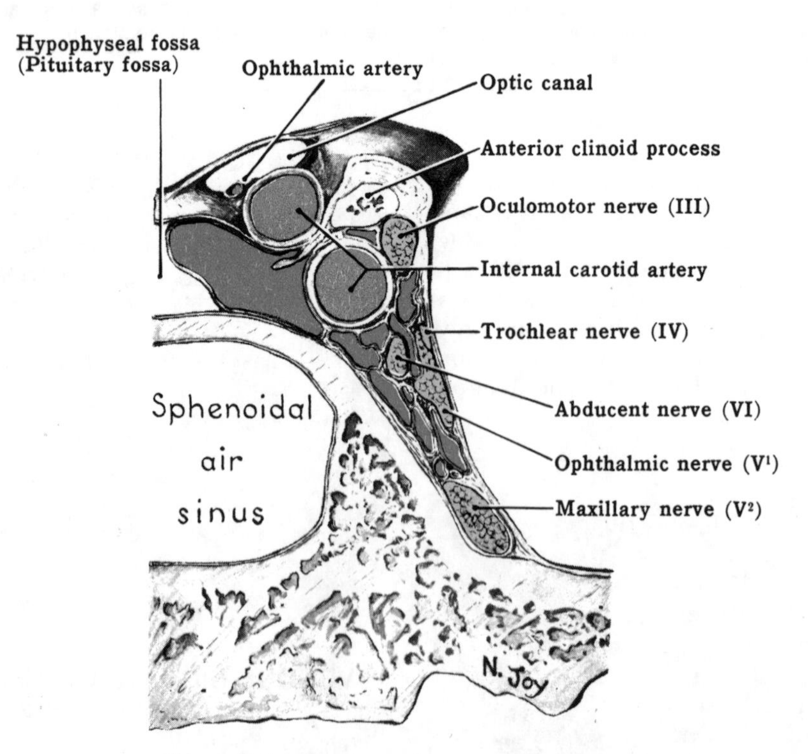

Figure 39.7. Cavernous sinus, coronal section.

transverse sinus intercommunicate, these infections can readily become bilateral and affect both the right and left side of the intracranial cavity and associated cranial nerves.

Cranial Nerves

There are 12 cranial nerves in man. They arise and are "numbered" sequentially from I–XII (1–12) in the anterior, middle, and posterior cranial fossa (*figs. 39.8 and 39.9*).

The **olfactory nerve** (I) arises in the nasal cavities and enters the anterior cranial fossae through the cribriform plate of the ethmoid bone (*see fig. 39.14*). The **olfactory bulb** and **olfactory tract** convey the sensory pathways for smell through the floor of the anterior cranial fossa to the brain (*figs. 39.5 and 39.8*).

The **optic nerve** (II) leaves the orbit and enters the middle cranial fossa through the **optic canal** within the lesser wings of the sphenoid bone (*figs. 39.5 and 39.8*). The portions of the optic nerves from the nasal one-half of the retina cross the midline in the **optic (chiasmic) groove** on the sphenoid bone to form the **optic chiasm**. Posterior to the optic chiasm, the visual fibers project to the brain as the **optic tracts**. Due to the optic chiasm, the **right optic tract** will contain fibers from the nasal retina of the left eye and the temporal retina of the right eye. The right side of the brain (the right occipital lobe) will therefore receive visual stimuli from the left visual field of the patient.

The **occulomotor nerve** (III) is the highest cranial nerve to arise from the brainstem at the midbrain level. It passes anteriorly in the subarachnoid space of the posterior cranial fossa between the posterior cerebral and superior cerebellar arteries to enter the dura between the attachments of the tentorium cerebelli to the posterior and anterior clinoid processes (*fig. 39.5*). It has an extensive course through the superior aspect of the lateral wall of the cavernous sinus to enter the orbit via the **superior orbital fissure** of the sphenoid bone (*figs. 39.7, 39.10, and 39.11*). The cranial nerve III provides voluntary motor fibers to four extrinsic skeletal muscles of the eye: medial rectus, superior rectus, inferior rectus, and inferior oblique muscles; and a skeletal muscle of the eyelid, **the levator palpebrae superioris** muscle. It also supplies the preganglionic **parasympathetic** neurons to the pupillary constrictor muscle and ciliary body muscle within the eye. The latter two muscles are smooth muscles and require an innervation by autonomic neurons.

The **trochlear nerve** (IV) is also motor to an extrinsic eye muscle; the **superior oblique muscle**. The trochlear nerve leaves the posterior aspect of the midbrain to traverse the subarachnoid space of the posterior cranial fossa and enter the free edge of the tentorial dura in the same triangle as the occulomotor nerve (III) (*figs. 39.5 and 39.9*). Cranial nerve IV lies in the lateral wall of the cavernous sinus as it passes forward to enter the superior orbital fissure (*figs. 39.7 and 39.10*).

The **trigeminal nerve** (V) is the great sensory nerve to the skin of the face as well as a motor nerve to the muscles that move the mandible (**muscles of mastication**). It arises from the pons region of the brainstem and crosses the subarachnoid space to enter the dura at the most medial aspect of the superior border of the petrous bone (*figs. 39.8 and 39.9*). The trigeminal ganglion lies within the dura of the middle cranial fossa lateral to the cavernous sinus (*fig. 39.11*). The ganglion is a swelling on V formed by the presence of the sensory pseudo-unipolar neuron cell bodies. No synapses occur in this or any other peripheral sensory ganglion. It is simply a site for the cell bodies of the peripheral sensory nerves of V. The 3 branches from the ganglion are the **ophthalmic (V^1), maxillary (V^2)** and **mandibular (V^3)** divisions. V^1 exits the middle cranial fossa via the **superior orbital fissure** with cranial nerves III, IV, and VI. The maxillary division (V^2) exits the middle cranial fossa through the **foramen rotundum** of the sphenoid bone, while **V^3** exits through the **foramen ovale** of the sphenoid bone (*fig. 39.11*).

The **abducens nerve (VI)** innervates the lateral rectus (abductor) muscle of the eye. It has the longest intracranial course and is frequently a sensitive monitor to changes in intracranial pressure and disease. **VI** arises from the pons-medullary junction of the brainstem and passes forward in the subarachnoid space to enter the dura of the posterior cranial fossa that covers the posterior wall of the cavernous sinus (*figs. 39.8 and 39.9*). The 6th cranial nerve then enters the cavernous sinus to lie lateral to the internal carotid artery (*figs. 39.7 and 39.10*) as it proceeds anteriorly to the superior orbital fissure. The unique positioning of VI *within* the cavernous sinus forms the basis for observing difficulties in eye abduction as an initial sign in cavernous sinus thrombosis.

The **facial nerve (VII)** and its accompanying component, the **nervus intermedius,** leave the brainstem at the pons-medullary junction and pass through the subarachnoid space to enter the dura at the **internal acoustic (auditory) meatus** (*figs. 39.8 and 39.9*). Cranial nerve VII passes through the internal auditory meatus of the petrous temporal bone to reach the middle ear. An enlarged **geniculate ganglion** for taste is found on the facial nerve as this point and marks the bifurcation point for the peripheral distribution of the nerve (*fig. 39.11*). The **greater (superficial) petrosal branch** passes anteromedially to enter the floor of the middle cranial fossa, courses though the pterygoid canal, and eventually reaches the pterygopalatine fossa behind the nasal cavity. The greater (superficial) petrosal nerve is part of the nervus intermedius that will carry taste sensations from the palate and secretomotor (parasympathetic) innervation to the glands of the palate, nose, and orbit.

The remaining taste and secretomotor fibers of the

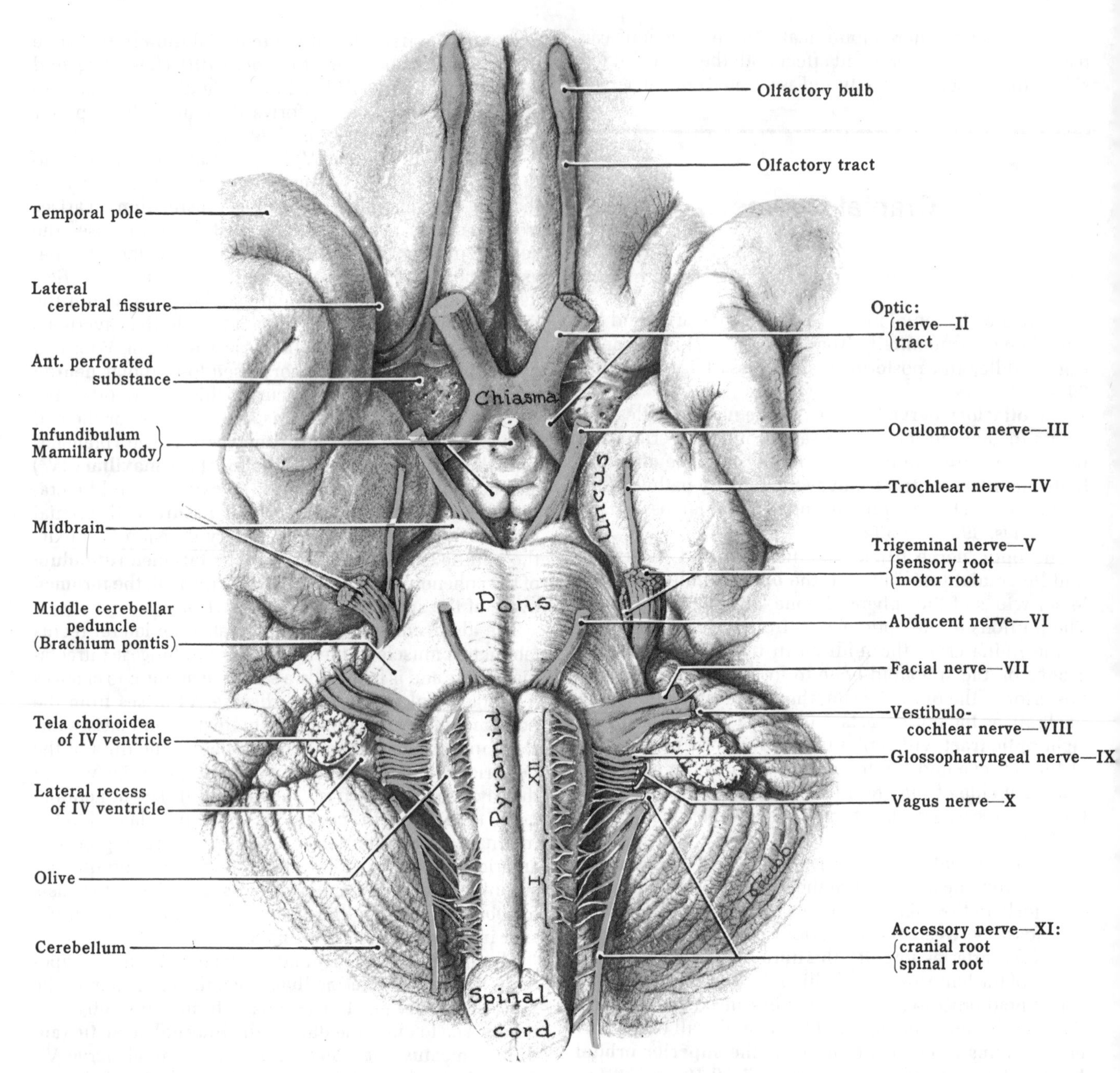

Figure 39.8. Base of the brain: The superficial origins of the cranial nerves.

Note:

1. The olfactory bulb, in which the olfactory (cranial I) nerves (not shown) end.
2. The superficial origin of the trochlear (cranial IV) nerve is from the dorsal aspect of the brainstem and cannot be seen in this figure.
3. The slender nervus intermedius, or so-called sensory root of the facial nerve (not labeled) between the facial (VII) and vestibulo-cochlear (VIII) nerves.
4. The fila of the hypoglossal (XII) nerve, arising between the pyramid and the olive, and in line with the ventral root of the 1st cervical nerve.

(From Anderson, J.E.: *Grant's Atlas*, 8th edition, Williams & Wilkins, Baltimore, 1983.)

nervus intermedius form the **chorda tympani** branch of VII. They leave the middle ear through the **petrotympanic fissure** and go to the floor of the mouth. Cranial nerve VII then exits the middle ear cavity via the stylomastoid foramen in the base of the skull just posterior to the **external acoustic (auditory) meatus** (external ear canal). The facial nerve is motor to the facial muscles that surround the cranial orifices (ears, mouth, nose, and eyes) as well as the stapedius, posterior digastric, and stylohyoid muscles. The trunk of VII continues anteriorly to

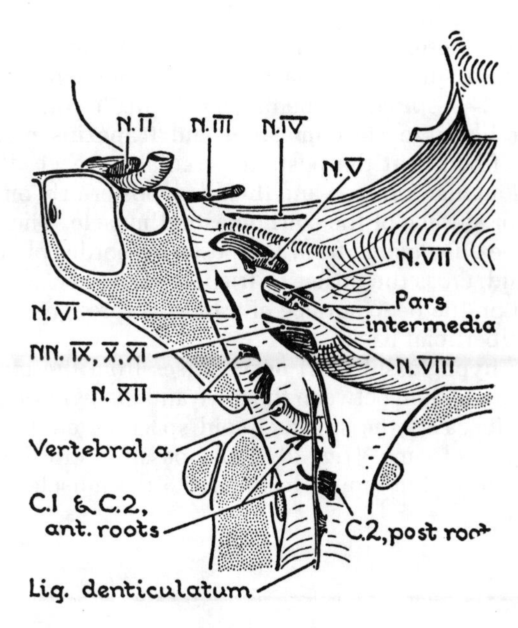

Figure 39.9. Cranial nerves, piercing dura mater.

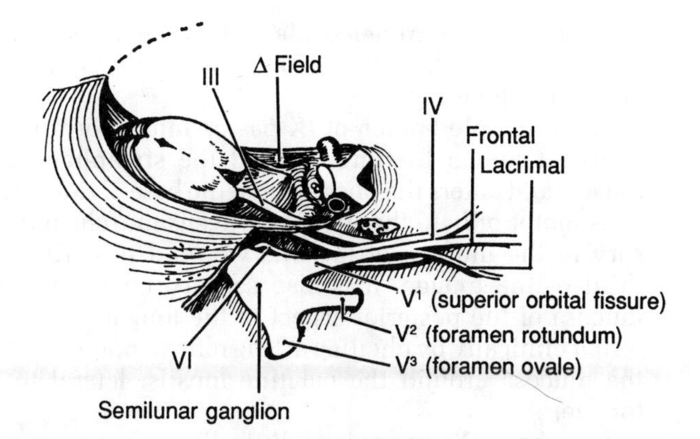

Figure 39.11. The 3 divisions of the trigeminal nerve (N. V.). *(Ganglion: semilunar = trigeminal.)*

enter the substance of the parotid gland. A plexus of nerves emerges from the parotid as the terminal branches of VII to the **muscles of facial expression**. Facial paralysis is the most common feature of disorders in VII, but additional signs and symptoms related to the nervus (pars) intermedius components can be seen in lesions of VII that occur within the temporal bone or the posterior cranial fossa.

The 8th cranial nerve (VIII) can be named the **auditory** or **vestibulocochlear nerve**. It has the same origin and pathway as the VII from the brainstem, through the subarachnoid space, and into the internal acoustic meatus (*fig. 39.9*). Cranial nerve VIII innervates the organs for hearing (the **cochlea**) and balance (the **semicircular canals, utricle** and **saccule**) within the petrous portion of the temporal bone.

The **glossopharyngeal nerve** (IX) is a mixed motor and sensory nerve to the tongue and pharynx. It arises with the Vagus (X) and cranial component of XI from the medulla of the brainstem and exits the skull via the jugular foramen (*fig. 39.9*). As IX passes inferiorly through the jugular foramen, it also sends sensory and secretomotor fibers through the lateral wall of the jugular foramen to enter the middle ear. The secretomotor (parasympathetic) fibers exit from the middle ear cavity as the *lesser petrosal* to pass through the middle cranial fossa and leave the

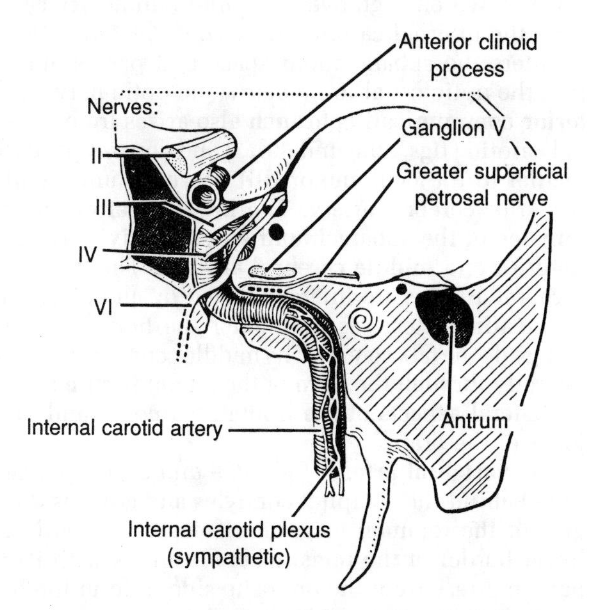

Figure 39.10. The course and relations of the intrapetrous and intracranial parts of the internal carotid artery.

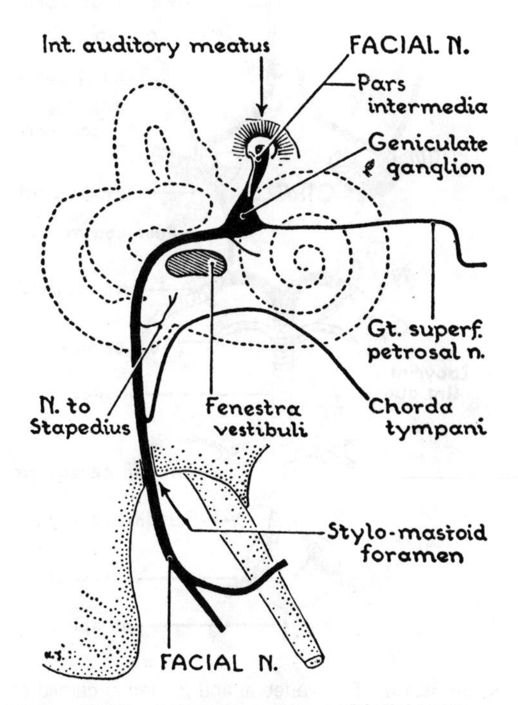

Figure 39.12. Intrapetrous course of facial nerve.

cranium via the **foramen ovale** (with V^3) to innervate the parotid gland. The sensory fibers innervate the mucosa of the middle ear.

The principle branch of IX passes inferiorly from the jugular foramen to spiral around the **stylopharyngeus muscle** and enters the pharynx and the base of the tongue. IX is motor *only* to the stylopharyngeus muscle and sensory to the mucosa of the pharynx and posterior one-third of the tongue. It is also sensory for taste in the mucosa of the posterior aspect of the tongue. **IX** can be tested clinically by eliciting a "gag reflex" upon touching the mucosa around the palatine tonsils, lateral to the tongue.

The **vagus (X) nerve** is a clinically important cranial nerve that supplies the head, neck, thorax, and abdomen. It contains a large portion of the parasympathetic fibers within the body. X is also motor to skeletal muscles in the head and neck and contain sensory neurons to skin in the external ear and mucous membranes of the gastrointestinal and respiratory systems. **X** arises from the medulla and exits the skull through the jugular foramen (*figs. 39.8 and 39.9*). Like IX, it sends fibers into the ear via the canals in the lateral wall of the jugular foramen. These fibers are sensory to the epithelium lining the external auditory meatus and the tympanic membrane. Below the jugular foramen, the vagus descends within the carotid sheath through the neck. It gives motor and sensory branches to the palate, pharynx, and larynx as it descends. It can be readily assessed in the head and neck by testing for pain in the ear, movement of the palate, and production of normal voice in the larynx.

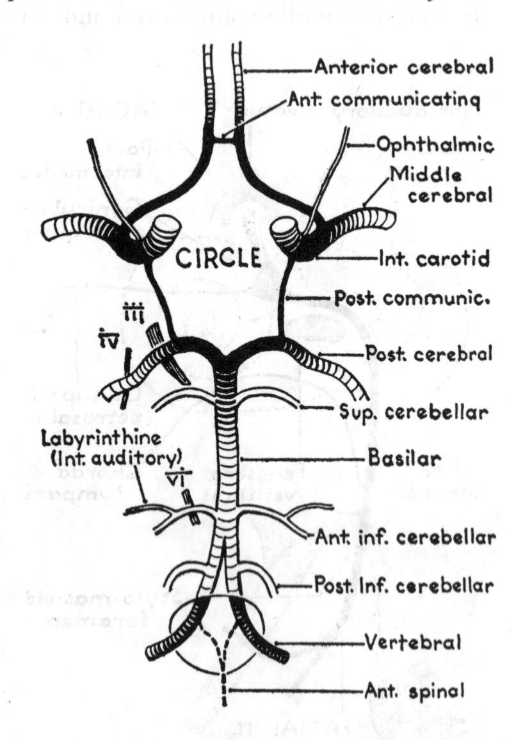

Figure 39.13. Two vertebral and 2 internal carotid arteries supply the brain and form an arterial circle.

The **accessory (XI) nerve** arises with IX and X as well as from a spinal root that enters the foramen magnum from C2–4 spinal segments (*fig. 39.8*). It contains the motor fibers to **sternomastoid** and **trapezius muscles**. These fibers exit the posterior cranial fossa via the internal jugular foramen and then pass posteriorly onto the inferior surface of the **sternomastoid muscle**. The fibers to trapezius emerge from the posterior border of sternomastoid, cross the posterior triangle of the neck, and descend on the deep surface of trapezius through the neck and superficial back.

The **hypoglossal (XII) nerve** arises from the medulla of the brainstem between the olive and the pyramids (*fig. 39.8*). It leaves the subarachnoid space by entering the hypoglossal canal (*fig. 39.9*) and passing into the neck and floor of the mouth to innervate the muscles of the tongue.

Cerebral Arteries in Relation to Cranial Nerves

The **internal carotid** and **vertebral** arteries arise bilaterally in the neck and ascend to the cranial cavity to supply the brain.

The **internal carotid artery** (*fig. 39.10*) enters the skull through the **carotid canal** and lies within the cavernous sinus. It forms a "hairpin loop" within the cavernous sinus (the "carotid siphon") (*figs. 39.7 and 39.13*) and then pierces the dura mater medial to the anterior clinoid process, which it grooves. The **ophthalmic artery** arises from the internal carotid artery (*figs. 39.7 and 39.13*) as it enters the subarachnoid space and passes anteriorly into the optic canal to lie below the optic nerve. A **posterior communicating branch** also arises from the internal carotid (*figs. 39.5 and 39.13*) and courses posteriorly, medial to the oculomotor (III) nerve, to unite with the posterior cerebral artery. The internal carotid then terminates in the subarachnoid space by dividing into the **anterior** and **middle cerebral arteries**. The two anterior cerebral arteries are interconnected by an **anterior communicating** artery, as they enter the horizontal fissure alongside the falx cerebri. The middle cerebral artery passes laterally through the stem of the lateral fissure to supply the lateral aspects of the frontal, temporal, and parietal lobes of the brain.

The **vertebral artery** (*figs. 39.9 and 39.13*) pierces the dura behind the occipital condyles and grooves the margins of the foramen magnum. It passes forward on the lower border of the pons, where it unites with its companion artery from the opposite side to form the basilar artery (*fig. 39.13*). In its intracranial course, the vertebral artery supplies the medulla, cerebellum, and spinal cord.

The **posterior inferior cerebellar arteries** arise from each vertebral artery in the posterior cranial fossa and are important clinical branches that loop over the rootlets of the IX, X and XI cranial nerves.

The **basilar artery** ascends on the bony slope of the basioccipital bone beneath the pons and midbrain (*fig. 39.5*). It bifurcates at the midbrain to **right** and **left posterior cerebral arteries** that pass through the tentorial notch and supply the medial aspects of the occipital and temporal lobes of the brain. The posterior cerebral arteries are closely related to the III and IV cranial nerves (*fig. 39.13*) in the posterior cranial fossa. **Superior** and **anterior inferior cerebellar arteries** are major branches of the basilar artery as it courses over the undersurface of the pons.

The resultant anastomosis of the internal carotid and the posterior cerebral branches of the basilar artery is termed the "Circle of Willis" (*fig. 39.13*). It provides the brain with the potential for alternate blood flow should a major carotid or vertebral vessel(s) become occluded or constricted by disease.

Floor or Base of the Skull

THREE CRANIAL FOSSAE

Boundaries (*fig. 39.14*)

The interior of the skull has three terraces, or levels, called fossae—an **anterior**, a **middle** and a **posterior**.

The anterior cranial fossa is sharply marked off from the middle cranial fossa by three free concave crests: one median and two lateral. The median concaved crest connects the **anterior clinoid processes** to which the free border of the tentorium cerebelli attach. The processes are components of the sphenoid bone and their interconnecting median crest overlies the **optic canals** and their openings, the **optic foramina** (singular: **optic foramen**). Each lateral crest is formed by the **lesser wing of the sphenoid**. It passes laterally from the anterior clinoid process over the superior orbital fissure to the **pterion** and projects posteriorly above the tips of the temporal lobes into the lateral sulcus of the brain.

The middle cranial fossa is demarcated from the posterior cranial fossa by a median rectangular plate, the **dorsum sellae**. The superior free angles of the dorsum sellae form the **posterior clinoid processes**, to which attach the most anterior peripheral components of the tentorium cerebelli. The lateral boundaries that separate the middle and posterior cranial fossae are the crests or superior borders of the **petrous part of the temporal bone**. This bony crest also serves to attach peripheral components of the tentorium cerebelli and mark the sites of the intradural **superior petrosal sinuses** (*fig. 39.14*).

CONTENTS

Anterior Cranial Fossa (*figs. 39.14 and 39.15*)

The floor of the anterior cranial fossa is mainly formed by the **orbital plates of the frontal bone** and the **jugum** (yoke) and lesser wings of the sphenoid bone. A central rectangular defect between these bones is filled by the ethmoid bone that separates the anterior cranial fossa from the nasal cavities inferiorly. The ethmoid component of the anterior cranial fossa contains the midline **crista galli** (L. cock's comb) which is an anterior attachment point for the falx cerebri.

Flanking the crista galli on each side is a sieve-like bony plate, the **cribriform plate** of the ethmoid bone. These bony perforations allow the **Olfactory Nerve (I)** to transmit its peripheral nerves from the mucosa of the nasal cavity to the **Olfactory bulb** that lies on the superior aspect of the cribriform plate. The olfactory bulb is connected to the brain tissue overlying the lesser wings of the sphenoid bone via the **olfactory tract**, which lies on the superior aspect of the cribiform plate and the **jugum** of the sphenoid bone.

Anteriorly, between the crista galli and the **crest of the frontal bone** is the **foramen cecum**. This bony channel may transmit an emissary vein from the nasal cavity to the superior sagittal sinus and serve as a potential route for nasal infections to spread to the meninges of the intracranial cavity.

Middle Cranial Fossa (*fig. 39.16*)

The middle cranial fossa is associated with six cranial nerves (II–VII), the internal carotid and middle meningeal arteries, the cavernous sinus, the hypothalamus, the pituitary gland, and the temporal lobes of the cerebral cortex. It is a very important area within the cranial cavity and highly relevant to clinical diagnosis of intracranial disease. A more detailed discussion of its contents will be given at the end of this chapter. The middle cranial fossa is shaped like a butterfly (*fig. 39.14*), with a median component and two expanded lateral components. Its bony floor is formed by parts of the sphenoid and temporal bones.

Median Part

The median part is likened to a bed with four clinoid processes representing the bedposts (Gr. kline = a bed). It lies above the body of the sphenoid, which is "inflated" by the sphenoid air sinuses contained within it. The space between the superior aspect of the body of the sphenoid

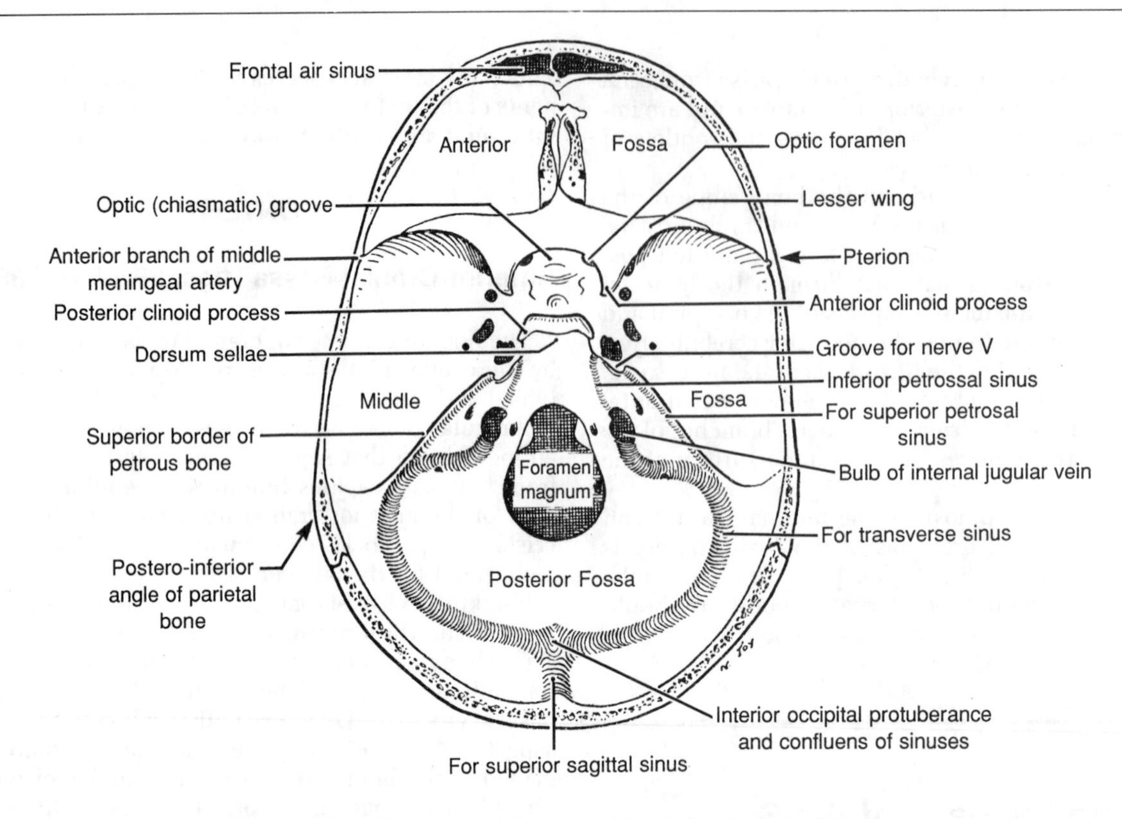

Figure 39.14. The interior of the base of the skull—the 3 cranial fossae.

and the clinoid processes is the **sellae turcica** (Turkish Saddle). It has a horizontal dural partition, the diaphragma sellae, that separates the **hypothalamus cerebri**, or pituitary gland, from the base of the brain (hypothalamus). The "seat" of the saddle is also called the **pituitary (hypophyseal) fossa** (*fig. 39.16*).

The **optic canals** arise in the anterior aspect of the medial part of the middle cranial fossa and run anterolaterally through the lesser wing of the sphenoid bone to open into the orbits. Each optic canal contains an **optic nerve** (II) and an **ophthalmic artery** inferior to the optic nerve. The **optic (chiasmatic) groove** is located between the optic foramina and above the **tuberculum sellae**. This groove is a bony landmark only and **does not** actually contain the optic chiasm, where the nerves from the two nasal parts of the retina cross the midline to enter the optic tracts on the opposite side of the head.

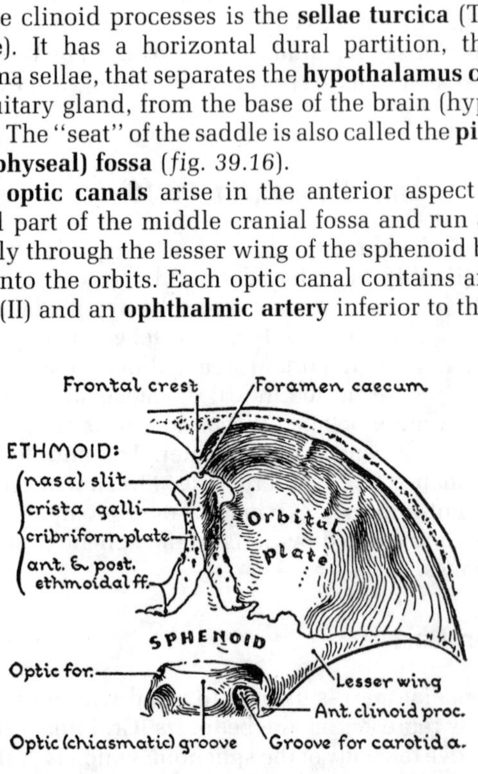

Figure 39.15. The anterior cranial fossa. (*Optic foramen and optic groove = optic canal and chiasmatic sulcus.*)

The sellae turcica lies behind the optic groove and has three parts—the **tuberculum sellae**, the **hypophyseal fossa**, and the **dorsum sellae** (back of saddle).

A ragged-edged foramen, the **foramen lacerum** (*fig. 39.14*), is readily observable in the middle cranial fossa of dried skull preparations or basal view radiographs of the skull. In life, it is filled by fibrocartilage and forms part of the floor of the middle cranial fossa between the median part of the fossa and **apex of the petrous temporal bone**. The carotid artery passes above the cartilage-filled foramen lacerum, as it leaves the carotid canal and enters the cavernous sinus on the lateral aspects of the body of the sphenoid bone.

Lateral Part

Each lateral part is limited anteriorly by the lesser wing of the sphenoid and posteriorly by the superior border of the petrous bone. The floor and lateral walls include the greater wing of the sphenoid and the petrous and squamous parts of the temporal bone.

Clinically significant features associated with the greater

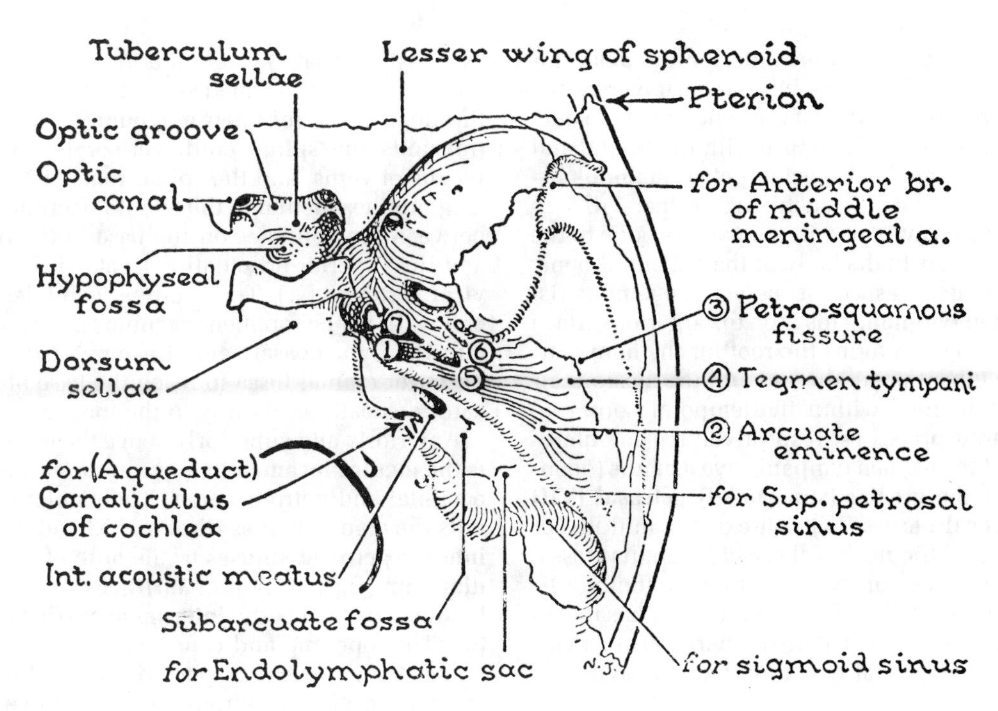

Figure 39.16. Middle cranial fossa. Seven details of petrous bone: (*1*) depression for trigeminal ganglion; (*2*) elevation for anterior semicircular canal; (*3*) remains of petrosquamous fissure; (*4*) tegmen tympani; (*5*) hiatus for greater petrosal nerve; (*6*) hiatus for lesser petrosal nerve; (*7*) roof of carotid canal, commonly membranous.

wing of the sphenoid, include four foramina: the **superior orbital fissure, foramen rotundum, foramen ovale** and **foramen spinosum**. These bony apertures are illustrated from a superior view in *Figures 39.14 and 39.16* and in schematic form in *Figure 39.17*.

The **superior orbital fissure** is formed between the lesser and greater wings of the sphenoid near their points of origin. It is readily visible radiographically as a crescent-shaped radiolucency in an anterior-posterior view of the orbit. It transmits the III, IV, V^1, VI cranial nerves and the superior ophthalmic veins, as they pass between the orbit and the middle cranial fossa.

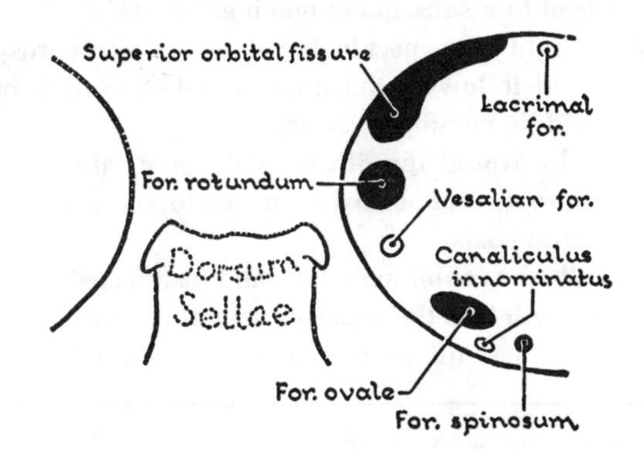

Figure 39.17. The crescent of foramina within the greater wing. Of these foramina, 4 are constant and 3 (lacrimal; Vesalian, for an emissary v.; and innominatus, for a small nerve) are not.

The **foramen rotundum** connects the middle cranial fossa to the pterygopalatine fossa that lies deep within the face lateral to the nasal cavities. V^2 enters the middle cranial fossa via the foramen rotundum with sensory fibers from the face, over the maxilla, the nasal cavity, and the palatal aspect of the oral cavity. The foramen rotundum is visible in the anteroposterior radiograph of the head as a round radiolucency inferior to the superior orbital fissure. It is commonly confused with the infraorbital foramen in radiographs. However, the two foramina are oriented in different planes and need different x-ray projections to reveal their bony location in skull films.

The **foramen ovale** is a horizontally oriented aperture on the floor of the medial cranial fossa. It transmits V^3 and the lesser petrosal branch of the IX from the cranial cavity to the infratemporal fossa inferiorly. It may also contain an **accessory meningeal artery**, which can arise from the maxillary artery below the base of the skull.

The **foramen spinosum** lies posterior to the foramen ovale and transmits the **middle meningeal artery** from the maxillary artery in the infratemporal fossa to the dura of the middle cranial fossa. The course of the middle meningeal artery in the middle cranial fossa is readily visible in dissection, skeletal preparations, and skull x-rays. The foramen ovale and foramen spinosum are also easily visualized as radiolucencies in the basal view of the skull.

The **squamous portion** of the temporal bone forms the lateral wall of the middle cranial fossa and overlies the dura covering aspects of the temporal lobe of the brain. The inner surface of the squama is grooved by the **middle**

meningeal artery and its terminal **anterior** and **posterior branches** (*fig. 39.16*). The anterior branch traverses the **pterion** where the greater wing of the sphenoid and squamous part of the temporal bone fuse with the frontal and parietal bones. The medial aspect of **petrosquamous fissure** overlies the auditory (Eustachian) tube that will connect the middle ear cavity within the petrous bone to the nasal cavities inferior to the body of the sphenoid bone. Posterior to the lateral aspect of the petrosquamous fissure is the **tegmen tympani**. This is on the anterior surface of the petrous bone and forms the roof for the bony part of the auditory tube, the middle ear, and the mastoid air cells that are contained within the temporal bone (petrous and mastoid processes, respectively). The anteromedial aspect of the tegmen tympani have a hiatus (opening) for each of the **greater (superficial) petrosal (VII)** and **lesser petrosal nerves (IX)** that exit the middle ear cavity and run over the floor of the middle cranial fossa. These nerves carry the parasympathetic secretomotor fibers to glands associated with the oral cavity, nasal cavities, and orbits. The greater petrosal nerve also carries taste fibers from the palatal region of the oral cavity.

Posterior Cranial Fossa (*figs. 39.14 and 39.4*)

This fossa lodges the hindbrain, which is comprised of the medulla (oblongata), pons, midbrain, and overlying cerebellum. The fossa is roofed by the tentorium cerebelli except at the tentorial notch, where the midbrain passes to connect with the forebrain in the middle cranial fossa.

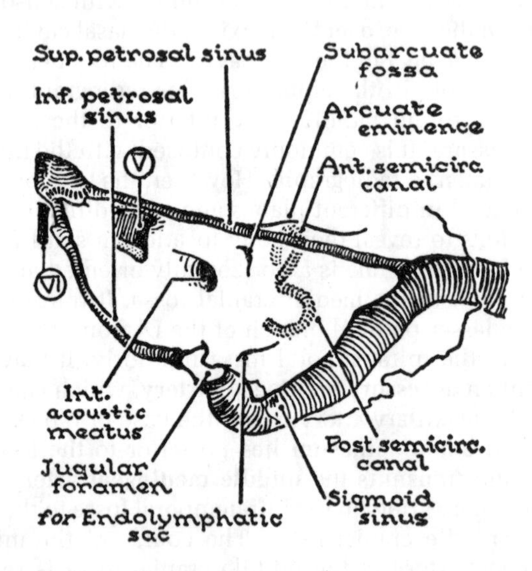

Figure 39.18. The posterior surface of the petrous bone is bounded by venous sinuses.

The posterior, lateral, and inferior aspects of the posterior cranial fossa are formed by the occipital bone. Within the floor is the **foramen magnum**, a large opening that transmits the spinal cord, vertebral arteries, vertebral plexus of veins, and the spinal roots of XI. The foramen magnum lies medial to the atlanto-occipital articulations between the condyles on the base of the occipital bone and the superior articulating facets of the **atlas** (1st cervical vertebra, C1). The occipital condyles form the lateral walls of the foramen magnum (*fig. 39.14*). They contain the hypoglossal canal for nerve XII, as it exits the posterior cranial fossa to reach the base of the skull and carotid sheath on its way to the tongue.

Anteriorly and superiorly above the hypoglossal canal, is the **jugular foramen** that lies in the suture between the occipital and petrous bones. The jugular foramen transmits the cranial nerves IX, X and XI and the **sigmoid** and **inferior petrosal sinuses** to the **bulb** of the **internal jugular vein** (*figs. 39.14 and 39.18*). Superior to the jugular foramen, one finds the **internal acoustic (auditory) meatus**. This opening and canal transmits the **VII** and **VIII** cranial nerves from the posterior cranial fossa to the internal aspect of the petrous bone. The **inner, middle,** and bony parts of the **external ear cavities** are contained in the petrous bone and are closely related to cranial nerves VII and VIII.

A number of emissary veins are also associated with the posterior cranial fossa. Most consistently, one can find major emissary veins in the posterior aspects of the occipital condyles and mastoid processes of the temporal bone. These emissary veins lack valves and can transmit venous blood from the scalp to the venous sinuses within the dural reflections of the posterior cranial fossa. The emissary veins are therefore potential routes by which infections of the scalp "spread" to the meninges.

Clinical Mini-Problems

1. **How could a scalp infection over the parietal bones lead to a subsequent meningitis?**
2. **Would you expect to find blood in the cerebrospinal fluid following the intracranial laceration of the middle meningeal artery?**
3. **Why would the 6th cranial nerve (abducens) be more susceptible to dysfunction in a cavernous sinus thrombosis?**
4. **Which cranial nerve(s) could be affected by tumor growth into the superior orbital fissure?**

(Answers to these questions can be found on p. 587.)

40

Ear

Clinical Case 40.1

Patient Bobby F. This 5-year-old preschooler has had a history of recurring upper respiratory infections with accompanying middle ear infections (otitis media). The episodes were treated by antibiotic therapy, but the child's mother had to miss a number of days at work while caring for the child during the treatment. When you examined the nasopharynx you could see an abundant amount of lymphoid tissue ("adenoids") that surrounded the opening of the auditory tube. You assume that your clinical supervisor (who is an otolaryngologist) will remove this tissue to facilitate the drainage capacity of the middle ear. However she elects to install a small tube in each eardrum to drain the middle ear into the external ear. The tubes are placed in the inferior posterior quadrant of each tympanic membrane. *Followup:* Subsequent upper respiratory infections during the next year are not accompanied by the severely painful earaches. The pediatrician intends to remove the ear tubes when the child's facial growth permits the natural drainage through the auditory tube—i.e., if they have not dropped out accidentally meanwhile.

External Ear

The external ear consists of an auricle and an external acoustic meatus (*fig. 40.1*).

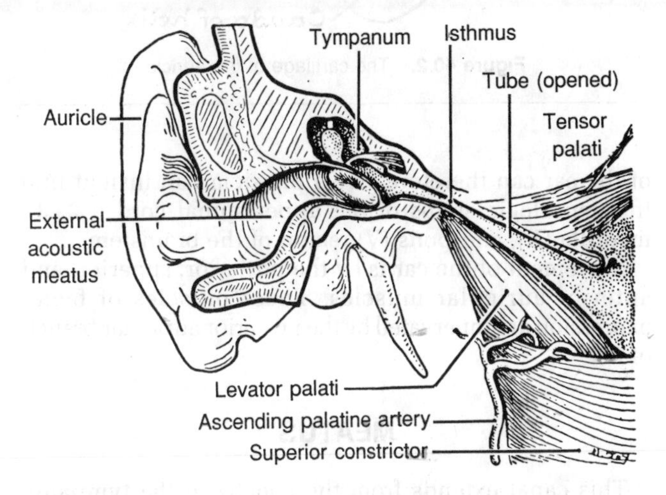

Figure 40.1. The external and middle ears and the auditory tube, opened throughout.

AURICLE

The framework of the auricle is made of a single piece of elastic cartilage (*fig. 40.2*) except at its most inferior portion, **the lobule**. The lobule, or ear lobe, is made of fibroareolar tissue. The auricular cartilage is continuous with the cartilage of the external acoustic meatus (L. meatus = canal).

The skin covering the cartilage and fibroareolar lobule is closely adherent to the underlying connective tissue skeleton of the ear. The surface anatomy of the skin of the ear is illustrated in *Figure 40.3*. The sensory supply to the skin of the ear is of significance in the clinical examination. The superior portion of the ear is innervated by V^3 via the auriculotemporal nerve. The inferior portion of the ear, particularly the lobule, is innervated by fibers of the greater auricular nerve from the cervical plexus (C2, 3). The external (acoustic) meatus and the skin surrounding its opening, **the concha**, intervenes between these two surface areas. They are innervated by nerve X for pain, temperature, touch, and pressure. Some touch may also be carried by nerve VII from the external auditory canal. A neurological examination of the skin

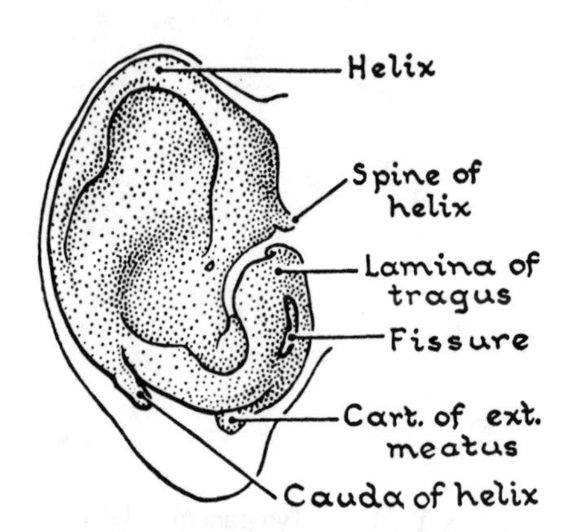

Figure 40.2. The cartilage of the auricle.

of the ear can therefore give a physician an insight into the neurologic status of the upper spinal cord (C2, 3), medulla (X), and pons (V) region of the brainstem.

Movement of the ear is by the posterior, superior, and anterior auricular muscles. These muscles of facial expression are innervated by the posterior auricular branch of VII.

MEATUS

This canal extends from the concha to the **tympanic membrane** or eardrum (*fig. 40.1*).

The lateral one-third of the meatus is chiefly cartilaginous and is attached to the outer edge of the bony meatus, which forms the medial two-thirds of the canal.

The meatus is sinuous, being convex superiorly and convex posteriorly. One must therefore pull the ear upward, backward, and laterally to straighten out the external auditory meatus, and then one can directly inspect the tympanic membrane with an ear speculum.

The cartilaginous portion of the meatus is lined with skin in which there are hairs, sebaceous glands, and modified sweat glands, which secrete cerumen or ear wax. Infections (boils) may occur here and, because of the close adherence of the skin to the cartilage, can cause great discomfort. The bony meatus is lined by a thin stratified squamous epithelium that also lines the external surface of the tympanic membrane. The *auricular branch of the vagus nerve* is the principal sensory supply to the external acoustic meatus.

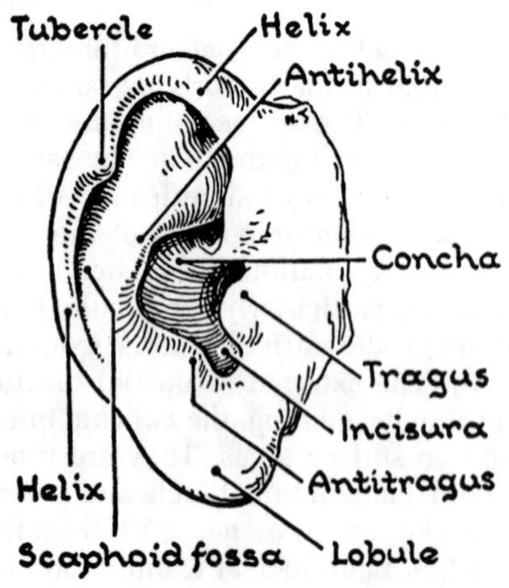

Figure 40.3. The auricle.

Middle Ear or Tympanum

Sound waves that enter the external acoustic meatus create a pressure on the tympanic membrane. The subsequent vibration of the tympanic membrane moves the three bony ossicles (**malleus, incus,** and **stapes**) within the middle ear. This bony movement is turn compresses the fluid in the inner ear by vibrating the oval window, the **fenestra vestibuli**, on the medial wall of the middle ear (*fig. 40.4*). This conversion of airwaves, to vibrations in a solid, to fluid vibration, is a highly efficient means of amplifying the sound waves that are entering the external acoustic meatus. *Figure 40.4* also depicts the difference in the size of the larger tympanic membrane and the much smaller oval window membrane, which is attached to the stapes. This concentration of vibrating surface area is also an amplifier of sound energy. Disorders that block the efficient transmission of these energies will therefore create varying degrees of deafness.

The middle ear is actually a modified air sinus within the petrous temporal bone. It communicates with the mastoid air cells through the *aditus* and also with the nasopharynx through the **auditory tube** (*figs. 40.5 and 40.6*). The middle ear cavity (tympanum) is a vertically oriented cavity whose shape resembles that of a red blood cell. The width at the center is only 2 mm, while the roof is 6 mm wide and the floor is 4 mm wide. The vertical height is approximately 15 mm. One can appreciate that the size of this cavity is comparable to a dime standing on its edge.

TYMPANIC CAVITY AND ITS WALLS

The **roof**, or tegmen tympani (*figs. 40.7 and 40.8*), is a thin layer of petrous temporal bone separating the middle cranial fossa from the middle ear. Just below this roof, the middle ear cavity is expanded into the **epitympanic recess**. This area provides space for the articular joint between the **head of the malleus** and the **body of the incus**.

The **floor** of the tympanic cavity is also a thin layer of petrous bone that rests upon the **superior jugular bulb**.

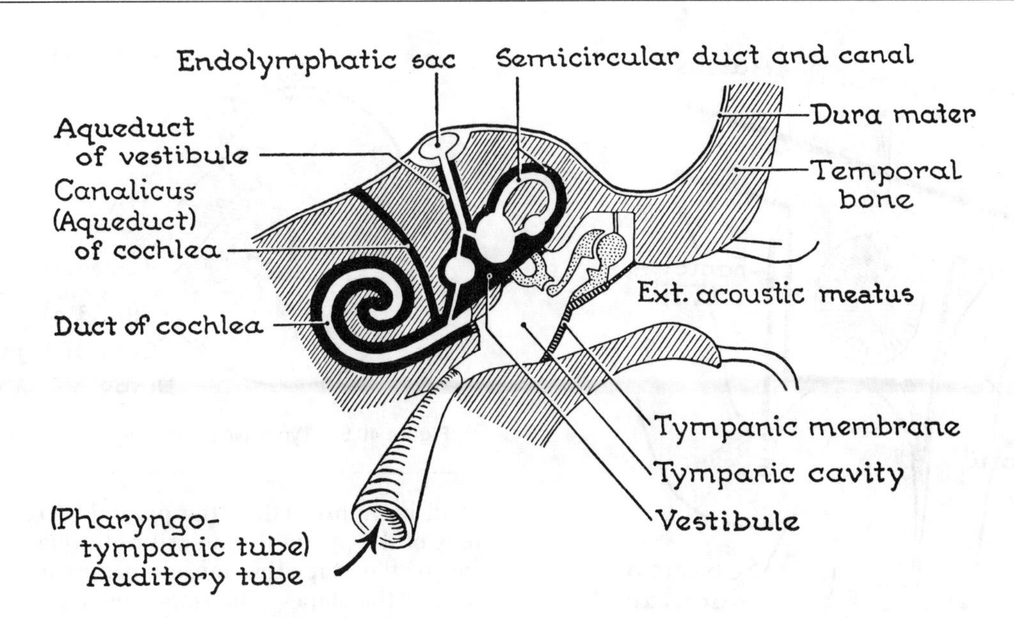

Figure 40.4. General plan of the 3 parts of the ear.

Figures 40.7 and 40.8 illustrate the relationship of the floor of the tympanic cavity and the superior aspect of the carotid sheath. The internal carotid artery and the internal jugular vein diverge from their close relationship in the carotid sheath at the floor of the tympanic cavity. The internal carotid artery courses anteriorly into the middle cranial fossa, while the internal jugular vein passes posteriorly into the posterior cranial fossa. The IX and X nerves that are within the carotid sheath also send their auricular branches into the tympanic cavity through foramina in its bony floor.

The **lateral wall** of the tympanic cavity is closed by the tympanic membrane (*figs. 40.8 and 40.9*). This wall is visualized by the clinician when the tympanic membrane is examined with an otoscope. The epitympanic recess rises a few millimeters about the tympanic membrane and contributes a bony component to the superior aspect of the lateral wall (*fig. 40.8*).

Anteriorly, the roof and floor of the tympanic cavity converge to form the bony component of the **auditory (Eustachian) tube**. The bony auditory tube is divided by a shelf of bone, the **processus cochleariformis** (*fig. 40.8*), into a superior compartment containing the **tensor tympani muscle** and a lower compartment that joins to the cartilaginous auditory tube. The free edge of the processus cochleariformis, which projects into the tympanic membrane, serves as a pulley for the tendon of the tensor tympani as it projects laterally to insert onto the handle of the malleus. The ascending carotid artery is also associated with the anterior wall of the tympanic cavity but is separated from the cavity by a thin layer of bone. In some clinical disorders, the patient can hear the pulsation of the blood in the carotid artery as it ascends through this portion of the anterior wall of the middle ear.

The **posterior wall** near the roof (*figs. 40.6 and 40.7*) has a tunnel, the **aditus**, that connects the cavity of the middle ear with the mastoid antrum. The aditus serves as a pathway for the fluids in the mastoid air cell to drain into the tympanum, then into the auditory tube and then into the nasopharynx. Upper respiratory infections that affect this drainage pathway can result in a fluid collection within the cavity of the middle ear. Below the aditus, the **facial nerve** (VII) enters the posterior wall of the mid-

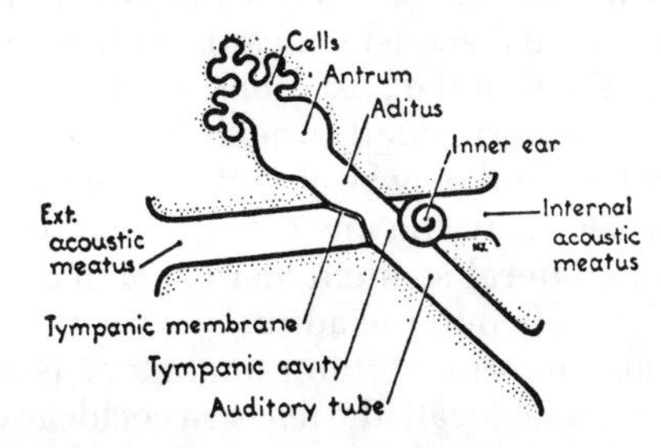

Figure 40.5. The 2 meatuses, which have blind ends, and the line of the airway (tube, cavity, aditus, and antrum), which passes between them, viewed from above.

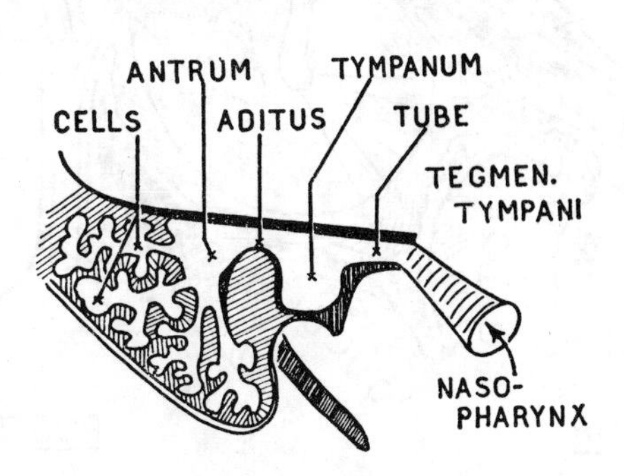

Figure 40.6. Tegmen tympani and passages it covers.

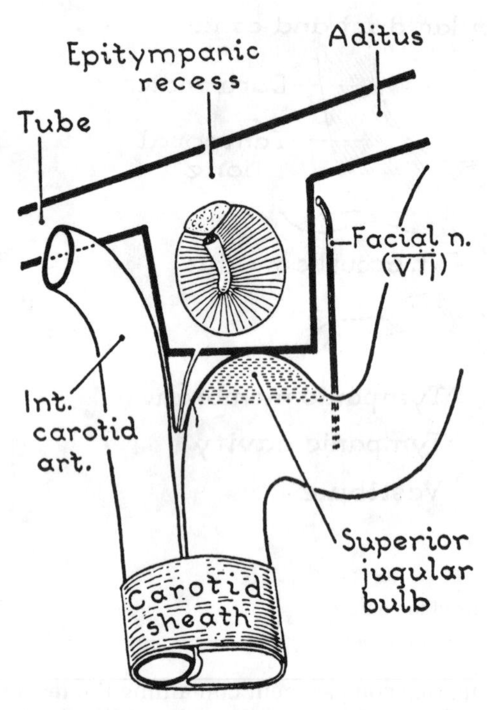

Figure 40.7. Tympanic cavity and its neighbors.

dle ear and descends to exit from the base of the temporal bone through the stylomastoid foramen (*figs. 40.8 and 40.10*). The **chorda tympani** arises from the facial nerve within the posterior wall of the middle ear, then courses along the lateral wall over the tympanic membrane and then exits into the infratemporal fossa via the **petrotympanic fissure** (*fig. 40.8*). The posterior wall also has a

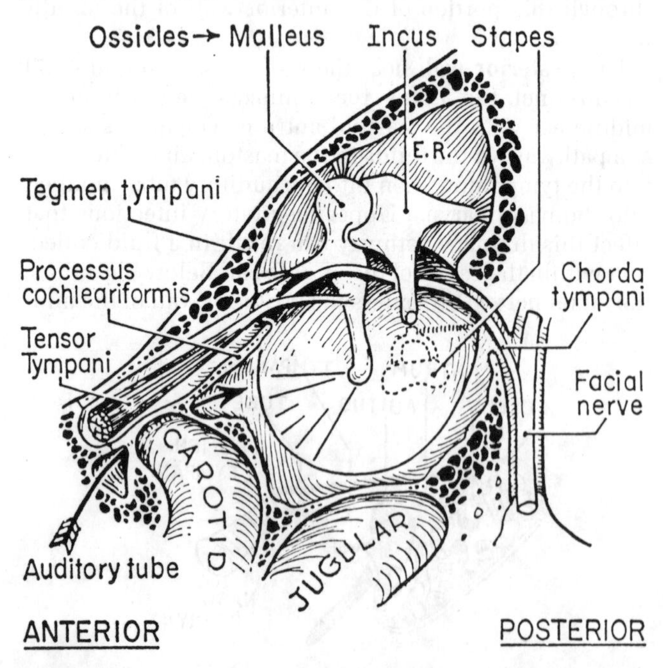

Figure 40.8. Lateral wall of the tympanic cavity and the ossicles. *ER*, epitympanic recess.

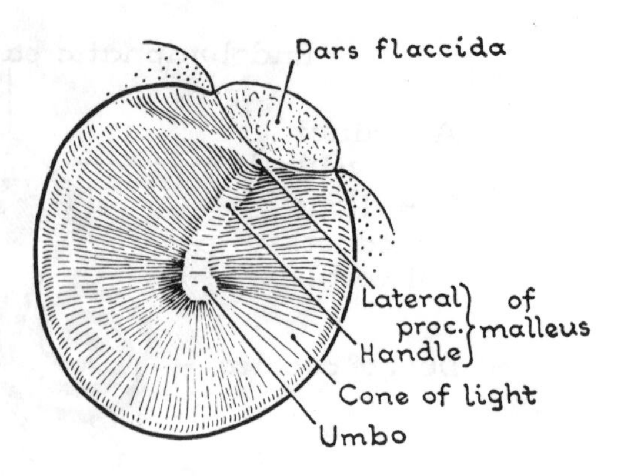

Figure 40.9. Tympanic membrane (lateral view).

small bony projection, the **pyramid** (*fig. 40.10*). At the apex of the pyramid is an orifice through which the tendon of the **stapedius** muscle passes to insert onto the neck of the stapes. The stapedius muscle acts to retract the footplate of the stapes away from the oval window (fenestra vestibuli) and thereby reflexively attenuates loud sounds by reducing the vibration to the fluid of the inner ear. The **stapedius muscle** receives its motor supply from VII, and in some cases of facial nerve palsy, patients complain of a sensitivity to loud sounds.

The **medial wall** of the tympanic cavity faces the portion of the petrous temporal bone, which contains the inner ear (*fig. 40.4*). In the center of the medial wall is a prominent bony swelling, the **promontory**, which overlies the first turn of the cochlea of the inner ear (*fig. 40.10*). Within the mucosa that covers the promontory is a plexus of nerves, the **tympanic plexus**. Branches of VII, IX, and X intermingle within this tympanic plexus and supply sensory fibers to the middle and external ear as well as preganglionic secretomotor fibers to the petrosal nerves.

Posterior and superior to the promontory on an oblique line projecting to the aditus are: (*a*) fenestra vestibuli; (*b*) the canal for the facial nerve and (*c*) the prominence formed by the lateral semicircular canal. The **fenestra vestibuli** is an **oval window** that is an opening between the vestibule of the inner ear and the tympanic cavity. This opening is approximately 3 mm across in its horizontal dimension and its form matches the surface of the footplate of the stapes, which overlies the oval window.

The **canal for the facial nerve** runs horizontally and connects the internal acoustic meatus with the descending facial canal within the posterior wall (*fig. 40.10*). The bone covering the horizontally oriented canal of the facial nerve is very thin and may be absent in some areas, leaving VII in contact with the mucosal lining of the middle ear cavity. The **lateral** semicircular canal also lies horizontally and bulges into the aditus.

Posterior and inferior to the promontory is a second opening in the medial wall, the **fenestra cochleae** or **round window**. It is between the cavity of the middle ear and the scala tympani of the inner ear. This bony opening is closed by a membrane.

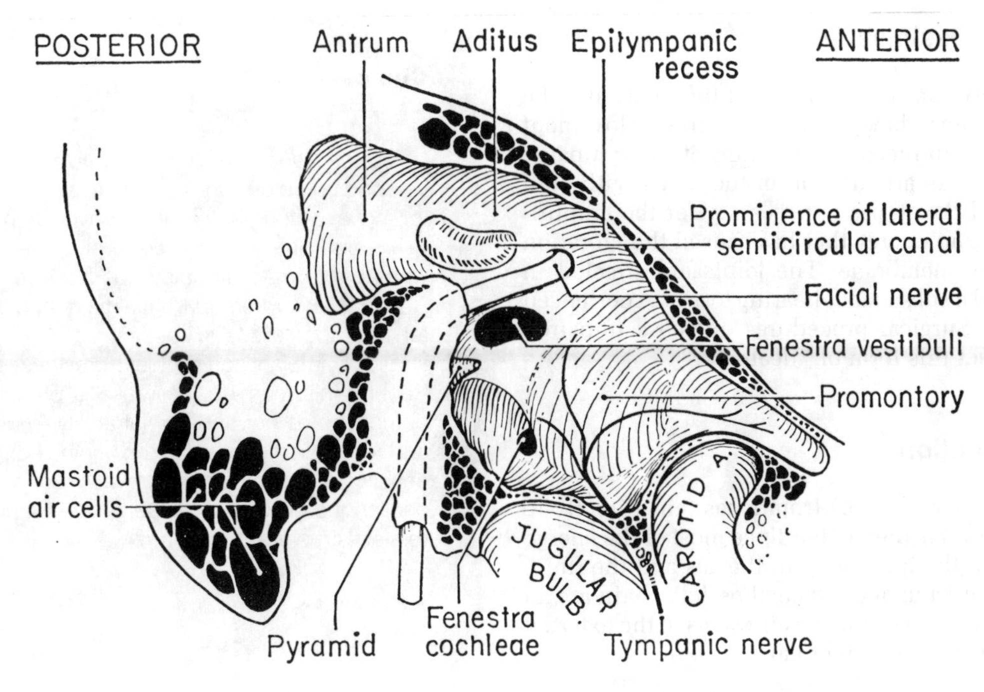

Figure 40.10. Medial wall of the tympanic cavity.

TYMPANIC MEMBRANE

This membrane or eardrum (*fig. 40.9*) is nearly circular, being 9 mm high and 8 mm wide. It is set in a sulcus of the tympanic bone. The tympanic membrane faces laterally, anteriorly, and inferiorly as though it were oriented to "catch" the sounds reflected from the ground as one advances. It is composed of circular and radial fibers and is lined by an ectodermally derived epidermis laterally and an endodermally derived mucous membrane medially. The handle of the malleus is attached to the fibrous tissue of the tympanic membrane and separated from the cavity of the middle ear by the overlying mucous membrane. The radial fibers of the tympanic membrane radiate from the inferior extent of the handle of the malleus. A relaxed superior portion of the tympanic membrane, the **pars flaccida**, attaches to the lateral process of the malleus. In "puncturing" an eardrum to drain the middle ear, one should avoid this pars flaccida region because of the underlying chorda tympani (*fig. 40.8*).

AUDITORY OSSICLES, JOINTS, AND MUSCLES

Ossicles

There are 3 ossicles—malleus (hammer), incus (anvil) and stapes (stirrup) (*figs. 40.4 and 40.8*). They intervene between the tympanic membrane and the oval window and have two synovial joints within their bony chain.

The **malleus** is approximately 8 mm long and vertically oriented within the tympanic cavity. It has a round **head** with an articulating facet for the incus facing posteriorly. The **neck** of the malleus joins the head with the long **manubrium** (handle) that extends inferiorly onto the upper half of the tympanic membrane. The membrane is attached to the manubrium from the **lateral process** superiorly to the inferior tip. An **anterior process** of the malleus is directed anteriorly and connected to a stabilizing ligament.

The **incus** is shaped like a molar tooth. It has a body and 2 diverging processes. The **short crus** (L. = leg-like process) is horizontal and has a ligament that attaches the incus to the posterior wall of the epitympanic recess (*fig. 40.8*). The **long crus** is vertically oriented and descends into the tympanic cavity. At the tip of the long crus is a medially projecting **lenticular process**. This articulates with the head of the stapes. The **body** of the incus articulates with the head of the malleus and has a prominent articular facet on its anterior aspect.

The **stapes** has a head with a concaved socket for articulation with the lentiform process of the incus. The head is supported by a short **neck** that intervenes between the two limbs (posterior and anterior) that attach to the medial **footplate** or **base**. The footplate covers the oval window and is attached to the margin of the oval window by an **annular ligament**.

Joints of the Ossicles

The joints between the malleus and the incus and the incus and the stapes have synovial cavities. Movement of the malleus is induced by vibration of the tympanic membrane. The free articulation of the incus with both the malleus and the stapes is necessary for the amplification of the sound energy that has caused the vibration of the tympanic membrane. The joints are affected in **otosclerosis** and a subsequent hearing loss is experienced by the patient. Surgical procedures on these impaired joints can correct this form of "deafness."

Sound Conduction

Transferring energy waves from a gas (air) directly to a fluid (perilymph) is ineffective since most of the energy is reflected from the interphase of the gas and the fluid. The vibrating eardrum and attached ossicle chain are an efficient means of converting the air waves in the external ear to fluid waves in the inner ear.

The effective vibratory area of the tympanic membrane is 55 mm^2 and the average size of the stapes footplate is 3.2 mm^2. Hence the hydraulic ratio (membrane to footplate) is 17:1.

The lever ratio of the malleus to the incus is 1.3:1.0, and the total transformer ratio is thus 22:1 (von Békésy).

Muscles of the Ossicles

There are two muscles that attach to the ossicles. Contraction of either muscle will result in an attenuation of sound by decreasing the movement of the ossicle to which they attach.

The **tensor tympani** is a muscle within the bony auditory tube. As its tendon leaves the bony canal, it uses the "lip" of the processus cochleariformis as a pulley (*fig. 40.8*). The tendon then courses laterally in the tympanic cavity to insert onto the handle of the malleus. Contraction of the tensor tympani pulls the tympanic membrane medially and thereby tenses the membrane. More energy is thereby needed in the vibrating air to move the membrane and attached malleus. The tensor tympani is innervated by a branch from the trunk of V^3 as it exits from the foramen ovale.

The **stapedius** is contained in the hollow **pyramidal process** of the posterior wall of the tympanic membrane (*fig. 40.10*). Its tendon leaves the foramen at the apex of the pyramid and passes anteriorly to insert onto the neck of the stapes. When it contracts, it will pull the footplate away from the oval window and "dampen" the sound energy, which is being relayed to the fluid in the vestibule of the inner ear. The stapedius is innervated by the facial nerve (VII). Paralysis of the facial nerve can therefore cause a sensitivity to loud sounds (hyperacusis) when the stapes is fully opposed to the oval window.

Clinical Case 40.2

Patient Nancy D. This 35-year-old woman developed complete inner-ear deafness bilaterally following a severe febrile illness when she was 11 years old. She was trained in sign language and was able to maintain fairly good speech patterns. Recently, she returned from a major audiology center that specializes in research and treatment by surgically implanting "cochlear prostheses." At ENT Rounds this morning, her case is presented. She demonstrates her substantial ability to understand and repeat sentences spoken behind her back. The ensuing lengthy discussion of hearing problems challenges your knowledge of the inner ear and related structures.

Internal Ear

The internal ear contains the specialized neurologic mechanisms and receptors for both hearing and balancing (vestibular) functions. It is contained within the petrous bone and has 2 parts: the **bony labyrinth** and the **membranous labyrinth**, which is within the bony labyrinth.

BONY LABYRINTH (*fig. 40.11*)

This has 3 parts—**cochlea, vestibule** and **semicircular canals**. They are related to the medial wall of the tympanic cavity. The cochlea lies deep to the promontory, the vestibule is associated with the oval window, and the lateral semicircular canal bulges into the aditus (*fig. 40.10*). The entire bony labyrinth is only 17 mm long.

The **cochlea** resembles a snail's shell with 2½ coils. It contains the neurological receptors for hearing. Vibration of the perilymph in the vestibule is transmitted to the fluids of the cochlea. The wave patterns in the fluids of the cochlea stimulate the hearing receptors in the inner ear, and the sound is perceived in the temporal lobe of the cerebral cortex.

The **vestibule** lies between the cochlea and the semicircular canals (*fig. 40.12*). It communicates with both of these chambers in the petrous bone. It opens laterally through the oval window into the tympanic cavity. The

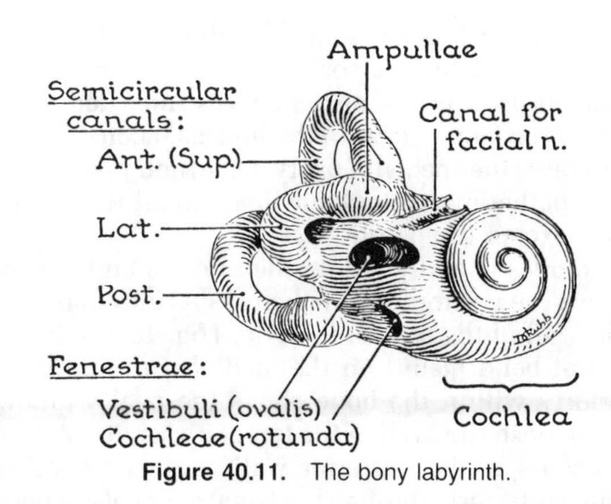

Figure 40.11. The bony labyrinth.

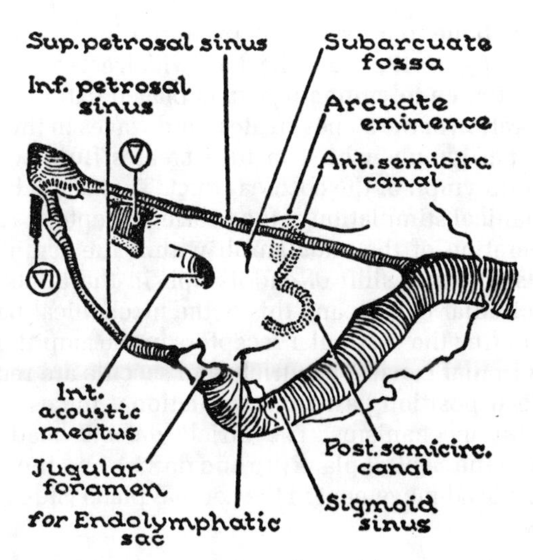

Figure 40.13. The posterior surface of the petrous bone is bounded by venous sinuses.

fluid (perilymph) within the vestibule contacts the footplate of the stapes in the oval window. The perilymph also surrounds the membranous labyrinth within the cochlea and the semicircular canals. While the perilymph has a composition similar to subarachnoid cerebrospinal fluid, it is not felt that the two spaces communicate on the surface of the petrous bone (Young).

The **three semicircular canals**—anterior (superior), posterior, and lateral—are set at right angles to each other and occupy 3 planes in space. The lateral canals of opposite sides of the ear are horizontal and lie in the same plane. The anterior canal of the one side lies in the same plane as the posterior canal of the opposite side and the two internal ears function together to maintain a proper equilibrium.

The lateral semicircular canal is horizontal and projects laterally forming the prominence in the medial wall of the aditus. The other anterior and posterior semicircular canals are vertically oriented. The anterior semicircular canal is oriented at a 90° angle to the transverse axis of the petrous bone, while the posterior semicircular canal is immediately deep to the posterior (cerebellar) surface of the petrous bone and parallel to this surface (*fig. 40.13*).

MEMBRANOUS LABYRINTH (*fig. 40.12*)

This has three components: (*a*) the cochlear duct; (*b*) the saccule and utricle; and (*c*) the 3 semicircular canal ducts. It is a closed system containing endolymph. A stalk, the **ductus endolymphaticus**, passes from the saccule and utricle through a canal (aqueduct of the vestibule) in the

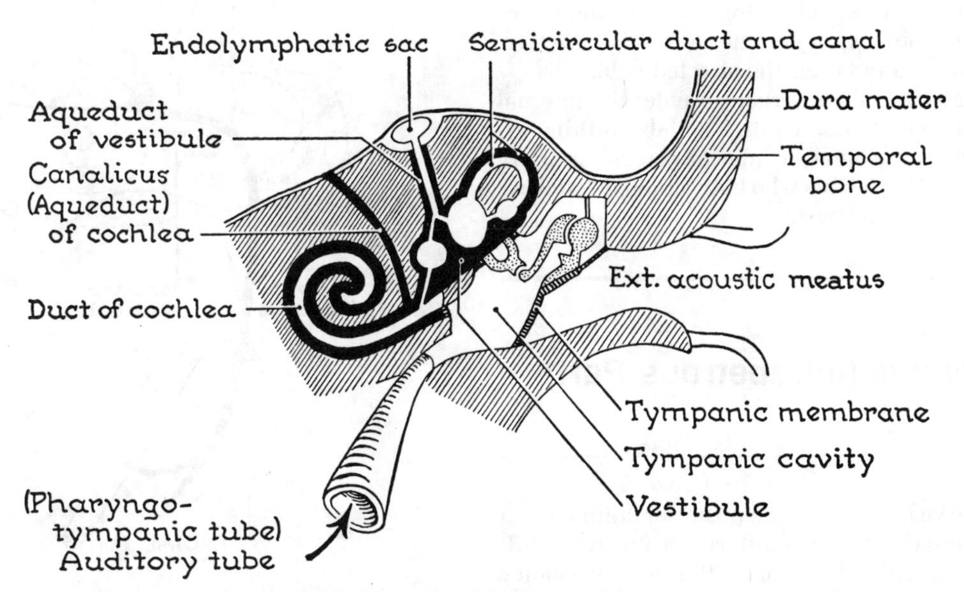

Figure 40.12. General plan of the 3 parts of the ear.

petrous bone to a fissure lateral to the internal acoustic meatus (*fig. 40.12*). There the duct, which acts as a "safety" expansion **endolymphatic sac**, is placed extradurally.

Vibration of the stapes creates fluid waves in the underlying perilymph, which in turn creates fluid waves in the endolymph of the cochlear duct. This is the basis for mechanical stimulation of the hearing receptors. Angular acceleration of the endolymph within the semicircular canals causes a shift of endolymph in the ducts of the semicircular canals, and this is the mechanical basis for stimulating the vestibular receptors in the ampulla of the semicircular canal. The utricle and saccule are receptors for head position based on gravitational forces on their receptor mechanisms. The utricle detects head movement in the sagittal plane (up and down), and the saccule detects head movement in the coronal plane (side to side).

NERVOUS AND ARTERIAL SUPPLY TO THE INTERNAL EAR

Nerve **VIII** supplies the special sensory receptors in the internal ear. The **auditory nerve** is better termed the **vestibulocochlear nerve** to depict both of its special functions of equilibrium and hearing. Clinical testing of nerve VIII should include both tests for hearing and balance. Hearing is tested with a "tuning fork" held in the air outside the external auditory meatus and then placed on the mastoid process. This tests air conduction through the tympanic membrane and ossicles and fluid vibration through the temporal bone. If the tympanic membrane or ossicles are impaired, the bone conduction should be heard normally when the tuning fork is placed on the mastoid process. Patients with neurologic damage in nerve VIII will also have hearing loss through bone conduction as well as air conduction.

Vestibular function may be tested by having the patients stand with their feet close together and their eyes closed. If the vestibular system or nerve VIII is defective, the patient will tend to fall to the affected side.

The blood vessels to the internal ear enter the internal acoustic meatus with VII and VIII. The **labyrinthine artery** is a branch of the anterior inferior cerebellar artery and can be affected in patients who have suffered "strokes" in their vertebral arterial system.

Facial Nerve (Intrapetrous Part)

The **facial nerve** (VII) leaves the posterior cranial fossa through the internal acoustic (auditory) meatus with VIII. VII lies above VIII within the canal. VII is mainly a motor nerve to the muscles derived from the second branchial (pharyngeal) arch of the embryo—the muscles of facial expression, stylohyoid, posterior belly of the digastric, and stapedius. There is also a **nervus intermedius** component of the facial nerve in the internal acoustic meatus that contains the special sensory (taste) and preganglionic parasympathetic secretomotor fibers to all the glands of the face except the parotid.

VII courses laterally within the internal auditory meatus above the vestibule of the bony labyrinth to reach the medial wall of the tympanic cavity (*fig. 40.14*). It makes an abrupt bend (genu) on the medial wall and courses posteriorly within the bone above the oval window to reach the posterior wall of the tympanic cavity. A second bend occurs at the posterior wall, and nerve VII then descends inferiorly through the facial canal of the petrous bone to exit the base of the skull through the **stylomastoid foramen**.

At the genu on the medial wall of the tympanic cavity is the **geniculate ganglion**. It contains the pseudo-unipolar cell bodies for the taste fibers from the anterior two-thirds of the tongue and the palate. **There are no synapses in the geniculate ganglion**. The **greater superficial petrosal nerve** branches from VII at the geniculate ganglion to pierce the anterior wall of the tympanic cavity and enter the middle cranial fossa. The superficial petrosal nerve contains taste fibers from the palate and secretomotor fibers for glands in the roof of the oral cavity, nasal cavity, and orbit.

The descending part of VII gives off the **motor branch to the stapedius muscle** and the **chorda tympani**. The chorda tympani courses between the handle of the malleus and vertical process of the incus to exit into the

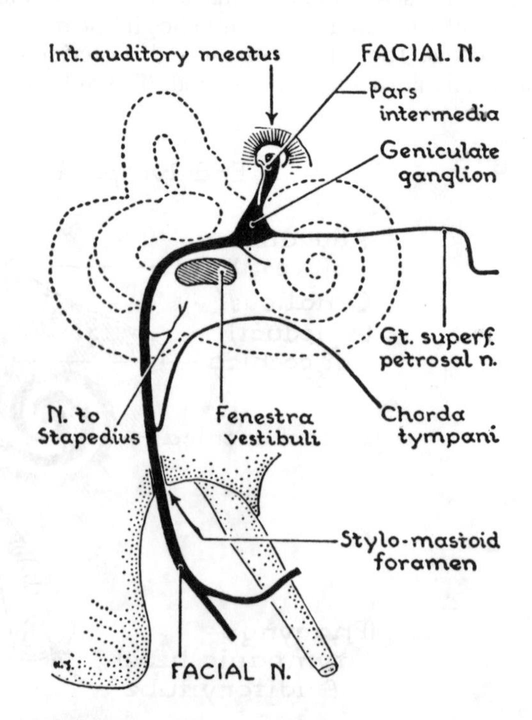

Figure 40.14. Intrapetrous course of facial nerve.

infratemporal fossa through the **petrotympanic fissure**. The chorda tympani contains taste fibers from the anterior two-thirds of the tongue and secretomotor fibers to the submandibular ganglion. The cell bodies for the taste fibers are in the geniculate ganglion on the medial wall of the tympanic cavity. Damage to nerve VII in the intrapetrous portion is tested by "back-tracking" the branches of VII above the stylomastoid foramen. If the patient has a paralysis of the facial muscles on one side of the face, the clinician would then test (*a*) taste in the anterior of the tongue (**chorda tympani**), (*b*) for signs of hyperacusis (**nerve to stapedius**) and (*c*) lacrimation on same side (**greater superficial petrosal nerve**). The lack of tearing on the side of facial paralysis presents a serious clinical situation. Since the lacrimal gland moistens the cornea, a lack of lacrimation causes desiccation of the cornea and can produce ulceration of the cornea and subsequent blindness. Corneal desiccation induces a painful stimulus, which is carried by V^1. Thus, a facial paralysis with associated corneal disorders may result from trauma to VII between the brainstem and the geniculate ganglion. The most common form of nerve VII paralysis is Bell's Palsy, and this usually only affects the branches of the nerve below the stylomastoid foramen.

Clinical Mini-Problems

1. **Severe earache and infection in the external auditory meatus may cause patients to vomit.**
   (a) Which nerve provides sensory innervation to the mucous membrane in the external auditory meatus?
   (b) Does this nerve have any sensory relationship with the gag reflex or gastrointestinal tract?
2. **Why would eruption of the molar teeth in children or adolescents sometimes "refer" a pain in the ear?**
3. **Which cranial nerve provides sensory and secretomotor innervation to the mucosa lining the middle ear?**
4. **Why would a clinician restrict the opening of the tympanic membrane to the lower one-half and particularly the postero-inferior quadrant?**
5. **Why would a nerve VII lesion in the internal acoustic meatus make the patient sensitive to loud noises?**
6. **Where in the temporal bone would one expect a lesion that gave signs of right-sided facial paralysis, dryness of the right eye, decreased auditory reception in the right ear, and a tendency to fall to the right side when standing with one's eyes closed?**

(Answers to these questions can be found on p. 587.)

41

Orbital Cavity and Contents

Clinical Case 41.1

Patient Frank S. This 63-year-old plumber with a 2-year history of intermittent headaches complained to his physician that he was experiencing "double vision." The physician examined his eyes and noted that the right eye deviated laterally and inferiorly at rest. The patient was unable to move the right eye medially (adduction), and the pupil was dilated and unresponsive to light, but no visual impairment was noted in the retina. The left eye was completely normal in its motor and sensory function. Now he is an inpatient at the University Hospital. Further tests reveal a meningioma (tumor of the dura) that is invading the upper aspects of the superior orbital fissure. At *combined neurology-ophthalmology ward rounds*, faculty and students discuss the alternatives for best treatment. You only have your anatomical knowledge to help you follow the debate.

Bony Orbital Cavity

Each orbital cavity is a pyramid-shaped compartment (*figs. 41.1 and 41.2*) with the base oriented toward the face and the apex directed into the middle cranial fossa. The roof of the orbit forms most of the floor of the anterior cranial fossa. The medial walls of each orbit parallel each other and are separated by the nasal cavities. The lateral walls of each orbit *do not* parallel each other. They diverge as they project anteriorly and are approximately at right angles to each other. The optic canal and the superior orbital fissure are located at the apex of each orbital cavity.

ORBITAL MARGIN

The **frontal, maxillary,** and **zygomatic** bones each constitute approximately one-third of the orbital margin (*fig. 41.2*). The orbital margin is subcutaneous and readily palpable in a physical examination. The inferior margin of the orbit continues medially as the **anterior lacrimal crest** of the **maxilla**. The superior orbital margin continues medially on the **frontal** bone to join the **posterior lacrimal crest** of the **lacrimal bone**. These lacrimal crests delineate the bony fossa that contain the **lacrimal sac** on the medial wall of the orbit. The orbital margin provides a suitable bony protection to the anterior of the eye except at the lateral margin. (Protective eye guards for squash or handball players are designed to complement the protective features of the orbital margin and still provide maximal peripheral vision.)

OPTIC CANAL

This bony channel is located in the sphenoid bone at the root of the lesser wing. It contains the **optic nerve (II)** and the **ophthalmic artery**, as they traverse between the orbit to the middle cranial fossa.

The length of the optic canal is 3–9 mm, and it parallels the plane of the lateral wall of the orbit. The canal is not visible in a standard head x-ray (anterior-posterior projection) (Plate 41.1). To demonstrate this canal in head films a special projection is used which requires that the x-rays parallel the long axis of the optic canal (*Plates 41.1 and 41.2*).

FISSURES AND SUTURES IN THE ORBITAL CAVITY

The **superior orbital fissure** is at the apex of the orbital cavity between the **lesser** and **greater wings** of the sphenoid bone. The superior orbital fissure contains connective tissue, cranial nerves **III, IV, V[1], VI,** and the **superior ophthalmic veins**. This non-bony component of the or-

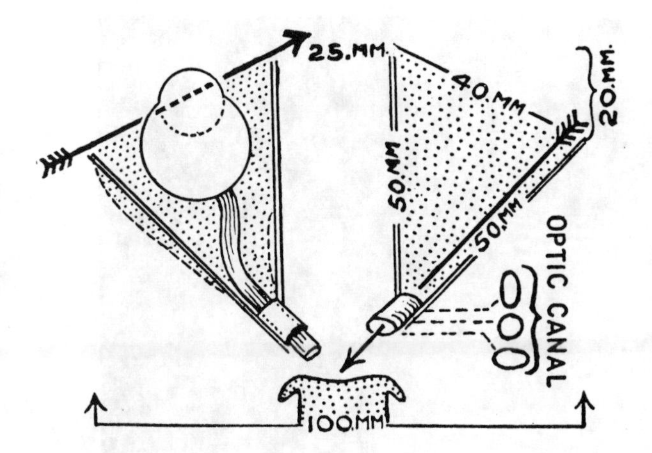

Figure 41.1. The orbital cavities on horizontal section, and their dimensions.

bital cavity appears as a radiolucent (dark) area on an x-ray film (*Plate 41.1*). The superior orbital fissure is lateral to the optic canal and inferiorly communicates with the **inferior orbital fissure**. The inferior orbital fissure lies between the greater wing of the sphenoid and the maxilla. It transmits the **infraorbital branch of V^2**, the **infraorbital vessels**, and the **inferior ophthalmic veins**.

The infraorbital groove on the orbital plate of the maxilla extends anteriorly from the inferior orbital fissure. It communicates with the face throughout the infraorbital foramen in the maxilla below the inferior margin of the orbit. The infraorbital nerve and vessels traverse this foramen and groove in their course between the face and the posterior aspect of the orbital cavity.

The sutural lines between the individual bones that make up the orbital cavity are shown in *Figure 41.2*. Six different skull bones: **frontal, maxilla, zygoma, sphenoid, ethmoid,** and **lacrimal;** fuse in sutural articulation to make a stable bony orbit.

WALLS OF THE ORBITAL CAVITY (*fig. 41.2*)

The medial wall is "paper-thin" and translucent in the dried skull. It separates the orbital cavity from the **ethmoid air cells** (sinuses) of the nasal cavity (*see fig. 41.5*). The **anterior** and **posterior ethmoid foramina** are openings in the wall that transmit nerves and vessels between the orbit and the ethmoid sinuses. The **lacrimal fossa** is also located anteriorly on the medial wall at the angle between the floor and the medial wall.

The inferior wall (floor) of the orbit is formed mainly by the **orbital process** of the **maxilla**. It contains the infraorbital groove, the infraorbital fissure, and the inferior part of the suture between the maxilla and zygoma.

The lateral wall is strong and functions as a major "force-transmitter" for the occlusal forces that project superiorly from the upper molar teeth. A **foramen for the zygomatic (V^2) nerve** is located in the lateral wall. A fossa for the **lacrimal gland** is also associated with the lateral wall at its angle with the frontal bone.

The superior wall (roof) of the orbit is formed by the **frontal** bone. The **supra-orbital notch** (sometimes a foramen) is a palpable defect on the superior margin of the orbit. The supra-orbital nerve (V^1) and vessels pass beneath the roof of the orbit, as they communicate with the skin and subcutaneous tissues over the superior margin of the orbit and forehead.

The bony orbit is lined internally by a periosteum (**the periorbita**). This tough connective tissue lining is easily detached from the roof and the medial wall during sur-

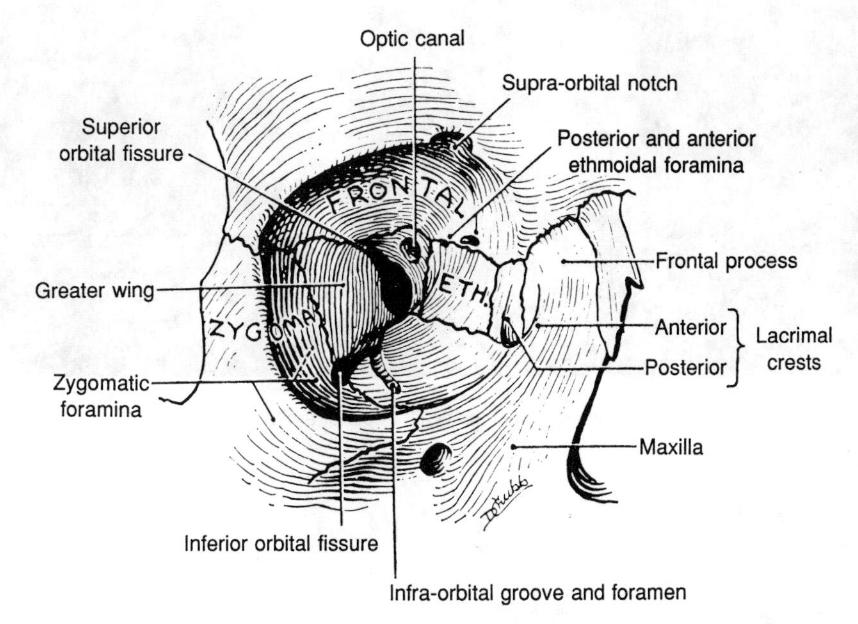

Figure 41.2. The bony walls of the orbit (orbital cavity).

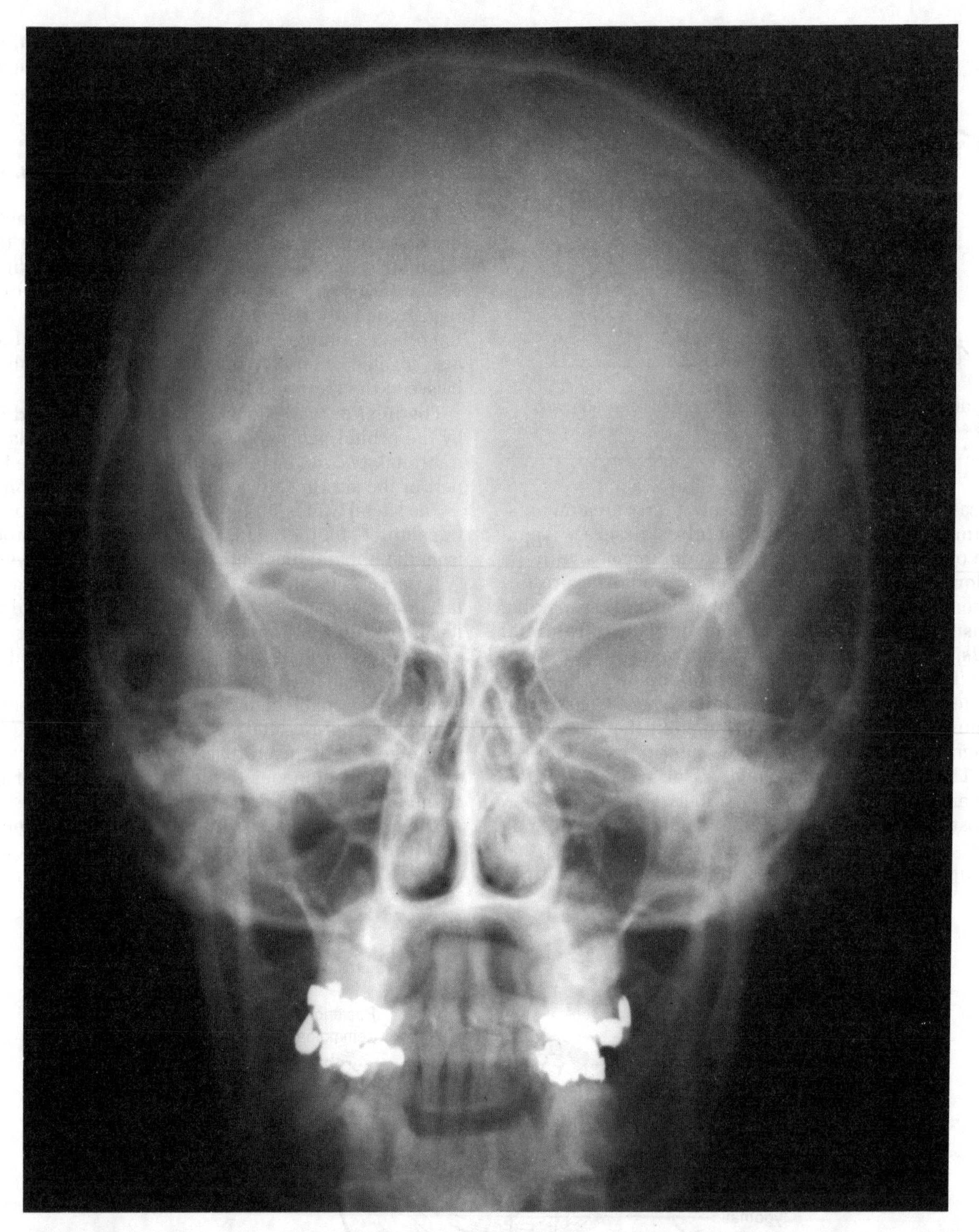

Plate 41.1. An anteroposterior projection of the orbital region of the skull. Superior orbital fissures appear as radiolucencies inferior to the lesser wing of the sphenoid.

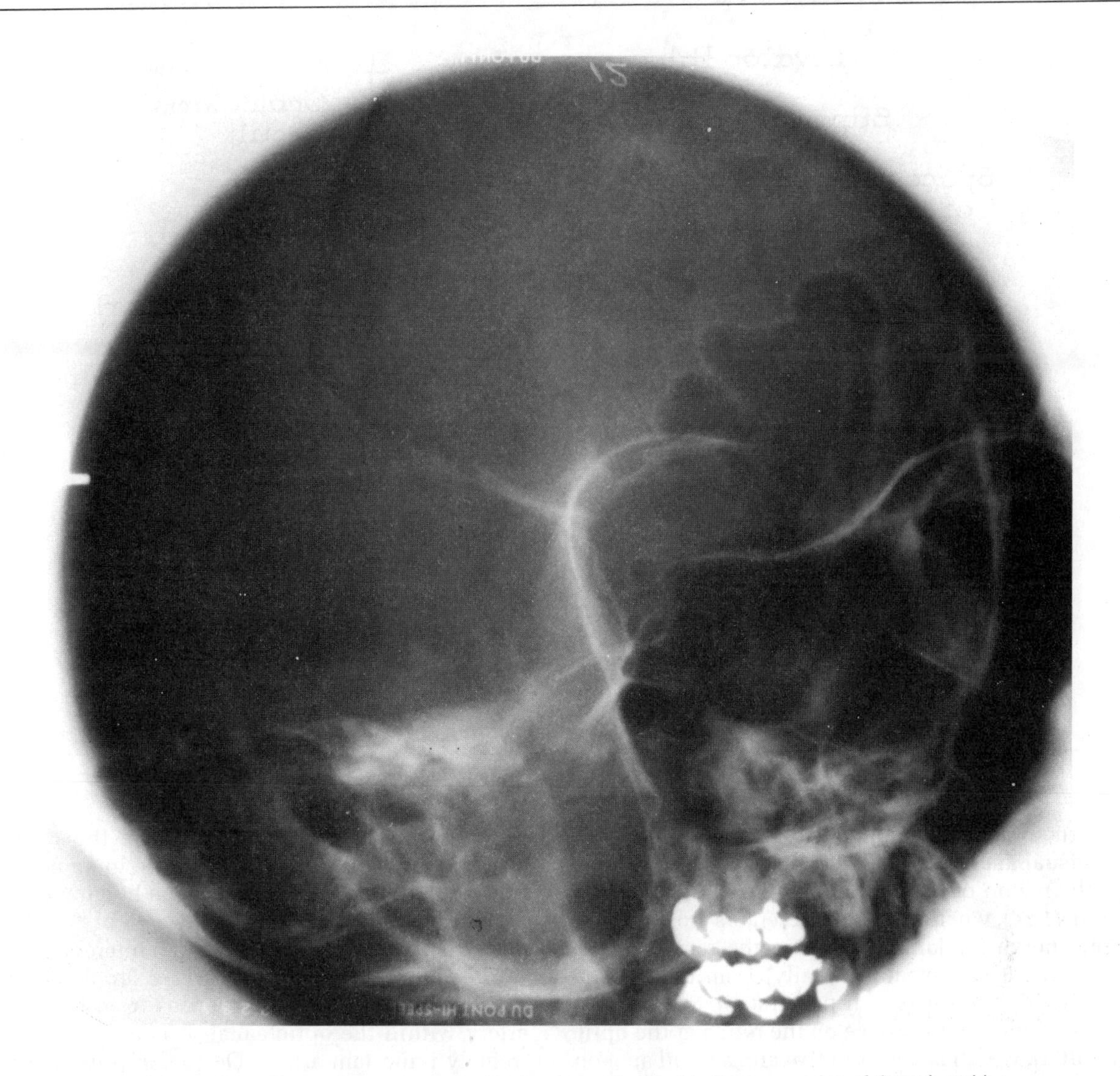

Plate 41.2. An oblique view of the orbit showing the optic canal in the lesser wing of the sphenoid.

gery. It is continuous with the **endosteal dura** of the middle cranial fossa at the superior orbital fissure.

Contents of the Orbital Cavity

EYEBALL (BULBUS OCULI)

The eye occupies the anterior half of the orbital cavity, while its associated muscles and nerves occupy the posterior half. Although the eye projects slightly beyond the bony orbital margin, the overlying soft tissues prevent a flat surface object, such as a book, from contacting the cornea when placed over the eye (*figs. 41.1 and 41.3*).

The posterior five-sixths of the eye is covered by a tough, white fibrous tissue (**the sclera**). The transparent anterior one-sixth is the **cornea**. The sclera and the cornea are continuous at the **corneoscleral junction**. The center of the cornea forms the **anterior pole** of the eye and the **posterior pole** is opposite.

A line joining the anterior and posterior poles is termed the **anteroposterior, sagittal,** or **optic axis**. This is a distance of approximately 24 mm. The eye reaches its adult size in early childhood (2–3 years), which correlates with the growth of the forebrain. The orbital and cranial growth in a young child is more prominent than facial growth in the lower half of the face. The latter region normally reaches its maximal growth after puberty in the late teens.

The **retina** lines the inner surface of the eye posterior

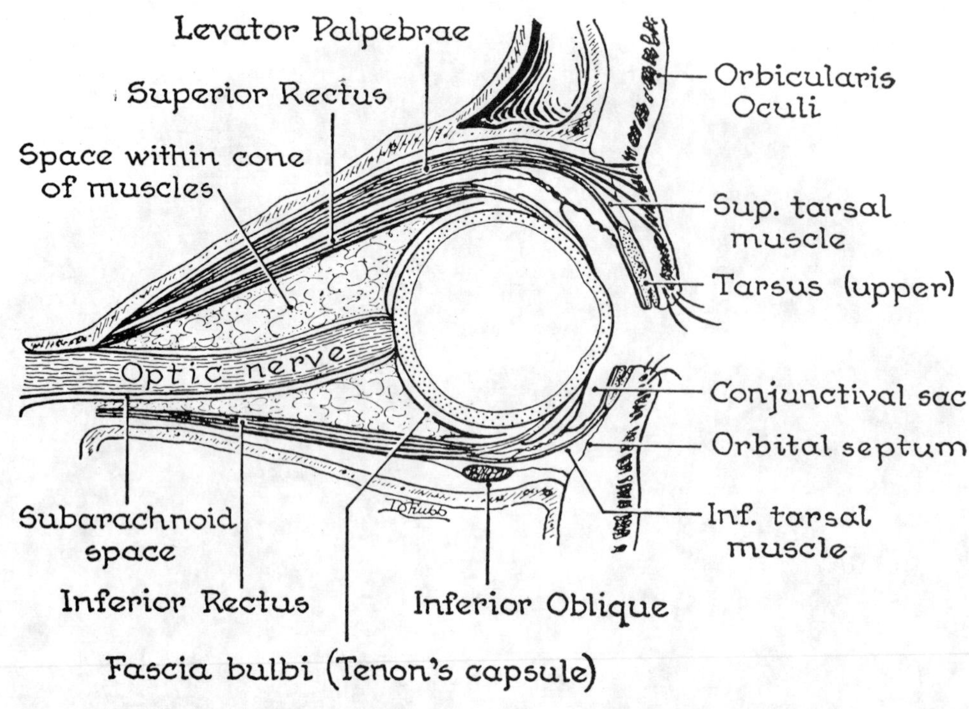

Figure 41.3. Diagram of the orbital cavity (on sagittal section).

to the corneoscleral junction. Vision is perceived most keenly (**visual acuity**) on the retina at the posterior pole. This is the region of the **macula** (yellow spot) (*see figs. 41.12 and 41.14*), where the retinal receptors are all cones. The **optic nerve** (II) joins the eye in the region of the posterior pole 3 mm medial (nasally) from the macula (*see fig. 41.14*). The fibers of II penetrate the entire posterior wall of the eye to surface on the retina at the **optic disc (papilla),** which is nonreceptive and a blind spot on the retina.

OPTIC NERVE (II)

The optic nerve carries the sensory impulses generated by stimulation of the receptors (rods and cones) in the retina. It passes posteriorly from the back of the eye, through the orbit to enter the optic canal in the lesser wing of the sphenoid bone. The optic nerves form the optic chiasm and optic tracts within the middle cranial fossa after they exit the optic canals via the optic foramen. The course of each optic nerve within the orbital cavity is somewhat sinuous (*fig. 41.1*). This permits a certain degree of unrestrained movement of the eye within the orbital cavity.

The optic nerve is a part of the brain and is therefore covered by meninges. *Figure 41.3* shows the *very* important relationship of the optic nerve and the dura mater and its underlying subarachnoid space. The dura is continuous with the sclera of the eye, and the subarachnoid space abuts against the posterior aspect of the retina around the optic disc. Increased pressure in the subarachnoid space is therefore readily detectable in an eye examination with an ophthalmoscope. The bulging of the retina around the optic disc creates the sign of **"papilledema,"** which is characteristic of increased intercranial pressure.

The optic nerve and the retina are supplied by a **central artery of the retina**, which arises from the ophthalmic artery within the optic canal. It is a very small but extremely important artery. Occlusion (thrombosis) of the central artery will cause blindness in the affected eye.

MUSCLES OF THE EYEBALL

There are six muscles that control the movement of each eyeball. Four of these muscles project in a straight line from the apex of the orbital cavity to the sclera, 6–8 mm behind the corneoscleral junction. They are called the rectus (straight) muscles. The two remaining muscles approach the eyeball from a medial and anterior direction, and they are termed oblique muscles (*figs. 41.4 and 41.5*).

The **superior rectus, inferior rectus, medial rectus,** and **lateral rectus** muscles arise from a fibrous, annular cuff at the apex of the orbital cavity (*fig. 41.6*). This annular arrangement of muscle origins encircles the opening of the optic canal and part of the superior orbital fissure as they communicate with the orbital cavity. The insertion of each rectus muscle into the sclera is just posterior to the conjunctiva that covers the corneoscleral junction on the anterior surface of the eye.

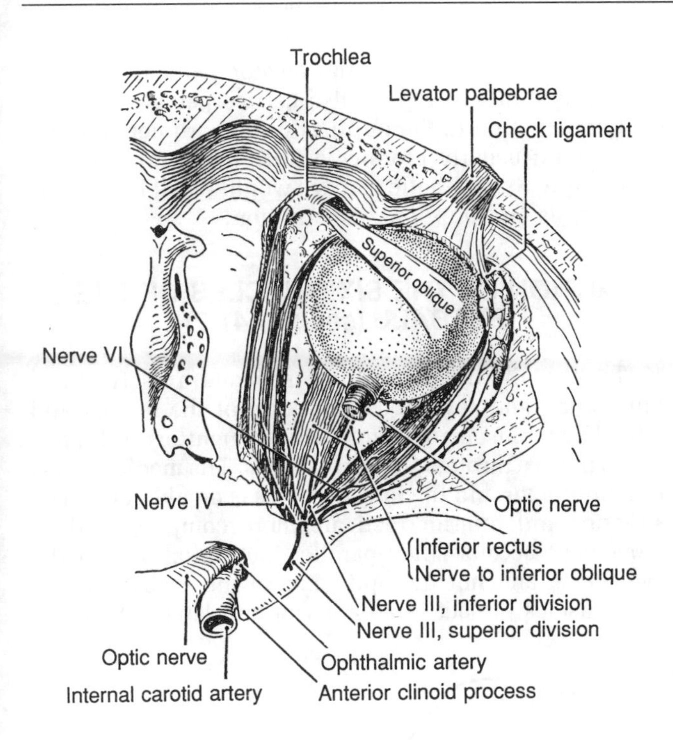

Figure 41.4. Dissection of orbit from above.

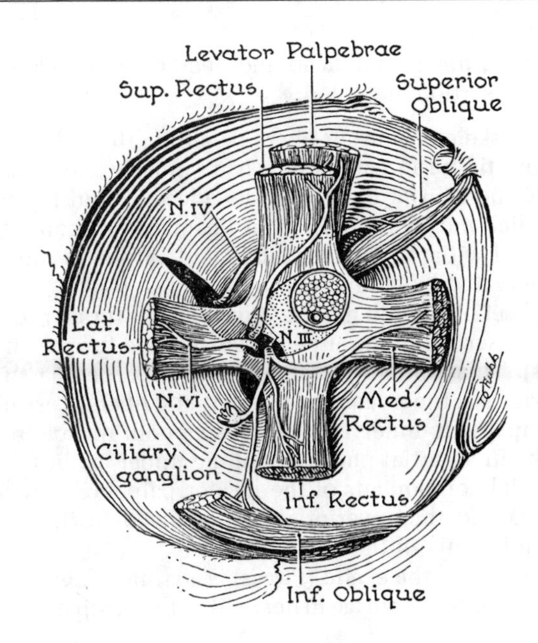

Figure 41.6. Sketch of the distribution of cranial nerves III, IV and VI.

The rectus muscles diverge to their respective quadrants on the eyeball. The rectus muscles form a cone of muscles in the posterior half of the orbit surrounding the optic nerve. The space within the muscular cone is filled with fatty tissue, which helps to support the contents of the orbit (*fig. 41.3*).

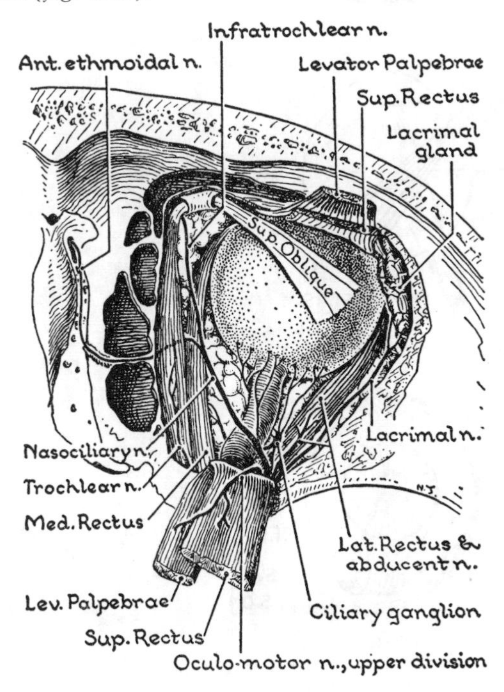

Figure 41.5. Dissection of orbital contents.

The **superior** and **inferior oblique muscles** are skeletal muscles that project from the medial wall of the orbit in a posterior and lateral direction into the sclera on the posterior half of the eye. The oblique muscle passes inferior to the associated rectus muscles as they cross the superior and inferior surfaces of the eyeball (*fig. 41.7*).

The superior oblique arises from the apex of the orbit in the medio-superior angle just outside the ring of fibrous tissues that serves as the origin for the recti muscles. The muscular part of the superior oblique projects anteriorly to a connective tissue pulley, the **trochlea**, which is located in the superomedial angle just inside the anterior margin of the orbit (*figs. 41.4 and 41.5*). The tendon of the superior oblique passes through the trochlea and then "fans" out as it projects to insert into the sclera of the posterolateral quadrant of the eye. The muscle therefore shortens along the medial wall of the orbit and its effective force is redirected by the trochlea to pull on the eye from a medial and anterior direction.

The inferior oblique arises from the periorbita on the medial wall on the inferomedial angle near the lacrimal fossa. It passes posteriorly and laterally beneath the inferior rectus to insert into the posterior lateral quadrant of the undersurface of the eyeball (*fig. 41.6*).

The **levator palpebrae superioris** muscle is delaminated from the superior portion of the superior rectus muscle. It shares a similar nerve supply with the superior rectus but inserts into the eyelid and **not** the eyeball. The muscles work congruently so that as the superior rectus rotates the eye superiorly, the levator palpebrae raises the eyelid to expand the visual field of the patient.

The details of the insertion of the levator palpebrae superioris are illustrated in *Figure 41.3* and are clinically important in diagnosing head and neck disorders. Four points of insertion are noted:

1. Most skeletal muscle fibers penetrate the subcutaneous tissue and orbicularis oculi muscle and insert into the skin of the eyelid. Its edges extend to the medial and lateral palpebral ligaments and are attached with them. This component of the muscle is innervated by cranial nerve III.
2. A sheet of **smooth muscle, the superior tarsal muscle**, inserts into the superior tarsal plate (upper tarsus). This component of the levator palpebrae superioris is innervated by sympathetic fibers. Loss of sympathetic innervation to this smooth muscle results in a partial ptosis (drooping upper eyelid).
3. Fascial components of the levator palpebrae superioris and the superior rectus attached to the superior fornix of the conjunctiva.
4. Drooping of the eyelid (*Ptosis*) is a clinical sign that reflects either damage to nerve III or the sympathetic fibers that innervate the levator palpebrae superioris. Since nerve III also carries parasympathetic innervation to the pupil as well, a 3rd nerve lesion would usually produce a full ptosis and a dilated pupil. Sympathetic damage would result in a partial ptosis and pupillary constriction.

ACTION OF THE SIX MUSCLES OF THE EYEBALL (*see fig. 41.7*)

The eyeball rotates on three separate axes (vertical, horizontal, and sagittal). The action of the medial and lateral recti muscles are simple and straightforward. They rotate the eye through the vertical axis. The **medial rectus** is an **adductor** and the **lateral rectus** is an **abductor**. The superior and inferior recti are more complex in their movement because they parallel the orbital axis of the eye socket and not the optic axis of the eyeball. These two muscles project in an anterior and lateral direction

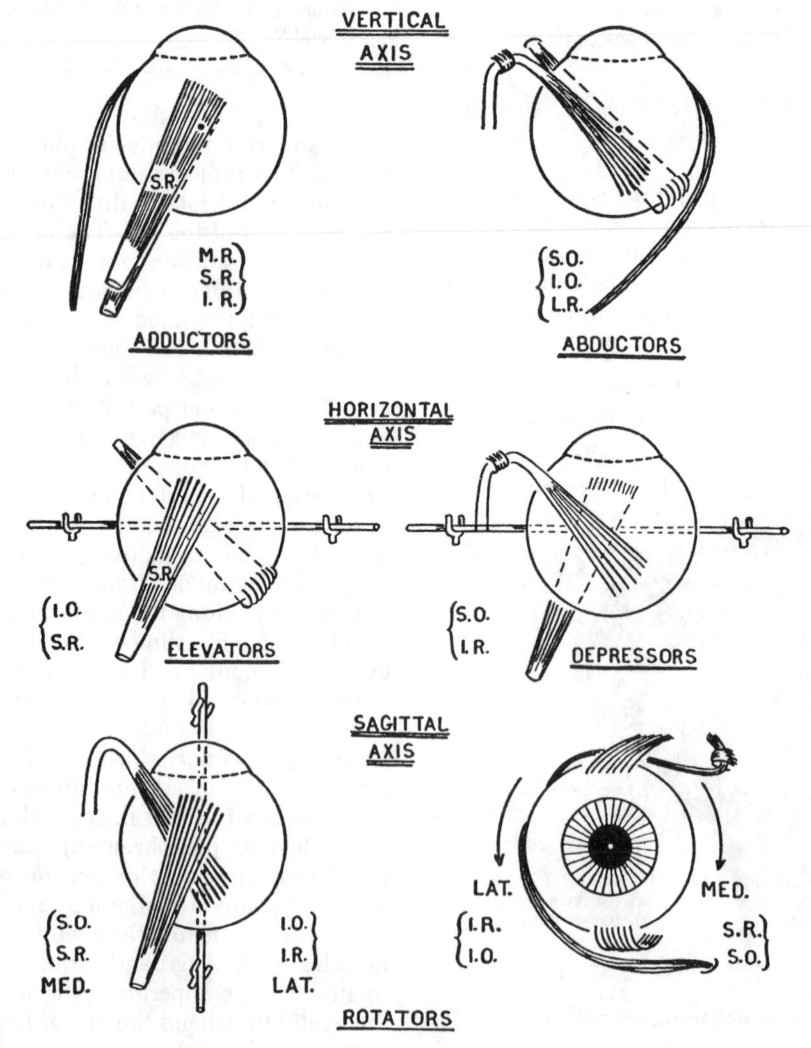

Figure 41.7. The actions of the 6 muscles of the right eyeball represented graphically. *MR*, medial rectus; *SR*, superior rectus; *IR*, inferior rectus; *SO*, superior oblique; *IO*, inferior oblique.

as they cross their respective surface of the eyeball. The major function of these muscles are through the horizontal axis. The **superior rectus** is an **elevator** and the **inferior rectus** is a **depressor**. Since both muscles lie medial to the center of the vertical axis, they are **both adductors** of the eyeball as well. The superior rectus and inferior rectus also approach the sagittal axis from the medial side and will therefore produce a "torque" through the sagittal axis. The **superior rectus** will effect a **clockwise medial rotation** (intortion) and the **inferior rectus** will produce a **counterclockwise lateral rotation** (extortion) of the eye.

The oblique muscles are also complex movers of the eye. Their effective direction of force is from medial to lateral in a posterior direction. These forces are nearly perpendicular to those of the superior and inferior recti muscles. The **superior and inferior oblique** muscles pass posterior to the center of the vertical axis and therefore act as **abductors** of the eye. In the horizontal axis, the **inferior oblique** acts as an **elevator** by pulling the posteroinferior surface forward. The **superior oblique** acts as a **depressor** by pulling the posterosuperior surface of the eye forward. The oblique muscles also cross the sagittal axis and therefore produce torque through the optic axis. The **superior oblique** produces a **clockwise medial rotation**, while the **inferior oblique** induces a **counterclockwise lateral rotation** of the eye. Movements of the eye are very important in clinical diagnosis. A test for examining the movement of each individual muscle will be described after the description of the nerve supply to these muscles.

The brain coordinates these complex movements to create smooth bilateral pursuit movements of the eyes. Dysfunction in these movements usually causes the symptom of "double vision" or **diplopia**. This symptom reflects a problem in the motor supply to the eye muscles and not the sensory supply to the retina, which would produce the symptom of blindness.

Nerves in the Orbital Cavity

Four major nerve groups enter the orbital cavity: (*a*) motor nerves to skeletal muscles; (*b*) special sensory nerves to the retina (II); (*c*) general sensory nerves for pain and temperature sensation; and (*d*) autonomic nerves from the sympathetic and parasympathetic divisions that will supply smooth muscle and glands.

THE THREE MOTOR NERVES TO SKELETAL MUSCLES OF THE ORBIT (*fig. 41.6*)

The **abducens nerve (VI)** enters the orbit through the superior orbital fissure. It is enclosed by the annular ring that serves as a point of origin for the recti and it **innervates the lateral rectus** on the surface within the muscular cone. **Nerve VI only** innervates this one eye muscle and therefore influences **abduction** of the eye.

The **trochlear nerve (IV)** also enters the orbit through the superior orbital fissure, but is outside of the fibrous annular ring. Nerve IV courses superiorly across the upper surface of the origins of the superior rectus and levator palpebral superioris muscles to reach the superior border of the superior oblique muscle. **Nerve IV innervates only the superior oblique muscle.**

The **oculomotor nerve (III)** innervates the superior rectus and levator palpebrae superioris by a superior division and the medial rectus, inferior rectus and inferior oblique muscles through an inferior division (*fig. 41.6*). Both divisions of III enter the orbit through the superior orbital fissure and are contained within the fibrous annular ring. The muscles are therefore supplied from the surface that is within the muscular cone of the orbit. The inferior division of III also carries preganglionic parasympathetic fibers to the ciliary ganglion.

A convenient memory aid for remembering the innervation of the eye muscles is $LR_6(SO_4)_3$—the **lateral rectus** by VI, **superior oblique** by IV, and all the rest by III.

A convenient memory aid for remembering how to test the individual muscles of the eye and their nerve supply is the letter H (*fig. 41.8*). Please note that the test is not the action of the muscle but a trick to isolate each muscle and test it in isolation! Therefore, to test the superior oblique muscle, ask the patient to look medially and down. The inferior movement of the eyeball from the medially adducted position is solely affected by the superior oblique. This tests IV. The abducens nerve (VI) is tested by contracting the lateral rectus muscle and observing the eye in abduction. All of the other eye muscles are innervated by the oculomotor nerve (III). Each muscle that is innervated by nerve III can be isolated from the others: the

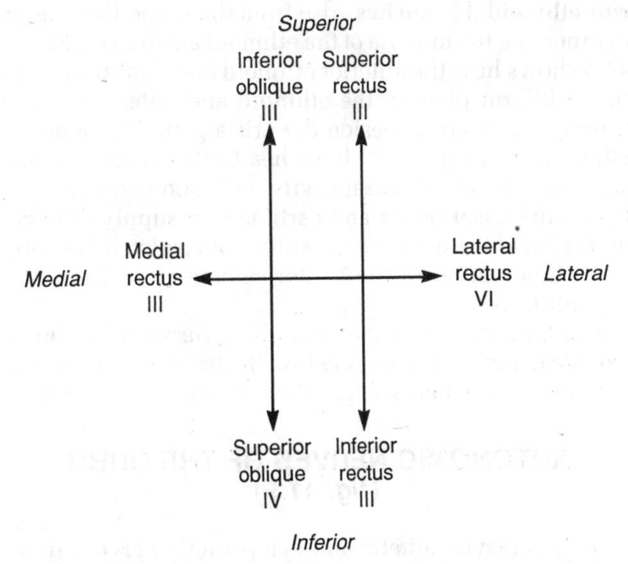

Figure 41.8. Diagram of the test movements for examining individual muscles that control eye movements.

medial rectus by adducting the eye, the inferior oblique by elevating the *adducted* eye, the inferior rectus by depressing the *abducted* eye, and the superior rectus by elevating the *abducted* eye.

SENSORY NERVES IN THE ORBIT (V^1)

The **frontal nerve** passes above the lateral rectus to lie between the levator palpebrae superioris muscle and the periorbita on the inferior surface of the orbital plates of the frontal bone. It divides into supraorbital and supratrochlear branches, which supply sensation to the skin of the upper eyelid, forehead, and scalp (*fig. 38.11*).

The **lacrimal nerve** follows the upper border of the lateral rectus muscle and ends in the lacrimal gland and the upper eyelid (*fig. 41.5*). It is *only* sensory to the lacrimal gland. The secretomotor supply to the gland is via the parasympathetic fibers from nerve VII through the pterygopalatine ganglion and branches of V^2.

The **nasociliary nerve** (*fig. 41.5*) is of utmost clinical importance. It enters the muscular cone of the orbit between the two heads of the lateral rectus within the fibrous annular ring. It courses over the optic nerve to the medial wall of the orbit and proceeds anteriorly between the superior oblique and medial rectus muscles.

Long ciliary nerves arise from the nasociliary nerve, as it crosses nerve II. The long ciliary nerves pass forward to enter the sclera on the anterior half of the eye. They will carry the clinically important sensory fibers required for the "corneal reflex."

Short ciliary nerves also arise from the nasociliary nerves and then pass forward to the sclera of the posterior half of the eye. Although these are sensory nerves to the eye, they also carry postganglionic parasympathetic fibers from the ciliary ganglion associated with nerve III to two **smooth muscles** in the eye: **pupillary constrictor** and **ciliary body** muscles.

On the medial wall of the orbit, the **posterior** and **anterior ethmoidal branches** arise from the nasociliary nerve to innervate the mucosa of the ethmoid air sinuses. *Figure 41.5* shows how the anterior ethmoid also continues onto the cribiform plate of the ethmoid and enters the nose through an opening beside the crista galli. The anterior ethmoid nerve gives V^1 branches to the mucosa of the superior part of the nasal cavity and then continues between the nasal bones and cartilages to supply the skin on the tip of the nose. This results in an isolated "island" of V^1 innervation in the V^2 dermatome area of the face (*fig. 38.11*).

The termination of the nasociliary nerve is the **infratrochlear nerve**. It passes below the trochlea and is sensory to the lacrimal sac and the mucosa surrounding it.

AUTONOMIC NERVES OF THE ORBIT (*fig. 41.9*)

Both parasympathetic and sympathetic nerves are required to innervate the smooth muscles and glands within the eye and orbit. The parasympathetic preganglionic fibers are contained in the **oculomotor nerve (III)**. They enter the orbit with III and proceed through the inferior division of nerve III to synapse in the **ciliary ganglion** on the cell bodies (neurons) of the postganglionic parasympathetic neurons. These join the short ciliary branches of the nasociliary nerve and enter the posterior aspect of the eye and pass anteriorly between the scleral and choriod coats of the eye to innervate two intrinsic smooth muscles: the **pupillary constrictor muscle** and the **ciliary body muscle**. The parasympathetic innervation of the pupil can be tested by shining a light in the eye and observing a pupillary constriction (**pupillary light reflex**). The parasympathetic innervation of the ciliary body is tested by asking the patient to read fine print. This is the **accommodation reflex**, which also requires pupillary constriction and medial convergence of both eyes by the medial rectus muscles. All of these functions are controlled by nerve III, even though the accommodation reflex involves both smooth and skeletal muscle action.

The parasympathetic innervation of the lacrimal gland will be described with the pterygopalatine ganglion (page 486).

The **preganglionic sympathetic fibers** of the orbit arise in the upper thoracic region of the spinal cord. They ascend in the sympathetic trunk to the superior cervical ganglion (*fig. 41.9*) to synapse on the cell bodies of the postganglionic neurons. The **postganglionic sympathetic neurons** then travel to the orbit on the plexuses of the internal carotid and ophthalmic arteries. They enter the orbit with the ophthalmic artery in the optic canal and proceed to the anterior half of the eye on either the long ciliary nerves or arteries. Within the eye, the sympathetics innervate the **dilator pupillae muscle**. These radial fibers of the iris cause pupillary dilatation when they contract. Pupillary dilatation can be tested when the lights

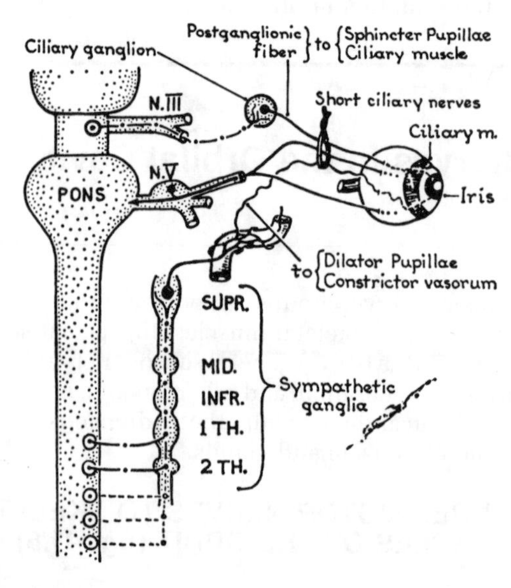

Figure 41.9. The autonomic nerve supply to the eyeball.

in a room are dimmed. Sympathetic fibers also innervate the blood vessels of the orbit to control the vasomotor tone of the blood vessels by active constriction of the vascular smooth muscle. Damage to the sympathetic system that supplies the orbit produces a **Horner's Syndrome**, which is characterized by 4 signs: (*a*) pupillary constriction (miosis) due to unopposed parasympathetic influence on the iris musculature; (*b*) partial drooping of the eyelid (ptosis) due to denervation of the smooth muscle component of levator palpebrae superioris; (*c*) lack of sweating on the skin around the orbit (anhydrosis); and (*d*) vasodilatation of the blood vessels of the skin around the orbit causing a "flushing" of the skin. A fifth sign of enopthalmos (retraction of the eye into the orbit) is described but not well explained.

VESSELS OF THE ORBITAL CAVITY

The **ophthalmic artery** (*fig. 41.10*) is the major blood supply to the orbit and its contents. It arises in the middle cranial fossa from the internal carotid artery within the cavernous sinus. The ophthalmic artery enters the orbit on the inferior surface of the optic nerve in the optic canal. Its initial branch, the **central artery of the retina**, is given off in the optic canal, and it enters the optic nerve to reach the retina through the optic disc. The central artery can be examined **in situ** with an ophthalmoscope and is a valuable clinical observation for many vascular and neurologic disorders. It is also the major blood supply to the sensory retina and occlusion of the vessel causes instantaneous blindness.

The ophthalmic artery enters the muscular cone of the orbit on the lateral aspect of nerve II. It ascends and crosses the superior aspect of the optic nerve and courses toward the medial wall of the orbit. The individual branches of the ophthalmic artery are complementary to the branches of V^1. If one knows the branches of the frontal, lacrimal, and nasociliary nerves, one can name the complementary branches of the ophthalmic artery. An important anastomosis between the ophthalmic artery and the facial artery occurs on the medial orbital margin. This provides a potential connection between the external and internal carotid arterial systems.

OPHTHALMIC VEINS

The **superior ophthalmic vein** anastomoses with the facial vein at the medial angle of the eye. Since there are no valves in these veins, blood can flow in either direction (*fig. 39.3*). This relation has the potential of having blood from infected areas of the face flowing in the reverse direction through the orbit, through the superior orbital fissure, and into the cavernous sinus. The danger of a facial infection causing a severe meningitis is possible by this anatomical route.

The **inferior ophthalmic veins** on the floor of the orbit communicate with the pterygoid plexus of veins in the deep face. The inferior ophthalmic vein drains through the inferior orbital fissure to reach the cavernous sinus. These venous channels also lack valves and can permit deep facial infections to "spread" to the meningeal sinuses.

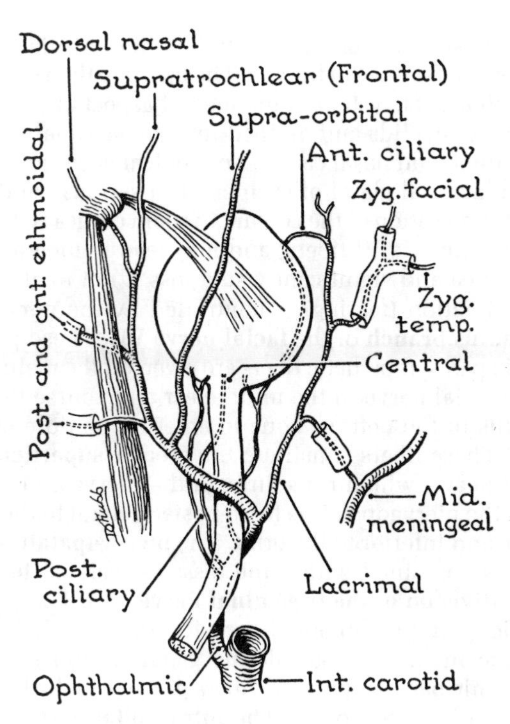

Figure 41.10. Ophthalmic artery and its branches.

THE LACRIMAL GLAND

This serous gland (*fig. 41.11*) lies behind the superolateral angle of the orbital margin, between the orbital plate of the frontal bone and the conjunctiva. It surrounds the lateral margin of the levator palpebrae muscle. A series of ducts leave the gland and pierce the conjunctiva, so that tears may spread over the conjunctiva and cornea

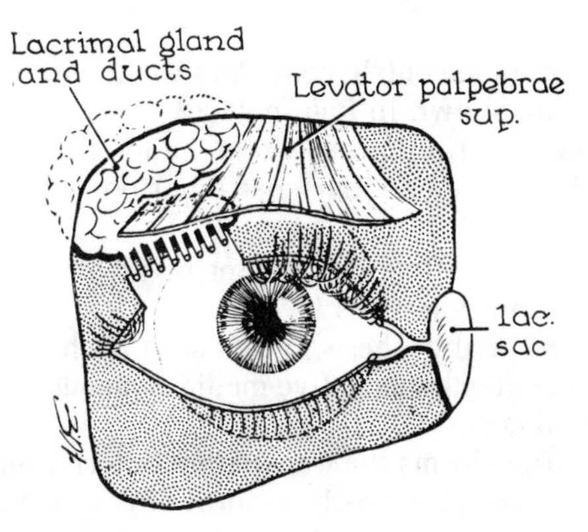

Figure 41.11. The lacrimal gland of the right orbit (schematic).

from the superolateral region to the inferomedial area of the anterior surface of the eye. The tears are then collected through a duct system in the medial aspect of the upper and lower eyelids and drained into the lacrimal sac, lacrimal duct, and nasal cavity. Loss of tear secretion by the **single** lacrimal gland of each eye is a serious condition. Rapid desiccation of the conjunctiva and cornea will cause severe pain (via V^1 fibers) and may lead to ulceration of the cornea and significant visual loss if not treated.

Lacrimation (tearing) is controlled by the **nervus intermedius branch** of the **facial nerve VII**. These preganglionic parasympathetic fibers are given off the main trunk of the facial nerve in the middle ear and course through a hiatus in the petrous bone to enter the middle cranial fossa. These fibers constitute the **greater superficial petrosal nerve**, which runs through the pterygoid canal to reach the pterygopalatine fossa posterolateral to the nasal cavity and inferior to the orbit. The **pterygopalatine ganglion** is contained within the fossa as well as the **maxillary division of the trigeminal nerve (V^2)**. The preganglionic parasympathetic fibers for the lacrimal gland synapse in the pterygopalatine ganglion and the postganglionic parasympathetic fibers proceed to the lacrimal gland on branches of V^2. The **infraorbital nerve** carries the postganglionic fibers into the orbit and passes them to the lateral wall of the orbit with its **zygomatic branch**. The zygomatic nerve enters the zygomatic foramen below the lacrimal gland, and the parasympathetic fibers ascend the short distance along the lateral orbital wall to enter the gland and innervate the secretomotor function of the gland. These terminal parasympathetic fibers may join the lacrimal nerve. The lacrimal nerve is a branch of V^1 that carries pain stimuli from the gland and adjacent areas of skin.

The Eyeball

The **three concentric coats** and their individual components are shown in *Figure 41.12*:

(a) Outer or fibrous coat (sclera and cornea)

(b) Middle or vascular coat (choriod, ciliary body & iris)

(c) Inner or retinal coat (outer pigmented epithelium and inner nervous layer)

Enclosed within these 3 coats are the chambers of the eye, containing the refractive media: aqueous humor, lens, and vitreous body.

The cornea forms the transparent anterior one-sixth of the outer coat of the eye. It is more convex than the scleral portion. The cornea is 0.5 mm–1.0 mm thick and is responsible for most of the refraction by the eye. It is avascular and can be readily transplanted due to its lack of blood supply and lymphatics. The cornea is supplied by sensory branches of V^1 and V^2. When the cornea is touched lightly, the afferent (sensory) fibers induce a corneal reflex, which is expressed by blinking the eyelids through the motor supply to the orbicularis oculi muscle via nerve VII.

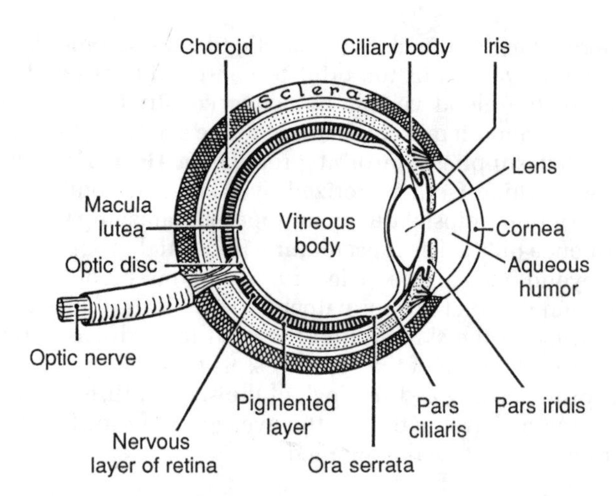

Figure 41.12. Scheme of an eyeball (sagittal section). *V*, vitreous body; *L*, lens; *A*, aqueous humor.

The **sclera** forms the posterior five-sixths of the outer coat of the eyeball. It is seen through the conjunctiva as the "white of the eye" and it is not transparent. The sclera serves as the insertion site for the extrinsic recti and oblique muscles of the eye. It is continuous with the dura mater of the optic nerve and is penetrated posteriorly by the neuronal fibers of nerve II. The **sclerocorneal junction** marks the union of the two portions of the outer coat of the eye. A pinpoint size canal, the **scleral sinus** (Canal of Schlemm) runs in a circular fashion adjacent to this sclerocorneal junction (*fig. 41.13*).

The space behind the cornea is the **anterior chamber**. It is continuous through the pupil with the **posterior chamber**. Both chambers contain aqueous humor. The **ciliary processes** of the ciliary body project into the posterior chamber and secrete the aqueous humor into the posterior chamber. The aqueous humor "circulates" through the pupil into the anterior chamber where it drains at the angle of the anterior chamber between the iris and the cornea (**iridocorneal angle**) through channels into the scleral sinus. The aqueous humor is then transported to the venous plexus within the sclera. Disruptions in the absorption of aqueous humor from the anterior chamber cause **glaucoma**.

The **ciliary body** is the anterior component of the middle or vascular coat of the eye. It is attached firmly to the outer coat at the sclerocorneal junction, and it also projects the **iris** and **ciliary processes** into the anterior and posterior chambers of the eye, respectively. The ciliary body is continuous posteriorly with the pigmented **cho-**

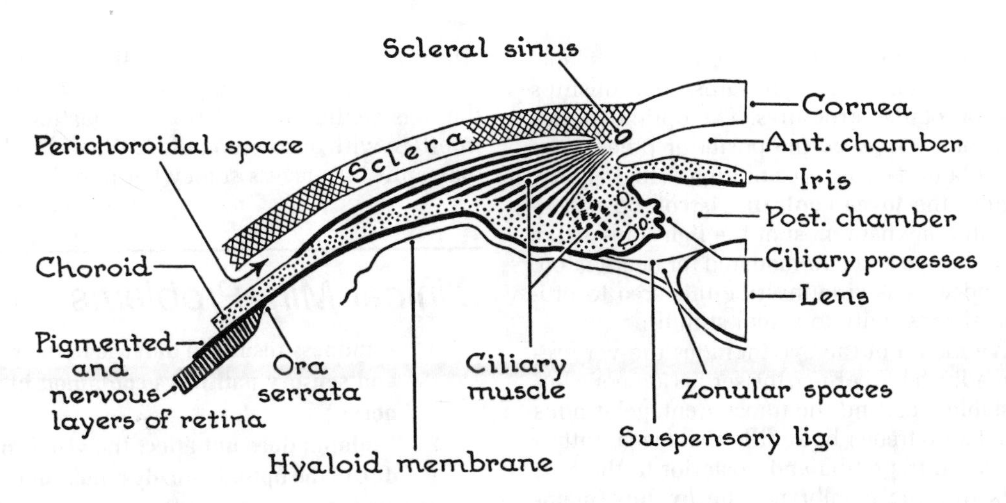

Figure 41.13. The ciliary region, enlarged. (After Cunningham's Anatomy.)

roid. The **ciliary muscle** is contained within the ciliary body. It is a smooth muscle that is innervated by parasympathetic fibers of the oculomotor nerve (III).

Contraction of this circular muscle decreases its diameter and relaxes the **suspensory ligaments** that extend from the ciliary body to the **lens**. The elastic nature of the lens allows it to "thicken" (become more rounded) when the suspensory ligaments are relaxed. The thicker lens is necessary for focusing the light rays on the retina in near vision. It therefore requires muscular energy to focus and do visual work at close range. Far vision is performed through a "thinner" lens, which occurs when the ciliary muscle relaxes and the diameter of the ciliary body expands. The tension on the suspensory ligaments pulls on the perimeter of the lens and makes it thinner and less refractive to the light rays entering the eye. The lens is therefore adjustable to some degree, while the cornea is a fixed refractive structure.

The colored **iris** is a circular diaphragm that lies anterior to the lens and has a central aperture, **the pupil**. The **pupillary constrictor** muscle (**sphincter pupillae**) is on the posterior aspect of the iris near the pupil. It is innervated by parasympathetic neurons in nerve III, which reach the eye via the short ciliary branches of V^1. The **pupillary dilator muscle (dilator pupillae)** is a radial array of smooth muscle fibers that extend like spokes of a wheel from the pupillary margin. The **pupillary dilator muscle** is innervated by **sympathetic fibers**.

The **retina** consists of pigmented and nervous layers on the posterior inner surface of the eyeball. The pigmented layer extends anteriorly on the inner surface of the choriod, ciliary body, and iris. The nervous portion, which contains the receptive rods and cones, is limited in its anterior extension and forms the **ora serrata** at its anterior limit in the eye. The retina, optic disc, macula lutea (yellow spot), and the central arteries and veins can

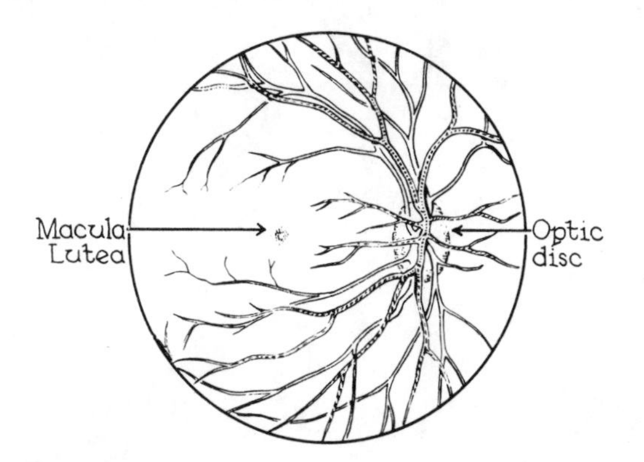

Figure 41.14. The retina and its arteries and veins as seen in the right eye through an ophthalmoscope.

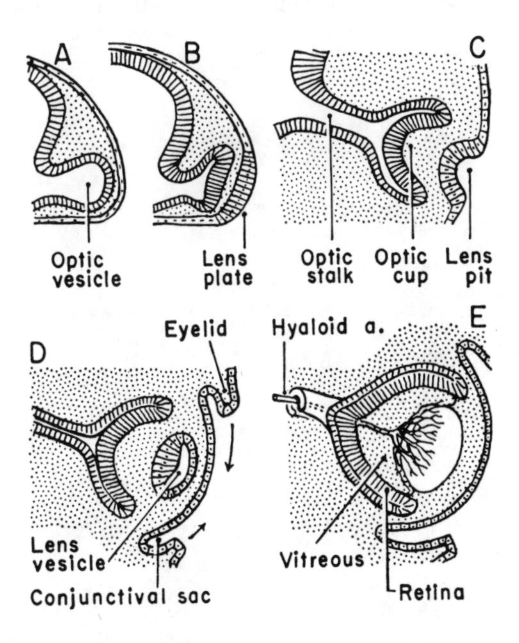

Figure 41.15. Stages in the development of retina, lens, and conjunctival sac. (After Mann.)

be visualized directly **in situ** with an ophthalmoscope. *Figure 41.14* is a diagram showing the gross relationships of these important ocular structures. The optic disc lies medial to the macula lutea at the posterior pole of the retina. The macula lutea is an area of high visual acuity. Its central region, the **fovea centralis**, is composed entirely of cones and is where most of the light rays strike the retina when the eye is accommodated for near vision. The retinal blood vessels circumscribe this area to provide the maximal sensitivity to visual stimuli.

The **refractive media** of the eye includes the very refractive but nonadjustable cornea, the serous aqueous humor, the adjustable lens, and the transparent, gelatinous **vitreous body**. The vitreous body fills the space within the globe internal to the retina and posterior to the lens (*fig. 41.13*). Its limiting membrane, the **hyaloid membrane**, abuts the inner surfaces of the ciliary body and lens (*fig. 41.13*).

DEVELOPMENT

The retina develops as an outgrowth of the brain (*fig. 41.15*) and is of ectodermal origin. In its initial stages, it resembles an inflated rubber balloon with a hollow stalk that is continuous with the cavity of the 3rd ventricle. At the same time, the lens is developing from the overlying skin ectoderm. The balloon-like **optic vesicle** collapses under the influence of the developing lens and the ensuing invaginated vesicle becomes cup-shaped. A two-layered retina is therefore formed.

The central artery originally passed through the hyaloid canal in the vitreous body and anastomosed with the capsule of the lens, which is highly vascular in the fetus. Before birth, the lens capsule atrophies and the artery degenerates back to the retina. The lens will harden and cornify with advancing age. The hardening begins at the lens center, and this loss of elasticity can gradually interfere with accommodation. When hardened, the lens may split into layers somewhat like those of an onion.

Clinical Mini-Problems

1. **Blindness results in damage to the visual receptors and sensory pathways contained in which cranial nerve?**
2. **Diplopia does not affect the visual pathway but is due to the optokinetic dysfunction resulting in paralyzed or weakened eye muscles. Which cranial nerves when damaged could cause diplopia?**
3. **Which cranial nerve would be damaged in a patient with a drooping upper right eyelid and an unresponsive dilated right pupil when exposed to bright light?**
4. **Which cranial nerve would be damaged in a patient with a drooping lower eyelid on the right side?**
5. **How would damage to the two structures in the optic canal cause blindness?**
6. **How would damage of cranial nerve VII cause pain in the eye?**
7. **Is the optic disc in the center of the posterior pole of the eyeball? When the blood vessels in the center of the optic disc radiate onto the retina are they medial or lateral to the macula lutea?**

(Answers to these questions can be found on p. 000.)

42

Posterior Triangle of the Neck

Clinical Case 42.1

Patient Mildred Y. This woman with chronic gallbladder disease now comes to the University Clinic complaining of pain over her right shoulder with each breath. You are able to explain that this peculiar symptom may be due to "referred pain" because the sensory nerves from the right dome of the diaphragm and the skin over the right shoulder enter the same level of the spinal cord.

The posterior triangle of the neck contains a number of important structures that enter the upper extremity. This region is therefore important in understanding the basis for signs and symptoms that occur in the upper extremity as well as the neck.

Boundaries

The posterior triangle is illustrated in *Figure 42.1*. The base of the triangle is formed by the middle one-third of the clavicle. The anterior side is formed by the posterior margin of the **sternomastoid muscle**, while the posterior side is formed by the anterior margin of the **trapezius muscle**. The apex of the triangle projects superiorly behind the ear to the level of the **superior nuchal line** of the **occipital** bone, where the sternomastoid and trapezius muscle meet.

STERNOMASTOID (STERNOCLEIDOMASTOID) AND TRAPEZIUS

These muscles arise from a single sheet of embryonic musculature that extended from the skull and neck to the shoulder girdle and sternum. They have a continuous superior attachment from the **mastoid process** to the **external occipital protuberance (inion)** along the superior nuchal line (*fig. 42.2*).

Inferiorly, their insertions are discontinuous across the clavicle. The sternomastoid muscle inserts into the medial third of the clavicle and the sternum, while the trapezius inserts into the lateral third of the clavicle, the acromion, and the spine of the scapula. The intervening separation of these muscles above the middle third of the clavicle forms the base of the posterior triangle.

Both muscles are innervated by a single motor nerve, **the accessory nerve**, cranial nerve XI. It lies on the deep surface of the sternomastoid muscle and traverses across the posterior triangle of the neck to reach the deep surface of the trapezius (*fig. 42.1*). It terminates in the trapezius musculature that arises from the spines of the thoracic vertebrae on the back.

Action

The sternomastoid muscles flex the head when they contract bilaterally. Individually, they rotate the head toward the opposite side of the body. They can also be used as accessory muscles of respiration in forced inspiration during exertion or certain diseased states.

The trapezius elevates the shoulder girdle and helps to support the weight of the upper limb. Its lower fibers draw the scapula toward the vertebral column. All of its movements can alter the position of the glenoid fossa of the scapula and permit a greater degree of circumduction in the shoulder joint.

Investment

The sternomastoid and trapezius muscles are invested by a tough dense cervical fascia. This deep **investing fascia** encircles the neck and covers the posterior triangle.

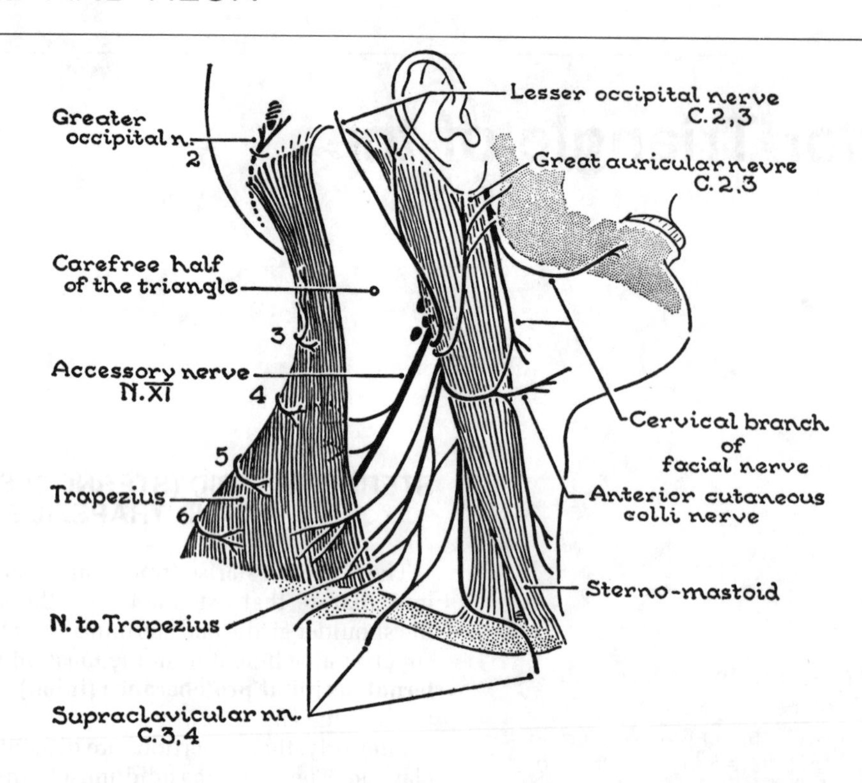

Figure 42.1. The superficial nerves of the neck. Of these, the facial and accessory are motor. (*Nerves of neck: anterior cutaneous = transverse.*)

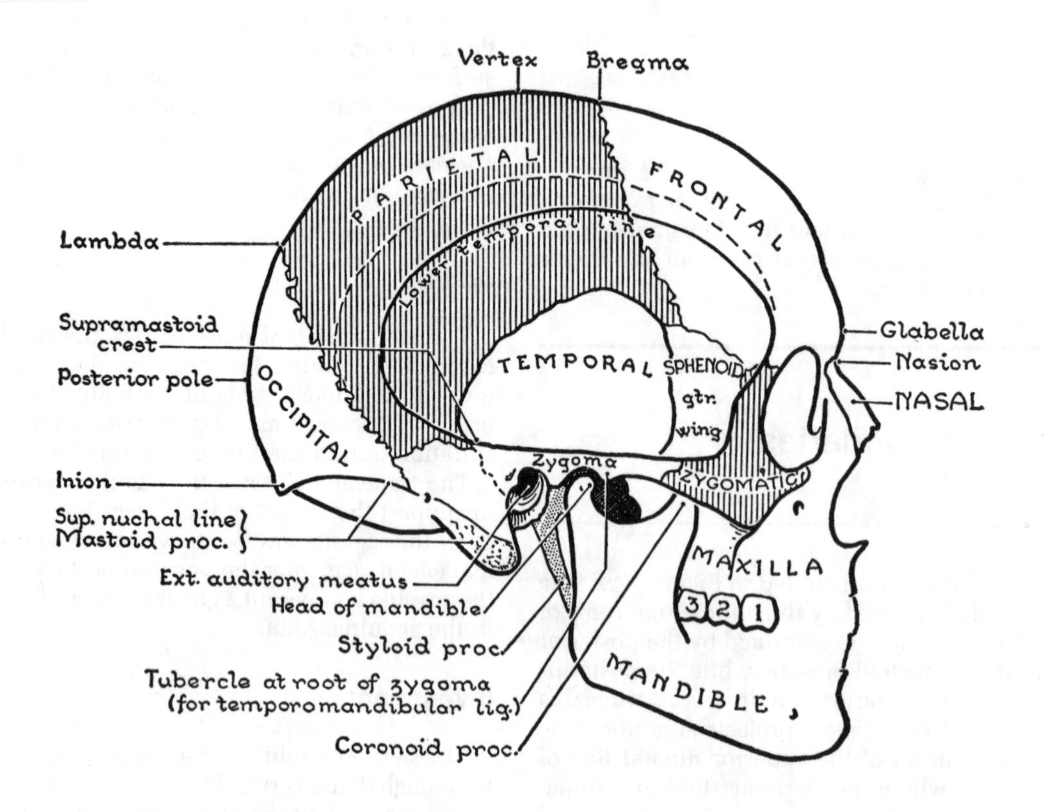

Figure 42.2. The skull, on side view (norma lateralis).

The accessory nerve is located in this fascial roof as it passes posteriorly and inferiorly from the sternomastoid muscle to the trapezius. The accessory nerve is vulnerable to damage in the roof of the posterior triangle. It lies deep to the skin and thin subcutaneous platysma muscle within the deep investing fascia.

CONTENTS OF THE POSTERIOR TRIANGLE

The roof of the posterior triangle contains the **accessory nerve (XI)**. The accessory nerve divides the fascial roof into a superior portion that contains no important structure (**the carefree part**) and a lower fascial roof, which contains the cutaneous branches of the cervical plexus where one must proceed with care not to injure or damage these structures (*fig. 42.1*). (Nerve XI is the most important structure to protect from injury.)

The cutaneous cervical nerves arise from the posterior border of the sternomastoid and penetrate the investing fascia over the posterior triangle. They then course through the overlying subcutaneous fascia and platysma to innervate the skin of the neck and shoulder region. These nerves are derived from the anterior rami of cervical nerves 2, 3, and 4. The **lesser occipital nerve** supplies the skin posterior to the ear and superficial to the mastoid process. The **greater auricular nerve** supplies the skin overlying the upper aspect of the sternomastoid, the earlobe, and the parotid gland. A small **transverse cervical nerve** of the neck crosses the sternomastoid in a horizontal plane to supply the skin of the neck overlying the "Adam's Apple" (laryngeal prominence). These three cutaneous nerves are derived from C2, 3 ventral rami. A very important group of cutaneous nerves are derived from the ventral rami of C3, 4 to exit from the posterior triangle through its roof and penetrate the subcutaneous fascia and platysma to innervate the skin that is superficial to the clavicle and scapula. These are the **supraclavicular nerves**. They overlap with cutaneous branches of the upper thoracic intercostal nerves on the chest wall above the 1st intercostal space. This is why the C4 dermatome approximates the T1 dermatome on the anterior chest wall. One must be sure to test the C5, C6, C7, and C8 dermatomes on the upper extremity in patients with suspected neurologic damage or you will be misled by this peculiar arrangement of overlapping high thoracic and midcervical dermatomes. The **supraclavicular nerves** also show a common neurologic origin with the **phrenic nerves** (C3, 4, 5). Sensations from pleura and peritoneum adhering to certain areas of the diaphragm are carried in the phrenic nerves and may be interpreted as pain from the skin supplied by the supraclavicular nerves, C3, 4 (a form of referred pain).

The posterior triangle also contains nerve fibers from the cervical plexus that enter the undersurface of the trapezius muscle. These are C3, 4 fibers that are thought to carry mainly sensory impulses from the muscle spindles of the trapezius back to the cervical regions of the cord. The sternomastoid has similar cervical branches from C2, 3, on its posterior surface. The unique way in which nerve XI arises from the cervical region of the cord and ascends into the foramen magnum and then exits the cranium via the jugular foramen is shown in *Figure 42.3*. Clinicians assess nerve XI as a peripheral nerve after it leaves the jugular foramen. Damage to XI will affect the motor function of the sternomastoid and trapezius muscle. Spinal cord damage to C2, 3, and 4 levels are frequently fatal due to the loss of phrenic nerve function. Patients on respiratory support units due to damage of the C3, 4, and 5 regions of the spinal cord may show weakness in the trapezius and sternomastoid muscles because of damage to some of the motor neurons that form nerve XI.

OMOHYOID (*fig. 42.4*)

The inferior belly of this neck strap muscle passes through the lower aspect of the posterior triangle as it inserts into the upper margin of the scapula. It runs one or two fingerbreadths above the clavicle. The inferior belly of the omohyoid is separated from its superior belly in the anterior triangle of the neck. The intervening tendon is bound to the clavicle by a fascial sling.

EXTERNAL JUGULAR VEIN (*fig. 42.4*)

This large vein descends subcutaneously across the sternomastoid muscle and pierces the deep fascia that

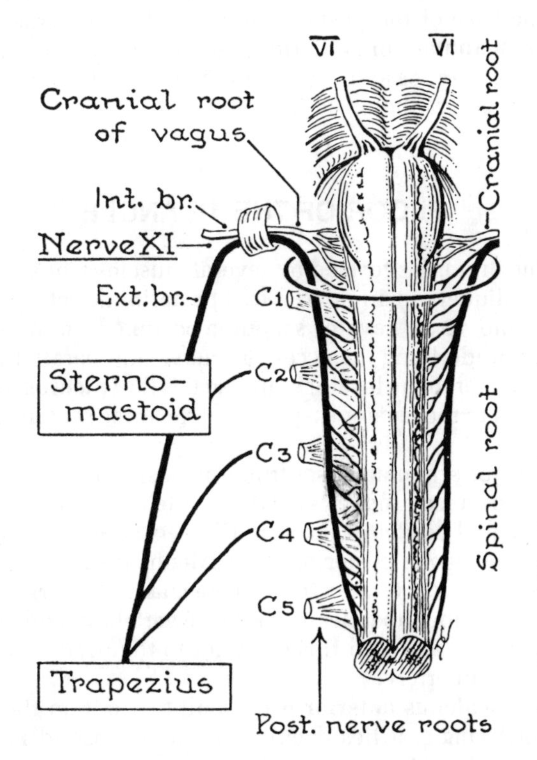

Figure 42.3. Origin and distribution of the accessory (XI) nerve. (Ventral nerve roots have been cut away.)

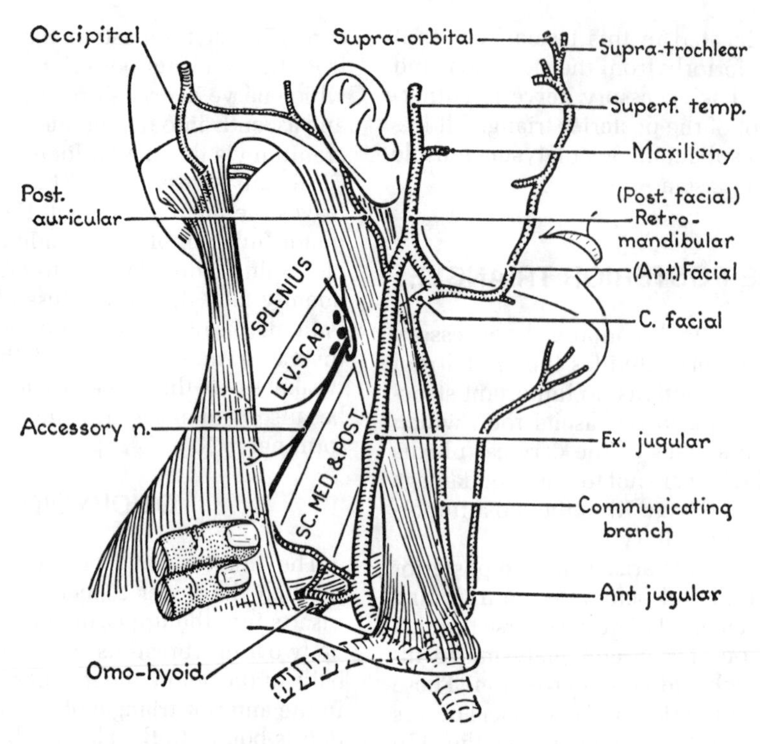

Figure 42.4. The superficial veins of the face and neck.

forms the roof of the posterior triangle. The vein terminates in the subclavian vein or the internal jugular vein at the base of the posterior triangle. The **traverse colli, suprascapular,** and **anterior jugular veins** communicate with each other and drain into the external jugular vein after it enters the posterior triangle.

FLOOR OF THE TRIANGLE

The "floor" is formed by several muscles whose fibers run obliquely in an inferior and posterior direction (*figs. 42.4 and 42.5*). **Levator scapulae** occupies a middle position underlying the accessory nerve and paralleling it.

Superior to the levator scapulae lies the **splenius,** while inferior to the levator scapulae are the three **scalene muscles.**

Levator scapulae arises from the posterior tubercles of the transverse processes of C1–4 vertebrae. It inserts into the medial border of the scapula at its superior angle.

The **scalenus posterior** and **scalenus medius** muscles are collectively termed the **scalene mass.** They arise from the posterior tubercles of cervical vertebrae and insert into the 1st and 2nd ribs posterior to the position of the subclavian artery.

The **scalenus anterior** is difficult to see from the posterior triangle. It lies deep to the sternomastoid muscle and is a *key* structure in understanding the visceral relationships in the neck. **Knowledge of the origin and insertion of this muscle and its key relationship is very important to easy understanding of the root of the neck.** The scalenus anterior arises from the **anterior** tubercles of the transverse processes of the cervical vertebrae C3–6. It inserts into the **scalene** tubercle of the 1st rib and separates the subclavian artery and vein at its insertion.

A small part of **semispinalis capitis** may appear at the apex of the posterior triangle, superior to the splenius (*fig. 42.5*).

FASCIAL CARPET (*fig. 42.6*)

The muscular floor of the posterior triangle is covered by a layer of deep fascia (**prevertebral fascia**). This fascia is part of the fascia that envelops the vertebral column and the prevertebral and postvertebral muscles of the neck.

This prevertebral fascia covers the subclavian vessels and the roots of the brachial plexus as they emerge between the **scalenus anterior** and **scalene mass** in the floor of the posterior triangle. The fascia of the prevertebral muscle is then drawn into the axilla on the brachial plexus and subclavian artery as the **axillary sheath** (*fig. 42.6*). This provides an anatomical course for infections within the posterior compartment of the neck to "track" into the axilla and produce upper extremity signs and symptoms. The space that lies between the investing fascia (roof of triangle) and the prevertebral fascia (floor of triangle) is a common site for head and neck fascial infections.

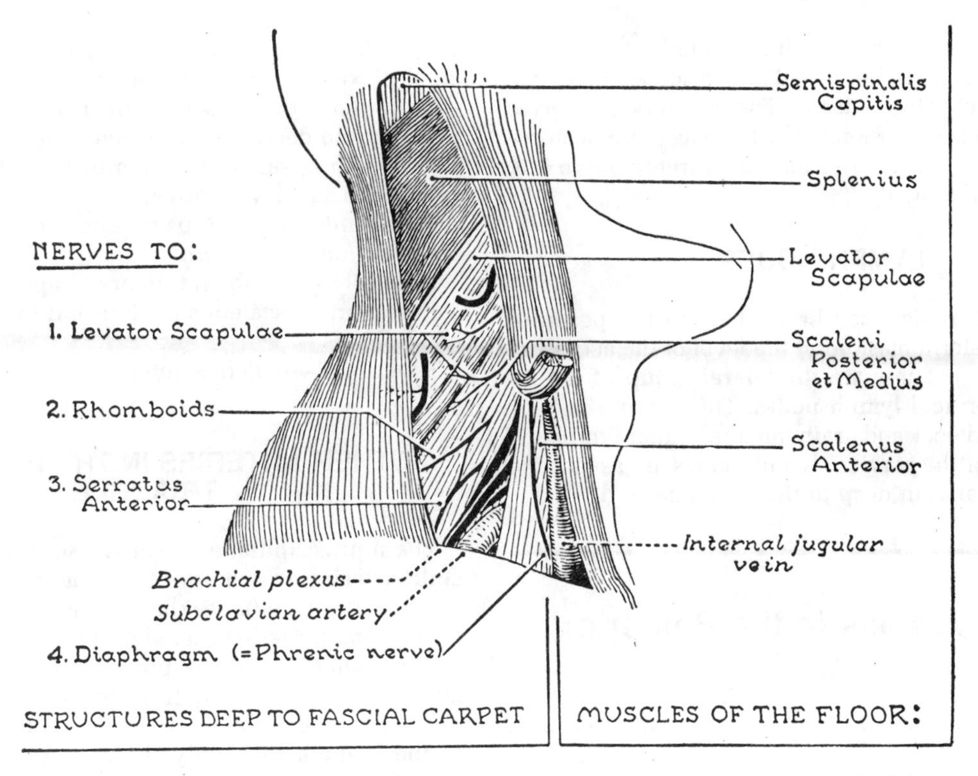

Figure 42.5. Floor of the posterior triangle and nerves that have no occasion to pierce its fascial carpet.

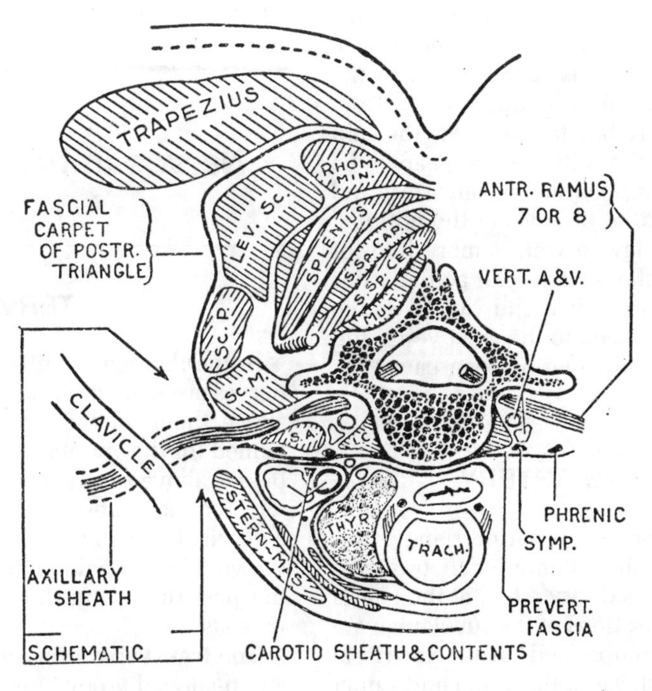

Figure 42.6. Transverse section of the neck showing the pre- and postvertebral fasciae prolonged into the axilla as a tubular covering (axillary sheath) for the brachial plexus and subclavian vessels. The nerves shown in *Figure 36.4* remain deep to this fascia.

The prevertebral muscles, which underlie the prevertebral fascia, also have important motor nerves which are deep and protected by the fascia. These include the nerves to levator scapulae and rhomboids, the **long thoracic nerve** (C5, 6, 7) to serratus anterior and the **phrenic nerve** (C3, 4, 5) to the diaphragm.

LYMPH NODES

Some lymph nodes may be evident on the posterior margin of the sternomastoid at the level of the accessory nerve (*fig. 42.4*). These are the **lateral group of the inferior deep cervical lymph nodes**. They drain the back of the scalp and neck and drain into the jugular lymphatic trunks. Most of the cervical lymph nodes are associated with the jugular vein deep to the sternomastoid.

Blood Vessels in the Posterior Triangle

THE SUBCLAVIAN VEIN

The subclavian vein is a continuation of the axillary vein. It begins at the lateral border of the first rib and ends medial to the scalenus anterior muscle where it joins the internal jugular to form the brachiocephalic vein. The subclavian vein is separated from the subclavian artery superior to the 1st rib by the insertion of the scalenus anterior muscle. Both vessels groove the upper surface of the 1st rib next to the **scalenus tubercle**. Since the subclavian vein is anterior to the scalenus anterior (prevertebral muscle), less prevertebral fascia covers the vein in the axillary sheath. Sometimes the vein appears to be separate from the densely wrapped subclavian artery and brachial plexus with the axilla. Because of the obliqueness of the 1st rib, the subclavian vein is more inferior in its position than the subclavian artery. It actually lies posterior and inferior to the clavicle and can be catheterized at this point to gain access to the great veins and right side of the heart via the superior vena cava (*fig. 42.4*).

THE SUBCLAVIAN ARTERY

The subclavian artery enters the posterior triangle posterior to the insertion of scalenus anterior. It traverses the triangle and becomes the axillary artery at the lateral aspect of the 1st rib. The pulsations of the subclavian are easily felt here. It can be compressed on the 1st rib at this point to control hemorrhage in the axilla and upper extremity. The portion of the subclavian artery that is in the posterior triangle may give off a **dorsal scapular artery**. This branch traverses through the **trunks** of the brachial plexus to supply the medial border of the scapula and the rhomboid muscles. This is more commonly supplied by the deep branch of the **transverse cervical artery**, which arises in the anterior triangle from the first part of the subclavian artery.

Important relationships around the subclavian artery are as follows (fig. 42.5):

Inferiorly—1st rib and pleura of apex of lung

Posteriorly—scalenus medius and lower trunk of brachial plexus

Anteriorly—scalenus anterior.

OTHER ARTERIES IN THE POSTERIOR TRIANGLE

The **suprascapular** and **transverse cervical (transverse colli) arteries**, which arise from the first part of the subclavian artery medial to the scalene anterior muscle, appear in the posterior triangle. They pass anterior to the scaleus anterior and the phrenic nerve, as they course laterally and posteriorly through the posterior triangle.

The **suprascapular** artery crosses the superior border of the scapula to supply the supraspinatus and infraspinatus fossae structures of the scapula.

The **transverse cervical artery** supplies the posterior surface of the trapezius through a superficial branch. Its deep branches, if present, would supply the rhomboids and medial border of the scapula. If the deep branch is absent then a **dorsal scapular artery** frequently arises from the 3rd part of the subclavian artery in the posterior triangle.

Nerves in the Posterior Triangle

TERMINOLOGY

In the thoracic, lumbar, and sacral regions, the spinal nerves are named numerically after the vertebrae that lie above them. In the cervical region, however, they are named for the vertebrae that lie inferior to the nerves. C1 passes above the 1st cervical vertebrae and sets the pattern for each subsequent cervical nerve. Since there are 8 cervical nerves and only 7 cervical vertebrae, C8 exits the vertebral canal below the 7th cervical vertebrae, and the pattern for the thoracic, lumbar, and sacral nerves persists.

The **transverse process** of the cervical vertebrae have a cup-shaped groove for the ventral rami of the cervical nerves. Each process has a posterior tubercle but only C3, 4, 5, and 6 vertebrae have anterior tubercles.

BRACHIAL PLEXUS

This plexus (*fig. 42.7*) is formed by the ventral rami of nerves C5, 6, 7, 8, and T1. The ventral ramus of C4 or the ventral ramus of T2 may also be included in some brachial plexuses. As illustrated, the ventral rami of C5 and C6 fuse in the posterior triangle to create the **upper trunk** of the brachial plexus. The ventral rami C7 continues into the posterior triangle as the **middle trunk** of the brachial plexus. C8 and T1 ventral rami combine to form the **lower trunk** of the brachial plexus. The lower trunk passes over the apex of the lung behind the subclavian artery to cross the superior aspect of the 1st rib in the posterior triangle.

Each of the three trunks divide into an **anterior** and a **posterior division** over the 1st rib. The three posterior divisions unite to form the **posterior cord** of the brachial plexus in the axilla. The anterior divisions of the upper and middle trunks unite to form the **lateral cord** and the anterior division of the lower trunk continues into the axilla in the **medial cord**.

Branches from the **roots** of the brachial plexus innervate the rhomboids (dorsal scapular nerve, C5), serratus anterior (long thoracic nerve C5, 6, 7) and diaphragm (phrenic nerve C3, 4, 5) (*fig. 42.5*). They also supply the adjacent prevertebral muscles. The roots of the brachial plexus also receive gray rami from the cervical sympathetic trunk that provides the sympathetic innervation of the skin supplied by the cutaneous branches of the brachial plexus.

The major nerve that arises from the **upper trunk** of brachial plexus in the posterior triangle is the **suprascapular nerve**. It arises from the upper trunk (C5, 6) and passes through the suprascapular notch of the scapula to innervate the supraspinatus and infraspinatus muscles. This branch can be tested by asking a patient to externally rotate their shoulder while palpating the infraspinatus.

A branch to the **subclavius** muscle also arises from the upper trunk in the posterior triangle. The nerve to the levator scapulae comes out of the cervical nerves C3 and 4 above the brachial plexus.

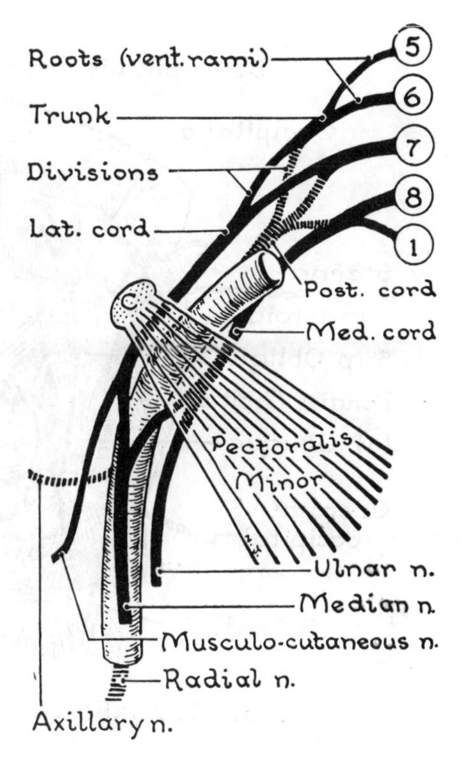

Figure 42.7. The brachial plexus.

Suboccipital Region

The back of the neck contains a large number of muscles that overlie the posterior aspects of the cervical vertebrae. The muscles of the suboccipital region are principally related to the 1st two cervical vertebrae, the **atlas** (C1) and **axis** (C2), and the occipital bone. *Figure 42.5* illustrates the superficial muscles in the apex of the posterior triangle, the **trapezius** and the **splenius**. Deep to the splenius are the true back muscles represented in the suboccipital region by the **semispinalis capitis**. This muscle overlies muscles that relate specifically to the atlas and the axis. The relation of the vertebral artery and the 1st and 2nd cervical nerves are of major importance in this region.

Contents of the Region

Four Muscles	—Two oblique muscles —Two straight (rectus) muscles
Two Nerves	—Greater occipital C2 —Suboccipital C1
Two Arteries	—Vertebral —Occipital

MUSCLES (*fig. 42.8*)

The **obliquus capitis inferior** (inferior oblique) passes from the spine of the axis (C2) in an oblique fashion to attach to the transverse process of the atlas (C1). The **greater occipital nerve (C2)** is inferior to this muscle and the **suboccipital nerve (C1)** is superior. The obliquus capitis inferior forms the inferior border of the suboccipital triangle.

The **obliquus capitis superior** (superior oblique) passes from the tip of the transverse process of the atlas (C1) in

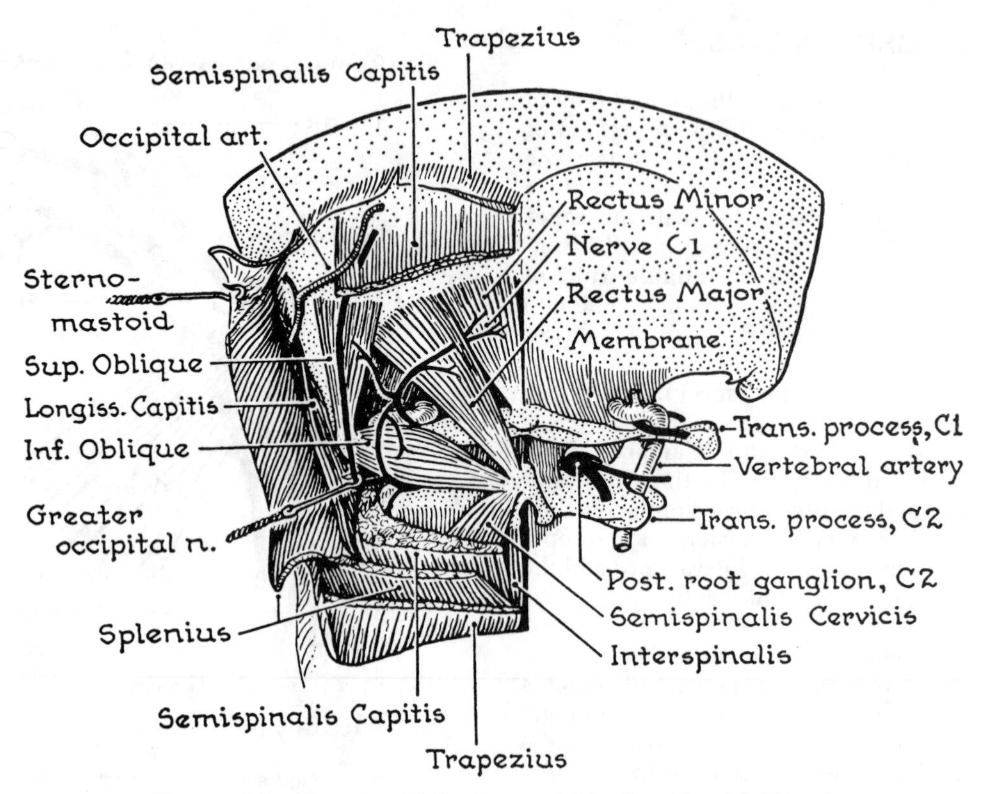

Figures 42.8. The suboccipital region, containing the suboccipital triangle.

an oblique fashion to attach to the occipital bone between the superior and inferior nuchal lines. It forms the lateral margin of the suboccipital triangle.

The **rectus posterior minor** attaches to the posterior tubercle of the atlas (C1) and the **rectus capitis posterior major** attaches to the spine of the axis (C2). These two muscles attach superiorly in an adjacent manner on the inferior nuchal line of the occipital bone just posterior to the foramen magnum. The lateral border of the **rectus capitis posterior major** forms the medial border of the suboccipital triangle.

Within the suboccipital triangle is the vertebral artery as it traverses medially above the arch of the atlas.

VERTEBRAL ARTERY

The vertebral artery arises from the first part of the subclavian artery in the root of the neck. Occasionally, the left vertebral arises directly from the arch of the aorta. The artery enters the foramen tranversarium of the 6th cervical vertebra and ascends through the successive foramina transversarii of vertebrae C5–C1. The artery assumes a near horizontal course above the posterior arch of the atlas and penetrates the posterior atlanto-occipital membrane and the dura mater to enter the subarachnoid space as it passes through the foramen magnum. The vertebral artery lies above the **suboccipital nerve** (C1), as it passes above the posterior arch of the atlas and posterior to its articulation with the occipital condyles.

SUBOCCIPITAL NERVE (C1 DORSAL RAMUS)

Supplies all four muscles of the suboccipital triangle. The C1 nerve is purely motor. There is no C1 dermatome.

GREATER OCCIPITAL NERVE (C2)

The dorsal rami of C2 exits the suboccipital triangle below the obliquus capitis inferior muscle. It is sensory to the scalp over the parietal and occipital bones, posterior to the vertex. The C2 dermatome overlaps with the V^1 dermatome on the back of the scalp. The greater occipital nerve accompanies the **occipital** artery in the scalp over the occipital bone.

DEEP CERVICAL ARTERY

A branch of the first part of the subclavian artery gives access to the dorsal aspect of the neck by passing between the 1st rib and the transverse process of the 7th cervical vertebra. It supplies blood to the postvertebral musculature and anastomoses with branches of the occipital and vertebral artery. This may augment supply to the brain when the carotid artery is occluded.

Clinical Mini-Problems

1. **How would one test the integrity of the accessory nerve (XI) at the base of the skull?**
2. **Which motor nerve(s) must be avoided when incising the investing fascia that forms the roof of the posterior triangle?**
3. **The axillary sheath is derived from prevertebral fascia of the neck. The structures that leave the neck between the anterior scalene muscles and scalene mass are tightly enclosed in the axillary fascia. Which structures would you expect to be enclosed securely in the axillary sheath?**
4. **When an anesthetist injects a local anesthetic into the brachial plexus in the neck, he palpates the subclavian artery over the 1st rib and injects the solution superior to the artery. This avoids traumatizing the artery and also prevents the chances of causing a pneumothorax by penetrating the apex of the lung. What relationship does the 1st rib have to the brachial plexus and the apex of the lung in the neck?**
5. **Which cervical spinal nerve lies inferior to the body of the 5th cervical vertebra?**
6. **How would one test the integrity of the upper trunk of the right brachial plexus in a patient with an injury to their right shoulder and neck?**

(Answers to these questions can be found on p. 587.)

43

Anterior Triangle of the Neck

Clinical Case 43.1

Patient Eleanor E. This elderly female nursing home resident aspirated some foreign substance (food) into the larynx. It induced a laryngospasm of the vocal cords. In a state of collapse she is rushed into the Emergency Department. The physician is unable to remove the aspirated debris and, with your help, immediately performs an emergency tracheostomy which re-establishes the airway. The tracheostomy is done through the upper tracheal rings superior to the isthmus of the thyroid gland.

Anterior Triangle and Median Line of the Neck

SUBDIVISIONS (*fig. 43.1*)

The anterior triangle of the neck is bounded by three borders: (a) median line of the neck from the chin to the manubrium; (b) the anterior margin of the sternomastoid muscle; and (c) the horizontal plane formed by the lower margin of the mandible. The anterior triangle is subdivided into three smaller triangles: **submandibular (digastric) triangle, carotid triangle**, and **muscular triangle**. The submandibular or digastric triangle is bounded by the posterior and anterior bellies of the digastric muscle and the inferior border of the mandible (*see fig. 43.8*). The carotid triangle is bounded by the posterior belly of the digastric, the superior belly of the omohyoid and the sternomastoid muscles. The muscular triangle overlies the strap muscles of the neck and is bounded by the superior belly of the omohyoid, the lower anterior margin of the sternomastoid, and the median line of the neck.

A **submental triangle** (*see fig. 43.8*) is formed by the anterior bellies of the right and left digastric muscles and a base above the hyoid bone. *Figure 43.1* also shows a parotid space beyond the superior margin of the anterior triangle in the area above the posterior belly of the digastric and posterior to the ramus of the mandible.

LANDMARKS (*figs. 43.2 and 43.3*)

The tips of the **transverse processes of the atlas (C1)** project more laterally than the transverse processes of the other higher cervical vertebrae. They can be palpated in the parotid space between the angle of the mandible and the mastoid process (*fig. 43.2*).

The **hyoid bone** can be felt in its subcutaneous relationship in the midline of the neck between the inferior border of the mandible and the upper border of the thyroid cartilage. The body of the hyoid bone is at the level of the 3rd cervical vertebra (C3) (*fig. 43.3*). It has a lesser and greater horn (cornu) that project posteriorly. The greater horns are palpable in the lateral aspect of the neck.

The **thyroid cartilage** is palpable in the midline of the neck at the levels of C4 and C5. The **laryngeal prominence** (Adam's Apple) is visible and most prominent in men. The vocal cords of the larynx attach anteriorly on the thyroid cartilage near the lower border of the laryngeal prominence. The thyroid cartilage possesses two lateral laminae that have superior and inferior horns (cornua). The inferior horns articulate with the cricoid cartilage.

The **cricoid cartilage** is at the level of C6 and its anterior arch can be palpated in the midline. This cartilage marks the upper limit of the trachea in the neck. The trachea and its cartilaginous rings are also palpable in the midline from the cricoid cartilage to the superior border of the manubrium. The lobes of the thyroid gland lie on the lateral aspects of the trachea in the neck (*fig. 43.4*).

Two important membranous structures are associated with the cartilages in the midline of the neck. The **thyrohyoid** membrane is between the thyroid cartilage and the hyoid bone. It is pierced by the **internal laryngeal nerve** and **vessels**. The **cricothyroid membrane** lies between the anterior arch of the cricoid cartilage and the

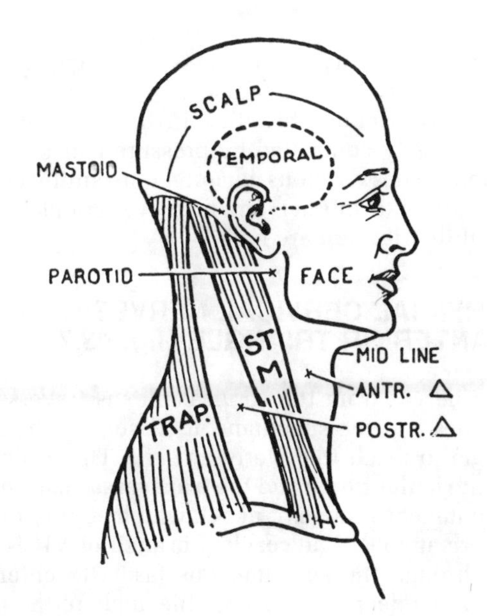

Figure 43.1. Superficial regions of head and neck. *STM*, sternomastoid.

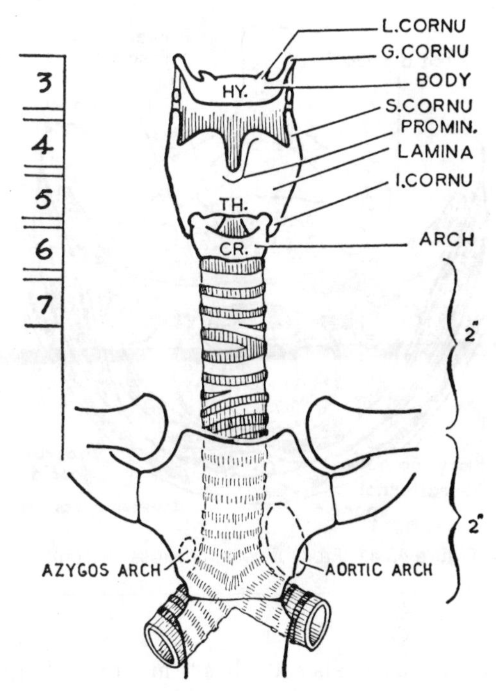

Figure 43.3. Landmarks and vertebral levels. *HY*, hyoid bone; *TH*, thyroid cartilage; *CR*, cricoid cartilage.

inferior borders of the laminae of the thyroid cartilage. It is just below the level of the vocal cords and may be penetrated to create a **high tracheostomy**. Since this has some possibilities for damaging the vocal cords, the preferred sight for a tracheostomy is below the cricoid cartilage but above the level of the isthmus of the thyroid gland (2nd–4th tracheal cartilages). At this point, the only tissues penetrated during a tracheostomy are skin, subcutaneous tissue, platysma fibers, investing fascia, the visceral fascia of the trachea, and the tracheal wall (*fig. 43.4*).

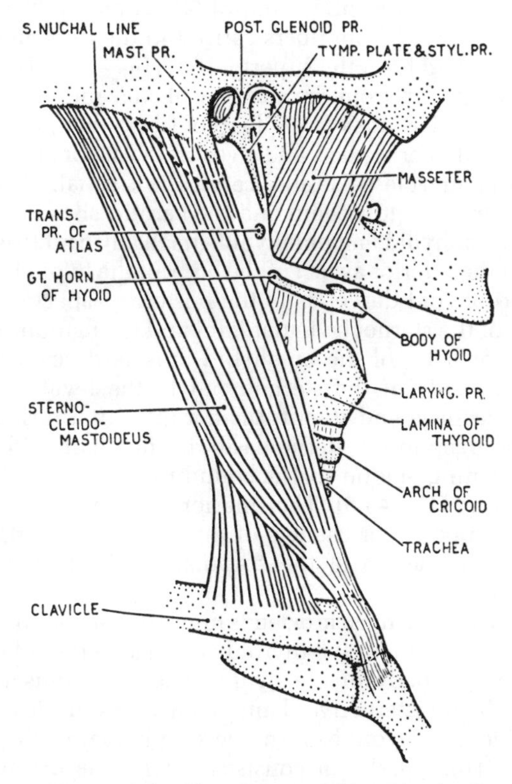

Figure 43.2. The sternomastoid and landmarks.

SUPERFICIAL STRUCTURES

Platysma

The platysma (Gk = a plate) is a muscle of facial expression that lies in the subcutaneous tissue. It extends from the face, lower lip, and lower margin of the mandible over the neck and clavicle to the skin over the second rib (*fig. 43.5*). It is supplied by the **cervical branch of the facial nerve (VII)**. Its action is to tense the subcutaneous tissue of the neck and assist in depressing the lower lip. Damage to the nerve supply and paralysis of the muscle creates an unpleasant cosmetic effect.

Deep Fascia

The fascia, which encloses the trapezius muscle in the posterior neck, continues anteriorly to form the tough roof of the posterior triangle and then invests the sternomastoid muscle. At the anterior borders of the sternomastoid, this **investing fascia** again forms a single-layer sheet that covers the anterior triangle and fuses with its opposite component in the midline of the neck (*fig. 43.4*).

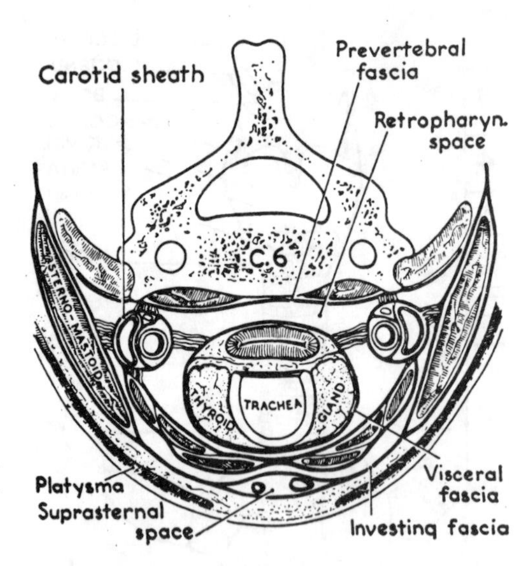

Figure 43.4. Front of neck (on cross-section).

The **investing fascia** also attaches firmly to the bony surfaces that it covers, i.e., the mandible superiorly and the clavicle and manubrium inferiorly.

Superficial Veins (*fig. 43.6*)

Superficial veins lie in the subcutaneous tissue external to and surrounding the investing fascia. **Anterior jugular veins** may be present near the midline. They arise in the submental triangle and descend near the midline to pierce the deep fascia above the manubrium. They usually pass between the posterior aspect of the sternomastoid and the upper border of the clavicle to enter the

Figure 43.5. The platysma muscle

external jugular veins in the posterior triangle. These veins may be damaged in a tracheostomy procedure and cause extensive venous bleeding in the surgical field. Bleeding is easily controlled by pressure above the operative field. These venous channels are quite variant from one patient to another, and many variations in their anatomical distribution are noted.

SUPERFICIAL CERVICAL NERVES OF THE ANTERIOR TRIANGLE (*fig. 43.7*)

Sensory nerves from the cervical plexus traverse the fascial planes and the subcutaneous tissues of the anterior triangle to reach the overlying skin. These include **the great auricular nerve** and the **transversus colli nerve** (anterior cutaneous nerve of neck), which supply the C2 and C3 dermatomes. The **cervical branch of VII** is also coursing through the subcutaneous fascia to enter the platysma on its deep surface near the angle of the mandible.

THE INFRAHYOID MUSCLES

These four paired muscles (*fig. 43.8*) are depressors of the larynx and the hyoid bone. Often referred to as "strap muscles," they lie between the layer of investing fascia and **visceral fascia**, which covers the thyroid gland, trachea, and esophagus (*fig. 43.4*). The infrahyoid muscles are innervated by the **ansa cervicalis**, which is a motor plexus formed by ventral rami of C1, 2, and 3. The C1 fibers are derived from fibers carried in the **hypoglossal nerve (XII)** and form the superior root (*fig. 43.9*). The C2 and C3 fibers are from the cervical plexus and form the inferior root. The **ansa** is the loop of fibers that joins the superior and inferior roots and gives off the major branches to the strap muscles. The ansa cervicalis is usually lateral to the internal jugular vein and the carotid sheath.

The **sternohyoid** and **omohyoid** attach adjacent to each other on the inferior aspect of the body of the hyoid bone. The sternohyoid descends to the posterior aspect of the capsule of the sternoclavicular joint and the manubrium. A superior belly of the omohyoid muscle descends on the lateral side of the sternohyoid to the level of the cricoid cartilage and then turns abruptly in a posterior direction deep to the sternomastoid. The omohyoid has an intervening tendon that is bound to the clavicle by a fascial sling (*fig. 43.8*). The inferior belly of omohyoid crosses the posterior triangle of the neck and inserts on the superior margin of the scapula, medial to the suprascapular notch.

The **thyrohyoid muscle** extends upward from the oblique line on the lamina of the thyroid cartilage to the inferior aspect of the body of the hyoid bone. This muscle is overlain by the sternohyoid and omohyoid muscles (*fig. 43.8*). The thyrohyoid has a nerve supply that is directly from XII (*fig. 43.9*) and consists of C1 anterior ramus fibers that were picked up by XII at the base of the skull.

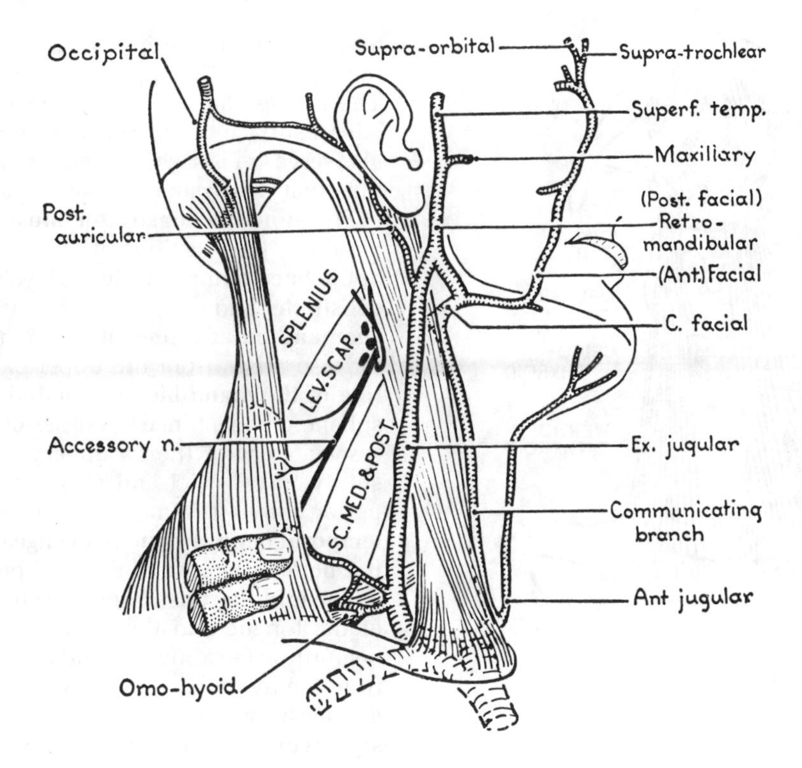

Figure 43.6. The superficial veins of the face and neck.

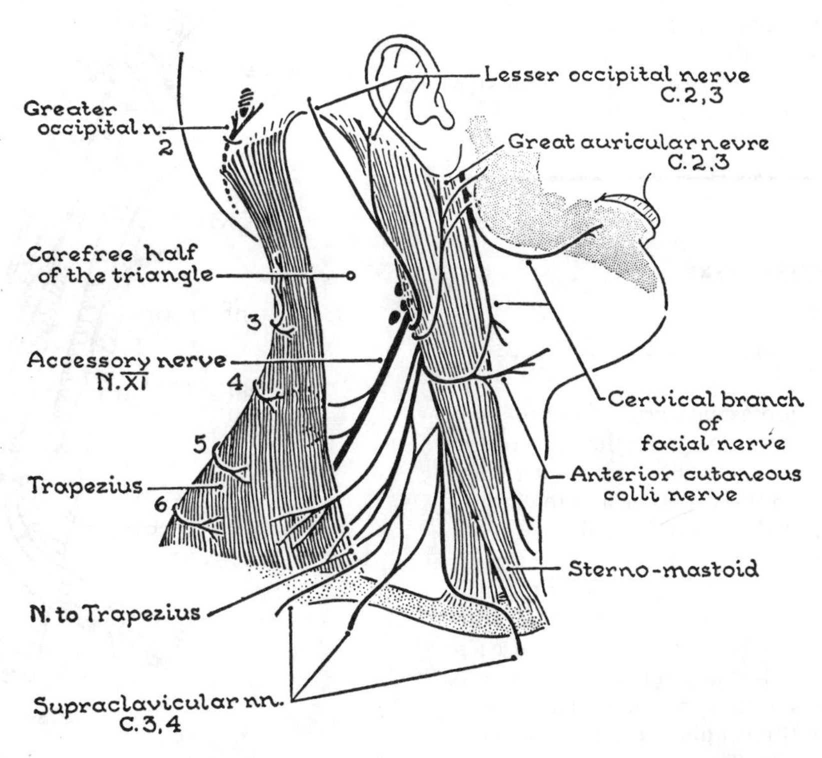

Figure 43.7. The superficial nerves of the neck. Of these, the facial and accessory are motor. (*Nerves of neck: anterior cutaneous = transverse.*)

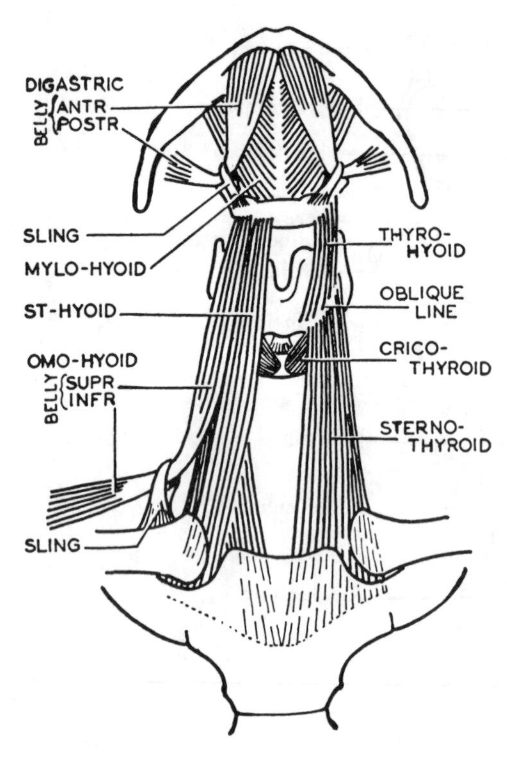

Figure 43.8. Muscles bounding midline of neck.

The **sternothyroid muscle** descends from its attachment on the oblique line of the thyroid cartilage to the posterior aspect of the manubrium. It is deep to the sternohyoid muscle. The sternothyroid receives its innervation from the ansa cervicalis like the overlying sternohyoid and omohyoid muscles (*fig. 43.9*).

Pharynx

The pharynx is a musculofascial derivitive of the embryonic foregut that is suspended from the base of the skull and extends to the level of the 6th cervical vertebra. It lies between the bodies of the vertebrae and the larynx at C4–C6. At the level of the cricoid cartilage (C6), it becomes continuous with the esophagus (*fig. 43.10*).

The lateral and posterior walls of the pharynx are formed mainly from three muscular **constrictors: superior, middle, and inferior**. Each muscle is fan-shaped, and each is fixed anteriorly by a smaller attachment than its posterior insertion. The expanding constrictors pass posteriorly and insert on a midline **raphe** with the same constrictor on the opposite side. The three constrictors are arranged in an "overlapping" fashion so that they appear to be stacked like flower pots from the inferior aspect of the pharynx to the upper extent behind the nasal cavities. The spaces that exist between the constrictors on the lateral surface of the pharynx are "closed" by fascia but also serve as passages for vessels, nerves, and the auditory tube that penetrate the pharyngeal wall.

The **superior constrictor muscle** lies posterior to the nasal and oral cavities. It is continuous with the plane of the **buccinator** muscle of the cheek. Both the superior constrictor and the buccinator arise in part from the **pterygomandibular raphe**. In addition, the superior constrictor also arises from the upper extent of the **mylohyoid line of the mandible** and **medial pterygoid plate** of the sphenoid bone. Since the superior constrictor is a muscle associated with the nasopharynx, it is restricted in its ability to constrict and close the upper airway. Its superior bony attachments are to the **apex of the petrous temporal bone** and the **pharyngeal tubercle of the occipital bone** as well as the medial pterygoid plates.

The **middle constrictor** muscle is posterior to the base of the tongue and the larynx. It plays a major role in deglutition (swallowing) and is a true constrictor. It arises from the hyoid bone at the angle between the **lesser and greater horns** (cornua) and from the inferior aspect of the **stylohyoid ligament** that attaches to the lesser horn. Its

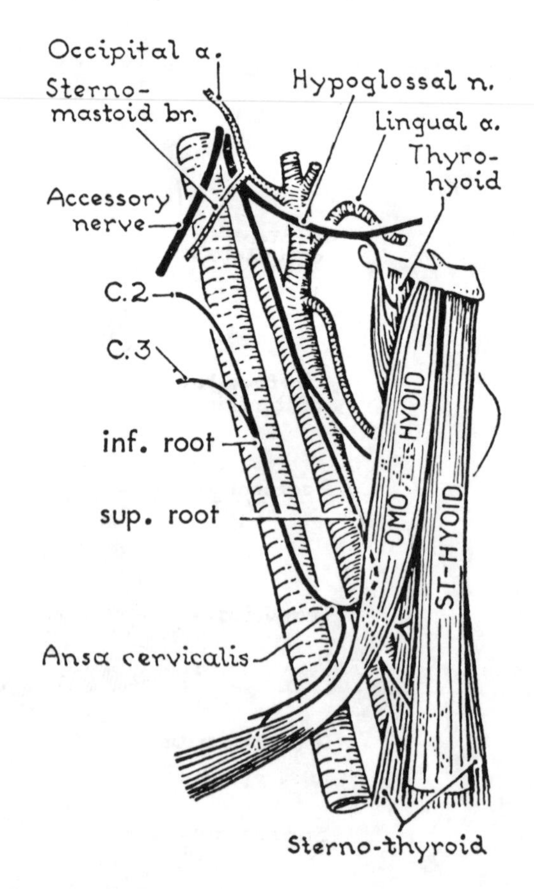

Figure 43.9. Ansa cervicalis: its roots and its branches to infrahyoid muscles.

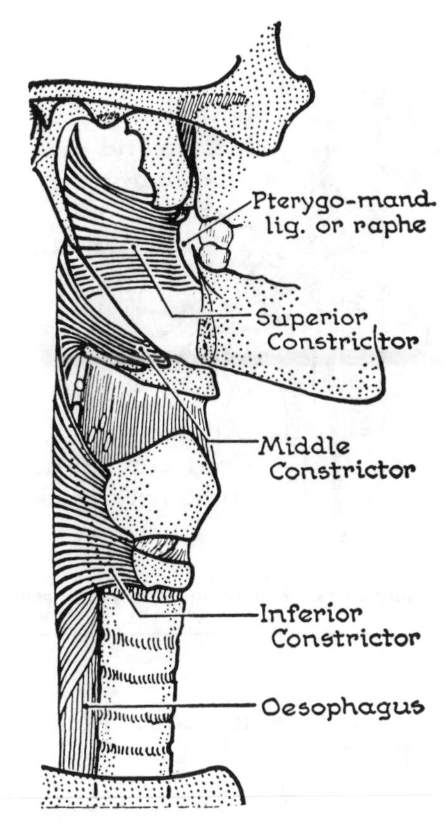

Figure 43.10. The 3 constrictors of the pharynx.

insertion is like the other constrictors into the middle raphe on the posterior aspect of the pharynx.

The **inferior constrictor muscle** arises from the **cricoid** and **thyroid cartilages**. The thyroid part is from the oblique line while the cricoid part is from the fascia overlying the cricothyroid muscle. The very inferior aspect of the inferior constrictor is termed the **cricopharyngeus** muscle and acts as a sphincter between the pharynx and the esophagus. The inferior constrictor and the cricopharyngeus are innervated by the recurrent laryngeal branch of the vagus (X). The superior and middle constrictor muscles are also skeletal muscles and under voluntary control via pharyngeal branches of the vagus nerve (X). The lower muscular fibers of the pharynx, however, are involved in the "swallowing reflex." The "orchestration" of this movement from skeletal muscles of the pharynx and upper esophagus to the smooth muscle of the thoracic esophagus is done by both the somatic (skeletal) motor fibers and the parasympathetic fibers of nerve X.

CAROTID SHEATH

The common and internal carotid arteries, the internal jugular vein, and the vagus nerve extend from the base of the skull to the thoracic cavity. They are all invested in a common loose areolar fascia, the **carotid sheath**. The artery lies deep and medial to the internal jugular vein, and the vagus nerve is posterior to the carotid artery. *Figure 43.4* shows the relationship of the carotid sheath to the three layers of deep fascia in the neck; **investing, visceral**, and **prevertebral fascia**. The carotid sheath is also anterior to the cervical sympathetic trunk, which lies on the **longus colli** and **longus capitis muscles** in front of the cervical vertebrae.

THYROID GLAND (*figs. 43.11 and 43.12A*)

The thyroid gland is a bilobular gland that lies lateral to the larynx and trachea in the lower neck. The two lobes are connected anteriorly by an **isthmus** of thyroid tissue that crosses the trachea at the level of the 2nd–4th tracheal cartilages. The lobes are overlapped by the sternothyroid muscles and related laterally to the carotid sheaths. The recurrent laryngeal branches of the vagus nerves lie deep to the thyroid lobes near the posterior aspect of the trachea. They are vulnerable to damage during thyroid surgery and can cause respiratory and vocal problems if damaged.

The thyroid gland may also have a **pyramidal lobe** arising from the isthmus and connected to the thyroid cartilage and hyoid bone. This is a reflection of its developmental origin in the region of the tongue and subsequent descent into the lower neck (*fig. 43.12B*). **Four parathyroid glands** are usually associated with the posterior surface of the thyroid gland. These two endocrine glands will share a common blood supply from **superior** and **inferior thyroid arteries**.

Carotid Triangle

The carotid triangle is bounded by the anterior border of the sternomastoid, the superior border of the omohyoid, and the inferior border of the posterior belly of the digastric.

POSTERIOR BELLY OF THE DIGASTRIC (*fig. 43.13*)

The posterior belly of the digastric is a key muscle for relationships in the upper neck. The muscle arises from the **digastric** groove on the medial aspect of the **mastoid process of the temporal bone**. It joins the anterior belly of the digastric by an intermediate tendon that is held down to the **body of the hyoid** by a fascial sling. The posterior and anterior bellies of the digastric area derived from separate branchiomeric (branchial or pharyngeal) arches in the head.

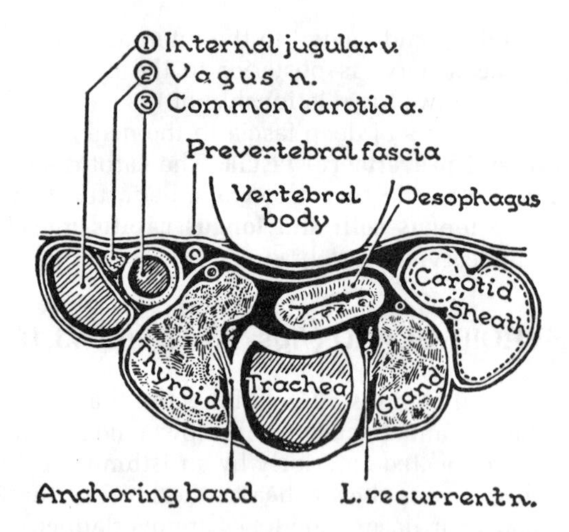

Figure 43.11. The thyroid gland and the carotid sheath (on cross-section).

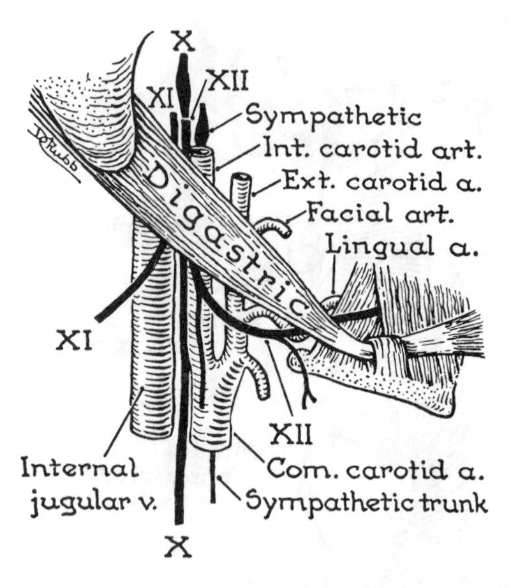

Figure 43.13. Key: posterior belly of digastric.

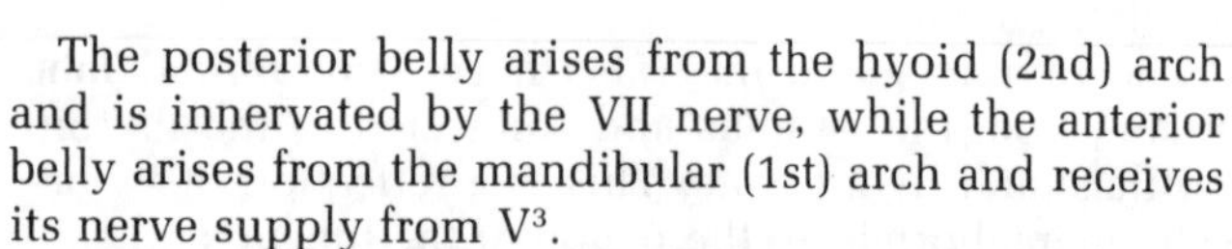

The posterior belly arises from the hyoid (2nd) arch and is innervated by the VII nerve, while the anterior belly arises from the mandibular (1st) arch and receives its nerve supply from V[3].

The posterior belly is also associated with the **stylohyoid muscle**. The VII nerve exits the stylomastoid canal on the base of the skull between these two muscles and innervates them before entering the substance of the parotid gland which lies superior to the posterior belly of the digastric muscle.

Important Relationships of the Posterior Belly of the Digastric

The parotid and the superficial lobe of the submandibular salivary glands contact the superior and lateral surfaces of the posterior belly of the digastric. Three superficial structures lie between the digastric and the skin of the upper neck. The cutaneous veins of the **external jugular** system, the cutaneous branches of the **great au-**

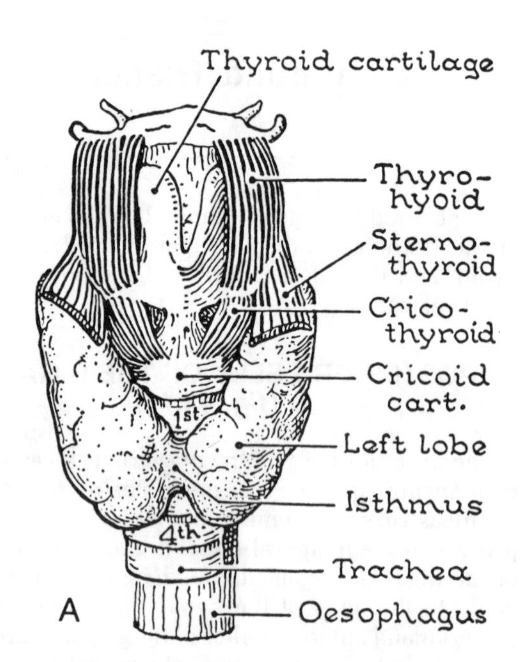

Figure 43.12A. The thyroid gland (front view).

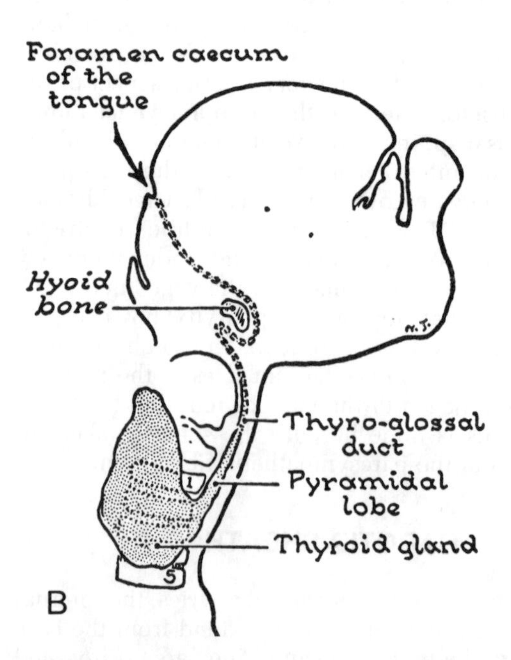

Figure 43.12B. The course of a developing thyroid gland. (From data by J. E. Frazer.)

ricular nerve and the important **cervical branch of the VII** nerve, which innervates the platysma (*figs. 43.6, 43.7*), all lie superficial to the posterior belly of the digastric muscle.

Deep to the posterior belly of the digastric (*fig. 43.13*), one can readily see 3 great vessels (**internal jugular vein, internal carotid artery** and **external carotid artery**) and 3 great cranial nerves (**vagus [X], accessory [XI]** and **hypoglossal [XII]**). In addition, deep to these structures are the **glossopharyngeal nerve (IX)** and the **sympathetic trunk**.

GENERAL DISPOSITION OF THE VESSELS AND NERVES

The internal jugular vein lies lateral and posterior to the internal and external carotid arteries when it descends deep to the posterior belly of the digastric. The jugular fossa lies posterior to the carotid canal in the base of the skull, and there is a gradual shift from this posterior to lateral relationship on the part of the vein as it descends into the lower neck region. The anterior aspect of the sternomastoid muscle overlies the internal jugular vein at the level of the posterior belly of the digastric muscle. The cranial nerves IX, X, XI, and XII descend together from the base of the skull in the carotid sheath to the level of the posterior belly of the digastric (C2). At this level, nerve IX passes medially to enter the pharynx between the superior and middle constrictors. The vagus nerve (X) continues inferiorly on the posterior aspect of the internal carotid artery; XI passes posteriorly to enter the sternomastoid muscle; and XII passes anteriorly, into the digastric triangle musculature above the hyoid bone.

Figures 43.13 and *43.14* depict the important relationships of the nerves to the vessels. Nerve XI lies superficial to the internal jugular vein in about 75% of the cases as it passes posteriorly to the sternomastoid muscle. As noted, X lies posterior to the internal carotid and the common carotid arteries as it descends. The pharyngeal branch of X passes anteriorly between the internal and external carotid arteries, while the superior laryngeal branch is deep to both of these arteries as it passes anteriorly to enter the laryngeal wall. The hypoglossal nerve remains external to both the external and internal carotid arteries as it passes anteriorly into the floor of the mouth to innervate the muscles of the tongue.

NERVES OF THE CAROTID TRIANGLE

Glossopharyngeal nerve IX is deep to the internal carotid artery and penetrates the lateral pharyngeal wall with the **stylopharyngeus muscle**. Nerve IX is motor to stylopharyngeus and sensory to the mucosa of the pharynx, posterior one-third of the tongue, palatine tonsil and part of the soft palate region. It will be described in detail in the chapter on the pharynx.

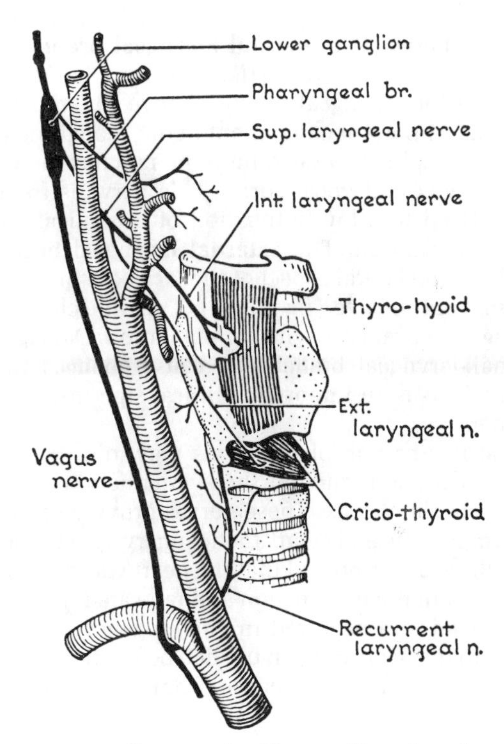

Figure 43.14. Pharyngeal and laryngeal branches of vagus.

Vagus Nerve (X) (*figs. 43.13 and 43.14*)

It descends through the entire length of the neck within the carotid sheath. As one of the most important nerves in the body, it is constantly assessed on examination of the patient. It maintains both voluntary motor control over skeletal muscles and involuntary control of smooth muscle, cardiac muscles, and glands. Its more superior branches are therefore voluntary motor nerves to muscles of the palate, pharynx, and larynx, and its lower cervical, thoracic, and abdominal branches are autonomic (parasympathetic) fibers. The vagus nerve also has extensive sensory fibers within it. Most are sensory to the visceral mucosa of the gastrointestinal and respiratory systems. The vagus has an **inferior (sensory) ganglion** in the jugular fossa (*fig. 43.14*). There are also some clinically important sensory fibers from the skin in the external auditory (ear) canal that have their sensory cell bodies located in a smaller superior ganglion in the jugular fossa.

The vagus innervates the skeletal muscles of the palate (except the tensor palati [V^3]), the pharynx (except stylopharyngeus, [IX]) and *all* the skeletal muscles of the larynx. The parasympathetic branches in the neck go to the mucous and serous glands of the endodermally derived foregut (pharynx, larynx, and trachea). Sensory branches arise from the same mucosa that contains these mucous glands as well as the skin of the external ear canal.

Branches of nerve X that relate specifically to the carotid triangle are two terminal branches of the superior laryngeal nerve. The **internal laryngeal branch** pierces the thyrohyoid membrane (*fig. 43.14*) and is sensory primarily to the laryngeal mucosa above the vocal cords. This is an important component in the "coughing reflex" when material is aspirated onto the mucous membranes of the larynx and vocal cords. This nerve is frequently anesthetized to allow intubation of a tracheal tube for general anesthesia. The **external laryngeal branch** descends on the lateral aspect of the larynx to innervate the **cricothyroid muscle** (*fig. 43.14*). This muscle assists in tensing the vocal cords during phonation. Damage to the external laryngeal branch can cause changes in voice quality. This is an important clinical sign in vagal nerve disorders.

A cardiac branch of the vagus may join the superior cervical cardiac branch of the sympathetic trunk to descend into the thorax. These nerves provide some of the parasympathetic and sympathetic innervation to the heart. It should be remembered that the heart was derived from mesoderm in the head region and relocated in the thorax as the head fold occurred in embryogenesis. The nerve supply to the heart was established before this embryonic folding and is therefore derived from the cervical region.

Accessory Nerve (XI)

This is a peripheral nerve that innervates two large muscles of the neck, **sternomastoid** and **trapezius**. It passes posteriorly from the inferior border of the posterior belly of the digastric, lies on the deep surfaces of these muscles, and is contained in the intervening (investing) fascia that forms the roof of the posterior triangle of the neck. This nerve can be tested in a patient by asking them to flex their head (chin placed on the chest). This requires the bilateral contraction of the sternomastoids. Unilateral contractions cause the head to rotate in the opposite direction. The lower part of the peripheral nerve is tested by asking the patient to "shrug" or elevate their shoulders.

Hypoglossal Nerve (N. XII)

This nerve is motor to the tongue musculature and passes anteriorly, deep to the middle tendon of the digastric bellies, to reach the digastric triangle. It then enters the tongue between the mylohyoid and hyoglossus muscles.

Two branches arise from nerve XII in the carotid triangle. These are actually C1 fibers that are "hitchhiking" on nerve XII as they pass anteriorly from the area of the foramen magnum on the base of the skull. The first branch leaves nerve XII on the lateral side of the carotid sheath inferior to the posterior belly of the digastric. This is the **superior root** of the **ansa cervicalis** (*figs. 43.9 and 43.13*). The second branch innervates the thyrohyoid muscle and leaves nerve XII just as it passes under the posterior belly of the digastric for the second time.

Ansa Cervicalis (*fig. 43.9*)

This is a loop of motor fibers derived from the anterior rami of C1, 2, and 3. C1 fibers are carried in the hypoglossal nerve to the carotid triangle and then leave nerve XII as a descending **superior root (descendens hypoglossi)**. This superior root joins an **inferior root (descendens cervicalis)** from a loop of the cervical plexus containing C2, 3 ventral rami. The **ansa** (Gr. handle) is usually on the lateral aspect of the internal jugular vein but may be deep to the vein. The two roots may at times not be joined by an "ansa" but proceed directly to the strap muscles which they innervate.

ARTERIES OF THE CAROTID TRIANGLE (*fig. 43.15*)

The common carotid bifurcates at the level of the hyoid bone (C3) to form the internal and external carotid arteries within the carotid triangle. The internal carotid artery is somewhat posterior and lateral to the external carotid artery at this point. The internal carotid artery does not give branches in the neck but ascends to the base of the skull to enter the cranial vault to supply the brain and orbit. A swelling in the internal carotid artery at its origin marks the position of the **carotid sinus**. This is a collection of pressure (baroreceptors) sensitive nerve endings associated with nerve IX. Changes in blood pressure are monitored here and conveyed to the brainstem

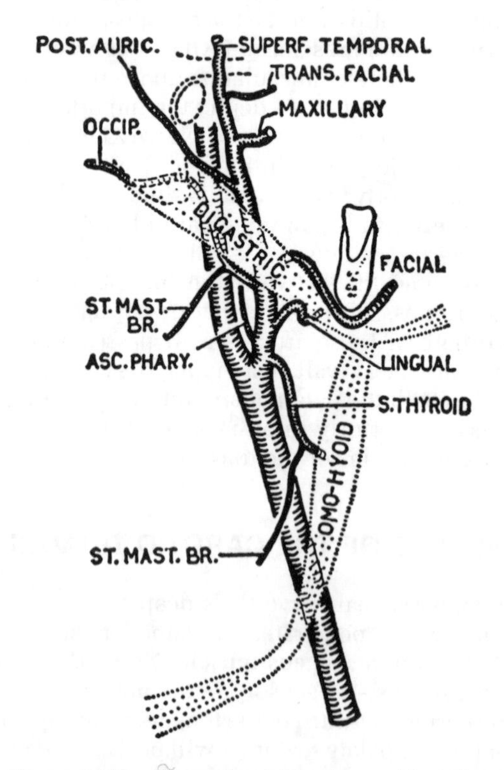

Figure 43.15. External carotid artery and branches.

for processing and maintenance of blood pressure. A **carotid body** is also present in the area of bifurcation of the common carotid artery. This is a microscopic structure containing neuronal chemoreceptors that are sensitive to oxygen content in the blood. They are also involved in mechanisms that control blood pressure and cardiac output.

The **external carotid artery** provides the major arterial branches to the extracranial structures of the head and upper neck. *Figure 43.15* shows the major branches of the external carotid artery that arise in the carotid triangle. The branches can be classified as being **inferior to, deep to** or **superior to the posterior belly of the digastric**.

Inferior to the posterior digastric muscle are three arteries. The **superior thyroid artery** arises from the base of the external carotid (sometimes from the common carotid artery) and descends to the superior pole of the thyroid gland at the level of the oblique line on the thyroid cartilage. Within the gland, branches ramify and anastomose with branches of the inferior thyroid artery as well as arterial branches from the opposite side through the isthmus of the gland. The superior thyroid arteries also give some vascular supply to the parathyroid glands within the thyroid. Three small branches of the superior thyroid artery come off as it descends to the thyroid gland. The **superior laryngeal artery** accompanies the internal laryngeal nerve as it pierces the thyrohyoid membrane to supply the larynx superior to the vocal cords. The **cricothyroid branch** accompanies the external laryngeal nerve to the cricothyroid muscle and enters the larynx to help supply the mucosa below the vocal cords. The superior thyroid artery also gives a **muscular branch** to the lower portion of the sternomastoid muscle. **The lingual artery** is the major blood supply to the tongue. It is a sinuous, S = shaped artery that projects anteriorly under the posterior belly of the digastric muscle. It accompanies the hypoglossal nerve into the digastric triangle. **The ascending pharyngeal artery** usually arises from the bifurcation of the common carotid artery and ascends deep to the posterior belly of the digastric muscle. It supplies the lateral wall of the pharynx and enters the nasopharynx with the auditory tube above the superior constrictor muscle.

Deep to the posterior belly of the digastric are the facial and occipital arteries. **The facial artery** arises from the external carotid and passes superiorly on the deep surface of the posterior belly of the digastric. It emerges on the superior border of the digastric and grooves the submandibular gland as it courses superficially to cross the inferior border of the mandible and masseter muscle (*fig. 43.16*). It can be palpated at this point as it rises to supply the angles of the mouth, lips, nose, and medial angles of eyes.

The **occipital artery** branches from the external carotid on its posterior surface (*fig. 43.15*). It lies on the deep surface of the posterior belly of the digastric. It is an important blood supply to the scalp overlying the occipital bone. It also has an important relationship to the hypoglossal nerve (*fig. 43.9*). The hypoglossal nerve passes inferiorly to the occipital artery as it passes anteriorly in the neck. The occipital artery also helps to supply the sternomastoid by a muscular branch and thereby assists the superior thyroid artery in supplying this muscle.

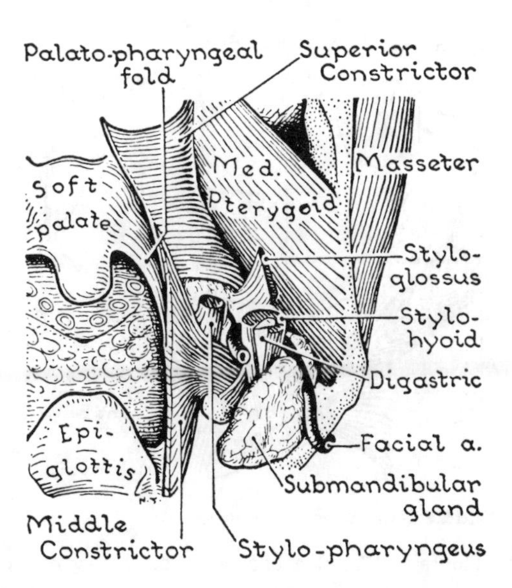

Figure 43.16. Details of S-shaped course of facial artery (posteroinferior view).

The **posterior auricular artery** arises from the superior border of the posterior belly of the digastric. It has an important **stylomastoid** branch, which supplies the facial nerve in the stylomastoid canal. Inflammation (neuritis) of VII can cause the nerve to "swell" and compress this artery. A sufficient lack of blood supply to nerve VII creates a unilateral palsy of the muscles of facial expression (Bell's Palsy). The posterior auricular artery also supplies the external auditory canal.

Additional branches of the external carotid artery arise within the parotid gland superior to the posterior belly of the digastric. They are the *maxillary, transverse facial* and *superficial temporal arteries*. They will be described in the chapter on the parotid gland (p. 518).

Digastric Triangle (Submandibular Triangle)

This area between the two bellies of the digastric muscles and the inferior border of the mandible contains the superficial lobe of the **submandibular (submaxillary) salivary gland** (*figs. 43.17 and 43.18*). The **floor** of the **di-**

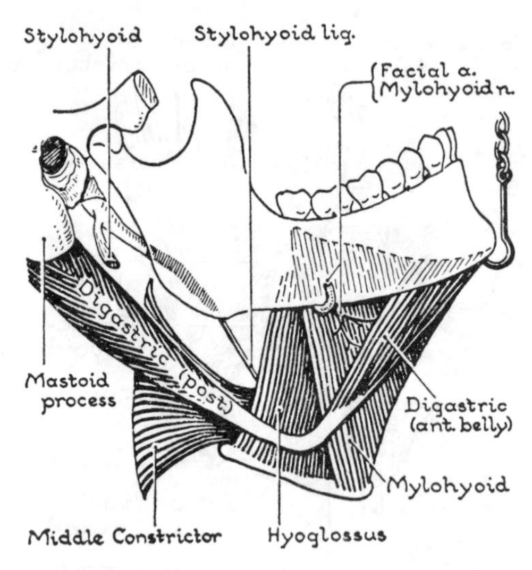

Figure 43.17. Floor of submandibular triangle.

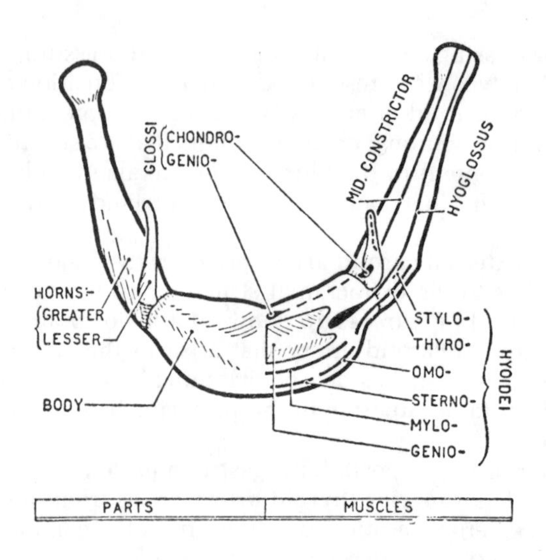

Figure 43.19. Hyoid and details of its muscles.

gastric triangle is formed by the fibers of 3 flat muscles: **mylohyoid, hyoglossus, and middle constrictor**.

The **mylohyoid** arises from the mylohyoid line on the medial aspect of the body of the mandible and inserts on the superior aspect of the body of the hyoid bone (*figs. 43.17 and 43.19*). It forms a "diaphragm" for the floor of the mouth and is continuous with the mylohyoid muscle on the opposite side through the midline raphe that extends from the mandible to the hyoid (*fig. 43.8*). Superficial to the mylohyoid is the anterior belly of the digastric muscle. Both the mylohyoid and anterior belly of the digastric are derived from the mandibular arch in the embryo, and they are both innervated by V^3 (**mylohyoid nerve**).

The hyoglossus arises from the greater cornu of the hyoid bone and inserts into the tongue. It has a very important relation with nerve XII on its external surface and the lingual artery on its internal surface in the digastric triangle (*fig. 43.13*).

The **middle constrictor** arises from the "V-shaped" union of the greater and lesser horns of the hyoid bone and from the **stylohyoid ligament** that attaches to the lesser cornu

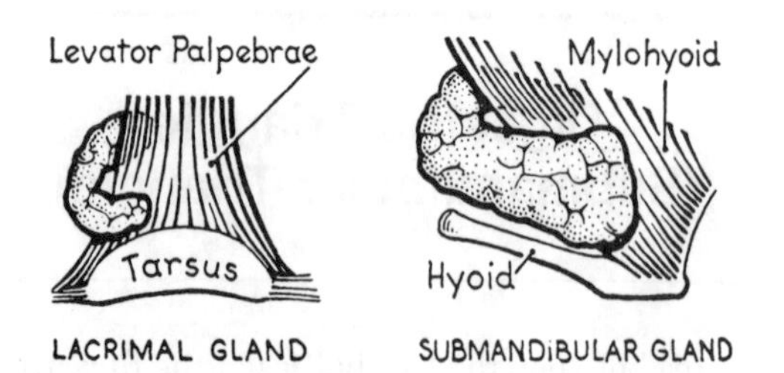

Figure 43.18. Two U-shaped glands and the two responsible muscles, compared.

(*fig. 43.10*). Its fibers pass posteriorly to the lateral wall of the pharynx.

CONTENTS OF THE DIGASTRIC TRIANGLE

1. The superficial lobe of the **submandibular gland** (submaxillary gland) fills this space and is separated from the parotid gland by an intervening fascial plane, **the stylomandibular ligament**. *Figure 43.18* shows how the submandibular gland forms a "U-shaped" gland around the posterior margin of the mylohyoid muscle. The deep part of the gland and the submandibular duct will be described in the floor of the mouth. The gland is innervated by parasympathetic secretomotor fibers in the chorda tympani and vasoconstrictive sympathetic fibers.
2. The **facial artery** and **facial vein** groove the superficial lobe of the submandibular gland as they traverse the digastric triangle (*fig. 43.16*).
3. The **hypoglossal nerve** and **lingual artery** enter the digastric triangle on the deep side of the posterior belly of the digastric and then become separated by the intervening hyoglossus muscle.
4. The mylohyoid nerve (V^3) and the submental artery and veins that arise from the facial artery and vein, respectively, lie on the superficial surface of the mylohyoid muscle. They are usually accompanied by lymphatics and a submental lymph node(s) is contained within the digastric triangle.

The **roof of the digastric triangle** is formed by **investing fascia** subcutaneous tissue, platysma muscle, and skin. Within the superficial fascia are the facial vein, cervical branch of VII, and branches of the transverse cervical nerve (C2, 3).

Clinical Mini-Problems

1. The hyoid bone is palpable at the level of the bifurcation of the common carotid artery. At which cervical vertebral level would one find the hyoid bone and carotid bifurcation?
2. How could an infection in the anterior triangle of the neck cause an inflammation of the pericardium (pericarditis)?
3. How could a disease of the thyroid gland cause a voice disorder in a patient?
4. A patient was given radioactive iodine to localize the thyroid tissue in a radiogram. The clinician found a large mass of thyroid tissue in the neck at the level of C6 and C7, but he also found some radioactivity in the posterior aspect of the tongue. How could there be two regions containing thyroid tissue?
5. Why would a local anesthetic deposited at the tip of each greater horn of the hyoid help prevent a patient from coughing when a bronchoscope is passed through the nasopharynx and larynx into the bronchial tree?
6. Which cranial nerve(s) are most susceptible to injury when the internal jugular vein is removed in radical neck surgery?

(Answers to these questions can be found on p. 587–588.)

44

Root of the Neck

Clinical Case 44.1

Patient Shirley V. This 46-year-old woman was thoroughly tested in the University Hospital and diagnosed as having a "goiter." She is scheduled for surgery and you are assigned to her case. The thyroidectomy is performed by successively clamping and ligating the terminal branches of the superior and inferior thyroid arteries from the posterior to the anterior aspect of the thyroid gland. Unfortunately, the left recurrent laryngeal nerve is damaged. When you examine her the next day, her voice quality is reduced to a whisper. This persists for 4 months following the surgery, and then her voice quality gradually improves to normal over the next 2 months. Meanwhile, her metabolism has normalized.

In the root of the neck the **key orienting structure** is the **scalenus anterior muscle**. It arises from the **anterior tubercles** of the cervical vertebrae C3–6 and inserts on the **scalene tubercle** on the superior aspect of the 1st rib. Virtually every important structure in the root of the neck approximates one of the surfaces of this important reference muscle.

The main vascular trunks to the head and neck and upper extremity pass posterior to the sternoclavicular joints and enter the root of the neck. On the right side, the brachiocephalic trunk divides into the right subclavian and right common carotid artery on the posterior aspect of the sternoclavicular joint. The left common carotid and left subclavian arteries arise from the arch of the aorta and ascend behind the left sternoclavicular joint to enter the root of the neck medial to the apex of the left lung.

Subclavian Artery

Arching over the apex of each lung (*fig. 44.1*), the subclavian artery passes **posterior to the scalenus anterior muscle** and superior to the 1st rib. At the lateral margin of the 1st rib, the subclavian artery becomes the axillary artery. As the subclavian passes through the root of the neck, it is divided into 3 parts by its relationship to the **scalenus anterior** muscle. The first part is medial to the scalenus anterior, the second part is directly posterior to the anterior scalene muscle, and the third part is lateral to the muscle in the posterior triangle of the neck above the 1st rib. The first part of the subclavian artery has the major arterial branches (*figs. 44.1 and 44.2*). A **dorsal scapular artery** frequently arises from the 2nd or 3rd part of the artery as the only other branch.

SURFACE ANATOMY

The artery can be traced by a curved line running from the sternoclavicular joint to 2–3 cm above the clavicle near its central point. The pulse in the subclavian artery can be felt against the 1st rib as it traverses its superior border above the level of the clavicle (*figs. 44.1 and 44.2*).

Subclavian Vein

The subclavian vein (*fig. 44.3*) has a similar course to the subclavian artery with the following 3 important differences: (*a*) the vein is inferior to the artery and within the arch formed by the artery as it arches over the apex of the lung; (*b*) the subclavian vein passes **anterior to**

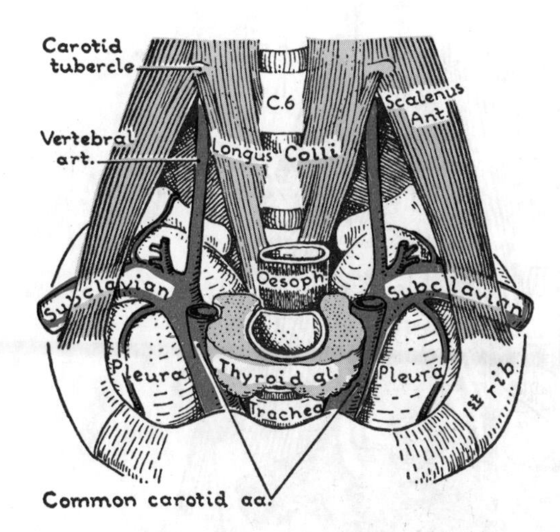

Figure 44.1. "The triangle of the vertebral artery." The "base" is the 1st part of the subclavian artery.

scalenus anterior at its insertion on the scalene tubercle; and (*c*) the subclavian vein is posterior and inferior to the clavicle and well protected by the clavicle. Access to the subclavian vein for intravenous catherization is done from the inferior aspect of the clavicle. The venous branches that join the subclavian vein are depicted in *Figure 44.4* and show a great deal of variation as compared to the comparable arterial branches of the subclavian artery that lie posterior and superior to the vein.

"Triangle of the Vertebral Artery" (*figs. 44.1, 44.3*)

The scalenus anterior forms the lateral side of this triangle, and the **longus colli muscle** forms the medial side. The base of the triangle is limited by the superior aspect of the subclavian artery.

CONTENTS

The **vertebral artery** and vein ascend to the apex of the triangle and enter the **foramen transversarum** of the transverse process of the 6th cervical vertebra (*figs. 44.1 and 44.2*). The **sympathetic trunk** and its associated **middle cervical ganglion** and **inferior cervical ganglion** (*fig. 44.5*) are within the triangle. The inferior cervical ganglion is associated with the posterior aspect of the vertebral artery at its origin from the subclavian, while the cervical sympathetic trunk is adherent to the anterior aspect of the longus colli muscle. The middle cervical ganglion is associated with the inferior thyroid artery.

The **common carotid artery** ascends through the anterior portion of the triangle to lie on the anterior aspect of the origins of the anterior scalene muscle. The carotid artery can be compressed against the transverse process of the 6th cervical vertebra. This area of the C6 transverse process is termed the **carotid tubercle** (*fig. 44.1*).

Surrounding the common carotid artery, the **internal jugular vein**, and the **vagus nerve** (*fig. 44.3*) is the carotid sheath. It descends through the neck on the medial border of the scalenus anterior. The right **recurrent** laryngeal nerve arises from the right vagus and loops under the right subclavian artery to ascend to the larynx between the trachea and esophagus.

The **phrenic nerve** enters the inferolateral corner of this triangle as it descends on the anterior surface of the **scalenus anterior**. It then crosses the anterior surface of the subclavian artery and apex of the lung to enter the superior aperture of the thorax within the internal borders of the 1st rib. On the left side, the phrenic nerve is crossed by the **thoracic duct** as it joins the venous system at the bifurcation of the **left brachiocephalic vein** (*fig. 44.3*). The **right lymphatic duct** joins the right brachiocephalic vein in the same manner, but it is small and difficult to demonstrate.

The **branches of the 1st part of the subclavian artery** are also located in this triangular area medial to the scalenus anterior muscle (*fig. 44.3*). They are:

1. Vertebral artery
2. Thyrocervical trunk
   Inferior thyroid artery
   Transverse cervical (transversa colli) artery
   Suprascapular artery

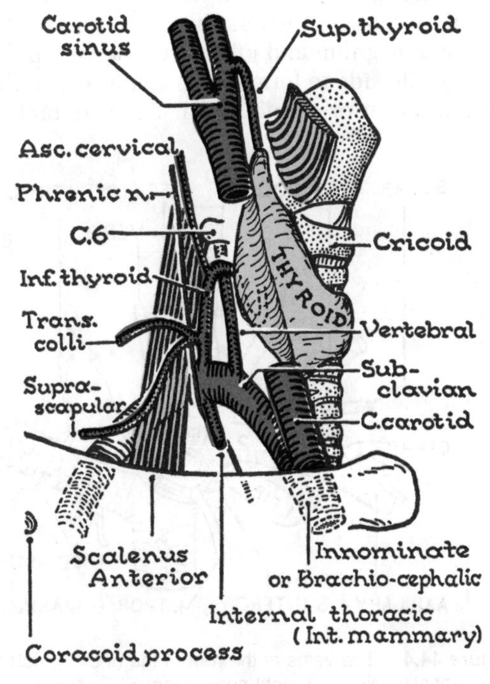

Figure 44.2. The arteries at the root of the neck.

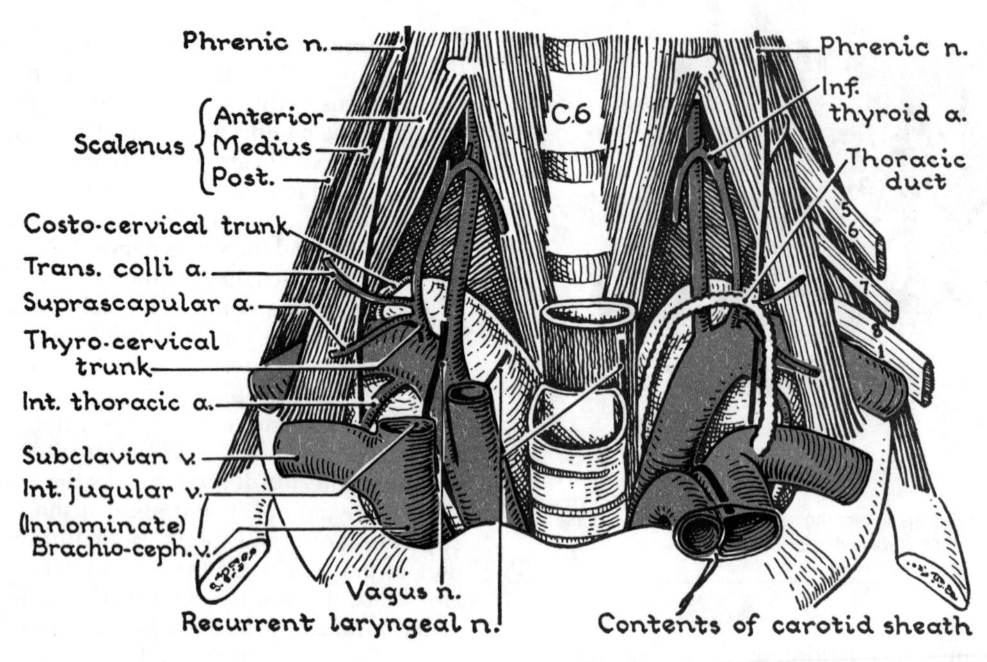

Figure 44.3. The root of the neck.

3. Internal thoracic (internal mammary) artery
4. Costocervical trunk
   - Deep cervical artery
   - Highest intercostal artery
     - 1st posterior intercostal artery
     - 2nd posterior intercostal artery

The **vertebral artery** is the largest branch and ascends to course through the foramen transversarii of the upper 6 cervical vertebrae. It enters the cranial cavity through the foramen magnum and joins with the vertebral artery of the opposite side to form the basilar artery in the midline of the posterior cranial fossa. The vertebral arteries are a major vascular supply to the brain along with the internal carotid arteries.

The **thyrocervical trunk** gives off three branches near its origin. The **suprascapular artery** and the **transverse cervical artery** pass anterior to the scalenus anterior and "clamp down" the phrenic nerve as they course into the posterior triangle of the neck. Both of these arteries supply muscles associated with the dorsal surface of the scapula. The **inferior thyroid artery** is the terminal branch of the thyrocervical trunk. It forms a loop at the level of C6, which is associated with the middle cervical ganglion of the sympathetic trunk (*figs. 44.3 and 44.5*). The inferior thyroid artery supplies the inferior pole of the thy-

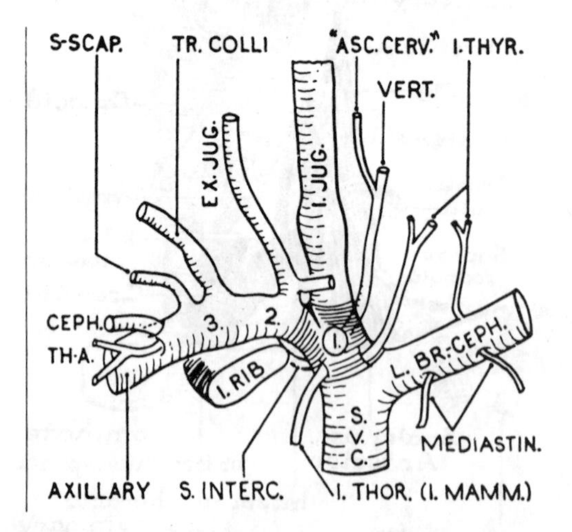

Figure 44.4. The veins at the root of the neck. 1, right brachiocephalic vein; 2, 3, right subclavian; *SVC*, superior vena cava; *THA*, thoracoacromial vein.

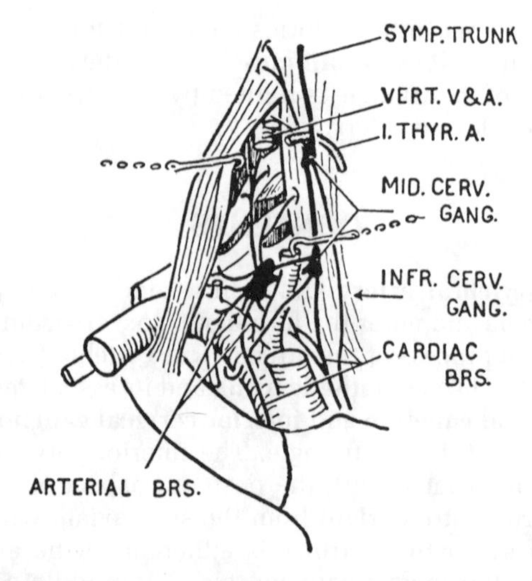

Figure 44.5. The sympathetic trunk, ganglia, and branches, at the root of the neck.

roid gland and anastomoses with the superior thyroid branch of the external carotid artery within the thyroid gland. A muscular branch of the inferior thyroid artery, the **ascending cervical artery** supplies the scalenus anterior and the longus colli muscles.

The **internal thoracic artery** descends on the pleura behind the subclavian vein and enters the thorax on the posterior aspect of the sternum (*figs. 44.1 and 44.3*).

The **costocervical trunk** passes posteriorly over the apex of the lung and the neck of the 1st rib (*figs. 44.1 and 44.3*). The **deep cervical branch** supplies the deep musculature of the back of the neck and the descending intercostal branch gives rise to the posterior intercostal arteries of the 1st and 2nd intercostal spaces.

Phrenic Nerve (*fig. 44.3*)

The phrenic nerve receives fibers from the anterior rami of C3–5. The nerve is formed at the superior lateral portion of the scalenus anterior and courses to the medial border of the muscle as it descends in the neck. It is contained in the prevertebral fascia that covers the anterior aspect of the scalenus anterior and is crossed superficially (anteriorly) by the suprascapular and transverse cervical arteries. The phrenic nerve is motor to the diaphragm muscle and sensory to the parietal pleura and peritoneum that cover the domes of the diaphragm. The diaphragm, like the heart, is largely derived from mesoderm that was first located in the head of the embryo. When the head fold places the diaphragm in the future thoracic area, its cervical nerve supply is carried from the cervical region to the thorax.

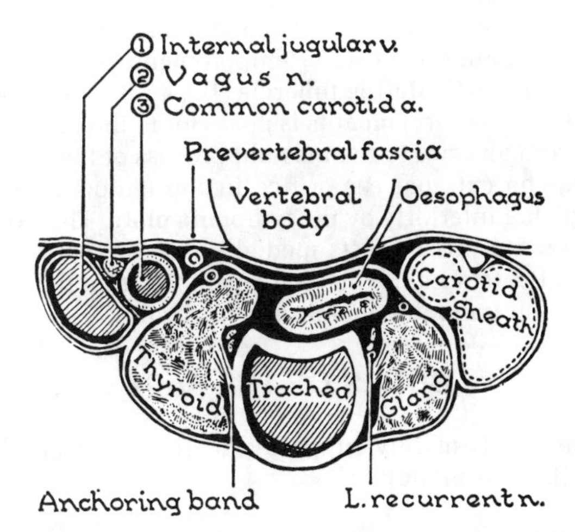

Figure 44.6. Trachea and esophagus (on cross-section). Note dense fibrous band attaching each lobe to trachea.

Recurrent Laryngeal Nerves

These important branches of the vagus are found in the root of the neck between the esophagus and trachea on the medial aspect of the thyroid glands (*figs. 44.3 and 44.6*). The **left recurrent laryngeal** is a branch of the left vagus nerve in the thorax, and the **right recurrent laryngeal nerve** arises from the right vagus at the base of the neck. The difference in these two origins is due to the persistence of the 6th aortic arch on the left side as the ductus arteriosus. On the right side, the most inferior aortic arch that is retained is the 4th, and it forms the initial segment of the right subclavian artery, which the right recurrent nerve loops around.

Trachea and Esophagus

The cervical portion of these two tubular organs exist in the root of the neck from C6 to the superior aperture of the thorax (T1, 1st rib, and manubrium). They are continuous with their thoracic components in the posterior mediastinum. *Figure 44.6* shows the trachea, the esophagus, and the thyroid gland invested by the visceral fascia of the neck. This visceral "column" is flanked by the carotid sheaths and lies anterior to the bodies of the vertebrae and the prevertebral fascia in the root of the neck.

Clinical Mini-Problems

1. **How would stimulation of the phrenic nerve produce shoulder pain in a patient?**
2. **How could disease in the apex of the right lung cause symptoms in the medial side of the hand?**
3. **How could disease in the apex of the lung cause a Horner's Syndrome in the right eye (pupillary constriction, partial ptosis, and lack of sweating around the eye)?**
4. **The carotid artery can be compressed against the transverse process of a cervical vertebra at the level of the cricoid cartilage.**
   (a) **Which vertebra is at this level?**
   (b) **Would one compress the vertebral artery in this maneuver?**

(Answers to these questions can be found on p. 588.)

45

Side of Skull, Parotid, Temporal and Infratemporal Regions

Clinical Case 45.1

Patient Peter L. This 48-year-old man has been diagnosed as having a benign left parotid tumor and scheduled for surgery by the otolaryngologist who is your clinical tutor. During the operation, the surgeon lays a skin flap in front of the left ear and exposes the parotid from the zygoma superiorly to the posterior belly of the digastric muscle inferiorly. The tumor is removed together with the superficial portion of the parotid gland. Special care is taken not to damage the facial nerve. The deep portion of the parotid is left intact around the retromandibular vein and external carotid artery. (At followup, the only postoperative symptom is some numbness over the left ear lobe.)

Lateral Aspect of the Skull

The view of the skull from the lateral aspect is termed **norma lateralis** (*fig. 45.1*). The skull can be divided into a superior cranial component and inferior facial component by a horizontal plane from the **external occipital protuberance (inion)** through the zygoma to the nasal cavity.

The cranial portion has an outer contour formed from the root of the nose by the frontal bone, the parietal bone, and the occipital bone. The highest point of this contour is reached on the parietal bone and termed the **vertex**. The **external occipital protuberance** is located at the base of this outer contour. An inferior surface of the occipital bone forms the roof of the suboccipital region, which is limited posteriorly by the **superior nuchal line**. This line is the bony attachment site on the skull of the trapezius muscle. The superior nuchal line continues anteriorly to blend with the **mastoid process** where the sternomastoid muscle arises from the skull.

Two prominent lines are formed on the lateral surfaces of the frontal and parietal bones. These are the **superior temporal line** and **inferior temporal line**. These mark the superior limit of the temporal fossa on the lateral aspect of the skull. The **superior temporal line** is the bony attachment site for the **temporal fascia** and the **inferior temporal line** is the attachment site of the **temporalis** muscle. The **pterion** is the H-shaped articulation site for the four bones that form the anterolateral portion of the temporal fossa. The parietal, frontal, greater wing of sphenoid, and the squamous part of the temporal bone form the margins of the pterion. This is an external landmark for the intracranial position of the **middle meningeal artery**.

The squamous portion of the temporal bone forms the central region of the temporal fossa. The inferior margin of the squamous portion of the temporal bone contains the **mandibular fossa** of the temporomandibular joint and its anterior **articulating tubercle** (*fig. 45.2*). The **external acoustic** (auditory) **meatus** is posterior to the mandibular fossa and anterior to the mastoid process of the temporal bone. The external ear orifice in the temporal bone is completed inferiorly by the **tympanic plate**. The external acoustic meatus projects medially into the petrous portion of the temporal bone. It is separated from the middle ear by the **tympanic membrane** (eardrum).

BONES OF THE FACE

The face is mainly supported by the **zygomatic bone, maxilla,** and **mandible**.

Zygomatic Bone

This "cheek bone" forms the lateral one-third of the orbital margin. It functions to disperse the forces generated in mastication over the frontal bone. Its major processes extend from the body of the bone to its articulating sutures. The zygomatic bone has a **frontal process**, a **maxillary process**, and a **temporal process**.

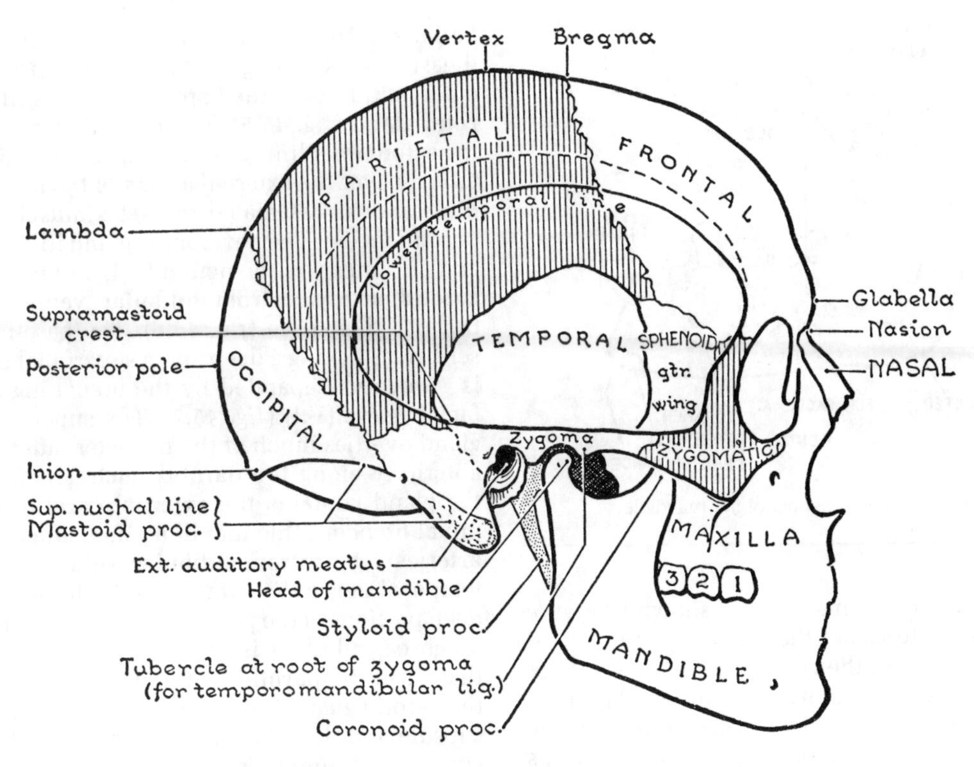

Figure 45.1. The skull, on side view (norma lateralis).

Mandible

The horseshoe-shaped mandible consists of a horizontally oriented **body** and two vertically oriented **rami**. The body of the mandible resulted from the fusion of a right and left component at 2 years of age. It has an **alveolar process** that surrounds the roots of the lower teeth and supports this dentition. Features on the lateral aspect of the body are: (*a*) the **oblique line** where the **depressor labii inferioris** and **depressor anguli oris** attach; (*b*) the **mental foramen** where the mental nerve (V^3) exits; and (*c*) the **mental protuberance**, which forms the "point" of the chin (*fig. 45.3*).

Each ramus has two major processes on its superior border. The **coronoid process** is a superior extension of the anterior margin of the ramus. It serves as an attachment for the strong **temporalis muscle**. The coronoid process lies deep to the zygomatic arch (*fig. 45.1*) and the tendon of the temporalis muscle passes medial to the zygomatic arch as it inserts onto the coronoid process and the anterior aspect of the ramus of the mandible. The **condylar process** of the mandible articulates with the mandibular fossa and articulating tubercle of the temporal bone (*figs. 45.1 and 45.2*). Each condyle has a **head** and **neck** (*fig. 45.3*). The head is covered by cartilage and is contained in the **temporomandibular joint**. The neck supports the head and also serves as an insertion site for the lateral pterygoid muscle on its anterior surface. The **mandibular notch** exits between the coronoid and condylar process. The mandibular notch is covered by the masseter muscle and transmits the masseter nerve and blood supply from the infratemporal fossa to the lateral aspect of the mandible. The **masseter** muscle arises from the zygomatic arch and inserts on the lateral aspect of the ramus from the **angle** to its junction with the body.

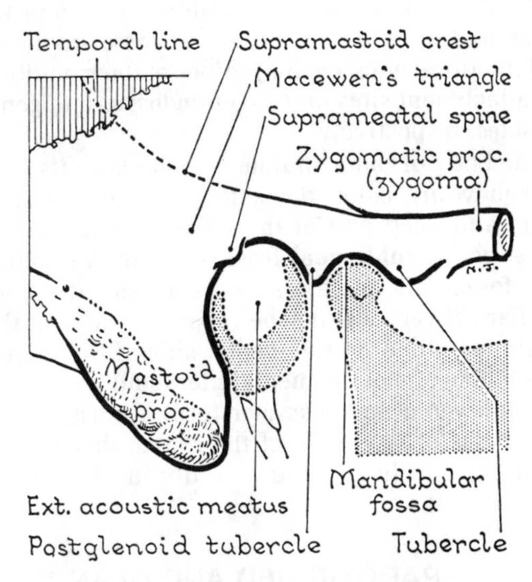

Figure 45.2. Twin depressions (external acoustic meatus and mandibular fossa) and postglenoid tubercle between. The suprameatal spine is the surgeon's guide to the mastoid antrum deep to the triangle behind.

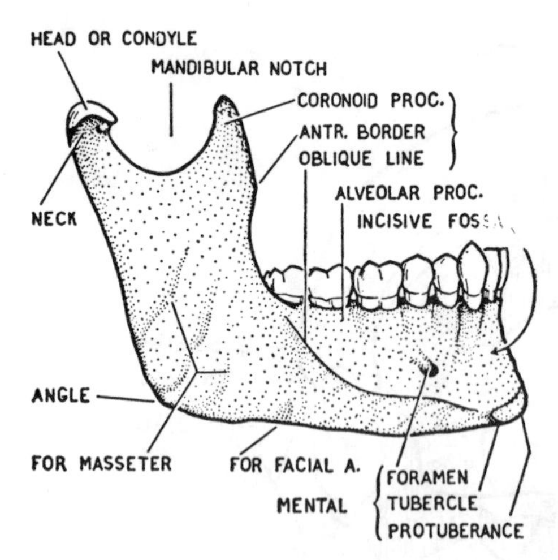

Figure 45.3. The lateral aspect of the mandible.

The medial aspect of the mandible is shown in *Figure 45.4*. The general features are the same as in the lateral view. The medial side of the ramus has an opening for the **mandibular canal**. This **mandibular foramen** permits the **inferior alveolar nerve** (V^3) and **vessels** to enter the ramus and body of the mandible and supply the pulps of the lower teeth. A small projection of bone extends above the anterior portion of the mandibular foramen. This is the **lingula** and is the attachment for the **sphenomandibular ligament**. A shallow **mylohyoid groove** extends inferiorly from the mandibular foramen. It contains the **mylohyoid nerve** and **vessels** that supply the **mylohyoid** and **anterior belly of the digastric** in the submental triangle of the neck. The mylohyoid muscle attaches to the **mylohyoid line** on the medial surface of the body of the mandible. The anterior belly of the digastric attaches to the **digastric fossa** on the inferior margin of the mandible. Just superior to the digastric fossa are the superior and inferior **mental (genial) spines** of the mandible. These are attachment sites for the **genioglossus** and **geniohyoid muscles**, respectively.

The body of the mandible is smooth on its medial aspect above and below the mylohyoid line. The sublingual gland and deep part of the submandibular gland lie in the superior **sublingual fossa**. The inferior **submandibular fossa** contains the superficial part of the submandibular salivary gland. The roughened area at the angle of the mandible on the medial side of the ramus is the attachment site of the **medial pterygoid** muscle. The **medial pterygoid** and **masseter** form a muscular sling on both sides of the angle of the ramus that supports the mandible and helps hold it firmly in the temporomandibular joint.

PAROTID BED AND GLAND

The parotid region (Gr. para = near, ous [otos] = the ear) lies between the ramus of the mandible and the external ear and is superior to the posterior belly of the digastric muscle (*fig. 43.1*). The region is a bony, muscular, and fascial-lined space, which is filled by the **parotid gland** (*fig. 45.5*). The parotid gland develops from the ectodermal lining of the oral cavity and migrates posteriorly over the external aspect of the ramus of the mandible and its associated **masseter muscle** to invade the parotid region. The developing gland fills the entire space and encompasses the neural (VII) and vascular (**external carotid artery, retromandibular vein**) structures that traverse this space (*fig. 45.6*). The parotid gland is described as having a deep and a superficial portion, which is "roughly" separated by the branching pattern of the **facial nerve** (VII) (*fig. 45.7*). The superficial part of the gland overlies much of the masseter muscle and extends anteriorly along the **parotid duct**. The deep portion of the gland contains the **external carotid artery**, its terminal branches (the **maxillary** and **superficial temporal arteries**), the **retromandibular vein**, and the **auriculotemporal nerve (V^3)** (*fig. 45.6*). Both the superficial and deep portions of the parotid gland are covered by a dense, tough capsule that is derived from the investing fascia that lines the **parotid bed** and the superficial aspect of the parotid gland. Swelling of the gland within this tough capsule and deep facial encasement produces a painful situation. "Mumps" is an example of parotid gland swelling induced by a virus.

The **parotid bed** is formed by 4 bony processes that are related to the base of the skull and their associated soft tissue elements. Posteriorly, the mastoid process and its two muscular origins for the sternomastoid and the posterior belly of the digastric form a limit for the parotid bed (*fig. 45.5*). The medial boundary of the parotid gland is formed by the **styloid process of the temporal bone** and the **stylohyoid muscle**. The **styloglossus** and **stylopharyngeus** muscles are medial to the parotid bed within the **lateral pharyngeal space**. The parotid fascia extends from the styloid process anteriorly to the **spine of the sphenoid bone** and then lateral to the posterior aspect of

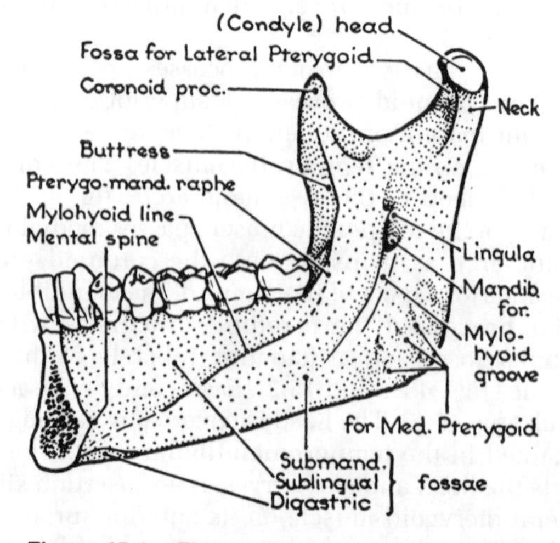

Figure 45.4. The medial aspect of the mandible.

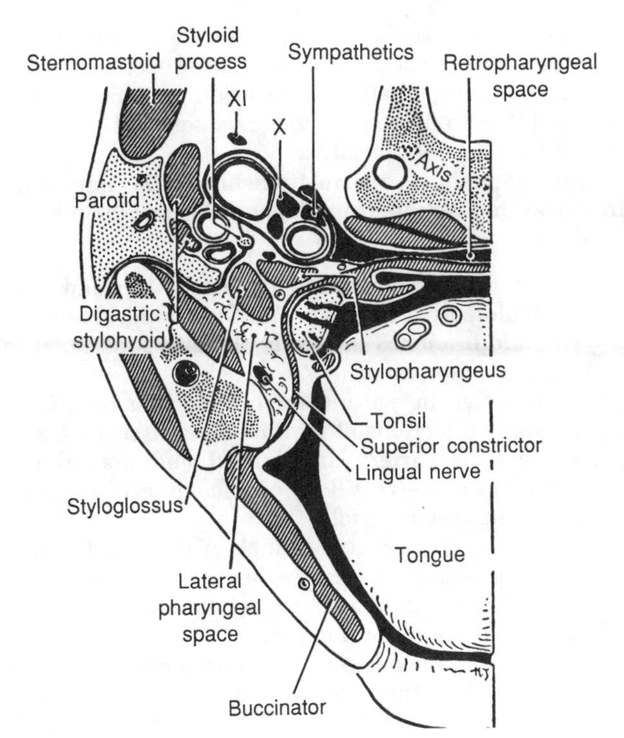

Figure 45.5. Cross-section at level of parotid gland and tonsil.

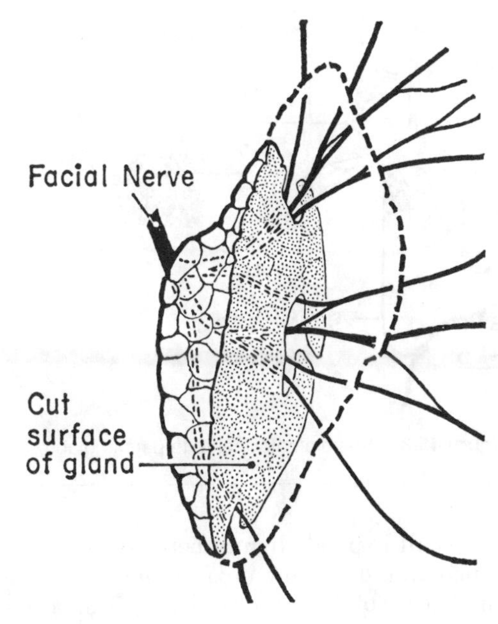

Figure 45.7. To show the relation of the facial nerve to the parotid gland. (After McKenzie.)

the ramus of the mandible. This parotid fascia blends with the **sphenomandibular** and **stylomandibular ligaments** as well as the fascia of the **medial pterygoid** and **masseter muscles** to form the anterior margin of the parotid bed. The superficial portion of the gland extends from the zygomatic arch superiorly to the posterior belly of the digastric inferiorly. The **investing fascia of the neck** that encircled the sternomastoid posteriorly also "invests" the entire parotid gland as it passes forward to attach to the mandible over the anterior triangle of the neck. The weakest part of this fascia is between the **styloid process** and the **spine** of the **sphenoid**. Therefore, infections that "break out" of the parotid fascia usually drain into the lateral pharyngeal space, which is in direct communication with the **retropharyngeal space** (*fig. 45.5*) between the pharynx and prevertebral musculature. These infections have the potential to "track" inferiorly through the neck and into the thorax along the course of the **carotid sheath** between the **visceral** and **prevertebral** fascia.

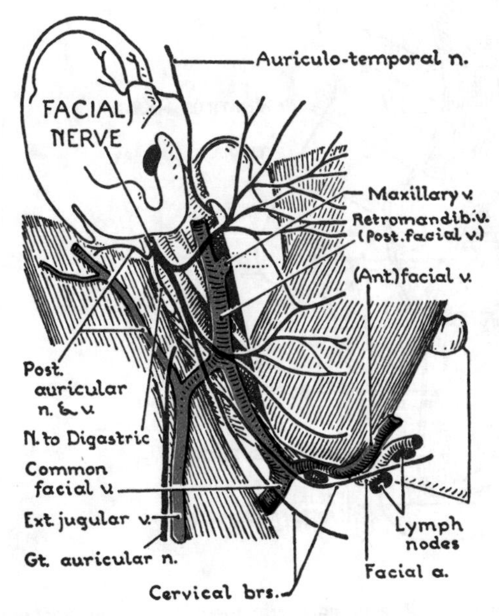

Figure 45.6. Facial nerve and veins in parotid region.

The parotid gland is a major salivary gland that is contained within the parotid fascia. The secretomotor control is via the **lesser petrosal branch** of the glossopharyngeal nerve (IX) (*fig. 45.8*). The **preganglionic fibers** arise from the tympanic plexus in the middle ear, enter the middle cranial fossa through a hiatus on the anterior aspect of the petrous bone and course through the periosteal dura to exit the middle cranial fossa with V^3 in the foramen ovale. The preganglionic parasympathetic fibers synapse on **postganglionic parasympathetic neurons** within the **otic ganglion** on the medial aspect of the trunk of V^3. These postganglionic fibers leave the otic ganglion and join the **auriculotemporal nerve** to be distributed to the parotid gland. Intraglandular branches of the auriculotemporal nerve carry both secretomotor fibers of nerve IX and sensory fibers of V^3. Pain sensation in the parotid gland in conditions such as "mumps" is carried in V^3. The parotid also receives **postganglionic sympathetic** innervation from the superior cervical ganglion via the arteries.

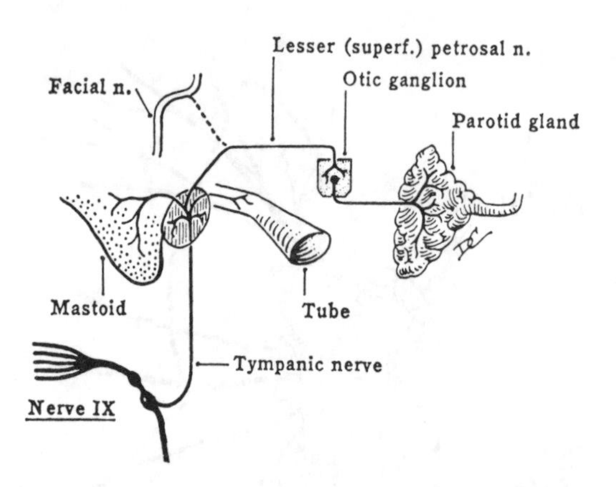

Figure 45.8. Secretomotor nerve to parotid gland.

The gland will expand slightly between meals as its cells produce and store new enzyme granules in their cytoplasm. Release of the extracellular zymogens at meal time is under parasympathetic control. Fluid content of saliva is controlled by the sympathetic innervation to the parotid vessels. The saliva is collected into a major single **parotid duct** that passes superficially over the masseter muscle and pierces the buccinator muscle to **open into the oral cavity** in the mucosa adjacent to the **upper 2nd molar tooth**. Saliva becomes thick during excitement due to a general sympathetic vasconstriction. This reduces the blood flow to the parotid gland vessels and decreases the fluid content of the saliva.

Figure 45.9 shows the relationship of the **important structures that radiate from the periphery of the parotid gland**.

1. Superiorly, the **superficial temporal artery and vein, auriculotemporal nerve**, and the **temporal** and **zygomatic branches of nerve VII** pass superficial to the zygomatic arch.
2. Anteriorly, the parotid duct (and occasionally accessory collections of parotid gland tissues), **transverse facial artery** and **buccal** and **mandibular branches** of nerve VII lie on the superficial aspect of the masseter muscle.
3. Inferiorly, the **cervical branch of nerve VII** to platysma passes under the angle of the mandible, and the external jugular vein and greater auricular nerve approach the parotid gland.
4. Posteriorly, the **occipital vein** and **posterior auricular** branch of the external carotid artery pass over the sternomastoid to supply the tissues of the scalp posterior to the ear.

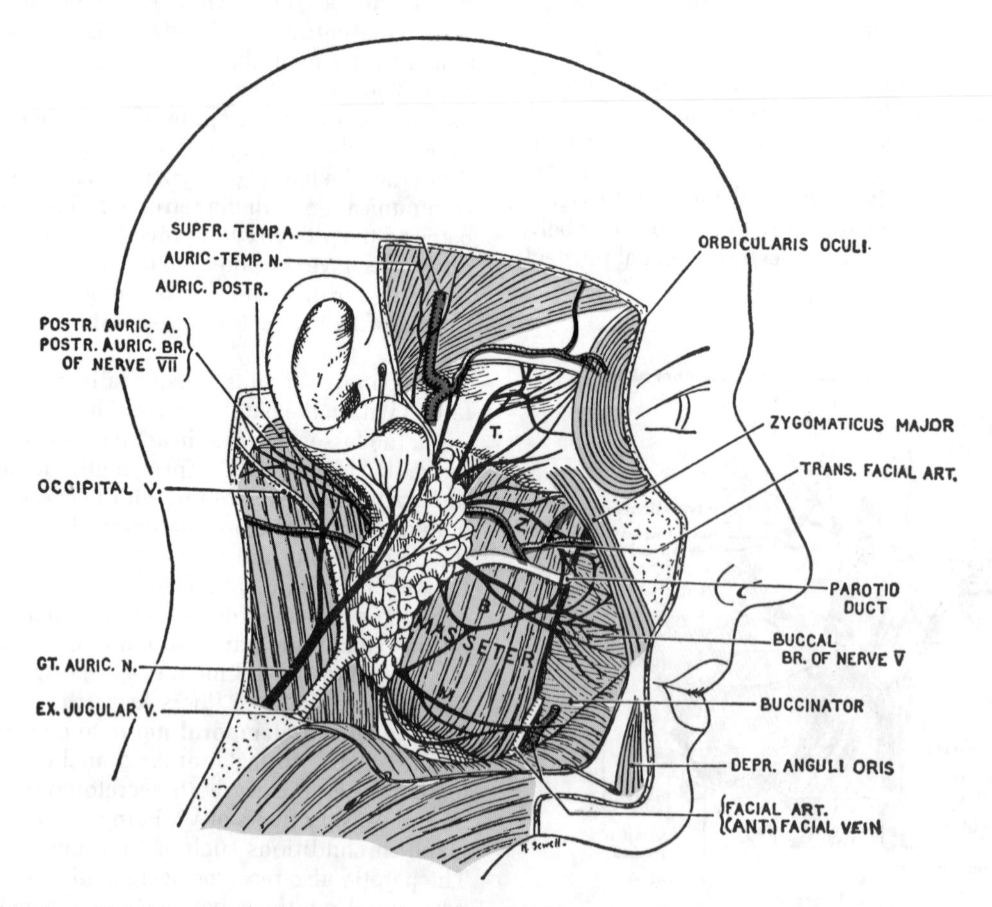

Figure 45.9. Dissection of the side of the face and facial nerve: *T*, temporal branches; *Z*, zygomatic branches; *B*, buccal branches; *M*, mandibular branch.

Structures within the parotid gland are shown in *Figure 45.6* after the parotid tissue has been removed.

1. The **facial nerve** is most superficial within the gland and passes lateral to the vascular elements between the superficial and deep portions of the parotid gland.
2. The **retromandibular vein** is formed within the deep portion of the parotid by the union of the **maxillary** and **superficial temporal veins**. It descends vertically to drain into both the **external** and **internal jugular veins** via the **common facial vein**.
3. The **external carotid artery** ascends in the deep portion of the parotid gland medial to the retromandibular vein and the facial nerve. It terminates in the gland by dividing into the **superficial temporal** and **maxillary arteries**.

The facial nerve is considered the most critical of these three structures when surgery is performed on the parotid gland. The veins and arteries can be tied off if necessary but damage to nerve VII may result in paralysis of the facial musculature with adverse cosmetic and functional effects.

THE FACIAL NERVE (VII) IN THE FACE

The facial nerve exits the base of the skull via the stylomastoid foramen as a purely motor nerve to the muscles of facial expression, the posterior belly of the digastric muscle, and the stylohyoid muscle. All of the sensory and parasympathetic fibers of VII have branched from the nerve within the middle ear and temporal bone.

Figure 45.7 shows how the facial nerve enters the posterior aspect of the deep portion of the parotid gland. As it traverses the gland, it becomes more superficial and leaves the periphery of the gland on the undersurface of the superficial part of the gland.

Branches of VII

The main trunk of nerve VII gives branches to the stylohyoid muscle and posterior belly of the digastric muscle that arise from the bone surrounding the stylomastoid foramen. A small posterior auricular branch also arises from the trunk to pass posteriorly into the scalp to innervate the **posterior auricular muscles** and **occipitalis muscle** behind the ear (*fig. 45.9*). The remaining branches to the muscles of facial expression arise within the parotid gland.

1. 3 **branches** of the temporofacial division (*figs. 45.7 and 45.9*).
   (a) **Temporal branches**—they supply mainly the muscles of the forehead and eyes (**anterior** and **superior auricular muscles, frontalis muscle, corrugator muscle**, and **orbicularis oculi muscle**) (*fig. 45.10*).
   (b) **Zygomatic branches**—they supply **orbicularis oculi** and **zygomaticus major muscles**.
   (c) **Buccal branches**—they supply **levator labii superioris** and the muscles of the nose and upper lip (*fig. 45.10*).

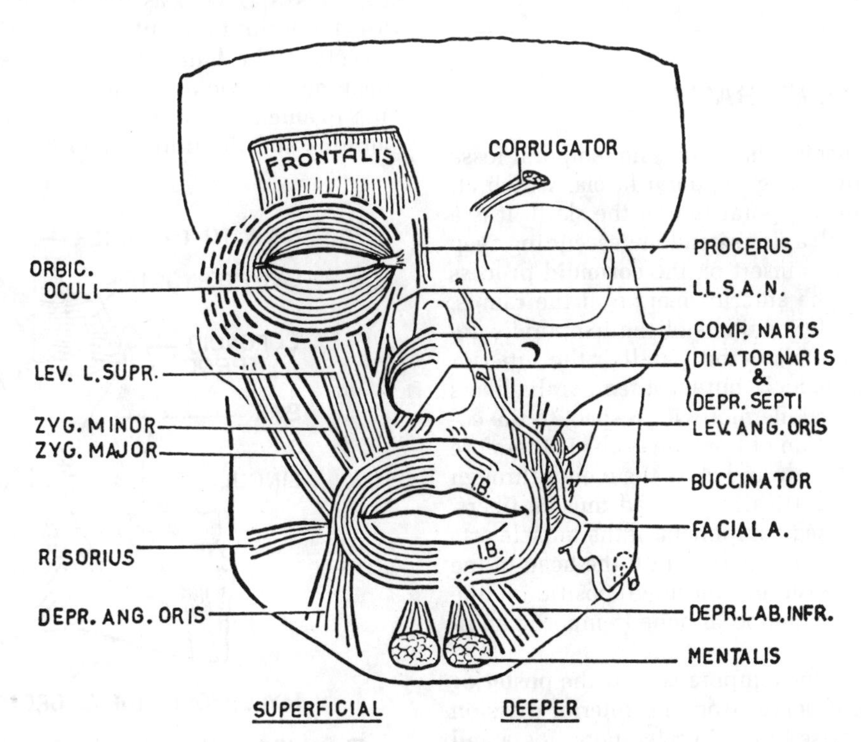

Figure 45.10. The muscles of the face. *LLSAN*, levator labii superioris alaeque nasi.

Clinical testing of the temporofacial division (upper half of the face) is done by asking patients to "wrinkle" their forehead or close their eyes against resistance.

2. **Branches of the cervicofacial division:**
   (a) **Buccal Branches**—to the muscles of the **cheek and mouth, buccinator, orbicularis oris,** and **levator anguli oris.**
   (b) **Mandibular branches**—to the muscles within the lower lip (**depressor anguli oris, depressor labii inferioris,** and **mentalis muscles**).
   (c) Cervical branch—to the **platysma** (*fig. 45.9*).

Clinical testing of the cervicofacial division is done by asking patients to show their teeth, purse their lips, or blow out on their cheeks. The lower face is sometimes paralyzed in patients who suffer "strokes" or ischemia to brain tissue. One must therefore test both upper and lower parts of nerve VII to distinguish between peripheral nerve problems that cause hemiparalysis of the face and certain diseases within the brain tissue. Since the muscles of facial expression are within the subcutaneous tissue of the face, the branches of the facial nerve are also vulnerable to damage in facial lacerations. One must therefore be able to test the terminal branches of the facial nerve as well.

Temporal and Infratemporal Regions

TEMPORALIS

The **temporalis muscle** arises from the temporal fossa (*fig. 45.1*) and the overlying temporal fascia, which attaches to the superior temporal line of the skull. It is a fan-shaped muscle with a single tendon descending deep to the zygomatic arch to insert on the **coronoid process** of the mandible and the anterior margin of the ramus. Some of the posteroinferior fibers of the temporalis are oriented in a horizontal plane that parallels the superior border of the zygomatic arch, but most temporalis fibers are oriented in a vertical manner with respect to the coronoid process. The action of the temporalis is therefore primarily to elevate the jaw and close the mouth through the contraction of its vertically oriented muscle fibers. The horizontally oriented component of the muscle acts to retract the mandible and thereby pull the head of the condyle into its most posterior (retruded) position in the mandibular fossa of the temporal bone (temporomandibular joint).

The nerve supply of the temporalis is by the **posterior** and **anterior temporal nerves** from the anterior division of V^3. These nerves arise in the infratemporal fossa and enter the temporal fossa by passing over the squamous part of the temporal bone (*see fig. 45.15*). They are therefore on the deep aspect of the temporal muscle.

INFRATEMPORAL REGION

This region lies inferior to the temporal fossa and zygomatic arch and deep to the ramus of the mandible. This area extends from the parotid fascia that attaches to the posterior aspect of the mandibular ramus anteriorly to the tuberosity of the maxilla.

Bony Boundaries

The **lateral wall** of the infratemporal fossa is formed by the medial aspect of the ramus of the mandible (*figs. 45.4 and 45.11*). The mandibular foramen of the ramus transmits the inferior alveolar nerve (V^3) and vessels into the substance of the mandible to supply the teeth and bone with their neural and vascular supply. This foramen is partially covered by a bony projection, the **lingula**, which is the inferior attachment of the **sphenomandibular ligament**. The superior attachment is on the spine of the sphenoid in the "roof" of the infratemporal fossa.

The **anterior wall** is formed by the body and tuberosity of the maxilla. This lies deep to the zygoma and zygomatic process of the maxilla that acts as the **buttress** for transmitting the forces of mastication from the upper dentition into the frontal bone of the cranial vault. In the medial aspect of the anterior wall is the **pterygomaxillary fissure**, which opens into the more medial **pterygopalatine fossa** as well as the **inferior orbital fissure** (*fig. 45.12*). Inferior to the **pterygomaxillary fissure** is a bony projection, the **hamulus**, which serves as the superior attachment of the **pterygomandibular raphe** (*fig. 43.10*). This ligamentous structure runs vertically between the hamulus and the upper **one-fifth of the mylohyoid line**

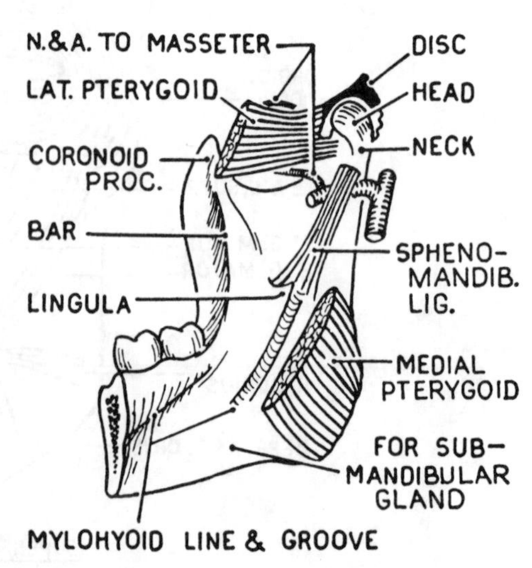

Figure 45.11. The lateral wall of the infratemporal fossa, i.e., the ramus of the jaw.

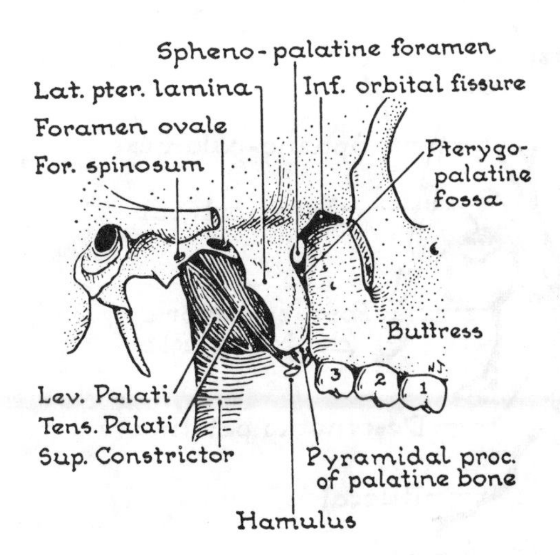

Figure 45.12. Anterior wall, medial wall, and roof of the infratemporal fossa.

on the mandible. It serves as a common site of origin for both the **buccinator muscle and the superior constrictor muscle**. The buccinator projects anteriorly to cross the mandible behind the molar teeth and enter the substance of the cheek (*fig. 45.5*). The superior constrictor projects posteriorly from the pterygomandibular raphe to form the upper part of the pharynx and part of the medial wall of the infratemporal fossa (*figs. 45.5 and 45.12*).

The **medial wall** of the infratemporal fossa is formed by the **lateral pterygoid plate** of the sphenoid bone, the **superior constrictor muscle**, and two small muscles arising from the roof of the fossa, the **levator and tensor palati muscles** (*fig. 45.12*). The lateral pterygoid plate serves as an attachment site for the pterygoid muscles, which act on the mandible. The medial pterygoid plate, its hamulus and pterygomandibular raphe are more medial and serve as attachments for the pharyngeal musculature.

The **roof** of the infratemporal fossa consists of two bones: the **greater wing of the sphenoid** anteriorly and the **squamous portion** of the temporal bone posteriorly. The **infratemporal crest** is on the anterior aspect of the undersurface of the greater wing of the sphenoid and serves as an attachment site for the upper head of the lateral pterygoid muscle. Posterior to this crest are two important foramina: (*a*) the **foramen ovale**, which transmits **V³** and the **lesser petrosal nerve (IX)** from the middle cranial fossa to the infratemporal fossa and (*b*) the **foramen spinosum**, which transmits the **middle meningeal artery** from the infratemporal fossa to the middle cranial fossa. The foramen spinosum is located on the sphenoid bone at the base of the spine (*fig. 45.12*).

Contents of the Infratemporal Fossa

The key orientation structure for the contents of the infratemporal fossa is the **lateral pterygoid muscle**.

The **lateral pterygoid muscle** arises from the infratemporal crest of the greater wing of the sphenoid and the lateral side of the *lateral pterygoid plate*. Its two heads are oriented horizontally below the zygomatic arch in the superior half of the infratemporal fossa (*fig. 45.13*). The superior head inserts into the capsule that is attached to the **interarticulating disc** of the temporomandibular joint (*fig. 45.14*). The inferior head is much larger and inserts into the **fovea** on the neck of the mandible. The action of this muscle is to protrude the mandible by pulling the head of the condyle onto the **articulating tubercle** of the temporal bone. The **articular disc** between the head of the condyle and the mandibular fossa of the temporal bone separates the temporomandibular joint into a lower hinge joint and upper sliding joint. The disc is always intervening between the mandibular head and the temporal bone. The disc is pulled anteriorly by the superior head of the lateral pterygoid muscle while the inferior head protrudes the mandible. **The protrusive actions of lateral pterygoid are used to clinically test the intactness of V³.** Bilateral contraction causes the mandible to protrude in the midline. Paralysis or severe weakness of either lateral pterygoid muscle will cause the mandible to deviate to the side of the damaged muscle or injured V³ nerve.

Lateral to the lateral pterygoid muscle is the maxillary artery and the ramus of the mandible (*fig. 45.13*). **Medial** and inferior to the lateral pterygoid muscle is the mandibular division of the trigeminal nerve V³ and the medial pterygoid muscle.

The **maxillary artery** (*fig. 45.13*) arises in the parotid gland and enters the posterior aspect of the infratemporal fossa by passing deep to the neck of the condyle of the mandible. It crosses the lateral side of the lateral pterygoid muscle and enters the pterygomaxillary fissure on the posterior aspect of the maxilla. The artery is divided into 3 parts by this lateral pterygoid muscle relationship: (*a*) 1st or mandibular part, (*b*) 2nd or pterygoid part, and (*c*) 3rd or pterygopalatine part. The mandibular and pterygoid parts are associated with the infratemporal fossa, and the pterygopalatine part is associated with the deep face and nasal region.

The mandibular portion of the maxillary artery has five branches. All of them enter a bony canal to supply their areas of distribution.

1. The **middle meningeal artery** is the largest and most important arterial branch. It ascends from the maxillary artery on the medial side of the lateral pterygoid muscle to pass through the foramen spinosum and enter the middle cranial fossa. It is the principal arterial supply to the periosteal dura of the cranial cavity.
2. The **inferior alveolar artery** enters the substance of the mandible through the mandibular foramen. It supplies the dental pulps of all the lower teeth as well as the mandible. The angle of the mandible has a poor blood supply and is a common site for **al-**

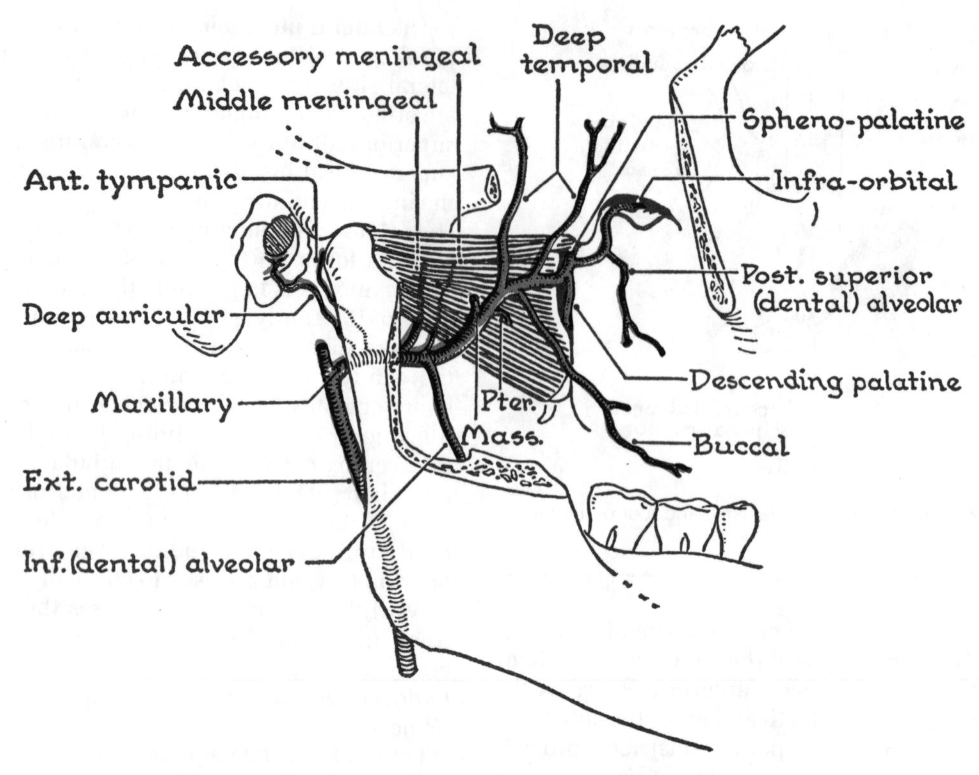

Figure 45.13. The lateral pterygoid muscle and the maxillary artery.

veolar osteitis (dry socket) following the extraction of the lower 3rd molars (wisdom teeth). Using an anesthetic with a vasoconstrictor (epinephrine) can induce an alveolar osteitis following dental surgery.

3. The **deep auricular artery** supplies the external auditory meatus and reaches the bony portion in a canal between the cartilaginous and bony canals.
4. The **anterior tympanic artery** accompanies the chorda tympani through the petrotympanic fissure to reach the middle ear.
5. The **accessory meningeal branch** is inconsistent, but when it is present, it enters the middle cranial fossa through the foramen ovale and supplies the trigeminal ganglion and the surrounding dura.

The pterygoid portion of the maxillary artery also has five branches and they are all muscular supplying the muscles of mastication in the pterygoid fossa. The two **deep temporal branches**, the **masseter, pterygoid**, and the **buccal** branch are all shown in *Figure 45.13*.

The **pterygoid plexus of veins** complements the maxillary artery in the infratemporal fossa. The veins are usually doubled (venae comitantes) and, in large part, lie lateral to the maxillary artery. These veins are extremely important because of their connections with the cavernous sinus via the deep facial, inferior ophthalmic, and emissary veins in the sphenoid bone. Since the veins of the head lack valves and constitute a very low-pressure system, infection and intravenously injected anesthetics can be forced in a retrograde fashion from the infratemporal region to the intracranial meninges. Facial infections can therefore lead to meningitis.

The **medial pterygoid** lies deep to the lateral pterygoid and arises from the medial side of the lateral pterygoid plate and the tuberosity of the maxilla. It projects in an inferior and posterior direction to insert on the medial aspect of the angle of the mandible (*figs. 45.11 and 45.15*). It largely complements the masseter muscle on the lateral side of the mandibular ramus and acts as a major elevator

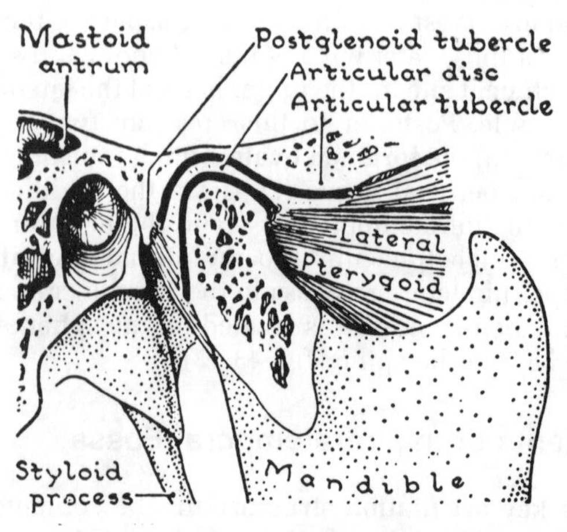

Figure 45.14. Mandibular joint (on sagittal section).

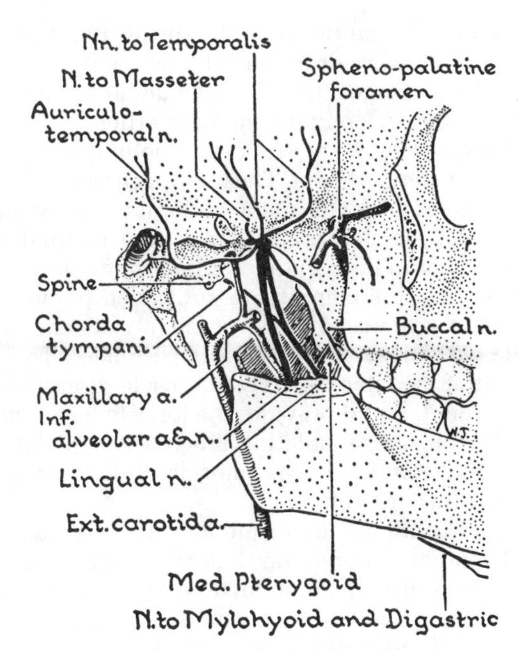

Figure 45.15. The mandibular nerve.

of the mandible along with the masseter and temporalis muscles. The fascia that invests the medial pterygoid muscle forms the floor of the infratemporal fossa. The mandibular division of the trigeminal nerve (V^3) lies between the lateral and medial pterygoid muscles.

The **mandibular nerve (V^3)** (*fig. 45.15*) leaves the trigeminal ganglion area in the middle cranial fossa and enters the infratemporal fossa through the **foramen ovale**. V^3 has a large component of sensory fibers whose cell bodies are located in the trigeminal ganglion. The smaller motor root has its cell bodies in the brainstem and not in the trigeminal ganglion.

In the infratemporal fossa, V^3 divides into two subsequent divisions: a smaller anterior division that is mainly motor to the muscles of mastication and a larger posterior division that contains mainly sensory nerves.

Branches of the Anterior Division of V^3

On the undersurface of the squamous portion of the temporal bone, branches of V^3 project to the **masseter muscle** by passing over the superior head of the lateral pterygoid, traversing the mandibular notch, and entering the deep surface of the masseter. Branches to the posterior and anterior portions of temporalis also ascend on the temporal bone above the superior head of the lateral pterygoid muscle to enter the undersurface of the temporalis muscle. The nerve to the lateral pterygoid is associated with the **buccal nerve** of V^3 that passes between the two heads of the lateral pterygoid muscle. The buccal nerve continues into the cheek on the lateral surface of the buccinator muscle. It is the terminal branch of the anterior division and is a **sensory nerve** to the mucosa of the inside of the cheek and lower gingiva (gums) around the molar teeth. The buccal nerve of V^3 must penetrate the buccinator muscle to supply the mucosa **but is not motor to buccinator**. The motor supply to the buccinator muscle is via the **buccal branches** of VII. The medial pterygoid muscle receives innervation by a branch directly from the trunk of V^3.

Three major nerves of the infratemporal fossa are associated with the larger posterior division of V^3. They are the **auriculotemporal, inferior alveolar (dental)**, and **lingual nerves**.

The Auriculotemporal Nerve (*fig. 45.15*)

The auriculotemporal nerve arises from V^3 just inferior to the foramen ovale and projects posteriorly in the infratemporal fossa parallel to the roof. The initial segment of the auriculotemporal nerve has two important features: (*a*) it encircles the middle meningeal artery as the artery ascends to enter the foramen spinosum; and (*b*) it receives the postganglionic parasympathetic fibers from the otic ganglion that are secretomotor to the parotid gland. The auriculotemporal nerve leaves the infratemporal fossa by passing medial to the head of the mandibular condyle and sending a sensory branch to the temporomandibular joint. The auriculotemporal nerve then enters the deep portion of the parotid gland. It gives sensory branches to the gland and its capsule as well as the secretomotor parasympathetic fibers that were carried from the otic ganglion (*fig. 45.8*). The terminal part of the auriculotemporal nerve accompanies the superficial temporal artery in its exit from the superior border of the parotid gland to the skin and superficial tissue of the upper half of the pinna of the ear and the temporal area (*fig. 45.6*). The auriculotemporal nerve is therefore sensory for pain and general sensation over the upper portion of the V^3 dermatome of the face (*fig. 38.11*).

The Inferior Alveolar (Dental) Nerve (*fig. 45.15*)

The inferior alveolar nerve descends from the foramen ovale to enter the mandibular foramen on the medial aspect of the ramus of the mandible. In its course through the infratemporal fossa, it lies between the medial and lateral pterygoid muscles and just posterior to the lingual nerve. Its only branch in the infratemporal fossa is a **motor** branch to the **mylohyoid muscle** and the **anterior belly of the digastric muscle**. This **mylohyoid nerve** arises from the inferior alveolar nerve just before it enters the mandibular foramen. The mylohyoid nerve then runs in the mylohyoid groove on the medial aspect of the ramus (*fig. 45.4*) and enters the submental triangle (*fig. 43.17*) on the inferior (superficial) aspect of the mylohyoid muscle.

The portion of the inferior alveolar nerve that enters the mandibular canal then courses through the ramus and body of the mandible. This portion is entirely sensory to the teeth, bone and mucosa of the lower lip and gingiva (gums) adjacent to the lower incisive teeth. *Figure 45.3* shows the mental foramen on the lateral aspect of the body, which allows the **mental nerve** branch of the inferior alveolar nerve to exit from the mandibular canal and reach the gingival and mucosal tissue in the vicinity of the lower lip.

The Lingual Nerve (*fig. 45.15*)

The lingual nerve has a similar course to the inferior alveolar nerve through the infratemporal fossa. It lies anterior to the inferior alveolar nerve, and it remains medial to the mandible throughout its entire course. The two nerves therefore become separated when the inferior alveolar nerve enters the mandibular canal. Within the infratemporal fossa, the lingual nerve **receives** a branch called the **chorda tympani**. These are preganglionic **parasympathetic secretomotor fibers of VII** from the tympanic plexus *and* **special sensory fibers for taste** that are running from the anterior two-thirds of the tongue to their cell bodies in the **geniculate ganglion of VII**. The chorda tympani passes through the **petrotympanic fissure** as it communicates between the middle ear cavity and the infratemporal fossa. This petrotympanic fissure is in the temporal bone but related to the medial aspect of the spine of the sphenoid bone, and therefore, the foramen spinosum and middle meningeal artery. The chorda tympani lies medial to the spine of the sphenoid, and the middle meningeal artery and the auriculotemporal nerve lie lateral to this structure in the infratemporal fossa (*fig. 45.15*).

The terminal distribution of the lingual nerve (V^3) and its associated fibers that are derived from VII are to the floor of the mouth and tongue. The lingual nerve is sensory for general sensation (pain, temperature, touch, and pressure) to the anterior two-thirds of the tongue and mucosa on the lingual surface of the mandible and floor of the mouth. Nerve VII is sensory for taste in this region and also secretomotor to *all* the glands in the floor of the mouth. The details of this innervation are discussed in Chapter 49.

Dentists manipulate the mandibular nerve division by placing anesthetics in the fascial compartment, defined by the fascia covering the medial pterygoid muscle and the medial aspect of the ramus of the mandible (*fig. 45.11*). The lipid-soluble anesthetic rapidly diffuses through this fat-filled space and enters the lipid (myelin) surrounding the inferior alveolar and lingual nerves. Anesthesia of the lower teeth, gingiva, lip, and mucosa of the tongue is thereby established in this **mandibular block technique**.

Temporomandibular Joint (TMJ)

The articulation of the mandible (jaw) with the temporal bone (base of the skull) occurs between the **head of the mandible** and the **mandibular fossa and articular tubercle of the temporal bone**. These bony surfaces are covered by cartilage and this synovial joint is somewhat unique in that it contains an intra-articular disc. The **articular disc** (*fig. 45.14*) divides the joint into a superior and inferior compartment. The inferior compartment permits a hinge rotation for the mandibular head, while the superior compartment is a sliding joint to permit the head to move either in the mandibular fossa or on the articular tubercle when the mandible is protruded (moved anteriorly). The articular disc is completely attached to the capsule of the joint throughout its periphery. The superior head of the lateral pterygoid muscle inserts into the capsule and disc and provides the force to move the disc anteriorly on the articulating tubercle when the lower head of the lateral pterygoid muscle pulls the mandible anteriorly during protrusion.

External to the joint capsule are three ligaments: the **lateral temporomandibular ligament** (simply a thickening of the capsule); the **stylomandibular ligament** (separating the parotid and submandibular glands); and the **sphenomandibular ligament** (*fig. 45.11*). These ligaments play a minor role in the stability and support of the joint. The **major supportive elements are the muscles of mastication**, which keep the mandibular head in contact with its articulating surfaces on the temporal bone. The masseter and medial pterygoid muscles form a "sling" that supports the angle of the mandible and the temporalis supports the anterior aspect of the ramus. All three muscles act to elevate the mandible and secure it into the temporal fossa.

Movements of the Mandible through the TMJ

Elevation (closure of the mouth)—Masseter, medial pterygoid, and temporalis (vertical fibers).

Depression (opening of the mouth)—Gravity, mylohyoid, anterior belly of digastric, and lateral pterygoid (as it pulls the head over the descending slope of the articular tubercle).

Protrusion (Anterior Projection)—Lateral pterygoid (medial pterygoid fibers may also assist since they course anterosuperiorly).

Retraction (Posterior Movement)—Temporalis (horizontal fibers).

Chewing or masticatory movements are complex intermixtures of these basic movements. Unilateral movements of the TMJ occur when one TMJ is stabilized in the mandibular fossa and, protrusive and depressive forces are applied to the opposite side of the mandible.

Clinical testing of the muscles of mastication and V^3 motor intervention can be done in 2 ways: (1) Superficial palpation of the temporalis and masseter is done while the patient clenches the teeth. (2) The patient is asked to protrude the jaw. A midline projection indicates a balanced protrusion by the right and left lateral pterygoid muscles. Weakness or paralysis of either muscle would produce a deviation of the protruded mandible toward the side of weakness.

DISLOCATION OF THE MANDIBLE (SUBLUXATION)

The head of the mandible may be dislocated from its articulating surface of the temporal bone in cases of extreme protrusion. Such movement usually occurs if the muscles are unresponsive to the impulses from the muscle stretch receptors, which normally prevent the excessive forces that would cause dislocation. Excessive yawning, having the mouth open continuously for extended periods during anesthesia or dental procedures, or having the anterior division of V^3 anesthetized by a local nerve block can interfere with the feedback influences that prevent dislocation.

When dislocation does occur (*fig. 45.16*), the head of the mandible "rides" over the crest of the articular tubercle and contacts the base of the skull, which is formed by the greater wing of the sphenoid bone. One must then depress the dislocated head and retract it into the mandibular fossa. Frequently muscle spasms of the masseter and temporalis are too great to do this. One then anesthetizes the nerves to these muscles by placing an anesthetic in the temporal fossa or notch of the mandible. This relaxes the muscles and allows reduction of the dislocation. The mandible must then be supported by a sling or by wiring the teeth together in an occluded position. This will assist the stability of the TMJ, while the muscles and ligaments heal and restore their support of the joint.

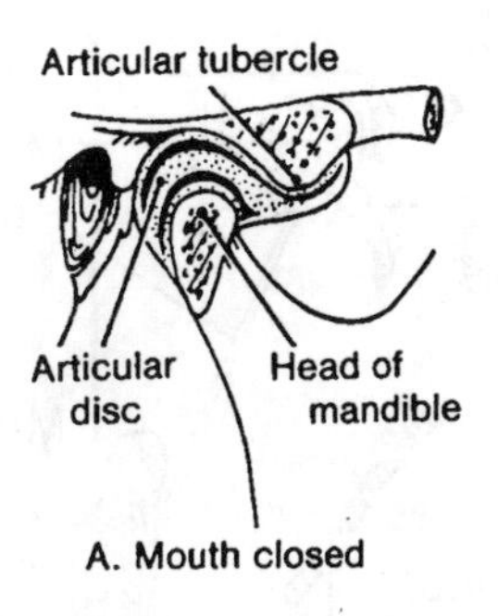

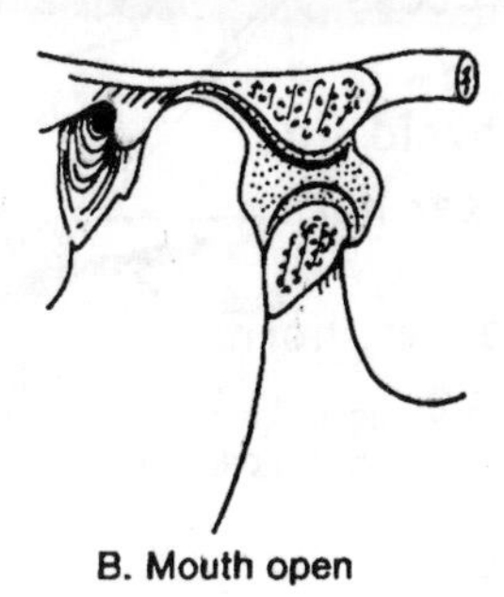

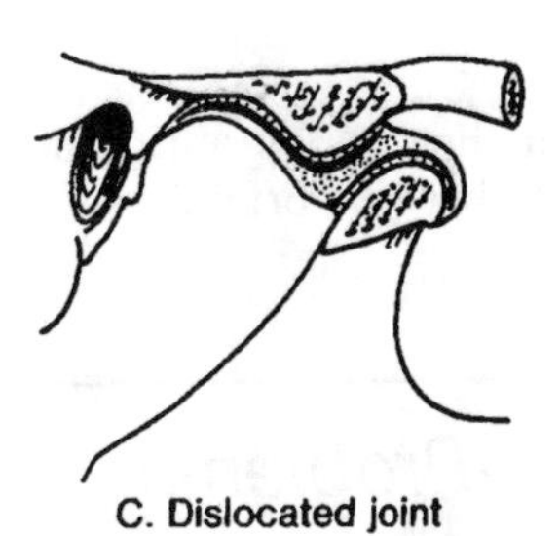

Figure 45.16. Diagrams illustrating the changing position of the head of the mandible and the temporal bone when *A*, the mouth is closed, *B*, the mouth is open, and *C*, the temporomandibular joint is dislocated. Note that the head of the mandible and the articular disc slide forward over the articular tubercle as the head rotates on the disc. (From Basmajian.)

EMBRYOLOGICAL NOTES

The mandible and its associated structures are derived from the mesoderm that surround the first aortic arch in the embryo (**1st branchial or 1st pharyngeal arch**). The mesoderm gives rise to cartilage, bone, muscle, and connective tissue elements. Since the 1st arch is innervated by the nerve V, the adult derivatives of this tissue will also have V innervation.

All the muscles that insert onto and move the mandible through the TMJ are innervated by V^3 (masseter, temporalis, medial pterygoid, lateral pterygoid, mylohyoid and anterior belly of the digastric muscles).

Meckel's cartilage in the embryo induces the formation of the mandible from mesodermal tissue in the pharyngeal arch. Structures that are derived from Meckel's cartilage directly are shown in *Figure 45.17*. They are the **sphenomandibular ligament, anterior ligament of the malleus, malleus,** and **incus**. The malleus and incus are two of the three middle ear bones (ossicles). The malleus is moved by the tympanic membrane (eardrum) vibration and a muscle (the **tensor tympani**). Since the **tensor tympani** muscle acts on a bone derived from the 1st pharyngeal arch, it will also be innervated by a branch from V^3. The V^3 branch to **tensor tympani** arises with the branch to the medial pterygoid just as the V^3 trunk exits the foramen ovale.

This embryonic relationship between the mandibular structures and the ear also explains, in part, how pain is "referred" from the mandible to the ear and **vice versa**. Dental pain can frequently be interpreted as an "earache." Temporomandibular joint pain and TMJ disorders may also exhibit earache and ear symptoms. Both are

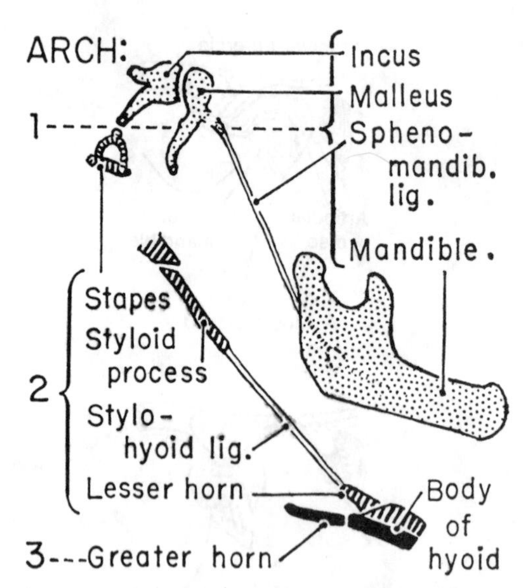

Figure 45.17. Two vestigial ligaments (sphenomandibular and stylohyoid) derived from the cartilages of the 1st and 2nd pharyngeal arches.

innervated by branches of the auriculotemporal nerve and have similar embryonic origins.

Clinical Mini-Problems

1. **Which major nerve(s) and vessel(s) would a surgeon need to avoid in removing the deep part of the parotid gland? What are the lateral to medial relationships of these intraglandular structures?**
2. **Why is pain in the parotid gland more intense prior to meal times and somewhat relieved after eating?**
3. **Why did a physician suspect that a child had "mumps" when the child presented with a fever, pain in the ear, and a sensitive and inflamed oral mucous membrane opposite to the upper second molar?**
4. **Where would one expect to find a lesion (disorder) in nerve VII of a patient who has paralysis of all the facial muscles on the right side but has normal parasympathetic function in the glands on the right side of the face?**
5. **If nerve axons regenerate by growing approximately 1 mm per day (1 inch per month), how would you advise the patient in problem 4 on the approximate time for recovery of this type of nerve damage?**
6. **Headache in the temporal region may be caused by stimulation of general pain receptors of the auriculotemporal nerve V^3, by vascular disorders (migraine) or by muscle spasms in the temporalis muscle. Where would one place a local anesthetic to temporarily "relax" the temporalis muscle in differentially diagnosing the possible cause for a patient's headache?**
7. **A 14-year-old girl with severely prognathic (anteriorly protruding) mandible was scheduled for a mandibular resection of the rami in order to shift the mandible posteriorly before orthodontic therapy. Why would the oral surgeon place the horizontal cuts through the rami above the level of the lingulae?**
8. **Which cranial nerve would be damaged if a patient's mandible deviated to the left side during protrusive movement?**
9. **"Shingles" is a painful skin disorder that results from a viral infection in the cell bodies of sensory neurons. If a patient presented with a painful rash and tiny raised skin blisters over their left jaw, upper ear and temporal region, where would you expect the viral infection to be located?**
10. **(a) When a dentist gives an anesthetic in the infratemporal fossa for anesthetizing the lower teeth, why are the tongue and skin of the same side also "numbed"?**
    (b) Would general sensation, taste sensation, or both be absent in the anesthetized tongue?

(Answers to these questions can be found on p. 588.)

46

Cervical Vertebrae, Prevertebral Region, and Exterior of Base of the Skull

Clinical Case 46.1

Patient Andy G. This 17-year-old high school senior injured his neck diving into a stream during a pregraduation party. He was rushed to the emergency room where he was placed in a stabilizing brace to prevent any neck movement. The spinal cord injury management team finds he has no sensation from his paralyzed lower extremities, abdomen, and chest to the level of the sternal angle (T2 dermatome). He has some sensation over his shoulders and some bicep function when tested reflexively. There is no muscle or sensory function in his forearms or hands. Two days after the accident, he develops respiratory problems and must be intubated and placed on a respirator. X-rays show a fracture of the C5 vertebra.

Cervical Vertebrae

There are 7 cervical vertebrae. The 1st and 2nd are distinctively unique. The 3rd, 4th, 5th, and 6th are typical cervical vertebrae, and the 7th is transitory between the cervical and thoracic form of vertebrae. All cervical vertebrae have one distinguishing feature—they possess a foramen in their transverse processes (the **foramen transversarium**). The foramina transversarii of the upper six cervical vertebrae transmit the vertebral artery as it ascends to the posterior cranial fossa. One other major clinical and anatomical relationship regarding the cervical vertebrae is that the cervical spinal nerves pass superior to their related vertebrae as they exit the spinal column. Since there are 7 cervical vertebrae and 8 cervical spinal nerves, the 7th cervical vertebrae will have the 7th cervical nerve above and the 8th cervical nerve below. All other spinal nerves in the thoracic, lumbar, and sacral region therefore exit the spinal column below their associated vertebra.

TYPICAL CERVICAL VERTEBRA (*fig. 46.1*)

The body is rectangular and concaved on the superior aspect. It will articulate with the convex inferior surface of the vertebral body superior to it. The size of the vertebral bodies increases in the lower vertebrae to support the additional weight of the head, neck, and upper extremity on the spinal column.

The **pedicles** arise from the body and project in a posterolateral fashion to join the neural arch. The **vertebral foramen** is therefore triangular in shape in the cervical region. The **articular processes** are at the junction of the pedicles and the laminae. They form a column of bone laterally that articulates in concert with the bodies of the cervical vertebrae. The articulating facets are horizontally oriented superiorly but gradually assume an anterior oblique orientation as one descends in the cervical region. These joints allow flexion and extension of the cervical column.

The **laminae** form the posterolateral legs of the triangular vertebral foramen. They serve as bony attachments for the deep muscles of the neck and protect the underlying cervical spinal cord. A cervical **laminectomy** is a neurosurgical/orthopedic procedure that is done to gain access to the cervical spinal cord or to relieve pressure on the cervical cord following a severe neck injury.

The **spinous processes** of the cervical vertebrae are most posterior and unite the laminae. In the typical cervical vertebrae, the spinous process is bifid. The **ligamentum nuchae**, a thick midline septum of connective tissue, attaches to this bifid spinous process. The spinous process and the ligamentum nuchae are attachment sites for muscles of the back of the neck.

The **transverse processes** project anterolaterally from the vertebral body, pedicle and articular process. **Understanding the anatomy of the transverse processes of the typical vertebra is the key to understanding many of the important clinical points in the neck.** The transverse

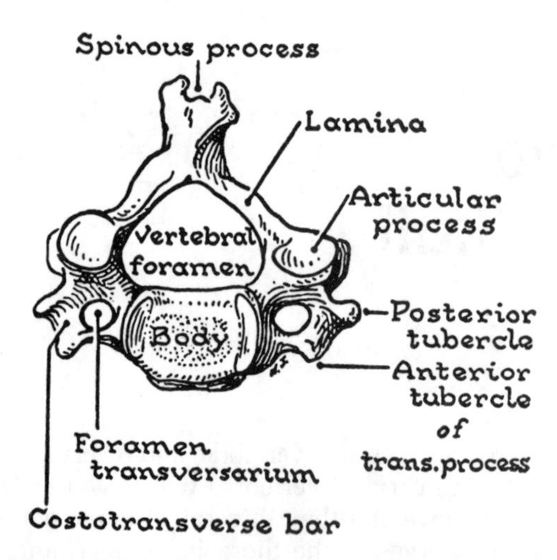

Figure 46.1. Typical cervical vertebra (from above).

process is "cup-shaped" on its superior surface and holds the emerging spinal nerve that passes from the intervertebral foramen. The **foramen transversarium** is situated in the anterior aspect of the base of each transverse process, and the vertebral artery contained within the foramen is therefore anterior to the emerging cervical nerve. **Posterior and anterior tubercles** exist at the tips of the transverse processes of C3–6 (the remaining cervical vertebrae have *only* posterior tubercles). The **posterior tubercles** give attachments to the muscles in the floor of the posterior triangle (scalene mass, levator scapulae, splenius, and semispinalis cervicis). The anterior tubercle gives attachment to *two* important reference muscles in the root of the neck—the **scalenus anterior** and the **longus colli** muscles. The emerging roots of the cervical and brachial plexuses are therefore found between the scalenus anterior and scalenus mass in the neck. The anterior tubercle is also equivalent to the bone mass that would form the ribs in the thoracic region. Cervical ribs are formed when the anterior tubercle is excessively extended into the neck. The **costotransverse bar** is a segment of bone in the transverse processes that connect the anterior and posterior tubercles distal to the foramen transversarium.

The **1st cervical vertebra** supports the skull and is therefore called the **atlas** (after the mythical Greek god who supported the globe). The atlas is unique in that it lacks a body and is composed of an **anterior** and **posterior arch** with laterally projecting **transverse processes** (*fig. 46.2*). The body of the atlas is the **dens** (odontoid process) of the 2nd cervical vertebra (*fig. 46.3*). The dens articulates with a posterior facet on the anterior arch of the atlas and a transverse ligament. This allows the atlas to rotate around the dens in a horizontal plane. This movement is demonstrated by asking patients to motion a "no" sign with their head.

The **superior articulating facets** of the **atlas** articulate with the **condyles** of the **occipital bone**. The movement in these **atlanto-occipital** joints is a rocking movement in the sagittal plane. This movement occurs when a patient nods the head to indicate a "yes" sign. The superior articulating facets are reniform (kidney-shaped) and match the form of the opposing occipital condyles. Within the concavity of the medial aspect of the superior articulating facets are the **tubercles** of the atlas. They serve to attach the **transverse ligament** (horizontal part of cruciate ligament) that articulates with the dens and prevents the dens from sliding posteriorly into the C1 vertebral foramen, which contains the spinal cord.

The transverse processes of the atlas are prominent and extend more laterally than the transverse processes of the underlying upper cervical vertebrae. The transverse processes of C1 can be felt behind the mandibular ramus (*fig. 43.2*). The foramina transversarii in the transverse processes of C1 are more lateral than the underlying foramina transversarii of the C2 vertebra. The vertebral artery therefore courses laterally as it ascends between the C2 and C1 vertebrae.

The **posterior arch** of the atlas overlies the cervical spinal cord just below the foramen magnum. It has **two** important physical markings that should be noted. First, a **groove for the vertebral artery** exists on the superior aspect of the posterior arch just behind the superior articulating facet. This indicates the course that the vertebral artery takes as it passes from the foramen transversarium of the atlas, to pierce the dura mater and to enter the posterior cranial fossa through the foramen magnum. Secondly, a sharp crest on the superior border of

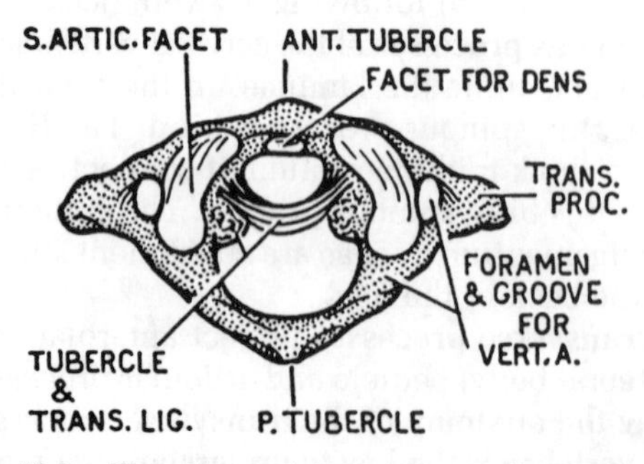

Figure 46.2. The atlas and the transverse ligament (posterosuperior view).

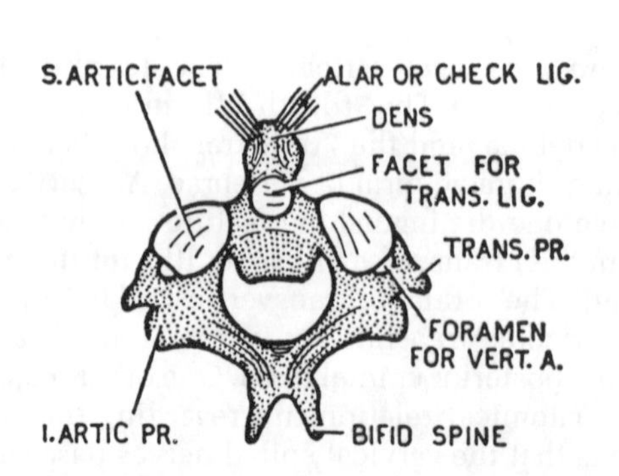

Figure 46.3. Axis (posterosuperior view).

the posterior arch of the atlas marks the inferior attachment site for the **posterior atlanto-occipital membrane**. This is a tough, elastic membrane (equivalent to **ligamentum flavum**) that attaches superiorly to the posterior margin of the foramen magnum. It may be pierced by a needle for a tap of the *cerebrospinal fluid* (CSF) in the cisterna magna of the posterior cranial fossa. The posterior atlanto-occipital membrane also bridges over the vertebral artery to fuse with the capsule of the atlanto-occipital joint.

The **2nd cervical vertebra or axis** (*fig. 46.3*) has a superior projecting process, **the dens**, which serves as the body of the overlying atlas. The **superior articulating facet of the axis** articulates with the **inferior articulating facet of the atlas**. This sliding joint permits a rotation of the atlas around the dens and over the atlanto-axial joint. The **transverse processes** of the axis are shorter in their lateral projection than the transverse processes of the atlas. Each transverse process has a **foramen transversarium** and a **posterior tubercle**. The laminae extend from the **articular processes** to the bifid spine in the same fashion as seen in a typical cervical vertebrae.

The **7th cervical vertebra** has a transitional form between a typical cervical and thoracic vertebra. The long **nonbifid spine** projects horizontally to be subcutaneous at the base of the neck. It is called the **vertebra prominens**. The **transverse processes** have a **foramen transversarium**, but it **usually does not contain the vertebral artery**. Veins are present in the foramen transversarium of the 7th cervical vertebra. An **anterior tubercle is not present** on the 7th cervical vertebra. Cervical ribs when they occur are most frequently found on the anterior aspect of the transverse process of the 7th cervical vertebra. The presence of a cervical rib in this location may produce compression of the lower trunk of the brachial plexus (C8, T1) and/or the subclavian artery as they cross the superior surface of the 1st rib.

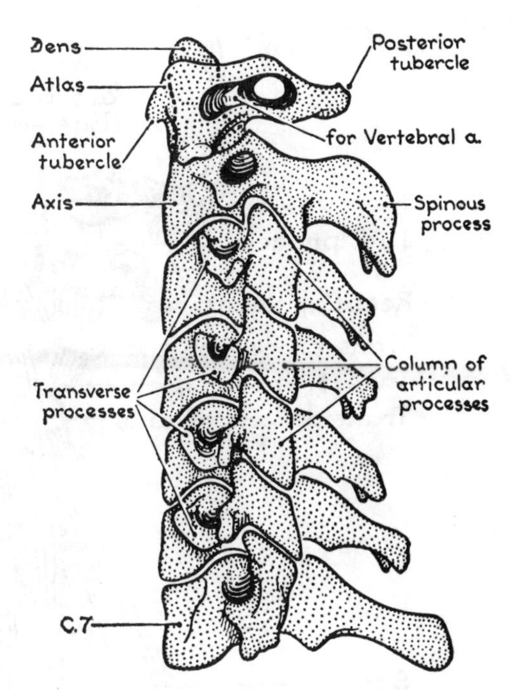

Figure 46.4. Cervical vertebrae on side view.

Deep Cervical Structures

Articulated cervical vertebrae (*fig. 46.4*), when clothed with the muscles attached to the anterior tubercles of the transverse processes, present a relatively flat prevertebral surface.

Deep anterior cervical muscles are grouped according to their relations to the roots of the cervical and brachial plexuses (*fig. 46.5*).

Muscles medial to the plexuses:

1. Rectus Capitus Anterior
2. Longus Colli (cervicis)
3. Longus Capitis
4. Scalenus Anterior

Muscles lateral to the plexuses:

1. Rectus Capitis Lateralis
2. Scalenus Medius and Posterior (Scalene Mass)
3. Levator Scapulae

The **longus colli** extends from the body of the 3rd thoracic vertebra to the anterior tubercle of the atlas and is attached to the bodies of the vertebrae in between. It is attached to the anterior tubercles of C3–6 and is a **landmark structure on** which the **cervical sympathetic trunk** ascends in the neck.

The **longus capitis** arises from the anterior tubercles of C3–6 and inserts superiorly on the **basiocciput** posterior to the pharyngeal tubercle. It is also related to the sympathetic chain in the upper neck.

The **scalenus anterior is the key muscle for relationships in the root of the neck** (Chapter 44). It attaches to the anterior tubercles of C3–6 and the scalene tubercle on the upper surface of the 1st rib. The brachial plexus and subclavian artery (2nd part) are posterior; the subclavian vein, phrenic nerve, suprascapular artery, and transverse cervical artery are anterior; and the carotid sheath, sympathetic chain, thyrocervical trunk, and vertebral artery are medial to the scalenus anterior.

The **rectus capitis anterior** and **rectus capitis lateralis** lie between the atlas and the base of the skull. The **hypoglossal nerve XII** and the **anterior ramus of C1** emerge from the prevertebral area at this point and join to course anteriorly to the strap muscles of the neck and floor of the mouth.

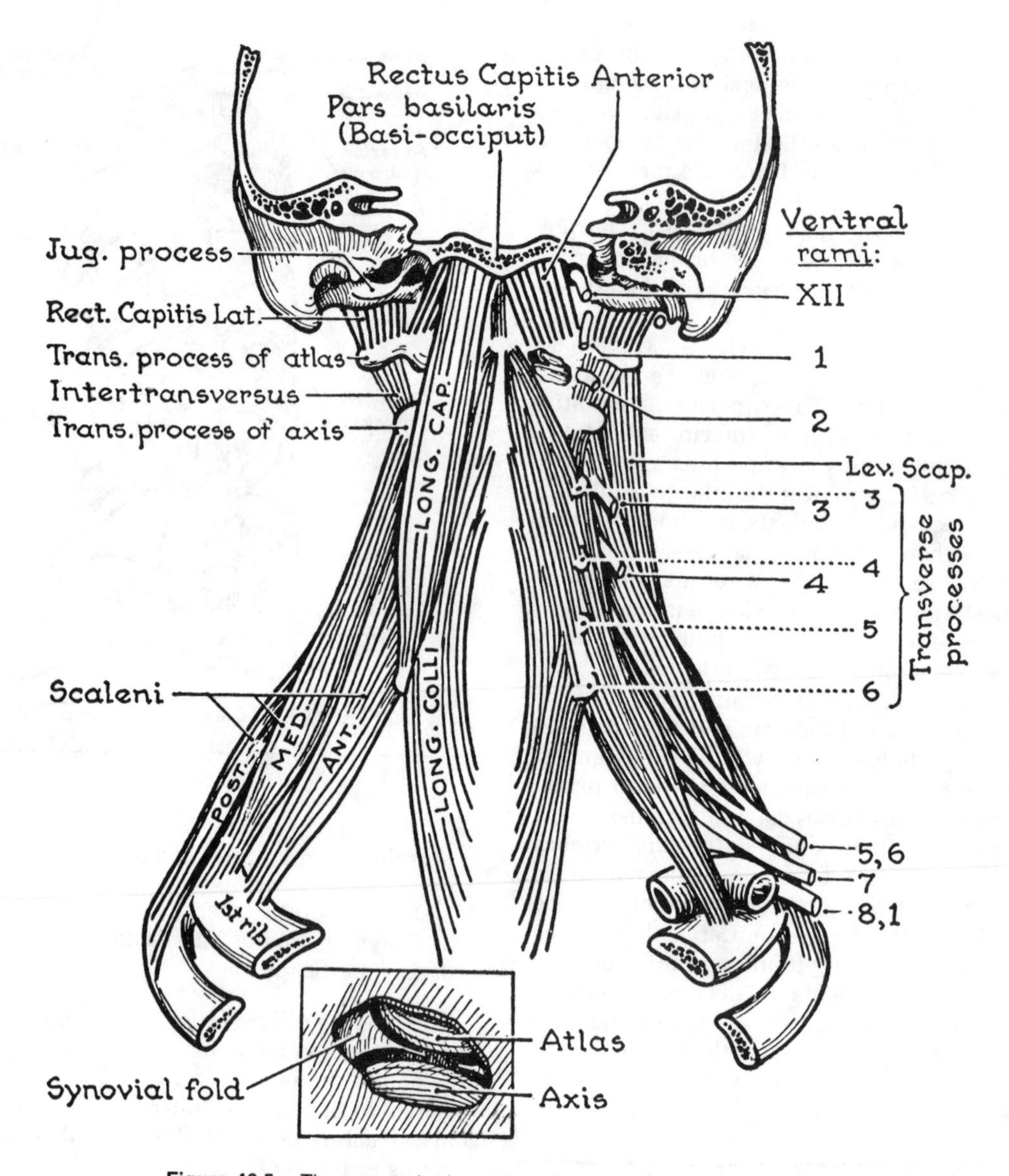

Figure 46.5. The prevertebral muscles: deep anterior cervical muscles.

The **scalenus medius, scalenus posterior**, and the **levator scapulae** (*fig. 46.5*) arise from the posterior tubercles of the cervical vertebrae and insert onto the 1st rib, 2nd rib and superior angle of the scapula, respectively. These prevertebral muscles can be torn in "whiplash" injuries and may be a source for the post-traumatic pain that is associated with this type of injury.

The **prevertebral fascia** covers the prevertebral muscles and forms the fascia on the floor of the posterior triangle of the neck. The brachial plexus and the subclavian artery carry this fascia into the axilla as the **axillary fascia**. The prevertebral fascia forms the posterior limit of the retropharyngeal space (*fig. 45.5*) and is a plane on which head and neck infections can spread inferiorly into the posterior mediastinum of the thorax.

VERTEBRAL ARTERY (*figs. 46.6 and 44.1*)

It arises from the first part of the subclavian artery and ascends between the scalenus anterior and longus colli to enter the foramen transversarium of C6. The subsequent course through the foramen transversarii of C5–C2 is deep to the prevertebral fascia. The artery courses laterally from C2 to enter the foramen transversarium of C1 and then turns medial to cross the posterior arch of the atlas (*fig. 46.6*). The vertebral artery then passes under the posterior atlanto-occipital membrane, pierces the dura, and enters the subarachnoid space. The two vertebral arteries then pass through the foramen magnum and unite to form a single **basilar artery** on the basiocciput of the posterior cranial fossa (*fig. 39.13*). The vertebral arteries

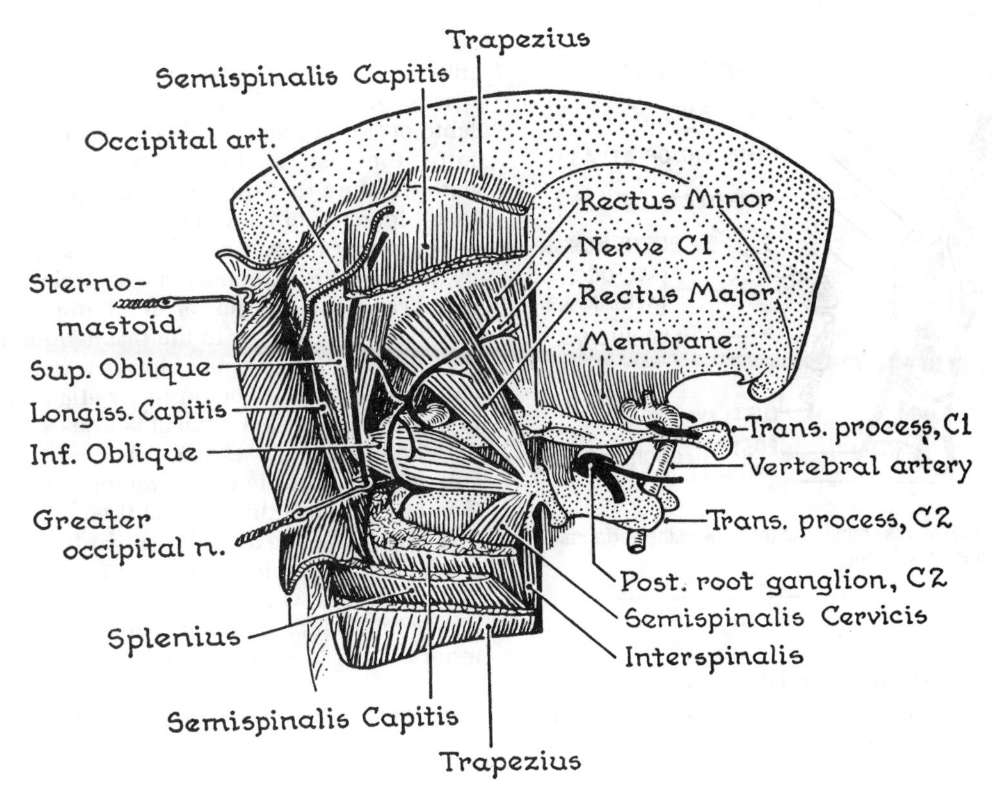

Figure 46.6. The suboccipital region, containing the suboccipital triangle.

provide a major blood supply to the cerebral cortex (medial and posterior surfaces), the brainstem and cerebellum, and the atlanto-occipital joints.

VENTRAL RAMI OF CERVICAL NERVES

The ventral ramus of C1 joins nerve XII and is distributed to the ansa cervicalis (superior root), the thyrohyoid muscle and the geniohyoid muscle by separate branches from XII as it courses to the tongue. C1 is a motor nerve. It has no dorsal root ganglion, and there is no C1 dermatome on the skin. The 2nd cervical nerve has a posterior ramus that gives rise to the **greater occipital nerve** in the suboccipital triangle (*fig. 46.6*). This nerve is the basis for the C2 dermatome that overlies the occipital bone and abuts the V^1 dermatome at the vertex of the skull. The anterior rami of C2 carry sensory fibers to the cervical plexus in the neck and motor fibers to the inferior root of the ansa cervicalis.

The cervical plexus is composed of sensory fibers in the ventral rami of C2–4, while the ansa cervicalis is composed of motor fibers from the ventral rami of C1–3. The ventral rami of C3–5 form the phrenic nerve that crosses the anterior aspect of the scalenus anterior and descends through the thorax to innervate the diaphragm and its adjacent parietal pleura and peritoneum. Preserving the motor supply to the diaphragm is of utmost importance in treating neck injuries.

Craniovertebral Joints

JOINTS BETWEEN SKULL, ATLAS, AND AXIS

Five synovial joints are involved between the skull, C1 vertebra, and C2 vertebra. The bilateral superior articulating facets on the axis (C2) articulate with the bilateral inferior articulating facets on the atlas (C1). These atlanto-axial joints and the joint between the dens (odontoid process) on the axis and the anterior arch and transverse ligament on the atlas (*fig. 46.7*) allow rotation of the head in the transverse plane. This allows the patients to turn their heads from side to side. The bilateral superior articulating facets on the atlas articulate with their corresponding occipital condyles on the base of the skull. This joint allows for a nodding movement of the skull in the sagittal plane.

The capsules of the bilateral atlanto-axial and atlanto-occipital joints enclose the synovial joints completely. Unlike other spinal column joints associated with the articulating processes of the vertebrae, the upper two cervical intervertebral joints are anterior to the emerging

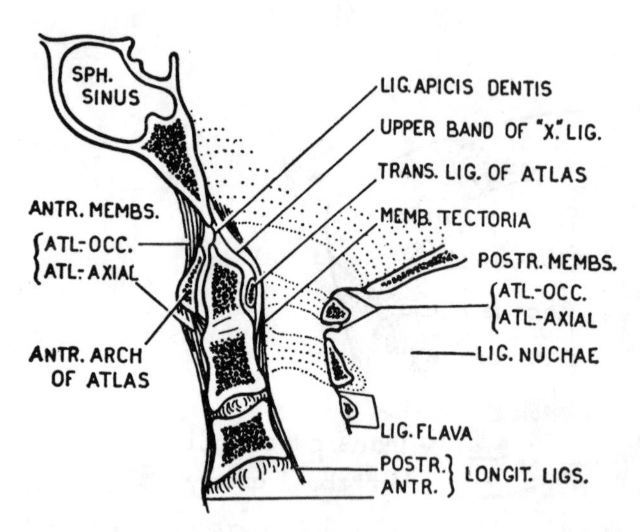

Figure 46.7. Ligaments connecting the skull to the vertebral column (paramedian section).

spinal nerves. The atlanto-occipital joint is also anterior to the horizontal component of the vertebral artery, which is crossing the posterior arch of C1 to enter the foramen magnum.

The midline atlanto-axial joint between the dens and the atlas has a very different form of capsular investment (*fig. 46.7*). The anterior arch of the atlas is attached in two connective tissue support ligaments: the **atlanto-axial ligament**, inferiorly, and the **anterior atlanto-occipital membrane**, superiorly. These two ligaments are comparable to the anterior longitudinal ligaments of the vertebrae in the lower elements of the spinal column. Posterior to this midline joint, the posterior longitudinal ligament of the spinal column is modified to span the interval from C2 to the foramen magnum. This connective tissue covering of the posterior aspect of the midline atlanto-occipital joint is called the **membrana tectoria**. It forms the posterior limit of the joint and is in contact with the dura that surrounds the spinal cord in the vertebral foramen of C1 and C2.

The posterior arches of C1 and C2 are also connected by dense elastic ligaments: the **posterior atlanto-occipital membrane** and the **posterior atlanto-axial membrane**. These are comparable to the ligamentum flava that connect the adjacent lamina of lower vertebrae in the spinal column (*fig. 46.7*). The posterior atlanto-occipital membrane is an important clinical landmark as well. Here the vertebral artery passes from its cervical relation to its subarachnoid position in the foramen magnum (*fig. 46.6*). One can also use this site between C1 and the foramen magnum for taking a cerebrospinal fluid "tap." The posterior atlanto-occipital membrane and its underlying dura and arachnoid would be penetrated by a needle to gain access to the subarachnoid cistern at the base of the posterior cranial fossa, the **cisterna magna**.

Exterior Base of the Skull

The undersurface of the skull can be divided into 3 anatomical regions—anterior, intermediate, and posterior—by two imaginary lines (*fig. 46.8*). Each line traverses an important set of foramina on the base of the skull

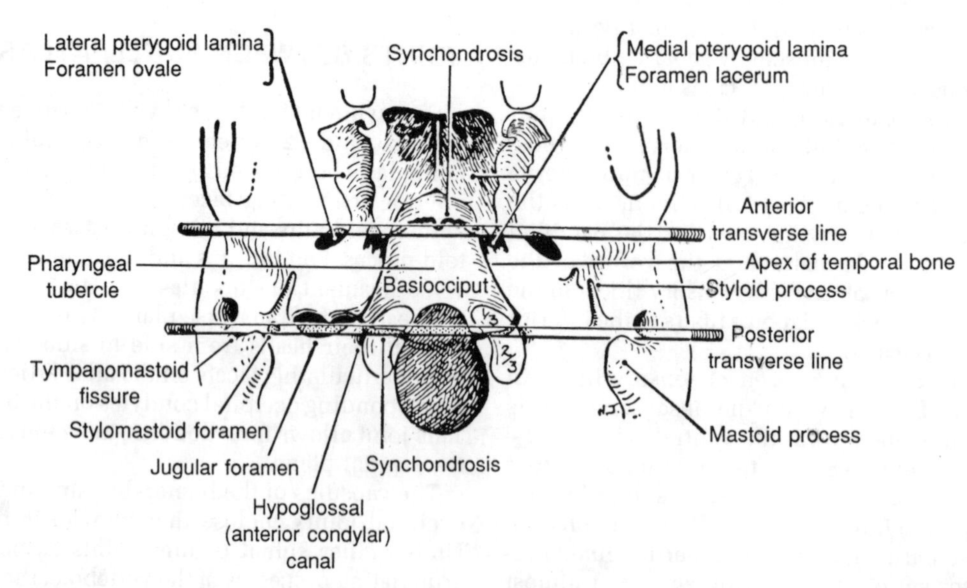

Figure 46.8. "Anterior transverse line" and "posterior transverse line" on exterior of the base of the skull.

and can serve as *keys* for understanding the relative relationships of major nerves and vessels at the base of the skull.

The **anterior transverse line** spans the base of the skull between the mandibular notches of each ramus of the mandible. This line traverses the **foramen ovale** at the base of the **lateral pterygoid plate** and the **foramen lacerum** at the base of the **medial pterygoid plate** of the sphenoid bone. It also demarks the site of fusion of the occipital and sphenoid bones on the basiocciput.

The **posterior transverse line** unites the anterior margins of the right and left mastoid processes of the temporal bones. This imaginary line crosses the **stylomastoid foramen**, the **jugular foramen**, the **hypoglossal canal**, the **occipital condyles**, and the **foramen magnum**.

Structures associated with the Anterior Transverse Line are:

1. **Masseter muscle** and its neurovascular supply in the mandibular notch.
2. **V³** exiting the foramen ovale with **middle meningeal artery** posteriorly entering the foramen spinosum (*fig. 46.9*).
3. **Lateral** and **medial pterygoid muscles** arising from the lateral and medial sides of the lateral pterygoid plates, respectively.
4. **Cartilaginous portion** of the **auditory (eustachian) tube** and the **tensor palati muscle** origin at the base of the medial pterygoid plate.
5. The **foramen lacerum** is filled with cartilage in life but has an inferior relationship to the **levator palati muscle** and **deep petrosal nerve** (sympathetic postganglionic fibers). The levator palati muscle arises from the tip of the petrous temporal bone at the margin of the foramen lacerum. The deep petrosal nerve courses anteriorly to enter the **pterygoid canal** of the sphenoid bone at the anterior margin of the foramen lacerum. In the canal, the deep petrosal nerve joins the **greater (superficial) petrosal nerve (VII)**, which passes over the intracranial surface of the foramen lacerum. The **internal carotid artery** is also superior to the foramen lacerum as it courses through the middle cranial fossa.
6. The **pharyngeal** tubercle on the basiocciput is just posterior to the anterior transverse line. It marks the superior attachment of the **midline raphe** of the

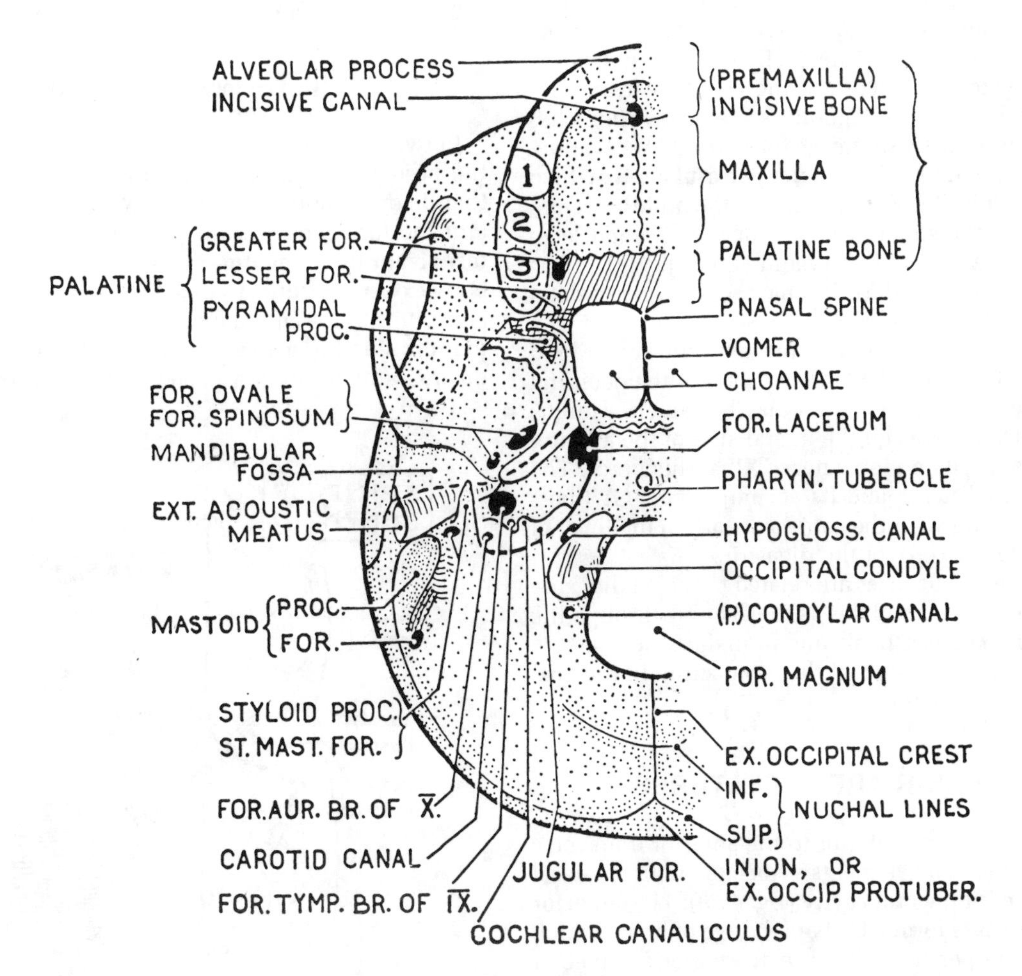

Figure 46.9. The exterior of the base of the skull.

pharynx. It also marks the separation between the visceral and prevertebral fascial planes at the base of the skull.

Structures on the Posterior Transverse Line include:

1. The origins of the **sternomastoid** and **posterior belly of the digastric muscle** are associated with the mastoid process.
2. **Nerve VII**, which is motor to the posterior belly of the digastric, the stylohyoid, and *all* **the muscles of facial expression**, exits the temporal bone from the stylomastoid foramen.
3. The **internal jugular vein** and cranial nerves **IX, X,** and **XI** exit the jugular foramen. They are enclosed in the carotid sheath with the internal carotid artery at the base of the skull. The **internal carotid artery** and the **nerve of the internal carotid artery** (sympathetic postganglionic fibers) enter the carotid canal anterior to the jugular foramen.
4. The **jugular fossa**, which is a bony space between the jugular foramen and the base of the skull, has small canaliculi in its lateral aspect. These small canals allow the auricular branches of IX and X to enter the middle ear and join VII to form the tympanic plexus. Nerve X will be sensory to the external auditory meatus, and nerve IX will given sensory branches to the mucosa of the middle ear and auditory tube, and preganglionic secretomotor branches to the **lesser petrosal nerve** for parotid innervation. Nerve VII will form the **greater petrosal nerve**, which exits the middle ear, courses over the middle cranial fossa and the superior aspect of the foramen lacerum, and enters the pterygoid canal. It will innervate the glands of the orbit, nasal cavity, and palatal region of the oral cavity.
5. The **hypoglossal canal** is the exit site for cranial nerve XII. It is on the lateral aspect of the occipital condyle on the base of the skull. The anterior ramus of C1 also courses on the lateral side of the atlanto-occipital joint and joins nerve XII at this point. The hypoglossal nerve and its accompanying C1 fibers then descend with the carotid sheath to the level of the posterior belly of the digastric.
6. The **vertebral arteries** are related to the medial side of the occipital condyles as they course through the foramen magnum to join and form the **basilar artery** on the superior aspect of the basiocciput in the posterior cranial fossa.

"ANTERIOR AREA" FEATURES

The basal skull area anterior to the anterior transverse line contains the bony elements associated with the nasal cavity and roof of the oral cavity (*fig. 46.9*). The **superior alveolar processes** form a U-shaped ridge of bone, which supports the upper teeth. Within the superior alveolar process, the **hard palate** arches superiorly to fuse with its opposite component in a midline suture. The hard palate is formed by two bony processes—the **palatine processes** of the **maxillae** anteriorly and the **horizontal processes** of the **palatine bones** posteriorly. An **incisive foramen** is present anteriorly in the midline between the palatine processes of the maxillae. This transmits the terminal components of the **nasopalatine nerve (V^2)** and the **greater palatine artery**. The horizontal process of the palatine bone contains the **greater** and **lesser palatine foramina** near the alveolar process. These foramina transmit the **greater and lesser palatine nerves (V^2) and vessels** to the hard and soft palate, respectively. The horizontal processes of the palatine bones fuse in the midline of the hard palate and form the **posterior nasal spine**.

The posterior openings of the nasal cavities, **the choanae**, are in the anterior area of the base of the skull. The choanae are superior to the soft, moveable palate in life. Each choana is separated by **the vomer**, a midline bone in the nasal cavity that extends along the midline suture of the palate from the posterior nasal spine to the region of the incisive foramen. The medial pterygoid plates form the lateral margins of choanae.

Lateral to the alveolar processes are the zygomatic bones and prominences of the "cheek bones" (zygomatic arch).

"INTERMEDIATE AREA"

Between the anterior and posterior transverse lines on the base of the skull are an important set of structures that lie on an oblique plane between the mastoid process and the foramen lacerum (*fig. 46.10*). Laterally, the **external auditory meatus** marks the lateral extent of the bony **external auditory canal**. The external auditory canal continues to the pinna by a cartilaginous extension. The **mandibular fossa** lies anterior to the external auditory meatus and forms the posterior articulating surface on

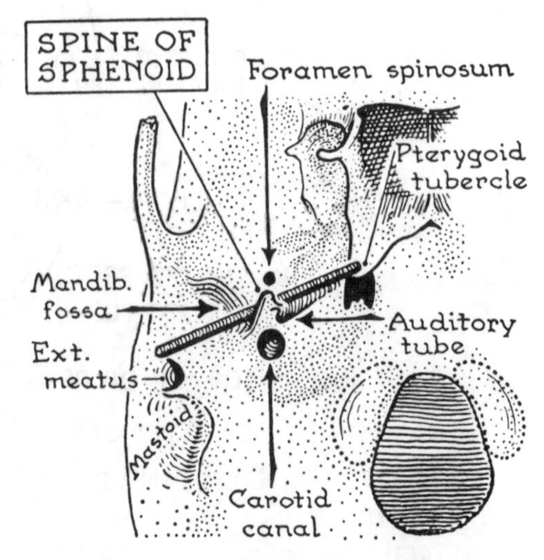

Figure 46.10. Oblique line on exterior of base of skull.

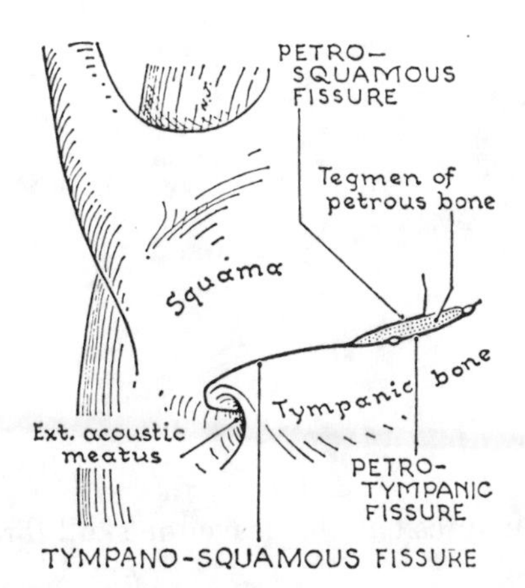

Figure 46.11. Explanation of terms. The tympanosquamous fissure is bifurcated by the tegmen tympani reaching the surface here from the roof (tegmen) of the middle ear. This is a detail of interest to ear surgeons.

the inferior surface of the squamous portion of the temporal bone for the **temporomandibular joint**. The articular tubercle of the temporal bone is anterior to the mandibular fossa and also serves as an articulating surface for the temporomandibular joint. Between the tympanic plate of the external acoustic meatus and the mandibular fossa is the **tympanosquamous fissure** (*fig. 46.11*). The medial extension of this fissure is the **petrotympanic fissure** where the **chorda tympani (VII)** exits from the middle ear to reach the infratemporal fossa. The chorda tympani courses on the medial aspect of the **spine of the sphenoid bone** (*fig. 46.10*). This bone projection is also related to the **foramen spinosum, middle meningeal artery**, and **auriculotemporal nerve (V³)**, which runs in a posterior oblique plane lateral to the spine of the sphenoid to enter the parotid gland between the external acoustic meatus and the temporomandibular joint.

The cartilaginous auditory tube lies medial to the spine of the sphenoid. It lies in the suture between the petrous temporal bone and the greater wing of the sphenoid. The auditory tube opens into the nasal cavity just posterior to the medial pterygoid plate. It connects the middle ear and the nasal cavity by a lateral osseous and medial cartilaginous tube and serves to equalize the pressure in the middle ear with the pressure in the nasal cavity (atmospheric pressure).

"POSTERIOR AREA"

This is the bony surface of the occipital bone that forms the roof of the suboccipital triangle. It is a site for muscle attachments and is described on page 343.

Clinical Mini-Problems

1. **(a) Why would a "cervical rib" on C7 create vascular and nerve disorders in the upper extremity?**
   (b) If you have studied the hand, would the neurological symptoms be on the thumb or little finger side of the hand?
2. **The lateral cervical radiograph of a patient with rheumatoid arthritis showed marked posterior displacement of the odontoid process (dens) of C2 into the vertebral canal in the superior aspect of the vertebral canal. Which specific ligament has been affected in this disease to produce this radiographic finding?**
3. **Three weeks following a rear-end car accident, a patient complained of constant neck pain but showed no radiologic evidence of bone injury in the cervical region. The physician diagnosed the disorders as "whiplash syndrome" and explained to the patient that some of the muscles on the anterior aspect of the cervical vertebral column may be injured.**
   (a) Which muscles lie on the anterior aspect of the cervical vertebral bodies?
   (b) Which major nerves are associated with these muscles in the neck?
4. **What nerves and vessels would be in jeopardy of being damaged if the spine of the sphenoid bone became necrotic (dissolved by infection)?**

(Answers to these questions can be found on p. 588.)

47

Great Vessels and Nerves of the Neck: Review and Summary

Clinical Case 47.1

Patient Doris F. This 55-year-old woman temporarily lost vision in her left eye. Greatly alarmed, she went immediately to see her physician. She described her "blindness" as being like someone had drawn a shade down over her left eye. The blindness lasted 30 minutes and then her sight was restored quite rapidly. The physician examined her retina with an ophthalmoscope and found no abnormality. He immediately got an appointment for her with your tutor, the chairman of ophthalmology. After hearing the history, he asks you not to look at the retina but to listen with a stethoscope to the left carotid artery over the carotid triangle. You hear a loud "bruit" over the artery and learn that this is due to an atheromatous plaque at the bifurcation of the common carotid. The blindness was due to platelet clots breaking off from the ulcerated area of the vessel and occluding the central artery of the retina in the left optic nerve. A consultation is arranged for a vascular surgeon to correct the arterial defect in the carotid system. You arrange to watch the operation.

General Disposition

The structures **deep to the parotid region and the posterior belly of the digastric muscle** are:

1. Internal jugular vein
2. Internal carotid artery
3. Cranial nerves IX, X, XI, and XII
4. Sympathetic trunk

These structures enter or leave the skull through 1 of 3 openings:

1. Jugular foramen—internal jugular vein and cranial nerves IX, X, XI.
2. Hypoglossal canal—nerve XII.
3. Carotid canal—internal carotid artery and nerve of the internal carotid artery.

The **jugular foramen** opens into the **jugular fossa**, which lies between the occipital and the temporal bones. The fossa contains the superior bulb of the internal jugular vein. This bulb receives the continuation of sigmoid sinus from the lateral aspect of the jugular foramen and the inferior petrosal sinus from the medial side of the jugular foramen. The three cranial nerves IX, X, and XI enter the jugular foramen alongside of the inferior petrosal sinus from the medial side adjacent to the brainstem.

The **hypoglossal canal** transmits the hypoglossal nerve (XII). It converges on the medial side of the inferior opening of the jugular fossa and joins the internal jugular vein IX, X, XI, and the internal carotid artery in the carotid sheath from the base of the skull to the level of the posterior belly of the digastric muscle (*figs. 47.1 and 47.2*).

The **carotid canal** is anterior to the jugular foramen and opens into the middle cranial fossa. The jugular foramen is within the posterior cranial fossa.

All four cranial nerves descend for a short distance between the internal jugular vein and the internal carotid artery (*figs. 47.1 and 47.2*). Cranial nerve X (**vagus nerve**) continues vertically through the neck within the carotid sheath between the internal jugular vein and the internal carotid and common carotid arteries. Cranial nerve IX (**glossopharyngeal nerve**) leaves the carotid sheath at the level of the posterior belly of the digastric to enter the pharynx with the **stylopharyngeus muscle** between the internal and external carotid arteries. Nerve IX innervates the stylopharyngeus as it spirals around this muscle to enter the pharynx between the superior and middle constrictor muscles. Nerve IX is then sensory to the posterior one-third of the tongue, tonsilar area, and pharynx. Cranial nerve XI (**accessory nerve**) leaves the carotid sheath below the posterior belly of the digastric and crosses the internal jugular vein to reach the sternomastoid muscle. The **hypoglossal nerve** (XII) passes anteriorly on the lateral surface of both the internal and external carotid arteries to reach the tongue.

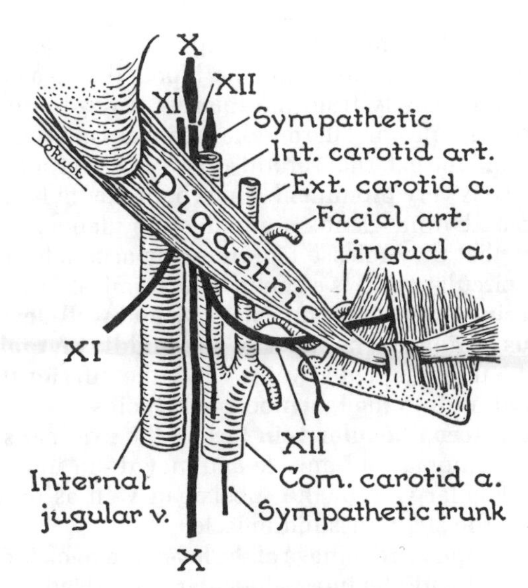

Figure 47.1. Key: posterior belly of digastric.

Common and Internal Carotid Arteries

The carotid system (*fig. 47.3*) ascends in the neck in the carotid sheath with the vagus nerve posterior and the internal jugular vein lateral to the arteries. The common carotid bifurcates at the level of the hyoid bone (C3 vertebral level), and the internal carotid artery remains associated with the carotid sheath to the base of the skull. The sympathetic trunk lies posterior to the carotid sheath throughout its course through the deep neck. The cervical sympathetic trunk is contained within a separate fascial investment between the carotid sheath and the prevertebral fascia that invests the longus colli and the longus capitis muscles.

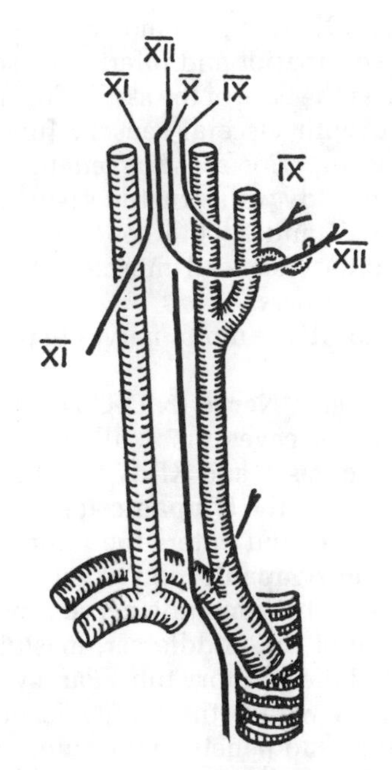

Figure 47.2. Relations of nerves IX–XII to great vessels.

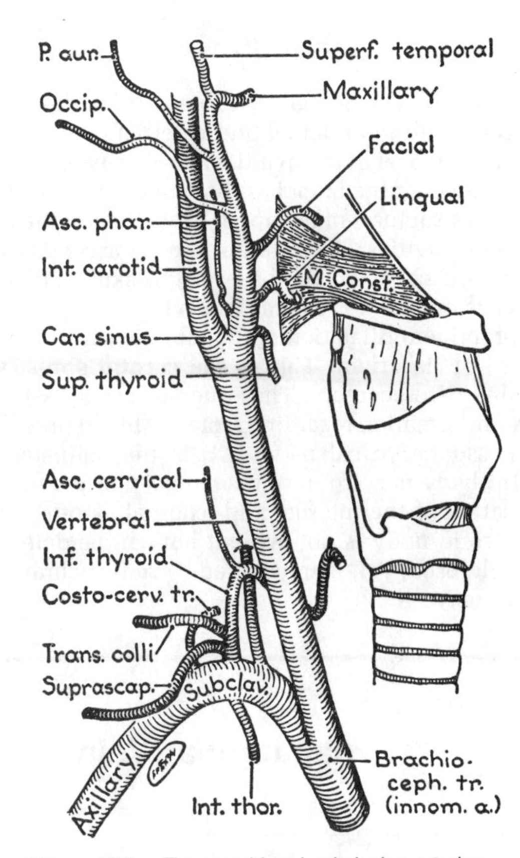

Figure 47.3. The carotid and subclavian arteries.

MEDIAL RELATIONSHIPS

The carotid system lies adjacent to the side of the digestive and respiratory passages in the neck. The narrowing of the pharynx inferiorly as it joins the esophagus brings the right and left common carotid arteries within 2 cm of each other at the root of the neck.

The thyroid gland, the trachea, and the thyroid cartilage lie medial to the common carotid arteries in the lower part of the neck. The **superior laryngeal nerve** lies on the medial aspect of **both** the internal and external carotid arteries. Its terminal branches, the **internal laryngeal nerve** and the **external laryngeal nerves** are medial to the common carotid artery as they descend to innervate structures (*fig. 43.14*).

BRANCHES

The common carotid gives no branches in the neck and bifurcates into the internal and external carotid arteries at about the level of the hyoid bone. The external carotid arteries give off the branches that supply the face, scalp, and deep structures of the upper neck, floor of the mouth, and nasal cavities. The internal carotid ascends to enter the carotid canal on the base of the skull and has **no branches outside** of the cranial cavity.

A prominent dilatation of the internal carotid artery is evident at its origin. This is the **carotid sinus**, which contains baroreceptors. These nerve receptors transmit neural information regarding systemic blood pressure via the glossopharyngeal nerve (IX) to the brainstem. The **carotid body** is a group of chemoreceptor organs at the bifurcation of the internal and external carotid arteries. The carotid body is not evident but can be detected by light microscopy. It is innervated by sensory branches of cranial nerve X.

Internal Jugular Vein

This continuation of the sigmoid sinus exits the posterior cranial fossa through the jugular foramen and ends in the root of the neck by uniting with the subclavian vein to form the brachiocephalic vein (*fig. 47.4*).

Bulbs. The internal jugular vein has a bulb at both ends. The **superior bulb** is a dilatation of the vein in the jugular fossa. The **inferior bulb** is a dilatation of the vein below a bicuspid valve. This valve serves to resist the "backflow" of blood into the neck when pressure increases in the right atrium and superior vena cava. The inferior bulb is located approximately 1 cm above the superior border of the sternoclavicular joint.

The **internal jugular vein** lies posterior to the internal carotid artery at the base of the skull. As it descends in the carotid sheath, the internal jugular vein moves laterally to lie on the external aspect of the carotid sheath in the neck. **Nerve XI and the ansa cervicalis** are usually lateral to the internal jugular vein below the posterior belly of the digastric muscle (*fig. 47.1*).

At the root of the neck, the internal jugular vein crosses anterior to the first part of the subclavian artery and its branches. It also receives a lymphatic trunk (thoracic duct and right lymphatic duct) at its point of union with the subclavian vein. The vein lies on the anterior aspect of the cupola of the lung as it enters the brachiocephalic vein.

The sigmoid sinuses and the inferior petrosal sinuses drain the posterior cranial fossa and the cavernous sinuses. They drain into the internal jugular vein through the jugular foramen. These intracranial sinuses and the internal jugular vein drain the blood that reaches the intracranial cavity from the internal carotid artery, the vertebral arteries, and the middle meningeal arteries, which arise from the external carotid artery. The thyroid venous drainage is very prominent. Since both the thyroid gland and parathyroid glands are endocrine glands, the hormones that they release require ready access to the systemic circulation. It is a common anatomical feature that **endocrine glands have an extensive and well-developed venous drainage.** The **superior** and **middle thyroid veins** join the internal jugular veins while the **inferior thyroid veins** drain into the brachiocephalic veins.

The **external jugular vein** lies on the external surface of the sternomastoid muscle and may drain into the internal jugular vein or the subclavian vein as they pass deep to the sternomastoid muscle.

The **lymphatic drainage of the head and neck** is closely associated with the internal jugular vein. Head and neck infections and diseases can stimulate lymph node enlargements around the internal jugular veins. These cervical lymph nodes then become palpable and painful at the margins of the sternomastoid muscles as they enlarge. Normally, these lymph nodes are not palpable as they surround the internal jugular vein deep to the sternomastoid muscle.

The Last Four Cranial Nerves (Extracranial Courses)

Nerves IX and X are mixed motor and sensory nerves. They both have superior and inferior sensory ganglia located on them at the base of the skull. The inferior ganglia are concerned with visceral sensory functions and are larger than the superior sensory ganglia, which receive sensations from body wall structures (somatic sensation). Cranial nerves IX and X both contain motor fibers that innervate voluntary skeletal muscle. They also contain parasympathetic fibers.

Nerves XI and XII are motor nerves that innervate skeletal muscles.

Glossopharyngeal Nerve (N.IX). This nerve (summarized in *fig. 47.5*) leaves the skull through the jugular foramen with nerves X and XI. Within the jugular fossa, an auricular branch, the **tympanic nerve**, passes through a bony canaliculus and enters the middle ear cavity to form part of the **tympanic plexus** on the **promontory**. **Sensory fibers** in the **tympanic nerve** innervate the mucous membrane of the middle ear, mastoid air sinuses, and bony part of the auditory tube. **Parasympathetic preganglionic fibers** within the tympanic nerve leave the tympanic plexus and penetrate the anterior aspect of the petrous bone. These fibers enter the middle cranial fossa

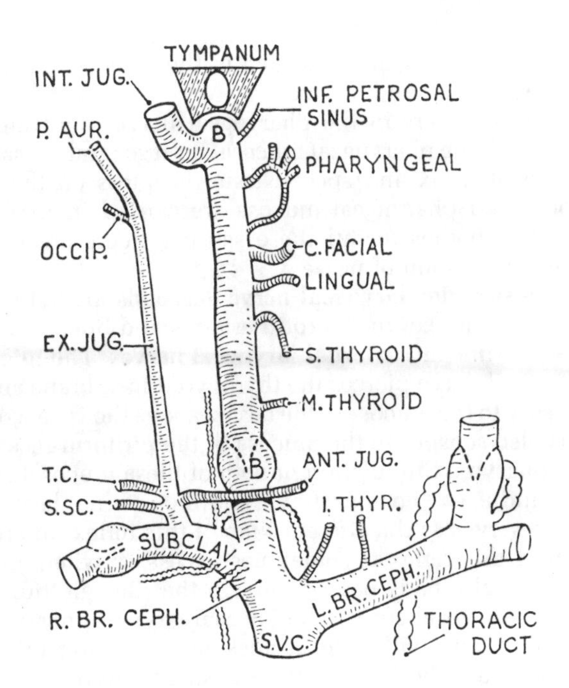

Figure 47.4. The internal jugular vein. *B*, jugular bulb; *SSC*, suprascapular; *SVC*, superior vena cava; *TC*, transversa colli.

through the **hiatus** of the **lesser petrosal nerve**. The **lesser petrosal nerve** courses through the dura of the middle cranial fossa to exit the cranium through the foramen ovale with V^3. The preganglionic parasympathetic fibers then enter the **otic ganglion**, which is associated with the medial aspect of V^3, and synapse on the postganglionic

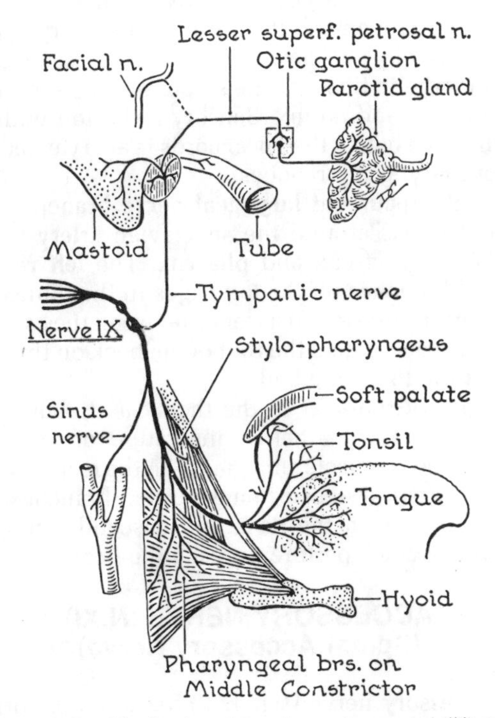

Figure 47.5. Distribution of the glossopharyngeal (IX) nerve.

parasympathetic neurons within the otic ganglion. Postganglionic fibers leave the ganglion, join the auriculotemporal nerve, and innervate the parotid gland.

The main trunk of the glossopharyngeal nerve descends in the carotid sheath to the posterior border of the stylopharyngeus muscle, which it joins to penetrate the wall of the pharynx. While winding around the stylopharyngeus in its descent, nerve IX innervates this muscle, which assists in elevation of the thyroid cartilage during swallowing. The **sensory distribution** of the glossopharyngeal nerve within the pharynx is into the mucosa on the middle constrictor (pharyngeal plexus), the mucosa of the posterior one-third of the tongue, the mucosa of the tonsilar region, and some of the mucous membrane of the soft palate. Nerve IX carries special taste sensation from the posterior one-third of the tongue and general sensation (pain and temperature) from the posterior one-third of the tongue and the other mucosal areas. The cell bodies for these visceral sensory receptors in the pharyngeal region are in the inferior ganglion of nerve IX in the jugular fossa.

The **sinus nerve** branch from nerve IX innervates the carotid sinus and carries afferent information regarding systemic blood pressure to the brainstem.

VAGUS NERVE (N.X)

This vagrant or wandering nerve supplies structures in the head, neck, thorax, and abdomen (*fig. 47.6*). The vagus nerve enters the jugular foramen and gives its first branches from the superior sensory ganglion. A **meningeal branch** re-enters the jugular foramen and innervates the dura of the posterior cranial fossa. An **auricular branch** enters the middle ear through a bony canaliculus in the lateral wall of the jugular fossa. It will leave the tympanic plexus and penetrate the tympanic membrane to innervate the **skin lining the external auditory meatus**. Nerve X can be tested clinically by stimulating this area of skin (X dermatome) with a "pinprick." Stimulation of this tissue in ear infections may induce nausea and vomiting, which is associated with the vagal innervation of the gastrointestinal tract. The dura and skin are associated with somatic (body wall) development and these sensory nerves have their cell bodies in the superior ganglion of nerve X.

The inferior ganglion of nerve X is the largest and it has direct branches to the carotid bifurcation (**sinus nerve**), the pharyngeal plexus (**pharyngeal branch**), and the larynx (**superior laryngeal nerve**). The sinus nerve innervates the carotid sinus and carotid body along with nerve IX. Its sensory fibers transmit information regarding blood pressure and oxygen tension in the blood to the cardiovascular regulating "centers" in the brainstem.

The pharyngeal nerve contains motor and sensory fibers to the palate and pharynx. The voluntary control of skeletal muscles in the palate and pharynx can be tested clinically by asking the patient to say "ah." If the vagi

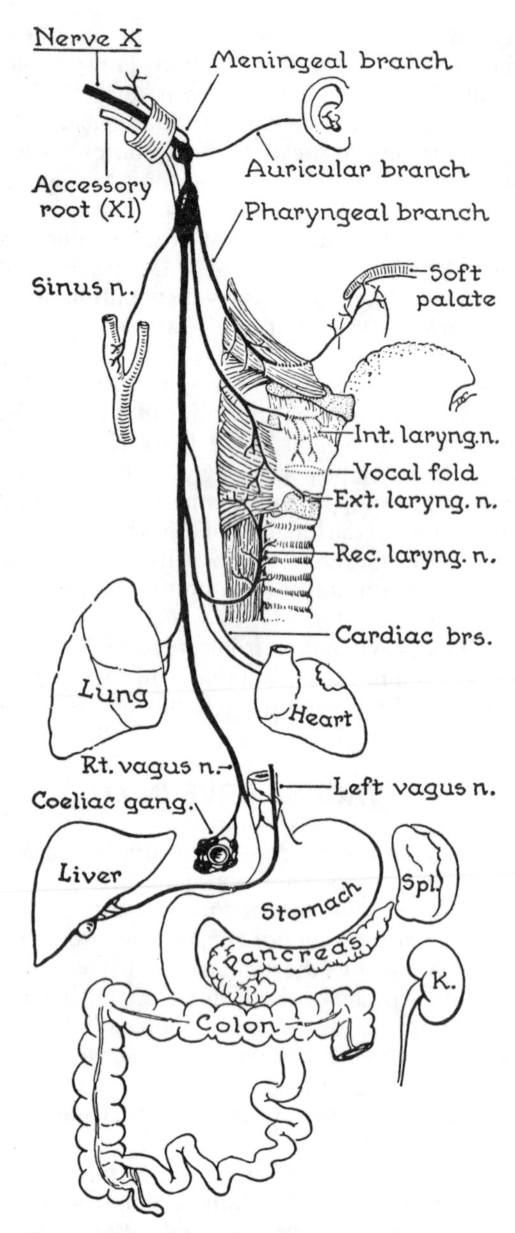

Figure 47.6. Distribution of the vagus (X) nerve.

and their pharyngeal branches to the palate are intact, the palate will rise in the midline. If one of the vagi or its pharyngeal branch is damaged, the palate will deviate away from the side of the affected nerve. **This is a very common clinical test for assessing the "intactness" of the vagus nerve at the base of the skull.** The major muscle that produces this clinical sign is the contraction of levator palati. The pharyngeal branch of nerve X also innervates the superior constrictor, middle constrictor, palatopharyngeus, and palatoglossal muscles. Secretomotor fibers in the vagus nerve are parasympathetic to the mucous and serous glands in the pharynx and larynx. These preganglionic parasympathetics run in the vagus nerve and synapse in ganglia located in the myenteric regions of the pharynx and larynx.

Sensory fibers in the pharyngeal nerve innervate receptors in the pharyngeal mucosa. Both general sensation to the pharynx and special taste receptors on the epiglottis and pharyngeal mucosa are carried in nerve X. The cell bodies for all these sensory fibers are in the inferior ganglion of nerve X.

The **superior laryngeal nerve** descends medial to the internal and external carotid arteries and branches into the **internal** and **external laryngeal nerves**. The internal laryngeal nerve pierces the thyrohyoid membrane and is sensory to the mucosa of the larynx above the vocal cords. It is also sensory to the mucosa of the piriform recess of the pharynx. Stimulation of both of these mucosal areas will induce a "cough reflex," and the internal laryngeal nerve serves as the afferent limb of this reflex. Interruption of this reflex is sometimes necessary when an endotracheal tube is passed through the rima glottidis between the vocal cords. The internal laryngeal nerve may be "blocked" with a local anesthetic solution injected subcutaneously at the tip of the greater horns of the hyoid bone.

The external laryngeal branch is motor to the cricothyroid muscle. This muscle assists in "tensing" the vocal cords. Damage to the external laryngeal can cause alterations in the production of voice quality.

The recurrent laryngeal nerve, which branch from the vagus lower in its course, are extremely important in the production of a normal voice. Damage to either the recurrent laryngeal nerve or the vagus above their origin usually produces a hoarse, raspy voice similar to a whisper. The recurrent laryngeal nerves are motor to the inferior constrictor and all the muscles of the larynx except the cricothyroid muscles. They also carry secretomotor parasympathetic fibers to the serous and mucous glands of the trachea and larynx. Sensory reception within the bronchial tree below the vocal cords is also carried in the recurrent laryngeal branches.

The **right recurrent laryngeal nerve** branches in the neck and hooks around the subclavian artery before it ascends to the larynx and pharynx. The **left recurrent laryngeal** branches off the left vagus in the thorax at the level of the aortic arch. It ascends between the esophagus and trachea and enters the root of the neck on the medial aspect of the thyroid gland.

Branches of the vagi in the thorax and abdomen are related to parasympathetic innervation of the cardiac muscle, smooth muscle, and glands in the respiratory and gastrointestinal systems. Some cardiac branches of the vagus may arise in the neck and descend through the superior thoracic aperture to reach the heart.

ACCESSORY NERVE (N.XI)
(Spinal Accessory Nerve)

The accessory nerve (*fig. 47.7*) has a double origin—**spinal and cranial**.

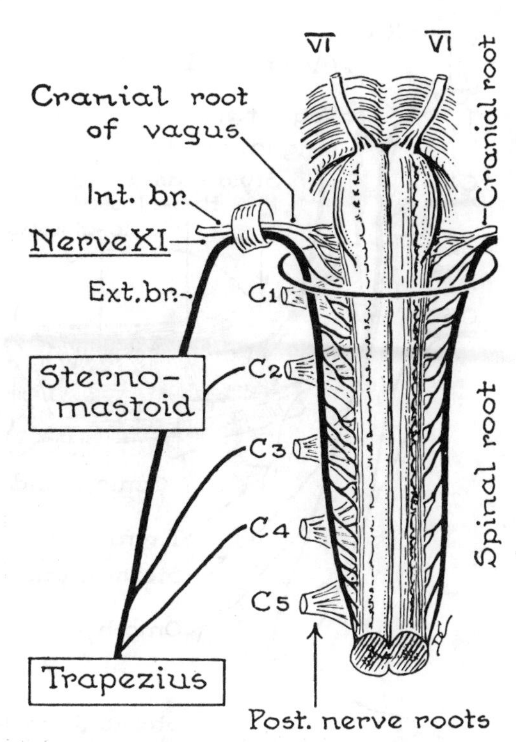

Figure 47.7. Origin and distribution of the accessory (XI) nerve. (Ventral nerve roots have been cut away.)

The spinal root arises from the anterior gray column (motor column) of the upper 4 or 5 segments of the spinal cord. The spinal root ascends in the subarachnoid space surrounding the spinal cord to enter the posterior cranial fossa through the foramen magnum. The **cranial root** arises from the medulla oblongata above the olive and is aggregated with the rootlets of the vagus nerve. The spinal and cranial roots of the accessory nerve fuse with the vagus nerve as they enter the jugular foramen and the spinal component of nerve XI separates from nerve X in the carotid sheath at the exit of the jugular fossa on the base of the skull.

The cranial component of nerve XI remains in the vagus nerve and in a clinical sense may be considered a part of the vagus nerve. Since nerve XI arises from the same motor nucleus of the brainstem (nucleus ambiguus) and goes to the same skeletal muscles in the head and neck—palate, pharynx, and larynx—as the vagus, the cranial root of nerve XI is tested with the vagus. The cranial root of XI is also affected by the same lesions in the brainstem and peripheral nerves that are considered to be vagal branches. It therefore follows, that the cranial root of XI cannot be distinguished clinically from the vagus and is thus included in the testing and function of X.

The spinal root of XI is a separate and distinguishable peripheral nerve. It separates from nerve X and descends in the carotid sheath to the level of the posterior belly of the digastric muscle. The accessory nerve then courses posteriorly across the superficial aspect of the jugular vein to enter the deep surface of the sternomastoid muscle. **After innervating the sternomastoid muscle** with motor fibers (C2, 3 also contribute proprioceptive innervation to the sternomastoid muscle), the accessory nerve enters the investing fascia that forms the thick dense roof of the posterior triangle of the neck. It courses from the midpoint of the sternomastoid to the undersurface of the trapezius at the posterior aspect of the posterior triangle. Nerve XI is then distributed with proprioceptive fibers of C3 and 4 to the undersurface of the trapezius throughout its neck and superficial back location.

Damage to the spinal root of nerve XI may occur at the base of the skull, within the carotid sheath or more likely in the posterior triangle of the neck. Functional "intactness" of XI is tested by asking patients to "shrug their shoulders." Since the muscle can also be palpated during this maneuver, the clinician can distinguish if the trapezius on either side is functional. Testing the sternomastoid for the intactness of the more proximal portions of XI is done by asking patients to flex their head (chin to chest position) against resistance. Individual sternomastoids can be tested by having patients rotate their head to one side while the clinician palpates the sternomastoid muscle on the opposite side of the head and neck.

It should be noted that it is the peripheral component of the spinal branch of nerve XI that is being examined in the physical examination. The motor cells of XI in the cervical cord region of C1–6 overlap the motor cell bodies from the phrenic nerve to the diaphragm. Spinal cord injuries that would paralyze the sternomastoid and trapezius muscles would likely paralyze the diaphragm as well. If the patient were not supported by a mechanical respirator, it would be very likely that a spinal cord injury that affected the spinal root of XI would be fatal.

HYPOGLOSSAL NERVE (N.XII) (*fig. 47.8*)

The hypoglossal nerve innervates the muscle of the first somite, which migrates into the tongue during its development in man. Even though the hypoglossal nerve has a prominent and well-defined nucleus in the lower brainstem and exits the posterior cranial fossa through the hypoglossal canal in the occipital condyles, it is intimately associated with the first cervical somite in development. In fact, **nerve XII and the anterior ramus of C1 unite** as the two nerves course between the medial and lateral anterior rectus capitis muscles at the base of the skull. Nerve XII and its accompanying C1 motor fibers then descend in the carotid sheath to the lower border of the posterior belly of the digastric muscle. At this point, C1 fibers forming the **superior root (descendens hypoglossi)** of the **ansa cervicalis** are given off to descend on the external aspect of the carotid sheath and innervate the long infrahyoid ("strap") muscles of the neck. Nerve XII loops around the occipital artery as it leaves the carotid sheath to course anteriorly to the lateral surface of the hyoglossus muscle (*fig. 47.8*). Here a small branch of **C1 fibers are given off to innervate the thyrohyoid muscle**, and the main trunk of cranial nerve XII passes deep to the intervening tendon of the digastric muscles.

As nerve XII ascends into the musculature of the tongue on the lateral aspect of the **hyoglossus** and **genioglossus muscles**, it gives off its final C1 branch to the **geniohyoid muscle** in the floor of the mouth. Nerve XII is actually motor to the **three intrinsic muscles of the tongue (longitudinal, transverse, and vertical muscles)** and the three extrinsic muscles of the tongue **(hyoglossus, genioglossus, and styloglossus muscles)** (*fig. 47.8*). Nerve XII innervates all the muscles in the head and neck that end in **glossus** *except* the palatoglossus which is innervated by X.

Testing nerve XII can be done by asking the patients to protrude their tongue (stick it out). A bilaterally innervated tongue would protrude in the midline. Damage to nerve XII would cause the tongue to deviate toward the side of the damaged nerve and weakened or paralyzed muscles.

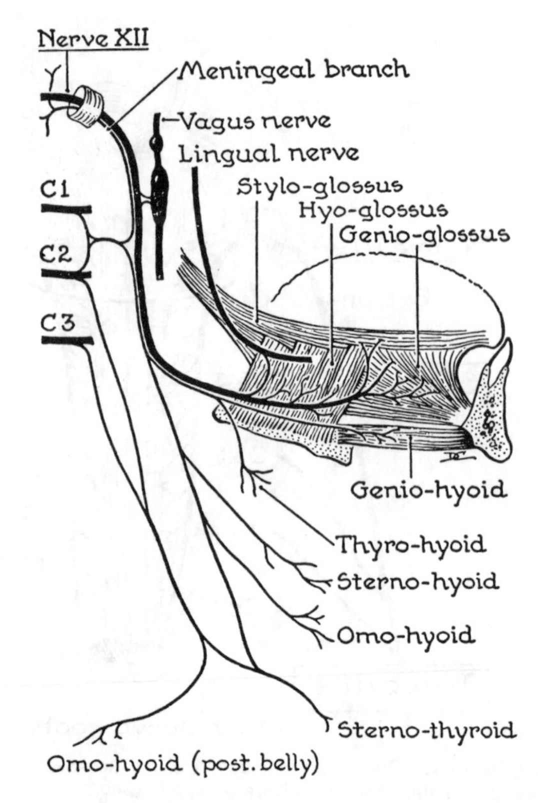

Figure 47.8. Distribution of hypoglossal (XII) nerve.

Sympathetic Trunk (Cervical Part) (*fig. 47.9*)

This part of the sympathetic trunk is an upper extension of the thoracic sympathetic trunk, which is located paravertebrally on the anterior aspect of the neck of the ribs. As the cervical sympathetic trunk ascends in the neck, it lies on the longus colli and longus capitis muscles, which arise from the anterior tubercles (costal equivalent) of the cervical vertebrae C3–6. The cervical sympathetic trunk is contained in a separate fascia (alar fascia), between the prevertebral fascia covering the longus colli and longus capitis muscles and the carotid sheath.

The sympathetic chain in the neck differs from the thoracic sympathetic chain in three major ways: (*a*) its location and relation to the vertebral column and prevertebral musculature; (*b*) the cervical sympathetic trunk has only 3 ganglia associated with 8 cervical nerves; and (*c*) there are *no* white rami communicantes associated with the cervical nerves.

The preganglionic parasympathetic fibers that ascend in the sympathetic trunk of the neck are from cell bodies in the intermediate lateral gray column of the T1–4 spinal cord segments. The preganglionic fibers will synapse in one of the 3 cervical ganglia—inferior, middle, and superior—on cell bodies of the postganglionic sympathetic neurons, which innervate the heart, upper extremity, head, and neck.

The **inferior cervical ganglion** lies anterior to the transverse process of the 7th cervical vertebra on the longus colli muscle. It lies posterior to the vertebral artery as the artery arises from the first part of the subclavian artery. The inferior cervical ganglion may be fused with the first thoracic ganglion to form an enlarged stellate ganglion. The sympathetic trunk can be blocked at this point with an injection of local anesthetic ("stellate block").

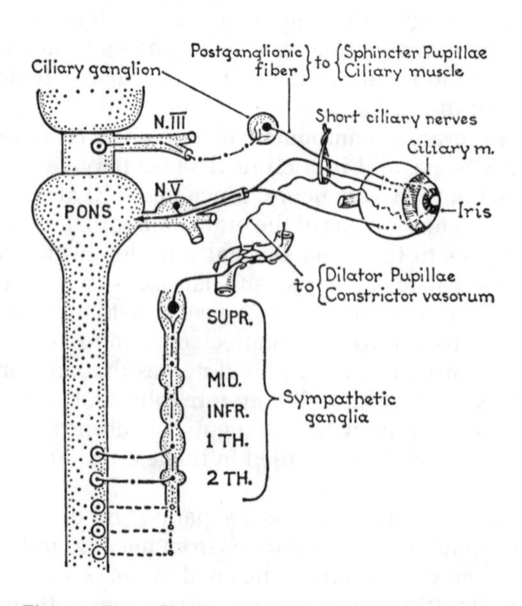

Figure 47.9. The nerve supply to the eyeball.

The inferior cervical ganglion has branches to the following structures: (*a*) **gray rami communicantes** to cervical nerves C7 and C8 of the branchial plexus; (*b*) **vascular branches** to the heart (which developed from mesoderm in the head region); (*c*) **vascular branches** to blood vessels arising from the first part of the subclavian artery; and (*d*) continuations of the sympathetic chain to the ganglia that lie subjacent (T1) and suprajacent (middle cervical ganglion). Many of the ascending fibers in the cervical chain that ascend from the inferior cervical ganglion "loop" anteriorly around the first part of the subclavian artery to form the **ansa subclavia**. Damage to the cervical sympathetic can occur if caution is not taken to avoid the subclavian artery in neck surgery.

The **middle cervical ganglion** is located at the height of the "loop" in the inferior thyroid artery (C6) on the anterior surface of the longus colli muscle. Postganglionic sympathetic neurons form the branches from the middle cervical ganglion: (*a*) the **gray rami communicantes** to cervical nerves 5 and 6, which contribute to the brachial plexus; (*b*) the cardiac branches to the heart; (*c*) arterial branches to vessels arising from the thyrocervical trunk; and (*d*) connections to the other two cervical ganglia. The superior cervical ganglion is the largest of the three cervical ganglia, and it lies on the longus capitis muscle at the vertebral levels of C2–3. It is the most superior of the sympathetic ganglia and the last point where ascending preganglionic sympathetic fibers can synapse on their respective postganglionic nerve cell bodies. The postganglionic sympathetic fibers leave the superior cervical ganglia as the following branches: (*a*) **gray rami communicantes** to the cervical nerves 2, 3, and 4 (C1 does not have a dorsal root and therefore would not transmit the postganglionic fibers to any associated skin dermatome); (*b*) cardiac branches to the heart; (*c*) vascular branches to the arteries arising from the external carotid in the neck; and (*d*) **the internal carotid nerve or plexus** (*fig. 47.9*).

The internal carotid nerve is the final termination of the sympathetic trunk superiorly. The fibers are postganglionic and course on the adventitia of the internal carotid artery. Just before the internal carotid artery enters the carotid canal in the base of the petrous temporal bone, the **deep petrosal nerve** leaves the arterial plexus to course anteriorly and enter the pterygoid canal of the sphenoid bone to fuse with the greater superficial petrosal nerve (VII). This mixed nerve of postganglionic sympathetic and preganglionic parasympathetic fibers is termed the **nerve of the pterygoid canal**, and it will enter into the pterygopalatine fossa. The postganglionic sympathetic fibers will associate with the terminal branches of the maxillary artery in the pterygoid fossa and travel to the arterioles of the vascular beds in the facial, nasal, and palatal regions served by these arteries.

The sympathetic fibers that continue into the middle cranial fossa on the internal carotid nerve go to vascular branches in the orbit on the ophthalmic artery. These postganglionic sympathetic fibers also contain the **motor fibers to the dilator pupillae muscles of the iris** and the **superior tarsal muscle (of Mueller) in the levator palpebrae superioris muscle**. Both of these muscles are smooth muscle and require an innervation from the autonomic nervous system. Damage to these sympathetic fibers will result in a **Horner's syndrome**, which is characterized by 3 prominent signs: (*a*) miosis (pupillary constriction due to unopposed parasympathetic action on the constrictor pupillae); (*b*) ptosis (partial drooping of the eyelid due to paralysis of the smooth muscle component in the levator palpebrae superioris muscle); and (*c*) anhydrosis (lack of sweating on the V^1 and V^2 dermatomes that surround the eye). A fourth sign of enophthalmos is obscure in its anatomical basis and may be illusory. Horner's syndrome is a very important and frequent clinical sign. It can occur from brainstem trauma to the sympathetic neurons from the brain that descend to the thoracic level. It may also occur in T1 nerve lesions, cervical sympathetic trunk and ganglion lesions, and damage to the nerve of the internal carotid artery.

The **testing of the sympathetic innervation in the head and neck** is best done by an examination of the pupillary reflexes. Dim light causes the sympathetic system to dilate the pupil by activating the pupillary dilator and inhibiting the parasympathetically innervated pupillary constrictor. The converse reaction of pupillary constriction is a test for the integrity of the parasympathetic fibers in III. **Tachycardia** (increased rate of heart beat) is another sympathetic clinical sign that is associated with the cardiac innervation from the cervical ganglia.

Clinical Mini-Problems

1. **Which cranial nerve is sensory to both the carotid sinus and the tonsillar region of the oral cavity? This nerve may be damaged in some forms of hypertension (high blood pressure), and the patient would also not have a gag reflex on the side of the damaged nerve.**
2. **Where does the thoracic duct drain into the venous system? It can be cannulated at this point in patients with lymphocytic leukemia.**
3. **Where would a clinician examine the lymph nodes that filter the lymph draining from the head?**
4. **How would one differentiate between a lesion in the vagus nerve in the posterior cranial fossa and a lesion in the vagus nerve at the level of the posterior belly of the digastric?**

(Answers to these questions can be found on p. 588.)

48

Pharynx and Palate

Clinical Case 48.1

Patient Jennifer L. This 6-year-old girl has a past history of multiple upper respiratory infections during her first year in school. She has become a "mouth breather," apparently due to increased lymphoid tissue in the upper pharyngeal region causing partial obstruction of her respiratory pathway. Her physician arranges with a specialist for a tonsillectomy at the Medical Center. You get permission to watch close-up. The child is given a general anesthetic through a nasotracheal tube. The enlarged tonsils are excised from the lateral pharyngeal wall. Some superior constrictor muscle is also inadvertently excised at the base of the tonsil, but the bleeding is controlled by cauterizing the severed blood vessels. The child is placed on antibiotics to prevent infection "and spread into the lateral pharyngeal space," an area whose importance you had not realized. *Followup:* There is considerable pain in her throat for 3 days following the surgery, and she also has some "gagging" when she swallows. Released from hospital after 3 days, she recovers without other postsurgical complications. Her general health improves during the next year and her breathing normalizes.

Exterior of Pharynx

The posterior wall of the pharynx is related to the prevertebral fascia posteriorly and the great vessels and nerves within the carotid sheath posterolaterally. This fasciomuscular wall is attached superiorly to the pharyngeal tubercle on the basiocciput (*see fig. 48.4*), the medial tip of the petrous bone, and the medial pterygoid plate of the sphenoid bone. It extends inferiorly to the level of the C6 vertebra, where it is continuous with the esophagus. The inferior union with the esophagus produces the narrowest and least dilatable part of the upper alimentary canal and is a site for foreign bodies to obstruct the canal.

PHARYNGEAL WALL

It has 4 distinct components (coats) from its exterior wall to its interior cavity: (*a*) investment of visceral fascia; (*b*) skeletal muscular wall; (*c*) fibrous internal muscular fascia; and (*d*) mucosal lining.

The outer facial lining is formed from the visceral fascia of the neck that also covers the external surface of the buccinator muscle. This **buccopharyngeal fascia** continues onto the external surface of the superior constrictor muscle, which shares a common origin with the buccinator—the **pterygomandibular raphe** (*fig. 48.1*). The pharyngeal plexus of veins and nerves lie deep to this fascia. The venous plexus drains the pharynx, including the soft palate and pharyngeal tonsil. This plexus communicates with the pterygoid plexus of veins and drains into the internal jugular vein above the posterior belly of the digastric (*fig. 48.2*). These veins lack valves, and infections of the tonsil, palate, and pharynx that gain access to this venous plexus (septicemia) can spread to the systemic circulation or in a retrograde fashion into the meningeal venous sinuses. The **nervous plexus** is formed by skeletal and parasympathetic motor branches of nerve X, sensory branches of nerve IX and vasomotor branches of the sympathetic system.

The **muscular layer** of the pharyngeal wall is composed of five paired skeletal (voluntary) muscles. Three form the "circular" outer layer of muscle: **the superior constrictor, the middle constrictor and the inferior constrictor**. Two are longitudinal muscles that run from the base of the skull and palate in a vertical manner to attach to the thyroid cartilage. These longitudinal muscles of the pharynx are the **stylopharyngeus** and **palatopharyngeus**.

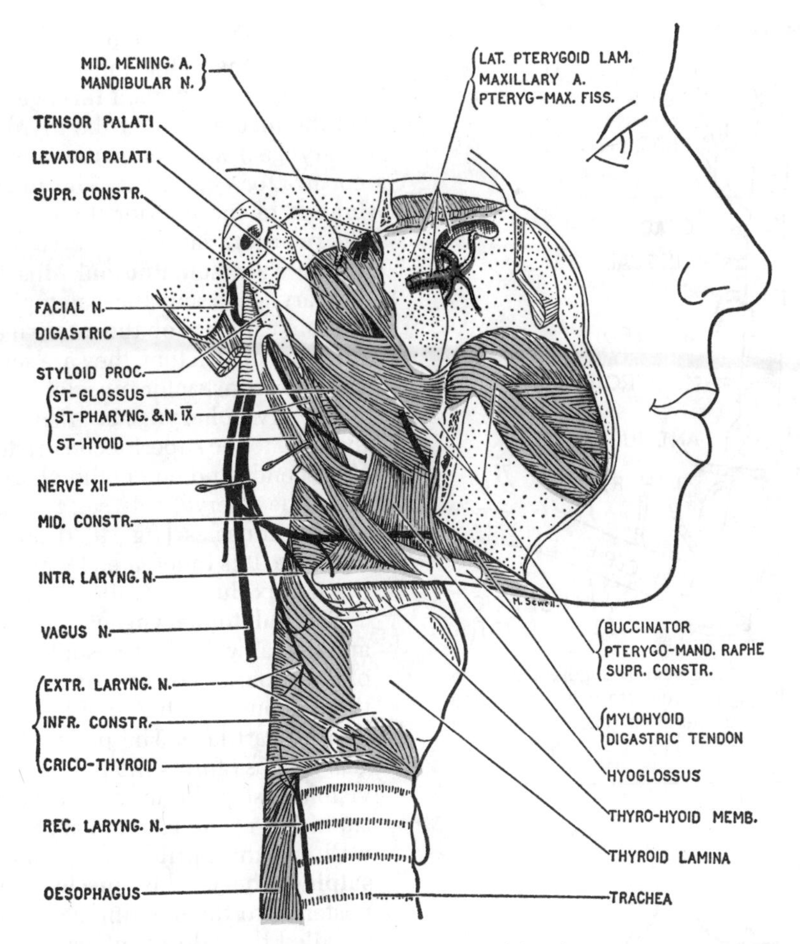

Figure 48.1. The lateral aspect of the pharynx.

THE THREE CONSTRICTORS (*fig. 48.3*)

The **superior constrictor** muscle arises from the medial pterygoid plate of the sphenoid bone, the pterygomandibular raphe and the upper 1/5th of the mylohyoid line on the mandible behind the 3rd molar tooth. The superior border of the muscle attaches to the petrous temporal bone apex and the pharyngeal tubercle through its associated investment of deep fascia. An interval exists between the base of the skull and the superior constrictor at the suture between the sphenoid and the petrous part of the temporal bone. This interval provides an entrance for the auditory tube to extend from the base of the skull to the nasopharynx. It also transmits the ascending pharyngeal artery from the external aspect of the superior constrictor to the internal surface of the pharynx (*fig. 48.4*). The inferior border of the superior constrictor is unattached and projects onto the inner surface of the middle constrictor to reach the midline pharyngeal raphe (*fig. 48.4*).

The **middle constrictor** muscle arises from the angle between the lesser and greater horns (cornua) of the hyoid bone (*figs. 48.3 and 48.5*). *Figure 48.5* shows how this origin is also associated with the lower aspect of the **stylohyoid ligament**. The insertion of this muscle into the midline raphe is such that the fibers of the middle constrictor overlap the superior constrictor superiorly but are internal to the fibers of the inferior constrictor, inferiorly (*figs. 48.3 and 48.4*). The interval between the superior and middle constrictors transmits the stylopharyngeus muscle and cranial nerve IX (*figs. 48.3 and 48.4*). The interval below the origin of the middle constrictor muscle is covered by the thyrohyoid membrane and transmits the internal laryngeal nerve and artery into the laryngeal cavity.

The **inferior constrictor** muscle has a broad area of origin from the oblique line of the thyroid cartilage and the fascia covering the cricothyroid muscle (*fig. 48.3*). It inserts into the midline raphe on the posterior aspect of the pharynx (*fig. 48.4*). The superior fibers of this muscle overlap the inferior fibers of the middle constrictor. The inferior border of the inferior constrictor blends with the superior end of the esophagus. A pronounced thickening of the inferior fibers of the inferior constrictor forms the **cricopharyngeus muscle**. This acts as a sphincter for the esophagus and prevents air from being sucked into

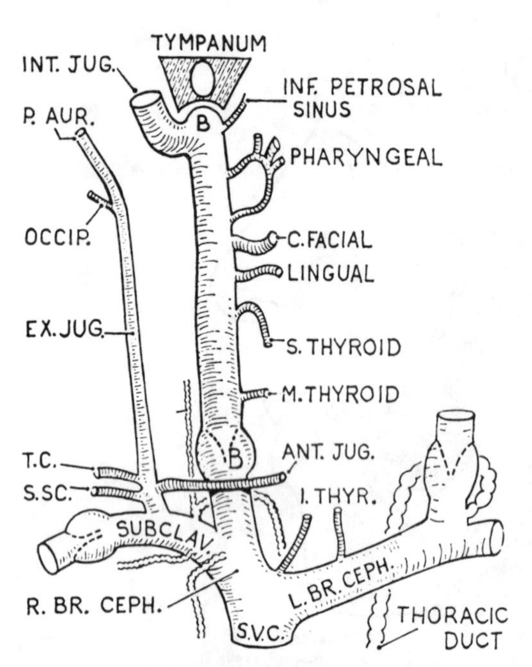

Figure 48.2. The internal jugular vein. *B*, jugular bulb; The vessels above the inferior jugular bulb are valveless. Head and neck venous drainage can therefore "back up" into the meningeal sinuses.

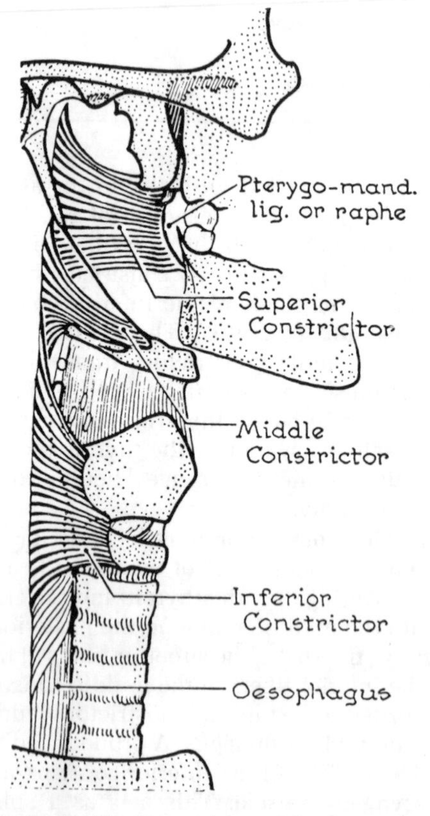

Figure 48.3. The 3 constrictors of the pharynx.

the alimentary system during inspiration (Negus). The inferior constrictor and the cricopharyngeus are innervated by the **recurrent laryngeal nerve (X)**, as it ascends on the medial side of the thyroid gland and pierces the pharyngeal wall below the inferior margin of the inferior constrictor (*fig. 48.4*). The **inferior laryngeal artery**, a branch of the inferior thyroid artery, also accompanies the recurrent laryngeal nerve into the laryngeal cavity.

The Two Longitudinal Muscles. While both the stylopharyngeus and the palatopharyngeus insert into the thyroid cartilage on the internal aspect of the middle and inferior constrictors, they are separated superiorly by the superior constrictor muscle.

The stylopharyngeus arises from the styloid process external to the superior constrictor at the base of the skull. It descends and enters the pharyngeal cavity with nerve IX in the interval between the superior and middle constrictor muscles (*fig. 48.4*) and inserts on the thyroid cartilage. It is innervated by nerve IX and helps to elevate the larynx during swallowing.

The **palatopharyngeus** muscle arises from the palate and lateral wall of the nasopharynx on the internal aspect of the superior constrictor muscle. It descends under the the mucous membrane of the pharynx to insert into the thyroid cartilage. The palatopharyngeal fold of pharyngeal mucosa forms the **palatopharyngeal arch**, which is visible just posterior to the palatine tonsil. A superior extension of the palatopharyngeus onto the lateral nasal wall and the cartilage of the auditory tube is called the **salpingopharyngeus muscle**. The overlying mucosal fold posterior to the opening of the auditory tube (*fig. 48.6*) is called the **salpingopharyngeal fold**.

Interior of the Pharynx and the Palate

Opening into the pharynx anteriorly are the orifices leading from the nasal cavity, oral cavity, and laryngeal cavity (*fig. 48.6*). Thus, the pharynx is divided into three parts: the **nasopharynx, oropharynx**, and **laryngopharynx**. The **soft palate** separates the nasopharynx from the oropharynx and acts as a flap-valve to allow continuity between these regions during respiration.

The **nasopharynx** lies superior to the soft palate and opens into each nasal cavity through the right and left **choanae** (posterior nasal apertures). The lateral and posterior walls of the nasopharynx are formed by mucous membrane lining the inner surface of the superior aspect of the **superior constrictor muscle** and the two palatal muscles that surround the opening of the auditory tube, **levator palati and tensor palati** (*figs. 48.1 and 48.7*). These muscular walls of the nasopharynx are attached to the bones on the base skull; medial pterygoid lamina

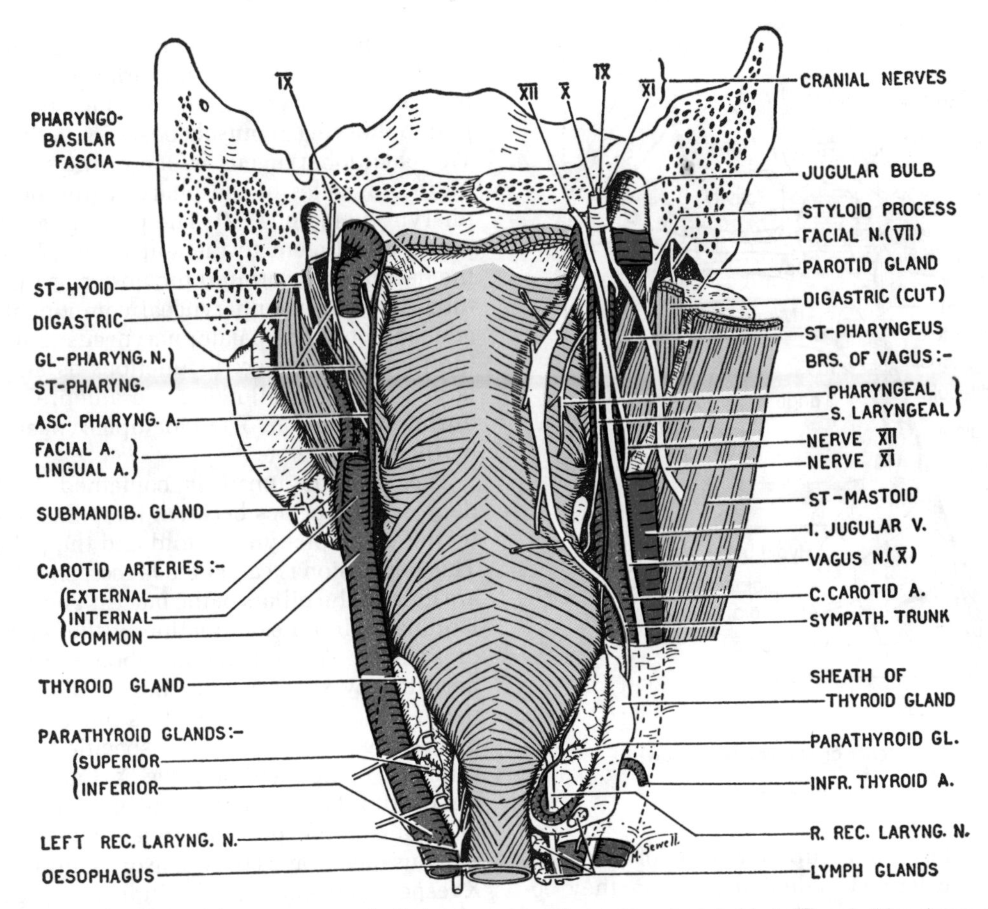

Figure 48.4. The pharynx, last 4 cranial nerves, sympathetic trunk, and great vessels—from behind. (The skull has been sectioned in the plane of 'the posterior transverse line.')

(plate), apex of the petrous part of the temporal bone, and the pharyngeal tubercle on the basiocciput (*fig. 48.8*). The firm superior attachment of the superior constrictor maintains an opened nasopharynx even when the muscle constricts. This "ensured" opening of the soft tissue respiratory passages posterior to the bony-lined nasal cavities is an anatomical feature that one sees throughout the descending bronchial tree. Cartilaginous plates and rings are present in the larynx, trachea, and bronchi to maintain the patent respiratory passages to the level of the bronchioles.

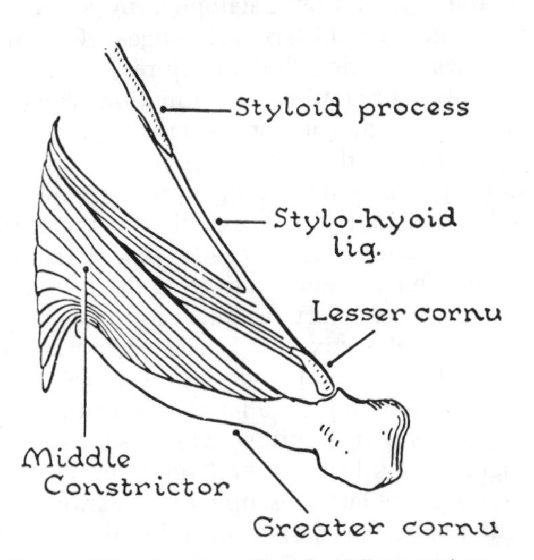

Figure 48.5. Angular origin of middle constrictor.

An important orifice on the lateral wall of the nasopharynx is the opening of the **auditory (Eustachian) tube** (*see figs. 48.6 and 48.12*). The mucous membrane of the nasopharynx bulges above the opening of the auditory tube. The superior and posterior aspects of this tubal elevation are most prominent due to the underlying cartilaginous extensions of the auditory tube. Extending inferiorly from the base of the posterior aspect of the tubal elevation is a fold of mucous membrane, the **salpingopharyngeal fold**, which overlies the descending salpingopharyngeal part of the palatopharyngeal muscle. Tonsillar tissue in the submucosal tissue of this region is referred to as the **tubular tonsil**. It may enlarge in upper respiratory infections and cause the closure of the pharyngeal opening of the auditory tube. This may lead to middle ear infections due to inadequate drainage of the middle ear and mastoid air cells.

The **(naso-)pharyngeal tonsils** lie in the submucosa

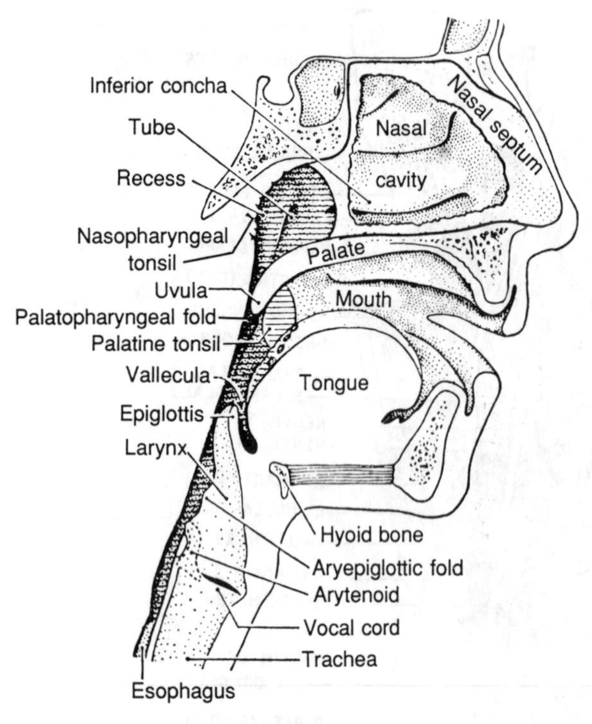

Figure 48.6. Interior of pharynx (side view).

that lines the posterior and superior walls of the nasopharynx. They are termed "adenoids," when they become enlarged or overgrown due to upper respiratory infections. Since they lie superior to the soft palate, severe enlargement may render flap-valve function of the palate ineffective during breathing. Children with prominent "adenoids" may become "mouth breathers" to bypass the obstruction in the respiratory passage between the nasopharynx and the oropharynx. Excessive mouth breathing in children can induce changes in the growth of the facial bones and produce undesirable hygienic and cosmetic effects. Removal of the "adenoids" is indicated in some patients with persistent mouth breathing habits. The **palatine tonsil** is what is commonly termed the "tonsil." The palatine tonsils are found in the oropharynx inferior to the soft palate.

The **Oropharynx** is situated inferior to the soft palate and superior and posterior to the root of the tongue (*fig. 48.6*). It is continuous with the oral cavity anteriorly and its demarcation from the oral cavity is marked by the presence of the **palatoglossal arches** (anterior pillars of the fauces) on the lateral walls. This is the site of the **buccopharyngeal membrane** in the embryo that separates the ectodermally lined stomodeum from the endodermally lined foregut.

The palatoglossal arches are folds of mucous membrane overlying the underlying **palatoglossal muscles**. These muscles are depressors of the soft palate and are innervated by the vagus nerve. They lie anterior to the palatine tonsil and attach to the lateral aspect of the tongue at the junction of its anterior two-thirds and posterior one-third segments. The posterior palatopharyngeal arch is a fold of pharyngeal mucous membrane overlying the **palatopharyngeal muscle** posterior to the palatine tonsil. The palatopharyngeus is one of the two internal longitudinal muscles of the pharynx (the other is the stylopharyngeus) (*fig. 48.7*). The palatopharyngeus has a superior attachment to the soft palate and lateral wall of the nasopharynx around the posterior lip of the auditory tube (the salpingopharyngeal component). The inferior attachment of the palatopharyngeus is its insertion into the thyroid cartilage. It functions as an elevator of the larynx during swallowing and a depressor of the palate during respiration. The palatopharyngeus is innervated by the vagus nerve.

The palatine tonsil is contained in the triangular depression that lies between the palatine arches formed by the palatopharyngeal fold and the palatoglossal fold. This depression is called the **fauces**, and the folds (arches) are termed the pillars of the fauces. They are very obvious landmarks in an examination of the lateral walls of the oral cavity. The mucous membrane of this region is innervated by sensory branches of the glossopharyngeal nerve (IX). Stimulation of this mucous membrane produces a "gag reflex" and is a specific test for the afferent (sensory) component of nerve IX. Caution must be taken not to stimulate the soft palate or lateral pharyngeal wall posterior to the palatopharyngeal fold. These areas of mucous membranes have sensory innervations by V^2 and X respectively and will also induce a "gag reflex" if stimulated. One must specifically stimulate the area of the "tonsillar bed" or fauces to isolate the clinical test for the sensory component of the right or left 9th nerve (IX). The skeletal motor component of nerve IX is to the stylopharyngeus. It is difficult to test this muscle in a unilateral fashion to assess the right or left 9th nerve (IX). The action of each stylopharyngeus is to assist in elevation of the larynx. Since this movement involves the bilateral action of the right and left stylopharyngeal muscles and the right and left palatopharyngeal muscles (innervated by nerve X), observing laryngeal elevation when a patient swallows does not clearly reflect the action of a single muscle or cranial nerve function. The parasympathetic secretomotor component of nerve IX may be tested by observing parotid secretion from either the right or left parotid papilla lateral to the upper 2nd molar in the oral cavity. This is the only exit site for the parotid secretion. A visible secretion of saliva can be readily stimulated before mealtime by activating the taste receptors in the mouth or olfactory receptors in the nose.

The **laryngopharynx** lies posterior to the superior opening of the laryngeal cavity at the vertebral level of C3–4. *Figure 48.6* depicts a lateral view of the superior laryngeal opening (the aditus), which is commonly seen in radiographs, while *Figure 48.7* shows the coronal view that is seen in the laryngoscopic examination.

The inlet of the larynx is oval and obliquely oriented. Anteriorly, the opening is bounded by the flexible epi-

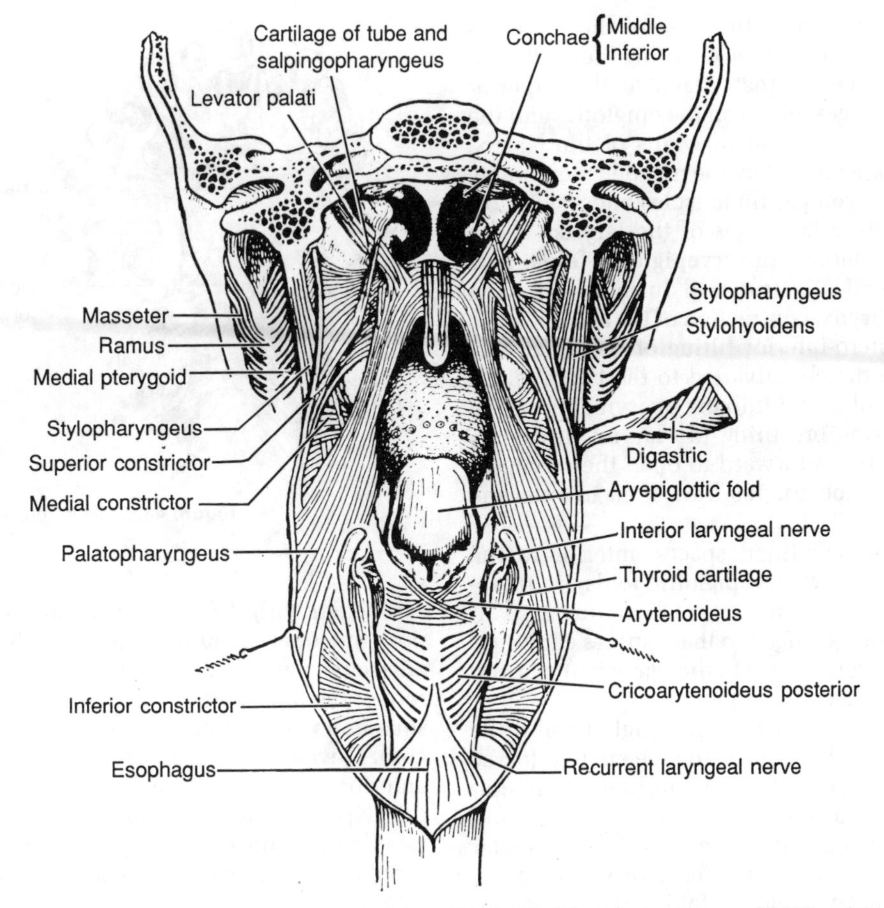

Figure 48.7. The muscles of the pharynx (from behind). *ST-PH*, stylopharyngeus; *ST-HY*, stylohyoid.

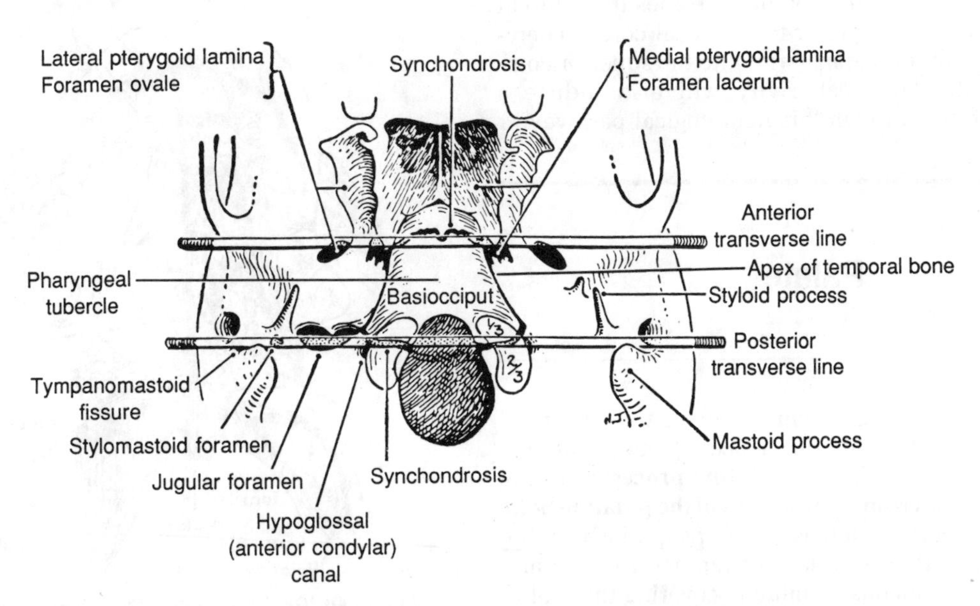

Figures 48.8. The anterior and posterior transverse lines on the exterior of the base of the skull.

glottis. The posterior limit of the inlet of the larynx is formed by the mucous membrane covering the arytenoid cartilages and the muscles that attach to the superior aspects of these cartilages. Between the epiglottis and the arytenoid cartilages are the lateral walls of the larynx. These lateral walls are formed by the mucous membrane, which overlies the aryepiglottic muscles and small corniculate and cuneiform cartilages of the larynx. These lateral walls are thus termed the **aryepiglottic folds**. This inlet to the larynx is effectively closed during swallowing by elevation of the larynx, contraction of the aryepiglottic muscles, and a postero-inferior tilting of the epiglottis. Food and drink are thereby diverted to the lateral sides of the aryepiglottic folds and into the pharynx below the laryngeal inlet. During breathing the larynx descends, and the tongue protrudes forward to open the laryngeal inlet and permit an unobstructed flow of air to enter the laryngeal cavity.

The mucous-membrane-lined spaces anterior to the epiglottis and lateral to the aryepiglottic folds are therefore important pathways for a bolus of food during swallowing. Food or particles lodged in these spaces can cause a "coughing reflex" in response to the persistent sensory stimulation of this mucosa.

Between the anterior surface of the epiglottis and the posterior surface of the base of the tongue are two fossae called the **valleculae** (*fig. 48.6*). A midline glosso-epiglottic fold of mucous membrane separates the right and left depressions that form the vallecula. The vallecular fossae communicate with the **piriform recesses** on the lateral aspect of the aryepiglottic folds. *Figure 48.7* illustrates the area of the piriform recesses with the mucous membrane removed. The important relationship that is shown in this illustration is the position of the underlying **internal laryngeal nerve** of X, which is the sensory (afferent) root of the "coughing reflex." This nerve also innervates the mucous membrane that lines the internal aspect of the larynx, superior to the vocal cords. Therefore, aspiration of foreign material into the superior compartment of the laryngeal cavity will also induce a "coughing reflex" through this neurological pathway.

Palate

The palate develops in the embryo to separate the nasal and oral cavities. The anterior aspect of the palate is a bony partition formed by the **palatine processes of the maxilla** and the **horizontal processes of the palatine bone** (*fig. 48.9*). This **hard palate** is covered superiorly by the mucous membrane of the floor of the nasal cavity and inferiorly by the mucous membrane covering the roof of the oral cavity. These mucous membranes are very closely associated with the bony palate and serve as a dual purpose **mucoperiosteum** to maintain the bone and line the adjacent cavities with epithelium. The mucosa on the roof of the oral cavity is markedly thickened by stratified squamous epithelium and serves as a masticatory mucosa. Elevated transverse ridges (rugae) are evident in the anterior one-third of the oral hard palate (*fig. 48.10*).

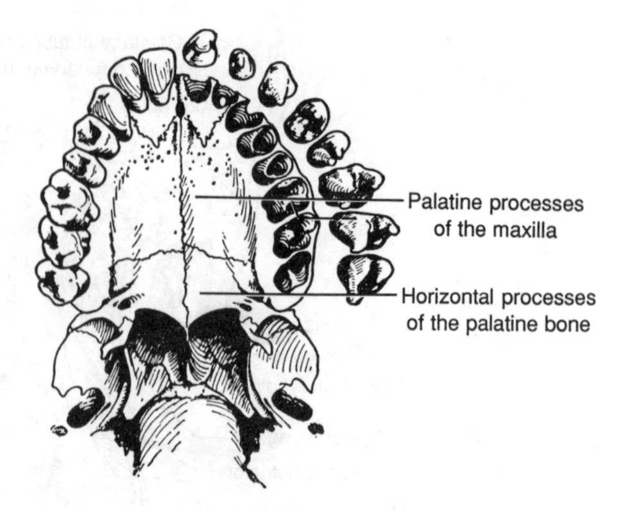

Figure 48.9. Hard palate.

The palate extends posteriorly from this bony partition as a flexible **soft palate**, separating the nasopharynx and oropharynx. The soft palate is composed of a central con-

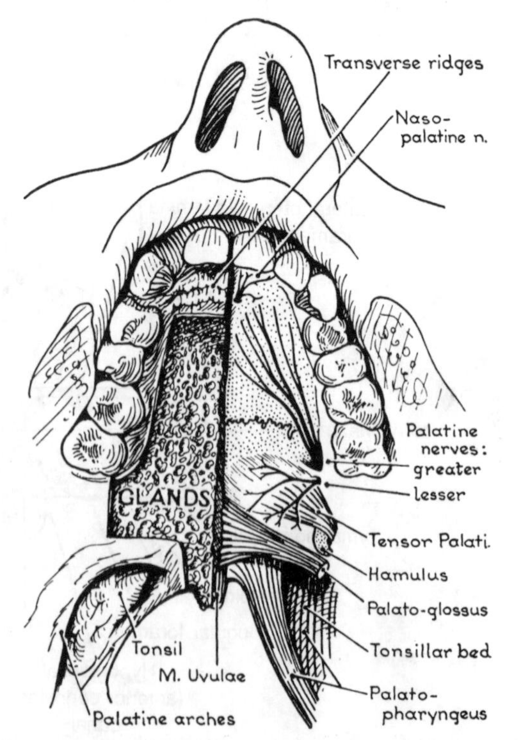

Figure 48.10. Dissection of the palate and palatine arches. The upper pole of the tonsil is deeply buried.

nective tissue aponeurosis, a number of skeletal muscles, and the overlying mucous membrane on the nasal and oral surfaces (*fig. 48.11*). The soft palate can be elevated during swallowing to close off the nasopharynx and divert the swallowed material inferiorly into the laryngopharynx and the esophagus. Depression of the soft palate occurs during respiration to open the respiratory passages from the nasopharynx to the inlet of the larynx.

The innervation of the mucous membrane of the palate is by the maxillary division of the trigeminal nerve (V^2). The mucosa on the oral surface of the anterior one-third of the hard palate is innervated by the **nasopalatine nerve** (*fig. 48.10*), which reaches the oral mucosa through the **incisive foramen** in the maxillary portion of the hard palate. This is the terminal branch of the nerve that also innervates the inferior aspect of the nasal septum and is described in the chapter on the nasal cavities. The posterior two-thirds of the hard palate is innervated by the **greater palatine nerve**, which branches from V^2 in the region of the pterygopalatine ganglion (*fig. 48.12*). The greater palatine nerve descends through the **greater palatine canal** of the palatine bone and enters the oral mucosa via the **greater palatine foramen** at the posterolateral aspect of the hard palate (*fig. 48.10*). The mucosa membrane of the soft palate received its sensory innervation through the **lesser palatine nerves**. These nerves parallel the course of the greater palatine nerves but open onto the oral mucosa of the palate via the **lesser palatine foramina** on the horizontal plates of the palatine bone (*figs. 48.10 and 48.12*).

Parasympathetic secretomotor fibers to the mucous and serous glands of the palate also accompany these sensory branches of V^2. The preganglionic secretomotor fibers are contained in the **greater (superficial) petrosal nerve (VII)**, which synapses on the postganglionic secretomotor neurons in the pterygopalatine ganglion (*fig. 48.12*). Taste fibers from palatal taste buds also course with these general sensory palatal neurons to the level of association of V^2 with the pterygopalatine ganglion in the **pterygopalatine fossa**. The taste fibers, like the preganglionic secretomotor fibers, are also associated with nerve VII, and they are contained in the greater superficial petrosal nerve.

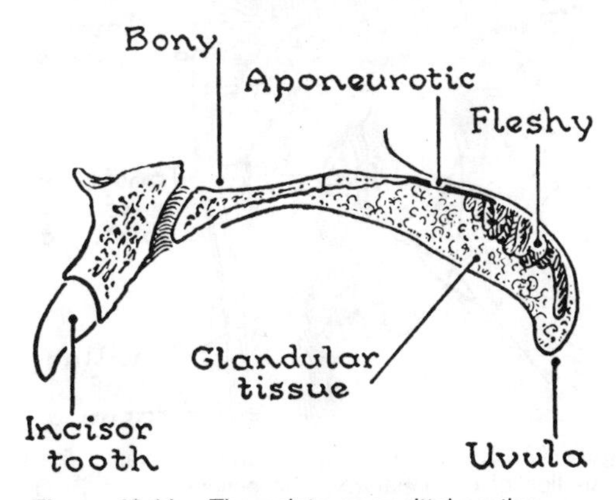

Figure 48.11. The palate on sagittal section.

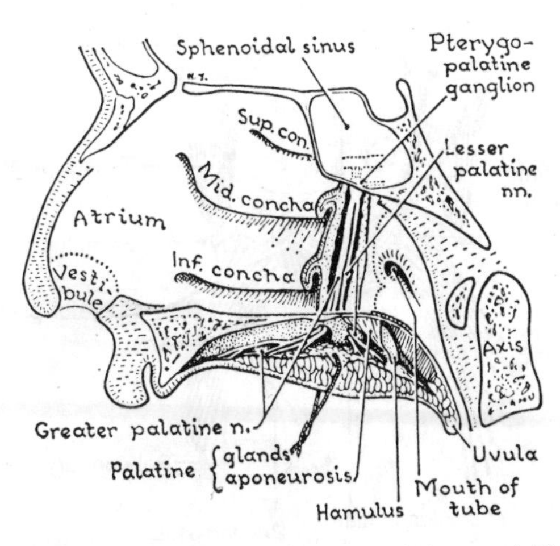

Figure 48.12. Exploration of greater palatine canal. Dissection of undersurface of palate.

The cell bodies of these taste fibers are located within the geniculate ganglion. The V^2 general sensory fibers enter the foramen rotundum to reach their cell bodies in the large trigeminal ganglion within the middle cranial fossa.

This detailed sensory innervation is particularly important to dentists and surgeons who wish to anesthetize these individual branches prior to commencing surgical procedures on the palatal soft tissues.

The arterial supply to the palate parallels the sensory innervation with one significant exception. The greater palatine artery supplies the oral mucosa of the entire hard palate and then ascends through the incisive foramen to supply a portion of the nasal septum as well. Thus, the nasopalatine and greater palatine *nerves* overlap within the mucous membrane on the oral surface of the hard palate while the **greater palatine artery** anastomoses with the **sphenopalatine artery** in the mucosa that covers the midline nasal septum in the nose. This interrelationship of the arteries in the nose is readily apparent because of the common occurrence of "nosebleeds" (epistaxis) at this point of arterial anastomosis.

MUSCLES OF THE SOFT PALATE

The muscles of the soft palate are depicted in *Figure 48.13*. They are associated with both the superior aspect and the inferior aspect of the palate. Their functions are to elevate and depress the palate as well as influence the size of the opening of the auditory tube. Opening the auditory tube will assist in maintaining an equilibrium between the air pressure within the middle ear and that in the nasal cavity (*fig. 48.14*). Testing the movement of the palate is also the way a clinician tests the pharyngeal branch of nerve X in a clinical examination. A clear understanding of these small muscles and their function is necessary for good and proper clinical practice.

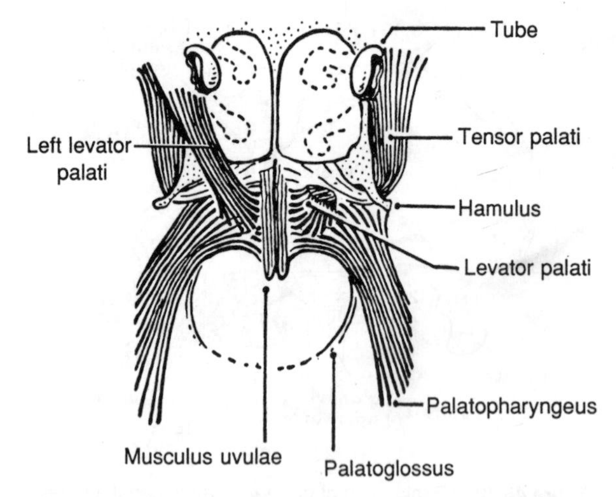

Figure 48.13. Five muscles of soft palate on each side.

Levator Palati and Tensor Palati Muscles (*figs. 48.13 and 48.15*)

These two palatal muscles arise from the base of the skull and insert into the palatal aponeurosis on the superior surface of the soft palate. The levator palati arises from the apex of the petrous bone and descends to the palate between the two cartilaginous laminae of the auditory tube. The levator palati raises a fold of mucous membrane that appears to be "poured out" of the ostium of the auditory tube as it opens into the nasopharynx. The levator palati is innervated by the pharyngeal branch of the vagus nerve and is the basis for accessing this branch of nerve X in the physical examination. Look at Figure *48.13*. If a patient were asked to say "Ah," both the right and left levator palati would contract, and normally, the palate would be elevated by the bilateral pull of the right and left levator palati muscles. The midline terminal projection of the soft palate (the **uvula** in *fig. 48.11*) would therefore rise in the midline of the oral cavity and could be visualized by the physician as one looked into the mouth. If the right levator palati were inactive (as shown by the cut muscle in *fig. 48.13*) due to a damaged right nerve X, the pull of the intact left levator palati would cause the uvula to shift to the left side. Therefore the **clinical test** for the pharyngeal branch of nerve X is to ask the patient to say "Ah" and observe if the palate rises in the midline (normal) or deviates away from the side of the damaged nerve X or its damaged pharyngeal branch. This test can assess the intactness of X at the level of the base of the skull, when the vagus exits the jugular foramen and enters the carotid sheath.

The **tensor palati** muscle also arises from the base of the skull but anterior to the levator palati and the anterior aspect of the cartilaginous auditory tube (*figs. 48.13 and 48.15*). The area of origin for the **tensor palati** is the **scaphoid fossa** at the base of the medial pterygoid plate on the sphenoid bone. The muscle descends vertically and its resulting tendon wraps around the hamulus of the medial pterygoid plate to insert into the soft palatal aponeurosis in the horizontal manner. Shortening of the muscle in the vertical plane therefore causes a "tensing" of the soft palate through this hamular "pulley mechanism" (*fig. 48.13*). The tensor palati is the *only* palatal muscle that is *not* innervated by X. The tensor palati is innervated by V³. The dual action of the levator palati moving the posterior cartilaginous plate of the auditory tube and the tensor palati moving the anterior cartilaginous plate of the auditory tube during swallowing effects an opening of the orifice of the auditory tube (*figs. 48.14 and 48.15*). This explains why one wishes to swallow, yawn, or move the palate when it is necessary to relieve

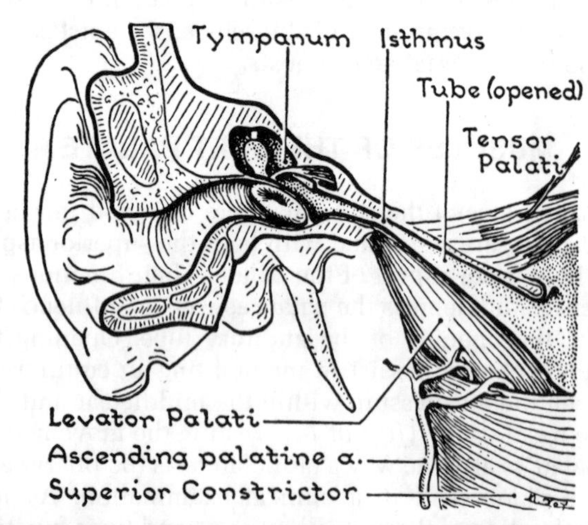

Figure 48.14. Auditory tube (pharyngotympanic tube), after removal of membranous and bony lateral wall.

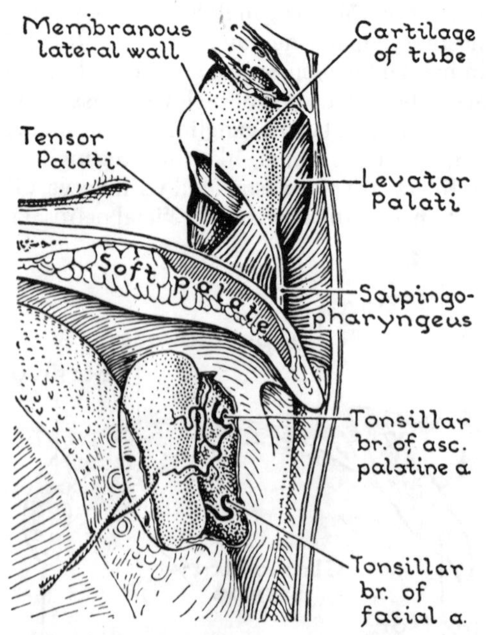

Figure 48.15. Second stage in the removal of the tonsil. Dissection of nasopharynx. (Dissections by Dr. P. G. Ashmore.)

a difference in pressure within the middle ear and the nasal cavity.

The **palatoglossus, palatopharyngeus,** and **stylopharyngeus** are related to the inferior aspect of the soft palate as they insert into the oropharynx and laryngopharynx. The **palatoglossus** muscle underlies the palatoglossal fold anterior to the tonsillar bed (*fig. 48.16*). It is innervated by the vagus nerve and assists in depressing the soft palate. The **palatopharyngeus** is a more prominent muscle and has two major components. The salpingopharyngeal part attaches superiorly to the posterior laminae of the cartilaginous portion of the auditory tube. It descends and fuses with the palatopharyngeal portion of the muscle that arises from the superior and inferior aspect of the soft palate (*fig. 48.17*). The palatopharyngeus muscle creates the palatopharyngeal fold (*fig. 48.16*), as it raises the mucous membranes of the oropharynx posterior to the tonsillar bed. It inserts onto the lamina of the thyroid cartilage and assists the **stylopharyngeus** in elevating the larynx during swallowing. The palatopharyngeus is innervated by X, while the stylopharyngeus is the only skeletal muscle innervated by IX.

The palatine tonsils are frequently removed in patients with persistent upper respiratory infections. The important blood vessels that supply the tonsils penetrate the lateral pharyngeal wall and enter the tonsil on its deep surface. *Figures 48.15* and *48.16* show two major blood vessels on the deep lateral aspect of the tonsil: the tonsillar branch of the ascending pharyngeal artery and the tonsillar branch of the facial artery. Tonsillar branches of the lingual artery may also enter the gland at its inferior pole. These vessels may be the source of extensive bleeding in tonsillectomies and are sometimes "tied off" or cauterized to control postsurgical hemorrhage.

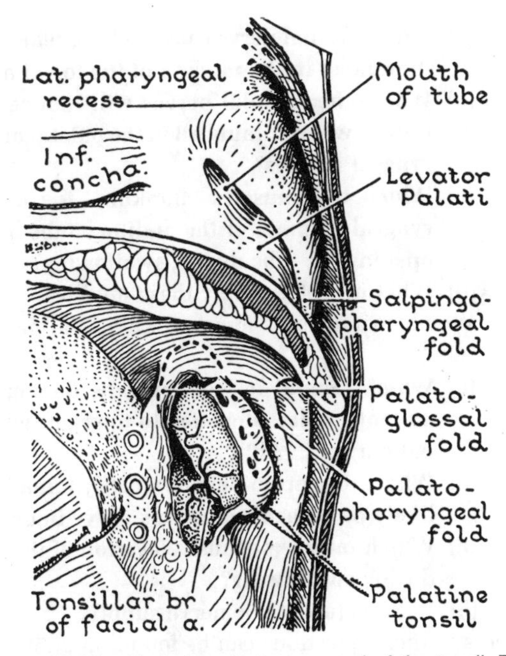

Figure 48.16. First stage in the removal of the tonsil. The side wall of the pharynx.

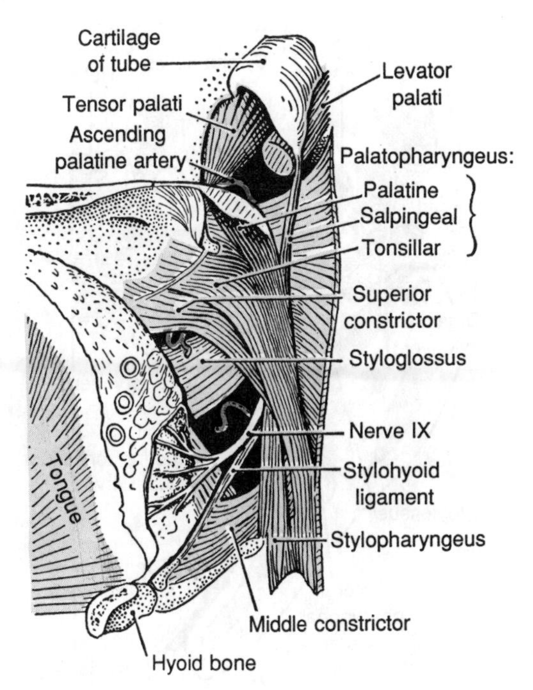

Figure 48.17. A stage in the dissection of the side wall of the pharynx from within, showing particularly the relations of the tonsil (by Dr. B. L. Guyatt).

The "tonsillar bed" in the lateral wall of the oropharynx lies in the submucosa of the pharynx. An internal fascia lining the constrictor muscles, the **pharyngobasilar fascia**, underlies the tonsil. The outer layer of fascia on the pharyngeal constrictor is the **buccopharyngeal** fascia, and this is a part of the visceral fascia of the neck. Infections of the tonsils or surgical procedures on the tonsils may compromise these fascial floors of the tonsillar bed and allow infections to spread into the **lateral pharyngeal space**, which is illustrated in *Figure 48.18*.

This space is lined with areolar fascia, filled with fat, muscles, and the structures contained within the carotid sheath. The space is continuous with the **retropharyngeal space**, which lies posterior to the pharynx and anterior to the **prevertebral fascia** that surround the spinal column and its associated muscles. Infections may spread in these fascial compartments by expansion (bacterial growth and exudate formation) and by gravity. They spread inferiorly into the triangles of the neck, deep to the investing fascia, and can eventually enter the upper extremity along the axillary sheath. They can also enter the thorax along the pathway of the carotid sheath. The latter can explain how pericarditis (inflammation of the pericardium) and infections of the posterior mediastinum can result from infections in the head and neck.

PALATAL DEVELOPMENT AND DEFECTS

Facial development occurs between the 5th and 8th weeks of embryonic life. The nasal and oral cavity are

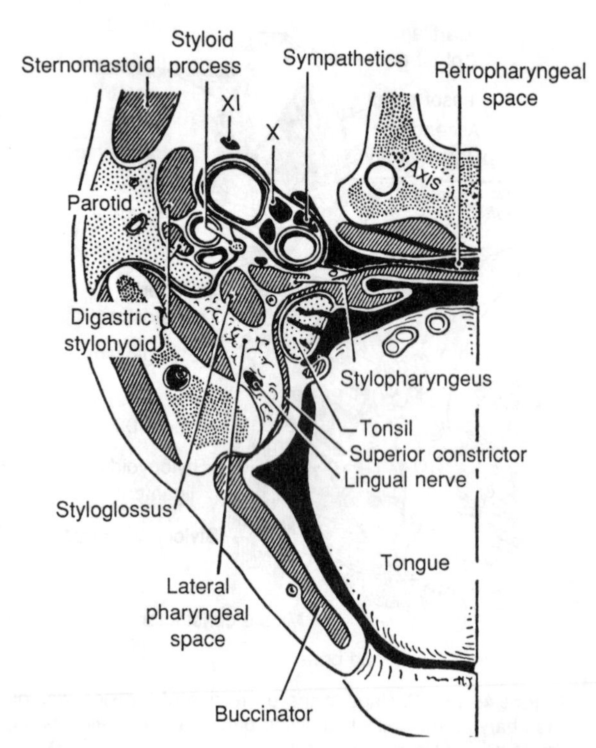

Figure 48.18. Cross-section of the head, at the level of the parotid gland and the tonsil.

separated by the development of the hard and soft palate. *Figure 48.19* shows the initial phases of palatal development prior to fusion (approximately the 7th week). The maxillary processes of the right and left side are joined anteriorly in the midline by the **intermaxillary segment**. This intermaxillary segment consists of the **globular processes** and the **premaxilla**. The globular processes give rise to the philtrum portion of the developing upper lip. The premaxilla forms the portion of the maxillary arch containing the four incisor teeth and the palatal bone between these teeth and the future incisive foramen. The secondary palate is formed from the development and

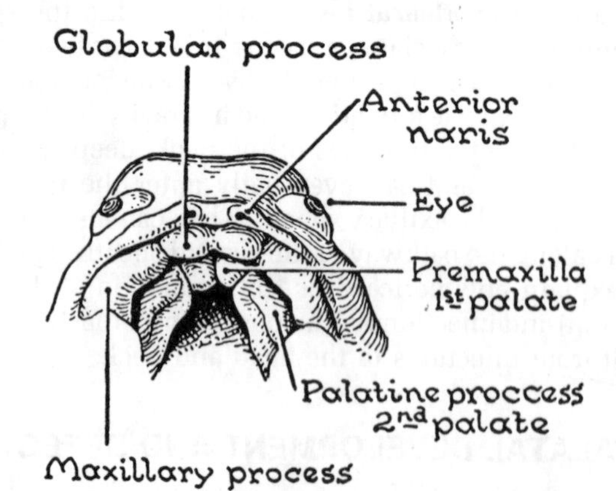

Figure 48.19. The palate develops from 3 shelves—anterior, right, and left.

fusion of the palatal shelves on the medial aspects of the maxillary processes. Fusion of these palatal shelves in the midline between the 8th and 11th week of embryonic life will complete the formation of the hard and soft palate.

Palatal fusion begins at the site of the future incisive foramen, which is at the junction of the anterior one-third and posterior two-thirds of the future hard palate. The resulting fusion then proceeds in two opposing directions from this initial site. Anteriorly, the premaxilla fuses first with the secondary palatal shelves of each side, and then the globular process of the intermaxillary segment fuses with the maxillary process. Failure of fusion in this process can result in a complete cleft from the lip to the incisive foramen or any segment of this interval from the lip toward the incisive foramen. Some newborns may have a normal fusion of the bony components and only demonstrate the cleft in the lip. Cleft lips may be bilateral or unilateral in their occurrence.

Posteriorly, the fusion proceeds from the incisive foramen region to the uvula. The entire process takes from the 8th to the 11th week. The hard palate fuses prior to the fusion of the soft palate. Complete or partial clefts of this fusion are also possible. The most severe cleft palate condition is a combination of the nonfusion of the two processes. This creates a cleft lip, maxillary arch, and a hard and soft palate defect. It produces an unsightly appearance and also creates problems for suckling and swallowing. It is a common congenital malformation and may be treated successfully by surgical correction.

Clinical Mini-Problems

1. **(a) Which cranial nerve causes a "gag reflex" when stimulated in the mucosa of the tonsillar bed?**
   (b) What other cranial nerves will induce a "gag reflex" when stimulated in the palatal and pharyngeal regions?
2. **(a) Which two muscles "flanking" the nasopharyngeal opening of the auditory tube assist in opening the tube during swallowing?**
   (b) What nerve(s) supply these muscles?
3. **Which two arteries anastomose on the nasal septum?**
4. **(a) Which cranial nerve would be damaged if a patient's palate deviated to the right when the patient said "Ah"?**
   (b) What other prominent clinical symptom would likely be recognized in this nerve injury?
5. **(a) Which muscle(s) cause elevation of the larynx during swallowing?**
   (b) What is (are) their nerve supply.

(Answers to these questions can be found on p. 588.)

49

Mouth, Tongue, and Teeth

Clinical Case 49.1

Patient Wally N. This 56-year-old seaman consulted a physician regarding a painful "canker sore" on the left edge of his tongue. A long-time user of chewing tobacco, he had noticed the condition for the previous month at sea. After examining the oral cavity and cervical lymph nodes, the physician requests a biopsy and advises the patient not to plan for early sailing. Carcinoma is the diagnosis and surgical removal is indicated. You assist in the operation. The lesion is excised from the left side of the tongue and the floor of the left side of the mouth. During the dissection, it is necessary to remove the left submandibular gland and duct. The lingual nerve on the left side is unavoidably traumatized by the surgical procedure, but it is carefully separated from the tumor where the latter invested it and surrounded the submandibular duct. Immediately following surgery, the left anterior two-thirds of the patient's tongue lacks all sensation, but sensation improves over the next 2 weeks and returns to normal following his release from hospital. He is scheduled for postoperative radiotherapy to augment the surgical resection.

Mouth

The direct visual inspection of your own mouth in a mirror or a fellow student's mouth will reveal the "living anatomy" described in this initial segment.

PARTS

The cavity of the mouth has two components—the **vestibule** and the **oral cavity proper**. These two parts are separated from each other by the teeth and their supportive tissues—**alveolar processes** and the overlying **gingival tissues**. The vestibule and oral cavity properly communicate with each other through the interdental spaces and through the area posterior to the last molar tooth and the anterior border of the ramus of the mandible.

The Vestibule

The vestibule is bounded externally by the lips and cheeks. It opens onto the skin of the face at the **aperture of the mouth**. The mucous membrane that lines the vestibule is a moist, nonkeratinized squamous epithelium that is transformed into keratinized skin at the vermilion border of the lip and keratinized gingiva on the alveolar processes. This mucosa may be raised in the midline by the connective tissue, which attaches the lips to the alveolar processes. These raised areas of mucosa are termed **frenula**. A prominent midline frenulum is present in both the superior and inferior vestibules of the mouth. The superior vestibule also contains the **parotid papilla** on its buccal (cheek) surface opposite the upper 2nd molar tooth. This parotid papilla is the constricted orifice of the parotid duct. It provides access into the parotid duct system for the injection of radiopaque dyes in radiological examinations, and it may be a site for obstruction of the duct by "stones" (sialoliths) that precipitate from salivary fluids in the parotid gland. Underlying the vestibular mucosa is a loose connective tissue that intervenes between the vestibule and the insertions of the muscle fibers of the orbicularis oris musculature onto the alveolar bone. The neurovascular supply to the vestibule lies in this loose connective tissue.

Blood Supply

Labial branches of the **facial artery** supply the tissues of the lips and vestibules. **Inferior** and **superior labial arteries** arise from the angles of the commissure where the upper and lower lips fuse. These arteries meet their counterparts from the other side of the face at the midline of the lip. Pulsations in the labial arteries can be felt by grasping the lip between one's fingers. Control of facial bleeding can be effected by compressing the lip in this manner if the lip has been lacerated.

Nerve Supply

The muscles of the lips and cheeks, the **orbicularis oris** and **buccinator**, are muscles of facial expression and therefore are innervated by the **facial nerve (VII)**. The sensory supply to the skin and mucosa of the lips and vestibule are by branches of the **trigeminal nerve (V)**. The skin and mucosa of the superior lip, cheek, and vestibule are supplied by V^2 branches: the **anterior, middle**, and **posterior superior alveolar nerves**. The skin and mucosa of the lower lip and adjacent anterior portion of the vestibule are innervated by the **mental nerve (V^3)**. The mucosa of the inferior vestibule adjacent to the cheek is innervated by the **long buccal (buccinator) nerve** from the anterior division of V^3. These subtle details in the differentiation of the nerve supplying this mucosa are important to dentists and surgeons. The detailed knowledge of this innervation is used in the techniques of locally anesthetizing this tissue prior to surgery.

Oral Cavity Proper

The oral cavity should be visually inspected in every head and neck examination. The **roof** of the oral cavity proper is formed by the hard and soft palates. The soft palate has a midline posterior projection, the **uvula**. It can be elevated by asking the patient to say "Ah." This function serves to assess the intactness of the pharyngeal branch of nerve **X** as well as to give the clinician a direct view of the posterior wall of the oropharynx.

Extending down the lateral walls of the oral cavity proper from the soft palate are two prominent folds or arches—the **palatoglossal arch** and the **palatopharyngeal arch**. The palatine tonsils are located between these arches.

The floor of the oral cavity proper is largely formed by the tongue. The tongue can be divided roughly into an anterior two-thirds component and a posterior one-third (root) component at the point where the palatoglossal arch reaches its lateral aspect. The anterior two-thirds and posterior one-third are also demarcated by a V-shaped groove, the **sulcus terminalis**, and the **circumvallate papillae** that lie anterior to this sulcus.

The undersurface of the tongue and the mucous membrane of the floor of the mouth under the tongue can be examined by asking patients to raise the tip of their tongue to their palate. A midline **lingual frenulum** is prominent on this undersurface of the tongue. It contains a component of the **genioglossus muscle** that is attached to the lingual surface of the mandible on the **superior genial (mental) tubercles**. On either side of the frenulum is the opening of the duct of the submandibular gland. Running posterolaterally from this orifice is the **plica sublingualis**, a fold of mucous membrane that overlies the upper border of the sublingual gland.

A final but important aspect of the examination of the tongue is done by grasping the tip of the tongue with a square gauze and pulling the tongue out of the mouth. This allows the clinician to examine the lateral aspects of the anterior two-thirds of the tongue. This is the most common site for cancer of the tongue to develop and should be assessed in every head and neck examination.

SUBLINGUAL REGION (FLOOR OF THE MOUTH)

Structures which are deep to the mucosa of the floor of the mouth include the **sublingual gland, lingual nerve**, the **submandibular gland**, and **duct** and the **hypoglossal nerve**. These structures all lie superior to the horizontally oriented **mylohyoid muscle** and lateral to the vertically oriented **hyoglossus muscle**.

The **sublingual gland** lies on the lingual aspect of the body of the mandible deep to the **plica sublingualis**. It has a row of short ducts (12–16) that empty into the floor of the mouth on the plica sublingualis. The duct of the submandibular gland and the lingual nerve lie on the medial surface of the sublingual gland as they course anteriorly in the floor of the mouth. The mylohyoid muscle lies inferior to the sublingual gland (*fig. 49.1*).

The sublingual gland is innervated by secretomotor postganglionic parasympathetic fibers that reach the gland through its sensory branch from the lingual nerve (V^3). The preganglionic parasympathetic nerves that control the salivary secretion of the sublingual glands are within the **chorda tympani (VII)** and synapse with the postganglionic neurons in the **submandibular ganglion** (*fig. 49.1*).

The **lingual nerve** (*fig. 49.1*) is a general sensory nerve for the anterior two-thirds of the tongue and the floor of the mouth. It also carries special taste fibers and secretomotor fibers of nerve VII that are in the chorda tympani when it joins the lingual nerve in the infratemporal fossa. The lingual nerve enters the floor of the mouth on the medial side of the mandible next to the 3rd molar tooth. (It is subject to damage at this point when the third molars [wisdom teeth] are being extracted.) Its initial branch in the floor of the mouth is to the submandibular ganglion. The preganglionic parasympathetic fibers to *all* of the salivary glands in the floor of the mouth leave the lingual nerve to synapse in the submandibular ganglion. Some postganglionic fibers, which leave the ganglion, **rejoin** the lingual nerve to reach the glands in the anterior aspect of the floor of the mouth. The lingual nerve courses into the floor of the mouth above the mylohyoid muscle. The lingual nerve has an important and consistent relation with the duct of the submandibular gland as it proceeds anteriorly. It is at first lateral to the **submandibular duct**

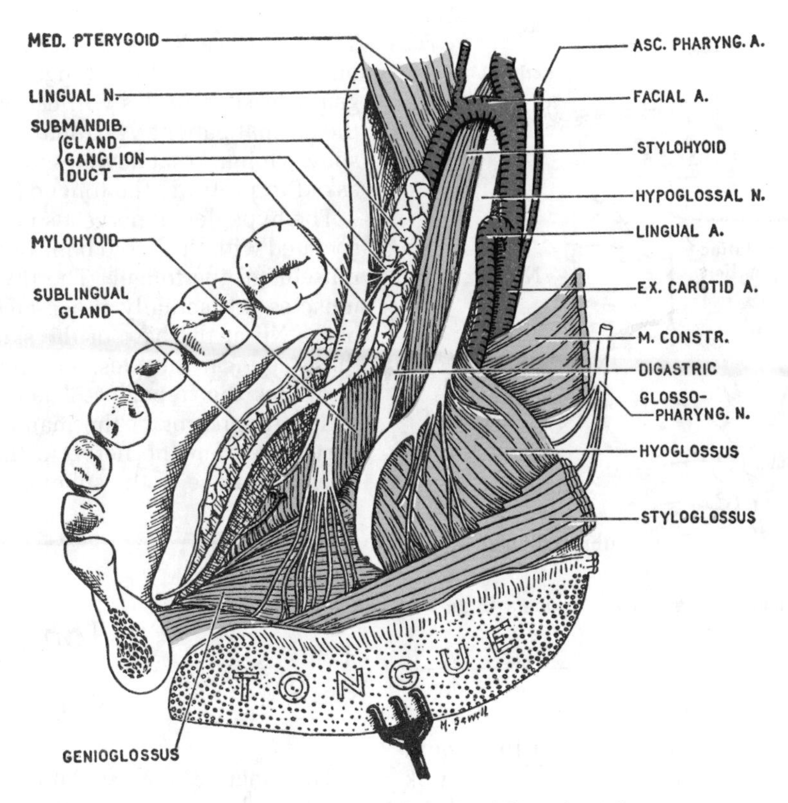

Figure 49.1. Dissection of right side of floor of mouth.

and medial to the **sublingual gland**. Subsequently in its course, the lingual nerve passes inferior to the submandibular duct and then ascends into the body of the tongue on the medial side of the submandibular duct. This relationship is illustrated, in part, in *Figures 49.1* and *49.2*. Since the tongue is being retracted in *Figure 49.1*, it is difficult to visualize the terminal ascent of the lingual nerve between the duct of the submandibular gland and the vertical muscles of the tongue proper (the **hyoglossus, styloglossus**, and **genioglossus**) (*fig. 49.2*). The lingual nerve transmits pain, temperature, touch, and pressure sensations in the mucosa on the lingual surface of the mandible and alveolar processes, the floor of the mouth, undersurface of the tongue, and the anterior two-thirds of the dorsum of the tongue.

The **chorda tympani** also contains special taste fibers for the anterior two-thirds of the tongue in addition to the preganglionic parasympathetic secretomotor fibers to the glands in the floor of the mouth. When the chorda tympani joins the lingual nerve in the infratemporal fossa, these special taste fibers remain associated with the general sensory fibers of V^3. They **do not** leave the lingual nerve to synapse in the submandibular ganglion (*fig. 49.3*). Since they are *sensory* they have a separate ganglion on nerve VII, the **geniculate ganglion**, which is located in the middle ear. The special sensory nerves of the chorda tympani will be associated with the taste buds in the mucosa of the anterior two-thirds of the tongue (*fig. 49.4*).

The **submandibular (submaxillary) ganglion** is a parasympathetic ganglion and the site of the synapses for the preganglionic fibers and postganglionic neurons that are secretomotor to the submaxillary, sublingual, and nu-

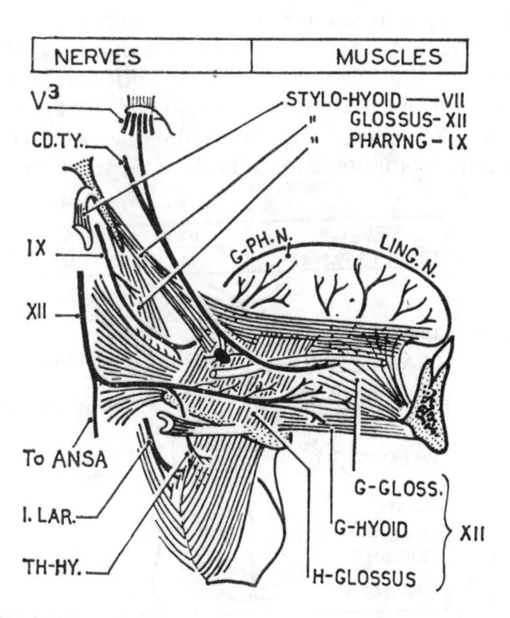

Figure 49.2. Nerves and muscles of tongue. *CD TY*, chorda tympani; *G-PHN*, glossopharyngeal nerve; *TH-HY*, thyrohyoid nerve and muscle.

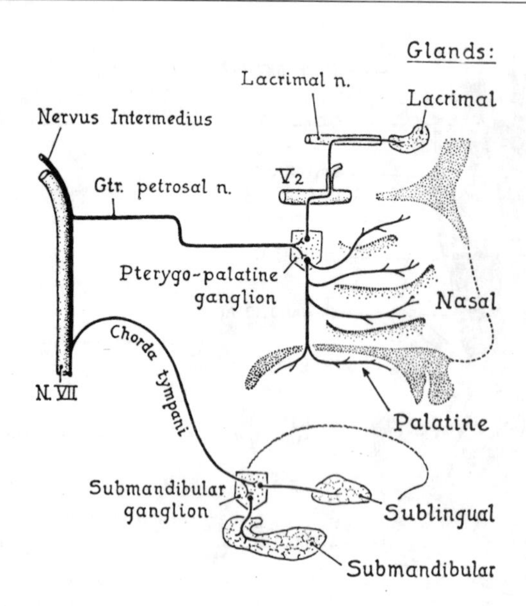

Figure 49.3. Distribution of parasympathetic fibers of the nervus intermedius.

merous small salivary glands in the floor of the mouth (*figs. 49.2 and 49.3*). It is suspended from the main trunk of the lingual nerve and adjacent to the deep part of the submandibular gland and the hyoglossus muscle. It is the most obvious of the parasympathetic ganglia of the head and neck and readily visualized in a dissection or surgical preparation. It has branches that extend directly to the submandibular gland and also a branch that rejoins the lingual nerve to reach the glands in the anterior region of the floor of the mouth.

The **hypoglossal nerve** courses anteriorly from the carotid sheath to enter the floor of the mouth on the lateral aspect of the hyoglossus muscle superior to the hyoid bone and the mylohyoid muscle (*fig. 49.2*). The hypoglossal (XII) nerve is inferior to the lingual nerve (V^3) and is a pure motor nerve to the skeletal muscles of the tongue.

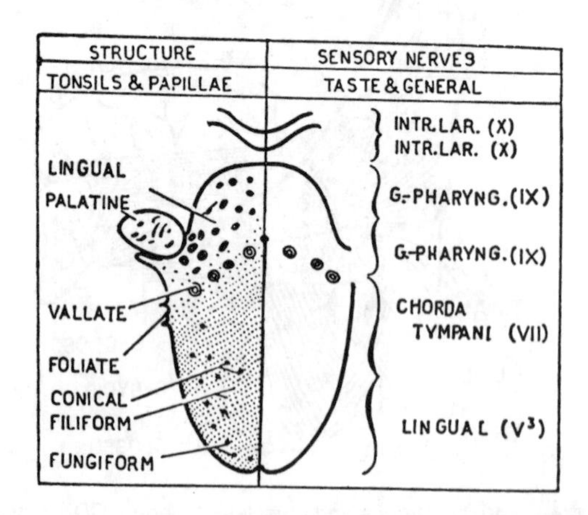

Figure 49.4. The dorsum of the tongue showing structure (*left*) and sensory nerve supply (*right*).

Testing nerve XII is done by asking patients to "stick out" or protrude their tongue. The bilateral pull of the genioglossal muscles will cause a midline protrusion in the normal patient. A patient with a XII nerve defect will show a tongue that deviates to the side of the lesion when asked to protrude the tongue (*fig. 49.5*).

The hypoglossal nerve also innervates two muscles associated with the hyoid bone but not considered extrinsic muscles of the tongue. The **thyrohyoid** and **geniohyoid muscles** are actually innervated by C1 fibers that join nerve XII at the base of the skull and "hitch a ride" to these cervical muscles. The mylohyoid muscle, which underlies the hypoglossal nerve, is a muscle of mastication (depressor of the mandible) and receives its innervation from the nerve to the mylohyoid (V^3) on its inferior surface in the submental triangle.

Tongue

The tongue is a muscular organ that functions in mastication (chewing), deglutition (swallowing), speech, and taste. Its two major components differ topographically, developmentally, structurally, functionally, neurologically and in their appearance. *Figure 49.4* summarizes these differences between the anterior two-thirds and the posterior one-third of the tongue (**L. lingua = Gk, glossa = tongue**).

The anterior two-thirds arises from the floor of the mouth and is covered by an ectodermally derived mucosa from the stomodeum. This is the **body** or **oral part** of the tongue. The posterior one-third is covered by endodermally derived mucosa from the rostral end of the primitive foregut. It lies in the anterior aspect of the oropharynx and is termed the **pharyngeal part** or **root** of the tongue. The boundary between these oral and pharyngeal parts of the

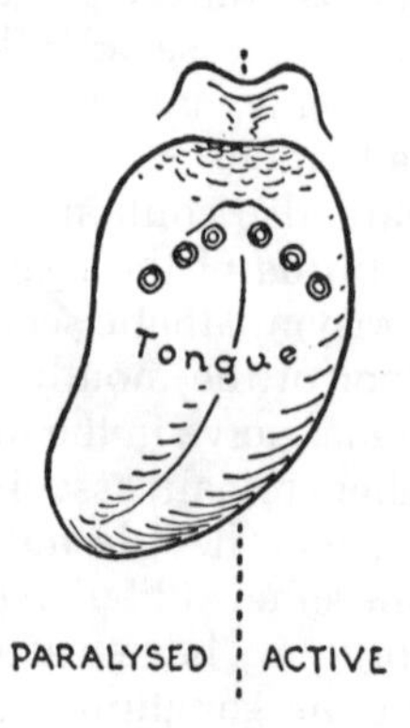

Figure 49.5. Unilateral action of the genioglossus projects the tongue to the contralateral side, i.e., inactive or paralyzed side.

tongue is the **sulcus terminalis** that lies posterior to the circumvallate papillae (*fig. 49.4*). The sulcus terminalis is V-shaped with its apex oriented in a posterior fashion in the midline of the tongue. The surface depression (pit) at this point is the **foramen cecum** of the tongue. This marks the site where the thyroid gland originated in its embryological formation. Thyroid tissue can be present in the foramen cecum in some individuals. Posterior to the sulcas terminalis are the **lingual tonsils** on the surface of the pharyngeal part of the tongue. These crypt-like structures are surrounded with lymphoid tissue and form part of the protective "lymphoid ring" of the upper respiratory tract (lingual tonsils in floor, palatine tonsils on lateral wall and pharyngeal tonsils on posterosuperior wall of the pharynx).

MUCOUS MEMBRANE OF TONGUE

The mucous membrane differs conspicuously on the two regions of the dorsum of the tongue (*fig. 49.4*). The anterior two-thirds is covered with papillae and has a "velvety" texture. The posterior one-third is studded with lymphoid tubercles and is smooth and glistening in appearance.

The **papillae** are of four varieties: filiform, fungiform, circumvallate, and foliate.

Filiform papillae are tapering and thread-like and are arranged in V-shaped rows that cover the dorsum of the body of the tongue. They contain sensory receptors for pain, temperature, touch, and pressure and have a scaly, cornified epithelium that helps to masticate food.

Fungiform papillae are small, rounded papillae that are dispersed among the filiform papillae. They are highly vascular and stand out visually due to their reddish color. They contain taste buds.

Circumvallate papillae are twelve or less circular structures approximately 2 mm in diameter. They are surrounded by a moat-like trough that is 2–3 mm deep. They lie on the anterior border of the sulcus terminalis and are prominent features on the dorsum of the tongue. The taste buds in the circumvallate papillae are present on the walls of the surrounding trough-like depression.

Foliate papillae are rudimentary in man. They are on the lateral aspect of the tongue anterior to the palatoglossal arch. They should be seen in surface anatomy as they occupy the area where cancerous lesions are most apt to occur on the tongue and should be distinguished from the mucosal changes that reflect precancerous conditions.

The **taste buds** of the anterior two-thirds of the tongue are innervated by nerve **VII** taste fibers in the chorda tympani. The taste buds on the posterior one-third of the tongue are innervated by nerve **IX**, while taste buds in the epiglottis and pharyngeal wall are innervated by nerve **X**. The palate also contains taste buds, and they are innervated by taste fibers of nerve **VII** that accompany the greater superficial petrosal nerve through the pterygoid canal and into the pterygopalatine fossa. They would then be distributed to the palate in the **greater** and **lesser palatine nerves (V^2).**

MUSCLES OF THE TONGUE

There are three extrinsic and three intrinsic muscles on each side of the tongue. *All* of these muscles are supplied by the hypoglossal nerve. The extrinsic muscles move the tongue in a bodily fashion, while the intrinsic muscles are mainly concerned with altering the shape of the tongue.

The **extrinsic muscles** (*fig. 49.6*) are the **genioglossus**, the **hyoglossus**, and the **styloglossus** muscles. The styloglossus will retract the tongue when it contracts, the hyoglossus will depress the body of the tongue, and the genioglossus will protrude the tongue. The genioglossus keeps the tongue forward in the mouth and prevents the posterior shift of the tongue that can interfere with respiration. Since the genioglossus is attached to the **superior genial tubercles** of the mandible, the anterior translation of the mandible in a patient receiving CPR (cardiopulmonary resuscitation) serves to pull the root of the tongue forward and away from the superior aperture of the larynx. The genioglossus is also the main tongue muscle used in clinically testing nerve XII (*fig. 49.5*). The patient is asked to protrude the tongue, and the clinician observes its path to determine if the hypoglossal nerves are functioning on both sides. The styloglossus and hyoglossus have an interrelation that resembles a meat cleaver (the styloglossus represents the handle of the cleaver, while the hyoglossus represents the blade—*fig. 49.6*). Both of these muscles are landmarks for finding the important lingual and hypoglossal nerves as they course

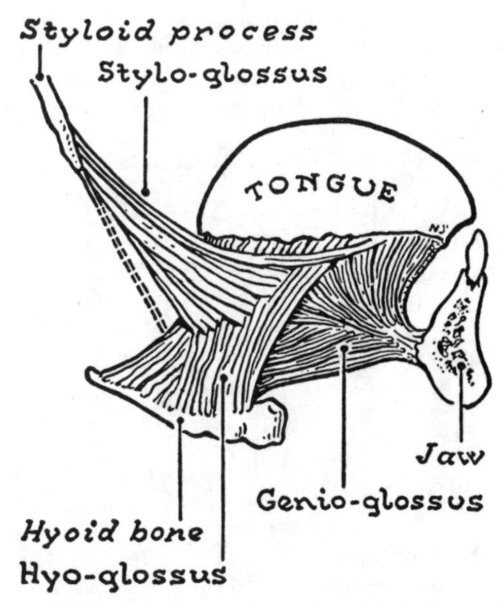

Figure 49.6. The 3 extrinsic muscles of the tongue and their 3 bony origins.

along the lateral aspects of these muscles as the nerves enter the floor of the mouth (*figs. 49.1 and 49.2*).

The **intrinsic muscles** are the longitudinal, vertical, and transverse skeletal muscle fibers, which decussate with each other and the insertions of the extrinsic muscles of the tongue.

VESSELS AND NERVES OF THE TONGUE

The **lingual artery** alone supplies the tongue. It arises from the external carotid artery at the level of the hyoid bone (*fig. 49.7*). It courses anteriorly on the middle constrictor parallel with nerve XII. It has a **key** relationship with the hyoglossus muscle. At the posterior edge of the hyoglossus, the lingual artery and nerve XII become separated by the vertically projecting hyoglossus. The lingual artery is the only major structure to pass on the medial aspect of the hyoglossus. All of the other major components in the floor of the mouth are lateral to this muscle (lingual veins, nerve XII, lingual nerve, submandibular gland and duct, and sublingual gland).

The lingual artery gives muscular branches, **dorsales lingulae arteries**, to the dorsum of the tongue before it emerges at the anterior border of the hyoglossus. Its large terminal branches, the **profunda lingulae arteries**, supply the body of the tongue. Other terminal branches supply the genioglossus muscle and the sublingual gland. The lingual artery is sinuous and contorted in its anterior course. This allows for an elongation of the arterial course when the tongue is protruded.

The **lymphatics** that drain the tongue, in general, parallel its arterial supply. Since the tongue is a midline structure, the lymphatics frequently drain to both the right and left jugular lymphatic trunks of the neck. Early detection of cancer of the tongue is therefore imperative. Surgical treatment of advanced cases of carcinoma of the tongue is a complicated and radical procedure. Because of these anatomical considerations, radiotherapy and chemotherapy are indicated and combined with more conservative surgical treatment.

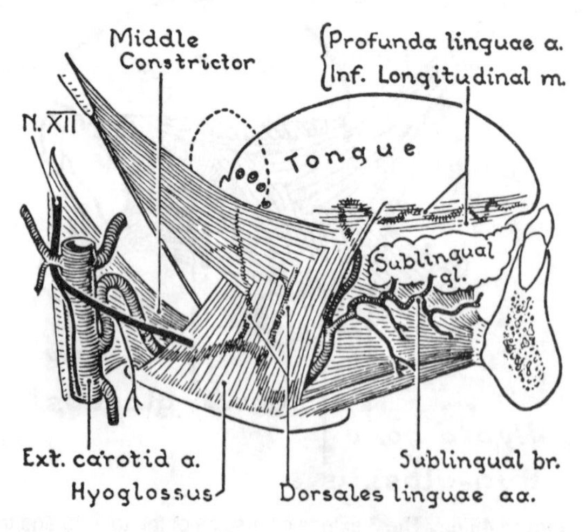

Figure 49.7. The lingual artery.

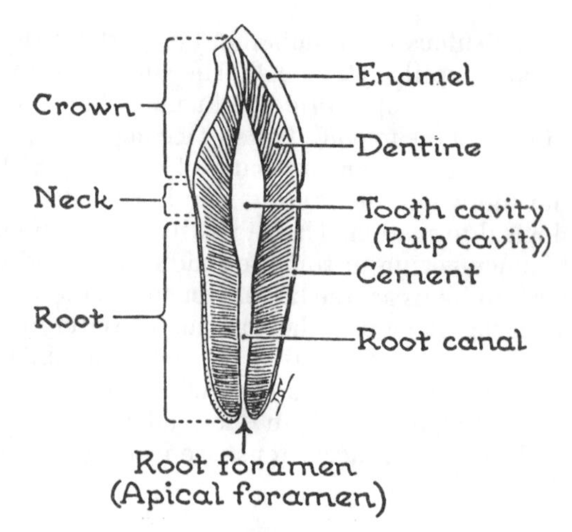

Figure 49.8. A tooth, on longitudinal section.

Teeth

PARTS (*fig. 49.8*)

The basic components of all teeth are the **crown, neck,** and **root(s)**. The root(s) are enclosed in the alveolar processes of the mandible and maxilla, while the crown of an erupted tooth projects into the oral cavity. The neck of each tooth is usually surrounded by the epithelial tissue of the oral cavity, **the gingiva**. Each tooth has a central **pulp cavity** within the crown portion of the tooth. The pulp cavity opens inferiorly through the **root canals** and **apical foramen** of each tooth root. These openings permit access for the sensory nerves and blood vessels to reach the **dental pulp** that fills the pulp cavity.

The crown of the tooth is covered by the white insensitive **enamel**. This is an **ectodermally derived** calcified material that is extremely hard and different from the **mesodermally derived** calcified tissues like dentin, cementum, cartilage, and bone. Underlying the enamel layer on the crown is a calcified layer of **dentin**. The dentin is a sensitive tissue that is intimately associated with the dental pulp. Irritation of the dentin by temperature or acids causes a painful sensation (toothache). The dentinal layer surrounds the entire pulpal cavity and root canal of the tooth and is not only confined to the crown portion. The dentin of the root portion is externally surrounded by **cementum**. This calcified tissue is similar to bone and acts, in part, as an anchoring site for the connective tissue, **periodontal ligament**, that supports the tooth in the alveolar socket. The periodontal ligament is therefore embedded in the bone of the alveolar process and the cementum of the tooth root. Its presence is noted in X-rays of the teeth by the radiolucent (dark) interval between the tooth root and the alveolar bone.

The pulp tissue of the tooth supports its nerves and vessels as well as the formative tissue for the dentinal layer. A viable and healthy pulp is extremely sensitive and responsive to stimuli that produce "toothache." The pulp responds to this stimuli by depositing more dentin between the pulp and the area of stimulus.

There are **32 permanent teeth**, 16 in the alveolar processes of the maxilla and 16 in the alveolar processes of the mandible. Within one side of each alveolar arch are 8 teeth, consisting of 2 incisors, 1 cuspid (canine), 2 bicuspids (premolars), and 3 molars. The dental formula for the permanent dentition is therefore designated as follows:

(Midline)

3	2	1	2	2	1	2	3	(maxilla)
3	2	1	2	2	1	2	3	(mandible)

A child between 6 months and 6 years of age has primary dentition. It consists of **20 deciduous teeth** (primary teeth). Each alveolar quadrant will therefore have 5 teeth: 2 incisors, 1 canine, and 2 molars. The dental formula for a child is therefore:

2	1	2	2	1	2
2	1	2	2	1	2

The primary dentition is usually fully erupted in the 2-year-old child. Children between 6 and 12 years of age have a mixed dentition of primary and secondary teeth.

ERUPTION

At birth, there are no erupted teeth, and the alveolar processes are covered by a smooth oral epithelium, suitable for grasping and suckling the nipple. At approximately 6 months of age, the central incisors begin to erupt through the oral epithelium. The lower incisors usually precede the upper incisors in their eruption. The entire primary dentition eruption should be completed by 24 months. Failure of any teeth in the primary dentition to erupt after 24 months of age should be assessed by radiographs.

Order and Time of Eruption of Deciduous Teeth

Teeth	Central Incisor	Lateral Incisors		1st Molar	Cuspid	2nd Molar
Months	6–8	8–10	10–12	12–16	16–20	20–24

The permanent dentition begins erupting in the 6th year of age when the 1st permanent molars erupt distal to the 2nd primary molars in both alveolar arches. The permanent dentition erupts on the lingual aspect of the primary dentition. The erupting permanent incisors create a pressure resorption of the roots of the primary incisors, and the primary teeth are loosened and exfoliated during the 6th year of life. The permanent incisors erupt during the 7th and 8th years. All of the primary teeth are eventually lost in a similar manner by 12 years of age and replaced by permanent teeth. The permanent 1st, 2nd, and 3rd molars erupt in the expanding alveolar arches behind the primary molars. The expanding size of the alveolar arches posterior to the arch occupied by the primary dentition is a component of the facial growth that occurs in prepubescent and adolescent children. Insufficient alveolar arch growth can be a factor in crowding, misalignment, and impaction of the teeth. The 3rd molars usually erupt between 15–21 years and are most commonly impacted if the alveolar arch space is insufficient. The permanent 1st, 2nd, and 3rd molars are frequently designated as the 6-year molars, 12-year molars, and wisdom teeth, respectively.

Order and Time of Eruption of Permanent Teeth

Teeth	1st Molars	Central Incisors	Lateral Incisors		1st Bicuspid	2nd Bicuspid & Cuspid	2nd Molar	3rd Molar
Years	6	7	8	9	10	11	12	18

DESCRIPTIVE TERMS

It is convenient to refer to the anterior surface of the incisors and cuspids as the **labial surface** (adjacent to the lips, L. = labia). Likewise, the anterior surface of the bicuspids and molars are termed the **buccal surface** since they oppose the cheek (L. = Bucca). The posterior surface of all teeth oppose the tongue and are therefore termed the **lingual surfaces**. The surfaces of the teeth that contact in the masticatory process are called **incisive surfaces** on the anterior teeth (incisors and cuspids). The bicuspids and molars have a flatter **occlusal surface** that is used for grinding and crushing the food bolus in mastication. The surfaces between the adjacent teeth are called the **proximal surfaces**. Those proximal surfaces on the side of the tooth that is toward the midline are called **mesial** proximal surfaces. Those on the opposite side of the tooth are the **distal** proximal surfaces. The distal surface of a given tooth will usually touch the mesial surface of the adjacent tooth in a given arch at the **contact point**. Teeth in the primary dentition are frequently spaced and lack contact points. Because of the angle of eruption of the permanent molars, the permanent dentition is usually compressed into a contact arrangement with adjacent teeth.

Crowns (*fig. 49.9*). The enamel-covered crowns have developed to form cutting components, the incisor edges and cusp points. The cuspid has a single cusp and is a transitional form between an incisor and molar tooth. The

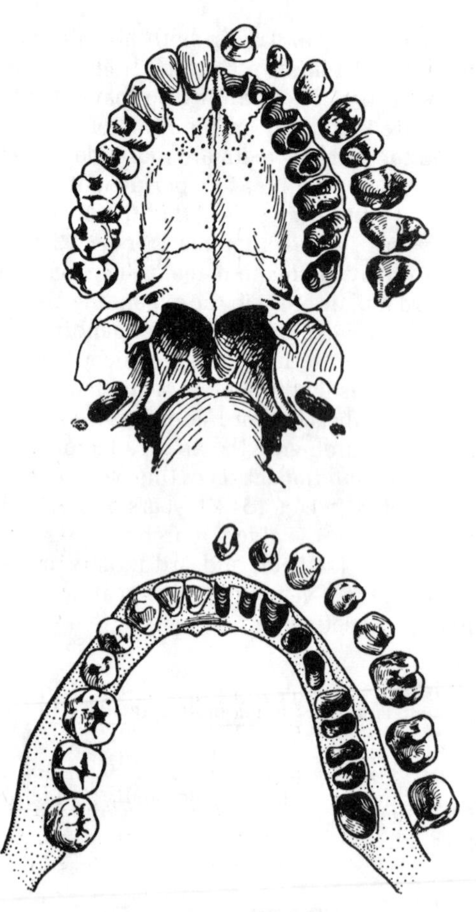

Figure 49.9. The upper and lower teeth and their sockets. (The 2nd upper premolar or bicuspid, as well as the 1st, happens to have 2 roots.)

bicuspids have 2 cusps as implied in their name, and the molars have 4–5 cusps. The cusps on the molars are on the periphery of the occlusal surface and are separated by grooves and fissures on the occlusal surfaces.

ROOTS (*fig. 49.9*)

The roots of the teeth are contained in the bony sockets in the alveolar processes. The incisors and cuspids have a single root with a single root canal. The bicuspids can vary between one and two root canals within their usually single root. The upper 1st bicuspid may have a double root in many cases. Molars have two or more roots. The lower molars have two flattened roots that are on the mesial and distal aspects of the tooth. The upper molars usually have two conical buccal roots and a larger cone-shaped lingual root. Each root has a root canal. Root formation is completed 3 years after the eruption of the tooth crown. The roots of the 3rd molar are therefore the last to complete their formation during the early twenties. They also show the greatest variation in their anatomical appearance.

OCCLUSION

The teeth of the upper arch project labially and buccally beyond the teeth of the lower arch (*fig. 49.10*). As a result, the buccal edges of the occlusal table on the lower teeth tend to wear down, while the lingual edges remain sharp. The opposite is true of the upper teeth, which have their lingual cusps occluding with the lower teeth. The upper incisors "overbite" the lower incisors and do not occlude in the same manner as the bicuspid and molar teeth.

Since the upper incisors are somewhat larger in their mesiodistal dimension than the corresponding lower incisors, the interdigitation of the upper and lower arch has the upper teeth slightly posterior to the corresponding tooth in the lower arch (*fig. 49.11*). This interdigitation enhances the masticatory process by having the intercuspation occur at the interproximal regions of the teeth above the contact points.

NERVE SUPPLY TO THE TEETH AND GINGIVAL TISSUE

The **maxillary nerve (V^2)** supplies the teeth and gingival tissue of the maxillary arch, while the **mandibular nerve (V^3)** supplies the teeth and gingiva of the mandibular arch (*fig. 49.11*).

The molar teeth of the maxillary arch are supplied by the **posterior superior alveolar nerve**. It branches from V^2 in the pterygopalatine fossa and descends on the tuberosity of the maxilla. The branches to the teeth enter the maxilla through the **posterior alveolar foramen or foramina** and course through the wall of the maxillary air sinus to reach the apical foramina of the upper molar roots. Branches to the buccal gingiva that surround the

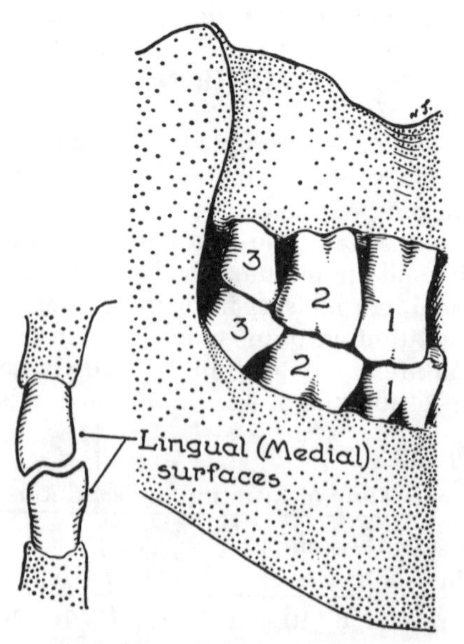

Figure 49.10. Right molar teeth in occlusion. Teeth, when in occlusion, bite on 2 teeth, except for the distal maxillary tooth and proximal mandibular tooth.

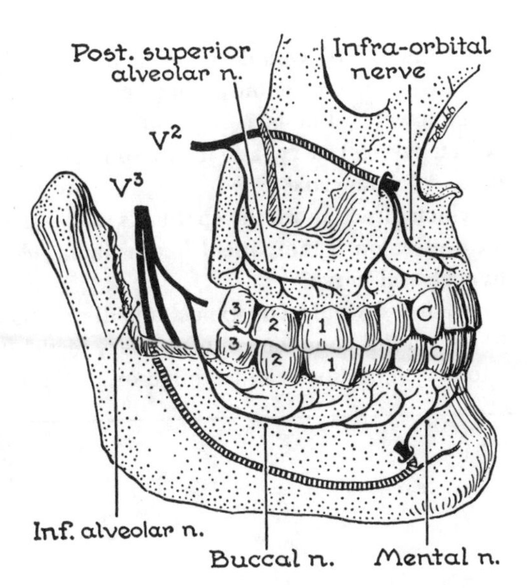

Figure 49.11. Nerve supply to the outer aspect of the gums. Teeth in occlusion. *C*, canine.

upper molar teeth and maxillary alveolar processes run over the lateral surface of the maxilla.

The bicuspids and incisors of the maxillary arch and their adjacent buccal and labial gingiva are innervated by branches of the **infraorbital nerve**. The **anterior superior alveolar nerves** descend on the maxilla, and those to the teeth enter the bone to reach the roots while the gingival branches course in the submucosa of the vestibular epithelium. The branches that go to the bicuspids are sometimes termed the **middle superior alveolar nerves**.

The lingual (palatal) gingiva on the maxillary arch is innervated by two other branches of V^2 (*fig. 49.12*). The **nasopalatine nerve** descends on the nasal septum and reaches the palate through the **incisive foramen**. The nasopalatine nerve supplies the general sensation to the gingiva on the palatal surface of the incisors. The gingiva on the palatal surface of the bicuspids and molars is supplied by the **greater palatine nerve**.

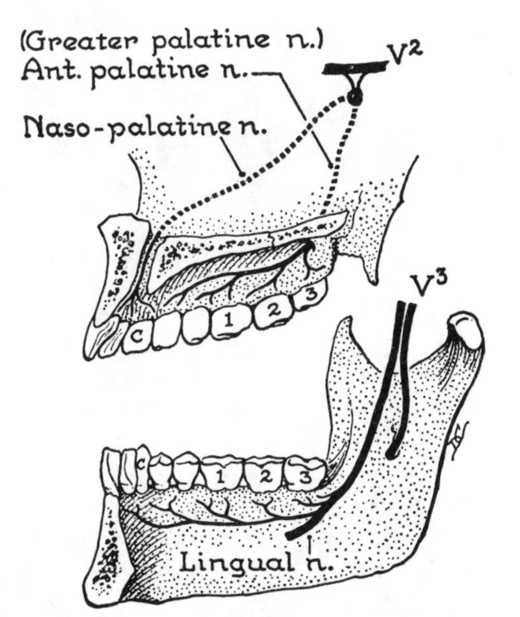

Figure 49.12. Nerve supply to the inner aspect of the gums. Branches also reach the gums between the tooth sockets (*fig. 46.1*). *C*, canine.

The teeth in the mandibular arch are innervated entirely by the **inferior alveolar (dental) nerve** as it courses through the mandibular canal within the mandible (*fig. 49.11*). The buccal gingiva, which surrounds the lower molars and bicuspids, is innervated by the **long buccal nerve**. This nerve arises from the anterior division of V^3 and leaves the infratemporal fossa by passing subcutaneously between the ramus of the mandible and the 3rd molar to reach the vestibular epithelium. The labial gingiva surrounding the lower incisors is innervated by the **mental nerve**, which exits the mandibular canal via the **mental foramen** near the root of the lower 1st bicuspid. The lingual gingiva of the mandibular arch is entirely supplied by branches from the lingual nerve (*fig. 49.12*).

DENTAL ANESTHESIA BY LOCAL INJECTIONS

The "mandibular block" is done by injecting an anesthetic into the infratemporal fossa between the lateral fascia of the medial pterygoid muscle and the medial surface of the ramus of the mandible. This injection will deposit the lipid-soluble anesthetic on the inferior alveolar nerve and the lingual nerve (*figs. 49.11 and 49.12*), which arise from the posterior division of V^3. The lower teeth, dorsum of the anterior two-thirds of the tongue, mucosa of the floor of the mouth, the lingual gingiva of the mandibular arch, the mucosa of the lower lip, and incisive gingiva are anesthetized. The remaining sensitive buccal gingiva around the molars and bicuspids is anesthetized by placing a small amount of local anesthetic in the "retromolar pad" area behind the lower 3rd molar. This blocks the sensory pathway through the long buccal nerve that courses this route to the gingiva.

Anesthetics to the maxillary arch require separate injections on the buccal and palatal surfaces of the maxillary process just distal to the tooth, which is to be anesthetized. Since the maxilla has a much thinner cortical layer of bone and its inner spongy bone allows rapid diffusion of anesthetic, this submucosal injection of local anesthetic is quite adequate. The mandible has a heavy cortical layer of bone on its outer surface. It is therefore necessary to anesthetize the inferior alveolar nerve at a site prior to entering the mandible (*fig. 49.12*) to get an adequate anesthetic effect.

LYMPH VESSELS

The dental pulps and the gingiva have a rich supply of lymphatics. These lymphatics follow the venous drainage of these structures and reach the deep cervical nodes that surround the internal jugular vein. Pulpal abscesses and gingival infections of the lower arch frequently stimulate the superficial nodes of the submental triangle and submandibular region before reaching the deep cervical lymph nodes.

Clinical Mini-Problems

1. **Where does the parotid duct open into the oral cavity?**
2. **Where does the submaxillary gland open into the oral cavity?**
3. **What function(s) would be lost if the left lingual nerve were severed at the point where it lies medial to the lower 3rd molar?**
4. **Which nerve(s) carry taste fibers from the oral cavity and oropharynx?**
5. **In which direction would a patient's tongue deviate on protrusion if the patient had a damaged right hypoglossal nerve?**
6. **At what age would you expect a pediatric patient to have a full complement of primary teeth?**

(Answers to these questions can be found on p. 588.)

50

Nose and Related Area

Clinical Case 50.1

Patient Rodney L. This accountant has suffered from spontaneous "nosebleeds" intermittently for 8 years. They have been difficult to stop with pressure and cold. On a recent occasion when nasal hemorrhage (epistaxis) was severe, he was taken to the emergency room and had to have his nasal cavities packed with epinephrine gauze to control the bleeding. He was then referred to the otolaryngologist, your clinical tutor. The mucosa of the left nasal septum was seen to be markedly engorged by the underlying vascular plexus. The otolaryngologist isolated the left sphenopalatine artery at the point where it reached the nasal septum posteriorly and ligated it. The patient has now been free of epistaxis for 3 months since the surgery.

The **external nose** is covered by thin skin with scant underlying subcutaneous tissue. The nerve supply to the skin on the nose is derived from both V^2 and V^1 branches (*fig. 50.1*). V^2 branches from the **infraorbital nerve** are the major sensory nerves to the skin of the nose (*fig. 50.2*). V^1 branches to the root of the nose are supplied by the **infratrochlear nerve**. A quite unique arrangement occurs in the innervation to the tip of the nose. The **anterior** ethmoidal branch of the nasociliary nerve (V^1) becomes the **external nasal nerve**, which descends from the root of the nose to the tip. The tip of the nose is therefore an "island of V^1 innervation in a sea of V^2 innervation." Loss of sensation to the tip of the nose may indicate an intracranial, intraorbital, or ethmoid air sinus disorder that affects the V^1 pathway from the trigeminal ganglion.

The skin of the external nose rests upon a framework of bone, cartilage, and fibroareolar tissue (*figs. 50.3 and 50.4*). The paired **nasal bones** are at the root of the nose, articulating with the frontal bone and the frontal processes of the maxilla. They may be fractured in facial trauma and surgery. The nasal cartilages underlie the central and lateral aspects of the nose inferior to the nasal bones (*fig. 50.3*). The central **septal cartilage** connects to the paired superior **lateral cartilages**, which articulate with the nasal bones. Inferiorly, the paired **alar cartilages** "wing out" from the midline septal cartilage to support the tip of the nose and the forward aspect of the medial, anterior, and lateral periphery of the nares (*figs. 50.3 and 50.4*). The alar cartilages are supported by the septal cartilage and the flexible and pliable fibroareolar tissue that forms the posterolateral aspect of the nares. The flexible nares can be expanded by facial muscles during breathing or expressive gestures.

Nasal Cavities

The **nares** open into the **right** and **left nasal cavities**. These cavities are superior to the hard palate and are separated from each other by the midline nasal septum. Each nasal cavity has an anterior and posterior aperture, a floor, a roof, a lateral wall, and a medial wall formed by the intervening midline nasal septum.

While each **naris** (nostril) forms an anterior opening for its respective nasal cavity, the **choana** forms the posterior nasal aperture. The two choanae are rigid, bony apertures that provide communicating passages between the nasal cavities and the nasopharynx. Each choana is an oval aperture that measures approximately 2 cm high and 1 cm wide. The choanae are oriented in the coronal plane of the head.

The **floor** of each nasal cavity is formed by the horizontally oriented hard palate, which is formed from the **palatine processes of the maxilla** and the **horizontal plates of the palatine bone** (*fig. 50.5*).

The **roof** of the nasal cavity is composed of three parts (*figs. 50.5 and 50.6*): (*a*) an **anterior part** whose slope corresponds to the slope of the inferior surface of the nasal bones at the bridge of the nose; (*b*) an **intermediate**

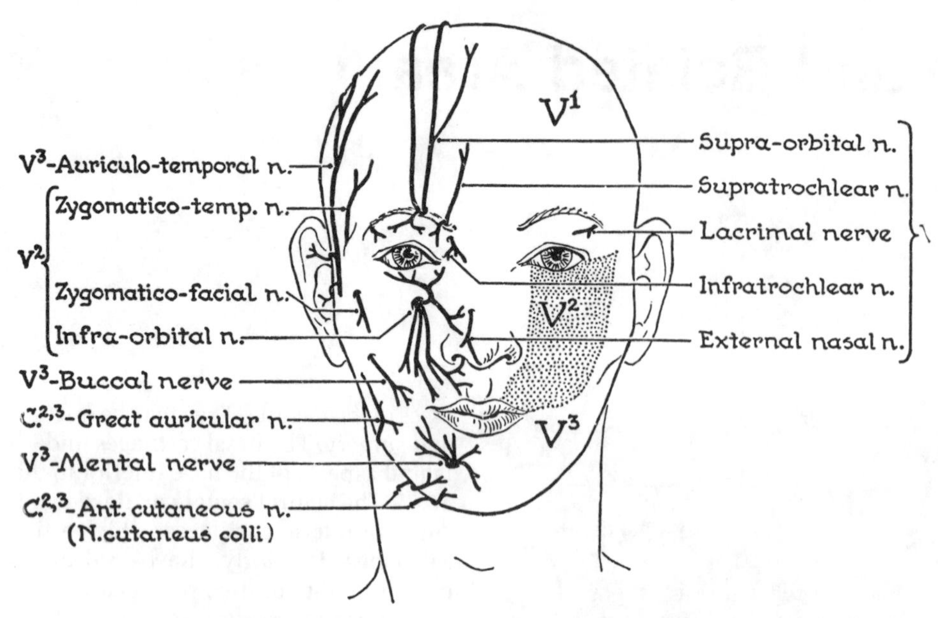

Figure 50.1. The sensory nerves of the face and front of the scalp.

part, formed by the horizontally oriented **cribriform plate of the ethmoid** bone, which provides the access for the olfactory nerve (I) into the nasal cavity; and (c) a **posterior part** of the roof is formed by the anterior and inferior aspects of the **body of the sphenoid bone**, which blends posteriorly with the roof of the nasopharynx.

The **nasal septum** is composed of bone and cartilage and is lined by a very thin and tightly adherent mucous membrane, which forms a mucoperiosteum and mucoperichondrium. The three major components of the nasal septum are illustrated in *Figure 50.5*. The two bony components, the **perpendicular plate of the ethmoid bone** and the **vomer** form the superior and inferior portions of the septum, respectively. They frequently articulate with each other in the posterior aspect of the nasal cavities but are widely separated anteriorly. The **septal cartilage** intervenes between the bony elements of the septum and is most extensive in the anterior part of the septum. The septal cartilage also forms the support for the midline ridge of the nose, the tip, and the **columella**, which separates the nares between the tip of the nose and the anterior nasal spine of the maxilla.

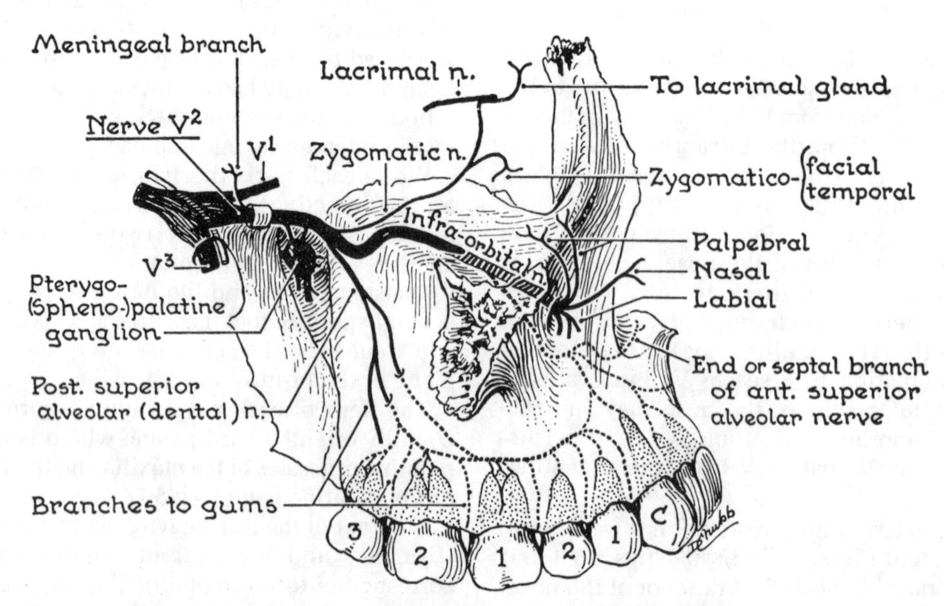

Figure 50.2. Distribution of the maxillary (V²) nerve. *C*, canine.

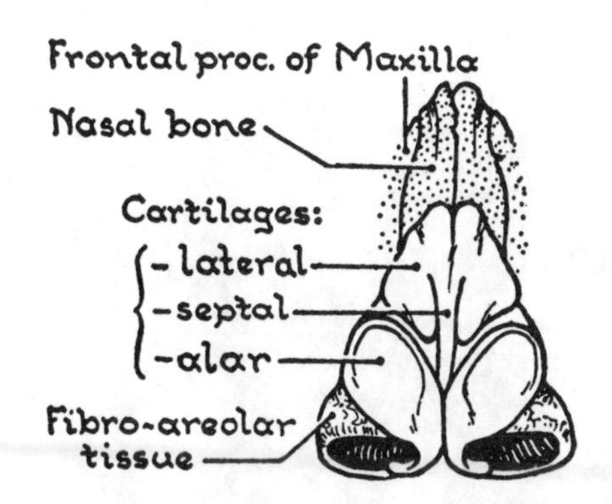

Figure 50.3 Framework of the external nose.

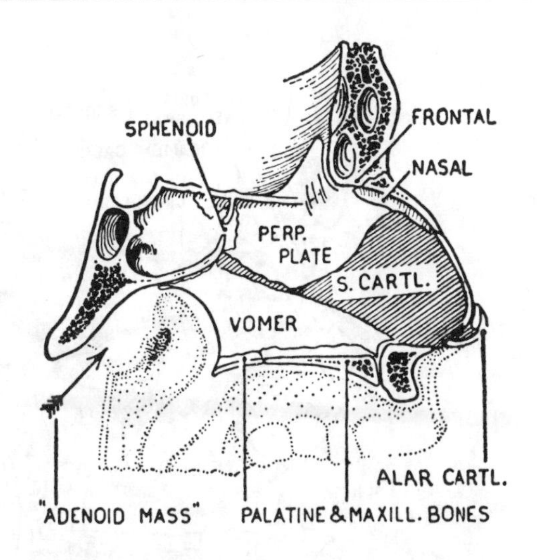

Figure 50.5. Bones and cartilages of nasal septum.

DEVIATED NASAL SEPTUM

The nasal septum is frequently not exactly in the midline and "deflects" or deviates to one side. This deflection usually occurs at the junction of the vomer and the septal cartilage. Extensive deviation may impair respiratory passage through one of the nasal cavities and require surgical correction.

The **lateral wall** of each nasal cavity presents an elaborate array of bone development, articulation, and interrelationship with the major facial bones in the upper face (*figs. 50.6 and 50.7*). A thorough understanding of the openings and mucous membrane relationships on the lateral wall of the nasal cavity is necessary for general clinical practice.

Three prominent inferiorly curved shelves, **inferior, middle, and superior conchae**, project from the lateral nasal wall into the nasal cavity (*figs. 50.6 and 50.7*). These conchae are scroll-like bones covered by a mucoperiosteum. The underlying spaces, which are inferior to each concha, are termed **meatus** (*fig. 50.7*). There is therefore an **inferior, middle**, and **superior meatus** on the lateral wall of each nasal cavity. In addition, the area on the lateral wall superior to the superior concha is termed the **spheno-ethmoid recess**.

The area between the conchae and the meatus and the nares is subdivided into two components. The **vestibule** is just inside each naris and subsequently communicates with the expanded **atrium** inferior to the nasal bones. The vestibule is lined by skin, which contains the nasal hairs, while the atrium is lined by glandular mucoperiosteum (*fig. 50.7*).

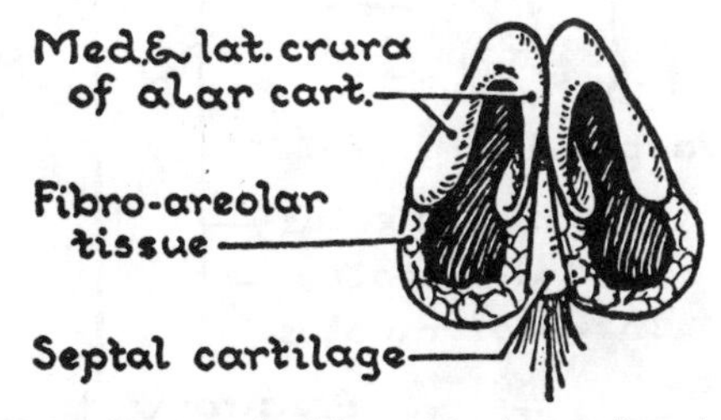

Figure 50.4. Framework of the external nose (from below).

The **inferior concha** is a separate bone that articulates with the maxilla, lacrimal, and palatine bones on the lateral wall of the nasal cavity. The underlying **inferior meatus** contains the opening for the **nasolacrimal duct**, which drains tears from the medial aspect of the orbit to the nasal cavity (*fig. 50.6B*). The posterior extent of the inferior concha and meatus is adjacent to the opening of the auditory tube into the nasopharynx. Examination of the ostium of the auditory tube may be done by observation with a nasal speculum placed in the naris and inferior aspect of the nasal cavity.

The **middle concha** is a process of the ethmoid bone. It overlies the middle meatus. Most of the paranasal air sinuses of the head open into this space and drain their seromucous secretions into the nasal cavity. Inflammation of the mucoperiosteum of the middle meatus in patients with upper respiratory infections and allergies can interfere with the normal drainage of these paranasal sinuses. *Figures 50.6B* and *50.8* illustrate the relationships of the middle meatus and the paranasal air sinuses. A crescent-shaped **hiatus semilunaris** opens on the wall of the middle meatus between the **unciform process of the ethmoid bone** and the **ethmoid bulla** (*fig. 50.6B*). Four major paranasal sinuses drain into the middle meatus through the openings that communicate with the hiatus semilunaris. The **frontal sinus** drains into the superior aspect of the hiatus semilunaris. The **anterior** and **middle ethmoidal air sinus** drains through openings on the ethmoid bulla on the superoposterior aspect of the hiatus semilunaris. The **maxillary air sinus** has its ostium directly inferior to the ethmoid bulla within the hiatus semilunaris.

The **superior concha** is also a process of the ethmoid bone. It overlies the **superior meatus**, which receives the opening of the **posterior ethmoidal air cells**. Superior to the superior concha is the **spheno-ethmoidal recess** where the **sphenoid air cells** drain into the nasal cavity (*figs. 50.6B* and *50.7*).

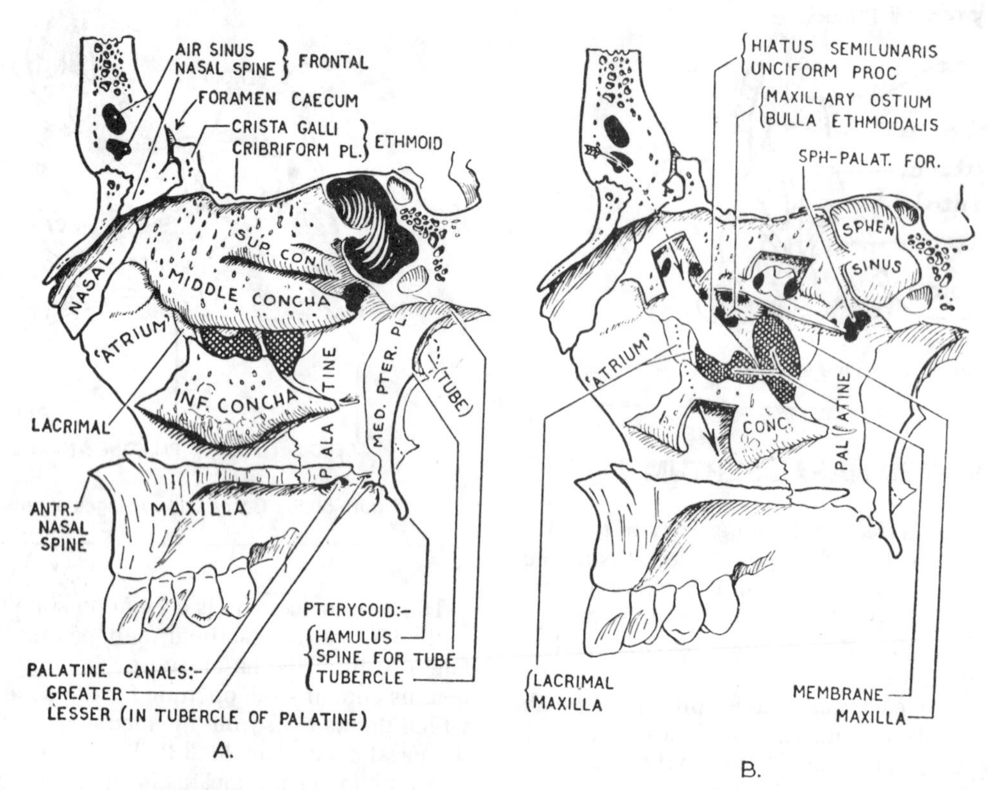

Figure 50.6. *A*, bones of lateral wall of nasal cavity; *B*, orifices of air sinuses and nasolacrimal canal, revealed by cutting away parts of the conchae. (*Arrows* lead from frontal sinus and nasolacrimal canal.)

NERVE SUPPLY TO THE NASAL CAVITY (*fig. 50.9*)

The mucosa of the nasal cavity is innervated by the **olfactory (I)** and the **trigeminal (V) nerves**. The olfactory nerve is a special sensory nerve that innervates the mucosa in the roof of the nasal cavity. Olfactory chemoreceptors detect odoriferous stimuli and transmit these sensations by way of the olfactory nerves through the cribriform plate of the ethmoid bone to the overlying olfactory bulbs. Olfactory pathways from the olfactory bulb to the brain pass through the olfactory tracts. Clinical testing of nerve **I** is difficult. Aromatic stimuli can be presented by each nasal cavity independently and identified by the patient. It is difficult to keep the stimuli confined in a unilateral manner and to keep receptors from being stimulated simultaneously. Taste and smell sensations are closely integrated and difficult to isolate in the head and neck examination.

Figure 50.9 also illustrates the route that the **anterior ethmoidal nerve (V^1)** takes to reach the mucosa of the atrium and the skin at the tip of the nose. These are sensory fibers that carry general sensation (pain, temperature, touch, and pressure). Most of the general sensation to the lateral wall and medial septum of the nose is carried in V^2. **V^2** (the maxillary division of V) leaves the trigeminal ganglion in the middle cranial fossa and enters the pterygopalatine fossa by way of the **foramen rotundum**. Within the pterygopalatine fossa, V^2 is associated with the pterygopalatine ganglion (*figs. 50.2 and 50.9*). This ganglion is the site of synapse between the preganglionic parasympathetic fibers in the **greater (su-**

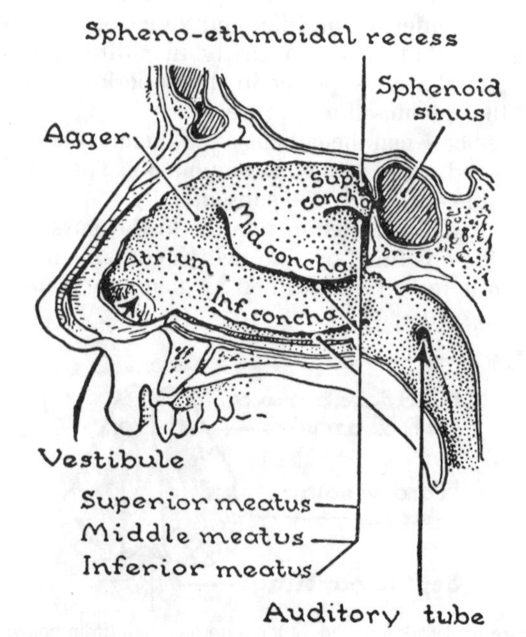

Figure 50.7. The lateral wall of the nasal cavity.

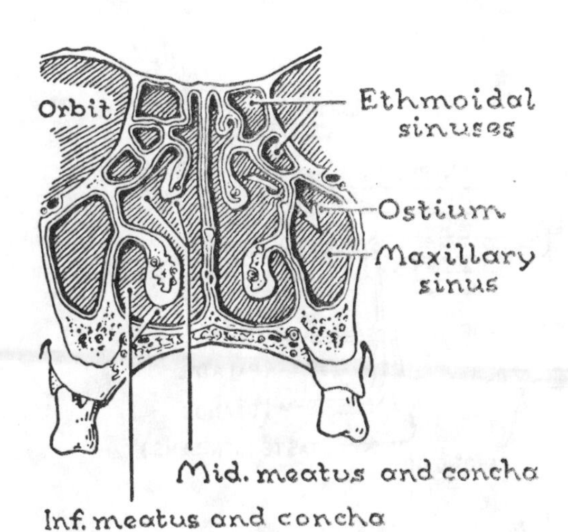

Figure 50.8. The nasal cavities and adjacent air sinuses, on coronal section.

perficial) petrosal nerve (VII) and the postganglionic neurons for the secretomotor fibers to glands above the floor of the mouth. The postganglionic fibers that leave the ganglion are then distributed to the glands of the nasal mucosa with the general sensory fibers of V^2. Terminal branches of the **infraorbital nerve** (*fig. 50.2*) enter the vestibule of the nose from the skin covering the nares. A small branch from the anterior superior alveolar nerve also innervates the anterior nasal mucosa of the inferior meatus. Most of the mucosa on the lateral nasal wall is innervated by branches of the descending **greater palatine nerve** that arises from the inferior pole of the pterygopalatine ganglion (*fig. 50.9*). **Superior** and **inferior posterior lateral nasal nerves** run in the mucoperiosteum covering the conchae and meatus.

The nasal septum mucosa is innervated by the nasopalatine nerve. It enters the nasal cavity from the pterygopalatine fossa through the **sphenopalatine foramen** (*fig. 50.6B*) and descends on the median nasal septum. The terminal branch of the nasopalatine nerve leaves the nasal cavity via the incisive foramen (*fig. 50.9*) to innervate the mucosa of the anterior one-third of the hard palate. Taste receptors on the palate parallel to the course of the general sensory fibers in the branches of V^2. The taste fibers, like the secretomotor fibers to the palatal and nasal glands, are branches of nerve **VII**. The taste fibers have their cell bodies in the geniculate ganglion of nerve VII within the petrous temporal bone (*fig. 50.10*).

Sympathetic innervation of the nasal cavity is from the branches of the superior cervical ganglion. These postganglionic sympathetic fibers reach the nasal cavity via the **nerve of the internal carotid artery** and the **deep petrosal nerve** of the pterygoid canal. Within the pterygopalatine fossa, the sympathetics join the terminal branches of the maxillary artery and control vasomotor "tone" in the blood vessels within the nasal cavity and palate.

BLOOD SUPPLY TO THE NASAL CAVITY

The **sphenopalatine artery** is a branch of the maxillary artery within the pterygopalatine fossa. It enters the nasal cavity with the nasopalatine nerve by passing through the sphenopalatine foramen (*fig. 50.6*). It supplies the upper two-thirds of the nasal septum and then anastomoses with the **greater palatine artery** that ascends through the incisive foramen to enter the nasal cavity on the inferior aspect of the nasal septum. The site of anastomosis of these two principal arteries is a common area of hemorrhage (epistaxis or nosebleed). The lateral

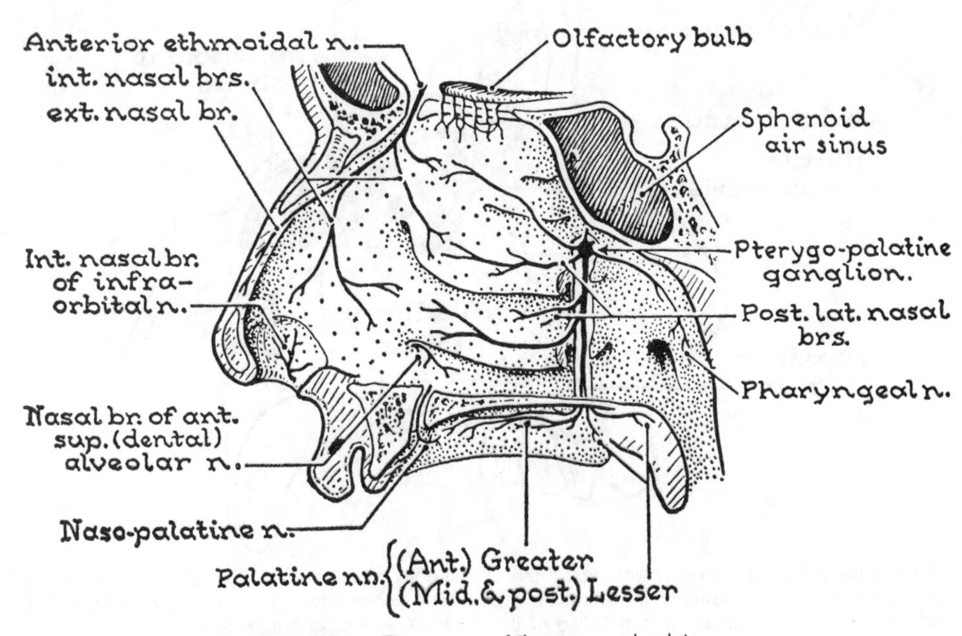

Figure 50.9. The nerves of the nose and palate.

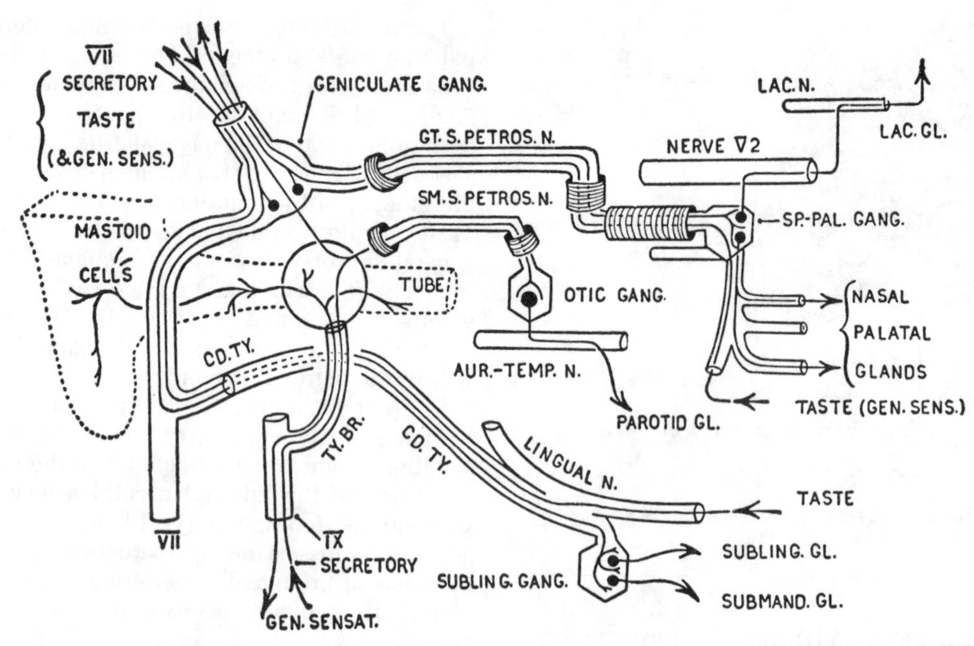

Figure 50.10. Parasympathetic ganglia on the branches of nerves VII and IX: pterygo(spheno)palatine, otic, and submandibular (sublingual). *CD TY*, Chorda tympani; *TY BR*, tympanic branch; *SP-PAL*, sphenopalatine.

walls of the nasal cavity are supplied by blood vessels that accompany the terminal branches of the anterior ethmoid nerve and the greater palatine nerve. The arteries of the lateral nasal wall have the same name as the accompanying sensory nerves.

Venous drainage of the nose parallels the arterial supply. The venous network of the nose forms a distensible cavernous tissue that principally overlies the inferior and middle concha. This tissue functions to humidify and warm the inspired air in the upper respiratory passage.

CLINICAL NOTES

Anesthesia of the nasal cavity can be readily effected by spraying a topical anesthetic on the nasal mucosa. V^2

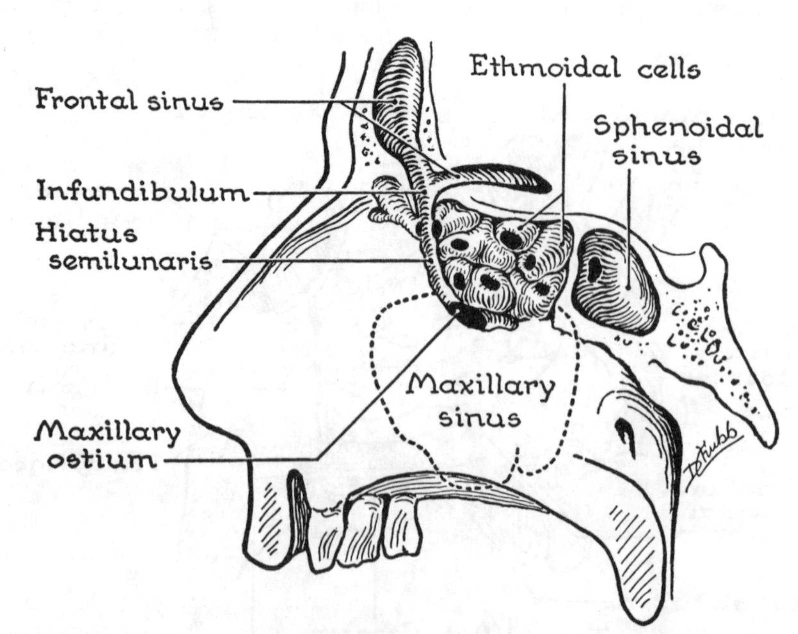

Figure 50.11. Diagram of the paranasal sinuses. Note that if there were fluid in the frontal sinus, it would trickle down the infundibulum, along the hiatus, and into the maxillary sinus, for the orifice of the frontal sinus is at its floor and that of the maxillary sinus at its roof. The sphenoidal ostium is in the upper half of its anterior wall ethmoidal ostia are variable.

nerve blocks via the greater palatine canal are also utilized to anesthetize the mucosa.

Hemorrhage control can be effected by packing the nasal cavity with a pressure gauze impregnated with a vasoconstrictor agent (epinephrine). Uncontrollable hemorrhages may require ligation of the sphenopalatine or greater palatine artery.

Paranasal Air Sinuses

The paranasal sinuses (**sphenoid, ethmoid, frontal**, and **maxillary**) are paired but asymmetrical (*figs. 50.8 and 50.11*). They are lined by respiratory epithelium, and they all drain into the nasal cavity. Chronic respiratory irritation, such as smoking, causes the mucosa to transform into a stratified squamous type of epithelium. The loss of the cilia can affect the drainage of these air sinuses and create conditions of chronic sinusitis.

SPHENOID SINUSES

The two sphenoid air sinuses are within the body of the sphenoid bone and are separated from each other by a midline bony partition (*fig. 50.12*). The sphenoid sinus drains through an ostium in its anterior wall into the spheno-ethmoidal recess of the nasal cavity. The **pituitary fossa** of the sphenoid bone is in the roof of the sphenoid sinuses and the pituitary gland is separated from the sinuses by the bony floor of the fossa. Surgical approaches to the pituitary are made through the nasal cavities and the sphenoid air sinuses.

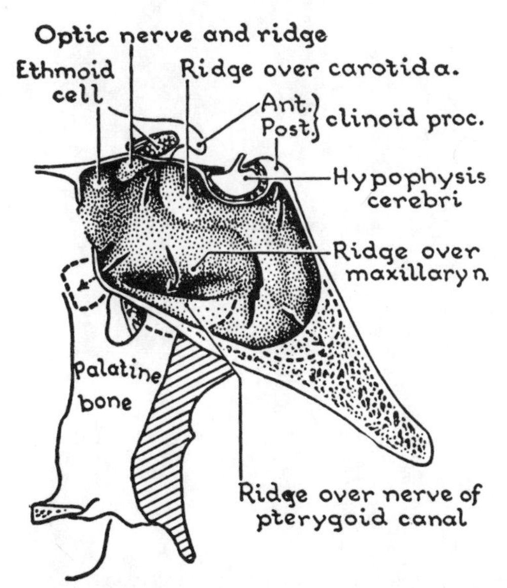

Figure 50.12. A large sphenoidal sinus with many diverticula. Note its important relationships.

ETHMOIDAL SINUSES (*figs. 50.8 and 50.11*)

The ethmoidal sinuses are likened to an oblong mass of 8–10 inflated balloons. They are divided into anterior, middle, and posterior ethmoid sinuses. The anterior and middle ethmoid air cells drain into the middle meatus of the nasal cavity through openings on the **ethmoid bulla**. The posterior ethmoidal air cells drain into the superior meatus.

FRONTAL SINUSES (*fig. 50.11*)

These sinuses invade the frontal bone above the superior margins of the orbits. The frontal sinus drains via its funnel-shaped **infundibulum** into the superior extension of the hiatus semilunaris.

MAXILLARY SINUS (ANTRUM) (*figs. 50.8 and 50.11*)

This large air sinus is within the maxilla posterior to the cuspid tooth region. Its floor is at the level of the hard palate, and the ostium of the maxillary sinus is near its roof on the medial wall. It drains into the middle meatus through the hiatus semilunaris. Because the ostium is highly positioned in this sinus, the maxillary sinus is prone to drainage disorders, particularly if the ciliated mucosa of the sinus is destroyed in states of chronic sinusitis. In persistent states of maxillary sinus congestion, a needle can be inserted through the thin bone below the inferior concha, and a saline solution can be injected to "flush" the sinus.

The floor of the maxillary sinus is closely associated with the roots of the superior bicuspid and molar teeth. Periapical dental infections may perforate the thin floor of the sinus and induce sinus infection. The nerve supply to the sinus mucosa is the same that supplies these teeth, the **posterior and middle alveolar nerves** (V^2).

The **mastoid air sinuses** are also similar to the paranasal sinuses, but they drain into the nasal region via the middle ear and auditory tube. All of the air sinuses function, in part, to lighten the bony masses that form the upper face. They develop as the facial bones expand to accompany the growth that occurs with the eruption of the permanent dentition. Most of the sinus growth is therefore seen during the years from 6 to 12.

Clinical Mini-Problems

1. **A patient had been to his dentist to have his two superior central incisors prepared for "porcelain**

caps." The dentist anesthetized both infraorbital nerves. The upper incisor teeth, labial gingiva, upper lip and lateral sides of the nose were numbed. Why did the patient still have sensation at the tip of his nose?

2. **What anatomical structures would guide you in finding the opening of the maxillary sinus in the nasal cavity.**
3. **Where does the nasolacrimal duct open in the nasal cavity?**
4. **Which nerve fibers synapse in the pterygopalatine ganglion?**
5. **Where are the cell bodies for the nerves that innervate the taste buds on the palate?**
6. **Why is the maxillary sinus in a smoker more susceptible to sinusitis?**

(Answers to these questions can be found on p. 588.)

51

Larynx

Clinical Case 51.1

Patient Florence T. This 40-year-old woman had her recurrent laryngeal nerves injured bilaterally during a thyroidectomy for "goiter" a year ago. When the anesthesiologist removed the nasotracheal tube following the surgery, she showed considerable respiratory distress. The surgeon therefore performed a tracheostomy and placed a endotracheal tube with an external opening in the midline of the lower neck. She was sent home after three days to convalesce, with no strenuous work allowed. She was scheduled for weekly visits to assess the regeneration of the damaged recurrent laryngeal nerves, but after one year both vocal cords remain paralyzed, and the patient is unable to abduct the vocal cords to expand the rima glottidis, as the otolaryngologist demonstrates to your class. He describes how he will create an "opened airway" through the trachea by suturing one of the paralyzed vocal cords to the lateral wall of the larynx. This should allow removal of the endotracheal tube and closure of the tracheostomy. The patient will have an adequate but diminished airway and a hoarse voice, which is expected to remain as a permanent condition.

The larynx is an important component of the respiratory passageway. It lies in the neck between the levels of the 4th and 6th cervical vertebrae, and it intervenes between the pharynx and the trachea. The larynx has a cartilaginous framework to maintain its patency (openness), but it also has an adjustable internal valve, the vocal cords, which can regulate the flow of air through the respiratory pathway within the larynx. The larynx functions mainly as an air passage and as a voice production organ.

External Structures

The **cricoid cartilage** (Gr. = like a ring) is a complete ring of cartilage at the lower level of the larynx (C6). It has a broad, flattened posterior component called the **lamina**. The anterior **arch** of the cricoid is much narrower, and it is attached superiorly to the inferior border of the thyroid cartilage by the **cricothyroid membrane**. The thickened anterior midline portion of this membrane is called the **cricothyroid ligament** (*fig. 51.1*).

The **thyroid cartilage** (Gr. = like a shield) consists of two quandrangular plates, the **right** and **left laminae**. These two laminae fuse anteriorly in the midline. The angle of this fusion is greater in females and prepubescent males. This angle becomes more acute in males after puberty creating a more defined **laryngeal prominence** (Adam's Apple) and a subsequent voice change (*fig. 51.2*). The posterior border of each lamina is prolonged into an **upper** and **lower horn** (*fig. 51.1*). An **oblique line** on the lateral aspect of the lamina is the attachment site for 3 muscles: **sternothyroid, thyrohyoid**, and **inferior constrictor**.

The upper border of the thyroid cartilage is attached to the hyoid bone by the **thyrohyoid membrane**. The membrane is pierced by the **internal laryngeal nerve (X)** and the **superior laryngeal artery**, a branch of the superior thyroid artery.

THE CRICOTHYROID JOINT AND MUSCLE

A facet on the tip of the inferior horn of the thyroid cartilage articulates with a corresponding facet on the lateral aspect of the cricoid cartilage (*fig. 51.1*). The **cricothyroid muscle** (*fig. 51.3*) arises from the upper border of the cricoid arch and inserts into the lower border of the thyroid lamina and inferior horn. Contraction of this muscle causes an anterior "tilting" of the thyroid cartilage, which results in a tensing of the vocal cords as

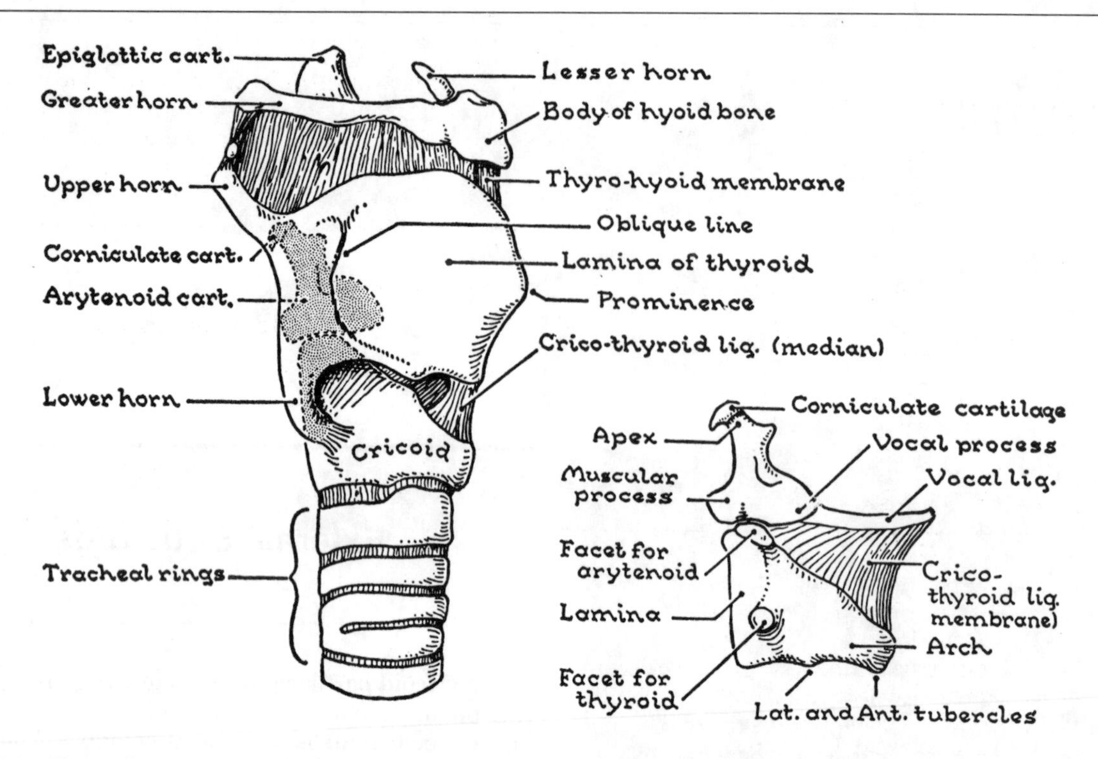

Figure 51.1. The cartilages of the larynx (side view). Three ligaments—median cricothyroid, cricothyroid, and vocal—constitute the conus elasticus.

illustrated in *Figure 51.3*. The cricothyroid muscle is innervated by the **external laryngeal nerve** (X), which branches from the superior laryngeal nerve in the carotid sheath of the upper neck.

Internal Framework

The paired **arytenoid cartilages** lie medial to the thyroid lamina and superior to the cricoid lamina (*fig. 51.1*). They are the most moveable of the laryngeal cartilages and serve as the principal mechanical basis for controlling the air passage between the vocal cords, the **rima glottidis** (*see fig. 51.9*). These pyramid-shaped cartilages have a **base** that articulates with the superior border of the cricoid lamina (*fig. 51.1*). Projecting anteriorly from the base of the arytenoid cartilage is a sharp **vocal process**, which attaches to the posterior aspect of the vocal ligament. A blunted **muscular process** projects laterally from the posterolateral aspect of the base of the arytenoid cartilage. This serves as a major insertion site for many of the intrinsic muscles of the larynx. The **apex** of the arytenoid cartilage projects superiorly and receives the aryepiglottic folds that border the superior laryngeal aperture. The small **corniculate cartilage** acts as an extension of the apex of the arytenoid cartilage within the aryepiglottic folds (*fig. 51.4*). A second small cartilage,

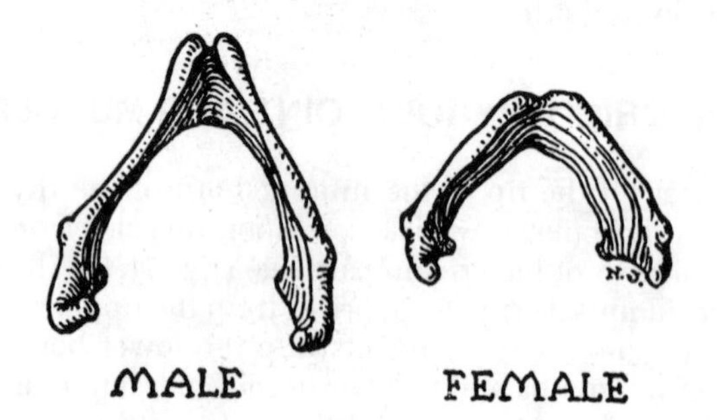

Figure 51.2. The angle at which the laminae of the thyroid cartilage meet varies with sex.

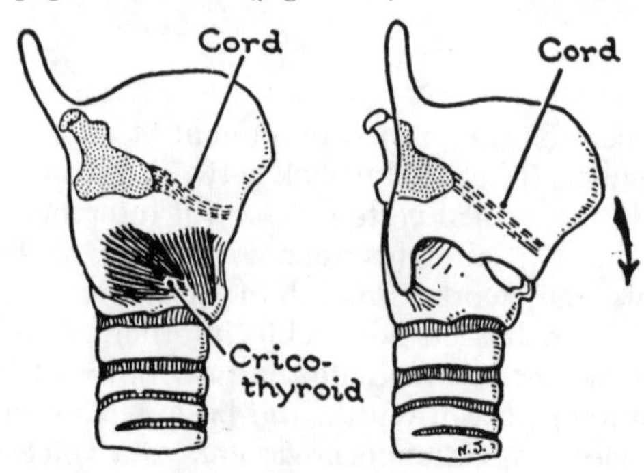

Figure 51.3. The cricothyroids render tense the vocal cords.

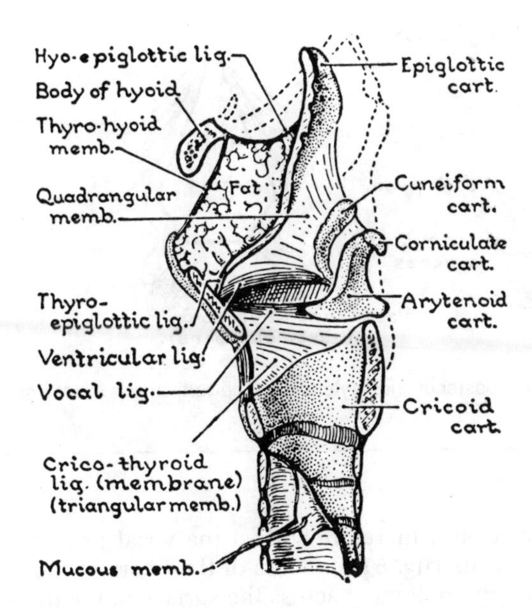

Figure 51.4. The cartilaginous and membranous skeleton of the larynx (on median section). (Cricothyroid lig. or membrane = cricovocal membrane = ½ conus elasticus.)

the **cuneiform cartilage** is also present in the aryepiglottic fold anterior to the arytenoid cartilage (*fig. 51.4*).

VOCAL AND CRICOTHYROID LIGAMENTS

The vocal ligament attaches to the vocal process of the arytenoid cartilage and runs anteriorly to attach to the internal aspect of the thyroid cartilage near its fusion angle. The vocal ligament is actually the anterior portion of the superior border of the elastic **cricothyroid membrane (conus elasticus)** (*figs. 51.1 and 51.4*). This is a triangular membrane that arises from the superior border of the lateral and anterior aspects of the cricoid cartilage. The midline portion of the cricothyroid membrane, **the cricothyroid ligament**, is sometimes penetrated in a tracheostomy to create an airway below the level of the vocal ligament. Damage to the cricothyroid membrane can cause changes in the vocal ligaments, which result in changes in the vocal quality of the voice.

The **epiglottic cartilage** is a single, leaf-shaped midline cartilage that borders the anterior part of the superior laryngeal aperture (*figs. 51.4 and 51.5*). Its inferior "stem" attaches by way of the thyro-epiglottic ligament to the angle of the thyroid laminae above the level of the vocal ligaments (*fig. 51.4*). The rounded superior border of the epiglottis rises above the level of the hyoid bone to approximate the surface of the posterior one-third of the tongue. A fibroelastic sheet, the **quadrangular membrane** attaches to the lateral borders of the epiglottis anteriorly and the arytenoid cartilages posteriorly (*fig. 51.4*). The free superior border of the quadrangular ligament is the **aryepiglottic ligament**, and the free inferior border of the quadrangular ligament forms the **vestibular ligament**. The aryepiglottic ligament is contained within the aryepiglottic fold of mucous membrane. The vestibular ligament is contained within the "false vocal cord."

Intrinsic Muscles

The intrinsic muscles of the larynx lie under the mucosa that covers the internal aspects of the larynx and they are all innervated by the recurrent laryngeal nerves (X). The intrinsic muscles of the larynx are positioned on the posterior and lateral aspects of the larynx (*see figs. 51.5 and 51.7*).

POSTERIOR MUSCLES (*fig. 51.5*)

Two prominent intrinsic muscles are evident from the posterior aspect.

The **posterior cricoarytenoid** is the most important muscle in the larynx to understand. It is the principal **abductor** of the vocal cords and is responsible for opening the airway during respiration (*fig. 51.6*). All other intrinsic muscles of the larynx cause closure of the **rima glottidis**, the opening between the vocal cords. The posterior cricoarytenoid arise from the posterior surface of the lamina of the cricoid cartilage and insert into the posterior

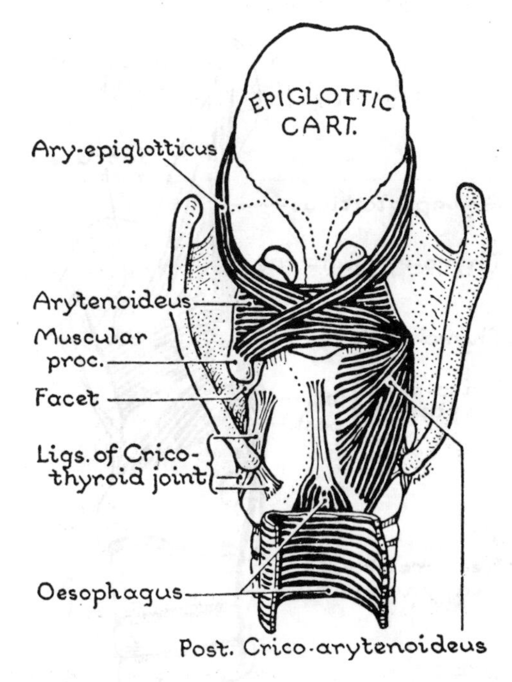

Figure 51.5. Intrinsic muscles of larynx (posterior view).

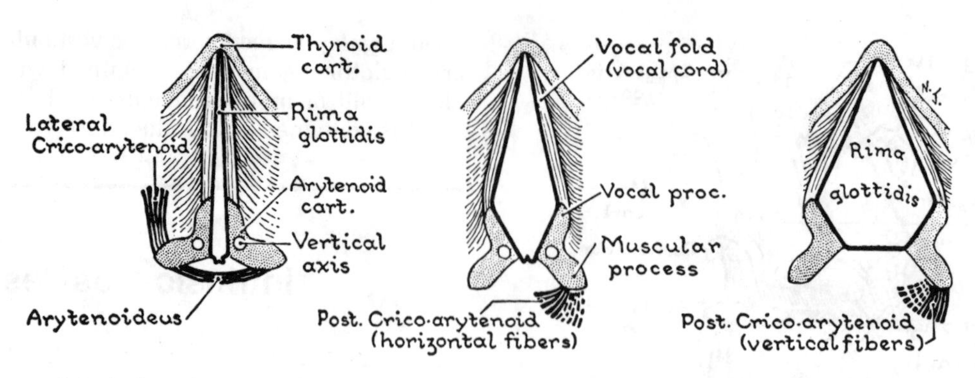

Figure 51.6. Scheme of glottis, from above, to explain the actions of the lateral and posterior cricoarytenoids. (The vertical axis is not a stationary one—see text.)

aspect of the muscular process of the arytenoid cartilage. The posterior "pull" on the muscular process of the arytenoid cartilage causes a lateral rotation of the anteriorly projecting vocal process of the arytenoid. Since the vocal cords are attached to the vocal process, this action by the posterior cricoarytenoid muscles "opens" the rima glottidis. Damage to the recurrent laryngeal nerve can result in paralysis of the posterior cricoarytenoid. Bilateral paralysis of the posterior cricoarytenoid muscles is a respiratory emergency and may require a **tracheostomy** to prevent suffocation in the affected patient.

The **arytenoideus** is a transverse muscle that attaches to the posterior aspects of the two arytenoid cartilages. Contraction of the arytenoideus approximates the two arytenoids by opposing their medial aspects. This muscle action would therefore adduct the vocal processes and vocal cords (*fig. 51.6*). Some of the fibers of the arytenoideus run obliquely across the surface of the transverse component of the muscle and form the **aryepiglotticus** that is in the aryepiglottic fold (*fig. 51.5*).

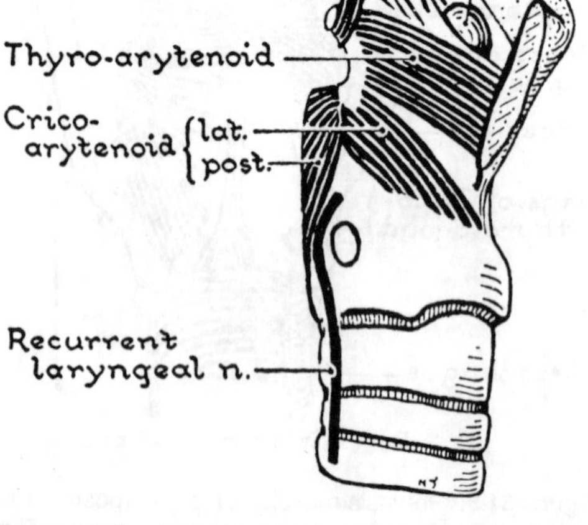

Figure 51.7. Intrinsic muscles of larynx (side view).

LATERAL MUSCLES (*fig. 51.7*)

All of these 5 muscles attach to the anterior aspect of the arytenoid cartilages. Tension in these muscles will cause the vocal processes to approximate each other and therefore adduct the vocal cords.

1. The **lateral cricoarytenoid muscle** arises from the superior aspect of the arch of the cricoid and inserts onto the anterior surface of the muscular process.
2. The **thyroarytenoideus** arises from the internal surface of the angle of the thyroid cartilage and inserts into the lateral aspect of the arytenoid cartilage.
3. The portion of the thyroarytenoideus that is adjacent to the vocal ligament is the **vocalis muscle**. The vocalis muscle is inserted onto the vocalis process of the arytenoid cartilage (von Leden and Moore).

Two muscles are associated with the quadrangular ligament above the level of the vocal cords.

4. The **aryepiglotticus** lies on the superior border of the quadrangular ligament within the aryepiglottic fold.
5. The **thyroepiglotticus** lies on the medial surface of the quadrangular ligament. It is a wisp-like muscle and not as evident as those associated with the arytenoid cartilages.

MUSCLE ACTIONS AND USES (*fig. 51.6*)

Respiration

The horizontal fibers of the posterior cricoarytenoids will abduct the vocal processes and open the rima glottidis. The vertical fibers of the posterior cricoarytenoid and the lateral cricoarytenoid will pull the arytenoid cartilages laterally to further open the rima glottidis.

Vocalization

The vocal cords are approximated to produce vocal sounds. A proper tension and a narrow airway opening in the rima glottidis is required to produce a normal voice. Paralysis of one of the posterior cricoarytenoids causes a "whisper" quality to the voice. Tension is affected by the vocalis muscle and the cricothyroid muscle as it tilts the thyroid cartilage (*fig. 51.3*).

Sphincter Function

Inspired air can be retained in the lungs by fully adducting the vocal cords and arytenoid cartilages. This function is used to increase pressure in the abdominal cavity during defecation or parturition (childbirth). The sphincter function also occurs in the "coughing reflex" to increase the force of expiration and cause a removal of the stimuli that initiates this reflex.

Swallowing

The aryepiglottic thyroarytenoid, and thyroepiglottic muscles close the vestibule, tilt the arytenoid cartilages anteriorly, and assist in depressing the epiglottis as the larynx is elevated during swallowing. This action functions to close the superior laryngeal aperture and divert the swallowed bolus into the piriform recesses that are lateral to the larynx in the laryngopharynx.

Interior of the Larynx

The **superior laryngeal aperture** or **aditus** is the superior entrance into the laryngeal cavity. The obliquely oriented aperture is bounded by the aryepiglottic folds and the mucosa on the posterior aspect of the epiglottis (*fig. 51.8*). The laryngeal cavity that exists between the

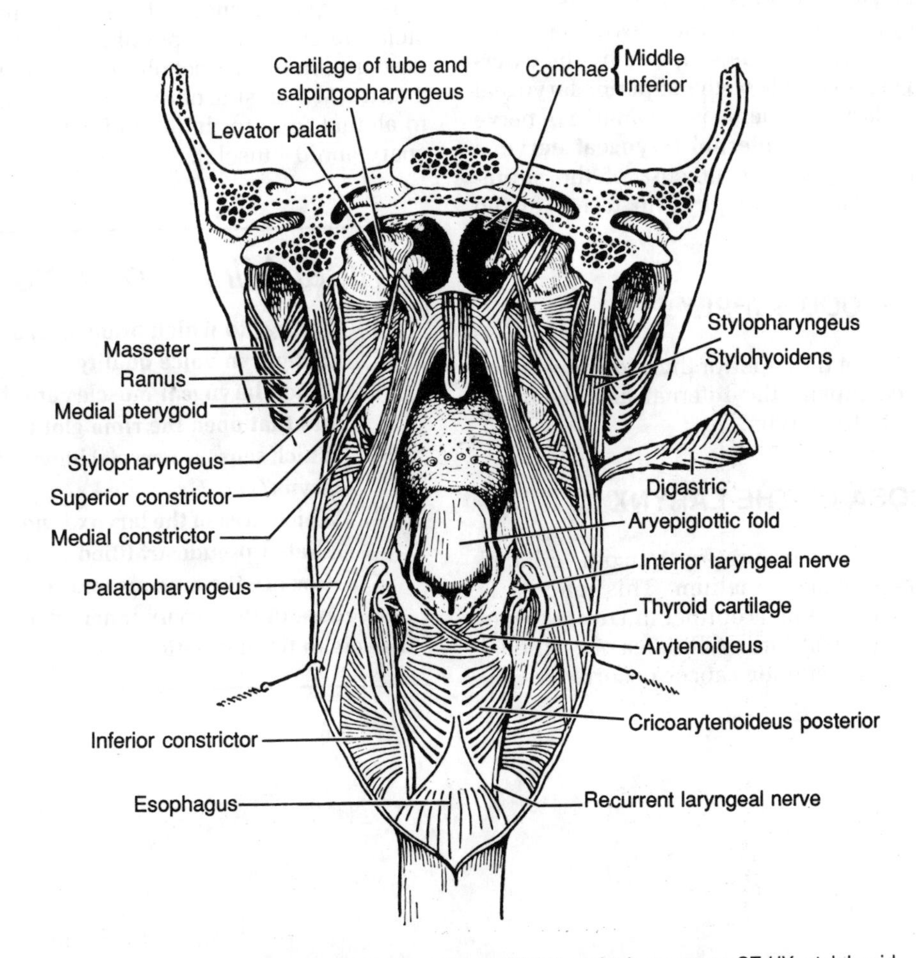

Figure 51.8. The muscles of the pharynx (from behind). *ST-PH*, stylopharyngeus; *ST-HY*, stylohyoid.

superior laryngeal aperture and the cricoid cartilage is divided into 3 parts. *Figure 51.9* depicts these divisions as they would appear in a coronal section of the larynx.

The **supraglottic cavity** is above the true vocal folds. There are two components to the supraglottic cavity, the **vestibule** and the **ventricles**. The **vestibule** is between the mucosa lining the two quadrangular membranes. It is limited superiorly by the aryepiglottic folds and inferiorly by the **vestibular folds** or "false vocal cords." Between the "true" and "false" vocal cords on each side of the larynx are diverticula called the **ventricles**. A cul-de-sac of mucous membrane, **the saccule**, extends superiorly from the anterior part of each ventricle. The saccule lies between the quadrangular membrane and the thyroarytenoid muscle (*fig. 51.7*). The saccule in man is vestigial and much larger in other primates.

The **infraglottic cavity** is that portion of the laryngeal cavity that is inferior to the vocal cords. It is continuous with the trachea through the cricoid cartilage lumen. The mucosa of the infraglottic cavity is innervated by sensory fibers in the **recurrent laryngeal nerves**. These nerves pierce the cricothyroid membrane to reach the mucosa in the internal laryngeal cavity. The laryngeal mucosa of the supraglottic cavity is innervated by the **internal laryngeal nerve**, which pierces the thyrohyoid membrane.

The **recurrent laryngeal nerves** are mixed sensory and motor nerves. They supply all of the muscles of the larynx except the cricothyroid, which is supplied by the **external laryngeal nerve**, a branch of the superior laryngeal nerve. The superior laryngeal nerve is the branch of nerve **X** that also gives rise to the **internal laryngeal nerve**, a sensory nerve that supplies the mucosa of the supraglottic cavity of the larynx and piriform recess of the pharynx.

BLOOD SUPPLY

Laryngeal branches of the superior thyroid and inferior thyroid arteries accompany the internal and recurrent laryngeal nerves into the larynx.

MUCOSA OF THE LARYNX

The larynx is lined by respiratory mucosa, ciliated pseudostratified columnar epithelium. This mucosa is converted to stratified squamous epithelium over the medial aspect of the true vocal cords. This area of epithelial transition is a potential site for cancer occurrence. Examination of the laryngeal cavity and vocal cords can be readily done by direct observation with a mirror placed in the oropharynx or with a laryngoscope.

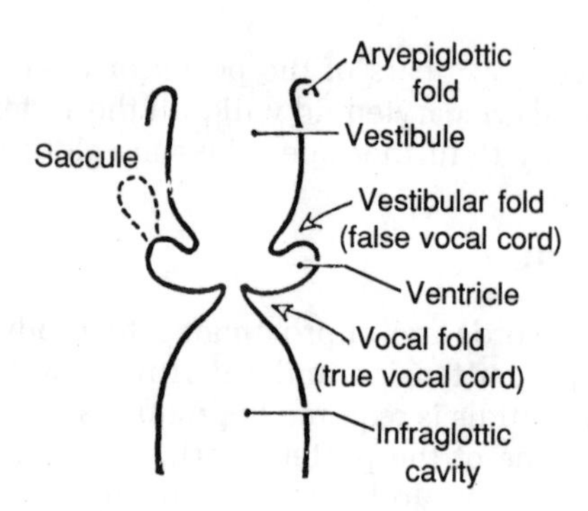

Figure 51.9. The 3 parts of the laryngeal cavity (coronal section).

TESTING

The quality of vocal tone is the most obvious sign in patients with laryngeal paralysis. Speaking with a whisper-like quality should be an obvious clue for the physician to suspect the possibility of a damaged recurrent laryngeal nerve. Direct observation of the vocal cords can confirm which side of the larynx is paralyzed and unable to abduct (due to the loss of action in the posterior cricoarytenoid muscle).

Clinical Mini-Problems

1. **Damage to which branch(es) of nerve X will cause changes in voice quality?**
2. **Which laryngeal muscles are the principal abductors that open the rima glottidis during breathing?**
3. **Which sensory nerve(s) innervate the mucosa of the larynx?**
4. **Which area of the larynx is *not* lined by respiratory ciliated pseudostratified columnar epithelium?**
5. **How are the mucous and serous glands in the mucosa of the larynx innervated?**

(Answers to these questions can be found on p. 589.)

52

Lymphatics of the Head and Neck

Clinical Case 52.1

Patient Sandra R. This 35-year-old office worker awoke one morning with a painful and tender lump in the submental triangle just deep to the skin. She waited for 3 days and then went to her physician because she was developing an additional painful and enlarged swelling at the "angle of her jaw" and in her neck. The physician examined the patient and concluded provisionally that the lumps were inflamed lymph nodes and that an infection was responsible for the stimulation of the lymph nodes. However, no open lesions or infective areas were apparent in an oral examination. He referred the patient to a dentist. The x-rays indicated that an abcess was present at the root tip of one of her lower central incisors. Root canal therapy was done to remove the infected pulpal tissue in the tooth and the patient was placed on antibiotic therapy for 5 days. The inflamed and enlarged lymph nodes were reduced in size and nonpainful 1 week after the antibiotic therapy ended.

The lymphatics of the head and neck are distributed in a systematic fashion just like the lymphatics in the rest of the body. The lymphatics that drain superficial structures of the head and neck will follow the venous channels that drain these structures. The lymphatics that drain the deeper structures within the head and neck will be associated with the arteries that supply these deep structures. All of the lymphatic channels in the head and neck will pass through a lymph node before draining into the junction of the jugular and subclavian veins in the root of the neck.

Lymph Nodes

The **main chain of lymph nodes** of the head and neck, called the *deep cervical nodes*, extends along the internal jugular vein from the base of the skull above to the clavicle below (*fig. 52.1*). Here it forms a *jugular lymph trunk*, which either opens independently into the angle between the internal jugular and subclavian veins or else joins the thoracic duct on the left side (right lymph duct on the right). Though the deep cervical nodes are largely covered by the obliquely set sternomastoid, a few of them spread forward into the upper part of the anterior triangle, and many spread backward beyond the posterior border of the sternomastoid into the posterior triangle.

All these nodes lie superficial to the prevertebral fascia and the roots of the cervical and brachial plexuses. The inferior belly of the omohyoid subdivides them into an upper and lower group. A few nodes of the *upper group*, which extend medially behind the nasopharynx, are called the *retropharyngeal nodes*. Their afferents come from the nasopharynx, soft palate, middle ear, and auditory tube.

Two nodes are especially to be noted (*fig. 52.2*): (1) the *jugulodigastric node*, which lies below the posterior belly of the digastric where the common facial vein enters the internal jugular; and (2) the *jugulo-omohyoid*, which lies above the inferior belly of the omohyoid where it crosses the internal jugular. The accessory nerve is surrounded by nodes both where it enters the sternomastoid and where it leaves it. The upper group of nodes drains into the lower group. The *lower group* (supraclavicular nodes) communicates with the nodes of the axilla and with the lymph vessels of the mamma.

All parts of the head and neck drain through the deep cervical chain. The chain has a few forward outposts in the neck, e.g., *infrahyoid* (on the thyrohyoid membrane), whose afferents follow the superior laryngeal artery and come from the larynx above the vocal cords; *prelaryngeal* (on the cricothyroid lig.) and *paratracheal* (in the groove between the trachea and esophagus), which follow the

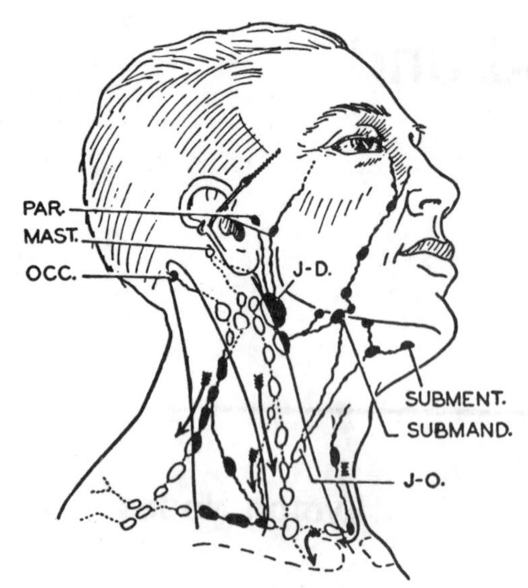

Figure 52.1. Lymphatics of head and neck. *J-D*, jugulo-digastric; *J-O*, jugulo-omohyoid. (Rouviere.)

inferior thyroid artery. These receive afferents from the larynx below the vocal cords and from the thyroid gland and adjacent parts.

A **horizontal series** of superficial nodes surrounds the junction of head and neck. These nodes are placed on the stem of named blood vessels, and they receive afferents from corresponding territories.

Thus, one or two *occipital nodes* lie on the trapezius where it is pierced by the occipital artery, 2 or 3 cm inferolateral to the inion. Their afferents are from the scalp; their efferents pass deep to the posterior border of the sternomastoid. They are notable for being palpable in German measles.

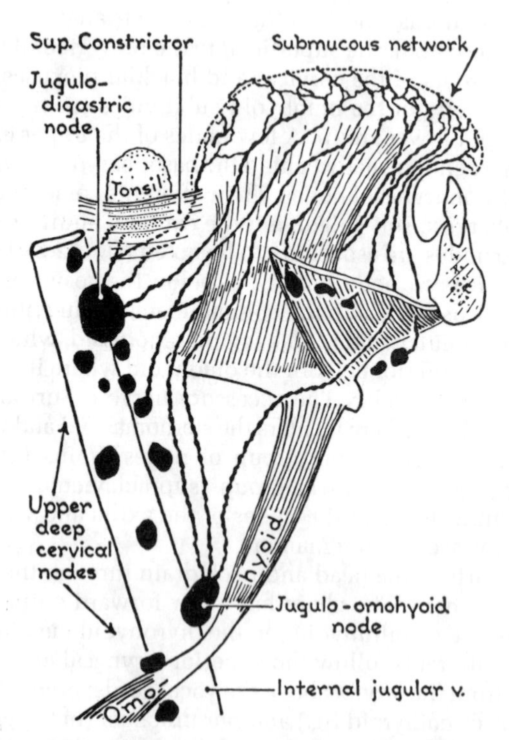

Figure 52.2. Lymphatics of tongue and tonsil. (After Jamieson and Dobson.)

One of two *retroauricular nodes* (mastoid nodes) lies on the mastoid with the posterior auricular artery. Their afferents come from the scalp and auricle; efferents pass to the deep cervical nodes.

Several *superficial parotid nodes* (preauricular nodes) lie superficial to the parotid fascia near the superficial temporal and transverse facial arteries. Their afferents come from the scalp, auricle, eyelids, and cheek. Their efferents pass to the deep parotid and superficial cervical nodes.

[The *deep parotid nodes* may be conveniently described now. Embedded in the parotid salivary gland, they receive afferents from the superficial parotid nodes and from the external acoustic meatus, tympanum, deep parts of the cheek, soft palate, and posterior part of the nasal cavity. Their efferents pass to the deep cervical nodes.]

The *superficial cervical nodes* are small and are placed beside the external jugular vein on the upper part of the sternomastoid. They are an offshoot of the superficial parotid nodes.

Half a dozen *submandibular nodes* lie on the surface of the submandibular salivary gland and also between it and the lower jaw, beside the facial artery. They have 2 extensions: (1) upward in the face along the course of the facial artery (the *facial nodes* are small and inconstant, except one or two at the lower border of the jaw); and (2) forward along the submental artery (the *submental nodes* lie on the mylohyoid below the symphysis menti). They receive afferents from the lower lip and chin and also from the tip of the tongue by vessels that pierce the mylohyoid in company with anastomotic branches of the sublingual artery. The efferents pass to the submandibular nodes and also to the jugulo-omohyoid nodes.

The *submandibular nodes* receive afferents from their 2 extensions and from the face, cheek, nose, upper lip, gums, and tongue. The efferents pass to the upper deep cervical nodes. To examine these nodes the subject should be told to drop his chin in order to slacken the cervical fascia. One index finger should then be placed below the tongue, and the fingers of the other hand should be placed below the jaw, and the structures between them palpated.

Lymphatics

The **lymph vessels of the tongue** (*fig. 52.2*) spring from the extensive submucous plexus, and all vessels drain ultimately into the deep cervical nodes alongside the internal jugular vein, between the levels of the digastric and the omohyoid, the uppermost node being the *jugulodigastric nodes*, the lowest the *jugulo-omohyoid node* (*fig. 52.1*).

Nearer the tip of the tongue the vessels arise, the lower the recipient node and the farther back, the higher node (*fig. 52.2*).

Course. The vessels from the apex of the tongue pierce the mylohyoid and are mostly intercepted by the submental nodes. The marginal or lateral vessels of the anterior two-thirds partly pierce the mylohyoid to end in the submandibular nodes and partly follow the blood vessels across both surfaces of the hyoglossus to the deep cervical nodes. The medial vessels, however, descend in (or near) the septum, between the genioglossi and, after either piercing or passing below that muscle, follow the lingual artery to the deep cervical nodes. The vessels from the posterior one-third pass through the pharyngeal wall below the tonsil.

Crossing in part to nodes of the opposite side are vessels near the median plane and also vessels leaving the submental nodes (*fig. 52.3*).

The **lymph vessels of the tonsil** pierce or run below the superior constrictor mainly to the jugulodigastric node.

The **lymph vessels of the upper teeth** pass through the infraorbital foramen and run with the facial artery to the submandibular nodes. Those from the **lower teeth** run through the mandibular canal to the deep cervical nodes.

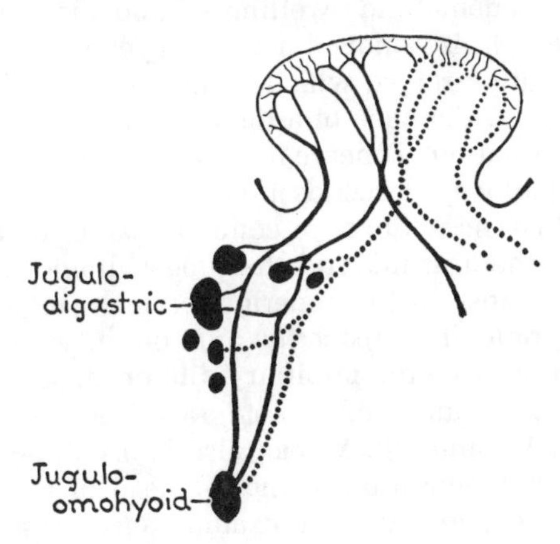

Figure 52.3. Lymphatics of the tongue decussating. (After Jamieson and Dobson.)

The vessels from the buccal surfaces of the **upper and lower gingiva (gums)** run to the submandibular nodes; those from the lingual part of the lower gums end in the submandibular and deep cervical nodes; those from the lingual part of the upper gums run dorsally with the palatine vessels to the deep cervical or retropharyngeal nodes.

The **lymph vessels of the larynx** *above the vocal cords* follow the superior laryngeal artery through the thyrohyoid membrane to the upper deep cervical nodes after partial interception by the infrahyoid nodes; those *below the vocal cords* pierce the cricothyroid and cricotracheal ligaments and pass to the deep cervical nodes after partial interception by the prelaryngeal and paratracheal nodes.

Vessels of the upper and lower parts of the larynx anastomose submucously in the posterior wall of the larynx but not in the region of the cords, which act as a barrier, comparable to the one erected at the pyloric sphincter.

THE LYMPH VESSELS OF THE EAR

Those of the auricle and external meatus pass to the retroauricular, upper cervical, and parotid nodes; those of the tympanic membrane and lateral wall of the tympanum pass to the parotid nodes; those of the auditory tube and medial wall of the tympanum pass to the retropharyngeal and deep cervical nodes.

The **lymph vessels of the nasal cavity** from the anterior part run with those of the external nose to the submandibular nodes; those from the posterior part pass to the retropharyngeal, deep parotid, and deep cervical nodes.

Clinical Mini-Problems

1. **In general, what kind of vascular channels are followed by lymphatics draining superficial structures?**
2. **In general, what kind of vascular channels are followed by lymphatics that drain deep structures?**

(Answers to these questions can be found on p. 589.)

53

The Normal Physical Examination of the Head and Neck

The physical examination of the head and neck is principally neurological but also includes the examination of the carotid arteries and the central artery of the retina, the lymphatics surrounding the internal jugular vein, the upper respiratory system, and the tympanic membrane. Basic understanding of head and neck anatomy is key to performing this physical examination.

Start from the top of the head, the orbital area, and descend to the neck. Use the cranial nerves I–XII as your guide, and assess the most obvious and relevant functions of these nerves. The intricate details of the neurological exam can be saved for the differential diagnosis of clinical disorders. The assessment for normality is simple, straightforward, and extremely valuable to the clinician.

The *olfactory nerve (I)* is usually not assessed in the general physical examination of the head and neck. It is closely associated with taste and difficult to isolate unilaterally. One therefore begins by examining the orbital region and its contents.

The Orbital Examination

Ask the patient to open the eyes widely and look at you. The *facial nerve (VII)* will create a bilateral wrinkling of the forehead due to the innervation of the **frontalis muscle**. The *oculomotor nerve (III)* will elevate the upper eyelid through its superior division branch to the **levator palpebrae superioris muscle**. If the patient can see you, the *optic nerve (II)* will be functioning to transmit the visual stimuli from the retina to the cortex.

Now ask the patient to move the eyes in the horizontal and vertical planes. Horizontal movement requires the *abducens nerve (VI)* and the *oculomotor nerve III* for lateral and medial movement of each eye. The oculomotor nerve (III) also elevates and depresses the eye by the action of the inferior oblique, superior rectus, and inferior rectus muscles. The *trochlear nerve (IV)* assists III in depressing the eye. If the eye can be moved fully in the horizontal and vertical planes, one can assume that nerves III, IV, and VI are functioning in a normal manner.

The final aspect of the orbital examination is done with an ophthalmoscope. A light is shined into the patient's eye, and one watches for pupillary constriction. This is a **parasympathetic** reflex that requires a sensitive **retina**, an intact nerve **II** for the afferent limb of the reflex to the brainstem and a normal nerve **III** as the efferent limb to the **pupillary constrictor muscle**. An examination of the central artery of the retina is also made at this time. This gives the clinician a unique opportunity to assess the vascular system *in vivo* in a non-invasive manner. It also allows one to examine the optic disc *in vivo* and confirm the presence of a normal intracranial pressure by the absence of edema and swelling around the optic disc. When the light is removed from the eye, one should observe a dilation reflex, which is initiated by the sympathetic system. The lack of a light response to the retina sends information via nerve II to the brainstem. The sympathetic pathway descends into the spinal cord to the level of T1. The white rami communicantes of T1 enter the sympathetic trunk and the preganglionic fibers ascend to synapse in the superior cervical ganglion. The postganglionic fibers reach the orbit on the vascular system and innervate the **pupillary dilator** muscle.

The orbital examination therefore assesses cranial nerves II, III, IV, VI, and VII. V^1 can also be examined by assessing touch sensation to the forehead or the corneal reflex. In addition, the eye examination assesses parasympathetic and sympathetic pathways in the head and neck, as well as the central artery of the retina and the intracranial pressure.

The Oral Examination

Ask the patient to open the mouth. Observe the bilateral movement of the upper and lower lips to assess the buccal branches of nerve VII. Also observe the protrusive pattern of the mandible when the mouth is opened. Bilateral contraction of the **lateral pterygoid** muscles (V^3) cause the jaw to remain in the midline when it protrudes. The sensory components of V^2 and V^3 can also be examined by assessing touch sensation to the upper and lower lips, respectively.

Examination of the posterior aspect of the oral cavity and the oropharynx is done by depressing the tongue and asking the patient to say "Ah." Elevation of the soft palate in the midline when the patient says "Ah" reveals a normal bilateral innervation of the vagi (X) to the **levator palati muscles**. Resistance to the manual depression of the tongue reveals an active contraction of the extrinsic muscles of the tongue and an intact hypoglossal nerve (XII). A gag reflex following the stimulation of the mucosa of the tonsillar bed is a test of the glossopharyngeal nerve (IX). The oral examination is completed by examining the lateral aspects of the tongue for normal tissue appearance. Ask the patient to protrude the tongue. This requires bilateral action of the genioglossi or the tongue will deviate from a midline protrusion. Holding the tongue in the protrusive position will assist in the examination of the lateral aspect as well as testing V^3 for touch sensation of the anterior mucosa. The oral examination therefore assesses nerves V, VII, IX, X, and XII and the mucosa of the tongue, tonsils, and oropharynx.

The Ear Examination

The clinician is mainly concerned with examining the external acoustic meatus and the tympanic membrane with an otoscope. Cutaneous sensation to the auricle and external auditory canal is carried by the auriculotemporal (V^3), **greater auricular** (C2, 3), and the vagus (X) nerves. Hearing and balance sensations are properties of the vestibulocochlear nerve (VIII). Patients are usually asked if they have had problems with hearing and balance because a thorough examination of these functions is complex and somewhat detailed.

The Neck Examination

The patient is asked to swallow and the clinician observes the midline movement of the thyroid cartilage. This is elevated by the **stylopharyngeus** (IX) and the **palatopharyngeus** (X) muscles. The thyroid gland is palpated below the thyroid cartilage in the root of the neck. The patient's voice is assessed for quality and volume. This is controlled by the vagi (X) primarily via the **recurrent laryngeal branches**. The carotid pulse is palpated within the carotid triangle anterior to the sternomastoid muscle. The blood flow through the carotid artery and its bifurcation may also be examined with a stethoscope at the carotid triangle.

The Lymphatic Examination

All lymphatic drainage of the head passes through the lymph nodes surrounding the internal jugular veins. One can palpate for this chain of lymph nodes along the anterior border of the **sternomastoid muscle**. These nodes are **normally** undetectable and insensitive but in infectious or cancerous states they react by enlarging. These enlarged nodes project beyond the anterior border of the sternomastoid and become palpable. They also become sensitive and painful when they are enlarged. Enlargement of the supraclavicular lymph nodes above the medial end of the clavicle can result from disorders in the head and neck, upper extremity, or chest wall (breast). They are palpated on the posterior border of the sternomastoid muscle near its insertion into the clavicle.

At the time one is palpating for lymphatics, nerve XI can be assessed for its normal innervation of the sternomastoid and trapezius muscle. The highly competent head and neck examination in the normal patient can be completed in a few minutes by one who has a basic knowledge of the regional anatomy. The complexities and details of this area should not dissuade the nonspecialist clinician from doing a thorough and simple examination in every patient.

Answer Section

Chapter 5

1. (a) Sternal angle
   (b) T2 dermatome
2. Right parasternal lymph nodes
3. C8 superior and T1 inferior
4. 5th posterior intercostal artery from aorta and 5th anterior intercostal artery from internal thoracic artery. They anastomose at the midclavicular line.

Chapter 6

1. Visceral layer
2. Parietal pleura that covers the diaphragm and adjacent costal wall
3. (a) 8th rib
   (b) 10th rib

Chapter 7

1. T4, 5 at intervertebral disc
2. Phrenic nerve
3. Between the right 4th and 6th ribs anterior to the midaxillary line
4. Right posterior basal
5. (a) Right phrenic nerve
   (b) Right 6th intercostal nerve

Chapter 8

1. 2nd intercostal space at the right margin of the sternum
2. 2nd intercostal space at the left margin of the sternum
3. Visceral afferents from the heart usually enter the spinal cord with the sympathetic fibers at T1
4. Septum primum or septum secundum

Chapter 9

1. T4, 5
2. By inducing an inflammation or compressing the left recurrent laryngeal nerve
3. The sympathetic outflow from T1–4 is disconnected from the brainstem, but the parasympathetic vagus remains intact.
4. The phrenic and supraclavicular nerves are both derived from spinal cord segment C3, 4. Stimulation of the sensory fibers in the phrenic nerve could be falsely preceived as stimuli from the skin over the shoulder by the brain.
5. At the junction of the left subclavian and jugular veins

Chapter 11

1. (a) T10
   (b) L1
2. The subclavian and external iliac arteries
3. Ilioinguinal nerve (L1)
4. Pectineal line of the pubis
5. (a) three (internal, cremasteric, and external fascia)
   (b) lateral

Chapter 12

1. 2nd part of duodenum
2. Common bile duct, hepatic artery proper, and the portal vein
3. Lower right quadrant
4. Via the epiploic foramen
5. Left 9–11 ribs

Chapter 13

1. Left gastric on the lesser curvature and right gastroepiploic artery on the greater curvature
2. Increased resistance would promote the anastomosis and enlargement of the veins in the rectum to allow the portal blood to flow into the systemic circulation via the inferior rectal veins.
3. L1, 2

Chapter 14

1. At the point where the middle colic artery branches end on the transverse mesocolon, usually the left colic flexure
2. Superior and inferior pancreaticoduodenal arteries
3. From the left colic flexure distal to the rectum (Parasympathetic fibers from S2–4 are affected in this injury.)
4. Retrograde flow of bile into the main pancreatic duct can obliterate pancreatic acinar tissue.

Chapter 15

1. At the bifurcation into the right external and internal iliac arteries.
2. (a) Vas deferens
   (b) Uterine artery
3. (a) L4
   (b) L5

Chapter 16

1. Middle and inferior rectal veins in the anal canal, left gastric and esophageal veins in thorax, and paraumbilical and thoracoabdominal veins surrounding the umbilicus
2. (a) Left umbilical vein
   (b) It becomes the ligamentum teres.
3. Postganglionic sympathetic fibers to the vascular bed in the lower limbs

Chapter 18

1. Pectineal ligament (Cooper's ligament)
2. Ventral rami of L4, 5, and S1–4
3. Pelvic splanchnic (parasympathetic) nerves
4. In the bulb of the corpus spongiosum penis
5. Venous route from prostatic plexus to vertebral plexus
6. S2–4

Chapter 19

1. Ureters, ovarian vessels, internal iliac vessels, branches from the inferior mesenteric vessels, and the sigmoid colon
2. The artery is superior to the ureter.
3. The visceral afferent fibers follow the ovarian artery back to the aorta at L2 and then course with the sympathetic fibers in the lesser splanchnic nerve. Therefore, the dermatome T10 is stimulated, and

this overlies the stomach region in the epigastric region.
4. Oblique. With the face oriented posteriorly, the fetus will be able to extend the neck as the birth canal bends anteriorly in the lower region of the pelvis.

Chapter 20

1. Sympathetic (L1, 2)
2. The L1, 2 spinal segments, which contain the preganglionic sympathetic neurons, are situated at the T12 vertebral level in the spinal column.
3. The male gonads require sympathetic innervation while the female gonads are under hormonal control from the pituitary.
4. External and internal iliac lymph nodes

Chapter 21

1. L1 and S3, 4
2. Pudendal nerve (S2–4)
3. Internal pudendal artery

Chapter 23

1. Femoral vein and femoral artery
2. Femoral artery
3. Anterior superior iliac spine laterally and pubic tubercle medially
4. Femoral artery, nerve to vastus medialis and saphenous nerve
5. Lumbar 2–4 in femoral nerve

Chapter 24

1. (a) Anterior inferior iliac spine and superior margin of the acetabulum are the superior attachments, and the intertrochanteric line is the inferior attachment.
   (b) Hip extension
2. Medial and lateral femoral circumflex, obturator, and gluteal arteries
3. The sciatic nerve lies at the midpoint of a line between the tip of the coccyx medially and the greater trochanter laterally.
4. Large muscle mass and relatively few neurovascular branches are subjected to the ensuing irritation.

Chapter 25

1. Biceps femoris—tibial and peroneal nerves
   Adductor magnus—obturator and tibial nerves
   Pectineus—femoral and obturator nerves
2. (a) The femoral vein is superficial to the femoral artery in the popliteal fossa.
   (b) The femoral vein is deep to the femoral artery in the adductor canal.
3. Biceps femoris
4. Superiorly—inferiorly; Gemelli with intervening obturator internus tendon, quadratus femoris, and adductor magnus.

Chapter 26

1. (a) Posterior cruciate ligament
   (b) Anterior cruciate ligament
2. Lateral collateral (fibular collateral) ligament and the tendon of biceps femoris
3. A more prominent lateral wall in the patellar groove of femur and the horizontally inserting fibers of the vastus medialis

Chapter 27

1. In the superficial fascia overlying the medial malleolus and subcutaneous portion of the tibia
2. By producing an anterior compartment syndrome and compressing the deep peroneal nerve
3. Peroneus longus and brevis tendons
4. Flexor digitorum longus and flexor hallucis longus

Chapter 28

1. Calcanonavicular ("Spring") ligament
2. 2nd metatarsal
3. Anterior talofibular ligament
4. (a) Between the first and second muscle layers
   (b) The lateral plantar nerve

Chapter 29

1. T1 and C7
2. (a) Trapezius
   (b) Ask the patient to "shrug" the shoulder against resistance.
3. The inferior lobe of the lung is examined from the back.

Chapter 30

1. The clavicle, because it articulates with the axial skeleton at the sternum and transmits the force from the upper extremity to the axial skeleton
2. Acromioclavicular joint
3. Lateral pectoral nerve
4. Nerve to the serratus anterior (long thoracic nerve)

Chapter 31

1. The axillary nerve and posterior humeral circumflex vessel
2. Test the function of the intraspinatus muscle by having the patient externally rotate his arm and palpate the muscle on the dorsum of the scapula. If the infraspinatus can contract, then the suprascapular nerve is intact.
3. The thyrocervical trunk of the first part of the subclavian gives off the suprascapular artery, which supplies the dorsal surface of the scapula. The suprascapular artery anastomoses with the scapular circumflex artery, a branch of the suprascapular artery off the third part of the axillary artery. This route could bypass most of the axillary artery.
4. (a) Infraspinatus and teres minor
   (b) Respectively innervated by suprascapular (C5, 6) and axillary (C5, 6) nerves

Chapter 32

1. (a) C4
   (b) T4
   (c) C6
   (d) C8
   (e) T2
2. Ulnar nerve (dorsal palmar branch)

Chapter 33

1. (a) Radial nerve and profunda brachii artery
   (b) Ulnar nerve

2. (a) C5–7
   (b) Musculocutaneous nerve
3. On the medial side of the biceps tendon
4. The long head of the triceps is innervated by a radial nerve branch that arises proximal to the spiral groove.

Chapter 34

1. Median nerve (innervates both pronator teres [directly] and pronator quadratus [via anterior interosseous branch])
2. The median nerve branches are mostly directed medially to the muscles arising from the medial epicondyle. The major nerve in the lateral aspect is the superficial branch of the nerve, a sensory nerve running on the deep surface of the brachioradialis in the proximal forearm.
3. Distal end of the radius.

Chapter 35

1. (a) scaphoid
   (b) lunate
2. Median nerve
3. Approximately 9 years of age (as a general rule, 1 carpal bone per year from age 1)

Chapter 36

1. Ulnar nerve (C8, T1)
2. The flexor pollicis longus is still intact due to its innervation by the anterior interosseous nerve in the forearm.
3. Have the patient pronate the right forearm against resistance. This requires an intact median nerve at the elbow but not at the wrist.
4. Scaphoid

Chapter 37

1. Deep radial nerve
2. Check for the "intactness" of the superficial radial nerve by testing sensation on the lateral side of the dorsum of the hand.
3. (a) Supinator and biceps brachii
   (b) Radial and musculocutaneous nerve, respectively
4. (a) Flexion of the metacarpophalangeal joints
   (b) Ulnar nerve

Chapter 38

1. (a) VII (facial palsy)
   (b) 2nd arch (hyoid arch)
2. Greater auricular nerve (C2,3); vagus (X) and auriculotemporal nerve (V^3)
3. Lower eyelid—(the muscles of upper eyelid are innervated in part by III).
4. Bleeding into the "areolar space" of the scalp under the parietal region and subsequent accumulation inferior to frontalis insertion. Gravity will assist this accumulation of blood in the periorbital region.
5. To "numb" the right supraorbital nerve (V^1), which supplies the scalp from the vertex to the supraorbital margin. There is also more subcutaneous tissue to receive the anesthetic in the supraorbital notch region, and the displacement of tissues by the injected anesthetic does not interfere with the site of tissue repair.

Chapter 39

1. Infection spread through an emissary vein to a dural sinus.
2. No—it is in the epidural space.
3. Nerve VI is within the sinus and not in the wall like III, IV and V^1. Therefore, VI is more subjected to initial pressure increases within the sinus.
4. III, IV, V^1 and VI

Chapter 40

1. (a) X
   (b) Yes
2. The common embryological original of mandible and ear ossicles (malleus and incus) and a common V^3 nerve innervation (although via different terminal branches of V^3), gives these areas a common basis for pain preception.
3. IX (glossopharyngeal).
4. To avoid the attachment to the malleus and injury to the chorda tympani
5. Paralysis of the stapedius and a decreased ability to attenuate sound at the oval window
6. In the right internal acoustic meatus

Chapter 41

1. II (Optic)
2. III, IV or VI (Oculomotor, Trochlear or Abducens)
3. Right III (Oculomotor Nerve) (Preganglionic or postganglionic fibers or neurons)
4. Right VII (Facial)
5. (a) By preventing impulses from travelling to the brain through the optic (II) nerve
   (b) By preventing the blood supply to the retina via the ophthalmic and central retinal arteries
6. If the lacrimal gland were denervated and corneal desiccation occurred. V^1 is intact and would transmit the pain sensation from the "dried" cornea.
7. No, the optic disc and its blood vessels are medial to the macula lutea, which is in the central portion of the posterior pole of the eyeball.

Chapter 42

1. Test flexion of head (sternomastoid muscles) and shrugging of the shoulders (trapezius muscles)
2. XI (Accessory Nerve).
3. Brachial Plexus and axillary artery. Since the subclavian vein lies anterior to the anterior scalene muscle in the neck, it is not heavily invested by prevertebral or axillary sheath fascia.
4. The brachial plexus is superior to the 1st rib and higher than the subclavian artery because of the obliquity of the 1st rib. The apex of the lung contacts the medial border of the rib and the superior aspect and lateral border of the 1st rib are "safe" for injecting anesthetic solutions.
5. C6.
6. Test right suprascapular nerve (C5, 6) by having the patient laterally rotate the arm against resistance.

Chapter 43

1. C3
2. By spreading inferiorly along the carotid sheath into the thorax. The carotid artery arises from the aortic

arch inferiorly, and the adventitia of the aortic arch is related to the fibrous pericardium.
3. By affecting the recurrent laryngeal nerves that ascend through the root of the neck on the posteromedial aspect of the thyroid lobes
4. The thyroid gland develops in the region of the foramen cecum in the tongue and then descends into the neck in embryogenesis.
5. It anesthetizes the internal laryngeal nerves, which innervate the laryngeal mucosa above the vocal cords.
6. IX, X, XI and XII (deep to the posterior belly of the digastric)

Chapter 44

1. Sensory impulses would enter the spinal cord at C3, 4, 5. This is the same segmental level of the spinal cord that the supraclavicular nerves (C3, 4) stimulate when the skin over the shoulder is stimulated.
2. By stimulating the T1 segments of the brachial plexus
3. By damaging the sympathetic trunk as it crosses the neck of the 1st rib.
4. (a) C6 (Its anterior tubercle is also called the carotid tubercle.)
   (b) No, the vertebral artery has entered the C6 foramen transversarium posterior to the carotid tubercle.

Chapter 45

1. Nerve VII, Retromandibular Vein, External Carotid Artery, and Auriculotemporal Nerve (V^3).
2. The parotid increases in size as the acinar cells store their secretory granules in their cytoplasm. These granules are released by parasympathetic stimulation during mastication.
3. The parotid duct opens into the oral cavity through the buccal mucosa opposite to the upper second molar. Inflammation of the gland also causes inflammation in the parotid duct. Pain in the ear is due to a common sensory nerve supply to the parotid and pinna through the auriculotemporal nerve (V^3).
4. Within the trunk of VII between the stylomastoid foramen and the parotid gland.
5. Measure the distance from the mastoid process to the eye and calculate 1mm per day for regeneration of the damaged nerve axons.
6. Directly on the bone above the zygomatic arch. The anterior and posterior deep temporal nerves which are motor to the temporalis muscle lie between the muscle and bone in this location.
7. To avoid damage to the inferior alveolar nerves and vessels that serve the mandible
8. Left V^3
9. Left trigeminal ganglion in the middle cranial fossa.
10. (a) The lingual and mental nerves are also anesthetized.
    (b) Both, because chorda tympani joins the lingual nerve within the infratemporal fossa.

Chapter 46

1. (a) By compressing the subclavian artery and brachial plexus against the superior aspect of the 1st rib
   (b) Little finger side (medial) due to compression of C8 and T1 below the cervical rib
2. Transverse ligament (horizontal part of cruciate ligament)
3. (a) Scalenus anterior and longus colli (capitis)
   (b) Brachial Plexus and phrenic nerves with scalenus anterior and cervical sympathetic trunk with longus colli (capitis)
4. Laterally—Auriculotemporal nerve (V^3) and middle meningeal artery Medially—Chorda tympani as it courses from the petrotympanic tissue to the lingual nerve in the infratemporal fossa

Chapter 47

1. IX (Glossopharyngeal)
2. Junction of left jugular and subclavian veins
3. Anterior to the sternomastoid. The lymph nodes are surrounding the underlying internal jugular vein.
4. Test for the presence of sensation in the external auditory meatus. If absent, the lesion is in the posterior cranial fossa. One could also test the pharyngeal branch of X, by asking the patient to say "Ah." If the palpate rises in the midline, the lesion is below the level of the digastric muscle.

Chapter 48

1. (a) IX (Glossopharyngeal)
   (b) V^2 (lesser palatine nerve) and X (pharyngeal plexus)
2. (a) Tensor palati muscle and levator palati muscle
   (b) V^3 and X, respectively
3. Greater palatine artery and sphenopalatine artery
4. (a) Left X
   (b) Hoarseness in the patient's voice
5. (a) Stylopharyngeus and palatopharyngeus
   (b) IX and X, respectively

Chapter 49

1. On the cheek mucosa opposite the upper 2nd molar
2. On the mucosa of the floor of the mouth adjacent to the lingual frenulum.
3. (a) General sensation to the left anterior two-thirds of tongue and floor of mouth
   (b) Taste sensation to the left anterior two-thirds of tongue
   (c) Secretomotor innervation to the glands in the floor of the left side of the mouth
4. VII, IX and X
5. To the right
6. 24 months

Chapter 50

1. External nasal branch of V^1 is still unanesthetized.
2. Middle concha—the ostium of the maxillary sinus opens inferior to it.
3. Inferior to the inferior concha
4. Parasympathetic preganglionic fibers from VII on their postganglionic neurons
5. In the geniculate ganglion on VII
6. The epithelium changes and the ciliated pseudostratified columnar epithelium is destroyed. The opening (maxillary ostium) is superior to the floor, and without the cilia, the sinus does not drain properly.

Chapter 51

1. Recurrent laryngeal nerve and external laryngeal nerve
2. Posterior cricoarytenoids
3. Internal laryngeal nerves to mucosa above vocal cords and recurrent laryngeal to the mucosa below the vocal cords
4. Mucosa of vocal cords that rim the rima glottidis.
5. Parasympathetic fibers in X (mainly recurrent laryngeal branches)

Chapter 52

1. Veins
2. Arteries

References

SECTION ONE—GENERAL CONSIDERATIONS

Barclay, A. E., Barcroft, J., Barron, D. H., Franklin, K. J., and Prichard, M. M. L.: Studies of the foetal circulation and of certain changes that take place after birth. Am. J. Anat., *69:* 383, 1941.

Barnett, C. H., Davies, D. V., and MacConaill, M. A.: *Synovial Joints.* Longmans Green & Co., London, 1960.

Basmajian, J. V.: *Primary Anatomy,* 7th edition. The Williams & Wilkins Co., Baltimore, 1976.

Basmajian, J. V.: *Muscles Alive: Their Functions Revealed by Electromyography,* 4th edition. The Williams & Wilkins Co., Baltimore, 1979.

Book, M. H.: The secreting area of the glomerulus. J. Anat., *71:* 91, 1936.

Brash, J. C.: Some problems in the growth and developmental mechanics of bone. Edinburgh Med. J., *41:* 305, 363, 1934.

Brash, J. C.: *Neuro-Vascular Hila of Limb Muscles.* E. & S. Livingstone Ltd., Edinburgh, 1955.

Bridgman, C.: Changes in intramuscular pressure during contraction (abstract). Anat. Rec., *148:* 263, 1964.

Brookes, M., Elkin, A. C., Harrison, R. G., and Heald, C. B.: A new concept of capillary circulation in bone cortex: some clinical applications. Lancet, *1:* 1078, 1961.

Charnley, J.: Articular cartilage. Brit. Med. J., *2:* 679, 1954.

Charnley, J.: How our joints are lubricated. Triangle, Sandoz J. Med. Sci., *4:* no. 5, 1960.

Charnley, J.: Athroplasty of the hip: a new operation. Lancet, *1:* 1129, 1961.

Clark, E. R., and Clark, E. L.: Further observations on living lymphatic vessels in the transparent chamber in the rabbit's ear. Am. J. Anat., *52:* 273, 1933.

Coventry, M. B., *et al.*: The intervertebral disc, etc. J. Bone Joint Surg., *27:* 105, 1945.

Cronkite, A. E.: The tensile strength of human tendons. Anat. Rec., *64:* 173, 1936.

Cox, H. T.: The cleavage lines of the skin. Br. J. Surg., *29:* 234, 1941.

Cummins, H., and Midlo, C.: *Finger Prints, Palms, and Soles.* Blakiston Co., New York, 1943.

Davies, D. V.: Observations on the volume, viscosity and nitrogen content of synovial fluid, etc. J. Anat., *78:* 68, 1944.

Dawson, B. H., and Hoyte, D. A. N.: Observations on premature fusion of the sutures of the cranial vault (abstracts). J. Anat., *91:* 590, 583 and 613, 1957.

Digby, K. H.: The measurement of diaphysial growth, etc. J. Anat., *50:* 187, 1916.

Dintenfass, L.: Lubrication in synovial joints. J. Bone Joint Surg., *45(A):* 1241, 1963.

Duchenne, G. B. A.: *Physiologie des mouvements.* (translated by Kaplan, E. B.; see below.) Paris, 1867.

Edwards, E. A.: The orientation of venous valves in relation to body surfaces. Anat. Rec., *64:* 369, 1936.

Ekholm, R.: Nutrition of articular cartilage. Acta Anat. (Basel), *24:* 329, 1955.

Gardner, E.: Physiology of movable joints. Physiol. Rev., *30:* 127, 1950.

Gardner, E.: The anatomy of the joints. American Academy of Orthopedic Surgeons, instruction course. vol. 9, p. 14. Edwards, Ann Arbor, Mich., 1952.

Girgis, F. G., and Pritchard, J. J.: Effects of skull damage on the development of sutural patterns in the rat. J. Anat., *92:* 39, 1958.

Haines, R. W.: On muscles of full and of short action. J. Anat., *69:* 20, 1934.

Haines, R. W.: The laws of muscle and tendon growth. J. Anat., *66:* 578, 1932.

Ham, A. W.: *Histology.* 5th edition. J. B. Lippincott Co., Philadelphia, 1965.

Harris, H. A.: *Bone Growth in Health and Disease.* Oxford University Press, London, 1933.

Hughes, H.: The factors determining the direction of the canal for the nutrient artery in the long bones of mammals and birds. Acta Anat., *15:* 261, 1952.

Inman, V. T., Saunders, J. B. deC. M., and Abbott, L. C.: Observations on the function of the shoulder joint. J. Bone Joint Surg., *26:* 1, 1944.

Inman, V. T., and Saunders, J. B. deC. M.: Anatomicophysiological aspects of injuries to the intervertebral disc. J. Bone Joint Surg., *29:* 461, 1947.

Kaplan, E. B.: *Physiology of Motion.* J. B. Lippincott Co., Philadelphia, 1949. (Translation of *Physiologie des mouvements* by G. B. A. Duchenne, see above.)

Keegan, J. J., and Garrett, F. D.: The segmental distribution of the cutaneous nerves in the limbs of man. Anat. Rec., *102:* 409, 1948.

Keith, A.: *Menders of the Maimed.* Froude, London, 1919.

Kimmel, D. L.: Innervation of spinal dura mater and dura mater of the posterior cranial fossa. Neurology, *11:* 800, 1961.

Kimmel, D. L.: The nerves of the cranial dura mater and their significance in dural headache and referred pain. Chicago Med. School Q., *22:* 16, 1961.

Langworthy, O. R., *et al.*: *Physiology of Micturition.* Williams & Wilkins Co., Baltimore, 1940.

Learmonth, J. E.: A contribution to the neurophysiology of the urinary bladder in man. Brain, *54:* 147, 1931.

Le Double, A. F.: *Traité des variations du système musculaire de l'homme.* Paris, 1897.

Le Gros Clark, W. E.: *The Tissues of the Body.* 2nd edition. Clarendon Press, Oxford, 1945.

Little, T. D., Freeman, M. A. R., and Swanson, S. A. V.: Experiments on friction in the human hip joint. In *Lubrication and Wear in Joints,* V. Wright (ed.), Lippincott, Philadelphia, 1969.

MacConaill, M. A.: The movements of bones and joints. The synovial fluid and its assistants. J. Bone Joint Surg., *32B:* 244, 1950.

MacConaill, M. A., and Basmajian, J. V.: *Muscles and Movements: A Basis for Human Kinesiology.* Williams & Wilkins Co., Baltimore, 1969.

MacDonald, I. B., *et al.*: Anterior rhizotomy. The accurate identification of motor roots at the lower end of the spinal cord. J. Neurosurg., *3:* 421, 1946.

Marondas, A.: Hyaluronic acid films. Proc. Inst. Mech. Engrs., *181* (part 3J): 122, 1967.

McCutcheon, C. W.: More on weeping lubrication. In *Lubrication and Wear in Joints,* V. Wright (ed.), J. B. Lippincott, Philadelphia, 1969.

McKern, T. W., and Stewart, T. D.: Skeletal age changes in young American males, analysed from the standpoint of age identification. Smithsonian Institute, Washington, D. C., 1957.

Mednick, L. W., and Washburn, S. L.: The role of the sutures in the growth of the braincase of the infant pig. Am. J. Phys. Anthropol., *14:* 175, 1956.

Mitchell, G. A. G.: *Anatomy of the Autonomic Nervous System.* E. & S. Livingstone, Ltd., Edinburgh, 1953.

Mortensen, O. A., and Guest, R. L.: The absorption of thorium dioxide by the reticuloendothelial system in the dog. Anat. Rec. *70:* 58, 1938.

Patten, B. M.: *Human Embryology.* edition 2. McGraw-Hill Book Co., Inc., New York, Toronto, London, 1953.

Petter, C. K.: Methods of measuring the pressure of the intervertebral disc. J. Bone Joint Surg., *15:* 365, 1933.

Phemister, D. B.: Bone growth and repair. Ann. Surg., *102:* 261, 1935.

Pressman, J. J., and Simon, M. B.: Experimental evidence of direct communications between lymph nodes and veins. Surg. Gynecol. Obstet., *113:* 537, 1961.

Pressman, J. J., Simon, M. B., Hand, K., and Miller, J.: Passage of fluids, cells, and bacteria via direct communications between lymph nodes and veins. Surg. Gynecol. Obstet., *115:* 207, 1962.
Radin, E. L., and Paul, I. L.: A consolidated concept of joint lubrication. J. Bone Joint Surg., *54A:* 607, 1972.
Rappaport, A. M., *et al.*: Subdivision of hexagonal liver lobules into a structural and functional unit. Anat. Rec., *119:* 11, 1954.
Rappaport, A. M.: *The Liver,* vol. 1. Academic Press, Inc. New York, 1963.
Rau, R. K.: Skull showing absence of coronal suture. J. Anat., *69:* 109, 1934.
Roofe, P. G.: Innervation of annulus fibrosus, etc. J. Neurol. Neurosurg. Psychiatry, *44:* 100, 1940.
Rusznyák, I., Földi, M., and Szabo, G.: *Lymphatics and Lymph Circulation.* Pergamon Press, London, 1960.
Smith, C. G.: Changes in length and position of the segments of the spinal cord, etc. Radiology, *66:* 259, 1956.
Smorto, M. P., and Basmajian, J. V.: *Clinical Electroneurography,* 2nd ed., Williams & Wilkins Co., Baltimore, 1979.
Sunderland, S.: Blood supply of the nerves of the upper limb in man. Arch. Neurol. Psychiatry, *53:* 91, 1945.
Todd, T. W., and Pyle, S. I.: A quantitative study of the vertebral column, etc. Am. J. Phys. Anthropol., *12:* 321, 1928.
Trueta, J., and Cavadias, A. X.: A study of the blood supply of the long bones. Surg. Gynecol. Obstet., *118:* 485, 1964.
Walls, E. W.: The fibre constitution of the human gastrocnemius and soleus muscles. J. Anat., *87:* 437, 1953.
Walmsley, T.: The articular mechanism of the diarthroses. J. Bone Joint Surg., *10:* 40, 1928.
Weinmann, J. P., and Sicher, H.: *Bone and Bones.* 2nd edition. C. V. Mosby Co., St. Louis, 1955.
White, J. C., and Smithwick, R. H.: *The Autonomic Nervous System.* The Macmillan Co., New York, 1946.
Whitnall, S. E.: *Anatomy of the Human Orbit,* 2nd edition. Oxford University Press, London, 1932.
Woodburne, R. T.: The sphincter mechanism of the urinary bladder and the urethra. Anat. Rec., *141:* 11, 1961.

SECTION TWO—THORAX

Arey, L. B.: *Developmental Anatomy,* 7th edition, W. B. Saunders, Philadelphia, 1965.
Batson, O. V.: The function of the vertebral veins and their role in the spread of metastases. Ann. Surg., *112:* 138, 1940.
Boyden, E. A.: *Segmental Anatomy of the Lungs.* McGraw-Hill Book Co., Inc., New York, 1955.
Bradley, W. F., *et al.*: Anatomic considerations of gastric neurectomy. JAMA, *133:* 459, 1947.
Brock, R. C.: *The Anatomy of the Bronchial Tree.* Oxford University Press, London, 1946.
Foster-Carter, A. F.: Broncho-pulmonary abnormalities. Br. J. Tuberc., Oct. 1946.
Gradwohl, R. B. H.: *Clinical Laboratory Methods and Diagnosis,* 4th edition, vol. 2. C. V. Mosby, St. Louis, 1948.
Grant, R. T.: Development of the cardiac coronary vessels in the rabbit. Heart, *13:* 261, 1926.
Gross, L.: *The Blood Supply to the Heart.* Paul B. Hoeber, Inc., New York, 1921.
Harper, W. F.: The blood supply of human heart valves. Br. Med. J., *2:* 305, 1941.
Hayek, H. von: *The Human Lung* (translated by V. E. Krahl. Illustration based on figure 216, by courtesy of Hafner Publishing Company, Inc., New York, 1960).
Jackson, C. L., and Huber, J. F.: Correlated applied anatomy of the bronchial tree and lungs with a system of nomenclature. Dis. Chest, *9:* 319, 1943.
James, T. N.: The arteries of the free ventricular walls in man. Anat. Rec., *136:* 371, 1960.
James, T. N.: Anatomy of the human sinus node. Anat. Rec., *141:* 109, 1961.
James, T. N., and Burch, G. E.: The atrial coronary arteries in man. Circulation *17:* 90, 1958.
Jones, D. S., Beargie, R. J., and Pauly, J. E.: An electromyographic study of some muscles of costal respiration in man. Anat. Rec., *117:* 17, 1953.
Krahl, V. E.: Translation of Hayek's *The Human Lung.* Hafner Publishing Company, Inc., New York, 1960.
Lachman, E.: The dynamic concept of thoracic topography, etc. Am. J. Roentgenol., *56:* 419, 1946.
Lachman, E.: A comparison of the posterior boundaries of lungs and pleura, etc. Anat. Rec., *83:* 521, 1942.
Macklin, C. C.: Bronchial length changes and other movements. Tubercle, Oct.–Nov. 1932.
Macklin, C. C.: The dynamic bronchial tree. Am. Rev. Tuberc., *25:* 393, 1932.
Mainland, D., and Gordon, E. J.: The position of organs determined from thoracic radiographs, etc. Am. J. Anat., *68:* 457, 1941.
Merklin, R. J.: Position and orientation of the heart valves, Anat. Rec., *125:* 375, 1969.
Miller, W. S.: *The Lung.* 2nd edition. Charles C Thomas, Springfield, Ill., 1921.
Mitchell, G. A. G.: *Anatomy of the Autonomic Nervouse System.* E. & S. Livingstone, Ltd., Edinburgh, 1953.
Mizeres, N. J.: The cardiac plexus in man. Am. J. Anat., *112:* 1963.
Morris, E. W. T.: Some features of the mitral valve. Thorax, *15:* 70, 1960.
Nathan, H.: Anatomical observations on the course of the azygos vein. Thorax, *15:* 229, 1960.
Nelson, H. P.: Postural drainage of the lungs. Br. Med. J., *2:* 251, 1934.
Reed, A. F.: The origins of the splanchnic nerves. Anat. Rec., *109:* 81, 1951.
Ross, J. K.: Review of the surgery of the thoracic duct. Thorax, *16:* 207, 1961.
Rouvière, H.: *Anatomie des lymphatiques de l'homme.* Masson & Cie, Paris, 1932.
Silvester, C. F.: On the presence of permanent communications between the lymphatic and the venous system at the level of the renal veins in South American monkeys. Am. J. Anat., *12:* 447, 1912.
Singer, R.: The coronary arteries of the Bantu heart. South African Med. J., *33:* 310, 1959.
Tobin, C. E.: The bronchial arteries and their connections with other vessels in the human lung. Surg. Gynecol. Obstet., *95:* 741, 1952.
Tobin, C. E.: Human pulmonic lymphatics. Anat. Rec., *127:* 611, 1957.
Tobin, C. E., and Zariquiey, M. O.: Arteriovenous shunts in the human lung. Proc. Soc. Exp. Biol. Med., *75:* 827, 1950.
Thoracic Society: The nomenclature of bronchopulmonary anatomy. Thorax, *5:* 222, 1950.
Trotter, M.: Synostosis between manubrium and body of the sternum in whites and negroes. Am. J. Phys. Anthropol., *18:* 439, 1934.
Truex, R. C., and Warshaw, L. J.: The incidence and size of the moderator band, etc. Anat. Rec., *82:* 361, 1942.
Walls, E. W.: Dissection of the atrio-ventricular node and bundle in the human heart. J. Anat., *79:* 45, 1945.
Walmsley, T.: The Heart, in *Quain's Anatomy,* 1929.
White, J. C., and Smithwick, R. H.: *The Autonomic Nervous System,* 2nd edition. The Macmillan Co., New York, 1946.
Woodburne, R. T.: The costomediastinal border of the left pleura in the precordial area. Anat. Rec., *97:* 197, 1947.
Zoll, P. M., Wessler, S., and Schlesinger, M. J.: Interarterial coronary anastomoses, etc. Circluation, *4:* 797, 1951.

SECTION THREE—ABDOMEN

Alvarez, W. C.: *An Introduction to Gastro-enterology.* Heinemann, London, 1940.

Anson, B. J., and McVay, C. B.: Inguinal hernia. The anatomy of the region. Surg. Gynecol. Obstet., *66:* 186, 1938.

Ayoub, S. F.: The anterior fibres of the levator ani muscle in man. J. Anat., *128:* 571, 1979.

Basmajian, J. V.: The marginal anastomoses of the arteries to the large intestine. Surg. Gynecol. Obstet., *99:* 614, 1954.

Basmajian, J. V.: The main arteries of the large intestine. Surg. Gynecol. Obstet., *101:* 585, 1955.

Benjamin, H. B., and Becker, A. B.: A vascular study of the small intestine. Surg. Gynecol. Obstet., *108:* 134, 1959.

Boyden, E. A.: The accessory gall-bladder. Am. J. Anat., *38:* 202, 1926.

Boyden, E. A.: The anatomy of the choledochoduodenal junction in man. Surg. Gynecol. Obstet., *104:* 641, 1957.

Boyden, E. A.: *Gallbladder. McGraw-Hill Encyclopedia of Science and Technology.* McGraw-Hill, New York, 1960.

Cullen, T. S.: *Embryology, Anatomy and Diseases of the Umbilicus.* W.B. Saunders Co., Philadelphia, 1916.

Curtis, G. M., and Movitz, D.: The surgical significance of the accessory spleen. Ann. Surg., *123:* 276, 1946.

Daseler, E. H., Anson, B. J., Hambley, W. C., and Reimann, A. F.: The cystic artery and constituents of the hepatic pedicle. A study of 500 specimens. Surg. Gynecol. Obstet., *85:* 45, 1947.

Dawson, W., and Langman, J.: An anatomical-radiological study on the pancreatic duct pattern in man. Anat. Rec., *139:* 59, 1961.

Doyle, J. F.: The superficial inguinal arch: a reassessment of what has been called the inguinal ligament. J. Anat., *108:* 297, 1971.

Drummond, H.: The arterial supply of the rectum and pelvic colon. Br. J. Surg., *1:* 677, 1914.

Edwards, E. A.: Functional anatomy of the portasystemic communications. Arch. Intern. Med., *88:* 137, 1951.

Farkas, L. G.: Basic morphological data of external genitals in 177 healthy Central European men. Am. J. Phys, Anthropol., *34:* 325, 1971.

Falconer, C. W. A., and Griffiths, E.: The anatomy of the blood-vessels in the region of the pancreas. Br. J. Surg., *37:* 334, 1950.

Finlayson, J.: Herophilus and Erasistratus. Glasgow Med. J., May, 1893.

Franklin, K. J.: *A Monograph on Veins.* Charles C Thomas, Springfield, Ill., 1937.

Frigerio, N. A., Stowe, R. R., and Howe, J. W.: Movement of sacroiliac joint. Clin. Orthop., *100:* 370, May, 1974.

Graves, F. T.: The anatomy of the intrarenal arteries and its application to segmental resection of the kidney. Br. J. Surg., *42:* 132, 1954.

Halbert, B., and Eaton, W. L.: Accessory spleens: a pilot study of 600 necropsies (abstract). Anat. Rec. *109:* 371, 1951.

Hanna, R. E., and Washburn, S. L.: The determination of the sex of skeltons as illustrated by a study of the Eskimo pelvis. Human Biol., *25:* 21, 1953.

Hardy, K. J.: Involuntary sphincter tone in the maintenance of continence. Aust. NZ J. Surg., *42:* 48, 1972.

Harrison, R. G.: The distribution of the vasal and cremasteric arteries to the testis, etc. J. Anat., *83:* 267, 1949.

Healey, J. E. and Schroy, P. C.: Anatomy of the biliary ducts within the human liver. Arch. Surg., *66:* 599, 1953.

Healey, J. E., Schroy, P. C., and Sorensen, R. J.: The intrahepatic distribution of the hepatic artery. J. Intern. Coll. Surgeons, *20:* 133, 1953.

Hjortsjo, C-H.: The topography of the intrahepatic duct systems (and of the portal vein). Acta Anat., *11:* 599, 1951.

Hjortsjo, C-H.: The intrahepatic ramifications of the portal vein. Lunds Universitets Arsskrift., *52:* 20, 1956.

Hyde, J. S, Swarts, C. L., Nicholas, E. E., Snead, C. R., and Strasser, N. F.: Superior mesenteric artery syndrome Am. J. Dis. Child., *106:* 25, 1963.

Jackson, A. J.: The spiral constrictor of the gastroesophageal junction. Am. J. Anat., *151:* 265, 1978.

Jamieson, J. K., and Dobson, J. F.: The lymphatic system of the stomach, and of the caecum and appendix. Lancet, April 20 and 27, 1907.

Jamieson, J. K., and Dobson, J. F.: The lymphatics of the testicle, Lancet, Feb. 19, 1910.

Jamieson, J. K., and Dobson, J. F.: The lymphatics of the colon. Proc. R. Soc. Med., March 1909.

Jay, G. D., III, *et al.*: Meckel's diverticulum: survey of 103 cases. Arch. Surg., *61:* 158, 1950.

Lofgren, F.: *Some Features in the Renal Morphogenesis and Anatomy with Practical Considerations.* Institute of Anatomy, University of Lund, Sweden, 1956.

Lofgren, F.: An attempt at homologizing different types of pyelus (renal pelvis). Urologia Intern., *5:* 1, 1956.

Lytle, W. J.: The internal inguinal ring. Br. J. Surg., *32:* 441, 1945.

Maisel, H.: The position of the human vermiform appendix. Anat. Rec., *136:* 385, 1960.

Mann, C. V., Greenwood, R. K., and Ellis, F. H., Jr.: The esophagogastric junction. Surg. Gynecol. Obstet., *118:* 853, 1964.

Michels, N. A.: *Blood Supply and Anatomy of the Upper Abdominal Organs.* J. B. Lippincott Co., Philadelphia, 1955.

Michels, N. A., Siddharth, P., Kornblith, P., and Parke, W. W.: The variant blood supply to the small and large intestine: its import in regional resections. J. Intern. Coll. Surgeons, *39:* 127, 1963.

Millbourn, E.: On the excretory ducts of the pancreas, etc. Acta Anat., *9:* 1, 1950.

Mills, R. W.: The relation of bodily habitus to visceral form, position, tonus and motility. Am. J. Roentgenol., *4:* 155, 1917.

Mitchell, G. A. G.: *Anatomy of the Autonomic Nervous System.* E. & S. Livingstone, Ltd., Edinburgh, 1953.

Moody, R. O., and Van Nuys, R. G.: Some results of a study of roentgenograms of the abdominal viscera. Am. J. Roentgenol., *20:* 348, 1928.

Moody, R. O., Van Nuys, R G., and Kidder, C. H.: The form and position of the empty stomach in healthy young adults. Anat. Rec., *43:* 359, 1929.

Moody, R. O., Van Nuys, R. G.: The position and mobility of the kidneys in healthy young men and women. Anat. Rec., *76:* 111, 1940.

Oh, C., and Kark, A. E.: Anatomy of the external anal sphincter. Br. J. Surg., *59:* 717, 1972.

Patey, D. H.: Some observations on the functional anatomy of inguinal hernia, etc. Br. J. Surg., *36:* 264, 1949.

Pierson, J. M.: The arterial blood supply of the pancreas. Surg. Gynecol. Obstet., *77:* 426, 1943.

Reeves, T. A.: A study of the arteries supplying the stomach and duodenum and their relation to ulcer. Surg. Gynecol. Obstet., *30:* 374, 1920.

Rienhoff, W. F., and Pickrell, K. L.: Pancreatitis. An anatomic study of the pancreatic and extrahepatic biliary systems. Arch. Surg., *51:* 205, 1945.

Roberts, W. H. B., and Taylor, W. H.: Inferior rectal nerve variations as it relates to pudendal block. Anat. Rec., *177:* 461, 1973.

Roche, M. B., and Rowe, G. G.: The incidence of separate neural arch, etc. J. Bone Joint Surg., *34A:* 491, 1952.

Ross, J. A.: Vascular patterns of small and large intestine compared. Br. J. Surg., *39:* 330, 1952.

Rowe, G. G., and Roche, M. B.: The etiology of separate neural arch. J. Bone Joint Surg., *35A:* 102, 1953.

Shah, M. A., and Shah, M.: The arterial supply of the vermiform appendix. Anat. Rec., *95:* 457, 1946.

Sheehan, D.: The afferent nerve supply of the mesentery, etc. J. Anat., *67:* 233, 1933.

Solanke, T. F.: The blood supply of the vermiform appendix in Nigerians J. Anat., *102:* 353, 1968.

Solanke, T. F.: The position, length and content of the vermiform appendix in Nigerians. Br. J. Surg., *57:* 100, 1970.

Stephens, F. D.: Nervous pathways in anorectal control. Aust. NZ J. Surg., *42:* 45, 1972.

Steward, J. A., and Rankin, F. W.: Blood supply of the large intestine: its surgical considerations. Arch. Surg., *26:* 843, 1933.

Stewart, T. D.: The age incidence of neural arch defects in Alaskan natives. J. Bone Joint Surg., *35A:* 937, 1953.

Tobin, C. E., and Benjamin, J. A.: Anatomic and clinical re-evaluation of Camper's, Scarpa's and Colles' fasciae. Surg. Gynecol. Obstet., *88:* 545, 1949.

Underhill, B. M. L.: Intestinal length in man. Br. Med. J., *2:* 1243, 1955.

Varma, K. K.: The role of the voluntary anal sphincter in the maintenance of faecal continence in normal and abnormal states. Aust. NZ J. Surg., *42:* 52, 1972.

Wakeley, C. P. G.: The position of the vermiform appendix, etc. J. Anat., *67:* 277, 1933.

Wells, L. J.: Descent of the testis: anatomical and hormonal considerations. Surgery, *14:* 436, 1943.

Wells L. J.: Observations on the development of the diaphragm in the human embryo. Anat. Rec., *100:* 778, 1948.

Wells, L. J.: Contributions to embryology. Carnegie Institute, *35:* 107, 1954.

Wharton, G. K.: The blood supply of the pancreas, etc. Anat. Rec., *53:* 55, 1932.

Wilkie, D. P. D.: The blood supply of the duodenum, etc. Surg. Gynecol. Obstet., *13:* 399, 1911.

Woodburne, R. T., and Olsen, L. L.: The arteries of the pancreas. Anat. Rec., *111:* 255, 1951.

SECTION FOUR—PERINEUM AND PELVIS

Blair, J. B., Holyoke, E., and Best, R. R.: A note on the lymphatics of the middle and lower rectum and anus. Anat. Rec., *108:* 635, 1950.

Braithwaite, J. L.: Vesiculo-deferential artery. Br. J. Urol., *24:* 64, 1952.

Braus, H.: *Anatomie des menchen.* 2nd edition. Springer-Verlag, Berlin, 1929.

Caldwell, W. E. and Moloy, H. C.: Anatomical variations in the female pelvis, etc. Am. J. Obstet. Gynecol., *26:* 479, 1933.

Derry, D. E: The innominate bone and the determination of sex. J. Anat., *43:* 266, 1908.

Frigerio, N. A., Stow, R. R., and Howe, J. W.: Movement of sacroiliac joint. Clin. Orthop., *100:* 370, 1974.

Greulich, W. W., and Thoms, H.: A study of pelvic type and its relationship to body build in white women. JAMA, *112:* 485, 1939.

Greulich, W. W., and Thomas, H.: The dimensions of the pelvic inlet of 789 white females. Anat. Rec., *72:* 45, 1938.

Keith, A.: *Human Embryology and Morphology.* 6th edition. Edward Arnold & Co., London, 1948.

Langworthy, O. R., *et al.*: *Physiology of Micturition.* Williams & Wilkins Co., Baltimore, 1940.

Leaf, C. H.: *The Lymphatics by Pouirier and Cuneo.* (translation) Constable, London, 1903.

Learmonth, J. E.: A contribution to the neurophysiology of the urinary bladder in man. Brain, *54:* 147, 1931.

McKern, T. W., and Stewart, T. D.: Skeletal age changes in young American males. Smithsonian Institute, Washington, D. C., 1957.

Mitchell, G. A. G.: *Anatomy of the Autonomic Nervous System.* E. & S. Livingstone, Ltd., Edinburgh, 1953.

Moloy, H. C.: *Evaluation of the Pelvis in Obstetrics.* W. B. Saunders Company, Philadelphia, 1951.

Oh, C., and Kark, A. E.: Anatomy of the external anal sphincter, Br. J. Surg., *59:* 717, 1972.

Phenice, T. W.: A newly developed visual method of sexing the os pubis. Am. J. Phys. Anthrop., *30:* 297, 1969.

Ricci, J. V. *et al.*: The female urethra: a histological study, etc. Am. J. Surg., N. S. *79:* 499, 1950.

Roberts, W. H. B., and Taylor, W. H.: Inferior rectal nerve variations as it relates to pudendal block. Anat. Rec. *177:* 461, 1973.

Rouvière, H.: Anatomie des lymphatiques de l'homme. Masson & Cie, Paris, 1932 (translated into English by Tobias, see below).

Shafik, A.: A new concept of the anatomy of the anal sphincter. Part V. Chir. Gastroent. (Surg. Gastroent.). *11:* 319, 1977.

Sheehan, D.: *Annual Review of Physiology*, Vol. 3, 1941.

Singh, S., and Potturi, B. R.: Greater sciatic notch in sex determination. J. Anat. (Lond.), *125:* 619, 1978.

Stopford, J. S. B.: The autonomic nerve supply of the distal colon. Brit. M. J., *1:* 572, 1934.

Thomas, J.: *Pelvimetry.* Paul B. Hoeber, Inc., 1956.

Tobias, M. J.: *Anatomy of the Human Lymphatic System.* Edwards, Ann Arbor, Mich., 1938 (translation of the work of Rouvière, see above).

Todd, T. W.: Mammalian pubic metamorphosis. Am. J. Phys. Anthropol., *4:* 407, 1921.

Washburn, S. L.: Sex differences in the pubic bone. Am. J. Phys. Anthropol., N. S. *6:* 199, 1948.

Wilde, R. F.: The anal intermuscular septum. Br. J. Surg., *36:* 279, 1949.

SECTION FIVE—LOWER LIMB

Basmajian, J. V.: The distribution of valves in the femoral, external iliac and common iliac veins, etc. Surg. Gynecol. Obstet., *95:* 357, 1952.

Basmajian, J. V., and Bentzon, J. W.: An electromyographic study of certain muscles of the leg and foot, etc. Surg. Gynecol. Obstet., *98:* 662, 1954.

Basmajian, J. V., and Lovejoy, J. F., Jr.: Functions of popliteus: a multifactorial electromyographic study. J. Bone Joint Surg., *53A:* 557, 1971.

Bing, R.: *Compendium of Regional Diagnosis in Lesions of the Brain and Spinal Cord*, Ed. 11, translated and edited by W. Haymaker. C. V. Mosby Company, St. Louis, 1940.

Doyle, J. F.: The superficial inguinal arch: a reassessment of what has been called the inguinal ligament. J. Anat., *108:* 297, 1971.

Gardner, E.: The innervation of the hip joint. Anat. Rec., *101:* 353, 1948.

Gardner, E.: The innervation of the knee joint. Anat. Rec., *101:* 109, 1948.

Harty, M.: Anatomic features of the lateral aspect of the knee joint. Surg. Gynecol. Obstet., *130:* 11, 1970.

Haxton, H.: The functions of the patella and the effects of its excision. Surg. Gynecol. Obstet., *80:* 389, 1945.

Haymaker, W., and Woodhall, B.: *Peripheral Nerve Injuries: Principles of Diagnosis*, Ed. 2. W. B. Saunders Company, Philadelphia and London, 1953.

Hicks, J. H.: The mechanics of the foot: the joints. J. Anat., *87:* 345, 1953. The plantar aponeurosis and the arch. J. Anat., *88:* 25, 1954. The foot as a support. Acta Anat., *25:* 34, 1955.

Hughston, J. C., and Eilers, A. F.: The role of the posterior oblique ligament in repairs of acute medial (collateral) ligament tears of the knee. J. Bone Joint Surg., *55A:* 923, 1973.

Jack, E. A.: Naviculocuneiform fusion in the treatment of flat foot. J. Bone Joint Surg., *35B:* 75, 1953.

Jones, F. W.: *The Foot, Structure and Function.* Ballière, Tindall & Cox, London, 1949.

Jones, R. L.: The human foot . . . the role of its muscles and ligaments in the support of the arch. Am. J. Anat., *68:* 1, 1941.

Keegan, J. J., and Garrett, F. D.: The segmental distribution of the cutaneous nerves in the limbs of man. Anat. Rec. *102:* 409, 1948.

Lambert, E. H.: The accessory deep peroneal nerve: a common variation in innervation of extensor digitorum brevis. Neurology, *19:* 1169, 1969.

Last, R. J.: The popliteus muscle and the lateral meniscus. J. Bone Joint Surg., N. S. *32B:* 93, 1950.

Lemont, H.: The branches of the superficial peroneal nerve and their clinical significance. J. Am. Podiatry Assoc., *65:* 310, 1975.

Lovejoy, J. F., Jr., and Harden, T. P.: Popliteus muscle in man. Anat. Rec., *169:* 727, 1971.

Marshall, J. L., Girgis, F. G., and Zelko, R. R.: The biceps femoris tendon and its functional significance. J. Bone Joint Surg., *54A:* 1444, 1972.

Mitchell, G. A. G.: *Anatomy of the Autonomic Nervous System.* E. & S. Livingstone, Ltd., Edinburgh, 1953.

Morton, D. J.: *The Human Foot.* Columbia University Press, New York, 1937.

Noyes, F. R., and Sonstegard, D. A.: Biomechanical function of the pes anserinus at the knee and the effect of its transplantation. J. Bone Joint Surg., *55A:* 1225, 1973.

O'Rahilly, R.: A survey of carpal and tarsal anomalies. J. Bone Joint Surg., *35A:* 626, 1953.

Robichon, J., and Romero, C.: The functional anatomy of the knee joint, with special reference to the medial collateral and anterior curciate ligaments. Can. J. Surg., *11:* 36, 1969.

Senior, H. D.: The development of the human femoral artery, a correction. Anat. Rec., *17:* 271, 1920.

Singer, C.: *The Evolution of Anatomy.* Kegan Paul, London, 1925.

Storton, C.: Personal communication.

Straus, W. L.: Human ilia: sex and stock. Am. J. Phys. Antrhopol., *11:* 1, 1927.

Trueta, J.: The normal vascular anatomy of the femoral head during growth. J. Bone Joint Surg., *39B:* 358, 1957.

Trueta, J., and Harrison, M. H. M.: The normal vascular anatomy of the femoral head in adult man. J. Bone Joint Surg., *35B:* 442, 1953.

Tucker, F. R.: Arterial supply to the femoral head and its clinical importance. J. Bone Joint Surg., *31B:* 82, 1949.

Walmsley, T.: The articular mechanism of the diarthroses, J. Bone Joint Surg., *10:* 40, 1928.

Weinert, C. R., Jr., McMaster, J. H., and Ferguson, R. J.: Dynamic fuction of the human fibula. Am. J. Anat., *138:* 145, 1973.

Wolcott, W. E.: The evolution of the circulation in the developing femoral head and neck. Surg. Gynecol. Obstet, *77:* 61, 1943.

SECTION SIX—UPPER LIMB AND BACK

Basmajian, J. V., and Bazant, F. J.: Factors preventing downward dislocation of the adducted shoulder joint. J. Bone Joint Surg., *41A:* 1182, 1959.

Basmajian, J. V., and Latif, A.: Integrated actions and functions of the chief flexors of the elbow. J. Bone Joint Surg., *39A:* 1106, 1957.

Batson, O. V.: The function of the vertebral veins and their role in the spread of metastases. Ann. Surg., *112:* 138, 1940.

Beevor, C. E.: Croonian lecture on muscular movements, etc. Br. Med. J., *1:* 1357, 1417, 1480; *2:* 12, 1903.

Bing, R.: *Compendium of Regional Diagnosis in Lesions of the Brain and Spinal Cord,* 11th edition (translated and edited by W. Haymaker). C. V. Mosby Co., St. Louis, 1940.

Braithwaite, F., *et al.*: The applied anatomy of the lumbrical and interosseous muscles of the hand. Guy's Hosp. Rep., *97:* 185, 1948.

Bunnell, S.: *Surgery of the Hand.* J. B. Lippincott Co., Philadelphia, 1944.

Corbin, K. B., and Harrison, F.: The sensory innervation of the spinal accessor and tonguye musculature in rhesus monkey. Brain, *62:* 191, 1939.

Cummins, H., and Midlo, C.: *Finger Prints, Palms, and Soles.* Blakiston Company, division of Doubleday & Co., New York, 1943.

Eckenhoff, J. E.: The physiologic significance of the vertebral venous plexus. Surg. Gynecol. Obstet., *131:* 72, 1970.

Fick, R.: *Handbuch der Anatomie und Mechanik der Gelenke,* vol. 3. Gustav Fischer, Jena, 1911.

Flecker, H. Time of appearance and fusion of ossification centers as observed by roentgenographic methods. Am. J. Roentgenol., *47:* 97, 1942. Also with similar title in J. Anat., *67:* 118, 1932.

Forrest, W. J.: Motor innervation of human thenar and hypothenar muscles in 25 hands: A study combining electromyography and percutaneous nerve stimulation. Can. J. Surg., *10:* 196, 1967.

Gardner, E.: The innervation of the elbow joint. Anat. Rec., *102:* 161, 1948.

Garn, S. M. *et al*: A rational approach to the assessment of skeletal maturation. Ann. Radiol. (Paris), V–VI, 1964.

George, R. K.: Personal communications.

Greulich, W. W., and Pyle, S. I.: Radiographic atlas of skeletal development of the hand and wrist, 2nd edition. Stanford University Press, Stanford, Calif., 1959.

Haines, R. W.: The mechanism of rotation at the first carpometacarpal joint. J. Anat., *78:* 44, 1944.

Haines, R. W.: The extensor apparatus of the finger. J. Anat., *85:* 251, 1951.

Halls, A. A., and Travill, A.: Transmission of pressures across the elbow joint. Anat. Rec., *150:* 243, 1964.

Harness, D., and Sekeles, E.: The double anastomotic innervation of thenar muscles. J. Anat., *109:* 461, 1971.

Haymaker, W., and Woodhall, B.: *Peripheral Nerve Injuries: Principles of Diagnosis.* 2nd edition. W. B. Saunders Co., Philadelphia and London, 1953.

Inman, V. T., Saunders, J. B. deC. M., and Abbott, L. C.: Observations on the functions of the shoulder joint. J. Bone Joint Surg., *26:* 1, 1944.

Johnson, G. F., Dorst, J. P., Kuhn, J. P., Roche, A. F., and Dávila, G. H.: Reliability of skeletal age assessments. Am. J. Roentgenol., *118:* 320, 1973.

Jones, F. W.: *The Principles of Anatomy as Seen in the Hand.* 2nd edition. Baillière, Tindal, & Cox, London, 1941.

Kanavel, A. B.: *Infections of the Hand.* 7th edition. Lea & Febiger, Philadelphia, 1939.

Keegan, J. J., and Garrett, F. D.: The segmental distribution of the cutaneous nerves in the limbs of man. Anat. Rec., *102:* 409, 1948.

Landsmeer, J. M. F.: The anatomy of the dorsal aponeurosis of the human finger, etc. Anat. Rec., *104:* 31, 1949.

McKern, T. W., and Stewart, T. D.: Skeletal age changes in young American males. Smithsonian Institute, Washington, D. C., 1957.

Poirer, P., and Charpy, A.: *Abrege d'anatomie.* Paris, Masson et Cie, 1908.

Rowntree, T.: Anomalous innervation of the hand muscles. J. Bone Joint Surg., *31B:* 505, 1949.

Salsbury, C. R.: The interosseous muscles of the hand. J. Anat., *71:* 395, 1937.

Stopford, J. S. B.: *Sensation and the Sensory Pathway.* Longmans Green & Co., Inc., London, 1930.

Sunderland, S.: The innervation of the first dorsal interosseous muscle of the hand. Anat. Rec., *95:* 7, 1946.

Sunderland, S.: Voluntary movements and the deceptive action of muscles in peripheral nerve lesions. Aust. NZ J. Surg., *13:* 160, 1944.

Testut, J. L.: *Traité d'anatomie Humaine,* 9th edition, revised by A. Latarget. Doin, Paris, 1948–1949.

Wilder, H. H.: *The History of the Human Body.* 2nd edition. Henry Holt & Co., New York, 1923.

Wookey, H.: Personal communication.

SECTION SEVEN—HEAD AND NECK

Basmajian, J. V., and Dutta, C. R.: Electromyography of the pharyngeal constrictors and levator palati in man. Anat. Rec., *139:* 561, 1961.

Batson, O. V.: The fuction of the vertebral veins and their role in the spread of metastases. Ann. Surg., *112:* 138, 1940.

Békésy, G. P. See von Békésy, G., *below.*

Bolz, E. A., and Lim, D. J.: Morphology of the stapediovestibular joint. Acta Otolaryngol., *73:* 10, 1972.

Browning, H.: The confluence of dural venous sinuses. Am. J. Anat., *93:* 307, 1953.

Carlsöö, S.: Nervous co-ordination and mechanical function of mandibular elevators. Acta Odontol. Scand., *10:* suppl. 11, 1952.

Cave, A. J. E.: A note on the origin of the m. scalenus medius. J. Anat., *67:* 480, 1933.

Cole, T. B., and Baylin, G.: Radiographic evaluation of the prevertebral space. Laryngoscope, *83:* 721, 1973.

Doran, G. A., and Baggett, H.: The genioglossus muscle: a reas-

sessment of its anatomy in some mammals, including man. Acta Anat., *83:* 403, 1972.
Eckenhoff, J. E.: The physiologic significance of the vertebral venous plexus. Surg. Gynecol. Obstet., *131:* 72, 1970.
Graves, G. O., and Edwards, L. F.: The eustachian tube. Arch. Otolaryngol., *39:* 359, 1944.
Hoshino, T., and Paparella, M. M.: Middle ear muscle anomalies. Arch. Otolaryngol., *94:* 235, 1971.
Jamieson, J. K., and Dobson, J. F.: The lymphatics of the tongue, etc. Br. J. Surg., *8:* 80, 1920.
Kimmel, D. L.: Innervation of spinal dura mater and dura mater of the posterior cranial fossa. Neurology, *11:* 800, 1961.
Kimmel, D. L.: The nerves of the cranial dura mater and their significance in dural headache and referred pain. Chicago Med. School Q., *22:* 16, 1961.
Latif, A.: An electromyographic study of the temporalis muscle, etc. Am. J. Orthod., *43:* 577, 1957.
Leden, H., and Moore, P.: See von Leden, H., and Moore, P., *below*.
Lewinsky, W., and Stewart, D.: An account of our present knowledge of the innervation of the teeth and their related tissues. Br. Dent. J., Dec. 1, 1938.
Lewis, G. F.: Personal communication.
Mann, I.: *The Development of the Human Eye*. Cambridge University Press, London, 1928.
McKenzie, J.: The parotid gland in relation to the facial nerve. J. Anat., *82:* 183, 1948.
McKern, T. W., and Stewart, T. D.: Skeletal age changes in young American males. Smithsonian Institute, Washington, D. C., 1957.
Mitchell, G. A. G.: *Anatomy of the Autonomic Nervous System*. E. & S. Livingston, Ltd., Edinburgh, 1953.
Mizeres, N. J.: The cardiac plexus in man. Am. J. Anat., *112:* 1963.
Moyers, R. E.: An electromyographic analysis of certain muscles involved in temporomandibular movement. Am. J. Orthodontics. *36:* 481, 1950.
Negus, V. E.: *The Comparative Anatomy and Physiology of the Larynx*. William Heinemann, Ltd., London, 1949.
Parkinson, D.: Collateral circulation of cavernous carotid artery: anatomy. Can. J. Surg., *7:* 251, 1964.
Pearson, A. A.: The hypoglossal nerve in human embryos. J. Comp. Neurol., *71:* 21, 1939.
Pearson, A. A., *et al.*: Cutaneous branches of the dorsal (primary) rami of the cervical nerves. Am. J. Anat., *112:* 169, 1963.
Powell, T. V., and Brodie, A. G.: Closure of sphenooccipital synchondrosis. Anat. Rec., *147:* 15, 1963.
Pressman, J. J., and Simon, M. B.: Experimental evidence of direct communications between lymph nodes and veins. Surg. Gynecol. Obstet., *113:* 537, 1961.
Rouvière, H.: *Anatomie des lymphatiques de l'homme*. Masson et Cie, Paris, 1932.
Saunders, J. B. deC. M., Davis, C., and Miller, E. R.: The mechanism of deglutition as revealed by cineradiography. Ann. Otol. Rhino. Laryngol., *60:* 897, 1951.
Schour, I., and Massler, M.: The development of the human dentition. J. Am. Dent. A., *28:* 1153, 1941.
Sellars, I. E., and Keen, E. N.: The anatomy and movements of the cricoarytenoid joint. Laryngoscope, *88:* 667, 1978.
Stewart, D., and Wilson, S. L.: Regional anaesthesia and innervation of the teeth. Lancet, *Oct. 20:* 809, 1928.
Stiles, H. J.: In *Cunningham's Text-Book of Anatomy*. Oxford University Press, London, 1913.
Sunderland, S.: The meningeal relations of the human hypophysis cerebri. J. Anat., *79:* 33, 1945.
Todd, T. W., and Lyon, D. W., Jr.: Endocranial suture closure, etc. Am. J. Phys. Anthropol, *7:* 325, 1924.
von Békésy, G.: The ear. Sci. Am., August 1957.
von Békésy, G.: *Experiments in Hearing* (translated and editied by E. G. Wever). McGraw-Hill Book Company, Inc., New York, 1960.
von Leden, H. and Moore, P.: The mechanics of the cricoarytenoid joint. Arch. Otolaryngol., *73:* 541, 1961.
Watt, J. C., and McKillop, A. N.: Relation of arteries to roots of nerves in posterior cranial fossa. Arch. Surg., *30:* 336, 1935.
Whitnall, S. E.: *Anatomy of the Human Orbit*, 2nd edition. Oxford University Press, London, 1932.
Young, M. W.: The termination of the perilymphatic duct. Anat. Rec., *112:* 404, 1952.

Index

Page numbers in *italics* denote figures; those followed by "t" denote tables.

Q

R

T

W

Z